Biopsychology

ELEVENTH EDITION

GLOBAL EDITION

John P. J. Pinel & Steven J. Barnes
University of British Columbia

Pearson

Harlow, England • London • New York • Boston • San Francisco • Toronto • Sydney • Dubai • Singapore • Hong Kong
Tokyo • Seoul • Taipei • New Delhi • Cape Town • São Paulo • Mexico City • Madrid • Amsterdam • Munich • Paris • Milan

John Pinel: *To Maggie, the love of my life.*

Steven Barnes: *To Behnaz and Mina, the loves of my life; to John Pinel, the best mentor one could ever hope to have; and to the countless students who have contributed to the evolution of Biopsychology.*

Please contact https://support.pearson.com/getsupport/s/contactsupport with any queries on this content.

Pearson Education Limited
KAO Two
KAO Park
Hockham Way
Harlow
Essex
CM17 9SR
United Kingdom

and Associated Companies throughout the world

Visit us on the World Wide Web at: www.pearsonglobaleditions.com

ISBN 10: 1-292-35193-4
ISBN 13: 978-1-292-35193-3

British Library Cataloguing-in-Publication Data
A catalogue record for this book is available from the British Library

1   21

Typeset in Palatino LT Pro 9.5/13 by SPi Global
Printed and bound by Neografia, Slovakia

Brief Contents

Contents

Preface

Welcome to the Eleventh Edition of *Biopsychology*! The Eleventh Edition of *Biopsychology* is a clear, engaging introduction to current biopsychological theory and research. It is intended for use as a primary course material in one- or two-semester courses in Biopsychology—variously titled Biopsychology, Physiological Psychology, Brain and Behavior, Psychobiology, Behavioral Neuroscience, or Behavioral Neurobiology.

The defining feature of *Biopsychology* is its unique combination of biopsychological science and personal, reader-oriented discourse. Instead of presenting the concepts of biopsychology in the usual fashion, the chapters address students directly and interweave the fundamentals of the field with clinical case studies, social issues, personal implications, useful metaphors, and memorable anecdotes.

Key Features in the Eleventh Edition

The following are features that have characterized recent editions of *Biopsychology* and have been maintained or expanded in this edition.

EMPHASIS ON BROAD THEMES   The emphasis of *Biopsychology* is "the big picture." Four broad themes are present throughout the chapters and a Themes Revisited section at the end of each chapter briefly summarizes how each theme was developed in that chapter. The four major themes provide excellent topics for essay assignments and exam questions.

EFFECTIVE USE OF CASE STUDIES   *Biopsychology* features many carefully selected case studies, which are highlighted in the chapters. These provocative cases stimulate interest, promote retention of the materials, and allow students to learn how biopsychological principles apply to the diagnosis and treatment of brain disorders.

REMARKABLE ILLUSTRATIONS   The illustrations in *Biopsychology* are special. Each one was conceptualized and meticulously designed to clarify and reinforce the chapter content by uniquely qualified scientists. John Pinel and his artist/designer wife, Maggie Edwards, created many of the original illustrations from previous editions.

FOCUS ON BEHAVIOR   In some biopsychological courseware, the coverage of neurophysiology, neurochemistry, and neuroanatomy subverts the coverage of behavioral research. *Biopsychology* gives top billing to behavior: It stresses that neuroscience is a team effort and that the unique contribution made by biopsychologists to this effort is their behavioral expertise.

EMPHASIS ON THE SCIENTIFIC METHOD   *Biopsychology* emphasizes the scientific method. It portrays the scientific method as a means of answering questions that is as applicable in daily life as in the laboratory. And *Biopsychology* emphasizes that being a scientist is fun.

DISCUSSION OF PERSONAL AND SOCIAL IMPLICATIONS   Several chapters of *Biopsychology*—particularly those on eating, sleeping, sex, and drug addiction—carry strong personal and social messages. In these chapters, students are encouraged to consider the relevance of biopsychological research to their lives outside the classroom.

ENGAGING, INSPIRING VOICES   Arguably the strongest pedagogical feature of *Biopsychology* is its personal tone. In the previous edition, Barnes and Pinel had addressed students directly and talked to them with warmth, enthusiasm, and good humor about recent advances in biopsychological science. This edition has not changed in this respect.

NEW! EMERGING THEMES   For this edition, Barnes and Pinel have identified and highlighted two "emerging themes" throughout the chapters: Themes that they feel are quickly emerging from the biopsychology literature. The Themes Revisited section at the end of each chapter briefly summarizes how each emerging theme was developed in that chapter. The two emerging themes provide excellent topics for essay assignments and exam questions.

New, Expanded, or Updated Coverage in the Eleventh Edition

Biopsychology remains one of the most rapidly progressing scientific fields. Like previous editions, the Eleventh Edition of *Biopsychology* has meticulously incorporated recent developments in the field—it contains more than 950 citations of articles or books that did not appear in the preceding edition. These recent developments have dictated changes to many parts of the chapters. The following list presents some of the content changes to this edition, organized by chapter.

CHAPTER 1: BIOPSYCHOLOGY AS A NEUROSCIENCE

- Introduction of emerging themes appearing in the chapters
- Five new citations

CHAPTER 2: EVOLUTION, GENETICS, AND EXPERIENCE

- Updated schematic illustration of how biopsychologists think about the biology of behavior
- Updated coverage and new key terms related to the topic of gene expression
- Expanded coverage of the topic of transgenerational epigenetics
- Simplified coverage of the evolution of humankind
- Three new key terms: *activators, repressors, hominins*
- Twenty new citations

CHAPTER 3: ANATOMY OF THE NERVOUS SYSTEM

- Updated and expanded coverage of the functions of glial cells
- Updated anatomical description of the basal ganglia
- Sixteen new citations

CHAPTER 4: NEURAL CONDUCTION AND SYNAPTIC TRANSMISSION

- Improved explanation and coverage of the action potential
- Coverage of the mechanical transmission of membrane potentials
- Two new key terms: *graded potentials, voltage-gated ion channels*
- Sixteen new citations

CHAPTER 5: THE RESEARCH METHODS OF BIOPSYCHOLOGY

- Expanded coverage of magnetic-field-based brain-imaging techniques
- Improved explanations of how MRI and fMRI work
- New section on ultrasound-based imaging techniques, such as functional ultrasound imaging
- Introduction of two new transcranial stimulation techniques: transcranial electrical stimulation and transcranial ultrasound stimulation
- Expanded coverage of magnetoencephalography
- Updated coverage of intracellular unit recording
- Expanded and comprehensive coverage of genetic methods, including coverage of gene-editing techniques like the CRISPR/Cas9 method
- Updated coverage on the various ways that fluorescent proteins are used in research
- New case study: The case of the vegetative patient

- New section on the study of functional connectivity
- Nine new key terms: *functional ultrasound imaging, transcranial electrical stimulation, transcranial ultrasound stimulation, gene knockin techniques, gene editing techniques, CRISPR/Cas9 method, resting-state fMRI, functional connectivity, functional connectome*
- Forty-two new citations

CHAPTER 6: THE VISUAL SYSTEM

- Updated and expanded coverage of modern research on visual system receptive fields
- Updated and expanded coverage of how the concept of a visual system receptive field is changing
- Updated coverage of research on the ventral and dorsal visual streams
- Updated and expanded coverage of the brain pathology associated with prosopagnosia
- One new key term: *occipital face area*
- Thirty-two new citations

CHAPTER 7: SENSORY SYSTEMS, PERCEPTION, AND ATTENTION

- New chapter title
- New chapter introduction, including coverage of some interesting exteroceptive senses only found in particular nonhuman species.
- Updated coverage of the subcortical auditory pathways
- Updated coverage of the organization and functions of the primary auditory cortex
- Updated coverage of the effects of auditory cortex damage
- Introduction of the thermal grid illusion—including a new figure
- Updated coverage of neuropathic pain
- Updated coverage of taste receptors
- Updated coverage of primary gustatory cortex organization
- New module on Perception
- Three new Check It Out features related to perception
- Updated coverage of the neural mechanisms of attention
- Twelve new key terms: *sensation, perception, periodotopy, thermal grid illusion, percept, perceptual decision making, bistable figures, phantom percepts, Charles Bonnet syndrome, binding problem, attentional gaze, frontal eye field*
- Sixty-one new citations

CHAPTER 8: THE SENSORIMOTOR SYSTEM

- Updated coverage of the primary motor cortex
- Updated coverage of the role of the cerebellum in sensorimotor function
- Updated and expanded coverage of the role of the basal ganglia in sensorimotor function
- More concise coverage of the descending motor pathways
- Updated coverage of the neuroplasticity associated with sensorimotor learning
- New key term: *movement vigor*
- Thirty-seven new citations

CHAPTER 9: DEVELOPMENT OF THE NERVOUS SYSTEM

- Updated coverage of the case of Genie
- Extensive updates to the coverage of stem cells and neurodevelopment
- New figure on the role of glia in neurodevelopment
- Updated coverage of the mechanisms of migration and aggregation of neurons
- Updated coverage of the chemoaffinity hypothesis
- Updated coverage of synapse formation
- Extensive updates to the module on early cerebral development in humans
- New case study written by a self-advocate with autism spectrum disorder
- New case study about the autistic savant Stephen Wiltshire, known by some as the "human camera"
- Coverage of the role of transcription-related errors in individuals with ASD
- Updated coverage of face processing in autism spectrum disorder
- Updated coverage of Williams syndrome, including coverage of face processing differences
- Four new key terms: *subventricular zone, radial glial cells, radial-glia-mediated migration, prenatal period*
- Eighty-three new citations

CHAPTER 10: BRAIN DAMAGE AND NEUROPLASTICITY

- Updated coverage of the mechanisms of ischemic stroke
- New section on traumatic brain injuries
- Coverage of mild traumatic brain injuries
- Updated coverage of chronic traumatic encephalopathy
- Updated discussion of causal factors in epilepsy

- Updated naming of the different types of seizures based on the new diagnostic criteria from the International League Against Epilepsy
- Extensive updates to the section on Parkinson's disease
- Updated and expanded coverage of Huntington's disease
- Updated and expanded coverage of multiple sclerosis
- Extensive updates to the section on Alzheimer's disease—including a new figure
- Five new key terms: *traumatic brain injury (TBI), closed-head TBI, subdural hematoma, mild TBI, alpha-synuclein*
- One hundred and forty-one new citations

CHAPTER 11: LEARNING, MEMORY, AND AMNESIA

- Updated coverage of H.M.
- Updated coverage of the amnesia of Korsakoff's syndrome
- New module: Amnesia after Traumatic Brain Injury: Evidence for Consolidation
- Updated coverage of the role of the hippocampus in consolidation
- Updated and improved coverage of the roles of grid cells
- Updated coverage of the relationship between place cells and grid cells
- New section: The hippocampus as a cognitive map
- Updated coverage of engram cells
- Coverage of the role of hippocampal-prefrontal connections in episodic memory
- Improved and updated coverage of long-term potentiation
- New section on nonsynaptic mechanisms of learning and memory
- Forty-eight new citations

CHAPTER 12: HUNGER, EATING, AND HEALTH

- New section: Evolution of Research on the Role of Hypothalamic Nuclei in Hunger and Satiety
- Updated and extended discussion of the role of hypothalamic circuits and gut peptides in hunger and eating
- Updated discussion of why some people gain weight, whereas others do not
- Updated coverage of leptin, insulin, and the arcuate melanocortin system
- Updated coverage of treatments for overeating
- New key term: *gut microbiome*
- Thirty new citations

CHAPTER 13: HORMONES AND SEX

- New module: Sexual development of brain and behavior
- Updated coverage of the aromatization hypothesis
- Extended and updated discussion of modern perspectives on sex differences in the brain
- Updated coverage of the role of gonadal hormones in female sexual behavior
- Extensive update to the module on sexual orientation and gender identity
- Four new key terms: *lesbian, transgender, gender identity, gender dysphoria*
- Forty-eight new citations

CHAPTER 14: SLEEP, DREAMING, AND CIRCADIAN RHYTHMS

- New module on dreaming
- Three new case studies directly related to the topic of dreaming
- Updated coverage of theories of dreaming
- Updated coverage of recuperation theories of sleep
- Updated coverage of the effects of sleep deprivation in humans
- Updated coverage of interventions for jet lag
- Updated coverage of the effect of shorter sleep times on health
- Two new figures
- One new key term: *lucid dreaming*
- One hundred and twenty-seven new citations

CHAPTER 15: DRUG USE, DRUG ADDICTION, AND THE BRAIN'S REWARD CIRCUITS

- Improved explanation of the relationship between drug withdrawal effects and conditioned compensatory responses
- Extensive update to coverage of nicotine
- Updated coverage of Korsakoff's syndrome
- Extensive update to coverage of marijuana
- Updated coverage of the history of cannabis use
- New discussion of the transgenerational epigenetic effects of drug taking
- Discussion of the current epidemic of opioid abuse
- Three new key terms: *smoking, vaping, drug craving*
- Eighty new citations

CHAPTER 16: LATERALIZATION, LANGUAGE, AND THE SPLIT BRAIN

- Updated coverage of sex differences in brain lateralization
- Updated coverage of anatomical asymmetries in the brain
- Updated coverage of the evolution of cerebral lateralization
- Updated coverage of the question of when cerebral lateralization evolved
- Twenty-seven new citations

CHAPTER 17: BIOPSYCHOLOGY OF EMOTION, STRESS, AND HEALTH

- Updated coverage of the facial feedback hypothesis
- Updated discussion of whether or not facial expressions are universal
- Thirty-two new citations

CHAPTER 18: BIOPSYCHOLOGY OF PSYCHIATRIC DISORDERS

- Major rewrite of this chapter
- Expanded coverage of all psychiatric disorders profiled in the chapter
- Coverage of the role of genetic, epigenetic, and neural factors for each psychiatric disorder
- Expanded and updated coverage of the discussion of the relative effectiveness of antidepressant medications
- Expanded coverage of theories of bipolar disorder
- Updated coverage of drug therapies for anxiety disorders
- Updated coverage of drug therapies for Tourette's disorder
- One hundred and seven new citations

Pedagogical Learning Aids

Biopsychology has several features expressly designed to help students learn and remember the material:

- **Scan Your Brain** study exercises appear within chapters at key transition points, where students can benefit most from pausing to consolidate material before continuing.
- **Check It Out** demonstrations apply biopsychological phenomena and concepts for students to experience themselves.
- **Themes Revisited** section at the end of each chapter summarizes the ways in which the book's four major themes, and its two emerging themes, relate to that chapter's subject matter.

- **Key Terms** appear in **boldface**, and other important terms of lesser significance appear in *italics*.
- **Appendixes** serve as convenient sources of additional information for students who want to expand their knowledge of selected biopsychology topics.

Ancillary Materials Available with *Biopsychology*

FOR INSTRUCTORS   Pearson Education is pleased to offer the following supplements to qualified adopters.

Test Bank (9781292352008)   The test bank for the Eleventh Edition of *Biopsychology* comprises more than 2,000 multiple-choice questions, including questions about accompanying brain images. Each item has answer justification, learning objective correlation, difficulty rating, and skill type designation, so that instructors can easily select appropriate questions for their tests.

Instructor's Manual (9781292351988)   The instructor's manual contains helpful teaching tools, including at-a-glance grids, activities and demonstrations for the classroom, handouts, lecture notes, chapter outlines, and other valuable course organization material for new and experienced instructors.

Video Embedded PowerPoint Slides (9781292401973) These slides, available in the Instructor's Resource Center, bring highlights of this edition of *Biopsychology* right into the classroom, drawing students into the lecture and providing engaging visuals, and include links to the videos referenced in each chapter.

Standard Lecture PowerPoint Slides (9781292351995) These accessible slides have a more traditional format, with excerpts of the chapter material and artwork, and are available online at www.pearsonglobaleditions.com.

Acknowledgments

Seven people deserve special credit for helping us create this edition of *Biopsychology*: Maggie Edwards, Linnea Ritland, Chandra Jade, Olivia Sorley, Natasha Au, Zamina Mithani, and Kim Nipp. Maggie is an artist/designer/writer/personal trainer who is John's partner in life. She is responsible for the original designs of most of the illustrations that appear in the chapters. Linnea, Chandra, Olivia, Natasha, Zamina, and Kim are six remarkable alumni of the University of British Columbia; Linnea helped with the drawing, editing, and voiceovers for the Chalk It Up Animations; Chandra helped with the editing of some of the Chalk It Up Animations; Olivia helped with the drawing of some of the Chalk It Up Animations; Natasha and Zamina helped with collecting articles as part of the research that went into this edition; and Kim co-designed the cover of the text with Barnes.

Pearson Education did a remarkable job of producing the original *Biopsychology*. They shared the dream of a solution that meets the highest standards of pedagogy but is also personal, attractive, and enjoyable. Now they have stepped up to support the conversion of *Biopsychology* to electronic format. Special thanks also go to Kelli Strieby, Matthew Summers, and Lisa Mafrici at Pearson; Marita Bley for her development, editing, and coordination; and Annemarie Franklin at SPi Global for coordinating the production—an excruciatingly difficult and often thankless job.

We thank the following instructors for providing us with reviews of various editions of *Biopsychology*. Their comments have contributed substantially to the evolution of this edition:

L. Joseph Acher, Baylor University
Nelson Adams, Winston-Salem State University
Marwa Azab, Golden West College
Michael Babcock, Montana State University–Bozeman
Ronald Baenninger, College of St. Benedict
Mark Basham, Regis University
Carol Batt, Sacred Heart University
Noel Jay Bean, Vassar College
Patricia Bellas, Irvine Valley College
Danny Benbasset, George Washington University
Thomas Bennett, Colorado State University
Linda Brannon, McNeese State University
Peter Brunjes, University of Virginia
John Bryant, Bowie State University
Michelle Butler, United States Air Force Academy
Donald Peter Cain, University of Western Ontario
Deborah A. Carroll, Southern Connecticut State University
John Conklin, Camosun College
Sherry Dingman, Marist College
Michael A. Dowdle, Mt. San Antonio College
Doug Engwall, Central Connecticut State University
Gregory Ervin, Brigham Young University
Robert B. Fischer, Ball State University
Allison Fox, University of Wollongong
Michael Foy, Loyola Marymount University
Ed Fox, Purdue University
Thomas Goettsche, SAS Institute, Inc.
Arnold M. Golub, California State University–Sacramento
Nakia Gordon, Marquette University
Mary Gotch, Solano College
Jeffrey Grimm, Western Washington University
Kenneth Guttman, Citrus College
Melody Smith Harrington, St. Gregory's University

Christopher Hayashi, Southwestern College
Theresa D. Hernandez, University of Colorado
Cindy Ellen Herzog, Frostburg State University
Peter Hickmott, University of California–Riverside
Michael Jarvinen, Emmanuel College
Tony Jelsma, Atlantic Baptist University
Roger Johnson, Ramapo College
Chris Jones, College of the Desert
John Jonides, University of Michigan
Jon Kahane, Springfield College
Craig Kinsley, University of Richmond
Ora Kofman, Ben-Gurion University of the Negev
Louis Koppel, Utah State University
Shannon Kundey, Hood College
Maria J. Lavooy, University of Central Florida
Victoria Littlefield, Augsburg College
Eric Littman, University of Cincinnati
Linda Lockwood, Metropolitan State College of Denver
Charles Malsbury, Memorial University
Michael R. Markham, Florida International University
Vincent Markowski, State University of New York–Geneseo
Michael P. Matthews, Drury College
Corinne McNamara, Kennesaw State University
Lin Meyers, California State University–Stanislaus
Maura Mitrushina, California State University, Northridge
Russ Morgan, Western Illinois University
Henry Morlock, SUNY–Plattsburgh
Caroline Olko, Nassau Community College
Lauretta Park, Clemson University
Ted Parsons, University of Wisconsin–Platteville
Jim H. Patton, Baylor University
Edison Perdorno, Minnesota State University
Michael Peters, University of Guelph

Michelle Pilati, Rio Hondo College
Joseph H. Porter, Virginia Commonwealth University
David Robbins, Ohio Wesleyan University
Dennis Rodriguez, Indiana University–South Bend
Margaret G. Ruddy, College of New Jersey
Jeanne P. Ryan, SUNY–Plattsburgh
Jerome Siegel, David Geffen School of Medicine, UCLA
Angela Sikorski, Texas A&M University–Texarkana
Patti Simone, Santa Clara University
Ken Sobel, University of Central Arkansas
David Soderquist, University of North Carolina at Greensboro
Michael Stoloff, James Madison University
Stuart Tousman, Rockford College
Dallas Treit, University of Alberta
Margaret Upchurch, Transylvania University
Dennis Vincenzi, University of Central Florida
Ashkat Vyas, Hunter College
Christine Wagner, University at Albany
Linda Walsh, University of Northern Iowa
Charles Weaver, Baylor University
David Widman, Juniata College
Jon Williams, Kenyon College
David Yager, University of Maryland
H.P. Ziegler, Hunter College

Global Edition Acknowledgments

Pearson would like to thank the following people for their work on the Global Edition:

Contributor
Antonia Ypsilanti, Sheffield Hallam University

Reviewers
Patrick Bourke, University of Lincoln
Kimberly Smith, University of Surrey

To the Student

We have tried to make *Biopsychology* different with content that includes clear, concise, and well-organized explanations of the key points but is still interesting to read—material from which you might suggest suitable sections to an interested friend or relative. To accomplish this goal, we thought about what kind of materials we would have liked when we were students, and we decided to avoid the stern formality and ponderous style of conventional science writing and to focus on ideas of relevance to your personal life.

We want *Biopsychology* to have a relaxed and personal style. In order to accomplish this, we imagined that we were chatting with you as we wrote and that we were telling you—usually over a glass of something—about the interesting things that go on in the field of biopsychology. Imagining these chats kept our writing from drifting back into conventional "textbookese," and it never let us forget that we were writing these materials for you.

As we write these words, we have finished work on this new edition, and now we are waiting with great excitement for *Biopsychology* to be released. There is more excitement around this edition than there has been since the first edition appeared in 1990—this time the excitement is about the release of *Biopsychology* in an online-only format and all the opportunities that it creates for effective teaching. We really hope that you will find this new format easy to use, interesting, and, most importantly, an effective learning tool.

We hope that *Biopsychology* teaches you much of relevance to your personal life and that reading it generates in you the same positive feelings that writing it did in us.

About the Authors

JOHN PINEL obtained his Ph.D. from McGill University in Montreal and worked briefly at the Massachusetts Institute of Technology before taking a faculty position at the University of British Columbia in Vancouver, where he is currently Professor Emeritus. Professor Pinel is an award-winning teacher and the author of more than 200 scientific papers. However, he feels that *Biopsychology* is his major career-related accomplishment: "It ties together everything I love about my job: students, teaching, writing, and research."

STEVEN BARNES obtained his Ph.D. from the University of British Columbia. He then worked as a postdoctoral fellow—first in the Department of Epileptology at the University of Bonn and then in the School of Interactive Arts and Technology at Simon Fraser University. He is currently an Associate Professor of Teaching, and Associate Head of Undergraduate Affairs, in the Department of Psychology at the University of British Columbia.

Steven is well-regarded for his work related to online learning technologies (e.g., the Tapestry Project; see tapestry-tool.com), student mental health and wellbeing, and bipolar disorder (BD). Steven co-directs the Collaborative RESearch Team to study psychosocial issues in BD (CREST. BD, see crestbd.ca), a BD research and knowledge exchange network, which received the 2018 Canadian Institutes for Health Research Gold Leaf Prize for Patient Engagement, Canada's most prestigious recognition for patient engagement in research across all health disciplines.

Steven is the recipient of multiple institutional awards for his teaching, including the prestigious Killam Teaching Prize and the 3M National Teaching Fellowship—the top national award given for teaching in any discipline in any postsecondary institution in Canada.

When he isn't teaching, writing, or doing research, he engages in the production of traditional pieces of visual art as well as interactive electronic artworks—some of which have been exhibited at prominent international venues. He sees his involvement in the creation of this new edition of *Biopsychology* as a complement to everything he loves to do: teaching, writing, visual and interactive art, and research.

Chapter 1
Biopsychology as a Neuroscience

What Is Biopsychology, Anyway?

Image Source/Alamy Stock Photo

 ## Chapter Overview and Learning Objectives

What Is Biopsychology?	**LO 1.1**	Define and discuss what is meant by *biopsychology*.
	LO 1.2	Discuss the origins of the field of biopsychology.
	LO 1.3	List the six fields of neuroscience that are particularly relevant to biopsychological inquiry.
What Types of Research Characterize the Biopsychological Approach?	**LO 1.4**	Compare the advantages and disadvantages of humans and nonhumans as subjects in biopsychological research.
	LO 1.5	Compare experiments, quasiexperimental studies, and case studies, emphasizing their utility in the study of causal effects.
	LO 1.6	Compare pure and applied research.
What Are the Divisions of Biopsychology?	**LO 1.7**	Describe the division of biopsychology known as physiological psychology.

LO 1.8   Describe the division of biopsychology known as psychopharmacology.

LO 1.9   Describe the division of biopsychology known as neuropsychology.

LO 1.10 Describe the division of biopsychology known as psychophysiology.

LO 1.11 Describe the division of biopsychology known as cognitive neuroscience.

LO 1.12 Describe the division of biopsychology known as comparative psychology.

How Do Biopsychologists Conduct Their Work?

LO 1.13 Explain how converging operations has contributed to the study of Korsakoff's syndrome.

LO 1.14 Explain scientific inference with reference to research on eye movements and the visual perception of motion.

Thinking Critically about Biopsychological Claims

LO 1.15 Define critical thinking and evaluate biopsychological claims.

The appearance of the human brain is far from impressive (see Figure 1.1). The human brain is a squishy, wrinkled, walnut-shaped hunk of tissue weighing about 1.3 kilograms. It looks more like something you might find washed up on a beach than one of the wonders of the world—which it surely is. Despite its disagreeable appearance, the human brain is an amazingly intricate network of **neurons** (cells that receive and transmit electrochemical signals) and many other cell types. Contemplate for a moment the complexity of your own brain's neural circuits. Consider the 90 billion neurons in complex array (Walløe, Pakkenberg & Fabricius, 2014), the estimated 100 trillion connections among them, and the almost infinite number of paths that neural signals can follow through this morass (Zimmer, 2011). The complexity of the human brain is hardly surprising, considering what it can do. An organ capable of creating a *Mona Lisa*, an artificial limb, and a supersonic aircraft; of traveling to the moon and to the depths of the sea; and of experiencing the wonders of an alpine sunset, a newborn infant, and a reverse slam dunk *must* be complex. Paradoxically, **neuroscience** (the scientific study of the nervous system) may prove to be the brain's ultimate challenge: Does the brain have the capacity to understand something as complex as itself (see Gazzaniga, 2010)?

Neuroscience comprises several related disciplines. The primary purpose of this chapter is to introduce you to one of them: biopsychology. Each of this chapter's five modules characterizes the neuroscience of biopsychology in a different way. However, before you proceed to the body of

this chapter, we would like to tell you about the case of Jimmie G. (Sacks, 1985), which will give you a taste of the interesting things that lie ahead.

Figure 1.1 The human brain: Appearances can be deceiving!

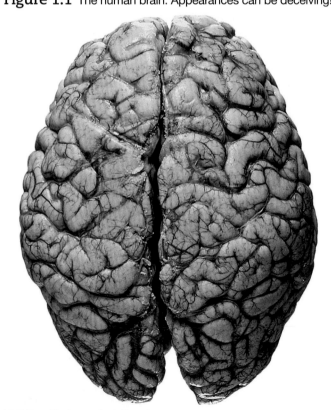

UHB Trust/The Image Bank/Getty Images

The Case of Jimmie G., the Man Frozen in Time

Jimmie G. was a friendly 49-year-old. He liked to chat about his school days and his time in the navy, both of which he could describe in remarkable detail. Jimmie was an intelligent man with superior abilities in math and science. So why was he a patient in a neurological ward?

When Jimmie talked about his past, there were hints of his problem. When he talked about his school days, he used the past tense; but when he recounted his early experiences in the navy, he switched to the present tense. More worrisome was that he never talked about anything that happened to him after his time in the navy.

Jimmie was tested by eminent neurologist Oliver Sacks, and a few simple questions revealed a curious fact: Jimmie believed he was 19. When asked to describe what he saw in a mirror, Jimmie became so frantic and confused that Dr. Sacks immediately took the mirror out of the room.

Returning a few minutes later, Dr. Sacks was greeted by a once-again cheerful Jimmie, who acted as if he had never seen Sacks before. Indeed, even when Sacks suggested they had met recently, Jimmie was certain they had not.

Then Dr. Sacks asked where Jimmie thought he was. Jimmie replied that all the beds and patients made him think that the place was a hospital. But he couldn't understand why he would be in a hospital. He was afraid that he might have been admitted because he was sick but didn't know it.

Further testing confirmed what Dr. Sacks feared. Although Jimmie had good sensory, motor, and cognitive abilities, he had one terrible problem: He forgot everything that was said or shown to him within a few seconds. Basically, Jimmie could not remember anything that had happened to him since his early 20s, and he was not going to remember anything that happened to him for the rest of his life. Dr. Sacks was stunned by the implications of Jimmie's condition.

Jimmie's situation was heart-wrenching. Unable to form new lasting memories, he was, in effect, a man frozen in time, a man without a recent past and no prospects for a future, stuck in a continuous present, lacking any context or meaning.

Remember Jimmie G.; you will encounter him again later in this chapter.

Four Major Themes of This Text

You will learn many new facts in this text—new findings, concepts, terms, and the like. But more importantly, many years from now, long after you have forgotten most of those facts, you will still be carrying with you productive new ways of thinking. We have selected four of these for special emphasis: Thinking Creatively, Clinical Implications, the Evolutionary Perspective, and Neuroplasticity.

THINKING CREATIVELY ABOUT BIOPSYCHOLOGY. We are all fed a steady diet of biopsychological information, misinformation, and opinion—by television, newspapers, the Internet, friends, relatives, teachers, and so on. As a result, you likely already hold strong views about many of the topics you will encounter in this text. Because these preconceptions are shared by many biopsychological researchers, they have often impeded scientific progress, and some of the most important advances in biopsychological science have been made by researchers who have managed to overcome the restrictive effects of conventional thinking and have taken creative new approaches. Indeed, **thinking creatively** (thinking in productive, unconventional ways) is the cornerstone of any science. In this text, we describe research that involves thinking "outside the box," we try to be creative in our analysis of the research we are presenting, or we encourage you to base your thinking on the evidence rather than on widely accepted views.

CLINICAL IMPLICATIONS. **Clinical** (pertaining to illness or treatment) considerations are woven through the fabric of biopsychology. There are two aspects to the clinical implications theme: (1) much of what biopsychologists learn about the functioning of a healthy brain comes from studying dysfunctional brains; and (2) many of the discoveries of biopsychologists have relevance for the treatment of brain dysfunction. One of our major focuses is on the interplay between brain dysfunction and biopsychological research.

THE EVOLUTIONARY PERSPECTIVE. Although the events that led to the evolution of the human species can never be determined with certainty, thinking of the environmental pressures that likely led to the evolution of our brains and behavior often leads to important biopsychological insights. This approach is called the **evolutionary perspective**. An important component of the evolutionary perspective is the comparative approach (trying to understand biological phenomena by comparing them in different species). Throughout this text, you will find that we humans have learned much about ourselves by studying species that are related to us through evolution. Indeed, the evolutionary approach has proven to be one of the cornerstones of modern biopsychological inquiry.

NEUROPLASTICITY. Until the early 1990s, most neuroscientists thought of the brain as a three-dimensional array of neural elements "wired" together in a massive network of circuits. The complexity of this "wiring diagram" of the brain was staggering, but it failed to capture one of the brain's most important features. In the past four decades, research has clearly demonstrated that the adult brain is not a static network of neurons: It is a plastic (changeable) organ that continuously grows and changes in response to an individual's environment and experiences. The discovery of **neuroplasticity** is arguably the single most influential discovery in modern neuroscience. As you will learn, it is a major component of many areas of biopsychological research.

You have probably heard of neuroplasticity. It is a hot topic in the popular media, where it is upheld as a panacea: A means of improving brain function or recovering from brain dysfunction. However, contrary to popular belief, the plasticity of the human brain is not always beneficial. For example, it also contributes to various forms of brain dysfunction (e.g., Tomaszcyk et al., 2014). Later on, you will see examples of both the positive and the negative sides of neuroplasticity.

Emerging Themes of This Text

As you read through this text you will start to see other themes in addition to the ones we outlined for you in the previous section. Many of them you will spot on your own. Here we highlight two "emerging" themes: themes that could become major themes in future editions of this text.

THINKING ABOUT EPIGENETICS. Most people believe their genes (see Chapter 2) control the characteristics they are born with, the person they become, and the qualities of their children and grandchildren. In this text, you will learn that genes are only a small part of what determines who you are. Instead, you are the product of ongoing interactions between your genes and your experiences—such interactions are at the core of a field of study known as **epigenetics**. But epigenetics isn't just about you: We now know that the experiences you have during your lifetime can be passed on to future generations. This is a fundamentally different way of thinking about who we are and how we are tied to both our ancestors and descendants. Epigenetics is currently having a major influence on biopsychological research.

CONSCIOUSNESS. As you will see, this text also examines different aspects of **consciousness** (the perception or awareness of some aspect of one's self or the world) from a biopsychological perspective. Indeed, one major goal of biopsychological research is to establish a better understanding of the neural correlates of consciousness (see Ward, 2013; Blackmore, 2018). To give you a taste of this emerging theme, you will soon appreciate that (1) we are not consciously aware of much of the information we receive from our environments, (2) there are many different states of consciousness, and (3) there can be dramatic alterations in consciousness as a result of brain dysfunction.

What Is Biopsychology?

This module introduces you to the discipline of biopsychology. We begin by exploring the definition and origins of biopsychology. Next, we examine how biopsychology is related to the various other disciplines of neuroscience.

Defining Biopsychology

LO 1.1 Define and discuss what is meant by
**   *biopsychology*.**

Biopsychology is the scientific study of the biology of behavior (see Dewsbury, 1991). Some refer to this field as *psychobiology, behavioral biology,* or *behavioral neuroscience;* but we prefer the term *biopsychology* because it denotes a biological approach to the study of psychology rather than a psychological approach to the study of biology: Psychology commands center stage in this text. *Psychology* is the scientific study of behavior—the scientific study of all overt activities of the organism as well as all the internal processes that are presumed to underlie them (e.g., learning, memory, motivation, perception, emotion).

What Are the Origins of Biopsychology?

LO 1.2 Discuss the origins of the field of
**   biopsychology.**

The study of the biology of behavior has a long history, but biopsychology did not develop into a major neuroscientific discipline until the 20th century. Although it is not possible to specify the exact date of biopsychology's birth, the publication of *The Organization of Behavior* in 1949 by Donald Hebb played a key role in its emergence (see Brown & Milner, 2003). In his book, Hebb developed the first comprehensive theory of how complex psychological phenomena, such as perceptions, emotions, thoughts, and memories, might be produced by brain activity. Hebb's theory did much to discredit the view that psychological functioning is too complex to have its roots in the physiology and chemistry of the brain. Hebb based his theory on experiments involving both human and nonhuman animals, on clinical case studies, and on logical arguments developed from his own insightful observations of daily life. This eclectic approach has become a hallmark of biopsychological inquiry.

In comparison to physics, chemistry, and biology, biopsychology is an infant—a healthy, rapidly growing infant, but an infant nonetheless. In this text, you will reap the benefits of biopsychology's youth. Because biopsychology does not have a long history, you will be able to move quickly to the excitement of modern research.

How Is Biopsychology Related to the Other Disciplines of Neuroscience?

LO 1.3   List the six fields of neuroscience that are particularly relevant to biopsychological inquiry.

Neuroscience is a team effort, and biopsychologists are important members of the team (see Albright, Kandel, & Posner, 2000; Kandel & Squire, 2000). Biopsychology can be further characterized by its relation to other neuroscientific disciplines.

Biopsychologists are neuroscientists who bring to their research a knowledge of behavior and of the methods of behavioral research. It is their behavioral orientation and expertise that make their contribution to neuroscience unique (see Cacioppo & Decety, 2009). You will be able to better appreciate the importance of this contribution if you consider that the ultimate purpose of the nervous system is to produce and control behavior (see Grillner & Dickinson, 2002). Think about it.

Biopsychology is an integrative discipline. Biopsychologists draw together knowledge from the other neuroscientific disciplines and apply it to the study of behavior. The following are a few of the disciplines of neuroscience that are particularly relevant to biopsychology:

- **Neuroanatomy.** The study of the structure of the nervous system (see Chapter 3).
- **Neurochemistry.** The study of the chemical bases of neural activity (see Chapters 4 and 15).
- **Neuroendocrinology.** The study of interactions between the nervous system and the endocrine system (see Chapters 13 and 17).
- **Neuropathology.** The study of nervous system dysfunction (see Chapters 10 and 18).
- **Neuropharmacology.** The study of the effects of drugs on neural activity (see Chapters 4, 15, and 18).
- **Neurophysiology.** The study of the functions and activities of the nervous system (see Chapter 4).

What Types of Research Characterize the Biopsychological Approach?

Biopsychology is broad and diverse. Biopsychologists study many different phenomena, and they approach their research in many different ways. This module discusses three major dimensions along which biopsychological research may vary: It can involve either human or nonhuman subjects, it can take the form of either formal experiments or nonexperimental studies, and it can be either pure or applied.

Human and Nonhuman Subjects

LO 1.4   Compare the advantages and disadvantages of humans and nonhumans as subjects in biopsychological research.

Both human and nonhuman animals are the subjects of biopsychological research. Of the nonhumans, mice and rats are the most common subjects; however, cats, dogs, and nonhuman primates are also commonly studied.

Humans have several advantages over other animals as experimental subjects of biopsychological research: They can follow instructions, they can report their subjective experiences, and their cages are easier to clean. Of course, we are joking about the cages, but the joke does serve to draw attention to one advantage humans have over other species of experimental subjects: Humans are often cheaper. Because only the highest standards of animal care are acceptable, the cost of maintaining an animal laboratory can be prohibitive for all but the most well-funded researchers.

Of course, the greatest advantage humans have as subjects in a field aimed at understanding the intricacies of human brain function is that they have human brains. In fact, you might wonder why biopsychologists would bother studying nonhuman subjects at all. The answer lies in the evolutionary continuity of the brain. The brains of humans are similar in fundamental ways to the brains of other mammals—they differ mainly in their overall size and the extent of their cortical development. In other words, the differences between the brains of humans and those of related species are more quantitative than qualitative, and thus many of the principles of human brain function can be clarified by the study of nonhumans (see Hofman, 2014; Katzner & Weigelt, 2013; Krubitzer & Stolzenberg, 2014).

One major difference between human and nonhuman subjects is that humans volunteer to be subjects. To emphasize this point, human subjects are more commonly referred to as *participants* or *volunteers*.

Nonhuman animals have three advantages over humans as subjects in biopsychological research. The first is that the brains and behavior of nonhuman subjects are simpler than those of human participants. Hence, the study of nonhuman species is often more likely to reveal fundamental brain–behavior interactions. The second advantage is that insights frequently arise from the **comparative approach**, the study of biological processes by comparing different species. For example, comparing the behavior of species that do not have a cerebral cortex with the behavior of species that do can provide valuable clues about cortical function. The third advantage is that it is possible to conduct research on laboratory animals that, for ethical reasons, is not possible with human participants. This is not to say that the study of nonhuman animals is not governed by a strict code of ethics (see Blakemore et al., 2012)—it is. However, there are fewer ethical constraints on the study of laboratory species than on the study of humans.

In our experience, most biopsychologists display considerable concern for their subjects, whether they are of their own species or not; however, ethical issues are not left to the discretion of the individual researcher. All biopsychological research, whether it involves human participants or nonhuman subjects, is regulated by independent committees according to strict ethical guidelines: "Researchers cannot escape the logic that if the animals we observe are reasonable models of our own most intricate actions, then they must be respected as we would respect our own sensibilities" (Ulrich, 1991, p. 197).

If you are concerned about the ethics of biopsychological research on nonhuman animals, you aren't alone. Both of us wrestle with various aspects of it. For example, a recurring concern we both have is whether the potential benefits of a research study outweigh the stress induced in the nonhuman subjects.

When people are asked for their opinion on nonhuman animal research, most fall into one of two camps: (1) Those in support of animal research—if and only if both the suffering of animals is minimized and the potential benefits to humankind cannot be obtained by other methods, or (2) those that are opposed to animal research—because it causes undue stress that is not outweighed by the potential benefits to humankind.

Journal Prompt 1.1

What are your initial feelings about biopsychological research on nonhuman animals? If you are sympathetic to one of the two aforementioned camps, explain your reasoning.

Because biopsychological research using nonhuman subjects is controversial, it first has to be approved by a panel of individuals from a variety of backgrounds and with different world views. These *nonhuman animal ethics committees* are tasked with very difficult decisions. Accordingly, it is usually the case that these committees will ask the researchers proposing a particular study to provide additional information or further justification before they approve their research.

Nonhuman animal ethics committees emphasize consideration of the so-called "three R's": Reduction, Refinement, and Replacement. Reduction refers to efforts to reduce the numbers of animals used in research. Refinement refers to refining research studies or the way animals are cared for, so as to reduce suffering. Providing animals with better living conditions is one example of refinement.

Finally, replacement refers to the replacing of studies using animal subjects with alternate techniques, such as experimenting on cell cultures or using computer models.

One of the earliest examples of replacement is the now ubiquitous crash-test dummy in the auto industry. Prior to the advent of the crash test dummy, live pigs were sometimes used as passengers in automobile crash tests. This example of replacement makes an important point about how notions of what is ethically acceptable in animal experimentation are in constant flux: Now that dummies are a viable alternative, nobody would be in favor of using pigs for crash tests. The recent development of complex computer models of nonhuman and human brains (see Frackowiak & Markram, 2015) might change the very nature of biopsychological research in your lifetime.

Experiments and Nonexperiments

LO 1.5   Compare experiments, quasiexperimental studies, and case studies, emphasizing their utility in the study of causal effects.

Biopsychological research involves both experiments and nonexperimental studies. Two common types of nonexperimental studies are quasiexperimental studies and case studies.

EXPERIMENTS.   The experiment is the method used by scientists to study causation, that is, to find out what causes what. As such, it has been almost single-handedly responsible for the knowledge that is the basis for our modern way of life. It is paradoxical that a method capable of such complex feats is so simple. To conduct an experiment involving living subjects, the experimenter first designs two or more conditions under which the subjects will be tested. Usually, a different group of subjects is tested under each condition (**between-subjects design**), but sometimes it is possible to test the same group of subjects under each condition (**within-subjects design**). The experimenter assigns the subjects to conditions, administers the treatments, and measures the outcome in such a way that there is only one relevant difference between the conditions being compared. This difference between the conditions is called the **independent variable**. The variable measured by the experimenter to assess the effect of the independent variable is called the **dependent variable**. If the experiment is done correctly, any differences in the dependent variable between the conditions must have been caused by the independent variable.

Why is it critical that there be no differences between conditions other than the independent variable? The reason is that when there is more than one difference that could affect the dependent variable, it is difficult to determine whether it was the independent variable or the unintended difference—called a **confounded variable**—that led to the observed effects on the dependent variable. Although the experimental method is conceptually simple, eliminating all confounded variables can be quite difficult. Readers of research papers must be constantly on the alert for confounded variables that have gone unnoticed by the experimenters.

An experiment by Lester and Gorzalka (1988) illustrates the prevention of confounded variables with good experimental design. The experiment was a demonstration of the **Coolidge effect** (see Lucio et al., 2014; Tlachi-López et al., 2012). The Coolidge effect is the fact that a copulating male who becomes incapable of continuing to copulate with one sex partner can often recommence copulating with a new sex partner (see Figure 1.2). Before your imagination

Figure 1.2 President Calvin Coolidge and Mrs. Grace Coolidge. Many students think the Coolidge effect is named after a biopsychologist named Coolidge. In fact, it is named after President Calvin Coolidge, of whom the following story is told. (If the story isn't true, it should be.)

During a tour of a poultry farm, Mrs. Coolidge inquired of the farmer how his farm managed to produce so many eggs with such a small number of roosters. The farmer proudly explained that his roosters performed their duty dozens of times each day.

"Perhaps you could point that out to Mr. Coolidge," replied the First Lady in a pointedly loud voice.

The President, overhearing the remark, asked the farmer, "Does each rooster service the same hen each time?"

"No," replied the farmer, "there are many hens for each rooster."

"Perhaps you could point that out to Mrs. Coolidge," replied the President.

Bettmann/Getty Images

starts running wild, we should mention that the subjects in Lester and Gorzalka's experiment were hamsters, not university students.

Lester and Gorzalka argued that the Coolidge effect had not been demonstrated in females because it is more difficult to conduct well-controlled Coolidge-effect experiments with females—not because females do not display a Coolidge effect. The confusion, according to Lester and Gorzalka, stemmed from the fact that the males of most mammalian species become sexually fatigued more readily than the females. As a result, attempts to demonstrate the Coolidge effect in females are almost always confounded by the fatigue of the males. When, in the midst of copulation, a female is provided with a new sex partner, the increase in her sexual receptivity could be either a legitimate Coolidge effect or a reaction to the greater vigor of the new male. Because female mammals usually display little sexual fatigue, this confounded variable is not a serious problem in demonstrations of the Coolidge effect in males.

Lester and Gorzalka devised a clever procedure to control for this confounded variable. At the same time a female subject was copulating with one male (the familiar male), the other male to be used in the test (the unfamiliar male) was copulating with another female. Then both males were given a rest while the female was copulating with a third male. Finally, the female subject was tested with either the familiar male or the unfamiliar male. The dependent variable was the amount of time that the female displayed **lordosis** (the arched-back, rump-up, tail-diverted posture of female rodent sexual receptivity) during each sex test. As Figure 1.3 illustrates, the females responded more vigorously to the unfamiliar males than they did to the familiar males during the third test, despite the fact that both the unfamiliar and familiar males were equally fatigued and both mounted the females with equal vigor. The purpose of this example—in case you have forgotten—is to illustrate the critical role played by good experimental design in eliminating confounded variables.

QUASIEXPERIMENTAL STUDIES. It is not possible for biopsychologists to bring the experimental method to bear on all problems of interest to them. Physical or ethical impediments frequently make it impossible to assign subjects to particular conditions or to administer particular conditions to the subjects who have been assigned to them. For example, experiments assessing whether frequent marijuana use causes brain dysfunction are not feasible because it would be unethical to assign a human to a condition that involves years of frequent marijuana use. (Some of you may be more concerned about the ethics of assigning humans to a control condition that involves many years of *not* getting high.) In such prohibitive situations, biopsychologists sometimes conduct **quasiexperimental studies**—studies of groups of subjects who have been exposed to the conditions of interest in

Figure 1.3 The experimental design and results of Lester and Gorzalka (1988). On the third test, the female hamsters were more sexually receptive to an unfamiliar male than they were to the male with which they had copulated on the first test.

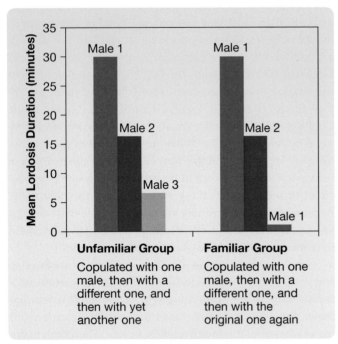

Based on Lester, G. L. L., & Gorzalka, B. B. (1988)

the real world. These studies have the appearance of experiments, but they are not true experiments because potential confounded variables have not been controlled—for example, by the random assignment of subjects to conditions.

In the popular press, quasiexperiments are often confused with experiments. Not a week goes by where one of us doesn't read a news article about how an "experiment" has shown something in human participants, when in reality the so-called experiment is actually a quasiexperiment.

Understanding the distinction between quasiexperiments and experiments is very important. Experiments can tell us whether an independent variable causes a change in a dependent variable (assuming that the experimenter has controlled for all confounding variables); quasiexperiments can tell us only that two variables are correlated with one another. For example, in interpreting experiments we can reach causal conclusions like "frequent alcohol consumption causes brain damage." In contrast, quasiexperimental studies can tell us only that "frequent alcohol use is associated with brain damage."

The importance of thinking clearly about quasiexperimental studies is illustrated by a study that compared 100 detoxified males who had previously been heavy drinkers of alcohol with 50 male nondrinkers (Acker et al., 1984). Overall, those who had been heavy drinkers performed more poorly on various tests of perceptual, motor, and cognitive ability, and their brain scans revealed extensive brain damage. Although this might seem like an experiment, it is not. It is a quasiexperimental study: Because the participants themselves decided which group they would be in—by drinking alcohol or not—the researchers had no means of ensuring that exposure to alcohol was the only variable that distinguished the two groups. Can you think of differences other than exposure to alcohol that could reasonably be expected to exist between a group of heavy drinkers and a group of abstainers—differences that could have contributed to the neuroanatomical or intellectual differences that were observed between them? There are several. For example, heavy drinkers as a group tend to be more poorly educated, more prone to accidental head injury, more likely to use other drugs, and more likely to have poor diets. Accordingly, although quasiexperimental studies have revealed that people who are heavy drinkers tend to have more brain damage than abstainers, such studies cannot prove that it was caused by the alcohol.

Have you forgotten the case of Jimmie G.? Jimmie's condition was a product of heavy alcohol consumption.

CASE STUDIES. Studies that focus on a single subject, or very small number of subjects, are called **case studies**. Such studies are rarely concerned with having control subjects. Rather, their focus is on providing a more in-depth picture than that provided by an experiment or a quasiexperimental study, and they are an excellent source of testable hypotheses. However, there is a major problem with all case studies: their **generalizability**—the degree to which their results can be applied to other cases. Because individuals differ from one another in both brain function and behavior, it is important to be skeptical of any biopsychological theory based entirely on a few case studies.

Pure and Applied Research

LO 1.6 Compare pure and applied research.

Biopsychological research can be either pure or applied. Pure research and applied research differ in a number of respects, but they are distinguished less by their own attributes than by the motives of the researchers involved in their pursuit. **Pure research** is motivated primarily by the curiosity of the researcher—it is done solely for the purpose of acquiring knowledge. In contrast, **applied research** is intended to bring about some direct benefit to humankind.

Many scientists believe that pure research will ultimately prove to be of more practical benefit than applied research. Their view is that applications flow readily from an understanding of basic principles and that attempts to move directly to application without first gaining a basic understanding are shortsighted. Of course, it is not necessary for a research project to be completely pure or completely applied; many research programs have elements of both approaches. Moreover, pure research often becomes the topic of **translational research**: research that aims to translate the findings of pure research into useful applications for humankind (see Howells, Sena, & Macleod, 2014).

One important difference between pure and applied research is that pure research is more vulnerable to the vagaries of political regulation because politicians and the voting public have difficulty understanding why research of no immediate practical benefit should be supported. If the decision were yours, would you be willing to grant millions of dollars to support the study of squid *motor neurons* (neurons that control muscles), learning in recently hatched geese, the activity of single nerve cells in the visual systems of monkeys, the hormones released by the *hypothalamus* (a small neural structure at the base of the brain) of pigs and sheep, or the functions of the *corpus callosum* (the large neural pathway that connects the left and right halves of the brain)? Which, if any, of these projects would you consider worthy of support? Each of these seemingly esoteric projects was supported, and each earned a Nobel Prize.

Table 1.1 provides a timeline of some of the Nobel Prizes awarded for research related to the brain and behavior. The purpose of this table is to give you a general sense of the official recognition that behavioral and brain research has received, not to have you memorize the list. You will learn later in the chapter that, when it comes to evaluating science, the Nobel Prize Committees have not been infallible.

What Are the Divisions of Biopsychology?

As you have just learned, biopsychologists conduct their research in a variety of fundamentally different ways. Biopsychologists who take the same approaches to their research tend to publish their research in the same journals, attend the same scientific meetings, and belong to the same professional societies. The particular approaches to biopsychology that have flourished and grown have gained wide recognition as separate divisions of biopsychological research. The purpose of this module is to give you a clearer sense of biopsychology and its diversity by describing six of its major divisions (see Figure 1.4): (1) physiological psychology, (2) psychopharmacology, (3) neuropsychology, (4) psychophysiology, (5) cognitive neuroscience, and (6) comparative psychology. For simplicity, they are presented as distinct approaches, but there is much overlap among them, and many biopsychologists regularly follow more than one approach.

Table 1.1 Nobel prizes specifically related to the nervous system or behavior.

Nobel Winner(s)	Date	Accomplishment
Ivan Pavlov	1904	Research on the physiology of digestion
Camillo Golgi and Santiago Ramón y Cajal	1906	Research on the structure of the nervous system
Charles Sherrington and Edgar Adrian	1932	Discoveries about the functions of neurons
Henry Dale and Otto Loewi	1936	Discoveries about the transmission of nerve impulses
Joseph Erlanger and Herbert Gasser	1944	Research on the functions of single nerve fibers
Walter Hess	1949	Research on the role of the brain in behavior
Egas Moniz	1949	Development of the prefrontal lobotomy
Georg von Békésy	1961	Research on the auditory system
John Eccles, Alan Hodgkin, and Andrew Huxley	1963	Research on the ionic basis of neural transmission
Ragnar Granit, Haldan Hartline, and George Wald	1967	Research on the chemistry and physiology of vision
Bernard Katz, Ulf von Euler, and Julius Axelrod	1970	Discoveries related to synaptic transmission
Karl Von Frisch, Konrad Lorenz, and Nikolaas Tinbergen	1973	Studies of animal behavior
Roger Guillemin and Andrew Schally	1977	Discoveries related to hormone production by the brain
Herbert Simon	1979	Research on human cognition
Roger Sperry	1981	Research on separation of the cerebral hemispheres
David Hubel and Torsten Wiesel	1981	Research on neurons of the visual system
Rita Levi-Montalcini and Stanley Cohen	1986	Discovery and study of nerve growth factors
Erwin Neher and Bert Sakmann	1991	Research on ion channels
Alfred Gilman and Martin Rodbell	1994	Discovery of G-protein–coupled receptors
Arvid Carlsson, Paul Greengard, and Eric Kandel	2000	Discoveries related to synaptic transmission
Linda Buck and Richard Axel	2004	Research on the olfactory system
John O'Keefe, May-Britt Moser, and Edvard Moser	2014	Research on the brain's system for recognizing locations
Jeffrey Hall, Michael Rosbach, and Michael Young	2017	Discoveries related to the molecular mechanisms controlling the circadian rhythm

Figure 1.4 The six major divisions of biopsychology.

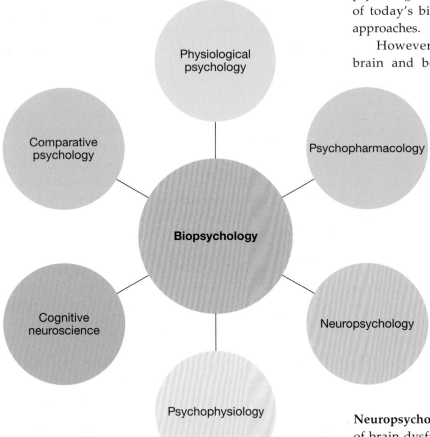

early psychopharmacologists were simply physiological psychologists who moved into drug research, and many of today's biopsychologists identify closely with both approaches.

However, the study of the effects of drugs on brain and behavior has become so specialized that psychopharmacology is regarded as a separate discipline. A substantial portion of psychopharmacological research is applied. Although drugs are sometimes used by psychopharmacologists to study the basic principles of brain–behavior interaction, the purpose of many psychopharmacological experiments is to develop therapeutic drugs (see Chapter 18) or to reduce drug abuse (see Chapter 15). Psychopharmacologists study the effects of drugs on laboratory species—and on humans, if the ethics of the situation permits it.

Neuropsychology

LO 1.9   Describe the division of biopsychology known as neuropsychology.

Neuropsychology is the study of the psychological effects of brain dysfunction in human patients. Because human volunteers cannot ethically be exposed to experimental treatments that endanger normal brain function, neuropsychology deals almost exclusively with case studies and quasiexperimental studies of patients with brain dysfunction resulting from disease, accident, or neurosurgery. The outer layer of the cerebral hemispheres— the **cerebral cortex**—is most likely to be damaged by accident or surgery; this is one reason why neuropsychology has focused on this important part of the human brain.

Neuropsychology is the most applied of the biopsychological subdisciplines; the neuropsychological assessment of human patients, even when part of a program of pure research, is always done with an eye toward benefiting them in some way. Neuropsychological tests facilitate diagnosis and thus help the attending physician prescribe effective treatments (see Benton, 1994). They can also be an important basis for patient care and counseling; Kolb and Whishaw (1990) described such an application in the case study of Mr. R.

Physiological Psychology

LO 1.7   Describe the division of biopsychology known as physiological psychology.

Physiological psychology is the division of biopsychology that studies the neural mechanisms of behavior through the direct manipulation and recording of the brain in controlled experiments—surgical and electrical methods are most common. The subjects of physiological psychology research are almost always laboratory animals because the focus on direct brain manipulation and controlled experiments precludes the use of human participants in most instances. There is also a tradition of pure research in physiological psychology; the emphasis is usually on research that contributes to the development of theories of the neural control of behavior rather than on research of immediate practical benefit.

Psychopharmacology

LO 1.8   Describe the division of biopsychology known as psychopharmacology.

Psychopharmacology is similar to physiological psychology except that it focuses on the manipulation of neural activity and behavior with drugs. In fact, many of the

The Case of Mr. R., the Student with a Brain Injury Who Switched to Architecture

Mr. R. was a 21-year-old honors student at a university. One day he was involved in a car accident in which he struck his head against the dashboard. Following the accident, Mr. R's grades began to decline; his once exceptional academic performance was now only average. He seemed to have particular trouble completing his term papers. Finally, after a year of struggling academically, he went for a neuropsychological assessment. The findings were striking.

Mr. R. turned out to be one of roughly one-third of left-handers whose language functions are represented in the right hemisphere of their brain, rather than in their left hemisphere. Furthermore, although Mr. R. had a superior IQ score, his verbal memory and reading speed were below average—something that is quite unusual for a person who had been so strong academically.

The neuropsychologists concluded that he may have suffered some damage to his right temporal lobe during the car accident, which would help explain his diminished language skills. The neuropsychologists also recommended that R. pursue a field that didn't require superior verbal memory skills. Following his exam and based on the recommendation of his neuropsychologists, Mr. R. switched majors and began studying architecture with substantial success.

Psychophysiology

LO 1.10 Describe the division of biopsychology known as psychophysiology.

Psychophysiology is the division of biopsychology that studies the relation between physiological activity and psychological processes in humans. Because the subjects of psychophysiological research are humans, psychophysiological recording procedures are typically noninvasive; that is, the physiological activity is recorded from the surface of the body. The usual measure of brain activity is the scalp **electroencephalogram (EEG)** (see Chapter 5). Other common psychophysiological measures are muscle tension, eye movement, and several indicators of autonomic nervous system activity (e.g., heart rate, blood pressure, pupil dilation, and electrical conductance of the skin). The **autonomic nervous system (ANS)** is the division of the nervous system that regulates the body's inner environment (see Chapter 3).

Most psychophysiological research focuses on understanding the physiology of psychological processes, such as attention, emotion, and information processing, but there have been some interesting clinical applications of the psychophysiological method. For example, psychophysiological experiments have indicated that people with schizophrenia have difficulty smoothly tracking a moving object with their eyes (see Meyhöfer et al., 2014)—see Figure 1.5.

> **Journal Prompt 1.2**
> What implications could the finding that people with schizophrenia have difficulty smoothly tracking moving objects have for the diagnosis of schizophrenia? (For a discussion of schizophrenia, see Chapter 18.)

Cognitive Neuroscience

LO 1.11 Describe the division of biopsychology known as cognitive neuroscience.

Cognitive neuroscience is the youngest division of biopsychology. Cognitive neuroscientists study the neural bases of **cognition**, a term that generally refers to higher intellectual processes such as thought, memory, attention, and complex perceptual processes (see Gutchess, 2014; Raichle, 2008). Because of its focus on cognition, most cognitive neuroscience research involves human participants, and because of its focus on human participants, its methods tend to be noninvasive, rather than involving penetration or direct manipulation of the brain.

The major method of cognitive neuroscience is *functional brain imaging*: recording images of the activity of the living human brain (see Chapter 5) while a participant is engaged in a particular mental activity. For example, Figure 1.6 shows that the visual areas of the left and right cerebral cortex at the back of the brain became active when the participant viewed a flashing light.

Because the theory and methods of cognitive neuroscience are so complex and pertinent to so many fields, cognitive neuroscience research often involves interdisciplinary

Figure 1.5 Visual tracking of a pendulum by a healthy control participant (top) and three participants with schizophrenia.

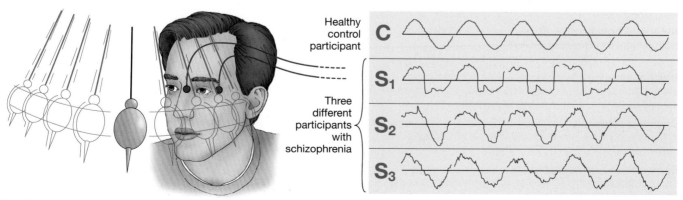

Based on Iacono, W. G., & Koenig, W. G. (1983).

Figure 1.6 Functional brain imaging is the major method of cognitive neuroscience. This image—taken from the top of the head with the participant lying on her back—reveals the locations of high levels of neural activity at one level of the brain as the participant viewed a flashing light. The red and yellow areas indicate high levels of activity in the visual cortex at the back of the brain. (Courtesy of Dr. Todd Handy, Department of Psychology, University of British Columbia.)

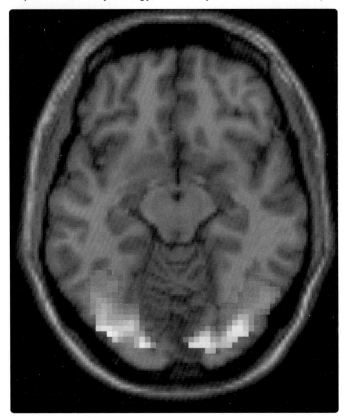

Todd C. Handy/University of British Columbia Department of Psychology

collaboration among many researchers with different types of training. Biopsychologists, cognitive psychologists, social psychologists, economists, computing and mathematics experts, and various types of neuroscientists commonly contribute to the field. Cognitive neuroscience research sometimes involves noninvasive electrophysiological recording, and it sometimes focuses on patients with brain dysfunction; in these cases, the boundaries between cognitive neuroscience and psychophysiology and neuropsychology, respectively, are blurred.

Comparative Psychology

LO 1.12 Describe the division of biopsychology known as comparative psychology.

Although most biopsychologists study the neural mechanisms of behavior, there is more to biopsychology than neural mechanisms. A biopsychologist should never lose sight of the fact that the purpose of their research is to understand the integrated behavior of the whole animal. The last division of biopsychology that we describe here is one that focuses on the behavior of animals in their natural environments. This division is **comparative psychology**.

Comparative psychologists compare the behavior of different species in order to understand the evolution, genetics, and adaptiveness of behavior. Some comparative psychologists study behavior in the laboratory; others engage in **ethological research**—the study of behavior in an animal's natural environment.

As a reminder, the purpose of this module was to demonstrate the diversity of biopsychology by describing six of its major divisions; these are summarized for you in Table 1.2. You will see all six of these divisions in action in subsequent chapters.

Table 1.2 The six major divisions of biopsychology with examples of how they have approached the study of memory.

Division of Biopsychology	Example from Memory Research
Physiological psychology: study of the neural mechanisms of behavior by manipulating the nervous systems of nonhuman animals in controlled experiments	Physiological psychologists have studied the contributions of one brain structure, the hippocampus, to memory by surgically removing it in rats and assessing their ability to perform various memory tasks.
Psychopharmacology: study of the effects of drugs on the brain and behavior	Psychopharmacologists have tried to improve the memory of Alzheimer's patients by administering drugs that alter brain chemistry.
Neuropsychology: study of the psychological effects of brain dysfunction in human patients	Neuropsychologists have shown that patients with damage to the hippocampus and surrounding structures are incapable of forming new long-term memories.
Psychophysiology: study of the relation between physiological activity and psychological processes in human volunteers by noninvasive physiological recording	Psychophysiologists have shown that familiar faces elicit the usual changes in autonomic nervous system activity even when patients with brain damage report that they do not recognize a face.
Cognitive neuroscience: study of the neural mechanisms of human cognition, largely through the use of functional brain imaging	Cognitive neuroscientists have used brain-imaging technology to observe the changes that occur in various parts of the brain while human volunteers perform memory tasks.
Comparative psychology: study of the evolution, genetics, and adaptiveness of behavior, largely through the use of the comparative method	Comparative psychologists have shown that species of birds that cache their seeds tend to have larger hippocampi, confirming that the hippocampus is involved in memory for location.

Scan Your Brain

To see if you are acquainted with the main premises of biopsychology and allied disciplines, fill in each of the following blanks with the most appropriate terms. The correct answers are provided at the end of the exercise. Before proceeding, review material related to your errors and omissions.

1. _____ is a branch of psychology that uses data from patients with brain damage to understand structure and function of the human brain.
2. Over the past few decades, researchers have realized that the adult brain connections are not static but changeable in response to the individual's genes and experiences. This is known as _____.

3. In a _____ design, participants are placed into different groups and exposed to different experimental conditions.
4. Studies that focus on a single participant rather than a group of participants are called _____.
5. The major method of cognitive neuroscience is _____, recording images of the activity of the living human brain.
6. _____ is a branch of biopsychology that studies genetic, evolutionary, and behavior differences across species.

Scan Your Brain answers: (1) Neuropsychology, (2) neuroplasticity, (3) between-subjects, (4) case studies, (5) functional brain imaging, (6) Evolutionary behavioral genetics.

How Do Biopsychologists Conduct Their Work?

This module explains how biopsychologists typically conduct their work. First, you will learn how biopsychologists collaborate with one another, and the importance of such collaboration in advancing a field of research. Second, you will learn about how biopsychologists make inferences about brain function that is not directly observable. These are important components of biopsychological research, and you will see in the next module what goes wrong when such collaboration and scientific inference are thrown by the wayside.

Converging Operations: How Do Biopsychologists Work Together?

LO 1.13 Explain how converging operations has contributed to the study of Korsakoff's syndrome.

Because each of the six biopsychological approaches to research has its own particular strengths and shortcomings and because the mechanisms by which the brain controls behavior are so complex, major biopsychological issues are rarely resolved by a single experiment or even by a series of experiments taking the same general approach. Progress is most likely when different approaches are focused on a single problem in such a way that the strengths of one approach compensate for the weaknesses of the others; this combined approach is called **converging operations** (see Thompson, 2005).

Consider, for example, the relative strengths and weaknesses of neuropsychology and physiological psychology in the study of the psychological effects of damage to the human cerebral cortex. In this instance, the strength of the neuropsychological approach is that it deals directly with human patients; its weakness is that its focus on human patients precludes experiments. In contrast, the strength of the physiological psychology approach is that it can use the power of experimental research on nonhuman animals; its weakness is that the relevance of research on laboratory animals to human brain damage is always open to question (see Couzin-Frankel, 2013; Reardon, 2016). Clearly these two approaches complement each other well; together they can answer questions that neither can answer individually.

To examine converging operations in action, let's return to the case of Jimmie G. The neuropsychological disorder from which Jimmie suffered was first described in the late 19th century by Sergei Korsakoff, a Russian physician, and subsequently became known as **Korsakoff's syndrome**. The primary symptom of Korsakoff's syndrome is severe memory loss, which is made all the more heartbreaking—as you have seen in Jimmie G.'s case—by the fact that its sufferers are often otherwise quite capable. Because Korsakoff's syndrome commonly occurs in heavy drinkers of alcohol, it was initially believed to be a direct consequence of the toxic effects of alcohol on the brain. This conclusion proved to be a good illustration of the inadvisability of inferring causality from the results of quasiexperimental studies. Subsequent research showed that Korsakoff's syndrome is largely caused by the brain damage associated with *thiamine* (vitamin B$_1$) deficiency.

Journal Prompt 1.3

Korsakoff's syndrome accounts for approximately 10 percent of adult dementias in the United States. Despite its relatively high prevalence, few people have heard of it. Why do you think this is the case?

The first support for the thiamine-deficiency interpretation of Korsakoff's syndrome came from the discovery of the syndrome in malnourished persons who consumed little or no alcohol. Additional support came from experiments in which thiamine-deficient rats were compared with otherwise identical groups of control rats. The thiamine-deficient rats displayed memory deficits and patterns of brain damage similar to those observed in many people who had been heavy drinkers of alcohol (Mumby, Cameli, & Glenn, 1999). Such people often develop Korsakoff's syndrome because most of their caloric intake comes in the form of alcohol, which lacks vitamins, and because alcohol interferes with the metabolism of what little thiamine they do consume. However, alcohol has been shown to accelerate the development of brain damage in thiamine-deficient rats, so it may have a direct toxic effect on the brain as well (Ridley, Draper, & Withall, 2013).

The point of this discussion of Korsakoff's syndrome is to show you that progress in biopsychology typically comes from converging operations—in this case, from the convergence of neuropsychological case studies (case studies of Korsakoff patients), quasiexperiments with human participants (comparisons of heavy drinkers with abstainers), and controlled experiments on laboratory animals (comparison of thiamine-deficient and control rats). The strength of biopsychology lies in the diversity of its methods and approaches. This means that, in evaluating biopsychological claims, it is rarely sufficient to consider the results of one study or even of one line of experiments using the same method or approach.

So what has all the research on Korsakoff's syndrome done for Jimmie G. and others like him? Today, heavy drinkers are counseled to stop drinking and are treated with large doses of thiamine. The thiamine limits the development of further brain damage and often leads to a slight improvement in the patient's condition; unfortunately, the acquired brain dysfunction is mostly irreversible.

Scientific Inference: How Do Biopsychologists Study the Unobservable Workings of the Brain?

LO 1.14   Explain scientific inference with reference to research on eye movements and the visual perception of motion.

Scientific inference is the fundamental method of biopsychology and of most other sciences—it is what makes being a scientist fun. This section provides further insight into the nature of biopsychology by defining, illustrating, and discussing scientific inference.

The scientific method is a system for finding things out by careful observation, but many of the processes studied by scientists cannot be observed. For example, scientists use empirical (observational) methods to study ice ages, gravity, evaporation, electricity, and nuclear fission—none of which can be directly observed; their effects can be observed, but the processes themselves cannot. Biopsychology is no different from other sciences in this respect. One of its main goals is to characterize, through empirical methods, the unobservable processes by which the nervous system controls behavior.

The empirical method that biopsychologists and other scientists use to study the unobservable is called **scientific inference**. Scientists carefully measure key events they can observe and then use these measures as a basis for logically inferring the nature of events they cannot observe. Like a detective carefully gathering clues from which to re-create an unwitnessed crime, a biopsychologist carefully gathers relevant measures of behavior and neural activity from which to infer the nature of the neural processes that regulate behavior. The fact that the neural mechanisms of behavior cannot be directly observed and must be studied through scientific inference is what makes biopsychological research such a challenge—and, as we said before, so much fun.

To illustrate scientific inference, we have selected a research project in which you can participate. By making a few simple observations about your own visual abilities under different conditions, you will be able to discover the principle by which your brain translates the movement of images on your retinas into perceptions of movement (see Figure 1.7). One feature of the mechanism is immediately obvious. Hold your hand in front of your face, and then move its image across your retinas by moving your eyes, by moving your hand, or by moving both at once. You will notice that only those movements of the retinal image produced by the movement of your hand are translated into the perception of motion; movements of the retinal image produced by your own eye movements are not. Obviously, there must be a part of your brain that monitors the movements of your retinal image and subtracts from the total those image movements produced by your own eye movements, leaving the remainder to be perceived as motion.

Now, let's try to characterize the nature of the information about your eye movements used by your brain in its perception of motion. Try the following. Shut one eye, then rotate your other eye slightly upward by gently pressing on your lower eyelid with your fingertip. What do you see? You see all of the objects in your visual field moving downward. Why? It seems that the brain mechanism responsible for the perception of motion does not consider eye movement per se. It considers only those eye movements that are actively produced by neural signals from the brain to the eye muscles, not those that are passively produced by

Figure 1.7 The perception of motion under four different conditions.

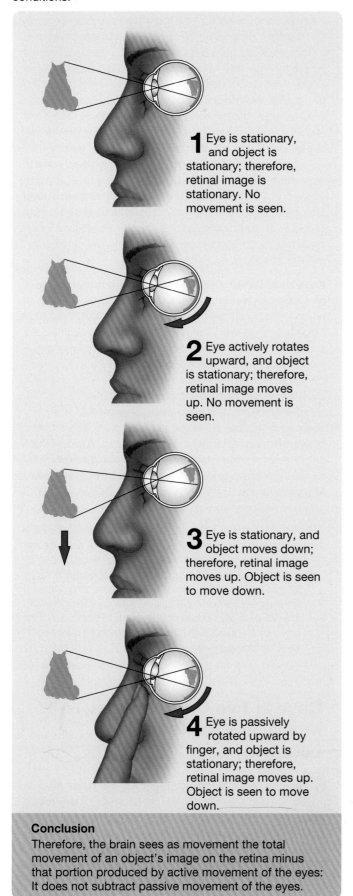

1 Eye is stationary, and object is stationary; therefore, retinal image is stationary. No movement is seen.

2 Eye actively rotates upward, and object is stationary; therefore, retinal image moves up. No movement is seen.

3 Eye is stationary, and object moves down; therefore, retinal image moves up. Object is seen to move down.

4 Eye is passively rotated upward by finger, and object is stationary; therefore, retinal image moves up. Object is seen to move down.

Conclusion
Therefore, the brain sees as movement the total movement of an object's image on the retina minus that portion produced by active movement of the eyes: It does not subtract passive movement of the eyes.

other means (e.g., by your finger). Thus, when your eye was moved passively, your brain assumed it had remained still and attributed the movement of your retinal image to the movement of objects in your visual field.

It is possible to trick the visual system in the opposite way; instead of the eyes being moved when no active signals have been sent to the eye muscles, the eyes can be held stationary despite the brain's attempts to move them. Because this experiment involves paralyzing the eye muscles, you cannot participate. Hammond, Merton, and Sutton (1956) injected a *paralytic* (movement-inhibiting) substance into the eye muscles of their participant—who was Merton himself. This paralytic substance was the active ingredient of *curare*, a drug with which some Indigenous people of South America coat their blow darts. What do you think Merton saw when he then tried to move his eyes? He saw the stationary visual world moving in the same direction as his attempted eye movements. If a visual object is focused on part of your retina, and it stays focused there despite the fact that you have moved your eyes to the right, it too must have moved to the right. Consequently, when Merton sent signals to his eye muscles to move his eyes to the right, his brain assumed the movement had been carried out, and it perceived stationary objects as moving to the right.

The point of the eye-movement example is that biopsychologists can learn much about the activities of the brain through scientific inference without directly observing them—and so can you. By the way, neuroscientists are still interested in the kind of feedback mechanisms inferred from the demonstrations of Hammond and colleagues, and they have refined our understanding of the mechanisms using modern neural recording techniques (e.g., Joiner et al., 2013; Wurtz et al., 2011).

Thinking Critically about Biopsychological Claims

We have all heard or read that we use only a small portion of our brains, that it is important to eat three meals a day, that intelligence is inherited, that everybody needs at least 8 hours of sleep per night, that there is a gene for schizophrenia, that heroin is a particularly dangerous (hard) drug, and that neurological diseases can now be cured by genetic engineering. These are but a few of the claims about biopsychological phenomena that have been widely disseminated (see Howard-Jones, 2014). You may believe many of these claims. But are they all true? How does one find out? And if they are not true, why do so many people believe them?

We hope that you will learn how to differentiate between flawed claims and exciting new discoveries. This, the final module of the chapter, begins teaching this lesson.

Evaluating Biopsychological Claims

LO 1.15 Define critical thinking and evaluate biopsychological claims.

As you have already learned, one of the major goals of this text is to teach you how to think creatively (to think in productive, unconventional ways) about biopsychological information. Often, the first step in creative thinking is spotting the weaknesses of existing ideas and the evidence on which they are based—the process by which these weaknesses are recognized is called **critical thinking**. The identification of weaknesses in existing beliefs is one of the major stimuli for scientists to adopt creative new approaches.

Journal Prompt 1.4

Do you think that improving your critical thinking abilities will impact your everyday life? Why or why not? (Suggestion: Revisit this journal prompt once you have finished this course!)

The purpose of this final module of the chapter is to develop your own critical thinking abilities by analyzing two claims that played major roles in the history of biopsychology. In both cases, the evidence proved to be grossly flawed. Notice that if you keep your wits about you, you do not have to be an expert to spot the weaknesses.

The first step in judging the validity of any scientific claim is to determine whether the claim and the research on which it is based were published in a reputable scientific journal. The reason is that, in order to be published in a reputable scientific journal, an article must first be reviewed by experts in the field—usually three or four of them—and judged to be of good quality. Indeed, the best scientific journals publish only a small proportion of the manuscripts submitted to them. You should be particularly skeptical of scientific claims that have not gone through this rigorous review process.

The first case that follows deals with an unpublished claim that was largely dispensed through the news media. The second deals with a claim that was initially supported by published research. Because both of these cases are part of the history of biopsychology, we have the advantage of 20/20 hindsight in evaluating their claims.

Case 1: José and the Bull

José Delgado, a particularly charismatic neuroscientist, demonstrated to a group of newspaper reporters a remarkable new procedure for controlling aggression. Delgado strode into a Spanish bullfighting ring carrying only a red cape and a small radio transmitter. With the transmitter, he could activate a battery-powered stimulator that had previously been mounted on the horns of the other inhabitant of the ring. As the raging bull charged, Delgado calmly activated the stimulator and sent a weak electrical current from the stimulator through an electrode that had been implanted in the caudate nucleus (see Chapter 3), a structure deep in the bull's brain. The bull immediately veered from its charge. After a few such interrupted charges, the bull stood tamely as Delgado swaggered about the ring. According to Delgado, this demonstration marked a significant scientific breakthrough—the discovery of a caudate taming center and the fact that stimulation of this structure could eliminate aggressive behavior, even in bulls specially bred for their ferocity.

To those present at this carefully orchestrated event—and to most of the millions who subsequently read about it—Delgado's conclusion was compelling. Surely, if caudate stimulation could stop the charge of a raging bull, the caudate must be a taming center. It was even suggested that caudate stimulation through implanted electrodes might be an effective treatment for human psychopathy. What do you think?

Analysis of Case 1 Delgado's demonstration provided little or no support for his conclusion. It should have been obvious to anyone who did not get caught up in the provocative nature of Delgado's media event that brain stimulation can abort a bull's charge in numerous ways, most of which are simpler, and thus more probable, than the one suggested by Delgado. For example, the stimulation may have simply rendered the bull confused, dizzy, nauseous, sleepy, or temporarily blind rather than nonaggressive; or the stimulation could have been painful. Clearly, any observation that can be interpreted in so many different ways provides little support for any one interpretation. When there are several possible interpretations for a behavioral observation, the rule is to give precedence to the simplest one; this rule is called **Morgan's Canon**. The following comments of Valenstein (1973) provide a reasoned view of Delgado's demonstration:

> Actually there is no good reason for believing that the stimulation had any direct effect on the bull's aggressive tendencies. An examination of the film record makes it apparent that the charging bull was stopped because as long as the stimulation was on it was forced to turn around in the same direction continuously. After examining the film, any scientist with knowledge in this field could conclude only that the stimulation had been activating a neural pathway controlling movement. (p. 98)

Case 2: Two Chimpanzees, Moniz, and the Prefrontal Lobotomy

In 1949, Dr. Egas Moniz was awarded the Nobel Prize in Physiology and Medicine for the development of **prefrontal lobotomy**—a surgical procedure in which the connections between the prefrontal lobes and the rest of the brain are cut as a treatment for mental illness. The **prefrontal lobes** are the large areas, left and right, at the very front of the brain (see Figure 1.8). Moniz's discovery was based on the report that two

Figure 1.8 The right and left prefrontal lobes, whose connections to the rest of the brain are disrupted by prefrontal lobotomy.

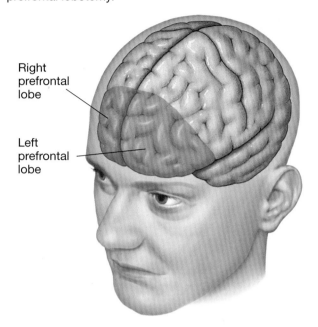

Right prefrontal lobe

Left prefrontal lobe

Figure 1.9 The prefrontal lobotomy procedure developed by Moniz and Lima.

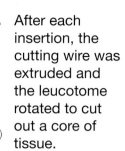

The leucotome was inserted six times into the patient's brain with the cutting wire retracted.

After each insertion, the cutting wire was extruded and the leucotome rotated to cut out a core of tissue.

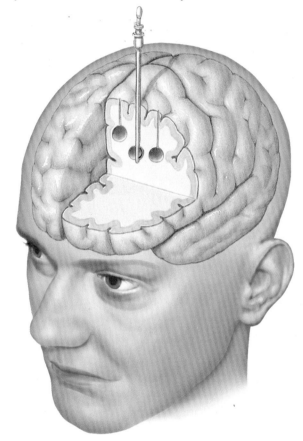

chimpanzees that frequently became upset when they made errors during the performance of a food-rewarded task, did not do so following the creation of a large *bilateral lesion* (an area of damage to both sides of the brain) of their prefrontal lobes. After witnessing a demonstration of this result at a scientific meeting in 1935, Moniz convinced neurosurgeon Almeida Lima to operate on a series of psychiatric patients (see Heller et al., 2006). Lima cut out six large cores of prefrontal tissue with a surgical device called a **leucotome** (see Figure 1.9).

Following Moniz's claims that prefrontal surgery was therapeutically useful, there was a rapid proliferation of various forms of prefrontal psychosurgery. One such variation was **transorbital lobotomy**, which was developed in Italy and then popularized in the United States by Walter Freeman in the late 1940s. It involved inserting an ice pick-like device under the eyelid, driving it through the orbit (the eye socket) with a few taps of a mallet, and pushing it into the prefrontal lobes, where it was waved back and forth to sever the connections between the prefrontal lobes and the rest of the brain (see Figure 1.10). This operation was frequently performed in doctors' offices.

Analysis of Case 2 Incredible as it may seem, Moniz's program of **psychosurgery** (any brain surgery, such as prefrontal lobotomy, performed for the treatment of a psychological problem) was largely based on the observation of two chimpanzees. Thus, Moniz displayed a lack of appreciation for the diversity of brain and behavior, both within and between species. No program of psychosurgery should ever be initiated without a thorough assessment of the effects of the surgery on a large sample of subjects from various nonhuman mammalian species. To do so is not only unwise, it is unethical.

A second major weakness in the scientific case for prefrontal lobotomy was the failure of Moniz and others to carefully

evaluate the consequences of the surgery in the first patients to undergo the operation (see Mashour, Walker, & Martuza, 2005; Singh, Hallmayer, & Illes, 2007). The early reports that the operation was therapeutically effective were based on the impressions of the individuals who were the least objective—the physicians who had prescribed the surgery and their colleagues. Patients were frequently judged as improved if they were more manageable, and little effort was made to evaluate more important aspects of their psychological adjustment or to document the existence of adverse side effects.

Eventually, it became clear that prefrontal lobotomies are of little therapeutic benefit and that they can produce a wide range of undesirable side effects, such as socially inappropriate behavior, lack of foresight, emotional unresponsiveness, epilepsy, and urinary incontinence. This led to the abandonment of prefrontal lobotomy in many parts of the world—but not before more

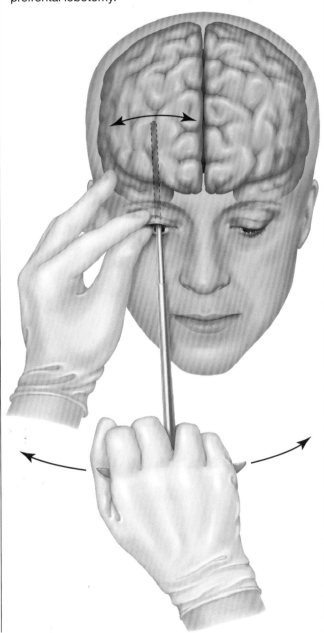

Figure 1.10 The transorbital procedure for performing prefrontal lobotomy.

than 40,000 patients had been lobotomized in the United States alone. And prefrontal lobotomies still continue to be performed in some countries.

A particularly troubling aspect of the use of prefrontal lobotomy is that not only informed, consenting adults received this "treatment." In his memoir, Howard Dully described how he had been lobotomized at the age of 12 (Dully & Fleming, 2007). The lobotomy was arranged by Dully's stepmother, agreed to by his father, and performed in 10 minutes by Walter Freeman. Dully spent most of the rest of his life in asylums, jails, and halfway houses, wondering what he had done to deserve the lobotomy and how much it had been responsible for his troubled life. Subsequent investigation of the case indicated that Dully was a normal child whose stepmother was obsessed by her hatred for him. Tragically, neither his father nor the medical profession intervened to protect him from Freeman's ice pick.

Some regard sound scientific methods as unnecessary obstacles in the paths of patients seeking treatment and therapists striving to provide it. However, the unforeseen consequences of prefrontal lobotomy should caution us against abandoning science for expediency. Only by observing the rules of science can scientists protect the public from bogus claims (see Rousseau & Gunia, 2016).

Thankfully, biopsychology has learned from the mistakes and faulty thinking of Delgado, Moniz, Freeman, and others. The practice of the scientific method and well-reasoned inference are nearly ubiquitous in modern biopsychology.

You are about to enter the amazing world of biopsychology. We hope your brain enjoys learning about itself.

Themes Revisited

The seeds of three of the major themes were planted in this chapter, but the thinking creatively theme predominated. First, you saw the creative approach that Lester and Gorzalka took in their research on the Coolidge effect in females. Then, you learned three important new ideas that will help you think about biopsychological claims: (1) the experimental method, (2) converging operations, and (3) scientific inference. Finally, you were introduced to two biopsychological claims that were once widely believed and saw how critical thinking identified their weaknesses and replaced them with creative new interpretations.

You also learned that two of the other major themes—clinical implications and the evolutionary perspective—tend to be associated with particular divisions of biopsychology. Clinical implications most commonly emerge from neuropsychological, psychopharmacological, and psychophysiological research; the evolutionary perspective is a defining feature of comparative psychology.

The two emerging themes, thinking about epigenetics and consciousness, will appear in later chapters.

Key Terms

Neurons, p. 26
Neuroscience, p. 26
Thinking creatively, p. 27
Clinical, p. 27
Evolutionary perspective, p. 27
Neuroplasticity, p. 28
Epigenetics, p. 28
Consciousness, p. 28

What Is Biopsychology?

Biopsychology, p. 28
Neuroanatomy, p. 29
Neurochemistry, p. 29
Neuroendocrinology, p. 29
Neuropathology, p. 29
Neuropharmacology, p. 29
Neurophysiology, p. 29

What Types of Research Characterize the Biopsychological Approach?

Comparative approach, p. 29
Between-subjects design, p. 30

Within-subjects design, p. 30
Independent variable, p. 30
Dependent variable, p. 30
Confounded variable, p. 30
Coolidge effect, p. 31
Lordosis, p. 31
Quasiexperimental studies, p. 31
Case studies, p. 32
Generalizability, p. 32
Pure research, p. 32
Applied research, p. 32
Translational research, p. 32

What Are the Divisions of Biopsychology?

Physiological psychology, p. 34
Psychopharmacology, p. 34
Neuropsychology, p. 34
Cerebral cortex, p. 34
Psychophysiology, p. 35
Electroencephalogram (EEG), p. 35

Autonomic nervous system (ANS), p. 35
Cognitive neuroscience, p. 35
Cognition, p. 35
Comparative psychology, p. 36
Ethological research, p. 36

How Do Biopsychologists Conduct Their Work?

Converging operations, p. 37
Korsakoff's syndrome, p. 37
Scientific inference, p. 38

Thinking Critically about Biopsychological Claims

Critical thinking, p. 40
Morgan's Canon, p. 40
Prefrontal lobotomy, p. 40
Prefrontal lobes, p. 40
Leucotome, p. 41
Transorbital lobotomy, p. 41
Psychosurgery, p. 41

Chapter 2
Evolution, Genetics, and Experience

Thinking about the Biology of Behavior

pyrozhenka/Shutterstock

 ## Chapter Overview and Learning Objectives

Thinking about the
Biology of Behavior: From
Dichotomies to Interactions

LO 2.1 Describe the origins of the physiological–psychological and nature–nurture ways of thinking.

LO 2.2 Explain why thinking about the biology of behavior in terms of traditional physiological–psychological and nature–nurture dichotomies is flawed.

Human Evolution

LO 2.3 Describe the origins of evolutionary theory.

LO 2.4 Explain the evolutionary significance of social dominance and courtship displays.

LO 2.5 Summarize the pathway of evolution from single-cell organisms to humans.

LO 2.6 Describe nine commonly misunderstood points about evolution.

LO 2.7 Describe how research on the evolution of the human brain has changed over time.

Fundamental Genetics

LO 2.8 Explain how Mendel's work with pea plants has informed us about the mechanisms of inheritance.

LO 2.9 Understand the structure and function of chromosomes.

LO 2.10 Describe the process of gene expression.

LO 2.11 Discuss several ways in which modern advances have changed our understanding of genetic processes.

LO 2.12 Define epigenetics, and explain how it has transformed our understanding of genetics.

Epigenetics of Behavioral Development: Interaction of Genetic Factors and Experience

LO 2.13 Discuss what insights into the genetics of behavior were gained from early research on selective breeding.

LO 2.14 Explain how classic research on phenylketonuria (PKU) has informed our understanding of the genetics of behavior.

Genetics of Human Psychological Differences

LO 2.15 Explain why it is important to distinguish between the development of individuals and the development of individual differences.

LO 2.16 Explain heritability estimates and how they are commonly misinterpreted.

LO 2.17 Describe two ways that twin studies can be used to study the interaction of genes and experience (i.e., nature and nurture).

We all tend to think about things in ways that have been ingrained in us by our **zeitgeist** (pronounced "TSYTE-gyste"), the general intellectual climate of our culture. That is why this is a particularly important chapter for you. You see, you are the intellectual product of a zeitgeist that promotes ways of thinking about the biological bases of behavior that are inconsistent with the facts. The primary purpose of this chapter is to help you bring your thinking about the biology of behavior in line with modern biopsychological science.

Thinking about the Biology of Behavior: From Dichotomies to Interactions

We tend to ignore the subtleties, inconsistencies, and complexities of our existence and to think in terms of simple, mutually exclusive dichotomies: right–wrong, good–bad, attractive–unattractive, and so on. The allure of this way of thinking is its simplicity.

The Origins of Dichotomous Thinking

LO 2.1 Describe the origins of the physiological–psychological and nature–nurture ways of thinking.

The tendency to think about behavior in terms of dichotomies is illustrated by two kinds of questions commonly asked about behavior: (1) Is it physiological, or is it psychological? (2) Is it inherited, or is it learned? Both questions have proved to be misguided, yet they are among the most common kinds of questions asked in biopsychology classrooms. That is why we are dwelling on them here.

IS IT PHYSIOLOGICAL, OR IS IT PSYCHOLOGICAL? The idea that human processes fall into one of two categories, physiological or psychological, has a long history in many cultures. For much of the history of Western cultures, truth was whatever the Church decreed to be true. Then, in about 1400, things started to change. The famines, plagues, and marauding armies that had repeatedly swept Europe during the Dark Ages subsided, and interest turned to art, commerce, and scholarship—this was the period of the Renaissance, or rebirth (1400–1700). Some Renaissance scholars were not content to follow the dictates of the Church; instead, they started to study things directly by observing them—and so it was that modern science was born.

Much of the scientific knowledge that accumulated during the Renaissance was at odds with Church dictates. However, the conflict was resolved by the prominent French philosopher René Descartes (pronounced "day-CART"). Descartes (1596–1650) advocated a philosophy that, in a sense, gave one part of the universe to science and the other part to the Church. He argued that the universe is composed of two elements: (1) physical matter, which behaves according to the laws of nature and is thus a suitable object of scientific investigation—the human body, including the brain, was assumed to be entirely physical, and so were nonhuman animals; and (2) the human mind (soul, self, or spirit), which lacks physical substance, controls human behavior, obeys no natural laws, and is thus the appropriate purview of the Church.

Cartesian dualism, as Descartes's philosophy became known, was sanctioned by the Roman Church, and so the idea that the human brain and the mind are separate entities became even more widely accepted. It has survived to this day, despite the intervening centuries of scientific progress. Most people now understand that human behavior has a physiological basis, but many still cling to the dualistic assumption that there is a category of human activity that somehow transcends the human brain.

IS IT INHERITED, OR IS IT LEARNED? The tendency to think in terms of dichotomies extends to the way people think about the development of behavioral capacities. For centuries, scholars have debated whether humans and other animals inherit their behavioral capacities or acquire them through learning. This debate is commonly referred to as the **nature–nurture issue**.

Most of the early North American experimental psychologists were totally committed to the nurture (learning) side of the nature–nurture issue. The degree of this commitment is illustrated by the oft-cited words of John B. Watson, the father of *behaviorism*:

> We have no real evidence of the inheritance of [behavioral] traits. I would feel perfectly confident in the ultimately favorable outcome of careful upbringing of a healthy, well-formed baby born of a long line of crooks, murderers and thieves, and prostitutes. Who has any evidence to the contrary?

> . . . Give me a dozen healthy infants, well-formed, and my own specified world to bring them up in and I'll guarantee to take any one at random and train him to become any type of specialist I might select—doctor, lawyer, artist, merchant-chief and, yes even beggar-man and thief. (Watson, 1930, pp. 103–104)

At the same time experimental psychology was taking root in North America, **ethology** (the study of animal behavior in the wild) was becoming the dominant approach to the study of behavior in Europe. European ethology, in contrast to North American experimental psychology, focused on the study of **instinctive behaviors** (behaviors that occur in all like members of a species, even when there seems to have been no opportunity for them to have been learned), and it emphasized the role of nature, or inherited factors, in behavioral development. Because instinctive behaviors are not learned, the early ethologists assumed they are entirely inherited. They were wrong, but then so were the early experimental psychologists.

Problems with Thinking about the Biology of Behavior in Terms of Traditional Dichotomies

LO 2.2 Explain why thinking about the biology of behavior in terms of traditional physiological–psychological and nature–nurture dichotomies is flawed.

The physiological-or-psychological debate and the nature-or-nurture debate are based on incorrect ways of thinking about the biology of behavior, and a new generation of questions is directing the current boom in biopsychological research (see Churchland, 2002). What is wrong with these old ways of thinking about the biology of behavior, and what are the new ways?

PHYSIOLOGICAL-OR-PSYCHOLOGICAL THINKING RUNS INTO DIFFICULTY. Not long after Descartes's mind–brain dualism was officially sanctioned by the Roman Church, it started to come under public attack.

> In 1747, Julien Offray de la Mettrie anonymously published a pamphlet that scandalized Europe. . . . La Mettrie fled to Berlin, where he was forced to live in exile for the rest of his life. His crime? He had argued that thought was produced by the brain—a dangerous assault, in the eyes of his contemporaries. (Corsi, 1991, cover)

There are two lines of evidence against *physiological-or-psychological thinking* (the assumption that some aspects of human psychological functioning are so complex that they could not possibly be the product of a physical brain). The first line is composed of the many demonstrations that even the most complex psychological changes (e.g., changes in self-awareness, memory, or emotion) can be produced by damage to, or stimulation of, parts of the brain (see Farah & Murphy, 2009). The second line of evidence is composed of demonstrations that some nonhuman species, particularly *primate* species, possess some abilities (e.g., complex problem solving) that were once assumed to be purely psychological and thus purely human (see Bartal, Decety, & Mason, 2011). The following two cases illustrate these two kinds of evidence. Both cases deal with self-awareness, which is widely regarded as one hallmark of the human mind (see Apps & Tsakiris, 2014).

The first case is Oliver Sacks's (1985) account of "the man who fell out of bed." This patient was suffering from **asomatognosia**, a deficiency in the awareness of parts of one's own body. Asomatognosia typically involves the left side of the body and usually results from damage to the *right frontal and parietal lobes* (see Feinberg et al., 2010; Figure 2.1). The point here is that, although the changes in self-awareness displayed by the patient were very complex, they were clearly the result of brain damage: Indeed, the full range of human experience can be produced by manipulations of the brain.

The Case of the Man Who Fell Out of Bed

When he awoke, Dr. Sacks's patient felt fine—that is, until he touched the thing in bed next to him. It was a severed human leg, all hairy and still warm! At first, the patient was confused. Then he figured it out. One of the nurses must have taken it from the autopsy department and put it in his bed as a joke. Some joke; it was disgusting. So, he threw the leg out of the bed, but somehow he landed on the floor with it attached to him.

The patient became agitated and desperate, and Dr. Sacks tried to comfort him and help him back into the bed. Making one last effort to reduce the patient's confusion, Sacks asked him where his left leg was, if the one attached to him wasn't it. Turning pale and looking like he was about to pass out, the patient replied that he had no idea where his own leg was—it had disappeared.

The second case describes G. G. Gallup's research on self-awareness in chimpanzees (see Gallup, 1983; Parker, Mitchell, & Boccia, 1994). The point of this case is that even nonhumans, which are assumed by some people to have no mind, are capable of considerable psychological complexity—in this case, self-awareness. Although their brains are less complex than the brains of humans, some species are capable of high levels of psychological complexity (see Gomez-Marin & Mainen, 2016).

Figure 2.1 Asomatognosia often involves damage to the right frontal and parietal lobes.

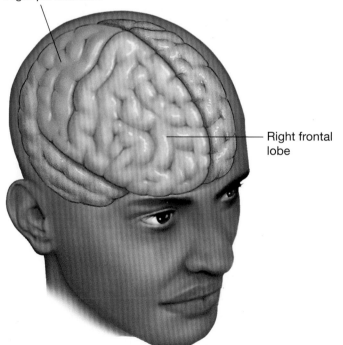

Right parietal lobe

Right frontal lobe

The Case of the Chimps with Mirrors*

One way of assessing an organism's self-awareness is to confront it with a mirror. Invariably, the first reaction of a chimpanzee to a mirror is to respond as if it were seeing another chimpanzee. However, after a day or two, it starts to act as if it were self-aware. It starts to use the mirror to groom itself, inspect parts of its body, and experiment with its reflection by making faces and assuming unusual postures while monitoring the results in the mirror.

In an attempt to provide even more convincing evidence of self-awareness, Gallup (1983) devised a clever test. Each chimpanzee was anesthetized, and its eyebrow was painted with a red, odorless, dye. Following recovery from anesthesia, the mirror was reintroduced. Upon seeing its painted eyebrow in the mirror, each chimpanzee repeatedly touched the marked area on its eyebrow while watching the image (see Figure 2.2.) Moreover, there was over a threefold increase in the time that the chimps spent looking in the mirror, and several kept touching their eyebrows and smelling their fingers. We suspect that you would respond pretty much the same way if you saw yourself in the mirror with a red spot on your face.

(continued)

* "Toward a Comparative Psychology of Mind" by G. G. Gallup, Jr., *American Journal of Primatology* 2:237–248, 1983. Copyright © 1983 John Wiley & Sons, Inc.

Figure 2.2 The reactions of chimpanzees to their *own* images suggest that they are self-aware. In this photo, the chimpanzee is reacting to the bright red, odorless dye that was painted on its eyebrow ridge while it was anesthetized.

The Povinelli Group LLC

Since Gallup's demonstration, many other species have passed what is now known as the *mirror self-recognition test*. These include Asian elephants, orangutans, and European magpies, to name a few. We humans pass the mirror self-recognition test only once we have reached 15 to 24 months of age.

NATURE-OR-NURTURE THINKING RUNS INTO DIFFICULTY. The history of nature-or-nurture thinking can be summed up by paraphrasing Mark Twain: "Reports of its death have been greatly exaggerated." Each time it has been discredited, it has resurfaced in a slightly modified form. First, factors other than genetics and learning were shown to influence behavioral development; factors such as the fetal environment, nutrition, stress, and sensory stimulation all proved to be influential. This led to a broadening of the concept of nurture to include a variety of experiential factors in addition to learning. In effect, it changed the nature-or-nurture dichotomy from "genetic factors or learning" to "genetic factors or experience."

Next, it was argued convincingly that behavior always develops under the combined control of both nature and nurture (see Johnston, 1987; Rutter, 1997), not under the control of one or the other. Faced with this point, many people merely substituted one kind of nature-or-nurture thinking for another. They stopped asking, "Is it genetic, or is it the result of experience?" and started asking, "How much of it is genetic, and how much of it is the result of experience?"

Like earlier versions of the nature-or-nurture question, the how-much-of-it-is-genetic-and-how-much-of-it-is-the-

result-of-experience version is fundamentally flawed. The problem is that it is based on the premise that genetic factors and experiential factors combine in an additive fashion—that a behavioral capacity, such as intelligence, is created by combining some amount of genetics with some amount of experience rather than through the interaction of genetics and experience. Once you learn more about how genetic factors and experience interact, you will better appreciate the folly of this assumption. For the time being, however, let us illustrate its weakness with a metaphor embedded in an anecdote.

The Case of the Thinking Student

One of my students told me (JP) she had read that intelligence was one-third genetic and two-thirds experience, and she wondered whether this was true. I responded by asking her the following question: "If I wanted to get a better understanding of music, would it be reasonable for me to begin by asking how much of it came from the musician and how much of it came from the instrument?"

"That would be dumb," she said. "The music comes from both; it makes no sense to ask how much comes from the musician and how much comes from the instrument. Somehow the music results from the interaction of the two together. You would have to ask about the interaction."

"That's exactly right," I said. "Now, do you see why..."

"Don't say any more," she interrupted. "I see what you're getting at. Intelligence is the product of the interaction of genes and experience, and it is dumb to try to find how much comes from genes and how much comes from experience."

"Yes!" I thought.

The point of this metaphor, in case you have forgotten, is to illustrate why it is inappropriate to try to understand interactions between two factors by asking how much each factor contributes. We would not ask how much a musician and how much her instrument contributes to producing music; we would not ask how much the water and how much the temperature contributes to evaporation; and we would not ask how much a male and how much a female contributes to reproduction. Similarly, we shouldn't ask how much genetic and how much experiential factors contribute to behavioral development. The answers to all these questions lie in understanding the nature of the interactions (see Sung et al., 2014; Uher, 2014). The importance of thinking about development in terms of interactions will become even more apparent later in this chapter.

A MODEL OF THE BIOLOGY OF BEHAVIOR. So far in this module, you have learned why people tend to think about the biology of behavior in terms of dichotomies, and you have learned some of the reasons why this way of thinking is inappropriate. Now, let's look at a way of thinking about the biology of behavior that has been adopted by most biopsychologists. It is illustrated in Figure 2.3. Like

Figure 2.3 A schematic illustration of the way in which most biopsychologists think about the biology of behavior.

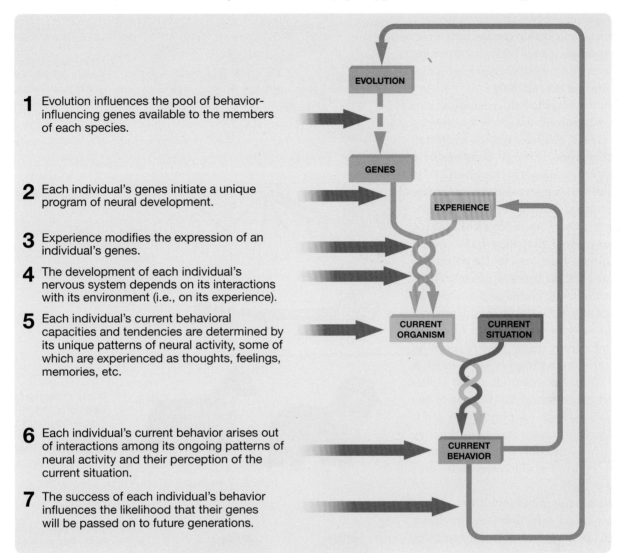

1 Evolution influences the pool of behavior-influencing genes available to the members of each species.

2 Each individual's genes initiate a unique program of neural development.

3 Experience modifies the expression of an individual's genes.

4 The development of each individual's nervous system depends on its interactions with its environment (i.e., on its experience).

5 Each individual's current behavioral capacities and tendencies are determined by its unique patterns of neural activity, some of which are experienced as thoughts, feelings, memories, etc.

6 Each individual's current behavior arises out of interactions among its ongoing patterns of neural activity and their perception of the current situation.

7 The success of each individual's behavior influences the likelihood that their genes will be passed on to future generations.

other powerful ideas, it is simple and logical. This model boils down to the single premise that all behavior is the product of interactions among three factors: (1) the organism's genetic endowment, which is a product of its evolution; (2) its experience; and (3) its perception of the current situation. Please examine the model carefully and consider its implications.

Journal Prompt 2.1

Imagine you are a biopsychology instructor. One of your students asks you whether depression is physiological or psychological. What would you say?

The next three modules of this chapter deal with three elements of this model of behavior: evolution, genetics, and the interaction of genetics and experience in behavioral development. The final module of the chapter deals with the genetics of human psychological differences.

Human Evolution

In this module, you will explore how brain and behavior have been shaped by evolutionary processes. As an entry point to the topic, and to provide some background, you will first learn about the history of the study of evolution. The module then builds upon that foundation by providing you with an overview of several key aspects of the role of evolution in brain and behavior. Moreover, you will learn about some of the most commonly misunderstood aspects about evolution.

Darwin's Theory of Evolution

LO 2.3   Describe the origins of evolutionary theory.

Modern biology began in 1859 with the publication of Charles Darwin's *On the Origin of Species*. In this monumental work, Darwin described his theory of evolution—the single most influential theory in the biological sciences. Darwin was not the first to suggest that species **evolve** (undergo systematic change) from preexisting species, but he was the first to amass a large body of supporting evidence and the first to suggest how evolution occurs (see Bowler, 2009).

Darwin presented three kinds of evidence to support his assertion that species evolve: (1) He documented the evolution of fossil records through progressively more recent geological layers. (2) He described striking structural similarities among living species (e.g., a human's hand, a bird's wing, and a cat's paw), which suggested that they had evolved from common ancestors. (3) He pointed to the major changes that had been brought about in domestic plants and animals by programs of selective breeding. However, the most convincing evidence of evolution comes from direct observations of rapid evolution in progress (see Barrick & Lenski, 2013). For example, Grant (1991) observed evolution of the finches of the Galápagos Islands—a population studied by Darwin himself (see Lamichhaney et al., 2015)—after only a single season of drought. Figure 2.4 illustrates these four kinds of evidence.

Darwin argued that evolution occurs through **natural selection** (see Pritchard, 2010). He pointed out that the members of each species vary greatly in their structure, physiology, and behavior and that the heritable traits associated with high rates of survival and reproduction are the most likely ones to be passed on to future generations (see Kingsley, 2009). He argued that natural selection, when repeated for generation after generation, leads to the evolution of species that are better adapted to surviving and reproducing in their particular environmental niche. Darwin called this process *natural selection* to emphasize its similarity to the artificial selective breeding practices employed by breeders of domestic animals. Just as horse breeders create faster horses by selectively breeding the fastest of their existing stock, nature creates fitter animals by "selectively" breeding the fittest. **Fitness**, in the Darwinian sense, is the ability of an organism to survive and contribute its genes to the next generation.

Darwin's theory of evolution was at odds with the various dogmatic views embedded in the 19th-century zeitgeist,

Figure 2.4 Four kinds of evidence supporting the theory that species evolve.

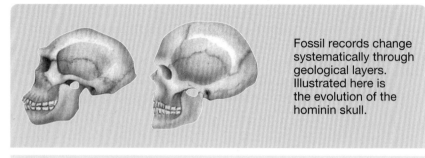

Fossil records change systematically through geological layers. Illustrated here is the evolution of the hominin skull.

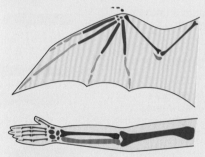

There are striking structural similarities among diverse living species (e.g., between a human arm and a bat's wing).

Major changes have been created in domestic plants and animals by programs of selective breeding.

Evolution has been observed in progress. For example, an 18-month drought on one of the Galápagos Islands left only large, difficult-to-eat seeds, which increased the chance that a bird with a long beak would survive and reproduce.

so it initially met with resistance. Although resistance still exists, virtually none comes from people who understand the evidence (see Short & Hawley, 2015).

Evolution is both a beautiful concept and an important one, more crucial nowadays to human welfare, to medical science, and to our understanding of the world than ever before [see Mindell, 2009]. It's also deeply persuasive—a theory you can take to the bank…the supporting evidence is abundant, various, ever increasing, and easily available in museums, popular books, textbooks, and a mountainous accumulation of scientific studies. No one needs to, and no one should, accept evolution merely as a matter of faith. (Quammen, 2004, p. 8)

Evolution and Behavior

LO 2.4 Explain the evolutionary significance of social dominance and courtship displays.

Some behaviors play an obvious role in evolution. For example, the ability to find food, avoid predation, or defend one's young obviously increases an animal's ability to pass on its genes to future generations. Other behaviors play a role that is less obvious but no less important—for example, social dominance and courtship displays, which are discussed here.

SOCIAL DOMINANCE. The males of many species establish a stable *hierarchy of social dominance* through combative encounters with other males (see Qu et al., 2017). In some species, these encounters often involve physical damage; in others, they involve mainly posturing and threatening until one of the two combatants backs down. The dominant male usually wins encounters with all other males of the group; the number two male usually wins encounters with all males except the dominant male; and so on down the line. Once a hierarchy is established, hostilities diminish because the lower-ranking males learn to avoid or quickly submit to the more dominant males. Because most of the fighting goes on between males competing for positions high in the social hierarchy, low-ranking males fight little, and the lower levels of the hierarchy tend to be only vaguely recognizable.

Why is social dominance an important factor in evolution? One reason is that in many species, dominant males copulate more than nondominant males and thus are more effective in passing on their characteristics to future generations. McCann (1981) studied the effect of social dominance on the rate of copulation in 10 bull elephant seals that cohabited the same breeding beach. These massive animals challenge each other by raising themselves to full height and pushing chest to chest. Usually, the smaller of the two backs down; if it does not, a vicious neck-biting battle ensues. McCann found that the dominant male accounted for about 37 percent of the copulations during the study, whereas poor number 10 accounted for only about 1 percent (see Figure 2.5).

Another reason why social dominance is an important factor in evolution is that in some species, dominant females are more likely to produce more and healthier offspring. For example, Pusey, Williams, and Goodall (1997) found that high-ranking female chimpanzees produced more offspring and that these offspring were more likely to survive to sexual maturity. They attributed these advantages to the fact that high-ranking female chimpanzees are more likely to maintain access to productive foraging areas (see Pusey & Schroepfer-Walker, 2013).

COURTSHIP DISPLAY. An intricate series of courtship displays precedes copulation in many species. The male approaches the female and signals his interest. His signal (which may be olfactory, visual, auditory, or tactual) may elicit a signal in the female, which may elicit another response in the male, and so on, until copulation ensues. But copulation is unlikely to occur if one of the pair fails to react appropriately to the signals of the other.

Courtship displays are thought to promote the evolution of new species. Let us explain. A **species** is a group of organisms reproductively isolated from other organisms; that is, the members of a species can produce fertile offspring only by mating with members of the same species (see de Knijff, 2014). A new species begins to branch off from an existing species when some barrier discourages breeding between a subpopulation of the existing species and the remainder of the species. Once such a reproductive barrier forms, the subpopulation evolves independently of the remainder of the species until cross-fertilization becomes impossible (see Arnegard et al., 2014; Roesti & Salzburger, 2014).

The reproductive barrier may be geographic; for example, a few birds may fly together to an isolated island, where many generations of their offspring breed among themselves and evolve into a separate species. Alternatively—to get back to the main point—the reproductive barrier may be behavioral. A few members of a species may develop

Figure 2.5 Two massive bull elephant seals challenge one another. Dominant bull elephant seals copulate more frequently than those lower in the dominance hierarchy.

Francois Gohier/Science Source

different courtship displays, and these may form a reproductive barrier between themselves and the rest of their **conspecifics** (members of the same species): Only the suitable exchange of displays between a courting couple will lead to reproduction.

Course of Human Evolution

LO 2.5   Summarize the pathway of evolution from single-cell organisms to humans.

By studying fossil records and comparing current species, we humans have looked back in time and pieced together the evolutionary history of our species—although some of the details are still controversial. The course of human evolution, as it is currently understood, is summarized in this section.

EVOLUTION OF VERTEBRATES.   Complex multicellular water-dwelling organisms first appeared on earth about 800 million years ago. About 250 million years later, the first chordates evolved (Satoh, 2016). **Chordates** (pronounced "KOR-dates") are animals with dorsal nerve cords (large nerves that run along the center of the back, or *dorsum*); they are 1 of the 40 or so large categories, or *phyla* (pronounced "FY-la"), into which zoologists group animal species (Zhang, 2013). The first chordates with spinal bones to protect their dorsal nerve cords evolved about 25 million years later. The spinal bones are called *vertebrae* (pronounced "VERT-eh-bray"), and the chordates that possess them are called **vertebrates**. The first vertebrates were primitive bony fishes (Shu et al., 1999). Today, there are seven classes of vertebrates: three classes of fishes, plus amphibians, reptiles, birds, and mammals.

EVOLUTION OF AMPHIBIANS.   About 410 million years ago, the first bony fishes started to venture out of the water (see Figure 2.6).

Fishes that could survive on land for brief periods of time had two great advantages: They could escape from stagnant pools to nearby fresh water, and they could take advantage of terrestrial food sources. The advantages of life on land were so great that, through the process of natural selection, the fins and gills of bony fishes transformed into legs and lungs, respectively, and so it was that the first **amphibians** evolved about 370 million years ago. Amphibians (e.g., frogs, toads, and salamanders) in their larval form must live in the water; only adult amphibians can survive on land.

EVOLUTION OF REPTILES.   About 315 million years ago, reptiles (e.g., lizards, snakes, and turtles) evolved from a branch of amphibians. Reptiles were the first vertebrates to lay shell-covered eggs and to be covered by dry scales. Both of these adaptations reduced the reliance of reptiles on watery habitats. A reptile does not have to spend the first stage of its life in the watery environment of a pond or lake; instead, it spends the first stage of its life in the watery environment of a shell-covered egg. And once hatched, a reptile can live far from water because its dry scales greatly reduce water loss through its water-permeable skin.

EVOLUTION OF MAMMALS.   About 225 million years ago, during the height of the age of dinosaurs, a new class of vertebrates evolved from one line of small reptiles. The females of this new class fed their young with secretions from special glands called *mammary glands*, and the members of the class are called **mammals** after these glands. Eventually, mammals stopped laying eggs; instead, the females nurtured their young in the watery environment of their bodies until the young were mature enough to be born. The duck-billed platypus is one surviving mammalian species that lays eggs.

Spending the first stage of life inside one's mother proved to have considerable survival value; it provided the long-term security and environmental stability necessary for complex programs of development to unfold. Today, most classification systems recognize about 26 different orders of mammals. The order to which we belong is the order **primates**. We humans—in our usual humble way—named our order using the Latin term *primus*, which means "first" or "foremost."

Primates have proven particularly difficult to categorize because there is no single characteristic possessed by all primates but no other animals. Still, most experts agree

Figure 2.6   A recently discovered fossil of a missing evolutionary link is shown on the right, and a reconstruction of the creature is shown on the left. It had scales, teeth, and gills like a fish and primitive wrist and finger bones similar to those of land animals.

Beth Rooney Photography

there are about 16 groups of primates. Species from five of them appear in Figure 2.7.

Apes (gibbons, orangutans, gorillas, and chimpanzees) are thought to have evolved from a line of Old World monkeys. Like Old World monkeys, apes have long arms and grasping hind feet that are specialized for arboreal (treetop) travel, and they have opposable thumbs that are not long enough to be of much use for precise manipulation (see Figure 2.8). Unlike Old World monkeys, though, apes have no tails and can walk upright for short distances. Chimpanzees are the closest living relatives of humans; almost 99 percent of genes are identical in the two species (see Rogers & Gibbs, 2014; but see Cohen, 2007); however, the actual ape ancestor of humans is likely long extinct (Jaeger & Marivaux, 2005).

EMERGENCE OF HUMANKIND. Primates of the same group that includes humans are known as **hominins** (see Figure 2.9). Hominins include six sub-groups including *Australopithecus* and *Homo*. Based on the fossil record, *Homo* is thought to be composed of at least eight species (see Wiedemann, 2014; Gibbons, 2015a); seven of which are now extinct. Perhaps you have heard of the Neanderthals (*Homo Neanderthalensis*)? They are one of those extinct *Homo* species. And we humans (*Homo Sapiens*) are the only one still kicking around.

It is difficult to reconstruct the events of human evolution because the evidence is so sparse. Only a few partial hominin fossils dating from the critical period have been discovered. However, three important hominin fossil discoveries have been particularly enlightening (see Harmon, 2013):

- An uncommonly complete fossil of a 3-year-old early *Australopithecus* girl in Ethiopia (see Figure 2.10; Gibbons, 2009; Suwa et al., 2009; White et al., 2009).

- Fossils indicating that a population of tiny hominins inhabited the Indonesian island of Flores as recently as 18,000 years ago (see Callaway, 2014; Stringer, 2014).

- Several early Australopithecine fossils with combinations of human and nonhuman characteristics in a pit in South Africa (Pickering et al., 2011; Wong, 2012).

Many experts believe that the Australopithecines evolved about 4 million years ago in Africa (see Krubitzer & Stolzenberg, 2014; Skinner et al., 2015; Wood, 2010) from a line of apes (*australo* means "southern," and *pithecus* means "ape"). Several species of Australopithecus are thought to have roamed the African plains for about 2 million years before becoming extinct. Australopithecines were only about 1.3 meters (4 feet) tall, and they had small brains, but analysis of their pelvis and leg bones indicates that their posture was upright. Any doubts about their upright posture were erased by the discovery of the fossilized footprints pictured in Figure 2.11 (see Raichlen et al., 2010).

The first *Homo* species are thought to have evolved from one

Figure 2.7 Species from five different groups of primates.

APE
Silver-Backed Lowland Gorilla

OLD WORLD MONKEY
Hussar Monkey

CEBID
Squirrel Monkey

TARSIER
Tarsier Monkey

HOMININ
Human

Clockwise from right corner: Kevin Schafer/Photolibrary/Getty Images (APE); Vladimir Sazonov/Shutterstock (HOMININ); Daniel Frauchiger, Switzerland/Moment/Getty Images (TARSIER); Michael Krabs/Alamy Stock Photo (CEBID); Anatoliy Lukich/Shutterstock (OLD WORLD MONKEY)

Figure 2.8 A comparison of the feet and hands of a human and a chimpanzee.

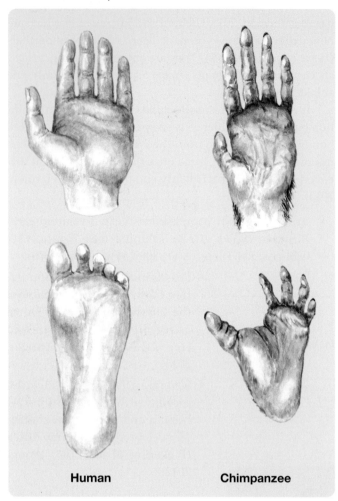

Human   Chimpanzee

Figure 2.9 A taxonomy of the human species.

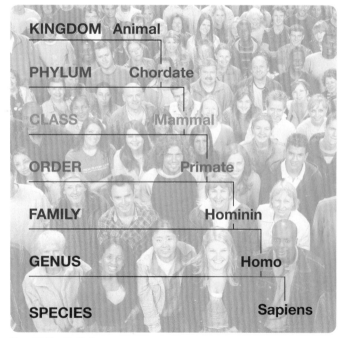

KINGDOM Animal

PHYLUM Chordate

CLASS Mammal

ORDER Primate

FAMILY Hominin

GENUS Homo

SPECIES Sapiens

Ryan McVay/Getty Images

Figure 2.10 The remarkably complete skull of a 3-year-old *Australopithecus* girl; the fossil is 3.3 million years old.

Lealisa Westerhoff/AFP/Getty Images

species of *Australopithecus* about 2 to 2.8 million years ago (see Antón, Potts, & Aiello, 2014; Dimaggio et al., 2015; Schroeder et al., 2014; Villmoare et al., 2015; but see Wiedemann, 2014; Wood, 2014). One distinctive feature of the early *Homo* species was the size of their brain cavity, larger than that of *Australopithecus* but smaller than that of modern humans. The early *Homo* species used fire and tools (see Orban & Caruana, 2014; Schwartz & Tattersall, 2015) and coexisted in Africa with various species of *Australopithecus* for about a half-million years, until the australopithecines died out. Early *Homo* species also lived outside of Africa for about 1.85 million years (see Lordkipanidze et al., 2013; Wood, 2011). Then, about 275,000 years ago (see Adler et al., 2014), early *Homo* species were gradually replaced in the fossil record by modern humans (*Homo sapiens*).

Paradoxically, although the big three human attributes—large brain, upright posture, and free hands with an opposable thumb—have been evident for hundreds of thousands of years, most human accomplishments are of recent origin. Artistic products (e.g., wall paintings and carvings) did not appear until about 40,000 years ago (see Krubitzer & Stolzenberg, 2014; Pringle, 2013), ranching and farming were not established until about 10,000 years ago (see Larson et al., 2014), and writing was not used until about 7,500 years ago.

Thinking about Human Evolution

LO 2.6   Describe nine commonly misunderstood points about evolution.

Figure 2.12 illustrates the main branches of vertebrate evolution. As you examine it, consider the following commonly misunderstood points about evolution. They should

Figure 2.11 Fossilized footprints of Australopithecine hominins who strode across African volcanic ash about 3.6 million years ago, leaving a 70-meter trail. There were two adults and a child; the child often walked in the footsteps of the adults.

John Reader/Science Source

provide you with a new perspective from which to consider your own origins.

- Evolution does not proceed in a single line. Although it is common to think of an evolutionary ladder or scale, a far better metaphor for evolution is a dense bush.

- We humans have little reason to claim evolutionary supremacy. We are the last surviving species of a group (i.e., *hominins*) that has existed for only a blip of evolutionary time.

- Evolution does not always proceed slowly and gradually. Rapid evolutionary changes (i.e., in a few generations) can be triggered by sudden changes in the environment or by adaptive genetic mutations. Whether human evolution occurred gradually or suddenly is still a matter of intense debate among *paleontologists* (those who scientifically study fossils).

- Few products of evolution have survived to the present day—only the tips of the branches of the evolutionary bush have survived. Fewer than 1 percent of all known species are still in existence.

- Evolution does not progress to preordained perfection—evolution is a tinkerer, not an architect. Increases in adaptation occur through changes to existing programs of development; and, although the results are improvements in their particular environmental context, they are never perfect designs. For example, the fact that mammalian sperm do not develop effectively at body temperature led to the evolution of the scrotum—hardly a perfect solution to any design problem.

- Not all existing behaviors or structures are adaptive. Evolution often occurs through changes in developmental programs that lead to several related characteristics, only one of which might be adaptive—the incidental nonadaptive evolutionary by-products are called **spandrels**. One example of a spandrel is the human belly button—it is a nonfunctional by-product of the umbilical cord. Also, behaviors or structures that were once adaptive might become nonadaptive, or even maladaptive, if the environment changes.

> **Journal Prompt 2.2**
>
> What might be an example of a behavior or structure that is currently adaptive but that might become non-adaptive, or even maladaptive, if our current environment were to change?

- Not all existing adaptive characteristics evolved to perform their current function. Some characteristics, called **exaptations**, evolved to serve one function and were later co-opted to serve another. For example, bird wings are exaptations—they are limbs that initially evolved for the purpose of walking.

- Similarities among species do not necessarily mean that the species have common evolutionary origins. Structures that are similar because they have a common evolutionary origin are termed **homologous**; structures that are similar but do not have a common evolutionary origin are termed **analogous**. The similarities between analogous structures result from **convergent evolution**, the evolution in unrelated species of similar solutions to the same environmental demands (see Stern, 2013). Deciding whether a

Figure 2.12 Hominin evolution.

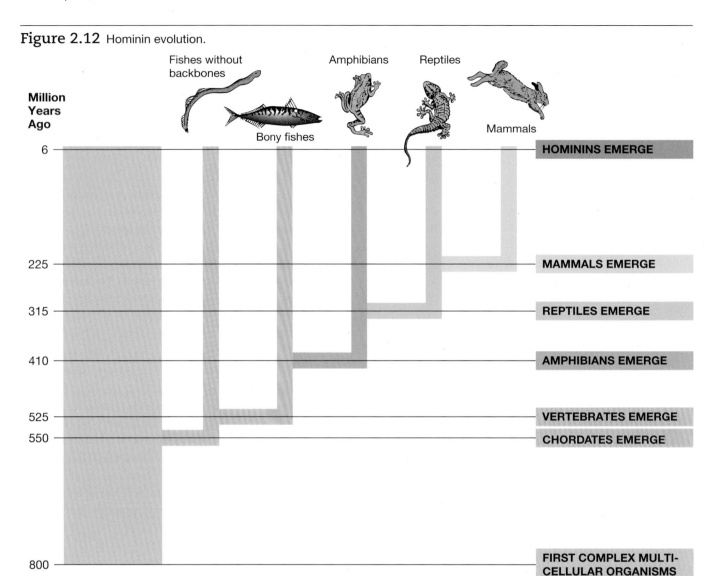

Evolution of the Human Brain

LO 2.7 Describe how research on the evolution of the human brain has changed over time.

structural similarity is analogous or homologous requires careful analysis of the similarity. For example, a bird's wing and a human's arm have a basic underlying commonality of skeletal structure that suggests a common ancestor; in contrast, a bird's wing and a bee's wing have few structural similarities, but they both evolved because of the common advantage of flight.

- There is now considerable evidence that *Homo sapiens* mated with other *Homo* species (e.g., Neanderthals) they encountered (see Dannemann & Racimo, 2018; Gibbons, 2014; Wong, 2015). The discovery of this pattern of mating changes the way we should view our origins: We are not the product of a single ancestral *Homo* population; rather, we are the combined offspring of many *Homo* populations that once coexisted and interacted.

Early research on the evolution of the human brain focused on size. This research was stimulated by the assumption that brain size and intellectual capacity are closely related—an assumption that quickly ran into two problems. First, it was shown that modern humans, whom we humans believe to be the most intelligent of all creatures, do not have the biggest brains. With brains weighing about 1,350 grams, humans rank far behind whales and elephants, whose brains weigh between 5,000 and 8,000 grams (Manger, 2013; Patzke et al., 2014). Second, the sizes of the brains of acclaimed intellectuals (e.g., Albert Einstein) were found to be unremarkable, certainly no match for their gigantic intellects. It is now clear that, although healthy adult human brains vary greatly in size—between about 1,000 and 2,000 grams—there is no clear relationship between overall human brain size and intelligence.

One obvious problem in relating brain size to intelligence is the fact that larger animals tend to have larger brains, presumably because larger bodies require more brain tissue to control and regulate them. Thus, the facts that large men tend to have larger brains than small men, that men tend to have larger brains than women, and that elephants have larger brains than humans do not suggest anything about the relative intelligence of these populations. This problem led to the proposal that brain weight expressed as a percentage of total body weight might be a better measure of intellectual capacity. This measure allows humans (2.33 percent) to take their rightful place ahead of elephants (0.20 percent), but it also allows both humans and elephants to be surpassed by that intellectual giant of the animal kingdom, the shrew (3.33 percent).

A more reasonable approach to the study of brain evolution has been to compare the evolution of different brain regions. For example, it has been informative to consider the evolution of the **brain stem** separately from the evolution of the **cerebrum** (cerebral hemispheres). In general, the brain stem regulates reflex activities that are critical for survival (e.g., heart rate, respiration, and blood glucose level), whereas the cerebrum is involved in more complex adaptive processes such as learning, perception, and motivation.

Figure 2.13 is a schematic representation of the relative size of the brain stems and cerebrums of several species that are living descendants of species from which humans evolved. This figure makes three important points about the evolution of the human brain:

- The brain has increased in size during evolution.

- Most of the increase in size has occurred in the cerebrum.

- An increase in the number of **convolutions**—folds on the cerebral surface—has greatly increased the surface area of the *cerebral cortex*, the outermost layer of cerebral tissue (see Geschwind & Rakic, 2013; Zilles, Palermo-Gallagher, & Amunts, 2013).

Although the brains of related species differ, there are fundamental similarities: All brains are constructed of many neurons, and the neural structures in the brains of one species can usually be found in the same locations in the brains of related species (see Goulas et al., 2014). For example, the brains of humans, monkeys, rats, and mice contain the same major structures connected in similar ways, and similar structures tend to perform similar functions (see Cole et al., 2009). The human brain appears to have evolved from the brains of our closest primate relatives (see Hofman, 2014; Matsuzawa, 2013).

Figure 2.13 The brains of animals of different evolutionary ages—cerebrums are shown in pink; brain stems are shown in orange.

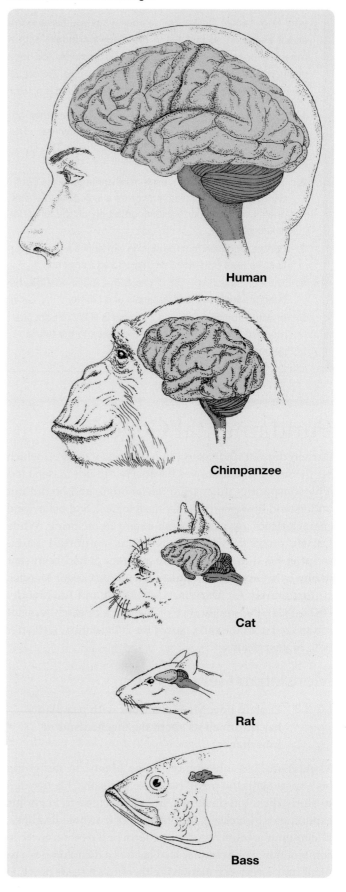

Human

Chimpanzee

Cat

Rat

Bass

Scan Your Brain

This is a good place to pause and scan your brain to check your knowledge. Do you remember what you have learned about evolution so far? Fill in the following blanks with the most appropriate terms from the first two modules. The correct answers are provided at the end of the exercise. Before proceeding, review material related to your errors and omissions.

1. There has been a long-standing debate on whether humans and other animals inherit their behavioral responses or acquire them through learning. This is called the _____ debate.

2. The condition that can result from damage to the right parietal lobe and typically involves a lack of awareness of one's own body parts (most commonly on the left side) is known as _____.

3. Darwin proposed that the striking similarities among living species were evidence that they shared a common _____.

4. Through selective _____ programs, major changes have been made to domestic animals and plants.

5. In some species, the _____ male is likely to copulate more and, therefore, pass on his genes to the future generations.

6. One distinctive feature of early *Homo* species was that they had brains _____ than Australopithecus but _____ than modern humans.

7. Incidental nonadaptive evolutionary by-products such as the belly button are called _____.

8. During the course of the vertebrate evolution, birds emerged approximately _____ years ago.

9. The overall human brain size does not predict _____.

10. Over millions of years, there has been a remarkable increase in the surface area of the _____, the outmost layer of the cerebral tissue in humans.

11. Evolutionary psychologists suggest that male–female _____ during copulation ensures that the offspring will survive, reproduce, and pass on their genes to the next generation.

12. _____ structures are similar because of convergent evolution.

Scan Your Brain answers: (1) nature–nurture, (2) asomatognosia, (3) ancestor, (4) breeding, (5) dominant, (6) larger, smaller, (7) spandrels, (8) 160 million, (9) intelligence, (10) cerebral cortex, (11) bonding, (12) Analogous.

Fundamental Genetics

Darwin did not understand two of the key facts on which his theory of evolution was based. He did not understand why conspecifics differ from one another, and he did not understand how anatomical, physiological, and behavioral characteristics are passed from parent to offspring. While Darwin puzzled over these questions, an unread manuscript in his files contained the answers. It had been sent to him by an unknown Augustinian monk, Gregor Mendel. Unfortunately for Darwin (1809–1882) and for Mendel (1822–1884), the significance of Mendel's research was not recognized until the early part of the 20th century, well after both of their deaths.

Mendelian Genetics

LO 2.8 **Explain how Mendel's work with pea plants has informed us about the mechanisms of inheritance.**

Mendel studied inheritance in pea plants. In designing his experiments, he made two wise decisions. He decided to study dichotomous traits, and he decided to begin his experiments by crossing the offspring of true-breeding lines. **Dichotomous traits** occur in one form or the other, never in combination. For example, seed color is a dichotomous pea plant trait: Every pea plant has either brown seeds or white seeds. **True-breeding lines** are breeding lines in which interbred members always produce offspring with the same trait (e.g., brown seeds), generation after generation.

In one of his early experiments, Mendel studied the inheritance of seed color: brown or white. He began by crossbreeding the offspring of a line of pea plants that had bred true for brown seeds with the offspring of a line of pea plants that had bred true for white seeds. The offspring of this cross all had brown seeds. Then, Mendel bred these first-generation offspring with one another, and he found that about three-quarters of the resulting second-generation offspring had brown seeds and about one-quarter had white seeds. Mendel repeated this experiment many times with various pairs of dichotomous pea plant traits, and each time the result was the same: One trait, which Mendel called the **dominant trait**, appeared in all of the first-generation offspring; the other trait, which he called the **recessive trait**, appeared in about one-quarter of the second-generation offspring. Mendel would have obtained a similar result if he had conducted an experiment with true-breeding lines of brown-eyed (dominant) and blue-eyed (recessive) humans.

The results of Mendel's experiment challenged the central premise on which all previous ideas about inheritance had rested: that offspring inherit the traits of their parents. Somehow, the recessive trait (white seeds) was passed on to one-quarter of the second-generation pea plants by first-generation pea plants that did not themselves possess it. An organism's observable traits are referred to as its **phenotype**; the traits that it can pass on to its offspring through its genetic material are referred to as its **genotype**.

Mendel devised a theory to explain his results. It comprised four central ideas. First, Mendel proposed that there are two kinds of inherited factors for each dichotomous trait—for example, that a brown-seed factor and a white-seed factor control seed color. Today, we call each inherited factor a **gene**. Second, Mendel proposed that each organism possesses two genes for each of its dichotomous traits; for example, each pea plant possesses either two brown-seed genes, two white-seed genes, or one of each. The two genes that control the same trait are called **alleles** (pronounced "a-LEELZ"). Organisms that possess two identical alleles (e.g., two white-seed alleles) are said to be **homozygous** for that trait; those that possess different alleles (e.g., one white-seed allele and one black-seed allele) for a trait are said to be **heterozygous** for that trait.

Third, Mendel proposed that one of the two kinds of genes for each dichotomous trait dominates the other in heterozygous organisms. For example, pea plants with a brown-seed gene and a white-seed gene always have brown seeds because the brown-seed gene always dominates the white-seed gene. And fourth, Mendel proposed that for each dichotomous trait, each organism randomly inherits one of its "father's" two factors and one of its "mother's" two factors. Figure 2.14 illustrates how Mendel's theory accounts for the result of his experiment on the inheritance of seed color in pea plants.

Chromosomes

LO 2.9   Understand the structure and function of chromosomes.

In this section, you will be presented with current knowledge related to two key aspects of chromosomal function: recombination and replication. The section ends with a discussion of the sex chromosomes and sex-linked traits.

REPRODUCTION AND RECOMBINATION.   It was not until the early 20th century that genes were found to be located on **chromosomes**—the threadlike structures in the nucleus of each cell (see Brenner, 2012). Chromosomes occur in matched pairs in virtually all multicellular organisms (see Sagi & Benvenisty, 2017), and each species has a characteristic number of pairs in each of its body cells (but see Sagi & Benvenisty, 2017); humans have 23 pairs. The two genes (alleles) that control each trait are situated at the same location, one on each chromosome of a particular pair.

The process of cell division that produces **gametes** (egg cells and sperm cells) is called **meiosis** (pronounced

Figure 2.14 How Mendel's theory accounts for the results of his experiment on the inheritance of seed color in pea plants.

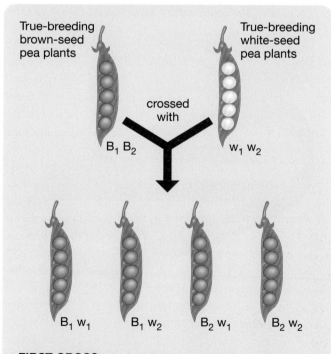

FIRST CROSS
One parent had two dominant brown-seed genes (B_1 B_2); the other had two recessive white-seed genes (w_1 w_2). Therefore, all offspring had one brown-seed gene and one white-seed gene (B_1 w_1, B_1 w_2, B_2 w_1, or B_2 w_2). Because the brown-seed gene is dominant, all had brown seeds.

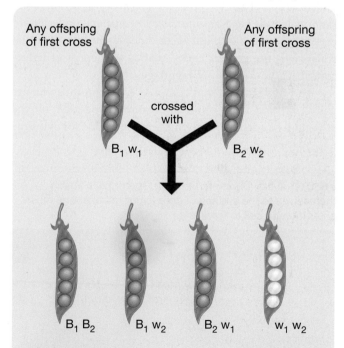

SECOND CROSS
Each parent had one brown-seed gene and one white-seed gene. Therefore, 25% of the offspring had two brown-seed genes (B_1 B_2), 50% had a brown-seed gene and a white-seed gene (B_1 w_2 or B_2 w_1), and 25% had two white-seed genes (w_1 w_2). Because the brown-seed gene is dominant, 75% had brown seeds.

"my-OH-sis"). In meiosis, the chromosomes divide, and one chromosome of each pair goes to each of the two gametes that result from the cell division. As a result, each gamete has only half the usual number of chromosomes (23 in humans); and when a sperm cell and an egg cell combine during fertilization (see Figure 2.15), a **zygote** (a fertilized egg cell) with the full complement of chromosomes (23 pairs in humans) is produced.

The random division of the pairs of chromosomes into two gametes is not the only way meiosis contributes to genetic diversity. Let us explain. During the first stage of meiosis, the chromosomes line up in their pairs. Then, the members of each pair cross over one another at random points, break apart at the points of contact, and exchange sections. As a result of this **genetic recombination**, each of the gametes that formed the zygote that developed into you contained chromosomes that were unique, spliced-together recombinations of chromosomes from your parents.

In contrast to the meiotic creation of the gametes, all other cell division in the body occurs by **mitosis** (pronounced "my-TOE-sis"). Just prior to mitotic division, the number of chromosomes doubles so that, when the cell divides, both daughter cells end up with the full complement of chromosomes (23 pairs in humans).

STRUCTURE AND REPLICATION. Each chromosome is a double-stranded molecule of **deoxyribonucleic acid (DNA)**. Each strand is a sequence of **nucleotide bases** attached to a chain of *phosphate* and *deoxyribose*; there are four nucleotide bases: *adenine, thymine, guanine,* and *cytosine*. It is the sequence of these bases on each chromosome that constitutes the genetic code—just as sequences of letters constitute the code of our language.

The two strands that compose each chromosome are coiled around each other and bonded together by the attraction of adenine for thymine and guanine for cytosine.

This specific bonding pattern has an important consequence: The two strands that compose each chromosome are exact complements of each other. For example, a sequence of adenine, guanine, thymine, cytosine, and guanine on one strand is always attached to a complementary sequence of thymine, cytosine, adenine, guanine, and cytosine on the other. Figure 2.16 illustrates the structure of DNA.

Replication is a critical process of the DNA molecule. Without it, mitotic cell division would not be possible. Figure 2.17 illustrates how *DNA replication* is thought to work. The two strands of DNA start to unwind. Then, the exposed nucleotide bases on each of the two strands attract their complementary bases, which are floating in the fluid

Figure 2.16 A schematic illustration of the structure of a DNA molecule. Notice the complementary pairings of nucleotide bases: thymine with adenine and guanine with cytosine.

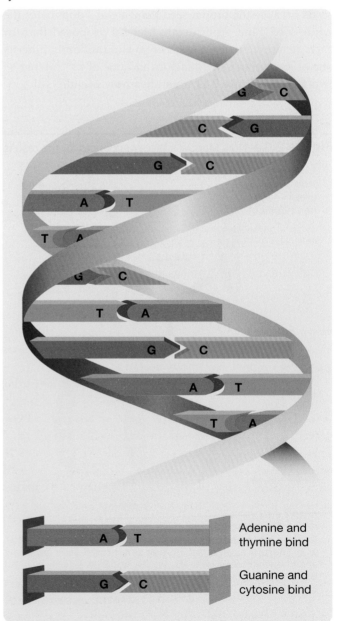

Adenine and thymine bind

Guanine and cytosine bind

Figure 2.15 During fertilization, sperm cells attach themselves to the surface of an egg cell; at least one must enter the egg cell to fertilize it.

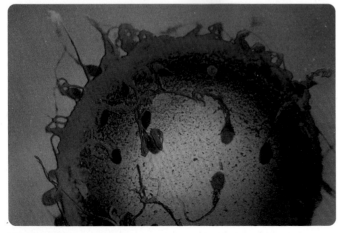

David M. Phillips/Science Source

Figure 2.17 DNA replication. As the two strands of the original DNA molecule unwind, the nucleotide bases on each strand attract free-floating complementary bases. Once the unwinding is complete, two DNA molecules, each identical to the first, will have been created.

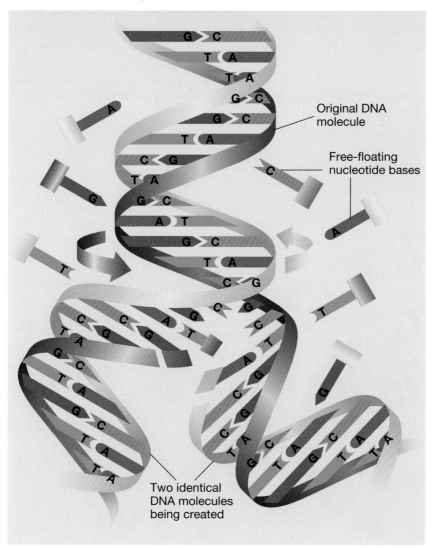

Original DNA molecule

Free-floating nucleotide bases

Two identical DNA molecules being created

in matched pairs, are called **autosomal chromosomes**; the one exception is the pair of **sex chromosomes**—the pair of chromosomes that determines an individual's sex. There are two types of sex chromosomes, X and Y, and the two look different and carry different genes. Females have two X chromosomes, and males have one X chromosome and one Y chromosome. Traits influenced by genes on the sex chromosomes are referred to as **sex-linked traits**. Virtually all sex-linked traits are controlled by genes on the X chromosome because the Y chromosome is small and carries few genes (see Maekawa et al., 2014).

Traits controlled by genes on the X chromosome occur more frequently in one sex than the other. If the trait is dominant, it occurs more frequently in females. Females have twice the chance of inheriting the dominant gene because they have twice the number of X chromosomes. In contrast, recessive sex-linked traits occur more frequently in males. The reason is that recessive sex-linked traits are manifested only in females who possess two of the recessive genes—one on each of their X chromosomes—whereas the traits are manifested in all males who possess the gene because they have only one X chromosome. The classic example of a recessive sex-linked trait is color blindness. Because the color-blindness gene is quite rare, females almost never inherit two of them and thus almost never possess the disorder; in contrast, every male who possesses one color-blindness gene is color blind.

of the nucleus. Thus, when the unwinding is complete, two double-stranded DNA molecules, both of which are identical to the original, have been created.

Chromosome replication does not always go according to plan; there may be errors. Sometimes, these errors can have significant clinical consequences. For example, in *Down syndrome*, which you will learn about in Chapter 10, there is an extra chromosome in each cell. But more commonly, errors in replication take the form of **mutations**—alterations of individual genes. In most cases, mutations disappear from the gene pool within a few generations because the organisms that inherit them are less fit. However, in some instances, mutations increase fitness and in so doing encourage rapid evolution.

SEX CHROMOSOMES AND SEX-LINKED TRAITS. There is one exception to the rule that chromosomes always come in matched pairs. The typical chromosomes, which come

Genetic Code and Gene Expression

LO 2.10 Describe the process of gene expression.

Structural genes contain the information necessary for the synthesis of proteins. **Proteins** are long chains of **amino acids**; they control the physiological activities of cells and are important components of cellular structure. All the cells in the body (e.g., brain cells, hair cells, and bone cells) contain exactly the same genes. How then do different kinds of cells develop? Part of the answer lies in those stretches of DNA that lack structural genes—indeed, although all genes were once assumed to be structural genes, those genes comprise only a small portion of each chromosome.

Although the stretches of DNA that lack structural genes are not well understood, it is clear that they include portions that do serve a function (see Hawkins, Al-attar, & Storey, 2018). These portions, called **promoters**, are stretches of DNA whose function is to determine whether or not particular structural genes are converted into proteins through a two-phase process known as **gene expression**. The control of gene expression by promoters is an important process because it heavily influences how a cell will develop and how it will function once it reaches maturity. Promoters are like switches because they can be regulated in two ways: They can be turned up, or they can be turned down.

Those proteins that bind to DNA and increase gene expression are called **activators**; whereas those that bind to DNA and decrease gene expression are called **repressors**.

The expression of a structural gene is illustrated in Figure 2.18. Gene expression involves two phases. In the first phase, known as **transcription**, the small section of the chromosome that contains the gene unravels, and

Figure 2.18 Gene expression. Transcription of a section of DNA into a complementary strand of messenger RNA (mRNA) is followed by the translation of the messenger RNA strand into a protein.

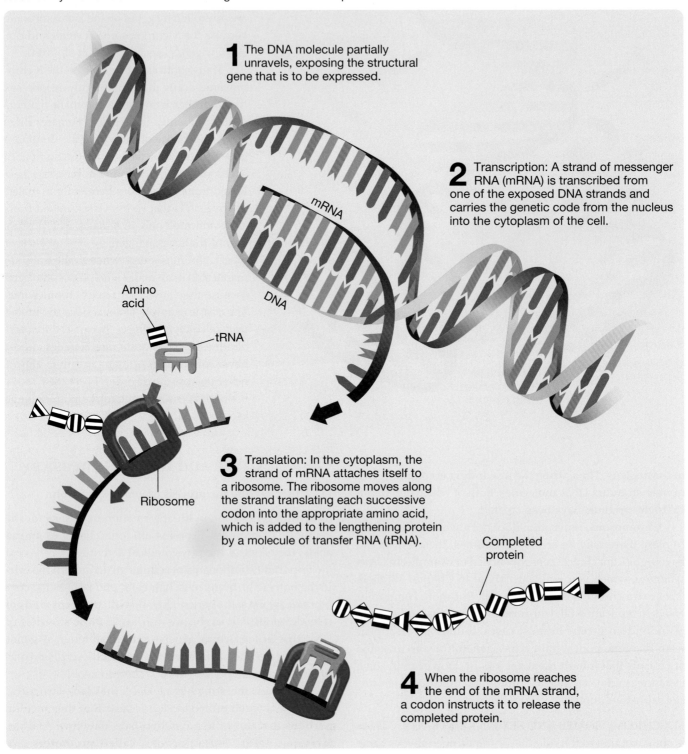

1 The DNA molecule partially unravels, exposing the structural gene that is to be expressed.

2 Transcription: A strand of messenger RNA (mRNA) is transcribed from one of the exposed DNA strands and carries the genetic code from the nucleus into the cytoplasm of the cell.

Amino acid

tRNA

mRNA

DNA

Ribosome

3 Translation: In the cytoplasm, the strand of mRNA attaches itself to a ribosome. The ribosome moves along the strand translating each successive codon into the appropriate amino acid, which is added to the lengthening protein by a molecule of transfer RNA (tRNA).

Completed protein

4 When the ribosome reaches the end of the mRNA strand, a codon instructs it to release the completed protein.

the unraveled section of one of the DNA strands serves as a template for the transcription of a short strand of **ribonucleic acid (RNA)**. RNA is like DNA except that it contains the nucleotide base uracil instead of thymine and has a phosphate and ribose backbone instead of a phosphate and deoxyribose backbone. The strand of transcribed RNA is called **messenger RNA** because it carries the genetic code out of the nucleus of the cell.

Once the messenger RNA leaves the nucleus, the second phase of gene expression, known as **translation**, begins. During translation, the messenger RNA attaches itself to any one of the many **ribosomes** present in the cell's *cytoplasm* (the clear fluid within the cell). The ribosome then moves along the strand of messenger RNA, translating the genetic code as it proceeds.

Each group of three consecutive nucleotide bases along the messenger RNA strand is called a **codon**. Each codon instructs the ribosome to add 1 of the 20 different kinds of amino acids to the protein it is constructing; for example, the sequence guanine-guanine-adenine instructs the ribosome to add the amino acid glycine. Each kind of amino acid is carried to the ribosome by molecules of **transfer RNA**; as the ribosome reads a codon, it attracts a transfer RNA molecule that is attached to the appropriate amino acid. The ribosome reads codon after codon and adds amino acid after amino acid until it reaches a codon that tells it the protein is complete, whereupon the completed protein is released into the cytoplasm.

In summary, the process of gene expression involves two phases. The first phase involves the transcription of the DNA base-sequence code to an RNA base-sequence code. The second phase involves the translation of the RNA base-sequence code into a protein.

Human Genome Project

LO 2.11 Discuss several ways in which modern advances have changed our understanding of genetic processes.

One of the most ambitious scientific projects of all time began in 1990. Known as the **Human Genome Project**, it was a loosely knit collaboration of major research institutions and individual research teams in several countries. Its purpose was to compile a map of the sequence of all 3 billion nucleotide bases that compose human chromosomes.

The Human Genome Project was motivated by potential medical applications. It was assumed that once the human genome was described, it would be a relatively straightforward matter to link variations in the genome to particular human diseases and then develop treatment and prevention programs tailored to individual patients.

However, more than two decades after the human genome was first described, these medical miracles have yet to be realized (see Hall, 2010). Be that as it may, The Human Genome Project has changed our understanding of ourselves and revolutionized the field of genetics. The following are three major contributions of the Human Genome Project:

- Many new techniques for studying DNA were developed during the Human Genome Project. Many things that were impossible before the Human Genome Project are now routine, and things that took months to accomplish before the Human Genome Project are now possible in only a few hours. Using this new technology, genomes have already been established for many species, including those of many long-extinct species (see Shapiro & Hofreiter, 2014), leading to important insights into evolution.

- The discovery that we humans, the most complex of all species, have relatively few structural genes surprised many scholars. Humans have about 21,000 structural genes; mice have about the same number, and corn has many more. Indeed, protein-encoding genes constitute only about 1 percent of human DNA. Researchers have now generated a nearly complete map of the entire set of proteins encoded for by our genes: the **human proteome** (see Ommen et al., 2018).

- Many variations in the human genome related to particular diseases have been identified. However, this has proven to be less useful than anticipated: So many genes have been linked to each disease that it has proven difficult to sort out the interactions among the numerous genes and experience (Hall, 2010). Compounding the problem is that even when many genes have been linked to a disease, all of them together often account for only a small portion of its heritability (Manolio et al., 2009). For example, 18 different gene variants have been linked to adult-onset diabetes, but these 18 variants account for only 6 percent of the heritability of the disease (see Stumvoll, Goldstein, & van Haeften, 2005).

Modern Genetics: Growth of Epigenetics

LO 2.12 Define epigenetics, and explain how it has transformed our understanding of genetics.

Since the discovery of genes in the 1960s, the structure and expression of protein-encoding genes had been the focus of genetics research and thinking (see Franklin & Mansuy, 2010; Zhang & Meaney, 2010). However, around the turn of the century, the field of genetics changed. Interest shifted away from genes and their expression to other possible mechanisms of inheritance. In particular, interest shifted to the mechanisms by which experience exerts its effects on development. This led to an explosion of interest in an area of genetics research that had been lingering in the background since 1942: epigenetics. **Epigenetics** is the study of all mechanisms of inheritance other than those mediated by changes to the gene sequence of DNA.

Why did epigenetics rise to prominence so quickly at the turn of the century? Four conditions set the stage. First, the Human Genome Project had generated an arsenal of new research techniques. Second, it was discovered that protein-coding genes constitute only about 1 percent of human DNA—it wasn't clear to researchers what the other 99% was doing (it was widely regarded as "junk DNA"). Third, it was found that the vast majority of RNA molecules were small—only 1.2 percent were of the large protein-encoding variety. This suggested that protein encoding is only a minor function of RNA (see Dolgin, 2015; Wilusz & Sharp, 2013). Finally, although there was a general consensus that development was the product of gene-experience interactions (see Figure 2.3), the mechanisms by which these critical interactions took place were unknown (see Qureshi & Mehler, 2012).

Stimulated by these four conditions, it was not long before a wave of research into epigenetics began to produce important discoveries. Genetics had just spent half a century focused exclusively on the genetic code as the mechanism of inheritance, and the new epigenetic research led to discoveries that challenged this narrow view.

Despite its relative youth, epigenetic research has already amassed an impressive array of discoveries. Here are five important ones:

- Epigenetic investigations of nongene DNA have identified many active areas. Many of these active areas seem to control the expression of nearby genes. Clearly, the belief that nongene DNA is junk DNA is no longer tenable (see Pennisi, 2014; Tragante, Moore, & Asselbergs, 2014).

- Many epigenetic mechanisms that can modulate gene expression have been discovered. Two of the most widely

studied are DNA methylation and histone remodeling (see Campbell & Wood, 2019; Cavalli & Heard, 2019; Schultz et al., 2015)—see Figure 2.19. **DNA methylation** is the reaction that occurs when a methyl group attaches to a DNA molecule, usually at cytosine sites in mammals (see Schübeler, 2012). **Histone remodeling** is the reaction that occurs when **histones** (proteins around which DNA is coiled) change their shape and in so doing influence the shape of the adjacent DNA—there are several different mechanisms by which this can occur. Both DNA methylation and histone remodeling can either decrease or increase expression (see Bintu et al., 2016; Keung & Khalil, 2016).

Figure 2.19 Two frequently studied epigenetic mechanisms. Histone remodeling involves modifications to a histone protein (around which DNA is coiled). DNA methylation involves the attachment of a methyl group to DNA. Both DNA methylation and histone remodeling can either decrease or increase gene expression.

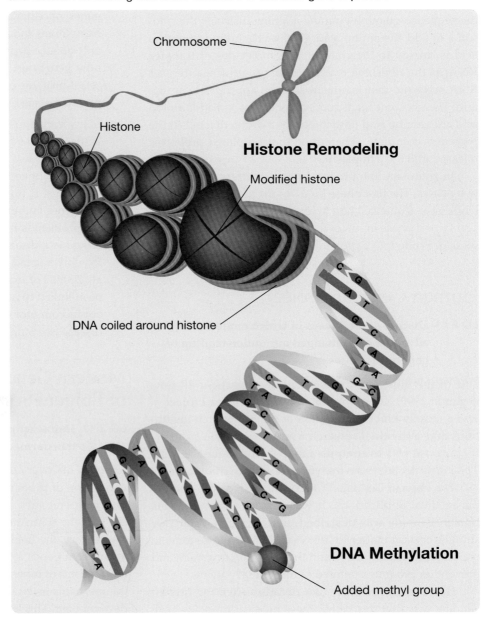

Chromosome

Histone

Histone Remodeling

Modified histone

DNA coiled around histone

DNA Methylation

Added methyl group

- So much interest has been generated by epigenetics research that a world-wide effort is now underway to catalogue the epigenome of each cell type. An **epigenome** represents a catalogue of all the modifications to DNA within a particular cell type other than changes to the nucleotide base sequence (see Romanoski et al., 2015). As of 2018, the epigenomes of over 600 cell types had been characterized (see Esteller, 2018).

- Some epigenetic effects involve post-transcription alterations to RNA that do not affect the RNA base sequence. This occurs in all RNA molecules that have been examined to date, though special attention has been paid to epigenetic modifications of messenger RNA and transfer RNA. The high prevalence of these RNA modifications has led to a new effort: The cataloguing of the so-called *epitranscriptomes* of various cell types. The **epitranscriptome** of a cell refers to all those modifications of RNA that occur after transcription—that do not involve modifications to the RNA base sequence (see Bludau et al., 2019; Helm & Motorin, 2017).

- Remarkably, epigenetic mechanisms (e.g., DNA methylation, histone remodeling) can be induced by particular experiences (e.g., neural activity, hormonal state, changes to the environment) and can last a lifetime (Campbell & Wood, 2019; Handel & Ramagopalan, 2010; Nadeau, 2009; Nelson & Nadeau, 2010; Riccio, 2010; Sweatt, 2013).

It is clear that epigenetic mechanisms can produce enduring changes in an individual. But can those experience-induced changes be passed on to future generations? That is, can the experiences of your mother and father be passed on to you and on to your children? Biologists first observed such transgenerational epigenetic effects in plants, but such effects have now been observed in mammals as well. **Transgenerational epigenetics** is a subfield of epigenetics that examines the transmission of experiences via epigenetic mechanisms across generations (see Hughes, 2014). For example, it has been shown that when mice experience an odor associated with a painful shock, the memory of that experience is passed on to subsequent generations through epigenetic mechanisms (see Dias et al., 2015; Dias & Ressler, 2014; Szyf, 2014; Welberg, 2014). There is growing evidence that inheritance via transgenerational epigenetic mechanisms can also occur in humans (see Yeshurun & Hannan, 2018). Indeed, it would be an evolutionary advantage to be able to rapidly pass on any new adaptations to a changing environment (see Lowdon, Jang, & Wang, 2016; Yeshurun & Hannan, 2018).

Before leaving this subsection on epigenetics, pause to consider the important implications of what you have just learned. It now seems likely that each person's genetic material changes through life as experiences accumulate, and there is evidence that these experience-induced changes can be passed on to future generations. These findings have revolutionized the field of genetics and have major implications for how we think about our ancestors, ourselves, and our descendants.

Journal Prompt 2.3

What implications might the study of epigenetics have for researchers who are trying to determine the genetic bases of a particular psychiatric disorder?

Scan Your Brain

Do you remember what you have just read about genetics so that you can move on to the next module with confidence? To find out, fill in the following blanks with the most appropriate terms. The correct answers are provided at the end of the exercise. Before proceeding, review material related to your errors and omissions.

1. According to Mendel's experiments, the dominant trait will result in about _____ of the offsprings in the second generation.

2. An offspring's observable traits are called its _____ while its genetic material is referred to as _____.

3. A single trait is controlled by two expressions of the same gene called _____.

4. Egg and sperm cells are also called _____, and each contains half the usual number of chromosomes.

5. During fertilization, the resulting _____ contains a full set of chromosomes.

6. The four nucleotide bases in DNA are adenine, thymine, cytosine, and _____.

7. Accidental errors in individual genes are called _____.

8. _____ are stretches of DNA that control the process of gene expression.

9. _____ mechanisms such as DNA methylation and histone remodeling may control gene.

10. The shape of DNA can be influenced by the change of shape of the adjacent _____.

11. All chromosomes except _____ chromosomes come in matched pairs.

12. The first phase of gene expression involves the transcription of DNA to RNA, and the second phase involves the _____ of RNA base-sequence code into a sequence of amino acids.

13. The Human Genome Project discovered that only _____ percent of human DNA contains protein-encoding genes.

Scan Your Brain answers: (1) three-quarters, (2) phenotype, genotype, (3) alleles, (4) gametes, (5) zygote, (6) guanine, (7) mutations, (8) Promoters, (9) Epigenetic, (10) histones, (11) sex, (12) translation, (13) 1.

Epigenetics of Behavioral Development: Interaction of Genetic Factors and Experience

This module comprises two classic examples of how genetic factors and experience interact to direct behavioral ontogeny. (**Ontogeny** is the development of individuals over their life span; **phylogeny**, in contrast, is the evolutionary development of species through the ages.) In each example, you will see that development is a product of the interaction of genetic and experiential factors, which we now know is likely mediated by epigenetic mechanisms (see Sweatt, 2013).

Selective Breeding of "Maze-Bright" and "Maze-Dull" Rats

LO 2.13 Discuss what insights into the genetics of behavior were gained from early research on selective breeding.

You have already learned in this chapter that most early psychologists assumed that behavior develops largely through learning. Tryon (1934) undermined this assumption by showing that behavioral traits can be selectively bred.

Tryon focused his selective-breeding experiments on the behavior that had been the focus of early psychologists in their investigations of learning: the maze running of laboratory rats. Tryon began by training a large heterogeneous group of laboratory rats to run a complex maze; the rats received a food reward when they reached the goal box. Tryon then mated the females and males that least frequently entered incorrect alleys during training—he referred to these rats as *maze-bright*. And he bred the females and males that most frequently entered incorrect alleys during training—he referred to these rats as *maze-dull*.

When the offspring of both the maze-bright and the maze-dull rats matured, their maze-learning performance was assessed. Then, the brightest of the maze-bright offspring were mated with one another, as were the dullest of the maze-dull offspring. This selective breeding procedure was continued for 21 generations. By the eighth generation, there was almost no overlap in the maze-learning performance of the two strains. With a few exceptions, the worst of the maze-bright strain made fewer errors than the best of the maze-dull strain (see Figure 2.20).

To control for the possibility that good maze-running performance was somehow being passed from parent to offspring through learning, Tryon used a *cross-fostering control procedure*: He tested maze-bright offspring that had been reared by maze-dull parents and maze-dull offspring that had been reared by maze-bright parents. However, the offspring of maze-bright rats made few errors even when they were reared by maze-dull rats, and the offspring of maze-dull rats made many errors even when they were reared by maze-bright rats.

Since Tryon's seminal selective-breeding experiments, many behavioral traits have been selectively bred. Indeed, it appears that any measurable behavioral trait that varies among members of a species can be selectively bred.

An important general point made by studies of selective breeding is that selective breeding based on one behavioral trait usually brings a host of other behavioral traits along with it. This indicates that the behavioral trait used as the criterion for selective breeding is not the only behavioral trait influenced by the genes segregated by the breeding. Indeed, Searle (1949) compared maze-dull and maze-bright rats on 30 different behavioral tests and found that they differed on many of them. The pattern of differences suggested that the maze-bright rats were superior maze learners not

Figure 2.20 Selective breeding of maze-bright and maze-dull strains of rats by Tryon (1934).

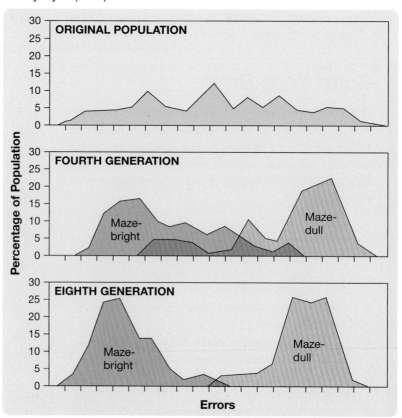

Data from Cooper, R. M., & Zubek, J. P. (1958)

because they were more intelligent but because they were less fearful—a trait that is not adaptive in many natural environments.

Selective-breeding studies have proved that genes influence the development of behavior. This conclusion in no way implies that experience does not. This point was driven home by Cooper and Zubek (1958) in a classic study of maze-bright and maze-dull rats. The researchers reared maze-bright and maze-dull rats in one of two environments: (1) an impoverished environment (a barren wire-mesh group cage) or (2) an enriched environment (a wire-mesh group cage that contained tunnels, ramps, visual displays, and other objects designed to stimulate interest). When the maze-dull rats reached maturity, they made significantly more errors than the maze-bright rats only if they had been reared in the impoverished environment (see Figure 2.21).

Phenylketonuria: A Single-Gene Metabolic Disorder

LO 2.14 Explain how classic research on phenylketonuria (PKU) has informed our understanding of the genetics of behavior.

It is often easier to understand the genetics of a behavioral disorder than it is to understand the genetics of typical behavior. The reason is that many genes influence the development of a typical behavioral trait, but it sometimes takes only one

Figure 2.21 Maze-dull rats did not make significantly more errors than maze-bright rats when both groups were reared in an enriched environment.

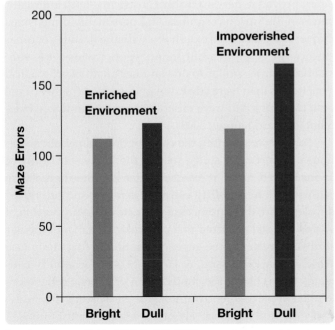

Data from Cooper, R. M., & Zubek, J. P. (1958)

abnormal gene to screw it up. A good example of this point is the neurological disorder **phenylketonuria (PKU)**.

PKU was discovered in 1934 when a Norwegian dentist, Asbjørn Følling, noticed a peculiar odor in the urine of his two intellectually disabled children. He correctly assumed that the odor was related to their disorder, and he had their urine analyzed. High levels of **phenylpyruvic acid** were found in both samples. Spurred on by his discovery, Følling identified other intellectually disabled children who had abnormally high levels of urinary phenylpyruvic acid, and he concluded that this subpopulation of intellectually disabled children was suffering from the same disorder. In addition to intellectual disability, the symptoms of PKU include vomiting, seizures, hyperactivity, irritability, and brain damage (Strisciuglio & Concolino, 2014).

The pattern of transmission of PKU through the family trees of afflicted individuals indicates that it is transmitted by a single gene mutation. About 1 in 100 people of European descent carry the PKU gene; but because the gene is recessive, PKU develops only in homozygous individuals (those who inherit a PKU gene from both their mother and their father). In the United States, about 1 in 16,000 infants is born with PKU (see Bilder et al., 2016). The incidence of PKU is lower in African Americans and Asian Americans than it is for Americans of European descent.

The biochemistry of PKU turned out to be reasonably straightforward. PKU homozygotes lack *phenylalanine hydroxylase*, an enzyme required for the conversion of the amino acid *phenylalanine* to *tyrosine*. As a result, phenylalanine accumulates in the body; and levels of *dopamine*, a neurotransmitter normally synthesized from tyrosine, are particularly low (see Boot et al., 2017). The consequence is abnormal brain development.

Like other behavioral traits, the behavioral symptoms of PKU result from an interaction between genetic and environmental factors: between the PKU gene and diet (see Rohde et al., 2014). Accordingly, in most modern hospitals, the blood of newborn infants is routinely screened for high phenylalanine levels (see Casey, 2013). If the level is high, the infant is immediately placed on a special phenylalanine-restricted diet; this diet reduces both the amount of phenylalanine in the blood and the development of intellectual disabilities—however, it does not prevent the development of subtle cognitive deficits (see Brown & Lichter-Konecki, 2016). The timing of this treatment is extremely important. The phenylalanine-restricted diet does not significantly reduce the development of intellectual disabilities in PKU homozygotes unless it is initiated within the first few weeks of life; conversely, the restriction of phenylalanine in the diet is often relaxed in adulthood, with few obvious adverse consequences to the patient. The period, usually early in life, during which a particular experience must occur to have a major effect on the development of a trait is the **sensitive period** for that trait.

Genetics of Human Psychological Differences

So far, this chapter has focused on three topics: human evolution, genetics, and the interaction of genetics and experience through epigenetic mechanisms. All three topics converge on one fundamental question: Why are we the way we are?

You have learned that each of us is a product of gene–experience interactions and that the effects of genes and experience on individual development are inseparable. This final module of the chapter continues to look at the effects of gene–experience interactions, but it focuses on a developmental issue that is fundamentally different from the ones we have been discussing: the development of individual differences rather than the development of individuals.

Development of Individuals versus Development of Differences among Individuals

LO 2.15 Explain why it is important to distinguish between the development of individuals and the development of individual differences.

This chapter has so far focused on the development of individuals. The remainder of the chapter deals with the development of differences among individuals. In the development of individuals, the effects of genes and experience are inseparable. In the development of differences among individuals, they are separable. This distinction is extremely important, but it confuses many people. Let's return to the musician metaphor to explain it.

The music of an individual musician is the product of the interaction of the musician and the instrument, and it doesn't make sense to ask what proportion of the music is produced by the musician and what proportion by the instrument. However, if we evaluated the playing of a large sample of musicians, each playing a different instrument, we could statistically estimate the degree to which the differences in the quality of the music they produced resulted from differences in the musicians themselves as opposed to differences in their instruments. For example, if we selected 100 people at random and had each one play a different professional-quality guitar, we would likely find that most of the variation in the quality of the music resulted from differences in the participants, some being experienced players and some never having played before. In the same way, researchers can select a group of volunteers and ask what proportion of the variation among them in some attribute (e.g., intelligence) results from genetic differences as opposed to experiential differences.

To assess the relative contributions of genes and experience to the development of differences in psychological attributes, behavioral geneticists study individuals of known genetic similarity. For example, they often compare **monozygotic twins**, who developed from the same zygote and thus are genetically similar, with **dizygotic twins**, who developed from two zygotes and thus are no more similar than any pair of *siblings* (brothers and sisters). Studies of pairs of monozygotic and dizygotic twins who have been separated at infancy by adoption are particularly informative about the relative contributions of genes and experience to differences in human psychological development. The most extensive of such adoption studies is the Minnesota Study of Twins Reared Apart (see Bouchard & Pedersen, 1999).

Heritability Estimates: Minnesota Study of Twins Reared Apart

LO 2.16 Explain heritability estimates and how they are commonly misinterpreted.

The Minnesota Study of Twins Reared Apart involved 59 pairs of monozygotic twins and 47 pairs of dizygotic twins who had been reared apart as well as many pairs of monozygotic and dizygotic twins who had been reared together. Their ages ranged from 19 to 68. Each twin was brought to the University of Minnesota for approximately 50 hours of testing, which focused on the assessment of intelligence and personality. Would the adult monozygotic twins reared apart prove to be similar because they were genetically similar, or would they prove to be different because they had been brought up in different environments?

The results of the Minnesota Study of Twins Reared Apart proved to be remarkably consistent—both internally, between the various cognitive and personality dimensions that were studied, and externally, with the findings of other studies. In general, adult monozygotic twins were substantially more similar to one another on all psychological dimensions than were adult dizygotic twins, whether or not both twins of a pair were raised in the same family environment (see Turkheimer, 2000).

In order to quantify the contributions of genetic variations in a particular study, such as the Minnesota Study of Twins Reared Apart, researchers often calculate heritability estimates. A **heritability estimate** is not about individual development; it is a numerical estimate of the proportion of variability that occurred in a particular trait in a particular study as a result of the genetic variation in that study (see Turkheimer, Pettersson, & Horn, 2014). Heritability estimates tell us about the contribution of genetic differences to phenotypic differences among the participants in a study; they have nothing to say about the relative contributions of genes and experience to the development of individuals.

The concept of heritability estimates can be quite confusing. We suggest that you pause here and carefully think about the definition before proceeding. The musician metaphor may help you here (see page 48). Recall that music is the product of an interaction between the musician and instrument. We cannot ask what proportion of that music is from the instrument and what proportion is from the musician. However, let's say we have many musicians and we listen to each make their music. We can now ask: What proportion of the differences among their music is due to their instruments and what proportion is due to their musical skills? Likewise, in the study of intelligence in many individuals, we can ask: What proportion of the differences in intelligence is due to differences in their environment and what proportion is due to differences in their genetics?

This analysis raises an important point: The magnitude of a study's heritability estimate depends on the amount of genetic and environmental variation from which it was calculated, and it cannot necessarily be generalized to other groups of individuals or other situations. For example, in the Minnesota study, there was relatively little environmental variation. All participants were raised in industrialized countries (Great Britain, Canada, and the United States) by parents who met the strict standards required for adoption. Accordingly, most of the variation in the subjects' intelligence and personality resulted from genetic variation. If the twins had each been separately adopted by vastly different parents (e.g., European royalty vs. a person living in extreme poverty), the resulting heritability estimates for IQ and personality would likely have been much lower.

Now that you understand the meaning of heritability estimates, let us tell you how big they tend to be for a variety of complex human traits and behaviors: for example, for intelligence, personality traits, aggression, divorce, religious beliefs, sports participation, psychiatric disorders, and television watching. The answer is simple because heritability estimates tend to be about the same regardless of the particular trait or behavior under consideration and regardless of the particular basis used to calculate them (i.e., twin, adoption, or family-tree studies). In the representative Western samples that have been studied, all complex traits and behaviors have substantial heritability estimates—most between 40 and 80 percent.

The discovery that genetic variability contributes substantially to individual differences in virtually all human traits and behaviors has led several eminent geneticists to argue that no more heritability estimate studies should be conducted (e.g., Johnson et al., 2009; Petronis, 2010). What could more heritability estimate studies possibly add? These geneticists are, however, excited about the potential of two other types of twin studies that have been increasingly reported. This chapter ends with them.

A Look into the Future: Two Kinds of Twin Studies

LO 2.17 Describe two ways that twin studies can be used to study the interaction of genes and experience (i.e., nature and nurture).

Two lines of research on twins have recently created considerable excitement among geneticists and other scholars. We hope you share their enthusiasm.

TWIN STUDIES OF EPIGENETIC EFFECTS. Most studies of epigenetic effects have focused on nonhuman species. In plants and nonhuman animals, it is quite clear that epigenetic changes can be triggered by experience, can last a lifetime, and can be passed on to future generations (see Szyf, 2014). To what extent do these amazing results apply to humans? Twin studies may provide a route to the answers (see Aguilera et al., 2010; Feil & Fraga, 2012).

The study of epigenetic effects in humans is difficult because experimental manipulation of human genetic material is not ethical. Monozygotic twins, however, provide a method of circumventing this difficulty. At conception monozygotic twins are genetically identical, and by repeatedly assessing their DNA one can document the development and survival of the many epigenetic differences that develop between them (see Bell & Saffery, 2012; Bell & Spector, 2011; Chatterjee & Morison, 2011; Silva et al., 2011). Moreover, by comparing monozygotic and dizygotic twins, it is possible to get a sense of the degree to which changes are caused by experiential as opposed to genetic factors—if epigenetic changes developed under genetic control, one would expect that the pattern of epigenetic changes would be more similar in monozygotic than dizygotic pairs.

The first systematic demonstration of epigenetic differences in human twins was published by Fraga and colleagues (2005). They took tissue samples (blood, skin, muscle) from 40 pairs of monozygotic twins, ranging in age from 3 to 74. Then, they screened the tissues for epigenetic alterations (e.g., DNA methylation, histone remodeling). They found that the twins were epigenetically indistinguishable early in life, but differences between them accumulated as they aged, each type of tissue displaying a different epigenetic profile (see Zong et al., 2012). As a result, the former assumption that monozygotic twins are genetically identical was disproven, and the common practice of referring to monozygotic twins as *identical twins* should be curtailed (see Figure 2.22).

In another study of epigenetic changes in twins, Wong and colleagues (2010) examined DNA methylation in *buccal cells* (cells of the lining of the mouth) scraped from 46 pairs of monozygotic twins and 45 pairs of dizygotic twins. They took samples from the twins at age 5 and again from the same twins at age 10. Then they assessed DNA methylation. Wong and colleagues found DNA methylation to be

Figure 2.22 Epigenetic research suggests that the common practice of referring to monozygotic twins as "identical twins" is no longer appropriate.

J. Lee/Getty Images

prominent in both groups of twins at both ages. Because the concordance rates of DNA methylation were the same between monozygotic twins and between dizygotic twins, they concluded that differences in DNA methylation are mainly a consequence of experiential factors.

The discovery of epigenetic differences in monozygotic twins raises the possibility that epigenetic differences may explain why one twin develops a disease and the other doesn't (Bell & Spector, 2011; Haque, Gottesman, & Wong, 2009). Once identified, such epigenetic differences would provide important clues to the causes and mechanisms of the disease. Bell and Spector (2011) suggest that *disease-discordant monozygotic twin studies* are a particularly powerful approach (see also Czyz et al., 2012). This kind of study begins with the identification of monozygotic twins who are discordant for a disease of interest. Then one searches each pair for epigenetic differences focusing on those areas of DNA that are thought

to be involved in the disorder. Large-scale studies in monozygotic twins across different ages, tissues, and epigenetic effects could greatly improve our understanding of human disease (see Bell & Spector, 2011; Tan et al., 2014).

TWIN STUDIES OF THE EFFECTS OF EXPERIENCE ON HERITABILITY. In thinking about heritability estimates, it is paramount to remember that heritability estimates depend on the particular conditions and subjects of a particular study. This point was driven home by the influential study of Turkheimer and colleagues (2003). Before the Turkheimer et al. study, all published studies of the heritability of intelligence were conducted on middle- to upperclass families, and the heritability estimates for intelligence tended to be about 75 percent.

Turkheimer and colleagues assessed heritability of intelligence in two samples of 7-year-old twins: those from families of low socioeconomic status (SES) and those from families of middle to high SES. The heritability estimates for intelligence in the middle- to high-SES twins was, as expected, about 70 percent. However, the heritability estimate for intelligence in the twins from low-SES families was only 10 percent. This effect was subsequently replicated and extended to other age groups: babies (Tucker-Drob et al., 2011) and adolescents (Harden, Turkheimer, & Loehlin, 2007).

One major implication of the study of Turkheimer et al. (2003) is that it forces us to think about intelligence as developing from the interaction of genes and experience, not from one or the other. It seems that one can inherit the potential to be of superior intelligence, but this potential is rarely realized in a poverty-stricken environment (see Nisbett et al., 2012).

This finding also has important implications for the development of programs to help the poor. Many politicians have argued against special programs for the poor because most heritability estimates of intelligence are high. They incorrectly argue that because intelligence is largely inherited (i.e., it has a high heritability estimate), special programs for the poor are a waste of money. However, the findings of Turkheimer and colleagues suggest otherwise: Reducing poverty would mean that all children would be able to reach their intellectual potential.

Themes Revisited

This chapter introduced the topics of evolution, genetics, and development, but two themes prevailed: thinking about epigenetics and thinking creatively about the biology of behavior. In terms of thinking about epigenetics, you were fully introduced to the field of epigenetics and you learned about how that field has huge implications for our understanding of behavior. This chapter also challenged you to think about important biopsychological phenomena in creative

new ways: the nature–nurture issue, the physiological-orpsychological dichotomy, the genetics of human psychological differences, the meaning of heritability estimates, and the important study of Turkheimer and colleagues (2003).

Two other themes also received coverage in this chapter: the evolutionary perspective and clinical implications. The evolutionary perspective was illustrated by comparative research on self-awareness in chimps and by consideration of the evolutionary significance of social dominance and courtship displays. The clinical implications theme was illustrated by the case of the man who fell out of bed, the discussion of phenylketonuria (PKU), and the discussion of disease-discordant twin studies.

This chapter was jam-packed with examples of one of the emerging themes: thinking about epigenetics. You were fully introduced to the field of epigenetics and you learned about how the field has huge implications for our understanding of behavior.

Key Terms

Zeitgeist, p. 45

Thinking about the Biology of Behavior: From Dichotomies to Interactions
Cartesian dualism, p. 46
Nature–nurture issue, p. 46
Ethology, p. 46
Instinctive behaviors, p. 46
Asomatognosia, p. 47

Human Evolution
Evolve, p. 50
Natural selection, p. 50
Fitness, p. 50
Species, p. 51
Conspecifics, p. 52
Chordates, p. 52
Vertebrates, p. 52
Amphibians, p. 52
Mammals, p. 52
Primates, p. 52
Hominins, p. 53
Spandrels, p. 55
Exaptations, p. 55
Homologous, p. 55
Analogous, p. 55
Convergent
  evolution, p. 55
Brain stem, p. 57
Cerebrum, p. 57
Convolutions, p. 57

Fundamental Genetics
Dichotomous traits, p. 58
True-breeding lines, p. 58
Dominant trait, p. 58
Recessive trait, p. 58
Phenotype, p. 58
Genotype, p. 58
Gene, p. 59
Alleles, p. 59
Homozygous, p. 59
Heterozygous, p. 59
Chromosomes, p. 59
Gametes, p. 59
Meiosis, p. 59
Zygote, p. 60
Genetic recombination, p. 60
Mitosis, p. 60
Deoxyribonucleic acid (DNA), p. 60
Nucleotide bases, p. 60
Replication, p. 60
Mutations, p. 61
Autosomal chromosomes, p. 61
Sex chromosomes, p. 61
Sex-linked traits, p. 61
Proteins, p. 61
Amino acids, p. 61
Promoters, p. 61
Gene expression, p. 62
Activators, p. 62
Repressors, p. 62
Transcription, p. 62
Ribonucleic acid (RNA), p. 63

Messenger RNA, p. 63
Translation, p. 63
Ribosomes, p. 63
Codon, p. 63
Transfer RNA, p. 63
Human Genome Project, p. 63
Human proteome, p. 63
Epigenetics, p. 64
DNA methylation, p. 64
Histone remodeling, p. 64
Histones, p. 64
Epigenome, p. 65
Epitranscriptome, p. 65
Transgenerational epigenetics, p. 65

Epigenetics of Behavioral Development: Interaction of Genetic Factors and Experience
Ontogeny, p. 66
Phylogeny, p. 66
Phenylketonuria (PKU), p. 67
Phenylpyruvic acid, p. 67
sensitive period, p. 67

Genetics of Human Psychological Differences
Monozygotic twins, p. 68
Dizygotic twins, p. 68

Heritability Estimates: Minnesota Study of Twins Reared Apart
Heritability estimate, p. 68

Chapter 3
Anatomy of the Nervous System

Systems, Structures, and Cells That Make Up Your Nervous System

Mike Kemp/Tetra Images/Alamy Stock Photo

 Chapter Overview and Learning Objectives

General Layout of the
Nervous System

LO 3.1 List and describe the major divisions of the nervous system.

LO 3.2 Describe the three meninges and explain their functional role.

LO 3.3 Explain where cerebrospinal fluid is produced and where it flows.

LO 3.4 Explain what the blood–brain barrier is and what functional role it serves.

Cells of the Nervous
System

LO 3.5 Draw, label, and define the major features of a multipolar neuron.

In order to understand what the brain does, it is first necessary to understand what it is—to know the names and locations of its major parts and how they are connected to one another. This chapter introduces you to these fundamentals of brain anatomy.

Before you begin this chapter, we want to apologize for the lack of foresight displayed by early neuroanatomists in their choice of names for neuroanatomical structures—but how could they have anticipated that Latin and Greek, universal languages of the educated in their day, would not be compulsory university fare in our time? To help you, we have provided the literal English meanings of many of the neuroanatomical terms, and we have kept this chapter as brief, clear, and to the point as possible, covering only the most important structures. The payoff for your effort will be a fundamental understanding of the structure of the human brain and a new vocabulary to discuss it.

General Layout of the Nervous System

In this module, we'll cover the general layout of the nervous system. We'll begin by discussing its two main divisions. Then, we'll look at the roles of meninges, ventricles, and cerebrospinal fluid. We'll conclude with a look at the blood–brain barrier.

Divisions of the Nervous System

LO 3.1 List and describe the major divisions of the nervous system.

The vertebrate nervous system is composed of two divisions: the central nervous system and the peripheral nervous system (see Figure 3.1). Roughly speaking, the **central nervous system (CNS)** is the division of the nervous system located within the skull and spine, and the **peripheral nervous system (PNS)** is the division located outside the skull and spine.

The central nervous system is composed of two divisions: the brain and the spinal cord. The *brain* is the part of the CNS located in the skull; the *spinal cord* is the part located in the spine.

The peripheral nervous system is also composed of two divisions: the somatic nervous system and the autonomic nervous system. The **somatic nervous system (SNS)** is the part of the PNS that interacts with the external environment. It is composed of **afferent nerves** that carry sensory signals from the skin, skeletal muscles, joints, eyes, ears, and so on, to the central nervous system and **efferent nerves** that carry motor signals from the central nervous system to the skeletal muscles. The **autonomic nervous system (ANS)** is the part of the peripheral nervous system that regulates the body's internal environment. It is composed of afferent nerves that carry sensory signals from internal organs to the CNS and efferent nerves that carry motor signals from the CNS to internal organs. You will not confuse the

Figure 3.1 The human central nervous system (CNS) and peripheral nervous system (PNS). The CNS is represented in red; the PNS in orange. Notice that even those portions of nerves that are within the spinal cord are considered to be part of the PNS.

Central nervous system

Peripheral nervous system

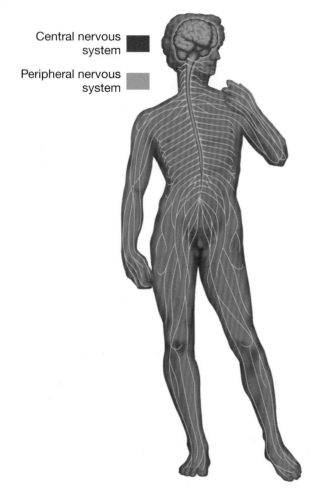

terms *afferent* and *efferent* if you remember that many words that involve the idea of going toward something—in this case, going toward the CNS—begin with an *a* (e.g., *advance*, *approach*, *arrive*) and that many words that involve the idea of going away from something begin with an *e* (e.g., *exit*, *embark*, *escape*).

The autonomic nervous system has two kinds of efferent nerves: sympathetic nerves and parasympathetic nerves. The **sympathetic nerves** are autonomic motor nerves that project from the CNS in the *lumbar* (small of the back) and *thoracic* (chest area) regions of the spinal cord. The **parasympathetic nerves** are those autonomic motor nerves that project from the brain and *sacral* (lower back) region of the spinal cord. See Appendix I. (Ask your instructor to specify the degree to which you are responsible for material in the appendices.) All sympathetic and parasympathetic nerves are two-stage neural paths: The sympathetic and parasympathetic neurons project from the CNS and go only part of the way to the target organs before they *synapse on*

other neurons (second-stage neurons) that carry the signals the rest of the way. However, the sympathetic and parasympathetic systems differ in that the sympathetic neurons project from the CNS synapse on second-stage neurons at a substantial distance from their target organs, whereas the parasympathetic neurons project from the CNS synapse near their target organs on very short second-stage neurons (see Appendix I).

The conventional view of the respective functions of the sympathetic and parasympathetic systems stresses three important principles: (1) sympathetic nerves stimulate, organize, and mobilize energy resources in threatening situations, whereas parasympathetic nerves act to conserve energy; (2) each autonomic target organ receives opposing sympathetic and parasympathetic input, and its activity is thus controlled by relative levels of sympathetic and parasympathetic activity; and (3) sympathetic changes are indicative of psychological arousal, whereas parasympathetic changes are indicative of psychological relaxation. Although these principles are generally correct, there are significant qualifications and exceptions to each of them (see Guyenet, 2006)—see Appendix II.

Most of the nerves of the peripheral nervous system project from the spinal cord, but there are 12 pairs of exceptions: the 12 pairs of **cranial nerves**, which project from the brain. They are numbered in sequence from front to back. The cranial nerves include purely sensory nerves such as the olfactory nerves (I) and the optic nerves (II), but most contain both sensory and motor fibers. The longest cranial nerves are the vagus nerves (X), which contain motor and sensory fibers traveling to and from the gut. The 12 pairs of cranial nerves and their targets are illustrated in Appendix III; the functions of these nerves are listed in Appendix IV. The autonomic motor fibers of the cranial nerves are parasympathetic.

The functions of the various cranial nerves are commonly assessed by neurologists as a basis for diagnosis. Because the functions and locations of the cranial nerves are specific, disruptions of particular cranial nerve functions provide excellent clues about the location and extent of tumors and other kinds of brain pathology.

Figure 3.2 summarizes the major divisions of the nervous system. Notice that the nervous system is a "system of twos."

Meninges

LO 3.2 Describe the three meninges and explain their functional role.

The brain and spinal cord (the CNS) are the most protected organs in the body. They are encased in bone and covered by three protective membranes, the three **meninges** (pronounced "men-IN-gees"; see Coles et al., 2017). The outer *meninx* (which, believe it or not, is the singular of *meninges*)

Figure 3.2 The major divisions of the nervous system.

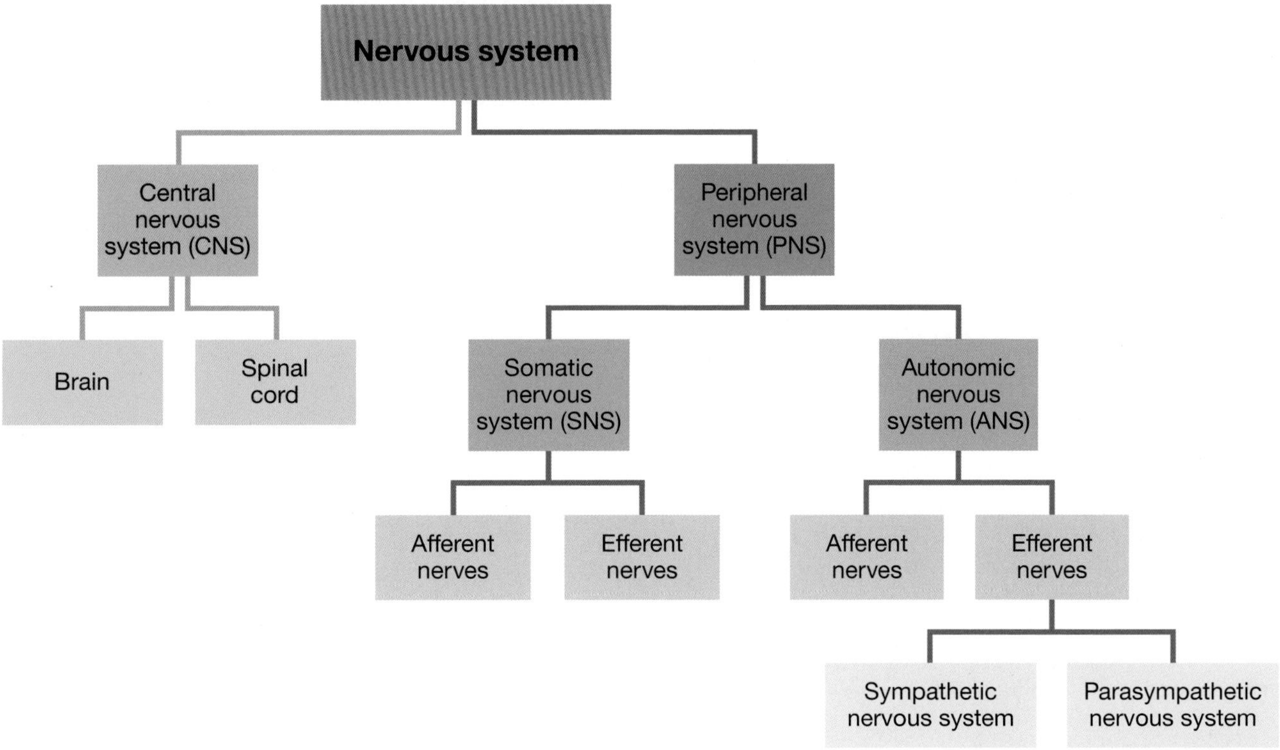

is a tough membrane called the **dura mater** (tough mother). Immediately inside the dura mater is the fine **arachnoid membrane** (spider-web-like membrane). Beneath the arachnoid membrane is a space called the **subarachnoid space**, which contains many large blood vessels and cerebrospinal fluid; then comes the innermost meninx, the delicate **pia mater** (pious mother), which adheres to the surface of the CNS.

Ventricles and Cerebrospinal Fluid

LO 3.3   Explain where cerebrospinal fluid is produced and where it flows.

Also protecting the CNS is the **cerebrospinal fluid (CSF)**, which fills the subarachnoid space, the central canal of the spinal cord, and the cerebral ventricles of the brain. The **central canal** is a small central channel that runs the length of the spinal cord; the **cerebral ventricles** are the four large internal chambers of the brain: the two lateral ventricles, the third ventricle, and the fourth ventricle (see Figure 3.3). The subarachnoid space, central canal, and cerebral ventricles are interconnected by a series of openings and thus form a single reservoir.

The cerebrospinal fluid supports and cushions the brain. Patients who have had some of their cerebrospinal fluid drained away often suffer raging headaches and experience stabbing pain each time they jerk their heads.

According to the traditional view, cerebrospinal fluid is produced by the **choroid plexuses** (networks of *capillaries*,

or small blood vessels that protrude into the ventricles from the pia mater), and the excess cerebrospinal fluid is continuously absorbed from the subarachnoid space into large blood-filled spaces, or *dural sinuses*, which run through the dura mater and drain into the large jugular veins of the neck. However, there is growing appreciation that cerebrospinal fluid production and absorption are more complex than was originally thought (see Brinker et al., 2014). Figure 3.4 illustrates the absorption of cerebrospinal fluid from the subarachnoid space into the large sinus that runs along the top of the brain between the two cerebral hemispheres.

Occasionally, the flow of cerebrospinal fluid is blocked by a tumor near one of the narrow channels that link the ventricles—for example, near the *cerebral aqueduct*, which connects the third and fourth ventricles. The resulting buildup of fluid in the ventricles causes the walls of the ventricles, and thus the entire brain, to expand, producing a condition called *hydrocephalus* (water head). Hydrocephalus is treated by draining the excess fluid from the ventricles and trying to remove the obstruction.

Journal Prompt 3.1

Hydrocephalus is often congenital (present from birth). What do you think might be some of the long-term effects of being born with hydrocephalus?

Figure 3.3 The cerebral ventricles and central canal.

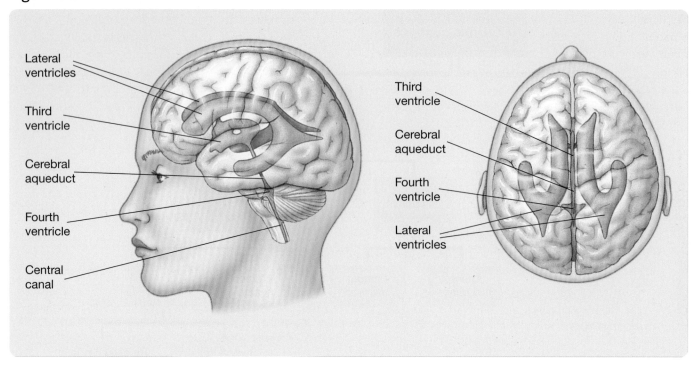

Figure 3.4 The absorption of cerebrospinal fluid (CSF) from the subarachnoid space (blue) into a major sinus. Note the three meninges.

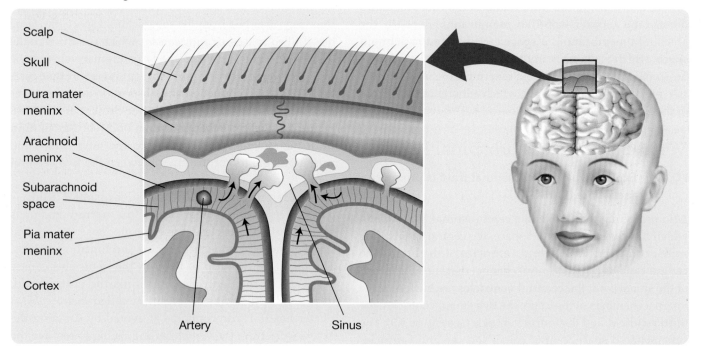

Blood–Brain Barrier

LO 3.4   Explain what the blood–brain barrier is and what functional role it serves.

The brain is a finely tuned electrochemical organ whose function can be severely disturbed by the introduction of certain kinds of chemicals. Fortunately, a mechanism impedes the passage of many toxic substances from the blood into the brain: the **blood–brain barrier**. This barrier is a consequence of the special structure of cerebral blood vessels. In the rest of the body, the cells that compose the walls of blood vessels are loosely packed; as a result, most molecules pass readily through them into surrounding tissue. In the brain, however, the cells of the blood vessel walls are tightly packed, thus forming a barrier to the passage of many molecules—particularly proteins and other large molecules (see Chow & Gu, 2015). The degree to which therapeutic or recreational drugs can influence brain activity depends on the ease with which they penetrate the blood–brain barrier (see Interlandi, 2013; Siegenthaler, Sohet, & Daneman, 2013).

The blood–brain barrier does not impede the passage of all large molecules. Some large molecules that are critical for normal brain function (e.g., glucose) are actively transported through cerebral blood vessel walls. Also, the blood vessel walls in some areas of the brain allow certain large molecules to pass through them unimpeded. Many CNS disorders are associated with impairment of the blood–brain barrier (see Bentivoglio & Kristensson, 2014; Jiang et al., 2018).

Scan Your Brain

This is a good place to pause and scan your brain to check your knowledge of the CNS. Fill in the following blanks with the most appropriate terms. The correct answers are provided at the end of the exercise. Before proceeding, review material related to your errors and omissions.

1. The somatic nervous system includes _____ nerves that carry motor signals from the central nervous system to the muscles.

2. The _____ is the part of the peripheral nervous system that regulates the body's internal environment.

3. The brain and the spinal cord are the only organs that are protected with three layers of protective membranes called _____.

4. _____ or "tough mother" is the outer meninx.

5. The _____ nervous system is activated when you encounter a threatening information such as a bear attacking you. This system is essential for the initiation of fight-or-flight responses.

6. Motor nerves that project from the brain and the lower region of the spine are called _____ nerves.

7. The _____ nerve is a purely sensory nerve that transfers visual information from the retina of the eye to the brain.

8. The _____ nerve is the nerve cell that extends directly from the brain to the gut.

9. The _____ is a channel that connects the third and fourth ventricles in the brain.

10. The ventricles of patients with a congenital condition called _____ build up fluid as a result of blocked channels in the brain.

11. Many toxic substances that are present in the bloodstream are prohibited from entering the brain by a mechanism called the _____ where cells of blood vessel walls are tightly packed, forming a barrier to the passage of large proteins.

12. Unlike large toxic molecules, _____, which is critical for the function of the brain, is actively transported through the vessel walls.

Scan Your Brain answers: (1) efferent, (2) ANS, (3) meninges, (4) Dura mater, (5) sympathetic, (6) parasympathetic, (7) optic, (8) vagus, (9) cerebral aqueduct, (10) hydrocephalus, (11) blood–brain barrier, (12) glucose.

Cells of the Nervous System

Most of the cells of the nervous system are of two fundamentally different types: neurons and glial cells. Their anatomy is discussed in the following two sections.

Anatomy of Neurons

LO 3.5 **Draw, label, and define the major features of a multipolar neuron.**

Recall that **neurons** are cells that are specialized for the reception, conduction, and transmission of electrochemical signals. They come in an incredible variety of shapes and sizes (see Sharpee, 2014; Shen, 2015; Underwood, 2015); however, many are similar to the one illustrated in Figures 3.5 and 3.6, which detail the major external and internal features of a neuron, respectively.

NEURON CELL MEMBRANE. The neuron cell membrane is composed of a *lipid bilayer*, or two layers of fat molecules (see Figure 3.7). Embedded in the lipid bilayer are numerous protein molecules that are the basis of many of the cell membrane's functional properties. Some membrane proteins are *channel proteins*, through which certain molecules can pass; others are *signal proteins*, which transfer a signal to the inside of the neuron when particular molecules bind to them on the outside of the membrane.

CLASSES OF NEURONS. Figure 3.8 illustrates a way of classifying neurons based on the number of processes (projections) emanating from their cell bodies. A neuron with more than two processes extending from its cell body is classified as a **multipolar neuron**; most neurons are multipolar. A neuron with one process extending from its cell body is classified as a **unipolar neuron**, and a neuron with two processes extending from its cell body is classified as a **bipolar neuron**. Neurons with a short axon or no axon at all are called **interneurons**; their function is to integrate neural

Figure 3.5 The major external features of a neuron.

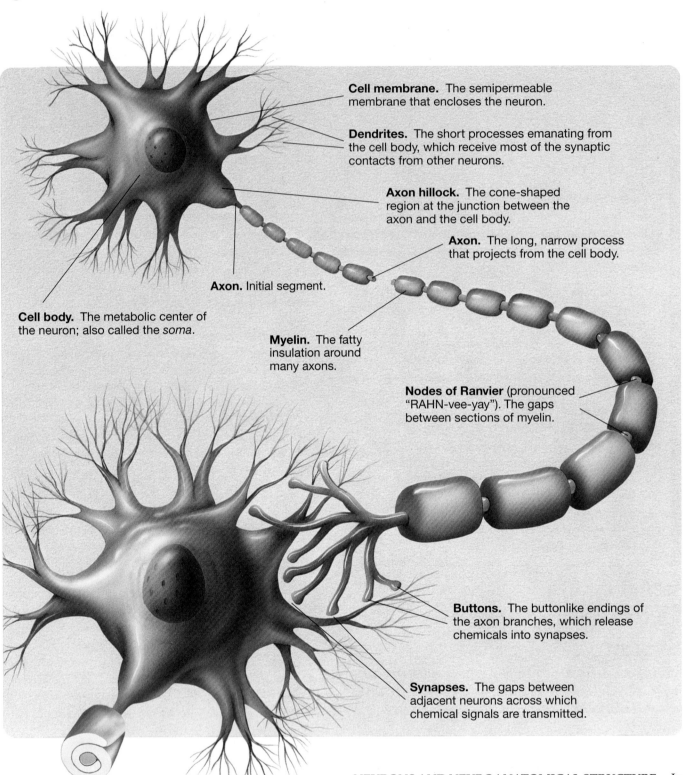

Cell membrane. The semipermeable membrane that encloses the neuron.

Dendrites. The short processes emanating from the cell body, which receive most of the synaptic contacts from other neurons.

Axon hillock. The cone-shaped region at the junction between the axon and the cell body.

Axon. The long, narrow process that projects from the cell body.

Axon. Initial segment.

Cell body. The metabolic center of the neuron; also called the *soma*.

Myelin. The fatty insulation around many axons.

Nodes of Ranvier (pronounced "RAHN-vee-yay"). The gaps between sections of myelin.

Buttons. The buttonlike endings of the axon branches, which release chemicals into synapses.

Synapses. The gaps between adjacent neurons across which chemical signals are transmitted.

activity within a single brain structure, not to conduct signals from one structure to another. Classifying neurons is a complex task, and neuroscientists still don't agree on the best method of classification (see Cembrowski & Menon, 2018; Wichterle, Gifford, & Mazzoni, 2013; Zeng & Sanes, 2017).

NEURONS AND NEUROANATOMICAL STRUCTURE. In general, there are two kinds of gross neural structures in the nervous system: those composed primarily of cell bodies and those composed primarily of axons. In the central nervous system, clusters of cell bodies are called **nuclei** (singular *nucleus*); in the peripheral nervous system, they are called **ganglia** (singular *ganglion*). (Note that the word *nucleus* has two different neuroanatomical meanings; it

Figure 3.6 The major internal features of a neuron.

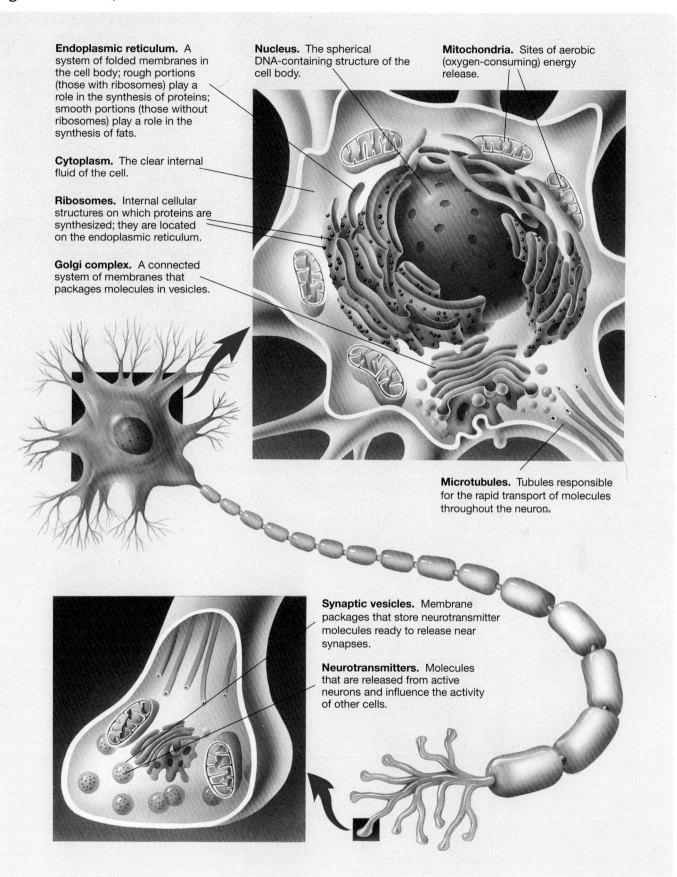

Endoplasmic reticulum. A system of folded membranes in the cell body; rough portions (those with ribosomes) play a role in the synthesis of proteins; smooth portions (those without ribosomes) play a role in the synthesis of fats.

Cytoplasm. The clear internal fluid of the cell.

Ribosomes. Internal cellular structures on which proteins are synthesized; they are located on the endoplasmic reticulum.

Golgi complex. A connected system of membranes that packages molecules in vesicles.

Nucleus. The spherical DNA-containing structure of the cell body.

Mitochondria. Sites of aerobic (oxygen-consuming) energy release.

Microtubules. Tubules responsible for the rapid transport of molecules throughout the neuron.

Synaptic vesicles. Membrane packages that store neurotransmitter molecules ready to release near synapses.

Neurotransmitters. Molecules that are released from active neurons and influence the activity of other cells.

Figure 3.7 The cell membrane is a lipid bilayer with signal proteins and channel proteins embedded in it.

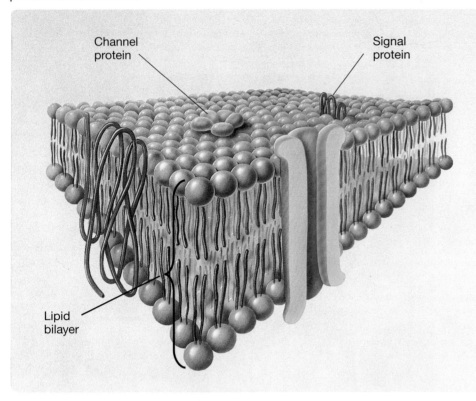

is a structure in the neuron cell body and a cluster of cell bodies in the CNS.) In the central nervous system, bundles of axons are called **tracts**; in the peripheral nervous system, they are called **nerves**.

Glia: The Forgotten Cells

LO 3.6 Briefly describe four kinds of glial cells.

Neurons are not the only cells in the nervous system; there are about as many **glial cells**, or *glia* (pronounced "GLEE-a"). It is commonly said that there are 10 times as many glia as neurons in the human brain, but this is incorrect: There are roughly two glia for every three neurons in your brain (see Nimmerjahn & Bergles, 2015; von Bartheld, 2017).

There are several kinds of glia. **Oligodendrocytes**, for example, are glial cells with extensions that wrap around the axons of some neurons of the central nervous system. These extensions are rich in **myelin**, a fatty insulating substance, and the **myelin sheaths** they form increase the speed of axonal conduction. A similar function is performed in the peripheral nervous system by **Schwann cells**, a second class of glia. Oligodendrocytes and Schwann cells are illustrated in Figure 3.9. Notice that each Schwann cell constitutes one myelin segment, whereas each oligodendrocyte provides several myelin segments, often on more than one axon. Another important difference between Schwann cells and oligodendrocytes is that only Schwann cells can guide axonal *regeneration* (regrowth) after damage. That is why effective axonal regeneration in the mammalian nervous system is restricted to the PNS.

Microglia make up a third class of glia. Microglia are smaller than

Figure 3.8 A unipolar neuron, a bipolar neuron, a multipolar neuron, and an interneuron.

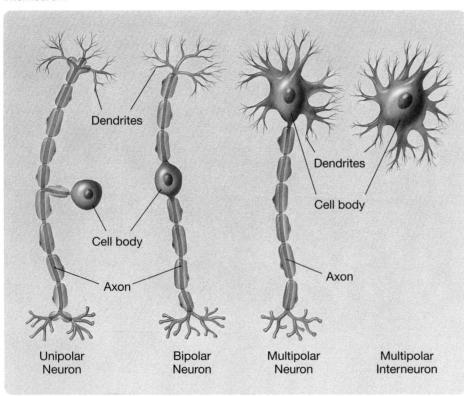

Figure 3.9 The myelination of CNS axons by an oligodendrocyte and the myelination of PNS axons by Schwann cells.

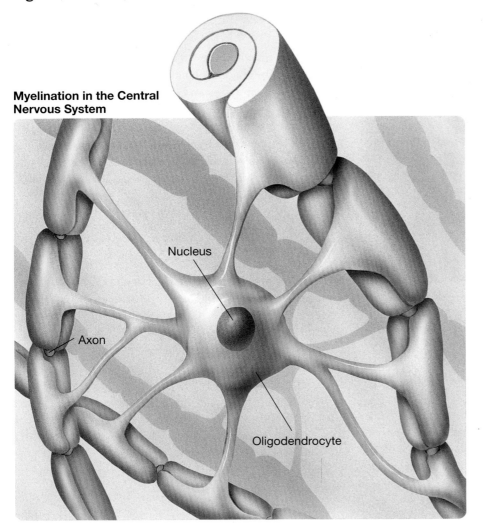

Myelination in the Central Nervous System

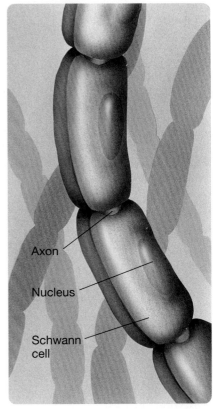

Myelination in the Peripheral Nervous System

other glial cells—thus their name. They respond to injury or disease by multiplying, engulfing cellular debris or even entire cells (see Brown & Neher, 2014), and triggering inflammatory responses (see Smith & Dragunow, 2014).

Astrocytes constitute a fourth class of glia. They are the largest glial cells, and they are so named because they are star-shaped (*astro* means "star"). The extensions of some astrocytes cover the outer surfaces of blood vessels that course through the brain; they also make contact with neurons (see Figure 3.10). These particular astrocytes appear to play a role in allowing the passage of some chemicals from the blood into CNS neurons and in blocking other chemicals (see Paixão & Klein, 2010), and they have the ability to contract or relax blood vessels based on the blood flow demands of particular brain regions (see Howarth, 2014; Mishra et al., 2016; Muoio, Persson, & Sendeski, 2014).

For decades, it was assumed that the function of glia was mainly to provide support for neurons—provide them with nutrition, clear waste, and form a physical matrix to hold neural circuits together (*glia* means "glue"). But this limited view of the role of glial cells has changed, thanks to a series of remarkable findings. For example, astrocytes, the most studied of the glial cells, have been shown to exchange chemical signals with neurons and other astrocytes (Araque et al., 2014; Montero & Orellana, 2015; Yoon & Lee, 2014), to control the establishment and maintenance of synapses between neurons (Baldwin & Eroglu, 2017), to modulate neural activity (Bouzier-Sore & Pellerin, 2013), to form functional networks with neurons and other astrocytes (Gittis & Brasier, 2015; Haim & Rowitch, 2017; Lee et al., 2014; Perea, Sur, & Araque, 2014), to control the blood–brain barrier (Alvarez, Katayama, & Prat, 2013; Cabezas et al., 2014), to respond to brain injury (Khakh & Sofroniew, 2015), and to play a role in certain forms of cognition (e.g., Dallérac & Rouach, 2016; Martin-Fernandez et al., 2017). Microglia have also been shown to play many more roles in brain function than had previously

Figure 3.10 Astrocytes (shown in pink) have an affinity for blood vessels (in red) and they also make contact with neurons (in blue).

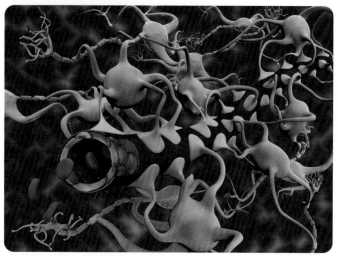

GUNILLA ELAM/Science Source

been thought (see Pósfai et al., 2018); for example, they have been shown to play a role in the regulation of cell death (Wake et al., 2013), synapse formation (Parkhurst et al., 2013; Welberg, 2014), and synapse elimination (Wake et al., 2012).

Research on the function of glia, although still in its early stages, is creating considerable excitement. There is now substantial evidence that the physiological effects of glia are both numerous and much more important than anyone might have imagined two decades ago. For example, some researchers have suggested that *glial networks* may be the dwelling places of thoughts (see Verkhratsky, Parpura, & Rodríguez, 2010). One final important discovery about glial cells is that they are much more varied than implied by the four types that we have just described: oligodendrocytes, Schwann cells, microglia, and astrocytes. For example, a new type of glial cell was recently discovered (see Fan & Agid, 2018); and at least fifteen different kinds of astrocytes have been identified, each with its own structure, physiology, and specific locations in the brain (Chai et al., 2017; Clarke & Liddelow, 2017; Lin et al., 2017). Sorting out the functions of each type is not going to be easy.

Neuroanatomical Techniques and Directions

This module first describes a few of the most common neuroanatomical techniques. Then, it explains the system of directions that neuroanatomists use to describe the location of structures in vertebrate nervous systems.

Neuroanatomical Techniques

LO 3.7 Compare several neuroanatomical research techniques.

The major problem in visualizing neurons is not that they are minute. The major problem is that neurons are so tightly packed and their axons and dendrites so intricately intertwined that looking through a microscope at unprepared neural tissue reveals almost nothing about them. The key to the study of neuroanatomy lies in preparing neural tissue in a variety of ways, each of which permits a clear view of a different aspect of neuronal structure, and then combining the knowledge obtained from each of the preparations. This point is illustrated by the following widely used neuroanatomical techniques.

GOLGI STAIN. The greatest blessing to befall neuroscience in its early years was the accidental discovery of the **Golgi stain** by Camillo Golgi (pronounced "GOLE-jee"), an Italian physician, in the early 1870s. Golgi was trying to stain the meninges, by exposing a block of neural tissue to potassium dichromate and silver nitrate, when he noticed an amazing thing. For some unknown reason, the silver chromate created by the chemical reaction of the two substances Golgi was using invaded a few neurons in each slice of tissue and stained each invaded neuron entirely black. This discovery made it possible to see individual neurons for the first time, although only in silhouette (see Figure 3.11). Golgi stains are commonly used to discover the overall shape of neurons.

NISSL STAIN. Although the Golgi stain permits an excellent view of the silhouettes of the few neurons that take up the stain, it provides no indication of the number of neurons in an area. The first neural staining procedure to overcome this shortcoming was the **Nissl stain**, which was developed by Franz Nissl, a German psychiatrist, in the 1880s. The most common dye used in the Nissl method is cresyl violet. Cresyl violet and other Nissl dyes penetrate all cells on a slide, but they bind to molecules (i.e., DNA and RNA) that are most prevalent in neuron cell bodies. Thus, they often are used to estimate the number of cell bodies in an area, by counting the number of Nissl-stained dots. Figure 3.12 is a photograph of a slice of brain tissue stained with cresyl violet. Notice that only the layers composed mainly of neuron cell bodies are densely stained.

ELECTRON MICROSCOPY. A neuroanatomical technique that provides information about the details of neuronal structure is **electron microscopy** (pronounced "my-CROSS-cuh-pee"). Because of the nature of light, the limit of magnification in light microscopy is about 1,500 times, a level of magnification insufficient to reveal

Figure 3.11 Neural tissue that has been stained by the Golgi method. Because only a few neurons take up the stain, their silhouettes are revealed in great detail, but their internal details are invisible.

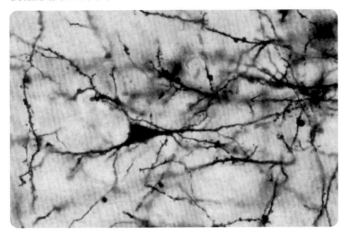

Martin M. Rotker/Science Source

the fine anatomical details of neurons. Greater detail can be obtained by first coating thin slices of neural tissue with an electron-absorbing substance that is taken up by different parts of neurons to different degrees, then passing a beam of electrons through the tissue onto a photographic film. The result is an *electron micrograph*, which captures neuronal structure in exquisite detail. A *scanning electron microscope* provides spectacular electron micrographs in three dimensions (see Figure 3.13), but it is not capable of as much magnification as conventional electron microscopy. The strength of electron microscopy is also a weakness: Because the images are so detailed, they can make it difficult to visualize general aspects of neuroanatomical structure.

NEUROANATOMICAL TRACING TECHNIQUES. Neuroanatomical tracing techniques are of two types: anterograde (forward) tracing methods and retrograde (backward) tracing methods (see Figure 3.14). *Anterograde tracing methods* are used when an investigator wants to trace the paths of axons projecting away from cell bodies located in a particular area. The investigator begins by injecting one of several chemicals commonly used for anterograde tracing into the cell body. It is then taken up by cell bodies and transported forward along their axons to their terminal buttons. Then, after a few days, the investigator removes the brain and slices it. Those slices are then treated to reveal the locations of the injected chemical.

Retrograde tracing methods work in the reverse manner; they are used when an investigator wants to trace the paths of axons projecting into a particular area. The investigator begins by injecting one of several chemicals commonly used for retrograde-tracing into an area of the brain. These chemicals are taken up by terminal buttons and then transported

Figure 3.12 The Nissl stain. Presented here is a Nissl-stained section through the rat hippocampus, at two levels of magnification to illustrate two uses of Nissl stains. Under low magnification (top panel), Nissl stains provide a gross indication of brain structure by selectively staining groups of neural cell bodies. Under higher magnification (bottom panel), one can distinguish individual neural cell bodies, and thus, count the number of neurons in various areas.

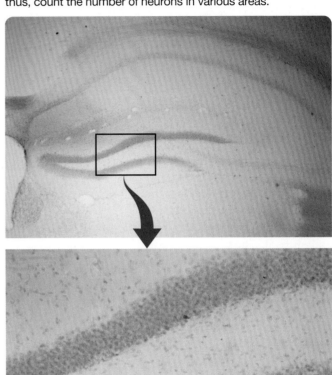

Carl Ernst/Brian Christie/University of British Columbia Department of Psychology

Figure 3.13 A color-enhanced scanning electron micrograph of a neuron cell body (green) studded with terminal buttons (orange). Each neuron receives numerous synaptic contacts.

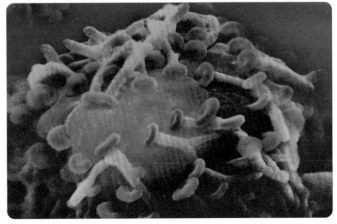

Photo Researchers/Science History Images/Alamy Stock Photo

Figure 3.14 One example of anterograde tracing (A) and one example of retrograde tracing (B).

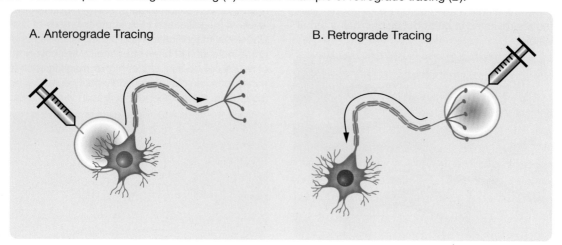

backward along their axons to their cell bodies. After a few days, the investigator removes the brain and slices it. Those slices are then treated to reveal the locations of the injected chemical.

Directions in the Vertebrate Nervous System

LO 3.8 Illustrate the neuroanatomical directions.

It would be difficult for you to develop an understanding of the layout of an unfamiliar city without a system of directional coordinates: north–south, east–west. The same goes for the nervous system. Thus, before introducing you to the locations of major nervous system structures, we will describe the three-dimensional system of directional coordinates used by neuroanatomists.

Directions in the vertebrate nervous system are described in relation to the orientation of the spinal cord. This system is straightforward for most vertebrates, as Figure 3.15a indicates. The vertebrate nervous system has three axes: anterior–posterior, dorsal–ventral, and medial–lateral. First,

Figure 3.15 (a) Anatomical directions in representative vertebrates, my (JP) cats Sambala and Rastaman. (b) Anatomical directions in a human. Notice that the directions in the cerebral hemispheres are rotated by 90° in comparison to those in the spinal cord and brain stem because of the unusual upright posture of humans.

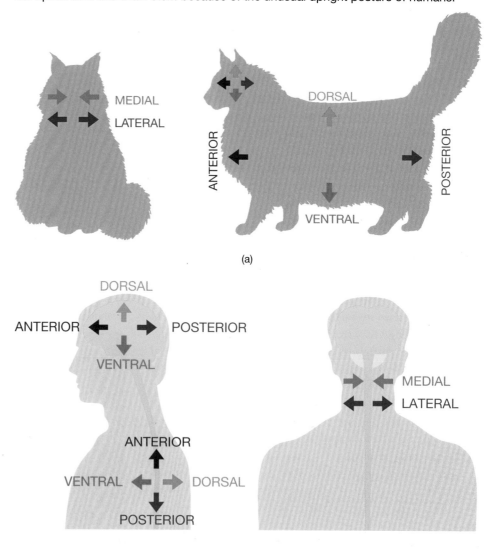

anterior means toward the nose end (the anterior end), and **posterior** means toward the tail end (the posterior end); these same directions are sometimes referred to as *rostral* and *caudal*, respectively. Second, **dorsal** means toward the surface of the back or the top of the head (the dorsal surface), and **ventral** means toward the surface of the chest or the bottom of the head (the ventral surface). Third, **medial** means toward the midline of the body, and **lateral** means away from the midline toward the body's lateral surfaces.

Humans complicate this simple three-axis (anterior–posterior, ventral–dorsal, medial–lateral) system of neuroanatomical directions by insisting on walking around on our hind legs. This changes the orientation of our cerebral hemispheres in relation to our spines and brain stems.

You can save yourself a lot of confusion if you remember that the system of vertebrate neuroanatomical directions was adapted for use in humans in such a way that the terms used to describe the positions of various body surfaces are the same in humans as they are in more typical, non-upright vertebrates. Specifically, notice that the top of the human head and the back of the human body are both referred to as *dorsal* even though they are in different directions, and the bottom of the human head and the front of the human body are both referred to as *ventral* even though they are in different directions (see Figure 3.15b). To circumvent this complication, the terms **superior** and **inferior** are often used to refer to the top and bottom of the primate head, respectively.

Proximal and *distal* are two other common directional terms. In general, **proximal** means "close," and **distal** means "far." Specifically, with regard to the peripheral

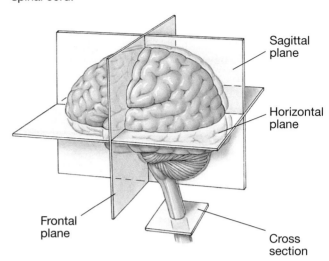

Figure 3.16 Horizontal, frontal (coronal), and sagittal planes in the human brain and a cross section of the human spinal cord.

Sagittal plane

Horizontal plane

Frontal plane

Cross section

nervous system, *proximal* means closer to the CNS, and *distal* means farther from the CNS. Your shoulders are proximal to your elbows, and your elbows are proximal to your fingers.

In the next module, you will see drawings of sections (slices) of the brain cut in one of three different planes: **horizontal sections**, **frontal sections** (also termed *coronal sections*), and **sagittal sections**. These three planes are illustrated in Figure 3.16. A section cut down the center of the brain, between the two hemispheres, is called a *midsagittal section*. A section cut at a right angle to any long, narrow structure, such as the spinal cord or a nerve, is called a **cross section**.

Scan Your Brain

This is a good place for you to pause to scan your brain. Are you ready to proceed to the structures of the brain and spinal cord? Test your grasp of the preceding modules of this chapter by drawing a line between each term in the left column and the appropriate word or phrase in the right column. The correct answers are provided at the end of the exercise. Before proceeding, review material related to your errors and omissions.

1. myelin	8. synaptic vesicles	a. gaps	h. protein synthesis
2. soma	9. astrocytes	b. cone-shaped region	i. the forgotten cells
3. axon hillock	10. ganglia	c. packaging membranes	j. CNS myelinators
4. Golgi complex	11. oligodendrocytes	d. fatty substance	k. black
5. ribosomes	12. Golgi stain	e. neurotransmitter storage	l. largest glial cells
6. synapses	13. dorsal	f. cell body	m. caudal
7. glial cells	14. posterior	g. PNS clusters of cell bodies	n. top of head

Scan Your Brain answers: (1) d, (2) f, (3) b, (4) c, (5) h, (6) a, (7) i, (8) e, (9) l, (10) g, (11) j, (12) k, (13) n, (14) m.

Anatomy of the Central Nervous System

In the first three modules of this chapter, you learned about the divisions of the nervous system, the cells that compose it, and some of the neuroanatomical techniques used to study it. This final module focuses exclusively on the anatomy of the CNS. Your ascent through the CNS will begin with a focus on the spinal cord, and then you will move up to the brain.

Spinal Cord

LO 3.9 Draw and label a cross section of the spinal cord.

In cross section, it is apparent that the spinal cord comprises two different areas (see Figure 3.17): an inner H-shaped core of gray matter and a surrounding area of white matter. **Gray matter** is composed largely of cell bodies and unmyelinated interneurons, whereas **white matter** is composed largely of myelinated axons. (It is the myelin that gives the white matter its glossy white sheen.) The two dorsal arms of the spinal gray matter are called the **dorsal horns**, and the two ventral arms are called the **ventral horns**.

Pairs of *spinal nerves* are attached to the spinal cord—one on the left and one on the right—at 31 different levels of the spine. Each of these 62 spinal nerves divides as it nears the cord (see Figure 3.17), and its axons are joined to the cord via one of two roots: the *dorsal root* or the *ventral root*.

All dorsal root axons, whether somatic or autonomic, are sensory (afferent) unipolar neurons with their cell bodies grouped together just outside the cord to form the **dorsal root ganglia** (see Figure 3.17). As you can see,

many of their synaptic terminals are in the dorsal horns of the spinal gray matter. In contrast, the neurons of the ventral root are motor (efferent) multipolar neurons with their cell bodies in the ventral horns. Those that are part of the somatic nervous system project to skeletal muscles; those that are part of the autonomic nervous system project to ganglia, where they synapse on neurons that in turn project to internal organs (heart, stomach, liver, etc.). See Appendix I.

Five Major Divisions of the Brain

LO 3.10 List and discuss the five major divisions of the human brain.

A necessary step in learning to live in an unfamiliar city is learning the names and locations of its major neighborhoods or districts. Those who possess this information can easily communicate the general location of any destination in the city. This section of the chapter introduces you to the five "neighborhoods," or divisions, of the brain—for much the same reason.

To understand why the brain is considered to be composed of five divisions, it is necessary to understand its early development. In the vertebrate embryo, the tissue that eventually develops into the CNS is recognizable as a fluid-filled tube (see Figure 3.18). The first indications of the developing brain are three swellings that occur at the anterior end of this tube. These three swellings eventually develop into the adult *forebrain, midbrain,* and *hindbrain.*

Before birth, the initial three swellings in the neural tube become five (see Figure 3.18). This occurs because the forebrain swelling grows into two different swellings, and so does the hindbrain swelling. From anterior to posterior, the five swellings that compose the developing brain at birth are the *telencephalon,* the *diencephalon,* the *mesencephalon* (or midbrain), the *metencephalon,* and the *myelencephalon* (*encephalon* means "within the head"). These swellings ultimately develop into the five divisions of the adult brain. As students, we memorized their order by remembering that the *t*elencephalon is on the *top* and the other four divisions are arrayed below it in alphabetical order.

Figure 3.19 illustrates the locations of the telencephalon, diencephalon, mesencephalon, metencephalon, and myelencephalon in the adult human brain. Notice that in humans, as in other higher vertebrates, the telencephalon (the left and right *cerebral hemispheres*) undergoes the greatest growth during development. The other four divisions of the brain are often referred to collectively as the **brain stem**—the stem on which the cerebral hemispheres sit. The myelencephalon is often referred to as the *medulla.*

Figure 3.17 A schematic cross section of the spinal cord, and the dorsal and ventral roots.

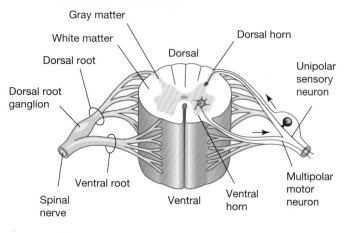

Gray matter
White matter
Dorsal root
Dorsal
Dorsal horn
Dorsal root ganglion
Unipolar sensory neuron
Ventral root
Spinal nerve
Ventral
Ventral horn
Multipolar motor neuron

Figure 3.18 The early development of the mammalian brain illustrated in schematic horizontal sections. Compare with the adult human brain in Figure 3.19.

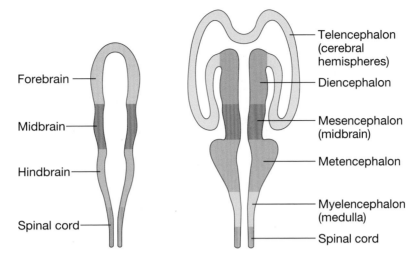

Forebrain

Midbrain

Hindbrain

Spinal cord

Telencephalon (cerebral hemispheres)

Diencephalon

Mesencephalon (midbrain)

Metencephalon

Myelencephalon (medulla)

Spinal cord

Figure 3.19 The five divisions of the adult human brain.

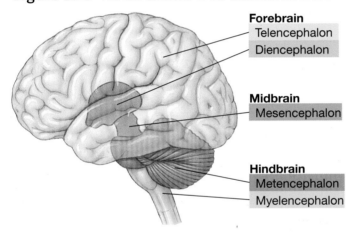

Forebrain
Telencephalon
Diencephalon

Midbrain
Mesencephalon

Hindbrain
Metencephalon
Myelencephalon

Now that you have learned the five major divisions of the brain, it is time to introduce you to their major structures. We begin our survey of brain structures in the myelencephalon, then ascend through the other divisions to the telencephalon.

Myelencephalon

LO 3.11 List and describe the components of the myelencephalon.

Not surprisingly, the **myelencephalon (or medulla)**, the most posterior division of the brain, is composed largely of tracts carrying signals between the rest of the brain and the body. An interesting part of the myelencephalon from a psychological perspective is the **reticular formation** (see Figure 3.20). It is a complex network of about 100 tiny nuclei that occupies the central core of the brain stem from the posterior boundary of the myelencephalon to the anterior boundary of the midbrain. It is so named because of its netlike appearance (*reticulum* means "little net"). Sometimes, the reticular formation is referred to as the *reticular activating system* because parts of it seem to play a role in arousal. However, the various nuclei of the reticular formation are involved in a variety of functions—including sleep, attention, movement, the maintenance of muscle tone, and various cardiac, circulatory, and respiratory reflexes. Accordingly, referring to this collection of nuclei as a *system* can be misleading.

Metencephalon

LO 3.12 List and describe the components of the metencephalon.

The **metencephalon**, like the myelencephalon, houses many ascending and descending tracts and part of the reticular formation. These structures create a bulge, called the **pons**, on the brain stem's ventral surface. The pons is one major division of the metencephalon; the other is the cerebellum (little brain)—see Figure 3.21. The **cerebellum** is the large, convoluted structure on the brain stem's dorsal surface. It is an important sensorimotor structure; cerebellar damage eliminates the ability to precisely control one's movements and to adapt them to changing conditions. However, the fact that cerebellar damage also produces a variety of cognitive deficits (e.g., deficits in decision making and in the use of language) suggests that the functions of the cerebellum are not restricted to sensorimotor control.

Figure 3.20 Structures of the human myelencephalon (medulla) and metencephalon.

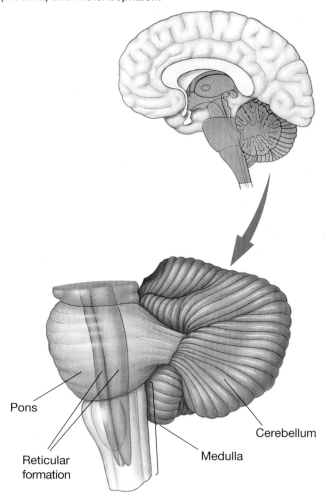

Pons

Reticular formation

Medulla

Cerebellum

Figure 3.21 The human mesencephalon (midbrain).

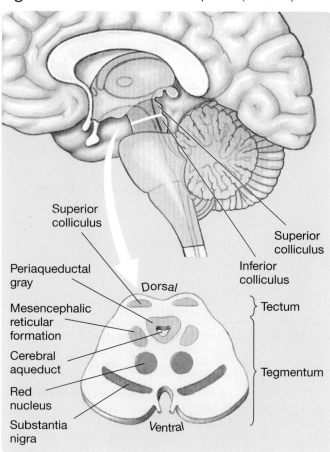

Superior colliculus

Periaqueductal gray

Mesencephalic reticular formation

Cerebral aqueduct

Red nucleus

Substantia nigra

Dorsal

Superior colliculus

Inferior colliculus

Tectum

Tegmentum

Ventral

Mesencephalon

LO 3.13 List and describe the components of the mesencephalon.

The **mesencephalon**, like the metencephalon, has two divisions. The two divisions of the mesencephalon are the tectum and the tegmentum (see Figure 3.21). The **tectum** (roof) is the dorsal surface of the midbrain. In mammals, the tectum is composed of two pairs of bumps, the *colliculi* (little hills). The posterior pair, called the **inferior colliculi**, have an auditory function. The anterior pair, called the **superior colliculi**, have a visual-motor function; more specifically, to direct the body's orientation toward or away from particular visual stimuli (see Gandhi & Katnani, 2011). In lower vertebrates, the function of the tectum is entirely visual-motor, and it is sometimes referred to as the *optic tectum*.

The **tegmentum** is the division of the mesencephalon ventral to the tectum. In addition to the reticular formation and tracts of passage, the tegmentum contains three

colorful structures of particular interest to biopsychologists: the periaqueductal gray, the substantia nigra, and the red nucleus (see Figure 3.21). The **periaqueductal gray** is the gray matter situated around the **cerebral aqueduct**, the duct connecting the third and fourth ventricles; it is of special interest because of its role in mediating the analgesic (pain-reducing) effects of opioid drugs. The **substantia nigra** (black substance) and the **red nucleus** are both important components of the sensorimotor system.

Diencephalon

LO 3.14 List and describe the components of the diencephalon.

The **diencephalon** is composed of two structures: the thalamus and the hypothalamus (see Figure 3.22). The **thalamus** is the large, two-lobed structure that constitutes the top of the brain stem. One lobe sits on each side of the third ventricle, and the two lobes are joined by the **massa intermedia**, which runs through the ventricle. Visible on the surface of the thalamus are white *lamina* (layers) that are composed of myelinated axons.

Figure 3.22 The human diencephalon.

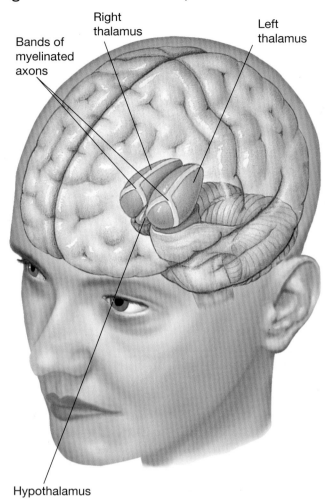

The thalamus comprises many different pairs of nuclei, most of which project to the cortex. The general organization of the thalamus is illustrated in Appendix V.

The most well-understood thalamic nuclei are the **sensory relay nuclei**—nuclei that receive signals from sensory receptors, process them, and then transmit them to the appropriate areas of sensory cortex. For example, the **lateral geniculate nuclei**, the **medial geniculate nuclei**, and the **ventral posterior nuclei** are important relay stations in the visual, auditory, and somatosensory systems, respectively. Sensory relay nuclei are not one-way streets; they all receive feedback signals from the very areas of cortex to which they project (Zembrzycki et al., 2013). Although less is known about the other thalamic nuclei, the majority of them receive input from areas of the cortex and project to other areas of the cortex (see Sherman, 2007).

The **hypothalamus** is located just below the anterior thalamus (*hypo* means "below")—see Figure 3.23. It plays an important role in the regulation of several motivated behaviors (e.g., eating, sleep, and sexual behavior). It exerts its effects in part by regulating the release of hormones from the **pituitary gland**, which dangles from it on the ventral surface of the brain. The literal meaning of *pituitary gland* is "snot gland"; it was first discovered in a gelatinous state behind the nose of a cadaver and was incorrectly assumed to be the main source of nasal mucus.

In addition to the pituitary gland, two other structures appear on the inferior surface of the hypothalamus: the optic chiasm and the mammillary bodies (see Figure 3.23). The **optic chiasm** is the point at which the *optic nerves* from each eye come together and then **decussate** (cross over to the other side of the brain) (see Chapter 6). The decussating fibers are said to be **contralateral** (projecting from one side of the body to the other), and the nondecussating fibers are said to be **ipsilateral** (staying on the same side of the body). The **mammillary bodies**,

Figure 3.23 The human hypothalamus (in color) in relation to the optic chiasm and the pituitary gland.

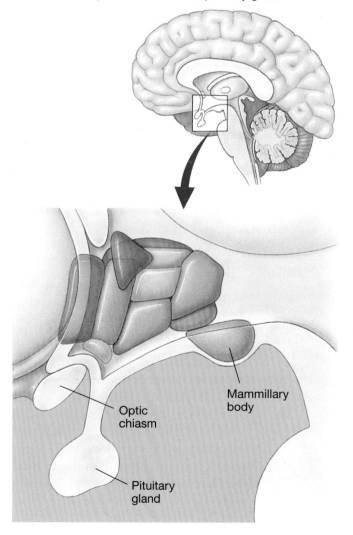

which are often considered to be part of the hypothalamus, are a pair of spherical nuclei located on the inferior surface of the hypothalamus, just behind the pituitary. The mammillary bodies and the other nuclei of the hypothalamus are illustrated in Appendix VI.

Telencephalon

LO 3.15 List and describe the components of the telencephalon.

The **telencephalon**, the largest division of the human brain, mediates the brain's most complex functions. It initiates voluntary movement, interprets sensory input, and mediates complex cognitive processes such as learning, speaking, and problem solving.

CEREBRAL CORTEX. The cerebral hemispheres are covered by a layer of tissue called the **cerebral cortex** (cerebral bark). Because the cerebral cortex is mainly composed of small, unmyelinated neurons, it is gray and is often referred to as the *gray matter*. In contrast, the layer beneath the cortex is mainly composed of large myelinated axons, which are white and often referred to as the *white matter*.

In humans, the cerebral cortex is deeply convoluted (furrowed)—see Figure 3.24. The *convolutions* have the effect of increasing the amount of cerebral cortex without increasing the overall volume of the brain. Not all mammals have convoluted cortexes; most mammals are *lissencephalic* (smooth-brained).

It was once believed that the number and size of cortical convolutions determined a species' intellectual capacities; however, the number and size of cortical convolutions appear to be related more to body size. Every large mammal has an extremely convoluted cortex.

> **Journal Prompt 3.2**
> Why do you think only large mammals have extremely convoluted cortices?

The large furrows in a convoluted cortex are called **fissures**, and the small ones are called **sulci** (singular *sulcus*). The ridges between fissures and sulci are called **gyri** (singular *gyrus*). It is apparent in Figure 3.24 that the cerebral hemispheres are almost completely separated by the largest of the fissures: the **longitudinal fissure**. The cerebral hemispheres are directly connected by a few tracts spanning the longitudinal fissure; these

Figure 3.24 The major fissures of the human cerebral cortex.

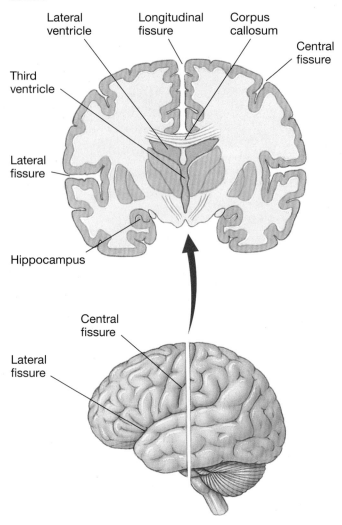

hemisphere-connecting tracts are called **cerebral commissures**. The largest cerebral commissure, the **corpus callosum**, is clearly visible in Figure 3.24.

As Figures 3.24 and 3.25 indicate, the two major landmarks on the lateral surface of each hemisphere are the **central fissure** and the **lateral fissure**. These fissures partially divide each hemisphere into four lobes: the **frontal lobe**, the **parietal lobe** (pronounced "pa-RYE-e-tal"), the **temporal lobe**, and the **occipital lobe** (pronounced "ok-SIP-i-tal"). Among the largest gyri are the **precentral gyri**, the **postcentral gyri**, and the **superior temporal gyri** in the frontal, parietal, and temporal lobes, respectively.

It is important to understand that the cerebral lobes are not functional units. It is best to think of the cerebral cortex as a flat sheet of cells that just happens to be divided into lobes because it folds in on itself at certain places during development. Thus, it is incorrect to think that a lobe is a functional unit, having one set of functions.

Figure 3.25 The lobes of the cerebral hemisphere.

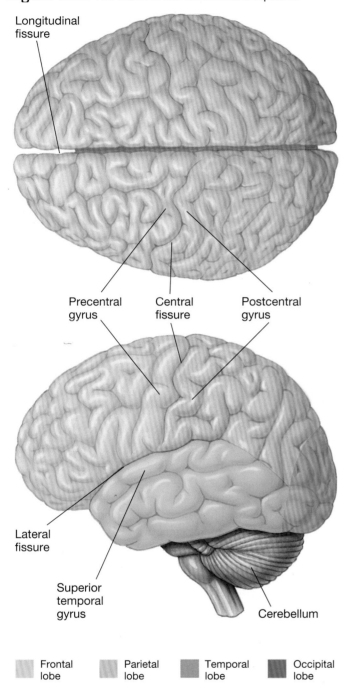

Longitudinal fissure

Precentral gyrus

Central fissure

Postcentral gyrus

Lateral fissure

Superior temporal gyrus

Cerebellum

Frontal lobe

Parietal lobe

Temporal lobe

Occipital lobe

Still, it is useful at this early stage of your biopsychological education to get a general idea of various functions of areas within each lobe. More thorough discussions of the cerebral localization of brain functions are presented in later chapters.

The main function of the occipital lobes is quite straightforward: We humans rely heavily on the analysis of visual input to guide our behavior, and the occipital cortex and large areas of adjacent cortex perform this function. There are two large functional areas in each parietal lobe: The postcentral gyrus analyzes sensations from the body (e.g., touch), whereas the remaining areas of cortex in the posterior parts of the parietal lobes play roles in perceiving the location of both objects and our own bodies and in directing our attention. The cortex of each temporal lobe has three general functional areas: The superior temporal gyrus is involved in hearing and language, the inferior temporal cortex identifies complex visual patterns, and the medial portion of temporal cortex (which is not visible from the usual side view) is important for certain kinds of memory. Lastly, each frontal lobe has two distinct functional areas: The precentral gyrus and adjacent frontal cortex have a motor function, whereas the frontal cortex anterior to motor cortex performs complex cognitive functions, such as planning response sequences, evaluating the outcomes of potential patterns of behavior, and assessing the significance of the behavior of others (see Euston, Gruber, & McNaughton, 2012; Isoda & Noritake, 2013; Pezzulo et al., 2014).

About 90 percent of human cerebral cortex is **neocortex** (new cortex), also known as *isocortex*. By convention, the layers of neocortex are numbered I through VI, starting at the surface. Figure 3.26 illustrates two adjacent sections of neocortex. One has been stained with a Nissl stain to reveal the number and shape of its cell bodies; the other has been stained with a Golgi stain to reveal the silhouettes of a small proportion of its neurons.

Three important characteristics of neocortical anatomy are apparent from the sections in Figure 3.26. First, it is apparent that many cortical neurons fall into one of two different categories: pyramidal (pyramid-shaped) cells and stellate (star-shaped) cells. **Pyramidal cells** are large multipolar neurons with pyramid-shaped cell bodies, a large dendrite called an *apical dendrite* that extends from the apex of the pyramid straight toward the cortex surface, and a very long axon (see Lodato, Shetty, & Arlotta, 2015). In contrast, **stellate cells** are small star-shaped interneurons (neurons with a short axon or no axon). Second, it is apparent that the six layers of neocortex differ from one another in terms of the size and density of their cell bodies and the relative proportion of pyramidal and stellate cell bodies that they contain. Third, it is apparent that many long axons and dendrites course vertically (i.e., at right angles to the cortical layers) through the neocortex. This vertical flow of information is the basis of the neocortex's **columnar organization**: Neurons in a given vertical column of neocortex often form a mini-circuit that performs a single function (see Rowland & Moser, 2014).

A fourth important characteristic of neocortical anatomy is not apparent in Figure 3.26: Although neocortex is six-layered, there are variations in the thickness of the respective layers from area to area (see Zilles & Amunts,

Figure 3.26 The six layers of neocortex. The thickness of the cell layers can give a clue as to the function of an area of neocortex. For example, the thickness of layer IV indicates that this is sensory neocortex.

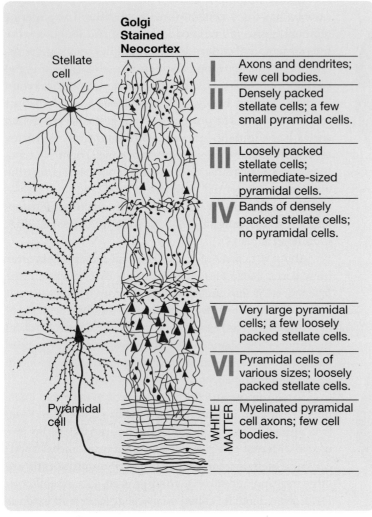

Golgi Stained Neocortex

Stellate cell

I	Axons and dendrites; few cell bodies.
II	Densely packed stellate cells; a few small pyramidal cells.
III	Loosely packed stellate cells; intermediate-sized pyramidal cells.
IV	Bands of densely packed stellate cells; no pyramidal cells.
V	Very large pyramidal cells; a few loosely packed stellate cells.
VI	Pyramidal cells of various sizes; loosely packed stellate cells.
WHITE MATTER	Myelinated pyramidal cell axons; few cell bodies.

Pyramidal cell

Based on Rakic, P. (1979)

2010). For example, because the stellate cells of layer IV are specialized for receiving sensory signals from the thalamus, layer IV is extremely thick in areas of sensory cortex. Conversely, because the pyramidal cells of layer V conduct signals from the neocortex to the brain stem and spinal cord, layer V is extremely thick in areas of motor cortex.

The **hippocampus** is one important area of cortex that is not neocortex—it has only three major layers (see Schultz & Engelhardt, 2014). The hippocampus is located at the medial edge of the cerebral cortex as it folds back on itself in the medial temporal lobe (see Figure 3.24). This folding produces a shape that is, in cross section, somewhat reminiscent of a seahorse (*hippocampus* means "sea horse"). The hippocampus plays a major role in some kinds of memory (see Chapter 11).

Limbic System and the Basal Ganglia

LO 3.16 List and describe the components of the limbic system and of the basal ganglia.

Although much of the subcortical portion of the telencephalon is taken up by axons projecting to and from the neocortex, there are several large subcortical nuclear groups. Some of them are considered part of either the *limbic system* or the *basal ganglia system*. Don't be misled by the word *system* in these contexts; it implies a level of certainty that is unwarranted. It is not entirely clear exactly what these hypothetical systems do, exactly which structures should be included in them, or even whether it is appropriate to view them as unitary systems. Nevertheless, if not taken too literally, the concepts of *limbic system* and *basal ganglia system* provide a useful means of conceptualizing the organization of several subcortical structures.

The **limbic system** is a circuit of midline structures that circle the thalamus (*limbic* means "ring"). The limbic system is involved in the regulation of motivated behaviors—including the four F's of motivation: fleeing, feeding, fighting, and sexual behavior. (This joke is as old as biopsychology itself, but it is a good one.) In addition to the structures about which you have already read (the mammillary bodies and the hippocampus), major structures of the limbic system include the amygdala, the fornix, the cingulate cortex, and the septum.

Let's begin tracing the limbic circuit (see Figure 3.27) at the **amygdala**—the almond-shaped nucleus in the anterior temporal lobe (*amygdala* means "almond" and is pronounced "a-MIG-dah-lah"). Posterior to the amygdala is the hippocampus, which runs beneath the thalamus in the medial temporal lobe. Next in the ring are the cingulate cortex and the fornix. The **cingulate cortex** is the large strip of cortex in the **cingulate gyrus** on the medial surface of the cerebral hemispheres, just superior to the corpus callosum; it encircles the dorsal thalamus (*cingulate* means "encircling"). The **fornix**, the major tract of the limbic system, also encircles the dorsal thalamus; it leaves the dorsal end of the hippocampus and sweeps forward in an arc coursing along the superior surface of the third ventricle and terminating in the septum and the mammillary bodies (*fornix* means "arc"). The **septum** is a midline nucleus located at the anterior tip of the cingulate cortex. Several tracts connect the septum and mammillary

Figure 3.27 The major structures of the limbic system: amygdala, hippocampus, cingulate cortex, fornix, septum, and mammillary body.

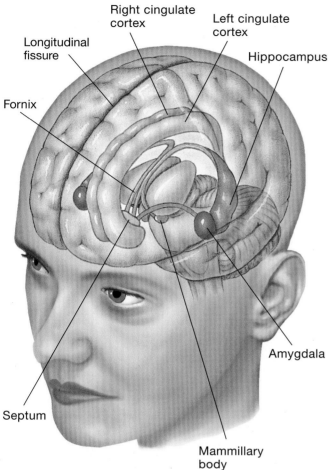

as the **globus pallidus** (pale globe). The globus pallidus is located medial to the putamen between the putamen and the thalamus.

The basal ganglia play a role in the performance of voluntary motor responses and decision making (see Hikosaka et al., 2014). Of particular interest is a pathway that projects to the striatum from the substantia nigra of the midbrain: *Parkinson's disease*, a disorder characterized by rigidity, tremors, and poverty of voluntary movement, is associated with the deterioration of this pathway. Another part of the basal ganglia that is of particular interest to biopsychologists is the *nucleus accumbens*, which is in the medial portion of the ventral striatum (see Figure 3.28). The nucleus accumbens is thought to play a role in the rewarding effects of addictive drugs and other reinforcers.

Figure 3.29 summarizes the major brain divisions and structures whose names have appeared in boldface in this section

Figure 3.28 The basal ganglia: striatum (caudate plus putamen), and globus pallidus. Notice that, in this view, the right globus pallidus is largely hidden behind the right thalamus and the left globus pallidus is totally hidden behind the left putamen.

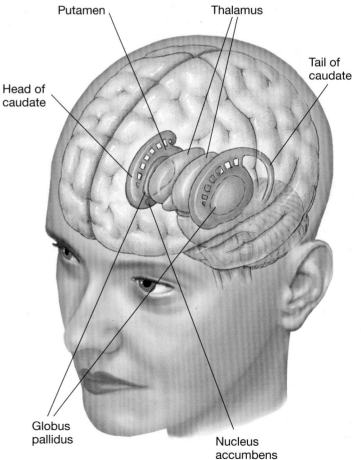

bodies with the amygdala and hippocampus, thereby completing the limbic ring.

The functions of the hippocampus, the hypothalamus and the amygdala have been investigated more than those of the other limbic structures. As stated previously, the hippocampus plays a role in certain forms of memory, and the hypothalamus is involved in a variety of motivated behaviors such as eating, sleep, and sexual behavior. The amygdala, on the other hand, is involved in emotion—particularly fear. You will learn much more about these structures in later chapters.

The **basal ganglia** are illustrated in Figure 3.28. The long tail-like **caudate** (*caudate* means "tail-like") and **putamen** (pronounced "pew-TAY-men") receive inputs from the neocortex (see Graybiel, 2000). Together, the caudate and putamen, which both have a striped appearance, are known as the **striatum** (striped structure). The striatum's major output is to a pale circular structure known

Figure 3.30 concludes this chapter, for reasons that too often get lost in the shuffle of neuroanatomical terms and technology. We have included it here to illustrate the beauty of the brain and the art of those who study its structure. We hope you are inspired by it. We wonder what thoughts its neural circuits once contained.

Figure 3.29 Summary of major brain structures.

Telencephalon	Cerebral cortex	Neocortex Hippocampus
	Major fissures	Central fissure Lateral fissure Longitudinal fissure
	Major gyri	Precentral gyrus Postcentral gyrus Superior temporal gyrus Cingulate gyrus
	Four lobes	Frontal lobe Temporal lobe Parietal lobe Occipital lobe
	Limbic system	Amygdala Hippocampus Fornix Cingulate cortex Septum Mammillary bodies
	Basal ganglia	Caudate Putamen } Striatum Globus pallidus
	Cerebral commissures	Corpus callosum
Diencephalon	Thalamus	Massa intermedia Lateral geniculate nuclei Medial geniculate nuclei Ventral posterior nuclei
	Hypothalamus	Mammillary bodies
	Optic chiasm	
	Pituitary gland	
Mesencephalon	Tectum	Superior colliculi Inferior colliculi
	Tegmentum	Reticular formation Cerebral aqueduct Periaqueductal gray Substantia nigra Red nucleus
Metencephalon	Reticular formation Pons Cerebellum	
Myelencephalon or Medulla	Reticular formation	

Figure 3.30 The art of neuroanatomical staining. This slide was stained with both a Golgi stain and a Nissl stain. Clearly visible on the Golgi-stained pyramidal neurons are the pyramid-shaped cell bodies, the large apical dendrites, and numerous dendritic spines. Less obvious here is the long, narrow axon that projects from each pyramidal cell body off the bottom of this slide.

National Institute of Mental Health

Scan Your Brain

If you have not previously studied the gross anatomy of the brain, your own brain is probably straining under the burden of new terms. To determine whether you are ready to proceed, scan your brain by labeling the following midsagittal view of a real human brain. You may find it challenging to switch from color-coded diagrams to a photograph of a real brain.

The correct answers are provided at the end of the exercise. Before proceeding, review material related to your errors and omissions. Notice that Figure 3.29 includes all the brain anatomy terms that have appeared in bold type in this module and thus is an excellent review tool.

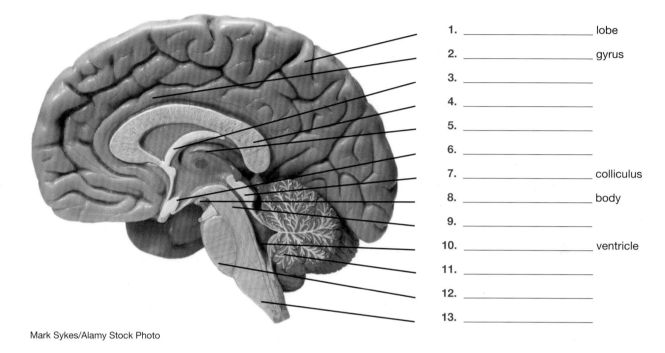

Mark Sykes/Alamy Stock Photo

1. _____ lobe
2. _____ gyrus
3. _____
4. _____
5. _____
6. _____
7. _____ colliculus
8. _____ body
9. _____
10. _____ ventricle
11. _____
12. _____
13. _____

Scan Your Brain answers: (1) parietal, (2) cingulate, (3) fornix, (4) corpus callosum, (5) thalamus, (6) hypothalamus, (7) inferior, (8) mammillary, (9) tegmentum, (10) fourth, (11) cerebellum, (12) pons, (13) medulla or myelencephalon.

Themes Revisited

This chapter contributed relatively little to the development of the text's themes. That development paused while you were being introduced to the key areas and structures of the human brain. A knowledge of fundamental neuroanatomy will serve as the foundation of discussions of brain function in subsequent chapters.

Key Terms

General Layout of the Nervous System

Central nervous system (CNS), p. 73
Peripheral nervous system (PNS), p. 73
Somatic nervous system (SNS), p. 73
Afferent nerves, p. 73
Efferent nerves, p. 73
Autonomic nervous system (ANS), p. 73
Sympathetic nerves, p. 74
Parasympathetic nerves, p. 74
Cranial nerves, p. 74

Meninges, p. 74
Dura mater, p. 75
Arachnoid membrane, p. 75
Subarachnoid space, p. 75
Pia mater, p. 75
Cerebrospinal fluid (CSF), p. 75
Central canal, p. 75
Cerebral ventricles, p. 75
Choroid plexuses, p. 75
Blood–brain barrier, p. 76

Cells of the Nervous System

Neurons, p. 77
Multipolar neuron, p. 77
Unipolar neuron, p. 77
Bipolar neuron, p. 77
Interneurons, p. 77
Nuclei, p. 78
Ganglia, p. 78
Tracts, p. 80
Nerves, p. 80

Chapter 4
Neural Conduction and Synaptic Transmission

How Neurons Send and Receive Signals

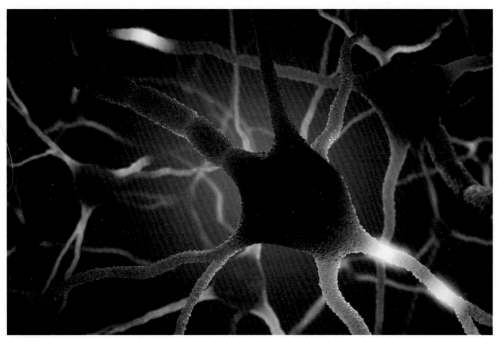

KTSDESIGN/SCIENCE PHOTO LIBRARY

⌄ Chapter Overview and Learning Objectives

Resting Membrane Potential	**LO 4.1**	Describe how the membrane potential is recorded.
	LO 4.2	Describe the resting membrane potential and its ionic basis, and describe the three factors that influence the distribution of Na^+ and K^+ ions across the neural membrane.
Generation, Conduction, and Integration of Postsynaptic Potentials	**LO 4.3**	Describe the types of postsynaptic potentials and how they are conducted.
	LO 4.4	Describe how postsynaptic potentials summate and how action potentials are generated.
Conduction of Action Potentials	**LO 4.5**	Explain the ionic basis of an action potential.
	LO 4.6	Explain how the refractory period is responsible for two important characteristics of neural activity.

The Lizard: A Case of Parkinson's Disease*

"I have become a lizard," he began. "A great lizard frozen in a dark, cold, strange world."

His name was Roberto Garcia d'Orta. He was a tall thin man in his sixties, but like most patients with Parkinson's disease, he appeared to be much older than his actual age. Not many years before, he had been an active, vigorous businessman. Then it happened—not all at once, not suddenly, but slowly, subtly, insidiously. Now he turned like a piece of granite, walked in slow shuffling steps, and spoke in a monotonous whisper.

What had been his first symptom?

A tremor.

Had his tremor been disabling?

"No," he said. "My hands shake worse when they are doing nothing at all"—a symptom called *tremor-at-rest*.

The other symptoms of Parkinson's disease are not quite so benign. They can change a vigorous man into a lizard. These include rigid muscles, a marked poverty of spontaneous movements, difficulty in starting to move, and slowness in executing voluntary movements once they have been initiated.

The term *reptilian stare* is often used to describe the characteristic lack of blinking and the widely opened eyes gazing out of a motionless face, a set of features that seems more reptilian than human. Truly a lizard in the eyes of the world.

What was happening in Mr. d'Orta's brain? A group of neurons called the *substantia nigra* (black substance) were unaccountably dying. These neurons make a particular chemical called dopamine, which they deliver to another part of the brain, known as the *striatum*. As the cells of the substantia nigra die, the amount of dopamine they can deliver to the cells in the striatum goes down. The striatum helps control movement, and to do that normally, it needs dopamine.

*Based on *Newton's Madness* by Harold Klawans (Harper & Row, 1990). Reprinted by permission of Jet Literary Associates, Inc.

Although dopamine levels are low in Parkinson's disease, dopamine is not an effective treatment because it does not readily penetrate the blood–brain barrier. However, knowledge of dopaminergic transmission has led to the development of an effective treatment: *L-dopa*, the chemical precursor of dopamine, which readily penetrates the blood–brain barrier and is converted to dopamine once inside the brain.

Mr. d'Orta's neurologist prescribed *L-dopa*, and it worked. He still had a bit of tremor, but his voice became stronger, his feet no longer shuffled, his reptilian stare faded away, and he was once again able to perform with ease many of the activities of daily life (e.g., eating, bathing, writing, speaking, and even having sex with his wife). Mr. d'Orta had been destined to spend the rest of his life trapped inside a body that was becoming increasingly difficult to control, but his life sentence was repealed—at least temporarily.

Mr. d'Orta's story does not end here. For the purposes of this chapter, his case illustrates why knowledge of the fundamentals of neural conduction and synaptic transmission is a must for any biopsychologist (see Südhof, 2017).

This chapter is about neural communication: How signals are sent from cell to cell, within networks of cells. This is not unlike what happens in a social network: Twitter is an illustrative example.

When a person in a social network tweets a message, that message is carried to other people. If the message isn't compelling enough, it gets lost in the void. If it is compelling, then some will retweet the message, propagating it to more people in their own social networks, and so on. Likewise, when a cell in our brain sends a message to the cells in its network, if the message is strong enough, some of those cells will propagate the message to other cells in their networks, and so on. Conversely, if the message isn't strong enough, the message will be lost.

Resting Membrane Potential

In order to understand how a message is conducted within neurons or transmitted from one neuron to another, you have to first learn about the **membrane potential**: the difference in electrical charge between the inside and the outside of a cell.

Recording the Membrane Potential

LO 4.1 Describe how the membrane potential is recorded.

To record a neuron's membrane potential, it is necessary to position the tip of one electrode inside the neuron and the tip of another electrode outside the neuron in the extracellular fluid. Although the size of the extracellular electrode is not critical, the tip of the intracellular electrode must be fine enough to pierce the neural membrane without damaging it. The intracellular electrodes are called **microelectrodes**; their tips are less than one-thousandth of a millimeter in diameter—much too small to be seen by the naked eye.

When both electrode tips are in the extracellular fluid, the voltage difference between them is zero. However, when the tip of the intracellular electrode is inserted into a neuron that is *at rest* (not receiving signals from other cells), a steady potential of about −70 millivolts (mV) is recorded. This indicates that the potential inside the resting neuron is about 70 mV less than that outside the neuron. This steady membrane potential of about −70 mV is called the neuron's **resting potential**. In its resting state, with the −70 mV charge built up across its membrane, a neuron is said to be **polarized** (it has a membrane potential that is not zero).

Ionic Basis of the Resting Potential

LO 4.2 Describe the resting membrane potential and its ionic basis, and describe the three factors that influence the distribution of Na^+ and K^+ ions across the neural membrane.

Like all salts in solution, the salts in neural tissue separate into positively and negatively charged particles called **ions**. There are many different kinds of ions in neurons, but this discussion focuses on only two of them: sodium ions and potassium ions. The abbreviations for sodium ions (Na^+) and potassium ions (K^+) are derived from their Latin names: *natrium* (Na) and *kalium* (K). The plus signs indicate that each Na^+ and K^+ ion carries a single positive charge.

In resting neurons, there are more Na^+ ions outside the cell than inside and more K^+ ions inside than outside. These unequal distributions of Na^+ and K^+ ions are maintained even though there are specialized pores in the neural membrane, called **ion channels**. Each type of ion channel is specialized for the passage of particular ions (e.g., Na^+ or K^+). For example, some ion channels are specialized for the passage of Na^+ ions, K^+ ions, or other ions.

There is substantial pressure on Na^+ ions to enter the resting neurons. This pressure is of two types. First is the *electrostatic pressure* from the resting membrane potential: Because opposite charges attract, the positively charged Na^+ ions are attracted to the −70 mV charge inside resting neurons. Second is the pressure from *random motion* for Na^+ ions to move down their *concentration gradient*. Let us explain. Like all ions in solution, the ions in neural tissue

Figure 4.1 Three factors that influence the distribution of Na^+ and K^+ ions across neural membranes, illustrated in a resting neuron.

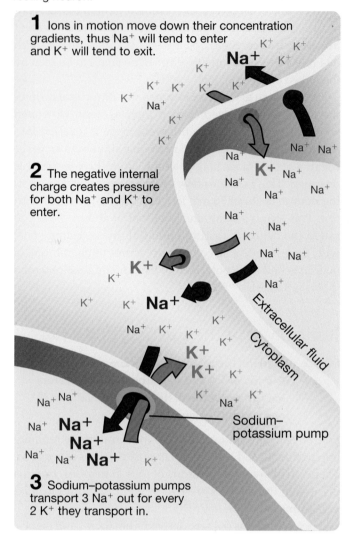

1 Ions in motion move down their concentration gradients, thus Na^+ will tend to enter and K^+ will tend to exit.

2 The negative internal charge creates pressure for both Na^+ and K^+ to enter.

3 Sodium–potassium pumps transport 3 Na^+ out for every 2 K^+ they transport in.

are in constant random motion, and particles in random motion tend to become evenly distributed because they are more likely to move down their *concentration gradients* than up them; that is, they are more likely to move from areas of high concentration to areas of low concentration than vice versa. Likewise, a drop of red ink placed in a bathtub full of water will move outwards to areas where there is no ink, and the water will gradually turn pink.

So, why then do Na^+ ions under electrostatic pressure and pressure from random movement not come rushing into neurons, thus reducing the resting membrane potential? The answer is simple: The sodium ion channels in resting neurons are closed, thus greatly reducing the flow of Na^+ ions into the neuron. In contrast, the potassium ion channels are open in resting neurons, but only a few K^+ ions exit because the electrostatic pressure that results from the negative resting membrane potential largely holds them inside.

In the 1950s, Alan Hodgkin and Andrew Huxley became interested in the stability of the resting membrane potential. Some Na^+ ions do manage to enter resting neurons despite the closed sodium channels and some K^+ ions do exit; then why does the resting membrane potential stay fixed? In a series of clever experiments, for which they were awarded Nobel Prizes, Hodgkin and Huxley discovered the answer. At the same rate that Na^+ ions leaked into resting neurons, other Na^+ ions were actively transported out; and at the same rate that K^+ ions leaked out of resting neurons, other K^+ ions were actively transported in. Such ion transport is performed by mechanisms in the cell membrane that continually exchange three Na^+ ions inside the neuron for two K^+ ions outside. These transporters are commonly referred to as **sodium–potassium pumps**.

Since the discovery of sodium–potassium pumps, several other classes of **transporters** (mechanisms in the membrane of a cell that actively transport ions or molecules across the membrane) have been discovered (see Kaila et al., 2014). You will encounter more of them later in this chapter.

Figure 4.1 summarizes the status of Na^+ and K^+ ions in a resting neuron. Now that you understand the basic properties of resting neurons, you are prepared to consider how neurons respond to input signals.

Generation, Conduction, and Integration of Postsynaptic Potentials

What happens when a resting membrane potential is disturbed? Typically, disturbances of the membrane potential occur as a result of input from cells that synapse on a neuron. For that reason, such disturbances of the resting membrane potential are termed **postsynaptic potentials (PSPs)**. In this module, you will learn how PSPs are generated by input to a neuron, how they are subsequently conducted to different parts of a neuron, and how they can cause a neuron to fire (produce an action potential).

Generation and Conduction of Postsynaptic Potentials

LO 4.3   Describe the types of postsynaptic potentials and how they are conducted.

When neurons fire, they release from their terminal buttons chemicals called *neurotransmitters*, which diffuse across the synaptic clefts and interact with specialized receptor molecules on the receptive membranes of the next neuron in the circuit. When neurotransmitter molecules bind to postsynaptic receptors, they typically have one of two effects, depending on the neurotransmitter, receptor, and postsynaptic neuron in question. They may **depolarize** the receptive membrane (decrease the resting membrane potential, from −70 to −67 mV, for example), or they may **hyperpolarize** it (increase the resting membrane potential, from −70 to −72 mV, for example).

Postsynaptic depolarizations are called **excitatory postsynaptic potentials (EPSPs)** because, as you will soon learn, they increase the likelihood that the neuron will fire. Postsynaptic hyperpolarizations are called **inhibitory postsynaptic potentials (IPSPs)** because they decrease the likelihood that the neuron will fire.

All PSPs, both EPSPs and IPSPs, are **graded potentials**. This means that the amplitudes of PSPs are proportional to the intensity of the signals that elicit them: Weak signals elicit small PSPs, and strong signals elicit large ones.

EPSPs and IPSPs travel passively from their sites of generation at synapses, usually on the dendrites or cell body, in much the same way that electrical signals travel through a cable. Accordingly, the transmission of PSPs has two important characteristics. First, it is *rapid*—so rapid that it can be assumed to be instantaneous for most purposes. It is important not to confuse the duration of PSPs with their rate of transmission; although the duration of PSPs varies considerably, all PSPs, whether brief or enduring, are transmitted almost instantaneously. Second, the transmission of PSPs is *decremental*: They decrease in amplitude as they travel through the neuron, just as a ripple on a pond gradually disappears as it travels outward. Most PSPs do not travel more than a couple of millimeters from their site of generation before they fade out completely.

Integration of Postsynaptic Potentials and Generation of Action Potentials

LO 4.4 Describe how postsynaptic potentials summate and how action potentials are generated.

The PSPs created at a single synapse typically have little effect on the firing of the postsynaptic neuron. The receptive areas of most neurons are covered with thousands of synapses, and whether a neuron fires is determined by the net effect of their activity. More specifically, whether a neuron fires depends on the balance between the excitatory and inhibitory signals reaching its axon. It was once believed that action potentials were generated at the **axon hillock** (the conical structure at the junction between the cell body and the axon), but they are actually generated in the adjacent section of the axon, called the **axon initial segment** (see Kuba, Adachi, & Ohmori, 2014; Tian et al., 2014).

The graded EPSPs and IPSPs created by the action of neurotransmitters at particular receptive sites on a neuron's membrane are conducted instantly and decrementally to the axon initial segment. If the sum of the depolarizations and hyperpolarizations reaching the axon initial segment at any time is sufficient to depolarize the membrane to a level referred to as its **threshold of excitation**—usually about −65 mV—an action potential is generated. The **action potential (AP)** is a massive but momentary—lasting for 1 millisecond—reversal of the membrane potential from about −70 to about +50 mV. Unlike PSPs, APs are not graded responses: Their magnitude is not related in any way to the intensity of the stimuli that elicit them. To the contrary, they are **all-or-none responses**; that is, they either occur to their full extent or do not occur at all. See Figure 4.2 for an

Figure 4.2 An EPSP, an IPSP, and an EPSP followed by an AP.

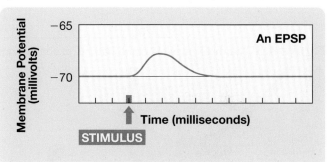

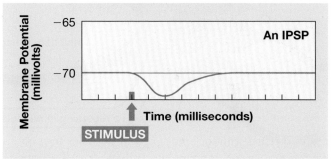

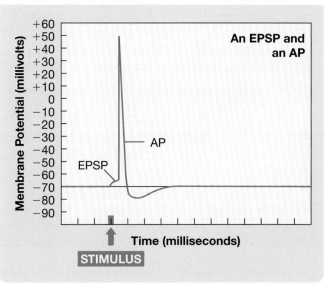

illustration of an EPSP, an IPSP, and an AP. Although many neurons display APs of the type illustrated in Figure 4.2, others do not—for example, some neurons display APs that have a longer duration, have a lower amplitude, or involve multiple spikes.

In effect, each neuron adds together all the graded excitatory and inhibitory PSPs reaching its axon initial segment and decides to fire or not to fire on the basis of their sum. The summation of PSPs occurs in two ways: over space and over time.

Figure 4.3 shows the three possible combinations of **spatial summation**. It shows how local EPSPs that are produced simultaneously on different parts of the receptive membrane sum to form a greater EPSP, how simultaneous IPSPs sum to form a greater IPSP, and how simultaneous EPSPs and IPSPs sum to cancel each other out.

Figure 4.4 illustrates **temporal summation**. It shows how PSPs produced in rapid succession at the same synapse

sum to form a greater signal. The reason that successive stimulations of a neuron can add together over time is that the PSPs they produce often outlast them. Thus, if a particular synapse is activated and then activated again before the original PSP has completely dissipated, the effect of the second stimulation will be superimposed on the lingering PSP produced by the first. Accordingly, it is possible for a brief subthreshold excitatory stimulus to fire a neuron if it is administered twice in rapid succession. In the same way, an inhibitory synapse activated twice in rapid succession can produce a greater IPSP than that produced by a single stimulation.

PSPs continuously summate over both time and space as a neuron is continually bombarded with stimuli from thousands of synapses. Although schematic diagrams of neural circuitry rarely show neurons with more than a few representative synaptic contacts, most neurons have thousands of synaptic contacts covering their dendrites and cell body. To better understand the summation of PSPs, consider what happens to the many ripples on the surface of a pond: The ripples are continuously interacting with each other to generate larger or smaller ripples.

The location of a synapse on a neuron's membrane had long been assumed to be an important factor in determining its potential to influence the neuron's firing. Because PSPs are transmitted decrementally, synapses near the axon had been assumed to have the most influence on the firing of the neuron. However, it has been demonstrated that some neurons have a mechanism for amplifying dendritic signals that originate far from their axon (see Adrian et al., 2014; Araya, 2014).

In some ways, the firing of a neuron is like the firing of a gun. Both reactions are triggered by graded responses. As a trigger is squeezed, it gradually moves back until it causes the gun to fire; as a neuron is stimulated, it becomes less polarized until the threshold of excitation is reached and firing occurs. Furthermore, the firing of a gun and neural firing are both all-or-none events. Just as squeezing a trigger harder does not make the bullet travel faster or farther, stimulating a neuron more intensely does not increase the speed or amplitude of the resulting action potential.

Figure 4.3 The three possible combinations of spatial summation.

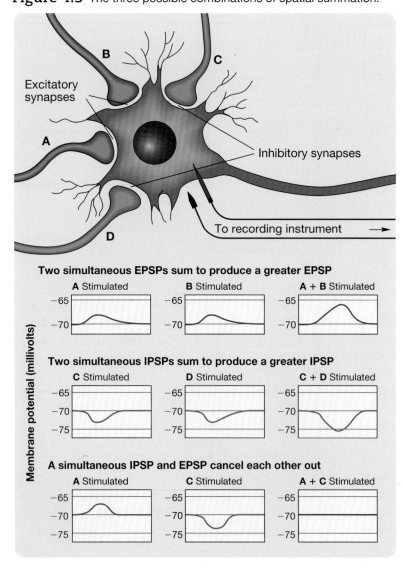

Journal Prompt 4.1

Can you think of a metaphor, other than the firing of a gun, that might serve as an accurate description of the firing of a neuron?

Figure 4.4 The two possible combinations of temporal summation.

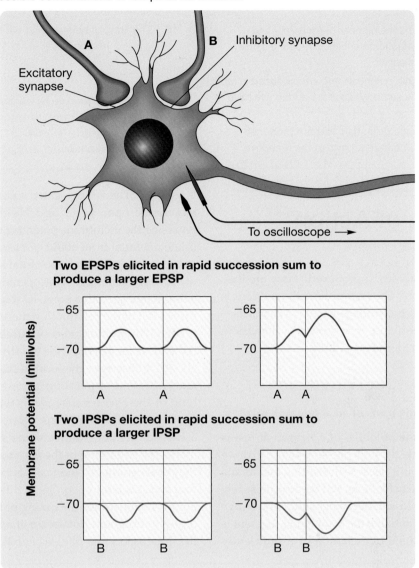

Scan Your Brain

This is a good place to pause and scan your brain to check your knowledge on synaptic transmission. Fill in the following blanks with the most appropriate terms. The correct answers are provided at the end of the exercise. Before proceeding, review material related to your errors and omissions.

1. _____ is a common chemical used to alleviate symptoms in people living with Parkinson's disease.

2. The difference in electrical charge between the inside and outside of a nerve cell is called _____ and is recorded using microelectrodes.

3. The resting potential inside the neuron is approximately _____ mV less than that outside the cell. This is called polarization.

4. Sodium and potassium ions are both _____ charged.

5. In a resting neuron, there are more _____ ions outside the cell and more _____ ions inside the cell.

6. K^+ ions are largely held inside the cell because of the membrane's _____ resting potential.

7. _____ pumps ensure that at resting potential, three Na^+ ions move inside the cell and two K^+ ions move outside the cell.

8. _____ are released into the synaptic cleft, and they attach to receptor molecules on the postsynaptic membrane of the next cell.

9. The neurotransmitters may _____ the postsynaptic receptive membrane, which implies that the resting membrane potential will increase.

10. _____ postsynaptic potentials increase the likelihood that a neuron will fire.

11. Postsynaptic potentials _____ in amplitude as they travel through the neuron.

12. A momentary increase of membrane potential to about +50 mV is called _____.

13. Each neuron sums the number of excitatory and inhibitory postsynaptic potentials to create a single signal, a process called _____.

14. When postsynaptic potentials that are produced in rapid succession at the same synapse are added, we have _____ summation.

15. When postsynaptic potentials that are produced simultaneously in different parts of the receptive membrane are added, we have _____ summation.

16. The firing of neurons and the firing of a gun are both _____ responses.

Scan Your Brain answers: (1) *L-dopa*, (2) membrane potential, (3) 70, (4) positively, (5) Na⁺, (6) negative, (7) Sodium–potassium, (8) Neurotransmitters, (9) hyperpolarize, (10) Excitatory, (11) decrease, (12) action potential, (13) integration, (14) temporal, (15) spatial, (16) all-or-none.

Conduction of Action Potentials

How are action potentials (APs) produced? How are they conducted along the axon? The answer to both questions is the same: through the action of **voltage-gated ion channels**—ion channels that open or close in response to changes in membrane potential (see Moran et al., 2015).

Ionic Basis of Action Potentials

LO 4.5 **Explain the ionic basis of an action potential.**

Recall that the membrane potential of a neuron at rest is relatively constant despite the high pressure acting to drive Na⁺ ions into the cell. This is because the resting membrane is relatively impermeable to Na⁺ ions and because those few that do pass in are pumped out. But things suddenly change when the membrane potential of the axon initial segment is depolarized to the threshold of excitation by a sufficiently large EPSP. The voltage-gated sodium channels in the axon membrane open wide, and Na⁺ ions rush in, suddenly reversing the membrane potential; that is, driving the membrane potential from about −70 to about +50 mV. The rapid change in the membrane potential associated with the *influx* of Na⁺ ions then triggers the opening of voltage-gated potassium channels. At this point, K⁺ ions near the membrane are driven out of the cell through these channels—first by their relatively high internal concentration and then, when the AP is near its peak, by the positive internal charge. After about 1 millisecond, the sodium channels close. This closure marks the end of the *rising phase* of the AP and the beginning of the *repolarization phase*, which is the result of the continued efflux of K⁺ ions. Once repolarization has been achieved, the potassium channels gradually close, which marks the beginning of the *hyperpolarization phase*. Because they close gradually, too many K⁺ ions flow out of the neuron, and it is left hyperpolarized for a brief period of time. Figure 4.5 illustrates the timing of the opening and closing of the sodium and potassium channels during an AP, and the three associated phases of an AP.

Figure 4.5 The opening and closing of voltage-gated sodium and potassium channels during an AP, and the three associated phases of an AP.

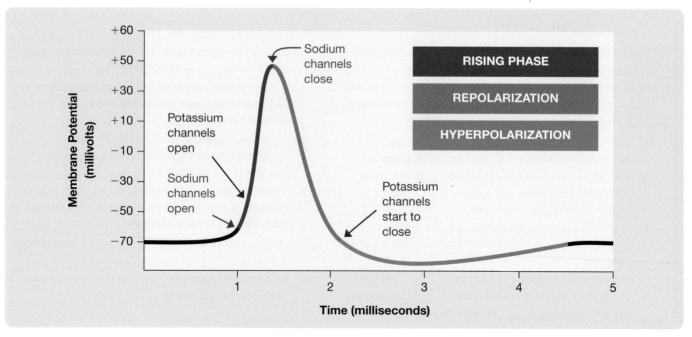

The number of ions that flow through the membrane during an AP is extremely small in relation to the total number inside and around the neuron. The AP involves only those ions right next to the membrane. Therefore, a single AP has little effect on the relative concentrations of various ions inside and outside the neuron, and the resting ion concentrations next to the membrane are rapidly reestablished by the random movement of ions. The sodium–potassium pumps play only a minor role in the reestablishment of the resting potential.

Refractory Periods

LO 4.6 **Explain how the refractory period is responsible for two important characteristics of neural activity.**

There is a brief period of about 1 to 2 milliseconds after the initiation of an AP during which it is impossible to elicit a second AP. This period is called the **absolute refractory period**. The absolute refractory period is followed by the **relative refractory period**—the period during which it is possible to fire the neuron again but only by applying higher-than-normal levels of stimulation. The end of the relative refractory period is the point at which the amount of stimulation necessary to fire the neuron returns to baseline.

Refractory periods are responsible for two important characteristics of neural activity. First, they are responsible for the fact that APs normally travel along axons in only one direction. Because the portions of an axon over which an AP has just traveled are left momentarily refractory, an AP cannot reverse direction. Second, refractory periods are responsible for the fact that the rate of neural firing is related to the intensity of the stimulation. If a neuron is subjected to continual high-intensity stimulation, it fires and then fires again as soon as its absolute refractory period is over—a maximum of about 1,000 times per second. However, if the level of continuous stimulation is of an intensity just sufficient to fire the neuron when it is at rest, the neuron does not fire again until both the absolute and the relative refractory periods have run their course. Intermediate intensities of continuous stimulation produce intermediate rates of neural firing.

Axonal Conduction of Action Potentials

LO 4.7 **Describe how action potentials are conducted along axons—both myelinated and unmyelinated.**

The conduction of APs along an axon differs from the conduction of PSPs in two important ways. First, the conduction of APs along an axon is typically *nondecremental*; APs do not grow weaker as they travel along the axonal membrane. Second, APs are conducted more slowly than PSPs.

These two differences are the result of the important role played by voltage-gated sodium channels in AP conduction. Once an AP has been generated, it travels along the axon as a graded potential; that is, it travels rapidly and decrementally. However, when that graded potential reaches the next voltage-gated sodium channel along the axon, and if it is sufficiently large (i.e., it exceeds the threshold of excitation), then those channels open and Na^+ ions rush into the axon and generate another full-blown AP. In essence, the AP is continually regenerated at each sodium channel along the length of the axon, again and again until a full-blown AP is triggered as the axon terminal buttons.

The following analogy may help you appreciate the major characteristics of axonal conduction. Consider a row of mouse traps on a wobbly shelf, all of them set and ready to be triggered. Each trap stores energy by holding back its striker against the pressure of the spring, in the same way that each voltage-gated sodium channel stores energy by holding back Na^+ ions, which are under pressure to move down their concentration and electrostatic gradients into the neuron. When the first trap in the row is triggered, the vibration is transmitted through the shelf rapidly and decrementally. When the vibration reaches the next trap and it is sufficiently large, then that trap is sprung—and so on down the line. Likewise, when a sodium channel at the axon initial segment is opened by an EPSP, an AP is generated and then that electrical signal travels instantly and decrementally (i.e., as a graded potential) to the next sodium channel along the axon. Then, that sodium channel opens to generate an AP, and so on down the length of the axon.

> **Journal Prompt 4.2**
> Can you think of a better analogy to describe axonal conduction?

The nondecremental nature of AP conduction is readily apparent from this analogy; the last trap on the shelf strikes with no less intensity than did the first. This analogy also illustrates another important point: The row of traps can transmit in either direction, just like an axon. If electrical stimulation of sufficient intensity is applied to a midpoint of an axon, two APs will be generated: One AP will travel along the axon back to the cell body—this is called **antidromic conduction**; the second AP will travel along the axon towards the terminal buttons—this is called **orthodromic conduction**. The elicitation of an AP and its orthodromic conduction are illustrated in Figure 4.6.

The generation of an AP at the axon initial segment also spreads back through the cell body and dendrites of the neuron as a large graded potential. It is believed that these antidromic (*backpropagating*) potentials play a role in certain forms of synaptic plasticity (see Stuart & Spruston, 2015).

CONDUCTION IN MYELINATED AXONS. Recall that the axons of many neurons are insulated from the extracellular fluid by segments of fatty tissue called *myelin*. In myelinated axons, ions can pass through the axonal membrane only at the **nodes of Ranvier**—the gaps between adjacent myelin segments.

Figure 4.6 The usual direction of signals conducted through a multipolar neuron (i.e., orthodromic conduction).

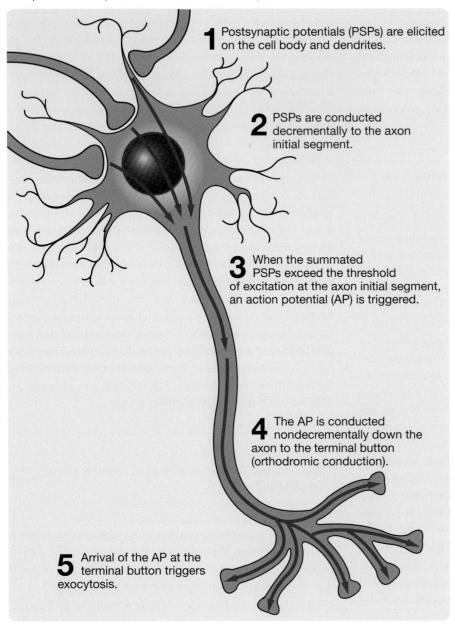

1 Postsynaptic potentials (PSPs) are elicited on the cell body and dendrites.

2 PSPs are conducted decrementally to the axon initial segment.

3 When the summated PSPs exceed the threshold of excitation at the axon initial segment, an action potential (AP) is triggered.

4 The AP is conducted nondecrementally down the axon to the terminal button (orthodromic conduction).

5 Arrival of the AP at the terminal button triggers exocytosis.

Indeed, in myelinated axons, axonal voltage-gated sodium channels are concentrated at the nodes of Ranvier. How, then, are APs transmitted in myelinated axons?

If we consider the mouse trap metaphor again, the answer is quite simple: It is just as if the mouse traps were placed further apart along the wobbly shelf. That is, because the sodium channels are concentrated at some distance from one another (at the nodes of Ranvier), the electrical signal generated at the sodium channels at one node of Ranvier travels instantly and decrementally (i.e., it is a graded potential) to the sodium channels at the next node, and so on down the length of the myelinated axon.

Myelination increases the speed of axonal conduction. Because conduction along the myelinated segments of the axon is instantaneous (i.e., it is a graded potential),

the signal "jumps" along the axon from one node of Ranvier to the next. There is, of course, a slight delay at each node while the AP is regenerated, but conduction is still much faster in myelinated axons than in unmyelinated axons. The transmission of APs in myelinated axons is called **saltatory conduction** (*saltare* means "to skip or jump"). Given the important role of myelin in neural conduction, it is hardly surprising that diseases that damage the nervous system by attacking myelin, like multiple sclerosis, have devastating effects on neural activity and behavior.

THE VELOCITY OF AXONAL CONDUCTION. At what speed are APs conducted along an axon? The answer to this question depends on two properties of the axon. Conduction is faster in large-diameter axons, and—as you have just learned—it is faster in those that are myelinated. Human *motor neurons* (neurons that synapse on skeletal muscles) are large and myelinated; thus, some can conduct at speeds up to 60 meters per second (about 134 miles per hour). In contrast, small, unmyelinated axons conduct APs at about 1 meter per second.

CONDUCTION IN NEURONS WITHOUT AXONS. APs are the means by which axons conduct all-or-none signals nondecrementally over relatively long distances. Thus, to keep what you have just learned about APs in perspective, it is important for you to remember that most neurons in mammalian brains either do not have axons or have very short ones, and many of these neurons do not normally display APs. Conduction in these *interneurons* is typically only through graded potentials.

The Hodgkin-Huxley Model in Perspective

LO 4.8 **Explain the shortcomings of the Hodgkin-Huxley model when applied to neurons in the mammalian brain.**

The preceding account of neural conduction is based largely on the *Hodgkin-Huxley model*, the theory first proposed by Hodgkin and Huxley in the early 1950s (see Catterall et al., 2012). Perhaps you have previously encountered some of this information about neural conduction in introductory biology or psychology courses, where it is often presented as a factual account of neural conduction and its mechanisms, rather than as a theory. To be fair, the Hodgkin-Huxley model was a major advance in our understanding of neural conduction (Catterall et al., 2012). Fully deserving of the 1963 Nobel Prize, the model provided a simple, effective introduction to what we now understand about the general ways in which neurons conduct signals. The problem is that the simple neurons and mechanisms of the Hodgkin-Huxley model are not representative of the variety, complexity, and plasticity of many of the neurons in the mammalian brain.

The Hodgkin-Huxley model was based on the study of squid motor neurons. Motor neurons are simple, large, and readily accessible in the PNS—squid motor neurons are particularly large. The simplicity, size, and accessibility of squid motor neurons contributed to the initial success of Hodgkin's and Huxley's research, but these same properties make it difficult to apply the model directly to the mammalian brain. Hundreds of different kinds of neurons are found in the mammalian brain, and many of these have actions not found in motor neurons.

Moreover, there is mounting evidence that neural conduction is not merely due to electrical impulses (see Holland, De Regt, & Drukarch, 2019). For example, there is evidence that electrical impulses, like APs or PSPs, are accompanied by mechanical impulses: travelling waves of expansion and contraction of the neural membrane—exactly like ripples on a pond (see Fox, 2018). In summary, the Hodgkin-Huxley model should be applied to cerebral neurons with great caution.

Synaptic Transmission: From Electrical Signals to Chemical Signals

Now that you have learned about how communication occurs within a single neuron—through postsynaptic potentials (PSPs) and action potentials (APs)—you are ready to learn how neurons communicate with other cells. In the remaining modules of this chapter, you will learn how APs arriving at terminal buttons trigger the release of neurotransmitters into synapses and how neurotransmitters carry signals to other cells. This module provides an overview of five aspects of synaptic transmission: (1) the structure of synapses; (2) the synthesis, packaging, and transport of neurotransmitter molecules; (3) the release of neurotransmitter molecules; (4) the activation of receptors by neurotransmitter molecules; and (5) the reuptake, enzymatic degradation, and recycling of neurotransmitter molecules.

Structure of Synapses

LO 4.9 Describe the structure of different types of synapses.

Some communication among neurons occurs across synapses such as the one illustrated in Figure 4.7. At such synapses, neurotransmitter molecules are released from specialized sites on buttons into synaptic clefts, where they induce EPSPs or IPSPs in other neurons by binding to receptors on their postsynaptic membranes. The synapse featured in Figure 4.7 is an *axodendritic synapse*—a synapse of an axon terminal button onto a dendrite. Also common are *axosomatic synapses*—synapses of axon terminal buttons on *somas* (cell bodies). Notice in Figure 4.7 that many axodendritic synapses terminate on **dendritic spines** (nodules of various shapes that are located on the surfaces of many dendrites)—see Figure 3.30. Also notice in Figure 4.7 that an astrocyte is situated at the synapse. Most synapses in the brain form a **tripartite synapse**: a synapse that involves two neurons and

Figure 4.7 Anatomy of a typical synapse.

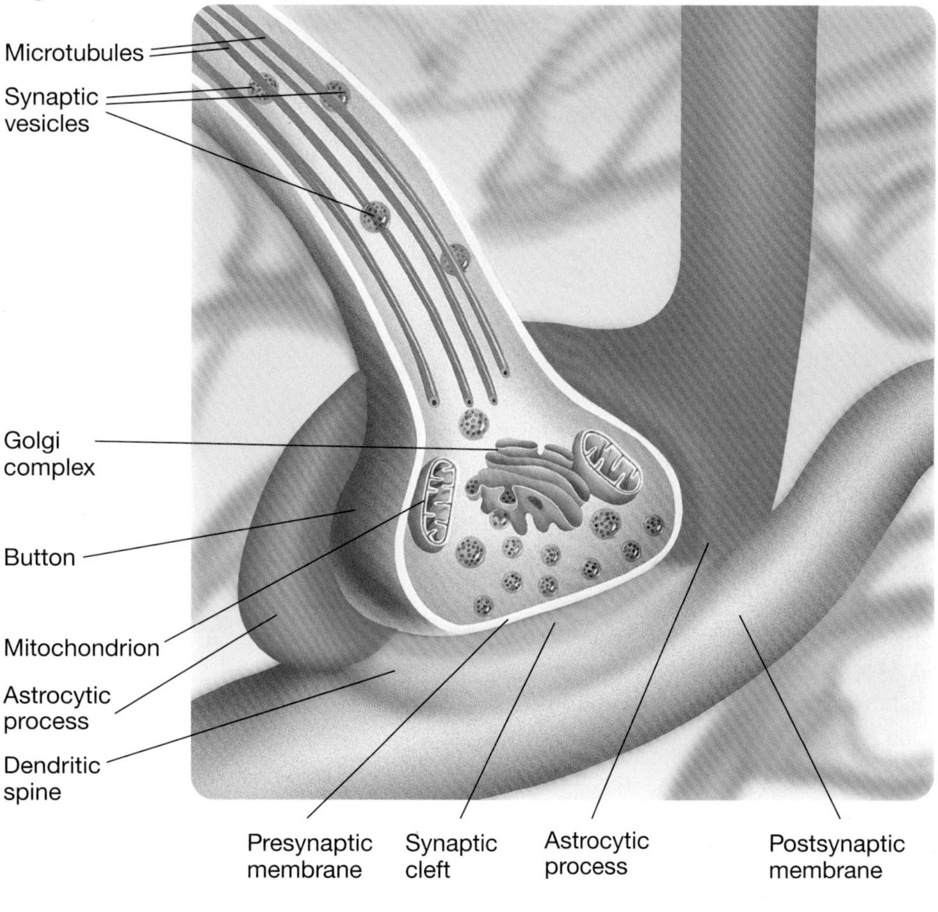

Microtubules, Synaptic vesicles, Golgi complex, Button, Mitochondrion, Astrocytic process, Dendritic spine, Presynaptic membrane, Synaptic cleft, Astrocytic process, Postsynaptic membrane

an astroglial cell (see Allen & Eroglu, 2017; Grosche & Reichenbach, 2013; Navarrete & Araque, 2014; Sun et al., 2013). All three cells communicate with one another through synaptic transmission.

Although axodendritic and axosomatic synapses are the most common synaptic arrangements, there are many others (see Matthews & Fuchs, 2010). For example, there are *dendrodendritic synapses*, which are interesting because they are often capable of transmission in either direction (see Urban & Castro, 2010). There are also *axoaxonic synapses*; these are particularly important because they can mediate *presynaptic facilitation and inhibition*. As illustrated in Figure 4.8, an axoaxonic synapse on or near a terminal button can selectively facilitate or inhibit the effects of that button on the postsynaptic neuron. The advantage of presynaptic facilitation and inhibition (compared to PSPs) is that they can selectively influence one particular synapse rather than the entire presynaptic neuron. Finally, in the central nervous system, there are also *axomyelenic synapses*, where an axon synapses on the myelin sheath of an oligodendrocyte. This newly discovered type of synapse represents yet another form of neuron–glia communication (see Dimou & Simons, 2017; Micu et al., 2018).

The synapses depicted in Figures 4.7 and 4.8 are **directed synapses**—synapses at which the site of neurotransmitter release and the site of neurotransmitter reception are in close proximity. This is a common arrangement, but there are also many nondirected synapses in the mammalian nervous system. **Nondirected synapses** are synapses at which the site of release is at some distance from the site of reception. One type of nondirected synapse is depicted in Figure 4.9. In this type of arrangement, neurotransmitter

Figure 4.8 Presynaptic facilitation and inhibition.

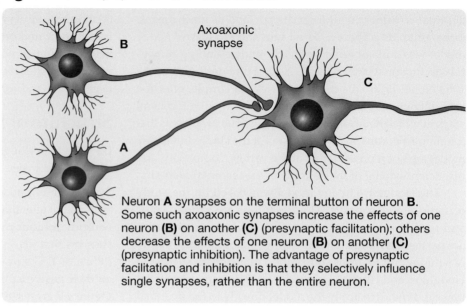

Neuron **A** synapses on the terminal button of neuron **B**. Some such axoaxonic synapses increase the effects of one neuron **(B)** on another **(C)** (presynaptic facilitation); others decrease the effects of one neuron **(B)** on another **(C)** (presynaptic inhibition). The advantage of presynaptic facilitation and inhibition is that they selectively influence single synapses, rather than the entire neuron.

Figure 4.9 One example of nondirected neurotransmitter release. Some neurons release neurotransmitter molecules diffusely from varicosities along the axon and its branches.

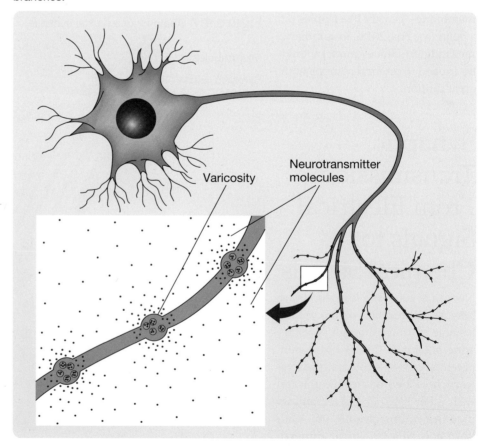

molecules are released from a series of *varicosities* (bulges or swellings) along the axon and its branches and thus are widely dispersed to surrounding targets. Because of their appearance, these synapses are often referred to as *string-of-beads synapses*.

Synthesis, Packaging, and Transport of Neurotransmitter Molecules

LO 4.10 Describe how neurotransmitter molecules are synthesized and packaged in vesicles.

There are two basic categories of neurotransmitter molecules: large and small.

All large neurotransmitters are neuropeptides. **Neuropeptides** are short amino acid chains composed of between 3 and 36 amino acids; in effect, they are short proteins.

Small-molecule neurotransmitters are typically synthesized in the cytoplasm of the terminal button and packaged in **synaptic vesicles** by the button's **Golgi complex**. (This may be a good point at which to review the internal structures of neurons in Figure 3.6.) Once filled with neurotransmitter, the vesicles are stored in clusters next to the presynaptic membrane. In contrast, neuropeptides, like other proteins, are assembled in the cytoplasm of the cell body on *ribosomes*; they are then packaged in vesicles by the cell body's Golgi complex and transported by *microtubules* to the terminal buttons at a rate of about 40 centimeters (about 16 inches) per day. The vesicles that contain neuropeptides are usually larger than those that contain small-molecule neurotransmitters, and they do not usually congregate as closely to the presynaptic membrane as the other vesicles do.

It was once believed that each neuron synthesizes and releases only one neurotransmitter, but it has been clear for some time that many neurons contain two neurotransmitters—a situation generally referred to as **coexistence**. It may have escaped your notice that the button illustrated in Figure 4.7 contains synaptic vesicles of two sizes. This suggests that it contains two neurotransmitters: a neuropeptide in the larger vesicles and a small-molecule neurotransmitter in the smaller vesicles. Although this type of coexistence was the first to be discovered, we now know that there is also coexistence of multiple small-molecule neurotransmitters in the same neuron (see Granger, Wallace, & Sabatini, 2017). Adding to the complexity is the fact that neurons can change the types of neurotransmitters they release over their lifespan (see Spitzer, 2017).

Release of Neurotransmitter Molecules

LO 4.11 Explain the process of neurotransmitter exocytosis.

Exocytosis—the process of neurotransmitter release—is illustrated in Figure 4.10 (see Shin, 2014). When a neuron

is at rest, synaptic vesicles that contain small-molecule neurotransmitters tend to congregate near sections of the presynaptic membrane that are particularly rich in *voltage-gated calcium channels* (see Simms & Zamponi, 2014). When stimulated by APs, these channels open, and Ca^{2+} (*calcium*) ions enter the button. The entry of the Ca^{2+} ions triggers a chain reaction that ultimately causes synaptic vesicles to fuse with the presynaptic membrane and empty their contents into the synaptic cleft (see Zhou et al., 2017).

The release of small-molecule neurotransmitters differs from the release of neuropeptides. Small-molecule neurotransmitters are typically released in a pulse each time an AP triggers a momentary influx of Ca^{2+} ions into the presynaptic membrane; in contrast, neuropeptides are typically released gradually in response to general increases in the level of intracellular Ca^{2+} ions, such as might occur during a general increase in the rate of neuron firing.

It is important to note that not all vesicles fuse with the presynaptic membrane. Some vesicles are released as intact packages into the extracellular space. These *extracellular vesicles* often carry larger molecules (e.g., proteins, RNA molecules) between different neurons and glia in the central nervous system (see Holm, Kaiser, & Schwab, 2018; Paolicelli, Bergamini, & Rajendran, 2018). Some of these transmitted molecules can induce persistent changes in the expression of genes through epigenetic mechanisms (see Bakhshandeh, Kamaleddin, & Aalishah, 2017).

Activation of Receptors by Neurotransmitter Molecules

LO 4.12 Describe the differences between ionotropic and metabotropic receptors.

Once released, neurotransmitter molecules produce signals in postsynaptic neurons by binding to **receptors** in the postsynaptic membrane. Each receptor is a protein that contains binding sites for only particular neurotransmitters; thus, a neurotransmitter can influence only those cells that have receptors for it. Any molecule that binds to another is referred to as its **ligand**, and a neurotransmitter is thus said to be a ligand of its receptor.

It was initially assumed that there is only one type of receptor for each neurotransmitter, but this has not proved to be the case. As more receptors have been identified, it has become clear that most neurotransmitters bind to several different types of receptors. The different types of receptors to which a particular neurotransmitter can bind are called the **receptor subtypes** for that neurotransmitter. The various receptor subtypes for a neurotransmitter are typically located in different brain areas, and they typically respond to the neurotransmitter in different ways.

Figure 4.10 Schematic illustration of exocytosis.

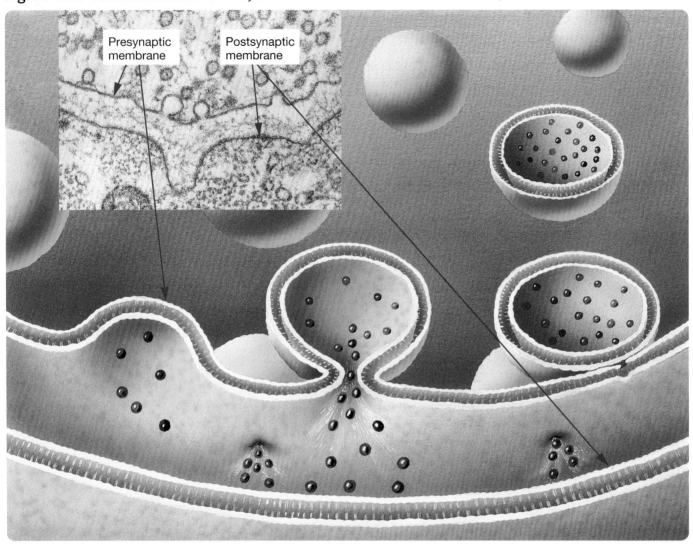

Don W. Fawcett/Science Source

Thus, one advantage of receptor subtypes is that they enable one neurotransmitter to transmit different kinds of messages to different parts of the brain.

The binding of a neurotransmitter to one of its receptor subtypes can influence a postsynaptic neuron in one of two fundamentally different ways, depending on whether the receptor is ionotropic or metabotropic. **Ionotropic receptors** are associated with ligand-activated ion channels; **metabotropic receptors** are typically associated with signal proteins and **G proteins** (*guanosine-triphosphate–sensitive proteins*); see Figure 4.11.

When a neurotransmitter molecule binds to an ionotropic receptor, the associated ion channel usually opens or closes immediately, thereby inducing an immediate postsynaptic potential. For example, in some neurons, EPSPs (depolarizations) occur because the neurotransmitter opens sodium channels, thereby increasing the flow of Na^+ ions into the neuron. In contrast, IPSPs (hyperpolarizations) often occur because the neurotransmitter opens potassium channels or chloride (Cl^-) channels, thereby increasing the

flow of K^+ ions out of the neuron or the flow of Cl^- ions into it, respectively.

Metabotropic receptors are more prevalent than ionotropic receptors, and their effects are slower to develop, longer-lasting, more diffuse, and more varied. There are many different kinds of metabotropic receptors, but each is attached to a serpentine signal protein that winds its way back and forth through the cell membrane seven times. The metabotropic receptor is attached to a portion of the signal protein outside the neuron; the G protein is attached to a portion of the signal protein inside the neuron.

When a neurotransmitter binds to a metabotropic receptor, a subunit of the associated G protein breaks away. Then, one of two things happen, depending on the particular G protein. The subunit may move along the inside surface of the membrane and bind to a nearby ion channel, thereby inducing an EPSP or IPSP; or it may trigger the synthesis of a chemical called a **second messenger** (neurotransmitters are considered to be the *first messengers*). Once created, a second messenger diffuses through the cytoplasm and

Figure 4.11 Ionotropic and metabotropic receptors.

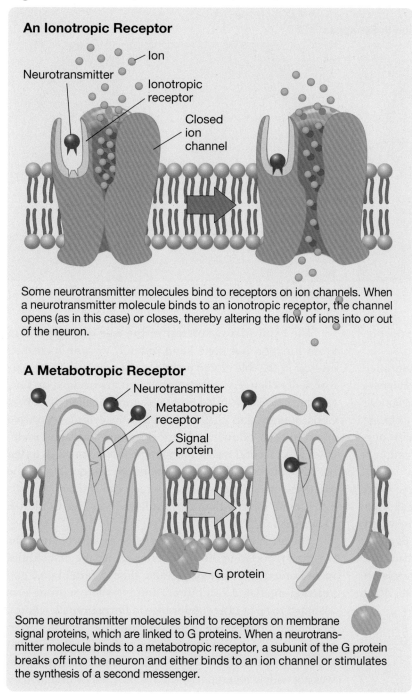

An Ionotropic Receptor

Neurotransmitter

Ion

Ionotropic receptor

Closed ion channel

Some neurotransmitter molecules bind to receptors on ion channels. When a neurotransmitter molecule binds to an ionotropic receptor, the channel opens (as in this case) or closes, thereby altering the flow of ions into or out of the neuron.

A Metabotropic Receptor

Neurotransmitter

Metabotropic receptor

Signal protein

G protein

Some neurotransmitter molecules bind to receptors on membrane signal proteins, which are linked to G proteins. When a neurotransmitter molecule binds to a metabotropic receptor, a subunit of the G protein breaks off into the neuron and either binds to an ion channel or stimulates the synthesis of a second messenger.

may influence the activities of the neuron in a variety of ways (Lyon, Taylor, & Tesmer, 2014)—for example, it may enter the nucleus and bind to the DNA, thereby influencing genetic expression. Thus, a neurotransmitter's binding to a metabotropic receptor can have radical, long-lasting effects. Furthermore, there is now evidence that ionotropic receptors can also produce second messengers that can have enduring effects (see Valbuena & Lerma, 2016; Reiner & Levitz, 2018).

Epigenetic mechanisms (see Chapter 2) that act on both ionotropic and metabotropic receptors are of increasing

interest to researchers. For example, there is strong evidence that the structures of both types of receptors (and thus their functionality) can be altered through epigenetic mechanisms (see Fomsgaard et al., 2018). Moreover, certain disorders may be the result of modifications to receptor structure via epigenetic mechanisms (see Matosin et al., 2017).

One type of metabotropic receptor—autoreceptors—warrants special mention. **Autoreceptors** are metabotropic receptors that have two unconventional characteristics: They bind to their neuron's own neurotransmitter molecules, and they are located on the presynaptic, rather than the postsynaptic, membrane. Their usual function is to monitor the number of neurotransmitter molecules in the synapse, to reduce subsequent release when the levels are high, and to increase subsequent release when they are low.

Differences between small-molecule and peptide neurotransmitters in patterns of release and receptor binding suggest that they serve different functions. Small-molecule neurotransmitters tend to be released into directed synapses and to activate either ionotropic receptors or metabotropic receptors that act directly on ion channels. In contrast, neuropeptides tend to be released diffusely, and virtually all bind to metabotropic receptors that act through second messengers. Consequently, the function of small-molecule neurotransmitters appears to be the transmission of rapid, brief excitatory or inhibitory signals to adjacent cells; and the function of neuropeptides appears to be the transmission of slow, diffuse, long-lasting signals.

Reuptake, Enzymatic Degradation, and Recycling

LO 4.13 Explain how neurotransmitters are removed from a synapse.

If nothing intervened, a neurotransmitter molecule would remain active in the synapse, in effect clogging that channel of communication. However, two mechanisms terminate synaptic messages and keep that from happening. These two message-terminating mechanisms are **reuptake** by transporters and **enzymatic degradation** (see Figure 4.12).

Reuptake is the more common of the two deactivating mechanisms. The majority of neurotransmitters, once

Figure 4.12 The two mechanisms for terminating neurotransmitter action in the synapse: reuptake and enzymatic degradation.

Two Mechanisms of Neurotransmitter Deactivation in Synapses

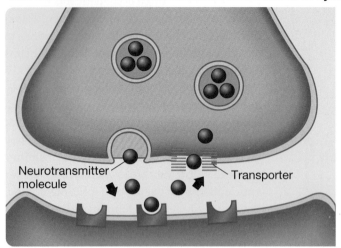

Reuptake

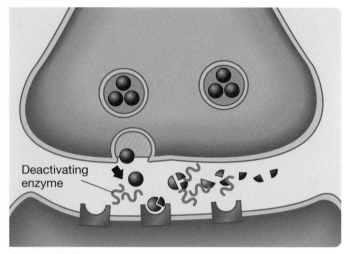

Enzymatic Degradation

released, are almost immediately drawn back into the presynaptic buttons by transporter mechanisms.

In contrast, other neurotransmitters are degraded (broken apart) in the synapse by the action of **enzymes**—proteins that stimulate or inhibit biochemical reactions without being affected by them. For example, *acetylcholine*, one of the few neurotransmitters for which enzymatic degradation is the main mechanism of synaptic deactivation, is broken down by the enzyme **acetylcholinesterase**.

Terminal buttons are models of efficiency. Once released, neurotransmitter molecules or their breakdown products are drawn back into the button and recycled, regardless of the mechanism of their deactivation. Even the vesicles, once they have done their job, are drawn back into the neuron from the presynaptic membrane and are used to create new vesicles (see Soykan, Maritzen, & Haucke, 2016).

Glia, Gap Junctions, and Synaptic Transmission

LO 4.14 Describe the roles of glia and gap junctions in synaptic transmission.

Glial cells, once overlooked as playing merely supportive roles in the nervous system, have been thrust to center stage by a wave of remarkable findings. For example, astrocytes have been shown to release chemical transmitters, to contain receptors for neurotransmitters, to conduct signals, and to influence synaptic transmission between neurons (see Bazargani & Attwell, 2016; Martín et al., 2015). Indeed, it is now inappropriate to think of brain function solely in terms of neuron–neuron connections. Neurons are only part of the story. Will *neuroscience* prove to be a misnomer? Anybody for "gliascience"?

The explosion of interest in the role of glial cells in brain function has gone hand in hand with an increased interest in the role of gap junctions. **Gap junctions** are narrow spaces between adjacent cells that are bridged by fine, tubular, cytoplasm-filled protein channels, called *connexins*. Consequently, gap junctions connect the cytoplasm of two adjacent cells, allowing electrical signals and small molecules (e.g., second messengers) to pass from one cell to the next (see Figure 4.13). Gap junctions are responsible for the existence of *electrical synapses*, which can transmit signals much more rapidly than chemical synapses (see Coulon & Landisman, 2017).

The presence of gap junctions between adjacent neurons was first reported in the 1950s, but because the first studies were limited to invertebrates and simple vertebrates, gap junction-mediated communication between neurons was assumed to be of little significance in the mammalian brain. Even after the presence of gap junctions was established in mammalian (i.e., rodent) brains in the early 1970s, the idea that gap junctions could play a major role in human brain function was not widely entertained. Then in the 1990s, stimulated by several important technical developments and the identification of the gap junction gene, gap junctions became the focus of neuroscientific research (see McCracken & Roberts, 2006). It is now firmly established that glial cells and gap junctions play major roles in brain function (see Rusakov et al., 2014; Coulon & Landisman, 2017; Szczupak, 2016).

The principles according to which astrocytes and gap junctions are distributed in the mammalian brain provide some of the best clues about their function. First, let's consider cerebral gap junctions. Cerebral gap junctions occur between all classes of cerebral cells; however, the majority of them seem to occur between cells of the same kind. For example, many gap junctions link astrocytes together

Figure 4.13 Gap junctions connect the cytoplasm of two adjacent cells. In the mammalian brain, there are many gap junctions between glial cells, between neurons, and between neurons and glia cells.

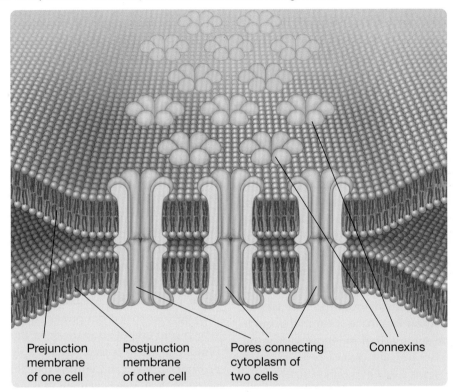

Prejunction membrane of one cell   Postjunction membrane of other cell   Pores connecting cytoplasm of two cells   Connexins

into networks of glial cells. Also, gap junctions between neurons are particularly prevalent between inhibitory interneurons of the same type (e.g., Lee et al., 2014). Accordingly, one function of gap junctions appears to be to synchronize the activities of like cells in a particular area.

One aspect of astrocytic organization suggests that they too play a role of synchronizing activities of like cells in a particular area. Unlike neurons, astrocytes are distributed evenly throughout a particular area, with only one astrocyte per location and little overlap between the projections of adjacent astrocytes. This suggests that each astrocyte coordinates the activity of neurons in its domain, and with as many as 40,000 processes, each astrocyte has a great potential to coordinate activity (see Pannasch & Rouach, 2013). Gap junctions on astrocytes tend to occur at the end of each process, where it comes in contact with processes from adjacent astrocytes.

Scan Your Brain

Before moving on to the discussion of specific neurotransmitters, review the general principles of axon conduction and synaptic transmission. Draw a line to connect each term in the left column with the appropriate word or phrase in the right column. The correct answers are provided at the end of the exercise. Before proceeding, review material related to your errors and omissions.

1. fatty
2. sclerosis
3. cell bodies
4. nondecremental
5. presynaptic facilitation
6. nondirected synapses
7. synaptic vesicles
8. from cell body to terminal buttons
9. acetylcholinesterase
10. short amino acid chains
11. saltatory
12. metabotropic receptors
13. electrical synapses
14. spines

a. axonal conduction of action potentials
b. orthodromic
c. myelin
d. nodes of Ranvier
e. multiple
f. dendritic
g. somas
h. axoaxonic synapses
i. string-of-beads
j. neuropeptides
k. store neurotransmitters
l. G proteins
m. enzymatic degradation
n. gap junctions

Scan Your Brain answers: (1) c, (2) e, (3) g, (4) a, (5) h, (6) i, (7) k, (8) b, (9) m, (10) j, (11) d, (12) l, (13) n, (14) f.

Neurotransmitters

Now that you understand the basics of neurotransmitter function, let's take a closer look at a select few of the well over 100 neurotransmitters that have been identified.

Overview of the Neurotransmitter Classes

LO 4.15 Name the major classes of neurotransmitters.

The following are three classes of conventional small-molecule neurotransmitters: the *amino acids*, the *monoamines*, and *acetylcholine*. A fourth group of various small-molecule neurotransmitters are often referred to as *unconventional neurotransmitters* because their mechanisms of action are unusual. In contrast to the small-molecule neurotransmitters, there is only one class of large-molecule neurotransmitters: the *neuropeptides*. All of the neurotransmitter classes and individual neurotransmitters that appear in this module in boldface type are summarized in Figure 4.16 at the end of this module.

The Roles and Functions of Neurotransmitters

LO 4.16 Identify the class, and discuss at least one function, of each of the neurotransmitters discussed in this section.

In this section, we examine some of the many neurotransmitters that enable intercellular communication within our nervous system. This section is organized according to the four classes of neurotransmitters discussed in the previous section.

AMINO ACID NEUROTRANSMITTERS. The neurotransmitters in the vast majority of fast-acting, directed synapses in the central nervous system are amino acids—the molecular building blocks of proteins. The four most widely studied **amino acid neurotransmitters** are **glutamate, aspartate, glycine,** and **gamma-aminobutyric acid (GABA)**. The first three are common in the proteins we consume, whereas GABA is synthesized by a simple modification of the structure of glutamate. Glutamate is the most prevalent excitatory neurotransmitter in the mammalian central nervous system. GABA is the most prevalent inhibitory neurotransmitter; however, it has excitatory effects at some synapses (see Watanabe, Fukuda, & Nabekura, 2014).

MONOAMINE NEUROTRANSMITTERS. Monoamines are another class of small-molecule neurotransmitters. Each is synthesized from a single amino acid—hence the name *monoamine* (one amine). **Monoamine neurotransmitters** are slightly larger than amino acid neurotransmitters, and their effects tend to be more diffuse. The monoamines are present in small groups of neurons whose cell bodies are, for the most part, located in the brain stem. These neurons often have highly branched axons with many varicosities (string-of-beads synapses), from which monoamine neurotransmitters are diffusely released into the extracellular fluid (see Figures 4.9 and 4.14).

There are four monoamine neurotransmitters: **dopamine, epinephrine, norepinephrine,** and **serotonin**. They are subdivided into two groups, **catecholamines** and **indolamines**, on the basis of their structures. Dopamine, norepinephrine, and epinephrine are catecholamines. Each is synthesized from the amino acid *tyrosine*. Tyrosine is converted to L-*dopa*, which in turn is converted to dopamine. Neurons that release norepinephrine have an extra enzyme (one that is not present in dopaminergic neurons), which converts the dopamine in them to norepinephrine. Similarly, neurons that release epinephrine have all the enzymes present in neurons that release norepinephrine, along with an extra enzyme that converts norepinephrine to epinephrine (see Figure 4.15). In contrast to the other monoamines, serotonin (also called *5-hydroxytryptamine*, or *5-HT*) is synthesized from the amino acid *tryptophan* and is classified as an indolamine.

Neurons that release norepinephrine are called *noradrenergic*; those that release epinephrine are called *adrenergic*. There are two reasons for this naming. One is that epinephrine and norepinephrine used to be called *adrenaline* and *noradrenaline*, respectively, by many scientists, until a drug company registered *Adrenalin* as a brand name. The other reason will become apparent if you try to say *norepinephrinergic*.

ACETYLCHOLINE. Acetylcholine (abbreviated *Ach*) is a small-molecule neurotransmitter that is, in one major respect, like a professor who is late for a lecture: It is in a class by itself. It is created by adding an *acetyl* group to a *choline* molecule. Acetylcholine is the neurotransmitter at neuromuscular junctions, at many of the synapses in the autonomic nervous system, and at synapses in several parts of the central nervous system. Recall that acetylcholine is broken down in the synapse by the enzyme *acetylcholinesterase*. Neurons that release acetylcholine are said to be *cholinergic*.

UNCONVENTIONAL NEUROTRANSMITTERS. The unconventional neurotransmitters act in ways that are

Figure 4.14 String-of-beads noradrenergic nerve fibers. The bright, beaded structures represent sites in these axons where the monoamine neurotransmitter norepinephrine is stored and released into the surrounding extracellular fluid.

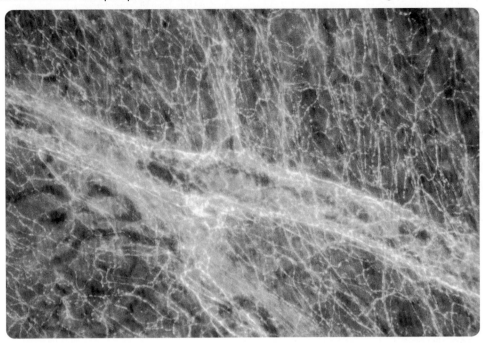

Dr. David Jacobwitz/SCIENCE PHOTO LIBRARY/Getty Images

different from those that neuroscientists have come to think of as typical for such substances. One class of unconventional neurotransmitters, the **soluble-gas neurotransmitters**, includes **nitric oxide** and **carbon monoxide**. These neurotransmitters are produced in the neural cytoplasm and immediately diffuse through the cell membrane into the extracellular fluid and then into

Figure 4.15 The steps in the synthesis of catecholamines from tyrosine.

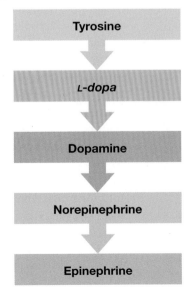

nearby cells. They easily pass through cell membranes because they are soluble in lipids. Once inside another cell, they stimulate the production of a second messenger and in a few seconds are deactivated by being converted to other molecules. They are difficult to study because they exist for only a few seconds.

Soluble-gas neurotransmitters have been shown to be involved in *retrograde transmission*. At some synapses, they transmit feedback signals from the postsynaptic neuron back to the presynaptic neuron. The function of retrograde transmission seems to be to regulate the activity of pre-synaptic neurons (see Iremonger, Wamsteeker Cusulin, & Bains, 2013).

Another class of unconventional neurotransmitters is the endocannabinoids. **Endocannabinoids** are neurotransmitters that are similar to *delta-9-tetrahydrocannabinol* (THC), the main *psychoactive* (producing psychological effects) constituent of marijuana. So far, two endocannabinoids have been discovered (see Di Marzo, Stella, & Zimmer, 2015). The most widely studied is **anandamide** (from the Sanskrit word *ananda*, which means "eternal bliss"). Like the soluble gases, the endocannabinoids are produced immediately before they are released. Endocannabinoids are synthesized from fatty compounds in the cell membrane; they tend to be released from the dendrites and cell body; and they tend to have most of their effects on presynaptic neurons, inhibiting subsequent synaptic transmission (see Ohno-Shosaku & Kano, 2014; Younts & Castillo, 2014).

NEUROPEPTIDES. About 100 neuropeptides have been identified. The actions of each neuropeptide depend on its amino acid sequence.

It is usual to loosely group **neuropeptide transmitters** into five categories. Three of these categories acknowledge that neuropeptides often function in multiple capacities, not just as neurotransmitters: One category (**pituitary peptides**) contains neuropeptides that were first identified as hormones released by the pituitary; a second category (**hypothalamic peptides**) contains neuropeptides that were first identified as hormones released by the hypothalamus; and a third category (**brain–gut peptides**) contains neuropeptides that were first discovered in the gut. The fourth category (**opioid peptides**) contains neuropeptides that are similar in structure to the active ingredients of opium, and the fifth (**miscellaneous peptides**) is a catch-all category that contains all of the neuropeptide transmitters that do not fit into one of the other four categories.

Figure 4.16 summarizes all the neurotransmitters that were introduced in this module. If it has not already occurred to you, this table should be very useful for reviewing the material in this module.

Pharmacology of Synaptic Transmission and Behavior

In case you have forgotten, the reason we have asked you to invest so much effort in learning about the neurotransmitters is that they play a key role in how the brain works. This chapter began on a behavioral note by considering the pathological behavior of Roberto Garcia d'Orta, which resulted from a Parkinson's disease-related disruption of his dopamine function. Now, let's return to behavior.

Most of the methods that biopsychologists use to study the behavioral effects of neurotransmitters are *pharmacological* (involving drugs). To study neurotransmitters and behavior, researchers administer to human or nonhuman subjects drugs that have particular effects on particular neurotransmitters and then assess the effects of the drugs on behavior.

Drugs have two fundamentally different kinds of effects on synaptic transmission: They facilitate it or they inhibit it. Drugs that facilitate the effects of a particular neurotransmitter are said to be **agonists** of that neurotransmitter. Drugs that inhibit the effects of a particular neurotransmitter are said to be its **antagonists**.

Figure 4.16 Classes of neurotransmitters and the particular neurotransmitters that were discussed (and appeared in boldface) in this module.

Small-Molecule Neurotransmitters

Amino acids		Glutamate Aspartate Glycine GABA
Monoamines	Catecholamines	Dopamine Epinephrine Norepinephrine
	Indolamines	Serotonin
Acetylcholine		Acetylcholine
Unconventional neurotransmitters	Soluble gases	Nitric oxide Carbon monoxide
	Endocannabinoids	Anandamide

Large-Molecule Neurotransmitters

Neuropeptides	Pituitary peptides Hypothalamic peptides Brain–gut peptides Opioid peptides Miscellaneous peptides

How Drugs Influence Synaptic Transmission

LO 4.17 Provide a general overview of how drugs influence synaptic transmission.

Although synthesis, release, and action vary from neurotransmitter to neurotransmitter, the following seven general steps are common to most neurotransmitters: (1) synthesis of the neurotransmitter, (2) storage in vesicles, (3) breakdown in the cytoplasm of any neurotransmitter that leaks from the vesicles, (4) exocytosis, (5) inhibitory feedback via autoreceptors, (6) activation of postsynaptic receptors, and (7) deactivation. Figure 4.17 illustrates these seven steps, and Figure 4.18 illustrates some ways that agonistic and antagonistic drugs influence them. For example, some agonists of a particular neurotransmitter bind to postsynaptic

Figure 4.17 Seven steps in neurotransmitter action: (1) synthesis, (2) storage in vesicles, (3) breakdown of any neurotransmitter leaking from the vesicles, (4) exocytosis, (5) inhibitory feedback via autoreceptors, (6) activation of postsynaptic receptors, and (7) deactivation.

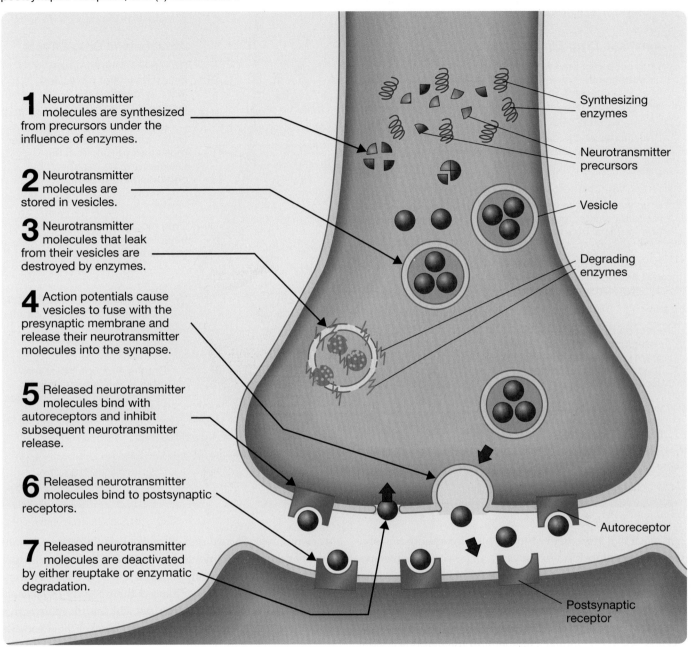

1 Neurotransmitter molecules are synthesized from precursors under the influence of enzymes.

2 Neurotransmitter molecules are stored in vesicles.

3 Neurotransmitter molecules that leak from their vesicles are destroyed by enzymes.

4 Action potentials cause vesicles to fuse with the presynaptic membrane and release their neurotransmitter molecules into the synapse.

5 Released neurotransmitter molecules bind with autoreceptors and inhibit subsequent neurotransmitter release.

6 Released neurotransmitter molecules bind to postsynaptic receptors.

7 Released neurotransmitter molecules are deactivated by either reuptake or enzymatic degradation.

Synthesizing enzymes

Neurotransmitter precursors

Vesicle

Degrading enzymes

Autoreceptor

Postsynaptic receptor

receptors and activate them, whereas some antagonistic drugs, called **receptor blockers**, bind to postsynaptic receptors without activating them and, in so doing, block the access of the usual neurotransmitter.

Behavioral Pharmacology: Three Influential Lines of Research

LO 4.18 Describe three examples of how drugs have been used to influence neurotransmission.

You will encounter discussions of the *putative* (hypothetical) behavioral functions of various neurotransmitters in subsequent chapters. However, this chapter ends with descriptions of three particularly influential lines of research on neurotransmitters and behavior. Each line of research led to the discovery of an important principle of neurotransmitter function, and each illustrates how drugs are used to study the nervous system and behavior.

WRINKLES AND DARTS: DISCOVERY OF RECEPTOR SUBTYPES. It was originally assumed that there was one kind of receptor for each neurotransmitter, but this notion was dispelled by research on acetylcholine receptors (see Changeux, 2013; Papke, 2014). Some acetylcholine receptors bind to *nicotine* (a CNS stimulant and the major

Figure 4.18 Some mechanisms of agonistic and antagonistic drug effects.

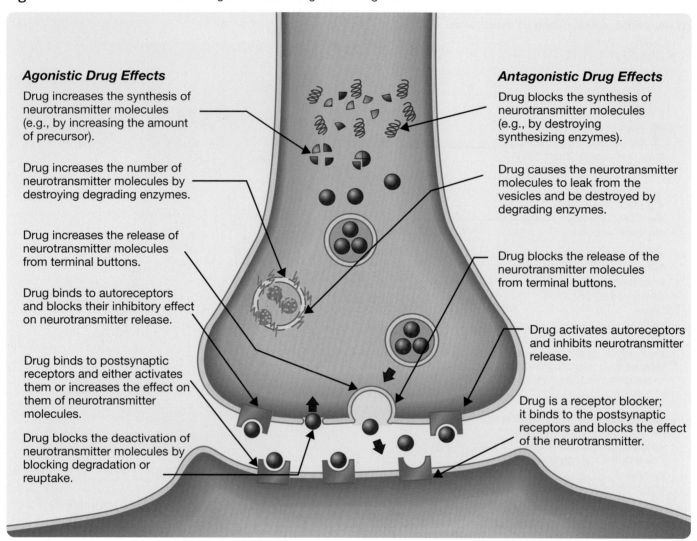

Agonistic Drug Effects

Drug increases the synthesis of neurotransmitter molecules (e.g., by increasing the amount of precursor).

Drug increases the number of neurotransmitter molecules by destroying degrading enzymes.

Drug increases the release of neurotransmitter molecules from terminal buttons.

Drug binds to autoreceptors and blocks their inhibitory effect on neurotransmitter release.

Drug binds to postsynaptic receptors and either activates them or increases the effect on them of neurotransmitter molecules.

Drug blocks the deactivation of neurotransmitter molecules by blocking degradation or reuptake.

Antagonistic Drug Effects

Drug blocks the synthesis of neurotransmitter molecules (e.g., by destroying synthesizing enzymes).

Drug causes the neurotransmitter molecules to leak from the vesicles and be destroyed by degrading enzymes.

Drug blocks the release of the neurotransmitter molecules from terminal buttons.

Drug activates autoreceptors and inhibits neurotransmitter release.

Drug is a receptor blocker; it binds to the postsynaptic receptors and blocks the effect of the neurotransmitter.

psychoactive ingredient of tobacco), whereas other acetylcholine receptors bind to *muscarine* (a poisonous substance found in some mushrooms). These two kinds of acetylcholine receptors thus became known as *nicotinic receptors* and *muscarinic receptors*.

Next, it was discovered that nicotinic and muscarinic receptors are distributed differently in the nervous system, have different modes of action, and consequently have different behavioral effects. Both nicotinic and muscarinic receptors are found in the CNS and the PNS. In the PNS, many nicotinic receptors occur at the junctions between motor neurons and muscle fibers, whereas many muscarinic receptors are located in the autonomic nervous system (ANS). Nicotinic and muscarinic receptors are ionotropic and metabotropic, respectively.

Many of the drugs used in research and medicine are extracts of plants that have long been used for medicinal and recreational purposes. The cholinergic agonists and antagonists illustrate this point well. For example, the ancient Greeks consumed extracts of the belladonna plant to treat stomach ailments and to make themselves more attractive. Greek women believed that the pupil-dilating effects of these extracts enhanced their beauty (*belladonna* means "beautiful lady"). **Atropine**, which is the main active ingredient of belladonna, is a receptor blocker that exerts its antagonist effect by binding to muscarinic receptors, thereby blocking the effects of acetylcholine on them. The pupil-dilating effects of atropine are mediated by its antagonist actions on muscarinic receptors in the ANS. In contrast, the disruptive effects of large doses of atropine on memory are mediated by its antagonistic effect on muscarinic receptors in the CNS. The disruptive effect of high doses of atropine on memory was one of the earliest clues that cholinergic mechanisms may play a role in memory (see Chapter 11).

South Americans have long used *curare*—an extract of a certain class of woody vines—on the tips of darts they use to kill their game. Like atropine, curare is a receptor blocker at cholinergic synapses, but it acts at nicotinic receptors. By binding to nicotinic receptors, curare blocks transmission at neuromuscular junctions, thus paralyzing its recipients and killing them by blocking their respiration. You may be surprised, then, to learn that the active ingredient of curare is sometimes administered to human patients during surgery to ensure that their muscles do not contract during an incision. When curare is used for this purpose, the patient's breathing must be artificially maintained by a respirator.

Botox (short for *Botulinium toxin*), a neurotoxin released by a bacterium often found in spoiled food, is another nicotinic antagonist, but its mechanism of action is different: It blocks the release of acetylcholine at neuromuscular junctions and is thus a deadly poison. However, injected in minute doses at specific sites, it has applications in medicine (e.g., reduction of tremors) and cosmetics (e.g., reduction of wrinkles; see Figure 4.19).

PLEASURE AND PAIN: DISCOVERY OF ENDOGENOUS OPIOIDS.

Opium, the sticky resin obtained from the seed pods of the opium poppy, has been used by humans since prehistoric times for its pleasurable effects. Morphine, its major psychoactive ingredient, is addictive. But morphine also has its good side: It is an effective *analgesic* (painkiller)—see Chapters 7 and 15.

In the 1970s, it was discovered that opioid drugs such as morphine bind effectively to receptors in the brain. These receptors were generally found in the hypothalamus and other limbic areas, but they were most concentrated in the area of the brain stem around the cerebral aqueduct, which connects the third and fourth ventricles; this part of the brain stem is called the **periaqueductal gray (PAG)**. Microinjection of morphine into the PAG, or even electrical stimulation of the PAG, produces strong analgesia.

Figure 4.19 Receiving cosmetic Botox injections.

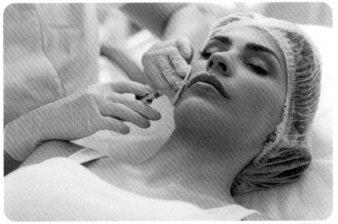

Viacheslav Iakobchuk/Alamy Stock Photo

The existence of selective opioid receptors in the brain raised an interesting question: Why are they there? They are certainly not there so that once humans discovered opium, opioids would have a place to bind. The existence of opioid receptors suggested that *opioid* chemicals occur naturally in the brain, and that possibility triggered an intensive search for them.

Several families of **endogenous** (occurring naturally within the body) opioids have been discovered. First discovered were the **enkephalins** (meaning "in the head"). Another major family of endogenous opioids are the **endorphins** (a contraction of "endogenous morphine"). All endogenous opioid neurotransmitters are neuropeptides, and their receptors are metabotropic.

TREMORS AND MENTAL ILLNESS: DISCOVERY OF ANTIPSYCHOTIC DRUGS.

Arguably, the most important event in the treatment of mental illness has been the development of drugs for the treatment of schizophrenia. Surprisingly, Parkinson's disease, the disease from which Roberto Garcia d'Orta suffered, played a major role in their discovery.

In the 1950s, largely by chance, two drugs were found to have antipsychotic effects; that is, they reduced the severity of psychosis—the major symptom of schizophrenia. Although these two drugs were not related structurally, they both produced a curious pattern of effects: Neither drug appeared to have any antipsychotic activity until patients had been taking it for about 3 weeks, at which point the drug also started to produce mild Parkinsonian symptoms (e.g., tremor-at-rest). Researchers put this result together with two other findings: (1) Parkinson's disease is associated with the degeneration of a main *dopamine* pathway in the brain, and (2) dopamine agonists (e.g., *cocaine* and *amphetamines*) produce a transient condition that resembles schizophrenia. Together, these findings suggested that schizophrenia might be caused by excessive activity at dopamine synapses and thus that potent dopamine antagonists would be effective in its treatment.

Journal Prompt 4.3

Why do biopsychologists need a solid understanding of neural conduction and synaptic transmission?

It would be a mistake to think that antipsychotic drugs cure schizophrenia or that they help in every case. However, they help improve the quality of life of many individuals with schizophrenia. You will learn much more about this important line of research in Chapter 18.

Themes Revisited

The function of the nervous system, like the function of any circuit, depends on how signals travel through it. The primary purpose of this chapter was to introduce you to neural conduction and synaptic transmission. This introduction touched on four of the text's five main themes.

The clinical implications theme was illustrated by the opening case of the Lizard, Roberto Garcia d'Orta. Then this theme was picked up again at the end of the chapter during discussions of curare, Botox, endogenous opioids, and antipsychotic drugs.

The thinking creatively theme arose in four metaphors: the social network metaphor of neural network communication, the ripples-on-the-pond metaphor of the summation of PSPs, the firing-gun metaphor of action potentials, and the mouse-traps-on-a-wobbly-shelf metaphor of axonal conduction. Metaphors are useful in teaching, and scientists find them useful for thinking about the phenomena they study.

Finally, the evolutionary perspective theme was implicit throughout the entire chapter because almost all neurophysiological research is conducted on the neurons and synapses of nonhuman subjects.

We also caught a glimpse of one of our two emerging themes in this chapter: thinking about epigenetics. In particular, this theme arose in the context of extracellular vesicles and ionotropic and metabotropic receptors.

Key Terms

Chapter 5
The Research Methods of Biopsychology

Understanding What Biopsychologists Do

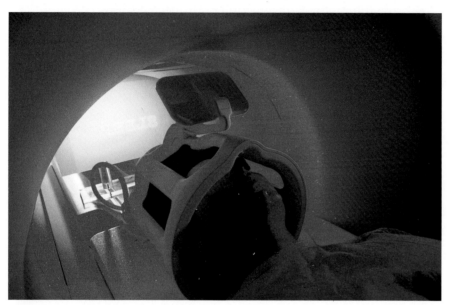

The Photolibrary Wales/Alamy Stock Photo

 ## Chapter Overview and Learning Objectives

PART ONE Methods of Studying the Nervous System

Methods of Visualizing and Stimulating the Living Human Brain	**LO 5.1**	Describe two x-ray-based techniques for visualizing the living human brain.
	LO 5.2	Describe the positron emission tomography (PET) technique.
	LO 5.3	Describe three magnetic-field-based techniques for imaging the living human brain.
	LO 5.4	Describe an ultrasound-based technique for imaging the living human brain.
	LO 5.5	Describe three transcranial stimulation techniques.
Recording Human Psychophysiological Activity	**LO 5.6**	Describe two psychophysiological measures of brain activity.
	LO 5.7	Describe two psychophysiological measures of somatic nervous system activity.

PART TWO Behavioral Research Methods of Biopsychology

Biopsychological
Paradigms of Animal
Behavior

LO 5.29 Describe three behavioral paradigms used to study species-common behaviors.

LO 5.30 Describe the Pavlovian conditioning paradigm and the operant conditioning paradigm.

LO 5.31 Describe four seminatural animal learning paradigms.

LO 5.32 Explain why multiple techniques should be used when trying to answer a specific question.

The Ironic Case of Professor P.

Two weeks before his brain surgery, Professor P. reported to the hospital for a series of tests. What amazed Professor P. most about these tests was how familiar they seemed. No, Professor P. was not a psychic; he was a biopsychologist, and he was struck by how similar the tests performed on him were to the tests he had encountered in his work.

Professor P. had a brain tumor on his right auditory-vestibular cranial nerve (cranial nerve VIII; see Appendices III and IV), and he had to have it *excised* (cut out). First, Professor P.'s auditory abilities were assessed by measuring his ability to detect sounds of various volumes and pitches and then by measuring the magnitude of the EEG signals evoked in his auditory cortex by clicks in his right ear.

Next, Professor P.'s vestibular function (balance) was tested by injecting cold water into his ear.

"Do you feel anything, Professor P.?"

"Well, a cold ear."

"Nothing else?"

"No."

So colder and colder water was tried with no effect until the final, coldest test was conducted. "Ah, that feels weird," said Professor P. "It's kind of like the bed is tipping."

The results of the tests were bad, or good, depending on your perspective—they certainly revealed his deficits. Professor P.'s hearing in his right ear was poor, and his right vestibular nerve was barely functioning. "At the temperatures we flushed down there, most people would have been on their hands and knees puking their guts out," said the medical technician. Professor P. smiled at her technical terminology.

Of course, he was upset that his brain had deteriorated so badly, but he sensed that his neurosurgeon was secretly pleased: "We won't have to try to save the nerve; we'll just cut it."

There was one last test. The skin of his right cheek was lightly pricked while the EEG responses of his somatosensory cortex were recorded from his scalp. "This is just to establish a baseline for the surgery," it was explained. "One main risk of removing tumors on the auditory-vestibular cranial nerve (VIII) is damaging the facial cranial nerve (VII), and that would make the right side of your face sag. So during the surgery, electrodes will be inserted in your cheek, and your cheek will be repeatedly stimulated with tiny electrical pulses. The cortical responses will be recorded and fed into a loudspeaker so that the surgeon can immediately hear changes in the activity if his scalpel starts to stray into the area."

As Professor P. was driving home from his pre-surgery tests, his mind wandered from his current plight to his day at the hospital. "Quite interesting," he thought to himself. There were biopsychologists everywhere, doing biopsychological things. In all three labs he had visited, there were people who had begun their training as biopsychologists.

Two weeks later, Professor P. was rolled into the preparation room. "Sorry to do this, Professor P., you were my favorite instructor," the nurse said, as she inserted a large needle into Professor P.'s face and left it there.

Professor P. didn't mind; he was barely conscious. He did not know that he wouldn't regain consciousness for several days—at which point he would be incapable of talking, eating, or even breathing. But more about that later.

Don't forget Professor P.; you will learn more about him in Chapter 10. For the time being, this case demonstrates that many of the fundamental research methods of biopsychology are also used in clinical settings. Let's move on to the methods themselves.

PART ONE Methods of Studying the Nervous System

This is the first of the two parts that compose this chapter. In this part, we present the methods used by biopsychologists to study the nervous system. As you will soon see, the methods are extremely diverse.

Methods of Visualizing and Stimulating the Living Human Brain

This module first describes four different sorts of methods for visualizing the living human brain: x-ray-based techniques, a radioactivity-based technique, magnetic-field-based techniques, and an ultrasound-based technique. Next, it presents three techniques for noninvasively stimulating the living human brain.

Figure 5.1 A cerebral angiogram of a healthy human.

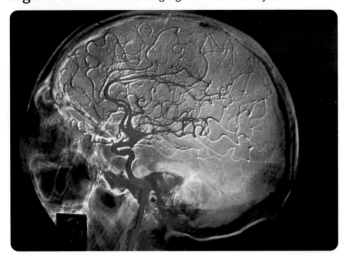

CNRI/Science Source

X-Ray-Based Techniques

LO 5.1 Describe two x-ray-based techniques for visualizing the living human brain.

Prior to the early 1970s, biopsychological research was impeded by the inability to obtain images of the organ of primary interest: the living human brain. Conventional x-ray photography is useless for this purpose. When an x-ray photograph is taken, an x-ray beam is passed through an object and then onto a photographic plate. Each molecule the beam has passed through absorbs some of the radiation; thus, only the unabsorbed portions of the beam reach the photographic plate. This makes x-ray photography effective in characterizing internal structures that absorb x-rays differently than their surroundings— just like a revolver in a suitcase full of clothes or a bone surrounded by flesh. However, by the time an x-ray beam has passed through the numerous overlapping structures of the brain, which differ only slightly in their ability to absorb x-rays, it carries little information about the structures through which it has passed.

CONTRAST X-RAYS. Although conventional x-ray photography is not useful for visualizing the brain, contrast x-ray techniques are. **Contrast x-ray techniques** involve injecting into one compartment of the body a substance that absorbs x-rays either less than or more than the surrounding tissue. The injected substance then heightens the contrast between the compartment and the surrounding tissue during x-ray photography.

One contrast x-ray technique, **cerebral angiography**, uses the infusion of a radio-opaque dye into a cerebral artery to visualize the cerebral circulatory system during x-ray photography (see Figure 5.1). Cerebral angiograms are most useful for localizing vascular damage, but the displacement of blood vessels from their normal position also can indicate the location of a tumor.

> **Journal Prompt 5.1**
>
> Egas Moniz, the inventor of the lobotomy, was also the pioneer of cerebral angiography. Some have argued that Moniz's Nobel Prize for the lobotomy should have been revoked. However, others have argued that he would have won it anyway for his important work on cerebral angiography. Do you think Moniz deserved to win the Nobel Prize? Why or why not?

COMPUTED TOMOGRAPHY. In the early 1970s, the introduction of computed tomography revolutionized the study of the living human brain. **Computed tomography (CT)** is a computer-assisted x-ray procedure that can be used to visualize the brain and other internal structures of the living body. During cerebral computed tomography, the neurological patient lies with his or her head positioned in the center of a large cylinder, as depicted in Figure 5.2.

Figure 5.2 Computed tomography (CT) uses x-rays to create a brain scan.

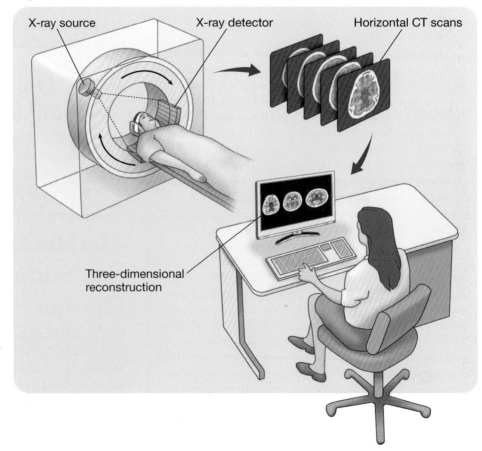

On one side of the cylinder is an x-ray tube that projects an x-ray beam through the head to an x-ray detector mounted on the other side. The x-ray tube and detector rotate rapidly around the head of the patient at one level of the brain, taking many individual x-ray photographs as they rotate. The meager information in each x-ray photograph is combined by a computer to generate a CT scan of one horizontal section of the brain. Then the x-ray tube and detector are moved along the axis of the patient's body to another level of the brain, and the process is repeated. Scans of eight or nine horizontal brain sections are typically obtained from a patient. When combined, these images provide three-dimensional representations of the brain.

Radioactivity-Based Techniques

LO 5.2   **Describe the positron emission tomography (PET) technique.**

Positron emission tomography (PET) was the first brain-imaging technique to provide images of brain activity (*functional brain images*) rather than images of brain structure (*structural brain images*). In one common version of PET, radioactive **fluorodeoxyglucose (FDG)** is injected into the patient's *carotid artery* (an artery of the neck that feeds the ipsilateral cerebral hemisphere). Because of its similarity to glucose, the primary metabolic fuel of the brain, fluorodeoxyglucose is rapidly taken up by active (energy-consuming) cells. However, unlike glucose, fluorodeoxyglucose cannot be metabolized; it therefore accumulates in active neurons and astrocytes until it is gradually broken down (see Zimmer et al., 2017). Each PET scan is an image of the levels of radioactivity (indicated by color coding) in various parts of one horizontal level of the brain. Thus, if a PET scan is taken of a patient who engages in an activity such as reading for about 30 seconds after the FDG injection, the resulting scan will indicate the areas of the target brain level that were most active during the 30 seconds (see Figure 5.3).

Notice from Figure 5.3 that PET scans are not really images of the brain. Each PET scan is merely a colored map of the amount of radioactivity in each of the tiny cubic voxels (volume pixels) that compose the scan. Exactly how each voxel maps onto a particular brain structure can be estimated only by superimposing the scan on a brain image.

The most significant current application of PET technology is its use in identifying the distribution of particular molecules (e.g., neurotransmitters, receptors, transporters) in the brain (see Camardese et al., 2014). This is accomplished by injecting volunteers with radioactively labeled **ligands** (ions or molecules that bind to other molecules). Then, PET scans can document the distribution of radioactivity in the brain.

Magnetic-Field-Based Techniques

LO 5.3   **Describe three magnetic-field-based techniques for imaging the living human brain.**

MAGNETIC RESONANCE IMAGING.   **Magnetic resonance imaging (MRI)** is a structural brain-imaging procedure in which high-resolution images are constructed from the measurement of radio-frequency waves that hydrogen atoms emit as they align with a powerful magnetic field. Such imaging is possible because: (1) water contains two hydrogen atoms (H_2O) and (2) different brain structures contain different amounts of water. This, in turn, means that the number of hydrogen atoms differs between brain structures, and, therefore, the radio-frequency waves emitted by a particular brain structure will be different from its neighboring brain structures. MRI provides clearer images of the brain than does CT (see Lerch et al., 2017). A two-dimensional MRI scan of the midsagittal plane of the brain is presented in Figure 5.4.

In addition to providing relatively high **spatial resolution** (the ability to detect and represent differences in spatial location), MRI can produce images in three

Figure 5.3 A pair of PET scans. A scan was done when the volunteer's eyes were either open (left) or closed (right). Areas of high activity are indicated by reds and yellows. Notice the high level of activity in the visual cortex of the occipital lobe when the volunteer's eyes were open.

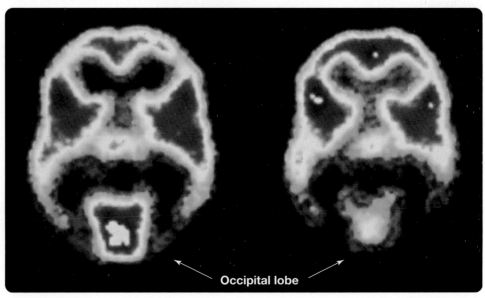

Occipital lobe

NIH/Science Source

Figure 5.4 A color-enhanced midsagittal MRI scan.

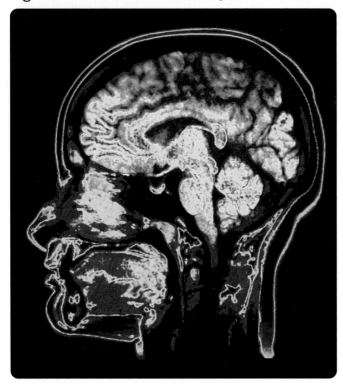

Scott Camazine/Science Source

dimensions. Figure 5.5 shows a three-dimensional MRI scan of a patient with a growing tumor.

DIFFUSION TENSOR MRI. Many variations of MRI have been developed. Arguably, one of the most innovative of

Figure 5.5 MRI of a growing tumor. The tumor is colored red.

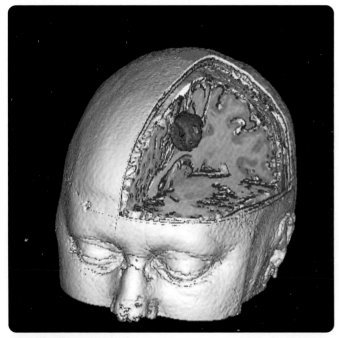

Simon Fraser/Science Source

these new MRI techniques has been diffusion tensor MRI. **Diffusion tensor MRI** is a method of identifying those pathways along which water molecules rapidly diffuse (see Jbadi et al., 2015). Because *tracts* (bundles of axons) are the major routes of rapid water diffusion in the brain, diffusion tensor imaging provides an image of major tracts—see Figure 5.6.

Most brain research focuses on the structures of the brain. However, in order to understand how the brain works, it is imperative to understand the connections among those structures—the so-called *connectome* (see Park & Friston, 2013; Glasser et al., 2016; Swanson & Lichtman, 2016). This is why diffusion tensor images have become a focus of neuroscientific research. Complete descriptions of connectomes already exist for some organisms, including the nematode *C. elegans* and the mouse (see Oh et al., 2014). Work on the so-called *Human Connectome Project* is well underway.

FUNCTIONAL MRI. MRI technology has been used to produce functional images of the brain. Indeed, functional MRI has become the most influential tool of cognitive neuroscience. It is often used to determine if a brain is dysfunctional, but it is also used for a variety of other purposes; for example, to infer the content of an individual's dreams (see Horikawa et al., 2013; Underwood, 2013).

Functional MRI (fMRI) produces images representing the increase in oxygenated blood flow to active areas of the brain. Functional MRI is possible because of two attributes of oxygenated blood. First, active areas of the brain take up more oxygenated blood than they need for their energy requirements, and thus oxygenated blood accumulates in active areas of the brain (see

Figure 5.6 Diffusion tensor MRI. This three-dimensional image shows the major tracts of the brain.

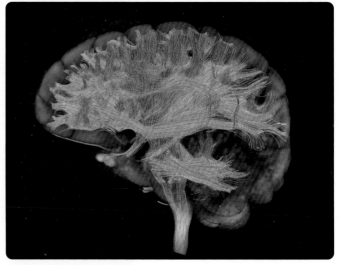

Image Source/Alamy Stock Photo

Hillman, 2014). Second, oxygenated blood has different magnetic properties than does deoxygenated blood, and this difference influences the radio-frequency waves emitted by hydrogen atoms in an MRI. The signal recorded by fMRI is called the **BOLD signal** (the blood-oxygen-level-dependent signal). The BOLD signal indicates the parts of the brain that are active or inactive during a cognitive or behavioral test, and thus it suggests the types of analyses the brain is performing. Because the BOLD signal is the result of blood flow through the brain, it is important to remember that it is not directly measuring the electrical activity of the brain.

Functional MRI has three advantages over PET: (1) nothing has to be injected into the volunteer; (2) it provides both structural and functional information in the same image; and (3) its spatial resolution is better. A functional MRI is shown in Figure 5.7.

It is important not to be unduly swayed by the impressiveness of fMRI images and technology. The images are often presented—particularly in the popular press or general textbooks—as if they are actual pictures of human neural activity. They aren't: They are images of the BOLD signal, and the relation between the BOLD signal and neural activity is complex (see Hillman, 2014). Furthermore, fMRI technology has poor **temporal resolution**, that is, it is poor at specifying the timing of neural events. Indeed, it takes 2 or 3 seconds to measure the BOLD signal, and

Figure 5.7 Functional magnetic resonance image (fMRI). This image illustrates the areas of cortex that became more active when the volunteers observed strings of letters and were required to specify which strings were words—in the control condition, volunteers viewed strings of asterisks (Kiehl et al., 1999). This fMRI illustrates surface activity; but images of sections through the brain can also be displayed.

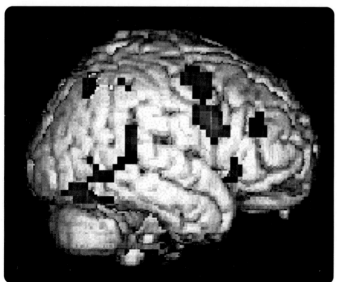

Kent Kiehl/Peter Liddle/University of British Columbia Department of Psychiatry

many neural responses, such as action potentials, occur in the millisecond range.

Ultrasound-Based Techniques

LO 5.4 Describe an ultrasound-based technique for imaging the living human brain.

Functional ultrasound imaging (fUS) is a new imaging technique that uses *ultrasound* (sound waves of a higher frequency than we can hear) to measure changes in blood volume in particular brain regions. When a brain region becomes active, blood levels increase there, and alter the passage of ultrasound through that brain region.

As a functional brain imaging method, fUS has three key advantages over PET and fMRI: (1) it is cheap, (2) highly portable; and (3) can be used for imaging some individuals, such as human infants, who cannot undergo PET or fMRI (see Deffieux et al., 2018).

Transcranial Stimulation

LO 5.5 Describe three transcranial stimulation techniques.

PET, fMRI, and fUS have allowed cognitive neuroscientists to create images of brain activity while volunteers are engaging in particular cognitive activities. Although technically impressive, these kinds of studies of brain activity and cognition all have the same shortcoming: They can be used to show a *correlation* between brain activity and cognitive activity, but they can't prove that the brain activity *caused* the cognitive activity (Sack, 2006). For example, a brain-imaging technique may show that the cingulate cortex becomes active when volunteers view disturbing photographs, but it can't prove that the cingulate activity causes the emotional experience—there are many other explanations. There are two obvious ways of supporting the hypothesis that the cingulate cortex is an area for emotional experience. One way would be to assess emotional experience in people lacking a functional cingulate cortex. This can be accomplished by studying patients with cingulate damage or by "turning off" the cingulate cortex of healthy patients—transcranial magnetic stimulation is a way of turning off particular areas of cortex. A second way would be to assess emotional experiences of volunteers after "turning on" their cingulate cortex—transcranial electrical stimulation and transcranial ultrasound stimulation are ways of turning on areas of cortex.

Let us briefly introduce you to transcranial magnetic stimulation and transcranial electrical stimulation, which are currently playing a major role in establishing the causal effects of human cortical activity on cognition and behavior. **Transcranial magnetic stimulation (TMS)** is a technique that can be used to turn off an area of human cortex by creating a magnetic field under a coil positioned next to the skull (e.g., Candidi et al., 2015). The magnetic stimulation temporarily turns off part of the brain while the effects of the disruption on cognition and behavior are assessed. Although there are still fundamental questions about safety, depth of effect, and mechanisms of neural disruption (see Polanía, Nitsche, & Ruff, 2018; Romei, Thut, & Silvanto, 2016), TMS is often employed to circumvent the difficulty that brain-imaging studies have in determining causation. Using different stimulation parameters, TMS can also be used to "turn on" an area of cortex (see Rossini et al., 2015).

Transcranial electrical stimulation (tES) is a technique that can be used to stimulate ("turn on") an area of the cortex by applying an electrical current through two electrodes placed directly on the scalp. The electrical stimulation temporarily increases activity in part of the brain while the effects of the stimulation on cognition and behavior are assessed (see Polanía, Nitsche, & Ruff, 2018).

The use of tES for its putative cognitive enhancement effects has become popular, and there are many relatively inexpensive tES systems available for purchase online (see Bourzac, 2016). However, there is conflicting evidence about whether tES has beneficial effects on cognition; some studies have even reported detrimental effects. Differing stimulation protocols might account for some of the discrepant findings (see Sellers et al., 2015).

Transcranial ultrasound stimulation (tUS) is a technique that, like tES and TMS, can be used to activate particular brain structures. However, unlike tES and TMS, which can only be used to stimulate cortical structures, tUS can also be used to activate subcortical structures.

To activate a brain structure using tUS, multiple sources of low-amplitude ultrasonic sound waves are placed around the head of the individual. Then, each of those sound sources is directed at the target brain structure. When the ultrasonic sound waves from each of those sources reach the target structure they sum together, such that the amplitude of the sound waves at the target brain structure is sufficiently large to stimulate activity in the cells there (see Tyler, Lani, & Hwang, 2018).

The tUS technique can also be used to make small permanent lesions to a brain structure. The procedure is the same as that for stimulation via tUS, except that the amplitude of each ultrasound source is larger, leading to a larger amplitude waveform that is sufficient to create a small (e.g., the size of a grain of rice) permanent lesion. This tUS-based lesion method has been used to treat several conditions (e.g., lesioning a thalamic nucleus to treat essential tremor)—all without having to make an incision. Accordingly, the tUS lesion technique is revolutionizing neurosurgery (see Landhuis, 2017).

Recording Human Psychophysiological Activity

The preceding module introduced you to structural and functional brain imaging. This module deals with *psychophysiological recording methods* (methods of recording physiological activity from the surface of the human body). Six of the most widely studied psychophysiological measures are described: two measures of brain activity (the scalp EEG and magnetoencephalography), two measures of somatic nervous system activity (muscle tension and eye movement), and two measures of autonomic nervous system activity (skin conductance and cardiovascular activity).

Psychophysiological Measures of Brain Activity

LO 5.6 Describe two psychophysiological measures of brain activity.

SCALP ELECTROENCEPHALOGRAPHY. The *electroencephalogram* (*EEG*) is a measure of the gross electrical activity of the brain. It is recorded through large electrodes by a device called an *electroencephalograph* (*EEG machine*), and the technique is called **electroencephalography**. In EEG studies of human participants, each channel of EEG activity is usually recorded from disk-shaped electrodes, about half the size of a dime, which are attached to the scalp.

The scalp EEG signal reflects the sum of electrical events throughout the head. These events include action potentials and postsynaptic potentials as well as electrical signals from the skin, muscles, blood, and eyes.

Thus, the utility of the scalp EEG does not lie in its ability to provide an unclouded view of neural activity. Its value as a research and diagnostic tool rests on the fact that some EEG wave forms are associated with particular states of consciousness or particular types of cerebral pathology (e.g., epilepsy). For example, **alpha waves** are regular, 8- to 12-per-second, high-amplitude waves that are associated with relaxed wakefulness. A few examples of EEG wave forms and their psychological correlates are presented in Figure 5.8.

Because EEG signals decrease in amplitude as they spread from their source, a comparison of signals recorded from various sites on the scalp can sometimes indicate the origin of particular waves (see Cohen, 2017). This is why it is usual to record EEG activity from many sites simultaneously.

Figure 5.8 Some typical electroencephalograms and their psychological correlates.

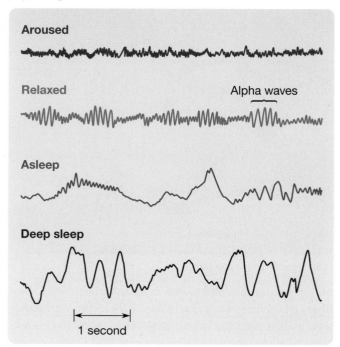

A method used to reduce the noise of the background EEG is **signal averaging**. First, a subject's response to a stimulus, such as a click, is recorded many—let's say 1,000—times. Then, a computer identifies the millivolt value of each of the 1,000 traces at its starting point (i.e., at the click) and calculates the mean of these 1,000 scores. Next, it considers the value of each of the 1,000 traces 1 millisecond (msec) from its start, for example, and calculates the mean of these values. It repeats this process at the 2-msec mark, the 3-msec mark, and so on. When these averages are plotted, the average response evoked by the click is more apparent because the random background EEG is canceled out by the averaging. See Figure 5.9, which illustrates the averaging of an auditory evoked potential.

The analysis of *average evoked potentials* (*AEPs*) focuses on the various waves in the averaged signal. Each wave is characterized by its direction, positive or negative, and by its latency. For example, the **P300 wave** illustrated in Figure 5.10 is the positive wave that occurs about 300 milliseconds after a momentary stimulus that has meaning for the participant (e.g., a stimulus to which the

Psychophysiologists are often more interested in the EEG waves that accompany certain psychological events than in the background EEG signal. These accompanying EEG waves are generally referred to as **event-related potentials (ERPs)**. One commonly studied type of event-related potential is the **sensory evoked potential**—the change in the cortical EEG signal elicited by the momentary presentation of a sensory stimulus. As Figure 5.9 illustrates, the cortical EEG that follows a sensory stimulus has two components: the response to the stimulus (the signal) and the ongoing background EEG activity (the noise). The *signal* is the part of any recording that is of interest; the *noise* is the part that isn't. The problem in recording sensory evoked potentials is that the noise of the background EEG is often so great that the sensory evoked potential is masked. Measuring a sensory evoked potential can be like detecting a whisper at a rock concert.

Figure 5.9 Signal averaging: Averaging of the background EEG (left) and of auditory evoked potentials (right). Averaging increases the signal-to-noise ratio.

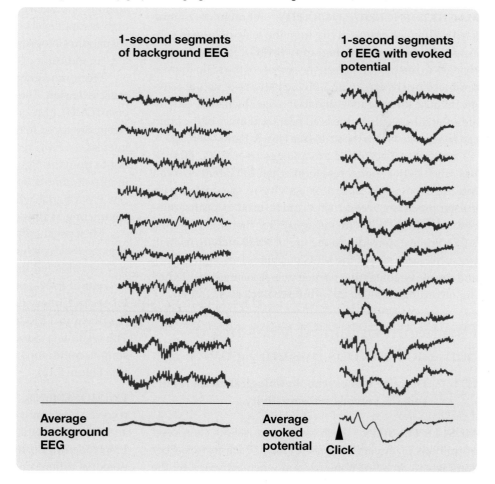

Figure 5.10 An average auditory evoked potential. Notice the P300 wave. This wave occurs only if the stimulus has meaning for the participant; in this case, the 'click' sound signals the imminent delivery of a reward. By convention, positive EEG waves are always shown as downward deflections.

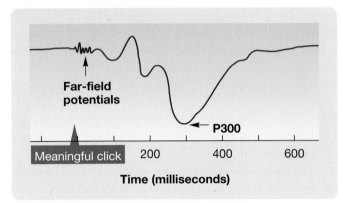

Figure 5.11 A magnetoencephalography (MEG) machine. Stylish in any home!

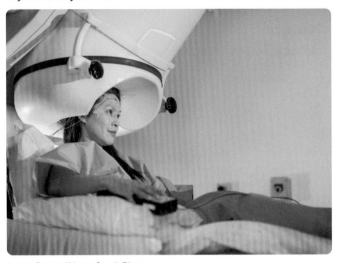

Image Source/Alamy Stock Photo

participant must respond). In contrast, the portions of an evoked potential recorded in the first few milliseconds after a stimulus are not influenced by the meaning of the stimulus for the participant. These small waves are called **far-field potentials** because, although they are recorded from the scalp, they originate far away in the sensory nuclei of the brain stem.

MAGNETOENCEPHALOGRAPHY. Another technique used to monitor brain activity from the scalp of human subjects is **magnetoencephalography (MEG)**. MEG measures changes in magnetic fields on the surface of the scalp that are produced by changes in underlying patterns of neural activity. Because the magnetic signals induced by neural activity are so small, only those induced near the surface of the brain can be recorded from the scalp (see Hari & Parkkonen, 2015).

MEG has two major advantages over EEG. First, it has much better spatial resolution than EEG; that is, it can localize changes in electrical activity in the cortex with greater precision. Second, MEG can be used to localize subcortical activity with greater reliability than EEG (Baillet, 2017). Some downsides to the use of MEG include its high price, the large size of the MEG machines (see Figure 5.11), and the requirement that participants remain very still during recordings (Baillet, 2017; but see Boto et al., 2018).

Psychophysiological Measures of Somatic Nervous System Activity

LO 5.7 Describe two psychophysiological measures of somatic nervous system activity.

MUSCLE TENSION. Each skeletal muscle is composed of millions of threadlike muscle fibers. Each muscle fiber contracts in an all-or-none fashion when activated by the

motor neuron that innervates it. At any given time, a few fibers in each resting muscle are likely to be contracting, thus maintaining the overall tone (tension) of the muscle. Movement results when a large number of fibers contract at the same time.

In everyday language, anxious people are commonly referred to as "tense." This usage acknowledges the fact that anxious or otherwise aroused individuals typically display high resting levels of tension in their muscles. This is why psychophysiologists are interested in this measure; they use it as an indicator of psychological arousal.

Electromyography is the usual procedure for measuring muscle tension. The resulting record is called an *electromyogram (EMG)*. EMG activity is usually recorded between two electrodes taped to the surface of the skin over the muscle of interest. An EMG record is presented in Figure 5.12. You will notice from this figure that the main correlate of an increase in muscle contraction is an increase in the amplitude of the raw EMG signal, which reflects the number of muscle fibers contracting at any one time.

Most psychophysiologists do not work with raw EMG signals; they convert them to a more workable form. The raw signal is fed into a computer that calculates the total amount of EMG spiking per unit of time—in consecutive 0.1-second intervals, for example. The integrated signal (i.e., the total EMG activity per unit of time) is then plotted. The result is a smooth curve, the amplitude of which is a simple, continuous measure of the level of muscle tension (see Figure 5.12).

EYE MOVEMENT. The electrophysiological technique for recording eye movements is called **electrooculography**, and the resulting record is called an *electrooculogram (EOG)*. Electrooculography is based on the fact that a steady potential difference exists between the front (positive) and

Figure 5.12 The relation between a raw EMG signal and its integrated version. The volunteer tensed her muscle beneath the electrodes and then gradually relaxed it.

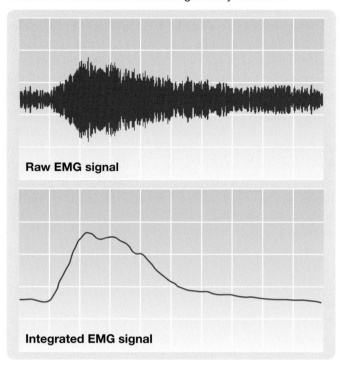

Raw EMG signal

Integrated EMG signal

back (negative) of the eyeball. Because of this steady potential, when the eye moves, a change in the electrical potential between electrodes placed around the eye can be recorded. It is usual to record EOG activity between two electrodes placed on each side of the eye to measure its horizontal movements and between two electrodes placed above and below the eye to measure its vertical movements (see Figure 5.13).

Figure 5.13 The typical placement of electrodes around the eye for electrooculography. The two electrooculogram traces were recorded as the volunteer scanned a circle.

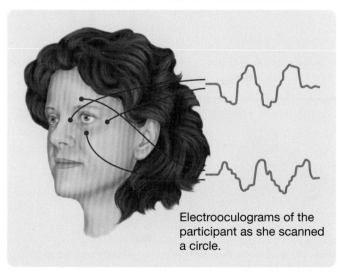

Electrooculograms of the participant as she scanned a circle.

Psychophysiological Measures of Autonomic Nervous System Activity

LO 5.8 **Describe two psychophysiological measures of autonomic nervous system activity.**

SKIN CONDUCTANCE.   Emotional thoughts and experiences are associated with increases in the ability of the skin to conduct electricity. The two most commonly employed indexes of *electrodermal activity* are the **skin conductance level (SCL)** and the **skin conductance response (SCR)**. The SCL is a measure of the background level of skin conductance that is associated with a particular situation, whereas the SCR is a measure of the transient changes in skin conductance that are associated with discrete experiences.

The physiological bases of skin conductance changes are not fully understood, but there is considerable evidence implicating the sweat glands. Although the main function of sweat glands is to cool the body, these glands tend to become active in emotional situations, causing the release of sweat that in turn increases the electrical conductivity of the skin (see Green et al., 2014). Sweat glands are distributed over most of the body surface, but, as you are almost certainly aware, those of the hands, feet, armpits, and forehead are particularly responsive to emotional stimuli.

CARDIOVASCULAR ACTIVITY.   The presence in our language of phrases such as *white with fear* and *blushing bride* indicates that modern psychophysiologists were not the first to recognize the relationship between *cardiovascular activity* and emotion. The cardiovascular system has two parts: the blood vessels and the heart. It is a system for distributing oxygen and nutrients to the tissues of the body, removing metabolic wastes, and transmitting chemical messages. Three different measures of cardiovascular activity are frequently employed in psychophysiological research: heart rate, arterial blood pressure, and local blood volume.

Heart Rate   The electrical signal associated with each heartbeat can be recorded through electrodes placed on the chest. The recording is called an **electrocardiogram** (abbreviated either **ECG**, for obvious reasons, or **EKG**, from the original German). The average resting heart rate of a healthy adult is about 70 beats per minute, but it increases abruptly at the sound, or thought, of a dental drill.

Blood Pressure   Measuring arterial blood pressure involves two independent measurements: a measurement of the peak pressure during the periods of heart contraction, the *systoles*, and a measurement of the minimum pressure during the periods of relaxation, the *diastoles*. Blood pressure is usually expressed as a ratio of systolic over diastolic blood pressure in millimeters of mercury (mmHg). The normal resting blood pressure for an adult is about 130/70 mmHg. A chronic blood pressure of more than 140/90 mmHg is viewed as a serious health hazard and is called **hypertension**.

You have likely had your blood pressure measured with a *sphygmomanometer*—a crude device composed of a hollow cuff, a rubber bulb for inflating it, and a pressure gauge for measuring the pressure in the cuff (*sphygmos* means "pulse"). More reliable, fully automated methods are used in research.

Blood Volume   Changes in the volume of blood in particular parts of the body are associated with psychological events. The best-known example of such a change is the engorgement of the genitals associated with sexual arousal in both males and females. **Plethysmography** refers to the various techniques for measuring changes in the volume of blood in a particular part of the body (*plethysmos* means "an enlargement").

One method of measuring these changes is to record the volume of the target tissue by wrapping a strain gauge around it. Although this method has utility in measuring blood flow in fingers or similarly shaped organs, the possibilities for employing it are somewhat limited. Another plethysmographic method is to shine a light through the tissue under investigation and to measure the amount of light absorbed by it. The more blood there is in a structure, the more light it will absorb.

Invasive Physiological Research Methods

We turn now from a consideration of the noninvasive techniques employed in research on living human brains to a consideration of more direct techniques, which are commonly employed in biopsychological studies of nonhuman animals. Most physiological techniques used in biopsychological research on nonhuman animals fall into one of three categories: lesion methods, electrical stimulation methods, and invasive recording methods. Each of these three methods is discussed in this module, but we begin with a description of *stereotaxic surgery* because each of these methods involves the use of stereotaxic surgery.

Stereotaxic Surgery

LO 5.9   Describe the process of stereotaxic surgery.

Stereotaxic surgery is the first step in many biopsychological experiments. *Stereotaxic surgery* is the means by which experimental devices are precisely positioned in the depths of the brain. Two things are required in stereotaxic surgery: an atlas to provide directions to the target site and an instrument for getting there.

The **stereotaxic atlas** is used to locate brain structures in much the same way that a geographic atlas is used to locate geographic landmarks. There is, however, one important difference. In contrast to the surface of the earth, which has only two dimensions, the brain has three. Accordingly, the brain is represented in a stereotaxic atlas by a series of individual maps, one per page, each representing the structure of a single, two-dimensional frontal brain slice. In stereotaxic atlases, all distances are given in millimeters from a designated reference point. In most rat atlases, the reference point is **bregma**—the point on the top of the skull where two of the major *sutures* (seams in the skull) intersect.

The **stereotaxic instrument** (see Figure 5.14) has two parts: a *head holder*, which firmly holds each subject's brain in the prescribed position and orientation; and an *electrode holder*, which holds the device to be inserted. A system of precision gears allows the electrode holder to be moved in three dimensions: anterior–posterior, dorsal–ventral, and lateral–medial. The implantation by stereotaxic surgery of an electrode in the amygdala of a rat is illustrated in Figure 5.15.

Figure 5.14 A stereotaxic instrument. This one is meant for surgery on rodents.

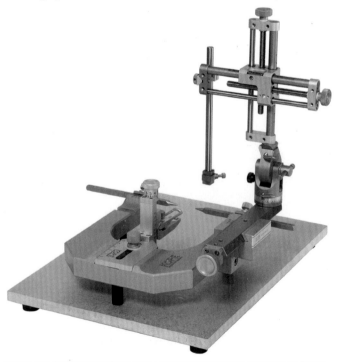

Model 900 Small Animal Stereotaxic Instrument originally designed by David Kopf Instruments in 1963.

Figure 5.15 Stereotaxic surgery: Implanting an electrode in the rat amygdala.

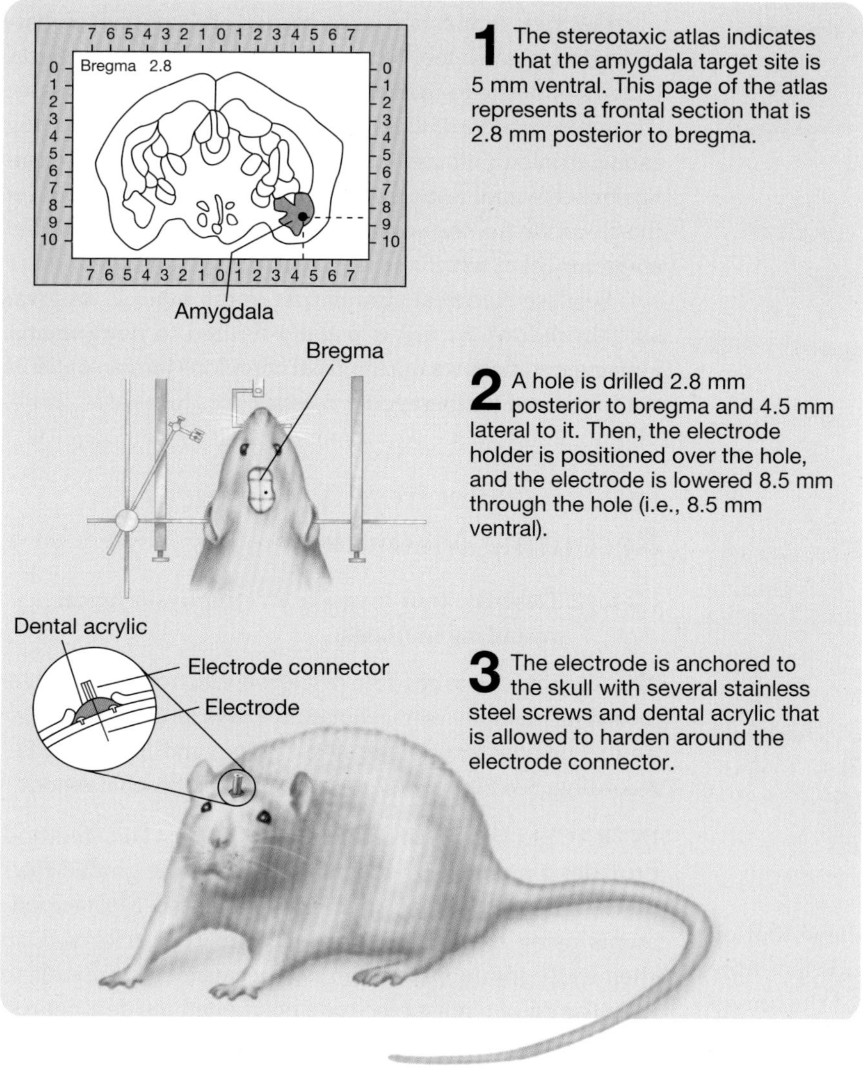

1 The stereotaxic atlas indicates that the amygdala target site is 5 mm ventral. This page of the atlas represents a frontal section that is 2.8 mm posterior to bregma.

2 A hole is drilled 2.8 mm posterior to bregma and 4.5 mm lateral to it. Then, the electrode holder is positioned over the hole, and the electrode is lowered 8.5 mm through the hole (i.e., 8.5 mm ventral).

3 The electrode is anchored to the skull with several stainless steel screws and dental acrylic that is allowed to harden around the electrode connector.

Lesion Methods

LO 5.10 Describe four types of lesion methods and explain why it is important to be cautious when interpreting the effects of lesions.

Those of you with an unrelenting drive to dismantle objects to see how they work will appreciate the lesion methods. In those methods, a part of the brain is damaged, destroyed, or inactivated; then the behavior of the subject is carefully assessed in an effort to determine the functions of the lesioned structure. Four types of lesions are discussed here: aspiration lesions, radio-frequency lesions, knife cuts, and reversible lesions.

ASPIRATION LESIONS. When a lesion is to be made in an area of cortical tissue that is accessible to the eyes and instruments of the surgeon, **aspiration** is frequently the method of choice. The cortical tissue is drawn off by suction through a fine-tipped handheld glass pipette. Because the underlying white matter is slightly more resistant to suction than the cortical tissue itself, a skilled surgeon can delicately peel off the layers of cortical tissue from the surface of the brain, leaving the underlying white matter and major blood vessels undamaged.

RADIO-FREQUENCY LESIONS. Small subcortical lesions are commonly made by passing *radio-frequency current* (high-frequency current) through the target tissue from the tip of a stereotaxically positioned electrode. The heat from the current destroys the tissue. The size and shape of the lesion are determined by the duration and intensity of the current and the configuration of the electrode tip.

KNIFE CUTS. *Sectioning* (cutting) is used to eliminate conduction in a nerve or tract. A tiny, well-placed cut can unambiguously accomplish this task without producing extensive damage to surrounding tissue. How does one insert a knife into the brain to make a cut without severely damaging the overlying tissue? One method is depicted in Figure 5.16.

REVERSIBLE LESIONS. Reversible lesions are useful alternatives to *destructive lesions*. **Reversible lesions** are methods for temporarily eliminating the activity in a particular area of the brain while tests are being conducted. The advantage of reversible lesions is that the same subjects can be repeatedly tested in both the lesion and control conditions. The two most common methods of producing a reversible lesion are by temporarily cooling the target structure or by injecting an anesthetic (e.g., *lidocaine*) into it.

INTERPRETING LESION EFFECTS. Before you leave this section on lesions, a word of caution is in order. Lesion effects are deceptively difficult to interpret. Because the structures of the brain are small, convoluted, and tightly packed together, even a highly skilled surgeon cannot completely destroy a structure without producing significant damage to adjacent structures. There is, however, an unfortunate tendency to lose sight of this fact. For example, a lesion that leaves major portions of the amygdala intact and damages an assortment of neighboring structures comes to be thought of simplistically as an *amygdala lesion*. Such an apparently harmless abstraction can be misleading in two ways. If you believe that all lesions referred to as "amygdala lesions" include damage to no other brain structure, you may incorrectly attribute all of their behavioral effects

Figure 5.16 A device for performing subcortical knife cuts. The device is stereotaxically positioned in the brain; then the blade swings out to make the cut. Here, the anterior commissure is being sectioned.

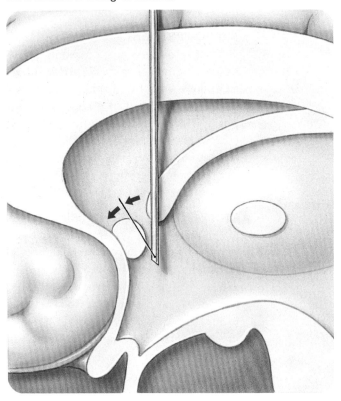

to amygdala damage; conversely, if you believe that all lesions referred to as "amygdala lesions" include the entire amygdala, you may incorrectly conclude that the amygdala does not participate in behaviors uninfluenced by the lesion.

BILATERAL AND UNILATERAL LESIONS. As a general principle—but one with several notable exceptions—the behavioral effects of *unilateral lesions* (lesions restricted to one half of the brain) are much milder than those of symmetrical *bilateral lesions* (lesions involving both sides of the brain), particularly in nonhuman species. Indeed, behavioral effects of unilateral lesions to some brain structures can be difficult to detect. As a result, most experimental studies of lesion effects are studies of bilateral, rather than unilateral, lesions.

Electrical Stimulation

LO 5.11 Describe the technique of electrical brain stimulation.

Clues about the function of a neural structure can be obtained by stimulating it electrically. Electrical brain stimulation is usually delivered across the two tips of a *bipolar electrode*—two insulated wires wound tightly together and cut at the end. Weak pulses of current produce an immediate increase in the firing of neurons near the tip of the electrode.

Electrical stimulation of the brain is an important biopsychological research tool because it often has behavioral effects, usually opposite to those produced by a lesion to the same site. It can elicit a number of behavioral sequences, including eating, drinking, attacking, copulating, and sleeping. The particular behavioral response elicited depends on the location of the electrode tip, the parameters of the current, and the test environment in which the stimulation is administered.

Because electrical stimulation of the brain is an invasive procedure, its use is usually limited to nonhumans. However, sometimes the brains of conscious human patients are stimulated for therapeutic reasons (e.g., Jonas et al., 2014).

Invasive Electrophysiological Recording Methods

LO 5.12 Describe four invasive electrophysiological recording methods.

This section describes four invasive electrophysiological recording methods: intracellular unit recording, extracellular unit recording, multiple-unit recording, and invasive EEG recording. See Figure 5.17 for an example of each method.

INTRACELLULAR UNIT RECORDING. This method provides a moment-by-moment record of the graded fluctuations in one neuron's membrane potential. Most experiments using this recording procedure are performed on chemically immobilized animals because it is difficult to keep the tip of a microelectrode positioned inside a neuron of a freely moving animal (see Long & Lee, 2012). However, special electrodes are now being developed that can allow researchers to do intracellular recordings in a freely moving animal (see Lee & Brecht, 2018).

EXTRACELLULAR UNIT RECORDING. With extracellular unit recording, it is possible to record the activity of a neuron through a microelectrode whose tip is positioned in the extracellular fluid next to it—each time the neuron fires, there is an electrical disturbance and a blip is recorded at the electrode tip. Accordingly, *extracellular unit recording* provides a record of the firing of a neuron but no information about the neuron's membrane potential. It is difficult to record extracellularly from a single neuron in a freely moving animal without the electrode tip shifting away from that neuron, but it can be accomplished with special flexible microelectrodes that can shift slightly with the brain. Initially, extracellular unit recording involved recording from one neuron at a time, each at the tip of a separately implanted electrode. However, it is now possible to simultaneously record extracellular signals from up to about 1,000 neurons by analyzing the correlations among the signals picked up

Figure 5.17 Four methods of recording electrical activity of the nervous system.

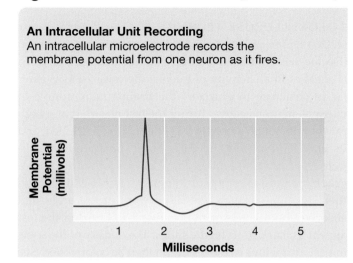

An Intracellular Unit Recording
An intracellular microelectrode records the membrane potential from one neuron as it fires.

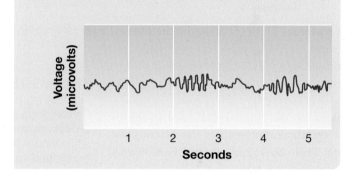

A Multiple-Unit Recording
A small electrode records the action potentials of many nearby neurons. These are added up and plotted. In this example, firing in the area of the electrode tip gradually declined and then suddenly increased.

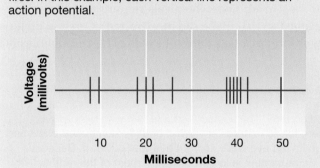

An Extracellular Unit Recording
An extracellular microelectrode records the electrical disturbance that is created each time an adjacent neuron fires. In this example, each vertical line represents an action potential.

An Invasive EEG Recording
A large implanted electrode picks up general changes in electrical brain activity. The EEG signal is not related to neural firing in any obvious way.

through several different electrodes implanted in the same general area (see Callaway & Garg, 2017; Harris et al., 2016; Jun et al., 2017).

MULTIPLE-UNIT RECORDING. In *multiple-unit recording*, the electrode tip is much larger than that of a microelectrode; thus, it picks up signals from many neurons, and slight shifts in its position due to movement of the subject have little effect on the overall signal. The many action potentials picked up by the electrode are fed into an integrating circuit, which adds them together. A multiple-unit recording is a graph of the total number of recorded action potentials per unit of time (e.g., per 0.1 second).

INVASIVE EEG RECORDING. In nonhuman animals, and sometimes in human patients (see Fox et al., 2018), EEG signals can be recorded through implanted electrodes rather than through scalp electrodes. In nonhuman animals, cortical EEG signals are frequently recorded through stainless steel skull screws, whereas subcortical EEG signals are typically recorded through implanted wire electrodes.

Pharmacological Research Methods

In the preceding module, you learned how physiological psychologists study the brain by manipulating it and recording from it using surgical and electrical methods. In this module, you will learn how psychopharmacologists manipulate the brain and record from it using chemical methods.

The major research strategy of psychopharmacology is to administer drugs that either increase or decrease the effects of particular neurotransmitters and to observe the behavioral consequences. Described here are routes of drug administration, methods of using chemicals to make selective brain lesions, methods of measuring the chemical activity of the brain that are particularly useful in biopsychological research, and methods for locating neurotransmitter systems.

Routes of Drug Administration

LO 5.13 Describe the various methods of drug administration.

In most psychopharmacological experiments, drugs are administered in one of the following ways: (1) they are fed to the subject; (2) they are injected through a tube into the stomach (*intragastrically*); or (3) they are injected hypodermically into the peritoneal cavity of the abdomen (*intraperitoneally, IP*), into a large muscle (*intramuscularly, IM*), into the fatty tissue beneath the skin (*subcutaneously, SC*), or into a large surface vein (*intravenously, IV*). A problem with these peripheral routes of administration is that many drugs do not readily pass through the blood–brain barrier. To overcome this problem, drugs can be administered in small amounts through a fine, hollow tube, called a **cannula**, that has been stereotaxically implanted in the brain.

Selective Chemical Lesions

LO 5.14 Describe the method of selective neurotoxic lesions.

The effects of surgical, radio-frequency, and reversible lesions are frequently difficult to interpret because they affect all neurons in the target area. In some cases, it is possible to make more selective lesions by injecting **neurotoxins** (neural poisons) that have an affinity for certain components of the nervous system. There are many selective neurotoxins. For example, when either *kainic acid* or *ibotenic acid* is administered by microinjection, it is preferentially taken up by cell bodies at the tip of the cannula and destroys those neurons, while leaving neurons with axons passing through the area largely unscathed.

Another selective neurotoxin that has been widely used is *6-hydroxydopamine (6-OHDA)*. It is taken up by only those neurons that release the neurotransmitter *norepinephrine* or *dopamine*, and it leaves other neurons at the injection site undamaged.

Measuring Chemical Activity of the Brain

LO 5.15 Describe two techniques for measuring chemical activity in the brain.

There are many procedures for measuring the chemical activity of the brains of laboratory animals. Two techniques that have

proved particularly useful in biopsychological research are the 2-deoxyglucose technique and cerebral dialysis.

2-DEOXYGLUCOSE TECHNIQUE. The *2-deoxyglucose (2-DG) technique* entails placing an animal that has been injected with radioactive 2-DG in a test situation in which it engages in an activity of interest. Because 2-DG is similar in structure to glucose—the brain's main source of energy—neurons active during the test absorb it at a high rate but do not metabolize it. Then the subject is killed, and its brain is removed and sliced. The slices are then subjected to **autoradiography**: They are coated with a photographic emulsion, stored in the dark for a few days, and then developed much like film. Areas of the brain that absorbed high levels of radioactive 2-DG during the test appear as black spots on the slides. The density of the spots in various regions of the brain can then be color-coded (see Figure 5.18).

CEREBRAL DIALYSIS. **Cerebral dialysis** is a method of measuring the extracellular concentration of specific neurochemicals in behaving animals (most other techniques for measuring neurochemicals require that the subjects be killed so that tissue can be extracted). Cerebral dialysis involves implanting a fine tube with a short semipermeable

Figure 5.18 The 2-deoxyglucose technique. The accumulation of radioactivity is shown in three frontal sections taken from the brain of a Richardson's ground squirrel. The subject was injected with radioactive 2-deoxyglucose; then, for 45 minutes, it viewed brightly illuminated black and white stripes through its left eye while its right eye was covered. Because the ground squirrel visual system is largely crossed, most of the radioactivity accumulated in the visual structures of the right hemisphere.

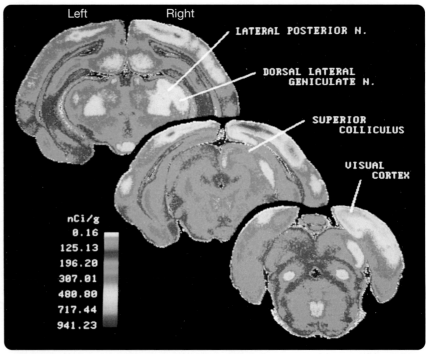

Rod Cooper/University of Calgary Department of Psychology

section into the brain. The semipermeable section is positioned in the brain structure of interest so that extracellular chemicals from the structure will diffuse into the tube. Once in the tube, they can be collected for freezing, storage, and later analysis; or they can be carried in solution directly to a *chromatograph* (a device for measuring the chemical constituents of liquids or gases).

Locating Neurotransmitters and Receptors in the Brain

LO 5.16 Describe two techniques for locating particular neurotransmitters or receptors in the brain.

A key step in trying to understand the psychological function of a particular neurotransmitter or receptor is finding out where it is located in the brain. Two of the techniques available for this purpose are immunocytochemistry and in situ hybridization. Each involves exposing brain slices to a labeled *ligand* of the molecule under investigation (the ligand of a molecule is another molecule that binds to it).

IMMUNOCYTOCHEMISTRY. When a foreign protein (an *antigen*) is injected into an animal, the animal's body creates *antibodies* that bind to it and help the body remove or destroy it; this is known as the body's *immune reaction*. Neurochemists have created stocks of antibodies for the brain's peptide neurotransmitters (neuropeptides) and their receptors. **Immunocytochemistry** is a procedure for locating particular neuroproteins in the brain by labeling their antibodies with a dye or radioactive element and then exposing slices of brain tissue to the labeled antibodies. Regions of dye or radioactivity accumulation in the brain slices mark the locations of the target neuroprotein.

Because all enzymes are proteins and because only those neurons that release a particular neurotransmitter are likely to contain all the enzymes required for its synthesis, immunocytochemistry can be used to locate neurotransmitters by binding to their enzymes. This is done by exposing brain slices to labeled antibodies that bind to enzymes located in only those neurons that contain the neurotransmitter of interest (see Figure 5.19).

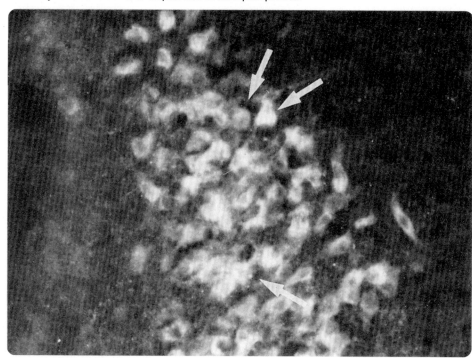

Figure 5.19 Immunocytochemistry. This section through a rat's pons reveals noradrenergic neurons that have attracted the antibody for dopamine-beta-hydroxylase, the enzyme that converts dopamine to norepinephrine.

Richard Mooney, University of Toledo College of Medicine, Department of Neurosciences

IN SITU HYBRIDIZATION. Another technique for locating peptides and other proteins in the brain is **in situ hybridization**. This technique takes advantage of the fact that all peptides and proteins are transcribed from sequences of nucleotide bases on strands of messenger RNA (mRNA). The nucleotide base sequences that direct the synthesis of many neuroproteins have been identified, and hybrid strands of mRNA with the complementary base sequences have been artificially created. In situ hybridization involves the following steps. First, hybrid RNA strands with the base sequence complementary to that of the mRNA that directs the synthesis of the target neuroprotein are obtained. Next, the hybrid RNA strands are labeled with a dye or radioactive element. Finally, the brain slices are exposed to the labeled hybrid RNA strands; they bind to the complementary mRNA strands, marking the location of neurons that release the target neuroprotein.

Genetic Methods

Genetics is a science that has made amazing progress, and biopsychologists are reaping the benefits. Modern genetic methods are now widely used in biopsychological research, which just a few decades ago would have seemed like science fiction. These genetic methods allow for adding, removing, and altering specific genes. There are three categories of genetic methods: (1) gene knockout techniques, (2) gene knockin techniques, and (3) gene editing techniques.

Gene Knockout Techniques

LO 5.17 Explain the gene knockout technique by describing an experiment that employed the technique.

Gene knockout techniques are procedures for creating organisms that lack a particular gene under investigation (e.g., Gingras et al., 2014). Mice (the favored mammalian subjects of genetic research) that are the products of gene knockout techniques are referred to as *knockout mice*. (This term often makes us smile, as images of little mice with boxing gloves flit through our minds.)

Many gene knockout studies have been conducted to clarify the neural mechanisms of behavior. For example, Ruby and colleagues (2002) and Hattar and colleagues (2003) used *melanopsin knockout mice* (mice in whom the gene for the synthesis of melanopsin has been deleted) to study the role of melanopsin in regulating the light–dark cycles that control circadian (about 24 hours) rhythms of bodily function—for example, daily cycles of sleep, eating, and body temperature. *Melanopsin* is a protein found in some neurons in the mammalian *retina* (the receptive layer of the eye), and it had been implicated in the control of circadian rhythms by light. Knockout of the gene for synthesizing melanopsin impaired, but did not eliminate, the ability of mice to adjust their circadian rhythms in response to changes in the light–dark cycle. Thus, melanopsin appears to contribute to the control of circadian rhythms by light, but it is not the only factor.

This type of result is typical of gene knockout studies of behavior: Many genes have been discovered that contribute to particular behaviors, but invariably other mechanisms are involved. It may be tempting to think that each behavior is controlled by a single gene, but the reality is much more complex. Each behavior is controlled by many genes interacting with one another and with experience through *epigenetic mechanisms* (see Chapter 2).

Gene Knockin Techniques

LO 5.18 Explain the gene knockin technique by describing an experiment that employed the technique.

It is now possible to replace one gene with another or add a gene that doesn't exist in an organism. Such **gene knockin techniques** have created interesting possibilities for research and therapy. Pathological genes from human cells can be inserted in other animals such as mice—mice that contain the genetic material of another species are called **transgenic mice**. For example, Shen and colleagues (2008) created transgenic mice by inserting a defective human gene that had been found to be associated with schizophrenia in a Scottish family with a particularly high incidence of the condition. The transgenic mice displayed a variety of cerebral abnormalities (e.g., reduced cerebral cortex and enlarged ventricles) and atypical behaviors reminiscent of human schizophrenia. Treating neurological disease by replacing faulty genes in patients suffering from genetic disorders is an exciting, but as yet unrealized, goal.

Gene Editing Techniques

LO 5.19 Describe how modern gene editing techniques, such as the CRISP-Cas9 method, can provide better ways of assessing the role of a gene in behavior.

Many genes that contribute to particular behaviors have been discovered, but invariably other mechanisms are involved. It may be tempting to think that each behavior is controlled by a single gene, but the reality is much more complex. Each behavior is controlled by many genes interacting with one another and with experience over the course of development (i.e., through epigenetic mechanisms; see Chapter 2).

As a means of controlling for the interaction of genes with experience, many researchers now use modern **gene editing techniques**. These new gene editing techniques allow researchers to edit genes at a particular time during development (see Heidenreich & Zhang, 2016; Khan, 2019).

Of the many gene editing techniques currently available, the **CRISPR/Cas9 method** is generating the most excitement (see Adli, 2018; Luo, Callaway, & Svoboda, 2018). In the most widely used version of the CRISPR/Cas9 method, Cas9 (a protein) is linked to a strand of RNA called the *guide-RNA*. The guide-RNA is made up of a sequence of nucleotide bases that are complementary to one or more strands of DNA. Once the guide-RNA and Cas9 are linked, it can be integrated into a virus. The virus can then be injected into an animal—either peripherally if one wants to edit the genome of the whole organism, or intracranially into a specific brain region if one wants to observe the focal effects of editing a gene in cells in that brain region.

Once the virus enters a cell, and the guide-RNA lines up with a complementary strand of DNA in the organism, Cas9 can either inhibit or activate the genes through various mechanisms. Furthermore, Cas9 can be regulated in various ways so that a researcher can control the effects of Cas9 on *gene expression* (see Chapter 2); thus, a researcher can reversibly alter the expression of a gene (or set of genes) in a particular brain region and examine the effects on behavior. Cas9 can also be used to alter an organism's *epigenome* (see Chapter 2)—see Gomez, Beitnere, and Segal, 2019.

Fantastic Fluorescence and the Brainbow

LO 5.20 Explain how green fluorescent protein has been used as a research tool in the neurosciences.

Green fluorescent protein (GFP) is a protein that exhibits bright green fluorescence when exposed to blue light. First isolated by Shimomura, Johnson, and Saiga (1962), from a species of jellyfish found off the west coast of North America, GFP is currently stimulating advances in many fields of biological research. Martin Chalfie, Osamu Shimomura, and Roger Tsien were awarded the 2008 Nobel Prize in chemistry for its discovery and study.

The utility of GFP as a research tool in the biological sciences could not be realized until its gene was identified and cloned in the early 1990s. The general strategy is to activate the GFP gene in only the particular cells under investigation so that they can be readily visualized. This can be accomplished in two ways: by inserting the GFP gene in only the target cells or by introducing the GFP gene in all cells of the subject but expressing the gene in only the target cells. Chalfie and colleagues (1994) were the first to use GFP to visualize neurons. They introduced the GFP gene into a small transparent roundworm, *Caenorhabditis elegans*, in an area of its chromosomes that controls the development of touch receptor neurons. Figure 5.20 shows the glowing touch receptor neurons. The GFP gene has now been expressed in the cells of many plant and animal species, including humans.

Livet and colleagues (2007) took the very useful GFP technique one step further—one big step. First, Tsien (1998) found that making minor alterations to the GFP gene resulted in the synthesis of proteins that fluoresced in different colors. Livet and colleagues (2007) then introduced the mutated genes for cyan, yellow, and blue fluorescent proteins into the genomes of developing mice in such a way that they were expressed in developing neurons. Each neuron produced different amounts of the three proteins, giving it a distinctive color—in the same way that a color printer can make any color by mixing only three colored inks in differing proportions. Because each neuron was labeled with its own distinctive color, the pathways of neural axons could be traced to their destinations through the cellular morass. This technique has been dubbed **brainbow** for obvious reasons—see Figure 5.21.

In addition to making brainbows, fluorescent proteins have allowed researchers to (1) label specific

Figure 5.20 Touch receptor neurons of the transparent *Caenorhabditis elegans* labeled by green fluorescent protein.

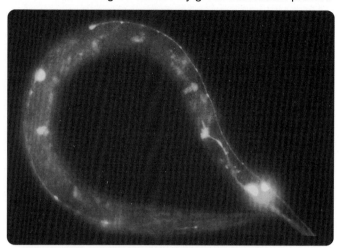

Chalfie, M., et. al. (1994). Green fluorescent protein as a marker for gene expression. Science, 263(5148), 802-805. Used with permission from American Association for the Advancement of Science (AAAS).

neurotransmitters so their activity can be observed (see Wang, Jing, & Li, 2018), (2) label synaptic vesicle proteins in order to observe the fusion of synaptic vesicles with the presynaptic membrane (see Deo & Lavis, 2018), (3) visualize postsynaptic potentials (see Chapter 4) by using fluorescent proteins that light up during membrane hyperpolarizations or depolarizations, and (4) observe the binding of neurotransmitters to receptors by creating receptors that light up when they bind their transmitter (see Lin & Schnitzer, 2016; Storace et al., 2016).

Optogenetics: A Neural Light Switch

LO 5.21 Explain how opsins have been used as a research tool in the neurosciences.

Opsins are light-sensitive ion channels that are found in the cell membranes of certain bacteria and algae (see Boyden, 2014). When opsins are illuminated with light, they open and allow ions to enter the cell. Depending on the particular opsin, light can either hyperpolarize or depolarize the cell membrane they are embedded in. The use of opsins is currently revolutionizing how neuroscientists study the brain (see Boyden, 2015; Kim, Adhikari, & Deisseroth, 2017; Paoletti, Ellis-Davies, & Mourot, 2019).

The utility of opsins for neuroscience research couldn't be realized until their gene was identified and rendered into a form that was expressible within a mammalian cell—a feat that was first accomplished in 2003 (see Boyden, 2014). Soon thereafter, neuroscientists started to use genetic engineering techniques to insert the opsin gene, or variants of the opsin gene, into particular types of neurons. In effect, by inserting an opsin gene into a particular type of neuron, a neuroscientist could use light to hyperpolarize or depolarize

Figure 5.21 With the research technique called *brainbow*, each neuron is labeled with a different color, facilitating the tracing of neural axons.

Jeff W. Lichtman, Harvard University, Department of Molecular and Cellular Biology

neurons. This novel method is known as **optogenetics** (see Goshen, 2014; Paoletti, Ellis-Davies, & Mourot, 2019; Yuste & Church, 2014), and it is increasingly being used by many neuroscientists (see Berndt & Deisseroth, 2015). For example, it can be used in living animals by injecting the animal with a virus carrying an opsin gene that targets a particular type of neuron (e.g., dopaminergic neurons; see Chang et al., 2016). An optical fiber can then be implanted in the animal and light can be shone through the fiber (see Paoletti, Ellis-Davies, & Mourot, 2019; Pisanello et al., 2017) to activate the opsin ion channels—causing the activity of only specific neurons to either be increased or suppressed (see Wiegert et al., 2017). Recent advances have allowed for the use of optogenetics in freely moving animals through the use of wireless technology (see Gutruf & Rogers, 2018).

Scan Your Brain

The research methods of biopsychology illustrate a psychological disorder suffered by many scientists. We call it "unabbreviaphobia"— the fear of leaving any term unabbreviated. To determine whether you have mastered Part One of this chapter and are ready for Part Two, supply the full term for each of the following abbreviations. The correct answers are provided at the end of the exercise. Before proceeding, review material related to your incorrect answers and omissions.

1. CT: _____
2. MRI: _____
3. PET: _____
4. 2-DG: _____
5. fMRI: _____
6. MEG: _____

7. TMS: _____
8. EEG: _____
9. ERP: _____
10. AEP: _____
11. EMG: _____
12. EOG: _____

13. SCL: _____
14. SCR: _____
15. ECG: _____
16. EKG: _____
17. IP: _____
18. IM: _____

19. IV: _____
20. SC: _____
21. 6-OHDA: _____
22. GFP: _____
23. fUS: _____
24. tES: _____

PART TWO Behavioral Research Methods of Biopsychology

We turn now from methods used by biopsychologists to study the nervous system to those that deal with the behavioral side of biopsychology. Because of the inherent invisibility of neural activity, the primary objective of the methods used in its investigation is to render the unobservable observable. In contrast, the major objectives of behavioral research methods are to control, to simplify, and to objectify.

A single set of procedures developed for the investigation of a particular behavioral phenomenon is commonly referred to as a **behavioral paradigm**. Each behavioral paradigm normally comprises a method for producing the behavioral phenomenon under investigation and a method for objectively measuring it.

Neuropsychological Testing

A patient suspected of suffering from some sort of nervous system dysfunction is usually referred to a *neurologist*, who assesses simple sensory and motor functions. More subtle changes in perceptual, emotional, motivational, or cognitive functions are the domain of the *neuropsychologist*.

Because neuropsychological testing is so time consuming, it is typically prescribed for only a small portion of brain-damaged patients. This is unfortunate; the results of neuropsychological testing can help brain-damaged patients in three important ways: (1) by assisting in the diagnosis of neural disorders, particularly in cases in which brain imaging, EEG, and neurological testing have proved equivocal; (2) by serving as a basis for counseling and caring for the patients; and (3) by providing a basis for objectively evaluating the effectiveness of a treatment or the seriousness of its side effects.

Modern Approach to Neuropsychological Testing

LO 5.22 Describe three approaches to neuropsychological testing.

The nature of neuropsychological testing has changed radically since the 1950s. Indeed, the dominant approach to psychological testing has evolved through three distinct phases: the *single-test approach*, the *standardized-test-battery approach*, and the modern *customized-test-battery approach*.

THE SINGLE-TEST APPROACH. Before the 1950s, the few existing neuropsychological tests were designed to detect the presence of brain damage; in particular, the goal of these early tests was to discriminate between patients with psychological problems resulting from structural brain damage and those with psychological problems resulting from functional, rather than structural, changes to the brain. This approach proved unsuccessful, in large part because no single test could be developed that would be sensitive to all the varied and complex psychological symptoms that could potentially occur in a brain-damaged patient.

THE STANDARDIZED-TEST-BATTERY APPROACH. The standardized-test-battery approach to neuropsychological testing grew out of the failures of the single-test approach, and by the 1960s, it became predominant in North America. The objective stayed the same—to identify brain-damaged patients—but the testing involved *standardized batteries* (sets) of tests rather than a single test. The most widely used standardized test battery has been the *Halstead-Reitan Neuropsychological Test Battery*. The Halstead-Reitan is a set of tests that tend to be performed poorly by brain-damaged patients in relation to other patients or healthy controls; the scores on each test are added together to form a single aggregate score. An aggregate score below the designated cutoff leads to a diagnosis of brain damage. The standardized-test-battery approach proved only marginally successful; standardized test batteries discriminate effectively between neurological patients and healthy individuals, but they are not so good at discriminating between neurological patients and psychiatric patients.

THE CUSTOMIZED-TEST-BATTERY APPROACH. The customized-test-battery approach—an approach largely developed by Luria and other Soviet Union neuropsychologists (see Ardila, 1992; Luria & Majovski, 1977)—began to be used routinely in a few American neuropsychological research institutions in the 1960s. This approach proved highly successful in research, and it soon spread to clinical practice. It now predominates in both the research laboratory and the neurological ward.

The objective of current neuropsychological testing is not merely to identify patients with brain damage; the objective is to characterize the nature of the psychological deficits of each brain-damaged patient. So how does the customized-test-battery approach to neuropsychological testing work? It usually begins in the same way for all patients: with a common battery of tests selected by the neuropsychologist to provide an indication of the general nature of the neuropsychological symptoms. Then, depending on the results of the common test battery, the neuropsychologist selects a series of tests customized to each patient in an effort to characterize in more detail the general symptoms revealed by the common battery. For example, if the results of the test battery indicated that a patient had a memory problem, subsequent tests would include those designed to reveal the specific nature of the memory problem.

The tests used in the customized-test-battery approach differ in three respects from earlier approaches. First, the newer tests are specifically designed to measure aspects of psychological function that have been spotlighted by modern theories and data. For example, modern theories, and the evidence on which they are based, suggest that the mechanisms of short-term and long-term memory are totally different; thus, the testing of patients with memory problems virtually always involves specific tests of both short-term and long-term memory. Second, the interpretation of the test results often does not rest entirely on how well the patient does; unlike early neuropsychological tests, currently used tests often require the neuropsychologist to assess the cognitive strategy that the patient employs in performing the test. Third, the customized-test-battery approach requires more skill and knowledge on the part of the neuropsychologist to select just the right battery of tests to expose a particular patient's deficits and to identify qualitative differences in cognitive strategy.

Tests of the Common Neuropsychological Test Battery

LO 5.23 Describe those tests that are often administered as part of an initial common neuropsychological test battery.

Because the customized-test-battery approach to neuropsychological testing typically involves two phases—a battery of general tests given to all patients followed by a series of specific tests customized to each patient—we'll cover examples of these neuropsychological tests in two sections. In this section, we'll look at some tests that are often administered as part of the initial common test battery.

INTELLIGENCE. Although the overall *intelligence quotient* (*IQ*) is a notoriously poor measure of brain damage, a test of general intelligence is nearly always included in the battery of neuropsychological tests routinely given to all patients. Many neuropsychological assessments begin with the **Wechsler Adult Intelligence Scale (WAIS)**, first published in 1955 and standardized in 1981 on a sample of 1,880 U.S. citizens between 16 and 71. The WAIS is composed of many subtests and is often the first test administered, because knowing a patient's IQ can help a neuropsychologist interpret the results of subsequent tests. Also, a skilled neuropsychologist can sometimes draw inferences about a patient's neuropsychological dysfunction from the pattern of deficits on the subtests of the WAIS. For example, low scores on subtests of verbal comprehension tend to be associated with left hemisphere damage.

MEMORY. One weakness of the WAIS is that it often fails to detect memory deficits, despite including subtests specifically designed to test memory function. For example, the information subtest of the WAIS assesses memory for general knowledge (e.g., "Who is Queen Elizabeth?"), and the **digit span** subtest (the most widely used test of short-term memory) identifies the longest sequence of random digits that a patient can repeat correctly 50 percent of the time; most people have a digit span of 7. However, these two forms of memory are among the least likely to be disrupted by brain damage—patients with seriously disturbed memory function often show no deficits on either the information or the digit span subtest. Be that as it may, memory problems rarely escape unnoticed because they are usually reported by the patient or the family of the patient.

LANGUAGE. If a neuropsychological patient has taken the WAIS, deficits in the use of language can be inferred from a low aggregate score on the verbal subtests. A patient who has not taken the WAIS can be quickly screened for language-related deficits with the **token test**. Twenty tokens of two different shapes (squares and circles), two different sizes (large and small), and five different colors (white, black, yellow, green, and red) are placed on a table in front of the patient. The test begins with the examiner reading simple instructions—for example, "Touch a red square"—and the patient trying to follow them. Then the test progresses to more difficult instructions, such as "Touch the small, red circle and then the large, green square." Finally, the patient is asked to read the instructions aloud and follow them.

LANGUAGE LATERALIZATION. It is usual for one hemisphere to participate more than the other in language-related activities. In most people, the left hemisphere is dominant for language, but in some, the right hemisphere is dominant. A test of language lateralization is often included in the common test battery because knowing which hemisphere is dominant for language is often useful in interpreting the results of other tests. Furthermore, a test of language lateralization is virtually always given to patients before any surgery that might encroach on the cortical language areas. The results are used to plan the surgery, trying to avoid the language areas if possible.

There are two widely used tests of language lateralization. The sodium amytal test (Wada, 1949) is one and the dichotic listening test (Kimura, 1973) is the other.

The **sodium amytal test** involves injecting the anesthetic *sodium amytal* into either the left or right carotid artery in the neck. This temporarily anesthetizes the *ipsilateral* (same-side) hemisphere while leaving the *contralateral* (opposite-side) hemisphere largely unaffected. Several tests of language function are quickly administered while the ipsilateral hemisphere is anesthetized. Later, the process is repeated for the other side of the brain. When the injection is on the side dominant for language, the patient is completely mute for about 2 minutes. When the injection is on the nondominant side, there are only a few minor speech problems. Because the sodium amytal test is invasive, it can be administered only for medical reasons—-usually to determine the dominant language hemisphere prior to brain surgery.

In the standard version of the **dichotic listening test**, sequences of spoken digits are presented to volunteers through stereo headphones. Three digits are presented to one ear at the same time that three different digits are presented to the other ear. Then, they are asked to report as many of the six digits as they can. Kimura (1973) found that patients correctly report more of the digits heard by the ear contralateral to their dominant hemisphere for language, as determined by the sodium amytal test.

Tests of Specific Neuropsychological Function

LO 5.24 Describe tests that might be used by a neuropsychologist to investigate in more depth general problems revealed by a common neuropsychological test battery.

Following analysis of the results of a neuropsychological patient's performance on a common test battery, the neuropsychologist selects a series of specific tests to clarify the nature of the general problems exposed by the common battery. There are thousands of tests that might be selected from. This section describes a few of them and mentions some of the considerations that might influence their selection.

Journal Prompt 5.2

What are some of the clinical implications of this two-stage approach to neuropsychological testing?

MEMORY. Following the discovery of memory impairment by the common test battery, at least four fundamental questions about the memory impairment must be answered (see Chapter 11): (1) Does the memory impairment involve *short-term memory, long-term memory,* or both? (2) Are any deficits in long-term memory *anterograde* (affecting the retention of things learned after the damage), *retrograde* (affecting the retention of things learned before the damage), or both? (3) Do any deficits in long-term memory involve *semantic memory* (memory for knowledge of the world) or *episodic memory* (memory for personal experiences)? (4) Are any deficits in long-term memory deficits of *explicit memory* (memories of which the patient is aware and can thus express verbally), *implicit memory* (memories demonstrated by the improved performance of the patient without the patient being conscious of them), or both?

Many amnesic patients display severe deficits in explicit memory with no deficits at all in implicit memory (see Squire & Dede, 2015). **Repetition priming tests** have proven instrumental in the assessment and study of this pattern. Patients are first shown a list of words and asked to study them; they are not asked to remember them. Then, at a later time, they are asked to complete a list of word fragments, many of which are fragments of words from the initial list. For example, if "purple" had been in the initial test, "pu_p_ _" could be one of the test word fragments. Amnesic patients often complete the fragments as accurately as healthy control subjects. But—and this is the really important part—they often have no conscious memory of any of the words in the initial list or even of ever having seen the list. In other words, they display good implicit memory of experiences without explicit memories of them.

LANGUAGE. If a neuropsychological patient turns out to have language-related deficits on the common test battery, a complex series of tests is administered to clarify the nature of the problem. For example, if a patient has a speech problem, it may be one of three fundamentally different problems: problems of *phonology* (the rules governing the sounds of the language), problems of *syntax* (the grammar of the language), or problems of *semantics* (the meaning of the language). Because brain-damaged patients may have one of these problems but not the others, it is imperative that the testing of all neuropsychological patients with speech problems include tests of each of these three capacities.

Reading aloud can be disrupted in different ways by brain damage, and follow-up tests must be employed that can differentiate between the different patterns of disruption. Some *dyslexic* patients (those with reading problems) remember the rules of pronunciation but have difficulties pronouncing words that do not follow these rules, words such as *come* and *tongue*, whose pronunciation must be remembered. Other dyslexic patients pronounce simple familiar words based on memory but have lost the ability to apply the rules of pronunciation—they cannot pronounce nonwords such as *trapple* or *fleeming*.

Behavioral Methods of Cognitive Neuroscience

The Case of the Vegetative Patient

What if a patient in a vegetative state was in fact completely conscious but simply unable to generate an observable response to stimuli? Owen and colleagues (2014) asked this precise question. They set about studying patients in a *vegetative state*. While in an MRI scanner, they instructed one of their patients to visualize each of two things: (1) playing tennis and (2) navigating through her home, room-to-room; each of these tasks was known to activate a distinct set of brain regions in healthy control participants. Amazingly, she displayed patterns of activation on her fMRI that were remarkably similar to what had been observed in controls. Shocked by this finding, Owen and colleagues wanted to see if they could communicate with the patient. To do so, while undergoing fMRI they instructed her to visualize playing tennis to say "yes" and to visualize walking around her home to say "no." Then, they asked her some questions for which they knew the answers (e.g., "Is your father's name Karlis?"). She responded with the correct answer to each of their questions—faint whispers from a trapped mind.

This case study is just one example of the powerful things that functional imaging is capable of. In this module you will learn about the techniques that *cognitive neuroscientists* use to study relationships between brain and cognition.

As we warned you earlier in this chapter, it is important to think critically about the results of functional brain imaging studies (see Raz, 2012). Because fMRI images are so compelling, it is particularly important to be an informed consumer of them; to understand the assumptions and research methods of **cognitive neuroscience:** a division of biopsychology that focuses on understanding cognition.

Before we present the behavioral methods of cognitive neuroscience, let's first discuss two key assumptions that are common in cognitive neuroscience. The first assumption is that each complex cognitive process results from the combined activity of simple cognitive processes called **constituent cognitive processes**. The second assumption is that each constituent cognitive process is mediated by neural activity within a particular brain region or across a set of brain regions (see Sporns & Betzel, 2016). Accordingly, the main goal of cognitive neuroscience is to identify the parts of the brain that mediate various constituent cognitive processes.

Paired-Image Subtraction Technique

LO 5.25 Describe the paired-image subtraction technique.

With the central role played by PET and fMRI in cognitive neuroscience research, the **paired-image subtraction technique** has become one of the key behavioral research methods in such research (see Kriegeskorte, 2010; Posner & Raichle, 1994). Let us illustrate this technique with the classic PET study of single-word processing by Petersen and colleagues (1988). Petersen and his colleagues were interested in locating the parts of the brain that enable a person to make a word association (to respond to a printed word by saying a related word). You might think this would be an easy task to accomplish by having a volunteer perform a word-association task while a PET image of the volunteer's brain is recorded. The problem with this approach is that many parts of the brain that would be active during the test period would have nothing to do with the constituent cognitive process of forming a word association; much of the activity recorded would be associated with other processes such as seeing the words, reading the words, and speaking. The paired-image subtraction technique was developed to deal with this problem.

The paired-image subtraction technique involves obtaining functional brain images during several different cognitive tasks. Ideally, the tasks are designed so that pairs of them differ from each other in terms of only a single constituent cognitive process. Then the brain activity associated with that process can be estimated by subtracting the activity in the image associated with one of the two tasks from the activity in the image associated with the other. For example, in one of the tasks in the study by Petersen and colleagues, volunteers spent a minute reading aloud printed nouns as they appeared on a screen; in another, they observed the same nouns on the screen but responded to each of them by saying aloud an associated verb (e.g., *truck—drive*). Then Petersen and his colleagues subtracted the activity in the images they recorded during the two tasks to obtain a *difference image*. The difference image illustrated the areas of the brain specifically involved in the constituent cognitive process of forming the word association; the activity associated with fixating on the screen, seeing the nouns, saying the words, and so on, was eliminated by the subtraction.

Default Mode Network

LO 5.26 Understand the default mode network and know the structures that are part of that network.

Interpretation of difference images is complicated by the fact that there is substantial brain activity when humans sit quietly and let their minds wander—this level of activity has been termed the brain's **default mode** (Raichle, 2010). Brain structures typically active in the default mode but less active during cognitive or behavioral tasks are collectively referred to as the **default mode network,** and their pattern of activity is known as the **resting state-fMRI (R-fMRI).** The default mode network comprises many structures (see Fox et al., 2015) including the following four cortical areas: medial parietal cortex, lateral parietal cortex, medial prefrontal cortex, and lateral temporal cortex. See Figure 5.22.

Mean Difference Images

LO 5.27 Explain what a mean difference image is.

Another difficulty in using PET and fMRI to locate constituent cognitive processes results from the *noise* associated with random cerebral events that occur during the test—for example, thinking about a sudden pang of hunger, noticing a fly on the screen, or wondering whether the test will last much longer (see Mason et al., 2007). The noise created by such events can be significantly reduced with a technique discussed earlier in this chapter: *signal averaging*. By averaging the difference images obtained from repetitions of the same tests, the researchers can greatly increase the *signal-to-noise ratio*. It is standard practice to average the images obtained from several volunteers; the resulting **mean** (averaged) **difference image** emphasizes areas of activity that are common to many volunteers and de-emphasizes areas of activity that are peculiar to a few of them. However, this averaging procedure can lead to at least two serious problems. First, if two volunteers had specific but different patterns of cortical activity, the average image derived from the two would reveal little about either. Because people differ substantially from one another in the cortical localization of cognitive abilities, this is a serious problem (see Braver, Cole, & Yarkoni, 2010; Kanai & Rees, 2011; Lichtman & Denk, 2011). Second, the area of cortex that controls a particular ability could change in an individual as a result of experience.

Journal Prompt 5.3

If an area of cortex that controls a particular ability can change in an individual based on their experiences, what implications, if any, does this neuroplasticity have for the reliability and validity of mean difference images?

Figure 5.22 The default mode network: areas of the brain in which activity is commonly recorded by functional brain-imaging techniques when the mind wanders, but not when it is actively engaged.

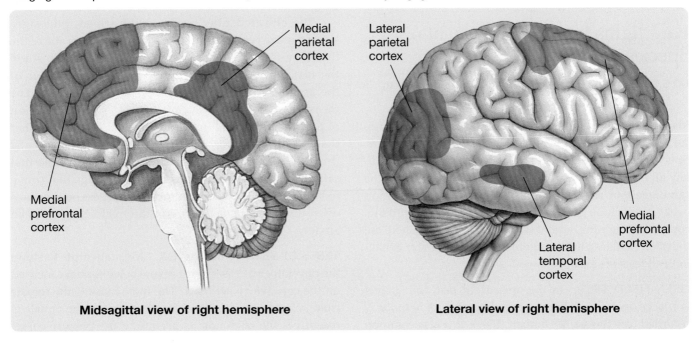

Midsagittal view of right hemisphere

Lateral view of right hemisphere

Functional Connectivity

LO 5.28 Explain the concept of functional connectivity.

In addition to being interested in which brain regions are active during particular cognitive tasks, cognitive neuroscientists are also eager to understand how network activity across multiple brain regions is related to a particular cognitive task. This approach is referred to as the study of **functional connectivity (FC)**. To measure functional connectivity, a cognitive neuroscientist examines which brain regions have parallel patterns of activity over time.

When a cognitive neuroscientist studies changes in FC with the presentation of a stimulus, or during the performance of a task, they are studying *extrinsic FC*. This is in contrast to *intrinsic FC*, which is FC that is present during the R-fMRI (see Kelly & Castellanos, 2014). Collectively, the task of characterizing the FC associated with each behavior and cognitive process is known as the study of the **functional connectome** (see Matthews & Hampshire, 2016).

Biopsychological Paradigms of Animal Behavior

Noteworthy examples of the behavioral paradigms used to study the biopsychology of laboratory species are provided here under three headings: (1) paradigms for the assessment of species-common behaviors, (2) traditional conditioning paradigms, and (3) seminatural animal learning paradigms. In each case, the focus is on methods used to study the behavior of the laboratory rat, one of the most common subjects of biopsychological research.

Paradigms for the Assessment of Species-Common Behaviors

LO 5.29 Describe three behavioral paradigms used to study species-common behaviors.

Many of the behavioral paradigms used in biopsychological research are used to study species-common behaviors. **Species-common behaviors** are those displayed by virtually all members of a species, or at least by all those of the same age and sex. Commonly studied species-common behaviors include grooming, swimming, eating, drinking, copulating, fighting, and nest building. Described here are the open-field test, tests of aggressive and defensive behavior, and tests of sexual behavior.

OPEN-FIELD TEST. In the **open-field test**, the subject is placed in a large, barren chamber, and its activity is recorded (see Brooks & Dunnett, 2009). It is also common in the open-field test to count the number of *boluses* (pieces of excrement) that were dropped by an animal during the test. Low activity scores and high bolus counts are frequently used as indicators of fearfulness. Fearful rats are also highly **thigmotaxic**; that is, they rarely venture away from the walls of the test chamber and rarely engage in such activities as rearing and grooming. Rats are often fearful when they are first placed in a strange open field, but this fearfulness usually declines with repeated exposure to the same open field.

TESTS OF AGGRESSIVE AND DEFENSIVE BEHAVIOR. Typical patterns of aggressive and defensive behavior can be observed and measured during combative encounters between the dominant male rat of an established colony and a smaller male intruder (see Blanchard & Blanchard, 1988). This is called the **colony-intruder paradigm**. The behaviors of the dominant male are considered to be aggressive and those of the hapless intruder defensive. The dominant male of the colony (the *alpha male*) moves sideways toward the intruder, with its hair erect. When it nears the intruder, it tries to push the intruder off balance and to deliver bites to its back and flanks. The defender tries to protect its back and flanks by rearing up on its hind legs and pushing the attacker away with its forepaws or by rolling onto its back. Thus, piloerection, lateral approach, and flank- and back-biting indicate conspecific aggression in the rat; freezing, boxing (rearing and pushing away), and rolling over indicate defensiveness.

Some tests of rat defensive behavior assess reactivity to the experimenter rather than to another rat. For example, it is common to rate the resistance of a rat to being picked up—no resistance being the lowest category and biting the highest—and to use the score as one measure of defensiveness.

The **elevated plus maze**, a four-armed, plus-sign-shaped maze typically mounted 50 centimeters above the floor, is a test of defensiveness commonly used to study the *anxiolytic* (anxiety-reducing) effects of drugs. Two of the arms of the maze have sides, and two do not. The measure of defensiveness, or anxiety, is the proportion of time the rats spend in the protected closed arms rather than on the exposed arms. Many established anxiolytic drugs significantly increase the proportion of time that rats spend on the open arms, and new drugs that prove to be effective in reducing rats' defensiveness on the maze often turn out to be effective in the treatment of human anxiety.

TESTS OF SEXUAL BEHAVIOR. Most attempts to study the physiological bases of rat sexual behavior have focused on the copulatory act itself. The male mounts the female from behind and clasps her hindquarters. If the female is receptive, she responds by assuming the posture called

lordosis; that is, she sticks her hindquarters in the air, she bends her back in a U, and she deflects her tail to the side. During some mounts, the male inserts his penis into the female's vagina; this act is called **intromission**. After intromission, the male dismounts by jumping backward. He then returns a few seconds later to mount and intromit once again. Following about 10 such cycles of mounting, intromitting, and dismounting, the male mounts, intromits, and **ejaculates** (ejects his sperm).

Three common measures of male rat sexual behavior are the number of mounts required to achieve intromission, the number of intromissions required to achieve ejaculation, and the interval between ejaculation and the reinitiation of mounting. The most common measure of female rat sexual behavior is the **lordosis quotient** (the proportion of mounts that elicit lordosis).

Traditional Conditioning Paradigms

LO 5.30 Describe the Pavlovian conditioning paradigm and the operant conditioning paradigm.

Learning paradigms play a major role in biopsychological research for three reasons. The first is that learning is a phenomenon of primary interest to psychologists. The second is that learning paradigms provide an effective technology for producing and controlling animal behavior. Because animals cannot follow instructions from the experimenter, it is often necessary to train them to behave in a fashion consistent with the goals of the experiment. The third reason is that it is possible to infer much about the sensory, motor, motivational, and cognitive state of an animal from its ability to learn and perform various responses.

If you have taken a previous course in psychology, you will likely be familiar with the Pavlovian and operant conditioning paradigms. In the **Pavlovian conditioning paradigm**, the experimenter pairs an initially neutral stimulus called a *conditional stimulus* (e.g., a tone or a light) with an *unconditional stimulus* (e.g., meat powder)—a stimulus that elicits an *unconditional* (reflexive) *response* (e.g., salivation). As a result of these pairings, the conditional stimulus eventually acquires the capacity, when administered alone, to elicit a *conditional response* (e.g., salivation)—a response that is often, but not always, similar to the unconditional response.

In the **operant conditioning paradigm**, the rate at which a particular voluntary response (such as a lever press) is emitted is increased by *reinforcement* or decreased by *punishment*. One widely used operant conditioning paradigm in biopsychology is the self-stimulation paradigm. In the **self-stimulation paradigm**, animals press a lever to deliver electrical stimulation to particular sites in their own brains; those structures in the brain that support self-stimulation have often been called *pleasure centers*.

Seminatural Animal Learning Paradigms

LO 5.31 Describe four seminatural animal learning paradigms.

In addition to Pavlovian and operant conditioning paradigms, biopsychologists use animal learning paradigms that have been specifically designed to mimic situations that an animal might encounter in its natural environment. Development of these paradigms stemmed in part from the reasonable assumption that forms of learning tending to benefit an animal's survival in the wild are likely to be more highly developed and more directly related to innate neural mechanisms. The following are four common seminatural learning paradigms: conditioned taste aversion, radial arm maze, Morris water maze, and conditioned defensive burying.

CONDITIONED TASTE AVERSION. A **conditioned taste aversion** is the avoidance response that develops to tastes of food whose consumption has been followed by illness (see Garcia & Koelling, 1966; Lin, Arthurs, & Reilly, 2014). In the standard conditioned taste aversion experiment, rats receive an *emetic* (a nausea-inducing drug) after they consume a food with an unfamiliar taste. On the basis of this single conditioning trial, the rats learn to avoid the taste.

The ability of rats to readily learn the relationship between a particular taste and subsequent illness unquestionably increases their chances of survival in their natural environment, where potentially edible substances are not routinely screened by government agencies. Rats and many other animals are *neophobic* (afraid of new things); thus, when they first encounter a new food, they consume it in only small quantities. If they subsequently become ill, they will not consume it again. Conditioned aversions also develop to familiar tastes, but these typically require more than a single trial to be learned.

Humans also develop conditioned taste aversions. Cancer patients have been reported to develop aversions to foods consumed before nausea-inducing chemotherapy (Bernstein & Webster, 1980). Many of you will be able to testify on the basis of personal experience about the effectiveness of conditioned taste aversions. I (JP) still have vivid memories of a batch of red laboratory punch that I overzealously consumed after eating two pieces of blueberry pie. But that is another story—albeit a particularly colorful one.

The discovery of conditioned taste aversion challenged three widely accepted principles of learning (see Revusky & Garcia, 1970) that had grown out of research on traditional operant and Pavlovian conditioning paradigms. First, it challenged the view that animal conditioning is always a gradual step-by-step process; robust taste aversions can be established in only a single trial. Second, it showed that *temporal contiguity* is not essential for conditioning; rats acquire taste aversions even when they do not become ill until several hours after eating. Third, it challenged the *principle of equipotentiality*—the view that conditioning proceeds in basically the same manner regardless of the particular stimuli and responses under investigation. Rats appear to have evolved to readily learn associations between tastes and illness; it is only with great difficulty that they learn relations between the color of food and nausea or between taste and footshock.

RADIAL ARM MAZE. The radial arm maze taps into the well-developed spatial abilities of rodents. The survival of rats in the wild depends on their ability to navigate quickly and accurately through their environment and to learn which locations in it are likely to contain food and water. This task is much more complex for a rodent than it is for us. Most of us obtain food from locations where the supply is continually replenished; we go to the market confident that we will find enough food to satisfy our needs. In contrast, the foraging rat must learn and retain a complex pattern of spatially coded details. It must not only learn where morsels of food are likely to be found but must also remember which of these sites it has recently stripped of their booty so as not to revisit them too soon. Designed by Olton and Samuelson (1976) to study these spatial abilities, the **radial arm maze** (see Figure 5.23) is an array of arms—usually eight or more—radiating from a central starting area. At the end of each arm is a food cup, which may or may not be baited, depending on the purpose of the experiment.

In one version of the radial arm maze paradigm, rats are placed each day in a maze that has the same arms baited each day. After a few days of experience, rats rarely visit unbaited arms at all, and they rarely visit baited arms more than once in the same day—even when control procedures make it impossible for them to recognize odors left

Figure 5.23 A radial arm maze.

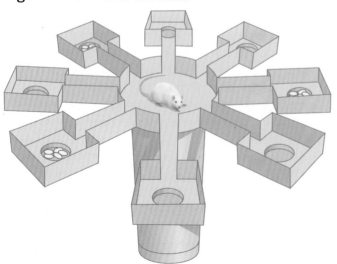

during previous visits to an arm or to make their visits in a systematic sequence. Because the arms are identical, rats must orient themselves in the maze with reference to external room cues; thus, their performance can be disrupted by rotation of the maze or by changes in the appearance of the room.

MORRIS WATER MAZE. Another seminatural learning paradigm that has been designed to study the spatial abilities of rats is the **Morris water maze** (Morris, 1981). The rats are placed in a circular, featureless pool of cool milky water in which they must swim until they discover the escape platform—which is invisible just beneath the surface of the water. The rats are allowed to rest on the platform before being returned to the water for another trial. Despite the fact that the starting point is varied from trial to trial, the rats learn after only a few trials to swim directly to the platform, presumably by using spatial cues from the room as a reference. The Morris water maze is useful for assessing the navigational skills of brain-lesioned or drugged animals.

CONDITIONED DEFENSIVE BURYING. Yet another seminatural learning paradigm useful in biopsychological research is conditioned defensive burying (e.g., Pinel & Mana, 1989; Pinel & Treit, 1978). In studies of **conditioned defensive burying**, rats receive a single aversive stimulus (e.g., a shock, air blast, or noxious odor) from an object mounted on the wall of the chamber just above the floor, which is littered with bedding material. After a single trial, almost every rat learns that the test object is a threat and responds by flinging bedding material at the test object with its head and forepaws (see Figure 5.24). Antianxiety drugs reduce the amount of conditioned defensive burying, and thus the paradigm is used to study the neurochemistry of anxiety (see Steimer, 2011).

Thinking Creatively About Biopsychological Research

LO 5.32 Explain why multiple techniques should be used when trying to answer a specific question.

Before moving on to the next chapter, you need to appreciate that multiple research methods almost always need to be used to answer a question; seldom, if ever, is an important biopsychological issue resolved by use of a single method. The reason for this is that neither the methods used to manipulate the brain nor the methods used to assess the behavioral consequences of these manipulations are totally selective; there are no methods of manipulating the brain that change only a single aspect of brain function, and there are no measures of behavior that reflect only a single psychological process. Accordingly, lines of research that use a single method can usually be

Figure 5.24 These photos show a rat burying a test object from which it has just received a single mild shock.

John Pinel

interpreted in more than one way and thus cannot provide unequivocal evidence for any one interpretation. Typically, important research questions are resolved only when several methods are brought to bear on a single problem. This general approach, as you may recall, is called *converging operations*.

Journal Prompt 5.4

Think of a research question that would require converging operations (i.e., using several methods to address a single problem) among two or more of the research methods described in this chapter.

Themes Revisited

This chapter introduced you to the two kinds of research methods used by biopsychologists: methods of studying the brain and methods of studying behavior. In the descriptions of these methods, all five of the main themes of the text were apparent.

The chapter-opening case of Professor P. alerted you to the fact that many of the methods used by biopsychologists to study the human brain are also used clinically, in either diagnosis or treatment. The clinical implications theme came up again during discussions of brain imaging,

genetic engineering, neuropsychological testing, and use of the elevated plus maze to test anxiolytic drugs.

The thinking creatively theme was implicit throughout this entire chapter. That is, a major purpose of this chapter was to help you better understand the research methods of biopsychology so you can be an informed consumer of biopsychological research. Moreover, the development of new research methods often requires considerable creativity.

The neuroplasticity theme arose during the discussion of the methods of cognitive neuroscience. Experience can produce changes in brain organization that can complicate the interpretation of functional brain images.

The evolutionary perspective theme arose in the discussion of green fluorescent protein, first isolated from jellyfish, and again during the discussion of the rationale for using seminatural animal learning paradigms, which assess animal behavior in environments similar to those in which it evolved.

Only one of the emerging themes appeared in this chapter. The thinking about epigenetics theme appeared only briefly during the discussion of modern tools for editing an animal's genome.

Key Terms

Chapter 6
The Visual System

How We See

Indiapicture/Alamy Stock Photo

 Chapter Overview and Learning Objectives

Light Enters the Eye and Reaches the Retina

LO 6.1 Explain how the pupil and the lens can affect the image that falls on the retina.

LO 6.2 Explain why some vertebrates have one eye on each side of their head, whereas other vertebrates have their eyes mounted side-by-side on the front of their heads. Also, explain the importance of binocular disparity.

The Retina and Translation of Light into Neural Signals

LO 6.3 Describe the structure of the retina and name the cell types that make up the retina.

LO 6.4 Describe the duplexity theory of vision and explain the differences between the photopic and scotopic systems.

LO 6.5 Explain the difference between the photopic and scotopic spectral sensitivity curves and explain how that difference can account for the Purkinje effect.

LO 6.6 Describe the three types of involuntary fixational eye movements and explain what happens when all eye movements are blocked.

LO 6.7 Describe the process of visual transduction.

From Retina to Primary Visual Cortex

LO 6.8 Describe the components and layout of the retina-geniculate-striate system.

LO 6.9 In the context of the retina-geniculate-striate system, explain what is meant by retinotopic.

LO 6.10 Describe the M and P channels.

Seeing Edges

LO 6.11 Describe contrast enhancement.

LO 6.12 Define the term *receptive field* and describe the methods used by David Hubel and Torsten Wiesel to map the receptive fields of visual system neurons.

LO 6.13 Describe the work of Hubel & Wiesel that helped to characterize the receptive fields of retinal ganglion cells, lateral geniculate neurons, and striate neurons of lower layer IV.

LO 6.14 Describe the work of Hubel & Wiesel that characterized the receptive fields of simple and complex cells in the primary visual cortex.

LO 6.15 Describe the organization of the primary visual cortex.

LO 6.16 Describe how views about the receptive fields of retinal ganglion cells and lateral geniculate neurons have recently changed.

LO 6.17 Describe the changing view of visual system receptive fields.

Seeing Color

LO 6.18 Describe the component and opponent-process theories of color vision.

LO 6.19 Describe Land's demonstration of color constancy and explain his retinex theory.

Cortical Mechanisms of Vision and Conscious Awareness

LO 6.20 Describe the three classes of visual cortex and identify their locations in the brain.

LO 6.21 Explain what happens when an area of primary visual cortex is damaged.

LO 6.22 Describe the areas of secondary visual cortex and association cortex involved in vision.

LO 6.23 Explain the difference between the dorsal and ventral streams and the functions that have been attributed to each stream by different theories.

LO 6.24 Describe the phenomenon of prosopagnosia and discuss the associated theoretical issues.

LO 6.25 Describe the phenomenon of akinetopsia and discuss the associated theoretical issues.

This chapter is about your visual system. Most people think their visual system has evolved to respond as accurately as possible to the patterns of light that enter their eyes. They recognize the obvious limitations in the accuracy of their visual system, of course; and they appreciate those curious instances, termed *visual illusions*, in which it is "tricked" into seeing things the way they aren't. But such shortcomings are generally regarded as minor imperfections in a system that responds as faithfully as possible to the external world.

But, despite the intuitive appeal of thinking about it in this way, this is not how the visual system works. The visual system does not produce an accurate internal copy of the external world. It does much more. From the tiny, distorted, upside-down, two-dimensional retinal images projected on the visual receptors that line the backs of the eyes, the visual system creates an accurate, richly detailed, three-dimensional perception that is—and this is the really important part—in some respects even better than the external reality from which it was created. Our primary goal in this chapter is to help you appreciate the inherent creativity of your own visual system.

You will learn in this chapter that understanding the visual system requires the integration of two types of research: (1) research that probes the visual system with sophisticated neuroanatomical, neurochemical, and neurophysiological techniques; and (2) research that focuses on the assessment of what we see. Both types of research receive substantial coverage in this chapter, but it is the second type that provides you with a unique educational opportunity: the opportunity to participate in the very research you are studying. Throughout this chapter, you will be encouraged to participate in a series of Check It Out demonstrations designed to illustrate the relevance of what you are learning in this text to life outside its pages.

This chapter is composed of six modules. The first three take you on a journey from the external visual world to the visual receptors of the retina and from there over the major visual pathway to the primary visual cortex. The next two modules describe how the neurons of this visual pathway mediate the perception of two particularly important features of the visual world: edges and color. The final module deals with the flow of visual signals from the primary visual cortex to other parts of the cortex that participate in the complex process of vision.

Before you begin the first module of the chapter, we'd like you to consider an interesting clinical case. Have you ever wondered whether one person's subjective experiences are like those of others? This case provides evidence that at least some of them are. It was reported by Whitman Richards (1971), and his participant was his wife. Mrs. Richards suffered from migraine headaches (see Goadsby, 2015), and like 20 percent of migraine sufferers, she often experienced visual displays, called *fortification illusions*, prior to her attacks (see Charles & Baca, 2013; Thissen et al., 2014).

The Case of Mrs. Richards: Fortification Illusions and the Astronomer

Each fortification illusion began with a gray area of blindness near the center of her visual field—see Figure 6.1. During the next few minutes, the gray area would begin to expand into a horseshoe shape, with a zigzag pattern of flickering lines at its advancing edge (this pattern reminded people of the plans for a fortification, hence the name of the illusions).

It normally took about 20 minutes for the lines and the trailing area of blindness to reach the periphery of her visual field. At this point, her headache would usually begin.

Because the illusion expanded so slowly, Mrs. Richards was able to stare at a point on the center of a blank sheet of paper and periodically trace on the sheet the details of her illusion. This method made it apparent that the lines became thicker and the expansion of the area of blindness occurred faster as the illusion spread into the periphery.

Interestingly, Dr. Richards discovered that a similar set of drawings was published in 1870 by the famous British astronomer George Biddell Airy. They were virtually identical to those done by Mrs. Richards.

Figure 6.1 The fortification illusions associated with migraine headaches.

1 An attack begins, often when reading, as a gray area of blindness near the center of the visual field.

2 Over the next 20 minutes, the gray area assumes a horseshoe shape and expands into the periphery, at which point the headache begins.

We will return to fortification illusions after you have learned a bit about the visual system. At that point, you will be better able to appreciate their significance.

Light Enters the Eye and Reaches the Retina

Everybody knows that cats, owls, and other nocturnal animals can see in the dark. Right? Wrong! Some animals have special adaptations that allow them to see under very dim illumination, but no animal can see in complete darkness. The light reflected into your eyes from the objects around you is the basis for your ability to see them; if there is no light, there is no vision.

You may recall from high-school physics that light can be thought of in two different ways: as discrete particles of energy, called *photons*, traveling through space at about 300,000 kilometers (186,000 miles) per second, or as waves of energy. Both theories are useful; in some ways, light behaves like particles; and in others, it behaves like waves. Physicists have learned to live with this nagging inconsistency, and we must do the same.

Light is sometimes defined as waves of electromagnetic energy between 380 and 760 *nanometers* (billionths of a meter) in length (see Figure 6.2). There is nothing special about these wavelengths except that the human visual system responds to them. In fact, some animals can see wavelengths that we cannot (see Gehring, 2014). For example, rattlesnakes can see *infrared waves*, which are too long for humans to see; as a result, they can see warm-blooded prey in what for us would be complete darkness. So, if we were writing this text for rattlesnakes, we would be forced to provide a different definition of light for them.

Wavelength and intensity are two properties of light that are of particular interest—wavelength because it plays an important role in the perception of color, and intensity because it plays an important role in the perception of brightness. In everyday language, the concepts of *wavelength* and *color* are often used interchangeably, as are *intensity* and *brightness*. For example, we commonly refer to an intense light with a wavelength of 700 nanometers as being a bright red light (see Figure 6.2), when in fact it is our perception of the light, not the light itself, that is bright and red. We know that these distinctions may seem trivial to you now, but by the end of the chapter you will appreciate their importance.

Figure 6.2 The electromagnetic spectrum and the colors that had been associated with wavelengths visible to humans.

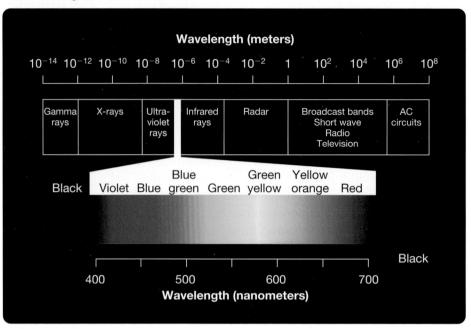

Pupil and Lens

LO 6.1 Explain how the pupil and the lens can affect the image that falls on the retina.

The amount of light reaching the *retinas* is regulated by the donut-shaped bands of contractile tissue, the *irises*, which give our eyes their characteristic color (see Figure 6.3). Light enters the eye through the *pupil*, the hole in the iris. The adjustment of pupil size in response to changes in illumination represents a compromise between

Figure 6.3 The human eye. Light enters the eye through the pupil, whose size is regulated by the iris. The iris gives the eye its characteristic color—blue, brown, or other.

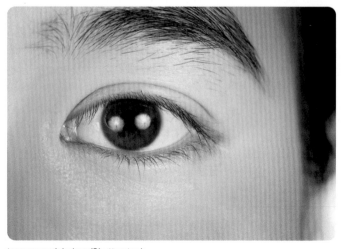

tarapong srichaiyos/Shutterstock

sensitivity (the ability to detect the presence of dimly lit objects) and **acuity** (the ability to see the details of objects). When the level of illumination is high and sensitivity is thus not important, the visual system takes advantage of the situation by constricting the pupils. When the pupils are constricted, the image falling on each retina is sharper and there is a greater *depth of focus*; that is, a greater range of depths is simultaneously kept in focus on the retinas. However, when the level of illumination is too low to adequately activate the receptors, the pupils dilate to let in more light, thereby sacrificing acuity and depth of focus.

Behind each pupil is a *lens*, which focuses incoming light on the retina (see Figure 6.4). When we direct our gaze at something near, the tension on the ligaments holding each lens in place is adjusted by the **ciliary muscles**, and the lens assumes its natural cylindrical shape. This increases the ability of the lens to *refract* (bend) light and thus brings close objects into sharp focus. When we focus on a distant object, the lens is flattened. The process of adjusting the configuration of the lenses to bring images into focus on the retina is called **accommodation**.

Eye Position and Binocular Disparity

LO 6.2 **Explain why some vertebrates have one eye on each side of their head, whereas other vertebrates have their eyes mounted side-by-side on the front of their heads. Also, explain the importance of binocular disparity.**

No description of the eyes of vertebrates would be complete without a discussion of their most obvious feature: the fact that they come in pairs. One reason vertebrates have two eyes is that vertebrates have two sides: left and right. By having one eye on each side, which is by far the most common arrangement, vertebrates can see in almost every direction without moving their heads. But then why do some vertebrates, including humans, have their eyes mounted side-by-side on the front of their heads? (See the first Check It Out demonstration on the next page.) This arrangement sacrifices the ability to see behind so that what is in front can be viewed through both eyes simultaneously—an arrangement that is an important basis for our visual system's ability to create three-dimensional perceptions (to see depth) from two-dimensional retinal images (see Baden, Euler, & Berens, 2020).

Figure 6.4 The human eye, a product of approximately 600 million years of evolution.

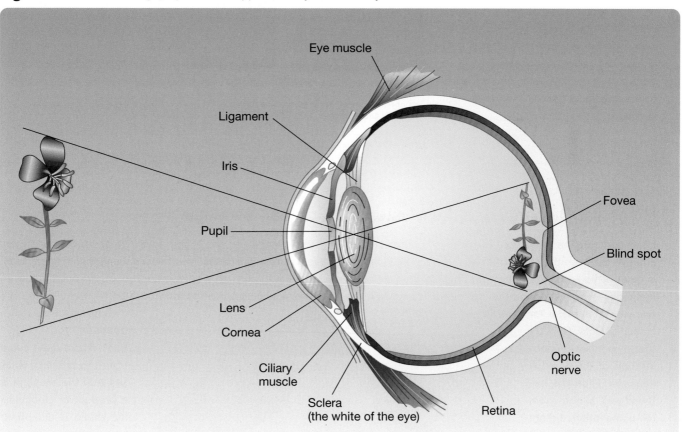

Based on Lamb, T. D., Collin, S. P., & Pugh, E. N. (2007). Evolution of the vertebrate eye: Opsins, photoreceptors, retina and eye cup. *Nature Reviews Neuroscience*, 8, 960–975.

Journal Prompt 6.1

Why do you think the two-eyes-on-the-front arrangement has evolved in some species but not in others? (After you've written your answer, see the first Check It Out demonstration below for more on this issue.)

The movements of your eyes are coordinated so that each point in your visual world is projected to corresponding points on your two retinas. To accomplish this, your eyes must *converge* (turn slightly inward); convergence is greatest when you are inspecting things that are close. But the positions of the images on your two retinas can never correspond exactly because your two eyes do not view the world from exactly the same position. **Binocular disparity**—the difference in the position of the same image on the two retinas—is greater for close objects than for distant objects; therefore, your visual system can use the degree of binocular disparity to construct one three-dimensional perception from two two-dimensional retinal images (see Lappin, 2014). (Look at the second Check It Out demonstration below.)

Check It Out
The Position of Eyes

Here you see three animals whose eyes are on the front of their heads (a human, an owl, and a lion) and three whose eyes are on the sides of their heads (an antelope, a canary, and a squirrel). Why do a few vertebrate species have their eyes side-by-side on the front of the head while most species have one eye on each side?

In general, predators tend to have the front-facing eyes because this enables them to accurately perceive how far away prey animals are; prey animals tend to have side-facing eyes because this gives them a larger field of vision and the ability to see predators approaching from most directions.

Top row from left: Guiziou Franck/Hemis/Alamy Stock Photo; Matthew Cuda/Alamy Stock Photo; C.K. Lorenz/Science Source
Bottom row from left: Naomi Engela Le Roux/123RF; Vasiliy Vishnevskiy/123RF; Colin Varndell/Nature Picture Library

Check It Out
Binocular Disparity and the Mysterious Cocktail Sausage

If you compare the views from each eye (by quickly closing one eye and then the other) of objects at various distances in front of you—for example, your finger held at different distances—you will notice that the disparity between the two views is greater for closer objects. Now try the mysterious demonstration of the cocktail sausage. Face the farthest wall in the room (or some other distant object) and bring the tips of your two pointing fingers together at arm's length in front of you—with the backs of your fingers away from you (unless you prefer sausages with fingernails). Now, with both eyes open, look through the notch between your touching fingertips, but focus on the wall. Do you see the cocktail sausage between your fingertips? Where did it come from? To prove to yourself that the sausage is a product of binocularity, make it disappear by shutting one eye. Warning: Do not eat this sausage.

The Retina and Translation of Light into Neural Signals

After light passes through the pupil and the lens, it reaches the retina. The retina converts light to neural signals, conducts them toward the CNS, and participates in the processing of the signals (Hoon et al., 2014; Seung & Sümbül, 2014).

Structure of the Retina

LO 6.3 **Describe the structure of the retina and name the cell types that make up the retina.**

Figure 6.5 illustrates the fundamental cellular structure of the retina. The retina is composed of five different types of neurons: **receptors**, **horizontal cells**, **bipolar cells**, **amacrine cells**, and **retinal ganglion cells**. Each of these five types of retinal neurons comes in a variety of subtypes: More than 60 different kinds of retinal neurons have been identified (see Cepko, 2015; Seung & Sümbül, 2014), including about 30 different retinal ganglion cells (see Baden et al., 2016). Notice that the amacrine cells and the horizontal cells are specialized for *lateral communication* (communication across the major channels of sensory input). Retinal neurons communicate both chemically via synapses and electrically via gap junctions (see Pereda, 2014).

Also notice in Figure 6.5 that the retina is in a sense inside-out: Light reaches the receptor layer only after passing through the other layers. Then, once the receptors have been activated, the neural message is transmitted back out through the retinal layers to the retinal ganglion cells, whose axons project across the outside of the retina before gathering together in a bundle and exiting the eyeball. This inside-out arrangement creates two visual problems: One is that the incoming light is distorted by the retinal tissue through which it must pass before reaching the receptors. The other is that for the bundle of retinal ganglion cell axons to leave the eye, there must be a gap in the receptor layer; this gap is called the **blind spot**.

The first of these two problems is minimized by the fovea (see Figure 6.6). The **fovea** is an indentation, about 0.33 centimeter in diameter, at the center of the retina;

Figure 6.5 The cellular structure of the mammalian retina.

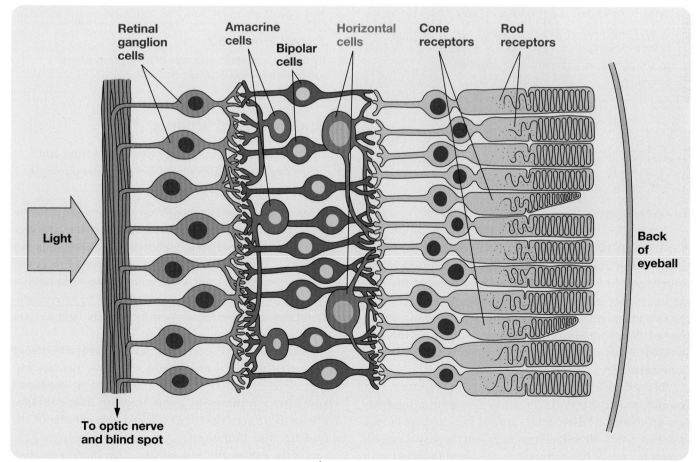

Figure 6.6 A section of the retina. The fovea is the indentation at the center of the retina; it is specialized for high-acuity vision.

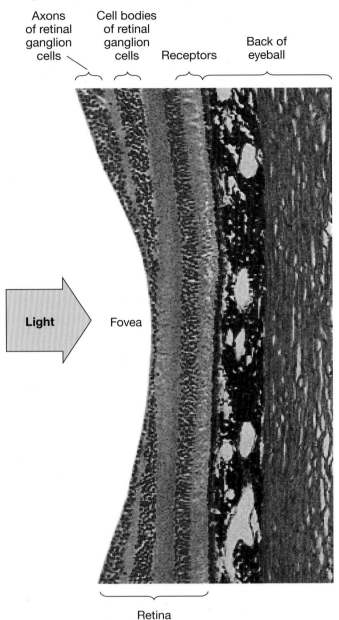

Ralph C. Eagle, Jr./Science Source

it is the area of the retina that is specialized for high-acuity vision (for seeing fine details). The thinning of the retinal ganglion cell layer at the fovea reduces the distortion of incoming light. The blind spot, the second of the two visual problems created by the inside-out structure of the retina, requires a more creative solution—which is illustrated in the accompanying Check It Out demonstration.

In the Check It Out demonstration, you will experience **completion** (or *filling in*). The visual system uses information provided by the receptors around the blind spot to fill in the gaps in your retinal images. When the visual system detects a straight bar going into one side of the blind spot and another straight bar leaving the other side, it fills in the missing bit for you; and what you see is a continuous straight bar, regardless of what is actually there. The completion phenomenon is one of the most compelling demonstrations that the visual system does much more than make a faithful copy of the external world.

It is a mistake to think that completion is merely a response to blind spots. Indeed, completion plays an important role in normal vision (see Murray & Herrmann, 2013; Weil & Rees, 2011). When you look at an object, your visual system does not conduct an image of that object from your retina to your cortex. Instead, it extracts key information about the object—primarily information about its edges and their location—and conducts that information to the cortex, where a perception of the entire object is created from that partial information. For example, the color and brightness of large unpatterned surfaces are not perceived directly but are filled in (completed) by a completion process called **surface interpolation** (the process by which we perceive surfaces; the visual system extracts information about edges and from it infers the appearance of large surfaces). The central role of surface interpolation in vision is an extremely important but counterintuitive concept. We suggest you read this paragraph again and think about it. Are your creative thinking skills developed enough to feel comfortable with this new way of thinking about your own visual system?

Journal Prompt 6.2
Try to give a specific example of a situation where surface interpolation would occur.

Cone and Rod Vision

LO 6.4 Describe the duplexity theory of vision and explain the differences between the photopic and scotopic systems.

You likely noticed in Figure 6.5 that there are two different types of receptors in the human retina: cone-shaped receptors called **cones** and rod-shaped receptors called **rods** (see Figure 6.7). The existence of these two types of receptors puzzled researchers until 1866, when it was first noticed that species active only in the day tend to have cone-only retinas, and species active only at night tend to have rod-only retinas.

From this observation emerged the **duplexity theory of vision**—the theory that cones and rods mediate different kinds of vision. **Photopic vision** (cone-mediated vision) predominates in good lighting and provides high-acuity (finely detailed) colored perceptions of the world. In dim illumination, there is not enough light to reliably excite the cones, and the more sensitive

Check It Out
Your Blind Spot and Completion

First, prove to yourself that you do have areas of blindness that correspond to your retinal blind spots. Close your left eye and stare directly at the A below, trying as hard as you can to not shift your gaze. While keeping the gaze of your right eye fixed on the A, hold the text at different distances from you until the black dot to the right of the A becomes focused on your blind spot and disappears at about 13 centimeters (5 inches).

If each eye has a blind spot, why is there not a black hole in your perception of the world when you look at it with one eye? You will discover the answer by focusing on B with your right eye while holding the text at the same distance as before. Suddenly, the broken line to the right of B will become whole. Now focus on C at the same distance with your right eye. What do you see?

Figure 6.7 Cones and rods. The red colored cells are cones; the blue colored cells are rods.

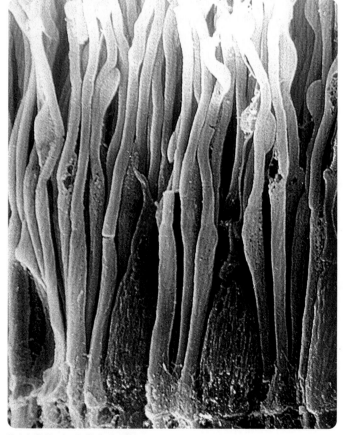

Ralph C. Eagle, Jr./Science Source

scotopic vision (rod-mediated vision) predominates. However, the sensitivity of scotopic vision is not achieved without cost: Scotopic vision lacks both the detail and the color of photopic vision.

The differences between photopic and scotopic vision result in part from a difference in the way the two systems are "wired." As Figure 6.8 illustrates, there is a large difference in convergence between the two systems. In the scotopic system, the output of several hundred rods converges on a single retinal ganglion cell, whereas in the photopic system, only a few cones converge on each retinal ganglion cell. As a result, when dim light stimulates many rods simultaneously, the outputs of this stimulation converge and *summate (add)* on the retinal ganglion cell. On the other hand, the effects of the same dim light applied to a sheet of cones cannot summate to the same degree, and the retinal ganglion cells may not respond at all to the light.

The convergent scotopic system pays for its high degree of sensitivity with a low level of acuity. When a retinal ganglion cell that receives input from hundreds of rods changes its firing, the brain has no way of knowing which portion of the rods contributed to the change. Although a more intense light is required to change the firing of a retinal ganglion cell that receives signals from cones, when such a retinal ganglion cell does react, there is less ambiguity about the location of the stimulus that triggered the reaction.

Figure 6.8 Schematic representations of the convergence of cones or rods on a retinal ganglion cell. There is a low degree of convergence in cone-fed pathways and a high degree of convergence in rod-fed pathways.

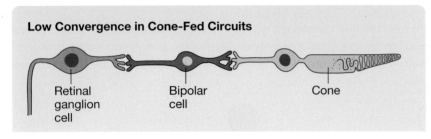

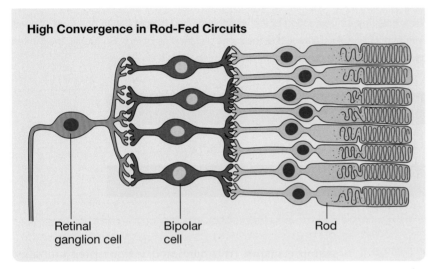

Figure 6.9 The distribution of cones and rods over the human retina. The figure illustrates the number of cones and rods per square millimeter as a function of distance from the center of the fovea.

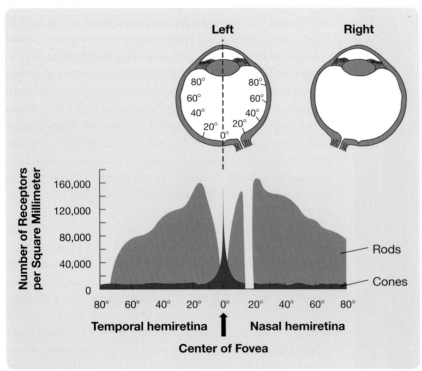

Based on Lindsay, P. H., & Norman, D. A. (1977). *Human Information Processing* (2nd ed.). New York, NY: Academic Press.

Cones and rods differ in their distribution on the retina. As Figure 6.9 illustrates, there are no rods at all in the fovea, only cones. At the boundaries of the foveal indentation, the proportion of cones declines markedly, and there is an increase in the number of rods. The density of rods reaches a maximum at 20 degrees from the center of the fovea. Notice that there are more rods in the *nasal hemiretina* (the half of each retina next to the nose) than in the *temporal hemiretina* (the half of each retina next to the temples).

Spectral Sensitivity

LO 6.5 **Explain the difference between the photopic and scotopic spectral sensitivity curves and explain how that difference can account for the Purkinje effect.**

Generally speaking, more intense lights appear brighter. However, wavelength also has a substantial effect on the perception of brightness. Because our visual systems are not equally sensitive to all wavelengths in the visible spectrum, lights of the same intensity but of different wavelengths can differ markedly in brightness. A graph of the relative brightness of lights of the same intensity presented at different wavelengths is called a *spectral sensitivity curve*.

By far the most important thing to remember about spectral sensitivity curves is that humans and other animals with both cones and rods have two of them: a **photopic spectral sensitivity curve** and a **scotopic spectral sensitivity curve**. The photopic spectral sensitivity of humans can be determined by having subjects judge the relative brightness of different wavelengths of light shone on the fovea. Their scotopic spectral sensitivity can be determined by asking subjects to judge the relative brightness of different wavelengths of light shone on the periphery of the retina at an intensity too low to activate the few peripheral cones located there.

The photopic and scotopic spectral sensitivity curves of human subjects are plotted in Figure 6.10. Under photopic conditions, notice that the visual system is maximally sensitive to wavelengths of about 560 nanometers; thus, under photopic conditions, a light at 500 nanometers would have to be much more intense than one at 560 nanometers to be seen as equally bright. In contrast, under scotopic conditions, the visual system is maximally sensitive to wavelengths of about 500 nanometers; thus, under scotopic conditions, a light of 560 nanometers would have to be much more intense than one at 500 nanometers to be seen as equally bright.

Because of the difference in photopic and scotopic spectral sensitivity, an interesting visual effect can be observed during the transition from photopic to scotopic vision. In 1825, Jan Purkinje described the following occurrence, which has become known as the **Purkinje effect** (pronounced "pur-KIN-jee"). One evening, just before dusk, while Purkinje was walking in his garden, he noticed how his yellow and red flowers appeared brighter in relation to his blue ones. What amazed him was that just a few minutes later, as the sun went down, the relative brightness of his flowers had somehow been reversed; the entire scene, when viewed at night, appeared completely in shades of gray, but most of the blue flowers appeared as brighter shades of gray than the yellow and red ones. Can you explain this shift in relative brightness by referring to the photopic and scotopic spectral sensitivity curves in Figure 6.10?

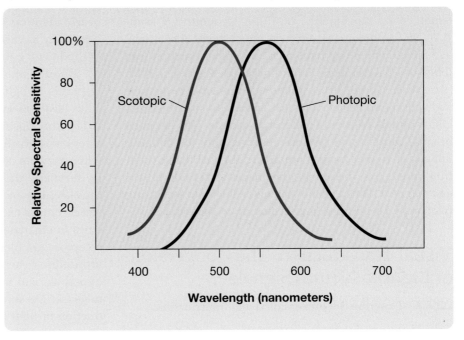

Figure 6.10 Human photopic (cone) and scotopic (rod) spectral sensitivity curves. The peak of each curve has been arbitrarily set at 100 percent.

Eye Movement

LO 6.6 Describe the three types of involuntary fixational eye movements and explain what happens when all eye movements are blocked.

If cones are responsible for mediating high-acuity color vision under photopic conditions, how can they accomplish their task when most of them are crammed into the fovea? (See Figure 6.9.) Look around you. What you see is not a few colored details at the center of a grayish scene. You seem to see an expansive, richly detailed, lavishly colored world. How can such a perception be the product of a photopic system that, for the most part, is restricted to a few degrees in the center of your *visual field* (the entire area that you can see at a particular moment)? The Check It Out demonstration provides a clue. It shows that what we see is determined not just by what is projected on the retina at that instant. Although we are not aware of it, the eyes continually scan the visual field, and our visual perception at any instant is a summation of recent visual information. It is because of this *temporal integration* that the world does not vanish momentarily each time we blink.

Check It Out
Periphery of Your Retina Does Not Mediate the Perception of Detail or Color

Close your left eye, and with your right eye stare at the fixation point (+) at a distance of about 13 centimeters (5 inches) from the page. Be very careful that your gaze does not shift. You will notice when your gaze is totally fixed that it is difficult to see detail and color at 20 degrees or more from the fixation point because there are so few cones there. Now look at the page again with your right eye, but this time without fixing your gaze. Notice the difference that eye movement makes to your vision.

W   F   D M E A ✛
50°   40°   30° 20° 10° 5° 0°

Our eyes continuously move even when we try to keep them still (i.e., fixated). Involuntary **fixational eye movements** are of three kinds: tremor, drifts, and **saccades** (small jerky movements, or flicks; pronounced "sah-KAHDS"). Although we are normally unaware of fixational eye movements, they have a critical visual function (see Ibbotson & Krekelberg, 2011; Spering & Carrasco, 2015; Zirnsak & Moore, 2014). When eye movements or their main effect (movement of images on the retina) are blocked, visual objects begin to fade and disappear. This happens because most visual neurons respond only to changing images; if retinal images are artificially stabilized (kept from moving on the retina), the images start to disappear and reappear. Thus, eye movements enable us to see during fixation by keeping the images moving on the retina.

Visual Transduction: The Conversion of Light to Neural Signals

LO 6.7   Describe the process of visual transduction.

Transduction is the conversion of one form of energy to another. *Visual transduction* is the conversion of light to neural signals by the visual receptors. A breakthrough in the study of visual transduction came in 1876 when a red *pigment* (a pigment is any substance that absorbs light) was extracted from rods. This pigment had a curious property. When the pigment—which became known as **rhodopsin**—was exposed to continuous intense light, it was *bleached* (lost its color) and lost its ability to absorb light, but when it was returned to the dark, it regained both its redness and its light-absorbing capacity.

It is now clear that rhodopsin's absorption of light (and the accompanying bleaching) is the first step in rod-mediated vision. Evidence comes from demonstrations that the degree to which rhodopsin absorbs light in various situations predicts how humans see under the very same conditions. For example, it has been shown that the degree to which rhodopsin absorbs lights of different wavelengths is related to the ability of humans and other animals with rods to detect the presence of different wavelengths of light under scotopic conditions.

Figure 6.11 illustrates the relationship between the **absorption spectrum** of rhodopsin and the human scotopic spectral sensitivity curve. The fact that the two curves are nearly identical leaves little doubt that, in dim light, our sensitivity to various wavelengths is a direct consequence of rhodopsin's ability to absorb them.

Rhodopsin is a G-protein–coupled receptor that responds to light rather than to neurotransmitter molecules (see Krishnan & Schiöth, 2015; Manglik & Kobilka, 2014). Rhodopsin receptors, like other G-protein–coupled receptors, initiate a cascade of intracellular chemical events when they are activated (see Figure 6.12). When rods are in darkness, their sodium channels are partially open, thus keeping the rods slightly depolarized and allowing a steady flow of excitatory glutamate neurotransmitter molecules to emanate from them. However, when rhodopsin receptors are bleached by light, the resulting cascade of intracellular chemical events closes the sodium channels, hyperpolarizes the rods, and reduces the release of glutamate (see Oesch, Kothmann, & Diamond, 2011). The transduction of light by rods exemplifies an important point: Signals are often transmitted through neural systems by decreases in activity.

Figure 6.11 The absorption spectrum of rhodopsin compared with the human scotopic spectral sensitivity curve.

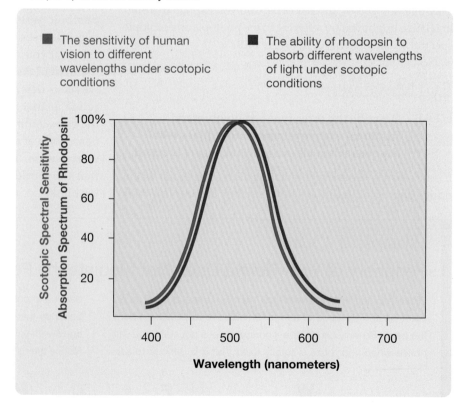

Figure 6.12 The inhibitory response of rods to light. When light bleaches rhodopsin molecules, the rods' sodium channels close; as a result, the rods become hyperpolarized and release less glutamate.

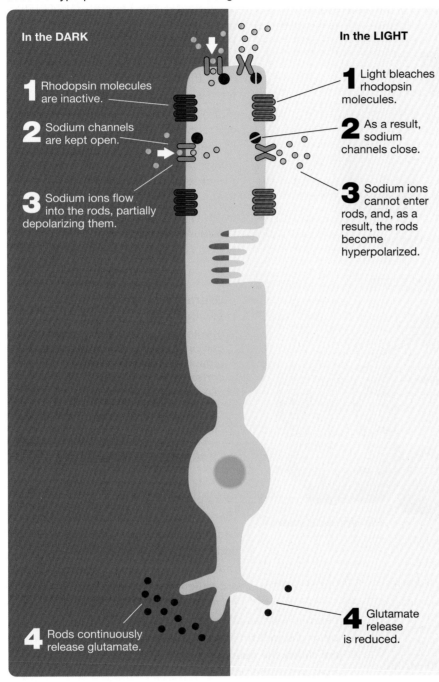

In the DARK

1 Rhodopsin molecules are inactive.

2 Sodium channels are kept open.

3 Sodium ions flow into the rods, partially depolarizing them.

4 Rods continuously release glutamate.

In the LIGHT

1 Light bleaches rhodopsin molecules.

2 As a result, sodium channels close.

3 Sodium ions cannot enter rods, and, as a result, the rods become hyperpolarized.

4 Glutamate release is reduced.

From Retina to Primary Visual Cortex

Many pathways in the brain carry visual information. By far the largest and most thoroughly studied visual pathways are the **retina-geniculate-striate pathways**, which conduct signals from each retina to the **primary visual cortex** (also known as *striate cortex* or *V1*) via the **lateral geniculate nuclei** of the thalamus.

Retina-Geniculate-Striate System

LO 6.8   Describe the components and layout of the retina-geniculate-striate system.

About 90 percent of axons of retinal ganglion cells become part of the retina-geniculate-striate pathways (see Tong, 2003). No other sensory system has such a predominant pair (left and right) of pathways to the cortex. The organization of these visual pathways is illustrated in Figure 6.13. Examine it carefully.

The main idea to take away from Figure 6.13 is that all signals from the left visual field reach the right primary visual cortex, either ipsilaterally from the *temporal hemiretina* of the right eye or contralaterally (via the *optic chiasm*) from the *nasal hemiretina* of the left eye—and that the opposite is true of all signals from the right visual field. Each lateral geniculate nucleus has six layers, and each layer receives input from all parts of the contralateral visual field of one eye. In other words, each lateral geniculate nucleus receives visual input only from the contralateral visual field; three layers receive input from one eye, and three receive input from the other. Most of the lateral geniculate neurons that project to the primary visual cortex terminate in the lower part of cortical layer IV (see Muckli & Petro, 2013), producing a characteristic stripe, or striation, when viewed in cross section—hence, primary visual cortex is often referred to as *striate cortex*. Note: Figure 6.13 depicts only the axonal projections from the lateral geniculate nuclei to the primary visual cortex, but there are just as many projections from the primary visual cortex to the lateral geniculate nuclei.

Figure 6.13 The retina-geniculate-striate system: the neural projections from the retinas through the lateral geniculate nuclei to the left and right primary visual cortex (striate cortex). The colors indicate the flow of information from various parts of the visual fields of each eye to various parts of the visual system.

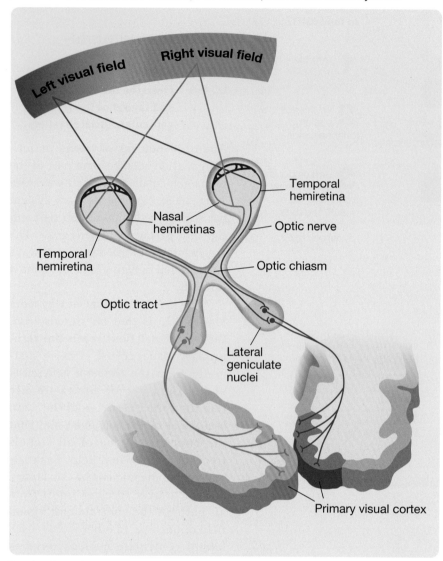

Based on Netter, F. H. (1962). *The CIBA Collection of Medical Illustrations. Vol. 1, The Nervous System.* New York, NY: CIBA.

Retinotopic Organization

LO 6.9   In the context of the retina-geniculate-striate system, explain what is meant by retinotopic.

The retina-geniculate-striate system is **retinotopic**; each level of the system is organized like a map of the retina. This means two stimuli presented to adjacent areas of the retina excite adjacent neurons at all levels of the system (see Kremkow et al., 2016). The retinotopic layout of the primary visual cortex has a disproportionately large representation of the fovea; although the fovea is only a small part of the retina, a relatively large proportion of the primary visual cortex (about 25 percent) is dedicated to the analysis of its input.

A dramatic demonstration of the retinotopic organization of the primary visual cortex was provided by Dobelle, Mladejovsky, and Girvin (1974). They implanted an array of electrodes in the primary visual cortex of patients who were blind because of damage to their eyes. If electrical current was administered simultaneously through an array of electrodes forming a shape, such as a cross, on the surface of a patient's cortex, the patient reported "seeing" a glowing image of that shape. This finding and recent research on retinal implants (see Roska & Sahel, 2018; Wood, 2018) could be the basis for the development of visual prostheses that could benefit many blind people (see Shepherd et al., 2013).

Journal Prompt 6.3

How does the prosthesis developed by Dobelle et al. (1974) demonstrate the retinotopic organization of the primary visual cortex?

The M and P Channels

LO 6.10   Describe the M and P channels.

Not apparent in Figure 6.13 is the fact that at least two parallel channels of communication flow through each lateral geniculate nucleus. One channel runs through the top four layers. These layers are called the **parvocellular layers** (or *P layers*) because they are composed of neurons with small cell bodies (*parvo* means "small"). The other channel runs through the bottom two layers, which are called the **magnocellular layers** (or *M layers*) because they are composed of neurons with large cell bodies (*magno* means "large").

The parvocellular neurons are particularly responsive to color, fine pattern details, and stationary or slowly moving objects. In contrast, the magnocellular neurons are particularly responsive to movement. Cones provide the majority of the input to the P layers, whereas rods provide the majority of the input to the M layers.

The parvocellular and magnocellular neurons project to different areas in the lower part of layer IV of the striate cortex. In turn, these M and P areas of lower layer IV project to different areas of visual cortex.

Scan Your Brain

This is a good place to pause and scan your brain to check your knowledge on the basics of the visual process before you move on to studying how we perceive edges and color. Fill in the following blanks with the most appropriate terms. The correct answers are provided at the end of the exercise. Before proceeding, review material related to your errors and omissions.

1. Light reflected from objects enters the eye though the _____.

2. Depending on how close or far away an object is, the lens is adjusted using the _____ muscles.

3. About 25 percent of the primary visual cortex is dedicated to analyzing input from the _____.

4. _____ is the process by which the lenses adjust their shape to bring images to focus on the retina.

5. The difference in the position of the same image on the two retinas is called _____.

6. Nasal hemiretina of the right visual field project to the _____ hemisphere, while temporal hemiretina project to the _____ hemisphere.

7. The location on the retina where a bundle of cell axons leaves the eye is called the _____.

8. The theory that the two types of retinal receptors, rods and cones, mediate different kinds of vision is called the _____ theory.

9. There are two _____ channels that run through the lateral geniculate nucleus called M and P channels.

Scan Your Brain answers: (1) pupil, (2) ciliary, (3) fovea, (4) Accommodation, (5) binocular disparity, (6) ipsilateral, contralateral, (7) blind spot, (8) duplexity, (9) parallel.

Seeing Edges

Edge perception (seeing edges) does not sound like a particularly important topic, but it is. Edges are the most informative features of any visual display because they define the extent and position of the various objects in it. Given the importance of perceiving visual edges and the unrelenting pressure of natural selection, it is not surprising that the visual systems of many species are particularly good at edge perception.

Before considering the visual mechanisms underlying edge perception, it is important to appreciate exactly what a visual edge is. In a sense, a visual edge is nothing: It is merely the place where two different areas of a visual image meet. Accordingly, the perception of an edge is really the perception of a contrast between two adjacent areas of the visual field. This module reviews the perception of edges (the perception of contrast) between areas that differ from one another in brightness (i.e., that show brightness contrast).

Contrast Enhancement

LO 6.11 Describe contrast enhancement.

Carefully examine the stripes in Figure 6.14. The intensity graph in the figure indicates what is there—a series of homogeneous stripes of different intensity. But this is not exactly what you see, is it? What you see is indicated in the brightness graph. Adjacent to each edge, the brighter stripe looks brighter than it really is and the darker stripe looks darker than it really is. The nonexistent stripes of brightness and darkness running adjacent to the edges are called *Mach bands*; they enhance the contrast at each edge and make the edge easier to see.

It is important to appreciate that **contrast enhancement** is not something that occurs just in books. Although we are normally unaware of it, every edge we look at is highlighted for us by the contrast-enhancing mechanisms of our nervous systems. In effect, our perception of edges is better than the real thing (as determined by measurements of the physical properties of the light entering our eyes).

Figure 6.14 The illusory bands visible in this figure are often called Mach bands, although Mach used a different figure to generate them in his studies (see Eagleman, 2001).

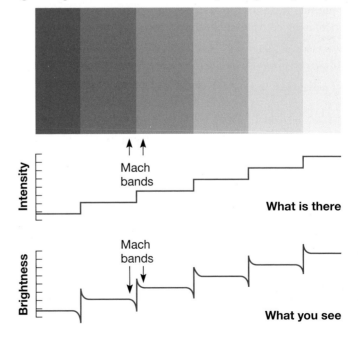

Receptive Fields of Visual Neurons: Hubel & Wiesel

LO 6.12 Define the term *receptive field* and describe the methods used by David Hubel and Torsten Wiesel to map the receptive fields of visual system neurons.

The Nobel Prize–winning research of David Hubel and Torsten Wiesel (see Hubel & Wiesel, 2004) is the fitting focus of this discussion of seeing edges. Their research revealed much about the neural mechanisms of vision, and their method has been adopted by subsequent generations of sensory neurophysiologists.

Hubel and Wiesel's influential method is a technique for studying single neurons in the visual systems of laboratory animals—their research subjects were cats and monkeys. First, the tip of a microelectrode is positioned near a single neuron in the part of the visual system under investigation. During testing, eye movements are blocked by paralyzing the eye muscles, and the images on a screen in front of the subject are focused sharply on the retina by an adjustable lens. The next step in the procedure is to identify the receptive field of the neuron. The **receptive field** of a visual neuron is the area of the visual field within which it is possible for a visual stimulus to influence the firing of that neuron. The final step in the method is to record the responses of the neuron to various simple stimuli within its receptive field in order to characterize the types of stimuli that most influence its activity. Then the electrode is advanced slightly, and the entire process of identifying and characterizing the receptive field properties is repeated for another neuron, and then for another, and another, and so on. The general strategy is to begin by studying neurons near the receptors and gradually work up through "higher" and "higher" levels of the system in an effort to understand the increasing complexity of the neural responses at each level.

Receptive Fields of the Retina-Geniculate-Striate System: Hubel & Wiesel

LO 6.13 Describe the work of Hubel & Wiesel that helped to characterize the receptive fields of retinal ganglion cells, lateral geniculate neurons, and striate neurons of lower layer IV.

Hubel and Wiesel (1979) began their studies of visual system neurons by recording from the three levels of the retina-geniculate-striate system: first from retinal ganglion cells, then from lateral geniculate neurons, and finally from the striate neurons of lower layer IV. They tested the neurons with stationary spots of *achromatic* (uncolored) light shone on the retina. They found little change in the receptive fields as they worked through the levels.

When Hubel and Wiesel compared the receptive fields recorded from retinal ganglion cells, lateral geniculate nuclei, and lower layer IV striate neurons, four commonalties were readily apparent:

- At each level, the receptive fields in the foveal area of the retina were smaller than those at the periphery; this is consistent with the fact that the fovea mediates fine-grained (high-acuity) vision.

- All the neurons (retinal ganglion cells, lateral geniculate neurons, and lower layer IV neurons) had receptive fields that were circular.

- All the neurons were **monocular**; that is, each neuron had a receptive field in one eye but not the other.

- Many neurons at each of the three levels of the retina-geniculate-striate system had receptive fields that comprised an excitatory area and an inhibitory area separated by a circular boundary.

Let us explain this last point—it is important. When Hubel and Wiesel shone a spot of achromatic light onto the various parts of the receptive fields of a neuron in the retina-geniculate-striate pathway, they discovered two different responses. The neuron responded with either "on" firing or "off" firing, depending on the location of the spot of light in the receptive field. That is, the neuron either displayed a burst of firing when the light was turned on (*"on" firing*), or it displayed an inhibition of firing when the light was turned on and a burst of firing when it was turned off (*"off" firing*).

For most of the retinal ganglion cells, lateral geniculate nuclei, and lower layer IV striate neurons, the reaction—"on" firing or "off" firing—to a light in a particular part of the receptive field was quite predictable. It depended on whether they were on-center cells or off-center cells, as illustrated in Figure 6.15.

On-center cells respond to lights shone in the central region of their receptive fields with "on" firing and to lights shone in the periphery of their receptive fields with inhibition, followed by "off" firing when the light is turned off. **Off-center cells** display the opposite pattern: They respond with inhibition and "off" firing in response to lights in the center of their receptive fields and with "on" firing to lights in the periphery of their receptive fields.

In effect, on-center and off-center cells respond best to contrast. Figure 6.16 illustrates this point. The most effective way to influence the firing rate of an on-center or off-center cell is to maximize the contrast between the center and the periphery of its receptive field by illuminating either the entire center or the entire surround (periphery) while leaving the other region completely dark. Diffusely illuminating the entire receptive field has little effect on firing. Hubel and Wiesel thus concluded that one function of many of the neurons in the retina-geniculate-striate system is to respond to the degree of brightness contrast between

Figure 6.15 The receptive fields of an on-center cell and an off-center cell.

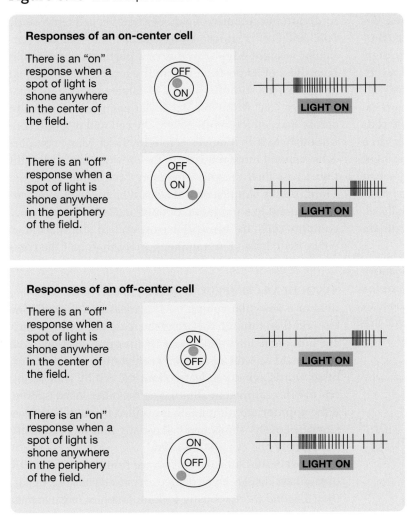

Responses of an on-center cell

There is an "on" response when a spot of light is shone anywhere in the center of the field.

There is an "off" response when a spot of light is shone anywhere in the periphery of the field.

Responses of an off-center cell

There is an "off" response when a spot of light is shone anywhere in the center of the field.

There is an "on" response when a spot of light is shone anywhere in the periphery of the field.

the two areas of their receptive fields (see Livingstone & Hubel, 1988).

Before moving on, notice one important thing from Figures 6.15 and 6.16 about visual system neurons: Most are continually active, even when there is no visual input (see Lee et al., 2013). Indeed, spontaneous activity is characteristic of most cerebral neurons, and responses to external stimuli consume only a small portion of the energy required for ongoing brain activity (see Zhang & Raichle, 2010).

Receptive Fields of Primary Visual Cortex Neurons: Hubel & Wiesel

LO 6.14 Describe the work of Hubel & Wiesel that characterized the receptive fields of simple and complex cells in the primary visual cortex.

The striate cortex neurons you just read about—that is, the neurons of lower layer IV—have receptive fields unlike those of the vast majority of striate neurons. The receptive fields of most primary visual cortex neurons fall into one of two classes: simple or complex. Neither of these classes includes the neurons of lower layer IV.

Figure 6.16 The responses of an on-center cell to contrast.

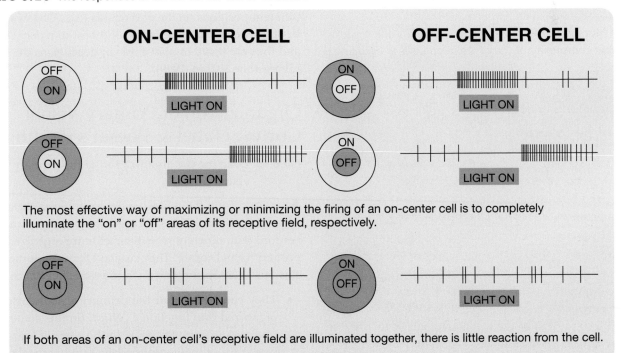

The most effective way of maximizing or minimizing the firing of an on-center cell is to completely illuminate the "on" or "off" areas of its receptive field, respectively.

If both areas of an on-center cell's receptive field are illuminated together, there is little reaction from the cell.

SIMPLE STRIATE CELLS. **Simple cells**, like lower layer IV neurons, have receptive fields that can be divided into antagonistic "on" and "off" regions and are thus unresponsive to diffuse light. And like lower layer IV neurons, they are all monocular. The main difference is that the borders between the "on" and "off" regions of the cortical receptive fields of simple cells are straight lines rather than circles. Several examples of receptive fields of simple cortical cells are presented in Figure 6.17. Notice that simple cells respond best to bars of light in a dark field, dark bars in a light field, or single straight edges between dark and light areas; that each simple cell responds maximally only when its preferred straight-edge stimulus is in a particular position and in a particular orientation (see Vidyasagar & Eysel, 2015); and that the receptive fields of simple cortical cells are rectangular rather than circular.

COMPLEX STRIATE CELLS. **Complex cells** are more numerous than simple cells. Like simple cells, complex cells have rectangular receptive fields, respond best to straight-line stimuli in a specific orientation, and are unresponsive to diffuse light. However, complex cells differ from simple cells in three important ways. First, they have larger receptive fields. Second, it is not possible to divide the receptive fields of complex cells into static "on" and "off" regions: A complex cell responds to a particular straight-edge stimulus of a particular orientation regardless of its position within the receptive field of that cell. Thus, if a stimulus (e.g., a 45-degree bar of light) that produces "on" firing in a particular complex cell is swept across that cell's receptive field, the cell will respond continuously to it as it moves across the field. Many complex cells respond more robustly to the movement of a straight line across their receptive fields in a particular direction. Third, unlike simple cortical cells, which are all monocular (respond to stimulation of only one of the eyes), many complex cells are **binocular** (respond to stimulation of either eye). Indeed, in monkeys, more than half the complex cortical cells are binocular.

BINOCULAR COMPLEX STRIATE CELLS. If the receptive field of a binocular complex cell is measured through one eye and then through the other, the receptive fields in each eye turn out to have almost exactly the same position in the visual field as well as the same orientation preference. In other words, what you learn about the cell by stimulating one eye is confirmed by stimulating the other. What is more, if the appropriate stimulation is applied through both eyes simultaneously, a binocular cell usually fires more robustly than if only one eye is stimulated.

Most of the binocular cells in the primary visual cortex of monkeys display some degree of *ocular dominance*; that is, they respond more robustly to stimulation of one eye than they do to the same stimulation of the other. In addition, some binocular cells fire best when the preferred stimulus is presented to both eyes at the same time but in slightly different positions on the two retinas (e.g., Ohzawa, 1998). In other words, these cells respond best to *retinal disparity* and thus are likely to play a role in depth perception (e.g., Livingstone & Tsao, 1999).

Organization of Primary Visual Cortex: Hubel & Wiesel's Findings

LO 6.15 Describe the organization of the primary visual cortex.

After describing the receptive fields of visual cortex neurons, Hubel and Wiesel focused their analyses on how neurons with different receptive fields are organized in the primary visual cortex. They reached three important conclusions about the organization of primate visual cortex:

- They concluded that the primary visual cortex was organized into functional *vertical* (in this context, *vertical* means at right angles to the cortical layers) columns: All of the neurons in the same vertical column

Figure 6.17 Examples of receptive fields of simple striate cells.

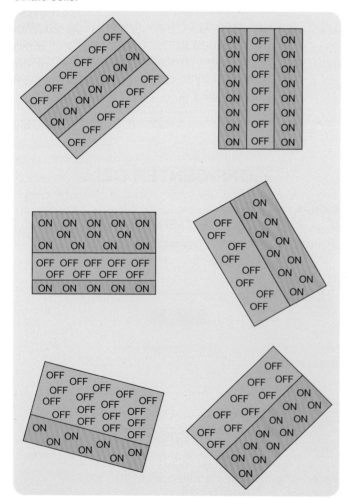

respond to stimuli applied to the same area of the retina, are dominated by the same eye (if they display dominance or monocularity), and "prefer" the same straight-line angles (if they display a preference for straight-line stimuli).

- They found that the location of various functional columns in primary visual cortex is influenced by the location on the retina of the column's visual fields, by the dominant eye of the column, and by the column's preferred straight-line angle. Hubel and Wiesel concluded that all of the functional columns in the primary visual cortex that analyze input from one area of the retina are clustered together, that half of a cluster receives input from the left eye and the other half receives input from the right eye, and that each cluster includes neurons with preferences for straight-line stimuli of various orientations.

- As Hubel and Wiesel's studies progressed from retina, to thalamus, to lower layer IV of visual cortex, to simple cortical cells, to complex cortical cells, the "preferences" of the neurons became more complex. Hubel and Wiesel concluded that this occurred because neurons with simpler preferences converged on neurons with more complex preferences.

Now that you know a bit about how the visual cortex is organized, you are in a better position to think constructively about Mrs. Richards's fortification illusions.

The Case of Mrs. Richards, Revisited

There was obviously a disturbance in Mrs. Richards's visual system. But where? And what kind of disturbance? And why the straight lines? A simple test located the disturbance. Mrs. Richards was asked to shut one eye and then the other and to report what happened to her illusion when she changed eyes. The answer was "Nothing." This suggested that the disturbance was cortical because the visual cortex is the first part of the retina-geniculate-striate system that contains neurons that receive input from both eyes.

This hypothesis was confirmed by a few simple calculations: The gradual acceleration of the illusion as it spread out to the periphery is consistent with a wave of disturbance expanding from the "foveal area" of the primary visual cortex to its boundaries at a constant rate of about 3 millimeters per minute—the illusion accelerated because proportionally less visual cortex is dedicated to receiving signals from the periphery of the visual field.

And why the lines? Would you expect anything else from an area of the cortex whose elements appear to have a preference for straight-line stimuli?

Changing Concept of the Characteristics of Visual Receptive Fields

LO 6.16 Describe how views about the receptive fields of retinal ganglion cells and lateral geniculate neurons have recently changed.

Since the seminal work of Hubel and Wiesel, there has been a massive amount of research focused on further characterizing the receptive fields of cells in the retina-geniculate-striate system. In general, these studies have discovered that receptive fields are much more complex than was originally recognized.

In this section, we will explain some of the important new findings about the receptive fields of retinal ganglion cells and lateral geniculate neurons—cells that were once believed to have just on-center and off-center receptive fields. As you will see, the characteristics described by Hubel and Wiesel, and others, turned out to be much simpler than what is currently known to be the case.

RETINAL GANGLION CELLS. Recent analyses of the visual processing of stimuli in primates and mice have shown that they have about 20 and 40 distinct sorts of retinal ganglion cells, respectively—each with its own sort of receptive field. In addition to the on-center and off-center receptive fields documented by Hubel and Wiesel, there are also retinal ganglion cells with receptive fields that are selective to one or more of the following: (1) uniform illumination, (2) orientation, (3) motion, and (4) direction of motion (see Baden et al., 2016; Ding et al., 2016; Hillier et al., 2017; Mauss et al., 2017; Morrie & Feller, 2016).

LATERAL GENICULATE CELLS. There has also been recent attention to the functional roles of lateral geniculate cells. Analyses have shown that some cells in the lateral geniculate nucleus have receptive fields that are sensitive to more than just contrast (i.e., on-center and off-center receptive fields). Indeed, those cells have receptive fields that are sensitive to one or more of the following: (1) orientation, (2) motion, and (3) direction of motion (see Sun, Tan, & Ji, 2016). As you may have recognized already, these receptive fields are similar to those of retinal ganglion cells (see Ghodrati, Khaligh-Razavi, & Lehky, 2017).

Changing Concept of Visual Receptive Fields: Contextual Influences in Visual Processing

LO 6.17 Describe the changing view of visual system receptive fields.

Most investigations of the responsiveness of visual system neurons have been based on two implicit assumptions. The first is that the mechanisms of visual processing can be best identified by studies using simplified, controllable, artificial stimuli (see Einhäuser & König, 2010). The second is that the receptive field properties of each neuron are static, unchanging properties of that neuron. Research that has employed video clips of real scenes involving natural movement suggests that neither of these assumptions is correct (see Haslinger et al., 2012) at any of the three levels of the retina-geniculate-striate system (see Rivlin-Etzion, Grimes, & Rieke, 2018; Rose & Bonhoeffer, 2018).

Studies of the responses of visual cortex to natural scenes—just the type of scenes the visual system has evolved to perceive—indicate that the response of a visual cortex neuron depends not only on the stimuli in its receptive field but also on the larger scene in which these stimuli are embedded (see Baden, Euler, & Berens, 2020; Coen-Cagli, Kohn, & Schwartz, 2015; Dumoulin & Knapen, 2018; Iacaruso, Gasler, & Hofer, 2017). The influences on a visual neuron's activity that are caused by stimuli outside the neuron's receptive field are generally referred to as *contextual influences* (see Gilbert & Li, 2013). Contextual influences can take many forms depending on the exact timing, location, and shape of the visual stimuli under investigation and on the ambient light levels (e.g., Iacaruso, Gasler, & Hofer, 2017; Roth et al., 2016; Tikidji-Hamburyan et al., 2015); and the nature of such contextual influences is dependent on prior exposure to them (see Khan et al., 2018; Thompson et al., 2017). Moreover, contextual signals associated with particular actions or states—such as engaging in locomotion or being exposed to stimuli that have biological relevance (e.g., a stimulus that has previously been associated with the presentation of food)—can help shape the properties of a receptive field (see Fiser et al., 2016; Hrvatin et al., 2018; Khan & Hofer, 2018).

Think for a moment about the implications that the discovery of contextual influences has for understanding how visual receptive fields function. A visual neuron's receptive field was initially assumed to be a property of the neuron resulting from the hardwired convergence of neural circuits; now, a neuron's receptive field is viewed as a plastic property of the neuron that is continually fine-tuned on the basis of prior experience and current signals from the animal's environment.

Journal Prompt 6.4

Why should natural scenes be used to study visual system neurons?

Seeing Color

Color is one of the most obvious qualities of human visual experience. So far in this chapter, we have largely limited our discussion of vision to black, white, and gray. Black is experienced when there is an absence of light; the perception of white is produced by an intense mixture of a wide range of wavelengths in roughly equal proportions; and the perception of gray is produced by the same mixture at lower intensities. In this module, we deal with the perception of colors such as blue, green, and yellow. The correct term for colors is *hues*, but in everyday language they are referred to as colors; and for the sake of simplicity, we will do the same.

What is there about a visual stimulus that determines the color we perceive? To a large degree, the perception of an object's color depends on the wavelengths of light that it reflects into the eye. Figure 6.2 is an illustration of the colors associated with individual wavelengths; however, outside the laboratory, one never encounters objects that reflect single wavelengths. Sunlight and most sources of artificial light contain complex mixtures of most visible wavelengths. Most objects absorb the different wavelengths of light that strike them to varying degrees and reflect the rest. The mixture of wavelengths that objects reflect influences our perception of their color, but it is not the entire story—as you are about to learn.

Component and Opponent Processing

LO 6.18   Describe the component and opponent-process theories of color vision.

The **component theory** (*trichromatic theory*) of color vision was proposed by Thomas Young in 1802 and refined by Hermann von Helmholtz in 1852. According to this theory, there are three different kinds of color receptors (cones), each with a different spectral sensitivity, and the color of a particular stimulus is presumed to be encoded by the ratio of activity in the three kinds of receptors. Young and Helmholtz derived their theory from the observation that any color of the visible spectrum can be matched by a mixing together of three different wavelengths of light in different proportions. This can be accomplished with any three wavelengths, provided that the color of any one of them cannot be matched by a mixing of the other two. The fact that three is normally the minimum number of different wavelengths necessary to match every color suggested that there were three types of receptors.

Another theory of color vision, the **opponent-process theory** of color vision, was proposed by Ewald Hering in 1878. He suggested that there are two different classes of cells in the visual system for encoding color and another class for encoding brightness. Hering hypothesized that each of the three classes of cells encoded two complementary color perceptions. One class of color-coding cells signaled red by changing its activity in one direction (e.g., hyperpolarization) and signaled red's complementary color, green, by changing its activity in the other direction (e.g., depolarization). Another class of color-coding cells was hypothesized to signal blue and its complement, yellow, in the same opponent fashion; and a class of brightness-coding cells was hypothesized to similarly signal both black and white. **Complementary colors** are pairs of colors (e.g., green light and red light) that produce white or gray when combined in equal measure.

Hering based his opponent-process theory of color vision on several behavioral observations. One was that complementary colors cannot exist together: There is no such thing as bluish yellow or reddish green (see Billock & Tsou, 2010). Another was that the afterimage produced by staring at red is green and vice versa, and the afterimage produced by staring at yellow is blue and vice versa (try the Check It Out demonstration).

Check It Out
Complementary Afterimages

Have you ever noticed complementary afterimages? To see them, stare at the fixation point (x) in the left panel for 1 minute without moving your eyes, then quickly shift your gaze to the fixation point in the right panel. In the right panel, you will see four squares whose colors are complementary to those in the left panel.

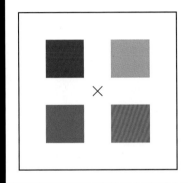
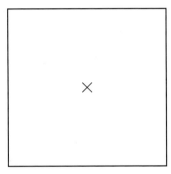

are most sensitive to medium wavelengths, and some are most sensitive to long wavelengths (see Shevell & Kingdom, 2008), but all three respond to most of the wavelengths of the visible spectrum.

Although the coding of color by cones seems to operate on a purely component basis (see Jameson, Highnote, & Wasserman, 2001), there is evidence of opponent processing of color at all subsequent levels of the retina-geniculate-striate system. That is, at all subsequent levels, there are cells that respond in one direction (e.g., increased firing) to one color and in the opposite direction (e.g., decreased firing) to its complementary color (see Joesch & Meister, 2016).

Most primates are *trichromats* (possessing three color vision photopigments). Most other mammals are *dichromats* (possessing two color vision photopigments)—they lack the photopigment sensitive to long wavelengths and thus have difficulty seeing light at the red end of the visible spectrum (see Figure 6.2). In contrast, some birds, fish, and reptiles have four photopigments, and some insects have five or more photopigments—the dragonfly wins first place with ten (see Kelber, 2016).

In a remarkable study, Jacobs and colleagues (2007) inserted into mice a gene for a third photopigment, thus converting them from dichromats to trichromats. Behavioral tests indicated that the transgenic mice had acquired the ability to see additional wavelengths of light.

A somewhat misguided debate raged for many years between supporters of the component and opponent theories of color vision. We say "misguided" because it was fueled more by the adversarial predisposition of scientists than by the incompatibility of the two theories. In fact, research subsequently proved that both color-coding mechanisms coexist in our visual systems (see DeValois et al., 2000).

It was the development in the early 1960s of a technique for measuring the absorption spectrum of the photopigment contained in a single cone that allowed researchers (e.g., Wald, 1964) to confirm the conclusion that Young had reached more than a century and a half before. They found that there are indeed three different kinds of cones in the retinas of those vertebrates with good color vision, and they found that each of the three has a different photopigment with its own characteristic absorption spectrum. As Figure 6.18 illustrates, some cones are most sensitive to short wavelengths, some

Figure 6.18 The absorption spectra of the three classes of cones.

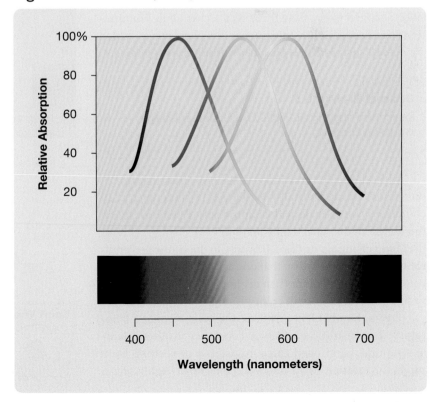

Color Constancy and the Retinex Theory

LO 6.19 Describe Land's demonstration of color constancy and explain his retinex theory.

Neither component nor opponent processing can account for the single most important characteristic of color vision: color constancy. **Color constancy** refers to the fact that the perceived color of an object is not a simple function of the wavelengths reflected by it.

Color constancy is an important—but much misunderstood—concept. Let us explain it with an example. As I (SB) write this at 6:15 on a January morning, it is dark outside, and I am working in my office by the light of a tiny incandescent desk lamp. Later in the morning, when students start to arrive, I will turn on my nasty fluorescent office lights; and then, in the afternoon, when the sun is brighter on my side of the building, I will turn off the lights and work by natural light. The point is that because these light sources differ markedly in the wavelengths they emit, the wavelengths reflected by various objects in my office—my blue shirt, for example—change substantially during the course of the day. However, although the wavelengths reflected by my shirt change markedly, its color does not—my shirt will be just as blue in mid-morning and in late afternoon as it is now. Color constancy is the tendency for an object to stay the same color despite major changes in the wavelengths of light that it reflects.

Although the phenomenon of color constancy is counterintuitive, its advantage is obvious. Color constancy improves our ability to tell objects apart in a memorable way so that we can respond appropriately to them; our ability to recognize objects would be greatly lessened if their color changed every time there was a change in illumination (see Foster, 2011). In essence, if it were not for color constancy, color vision would have little survival value.

Journal Prompt 6.5

Describe a specific example where color constancy would be adaptive.

Although color constancy is an important feature of our vision, we are normally unaware of it. Under everyday conditions, we have no way of appreciating just how much the wavelengths reflected by an object can change without the object changing its color. It is only in the controlled environment of the laboratory that one can fully appreciate that color constancy is more than an important factor in color vision: It is the essence of color vision.

Edwin Land (1977) developed several dramatic laboratory demonstrations of color constancy. In these demonstrations, Land used three adjustable projectors. Each projector emitted only one wavelength of light: one a short-wavelength light, one a medium-wavelength light, and one a long-wavelength light. Thus, it was clear that only three wavelengths of light were involved in the demonstrations. Land shone the three projectors on a test display like the one in Figure 6.19. (These displays are called *Mondrians* because they resemble the paintings of the Dutch artist Piet Mondrian.)

Land found that adjusting the amount of light emitted from each projector—and thus, the amount of light of each wavelength being reflected by the Mondrian—had no effect at all on the perception of its colors. For example, in one demonstration Land used a photometer to measure the amounts of the three wavelengths reflected by a rectangle judged to be pure blue by his participants. He then adjusted the emittance of the projectors, and he

Figure 6.19 The method of Land's (1977) color-vision experiments. Participants viewed Mondrians illuminated by various proportions of three different wavelengths: a short wavelength, a middle wavelength, and a long wavelength.

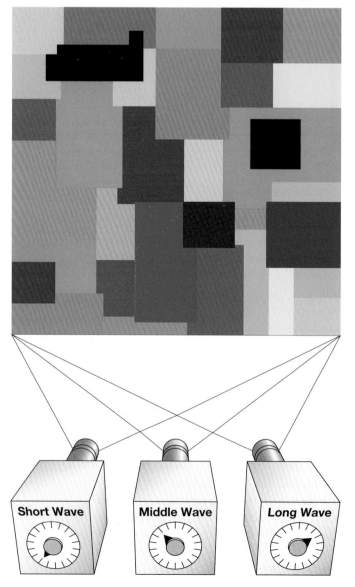

measured the wavelengths reflected by a red rectangle on a different Mondrian, until the wavelengths were exactly the same as those that had been reflected by the blue rectangle on the original. When he showed this new Mondrian to his participants, the red rectangle looked—you guessed it—red, even though it reflected exactly the same wavelengths as had the blue rectangle on the original Mondrian.

The point of Land's demonstration is that blue objects stay blue, green objects stay green, and so forth, regardless of the wavelengths they reflect. This color constancy occurs as long as the object is illuminated with light that contains some short, medium, and long wavelengths (such as daylight, firelight, and virtually all manufactured lighting) and as long as the object is viewed as part of a scene, not in isolation.

According to Land's **retinex theory** of color vision, the color of an object is determined by its *reflectance*—the proportion of light of different wavelengths that a surface reflects. Although the wavelengths of light reflected by a surface change dramatically with changes in illumination, the efficiency with which a surface absorbs each wavelength and reflects the unabsorbed portion does not change. According to the retinex theory, the visual system calculates the reflectance of surfaces, and thus, perceives their colors, by comparing the light reflected by adjacent surfaces in at least three different wavelength bands (short, medium, and long). You learned in the previous module that the context plays an important role in the processing of spatial contrast (i.e., edges), and the retinex theory suggests that the context plays a similarly important role in processing color (Shevell & Kingdom, 2008).

Scan Your Brain

The striate cortex is the main entrance point of visual signals to the cortex. In the upcoming module, we will follow visual signals to other parts of the cortex. This is a good point to pause and review what you have learned. Draw a line to connect each term in the first column with the closely related word or phrase in the second column. Each term should be linked to only one item in the second column. The correct answers are provided at the end of this exercise. Before proceeding, review material related to your errors and omissions.

1. contrast enhancement
2. simple cortical cells
3. complex cortical cells
4. ocular dominance columns
5. component
6. opponent
7. retinex

   a. many are binocular
   b. complementary afterimages
   c. striate cortex
   d. reflectance
   e. static on and off areas
   f. three
   g. Mach bands

Scan Your Brain answers: (1) g, (2) e, (3) a, (4) c, (5) f, (6) b, (7) d.

Cortical Mechanisms of Vision and Conscious Awareness

So far, you have followed the major visual pathways from the eyes to the primary visual cortex, but there is much more to the human visual system—we are visual animals. The entire occipital cortex as well as large areas of temporal cortex and parietal cortex are involved in vision (see Figure 6.20).

Figure 6.20 The visual areas of the human cerebral cortex.

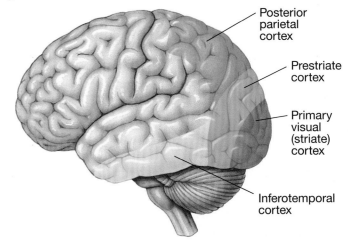

Posterior parietal cortex

Prestriate cortex

Primary visual (striate) cortex

Inferotemporal cortex

Three Different Classes of Visual Cortex

LO 6.20 Describe the three classes of visual cortex and identify their locations in the brain.

Visual cortex is often considered to be of three different classes. *Primary visual cortex*, as you have learned, is that area of cortex that receives most of its input from the visual relay nuclei of the thalamus (i.e., from the lateral geniculate nuclei). Areas of **secondary visual cortex** are those that receive most of their input from the primary visual cortex, and areas of **visual association cortex** are those that receive input from areas of secondary visual cortex as well as from the secondary areas of other sensory systems.

The primary visual cortex is located in the posterior region of the occipital lobes, much of it hidden from view in the longitudinal fissure. Most areas of secondary visual cortex are located in two general regions: in the prestriate cortex and in the inferotemporal cortex. The **prestriate cortex** is the band of tissue in the occipital lobe that surrounds the primary visual cortex. The **inferotemporal cortex** is the cortex of the inferior temporal lobe. Areas of association cortex that receive visual input are located in several parts of the cerebral cortex, but the largest single area is in the **posterior parietal cortex**.

The major flow of visual information in the cortex is from primary visual cortex to the various areas of secondary visual cortex to the areas of association cortex. As one moves up this visual hierarchy, the neurons have larger receptive fields and the stimuli to which the neurons respond are more specific and more complex.

Damage to Primary Visual Cortex: Scotomas and Completion

LO 6.21 Explain what happens when an area of primary visual cortex is damaged.

Damage to an area of the primary visual cortex produces a **scotoma**—an area of blindness—in the corresponding area of the contralateral visual field of both eyes (see Figure 6.13). Neurological patients with suspected damage to the primary visual cortex are usually given a **perimetry test**. While the patient's head is held motionless on a chin rest, the patient stares with one eye at a fixation point on a screen. A small dot of light is then flashed on various parts of the screen, and the patient presses a button to record when the dot is seen. Then, the entire process is repeated for the other eye. The result is a map of the right and left visual field of each eye, which indicates any areas of blindness. Figure 6.21 illustrates the perimetric maps of each eye of a man with a bullet wound in his left primary visual

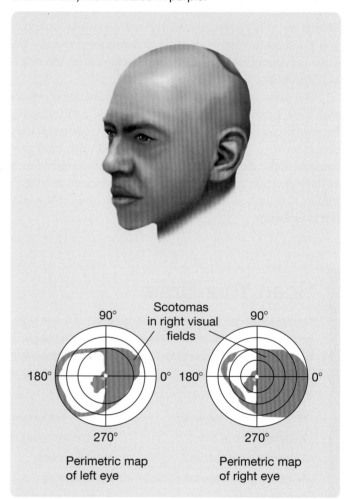

Figure 6.21 The perimetric maps of a man with a bullet wound in his left primary visual cortex. The scotomas (areas of blindness) are indicated in purple.

Based on Teuber, H.-L., Battersby, W. S., & Bender, M. B. (1960). Recovery of function after brain injury in man. In *Outcomes of severe damage to the nervous system. CIBA Foundation Symposium 34.* Amsterdam, Netherlands: Elsevier North-Holland.

cortex. Notice the massive scotoma in the right visual field of each eye.

Many patients with scotomas are not consciously aware of their deficits. One factor that contributes to this lack of awareness is completion. A patient with a scotoma who looks at a complex figure, part of which lies in the scotoma, often reports seeing a complete image (see Silvanto, 2014). In some cases, this completion may depend on residual visual capacities in the scotoma; however, completion also occurs in cases in which this explanation can be ruled out. For example, patients who are **hemianopsic** (having a scotoma covering half of the visual field) may see an entire face when they focus on a person's nose, even when the side of the face in the scotoma has been covered by a blank card.

Consider the completion phenomenon experienced by the esteemed physiological psychologist Karl Lashley (1941). He often developed a large scotoma next to his fovea during migraine attacks (see Figure 6.22).

Figure 6.22 The completion of a migraine-induced scotoma, as described by Karl Lashley (1941).

Lashley's scotoma   What Lashley saw

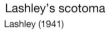
Lashley (1941)

Blindsight is the ability to respond to visual stimuli in a scotoma with no conscious awareness of them (de Gelder, 2010; Leopold, 2012; Silvanto, 2014). Of all visual abilities, perception of motion is most likely to survive damage to primary visual cortex (see Schmid & Maier, 2015). For example, a patient with blindsight might reach out and grab a moving object in her scotoma, all the while claiming not to see the object.

If blindsight confuses you, imagine how it confuses people who experience it. Consider, for example, the reactions to blindsight of D.B., a patient who was blind in his left visual field following surgical removal of his right occipital lobe (Weiskrantz et al., 1974).

The Physiological Psychologist Who Made Faces Disappear

Talking with a friend, I [Karl Lashley, an early leader in brain-behavior research] glanced just to the right of his face wherein his head disappeared. His shoulders and necktie were still visible but the vertical stripes on the wallpaper behind him seemed to extend down to the necktie. It was impossible to see this as a blank area when projected on the striped wallpaper of uniformly patterned surface, although any intervening object failed to be seen. (Lashley, 1941, p. 338)

Journal Prompt 6.6

Earlier you learned about surface interpolation. How is surface interpolation at work in this case study? (Hint: Take a close look at Figure 6.22.)

You probably equate perception with **conscious awareness**; that is, you might assume that if a person sees something, he or she will be consciously aware of seeing it. In everyday thinking, perceiving and being aware are inseparable processes: We assume that someone who has seen something will be able to acknowledge that he or she has seen it and be able to describe it. In the following pages, you will encounter examples of phenomena for which this is not the case: people who see things but have no conscious awareness of them. Blindsight is the first example.

Blindsight is sometimes displayed by patients with scotomas resulting from damage to primary visual cortex.

The Case of D.B., the Man Confused by His Own Blindsight

D.B. had no awareness of "seeing" in his blind left field. Despite this apparent left-field blindness, he could accurately reach for visual stimuli in his left field and could accurately differentiate between a horizontal or diagonal line in his left field if forced to "guess." When he was questioned about his vision in his left field, his most usual response was that he saw nothing. When he was shown a video of his accurate left-field performance through his good, right field, he was astonished and insisted he was just guessing.

Two neurological interpretations of blindsight have been proposed. One is that the striate cortex is not completely destroyed and the remaining islands of functional cells are capable of mediating some visual abilities in the absence of conscious awareness (see Wüst, Kasten, & Sabel, 2002). The other is that those visual pathways that ascend directly to the secondary visual cortex from subcortical visual structures without passing through the primary visual cortex are capable of maintaining some visual abilities in the absence of cognitive awareness (see Schmid & Maier, 2015). There is some support for both theories, but it is far from conclusive in either case (see Gross, Moore, & Rodman, 2004; Rosa, Tweedale, & Elston, 2000; Schärli, Harman, & Hogben, 1999a, 1999b). Indeed, it is possible that both mechanisms contribute to the phenomenon.

Functional Areas of Secondary and Association Visual Cortex

LO 6.22 Describe the areas of secondary visual cortex and association cortex involved in vision.

Secondary visual cortex and the portions of association cortex involved in visual analysis are both composed of many different areas, each specialized for a particular type of visual analysis. For example, in the macaque monkey, whose visual cortex has been most thoroughly mapped, there are 32 different functional areas of visual cortex; in addition to primary visual cortex, 24 areas of secondary visual cortex and 7 areas of association visual cortex have been identified. The neurons in each functional area respond most vigorously to different aspects of visual stimuli (e.g., to their color, movement, or shape); selective lesions to the different areas produce different visual losses; and there are anatomical and organizational differences among the areas (see Patel et al., 2014).

The various functional areas of secondary and association visual cortex in the macaque are prodigiously interconnected. Anterograde and retrograde tracing studies have identified more than 300 interconnecting pathways (see Markov & Kennedy, 2013). Connections between areas are virtually always reciprocal (see Gilbert & Li, 2013).

PET, fMRI, and evoked potentials (see Chapter 5) have been used to identify various areas of visual cortex in humans. The activity of volunteers' brains has been monitored while they inspect various types of visual stimuli. By identifying the areas of activation associated with various visual properties (e.g., movement or color), researchers have so far delineated about a dozen different functional areas of human visual cortex (see Grill-Spector & Mallach, 2004). A map of some of these areas is shown in Figure 6.23. Most are similar in terms of location, anatomical characteristics, and function to areas already identified in the macaque.

Dorsal and Ventral Streams

LO 6.23 Explain the difference between the dorsal and ventral streams and the functions that have been attributed to each stream by different theories.

As you have already learned, most visual information enters the primary visual cortex via the lateral geniculate nuclei. The information from the two lateral geniculate nuclei is received in the primary visual cortex, combined, and then segregated into multiple pathways that project separately to the various functional areas of secondary, and then association, visual cortex (see Horton & Sincich, 2004).

Figure 6.23 Some of the visual areas that have been identified in the human brain.

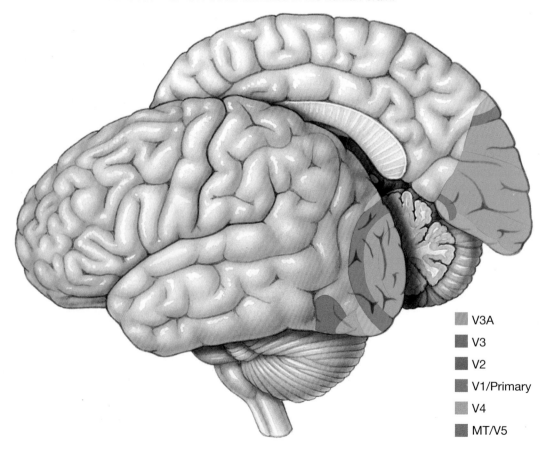

- ■ V3A
- ■ V3
- ■ V2
- ■ V1/Primary
- ■ V4
- ■ MT/V5

Many pathways that conduct information from the primary visual cortex through various specialized areas of secondary and association cortex can be thought of as components of two major streams: the dorsal stream and the ventral stream (Ungerleider & Mishkin, 1982). The **dorsal stream** flows from the primary visual cortex to the dorsal prestriate cortex to the posterior parietal cortex, and the **ventral stream** flows from the primary visual cortex to the ventral prestriate cortex to the inferotemporal cortex—see Figure 6.24.

Most visual cortex neurons in the dorsal stream respond most robustly to spatial stimuli, such as those indicating the location of objects or their direction of movement. In contrast, most neurons in the ventral stream respond to the characteristics of objects, such as color and shape (see Tompa & Sáry, 2010). Indeed, there are clusters of visual neurons in the ventral stream, and each cluster responds specifically to a particular class of objects—for example, most neurons in a particular cluster may respond to faces, whereas most neurons in another cluster might respond to animals (Haxby, 2006; Reddy & Kanwisher, 2006). Accordingly, Ungerleider and Mishkin (1982) proposed that the dorsal and ventral visual streams perform different visual functions. They suggested that the dorsal stream is involved in the perception of "where" objects are and the ventral stream is involved in the perception of "what" objects are.

A major implication of the **"where" versus "what" theory** of vision is that damage to some areas of cortex may abolish certain aspects of vision while leaving others unaffected. Indeed, the most convincing support for the influential "where" versus "what" theory has come from the comparison of the specific effects of damage to the dorsal and ventral streams (see Ungerleider & Haxby, 1994). Patients with damage to the posterior parietal cortex often have difficulty reaching accurately for objects they have no difficulty describing; conversely, patients with damage to the inferotemporal cortex often have no difficulty reaching accurately for objects they have difficulty describing.

Although the "where" versus "what" theory is widely accepted, there is an alternative interpretation for the same evidence (de Haan & Cowey, 2011; Haque et al., 2018; O'Reilly, 2010). Goodale and Milner (1992) argued that the primary difference between the dorsal and ventral streams is not the kinds of information they carry but the use to which that information is put. They suggested that the primary function of the dorsal stream is to direct behavioral interactions with objects, whereas the primary function of the ventral stream is to mediate the conscious perception of objects. Goodale and Milner's assertion has been termed the **"control of behavior" versus "conscious perception" theory** (see Logothetis & Sheinberg, 1996). One of the most interesting aspects of this theory is its evolutionary implication: Goodale (2004) suggested that the conscious awareness mediated by the ventral stream is one thing that distinguishes humans and their close relatives from their evolutionary ancestors.

The "control of behavior" versus "conscious perception" theory can readily explain the two major neuropsychological findings that are the foundation of the "where" versus "what" theory. Namely, the "control of behavior" versus "conscious perception" theory suggests that patients with dorsal stream damage may do poorly on tests of location and movement because most tests of location and movement involve performance measures, and that patients with ventral stream damage may do poorly on tests of visual recognition because most tests of visual recognition involve verbal responses, and thus, conscious awareness.

The major support for the "control of behavior" versus "conscious perception" theory is the confirmation of its two primary predictions: (1) that some patients with bilateral lesions to the ventral stream may have no conscious experience of seeing and yet be able to interact with objects under visual guidance, and (2) that some patients with bilateral lesions to the dorsal stream may consciously see objects but be unable to interact with them under visual guidance (see Figure 6.25). Following are two such cases.

Figure 6.24 Information about particular aspects of a visual display flow out of the primary visual cortex over many pathways. The pathways can be grouped into two general streams: dorsal and ventral.

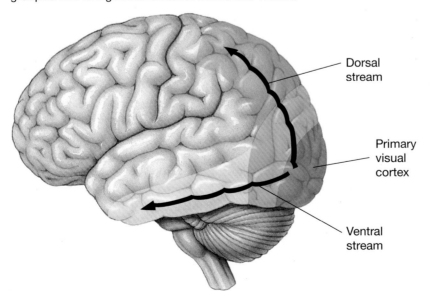

Dorsal stream

Primary visual cortex

Ventral stream

Figure 6.25 The "where" versus "what" and the "control of behavior" versus "conscious perception" theories make different predictions.

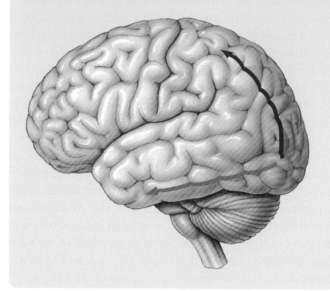

Dorsal and Ventral Streams: Two Theories and What They Predict

"Where" vs. "What" Theory

Dorsal stream specializes in visual spatial perception
Ventral stream specializes in visual pattern recognition

Predicts
- Damage to **dorsal stream** disrupts visual spatial perception
- Damage to **ventral stream** disrupts visual pattern recognition

"Control of Behavior" vs. "Conscious Perception" Theory

Dorsal stream specializes in visually guided behavior
Ventral stream specializes in conscious visual perception

Predicts
- Damage to **dorsal stream** disrupts visually guided behavior but not conscious visual perception
- Damage to **ventral stream** disrupts conscious visual perception but not visually guided behavior

D.F., the Woman Who Could Grasp Objects She Did Not Consciously See

D.F. has bilateral damage to her ventral prestriate cortex, thus interrupting the flow of the ventral stream; her case is described by Goodale and Milner (2004). Amazingly, she can respond accurately to visual stimuli that she does not consciously see.

Despite her inability to consciously recognize the size, shape and orientation of visual objects, D.F. displayed accurate hand movements directed at the same objects. For example, when she was asked to indicate the width of blocks with her index finger and thumb, her matches were variable and unrelated to the actual size of the blocks. However, when she was asked to pick up blocks of different sizes, the distance between her index finger and thumb changed appropriately with the size of the object. In other words, D.F. adjusted her hand to the size of objects she was about to pick up, even though she did not consciously perceive their size.

A similar dissociation occurred in her responses to the orientation of stimuli. When presented with a large slanted slot, she could not indicate the orientation of the slot either verbally or manually. However, she was as good as healthy volunteers at quickly placing a card in the slot, orienting her hand appropriately from the start of the movement.

A.T., the Woman Who Could Not Accurately Grasp Unfamiliar Objects That She Saw

The case of A.T. is in major respects complementary to that of D.F. A.T. is a woman with a lesion of the occipitoparietal region, which likely interrupts her dorsal route (Jeannerod et al., 1995).

A.T. was able to recognize objects and demonstrate their size with her fingers. In contrast, the preshape of her hand during object-directed movements was incorrect. As a consequence, she could not pick up objects between her fingertips—instead, the patient made awkward palmar grasps. Although A.T. could not preshape her hand to pick up neutral objects like blocks, when presented with a familiar object of standard size, like a lipstick, she grasped it with reasonable accuracy.

The widely held view that the dorsal stream controls visually guided behavior, whereas the ventral stream controls visual perception is currently under considerable scrutiny. For example, recent human and nonhuman studies have found that the dorsal stream carries information about objects (see Bi, Wang, & Caramazza, 2016; Freud, Plaut, & Behrmann, 2016) and that the ventral stream is active during interactions with objects (see Connor & Knierim, 2017). These findings are the exact opposite of the predictions of the "control of behavior" versus "conscious perception" theory and instead suggest that both visual streams carry information about what an object is and how to interact with it. Accordingly, the dichotomy of two functional streams within the higher visual areas may no longer be tenable.

The remainder of this chapter focuses on two neuropsychological conditions, *prosopagnosia* and *akinetopsia,* and the damage to visual cortical areas associated with each of them. Prosopagnosia refers to a difficulty in recognizing faces; akinetopsia to a difficulty in perceiving visual motion. Damage to the *fusiform face area* or the *occipital face area* has been linked to *prosopagnosia*, whereas damage to area *MT* (middle temporal) has been linked to *akinetopsia*.

Prosopagnosia

LO 6.24 Describe the phenomenon of prosopagnosia and discuss the associated theoretical issues.

Prosopagnosia, briefly put, is a visual agnosia for faces (see DeGutis et al., 2014) that can be acquired either during development (*developmental prosopagnosia*) or as a result of brain injury (*acquired prosopagnosia*; see Susilo & Duchaine, 2013). Let us explain. **Agnosia** is a failure of recognition (*gnosis* means "to know") that is not attributable to a sensory deficit or to verbal or intellectual impairment. A **visual agnosia** is a specific agnosia for visual stimuli. In other words, visual agnosics can see things, but they don't know what they are (see Albonico & Barton, 2019).

Visual agnosias are often specific to a particular aspect of visual input and are named accordingly; for example, *movement agnosia, object agnosia*, and *color agnosia* are difficulties in recognizing movement, objects, and color, respectively.

Prosopagnosics are visual agnosics with a specific difficulty in recognizing faces. They can recognize a face as a face, but they have problems recognizing whose face it is. In extreme cases, prosopagnosics cannot recognize themselves in photos.

IS PROSOPAGNOSIA SPECIFIC TO FACES?   The belief that prosopagnosia is a deficit specific to the recognition of faces has been challenged. To understand this challenge, you need to know that the diagnosis of prosopagnosia is typically applied to neuropsychological patients who have difficulty recognizing particular faces but can readily identify other test objects (e.g., a chair, a dog, or a tree). Surely, this is powerful evidence that prosopagnosics have recognition difficulties specific to faces. Not so. Pause for a moment and think about this evidence: It is seriously flawed.

Because prosopagnosics have no difficulty recognizing faces as faces, the fact that they can recognize chairs as chairs, pencils as pencils, and doors as doors is not relevant. The critical question is whether they can recognize which chair, which pencil, and which door. Careful testing of this sort usually reveals that their recognition deficits are not restricted to faces: For example, a farmer lost his ability to recognize particular cows when he lost his ability to recognize faces. This suggests that some prosopagnosic patients have a general problem recognizing specific objects that belong to complex classes of objects (e.g., particular automobiles or particular houses), not a specific problem recognizing faces (see Behrmann et al., 2005)—although in daily life the facial-recognition problems are likely to be the most problematic and, thus, obvious. Still, it is difficult to rule out the possibility that at least a few prosopagnosic patients have recognition deficits limited to faces. Indeed, several thorough case studies of prosopagnosia have failed to detect recognition deficits unrelated to faces (Duchaine & Nakayama, 2005). It seems likely that prosopagnosia is not a unitary disorder (Duchaine & Nakayama, 2006), and it appears that only some patients with this diagnosis have pattern-recognition deficits restricted to facial recognition (see Albonico & Barton, 2019; Robotham & Starrfelt, 2018).

R.P., a Typical Prosopagnosic

With routine testing, R.P. displayed a severe deficit in recognizing faces and in identifying facial expressions (Laeng & Caviness, 2001) but no other obvious recognition problems. If testing had stopped there, as it often does, it would have been concluded that R.P. is an agnosic with recognition problems specific to human faces. However, more thorough testing indicated that R.P. is deficient in recognizing all objects with complex curved surfaces, not just faces.

WHAT BRAIN PATHOLOGY IS ASSOCIATED WITH PROSOPAGNOSIA?   The diagnosis of acquired prosopagnosia is usually associated with damage to either or both of the *fusiform face area* and the *occipital face area* (see Haque et al., 2018). The **fusiform face area (FFA)** is located on the ventral surface of the boundary between the occipital and temporal lobes (see Figure 6.26). The FFA has been implicated in face identification because parts of it are selectively activated by human faces (see Collins & Olson, 2014; van den Hurk et al., 2015) and because electrical stimulation of this brain area in humans can metamorphose a viewed face into a completely different face (see Rangarajan et al., 2014). The **occipital face area (OFA)** is located on the ventral surface of the occipital lobe (see Figure 6.26). Reversible inactivation of the OFA by transcranial magnetic stimulation selectively disrupts the ability to discriminate between faces (see Freiwald, Duchaine, & Yovel, 2016). In addition to the FFA and OFA, recent evidence also

Figure 6.26 The location of the fusiform face area (FFA), the occipital face area (OFA), and area MT. Damage to the FFA or OFA is associated with prosopagnosia. Damage to area MT is associated with akinetopsia. The FFA and OFA are not visible in this figure; they lie on the ventral surface of the temporal lobe and occipital lobe, respectively.

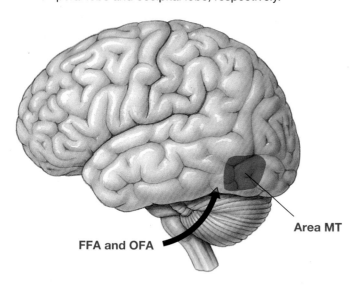

FFA and OFA

Area MT

points to an important role of the *lateral prefrontal cortex* in face identification (see Kornblith & Tsao, 2017). It makes sense that specialized mechanisms to perceive faces have evolved in the human brain because face perception plays such a major role in human social behavior (see Freiwald, Duchaine, & Yovel, 2016; Jack & Schyns, 2017; Powell, Kosakowski, & Saxe, 2018).

CAN PROSOPAGNOSICS PERCEIVE FACES IN THE ABSENCE OF CONSCIOUS AWARENESS? Tranel and Damasio (1985) were the first to demonstrate that prosopagnosics can recognize faces in the absence of conscious awareness. They presented a series of photographs to several patients, some familiar to the patients, some not. The patients claimed not to recognize any of the faces. However, when familiar faces were presented, the subjects displayed a large skin conductance response, which did not occur with unfamiliar faces, thus indicating that the faces were being unconsciously recognized by undamaged portions of the brain.

Akinetopsia

LO 6.25 Describe the phenomenon of akinetopsia and discuss the associated theoretical issues.

Akinetopsia is a deficiency in the ability to see movement progress in a normal smooth fashion—individuals affected by it only see periodic snapshots of the world. Akinetopsia can be either a permanent result of brain damage, or it can be the transient result of taking high doses of certain antidepressants (Haque et al., 2018; Horton, 2009).

Two Cases of Drug-Induced Akinetopsia

A 47-year-old depressed male receiving 100 mg of nefazodone twice daily reported a bizarre derangement of motion perception. Each moving object was followed by a trail of multiple freeze-frame images, which disappeared once the motion ceased. A 48-year-old female receiving 400 mg of nefazodone once daily at bedtime reported similar symptoms, with persistent multiple strobelike trails following moving objects. In both cases, stationary elements were perceived normally, indicating a selective impairment of the visual perception of motion. Vision returned to normal in both patients once the dosage was reduced.

When akinetopsia is the result of an acquired brain injury, it is often associated with damage to **area MT** (middle temporal area) of the cortex. The location of MT—near the junction of the temporal, parietal, and occipital lobes—is illustrated in Figure 6.26.

The function of MT appears to be the perception of motion. Given the importance of the perception of motion in primate survival, it is reasonable that an area of the visual system is dedicated to it. Some neurons at lower levels of the visual hierarchy (e.g., in the primary visual cortex) respond to movement as well as color and shape; however, they provide little information about the direction of movement because their receptive fields are so small. In contrast, 95 percent of the neurons of MT respond to specific directions of movement and little else. Also, each MT neuron has a large binocular receptive field, allowing it to track movement over a wide range.

The following four lines of research implicate MT in the visual perception of motion and damage to MT as a cause of akinetopsia:

- Patients with akinetopsia tend to have unilateral or bilateral damage to MT (Cooper et al., 2012; Haque et al., 2018).

- As measured by fMRI, activity in MT increases when humans view movement (see Haque et al., 2018; Zeki, 2015).

- Blocking activity in MT with transcranial magnetic stimulation (TMS) produces motion blindness (see Haque et al., 2018; Vetter, Grosbras, & Muckli, 2015).

- Electrical stimulation of MT in human patients induces the visual perception of motion (Blanke et al., 2002).

Themes Revisited

Vision is a creative process. Your visual system does not transmit complete and intact visual images of the world to the cortex. It carries information about a few critical features of the visual field—for example, information about location, movement, brightness contrast, and color contrast—and from these bits of information, it creates a perception far better than the retinal image in all respects and better than the external reality in some. Another main point is that your visual system can perceive things without your conscious awareness of them.

The Check It Out demonstrations in this chapter offered you many opportunities to experience firsthand important aspects of the visual process. We hope you checked them out and your experience made you more aware of the amazing abilities of your own visual system and the relevance of what you have learned in this chapter to your everyday life.

This chapter developed all four of the major themes. First, the evolutionary perspective theme was emphasized, largely because the majority of research on the neural mechanisms of human vision has been comparative and because thinking about the adaptiveness of various aspects of vision (e.g., color vision) has led to important insights.

Second, the thinking creatively theme was emphasized because the main point of the chapter was that we tend to think about our own visual systems in a way that is fundamentally incorrect: The visual system does not passively provide images of the external world; it extracts some features of the external world, and from these it creates our visual perceptions. Once you learn to think in this unconventional way, you will be able to better appreciate the amazingness of your own visual system.

Third, the clinical implications theme was developed through a series of clinical case studies: Mrs. Richards, who experienced fortification illusions before her migraine attacks; Karl Lashley, the physiological psychologist who used his scotoma to turn a friend's head into a wallpaper pattern; D.B., the man with blindsight; D.F., who showed by her accurate reaching that she detected the size, shape, and orientation of objects that she could not describe; A.T., who could describe the size and shape of objects she could not accurately reach for; R.P., a typical prosopagnosic; and two patients with akinetopsia induced by a particular antidepressant.

Fourth, this chapter touched on the neuroplasticity theme. The study of the visual system has focused on the receptive field properties of neurons in response to simple stimuli, and receptive fields have been assumed to be static. However, when natural visual scenes have been used in such studies, it has become apparent that each neuron's receptive field changes depending on the visual context.

This chapter was also filled with materials related to one of the emerging themes: consciousness. Two examples are the discussion of the distinction between the dorsal and ventral visual streams, and the unconscious processing of visual stimuli in cases of scotomas, blindsight, and prosopagnosia.

Key Terms

Light Enters the Eye and Reaches the Retina

Sensitivity, p. 155
Acuity, p. 155
Ciliary muscles, p. 155
Accommodation, p. 155
Binocular disparity, p. 156

The Retina and Translation of Light into Neural Signals

Receptors, p. 157
Horizontal cells, p. 157
Bipolar cells, p. 157
Amacrine cells, p. 157
Retinal ganglion cells, p. 157
Blind spot, p. 157
Fovea, p. 157

Completion, p. 158
Surface interpolation, p. 158
Cones, p. 158
Rods, p. 158
Duplexity theory, p. 158
Photopic vision, p. 158
Scotopic vision, p. 159
Photopic spectral sensitivity curve, p. 160
Scotopic spectral sensitivity curve, p. 160
Purkinje effect, p. 161
Fixational eye movements, p. 162
Saccades, p. 162
Transduction, p. 162
Rhodopsin, p. 162
Absorption spectrum, p. 162

From Retina to Primary Visual Cortex

Retina-geniculate-striate pathways, p. 163
Primary visual cortex, p. 163
Lateral geniculate nuclei, p. 163
Retinotopic, p. 164
Parvocellular layers, p. 164
Magnocellular layers, p. 164

Seeing Edges

Contrast enhancement, p. 165
Receptive field, p. 166
Monocular, p. 166
On-center cells, p. 166
Off-center cells, p. 166
Simple cells, p. 168

Chapter 7
Sensory Systems, Perception, and Attention

How You Know the World

Barry Diomede/Alamy Stock Photo

⌄ Chapter Overview and Learning Objectives

Principles of Sensory System Organization	**LO 7.1**	Name and define the three types of sensory cortex.
	LO 7.2	In the context of sensory system organization, explain what is meant by each of the following terms: *hierarchical organization*, *functional segregation*, and *parallel processing*. Summarize the current model of sensory system organization.
Auditory System	**LO 7.3**	Explain the relationship between the physical and perceptual dimensions of sound.
	LO 7.4	Describe the components of the human ear, and explain how sound is processed within its various structures.
	LO 7.5	Describe the major pathways that lead from the ear to the primary auditory cortex.

LO 7.6 Describe the organization of auditory cortex.

LO 7.7 Describe the effects of damage to the auditory system.

Somatosensory System: Touch and Pain

LO 7.8 Name some of the cutaneous receptors and explain the functional significance of fast versus slow receptor adaptation.

LO 7.9 Describe the two major somatosensory pathways.

LO 7.10 Describe the cortical somatosensory areas and their somatotopic layout.

LO 7.11 Name the areas of association cortex that somatosensory signals are sent to, and describe the functional properties of one of those areas.

LO 7.12 Describe the two major types of somatosensory agnosia.

LO 7.13 Describe the rubber-hand illusion and its neural mechanisms.

LO 7.14 Explain why the perception of pain is said to be paradoxical.

LO 7.15 Define neuropathic pain and describe some of its putative neural mechanisms.

Chemical Senses: Smell and Taste

LO 7.16 Describe two adaptive roles for the chemical senses.

LO 7.17 Describe the olfactory system.

LO 7.18 Describe the gustatory system.

LO 7.19 Explain the potential effects of brain damage on the chemical senses.

Perception

LO 7.20 Use examples to illustrate the role of experience in perception.

LO 7.21 Explain perceptual decision making, using some examples of phantom percepts to illustrate.

LO 7.22 Explain the binding problem and describe two potential solutions to it.

Selective Attention

LO 7.23 Describe the two characteristics of selective attention and explain what is meant by exogenous versus endogenous attention.

LO 7.24 Describe the phenomenon of change blindness.

LO 7.25 Describe the neural mechanisms of attention.

LO 7.26 Describe the disorder of attention known as simultanagnosia.

Two chapters in this text focus on the five human extero-ceptive sensory systems: Chapter 6 and this one. Whereas Chapter 6 introduced the visual system, this chapter focuses on the remaining four of our five **exteroceptive sensory systems** (sensory systems that detect stimuli outside of our bodies): the *auditory* (hearing), *somatosensory* (touch), *olfactory* (smell), and *gustatory* (taste) systems.

Although we focus on the five human exteroceptive senses in this book, it is important to realize that there are other sorts of exteroceptive senses that we don't have. For example, many species can sense the earth's magnetic field and use that information to navigate (see Nordmann, Hochstoeger, & Keays, 2017). In addition, sharks, electric fish, and many amphibians detect minute electrical signals, and use them for navigation, hunting, and communication (see Bellono, Leitch, & Julius, 2017).

In addition to covering the mechanisms of various sorts of **sensation** (the process of detecting the presence of stimuli), this chapter also discusses mechanisms of **perception**: the higher-order process of integrating, recognizing, and interpreting patterns of sensations. Although you will encounter examples of perception in the second, third, and fourth modules, the topic of perception is the focus of the fifth module. The chapter ends with an overview of the mechanisms of attention: how our brains manage to attend to a select few sensory stimuli despite being continuously bombarded by thousands of them.

Before you begin the first module of this chapter, consider the following case (Williams, 1970). As you read the chapter, think about this patient, the nature of his deficit, and the likely location of his brain damage. By the time you have reached the final module of this chapter, you will better understand this patient's problem.

The Case of the Man Who Could See Only One Thing at a Time

A 68-year-old patient was referred because he had difficulty finding his way around—even around his own home. The patient attributed his problems to his "inability to see properly." It was found that if two objects (e.g., two pencils) were held in front of him at the same time, he could see only one of them, whether they were held side by side, one above the other, or even one partially behind the other. Pictures of single objects or faces could be identified, even when quite complex; but if a picture included two objects, only one object could be identified at a time—he would perceive the first object, after which it would be replaced by a perception of the second object, which would then be replaced by a perception of the first object, and so on. If the patient was shown overlapping drawings (i.e., one drawn on top of another), he would see one but deny the existence of the other.

Principles of Sensory System Organization

The visual system is by far the most thoroughly studied sensory system. As a result, it is also the best understood. However, as more has been discovered about the other sensory systems, it has become apparent that each is organized like the visual system in fundamental ways.

Types of Sensory Areas of Cortex

LO 7.1   **Name and define the three types of sensory cortex.**

The sensory areas of the cortex are, by convention, considered to be of three fundamentally different types: primary, secondary, and association. The **primary sensory cortex** of a system is the area of sensory cortex that receives most of its input directly from the thalamic relay nuclei of that system. For example, as you learned in Chapter 6, the primary visual cortex is the area of the cerebral cortex that receives most of its input from the lateral geniculate nucleus of the thalamus. The **secondary sensory cortex** of a system comprises the areas of the sensory cortex that receive most of their input from the primary sensory cortex of that system or from other areas of secondary sensory cortex of the same system. **Association cortex** is any area of cortex that receives input from more than one sensory system. Most input to areas of association cortex comes via areas of secondary sensory cortex.

The interactions among these three types of sensory cortex and among other sensory structures are characterized by three major principles: hierarchical organization, functional segregation, and parallel processing.

Features of Sensory System Organization

LO 7.2   **In the context of sensory system organization, explain what is meant by each of the following terms: *hierarchical organization, functional segregation,* and *parallel processing.* Summarize the current model of sensory system organization.**

Sensory systems are characterized by **hierarchical organization**. A hierarchy is a system whose members can be assigned to specific levels or ranks in relation to one another. For example, an army is a hierarchical system because all soldiers are ranked with respect to their authority. In the same way, sensory structures are organized in a hierarchy on the basis of the specificity and complexity of their function. As one moves through a sensory system from receptors, to thalamic nuclei, to primary sensory cortex, to secondary sensory cortex, to association cortex, one finds neurons that respond optimally to stimuli of greater and greater specificity and complexity. Each level of a sensory hierarchy receives much of its input from lower levels and adds another layer of analysis before passing it on up the hierarchy (see Rees, Kreiman, & Koch, 2002).

The hierarchical organization of sensory systems is apparent from a comparison of the effects of damage to various levels: The higher the level of damage, the more specific and complex the deficit. For example, destruction of a sensory system's receptors produces a complete loss of ability to perceive in that sensory modality (e.g., total blindness or deafness); in contrast, destruction of an area of association or secondary sensory cortex typically produces complex and specific sensory deficits, while leaving fundamental sensory abilities intact. Dr. P., the man who mistook his wife for a hat (Sacks, 1985), displayed such a pattern of deficits.

Case of the Man Who Mistook His Wife for a Hat*

Dr. P. was a highly respected musician and teacher—a charming and intelligent man. He had been referred to the eminent neurologist Oliver Sacks for help with a vision problem. At least, as Dr. P. explained to the neurologist, other people seemed to think that he had a vision problem, and he did admit that he sometimes made odd errors.

Dr. Sacks tested Dr. P.'s vision and found his visual acuity to be excellent—Dr. P. could easily spot a pin on the floor. The first sign of a problem appeared when Dr. P. needed to put his shoe back on following a standard reflex test. Gazing at his foot, he asked Sacks if it was his shoe.

Continuing the examination, Dr. Sacks showed Dr. P. a glove and asked him what it was. Taking the glove and puzzling over it, Dr. P. could only guess that it was a container divided into five compartments for some reason. Even when Sacks asked whether the glove might fit on some part of the body, Dr. P. displayed no signs of recognition.

At that point, Dr. P. seemed to conclude that the examination was over and, from the expression on his face, that he had done rather well. Preparing to leave, he turned and grasped his wife's head and tried to put it on his own. Apparently, he thought it was his hat.

Mrs. P. showed little surprise. That kind of thing happened a lot.

FUNCTIONAL SEGREGATION. It was once assumed that the primary, secondary, and association areas of a sensory system were each *functionally homogeneous*. That is, it was assumed that all areas of cortex at any given level of a sensory hierarchy acted together to perform the same function. However, research has shown that **functional segregation**, rather than functional homogeneity, characterizes the organization of sensory systems. It is now clear that each of the three levels of cerebral cortex—primary, secondary, and association—in each sensory system contains functionally distinct areas that specialize in different kinds of analysis.

PARALLEL PROCESSING. It was once believed that the different levels of a sensory hierarchy were connected in a serial fashion. In a *serial system*, information flows among the components over just one pathway, like a string through a strand of beads. However, we now know that sensory systems are *parallel systems* in which information flows through the components over multiple pathways (see Lleras et al., 2017). Parallel systems feature **parallel processing**—the simultaneous analysis of a signal in different ways by the multiple parallel pathways of a neural network.

SUMMARY MODEL OF SENSORY SYSTEM ORGANIZATION. Figure 7.1 summarizes the information in this module by illustrating how thinking about the organization of sensory systems has changed. In the 1960s, sensory systems were believed to be hierarchical, functionally homogeneous,

Figure 7.1 Two models of sensory system organization: The former model was hierarchical, functionally homogeneous, and serial; the current model, which is more consistent with the evidence, is hierarchical, functionally segregated, and parallel. Not shown in the current model are the many descending pathways—one means by which higher levels of sensory systems can influence sensory input.

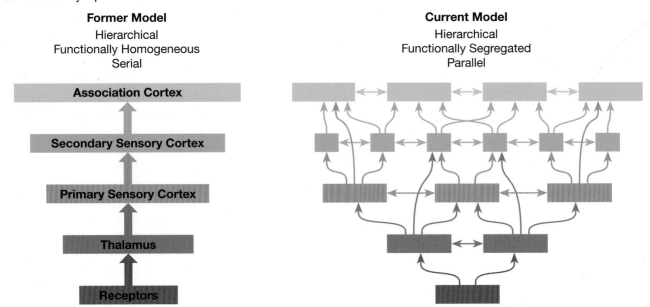

Former Model
Hierarchical
Functionally Homogeneous
Serial

Current Model
Hierarchical
Functionally Segregated
Parallel

Association Cortex

Secondary Sensory Cortex

Primary Sensory Cortex

Thalamus

Receptors

*Based on *The Man Who Mistook His Wife for a Hat and Other Clinical Tales* by Oliver Sacks. Copyright © 1970, 1981, 1983, 1984, 1986 by Oliver Sacks.

Figure 7.2 The relation between the physical and perceptual dimensions of sound.

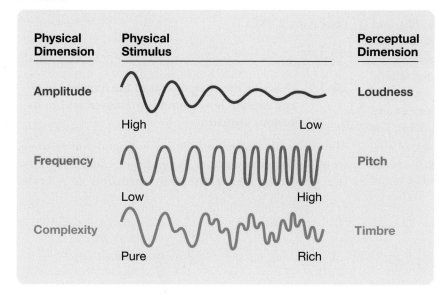

Figure 7.3 The breaking down of a sound—in this case, the sound of a clarinet—into its component sine waves by Fourier analysis. When added together, the component sine waves produce the complex sound wave.

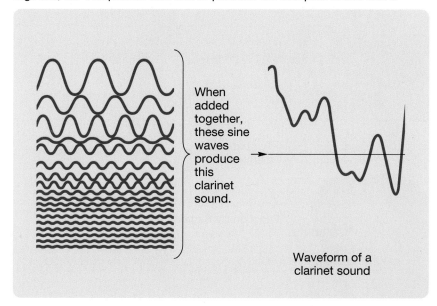

and serial. However, subsequent research has established that sensory systems are hierarchical, functionally segregated, and parallel (see Rauschecker, 2015).

Not shown in Figure 7.1 are the many neurons that descend through the sensory hierarchies. Although sensory systems carry information from lower to higher levels of their respective hierarchies, they also conduct information in the opposite direction (from higher to lower levels). These are known as *top-down signals* (see Bressler & Richter, 2015; Marques et al., 2018; Ruff, 2013).

Now that you have an understanding of the general principles of sensory system organization, let's take a look

in sequence at the auditory system, the somatosensory system, and the chemical sensory systems (smell and taste).

Auditory System

The function of the auditory system is the perception of sound. Sounds are vibrations of air molecules that stimulate the auditory system; humans hear only those molecular vibrations between about 20 and 20,000 *hertz* (cycles per second).

Physical and Perceptual Dimensions of Sound

LO 7.3 **Explain the relationship between the physical and perceptual dimensions of sound.**

Figure 7.2 illustrates how sounds are commonly recorded in the form of waves and the relation between the physical dimensions of sound vibrations and our perceptions of them. The *amplitude, frequency*, and *complexity* of the molecular vibrations are most closely linked to perceptions of *loudness, pitch*, and *timbre*, respectively.

Pure tones (sine wave vibrations) exist only in laboratories and sound recording studios; in real life, sound is always associated with complex patterns of vibrations. For example, Figure 7.3 illustrates the complex sound wave associated with one note of a clarinet. The figure also illustrates that any complex sound wave can be broken down mathematically into a series of sine waves of various frequencies and amplitudes; these component sine waves produce the original sound when they are added together. **Fourier analysis** is the mathematical procedure for breaking down complex waves into their component sine waves. One theory of audition is that the auditory system performs a Fourier-like analysis of complex sounds in terms of their component sine waves.

For any pure tone, there is a close relationship between the frequency of the tone and its perceived pitch; however, the relation between the frequencies that make up natural sounds (which are always composed of a mixture of frequencies) and their perceived pitch is complex (see Bidelman & Grall, 2014): The pitch of such sounds is related to their *fundamental frequency*: the frequency that is the highest common *divisor* (a number that divides another number) for the various component frequencies. For example, a sound that is a mixture of 100, 200, and 300 Hz frequencies normally has a pitch related to 100 Hz because 100 Hz is the highest common divisor of the three components. An extremely important characteristic of pitch perception is the fact that the pitch of a complex sound may not be directly related to the frequency of any of the sound's components (see Lau & Werner, 2014). For example, a mixture of pure tones with frequencies of 200, 300, and 400 Hz would be perceived as having the same pitch as a pure tone of 100 Hz—because 100 Hz is the

fundamental frequency (i.e., the highest common divisor) of 200, 300, and 400 Hz. This important aspect of pitch perception is referred to as the *missing fundamental* (see Oxenham, 2018).

The Ear

LO 7.4   Describe the components of the human ear, and explain how sound is processed within its various structures.

The ear is illustrated in Figure 7.4. Sound waves travel from the outer ear down the auditory canal and cause the **tympanic membrane** (the eardrum) to vibrate.

Figure 7.4 Anatomy of the ear.

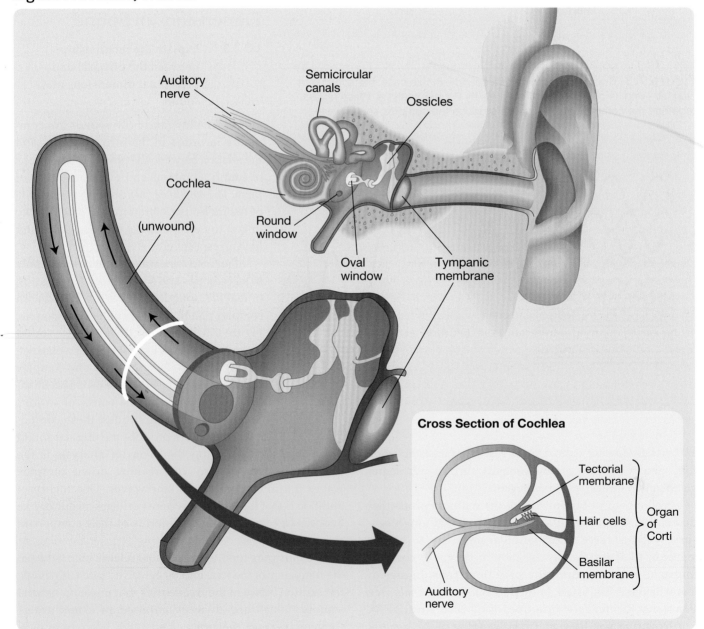

These vibrations are then transferred to the three **ossicles**—the small bones of the middle ear: the *malleus* (the hammer), the *incus* (the anvil), and the *stapes* (the stirrup). The vibrations of the stapes trigger vibrations of the membrane called the **oval window**, which in turn transfers the vibrations to the fluid of the snail-shaped **cochlea** (*kokhlos* means "land snail"). The cochlea is a long, coiled tube with an internal structure running almost to its tip. This internal structure is the auditory receptor organ, the **organ of Corti**.

Each pressure change at the oval window travels along the organ of Corti as a wave. The organ of Corti is composed of several membranes; we will focus on two of them: the basilar membrane and the tectorial membrane. The auditory receptors, the **hair cells**, are mounted in the **basilar membrane**, and the **tectorial membrane** rests on the hair cells. Accordingly, a deflection of the organ of Corti at any point along its length produces a shearing force on the hair cells at the same point. This force stimulates the hair cells, which in turn increase firing in axons of the **auditory nerve** (see Wu et al., 2017)—a branch of the *auditory-vestibular nerve* (one of the 12 cranial nerves). The vibrations of the cochlear fluid are ultimately dissipated by the *round window*, an elastic membrane in the cochlear wall.

The cochlea is remarkably sensitive (see Hudspeth, 2014). Humans can hear differences in pure tones that differ in frequency by only 0.2 percent. The major principle of cochlear coding is that different frequencies produce maximal stimulation of hair cells at different points along the basilar membrane—with higher frequencies producing greater activation closer to the windows and lower frequencies producing greater activation at the tip of the basilar membrane. Thus, the many component frequencies that compose each complex sound activate hair cells at many different points along the basilar membrane, and the many signals created by a single complex sound are carried out of the ear by many different auditory neurons. Like the cochlea, most other structures of the auditory system are arrayed according to frequency. Thus, in the same way that the organization of the visual system is largely **retinotopic**, the organization of the auditory system is largely **tonotopic** (see Schreiner & Polley, 2014).

This brings us to the major unsolved mystery of auditory processing. Imagine yourself in a complex acoustic environment such as a party. The music is playing; people are dancing, eating, and drinking; and numerous conversations are going on around you. Because the component frequencies in each individual sound activate many sites along your basilar membrane, the number of sites simultaneously activated at any one time by the party noises is enormous. But somehow your auditory system manages to sort these individual frequency messages into separate categories and combine them so that you hear each source of complex sounds independently (see Bremen &

Middlebrooks, 2013; Christison-Lagay & Cohen, 2014; Christison-Lagay, Gifford, & Cohen, 2015). For example, you hear the speech of the person standing next to you as a separate sequence of sounds, despite the fact that it contains many of the same component frequencies coming from other sources. The mechanism underlying this important ability has yet to be identified, but one theory is that it is due to the synchronous relationship over time of the frequency elements of each sound source (see Oxenham, 2018).

Figure 7.4 also shows the **semicircular canals**—the receptive organs of the **vestibular system**. The vestibular system carries information about the direction and intensity of head movements, which helps us maintain our balance (see Brandt & Dieterich, 2017; Gu, 2018).

From the Ear to the Primary Auditory Cortex

LO 7.5 Describe the major pathways that lead from the ear to the primary auditory cortex.

There is no major auditory pathway to the cortex comparable to the visual system's retina-geniculate-striate pathway. Instead, there is a network of auditory pathways, some of which are illustrated in Figure 7.5. The axons of each *auditory nerve* synapse in the ipsilateral *cochlear nuclei*, from which many projections lead to the **superior olives** on both sides of the brain stem at the same level. The axons of the olivary neurons project via the *lateral lemniscus* to the **inferior colliculi**, where they synapse on neurons that project to the **medial geniculate nuclei** of the thalamus, which in turn project to the *primary auditory cortex*. Notice that signals from each ear are combined at a very low level (in the superior olives) and are transmitted to both ipsilateral and contralateral auditory cortex.

The subcortical pathways of the auditory system are inherently complex, and they have many more synapses than the other senses (see Jasmin, Lima, & Scott, 2019; Wang, 2018). Some researchers believe that the complex subcortical organization of the auditory system is related to the complexity of the analyses that the auditory system has to perform (see Wang, 2018).

Auditory Cortex

LO 7.6 Describe the organization of auditory cortex.

Recent progress in the study of human auditory cortex has resulted from the convergence of functional brain-imaging studies in humans and invasive neural recording studies in monkeys (see Saenz & Langers, 2014). Still, primate auditory cortex is far from being well understood—for example, our understanding of it lags far behind our current understanding of the visual cortex.

Figure 7.5 Some of the pathways of the auditory system that lead from one ear to the cortex.

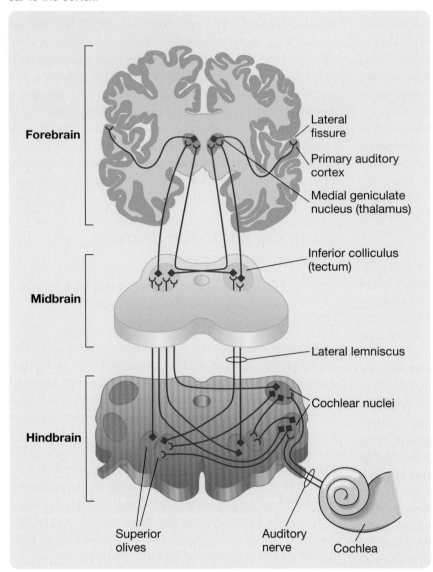

identified. First, like the primary visual cortex, the primary auditory cortex is organized in functional columns (see Mizrahi, Shalev, & Nelken, 2014): All of the neurons encountered during a vertical microelectrode penetration of primary auditory cortex (i.e., a penetration at right angles to the cortical layers) tend to respond optimally to sounds in the same frequency range. Second, like the cochlea, auditory cortex has a tonotopic organization (see Jasmin, Lima, & Scott, 2019; Schreiner & Polley, 2014): Each area of auditory cortex appears to have a gradient of frequencies from low to high along its length. Third, auditory cortex is also organized according to the temporal components of sound; that is, variations in the amplitude of particular sound frequencies over time. For example, our auditory environments almost never consist of sounds that do not vary in their intensity over time. It seems that auditory cortex is sensitive to such fluctuations. This third organizing principle of auditory cortex is known as **periodotopy** (see Brewer & Barton, 2016).

WHAT SOUNDS SHOULD BE USED TO STUDY AUDITORY CORTEX? Why has research on auditory cortex lagged behind research on visual cortex? There are several reasons, but a major one is a lack of clear understanding of the dimensions along which auditory cortex evaluates

In primates, the primary auditory cortex, which receives the majority of its input from the medial geniculate nucleus, is located in the temporal lobe, hidden from view within the lateral fissure (see Figure 7.6). Primate primary auditory cortex comprises three adjacent areas (see Moerel, De Martino, & Formisano, 2014): Together these three areas are referred to as the *core region*. Surrounding the core region is a band—often called the *belt*—of areas of secondary auditory cortex. Areas of secondary auditory cortex outside the belt are called *parabelt areas* (Jasmin, Lima, & Scott, 2019). In total, there seem to be about 13 separate areas of auditory cortex in primates (see Brewer & Barton, 2016).

ORGANIZATION OF PRIMATE AUDITORY CORTEX. Three important principles of organization of primary auditory cortex have been

Figure 7.6 General location of the primary auditory cortex and areas of secondary auditory cortex. Most auditory cortex is hidden from view in the lateral fissure.

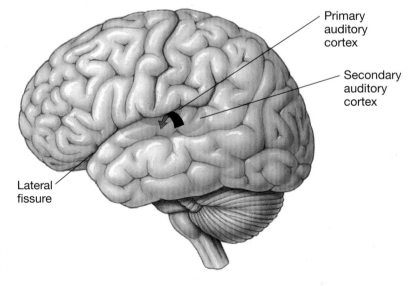

sound (Sharpee, Atencio, & Schreiner, 2011). You may recall that research on the visual cortex did not start to progress rapidly until it was discovered that most visual neurons respond to contrast. There is clear evidence of a hierarchical organization in auditory cortex—the neural responses of secondary auditory cortex tend to be more complex and varied than those of primary auditory cortex (see Jasmin, Lima, & Scott, 2019).

Many neurons in auditory cortex respond only weakly to simple stimuli such as pure tones, which have been widely employed in electrophysiological studies of auditory cortex. This practice is changing, however, partly in response to the discovery that natural sounds, in general, are better at eliciting responses from neurons in mammalian auditory cortex (see Gervain & Geffen, 2019; Kopp-Scheinpflug, Sinclair, & Linden, 2019).

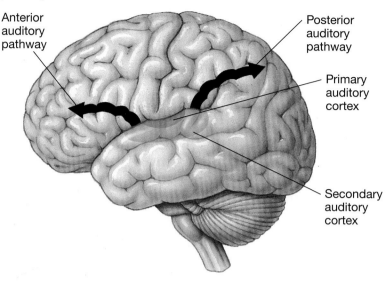

Figure 7.7 The hypothesized anterior and posterior auditory pathways.

WHAT ANALYSES DOES THE AUDITORY CORTEX PERFORM? We now know that calculations by the auditory cortex produce signals that are not faithful representation of sounds (see Tsunada et al., 2016; Wang, 2018). More specifically, auditory cortex is now known to integrate information about the current perceptions and behaviors of an animal in order to produce auditory signals that are relevant to the animal's current situation (see Kuchibhotla & Bathellier, 2018; Lima, Krishnan, & Scott, 2016; Schneider & Mooney, 2018).

One example of an output signal from auditory cortex that is particularly relevant to an animal's current situation is the creation of representations of *auditory objects*. For example, it is believed that the auditory cortex can take the complex mixture of frequencies produced by a piano and convert it into a sound representation that allows us to say "That's the sound of a piano!" (see Angeloni & Geffen, 2018; Kuchibhotla & Bathellier, 2018; Tsunada et al., 2016).

TWO STREAMS OF AUDITORY CORTEX. Thinking about the general organization of auditory cortex has been inspired by research on visual cortex. Researchers have proposed that, just as there are two main cortical streams of visual analysis (dorsal and ventral), there are two main cortical streams of auditory analysis. Auditory signals are ultimately conducted to two large areas of association cortex: prefrontal cortex and posterior parietal cortex. There is good evidence that the *anterior auditory pathway* is more involved in identifying sounds (what), whereas the *posterior auditory pathway* is more involved in locating sounds (where)—see Jasmin, Lima, & Scott (2019) and van der Heijden et al. (2019). These pathways are illustrated in Figure 7.7.

AUDITORY–VISUAL INTERACTIONS. Sensory systems have traditionally been assumed to interact in association cortex. Indeed, as you have already learned, association cortex is usually defined as areas of cortex where such interactions, or associations, take place. Much of the research on sensory system interactions has focused on interactions between the auditory and visual systems, particularly on those that occur in the posterior parietal cortex (see Brang et al., 2013; Cohen, 2009). In one study of monkeys (Mullette-Gillman, Cohen, & Groh, 2005), some posterior parietal neurons were found to have visual receptive fields, some were found to have auditory receptive fields, and some were found to have both.

Functional brain imaging is widely used to investigate sensory system interactions. One advantage of functional brain imaging is that it does not focus on any one part of the brain; it records activity throughout the brain. Functional brain-imaging studies have confirmed that sensory interactions do occur in association cortex, but more importantly, they have repeatedly found evidence of sensory interactions at the lowest level of the sensory cortex hierarchy, in areas of primary sensory cortex (see Man et al., 2013; Smith & Goodale, 2015). This discovery is changing how we think about the interaction of sensory systems: Sensory system interaction is not merely tagged on after *unimodal* (involving one system) analyses are complete; sensory system interactions seem to be an early and integral part of sensory processing.

WHERE DOES THE PERCEPTION OF PITCH OCCUR? Recent research has answered one fundamental question about auditory cortex: Where does the perception of pitch likely occur? This seemed like a simple question to answer because most areas of auditory cortex have a clear tonotopic organization. However, when experimenters used sound stimuli in which frequency and pitch were different—for example, by using the missing fundamental technique—most auditory neurons responded to changes in frequency rather than pitch. This information led Bendor and Wang (2005) to probe primary and secondary areas of monkey auditory cortex with microelectrodes to assess the responses of individual neurons to missing fundamental stimuli. They discovered one small area just anterior to primary auditory cortex that contained many neurons that responded to pitch rather than frequency, regardless of the quality of the sound. The same small area also contained neurons that responded to frequency, and Bendor and Wang suggested that this area was likely the place where frequencies of sound were converted to the perception of pitch. A comparable pitch area has been identified by fMRI studies in a similar location in the human brain.

Effects of Damage to the Auditory System

LO 7.7 Describe the effects of damage to the auditory system.

The study of damage to the auditory system is important for two reasons. First, it provides information about how the auditory system works. Second, it can serve as a source of information about the causes and treatment of clinical deafness.

AUDITORY CORTEX DAMAGE. Following bilateral lesions to the primary auditory cortex, there is often a complete loss of hearing, which presumably results from the shock of the lesion because hearing recovers in the ensuing weeks. The major permanent effects are loss of the ability to process the structural aspects of sounds—an ability that is necessary for the processing of speech sounds. Accordingly, patients with bilateral auditory cortex lesions are often said to be "word deaf" (see Jasmin, Lima, & Scott, 2019).

Consistent with the definitions of the two cortical auditory pathways, patients with damage to the anterior auditory cortex pathway (the *what pathway*) have trouble identifying sounds. Whereas, patients with damage to the posterior auditory cortex pathway (the *where pathway*) have difficulty localizing sounds (see Jasmin, Lima, & Scott, 2019).

DEAFNESS IN HUMANS. Deafness is one of the most prevalent human disabilities: An estimated 360 million people currently suffer from disabling hearing impairments (Lesica, 2018). Hearing impairment affects more than one's ability to detect sounds: it can lead to feelings of social isolation and has been associated with an increased risk for dementia (see Lesica, 2018; Peelle & Wingfield,

2016). Total deafness is rare, occurring in only 1 percent of hearing-impaired individuals.

Severe hearing problems typically result from damage to the inner ear or the middle ear or to the nerves leading from them rather than from more central damage. There are two common classes of hearing impairments: those associated with damage to the ossicles (*conductive deafness*) and those associated with damage to the cochlea or auditory nerve (*nerve deafness*). The major cause of nerve deafness is a loss of hair cell receptors (see Wallis, 2018; Wong & Ryan, 2015).

If only part of the cochlea is damaged, individuals may have nerve deafness for some frequencies but not others. For example, age-related hearing loss features a specific deficit in hearing high frequencies. That is why elderly people often have difficulty distinguishing "s," "f," and "t" sounds: They can hear people speaking to them but often have difficulty understanding what people are saying. Often, relatives and friends do not realize that much of the confusion displayed by the elderly stems from difficulty discriminating sounds (see Wingfield, Tun, & McCoy, 2005). Unfortunately, hearing aids often do not help with the processing of speech (see Lesica, 2018; Peelle & Wingfield, 2016).

Hearing loss is sometimes associated with **tinnitus** (ringing of the ears). When only one ear is damaged, the ringing is perceived as coming from that ear; however, cutting the nerve from the ringing ear has no effect on the ringing. This suggests that neuroplastic changes to the auditory system resulting from deafness are the cause of tinnitus (see Eggermont & Tass, 2015; Elgoyhen et al., 2015; Shore, Roberts, & Langguth, 2016; Sedley et al., 2016).

Journal Prompt 7.1

Why do you think tinnitus is associated with deafness? [Hint: Sensory neurons don't stop firing in the absence of sensory input.]

Some people with nerve deafness benefit from cochlear implants (see Figure 7.8). *Cochlear implants* bypass damage to the auditory hair cells by converting sounds picked up by a microphone on the patient's ear to electrical signals, which are then carried into the cochlea by a bundle of electrodes. These signals excite the auditory nerve. Although cochlear implants can provide major benefits, they do not restore normal hearing. The sooner a person receives a cochlear implant after becoming deaf, the more likely they are to benefit, because disuse leads to alterations of the auditory neural pathways (see Kral & Sharma, 2012).

Figure 7.8 Cochlear implant: The surgical implantation is shown on the left, and a child with an implant is shown on the right.

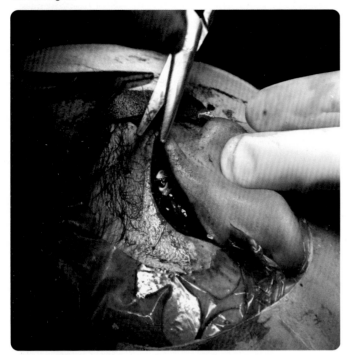

AJPhoto/Science Source

Gene J. Puskar/AP Images

Scan Your Brain

Before we go on to discuss the other sensory systems, pause and scan your brain to check your knowledge of what you have learned in this chapter so far. Fill in the following blanks with the most appropriate terms. The correct answers are provided at the end of the exercise. Before proceeding, review material related to your errors and omissions.

1. An area of the cortex that receives input from multiple sensory systems and integrates sensory information is called the _____ cortex.

2. Sensation is the process of detecting the presence of stimuli, and the higher-order process called _____ allows the interpretation of sensory patterns.

3. The simultaneous analysis of signals, also known as _____ processing, allows for information to flow through multiple pathways at the same time.

4. In the 1960s, the sensory organization was believed to be hierarchical, _____ _____, and serial.

5. The frequency of sound vibrations is linked to perceptions of _____.

6. Sound waves travel from the external environment to the outer ear, and through the auditory canal where they reach the _____ membrane.

7. The three smallest bones in the human body are the malleus, the incus, and the _____.

8. The _____ are auditory receptors located in the cochlea on the basilar membrane, and they increase firing in axons of the auditory nerve.

9. Axons from olivary neurons project to the _____ _____ via the lateral lemniscus.

10. The cochlea and the primary auditory cortex are both organized _____ on the bases of sound frequencies.

11. Damage to the ossicles is associated with _____ deafness, while damage to the cochlea is associated with _____ deafness.

12. The anterior auditory pathway is more involved in identifying sounds (what), whereas the _____ auditory pathway is more involved in locating sounds (where).

Scan Your Brain answers: (1) association, (2) perception, (3) parallel, (4) functionally homogeneous, (5) pitch, (6) tympanic, (7) stapes, (8) hair cells, (9) inferior colliculi, (10) tonotopically, (11) conductive, nerve, (12) posterior.

Somatosensory System: Touch and Pain

Sensations from your body are referred to as *somatosensations*. The system that mediates these bodily sensations—the *somatosensory system*—is composed of three separate but interacting systems: (1) an *exteroceptive system*, which senses external stimuli that are applied to the skin; (2) a *proprioceptive system*, which monitors information about the position of the body that comes from receptors in the muscles, joints, and organs of balance; and (3) an *interoceptive system*, which provides general information about conditions within the body (e.g., temperature and blood pressure). This module deals almost exclusively with the exteroceptive system, which itself comprises three somewhat distinct divisions: a division for perceiving *mechanical stimuli* (touch), one for *thermal stimuli* (temperature), and one for *nociceptive stimuli* (pain).

Cutaneous Receptors

LO 7.8 **Name some of the cutaneous receptors and explain the functional significance of fast versus slow receptor adaptation.**

There are many kinds of receptors in the skin (see Owens & Lumpkin, 2014; Zimmerman, Bai, & Ginty, 2014). Figure 7.9 illustrates four of them. The simplest cutaneous receptors are the *free nerve endings* (neuron endings with no specialized structures on them), which are particularly sensitive to temperature change and pain. The largest and deepest cutaneous receptors are the onion-like *Pacinian corpuscles*; because they adapt rapidly, they respond to sudden displacements of the skin but not to constant pressure. In contrast, *Merkel's disks* and *Ruffini endings* both adapt slowly and respond to gradual skin indentation and skin stretch, respectively.

To appreciate the functional significance of fast and slow receptor adaptation, consider what happens when a constant pressure is applied to the skin. The pressure evokes a burst of firing in all receptors, which corresponds to the sensation of being touched; however, after a few hundred milliseconds, only the slowly adapting receptors remain active, and the quality of the sensation changes. In fact, you are often totally unaware of constant skin pressure; for example, you are usually unaware of the feeling of your clothes against your body until you focus attention on it. As a consequence, when you try to engage in **stereognosis** (identifying objects by touch), you manipulate the object in your hands so that the pattern of stimulation continually changes. Having some receptors that adapt quickly and some that adapt slowly provides information about both the dynamic and static qualities of tactual stimuli.

The structure and physiology of each type of somatosensory receptor seem to be specialized for a different

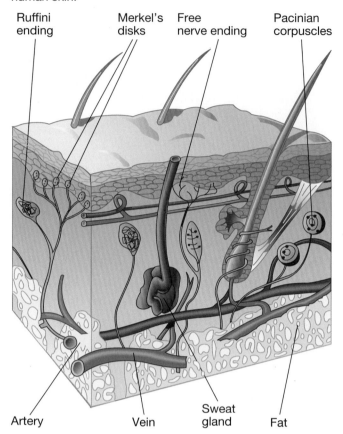

Figure 7.9 Four cutaneous receptors that occur in human skin.

Ruffini ending   Merkel's disks   Free nerve ending   Pacinian corpuscles

Artery   Vein   Sweat gland   Fat

function. However, in general, the various receptors tend to function in the same way: Stimuli applied to the skin deform or change the chemistry of the receptor, and this in turn changes the permeability of the receptor cell membrane to various ions (see Delmas, Hao, & Rodat-Despoix, 2011). The result is a neural signal.

Initially, it was assumed that each type of receptor located in the skin (see Figure 7.9) mediates a different tactile sensation (e.g., touch, pain, heat), but this has not proven to be the case. Each tactile sensation appears to be produced by the interaction of multiple receptor mechanisms, and each receptor mechanism appears to contribute to multiple sensations (see Hollins, 2010; Lumpkin & Caterina, 2007; McGlone & Reilly, 2010). In addition, skin cells that surround particular receptors also seem to play a role in the quality of the sensations produced by that receptor (see Zimmerman, Bai, & Ginty, 2014). Indeed, new forms of tactile sensation are still being discovered (see McGlone, Wessberg, & Olausson, 2014; Ran, Hoon, & Chen, 2016).

Two Major Somatosensory Pathways

LO 7.9 **Describe the two major somatosensory pathways.**

Somatosensory information ascends from each side of the body to the human cortex over several pathways, but there are two major ones: the dorsal-column medial-lemniscus system and the anterolateral system. The **dorsal-column medial-lemniscus system** tends to carry information about touch and proprioception, and the **anterolateral system** tends to carry information about pain and temperature (see Ran et al., 2016). The key words in the preceding sentence are "tends to": The separation of function in the two pathways is far from complete. Accordingly, lesions of the dorsal-column medial-lemniscus system do not eliminate touch perception or proprioception, and lesions of the anterolateral system do not eliminate perception of pain or temperature.

The dorsal-column medial-lemniscus system is illustrated in Figure 7.10. The sensory neurons of this system

enter the spinal cord via a dorsal root, ascend ipsilaterally in the **dorsal columns**, and synapse in the *dorsal column nuclei* of the medulla. The axons of dorsal column nuclei neurons *decussate* (cross over to the other side of the brain) and then ascend in the **medial lemniscus** to the contralateral **ventral posterior nucleus** of the thalamus. The ventral posterior nuclei also receive input via the three branches of the trigeminal nerve, which carry somatosensory information from the contralateral areas of the face. Most neurons of the ventral posterior nucleus project to the *primary somatosensory cortex (SI)*; others project to the *secondary somatosensory cortex (SII)* or the posterior parietal cortex. Neuroscience trivia buffs will almost certainly want to add to their collection the fact that the dorsal column neurons that originate in the toes are the longest neurons in the human body.

The anterolateral system is illustrated in Figure 7.11. Most dorsal root neurons of the anterolateral system synapse as soon as they enter the spinal cord. The axons of most of the second-order neurons decussate but then ascend to the brain in the contralateral anterolateral portion of the spinal cord; however, some do not decussate but ascend ipsilaterally. The anterolateral system comprises three different tracts: the *spinothalamic tract*, the *spinoreticular tract*, and the *spinotectal tract*. The three branches of the trigeminal nerve carry pain and temperature information from the face to the same thalamic sites. The pain and temperature information that reaches the thalamus is then distributed to somatosensory cortex and other parts of the brain.

If both ascending somatosensory paths are completely transected by a spinal injury, the patient can feel no body sensation from below the level of the cut. Clearly, when it comes to spinal injuries, lower is better.

Figure 7.10 The dorsal-column medial-lemniscus system. The pathways from only one side of the body are shown.

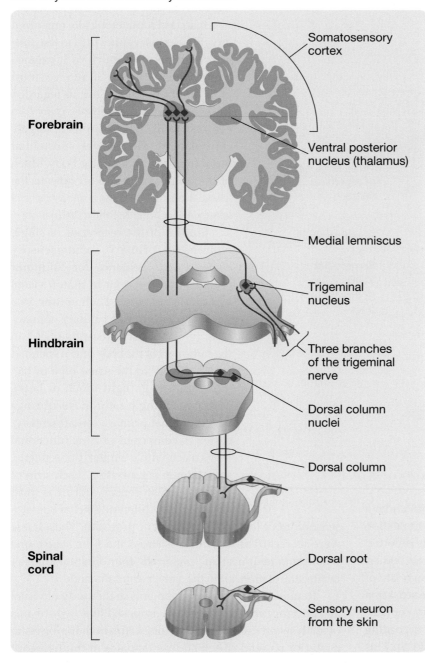

Somatosensory cortex

Forebrain

Ventral posterior nucleus (thalamus)

Medial lemniscus

Trigeminal nucleus

Hindbrain

Three branches of the trigeminal nerve

Dorsal column nuclei

Dorsal column

Spinal cord

Dorsal root

Sensory neuron from the skin

Cortical Areas of Somatosensation

LO 7.10 Describe the cortical somatosensory areas and their somatotopic layout.

In 1937, Penfield and his colleagues mapped the primary somatosensory cortex of patients during neurosurgery (see Figure 7.12). Penfield applied electrical stimulation to various sites on the cortical

Figure 7.11 The anterolateral system. The pathways from only one side of the body are shown.

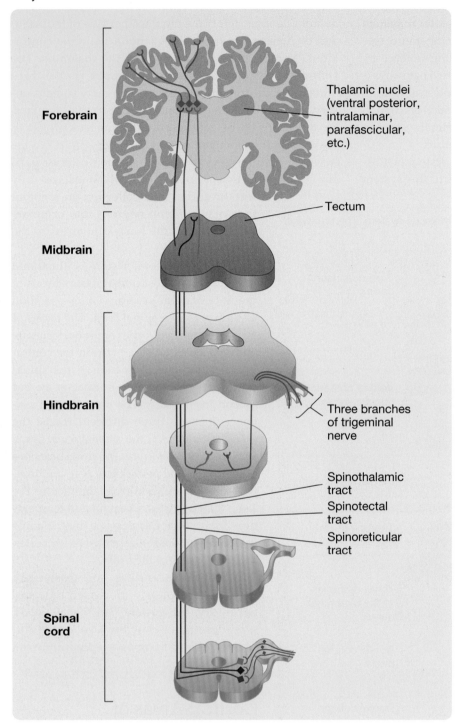

Forebrain

Thalamic nuclei
(ventral posterior,
intralaminar,
parafascicular,
etc.)

Tectum

Midbrain

Hindbrain

Three branches
of trigeminal
nerve

Spinothalamic
tract

Spinotectal
tract

Spinoreticular
tract

Spinal
cord

surface, and the patients, who were fully conscious under a local anesthetic, described what they felt. When stimulation was applied to the *postcentral gyrus*, the patients reported somatosensory sensations in various parts of their bodies. When Penfield mapped the relation between each site of stimulation and the part of the body in which the sensation was felt, he discovered that the human primary somatosensory cortex (SI) is **somatotopic**—organized according to a map of the body surface (see Chen et al., 2015). This

somatotopic map is commonly referred to as the **somatosensory homunculus** (*homunculus* means "little man").

Notice in Figure 7.12 that the somatosensory homunculus is distorted; the greatest proportion of SI is dedicated to receiving input from the parts of the body we use to make tactile discriminations (e.g., hands, lips, and tongue). In contrast, only small areas of SI receive input from large areas of the body, such as the back, that are not usually used to make somatosensory discriminations. The Check It Out demonstration on page 197 allows you to experience the impact this organization has on your ability to perceive touches.

A second somatotopically organized area, SII, lies just ventral to SI in the postcentral gyrus, and much of it extends into the lateral fissure. SII receives most of its input from SI and is thus regarded as secondary somatosensory cortex. In contrast to SI, whose input is largely contralateral, SII receives substantial input from both sides of the body. Much of the output of SI and SII goes to the association cortex of the *posterior parietal lobe* (see McGlone & Reilly, 2010).

Studies of the responses of single neurons in primary somatosensory cortex found evidence for columnar organization—similar to that in visual and auditory cortex. Each neuron in a particular column of primary somatosensory cortex had a receptive field on the same part of the body and responded most robustly to the same type of tactile stimuli (e.g., light touch or heat). Moreover, single-neuron recordings suggested that primary somatosensory-cortex is composed of four functional strips, each with a similar, but separate, somatotopic organization. Each strip of primary somatosensory cortex is most sensitive to a different kind of somatosensory input (e.g., to light touch or pressure). Thus, if one were to record from neurons across the four strips, one would find neurons that "preferred" four different kinds of tactile stimulation, all to the same part of the body.

Reminiscent of the developments in the study of visual and auditory cortex, it has been proposed that two streams of analysis proceed from SI: a dorsal stream that projects to posterior parietal cortex and participates in multisensory

Figure 7.12 The locations of human primary somatosensory cortex (SI) and one area of secondary somatosensory cortex (SII) with the conventional portrayal of the somatosensory homunculus. Something has always confused us about this portrayal of the somatosensory homunculus: The body is upside down, while the face is right side up. It now appears that this conventional portrayal is wrong. The results of an fMRI study suggest that the face representation is also inverted. (Based on Servos et al., 1999.)

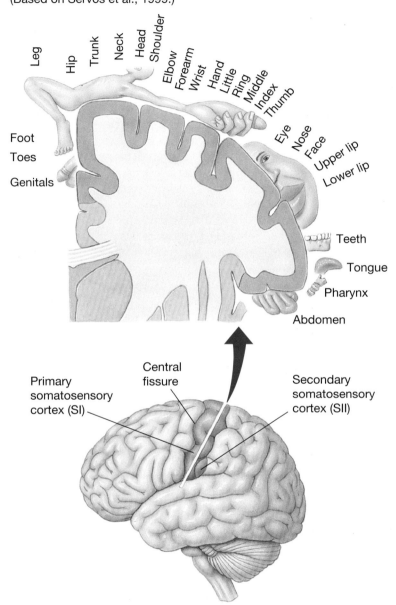

Based on Servos, P., Engel, S. A., Gati, J., & Menon, R. (1999). fMRI evidence for an inverted face representation in human somatosensory cortex. *Neuroreport, 10*(7), 1393–1395.

Check It Out
Touching a Back

Because only a small portion of human primary somatosensory cortex receives input from the entire back, people have difficulty recognizing objects that touch their backs. You may not have noticed this tactile deficiency—unless, of course, you often try to identify objects by feeling them with your back. You will need one thing to demonstrate the recognition deficiencies of the human back: a friend. Touch your friend on the back with one, two, or three fingers, and ask your friend how many fingers he or she feels. When using two or three fingers, be sure they touch the back simultaneously because temporal cues invalidate this test of tactile discrimination. Repeat the test many times, adjusting the distance between the touches on each trial. Record the results. What you should begin to notice is that the back is incapable of discriminating between separate touches unless the distance between the touches is considerable. In contrast, fingertips can distinguish the number of simultaneous touches even when the touches are very close.

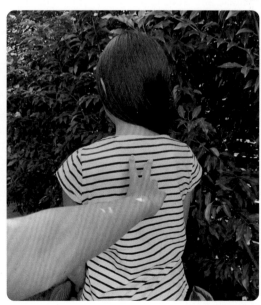

Steven J. Barnes

integration and direction of attention, and a ventral stream that projects to SII and participates in the perception of objects' shapes (Yau, Connor, & Hsiao, 2013).

EFFECTS OF DAMAGE TO THE PRIMARY SOMATOSENSORY CORTEX. The effects of damage to the primary somatosensory cortex are often remarkably mild—presumably because the somatosensory system features numerous parallel pathways. Corkin, Milner, and Rasmussen

(1970) assessed the somatosensory abilities of patients with epilepsy before and after a unilateral excision that included SI. Following the surgery, the patients displayed two minor contralateral deficits: a reduced ability to detect light touch and a reduced ability to identify objects by touch (i.e., a deficit in stereognosis). These deficits were bilateral only in those cases in which the unilateral lesion encroached on SII.

Somatosensory System and Association Cortex

LO 7.11 Name the areas of association cortex that somatosensory signals are sent to, and describe the functional properties of one of those areas.

Somatosensory signals are ultimately conducted to the highest level of the sensory hierarchy, to areas of association cortex in prefrontal and posterior parietal cortex.

Posterior parietal cortex contains *bimodal neurons* (neurons that respond to activation of two different sensory systems); some of these respond to both somatosensory and visual stimuli. The visual and somatosensory receptive fields of each neuron are spatially related; for example, if a neuron has a somatosensory receptive field centered in the left hand, its visual field is adjacent to the left hand (see Crochet, Lee, & Peterson, 2019). Remarkably, as the left hand moves, the visual receptive field of the neuron moves with it. The existence of these bimodal neurons motivated the following interesting case study by Schendel and Robertson (2004).

The Case of W.M., Who Reduced His Scotoma with His Hand

W.M. suffered a stroke in his right posterior cerebral artery. The stroke affected a large area of his right occipital and parietal lobes and left him with severe left *hemianopsia* (a condition in which a scotoma covers half the visual field). When tested with his left hand in his lap, W.M. detected 97.8 percent of the stimuli presented in his right visual field and only 13.6 percent of those presented in his left visual field. However, when he was tested with his left hand extended into his left visual field, his ability to detect stimuli in his left visual field improved significantly. Further analysis showed that this general improvement resulted from W.M.'s greatly improved ability to see those objects in the left visual field that were near his left hand. Remarkably, this area of improved performance around his left hand was expanded even further when he held a tennis racket in his extended left hand.

Somatosensory Agnosias

LO 7.12 Describe the two major types of somatosensory agnosia.

There are two major types of somatosensory agnosia. One is **astereognosia**—the inability to recognize objects by touch. Cases of pure astereognosia—those that occur in the absence of simple sensory deficits—are rare (Corkin, Milner, & Rasmussen, 1970). The other type of somatosensory agnosia is **asomatognosia**—the failure to recognize parts of one's own body. Asomatognosia is usually unilateral, affecting only the left side of the body, and it is usually associated with extensive damage to the right temporal and posterior parietal lobe (Feinberg et al., 2010). The case of Aunt Betty (Klawans, 1990) is an example.

The Case of Aunt Betty, Who Lost Half of Her Body*

Aunt Betty was my patient. She wasn't really my aunt, she was my mother's best friend.

As we walked to her hospital room, one of the medical students described the case. "Left hemiplegia [left-side paralysis], following a right-hemisphere stroke." I was told.

Aunt Betty was lying on her back with her head and eyes turned to the right. "Betty," I called out.

I approached her bed from the left, but Aunt Betty did not turn her head or even her eyes to look toward me.

"Hal," she called out. "Where are you?"

I turned her head gently toward me, and we talked. It was clear that she had no speech problems, no memory loss, and no confusion. She was as sharp as ever. But her eyes still looked to the right, as if the left side of her world did not exist.

I held her right hand in front of her eyes. "What's this?" I asked.

"My hand, of course," she said with an intonation that suggested what she thought of my question.

"Well then, what's this?" I said, as I held up her limp left hand where she could see it.

"A hand."

"Whose hand?"

"Your hand, I guess," she replied. She seemed puzzled. I placed her hand back on the bed.

"Why have you come to the hospital?" I asked.

"To see you," she replied hesitantly. I could tell that she didn't know why.

Aunt Betty was in trouble.

As in the case of Aunt Betty, asomatognosia is often accompanied by **anosognosia**—the failure of neuropsychological patients to recognize their own symptoms. Indeed, anosognosia is a common, but curious, symptom of many neurological disorders—many neurological patients with severe behavioral problems think that they are doing quite well.

Asomatognosia is commonly a component of **contralateral neglect**—the tendency not to respond to stimuli that are contralateral to a right-hemisphere injury. You will learn more about contralateral neglect in Chapter 8.

---

*Based on *Newton's Madness* by Harold Klawans (Harper & Row 1990).

Rubber-Hand Illusion

LO 7.13 Describe the rubber-hand illusion and its neural mechanisms.

We perceive ownership of our own body parts. Somesthetic sensation is so fundamental that it is taken for granted. This is why exceptions to it, such as asomatognosia, are so remarkable. In the past decade, another exception—one that is in some respects the opposite of asomatognosia—has been a focus of research. This exception is the **rubber-hand illusion** (the feeling that an extraneous object, in this case a rubber hand, is actually part of one's own body).

The rubber-hand illusion can be generated in a variety of ways, but it is usually induced in the following manner (see Kilteni et al., 2015; Moseley, Gallace, & Spence, 2012). A healthy volunteer's hand is hidden from view by a screen, and a rubber hand is placed next to the hidden hand but in clear sight. Then the experimenter repeatedly strokes the hidden hand and the rubber hand synchronously—see Figure 7.13. In less than a minute, many volunteers begin to feel that the rubber hand is part of their own body (see Blanke, Slater, & Serino, 2015). Interestingly, when this happens, the temperature in the hidden hand drops (Moseley et al., 2008).

Although the neural mechanisms for the rubber-hand illusion are unknown, functional imaging studies have suggested that association cortex in the posterior parietal and frontal lobes plays a role in its induction (Limanowski & Blankenburg, 2015; Tsakiris et al., 2007). It has been suggested that those frontal and parietal-*bimodal neurons* with

both visual and somatosensory fields play a critical role (Kilteni et al., 2015). Interestingly, the rubber-hand illusion is not limited to humans: rubber tails have been demonstrated in mice and rubber arms in monkeys (see Brecht et al., 2017).

Perception of Pain

LO 7.14 Explain why the perception of pain is said to be paradoxical.

A paradox is a logical contradiction. The perception of pain is paradoxical in three important respects, which are explained in the following three subsections.

PAIN IS ADAPTIVE. One paradox of pain is that an experience that seems in every respect to be so bad is in fact extremely important for our survival. There is no special stimulus for pain; it is a response to potentially harmful stimulation of any type. It warns us to stop engaging in potentially harmful activities or to seek treatment (see Navratilova & Porreca, 2014).

The value of pain is best illustrated by the cases of people, like Miss C., who experience no pain (Melzack & Wall, 1982).

Figure 7.13 One common induction method for the rubber-hand illusion. The participant's hand is hidden from view by a screen, and a rubber hand is placed next to their hidden hand but in clear sight. Then the experimenter repeatedly strokes the hidden hand and the rubber hand synchronously.

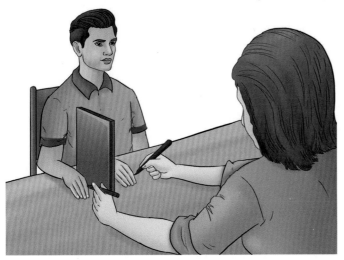

The Case of Miss C., the Woman Who Felt No Pain

Miss C., a university student, was very intelligent, and she was normal in every way except that she never felt pain. (Her condition is now referred to as *congenital insensitivity to pain*.)

She felt no pain when subjected to strong electric shock, burning hot water, or an ice bath. Equally astonishing was the fact that she showed no changes in blood pressure, heart rate, or respiration when these stimuli were presented. Furthermore, she did not sneeze, cough, or display corneal reflexes (blinking to protect the eyes). As a child, she had bitten off the tip of her tongue while chewing food and had suffered severe burns after kneeling on a radiator.

Miss C. exhibited pathological changes in her knees, hip, and spine because of the lack of protection to joints provided by pain sensation. She apparently failed to shift her weight when standing, to turn over in her sleep, or to avoid harmful postures.

Miss C. died at the age of 29 of massive infections and extensive skin and bone trauma.

Journal Prompt 7.2

Cases of congenital insensitivity to pain illustrate something important about the adaptive value of pain. Based on this case study, can you specify what that adaptive value might be?

Cox and colleagues (2006) studied six cases of congenital insensitivity to pain among members of a family from Pakistan. They were able to identify the genetic abnormality underlying the disorder in these six individuals: a gene that influences the synthesis of sodium ion channels. Indeed, knockout mice that are missing this sodium ion channel gene show a comparable indifference to pain (Gingras et al., 2014).

PAIN HAS NO CLEAR CORTICAL REPRESENTATION.
The second paradox of pain is that it has no obvious cortical representation (Rainville, 2002). Painful stimuli activate many areas of cortex including the thalamus, SI and SII, the insula, and the anterior cingulate cortex (see Figure 7.14)—see Navratilova and Porreca (2014). However, none of those areas seems necessary for the perception of pain. For example, painful stimuli usually elicit responses in SI and SII (see Zhuo, 2008). However, removal of SI and SII in humans is not associated with any change in the threshold for pain. Indeed, *hemispherectomized* patients (those with one cerebral hemisphere removed) can still perceive pain from both sides of their bodies.

The cortical area that has been most frequently linked to pain is the **anterior cingulate cortex** (see Figure 7.14). For example, Craig et al. (1996) examined the effect of the **thermal grid illusion:** the perception of pain that results from placing one's hand on a grid of metal rods that alternate between cool and warm (see Figure 7.15; Bokiniec et al., 2018). When participants experienced such illusory pain while undergoing fMRI, only the anterior cingulate cortex displayed a marked increase in activity (Craig et al., 1996).

PAIN IS MODULATED BY COGNITION AND EMOTION.
The third paradox of pain is that this most compelling of all sensory experiences can be so effectively suppressed by cognitive and emotional factors (see Bushnell, Čeko, & Low, 2013; Senkowski, Höfle, & Engel, 2014). For example, males participating in a certain religious ceremony suspend objects from hooks embedded in their backs with little evidence of pain (see also Figure 7.16); severe wounds suffered by soldiers in battle are often associated with little pain; and people injured in life-threatening situations frequently feel no pain until the threat is over.

Three discoveries led to the identification of a descending pain-control circuit. First was the discovery that electrical stimulation of the **periaqueductal gray (PAG)** has analgesic (pain-blocking) effects: Reynolds (1969) was able to perform surgery on rats with no analgesia other than that provided by PAG stimulation. Second was the discovery that the PAG and other areas of the brain contain specialized receptors for opioid analgesic drugs such as morphine. And third was the isolation of several endogenous (internally produced) opioid analgesics, the **endorphins**, which you learned about in Chapter 4. These three findings together suggested that analgesic drugs and psychological factors might block pain through an endorphin-sensitive circuit that descends from the PAG.

Figure 7.14 Location of the anterior cingulate cortex in the cingulate gyrus.

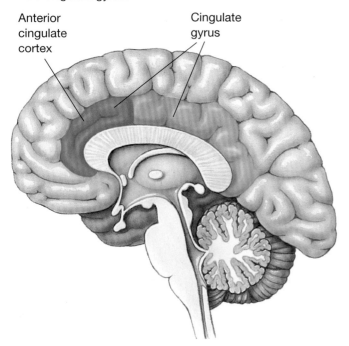

Figure 7.15 The thermal grid illusion. Pain is perceived when one's hand is placed on a grid of metal rods that alternate between cool and warm.

Figure 7.16 When experienced as part of a ritual, normally excruciating conditions (e.g., walking on hot coals) often produce little pain.

VanVang/Alamy Stock Photo

Figure 7.17 illustrates the descending analgesia circuit first hypothesized by Basbaum and Fields (1978). They proposed that the output of the PAG excites the serotonergic neurons of the *raphé nuclei* (a cluster of serotonergic nuclei in the core of the medulla), which in turn project down the dorsal columns of the spinal cord and excite interneurons that block incoming pain signals in the dorsal horn.

Descending analgesia pathways have been the subject of intensive investigation since the first model was proposed by Basbaum and Fields in 1978. In order to incorporate the mass of accumulated data, models of the descending analgesia circuits have grown much more complex (see Lau & Vaughan, 2014). For example, both the anterior cingulate cortex and prefrontal cortex are now believed to be important components of descending analgesia circuits (see Davis et al., 2017).

Neuropathic Pain

LO 7.15 Define neuropathic pain and describe some of its putative neural mechanisms.

In most cases, plasticity of the human nervous system helps it function more effectively. In the case of neuropathic pain,

just the opposite is true (see Luo, Kuner, & Kuner, 2014). **Neuropathic pain** is severe chronic pain in the absence of a recognizable pain stimulus. A typical case of neuropathic pain develops after an injury: Once the injury heals, there seems to be no reason for further pain, but the patient experiences chronic excruciating pain. In many cases, attacks of neuropathic pain can be triggered by an innocuous stimulus, such as a gentle touch.

Although neuropathic pain may be perceived to be in a limb—even in an amputated limb (see Chapter 10)—it is caused by abnormal activity in the CNS. Thus, cutting nerves from the perceived location of the pain often brings little or no comfort. And, unfortunately, medications that have been developed to treat the pain associated with injury are usually ineffective against neuropathic pain.

There are three promising lines of research into the neural mechanisms of neuropathic pain. First, recent research has implicated aberrant microglial activity in neuropathic

Figure 7.17 Basbaum and Fields's (1978) model of the descending analgesia circuit.

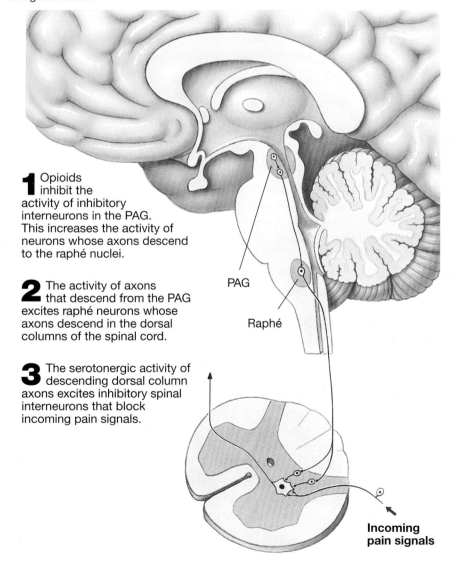

1 Opioids inhibit the activity of inhibitory interneurons in the PAG. This increases the activity of neurons whose axons descend to the raphé nuclei.

2 The activity of axons that descend from the PAG excites raphé neurons whose axons descend in the dorsal columns of the spinal cord.

3 The serotonergic activity of descending dorsal column axons excites inhibitory spinal interneurons that block incoming pain signals.

PAG

Raphé

Incoming pain signals

pain, including the induction of neuroplastic changes that lead to the persistence of pain long after the injury has healed (see Inoue & Tsuda, 2018; Kuner & Flor, 2017). Second, there is considerable evidence supporting the involvement of *epigenetic mechanisms* (see Chapter 2) in neuropathic pain (see Birklein et al., 2018; Zhang et al., 2016). Drugs are currently being developed to modify such epigenetic changes in order to treat neuropathic pain (see Neiderberger et al., 2017). Third, the aforementioned neuroplastic and epigenetic changes are most prominent in the anterior cingulate cortex (Bliss et al., 2016; Zhuo, 2016) and prefrontal cortex (see Peng et al., 2017)—structures that are both involved in descending analgesia pathways (see Davis et al., 2017).

Chemical Senses: Smell and Taste

Olfaction (smell) and *gustation* (taste) are referred to as the chemical senses because their function is to monitor the chemical content of the environment. Smell is the response of the olfactory system to airborne chemicals that are drawn by inhalation over receptors in the nasal passages, and taste is the response of the gustatory system to chemicals in solution in the oral cavity.

Adaptive Roles of the Chemical Senses

LO 7.16 Describe two adaptive roles for the chemical senses.

When we are eating, smell and taste act in concert. Molecules of food excite both smell and taste receptors and produce an integrated sensory impression termed **flavor**. The contribution of olfaction to flavor is often underestimated, but you won't make this mistake if you remember that people with no sense of smell have difficulty distinguishing the flavors of apples and onions. Flavor is also influenced by a number of other factors such as the temperature, texture, and appearance of the food and a person's level of satiety (see Chaudhauri & Roper, 2010; Rolls et al., 2010).

In humans, the main adaptive role of the chemical senses is the evaluation of potential foods (i.e., encouraging the consumption of sources of energy and nutrients while avoiding toxins) in natural environments, where potential foods do not come with labels (see Yarmolinsky, Zuker, & Ryba, 2009). However, in many other species, the chemical senses also play a major role in regulating social interactions. The members of many species release **pheromones**—chemicals that influence the physiology and behavior of *conspecifics* (members of the same species)—see Stowers and Kuo (2015); Leighton and Sternberg (2016).

For example, Murphy and Schneider (1970) showed that the sexual and aggressive behavior of hamsters is under pheromonal control. Normal male hamsters attack and kill unfamiliar males that are placed in their colonies, whereas they mount and impregnate unfamiliar sexually receptive females. However, male hamsters that are unable to smell the intruders engage in neither aggressive nor sexual behavior. Murphy and Schneider confirmed the olfactory basis of hamsters' aggressive and sexual behavior in a particularly devious fashion. They swabbed a male intruder with the vaginal secretions of a sexually receptive female before placing it in an unfamiliar colony; in so doing, they converted it from an object of hamster assassination to an object of hamster lust.

The possibility that humans may release sexual pheromones has received considerable attention because of its financial and recreational potential. However, there is little evidence that humans release or respond to sexual pheromones (see Wyatt, 2017).

Olfactory System

LO 7.17 Describe the olfactory system.

The olfactory system is illustrated in Figure 7.18. The olfactory receptor cells are located in the upper part of the nose, embedded in a layer of mucus-covered tissue called the **olfactory mucosa**. Their dendrites are located in the nasal passages, and their axons pass through a porous portion of the skull (the *cribriform plate*) and enter the **olfactory bulbs**, where they synapse on neurons that project via the *olfactory tracts* to the brain.

For decades, it was assumed that there were only a few types of olfactory receptors. Different profiles of activity in a small number of receptor types were thought to lead to the perception of various smells—in the same way that the profiles of activity in three types of cones were once thought to lead to the perception of colors. Then, at the turn of the 21st century, it was discovered that rats and mice have about 1,000 different kinds of olfactory receptor proteins and that humans have about 300 (see Grabska-Barwińska et al., 2017; Uchida, Poo, & Haddad, 2014).

In mammals, each olfactory receptor cell contains only one type of receptor protein molecule (see Giesel & Datta, 2014; Uchida, Poo, & Haddad, 2014). Olfactory receptor proteins are in the membranes of the dendrites of the olfactory receptor cells, where they can be stimulated by circulating airborne chemicals in the nasal passages. Researchers have attempted to discover the functional principle by which the various receptors are distributed through the olfactory mucosa. If there is such a principle, it has not yet been discovered: All types of olfactory receptors appear to be scattered throughout the mucosa, providing no clue about the organization of the system. Because each type of olfactory receptor responds in varying degrees to a wide variety of

Figure 7.18 The human olfactory system.

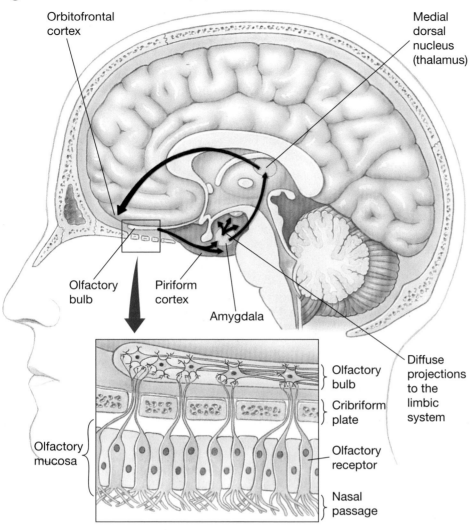

Orbitofrontal cortex

Medial dorsal nucleus (thalamus)

Olfactory bulb

Piriform cortex

Amygdala

Olfactory mucosa

Olfactory bulb

Cribriform plate

Olfactory receptor

Nasal passage

Diffuse projections to the limbic system

- There is mirror symmetry between the left and right olfactory bulbs—glomeruli sensitive to particular odors tend to be located at the same sites on the two bulbs.

- The glomeruli sensitive to particular odors are arrayed on the olfactory bulbs in the same way in different members of the same species.

- The layout of the glomeruli is similar in related species (i.e., rats and mice).

Although it is clear that the olfactory bulbs are organized topographically and that the layout is not random (see Wanner et al., 2016), the topographic principle according to which the glomeruli are arrayed has yet to be discovered (see Murthy, 2011). The poorly understood topographic organization of the olfactory bulbs has been termed a **chemotopic map** (see Falasconi et al., 2012).

New olfactory receptor cells are created throughout each individual's life to replace those that have deteriorated. Once created, the new receptor cells develop axons, which grow until they reach the appropriate target sites in the olfactory bulb. Each new olfactory receptor cell survives only a few weeks before being replaced. How the axons from newly formed receptors scattered about the nasal mucosa find their target glomeruli in the olfactory bulb remains a mystery (see Mori & Sakano, 2011).

Each olfactory bulb projects axons to several structures of the medial temporal lobes, including the amygdala and the **piriform cortex**—an area of medial temporal cortex adjacent to the amygdala (see Bekkers & Suzuki, 2013). The piriform cortex is considered to be primary olfactory cortex, but this designation is somewhat arbitrary (see Gottfried, 2010). The olfactory system is the only sensory system whose major sensory pathway reaches the cerebral cortex without first passing through the thalamus.

Two major olfactory pathways leave the amygdala-piriform area. One projects diffusely to the limbic system, and the other projects via the **medial dorsal nuclei** of the thalamus to the **orbitofrontal cortex**—the area of cortex on the inferior surface of the frontal lobes next to the *orbits* (eye sockets)—see Mainland et al. (2014). The limbic pathway

odors, each odor seems to be encoded by component processing—that is, by the pattern of activity across different receptor types (see Giesel & Datta, 2014).

The axons of olfactory receptors terminate in discrete clusters of neurons that lie near the surface of the olfactory bulbs—these clusters are called the **olfactory glomeruli**. Each glomerulus receives input from several thousand olfactory receptor cells, all with the same receptor protein (see Gupta, Albeanu, & Bhalla, 2015; Wanner et al., 2016; Tian et al., 2016). In mice, there are one or two glomeruli in each olfactory bulb for each receptor protein type (see Schoppa, 2009).

Because systematic topographic organization is apparent in other sensory systems (e.g., *retinotopic* and *tonotopic* layouts), researchers have been trying to discover whether glomeruli sensitive to particular odors are arrayed systematically on the surfaces of the olfactory bulbs. Indeed, three lines of evidence indicate that there is a systematic layout (see Cheetham & Belluscio, 2014; Tsai & Barnea, 2014):

is thought to mediate the emotional response to odors; the thalamic-orbitofrontal pathway is thought to mediate the conscious perception of odors.

Gustatory System

LO 7.18 Describe the gustatory system.

Taste receptor cells are found on the tongue and also throughout the gastrointestinal tract (see Nolden & Feeney, 2020). On the tongue, they typically occur in clusters of 50 to 100 called **taste buds** (see Barretto et al., 2015) that are often located around small protuberances called *papillae* (singular *papilla*). The relation between taste receptors, taste buds, and papillae is illustrated in Figure 7.19.

The 50 to 100 receptor cells that compose each taste bud is said to be one of three types: (1) cells that detect bitter,

sweet and *umami* (savory); (2) cells that detect sour; and (3) cells that detect salty (see Roper & Chaudhari, 2017). In each taste bud, only one of the receptor cells, the *presynaptic cell*, synapses onto the neuron carrying signals away from the bud; communication among the other cells of a taste bud appears to occur via gap junctions (see Dando & Roper, 2009). Like olfactory receptor cells, gustatory receptor cells survive only a few weeks before being replaced by new cells. The tastes that have been most studied are sweet, sour, bitter, salty, and *umami*, but a case can be made for others (see Liman, Zhang, & Montell, 2014; Vincis & Fontanini, 2016).

Taste transduction for sweet, umami, and bitter is mediated by *metabotropic* receptors. There are two metabotropic receptors for sweet, one for umami, and about 25 for bitter. By contrast, taste transduction for salty and sour is mediated by *ionotropic* receptors. Sour is transduced by three different ionotropic receptors, and salty is mediated by two (see Nolden & Feeney, 2020). As in the olfactory system, there appears to be only one type of receptor protein per each receptor cell.

The major pathways over which gustatory signals are conducted to the cortex are illustrated in Figure 7.20. Gustatory afferents leave the mouth as part of the *facial* (VII), *glossopharyngeal* (IX), and *vagus* (X) *cranial nerves*, which carry information from the front of the tongue, the back of the tongue, and the back of the oral cavity, respectively. These fibers all terminate in the **solitary nucleus** of the medulla, where they synapse on neurons that project to the *ventral posterior nucleus* of the thalamus. The gustatory axons of the ventral posterior nucleus project to the *primary gustatory cortex*, which is in the *insula*, an area of cortex hidden in the lateral fissure (see Linster & Fontanini, 2014). A different area of primary gustatory cortex represents each taste (see Peng et al., 2015). Secondary gustatory cortex is in the orbitofrontal cortex (see Figure 7.20). Unlike the projections of other sensory systems, the projections of the gustatory system are primarily ipsilateral.

Some evidence suggests that the primary gustatory cortex, like primary olfactory cortex, is chemotopically organized. Schoenfeld et al. (2004) measured fMRI responses to the five primary tastes and found that each primary taste produced

---

Figure 7.19 Taste receptors, taste buds, and papillae on the surface of the tongue, and a cross-section of a papilla that shows a taste bud and its taste receptors. Two sizes of papillae are visible in the photograph; only the larger papillae contain taste buds and receptors.

Surface of Tongue

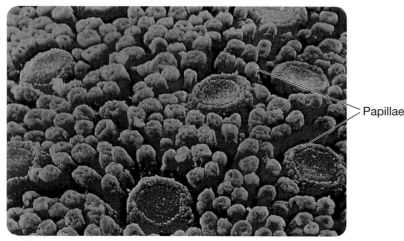

Omikron / Science Source

Cross Section of a Papilla

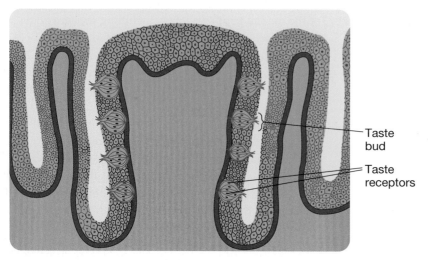

Figure 7.20 The human gustatory system.

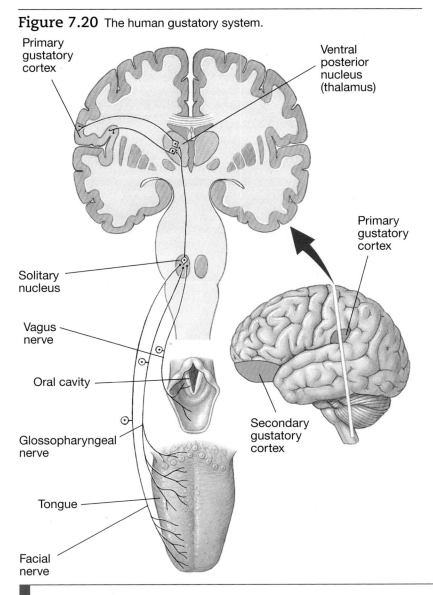

Primary gustatory cortex

Ventral posterior nucleus (thalamus)

Primary gustatory cortex

Solitary nucleus

Vagus nerve

Oral cavity

Glossopharyngeal nerve

Secondary gustatory cortex

Tongue

Facial nerve

activity in a different area of primary gustatory cortex. The chemotopic map was different in each volunteer, and there was considerable overlap of the five areas, but the map in each volunteer was stable over time. A similar finding has been reported in mice (Chen et al., 2011).

Brain Damage and the Chemical Senses

LO 7.19 **Explain the potential effects of brain damage on the chemical senses.**

The inability to smell is called **anosmia**; the inability to taste is called **ageusia**. The most common neurological cause of anosmia is a blow to the head that causes a displacement of the brain within the skull and shears the olfactory nerves where they pass through the cribriform plate. Less complete deficits in olfaction have been linked to a wide variety of neurological disorders including Alzheimer's disease, Down syndrome, epilepsy, multiple sclerosis, Korsakoff's syndrome, and Parkinson's disease (see Godoy et al., 2015).

Ageusia is rare, presumably because sensory signals from the mouth are carried via three separate pathways. However, partial ageusia, limited to the anterior two-thirds of the tongue on one side, is sometimes observed after damage to the ear on the same side of the body. This is because the branch of the facial nerve (VII) that carries gustatory information

---

Scan Your Brain

Now that you have reached the threshold of this chapter's final module, a module that focuses on attention, you should scan your brain to test your knowledge of the sensory systems covered in the preceding modules. Complete each sentence with terms related to the appropriate system. The correct answers are provided at the end of the exercise. Before proceeding, review material related to your incorrect answers and omissions.

1. Information from the receptors in the muscles, joints, and balance organs about the position of the body is monitored by the _____ system.

2. The largest and deepest receptors of the skin that respond to sudden displacements of the skin but not to constant pressure are called _____.

3. Some skin receptors such as the Ruffini endings and _____ adapt slowly, allowing the static properties of tactile stimuli to become unnoticeable.

4. Most of the output of the secondary somatosensory cortex goes to the association cortex of the _____

_____ lobe and participates in the perception of objects' shapes.

5. A rare neuropsychological condition known as _____ affects the ability to recognize one's own body parts.

6. Some of the areas of the brain implicated in pain perception are the thalamus, SI and SII, the anterior cingulate gyrus, and the _____.

7. Sexual and aggressive behavior in animals, such as hamsters, are controlled by _____.

8. Clusters of neurons that lie near the surface of the olfactory bulb are called olfactory _____.

9. The emotional response to odors is thought to be mediated by the _____ system.

10. Gustatory signals leave the back of the tongue reaching the solitary nucleus of the medulla and from there to the thalamus. The final destination is the primary gustatory cortex located in the _____.

Scan Your Brain answers: (1) proprioceptive, (2) Pacinian corpuscles, (3) Merkel's disks, (4) posterior parietal, (5) asomatognosia, (6) insula, (7) pheromones, (8) glomeruli, (9) limbic, (10) insula.

from the anterior two-thirds of the tongue passes through the middle ear.

Perception

The three previous modules examined four of our five sensory systems. However, knowing how sensory systems work leaves unanswered how we construct a rich multisensory **percept** (the outcome of perception) of the world from moment to moment. As you will see, our nervous system makes decisions about how we perceive incoming sensory information based on our prior experiences.

Role of Prior Experience in Perception

LO 7.20 Use examples to illustrate the role of experience in perception.

Prior knowledge has a major influence on how we perceive the world. For example, in our visual world, shadows usually appear beneath objects rather than above them, because the sun is our primary source of light. Moreover, our knowledge of the identity of objects affects our perception. Check It Out: How Knowledge Affects Perception on the next page illustrates each of these points (see also de Lange, Heilbron, & Kok, 2018).

Our prior knowledge about the temporal order of sensory events also affects our perception (see Nobre & van Ede, 2018). For example, through learning we come to know the order in which colored lights appear at a stoplight (e.g., green to yellow to red to green, etc.).

Perceptual Decision Making

LO 7.21 Explain perceptual decision making, using some examples of phantom percepts to illustrate.

Many researchers believe that we create mental models of the world based on predictable and recurring sensory events. According to this theory, we are "prediction machines" that actively construct models of ourselves and our world via our five senses (see de Lange, Heilbron, & Kok, 2018; Freedman & Assad, 2016; Whitney & Leib, 2018; Murray et al., 2016). That is, from moment to moment, our brain is making decisions about what we should perceive, and those decisions are based on prior experiences and current incoming sensory information. It appears that such ongoing **perceptual decision making** consumes a large proportion of the energy used by our brains (see Richmond & Zacks, 2017). The phenomenon of **bistable figures** (see Check It Out: Bistable Figures) illustrates perceptual decision making at work.

Check It Out
Bistable Figures

The image below is an example of a bistable figure: While viewing a bistable figure, our percept alternates between that of two faces and that of a chalice. Interestingly, this alternation between percepts occurs at a regular frequency, and that frequency is different from person to person (see Brascamp et al., 2018).

Even in the absence of sensory input, we still perceive. This is illustrated by the existence of **phantom percepts**: the products of perception when there is an absence of sensory input (see Mohan & Vanneste, 2017). One example of this is the phenomenon of **phantom limbs**, wherein amputees perceive the presence of their missing limb long after it has been lost to injury or amputation.

Another example is what happens when an individual is deprived of visual input later in life, such as is the case for individuals who develop *glaucoma* (a medical condition wherein there is irreversible damage to the optic nerve). Some of these individuals experience rich and complex hallucinations (e.g., people's faces, complex landscapes). This condition is known as **Charles Bonnet Syndrome** (see Mohan & Vanneste, 2017; Sacks, 2012). The characteristics of the phantom percepts in Charles Bonnet syndrome seem to be dependent on the person's experience. For example, some musicians with Charles Bonnet syndrome experience phantom percepts of musical notation (see Sacks, 2013).

The mechanisms by which we make perceptual decisions are not entirely clear, but there is mounting evidence that such decisions are mediated by several areas in the brain, including

Check It Out

How Knowledge Affects Perception

The image below illustrates that we perceive the world based on our knowledge of the how the world works. That is, you perceive each circle as either convex or concave, such that in all instances it looks like the shading is due to a light source coming from above. This is presumably because we know that our most common source of light, the sun, shines down on objects.

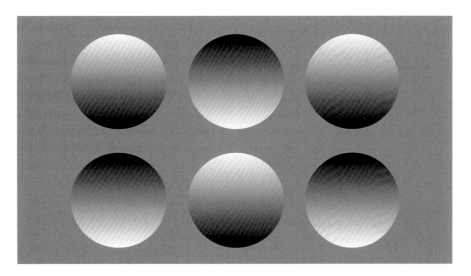

The next part of this Check It Out requires that you be the experimenter! Cover up the right image and then ask someone else what they see in the left image. Next, uncover the right image. Now, ask them what they see in the left image. The fact that they have a richer percept (i.e., they see the dalmation in it) of the left image after viewing the right image illustrates the effect of prior knowledge on perception.

Mala Iryna/Shutterstock

the dorsolateral prefrontal cortex and the posterior parietal cortex (see Brascamp et al., 2018; Fleming, van der Putten, & Daw, 2018). Interestingly, both structures are involved in the decision to initiate a physical movement (see Chapter 8). That is, those brain structures involved in action decision making are the same as those implicated in perceptual decision making.

The Binding Problem

LO 7.22 Explain the binding problem and describe two potential solutions to it.

Sensory systems are characterized by a division of labor: Multiple specialized areas, at multiple levels, are interconnected by multiple parallel pathways. For example, each area of the visual system is specialized for perceiving specific aspects of visual scenes (e.g., shape, color, movement). Yet, complex stimuli are normally perceived as integrated wholes, not as combinations of independent attributes. How does the brain combine individual sensory attributes to produce integrated perceptions? This is called the **binding problem** (see Bizley, Maddox, & Lee, 2016; Feldman, 2013; but see Di Lollo, 2012).

One possible solution to the binding problem is that there is a single area of the cortex at the top of the sensory hierarchy that receives signals from all other areas of the various sensory systems and puts them together to form a percept. One area of the brain that has received recent attention as the potential location for the binding of sensory information is the *claustrum* (see Figure 7.21), a structure that is made up of a fine sheet of neurons located just underneath the cortex towards the middle of the brain (see Goll, Atlan, & Citri, 2015; Tan et al., 2017).

An alternative solution to the binding problem is that there is no single area responsible for putting together perceptions. Rather, perceptions might be the result of multiple interactions at each of the cortical levels of the hierarchy.

Figure 7.21 The location of the claustrum.

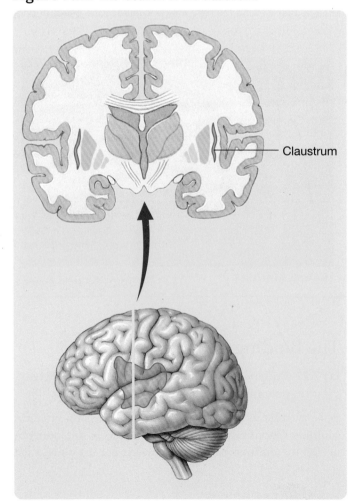

This solution is informed by the study of communication between sensory areas in the cortex. For example, we now know that, beginning at the level of the primary sensory cortices, there is integration of information from multiple senses (see Murray et al., 2016). Accordingly, it might be the case that binding emerges from the constant exchange of information between the many sensory cortices, both within and across sensory modalities (see Bizley, Maddox, & Lee, 2016). A role for subcortical structures in this information exchange has also been suggested (see Romo & Rossi-Pool, 2020).

Selective Attention

We consciously perceive only a small subset of the many stimuli that excite our sensory organs at any one time and largely ignore the rest (see Peelen & Kastner, 2014; Squire et al., 2013). The process by which this occurs is selective attention.

Characteristics of Selective Attention

LO 7.23 Describe the two characteristics of selective attention and explain what is meant by exogenous versus endogenous attention.

Selective attention has two characteristics: It improves the perception of the stimuli that are its focus, and it interferes with the perception of the stimuli that are not its focus (see Sprague, Saproo, & Serences, 2015). For example, if you focus your attention on a potentially important announcement in a noisy airport, your chances of understanding it increase, but your chances of understanding a simultaneous comment from a traveling companion decrease.

Attention can be focused in two different ways: by internal cognitive processes (*endogenous attention*) or by external events (*exogenous attention*)—see Chica, Bartolomeo, and Lupiáñez (2013), but see Macaluso and Doricchi (2013). For example, your attention can be focused on a tabletop because you are searching for your keys (endogenous attention), or it can be drawn there because your cat tipped over a lamp (exogenous attention). Endogenous attention is thought to be mediated by **top-down** (from higher to lower levels) neural mechanisms, whereas exogenous attention is thought to be mediated by **bottom-up** (from lower to higher levels) neural mechanisms (see Miller & Buschman, 2013).

Eye movements often play an important role in visual attention, but it is important to realize that visual attention can be shifted without shifting the direction of visual focus (see Krauzlis, Lovejoy, & Zénon, 2013). To prove this to yourself, look at the next Check It Out demonstration.

One other important characteristic of selective attention has been called the cocktail-party phenomenon (see Du et al., 2011). The **cocktail-party phenomenon** is the fact that even when you are focusing so intently on one conversation

Check It Out

Shifting Visual Attention Without Shifting Visual Focus

Fix your gaze on the +; concentrate on it. Next, shift your attention to one of the letters without shifting your gaze from +. Now, shift your attention to other letters, again without shifting your gaze from the +. You have experienced *covert attention*—a shift of visual attention without any corresponding eye movement. A change in visual attention that involves a shift in gaze is called *overt attention*.

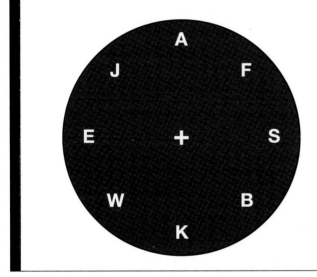

that you are totally unaware of the content of other conversations going on around you, the mention of your name in one of the other conversations will immediately gain access to your consciousness. This phenomenon suggests that your brain can block from conscious awareness all stimuli except those of a particular kind while still unconsciously monitoring the blocked-out stimuli just in case something comes up that requires your attention.

Change Blindness

LO 7.24 Describe the phenomenon of change blindness.

There is no better illustration of the importance of attention than the phenomenon of change blindness (Land, 2014). To study **change blindness**, a volunteer is shown a photo on a computer screen and asked to report any change in the image as soon as it is noticed. In fact, the image is composed of two images that alternate with a delay of less than 0.1 second between them. The two photographic images are identical except for one gross feature. For example, the two images in Figure 7.22 are identical except that the picture in the center of the wall is missing from one. You might think that any person would immediately notice the picture disappearing and reappearing. But this is not what happens. Most volunteers spend many seconds staring at the image—searching, as instructed, for some change—before they notice the disappearing and reappearing picture. When they finally notice it, they wonder in amazement why it took them so long.

Why does change blindness occur? It occurs because, contrary to our impression, when we view a scene, we have absolutely no memory for parts of the scene that are not the focus of our attention. When viewing the scene in Figure 7.22, most volunteers attend to the two people and do not notice when the picture disappears from the wall between them. Because they have no memory of the parts of the image to which they did not attend, they are not aware when those parts change.

The change blindness phenomenon does not occur without the brief (i.e., less than 0.1 second) intervals between

Figure 7.22 The change blindness phenomenon. These two illustrations were continually alternated, with a brief (less than 0.1 second) interval between each presentation, and the subjects were asked to report any changes they noticed. Amazingly, it took most of them many seconds to notice the disappearing and reappearing building in the distance, behind the person on the left.

Creativa Images/Shutterstock

Creativa Images/Shutterstock

images. Without the intervals, no memory is required and the changes are perceived immediately.

Neural Mechanisms of Attention

LO 7.25 Describe the neural mechanisms of attention.

Where do top-down attentional influences on sensory systems originate? There is a general consensus that both prefrontal cortex and posterior parietal cortex play major roles in directing top-down attention (see Baluch & Itti, 2011; Noudoost et al., 2010).

Moran and Desimone (1985) were the first to demonstrate the effects of attention on neural activity. They trained monkeys to stare at a fixation point on a screen while they recorded the activity of neurons in a prestriate area that was part of the ventral stream and particularly sensitive to color. In one experiment, they recorded from individual neurons that responded to either red or green bars of light in their receptive fields. When the monkey was trained to perform a task that required attention to the red cue, firing of the neurons in response to the red cue was increased, while the response to the green cue was reduced. The opposite happened when the monkey was required to attend to green.

Experiments paralleling those in monkeys have been conducted in humans using functional brain-imaging techniques. For example, Corbetta and colleagues (1990) presented a collection of moving, colored stimuli of various shapes and asked volunteers to discriminate among the stimuli based on their movement, color, or shape. Attention to shape or color produced increased activity in areas of the ventral stream; attention to movement produced increased activity in an area of the dorsal stream.

In another study of attention in human volunteers, Ungerleider and Haxby (1994) showed volunteers a series of faces. The volunteers were asked whether the faces belonged to the same person or whether they were located in the same position relative to the frame. When they were attending to identity, regions of the ventral stream were more active; when they were attending to position, regions of the dorsal stream were more active.

The preceding studies indicate the principle by which the neural mechanisms of selective attention work. Selective attention works by strengthening the neural responses to attended-to stimuli and by weakening the responses to others (see Buschman, 2015; Luo & Maunsell, 2015). This dual mechanism has been termed a *push–pull mechanism* (see Stevens & Bavelier, 2012).

Some neural mechanisms of attention involve a surprising degree of neural plasticity. For example, the location of the receptive fields of visual neurons, which had been assumed to be a static property of visual neurons, can be shifted by spatial attention (see Anton-Erxleben & Carrasco, 2013). Recording from neurons in an area of monkey secondary visual cortex in the dorsal stream, Wommelsdorf and colleagues (2006) found that the receptive fields of many of the neurons shifted toward points in the visual field to which the subjects were attending. Similarly, Rolls (2008) found that visual receptive fields of inferotemporal cortex neurons shrink to become little more than the size of objects on which they are focusing.

Covert attention is the process of attending to a sensory stimulus without fixing one's gaze (see the Check It Out on page 209). Covert attention is of great interest to biopsychologists because it allows them to study the neural mechanisms of shifts in **attentional gaze** (i.e., the shift in attention from one perceptual object to another) without worrying about the confounding effects of eye movements. Studies of the neural mechanisms of attentional gaze shifts have identified the **frontal eye field** (an area on the ventral surface of the frontal cortex) as important.

The frontal eye field, which is also active during eye movements, is active during shifts in attentional gaze. Moreover, stimulation of neurons in the frontal eye field, at intensities below the threshold necessary to elicit eye movements, leads to improvements in the detection of stimuli via covert attention (see Moore & Zirnsak, 2017). Thus, it appears that the frontal eye field mediates shifts in visual attention, though other brain areas have also been implicated (e.g., Schmitz & Duncan, 2018).

If you concluded from the foregoing discussion that most of the research on the neural mechanisms of selective attention has focused on visual attention, you would be correct. However, there are also some studies of attention to auditory (e.g., Shamma, Elhilali, & Micheyl, 2011), somatosensory (e.g., Fujiwara et al., 2002), gustatory (e.g., Stevenson, 2012; Veldhuizen, Gitelman, & Small, 2012), and olfactory (e.g., Veldhuizen & Small, 2011) stimuli.

Simultanagnosia

LO 7.26 Describe the disorder of attention known as simultanagnosia.

We have not forgotten that we asked you to think about the patient whose case opened this chapter. He could identify objects in any part of his visual field if they were presented individually; thus, he was not suffering from blindness or other visual field-defects. His was a disorder of attention called **simultanagnosia**. Specifically, he suffered from *visual-simultanagnosia*—a difficulty in attending visually to more than one object at a time. Because the dorsal stream (which includes the posterior parietal association cortex) is responsible for visually localizing objects in space, you may have hypothesized that the patient's problem was associated with damage to this area. If you did, you were correct. Simultanagnosia is usually associated with bilateral damage to the posterior parietal cortex.

Themes Revisited

The clinical implications theme was prominent in this chapter, but you saw it in a different light. Previous chapters discussed how biopsychological research is leading to the development of new treatments; this chapter focused exclusively on what particular clinical cases have revealed about the organization of healthy sensory systems. The following cases played a key role in this chapter: the patient with visual simultanagnosia; Dr. P., the visual agnosic who mistook his wife for a hat; Aunt Betty, the asomatognosic who lost the left side of her body; Miss C., the student who felt no pain and died as a result; and W.M., the man who reduced his scotoma with his hand.

The neuroplasticity theme was also developed in this chapter: the neuroplasticity theme and the evolutionary perspective theme. Although this chapter did not systematically discuss the plasticity of sensory systems, three important examples of sensory system plasticity were mentioned: the effects of tinnitus on the auditory system, partial recovery of vision in a scotoma by placing a hand in the scotoma, and the movement of the receptive fields of visual neurons toward a location that is the focus of attention.

The thinking creatively theme came up once. The case of Miss C. taught us that pain is a positive sensation that we can't live without.

The two emerging themes were also present in this chapter. The thinking about epigenetics theme arose when we saw that epigenetic mechanisms are important for neuropathic pain and that drug therapies are being developed to alter those epigenetic mechanisms. The consciousness theme was present throughout this chapter, though it was truly brought to the forefront in the modules on perception and attention.

Key Terms

Exteroceptive sensory systems, p. 184
Sensation, p. 185
Perception, p. 185

Principles of Sensory System Organization

Primary sensory cortex, p. 185
Secondary sensory cortex, p. 185
Association cortex, p. 185
Hierarchical organization, p. 185
Functional segregation, p. 186
Parallel processing, p. 186

Auditory System

Fourier analysis, p. 187
Tympanic membrane, p. 188
Ossicles, p. 189
Oval window, p. 189
Cochlea, p. 189
Organ of Corti, p. 189
Hair cells, p. 189
Basilar membrane, p. 189
Tectorial membrane, p. 189
Auditory nerve, p. 189
Retinotopic, p. 189
Tonotopic, p. 189
Semicircular canals, p. 189
Vestibular system, p. 189
Superior olives, p. 189
Inferior colliculi, p. 189

Medial geniculate nuclei, p. 189
Periodotopy, p. 190
Tinnitus, p. 192

Somatosensory System: Touch and Pain

Stereognosis, p. 194
Dorsal-column medial-lemniscus system, p. 195
Anterolateral system, p. 195
Dorsal columns, p. 195
Medial lemniscus, p. 195
Ventral posterior nucleus, p. 195
Somatotopic, p. 196
Somatosensory homunculus, p. 196
Astereognosia, p. 198
Asomatognosia, p. 198
Anosognosia, p. 198
Contralateral neglect, p. 198
Rubber-hand illusion, p. 199
Thermal grid illusion, p. 200
Anterior cingulate cortex, p. 200
Periaqueductal gray (PAG), p. 200
Endorphins, p. 200
Neuropathic pain, p. 201

Chemical Senses: Smell and Taste

Flavor, p. 202
Pheromones, p. 202
Olfactory mucosa, p. 202

Olfactory bulbs, p. 202
Olfactory glomeruli, p. 203
Chemotopic, p. 203
Piriform cortex, p. 203
Medial dorsal nuclei, p. 203
Orbitofrontal cortex, p. 203
Taste buds, p. 204
Solitary nucleus, p. 204
Anosmia, p. 205
Ageusia, p. 205

Perception

Percept, p. 206
Perceptual decision making, p. 206
Bistable figures, p. 206
Phantom percepts, p. 206
Phantom limbs, p. 206
Charles Bonnet syndrome, p. 206
Binding problem, p. 208

Selective Attention

Selective attention, p. 208
Top-down, p. 208
Bottom-up, p. 208
Cocktail-party phenomenon, p. 208
Change blindness, p. 209
Attentional gaze, p. 210
Frontal eye field, p. 210
Simultanagnosia, p. 210

Chapter 8
The Sensorimotor System

How You Move

Matteo Carta/Alamy Stock Photo

⌄ Chapter Overview and Learning Objectives

Three Principles of Sensorimotor Function	**LO 8.1**	In the context of the sensorimotor system, explain what *hierarchically organized* means.
	LO 8.2	Explain the important role of sensory input for motor output.
	LO 8.3	Describe how learning changes the nature and locus of sensorimotor control.
	LO 8.4	Describe and/or draw the general model of sensorimotor function.
Sensorimotor Association Cortex	**LO 8.5**	Explain the role of the posterior parietal cortex in sensorimotor function and describe what happens when it is damaged or stimulated.
	LO 8.6	Explain the role of the dorsolateral prefrontal association cortex in sensorimotor function and describe the response properties of neurons in this region of cortex.

Secondary Motor Cortex	**LO 8.7**	Explain the general role of areas of secondary motor cortex.
	LO 8.8	Describe the major features of mirror neurons and explain why they have received so much attention from neuroscientists.
Primary Motor Cortex	**LO 8.9**	Describe the conventional view of primary motor cortex function and the evidence upon which it was based.
	LO 8.10	Describe the current view of primary motor cortex function and the evidence upon which it is based.
Cerebellum and Basal Ganglia	**LO 8.11**	Describe the structure and connectivity of the cerebellum and explain the current view of cerebellar function.
	LO 8.12	Describe the anatomy of the basal ganglia and explain the current view of their function.
Descending Motor Pathways	**LO 8.13**	Compare and contrast the two dorsolateral motor pathways and the two ventromedial motor pathways.
Sensorimotor Spinal Circuits	**LO 8.14**	Describe the components of a motor unit and distinguish between the different types of muscles.
	LO 8.15	Describe the receptor organs of tendons and muscles.
	LO 8.16	Describe the stretch reflex and explain its mechanism.
	LO 8.17	Describe the withdrawal reflex and explain its mechanism.
	LO 8.18	Explain what is meant by *reciprocal innervation*.
	LO 8.19	Explain recurrent collateral inhibition.
	LO 8.20	Describe the phenomenon of walking and the degree to which it is controlled by spinal circuits.
Central Sensorimotor Programs and Learning	**LO 8.21**	Explain what is meant by a hierarchy of central sensorimotor programs and explain the importance of this arrangement for sensorimotor functioning.
	LO 8.22	Describe the various characteristics of central sensorimotor programs.
	LO 8.23	Explain how the classic Jenkins and colleagues PET study of simple motor learning summarizes the main points of this chapter.
	LO 8.24	Describe two examples of neuroplasticity—one at the cortical level and one at the subcortical level.

The evening before we started to write this chapter, I (JP) was standing in a checkout line at the local market. As I waited, I scanned the headlines on the prominently displayed magazines—WOMAN GIVES BIRTH TO CAT; FLYING SAUCER LANDS IN CLEVELAND SHOPPING MALL; HOW TO LOSE 20 POUNDS IN 2 DAYS. Then, my mind began to wander, and I started to think about beginning to write this chapter. That is when I began to watch Rhonelle's movements, and to wonder about the neural system that controlled them. Rhonelle is a cashier—the best in the place.

The Case of Rhonelle, the Dexterous Cashier

I was struck by the complexity of even Rhonelle's simplest movements. As she deftly transferred a bag of tomatoes to the scale, there was a coordinated adjustment in almost every part of her body. In addition to her obvious finger, hand, arm, and shoulder movements, coordinated movements of her head and eyes tracked her hand to the tomatoes; and there were adjustments in the muscles of her feet, legs, trunk, and other arm, which kept her from lurching forward. The accuracy of these responses suggested that they were guided in part by the patterns of visual, somatosensory, and vestibular changes they produced. (The term *sensorimotor* in the title of this chapter formally recognizes the critical contribution of sensory input to guiding motor output.)

As my purchases flowed through her left hand, Rhonelle scanned the items with her right hand and bantered with Rich, the bagger. I was intrigued by how little of what Rhonelle was doing appeared to be under her conscious control. She made general decisions about which items to pick up and where to put them, but she seemed to give no thought to the exact means by which these decisions were carried out. Each of her responses could have been made with an infinite number of different combinations of finger, wrist, elbow, shoulder, and body adjustments; but somehow she unconsciously picked one. The higher parts of her sensorimotor system—perhaps her cortex—seemed to issue conscious general commands to other parts of the system, which unconsciously produced a specific pattern of muscular responses that carried them out.

The automaticity of Rhonelle's performance was a far cry from the slow, effortful responses that had characterized her first days at the market. Somehow, experience had integrated her individual movements into smooth sequences, and it seemed to have transferred the movements' control from a mode that involved conscious effort to one that did not.

I was suddenly jarred from my contemplations by a voice. "Sir, excuse me, sir, that will be $78.65," Rhonelle said, with just a hint of delight at catching me mid-daydream. I hastily paid my bill, muttered "thank you," and scurried out of the market.

As we write this, I am smiling both at my own embarrassment and at the thought that Rhonelle has unknowingly introduced you to three principles of sensorimotor control that are the foundations of this chapter: (1) The sensorimotor system is hierarchically organized. (2) Motor output is guided by sensory input. (3) Learning can change the nature and the locus of sensorimotor control.

Three Principles of Sensorimotor Function

Before getting into the details of the sensorimotor system, let's take a closer look at the three principles of sensorimotor function introduced by Rhonelle. You will better appreciate these principles if you recognize that they also govern the operation of any large, efficient company—perhaps because that is another system for controlling output that has evolved in a competitive environment. You may find this metaphor useful in helping you understand the principles of sensorimotor system organization—many scientists find that metaphors help them think creatively about their subject matter.

The Sensorimotor System Is Hierarchically Organized

LO 8.1 In the context of the sensorimotor system, explain what *hierarchically organized* means.

The operation of both the sensorimotor system and a large, efficient company is directed by commands that cascade down through the levels of a hierarchy (see Sadnicka et al., 2017)—from the association cortex or the company president (the highest levels) to the muscles or the workers (the lowest levels). Like the orders issued from the office of a company president, the commands that emerge from the association cortex specify general goals rather than specific plans of action. Neither the association cortex nor the company president routinely gets involved in the details. The main advantage of this *hierarchical organization* is that the higher levels of the hierarchy are left free to perform more complex functions.

Both the sensorimotor system and a large, efficient company are parallel hierarchical systems; that is, they are hierarchical systems in which signals flow between levels over multiple paths. This parallel structure enables the association cortex or company president to exert control over the lower levels of the hierarchy in more than one way. For example, the association cortex can directly inhibit an eye blink reflex to allow the insertion of a contact lens, just as a company president can personally organize a delivery to an important customer (see McDougle, Ivry, & Taylor, 2016).

The sensorimotor and company hierarchies are also characterized by *functional segregation*. That is, each level of the sensorimotor and company hierarchies tends to be composed of different units (neural structures or departments), each of which performs a different function.

In summary, the sensorimotor system—like the sensory systems you read about in Chapter 7—is a parallel, functionally segregated, hierarchical system. The main difference between the sensory systems and the sensorimotor system is the primary direction of information flow. In sensory systems, information mainly flows up through the hierarchy; in the sensorimotor system, information mainly flows down.

Motor Output Is Guided by Sensory Input

LO 8.2 **Explain the important role of sensory input for motor output.**

Efficient companies are flexible. They continuously monitor the effects of their own activities, and they use this information to fine-tune their activities. The sensorimotor system does the same (see Azim, Fink, & Jessell, 2014; Danna & Velay, 2015; Scott, 2016). The eyes, the organs of balance, and the receptors in skin, muscles, and joints all monitor the body's responses, and they feed their information back into sensorimotor circuits. In most instances, this **sensory feedback** plays an important role in directing the continuation of the responses that produced it. The only responses that are not normally influenced by sensory feedback are *ballistic movements*—brief, all-or-none, high-speed movements, such as swatting a fly.

Behavior in the absence of just one kind of sensory feedback—the feedback carried by the somatosensory nerves of the arms—was studied in G.O., a former darts champion (Rothwell et al., 1982).

The Case of G.O., the Man with Too Little Feedback

An infection had selectively destroyed the somatosensory nerves of G.O.'s arms. He had great difficulty performing intricate responses such as doing up his buttons or picking up coins, even under visual guidance. Other difficulties resulted from his inability to adjust his motor output in light of unanticipated external disturbances; for example, he could not keep from spilling a cup of coffee if somebody brushed against him. However, G.O.'s greatest problem was his inability to maintain a constant level of muscle contraction.

The result of his infection was that even simple tasks requiring a constant motor output to the hand required continual visual monitoring. For example, when carrying a suitcase, he had to watch it to reassure himself that he had not dropped it. However, even visual feedback was of little use to him in tasks requiring a constant force, tasks such as grasping a pen while writing or holding a cup. In these cases, he had no indication of the pressure that he was exerting on the object; all he saw was the pen or cup slipping from his grasp.

Many adjustments in motor output that occur in response to sensory feedback are controlled unconsciously by the lower levels of the sensorimotor hierarchy without the involvement of the higher levels (see Deliagina, Zelenin, & Orlovsky, 2012). In the same way, large companies run more efficiently if the interns do not check with the company president each time they encounter a minor problem.

Learning Changes the Nature and Locus of Sensorimotor Control

LO 8.3 **Describe how learning changes the nature and locus of sensorimotor control.**

When a company is just starting up, each individual decision is made by the company president after careful consideration. However, as the company develops, many individual actions are coordinated into sequences of prescribed procedures routinely carried out by personnel at lower levels of the hierarchy.

Similar changes occur during sensorimotor learning (see Bassett et al., 2015). During the initial stages of motor learning, each individual response is performed under conscious control; then, after much practice, individual responses become organized into continuous integrated sequences of action that flow smoothly and are adjusted by sensory feedback without conscious regulation. If you think for a moment about the sensorimotor skills you have acquired (e.g., typing, swimming, knitting, basketball playing, dancing, piano playing), you will appreciate that the organization of individual responses into continuous motor programs and the transfer of their control to lower levels of the CNS characterize most sensorimotor learning.

General Model of Sensorimotor System Function

LO 8.4 **Describe and/or draw the general model of sensorimotor function.**

Figure 8.1 is a model that illustrates several principles of sensorimotor system organization; it is the framework of this chapter. Notice its hierarchical structure, the functional

Figure 8.1 A general model of the sensorimotor system. Notice its hierarchical structure, functional segregation, parallel descending pathways, and feedback circuits.

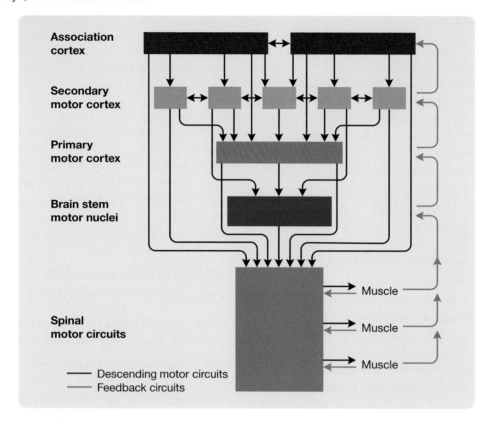

segregation of the levels (e.g., of secondary motor cortex), the parallel connections between levels, and the numerous feedback pathways.

This chapter focuses on the neural structures that play important roles in the control of voluntary behavior (e.g., picking up an apple). It begins at the level of association cortex and traces major motor signals as they descend the sensorimotor hierarchy to the skeletal muscles that ultimately perform the movements.

Sensorimotor Association Cortex

Association cortex is at the top of your sensorimotor hierarchy. There are two major areas of sensorimotor association cortex: the posterior parietal association cortex and the dorsolateral prefrontal association cortex. Posterior parietal cortex and the dorsolateral prefrontal cortex are each composed of several different areas, each with different functions (see Davare et al., 2011; Wilson et al., 2010). However, there is no general consensus on how best to divide either of them for analysis or even how comparable the areas are in humans, monkeys, and rats (see Teixeira et al., 2014; Turella & Lingnau, 2014).

Posterior Parietal Association Cortex

LO 8.5 **Explain the role of the posterior parietal cortex in sensorimotor function and describe what happens when it is damaged or stimulated.**

Before an effective movement can be initiated, certain information is required. The nervous system must know the original positions of the parts of the body that are to be moved, and it must know the positions of any external objects with which the body is going to interact. The **posterior parietal association cortex** (the portion of parietal neocortex posterior to the primary somatosensory cortex) plays an important role in integrating these two kinds of information, in directing behavior by providing spatial information, and in directing attention (Freedman & Ibos, 2018; Hutchinson et al., 2014; Kuang, Morel, & Gail, 2015; Wilber et al., 2014).

You learned in Chapter 7 that the posterior parietal cortex is classified as *association cortex* because it receives input from more than one sensory system. It receives information from the three sensory systems that play roles in the localization of the body and external objects in space: the visual system, the auditory system, and the somatosensory system (see Figure 8.2)—see Sereno and Huang (2014). In turn, much of the output of the posterior parietal cortex goes to areas of motor cortex, which are located in the frontal

Figure 8.2 The major cortical input and output pathways of the posterior parietal association cortex. Shown are the lateral surface of the left hemisphere and the medial surface of the right hemisphere.

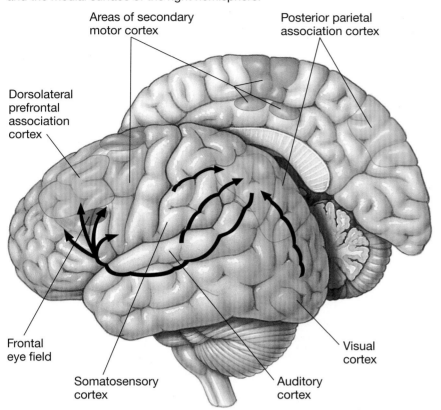

cortex: to the *dorsolateral prefrontal association cortex*, to the various areas of *secondary motor cortex*, and to the **frontal eye field**—a small area of prefrontal cortex that controls both eye movements and shifts in attention (see Moore & Zirnsak, 2015; Figure 8.2). Electrophysiological studies in macaque monkeys and functional magnetic resonance imaging (fMRI) and transcranial magnetic stimulation (TMS) studies in humans indicate that the posterior parietal cortex contains a mosaic of small areas, each specialized for guiding particular movements of eyes, head, arms, or hands (Man et al., 2015; Wang et al., 2015).

Desmurget and colleagues (2009) applied electrical stimulation to the inferior portions of the posterior parietal cortexes of conscious neurosurgical patients. At low current levels, the patients experienced an intention to perform a particular action, and, at higher current levels, they felt that they had actually performed it. However, in neither case did the action actually occur (see Desmurget & Sirigu, 2012).

Damage to the posterior parietal cortex can produce a variety of deficits, including deficits in the perception and memory of spatial relationships, in accurate reaching and grasping, in the control of eye movement, and in attention (see Andersen et al., 2014; Turella & Lingnau, 2014). However, apraxia and contralateral neglect are the two most striking consequences of posterior parietal cortex damage.

Apraxia is a disorder of voluntary movement that is not attributable to a simple motor deficit (e.g., not to paralysis or weakness) or to any deficit in comprehension or motivation (see Niessen, Fink, & Weiss, 2014). Remarkably, patients with apraxia have difficulty making specific movements when they are requested to do so, particularly when the movements are out of context; however, they can often readily perform the very same movements under natural conditions when they are not thinking about what they are doing. For example, a carpenter with apraxia who has no difficulty at all hammering a nail during the course of her work might not be able to demonstrate hammering movements when requested to make them, particularly in the absence of a hammer. Although its symptoms are bilateral, apraxia is often caused by unilateral damage to the left posterior parietal cortex or its connections (Hoeren et al., 2014; Niessen, Fink, & Weiss, 2014).

Contralateral neglect, the other striking consequence of posterior parietal cortex damage, is a disturbance of a patient's ability to respond to stimuli on the side of the body opposite (contralateral) to the side of a brain lesion in the absence of simple sensory or motor deficits. Most patients with contralateral neglect often behave as if the left side of their world does not exist, and they often fail to appreciate that they have a problem (see Li & Malhotra, 2015). The disturbance is often associated with large lesions of the right posterior parietal cortex, though damage to other brain regions has also been implicated (see Karnath & Otto, 2012). Mrs. S. suffered from contralateral neglect after a massive stroke to the posterior portions of her right hemisphere (Sacks, 1999).

The Case of Mrs. S., the Woman Who Turned in Circles

After her stroke, Mrs. S. could not respond to things on her left—including objects and parts of her own body. For example, she often put makeup on the right side of her face but ignored the left.

Mrs. S.'s left-side contralateral neglect created many problems for her, but a particularly bothersome one was that she had difficulty getting enough to eat. When a plate of food was put directly in front of her, she could see only the food on the right half of the plate, and she ate only that half, even if she was very hungry. However, Mrs. S. developed an effective way of getting more food. If she was still hungry after completing a meal, she turned her wheelchair to the right in a full circle until the remaining half of her meal appeared once more directly in front of her. Then, she ate that remaining food, or more precisely, she ate the right half of it. If she was still hungry after that, she turned once again in a circle to the right until the remaining quarter of her meal appeared, and she ate half of that…and so on.

Most patients with contralateral neglect have difficulty responding to things to the left. But to the left of what? For most patients with contralateral neglect, the deficits in responding occur for stimuli to the left of their own bodies, referred to as *egocentric left* (see Karnath, 2015). Egocentric left is partially defined by gravitational coordinates: When patients tilt their heads, their field of neglect is not normally tilted with it.

Some patients also tend not to respond to the left sides of objects, regardless of where the objects are in their visual fields (see Karnath, 2015). These patients, who are said to suffer from *object-based contralateral neglect,* fail to respond to the left side of objects (e.g., the left hand of a statue) even when the objects are presented horizontally or upside down (see Adair & Barrett, 2008).

As you will recall, failure to perceive an object consciously does not necessarily mean the object is not perceived. Indeed, two types of evidence suggest that information about objects that are not noticed by patients with contralateral neglect may be unconsciously perceived (see Jerath & Crawford, 2014). First, when objects were repeatedly presented in the same location to the left of patients with contralateral neglect, they tended to look more in that general direction on future trials, although they were unaware of the objects (Geng & Behrmann, 2002). Second, patients could more readily identify fragmented (partial) drawings viewed to their right if complete versions of the drawings had previously been presented to the left, where they were not consciously perceived (Vuilleumier et al., 2002).

Dorsolateral Prefrontal Association Cortex

LO 8.6 **Explain the role of the dorsolateral prefrontal association cortex in sensorimotor function and describe the response properties of neurons in this region of cortex.**

The other large area of association cortex that has important sensorimotor functions

is the **dorsolateral prefrontal association cortex** (see Kaller et al., 2011). It receives projections from the posterior parietal cortex, and it sends projections to areas of *secondary motor cortex*, to *primary motor cortex*, and to the *frontal eye field*. These projections are shown in Figure 8.3.

Several studies have characterized the activity of monkey dorsolateral prefrontal neurons while the monkeys identify and respond to objects (e.g., Rao, Rainer, & Miller, 1997). The activity of some neurons depends on the characteristics of objects; the activity of others depends on the locations of objects; and the activity of still others depends on a combination of both. The activity of other dorsolateral prefrontal neurons is related to the response rather than to the object. These neurons typically begin to fire before the response and continue to fire until the response is complete. Neurons in many cortical motor areas begin to fire in anticipation of a motor activity (see Rigato, Murakami, & Mainen, 2014; Siegel, Buschman, & Miller, 2015), but those in the dorsolateral prefrontal association cortex tend to fire first.

The response properties of dorsolateral prefrontal neurons suggest that decisions to initiate voluntary movements may be made in this area of cortex, but these decisions depend on critical interactions with posterior parietal cortex and other areas of frontal cortex (Lee, Seo, & Jung, 2012; Ptak, Schnider, & Fellrath, 2017).

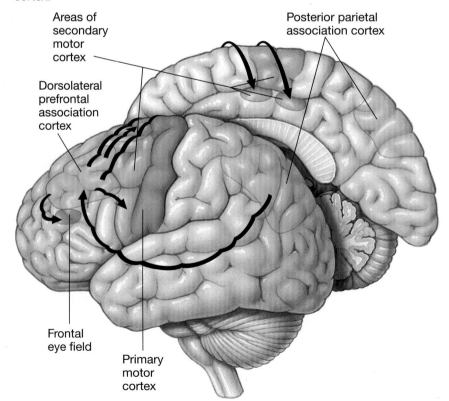

Figure 8.3 The major cortical input and output pathways of the dorsolateral prefrontal association cortex. Shown are the lateral surface of the left hemisphere and the medial surface of the right hemisphere. Not shown are the major projections back from dorsolateral prefrontal cortex to posterior parietal cortex.

Areas of secondary motor cortex

Posterior parietal association cortex

Dorsolateral prefrontal association cortex

Frontal eye field

Primary motor cortex

Secondary Motor Cortex

Areas of **secondary motor cortex** are those that receive much of their input from association cortex (i.e., posterior parietal cortex and dorsolateral prefrontal cortex) and send much of their output to primary motor cortex (see Figure 8.4). For many years, only two areas of secondary motor cortex were known: the supplementary motor area and the premotor cortex. Both of these large areas are clearly visible on the lateral surface of the frontal lobe, just anterior to the *primary motor cortex*. The **supplementary motor area** wraps over the top of the frontal lobe and extends down its medial surface into the longitudinal fissure, and the **premotor cortex** runs in a strip from the supplementary motor area to the lateral fissure.

Identifying the Areas of Secondary Motor Cortex

LO 8.7 Explain the general role of areas of secondary motor cortex.

The simple two-area conception of secondary motor cortex has become more complex. Neuroanatomical and neurophysiological research with monkeys has made a case for at least eight areas of secondary motor cortex in each hemisphere, each with its own subdivisions (Nachev, Kennard, & Husain, 2008). Although most of the research on secondary motor cortex has been done in monkeys, functional brain-imaging studies have suggested that human secondary motor cortex has a comparable organization (see Caminiti, Innocenti, & Battaglia-Mayer, 2015).

To qualify as secondary motor cortex, an area must be appropriately connected with association and secondary motor areas (see Figure 8.4). Electrical stimulation of an area of secondary motor cortex typically elicits complex movements, often involving both sides of the body. Neurons in an area of secondary motor cortex often become more active just prior to the initiation of a voluntary movement and continue to be active throughout the movement.

In general, areas of secondary motor cortex are thought to be involved in the programming of specific patterns of movements after taking general instructions from dorsolateral prefrontal cortex (see Pearce

& Moran, 2012). Evidence of such a function comes from brain-imaging studies in which the patterns of activity in areas of secondary motor cortex have been measured while a volunteer is either imagining his or her own performance of a particular series of movements or planning the performance of the same movements (see Olshansky et al., 2015; Park et al., 2015).

Mirror Neurons

LO 8.8 Describe the major features of mirror neurons and explain why they have received so much attention from neuroscientists.

Few discoveries have captured the interest of neuroscientists as much as the discovery of mirror neurons (see Rizzolatti & Fogassi, 2014). **Mirror neurons** are neurons that fire when an individual performs a particular goal-directed movement or when they observe the same goal-directed movement performed by another.

Mirror neurons were discovered in the early 1990s in the laboratory of Giacomo Rizzolatti (see Ferrari & Rizzolatti, 2014). Rizzolatti and his colleagues had been studying a class of macaque monkey ventral-premotor-area neurons that seemed to encode for particular goal objects;

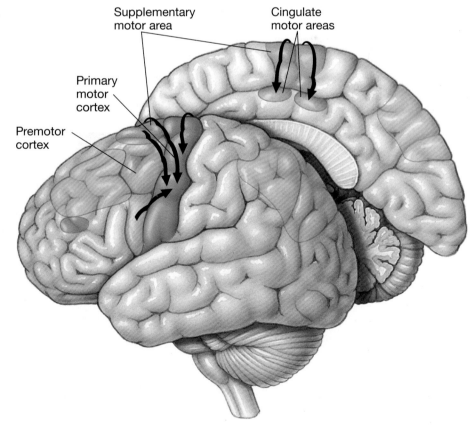

Figure 8.4 Three sorts of secondary motor cortex—supplementary motor area, premotor cortex, and cingulate motor areas—and their output to the primary motor cortex. Shown are the lateral surface of the left hemisphere and the medial surface of the right hemisphere.

Supplementary motor area

Cingulate motor areas

Primary motor cortex

Premotor cortex

that is, these neurons fired when the monkey reached for one object (e.g., a specific toy) but not when the monkey reached for another. Then, they noticed something strange: Some of these neurons, later termed *mirror neurons*, fired just as robustly when the monkey watched the experimenter pick up the same object but not others—see Figure 8.5.

Why did the discovery of mirror neurons in the ventral premotor area create such a stir? The reason is that they provide a possible mechanism for *social cognition* (knowledge of the perceptions, ideas, and intentions of others). Mapping the actions of others onto one's own action repertoire might facilitate social understanding,

cooperation, and imitation (see Farina, Borgnis, & Pozzo, 2020; Rizzolatti & Sinigaglia, 2016; Wood et al., 2016; but see Fitch, 2017; Kennedy-Constantini, 2017).

Support for the idea that mirror neurons play a role in social cognition has come from demonstrations that these neurons respond to the *understanding* of the purpose of an action, not to some superficial characteristic of the action itself (Rizzolatti & Sinigaglia, 2016; but see Churchland, 2014). For example, mirror neurons that reacted to the sight of an action that made a sound (e.g., cracking a peanut) were found to respond just as robustly to the sound alone—in other words, they responded fully to the particular action and its goal regardless of how it was detected. Indeed, many ventral premotor mirror neurons fire even when a monkey does not perceive the key action but just creates a mental representation of it.

Mirror neurons have been found in several areas of the macaque monkey frontal and parietal cortex (see Bonini & Ferrari, 2011; Giese & Rizzolatti, 2015). However, despite more than 600 published studies of "mirror systems" in humans, descriptions of individual mirror neurons in humans are rare (see Molenberghs, Cunnington, & Mattingley, 2012). Indeed, we know of only one: Mukamel and colleagues (2010). This is because there are few opportunities to record the firing of individual neurons in humans while conducting the required behavioral tests.

Most of the research on human mirror neuron mechanisms have been functional MRI studies. Many of these studies have found areas of human motor cortex that are active when a person performs, watches, or imagines a particular action (e.g., Farina, Borgnis, & Pozzo, 2020; Rizzolatti & Sinigaglia, 2016; Vogeley, 2017). There is no direct evidence that mirror neurons are responsible for these human findings—it is possible that different neurons in the same cortical areas contribute to the functional

Figure 8.5 Responses of a mirror neuron of a monkey.

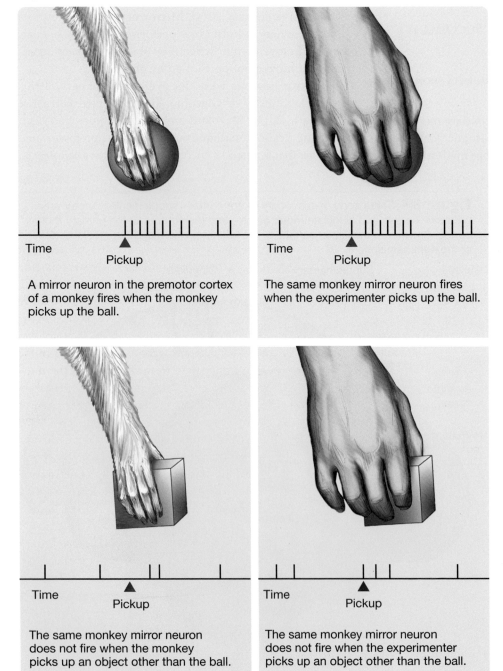

Time
Pickup

A mirror neuron in the premotor cortex of a monkey fires when the monkey picks up the ball.

Time
Pickup

The same monkey mirror neuron fires when the experimenter picks up the ball.

Time
Pickup

The same monkey mirror neuron does not fire when the monkey picks up an object other than the ball.

Time
Pickup

The same monkey mirror neuron does not fire when the experimenter picks up an object other than the ball.

MRI activity in these different conditions. However, the mirror mechanisms identified by functional MRI in humans tend to be in the same areas of cortex as those identified by single cell recording in macaques (Molenberghs et al., 2012).

Primary Motor Cortex

The **primary motor cortex** is located in the *precentral gyrus* of the frontal lobe (see Figures 8.3, 8.4, and 8.6). It is the major point of convergence of cortical sensorimotor signals, and it is the major, but not the only, point of departure of sensorimotor signals from the cerebral cortex. Understanding of the function of primary motor cortex has undergone radical changes over the past two decades—see Graziano (2016). The following two sections describe these changes.

Figure 8.6 The motor homunculus: the somatotopic map of the human primary motor cortex. Electrical stimulation of various sites in the primary motor cortex elicits simple movements in the indicated parts of the body.

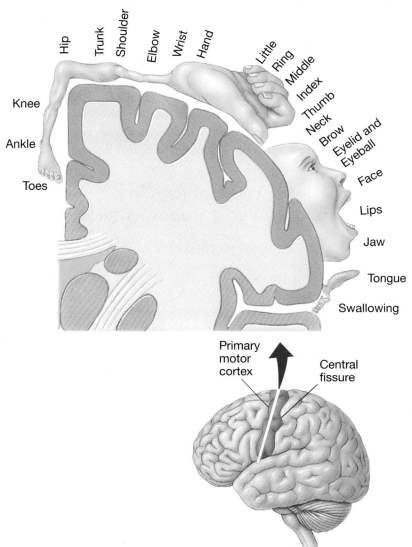

Based on Penfield, W., & Rasmussen, T. (1950). *The cerebral cortex of man: a clinical study of the localization of function*. New York, NY: Macmillan.

Conventional View of Primary Motor Cortex Function

LO 8.9 Describe the conventional view of primary motor cortex function and the evidence upon which it was based.

In 1937, Penfield and Boldrey mapped the primary motor cortex of conscious human patients during neurosurgery by applying brief, low-intensity electrical stimulations to various points on the cortical surface and noting which part of the body moved in response to each stimulation. They found that the stimulation of each particular cortical site activated a particular contralateral muscle and produced a simple movement. When they mapped out the relation between each cortical site and the muscle that was activated by its stimulation, they found that the primary motor cortex is organized somatotopically—that is, according to a map of the body. The **somatotopic** layout of the human primary motor cortex is commonly referred to as the **motor homunculus** (see Figure 8.6). Notice that most of the primary motor cortex is dedicated to controlling parts of the body that are capable of intricate movements, such as the hands and mouth.

It is important to appreciate that each site in the primary motor cortex receives sensory feedback from receptors in the muscles and joints that the site influences. One interesting exception to this general pattern of feedback has been described in monkeys: Monkeys have at least two different hand areas in the primary motor cortex of each hemisphere, and one receives input from receptors in the skin rather than from receptors in the muscles and joints. Presumably, this latter adaptation facilitates **stereognosis**—the process of identifying objects by touch. Close your eyes and explore an object with your hands; notice how stereognosis depends on a complex interplay between motor responses and the somatosensory stimulation produced by them (see Kappers, 2011).

What is the function of each primary motor cortex neuron? Until recently, each neuron was thought to encode the direction of movement. The main evidence for this was the finding that each neuron in the arm area of the primary motor cortex fires maximally when the arm reaches in a particular direction and that each neuron has a different preferred direction.

Current View of Primary Motor Cortex Function

LO 8.10 Describe the current view of primary motor cortex function and the evidence upon which it is based.

Recent efforts to map the primary motor cortex have used a new stimulation technique—see Graziano (2016). Rather than stimulating with brief pulses of current that are just above the threshold to produce a reaction, investigators have used longer bursts of current (e.g., 0.5 to 1 seconds; see Van Acker et al., 2014), which are more similar to the duration of a motor response. The results were amazing: Rather than eliciting the contractions of individual muscles, these currents elicited complex natural-looking response sequences. For example, stimulation at one site reliably produced a feeding response: The arm reached forward, the hand closed as if clasping some food, the closed hand was brought to the mouth, and finally the mouth opened. These recent studies have revealed a looser somatotopic organization than was previously thought. For example, although stimulations to the face area do tend to elicit facial movements, those movements are complex species-typical movements (e.g., an aggressive facial expression) rather than individual muscle contractions (see Ejaz, Hamada, & Diedrichsen, 2015). Also, sites that move a particular body part overlap greatly with sites that move other body parts (Sanes et al., 1995). Presumably that is why small lesions in the hand area of the primary motor cortex of humans (Scheiber, 1999) or monkeys (Scheiber & Poliakov, 1998) do not selectively disrupt the activity of a single finger.

The conventional view that many primary motor cortex neurons are tuned to movement in a particular direction has also been challenged. In the many studies that have supported this conventional view, the monkey subjects were trained to make arm movements from a central starting point so that the relation between neural firing and the direction of movement could be precisely assessed. In each case, each neuron fired only when the movements were made at a particular angle. However, an alternative to the idea that motor neurons are coded to particular angles of movement has come from the findings of studies in which the activity of individual primary motor cortex neurons is recorded as monkeys moved about freely (see Graziano, 2016; Harrison & Murphy, 2014)—rather than as they performed simple, learned arm movements from a set starting point. The firing of many primary motor cortex neurons in freely moving monkeys was often related to the particular end point of a movement, not to the direction of the movement. That is, if a monkey reached toward a particular location, primary motor cortex neurons sensitive to that target location tended to become active regardless of the direction of the movement that was needed to get from the starting point to the target.

The importance of the target of a movement, rather than the direction of a movement, for the function of primary motor cortex was also apparent in stimulation studies (see Graziano, 2016; Harrison & Murphy, 2014). For example, if stimulation of a particular motor cortex site caused a straight arm to bend at the elbow to a 90-degree angle, stimulation of the same site caused a tightly bent arm (i.e., bent past 90 degrees towards the animal's body) to straighten to the same 90-degree angle. In other words, the same stimulation of motor cortex can produce opposite movements depending on the starting position, but the end position of the movements remains the same. Stop for a moment and consider the implications of this finding—they are as important as they are counterintuitive. First, the finding means that the signals from every site in the primary motor cortex diverge greatly, so each particular site has the ability to get a body part (e.g., an arm) to a target location regardless of the starting position. Second, it means that the sensorimotor system is inherently plastic. Apparently, each location in the primary motor cortex can produce the innumerable patterns of muscle contraction and relaxation (Davidson et al., 2007) required to get a body part from any starting point to a specific target location. Accordingly, it has been suggested that the primary motor cortex contains an **action map** (see Graziano, 2016) in addition to a topographic map.

The neurons of the primary motor cortex play a major role in initiating body movements. With an appropriate interface, could they control the movements of a machine (see Georgopoulos & Carpenter, 2015)? Belle says, "Yes."

Belle: The Monkey That Controlled a Robot with Her Mind

In the laboratory of Miguel Nicolelis and John Chapin, a tiny owl monkey called Belle watched a series of lights on a control panel. Belle had learned that if she moved the joystick in her right hand in the direction of a light, she would be rewarded with a drop of fruit juice. On this particular day, as a light flashed on the panel, 100 microelectrodes recorded extracellular unit activity from neurons in Belle's primary motor cortex. This activity moved Belle's arm toward the light, but at the same time, the signals were analyzed by a computer, which fed the output to a laboratory several hundred kilometers away, at the Massachusetts Institute of Technology. At MIT, the signals from Belle's brain entered the circuits of a robotic arm. On each trial, the activity of Belle's primary motor cortex moved her arm toward the test light, and it moved the robotic arm in the same direction. Belle's neural signals were directing the activity of a robot.

Belle's remarkable feat raised a possibility that is starting to be realized. Indeed, there has been a recent flurry of technological advances involving *brain–computer interfaces* (i.e., direct communication between a computer and the brain—usually via an array of electrodes placed in the brain). For example, paralyzed patients have learned to control robotic arms with neural signals collected via multi-electrode arrays implanted in the primary motor cortex (Collinger et al., 2013; Golub et al., 2016; Pruszynski & Diedrichsen, 2015).

Brain-computer interfaces have also been used to mitigate the effects of spinal-cord damage. For example, in a study by Capogrosso and colleagues (2016), monkeys with transected spinal cords each had a wireless transmitter implanted in their motor cortex to record and transmit motor cortex activity. Another wireless receiver positioned on their spinal cord just below the site of transection converted that transmitted message into a pattern of stimulation that elicited movement of their paralyzed limb (Capogrosso et al., 2016). Such *brain–spine-computer interfaces* might someday allow for significant recovery from the effects of spinal cord damage (see Jackson, 2016).

EFFECTS OF PRIMARY MOTOR CORTEX LESIONS. Extensive damage to the human primary motor cortex has less effect than you might expect, given that this cortex is the major point of departure of motor fibers from the cerebral cortex. Large lesions to the primary motor cortex may disrupt a patient's ability to move one body part (e.g., one finger) independently of others (see Ebbesen & Brecht, 2017), may produce **astereognosia** (deficits in stereognosis), and may reduce the speed, accuracy, and force of a patient's movements. Such lesions do not, however, eliminate voluntary movement, presumably because there are parallel pathways that descend directly from secondary and association motor areas to subcortical motor circuits without passing through primary motor cortex.

Cerebellum and Basal Ganglia

The cerebellum and the basal ganglia (see Figures 3.20 and 3.28) are both important and highly interconnected sensorimotor structures (see Bostan & Strick, 2018), but neither is a major part of the pathway by which signals descend through the sensorimotor hierarchy. Instead, both the cerebellum and the basal ganglia interact with different levels of the sensorimotor hierarchy and, in so doing, coordinate and modulate its activities.

Cerebellum

LO 8.11 Describe the structure and connectivity of the cerebellum and explain the current view of cerebellar function.

The cerebellum's structure and complex connectivity with other brain structures suggest its functional complexity

(see Bostan & Strick, 2018; Buckner, 2013). And while it constitutes only 10 percent of the mass of the brain, the cerebellum contains more than half of the brain's neurons (Azevedo et al., 2009).

The cerebellum receives information from primary and secondary motor cortex, information about descending motor signals from brain-stem motor nuclei, and feedback from motor responses via the somatosensory and vestibular systems. The cerebellum is thought to compare these three sources of input and correct ongoing movements that deviate from their intended course (see Bastian, 2006; Bell, Han, & Sawtell, 2008; Herzfeld & Shadmehr, 2014). By performing this function, it is believed to play a major role in motor learning, particularly in the learning of sequences of movements in which timing is a critical factor (see Pritchett & Carey, 2014).

The effects of diffuse cerebellar damage on motor function are devastating. The patient loses the ability to accurately control the direction, force, velocity, and amplitude of movements and the ability to adapt patterns of motor output to changing conditions. It is particularly difficult to maintain steady postures (e.g., standing), and attempts to do so frequently lead to tremor. There are also severe disturbances in balance, gait, speech, and the control of eye movement. Learning new motor sequences is difficult (Thach & Bastian, 2004). These effects of cerebellar damage suggest that the cerebellum plays a major role in monitoring and adapting ongoing patterns of movement (see Peterburs & Desmond, 2016)

The functions of the cerebellum were once thought to be entirely sensorimotor, but this conventional view is no longer tenable (see Buckner, 2013; Koziol et al., 2014). Patients with cerebellar damage often display diverse sensory, cognitive, emotional, and memory deficits (see Sokolov, Miall, & Ivry, 2017). Also, healthy volunteers often display cerebellar activity during sensory, cognitive, or emotional activities. There are several competing theories of cerebellar function (see Koziol et al., 2014), but a popular one is that the cerebellum plays an important role in learning from one's errors and in the prediction of errors (see Herzfeld et al., 2018; Sokolov, Miall, & Ivry, 2017).

Basal Ganglia

LO 8.12 Describe the anatomy of the basal ganglia and explain the current view of their function.

The basal ganglia do not contain as many neurons as the cerebellum, but in one sense they are more complex. Unlike the cerebellum, which is organized systematically in lobes, columns, and layers, the basal ganglia are a complex heterogeneous collection of interconnected nuclei.

The anatomy of the basal ganglia suggests that, like the cerebellum, they perform a modulatory function (see Nelson & Kreitzer, 2014). They contribute few fibers to descending motor pathways; instead, they form neural loops via their numerous reciprocal connections with cortical areas and the cerebellum (Bostan & Strick, 2018; Nelson & Kreitzer, 2014; Oldenburg & Sabatini, 2015). Many of the cortical loops carry signals to and from the motor areas of the cortex (see Nambu, 2008).

Theories of basal ganglia function have changed in much the same way that theories of cerebellar function have changed. The traditional view of the basal ganglia was that they, like the cerebellum, play a role in the modulation of motor output. Now, the basal ganglia are thought to also be involved in a variety of cognitive functions (see Eisinger et al., 2018; Hikosaka et al., 2014; Lim, Fiez, & Holt, 2014; Rektor et al., 2015) and in many aspects of motivation (see Bostan & Strick, 2018). The basal ganglia have also been shown to participate in learning. For example, they play a role in *habit learning*, a type of learning that is usually acquired gradually, trial-by-trial (see Ashby, Turner, & Horovitz, 2010), and in classical conditioning (see Stephenson-Jones et al., 2016).

One theory of basal ganglia sensorimotor function is based on its known roles in both movement and motivation (see Averbeck & Costa, 2017; Schultz, 2016). This theory comprises two major assertions. The first states that the basal ganglia are responsible for **movement vigor** (see Dudman & Kraukauer, 2016): the control of the speed and amplitude of movement based on motivational factors. For example, the basal ganglia might enable a concert pianist to play a particular piece with more or less vigor. The second assertion is that movement not only involves the execution of actions but also requires that we actively suppress motor activity that would otherwise be inappropriate or unhealthy. For example, we have to suppress our tendency to display inappropriate yawning or scratching at social gatherings; we also need to suppress unwanted movements generated by the ongoing spontaneous activity of our muscles (e.g., tremor, twitching, coughing). When such movement inhibition fails, symptoms of a neurological or psychiatric disorder can emerge (see Duque et al., 2017).

Scan Your Brain

Now that you have learned about the sensorimotor pathways, this is a good place for you to pause to scan your brain to evaluate your knowledge by completing the following statements. The correct answers are provided at the end of the exercise. Before proceeding, review material related to your incorrect answers and omissions.

1. The _____ _____ association cortex provides important spatial information and helps direct attention to external stimuli.

2. _____ movements are normally not affected by sensory feedback.

3. Motor learning, such as riding a bike, begins with responses under conscious control, but with practice these are adjusted by _____ feedback without conscious regulation.

4. _____ is a disorder of voluntary movement that cannot be explained by paralysis; rather, it is attributed to the inability to perform motor movements when instructed to do so.

5. _____ _____, a striking consequence of posterior parietal cortex damage, is a disturbance of a patient's ability to respond to stimuli on the side of the body

opposite to the side of a brain lesion in the absence of simple sensory or motor deficits.

6. The supplementary motor area and premotor cortex are part of the _____ _____ cortex that is involved in programming patterns of movements.

7. _____ are neurons that fire when an individual performs a particular goal-directed hand movement or when they observe the same goal-directed movement performed by another.

8. The somatotopic layout of the human primary cortex is also known as the _____ _____.

9. The _____ is involved in motor learning and the temporal association of motor actions. As such, damage to this area can cause detrimental effects on posture, gait, speech, and balance.

10. Recent views of the function of the basal ganglia suggest that they are involved in various cognitive functions including _____ learning.

Scan Your Brain answers: (1) posterior parietal, (2) Ballistic, (3) sensory, (4) Apraxia, (5) Contralateral neglect, (6) secondary motor, (7) Mirror neurons, (8) motor homunculus, (9) cerebellum, (10) habit.

Descending Motor Pathways

Neural signals are conducted from the primary motor cortex to the motor neurons of the spinal cord over four different pathways. Two pathways descend in the *dorsolateral* region of the spinal cord—collectively known as the *dorsolateral motor*

pathways, and two descend in the *ventromedial* region of the spinal cord—collectively known as the *ventromedial motor pathways*. Signals conducted over these pathways act together in the control of voluntary movement (see Iwaniuk & Whishaw, 2000). Like a large company, the sensorimotor system does not work well unless there are good lines of communication from the executive level (the cortex) to the office personnel (the spinal motor circuits) and workers (the muscles).

The Two Dorsolateral Motor Pathways and the Two Ventromedial Motor Pathways

LO 8.13 Compare and contrast the two dorsolateral motor pathways and the two ventromedial motor pathways.

The descending dorsolateral and ventromedial motor pathways are similar in that each is composed of two major tracts, one whose axons descend directly to the spinal cord and another whose axons synapse in the brain stem on neurons that in turn descend to the spinal cord. However, the *dorsolateral tracts* differ from the *ventromedial tracts* in two major respects:

- The ventromedial tracts are much more diffuse. Many of their axons innervate interneurons on both sides of the spinal gray matter and in several different segments, whereas the axons of the dorsolateral tracts terminate in the contralateral half of one spinal cord segment, sometimes directly on a motor neuron.

- The motor neurons activated by the ventromedial tracts project to proximal muscles of the trunk and limbs (e.g., shoulder muscles), whereas the motor neurons activated by the dorsolateral tracts project to distal muscles (e.g., finger muscles).

Because all four of the descending motor tracts originate in the cerebral cortex, all are presumed to mediate voluntary movement; however, major differences in their routes and destinations suggest that they have different functions. This difference was first demonstrated in an experiment on monkeys by Lawrence and Kuypers.

In their experiment, Lawrence and Kuypers (1968) made complete transections of the dorsolateral tracts in monkeys. The monkeys could stand, walk, and climb after this transection, but when they were sitting, their arms hung limply by their sides (remember that monkeys normally use their arms for standing and walking). In those few instances in which the monkeys did use an arm for reaching, they used it like a rubber-handled rake—throwing it out from the shoulder and using it to draw small objects of interest back along the floor.

The other group of monkeys in their experiment had complete transections of their ventromedial tracts. In contrast to the first group, these subjects had severe postural abnormalities: They had great difficulty walking or sitting. If they did manage to sit or stand without clinging to the bars of their cages, the slightest disturbance, such as a loud noise, frequently made them fall.

What do these experiments tell us about the roles of the various descending sensorimotor tracts in the control of primate movement? They suggest that the ventromedial tracts are involved in the control of posture and whole-body movements (e.g., walking, climbing) and that they can exert control over the limb movements involved in such activities.

In contrast, the dorsolateral tracts control the movements of the limbs (see Ruder & Arber, 2019).

Sensorimotor Spinal Circuits

We have descended the sensorimotor hierarchy to its lowest level: the spinal circuits and the muscles they control. Psychologists, including us, tend to be brain-oriented, and they often think of the spinal cord motor circuits as mere cables that carry instructions from the brain to the muscles. If you think this way, you will be surprised: The motor circuits of the spinal cord show considerable complexity in their functioning, independent of signals from the brain (see Dasen, 2017; Giszter, 2015; Kiehn, 2016; Ruder & Arber, 2019). Again, the business metaphor helps put this in perspective: Can the office managers (spinal circuits) and workers (muscles) of a company function effectively when all of the executives are at a convention in Hawaii? Of course they can.

Muscles

LO 8.14 Describe the components of a motor unit and distinguish between the different types of muscles.

Motor units are the smallest units of motor activity. Each motor unit comprises a single motor neuron and all of the individual skeletal muscle fibers that it innervates (see Figure 8.7). When the motor neuron fires, all the muscle fibers of its unit contract together. Motor units differ appreciably in the number of muscle fibers they contain; the units with the fewest fibers—those of the fingers and face—permit the highest degree of selective motor control.

A skeletal muscle comprises hundreds of thousands of threadlike muscle fibers bound together in a tough membrane and attached to a bone by a *tendon*. *Acetylcholine*, which is released by motor neurons at *neuromuscular junctions*, activates the **motor end-plate** on each muscle fiber and causes the fiber to contract. Contraction is the only method that muscles have for generating force, thus, any muscle can generate force in only one direction. All of the motor neurons that innervate the fibers of a single muscle are called its **motor pool**.

Although it is an oversimplification (see Gollnick & Hodgson, 1986), skeletal muscle fibers are often considered to be of two basic types: fast and slow. *Fast muscle fibers*, as you might guess, are those that contract and relax quickly. Although they are capable of generating great force, they fatigue quickly because they are *poorly vascularized* (have few blood vessels, which gives them a pale color). In contrast, *slow muscle fibers*, although slower and weaker, are

Figure 8.7 An electron micrograph of a motor unit: a motor neuron (pink) and the muscle fibers it innervates.

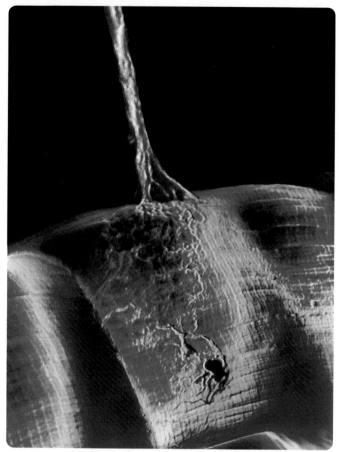

CNRI/Science Source

capable of more sustained contraction because they are more richly vascularized (and hence much redder). Each muscle has both fast and slow fibers—the fast muscle fibers participate in quick movements such as jumping, whereas the slow muscle fibers participate in gradual movements such as walking. Because each muscle can apply force in only one direction, joints that move in more than one direction must be controlled by more than one muscle. Many skeletal muscles belong unambiguously to one of two categories: flexors or extensors. **Flexors** act to bend or flex a joint, and **extensors** act to straighten or extend it. Figure 8.8 illustrates the *biceps* and *triceps*—the flexor and extensor, respectively, of the elbow joint. Any two muscles whose contraction produces the same movement, be it flexion or extension, are said to be **synergistic muscles**; those that act in opposition, like the biceps and the triceps, are said to be **antagonistic muscles**.

To understand how muscles work, it is important to realize that they are elastic, rather than inflexible and cablelike. If you think of an increase in muscle tension as analogous to an increase in the tension of an elastic band joining two bones, you will appreciate that muscle contraction can be of two types. Activation of a muscle can increase the tension that it exerts on two bones without shortening and pulling them together; this is termed **isometric contraction**. Or it can shorten and pull them together; this is termed **dynamic contraction**. The tension in a muscle can be increased by increasing the number of neurons in its motor pool that are firing, by increasing the firing rates of those already firing, or more commonly by a combination of these two changes.

Receptor Organs of Tendons and Muscles

LO 8.15 Describe the receptor organs of tendons and muscles.

The activity of skeletal muscles is monitored by two kinds of receptors: Golgi tendon organs and muscle spindles. **Golgi tendon organs** are embedded in the *tendons*, which connect each skeletal muscle to bone; **muscle spindles** are embedded in the muscle tissue itself. Because of their different locations, Golgi tendon organs and muscle spindles respond to different aspects of muscle contraction. Golgi tendon organs respond to increases in muscle tension (i.e., to the pull of the muscle on the tendon), but they are completely insensitive to changes in muscle length. In contrast, muscle spindles respond to changes in muscle length, but they do not respond to changes in muscle tension.

Under normal conditions, the function of Golgi tendon organs is to provide the central nervous system with

Figure 8.8 The biceps and triceps, which are the flexor and extensor muscles, respectively, of the elbow joint.

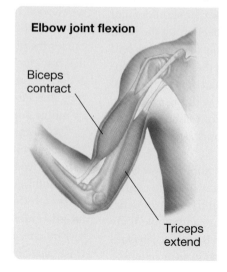
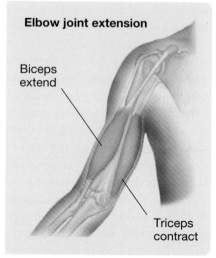

Elbow joint flexion

Biceps contract

Triceps extend

Elbow joint extension

Biceps extend

Triceps contract

information about muscle tension, but they also serve a protective function. When the contraction of a muscle is so extreme that there is a risk of damage, the Golgi tendon organs excite inhibitory interneurons in the spinal cord that cause the muscle to relax.

Figure 8.9 is a schematic diagram of the *muscle-spindle feedback circuit*. Examine it carefully. Notice that each muscle spindle has its own threadlike **intrafusal muscle**, which is innervated by its own **intrafusal motor neuron**. Why would a receptor have its own muscle and motor neuron? The reason becomes apparent when you consider what would happen to a muscle spindle without them. Without its intrafusal motor input, a muscle spindle would fall slack each time its **skeletal muscle (extrafusal muscle)** contracted. In this slack state, the muscle spindle could not do its job, which is to respond to slight changes in extrafusal muscle length. As Figure 8.10 illustrates, the intrafusal motor neuron solves this problem by shortening the intrafusal muscle each time the extrafusal muscle becomes shorter, thus keeping enough tension on the middle, stretch-sensitive portion of the muscle spindle to keep it responsive to slight changes in the length of the extrafusal muscle.

Figure 8.9 The muscle-spindle feedback circuit. There are many muscle spindles in each muscle; for clarity, only one much-enlarged muscle spindle is illustrated here.

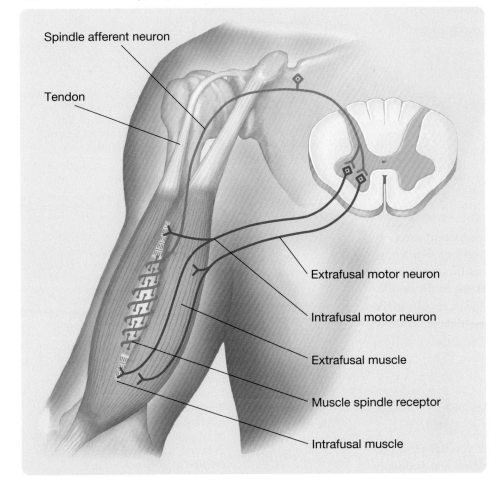

Spindle afferent neuron

Tendon

Extrafusal motor neuron

Intrafusal motor neuron

Extrafusal muscle

Muscle spindle receptor

Intrafusal muscle

Stretch Reflex

LO 8.16 Describe the stretch reflex and explain its mechanism.

When the word *reflex* is mentioned, many people think of themselves sitting on the edge of their doctor's examination table having their knees rapped with a little rubber-headed hammer. The resulting leg extension is called the **patellar tendon reflex** (*patella* means "knee"). This reflex is a **stretch reflex**—a reflex elicited by a sudden external stretching force on a muscle.

When your doctor strikes the tendon of your knee, the extensor muscle running along your thigh is stretched. This initiates the chain of events depicted in Figure 8.11. The sudden stretch of the thigh muscle stretches its muscle-spindle stretch receptors, which in turn initiate a volley of action potentials carried from the stretch receptors into the spinal cord by **spindle afferent neurons** via the *dorsal root*. This volley of action potentials excites motor neurons in the *ventral horn* of the spinal cord, which respond by sending action potentials back to the muscle whose stretch originally excited them (see Illert & Kummel, 1999). The arrival of these impulses back at the starting point results in a compensatory muscle contraction and a sudden leg extension.

The method by which the patellar tendon reflex is typically elicited in a doctor's office—that is, by a sharp blow to the tendon of a completely relaxed muscle—is designed to make the reflex readily observable. However, it does little to communicate its functional significance. In real-life situations, the function of the stretch reflex is to keep external forces from altering the intended position of the body. When an external force, such as a push on your arm while you are holding a cup of coffee, causes an unanticipated extrafusal muscle stretch, the muscle-spindle feedback circuit produces an immediate compensatory contraction of the muscle that counteracts the force and keeps you from spilling the coffee—unless, of course, you are wearing your best clothes.

The mechanism by which the stretch reflex maintains limb

Figure 8.10 The function of intrafusal motor neurons.

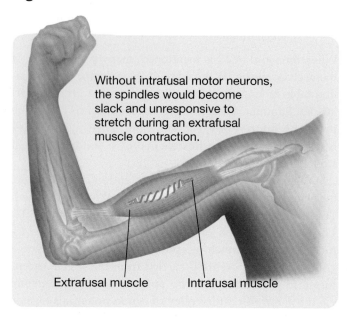

Without intrafusal motor neurons, the spindles would become slack and unresponsive to stretch during an extrafusal muscle contraction.

Extrafusal muscle   Intrafusal muscle

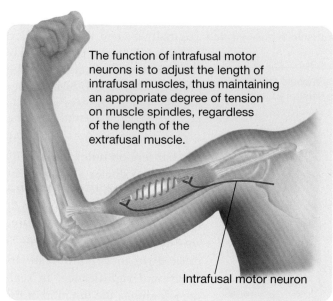

The function of intrafusal motor neurons is to adjust the length of intrafusal muscles, thus maintaining an appropriate degree of tension on muscle spindles, regardless of the length of the extrafusal muscle.

Intrafusal motor neuron

stability is illustrated in Figure 8.12. Examine it carefully because it illustrates two of the principles of sensorimotor system function that are the focus of this chapter: the important role played by sensory feedback in the regulation of motor output and the ability of lower circuits in the motor hierarchy to take care of "business details" without the involvement of higher levels.

Withdrawal Reflex

LO 8.17 Describe the withdrawal reflex and explain its mechanism.

We are sure that, at one time or another, you have touched something painful—a hot pot, for example—and suddenly pulled back your hand. This is a **withdrawal reflex**. Unlike

the stretch reflex, the withdrawal reflex is *not monosynaptic*. When a painful stimulus is applied to the hand, the first responses are recorded in the motor neurons of the arm flexor muscles about 1.6 milliseconds later, about the time it takes a neural signal to cross two synapses. Thus, the shortest route in the withdrawal-reflex circuit involves one interneuron. Other responses are recorded in the motor neurons of the arm flexor muscles after the initial volley; these responses are triggered by signals that have traveled over multisynaptic pathways—some involving the cortex. See Figure 8.13.

Reciprocal Innervation

LO 8.18 Explain what is meant by *reciprocal innervation*.

Reciprocal innervation is an important principle of spinal cord circuitry. It refers to the fact that antagonistic muscles

Figure 8.11 The elicitation of a stretch reflex. All of the muscle spindles in a muscle are activated during a stretch reflex, but only a single muscle spindle is depicted here.

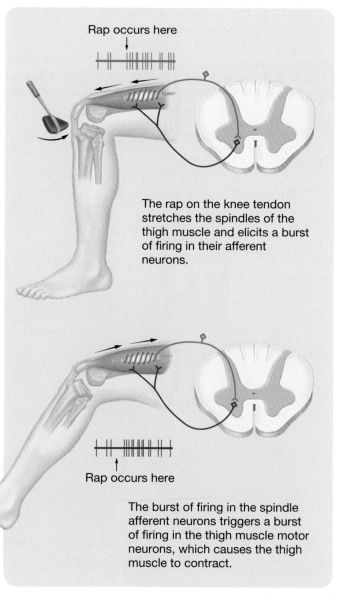

Rap occurs here

The rap on the knee tendon stretches the spindles of the thigh muscle and elicits a burst of firing in their afferent neurons.

Rap occurs here

The burst of firing in the spindle afferent neurons triggers a burst of firing in the thigh muscle motor neurons, which causes the thigh muscle to contract.

Figure 8.12 The automatic maintenance of limb position by the muscle-spindle feedback system.

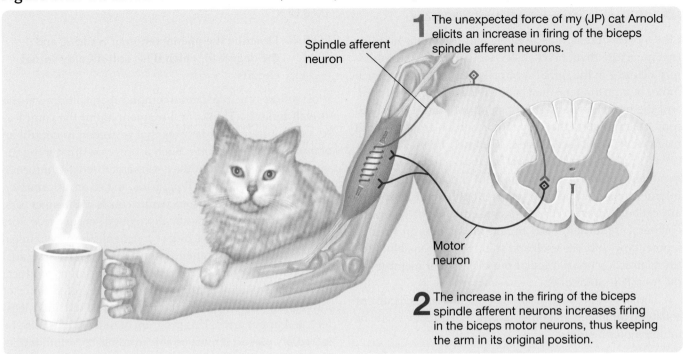

1 The unexpected force of my (JP) cat Arnold elicits an increase in firing of the biceps spindle afferent neurons.

Spindle afferent neuron

Motor neuron

2 The increase in the firing of the biceps spindle afferent neurons increases firing in the biceps motor neurons, thus keeping the arm in its original position.

---

are innervated in a way that permits a smooth, unimpeded motor response: When one is contracted, the other relaxes. Figure 8.13 illustrates the role of reciprocal innervation in the withdrawal reflex. "Bad news" of a sudden painful event in the hand arrives in the dorsal horn of the spinal cord and has two effects: The signals excite both excitatory and inhibitory interneurons. The excitatory interneurons excite the motor neurons of the elbow flexor; the inhibitory interneurons inhibit the motor neurons of the elbow extensor. Thus, a single sensory input produces a coordinated pattern of motor output; the activities of agonists and antagonists are automatically coordinated by the internal circuitry of the spinal cord (see Nielsen, 2016).

Movements are quickest when there is simultaneous excitation of all agonists and complete inhibition of all antagonists; however, this is not the way voluntary movement is normally produced. Most muscles are always contracted to some degree, and movements are produced by adjustment in the level of relative cocontraction between antagonists. Movements produced by **cocontraction** are smooth, and they can be stopped with precision by a slight increase in the contraction of the antagonistic muscles. Moreover, cocontraction insulates us from the effects of unexpected external forces.

Figure 8.13 The reciprocal innervation of antagonistic muscles in the arm. During a withdrawal reflex, elbow flexors are excited, and elbow extensors are inhibited.

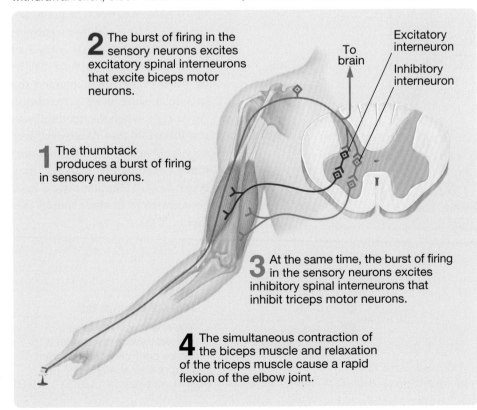

2 The burst of firing in the sensory neurons excites excitatory spinal interneurons that excite biceps motor neurons.

1 The thumbtack produces a burst of firing in sensory neurons.

To brain

Excitatory interneuron

Inhibitory interneuron

3 At the same time, the burst of firing in the sensory neurons excites inhibitory spinal interneurons that inhibit triceps motor neurons.

4 The simultaneous contraction of the biceps muscle and relaxation of the triceps muscle cause a rapid flexion of the elbow joint.

Recurrent Collateral Inhibition

LO 8.19 Explain recurrent collateral inhibition.

Like most workers, muscle fibers and the motor neurons that innervate them need an occasional break, and inhibitory neurons in the spinal cord make sure they get it. Each motor neuron branches just before it leaves the spinal cord, and the branch synapses on a small inhibitory interneuron, which inhibits the very motor neuron from which it receives its input (see Illert & Kummel, 1999). The inhibition produced by these local feedback circuits is called **recurrent collateral inhibition**, and the small inhibitory interneurons that mediate recurrent collateral inhibition are called *Renshaw cells*. As a consequence of recurrent collateral inhibition, each time a motor neuron fires, it momentarily inhibits itself and shifts the responsibility for the contraction of a particular muscle to other members of the muscle's motor pool.

Figure 8.14 provides a summary; it illustrates recurrent collateral inhibition and other factors that directly excite or inhibit motor neurons.

Walking: A Complex Sensorimotor Reflex

LO 8.20 Describe the phenomenon of walking and the degree to which it is controlled by spinal circuits.

Most reflexes are much more complex than withdrawal and stretch reflexes. Think for a moment about the complexity of the program of reflexes that is needed to control an activity such as walking. Such a program must integrate visual information from the eyes; somatosensory information from the feet, knees, hips, arms, and so on; information about balance from the semicircular canals of the inner ears; and information from newly discovered sensory receptors in the spinal cord that detect mechanical and chemical information from the cerebrospinal fluid in the central canal (see Böhm & Wyart, 2016). That program must produce, on the basis of all this information, an integrated series of movements that involves the muscles of the trunk, legs, feet, and upper arms. This program of reflexes must also be incredibly plastic; it must be able to adjust its output immediately to changes in the slope of the terrain, to instructions from the brain, or to sudden external forces. Remarkably, similar patterns of neural activity control walking in humans, other mammals, and birds (see Dominici et al., 2011; Grillner, 2011).

Grillner (1985) showed that the spinal cord, with no contribution whatsoever from the brain, can control walking. Grillner's subjects were cats whose spinal cords had been separated from their brains by transection. He suspended the cats in a sling over a treadmill. Amazingly, when the treadmill was started so that the cats received sensory feedback of the sort that normally accompanies walking, they began to walk. Similar results have been observed in other species (see Kiehn, 2016).

Figure 8.14 The excitatory and inhibitory signals that directly influence the activity of a motor neuron.

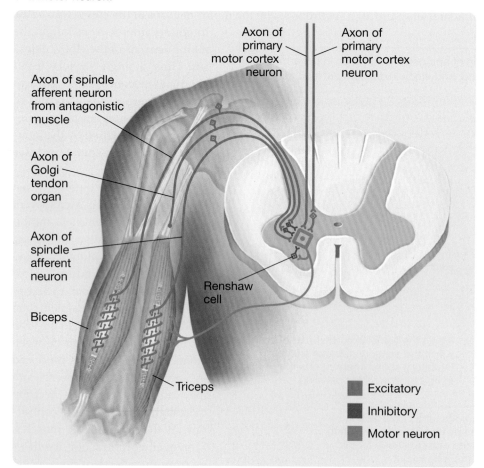

Axon of primary motor cortex neuron

Axon of primary motor cortex neuron

Axon of spindle afferent neuron from antagonistic muscle

Axon of Golgi tendon organ

Axon of spindle afferent neuron

Renshaw cell

Biceps

Triceps

■ Excitatory
■ Inhibitory
■ Motor neuron

Journal Prompt 8.2

Why might walking have evolved in such a way that it can be controlled independent of input from the brain?

Central Sensorimotor Programs and Learning

In this chapter, you have learned that the sensorimotor system is like the hierarchy of a large efficient company. You have learned how the executives—the dorsolateral prefrontal cortex and the secondary motor cortexes—issue commands based on information supplied to them in part by the posterior parietal cortex. And you have learned how these commands are forwarded to the director of operations (the primary motor cortex) for distribution over four main channels of communication (the two dorsolateral and the two ventromedial spinal motor pathways) to the metaphoric office managers of the sensorimotor hierarchy (the spinal sensorimotor circuits). Finally, you have learned how spinal sensorimotor circuits direct the activities of the workers (the muscles).

In this final module, you will learn about central sensorimotor programs. The module concludes with revisiting the case of Rhonelle the cashier.

A Hierarchy of Central Sensorimotor Programs

LO 8.21 Explain what is meant by a hierarchy of central sensorimotor programs and explain the importance of this arrangement for sensorimotor functioning.

One view of sensorimotor function is that the sensorimotor system comprises a hierarchy of **central sensorimotor programs**. According to this view, all but the highest levels of the sensorimotor system have certain patterns of activity programmed into them, and complex movements are produced by activating the appropriate combinations of these programs. For example, if you want to look at a magazine, your association cortex will activate high-level cortical programs that in turn will activate lower-level programs—perhaps in your brain stem—for walking, bending over, picking up, and thumbing through. These programs in turn will activate spinal programs that control the various elements of the sequences and cause your muscles to complete the objective (Grillner & Jessell, 2009).

Once activated, each level of the sensorimotor system is capable of operating on the basis of current sensory feedback without the direct control of higher levels. Thus, although the highest levels of your sensorimotor system retain the option of directly controlling your activities, most of the individual responses that you make are performed without direct cortical involvement, and you are often barely aware of them (see Custers & Aarts, 2010).

In much the same way, a company president who wishes to open a new branch office simply issues the command to one of the executives, and the executive responds in the usual fashion by issuing a series of commands to the appropriate people lower in the hierarchy, who in turn do the same. Each of the executives and workers of the company knows how to complete many different tasks and executes them in the light of current conditions when instructed to do so. Good companies have mechanisms for ensuring that the programs of action at different levels of the hierarchy are well coordinated and effective. In the sensorimotor system, these mechanisms seem to be the responsibility of the cerebellum and basal ganglia.

Characteristics of Central Sensorimotor Programs

LO 8.22 Describe the various characteristics of central sensorimotor programs.

CENTRAL SENSORIMOTOR PROGRAMS ARE CAPABLE OF MOTOR EQUIVALENCE. Like a large, efficient company, the sensorimotor system does not always accomplish a particular task in exactly the same way. The fact that the same basic movement can be carried out in different ways involving different muscles is called **motor equivalence**. For example, you have learned to sign your name with stereotypical finger and hand movements, yet if you wrote your name with your toe on a sandy beach, your signature would still retain many of its typical characteristics.

Motor equivalence illustrates the inherent plasticity of the sensorimotor system. It suggests that specific central sensorimotor programs for signing your name are not stored in the neural circuits that directly control your preferred hand; general programs are stored higher in your sensorimotor hierarchy and then are adapted to the situation as required. In an fMRI study, Rijntjes and others (1999) showed that the central sensorimotor programs for signing one's name seem to be stored in areas of secondary motor cortex that control the preferred hand. Remarkably, these same hand areas were also activated when the signature was made with a toe.

SENSORY INFORMATION THAT CONTROLS CENTRAL SENSORIMOTOR PROGRAMS IS NOT NECESSARILY CONSCIOUS. In Chapter 6, you learned that the neural mechanisms of conscious visual perception (ventral stream) are not necessarily the same as those that mediate the visual control of behavior (dorsal stream). Initial evidence for this theory came from neuropsychological patients who could respond to visual stimuli of which they had little conscious awareness and from others who could not effectively interact with objects that they consciously perceived.

Is there evidence for the separation of conscious perception and sensory control of behavior in intact humans? Haffenden and Goodale (1998) supplied such evidence (see also Ganel, Tanzer, & Goodale, 2008; Goodale & Westwood, 2004). They showed healthy volunteers a three-dimensional version of the visual illusion in Figure 8.15—notice that the two central disks appear to be different sizes, even though they are identical. Remarkably, when the volunteers were asked to indicate the size of each central disk with their right thumb and pointing finger, they judged the disk on the left to be bigger than the one on the right; however, when they were asked to reach out and pick up the disks with the same two digits, the preparatory gap between the digits was a function of the actual size of each disk rather than its perceived size.

CENTRAL SENSORIMOTOR PROGRAMS CAN DEVELOP WITHOUT PRACTICE. Although central sensorimotor programs for some behaviors can be established by practicing the behaviors, the central sensorimotor programs for many species-typical behaviors are established

---

Figure 8.15 The Ebbinghaus illusion. Notice that the central disk on the left appears larger than the one on the right. In fact, both central disks are exactly the same size. Haffenden and Goodale (1998) found that when volunteers reached out to pick up either of the central disks, the position of their fingers as they approached the disks indicated that their responses were being controlled by the actual sizes of the disks, not their consciously perceived sizes.

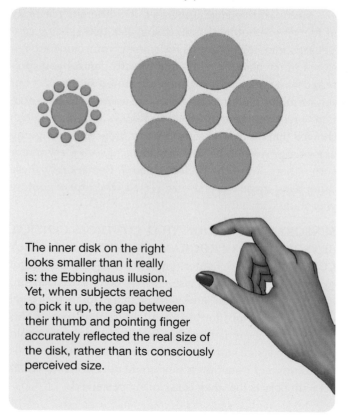

The inner disk on the right looks smaller than it really is: the Ebbinghaus illusion. Yet, when subjects reached to pick it up, the gap between their thumb and pointing finger accurately reflected the real size of the disk, rather than its consciously perceived size.

---

without explicit practice of the behaviors. This point was made clear by the classic study of Fentress (1973). Fentress showed that adult mice raised from birth without forelimbs still made the patterns of shoulder movements typical of grooming in their species—and that these movements were well coordinated with normal tongue, head, and eye movements. For example, the mice blinked each time they made the shoulder movements that would have swept their forepaws across their eyes. Fentress's study also demonstrated the importance of sensory feedback in the operation of central sensorimotor programs. The forelimbless mice, deprived of normal tongue–forepaw contact during face grooming, would often interrupt ostensible grooming sequences to lick a cage-mate or even the floor.

PRACTICE CAN CREATE CENTRAL SENSORIMOTOR PROGRAMS. Although central sensorimotor programs for many species-typical behaviors develop without practice, practice can generate or modify them. Theories of sensorimotor learning emphasize two kinds of processes that influence the learning of central sensorimotor programs: response chunking and shifting control to lower levels of the sensorimotor system.

Response Chunking According to the **response-chunking hypothesis**, practice combines the central sensorimotor programs that control individual responses into programs that control sequences (chunks) of behavior. In a novice typist, each response necessary to type a word is individually triggered and controlled; in a skilled typist, sequences of letters are activated as a unit, with a marked increase in speed and continuity.

An important principle of chunking is that chunks can themselves be combined into higher-order chunks. For example, the responses needed to type the individual letters and digits of one's address may be chunked into longer sequences necessary to produce the individual words and numbers, and these chunks may in turn be combined so that the entire address can be typed as a unit.

Shifting Control to Lower Levels In the process of learning a central sensorimotor program, control is shifted from higher levels of the sensorimotor hierarchy to lower levels (see Bassett et al., 2015; Kawai et al., 2015; Makino et al., 2016). Shifting the level of control to lower levels of the sensorimotor system during training has two advantages. One is that it frees up the higher levels of the system to deal with more esoteric aspects of performance. For example, skilled pianists can concentrate on interpreting a piece of music because they do not have to consciously focus on pressing the right keys. The other advantage of shifting the level of control is that it permits great speed because different circuits at the lower levels of the hierarchy can act simultaneously, without interfering with one another. It is possible to type 120 words per minute only because the circuits responsible for activating each individual key press can become active before the preceding response has been completed.

Functional Brain Imaging of Sensorimotor Learning

LO 8.23 Explain how the classic Jenkins and colleagues PET study of simple motor learning summarizes the main points of this chapter.

Functional brain-imaging techniques have provided opportunities for studying gross neural correlates of sensorimotor learning. By recording the brain activity of human volunteers as they learn to perform new motor sequences, researchers can develop hypotheses about the roles of various structures in sensorimotor learning. One of the first studies of this type was the PET study of Jenkins and colleagues (1994). These researchers recorded PET activity from human volunteers who performed two different sequences of key presses. There were four different keys, and each sequence was four presses long. The presses were performed with the right hand, one every 3 seconds, and tones indicated when to press and whether or not a press was correct. There were three conditions: (1) a rest control condition, (2) a condition in which the volunteers performed a newly learned sequence, and (3) a condition in which they performed a well-practiced sequence.

The results of Jenkins and colleagues are summarized in Figure 8.16. Notice two things. First, notice the involvement of the cortical sensorimotor areas that you were introduced to in this chapter. Second, notice how the involvement of association areas and the cerebellum diminished when sequences were well practiced.

The Jenkins and colleagues brain-imaging study of sensorimotor learning and subsequent studies like it (e.g., Bassett et al., 2015; Ostry & Gribble, 2015) have made important contributions by identifying where changes occur in the brain while volunteers learn sensorimotor tasks. But what is the nature of those changes? The next section addresses this question.

Figure 8.16 The activity recorded by PET scans during the performance of newly learned and well-practiced sequences of finger movements.

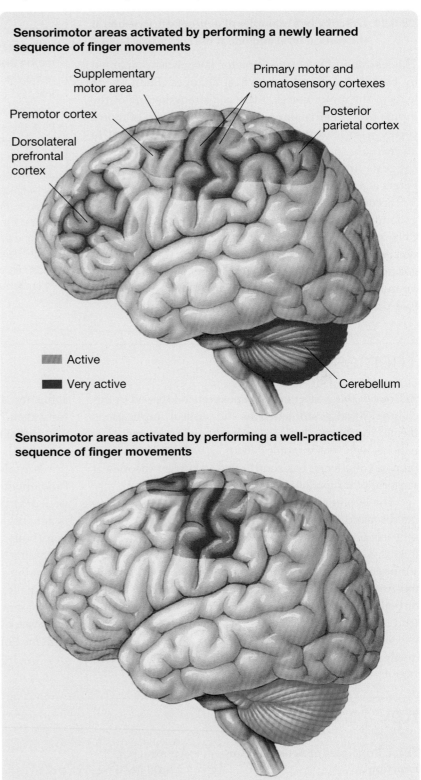

Based on Jenkins, I. H., Brooks, D. J., Bixon, P. D., Frackowiak, R. S. J., & Passingham, R. E. (1994). Motor sequence learning: A study with positron emission tomography. *Journal of Neuroscience, 14*(6), 3775–3790.

Neuroplasticity Associated with Sensorimotor Learning

LO 8.24 Describe two examples of neuroplasticity—one at the cortical level and one at the subcortical level.

The learning of new sensorimotor tasks is accompanied by both cortical and subcortical changes. First, at the level of the motor cortex, there is a strengthening of the inputs from the thalamus and from other areas of motor cortex with learning. Such strengthening is related to an increase in the number of dendritic spines—which suggests an increase in the number of synapses (see Peters et al., 2017).

Second, there is a large increase in the number of oligodendrocytes (glia that produce myelin sheaths in the CNS; see Chapter 4) in subcortical white matter just after sensorimotor learning. This increase in the number of oligodendrocytes is presumably due to an increased demand for myelination of new and/or existing axonal connections (see Xiao et al., 2016).

The Case of Rhonelle, Revisited

A few days after we finished writing this chapter, I (JP) stopped off to pick up a few fresh vegetables and some fish for dinner, and I once again found myself waiting in Rhonelle's line. It was the longest line, but I am a creature of habit. This time, I felt rather smug as I watched her. All of the reading and thinking that had gone into the preparation of this chapter had provided me with some new insights into what she was doing and how she was doing it. I wondered whether she appreciated her own finely tuned sensorimotor system as much as I did.

Then I hatched my plot—a little test of Rhonelle's muscle-spindle feedback system. How would Rhonelle's finely tuned sensorimotor system react to a bag that looked heavy but was in fact extremely light? Next time, I would get one of those paper bags at the mushroom counter, blow it up, drop one mushroom in it, and then fold the top so it looked completely full. I smiled at the thought. But I wasn't the only one smiling. My daydreaming ended abruptly, and the smile melted from my face as I noticed Rhonelle's amused grin. Will I never learn?

Themes Revisited

All four of this text's major themes were addressed in this chapter. Most prominent was the clinical implications theme. You learned how research with neuropsychological patients with sensorimotor deficits, as well as with normal human volunteers, has contributed to current theories of sensorimotor functioning.

The evolutionary perspective theme was evident in the discussion of several comparative experiments on the sensorimotor system, largely in nonhuman primates. An important point to keep in mind is that although the sensorimotor functions of nonhuman primates are similar to those of humans, they are not identical (e.g., monkeys walk on both hands and feet). Remarkably, programs for walking tend to be similar in humans, other mammals, and birds.

You learned how metaphors can be used to think productively about science—in particular, how a large, efficient company can serve as a useful metaphor for the sensorimotor system. You also learned how recent analyses have suggested that primary motor cortex encodes the end point of movements rather than the movements themselves.

You also learned that the sensorimotor system is fundamentally plastic. General commands to move are issued by cortical circuits, but exactly how a movement is actually completed depends on the current situation (e.g., body position). Moreover, the sensorimotor system maintains the ability to change itself in response to learning and practice.

Finally, one of the two emerging themes, consciousness, was present in the pervasive discussion of conscious versus unconscious control of movement, as can be demonstrated with the Ebbinghaus illusion.

Key Terms

Three Principles of Sensorimotor Function

Sensory feedback, p. 215

Sensorimotor Association Cortex

Posterior parietal association cortex, p. 216
Frontal eye field, p. 217

Apraxia, p. 217
Contralateral neglect, p. 217
Dorsolateral prefrontal association cortex, p. 218

Secondary Motor Cortex

Secondary motor cortex, p. 219
Supplementary motor area, p. 219

Premotor cortex, p. 219
Mirror neurons, p. 219

Primary Motor Cortex

Primary motor cortex, p. 221
Somatotopic, p. 221
Motor homunculus, p. 221
Stereognosis, p. 221

Chapter 9
Development of the Nervous System

From Fertilized Egg to You

Robbi Akbari Kamaruddin/Alamy Stock Photo

 ## Chapter Overview and Learning Objectives

Five Phases of Early Neurodevelopment

LO 9.1 Define the terms *totipotent, pluripotent, multipotent, unipotent* and *stem cell*, and identify the major sources of new cells in the developing nervous system.

LO 9.2 Describe the development of the neural plate into the neural tube.

LO 9.3 Describe the process of neural proliferation and identify the two organizer areas.

LO 9.4 Describe the processes of migration and aggregation.

Most of us tend to think of the brain as a three-dimensional array of neural elements "wired" together in a massive network of circuits. However, the brain is not a static network of interconnected elements. It is a *plastic* (changeable) living organ that continuously changes in response to your ongoing experiences.

This chapter focuses on the incredible process of *neurodevelopment* (development of the nervous system), which begins with a single fertilized egg cell and continues through to adulthood. Four general ideas are emphasized: (1) the amazing nature of neurodevelopment, (2) the important role of experience in neurodevelopment, (3) the plasticity of the adult brain; and (4) the consequences of "errors"

in neurodevelopment, such as occur in autism spectrum disorder and Williams disorder.

But first, a case study. Many of us are reared in similar circumstances—we live in warm, safe, stimulating environments with supportive families and communities and plenty to eat and drink. Because there is so little variation in most people's early experience, the critical role of experience in human cerebral and psychological development is not always obvious. In order to appreciate the critical role played by experience in neurodevelopment, it is important to consider cases in which children have been reared in grossly abnormal environments. Genie is such a case (Curtiss, 1977; Rymer, 1993).

The Case of Genie

When Genie was admitted to the hospital at the age of 13, she was only 1.35 meters (4 feet, 5 inches) tall and weighed only 28.1 kilograms (62 pounds). She could not stand erect, chew solid food, or control her bladder or bowels. Since the age of 20 months, Genie had spent most days tied to a potty in a small, dark, closed room. Her only clothing was a cloth harness, which kept her from moving anything other than her feet and hands. In the evening, Genie was transferred to a covered crib and a straitjacket. Her father was intolerant of noise, and he beat Genie if she made any sound whatsoever. According to her mother, who was almost totally blind, Genie's father and brother rarely spoke to Genie, although they sometimes barked at her like dogs. The mother was permitted only a few minutes with Genie each day, during which time she fed Genie cereal or baby food—Genie was allowed no solid food. Genie's severe childhood deprivation left her seriously scarred. When she was admitted to the hospital, she made almost no sounds and was totally incapable of speech.

After Genie was rescued from this horrific life at the age of 13, a major effort was made to get her development back on track and to document her problems and improvements. Genie received special care and training after her rescue, but her behavior never became typical. The following were a few of her continuing problems: She did not react to extremes of warmth and cold; she tended to have silent tantrums during which she would flail, spit, scratch, urinate, and rub her own "snot" on herself; she was easily terrified (e.g., of dogs and men wearing khaki); she could not chew; she could speak only short, poorly pronounced phrases.

It is believed that Genie is currently living in a home for intellectually disabled adults.

Genie's developmental issues were apparently the result of the severe abuse she experienced. Accordingly, this case study suggests the important role that experience plays in neurodevelopment. Reports of childhood adversity affecting neurodevelopment are by no means limited to Genie's case (see Teicher et al., 2016). For example, research has established that malnutrition associated with poverty is sufficient to negatively impact neurodevelopment (see Storrs, 2017). It is clear that neurodevelopment is an important and vulnerable process. Let's learn more.

Five Phases of Early Neurodevelopment

In the beginning, there is a **zygote**, a single cell formed by the amalgamation of an *ovum* (an egg) and a *sperm*. The zygote divides to form two daughter cells. These two divide to form four, the four divide to form eight, and so on, until

a mature organism is produced. Of course, there must be more to development than this; if there were not, each of us would have ended up like a bowl of rice pudding: an amorphous mass of homogeneous cells.

To save us from this fate, three things other than cell multiplication must occur. First, cells must *differentiate*; some must become muscle cells, some must become multipolar neurons, some must become glial cells, and so on. Second, cells must make their way to appropriate sites and align themselves with the cells around them to form particular structures. And third, cells must establish appropriate functional relations with other cells. This module describes how the developing nervous system accomplishes these three things in five phases: (1) induction of the neural plate, (2) neural proliferation, (3) migration and aggregation, (4) axon growth and synapse formation, and (5) neuron death and synapse rearrangement.

Stem Cells and Neurodevelopment

LO 9.1   **Define the terms *totipotent*, *pluripotent*, *multipotent*, *unipotent* and *stem cell*, and identify the major sources of new cells in the developing nervous system.**

A fertilized egg is **totipotent**, that is, the cell has the ability to develop into any class of cell in the body (e.g., bone, skin, neuron, or heart cells). However, soon after, generations of new cells start to be created by cell division; these newly created cells are not totipotent (see Boroviak & Nichols, 2014; Kohwi & Doe, 2013). At this stage, developing cells have the ability to develop into many, but not all, classes of body cells and are said to be **pluripotent**. As the embryo develops, new cells become more and more specialized, and eventually the new cells can develop into different cells of only one class (e.g., different kinds of blood cells). These new cells are said to be **multipotent**. Eventually, most developing cells are **unipotent**: that is, they can develop into only one type of cell (e.g., bipolar neurons).

The totipotent, pluripotent, and multipotent cells created during early development are all embryonic **stem cells** (see Figure 9.1). To understand nervous system development, it is necessary to understand two important properties of stem cells (see Morey, Santanach, & Di Croce, 2015). First, they have an almost unlimited capacity for self-renewal if maintained in an appropriate cell culture—for example, cultures of embryonic stem cells can be kept alive and multiplying for more than a year. This almost unlimited capacity of stem cells for self-renewal is a product of *asymmetric cell division*. The second property of stem cells that plays a critical role in nervous system development is the ability of each stem cell to develop into many different kinds of cells.

These two defining properties of embryonic stem cells appear to be related to the mechanism by which

Figure 9.1 Totipotent, pluripotent, and multipotent cells are all considered to be stem cells. However, their capacity to develop into the different cells of the body differs. Also depicted here is asymmetric cell division—which is required for a stem cell culture to be self-renewing.

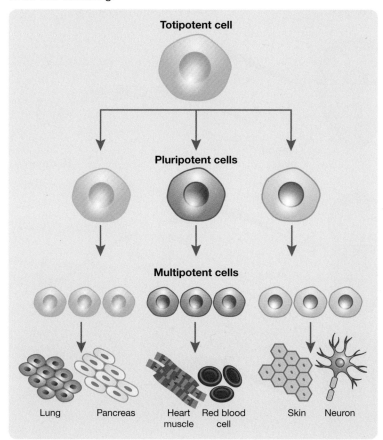

they multiply: asymmetric cell division (cell division that produces two daughter cells with different characteristics). When a stem cell divides into two daughter cells, one of the daughter cells is a stem cell, while the other daughter cell develops into a more specific cell type—see Ito and Suda (2014). But how does one embryonic stem cell develop into different cell types when every cell in a person's body has exactly the same DNA? What makes one cell develop into a skin cell and another into a neuron? The answer lies in the ability of cells to transcribe different sections of DNA depending on their experience. These mechanisms are generally referred to as *epigenetic* (see Bale, 2015; Yao et al., 2016). *Epigenetic mechanisms* were the focus of Chapter 2 and thinking about epigenetics is one of the emerging themes in this text.

The ability of scientists to study development has increased greatly by the development of biochemical tools for controlling the fates of developing cells. For example, with modern biochemical tools, a culture of stem cells can be induced to develop into one of many different brain cell types (see Tsunemoto et al., 2018). It is even possible to watch a miniature three-dimensional brain (a so-called *brain organoid*) develop in culture from stem cells (see

Di Lullo & Kriegstein, 2017; Kelava & Lancaster, 2016; Pasca, 2018; Shen, 2018).

Because of the ability of stem cells to develop into different types of mature cells, their therapeutic potential is under intensive investigation. Do stem cells injected into a damaged part of a mature brain develop into the appropriate brain structure and improve function? You will learn about the potential of stem-cell therapies in Chapter 10.

Induction of the Neural Plate

LO 9.2 Describe the development of the neural plate into the neural tube.

Three weeks after conception, the tissue that is destined to develop into the human nervous system becomes recognizable as the **neural plate**—a small patch of ectodermal tissue on the dorsal surface of the developing embryo. The ectoderm is the outermost of the three layers of embryonic cells: *ectoderm, mesoderm,* and *endoderm.* The development of the neural plate is the first major stage of neurodevelopment in all vertebrates (see Figure 9.2).

The development of the neural plate is *induced* by chemical signals from an area of the underlying **mesoderm layer**—an area consequently referred to as an *organizer* (see Araya et al., 2014; Kiecker & Lumsden, 2012). Indeed, tissue taken from the dorsal mesoderm of one embryo (i.e., the *donor*) and implanted beneath the ventral ectoderm of another embryo (i.e., the *host*) induces the development of an extra neural plate on the ventral surface of the host.

As Figure 9.3 illustrates, the growing neural plate folds to form the *neural groove,* and then the lips of the neural groove fuse to form the **neural tube**—*neural tube defects,* which develop into severe birth defects of the CNS, can

Figure 9.2 A cross section through the ectoderm, mesoderm, and endoderm in a developing embryo. The neural plate develops from some of the tissue in the endoderm.

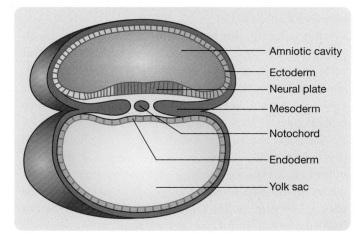

result from errors in this folding process (see Greene & Copp, 2014). The inside of the neural tube eventually becomes the *cerebral ventricles* and *spinal canal*. By 40 days after conception, three swellings are visible at the anterior end of the human neural tube.

Neural Proliferation

LO 9.3   Describe the process of neural proliferation and identify the two organizer areas.

Once the lips of the neural groove have fused to create the neural tube, the cells of the tube begin to *proliferate* (increase greatly in number). Remarkably, cells are generated at a rate of more than 4 million per hour (see Silbereis et al., 2016). This **neural proliferation** does not occur simultaneously or equally in all parts of the tube. Most cell division in the neural tube occurs in the **ventricular zone** and **subventricular zone**—the two regions adjacent to the *ventricle* (the fluid-filled center of the tube). In each species, the cells in different parts of the neural tube proliferate in a particular sequence that is responsible for the pattern of swelling and folding that gives the brain of each member of that species its characteristic shape. For example, in many animals, these swellings ultimately develop into the *forebrain, midbrain,* and *hindbrain* (see Figure 3.18). The complex pattern of proliferation is in part controlled by chemical signals from two organizer areas in the neural tube: the *floor plate,* which runs along the midline of the ventral surface of the tube, and the *roof plate,* which runs along the midline of the dorsal surface of the tube (see Kanold & Luhmann, 2010).

Remarkably, the stem cells created in the developing neural tube are virtually always **radial glial cells** (see Falk & Götz, 2017)—cells whose cell bodies lie either in the ventricular zone or subventricular zone and have a long process that extends to the outermost part of the developing neural tube. Figure 9.4 depicts how radial glial cells undergo asymmetric cell division to achieve the creation of neurons, glia, and other cells of the developing nervous system. Interestingly, in addition to being stem cells, radial glial cells play a key role in cell migration during development (see Figure 9.4).

Figure 9.3 How the neural plate develops into the neural tube during the third and fourth weeks of human embryological development.

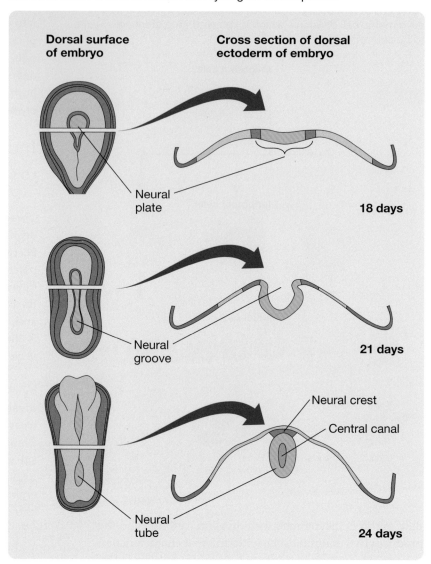

Based on Cowan, W. M. (1979, September). The development of the brain. *Scientific American, 241*, 113–133.

Migration and Aggregation

LO 9.4   Describe the processes of migration and aggregation.

MIGRATION.   Once cells have been created through cell division in the ventricular zone of the neural tube, they migrate to the appropriate target location. During this period of **migration**, the cells are still in an immature form, lacking the processes (i.e., axons and dendrites) that characterize mature neurons. Two major factors govern migration in the developing neural tube: time and location. In a given region of the tube, subtypes of neurons arise on a precise and predictable schedule and then migrate together to particular destinations (see Itoh, Tyssowski, & Gotoh, 2013; Kohwi & Doe, 2013).

Figure 9.4 Radial glial cells are the stem cells in the developing nervous system. Asymmetric cell division of radial glial cells leads to the production of neurons, glia, and other cells of the nervous system.

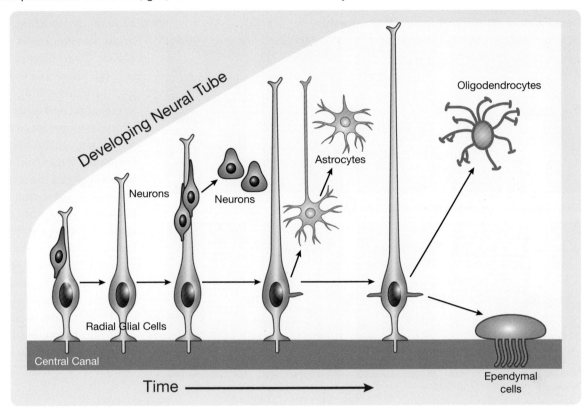

Based on Kriegstein, A., & Alvarez-Buylla, A. (2009).

Cell migration in the developing neural tube is considered to be of two kinds (see Figure 9.5): **Radial migration** proceeds from the ventricular zone in a straight line outward toward the outer wall of the tube; **tangential migration** occurs at a right angle to radial migration—that is, parallel to the tube's walls. Many cells engage in both radial and tangential migration to get from their point of origin in the ventricular zone to their target destination (see

Figure 9.5 Two types of neural migration: radial migration and tangential migration.

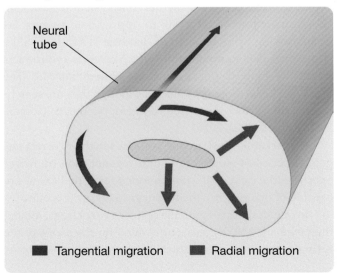

■ Tangential migration   ■ Radial migration

Budday, Steinmann, & Kuhl, 2015; Evsyukova, Plestant, & Anton, 2013).

There are two mechanisms by which developing cells migrate (see Figure 9.6). One is somal translocation. The second is radial-glia-mediated migration. In **somal translocation**, the developing cell has a process that extends from its cell body that seems to explore the immediate environment. Numerous chemicals guide the movement of these processes, by either attracting or repelling them (see Devreotes & Horwitz, 2015; Maeda, 2015). Once the process finds a suitable environment, the cell body moves into and along the process (see Cooper, 2013). Somal translocation allows a cell to migrate in either a radial or tangential fashion.

In **radial-glia-mediated migration,** the developing cell uses the long process that extends from each radial-glia cell as a sort of rope along which it pulls itself up and away from the ventricular zone (see Figure 9.6). Radial-glia-mediated migration allows a cell to migrate in only a radial fashion (see Ohshima, 2015).

Early research on migration in the developing neural tube focused on the cortex. Based on the results of that research, it was asserted that cells migrated in an orderly fashion, progressing from deeper to more superficial layers. Because each wave of cortical cells migrates through the already formed lower layers of cortex before reaching its destination, this radial pattern of cortical development is referred to as an **inside-out pattern** (see Greig et al., 2013).

Figure 9.6 Two methods by which cells migrate in the developing neural tube: somal translocation and radial-glia-mediated migration.

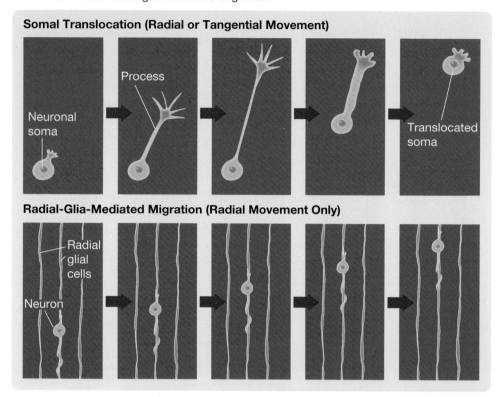

However, cortical migration patterns have turned out to be much more complex than originally thought: Many cortical cells engage in long tangential migrations to reach their final destinations, and the patterns of proliferation and migration are different for different areas of the developing cortex (see Anderson & Vanderhaeghen, 2014; Silbereis et al., 2016; Sun & Hevner, 2014).

The **neural crest** is a structure situated just dorsal to the neural tube (see Figure 9.3). It is formed from cells that break off from the neural tube as it is being formed. Neural crest cells develop into the neurons and glial cells of the peripheral nervous system as well as many other cell types in the body (see Buitrago-Delgado et al., 2015; Kuratani, Kusakabe, & Hirasawa, 2018).

AGGREGATION. Once developing neurons have migrated, they must align themselves with other developing neurons that have migrated to the same area to form the structures of the nervous system. This process is called **aggregation**.

Aggregation is thought to be mediated by at least three non-exclusive mechanisms. First, **cell-adhesion molecules (CAMs)**, which are located on the surfaces of neurons and other cells (see Mori et al., 2014; Sytnyk, Leshchyns'ka, & Schachner, 2017; Weledji & Assob, 2014), have the ability to recognize molecules on other cells and adhere to them. Elimination of just one type of CAM in a *knockout mouse*

(mice that lack a particular gene under investigation; see Chapter 5) has been shown to have a devastating effect on brain development (DiCicco-Bloom, 2006; Lien et al., 2006). Second, there is evidence that gap junctions play a role in aggregation and other aspects of neurodevelopment (see Belousov & Fontes, 2013; Niculescu & Lohmann, 2014). You may recall from Chapter 4 that *gap junctions* are points of communication between adjacent cells; the gaps are bridged by narrow tubes called *connexins* (see Figure 4.13). Third, the process of aggregation is also achieved through interactions between glial cells and neurons; for example, through interactions between microglia and neurons (see Thion & Garel, 2017).

Axon Growth and Synapse Formation

LO 9.5 Describe the processes of axon growth and synapse formation. Also, explain the chemoaffinity hypothesis and the topographic gradient hypothesis.

AXON GROWTH. Once neurons have migrated to their appropriate positions and aggregated into neural structures, axons and dendrites begin to grow from them (see Yogev & Shen, 2017). For the nervous system to function, these projections must grow to appropriate targets. At each growing tip of an axon or dendrite is an amoebalike structure called a **growth cone**, which extends and retracts fingerlike cytoplasmic extensions called *filopodia* (see Figure 9.7). The filopodia behave as though they are searching for the correct route (see Kerstein, Nichol, & Gomez, 2015; Kahn & Baas, 2016).

Remarkably, most growth cones reach their correct targets. A series of studies of neural regeneration by Roger Sperry in the early 1940s first demonstrated that axons are capable of precise growth and suggested how it occurs.

In one study, Sperry cut the optic nerves of frogs, rotated their eyeballs 180 degrees, and waited for the axons of the **retinal ganglion cells**, which compose the optic nerve,

Figure 9.7 Growth cones. The cytoplasmic extensions (the filopodia) of growth cones seem to search for the correct route.

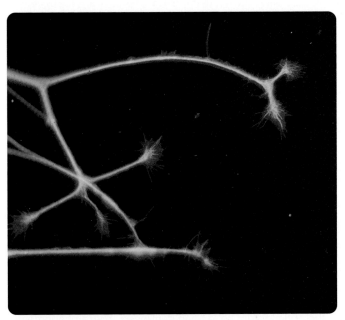

Courtesy of Naweed I. Syed, Ph.D., Departments of Anatomy and Medical Physiology, the University of Calgary.

to *regenerate* (grow again). (Frogs, unlike mammals, have retinal ganglion cells that regenerate.) Once regeneration was complete, Sperry used a convenient behavioral test to assess the frogs' visual capacities (see Figure 9.8). When he dangled a lure behind the frogs, they struck forward, thus indicating that their visual world, like their eyes, had been rotated 180 degrees. Frogs whose eyes had been rotated, but whose optic nerves had not been cut, responded in exactly the same way. This was strong behavioral evidence that each retinal ganglion cell had grown back to the same point of the **optic tectum** (called the *superior colliculus* in mammals) to which it had originally been connected. Neuroanatomical investigations have confirmed that this is exactly what happens (see Guo & Udin, 2000).

On the basis of his studies of regeneration, Sperry proposed the **chemoaffinity hypothesis** of axonal development (see Sperry, 1963). He hypothesized that each postsynaptic surface in the nervous system releases a specific chemical label and that each growing axon is attracted by the label to its postsynaptic target during both neural development and regeneration. In the time since Sperry proposed the chemoaffinity hypothesis, many guidance molecules for axon growth have been identified (see Dudanova & Klein, 2013; Onishi, Hollis, & Zou, 2014). Indeed, it is difficult to imagine another mechanism by which an axon growing out from a rotated eyeball could find its precise target on the optic tectum.

Although it generated a lot of research, the chemoaffinity hypothesis fails to account for the discovery that some growing axons follow the same circuitous route to reach

their target in every member of a species rather than growing directly to it. This discovery led to a revised notion of how growing axons reach their specific targets. According to this revised hypothesis, a growing axon is not attracted to its target by a single specific attractant released by the target, as Sperry thought. Instead, growth cones seem to be influenced by a series of chemical and physical signals along the route (see Goodhill, 2016; Koser et al., 2016; Squarzoni, Thion, & Garel, 2015; Tamariz & Varela-Echavarría, 2015); some attract and others repel the growing axons (see Seabrook et al., 2017).

Pioneer growth cones—the first growth cones to travel along a particular route in a developing nervous system—are believed to follow the correct trail by interacting with guidance molecules along the route. Then, subsequent

Figure 9.8 Sperry's classic study of eye rotation and regeneration.

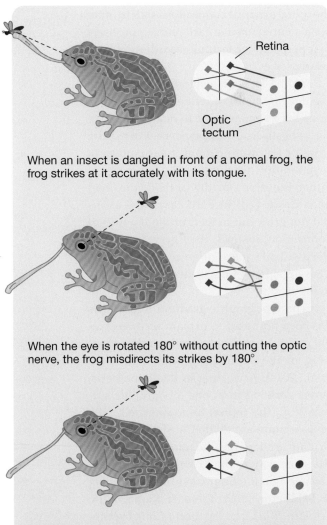

When an insect is dangled in front of a normal frog, the frog strikes at it accurately with its tongue.

When the eye is rotated 180° without cutting the optic nerve, the frog misdirects its strikes by 180°.

When the optic nerve is cut and the eye is rotated by 180°, at first the frog is blind; but once the optic nerve has regenerated, the frog misdirects its strikes by 180°. This is because the axons of the optic nerve, although rotated, grow back to their original synaptic sites.

growth cones embarking on the same journey are presumed to follow the routes blazed by the pioneers. The tendency of developing axons to grow along the paths established by preceding axons is called **fasciculation**.

Much of the axonal development in complex nervous systems involves growth from one topographic array of neurons to another. The neurons on one array project to another, maintaining the same topographic relation they had on the first array; for example, the topographic map of the retina is maintained on the optic tectum.

At first, it was assumed that the integrity of topographical relations in the developing nervous system was maintained by a point-to-point chemoaffinity, with each retinal ganglion cell growing toward a specific chemical label. However, evidence indicates that the mechanism must be more complex. In most species, the synaptic connections between retina and optic tectum are established long before either reaches full size. Then, as the retinas and the optic tectum grow at different rates, the initial synaptic connections shift to other tectal neurons so that each retina is precisely mapped onto the tectum, regardless of their relative sizes.

Studies of the regeneration (rather than the development) of retinal-tectum projections tell a similar story. In one informative series of studies, the optic nerves of mature frogs or fish were cut and their pattern of regeneration was assessed after parts of either the retina or the optic tectum had been destroyed. In both cases, the axons did not grow out to their original points of connection; instead, they grew out to fill the available space in an orderly fashion. These results are illustrated schematically in Figure 9.9.

The **topographic gradient hypothesis** has been proposed to explain accurate axonal growth involving topographic mapping in the developing brain (see Weth et al., 2014). According to this hypothesis, axons growing from one topographic surface (e.g., the retina) to another (e.g., the optic tectum) are guided to specific targets that are arranged on the terminal surface in the same way as the axons' cell bodies are arranged on the original surface (see Cang & Feldheim, 2013; Klein & Kania, 2014; Triplett, 2014). The key part of this hypothesis is that the growing axons are guided to their destinations by two intersecting signal gradients (e.g., an anterior–posterior

gradient and a medial–lateral gradient; see Seabrook et al., 2017; but see Goodhill, 2016). However, other mechanisms have also been shown to contribute to accurate topographic mapping (e.g., Quast et al., 2017), such as spontaneous neural activity (see Seabrook et al., 2017; Zhang et al., 2017) and neuron-astrocyte interactions (see López-Hidalgo & Schummers, 2014).

SYNAPSE FORMATION.   Once axons have reached their intended sites, they must establish an appropriate pattern of synapses. During neurodevelopment, synapses are formed at an astonishing rate of about 700,000 synapses per second (see Silbereis et al., 2016).

A single neuron can grow an axon on its own, but it takes coordinated activity in at least two neurons to create a synapse between them (see Andreae & Burrone, 2014). This is one reason why our understanding of how axons connect

Figure 9.9 The regeneration of the optic nerve of the frog after portions of either the retina or the optic tectum have been destroyed. These phenomena support the topographic gradient hypothesis.

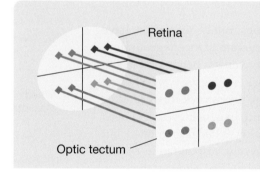

Retina

Optic tectum

Axons normally grow from the frog retina and terminate on the optic tectum in an orderly fashion. The assumption that this orderliness results from point-to-point chemoaffinity is challenged by the following two observations.

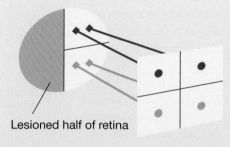

Lesioned half of retina

1 When half the retina was destroyed and the optic nerve cut, the retinal ganglion cells from the remaining half retina projected systematically over the entire optic tectum.

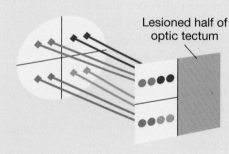

Lesioned half of optic tectum

2 When half the optic tectum was destroyed and the optic nerve cut, the retinal ganglion cells from the retina projected systematically over the remaining half of the optic tectum.

to their targets has lagged behind our understanding of how they reach them. Still, some exciting breakthroughs have been made: (1) some of the chemical signals that play a role in the location and formation of synapses have been identified (see Christensen, Shao, & Colón-Ramos, 2013; Inestrosa & Arenas, 2010; Koropouli & Kolodkin, 2014; Krueger et al., 2012); (2) it has been shown that spontaneous neurotransmitter release is important for synapse formation (see Andreae & Burrrone, 2018); and (3) cell surfaces have been shown to interact prior to synapse formation (see de Wit & Ghosh, 2016).

Perhaps the most exciting recent discovery about **synaptogenesis** (the formation of new synapses) is that it depends on the presence of glial cells, particularly astrocytes (see Bosworth & Allen, 2017) and microglia (see Stogsdill & Eroglu, 2017). Retinal ganglion cells maintained in culture formed seven times more synapses when astrocytes were present. Moreover, synapses formed in the presence of astrocytes were quickly lost when the astrocytes were removed. Early theories about the contribution of astrocytes to synaptogenesis emphasized a nutritional role: Developing neurons need high levels of cholesterol during synapse formation, and the extra cholesterol is supplied by astrocytes. However, current evidence suggests that astrocytes play a much more complex role in synaptogenesis, by processing, transferring, and storing information supplied by neurons (see Clarke & Barres, 2013).

Most current research on synaptogenesis is focused on determining the chemical signals that must be exchanged between presynaptic and postsynaptic neurons for a synapse to be created (see Ou & Shen, 2010; Sheffler-Collins & Dalva, 2012). One complication researchers face is the promiscuity that developing neurons display when it comes to synaptogenesis: In vitro studies show that any type of neuron can form synapses with any other type. However, once established, synapses that do not function appropriately tend to be eliminated (see Coulthard, Hawksworth, & Woodruff, 2018).

Neuron Death and Synapse Rearrangement

LO 9.6 **Describe the processes of neuron death and synapse rearrangement. Why is apoptosis safer than necrosis?**

Neuron death is a normal and important part of neurodevelopment. Many more neurons—about 50 percent more—are produced than are required, and waves of large-scale neuron death occur in various parts of the brain throughout development.

Journal Prompt 9.1
Why do you think the developing nervous system would produce 50 percent more neurons than are required?

Neuron death during development was initially assumed to be a passive process. It was assumed that developing neurons died when they failed to get adequate nutrition. However, it is now clear that cell death during development is usually active. Genetic programs inside neurons are triggered and cause them to actively commit suicide. Passive cell death is called **necrosis** ("ne-KROE-sis"); active cell death is called **apoptosis** ("A-poe-toe-sis").

Apoptosis is safer than necrosis. Necrotic cells break apart and spill their contents into the surrounding extracellular fluid, and the consequence is potentially harmful inflammation. In contrast, in apoptotic cell death, the internal structures of a cell are cleaved apart and packaged in membranes before the cell breaks apart. These membrane packages contain molecules that attract microglia who engulf and consume them.

Apoptosis removes excess neurons in a safe, neat, and orderly way. But apoptosis has a dark side as well: If genetic programs for apoptotic cell death are blocked, the consequence can be cancer (see Bardella et al., 2018); if the programs are inappropriately activated, the consequence can be neurodegenerative disease.

What triggers the genetic programs that cause apoptosis in developing neurons? There appear to be two kinds of triggers. First, some developing neurons appear to be genetically programmed for an early death—once they have fulfilled their functions, groups of neurons die together in the absence of any obvious external stimulus (see Dekkers & Barde, 2013; Underwood, 2013). Second, some developing neurons seem to die because they fail to obtain the life-preserving chemicals that are supplied by their targets (see Deppmann et al., 2008). Evidence that life-preserving chemicals are supplied to developing neurons by their postsynaptic targets comes from two kinds of observations: (1) Grafting an extra target structure (e.g., an extra limb) to an embryo before the period of synaptogenesis reduces the death of neurons growing into the area, and (2) destroying some of the neurons growing into an area before the period of cell death increases the survival rate of the remaining neurons.

Several life-preserving chemicals that are supplied to developing neurons by their targets have been identified. The most prominent class of these chemicals is the **neurotrophins. Nerve growth factor (NGF)** was the first neurotrophin to be isolated (see Levi-Montalcini, 1952, 1975); the second was **brain-derived neurotrophic factor (BDNF)**. The neurotrophins promote the growth and survival of neurons, function as axon guidance molecules, and stimulate synaptogenesis (Park & Poo, 2013).

SYNAPSE REARRANGEMENT.

During the period of cell death, neurons that have established incorrect connections are particularly likely to die. As they die, the space they leave vacant on postsynaptic membranes is filled by the sprouting axon terminals of surviving neurons. Thus, cell death results in a massive rearrangement of synaptic connections. This phase of synapse rearrangement also tends to focus the output of each neuron on a smaller number of postsynaptic cells, thus increasing the selectivity of transmission (see Figure 9.10). There is evidence that microglia play a role in synapse rearrangement (see Coulthard, Hawksworth, & Woodruff, 2018; Ueno & Yamashita, 2014).

Figure 9.10 The effect of synapse rearrangement on the selectivity of synaptic transmission. The synaptic contacts of each axon become focused on a smaller number of cells.

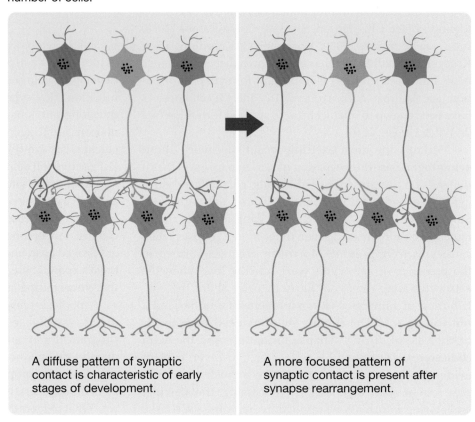

A diffuse pattern of synaptic contact is characteristic of early stages of development.

A more focused pattern of synaptic contact is present after synapse rearrangement.

Scan Your Brain

Are you ready to focus on the continuing development of the human brain after birth? To find out if you are prepared to proceed, scan your brain by filling in the blanks in the following *chronological list of stages of neurodevelopment*. The correct answers are provided at the end of the exercise. Before proceeding, review material related to your errors and omissions.

1. Induction of the neural _____
2. Formation of the _____ tube
3. Neural _____
4. Neural _____
5. _____ aggregation
6. Growth of neural _____
7. Formation of _____
8. Neuron _____ and synapse _____

Scan Your Brain answers: (1) plate, (2) neural, (3) proliferation, (4) migration, (5) Neural, (6) processes (axons and dendrites), (7) synapses, (8) death; rearrangement.

Early Cerebral Development in Humans

Much of our knowledge of the development of the human brain comes from the study of nonhuman species. This fact emphasizes the value of the evolutionary perspective. There is, however, one way in which the development of the human brain is unique: The human brain develops more slowly than those of other species, not achieving full maturity until late adolescence or early adulthood (see Crone & Dahl, 2012; Fuhrmann, Knoll, & Blakemore, 2015). This module deals with cerebral development that occurs during both the **prenatal period** (the period of development before birth) and the early **postnatal period** (the period of development after birth). This module places emphasis on the development of the *prefrontal cortex* (see Figure 1.8), because it is the last part of the human brain to reach maturity (see Giedd, 2015).

Prenatal Growth of the Human Brain

LO 9.7   Describe what has been discovered about human prenatal growth of the brain.

As you have already learned, much of our knowledge about neurodevelopment is inferred after examining development in nonhuman animals. This was especially the case for studies examining prenatal growth of the brain, because such procedures would be invasive or otherwise unfeasible to use in developing humans.

Journal Prompt 9.2

Much of our knowledge about human neurodevelopment is inferred from studies of non-human animals. Given that the human brain is unique in that it develops more slowly, what implications does this have for the validity of our knowledge of human neurodevelopment?

However, three major technical advances have allowed researchers to directly examine the development of human neural tissue. The first you have already learned about: The development of three-dimensional brain organoids in culture. Brain organoids are built from human cells and have been used to study the development of the human brain (see Arlotta, 2018). For example, Quadrato and colleagues (2017) induced the development of human photosensitive brain organoids and then used them to study the development of the visual system.

The second technical advance that has allowed us to more directly examine prenatal human development is the ability to image the brains of prenatal humans. For example, a small number of labs have recently imaged the *functional connectivity* (correlated activity between different brain regions over time) of the developing brain in utero. This line of work has already revealed the development of functional connections in 30- to 38-week-old fetuses (see Thomason et al., 2017).

The final technical advance that has furthered our understanding of the developing human brain is the characterization of cell-level *transcriptomes* (a catalogue of all the proteins transcribed in a particular cell). Transcriptomes have now been characterized for both the developing human prefrontal cortex (see Zhong et al., 2018), and even the entire human cerebral cortex (see Fan et al., 2018), in human embryos. Analysis of these transcriptomes will likely provide important insights into cell development in prenatal humans (see Bakken et al., 2016).

Postnatal Growth of the Human Brain

LO 9.8   Describe the various qualities of postnatal growth of the human brain.

The human brain grows substantially after birth, doubling in volume between birth and adulthood (see Gilmore, Knickmeyer, & Gao, 2018), with much of the growth occurring in the first year (Gilmore et al., 2012; Li et al., 2015) and continuing into the third year (see Silbereis et al., 2016). This increase in size does not result from the development of additional neurons. The postnatal growth of the human brain seems to result from three other kinds of growth: synaptogenesis, myelination of axons, and increased branching of dendrites.

There is a general increase in *synaptogenesis* in the human cortex shortly after birth, but there are differences among the cortical regions. For example, in the primary visual and auditory cortexes, there is a major burst of synaptogenesis in the fourth postnatal month, and maximum synapse density (150 percent of adult levels) is achieved in the seventh or eighth postnatal month. In contrast, synaptogenesis in the prefrontal cortex occurs at a relatively steady rate, reaching maximum synapse density in the second year.

Myelination increases the speed of axonal conduction, and the myelination of various areas of the human brain during development roughly parallels their functional development (see Purger, Gibson, & Monje, 2015). Myelination of sensory areas occurs in the first few months after birth, and myelination of the motor areas follows soon after that, whereas myelination of the prefrontal cortex continues into adulthood (Yap et al., 2013). Indeed, myelination appears to be an ongoing process that changes as a function of experience throughout one's lifespan (see Mount & Monje, 2017).

In general, the pattern of *dendritic branching* in the cortex duplicates the inside-out pattern of neural migration you have already learned about, in the sense that dendritic branching progresses from deeper to more superficial layers. Technical advances in imaging live neurons in culture are leading to insights into how dendrites can reconfigure themselves. Most surprising is the speed with which even mature *dendritic spines* (the small protuberances from dendrites, many of which form synapses with other cells) can change their shape in response to an animal's experiences and current environment (see Berry & Nedivi, 2017).

Early human brain development is not a one-way street; there are regressive changes as well as growth (see Jernigan et al., 2011). For example, once maximum synaptic density and gray matter volume have been achieved, there are periods of decline. Like periods of synaptogenesis, periods of synaptic and gray-matter loss occur at different times in different parts of the brain. For example, cortical thinning occurs first in primary sensory and motor areas, progresses to secondary areas, and culminates in association areas (see Jernigan et al., 2011). The achievement of the adult level of gray matter in a particular cortical area is correlated with that area's reaching functional maturity—sensory and motor areas reach functional maturity before association areas (see Purger, Gibson, & Monje, 2015).

As is true for studies of the prenatal human brain, new technologies have enhanced the rate at which researchers are learning about the developing postnatal human brain. For example, recent developments in infant magnetic resonance imaging (MRI) and functional MRI (see Chapter 5) have allowed for a survey of structural changes and functional changes, respectively, in the human brain across development (see Cao, Huang, & He, 2017; Gilmore, Knickmeyer, & Gao, 2018; Vijayakumar et al., 2018). That is, cognitive neuroscientists can now examine the neural correlates of particular cognitive capacities at different stages of early development (see Ellis & Turk-Browne, 2018). Another example comes from the development of methods for cataloguing the proteins in particular cell types: Such methods have allowed researchers to create a *proteome* (a catalogue of the proteins in different cell types) of the postnatal human brain (see Carlyle et al., 2017).

Development of the Prefrontal Cortex

LO 9.9   Describe the functions of the prefrontal cortex and what sorts of behaviors infants display prior to its development.

As you have just learned, the prefrontal cortex displays the most prolonged period of development of any brain region. Its development is believed to be largely responsible for the course of human cognitive development, which occurs over the same period (see Giedd, 2015).

Given the size, complexity, and heterogeneity of the prefrontal cortex, it is hardly surprising that no single theory can explain its function. Nevertheless, four types of cognitive functions have often been linked to this area in studies of adults with extensive prefrontal damage. Various parts of the adult prefrontal cortex seem to play roles in (1) *working memory*, that is, keeping relevant information accessible for short periods of time while a task is being completed; (2) planning and carrying out sequences of actions; (3) inhibiting responses that are inappropriate in the current context but not in others; and (4) following rules for social behavior (see Fareri & Delgado, 2014; Watanabe & Yamamoto, 2015). These cognitive functions seem to mature during the period of adolescence (often defined as 10 to 24 years of age; Patton et al., 2018) and appear to be associated with the growth of dopaminergic axons into the prefrontal cortex (see Hoops & Flores, 2017) and the maturation of the GABAergic system in the prefrontal cortex (see Caballero & Tseng, 2016).

One interesting line of research on prefrontal cortex development is based on Piaget's classic studies of psychological development in human babies. In his studies of 7-month-old children, Piaget noticed an intriguing error. A small toy was shown to an infant; then, as the child watched, it was placed behind one of two screens, left or right. After a brief delay, the infant was allowed to reach for the toy. Piaget found that almost all 7-month-old infants reached for the screen behind which they had seen the toy being placed. However, if, after being placed behind the same screen on several consecutive trials, the toy was placed behind the other screen (as the infant watched), most of the 7-month-old infants kept reaching for the previously correct screen rather than the screen that currently hid the toy. Children tend to make this *perseverative error* between about 7 and 12 months, but not thereafter (Diamond, 1985). **Perseveration** is the tendency to continue making a formerly correct response when it is currently incorrect.

Diamond (1991) hypothesized that the perseverative errors that occur in infants between 7 and 12 months are due to a lag in prefrontal cortex development. Current research supports this hypothesis. First, patients with damage to the prefrontal cortex display perseveration when switching tasks: They fail to suppress previously correct responses when the task switches such that those responses are now incorrect (see Shallice & Cipolloti, 2018). Second, the mammalian prefrontal cortex is known to be involved in the retention of information within working memory (see Cavanagh et al., 2018), which is required in order to perform well on Piaget's task. Finally, the number of synapses in the prefrontal cortex is not maximal until the second year of life.

Effects of Experience on Postnatal Development of Neural Circuits

Because the human brain develops so slowly, there are many opportunities for experience to influence its development. Indeed, as you will see in this module, the effects of experience on brain development are many and varied.

Critical Periods vs. Sensitive Periods

LO 9.10   Explain the difference between a "critical period" and a "sensitive period" of development.

An important feature of the effects of experience on development is that they are time-dependent: The effect of a given experience on development depends on when it occurs during development (see Makinodan et al., 2012). In most cases, there is a window of opportunity in which a particular experience can influence development. If it is absolutely essential (i.e., critical) for an experience to occur within a particular interval to influence development, the interval is called a **critical period**. If an experience has a great effect on development when it occurs during a particular interval but can still have weak effects outside the interval, the interval is called a **sensitive period**. Although the term *critical period* is widely used, the vast majority of experiential effects on development have been found to occur during *sensitive periods*.

Early Studies of Experience and Neurodevelopment: Deprivation and Enrichment

LO 9.11 Explain the different effects of deprivation and enrichment on neurodevelopment.

Most research on the effects of experience on the development of the brain has focused on sensory and motor systems—which lend themselves to experiential manipulation. Much of the early research focused on two general manipulations of experience: sensory deprivation and sensory enrichment.

The first studies of sensory deprivation assessed the effects of rearing animals in the dark. Rats reared from birth in the dark were found to have fewer synapses and fewer dendritic spines in their primary visual cortex, and as adults they were found to have deficits in depth and pattern vision. In contrast, the first studies of early exposure to *enriched environments* (e.g., environments that contain toys, running wheels, and other rats; see Sale, 2018) found that enrichment had beneficial effects. For example, rats that were raised in enriched (complex) group cages rather than by themselves in barren cages were found to have thicker cortex, with more dendritic spines and more synapses per neuron (see Berardi, Sale, & Maffei, 2015).

The effects of sensory deprivation have also been studied in human babies born with cataracts in both eyes, which render them nearly blind (Vavvas et al., 2018). When the cataracts were removed between 1 and 6 weeks after birth, their vision was comparable to that of a newborn (Maurer & Lewis, 2018). Thereafter, some aspects of vision improved quickly, but some visual deficits (e.g., deficits in face processing) persisted into adulthood (see Maurer, 2017).

Experience and Neurodevelopment

LO 9.12 Give two examples of the effects of experience on neurodevelopment.

Research on the effects of experience on brain development has progressed beyond simply assessing general sensory deprivation or enrichment. Manipulations of early experience have become more selective. Many of these selective manipulations of early experience have revealed a competitive aspect to the effects of experience on neurodevelopment. This competitive aspect is most clearly illustrated by the disruptive effects of monocular deprivation on the development of ocular dominance columns in

primary visual cortex (see Chapter 6) and by the development of topographic cortical maps in sensory systems.

OCULAR DOMINANCE COLUMNS. Depriving one eye of input for a few days early in life has a lasting adverse effect on vision in the deprived eye, but this does not happen if the other eye is also blindfolded. When only one eye is blindfolded, the ability of that eye to activate the visual cortex is reduced, whereas the ability of the other eye is increased. Both of these effects occur because early monocular deprivation changes the pattern of synaptic input into the primary visual cortex (see Jaepel et al., 2017).

In many species, ocular dominance columns in layer IV of the primary visual cortex are largely developed at birth. However, blindfolding one eye for just a few days during the first few months of life reorganizes the system: The width of the columns of input from the deprived eye is decreased, and the width of the columns of input from the nondeprived eye is increased (Hata & Stryker, 1994; Hubel, Wiesel, & LeVay, 1977). The exact timing of the sensitive period for this effect is specific to each species. Note that this is an example of a *sensitive period*, as opposed to a *critical period*, because during adulthood the effect still occurs but requires longer periods of monocular deprivation (see Levelt & Hübener, 2012).

Because the adverse effects of early monocular deprivation manifest themselves so quickly (i.e., in a few days), it was believed that they could not be mediated by structural changes. However, Antonini and Stryker (1993) found that a few days of monocular deprivation produced a massive decrease in branching of those axons that extend from the cell bodies of lateral geniculate nucleus neurons to lower layer IV of the primary visual cortex (see Figure 9.11).

Figure 9.11 The effect of a few days of early monocular deprivation on the structure of axons projecting from the lateral geniculate nucleus into lower layer IV of the primary visual cortex. Axons carrying information from the deprived eye displayed substantially less branching.

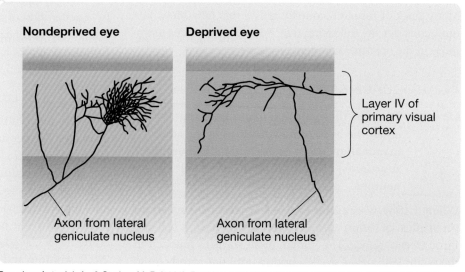

Based on Antonini, A., & Stryker, M. P. (1993). Rapid re-modeling of axonal arbors in the visual cortex. *Science, 260*, 1819–1821.

TOPOGRAPHIC SENSORY CORTEX MAPS. Some of the most remarkable demonstrations of the effects of experience on the organization of the nervous system come from research on sensory topographic maps. The following are two such demonstrations:

- Roe and colleagues (1990) surgically altered the course of developing axons of ferrets' retinal ganglion cells so that the axons synapsed in the medial geniculate nucleus of the auditory system instead of in the lateral geniculate nucleus of the visual system. Remarkably, the experience of visual input caused the auditory cortex of the ferrets to become organized *retinotopically* (laid out like a map of the retina). In general, surgically attaching the inputs of one sensory system to cortex that would normally develop into the primary cortex of another system leads that cortex to develop many, but not all, characteristics typical of the newly attached system (see Majewska & Sur, 2006).

- Several studies have shown that early music training influences the organization of human cortex (see Miendlarzewska & Trost, 2014). For example, early musical training expands the area of auditory cortex that responds to complex musical tones.

Neuroplasticity in Adults

If this text were a road trip we were taking together, at this point, we would spot the following highway sign: SLOW, IMPORTANT VIEWPOINT AHEAD. You see, you are about to encounter findings that have changed how neuroscientists think about the human brain.

Neuroplasticity was once thought to be restricted to the developmental period. Mature brains were considered to be set in their ways, incapable of substantial reorganization. Now, the accumulation of evidence has made clear that mature brains are continually changing and adapting. Many lines of research contributed to this new view. For now, consider the following two. You will encounter many more in the next two chapters.

Neurogenesis in Adult Mammals

LO 9.13 **Describe the evolution in our thinking about the birth of new neurons in the adult mammalian brain. Also, explain the possible function(s) of adult-born hippocampal neurons.**

When I (SB) was a student, I learned two important principles of brain development. The first I learned through experience: The human brain starts to function in the womb and never stops working until one stands up to speak in public. The second I learned in a course on brain development: **Neurogenesis** (the growth of new neurons) does not occur in adults. The first principle appears to be fundamentally correct, at least when applied to me, but the second has been proven wrong.

Prior to the early 1980s, brain development after the early developmental period was seen as a downhill slope: Neurons continually die throughout a person's life, and it was assumed that the lost cells are never replaced by new ones. Although researchers began to chip away at this misconception in the early 1980s, it persisted until the late 20th century in neuroscience, as one of the central principles of neurodevelopment. Nevertheless, the idea persists as the most widely held public misconception about brain function (see Yoo & Blackshaw, 2018).

The first serious challenge to the assumption that neurogenesis is restricted to early stages of development came with the discovery of the growth of new neurons in the brains of adult birds. Nottebohm and colleagues (e.g., Goldman & Nottebohm, 1983) found that brain structures involved in singing begin to grow in songbirds just before each mating season and that this growth results from an increase in the number of neurons.

Journal Prompt 9.3

As you will learn in this module, even the adult brain displays significant neuroplasticity. What do you think is the evolutionary significance of this adult neuroplasticity?

Then, in the 1990s, researchers, armed with new techniques for labelling newly born neurons, showed that adult neurogenesis occurs in the rat hippocampus (e.g., Cameron et al., 1993)—see Figure 9.12. And shortly thereafter, it was discovered that new neurons are also continually generated in the subventricular zone. (As you learned earlier, the subventricular zone is where neural proliferation occurs during early neurodevelopment.)

At first, reports of adult neurogenesis were not embraced by a generation of neuroscientists who had been trained to think of the adult brain as fixed, but acceptance grew as confirmatory reports accumulated. Particularly influential were reports that new neurons are added to the hippocampuses of primates (e.g., Kornack & Rakic, 1999), including humans (Erikkson et al., 1998), and that the number of new neurons added to the adult human hippocampus is substantial, an estimated 700 per day per hippocampus (see Kempermann, 2013; Kheirbek & Hen, 2013; Spalding et al., 2013).

In most nonhuman adult mammals, substantial neurogenesis seems to be restricted to the subventricular zone and hippocampus—although low levels have also been observed in the hypothalamus (see Sousa-Ferreira, de Almeida, & Cavadas, 2014), cortex (see Feliciano & Bordey, 2012), striatum and spinal cord (see Yoo & Blackshaw, 2018). In adult humans, neurogenesis has been observed in the

Figure 9.12 Adult neurogenesis. The top panel shows new cells in the dentate gyrus of the hippocampus that are labelled with different colors: the cell bodies of neurons are stained blue, mature glial cells are stained green, and new cells are stained red. The bottom panel shows the cells from the top panel under higher magnification, which makes it apparent that some of the cells have taken up both blue and red stain and are thus new neurons.

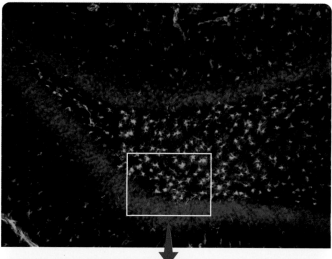

Courtesy of Carl Ernst and Brian Christie, Department of Psychology, University of British Columbia.

striatum (Ernst et al., 2014; Inta, Cameron, & Gass, 2015; Welberg, 2014a) and in the hippocampus (see Adams & Morshead, 2018; Kuhn, Toda, & Gage, 2018; Snyder, 2019; but see Sorrells et al., 2018), but not in the subventricular zone (see Bergmann & Frisén, 2013).

EFFECTS OF EXPERIENCE ON ADULT NEUROGENESIS. From a biopsychological perspective, evidence that experience can influence adult neurogenesis is particularly exciting. One line of research on adult neurogenesis began with a study of the effects on adult rodents living in enriched environments. It turned out that adult rats living in enriched environments produce more new hippocampal neurons than do adult rats living in nonenriched environments (Kempermann & Gage, 1999). This observed positive effect on neurogenesis in the adult rat hippocampus depends in part on the increases in exercise that typically occur in enriched environments (see DiFeo & Shors, 2017; Van Praag et al., 1999). However, exposure to enriched environments in the absence of increases in exercise can also effectively increase adult hippocampal neurogenesis (see Birch & Kelly, 2019). Taken together, these findings have a provocative implication: Because the hippocampus is involved in some kinds of memory, both enriching one's environment (e.g., by engaging in diverse cognitive activities) and engaging in regular exercise may reduce or delay memory problems (see Ma et al., 2017).

FUNCTIONS OF NEWLY BORN NEURONS IN THE ADULT BRAIN. What are the functions of neurons that are the products of adult neurogenesis? It is now established that neurons generated during adulthood survive, become integrated into neural circuits, and begin to conduct neural signals (see Kelsch, Sim, & Lois, 2010; Toni et al., 2008). Adult-generated olfactory bulb and striatal neurons become *interneurons* (see Brann & Firestein, 2014; Ernst et al., 2014; Sakamoto, Kageyama, & Imayoshi, 2015); and adult-generated hippocampal neurons become neurons that grow axons that form synapses on cells in other parts of the hippocampus (see Gonçalves, Schafer, & Gage, 2016). Although there has been some progress in understanding the anatomy and physiology of adult-generated neurons, understanding their function has proven more difficult.

Most research on the function of adult-generated neurons has focused on the hippocampus. Based on that research, a number of theories have been proposed over the past 25 years. Some researchers have proposed a role for these new hippocampal neurons in memory function (see Gonçalves, Schafer, & Gage, 2016; Miller & Sahay, 2019; Lieberwirth et al., 2016; Yagi & Galea, 2019)—including forgetting (see Akers et al., 2014; Mongiat & Schinder, 2014; Welberg, 2014b). One popular theory is that adult hippocampal neurogenesis is important for **pattern separation**: our ability to separate distinct percepts into individual memories for storage (see Aimone et al., 2014; Bergmann & Frisén, 2013; but see Becker, 2017). Other researchers have proposed that adult hippocampal neurogenesis serves a role in mood and anxiety regulation (see Cope & Gould, 2019) and resilience to stressful situations (Anacker et al., 2018).

Freund et al. (2013) took a naturalistic approach to studying the role of adult hippocampal neurogenesis: They monitored the exploratory behavior of a large number of genetically identical mice in a complex enriched environment over a 3-month period. They found that variation in individual behavior increased over time. As mentioned previously, physical activity promotes neurogenesis, so, as expected, those animals that were more active had greater amounts of hippocampal neurogenesis. However, Freund and colleagues also found that exploration had an even greater influence on adult hippocampal neurogenesis than physical activity; that is, those animals that explored over larger areas, and consequently were exposed to more cognitive challenges, showed the greatest amounts of newly born hippocampal neurons (see Bergmann & Frisén, 2013; Freund et al., 2013). Accordingly, one role of adult hippocampal neurogenesis might be to allow us to flexibly adapt to complex environments.

Effects of Experience on the Reorganization of the Adult Cortex

LO 9.14   Describe four examples of experience affecting the organization of the adult cortex.

We said we would consider two lines of current research on adult neuroplasticity, and you have just learned about one: adult neurogenesis. The second deals with the effects of experience on the reorganization of adult cortex.

Experience in adulthood can lead to reorganization of sensory and motor cortical maps. For example, Mühl-nickel and colleagues (1998) found that *tinnitus* (ringing in the ears) produces a major reorganization of primary auditory cortex; Elbert and colleagues (1995) showed that adult musicians who finger stringed instruments (e.g., the violin)

with their left hand have an enlarged hand-representation area in their right somatosensory cortex; and Rossini and colleagues (1994) showed that anesthetizing particular fingers reduced their representation in contralateral somatosensory cortex.

One study warrants special attention because it demonstrates an important aspect of adult neuroplasticity. Hofer and colleagues (2005) showed that the elimination of visual input to one eye of adult mice reduced the size of the ocular dominance columns for that eye in layer IV of primary visual cortex. More importantly, they showed that the reductions in size of the ocular dominance columns occurred more quickly and were more enduring if the adult mice had previously experienced visual deprivation in the same eye. Thus, once the brain has adapted to abnormal environmental conditions, it can more rapidly adapt if it encounters the same conditions again.

The discovery of adult neuroplasticity is changing the way that we humans think about ourselves. More importantly, for those with brain damage, it has suggested some promising new treatment options. You will learn about some of these in Chapter 10.

Scan Your Brain

Before you delve into the last module of the chapter on disorders of neurodevelopment, review what you have learned so far. Fill in each of the blanks with the most appropriate term. The correct answers are provided at the end of the exercise. Before proceeding, review material related to your errors and omissions.

1. Several areas of the prefrontal cortex play a role in executive functions, planning, following social rules, and ____.
2. Piaget found that 7-month-old infants reached for the same location to find a hidden object even after the object had been moved to a new location in front of them. This was termed as ____ ____.
3. If it is absolutely essential for an experience to occur within a particular interval to influence development, the interval is called a ____ ____.
4. Most experiential effects occur during the ____ period.

5. The effects of early stimulus deprivation, such as depriving visual input, can reduce the number of ____ projecting to the visual cortex from the lateral geniculate nucleus.
6. New olfactory bulb and striatal neurons are created from adult neural ____ cells at certain sites in the subventricular zone of the lateral ventricles.
7. Animal studies have shown that adult rats living in enriched environments produced 60 percent more neurons in the ____.
8. Adult-generated hippocampal neurons become ____ cells in the dentate gyrus.
9. ____, otherwise referred to as *ringing of the ears*, produces a major reorganization of the primary auditory cortex.

Scan Your Brain answers: (1) inhibition, (2) perseverative error, (3) critical period, (4) sensitive, (5) axons, (6) stem, (7) hippocampus, (8) granule, (9) Tinnitus.

Atypical Neurodevelopment: Autism Spectrum Disorder and Williams Syndrome

Like all complex processes, neurodevelopment is easily thrown off track. Because it can disrupt subsequent

stages, one tiny misstep can have far-reaching consequences. Ironically, much of what we have learned about typical development has come from studying developmental disorders—as you have already seen with the case of Genie. This final module of the chapter focuses on two forms of atypical neurodevelopment: autism spectrum disorder and Williams syndrome. It is informative to consider these two disorders together because, as you will soon learn, they are similar in some respects and opposite in others.

Autism Spectrum Disorder

LO 9.15   Describe autism spectrum disorder and attempts to identify its neural mechanisms.

Autism spectrum disorder (ASD) is a complex neurodevelopmental disorder. It is a difficult disorder to define because cases differ so greatly. Be that as it may, ASD is almost always apparent before the age of 3 and typically does not increase in severity after that age. Two symptoms are considered to be *core symptoms* because they are required for diagnosis: (1) a reduced capacity for social interaction (e.g., social avoidance, lack of interest in social interactions) and communication and (2) restricted and repetitive patterns of behavior, interests, or activities (American Psychiatric Association, 2013; Volkmar & McPartland, 2014). Other characteristics also tend to be associated with the disorder—about 66% to 75% percent of those with ASD are male (see Lai et al., 2017; Tordjman et al., 2014), many suffer from either an intellectual or learning disability, and they are more likely to suffer from epilepsy (see Lee, Smith, & Paciorkowski, 2015). Older mothers are more likely to give birth to a child with ASD, but the probability of a young mother (under 30) giving birth to a child with ASD increases if the father is over 40 (Shelton, Tancredi, & Hertz-Picciotto, 2010).

ASD IS A HETEROGENEOUS DISORDER.   ASD is *heterogeneous* in two senses. First, an affected individual with ASD might be severely impaired in some respects but typical, or even superior, in others. For example, individuals with ASD who have an accompanying intellectual disability often perform well on tests involving rote memory, jigsaw puzzles, music, and art. Second, there are substantial differences in the amount of impairment amongst individuals with ASD: some have many impairments, whereas others display few.

Because ASD has such diverse presentations, here we will present two cases for your consideration: one that lies on each end of the autism spectrum. It is important to note that the majority of individuals lie somewhere between these two extreme cases. Alex suffers from one extreme form of ASD, whereas S.D. lives with a form of ASD on the other end of the spectrum. Unlike the previous case studies in this text, S.D. has written their own story.

The Case of Alex: Are You Ready to Rock?

Alex cried so hard when he was a baby that he would sometimes vomit. It was the first sign of his relatively severe ASD. Alex is now 7, and he spends much of each day scampering around the house.

When Alex sees the delivery boy, he yells, "Are you ready to rock?" because the delivery boy once said that in passing to him. Alex is obsessed with spaghetti, chocolate ice cream, and trucks. He can spot trucks in a magazine that are so small that most people would not immediately recognize them, and he has all his toy trucks lined up in the main hallway of his house.

Despite his severe intellectual disability, Alex has little difficulty with computers. He recently went on the Internet and purchased a new toy truck.

Alex has echolalia; that is, he repeats almost everything that he hears. He recently told his mother that he loves her, but she doubts very much that he understands what this means—he is just repeating what she has said to him. He loves music and knows the words to many popular songs.

The Case of S.D: The Self-Advocate

I was diagnosed relatively late, at the age of 12, with ASD. I already knew I was different: I had struggled to relate to my peers in elementary school and didn't think I had any friends. At the same time, I excelled in my academic work.

I've always been an introspective person, wondering about why I think the way I do. Perhaps my ceaseless introspection is a sort of coping mechanism for living in a world where the majority of people have a fundamentally different nervous system than me. Regardless, such introspection is an important part of my experiences of living with ASD.

My first experience with self-advocacy was an in-class presentation on ASD in middle school, where I shared my personal experiences living with ASD. Then, as I got older, I began to give talks in the wider community, especially about ASD in educational settings. I wanted people to understand the challenges and opportunities of working with students like me. Simply telling my audience my story has had a powerful impact, both for myself and for the many other students living with ASD.

Most of my audience members are surprised by my diagnosis. Perhaps this is because many of the challenges I face living with ASD (e.g., issues with executive function and self-care) are not visible, whereas my talent for public speaking is.

Since I began attending university, the environment has been great for me as a student with ASD. I like the clear structure of the courses, the reduced social pressure (especially compared to high school), and being able to engage in my intellectual interests with other like-minded persons. I suppose my choice to major in Psychology was just a natural extension of my constant introspection.

There certainly still exist stereotypes and misconceptions of ASD in university, but I'm always happy to share my experiences and provide more information about ASD as a means of battling that. Most importantly, both my ASD and my self-advocacy have given me a sense of community.

Used with Permission of Emma Smith

ASD is common enough that everybody should be alert for its main early warning sign: a delay in the development of social interaction. For example, the following specific signs could be a cause for concern: a decline in eye contact from 2 months onwards (see Shultz, Klin, & Jones, 2018), no smiles or happy expressions by 9 months, and no communicative gestures such as pointing or waving by 12 months (see Constantino et al., 2017; Jones & Klin, 2014; Yoon & Vouloumanos, 2014).

ASD is the most prevalent neurodevelopmental disorder. Until the early 1990s, most *epidemiological studies* (studies of the incidence and distribution of disease in the general population) reported that ASD in the United States occurred in fewer than 1 in 1,000 births; however, the Centers for Disease Control and Prevention now estimate the incidence to be 1 in 68 births (see Biyani et al., 2015). This large increase in the incidence of ASD is cause for concern—although it may, in part, reflect recent broadening of the diagnostic criteria, increasing public awareness of the disorder, and improved methods of identifying cases (see Kim & Leventhal, 2015).

More severe forms of ASD, like that of Alex, are difficult to treat (see Walsh et al., 2011). Intensive behavioral therapy can improve the lives of some of those individuals (see Lange & McDougle, 2013), but it is still often difficult for them to live independently.

ASD SAVANTS. Perhaps the single most remarkable aspect of ASD is the tendency for some persons with ASD to be savants. **Savants** are persons with general intellectual disabilities who nevertheless display amazing and specific cognitive or artistic abilities (see Treffert, 2014a). Savant abilities can take many forms: feats of memory, naming the day of the week for any future or past date, identifying prime numbers (any number divisible only by itself and 1), drawing, and playing musical instruments (see Bonnel et al., 2003). Consider the following cases of savantism (Ramachandran & Blakeslee, 1998; Sacks, 1985).

Cases of Amazing Savant Abilities

- One savant could tell the time of day to the exact second without ever referring to his watch. Even when he was asleep, he would mumble the correct time.

- Tom was blind and could not tie his own shoes. He had never had any musical training, but he could play the most difficult piano piece after hearing it just once, even if he was playing with his back to the piano.

- One pair of ASD twins had difficulty doing simple addition and subtraction and could not even comprehend multiplication and division. Yet, if given any date in the last or next 40,000 years, they could specify the day of week that it fell on.

- Stephen Wiltshire, sometimes called the "human camera," has ASD and can draw an unbelievably detailed city skyline—down to the correct numbers of windows on particular buildings—after only a brief viewing of the city by helicopter. You can see Stephen Wiltshire's work at www.stephenwiltshire.co.uk.

Savant abilities remain a mystery. These abilities do not appear to develop through learning or practice; they seem to emerge spontaneously. Even ASD savants with good language abilities cannot explain their own feats. They seem to recognize patterns and relationships that escape others. Several investigators have speculated that somehow the atypical development of certain parts of their brains has led to compensatory responses in other parts. Indeed, savant-like abilities can emerge in otherwise healthy people following damage to the left anterior temporal lobe, or by transient inactivation (via transcranial magnetic stimulation; see Chapter 5) of the same structure (see Treffert, 2014b).

GENETIC MECHANISMS OF ASD. Genetic and epigenetic factors influence the development of ASD (see Bagni & Zukin, 2019; Iakoucheva, Muotri, & Sebat, 2019; Quesnel-Vallieres et al., 2019; Tremblay & Jiang, 2019). Siblings of people with ASD have about a 20 percent chance of being diagnosed with the disorder (see Gliga et al., 2014). This is well above the rate in the general population, but well below the 50 percent chance that would be expected if ASD were caused solely by a single dominant gene. Also, if one monozygotic twin is diagnosed with ASD, the other has a 60 percent chance of receiving the same diagnosis. These findings suggest that ASD is triggered by several genes interacting with the environment (see State & Levitt, 2011). Several dozen genes have already been implicated. In only about 1 to 5 percent of ASD cases can the disorder be attributed to a single gene mutation (see Mullins, Fishell, & Tsien, 2016; Sztainberg & Zoghbi, 2016; Takumi & Tamada, 2018).

Much better predictors of ASD are *transcription-related errors* (errors in the creation of RNA molecules from strands of DNA). Transcription-related errors are found in the brain cells of a majority of individuals with ASD (see Quesnel-Vallieres et al., 2019), and the number of those errors is positively correlated with the severity of ASD symptoms (Wu et al., 2016).

NEURAL MECHANISMS OF ASD. The heterogeneity of the symptoms of ASD—that is, deficits in some behavioral functions but not others—suggests underlying alterations to some neural structures but not others. In addition, marked differences in the symptoms displayed by various individuals with ASD suggest that similar variability exists for the underlying neural correlates (see Robertson & Baron-Cohen, 2017). Given this complex situation, it is clear that large, systematic studies will be required to identify the neural correlates of the various symptoms of ASD.

One line of research on the neural mechanisms of ASD has focused on the atypical reaction of individuals with ASD to faces: They spend less time than typical looking at faces, particularly at the eyes, and they remember faces less well (see Rigby, Stoesz, & Jakobson, 2018; but see Guillon et al., 2014). Subsequently, the *fusiform face area* (see Chapter 6) of adolescents with ASD, but not adults with ASD, was found to display less activity during fMRI than typical in response to the presentation of faces (see Nomi & Uddin, 2015; Lynn et al., 2018). This latter finding suggests a delay in the development of face processing circuitry in ASD (see Lynn et al., 2018).

It should be emphasized that the underlying mechanisms of ASD may not be entirely neural. Indeed, there is substantial interest in the role of glial activity in the development of ASD (see Welberg, 2014c; Zeidán-Chuliá et al., 2014). In particular, there is evidence of decreased synaptic pruning by glial cells during development in individuals with ASD, leading to larger than typical cerebral volumes during the first year of life (see Neniskyte & Gross, 2017).

Williams Syndrome

LO 9.16   Describe Williams syndrome and attempts to identify its neural mechanisms.

Williams syndrome, like ASD, is a neurodevelopmental disorder associated with intellectual disability and with a heterogeneous pattern of abilities and disabilities (see Van Herwegen, 2015). However, all individuals with Williams syndrome have an accompanying intellectual disability, whereas that is not necessarily the case for individuals with ASD. Moreover, in contrast to individuals with ASD who often have difficulties with social interaction, people with Williams syndrome are sociable, empathetic, and talkative. In many respects, ASD and Williams syndrome are opposites, which is one reason they can be fruitfully studied together (see Barak & Feng, 2016).

Journal Prompt 9.4
Autism spectrum disorder and Williams syndrome are similar, but they are also opposites in several ways. Thus, the study of their neural mechanisms can be complementary. Discuss.

Williams syndrome occurs in approximately 1 in 7,500 births (see Martens, 2013; Van Herwegen, 2015). Anne Louise McGarrah has Williams syndrome (Finn, 1991).

The Case of Anne Louise McGarrah: Uneven Abilities

Anne Louise McGarrah, 42, can't add 13 and 18. Yet, she is an enthusiastic reader. "I love to read," she says. "Biographies, fiction, novels, just about anything…"

McGarrah has difficulty telling left from right, and if asked to get several items, she often comes back with only one. But she plays the piano and loves classical music: "I love listening to music. I like Beethoven, but I love Mozart and Chopin and Bach. I like the way they develop their music—it's very light, very airy, and very cheerful…"

She's aware of her own condition. "One time I had a very weird experience. I was in the store, shopping, minding my own business. A woman came up to me and was staring. I felt really, really bewildered by it. I wanted to talk to her and ask her if she understood that I have a disability."

It is the language abilities of individuals with Williams syndrome that have attracted the most attention. Although they display both a delay in language development and language deficits in adulthood (see Niego & Benítez-Burraco, 2019), their language skills are remarkable considering their characteristically low IQs (which average around 55). For example, in one test, children with Williams syndrome were asked to name as many animals as they could in 60 seconds. Answers included koala, yak, ibex, condor, Chihuahua, brontosaurus, and hippopotamus. When asked to look at a picture and tell a story about it, children with Williams syndrome often produced an animated narrative. As they told the story, the children altered the pitch, volume, rhythm, and vocabulary of their speech to engage the audience. Sadly, the verbal and social skills of these children often lead teachers to overestimate their cognitive abilities, and they do not always receive the extra academic support they need.

The cognitive strengths of persons with Williams syndrome are not limited to language: They are often musically gifted (Niego & Benítez-Burraco, 2019). Although most cannot learn to read music, some have perfect or near-perfect pitch and an uncanny sense of rhythm. Many retain melodies for years, and some are professional musicians. As a group, people with Williams syndrome show more interest in, and emotional reaction to, music (see Järvinen et al., 2015) than does the general population. Another cognitive strength of individuals with Williams syndrome is their near-typical ability to recognize faces.

On the other hand, persons with Williams syndrome display several serious cognitive deficits. For example, they have severe attentional problems (see Lense, Key, & Dykens, 2011). Also, their spatial abilities are even worse than those of people with comparable IQs: They have difficulty remembering the locations of a few blocks placed on a test board, their space-related speech is poor, and their ability to draw objects is almost nonexistent (see Nagai, Inui, & Iwata, 2011; Rhodes et al., 2011).

A lot of attention has been paid to how persons with Williams syndrome respond to faces. Interestingly, they show an excessive interest in human faces. This is in marked contrast to what is seen in individuals with ASD. Moreover, persons with Williams syndrome have difficulty perceiving negative emotion in faces (e.g., they have difficulty identifying angry faces). They also rate unfamiliar faces as more friendly than do persons without Williams syndrome. Taken together, these face-processing biases may contribute to the hypersociability that is typical of persons with Williams syndrome (see Barak & Feng, 2016).

Williams syndrome is also associated with a variety of health problems, including several involving the heart. One heart disorder was found to result from a mutation in a gene on chromosome 7 that controls the synthesis of *elastin*, a protein that imparts elasticity to many organs and tissues, including the heart (see Cunniff et al., 2001). Aware that the same cardiac problem is prevalent in people with Williams syndrome, investigators have assessed the status of this gene in that group. They found that the gene was absent from one of the two copies of chromosome 7 in 95 percent of individuals with Williams syndrome (see Howald et al., 2006). It was missing through an accident of reproduction that deleted an entire segment of chromosome 7, a segment that included the elastin gene and about 26 others (see Chailangkarn et al., 2016; Järvinen, Korenberg, & Bellugi, 2013). Interestingly, individuals with extra copies of the same segment of chromosome 7 display symptoms of ASD.

Several brain differences have been reported in people with Williams syndrome (e.g., a decrease in basal ganglia volume; see Dennis & Thompson, 2013); however, most research on the neural correlates of Williams syndrome has focused on the cortex. Williams syndrome is associated with a general thinning of the cortex and underlying white matter (see Meyer-Lindenberg et al., 2006; Niego & Benítez-Burraco, 2019; Toga & Thompson, 2005). The cortical thinning is greatest in two areas: at the boundary of the parietal and occipital lobes and in the **orbitofrontal cortex** (the inferior area of frontal cortex near the

orbits, or eye sockets)—see Figure 9.13. Reduced cortical development in these two areas may be related to two of the major symptoms of Williams syndrome: profound impairment of spatial cognition and remarkable hypersociability, respectively. Conversely, the thickness of the cortex in one area in people with Williams syndrome is often typical: the **superior temporal gyrus**, which includes primary and secondary auditory cortex (refer to Figure 9.13). The typical nature of this area may be related to the relatively high levels of language and music processing in those with Williams syndrome (Niego & Benítez-Burraco, 2019).

Many cultures feature tales involving magical little people (pixies, elves, leprechauns, etc.). Descriptions of these creatures portray them as virtually identical to persons with Williams syndrome: short with small upturned noses, oval ears, broad mouths, full lips, puffy eyes, and small chins. Even the behavioral characteristics of elves—engaging storytellers, talented musicians, loving, trusting, and sensitive to the feelings of others—match those of individuals with Williams syndrome. These similarities suggest that folktales about elves may have originally been based on persons with Williams syndrome.

EPILOGUE. As a student, I (SB) was amazed to learn that—starting as one cell—my brain had constructed itself

Figure 9.13 Two areas of reduced cortical volume and one area of typical cortical volume observed in people with Williams syndrome.

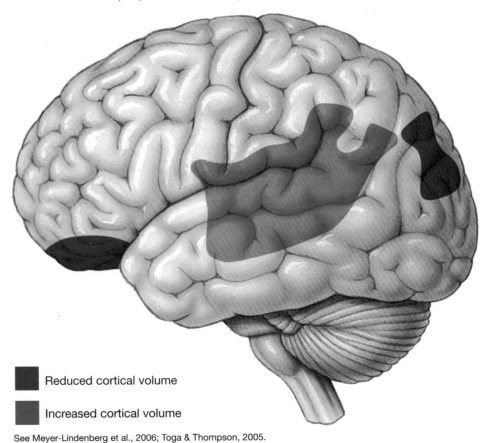

■ Reduced cortical volume

■ Increased cortical volume

See Meyer-Lindenberg et al., 2006; Toga & Thompson, 2005.

through the interaction of its genetic programs and experience. Later, I was astounded to learn that my adult brain is still plastic. We hope that this chapter has helped your brain appreciate its own development. Also, we hope that

the case of Genie, which began the chapter, and the cases of people with ASD and Williams syndrome, which completed it, have helped you appreciate what can happen when the complex process of neurodevelopment is disturbed.

Themes Revisited

This chapter was all about neurodevelopment, with a particular emphasis on how the brain continues to develop throughout an individual's life span. Accordingly, the neuroplasticity theme took center stage. We hope you now fully appreciate that your own brain is constantly changing in response to interactions between your genetic programs and experience.

The clinical implications and evolutionary perspective themes were also emphasized in this chapter. One of the best ways to understand the principles of typical neurodevelopment is to consider what happens when it goes "wrong": Accordingly, the chapter began with the tragic case of Genie and ended with a discussion of ASD and

Williams syndrome. The evolutionary perspective theme was emphasized because much of the information that we have about typical human neurodevelopment has come from studying other species.

The thinking creatively theme was also pervasive. Two general points were made. First, neurodevelopment always proceeds from gene-experience interactions rather than from either genetics or environment alone. Second, it is important to realize that "normal" experience plays an important role in fine-tuning the development of normal neural function.

Of the two emerging themes, only the thinking about epigenetics theme was present in this chapter: for example, in the discussion of the determinants of cell fate.

Key Terms

Five Phases of Early Neurodevelopment
Zygote, p. 238
Totipotent, p. 238
Pluripotent, p. 238
Multipotent, p. 238
Unipotent, p. 238
Stem cells, p. 238
Neural plate, p. 239
Mesoderm layer, p. 239
Neural tube, p. 239
Neural proliferation, p. 240
Ventricular zone, p. 240
Subventricular zone, p. 240
Radial glial cells, p. 240
Migration, p. 240
Radial migration, p. 241
Tangential migration, p. 241
Somal translocation, p. 241
Radial-glia-mediated migration, p. 241
Inside-out pattern, p. 241
Neural crest, p. 242
Aggregation, p. 242

Cell-adhesion molecules (CAMs), p. 242
Growth cone, p. 242
Retinal ganglion cells, p. 242
Optic tectum, p. 243
Chemoaffinity hypothesis, p. 243
Pioneer growth cones, p. 243
Fasciculation, p. 244
Topographic gradient hypothesis, p. 244
Synaptogenesis, p. 245
Necrosis, p. 245
Apoptosis, p. 245
Neurotrophins, p. 245
Nerve growth factor (NGF), p. 245
Brain-derived neurotrophic factor (BDNF), p. 245

Early Cerebral Development in Humans
Prenatal period, p. 246
Postnatal period, p. 246
Perseveration, p. 248

Effects of Experience on Postnatal Development of Neural Circuits
Critical period, p. 248
Sensitive period, p. 248

Neuroplasticity in Adults
Neurogenesis, p. 250
Pattern separation, p. 251

Atypical Disorders of Neurodevelopment: Autism Spectrum Disorder and Williams Syndrome
Autism spectrum disorder (ASD), p. 253
Savants, p. 254
Williams syndrome, p. 255
Orbitofrontal cortex, p. 256
Superior temporal gyrus, p. 256

Chapter 10
Brain Damage and Neuroplasticity

Can the Brain Recover from Damage?

Michael Ventura/Alamy Stock Photo

 ## Chapter Overview and Learning Objectives

Causes of Brain Damage

LO 10.1 Describe different types of brain tumors and explain the difference between an encapsulated and an infiltrating brain tumor.

LO 10.2 Describe differences between the two types of stroke: cerebral hemorrhage and cerebral ischemia.

LO 10.3 Describe the two sorts of closed-head traumatic brain injuries (TBIs).

LO 10.4 Describe two different types of infections of the brain.

LO 10.5 Describe three different types of neurotoxins.

LO 10.6 Discuss the symptoms of Down syndrome and what causes this disorder.

LO 10.7 Explain the difference between apoptosis and necrosis.

Neurological Diseases	**LO 10.8**	Define epilepsy; describe four common types of seizures; and discuss some treatments for epilepsy.
	LO 10.9	Describe the symptoms of Parkinson's disease and some treatments for this disorder.
	LO 10.10	Describe the symptoms of Huntington's disease and explain its genetic basis.
	LO 10.11	Describe the symptoms of multiple sclerosis (MS) and its risk factors.
	LO 10.12	Describe the symptoms of Alzheimer's disease and evaluate the amyloid hypothesis.
Animal Models of Human Neurological Diseases	**LO 10.13**	Describe the kindling model of epilepsy and explain the ways in which it models human epilepsy.
	LO 10.14	Describe the events that led to the discovery of the MPTP model of Parkinson's disease, and evaluate the utility of this animal model.
Responses to Nervous System Damage: Degeneration, Regeneration, Reorganization, and Recovery	**LO 10.15**	Explain the various types of neural degeneration that ensue following axotomy.
	LO 10.16	Compare neural regeneration within the CNS vs. the PNS.
	LO 10.17	Describe three examples of cortical reorganization following damage to the brain, and discuss the mechanisms that might underlie such reorganization.
	LO 10.18	Describe the concept of "cognitive reserve," and discuss the potential role of adult neurogenesis in recovery following CNS damage.
Neuroplasticity and the Treatment of CNS Damage	**LO 10.19**	Discuss early work on neurotransplantation for the treatment of CNS damage.
	LO 10.20	Discuss the methods and findings of modern research on neurotransplantation.
	LO 10.21	Discuss methods of promoting recovery from CNS damage through rehabilitative treatment.

The study of human brain damage serves two purposes: It increases our understanding of the healthy brain, and it serves as a basis for the development of new treatments. The first three modules of this chapter focus on brain damage itself, the fourth module focuses on the recovery and reorganization of the brain after damage, and the fifth discusses exciting new treatments that promote neuroplasticity. But first, the continuation of the ironic case of Professor P., which you first encountered in Chapter 5, provides a personal view of brain damage.

The Ironic Case of Professor P.

One night Professor P. sat at his desk staring at a drawing of the cranial nerves, much like the one in Appendix III of this book. As he mulled over the location and function of each cranial nerve (see Appendix IV), the painful truth became impossible for him to deny. The irony of the situation was that Professor P. was a neuroscientist, all too familiar with what he was experiencing.

His symptoms started subtly, with slight deficits in balance. Professor P. chalked up his occasional lurches to aging—after all, he thought to himself, he was past his prime. Similarly, his doctor didn't seem to think that it was a problem, but Professor P. monitored his symptoms nevertheless. Three years later, his balance problems unabated, Professor P. started to worry. He was trying to talk on the phone but was having trouble hearing until he changed the phone to his left ear. Professor P. was going deaf in his right ear.

Professor P. made an appointment with his doctor, who referred him to a specialist. After a cursory and poorly controlled hearing test, the specialist gave him good news. "You're fine, Professor P.; lots of people experience a little hearing loss when they reach middle age; don't worry about it." To this day, Professor P. regrets that he did not insist on a second opinion.

It was about a year later that Professor P. sat staring at the illustration of the cranial nerves. By then, he had begun to experience numbness on the right side of his mouth, he was having problems swallowing, and his right tear ducts were not releasing enough tears. He stared at the point where the auditory and vestibular nerves come together to form cranial nerve VIII (the auditory-vestibular nerve). He knew it was there, and he knew that it was large enough to be affecting cranial nerves V through X as well. It was something slow-growing, perhaps a tumor? Was he going to die? Was his death going to be terrible and lingering?

He didn't see his doctor right away. A friend of his was conducting a brain MRI study, and Professor P. volunteered to be a control subject, knowing that his problem would show up on the scan. It did: a large tumor sitting, as predicted, on the right cranial nerve VIII.

Then, MRI in hand, Professor P. went back to his doctor, who referred him to a neurologist, who in turn referred him to a neurosurgeon. Several stressful weeks later, Professor P. found himself on life support in the intensive care unit of his local hospital, tubes emanating seemingly from every part of his body. During the 6-hour surgery, Professor P. had stopped breathing.

In the intensive care unit, near death and hallucinating from the morphine, Professor P. thought he heard his wife, Maggie, calling for help, and he tried to go to her assistance. But one gentle morphine-steeped professor was no match for five nurses intent on saving his life. They quickly turned up his medication, and the next time he regained consciousness, he was tied to the bed.

Professor P.'s auditory-vestibular nerve was transected during his surgery, which has left him permanently deaf and without vestibular function on the right side. He was also left with partial hemifacial paralysis, including serious blinking and tearing problems.

Professor P. is still alive and much improved. Indeed, at the very moment that these words are being written, Professor P. is working on a forthcoming edition . . . If it has not yet occurred to you, I (JP) am Professor P., which is why this chapter has come to have special meaning for me.

Causes of Brain Damage

This module provides an introduction to six causes of brain damage: brain tumors, cerebrovascular disorders, closed-head injuries, infections of the brain, neurotoxins, and genetic factors. It concludes with a discussion of programmed cell death, which mediates many forms of brain damage.

Brain Tumors

LO 10.1 Describe different types of brain tumors and explain the difference between an encapsulated and an infiltrating brain tumor.

A **tumor**, or **neoplasm** (literally, "new growth"), is a mass of cells that grows independently of the rest of the body. About 20 percent of tumors found in the human brain are **meningiomas** (see Figure 10.1)—tumors that grow between the *meninges*, the three membranes that cover the central nervous system. All meningiomas are **encapsulated tumors**—tumors that grow within their own membrane. As a result, they are particularly easy to identify on a CT scan, they can influence the function of the brain only by the pressure they exert on surrounding tissue, and they are almost always **benign tumors**—tumors that are surgically removable with little risk of further growth in the body (see Barresi, Caffo, & Tuccari, 2016).

> **Journal Prompt 10.1**
> Were you surprised that one of the authors of your textbook suffered significant brain damage? Why or why not?

Unfortunately, encapsulation is the exception rather than the rule when it comes to brain tumors. Aside from meningiomas, most brain tumors are infiltrating. **Infiltrating tumors** are those that grow diffusely through surrounding tissue. As a result, they are usually **malignant tumors**; that is, it is difficult to remove or destroy them completely, and any cancerous tissue that remains after surgery usually continues to grow. **Gliomas** (brain tumors that develop from glial cells) are infiltrating, rapidly growing, and unfortunately the most common form of malignant brain

Figure 10.1 A meningioma.

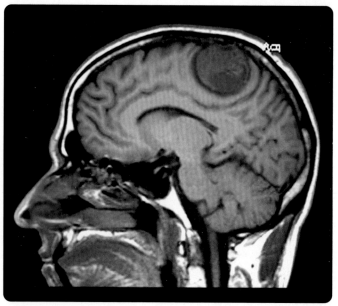

Living Art Enterprises/Science Source

tumors (see Broekman et al., 2018; Gusyatiner & Hegi, 2018; Laug, Glasgow, & Deneen, 2018).

About 10 percent of brain tumors do not originate in the brain. They grow from infiltrating cells that are carried to the brain by the bloodstream from some other part of the body. These tumors are called **metastatic tumors** (*metastasis* refers to the transmission of disease from one organ to another)—see Alarcón & Tavazoie (2016); Cheung and Ewald (2016). Many metastatic brain tumors originate as cancers of the lungs. Figure 10.2 illustrates the ravages of metastasis. Currently, the chance of recovering from a cancer that has already attacked two or more separate sites is slim.

Fortunately, my (JP) tumor was encapsulated. Encapsulated tumors that grow on cranial nerve VIII are referred to as *acoustic neuromas* (neuromas are tumors that grow on nerves or tracts). Figure 10.3 is an MRI scan of my acoustic neuroma, the same scan that I took to my doctor.

Strokes

LO 10.2 Describe differences between the two types of stroke: cerebral hemorrhage and cerebral ischemia.

Strokes are sudden-onset cerebrovascular disorders that cause brain damage. In the United States, stroke is the fifth leading cause of death, the major cause of neurological dysfunction, and a leading cause of adult disability (see

Figure 10.3 An MRI of Professor P.'s acoustic neuroma, the very one that he took to his doctor. The arrow indicates the tumor.

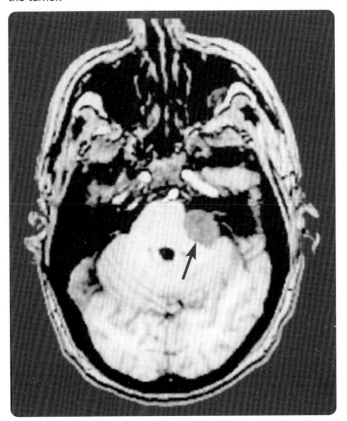

John Pinel

Feigin et al., 2016; Prabhakaran, Ruff, & Bernstein, 2015). The symptoms of a stroke depend on the area of the brain affected, but common consequences of stroke are amnesia, aphasia (language difficulties), psychiatric disorders, dementia, paralysis, and coma (see Ferro, Caeiro, & Figueira, 2016; Mok et al., 2017).

The area of dead or dying tissue produced by a stroke is called an *infarct*. Surrounding the infarct is a dysfunctional area called the **penumbra**. The tissue in the penumbra may recover or die in the ensuing days, depending on a variety of factors. Accordingly, the primary goal of treatment following stroke is to save the penumbra (see Baron, 2018; Evans et al., 2017).

There are two major types of strokes: those resulting from cerebral hemorrhage and those resulting from cerebral ischemia (pronounced "iss-KEEM-ee-a").

CEREBRAL HEMORRHAGE. Cerebral hemorrhage (bleeding in the brain) occurs when a cerebral blood vessel ruptures and blood seeps into the surrounding neural tissue and damages it (see Bulters et al., 2018; Carpenter et al., 2016). Bursting aneurysms are a common cause of intracerebral hemorrhage. An **aneurysm** is a pathological balloonlike dilation that forms in the wall of an artery at a point where the elasticity of the artery wall is defective (see Etminan & Rinkel, 2016). Although aneurysms of the

Figure 10.2 Multiple metastatic brain tumors. The colored areas indicate the location of the larger metastatic brain tumors in this patient.

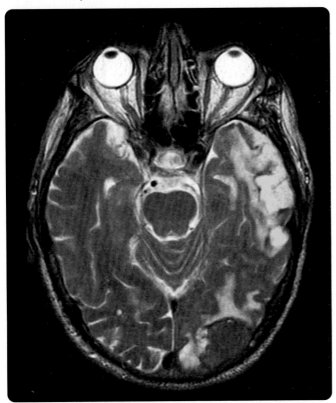

SMC Images/Photodisc/GettyImages

Figure 10.4 An angiogram that illustrates narrowing of one carotid artery (see arrow), a major pathway of blood to the brain.

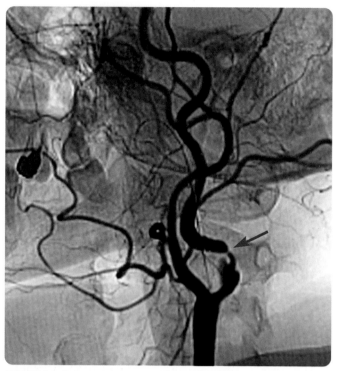

ZEPHYR/SPL/Science Photo Library/Alamy Stock Photo

brain are particularly problematic, aneurysms can occur in any part of the body. Aneurysms can be **congenital** (present at birth) or can result from exposure to vascular poisons or infection (see Caranci et al., 2013). Individuals at risk of aneurysms should make every effort to avoid cigarette smoking, alcohol consumption, and hypertension (see Brown & Broderick, 2014).

CEREBRAL ISCHEMIA. **Cerebral ischemia** is a disruption of the blood supply to an area of the brain. The three main causes of cerebral ischemia are thrombosis, embolism, and arteriosclerosis. In **thrombosis**, a plug called a *thrombus* is formed and blocks blood flow at the site of its formation. A thrombus may be composed of a blood clot, fat, oil, an air bubble, tumor cells, or any combination thereof. **Embolism** is similar, except that the plug, called an *embolus* in this case, is carried by the blood from a larger vessel, where it was formed, to a smaller one, where it becomes lodged; in essence, an embolus is just a thrombus that has taken a trip. In **arteriosclerosis**, the walls of blood vessels thicken and the channels narrow, usually as the result of fat deposits; this narrowing can eventually lead to complete blockage of the blood vessels. The *angiogram* in Figure 10.4 illustrates partial blockage of one carotid artery.

Ischemia-induced brain damage has two important properties. First, it takes a while to develop. Soon after a temporary cerebral ischemic episode, there usually is little or no evidence of brain damage; however, substantial neuron loss can often be detected a day or two later.

Second, ischemia-induced brain damage does not occur equally in all parts of the brain—particularly susceptible are neurons in certain areas of the hippocampus (see Schmidt-Kastner, 2015).

Paradoxically, **glutamate**, the brain's most prevalent excitatory neurotransmitter, plays a major role in ischemia-induced brain damage (see Chisholm & Sohrabji, 2015; Parsons & Raymond, 2014). Here is how this mechanism is thought to work (see Lai, Zhang, & Wang, 2014; Leng et al., 2014). After a blood vessel becomes blocked, many of the blood-deprived neurons become overactive and release excessive quantities of glutamate. The glutamate in turn overactivates glutamate receptors in the membranes of postsynaptic neurons; the glutamate receptors most involved in this reaction are the **NMDA (N-methyl-D-aspartate) receptors**. As a result, large numbers of Na^+ and $Ca2^+$ ions enter the postsynaptic neurons. The excessive internal concentrations of Na^+ and $Ca2^+$ ions in postsynaptic neurons affect them in two ways: They trigger the release of excessive amounts of glutamate from the neurons, thus spreading the toxic cascade to yet other neurons; and they trigger a sequence of internal reactions that ultimately kill the postsynaptic neurons. See Figure 10.5.

An implication of the discovery that excessive glutamate release causes much of the brain damage associated with stroke is the possibility of preventing stroke-related brain damage by blocking the glutaminergic cascade. Some clinical trials have shown that NMDA-receptor antagonists are effective following acute ischemic stroke. However, to be effective they need to be administered almost immediately after the stroke. This makes them impractical in most human clinical situations (see Leng et al., 2014). That being said, two other sorts of treatments have been shown to be effective for stroke (see Levy & Mokin, 2017): The administration of a *tissue plasminogen activator* (a drug that breaks down blood clots) or an *endovascular therapy* (the surgical removal of a thrombus or embolus from an artery). Administration of these treatments within a few hours after the onset of ischemic stroke can lead to better recovery (see Algra & Wermer, 2017; Dong et al., 2018; Ginsberg, 2016; Mokin, Rojas, & Levy, 2016; Law & Levine, 2015).

Traumatic Brain Injuries

LO 10.3 Describe the two sorts of closed-head traumatic brain injuries (TBIs).

About 50–60 million people experience some form of **traumatic brain injury (TBI)** each year; and about 50 percent of us will experience a TBI at least once in our lives (see Mollayeva, Mollayeva, & Colantonio, 2018). For the brain to be seriously damaged by a TBI, it is not necessary for the skull to be penetrated (i.e., a *penetrating TBI*). In fact, any blow to the head should be treated with extreme

Figure 10.5 The cascade of events by which the ischemia-induced release of glutamate kills neurons.

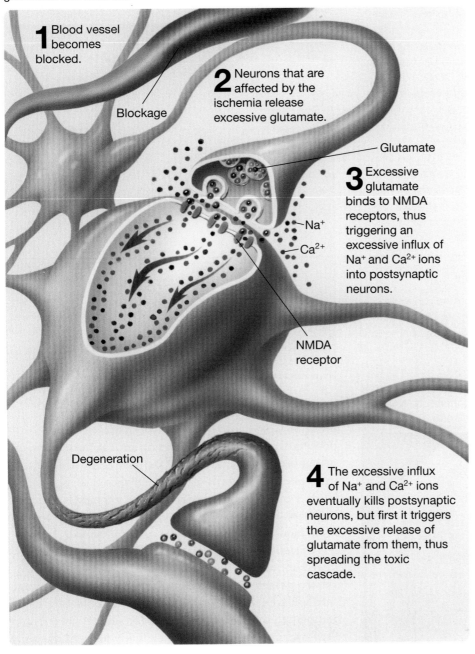

1 Blood vessel becomes blocked.

Blockage

2 Neurons that are affected by the ischemia release excessive glutamate.

Glutamate

3 Excessive glutamate binds to NMDA receptors, thus triggering an excessive influx of Na+ and Ca2+ ions into postsynaptic neurons.

Na+

Ca2+

NMDA receptor

Degeneration

4 The excessive influx of Na+ and Ca2+ ions eventually kills postsynaptic neurons, but first it triggers the excessive release of glutamate from them, thus spreading the toxic cascade.

occur when the brain slams against the inside of the skull. As Figure 10.6 illustrates, blood from such injuries can accumulate in the *subdural space*—the space between the dura mater and arachnoid membrane—and severely distort the surrounding neural tissue. Such a "puddle" of blood is known as a **subdural hematoma** (see Shi et al., 2019).

It may surprise you to learn that contusions frequently occur on the side of the brain opposite the side struck by a blow. The reason for such so-called **contrecoup injuries** is that the blow causes the brain to strike the inside of the skull on the other side of the head.

When there is a disturbance of consciousness following a blow to the head and there is no evidence of a contusion or other structural damage, the diagnosis is **mild TBI (mTBI)**. Most TBIs are mTBIs (see Mollayeva, Mollayeva, & Colantonio, 2018). mTBIs were once called *concussions*. However, the term "concussion" is no longer deemed appropriate because it was associated with the mistaken assumption that its effects involved no long-term damage (see Grasso & Landi, 2016; Moye & Pradhan, 2017; Zetterberg & Blennow, 2016). This has not turned out to be the case. For example, there is now substantial evidence that the effects of mTBIs can last many years and that the effects of repeated mTBIs can accumulate (see Carman et al., 2015).

caution, particularly when confusion, sensorimotor disturbances, or loss of consciousness ensues. Brain injuries produced by blows that do not penetrate the skull are called **closed-head TBIs**.

There are several types of closed-head TBIs. **Contusions** are closed-head TBIs that involve damage to the cerebral circulatory system. Such damage produces internal hemorrhaging, which in turn produces a localized collection of blood in the brain—in other words, a bruised brain.

It is paradoxical that the very hardness of the skull, which protects the brain from penetrating injuries, is the major factor in the development of contusions. Contusions

Chronic traumatic encephalopathy (CTE) is the **dementia** (general intellectual deterioration) and cerebral scarring often observed in boxers, rugby players, American football players (see Figure 10.7), and other individuals who have experienced repeated mTBIs (see Azad et al., 2016; Maroon et al., 2015; Smith et al., 2019; Underwood, 2015a). For example, there have been reports that most (approximately 90 percent) former American football players, and many former recreational American football players, meet the diagnostic criteria for chronic traumatic encephalopathy (Mez et al., 2017). The case of Junior Seau is particularly tragic (Azad et al., 2016).

The Case of Junior Seau

Junior Seau was an all-star linebacker in the National Football League (NFL) for 20 years. He was known for his hard-hitting, aggressive play. Although Seau never complained about head injuries to his coaches or medical staff, his family members reported that he had suffered many mTBIs. When he came home from games, he often experienced severe headaches and would go straight to his darkened bedroom. But, according to his ex-wife, "he always bounced back and kept on playing." He was a warrior.

After Seau's retirement from the NFL in 2010, his family and friends noticed several disturbing behavioral changes: heavy consumption of alcohol, reckless business and financial decisions, and gambling. Most disturbing were his frequent violent outbursts that were completely out of character for him and were often directed at friends and family—the very people who were trying to help him.

On May 2, 2012, at the age of 43, Junior Seau shot himself. He left no note—no explanation.

Seau's family donated his brain to the American National Institutes for Health (NIH) for study. A detailed autopsy of Seau's brain revealed that he met criteria for a diagnosis of chronic traumatic encephalopathy.

Figure 10.7 The NFL has acknowledged that there is a connection between playing football and chronic traumatic encephalopathy (CTE).

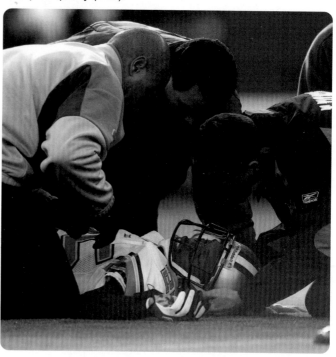

epa european pressphoto agency b.v./Alamy Stock Photo

Figure 10.6 A CT scan of a subdural hematoma. Notice that the hematoma has displaced the left lateral ventricle.

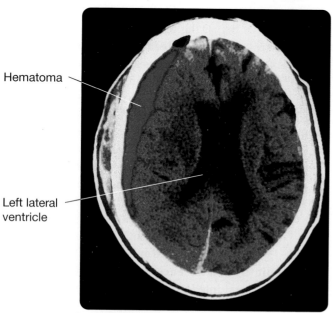

Hematoma

Left lateral ventricle

Scott Camazine/Alamy Images

Infections of the Brain

LO 10.4   Describe two different types of infections of the brain.

An invasion of the brain by microorganisms is a *brain infection*, and the resulting inflammation is called **encephalitis**. There are two common types of brain infections: bacterial infections and viral infections.

BACTERIAL INFECTIONS. When bacteria infect the brain, they often lead to the formation of *cerebral abscesses*—pockets of pus in the brain. Bacteria are also the major cause of **meningitis** (inflammation of the meninges), which is fatal in 30 percent of adults (see Castelblanco, Lee, & Hasbun, 2014). Penicillin and other antibiotics sometimes eliminate bacterial infections of the brain, but they cannot reverse brain damage that has already been produced.

Syphilis is one bacterial brain infection you have likely heard about (see Berger & Dean, 2014). Syphilis bacteria are passed from infected to noninfected individuals through contact with genital sores. The infecting bacteria then go into a dormant stage for several years before they become virulent and attack many parts of the body, including the brain. The syndrome of mental illness and dementia that results from a syphilitic infection is called **general paresis**.

VIRAL INFECTIONS. There are two types of viral infections of the nervous system: those that have a particular affinity for neural tissue and those that attack neural tissue but have no greater affinity for it than for other tissues.

Rabies, which is usually transmitted through the bite of a rabid animal, is a well-known example of a virus that has a particular affinity for the nervous system. The fits of rage caused by the virus's effects on the brain increase the probability that rabid animals that normally attack by biting (e.g., dogs, cats, raccoons, bats, and mice) will spread the disorder. Although the effects of the rabies virus on the brain are almost always lethal (see Schnell et al., 2010), the virus does have one redeeming feature: It does not usually attack the brain for at least a month after it has been contracted, thus allowing time for preventive vaccination.

The *mumps* and *herpes* viruses are common examples of viruses that can attack the nervous system but have no special affinity for it. Although these viruses sometimes spread into the brain, they typically attack other tissues of the body.

Viruses may play a far greater role in neuropsychological disorders than is currently thought. Their involvement in the *etiology* (cause) of disorders is often difficult to recognize because they can lie dormant for many years before producing symptoms (see Miller, Schnell, & Rall, 2016).

Neurotoxins

LO 10.5   Describe three different types of neurotoxins.

The nervous system can be damaged by exposure to any one of a variety of toxic chemicals—chemicals that can enter general circulation from the gastrointestinal tract, from the lungs, or through the skin. For example, heavy metals such as mercury and lead (see Chen, 2013a; Hare et al., 2015; Tshala-Katumbay et al., 2015) can accumulate in the brain and permanently damage it, producing a **toxic psychosis** (chronic mental illness produced by a neurotoxin). Have you ever wondered why *Alice in Wonderland*'s Mad Hatter was a "mad hatter" and not a "mad" something else? In 18th- and 19th-century England, hat makers commonly developed toxic psychosis from the mercury employed in the preparation of the felt used to make hats. In a similar vein, the word *crackpot* originally referred to the toxic psychosis observed in some people in England—primarily the poor—who steeped their tea in cracked ceramic pots with lead cores.

Sometimes, the very drugs used to treat neurological or psychiatric disorders prove to be toxic. For example, some of the antipsychotic drugs introduced in the early 1950s produced effects of distressing scope. By the late 1950s, millions of patients with schizophrenia were being maintained on these drugs. However, after several years of treatment, about 5 percent of the patients developed a motor disorder called **tardive dyskinesia (TD)**—see Cloud, Zutshi, and Factor (2014). Its primary symptoms are involuntary smacking and sucking movements of the lips, thrusting and rolling of the tongue, lateral jaw movements, and puffing of the cheeks.

Some neurotoxins are *endogenous* (produced by the patient's own body). For example, the body can produce antibodies that attack particular components of the nervous system (see Melzer, Meuth, & Wiendl, 2012). Stress hormones, such as *cortisol*, are also believed to produce neurotoxic effects (see Lupien et al., 2018). Likewise, as you have just learned from the discussion of glutamate and ischemic stroke, the excessive release of certain neurotransmitters can also damage the brain.

Genetic Factors

LO 10.6   Discuss the symptoms of Down syndrome and what causes this disorder.

Some neuropsychological diseases of genetic origin are caused by abnormal recessive genes that are passed from parent to offspring. (In Chapter 2, you learned about one such disorder, *phenylketonuria*, or *PKU*.) Inherited neuropsychological disorders are rarely associated with dominant genes because dominant genes that disturb neuropsychological function tend to be eliminated from the gene pool—individuals who carry one usually have major survival and reproductive disadvantages. In contrast, individuals who inherit one abnormal recessive gene do not develop the disorder, and the gene is passed on to future generations.

Genetic accident is another major cause of neuropsychological disorders of genetic origin. **Down syndrome**, which occurs in about 0.15 percent of births, is such a disorder. The genetic accident associated with Down syndrome occurs in the mother during ovulation, when an extra chromosome 21 is created in the egg. Thus, when the egg is fertilized, there are three chromosome 21s, rather than two, in the zygote (see Dekker et al., 2015). The consequences tend to be characteristic disfigurement, intellectual disability, early-onset *Alzheimer's disease* (a type of dementia), and other troublesome medical complications.

There was great optimism among professionals who study and treat neuropsychological disorders when the human genome was documented at the beginning of this century. Inherited factors play major roles in virtually all neuropsychological disorders, and it seemed that the offending genes would soon be identified and effective treatments developed to target them. This has not happened, for two reasons (see Maurano et al., 2012). First, numerous loci on human chromosomes have been associated with each disorder—not just one or two. Second, about 90 percent of the chromosomal loci involved in neuropsychological disorders did not involve protein-coding genes; rather, the loci were in poorly understood sections of the DNA.

Programmed Cell Death

LO 10.7   Explain the difference between apoptosis and necrosis.

Recall that neurons and other cells have genetic programs for destroying themselves by a process called **apoptosis** (pronounced "A-poe-toe-sis"). Apoptosis plays a critical role in early development by eliminating extra neurons. It also plays a role in brain damage. Indeed, all of the six causes of brain damage that have been discussed in this chapter (tumors, cerebrovascular disorders, closed-head TBIs, infections, toxins, and genetic factors) produce neural damage, in part, by activating apoptotic programs of self-destruction (see Viscomi & Molinari, 2014).

It was once assumed that the death of neurons follow-ing brain damage was totally necrotic—*necrosis* is passive cell death resulting from injury. It now seems that if cells are not damaged too severely, they will attempt to mar-shal enough resources to destroy themselves via apopto-sis. However, cell death is not an either–or situation: Some dying cells display signs of both necrosis and apoptosis (see Zhou & Yuan, 2014).

It is easy to understand why apoptotic mechanisms have evolved: Apoptosis is clearly more adaptive than necrosis. In necrosis, the damaged neuron swells and breaks apart, beginning in the axons and dendrites and ending in the cell body. This fragmentation leads to inflammation, which can damage other cells in the vicinity. Necrotic cell death is quick—it is typically complete in a few hours. In contrast, apoptotic cell death is slow, typically requiring a day or two. Apoptosis of a neuron proceeds gradually, start-ing with shrinkage of the cell body. Then, as parts of the neuron die, the resulting debris is packaged in vesicles—a process known as *blebbing* (one of my (SB) favorite words). As a result, there is no inflammation, and damage to nearby cells is kept to a minimum.

Neurological Diseases

The preceding module focused on some of the causes of human brain damage. This module considers five diseases associated with brain damage: epilepsy, Parkinson's disease, Huntington's disease, multiple sclerosis, and Alzheimer's disease.

Epilepsy

LO 10.8   **Define epilepsy; describe four common types of seizures; and discuss some treatments for epilepsy.**

The primary symptom of **epilepsy** is the epileptic *seizure*, but not all persons who suffer seizures are considered to have epilepsy. Sometimes, an otherwise healthy person may have one seizure and never have another—such a one-time seizure could be triggered by exposure to a con-vulsive toxin or by a high fever. The diagnosis of *epilepsy* is applied to only those patients whose seizures are repeatedly generated by their own chronic brain dysfunction. About 0.8 percent of the population are diagnosed with epilepsy at some point in their lives (see Fiest et al., 2017).

Because epilepsy is charac-terized by epileptic seizures— or, more accurately, by spontaneously recurring epileptic seizures—you might think that the task of diagnosing epi-lepsy would be an easy one. But you would be wrong. The task is made difficult by the diversity and complexity of epi-leptic seizures. You are probably familiar with seizures that take the form of **convulsions** (motor seizures); these often involve tremors (*clonus*), rigidity (*tonus*), and loss of both balance and consciousness. But most seizures do not take this form; instead, they involve subtle changes of thought, mood, or behavior that are not easily distinguishable from normal ongoing activity.

There are many causes of epilepsy. Viruses, neuro-toxins, tumors, and blows to the head can all cause epi-lepsy, and more than 30 different faulty genes have been linked to it, as well as many different *epigenetic mechanisms* (see Chapter 2; see Boison, 2016; Chen et al., 2017). Many cases of epilepsy are associated with faults at inhibitory synapses (e.g., GABAergic synapses)—synapses that are normally responsible for preventing excessive excitatory activity in the brain; this, in turn, leads to synchronous fir-ing of neurons and ultimately seizures (see Ben-Ari, 2017; Di Cristo et al., 2018; Moore et al., 2017; Trevelyan, 2016). Dysfunctional activity in astrocytes is also implicated in the development of seizures (see Almad & Maragakis, 2018; Bedner & Steinhäuser, 2016; Robel & Sontheimer, 2016). In other cases of epilepsy, inflammatory processes seem to be responsible for seizure activity (see Marchi, Granata, & Janigro, 2014).

The diagnosis of epilepsy rests heavily on evidence from electroencephalography (EEG). The value of scalp electroencephalography in confirming suspected cases of epilepsy stems from the fact that seizures are associ-ated with bursts of high-amplitude EEG spikes, which are often apparent in the scalp EEG during a seizure (see Figure 10.8), and from the fact that individual spikes often punctuate the scalp EEG between epileptic seizures (see Bragatti et al., 2014).

Some individuals with epilepsy experience pecu-liar psychological changes just before a seizure. These changes, called **epileptic auras**, may take many different forms—for example, a bad smell, a specific thought, a vague feeling of familiarity, a hallucination, or a tightness of the chest. Epileptic auras are important for two reasons. First, the nature of the auras provides clues concerning the

Figure 10.8 Cortical EEG recording of an epileptic seizure. Notice that the trace is characterized by epileptic spikes (sudden, high amplitude EEG signals that accom-pany epileptic seizures).

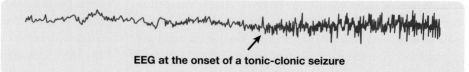

EEG at the onset of a tonic-clonic seizure

location of the epileptic focus. Second, epileptic auras can warn the patient of an impending convulsion (see Lohse et al., 2015).

Once an individual has been diagnosed with epilepsy, it is common to assign the type of seizures they experience to one of two general categories—*focal seizures* or *generalized seizures*—and then to one of their respective subcategories (see Falco-Walter, Scheffer, & Fisher, 2018). The various seizure types are so different from one another that epilepsy is best viewed not as a single disease but as a number of different, but related, diseases.

FOCAL SEIZURES. A **focal seizure** is a seizure that does not involve the entire brain. The epileptic neurons at a focus begin to discharge together in bursts, and it is this synchronous firing (see Figure 10.9) that produces epileptic spiking in the EEG. The synchronous activity tends to spread to other areas of the brain—but, in the case of focal seizures, not to the entire brain. The specific behavioral symptoms of a focal epileptic seizure depend on where the disruptive discharges begin and into what structures they spread. Because focal seizures do not involve the entire brain, they are often not accompanied by a total loss of consciousness or equilibrium.

There are many different categories of focal seizures, depending on where in the brain they start and where they spread to. Here we focus on two types of focal seizures: simple and complex. **Simple seizures** are focal seizures whose symptoms are primarily sensory or motor or both; they are sometimes called *Jacksonian seizures* after the famous 19th-century neurologist Hughlings Jackson. Simple seizures involve only one sort of sensory or motor symptom, and they are rarely accompanied by a loss of consciousness.

In contrast, **complex seizures** often begin in the temporal lobes and usually do not spread out of them. Accordingly, those who experience them are often said to have *temporal lobe epilepsy*. About half of all cases of epilepsy in adults are of the complex variety (see Bertram, 2014). During a complex seizure, the patient engages in compulsive, repetitive, simple behaviors commonly referred to as *automatisms* (e.g., doing and undoing a button) and in more complex behaviors that appear almost "normal." The diversity of complex seizures is illustrated by the following two cases (Lennox, 1960).

The Subtlety of Complex Seizures: Two Cases

A doctor received a call from his hospital informing him that he was needed to perform an emergency operation. A few hours after the surgery, he returned home feeling dazed and confused. He had performed the operation, a very difficult one, with his usual competence, but afterward he had said and done things that seemed peculiar to his colleagues. The next day he had no memory of the surgery or the related events.

While attending a concert, a young music teacher suddenly jumped up from his seat, walked down the aisle onto the stage, circled the piano twice, jumped to the floor, and hopped up the aisle out of the exit. He did not regain full consciousness until he was on his way home. This was not the first time that he had had such a seizure: He often found himself on a bus with no idea where he was going or how he got there.

In 2017, the International League Against Epilepsy (ILEA)—the group responsible for defining the diagnostic criteria for seizures and epilepsy—published new diagnostic guidelines. Amongst many changes, the new guidelines discourage the use of the categories "complex" and "simple." This stems from the finding that focal seizures, rather than merely being simple or complex, lie on a complexity continuum. Rather than rating a seizure in terms of the complexity of its behavioral symptoms, the ILEA recommends that focal seizures be classified in terms of the level of disruption of consciousness during the seizure, ranging from no disruption of consciousness (as is true for many simple seizures) to disrupted consciousness (as is true for many complex seizures) (see Zuberi & Brunklaus, 2018).

GENERALIZED SEIZURES. **Generalized seizures** involve the entire brain. Some begin as focal discharges that gradually spread through the entire brain. In other cases, the discharges seem to begin almost simultaneously in all parts of the brain. Such sudden-onset generalized seizures may result from diffuse pathology or may begin focally in a structure, such as the thalamus, that projects to many parts of the brain.

Like focal seizures, generalized seizures occur in many forms. One is the **tonic-clonic seizure**. The primary symptoms of a tonic-clonic seizure are loss of consciousness, loss of equilibrium, and a violent *tonic-clonic convulsion* (that is, a convulsion involving both tonus and clonus). Tongue biting, urinary incontinence, and *cyanosis* (turning blue from a lack of oxygen during a convulsion) are common manifestations of tonic-clonic convulsions. The **hypoxia** (a shortage of oxygen supply to a tissue, such as brain tissue) that accompanies a tonic-clonic seizure can itself cause brain damage.

Figure 10.9 The bursting of an epileptic neuron, recorded by extracellular unit recording.

A second type of generalized seizure is the **absence seizure**. Absence seizures are not associated with convulsions; their primary behavioral symptom is a loss of consciousness associated with a cessation of ongoing behavior, a vacant look, and sometimes fluttering eyelids. The EEG of an absence seizure is different from that of other seizures; it is a bilaterally symmetrical **3-per-second spike-and-wave discharge** (see Figure 10.10). Absence seizures are most common in children, and they frequently cease at puberty (see Guilhoto, 2017).

Although there is no cure for epilepsy, the frequency and severity of seizures can often be reduced by anticonvulsant medication (see Iyer & Marson, 2014; Rheims & Ryvlin, 2014). Unfortunately, these drugs often have adverse side effects (e.g., memory impairment), and they don't work for everyone (see Devinsky et al., 2013; Lowenstein, 2015). Other treatment options include stimulation of the vagus nerve (see Dugan & Devinsky, 2013; Vonck et al., 2014), transcranial magnetic stimulation (see Chapter 5) (see Chen et al., 2016), and the *ketogenic diet* (a diet consisting of high levels of fat, moderate levels of protein, and low levels of carbohydrates)—see Scharfman (2015). Brain surgery is sometimes used, but usually only in serious cases when other treatment options have been exhausted (see Jetté, Sander, & Keezer, 2016).

Parkinson's Disease

LO 10.9 Describe the symptoms of Parkinson's disease and some treatments for this disorder.

Parkinson's disease is a movement disorder of middle and old age that affects 1–2 percent of the population over the age of 65 (see Harris et al., 2020; Kalia & Lang, 2015). It is slightly more prevalent in males than in females (see Pringsheim et al., 2014).

Figure 10.10 The bilaterally symmetrical, 3-per-second spike-and-wave EEG discharge associated with absence seizures.

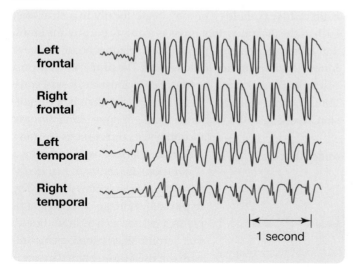

Left frontal

Right frontal

Left temporal

Right temporal

1 second

The initial symptoms of Parkinson's disease are mild—perhaps no more than a slight stiffness or tremor of the fingers—but they inevitably increase in severity with advancing years. The most common symptoms of the full-blown disorder are a tremor that is pronounced during inactivity but not during voluntary movement or sleep, muscular rigidity, difficulty initiating movement, slowness of movement, and a masklike face. Certain ailments often develop well before the motor symptoms of Parkinson's disease become apparent—up to a decade before in some cases. These include sleep disturbances, loss of the sense of smell, and depression (see Bourzac, 2016; Charvin et al., 2018; McGregor & Nelson, 2019; Potsuma & Berg, 2016).

Many Parkinson's patients display only mild cognitive deficits that don't interfere with their daily life (see Aarsland et al., 2017; Perugini et al., 2018). In essence, these patients are thinking people trapped inside bodies they cannot control. Do you remember the case of "The Lizard"—Roberto Garcia d'Orta—from Chapter 4?

However, some Parkinson's patients do experience more severe cognitive deficits. In general, there is a more rapid cognitive decline with age in Parkinson's patients than in the general population (see Aarsland et al., 2017; Yang, Tang, & Gou, 2016). Indeed, a majority of Parkinson's patients will display symptoms of dementia if they live for more than 10 years after their initial diagnosis (see Aarsland et al., 2017; Fyfe, 2018).

Like epilepsy, Parkinson's disease seems to have no single cause; faulty DNA, brain infections, strokes, tumors, TBI, and neurotoxins have all been implicated in specific cases (see Przedborski, 2017). However, in the majority of cases, there is no obvious cause, and no family history of the disorder (see Przedborski, 2017). Although numerous genes have been linked to Parkinson's disease (see Singleton & Hardy, 2016; Verstraetan, Theuns, & Van Broeckhoven, 2015), most cases of Parkinson's disease are likely the result of interactions between multiple genetic and environmental factors (see Przedborski, 2017).

Parkinson's disease is associated with widespread degeneration, but it is particularly severe in the **substantia nigra**—the midbrain nucleus whose neurons project via the **nigrostriatal pathway** to the **striatum** of the basal ganglia (see Harris et al., 2020; McGregor & Nelson, 2019; Michel, Hirsh, & Hunot, 2016). Although *dopamine* is normally the major neurotransmitter released by most neurons of the substantia nigra, there is little dopamine in the substantia nigra and striatum of long-term Parkinson's patients. Autopsy often reveals clumps of a protein called **alpha-synuclein** in the surviving dopaminergic neurons of the substantia nigra—these clumps are known as **Lewy bodies**, after the German pathologist who first reported them in 1912 (see Lashuel et al., 2013). Alpha-synuclein is currently the focus of intense research because it is believed to play an important role in the development and spread of pathology in the brains of Parkinson's patients (see Harris et al., 2020; Mor et al., 2017; Roy, 2017; Visanji et al., 2016; but see Surmeier, Obeso, & Halliday, 2017).

As you saw in the case of d'Orta (see Chapter 4), the symptoms of Parkinson's disease can be alleviated by injections of **L-dopa**—the chemical from which the body synthesizes dopamine. However, L-dopa is not a permanent solution; it typically becomes less and less effective with continued use, until its side effects (e.g., involuntary movements) outweigh its benefits (see Charvin et al., 2018; De Deurwaerdere, Di Giovanni, & Millan, 2017). This is what happened to d'Orta. L-dopa therapy gave him a 3-year respite from his disease, but ultimately it became ineffective. His prescription was then changed to another dopamine agonist, and again his condition improved—but again the improvement was only temporary.

There is currently no drug that will permanently block the progressive development of Parkinson's disease or permanently reduce the severity of its symptoms (see Charvin et al., 2018), though there are ongoing attempts to develop such drugs (see Elkouzi et al., 2019). Indeed, current evidence suggests that by the time the motor symptoms of Parkinson's disease become apparent, and a diagnosis is made, irreversible damage has already occurred (see Tison & Meissner, 2014). We will return to d'Orta's roller-coaster case later in this chapter.

When medication is not effective in the treatment of Parkinson's disease, **deep brain stimulation** is a treatment option. This entails applying low-intensity electrical stimulation continually to a particular area of the brain through a stereotaxically implanted electrode (see Faggiani & Benazzouz, 2017; Ligaard, Sannæs, & Pihlstrøm, 2019) (see Figure 10.11). The treatment of Parkinson's disease by this method usually involves chronic bilateral electrical

Figure 10.11 Deep brain stimulation for Parkinson's disease.

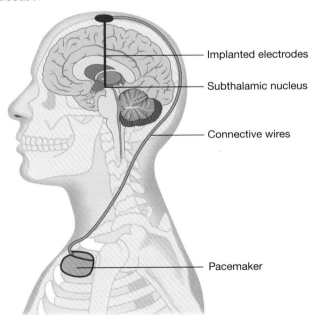

Implanted electrodes

Subthalamic nucleus

Connective wires

Pacemaker

stimulation of the **subthalamic nucleus,** a nucleus that lies just beneath the thalamus and is richly connected to the basal ganglia (see Faggiani & Benazzouz, 2017; Limousin & Foltynie, 2019; Stefani et al., 2017). High-frequency electrical stimulation is employed, and this blocks the function of the target structure, much as a lesion would (see Ashkan et al., 2017). Once the current is turned on, symptoms can be alleviated within seconds; and when the stimulation is turned off, the therapeutic improvements dissipate very quickly (see Lozano, Hutchison, & Kalia, 2017). Unfortunately, deep brain stimulation can cause side effects such as cognitive, speech, and gait problems (see Cossu & Pau, 2016; Moldovan et al., 2015), and it does not slow the progression of Parkinson's disease (Limousin & Foltynie, 2019).

Huntington's Disease

LO 10.10 Describe the symptoms of Huntington's disease and explain its genetic basis.

Like Parkinson's disease, **Huntington's disease** is a progressive motor disorder, but, unlike Parkinson's disease, it is rare (1 in 7,500), it has a simple genetic basis, and it is always associated with severe dementia.

The first clinical sign of Huntington's disease is often increased fidgetiness. As the disorder develops, rapid, complex, jerky movements of entire limbs (rather than individual muscles) begin to predominate. Also prominent are psychiatric symptoms and cognitive deficits (see Bachoud-Lévi et al., 2019). Eventually, motor and cognitive deterioration become so severe that sufferers are incapable of feeding themselves, controlling their bowels, or recognizing their friends and relatives. There is no cure; death typically occurs about 20 years after the appearance of the first symptoms.

Huntington's disease is passed from generation to generation by a single mutated dominant gene, called **huntingtin**. The protein it codes for is known as the **huntingtin protein**. Because the gene is dominant, all individuals carrying the gene develop the disorder, as do about half their offspring (see Essa et al., 2019). The huntingtin gene has remained in the gene pool because, once inherited, the first symptoms of the disease do not appear until after the peak reproductive years (at about age 40).

The mutated huntingtin protein seems to be "stickier" than normal huntingtin protein—leading to the accumulation of clumps of the mutated huntingtin protein within cells (see Tyebji & Hannan, 2017). It is thought that these clumps are toxic to cells (see Caron, Dorsey, & Hayden, 2018). Early on in the disease, this accumulation becomes particularly pronounced in cells in the striatum, where the cells begin to die (see Mattis & Svendsen, 2018). The cell death in the striatum destroys the connections between the striatum and the cortex, and it is this disconnection which is believed to result in the early symptoms of Huntington's disease. As the disease progresses, cells throughout the brain begin to die from the clumps of mutated huntingtin-protein (see Essa et al., 2019; Veldman & Yang, 2018).

If one of your parents were to develop Huntington's disease, the chance would be 50/50 that you too would develop it. If you were in such a situation, would you want to know whether you would suffer the same fate? Medical geneticists have developed a test that can tell relatives of Huntington's patients whether they are carrying the gene. Some choose to take the test, and some do not. One advantage of the test is that it permits the relatives of Huntington's patients who have not inherited the gene to have children without the fear of passing the disorder on to them. Another potential advantage is that it may permit preventative therapies for carriers of the gene in the future (see Caron, Dorsey, & Hayden, 2018).

There is currently no established treatment that slows the progression of Huntington's disease. However, a clinical trial in 2016 that used a novel medication generated some promising results (see Jankovic, 2017), and there are many other clinical trials for other candidate therapies in progress (see Caron, Dorsey, & Hayden, 2018).

Multiple Sclerosis

LO 10.11 Describe the symptoms of multiple sclerosis (MS) and its risk factors.

Multiple sclerosis (MS) is a progressive disease that attacks the myelin of axons in the CNS. It is particularly disturbing because the first symptoms typically appear in early adulthood (Filippi et al., 2018). First, there are microscopic areas of degeneration on the myelin sheaths provided by oligodendroglia (see Chapter 3) (see Lucchinetti et al., 2011; Filippi et al., 2018) but eventually damage to the myelin is so severe that the associated axons become dysfunctional and degenerate (see Faissner et al., 2019; Filippi et al., 2018). Ultimately, many areas of hard scar tissue develop in the CNS (*sclerosis* means "hardening"). Figure 10.12 illustrates degeneration of the white matter of a patient with MS.

Figure 10.12 Areas of sclerosis (see arrows) in the white matter of a patient with multiple sclerosis (MS).

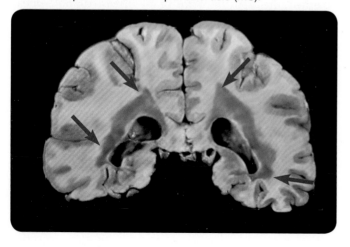

MS is often considered to be an *autoimmune disorder*—a disorder in which the body's immune system attacks part of the body as if it were a foreign substance (see Dong & Yong, 2019; Nave & Ehrenreich, 2017). In MS, the myelin sheath on axons is the focus of the faulty immune reaction. Indeed, an animal model of MS, termed *experimental autoimmune encephalomyelitis* (see Deshmukh et al., 2013; Dong & Yong, 2019), can be induced by injecting laboratory animals with myelin and a preparation that stimulates the immune system. However, it should be noted that in MS, damage to neurons occurs even without demyelination (see Friese, Schattling, & Fugger, 2014). Indeed, the general consensus is that the progression of MS is driven by an interaction between immune-system reactivity and neural degeneration (Faissner et al., 2019).

In MS, there is also a lack of *remyelination* of axons. Remyelination refers to the generation of new myelin sheaths on axons—a job taken care of by oligodendroglia in the CNS (see Faissner et al., 2019; Franklin & Ffrench-Constant, 2017). In healthy individuals, remyelination of axons by oligodendroglia and the generation of new oligodendroglia are both ongoing processes that occur throughout one's lifespan (see Jensen & Yong, 2016). In patients with MS, both processes appear to be disrupted. First, many newly born oligodendroglia fail to develop into cells that are capable of remyelinating axons. Second, those oligodendroglia that are functional do not remyelinate affected brain areas (see Nave & Ehrenreich, 2017). Accordingly, a major focus of current MS research is the development of treatments that encourage remyelination by oligodendroglia (see Filippi et al., 2018; Jensen & Yong, 2016; Kremer et al., 2016; Stangel et al., 2017).

Diagnosing MS is typically done with MRI (see Chapter 5) (see Matthews, 2019; Rotstein & Montalban, 2019), and it focuses on identifying the time course of development of white-matter lesions (see Geraldes et al., 2018; Tintore, Vidal-Jordana, & Sastre-Garriga, 2019). However, MRI-based diagnosis is typically complex because the nature and severity of MS lesions depends on a variety of factors, including their number, size, and location (see Rotstein & Montalban, 2019; Sati et al., 2016).

In the majority of cases of MS, there are periods of remission (up to 2 years) during which the patient displays no symptoms; however, these are usually just oases in the progression of MS, which eventually becomes continuous and severe (see Larochelle et al., 2016). Common symptoms of advanced MS are visual disturbances, muscular weakness, numbness, tremor, and **ataxia** (loss of motor coordination). In addition, cognitive deficits and emotional changes occur in some patients (see Filippi et al., 2018; Ransohoff, Hafler, & Lucchinetti, 2015; Rotstein & Montalban, 2019).

Epidemiological studies have revealed several puzzling features of MS (see Filippi et al., 2018). **Epidemiology** is the study of the various factors such as diet, geographic location, and age that influence the distribution of a disease in the general population. Genetic factors seem to play less of a causal role in MS than they do in other neurological disorders: The concordance rate is only 35 percent in monozygotic twins, compared with 6 percent in dizygotic twins. Also, the incidence of multiple sclerosis is substantially higher in females than in males (see Friese, Schattling, & Fugger, 2014; Ransohoff, Hafler, & Lucchinetti, 2015), and in Caucasians than in other groups (Filippi et al., 2018). Also, the incidence is higher in people who have lived in cold climates, particularly during their childhoods (Filippi et al., 2018).

Several established risk factors exist for MS. The most well-established ones include vitamin D deficiency, exposure to the *Epstein-Barr virus* (the most common cause of mononucleosis), and cigarette smoking (see Filippi et al., 2018; Hauser, Chan, & Oksenberg, 2013).

In the 1990s, *immunomodulatory drugs* were approved for the treatment of MS, and a large number of them are now available for MS treatment (see Chun et al., 2019). Although these drugs are still widely prescribed for MS, their benefits are only marginal, and they help only some MS patients (see Faissner et al., 2019; Marino & Cosentino, 2016). Still, this modest success has stimulated the current search for more effective drug treatments, such as those that encourage remyelination (see Stangel et al., 2017; Filippi et al., 2018).

Alzheimer's Disease

LO 10.12 Describe the symptoms of Alzheimer's disease and evaluate the amyloid hypothesis.

Alzheimer's disease is the most common cause of *dementia* in the elderly; it currently affects about 50 million people worldwide (see Drew & Ashour, 2018; Götz, Bodea, & Goedert, 2018). It sometimes appears in individuals as young as 40, but the likelihood of its development becomes greater with advancing years. About 10 percent of people over the age of 65 suffer from the disease (see Mielke, Vemuri, & Rocca, 2014).

Journal Prompt 10.2

Total dementia often creates less suffering than partial dementia for the individual with the dementia. Why do you think that is the case?

Alzheimer's disease is considered to progress through three stages (see Chiesa et al., 2017; Hickman, Faustin, & Wisniewski, 2016). The first stage, the *preclinical stage*, involves pathological changes in the brain without any behavioral or cognitive symptoms (see Brier et al., 2016; Dolgin, 2018; Polanco et al., 2019; Villemagne et al., 2018). The second stage of Alzheimer's disease, known as the *prodromal stage*, involves mild cognitive impairment. The cognitive impairment observed during this stage is not nearly as severe as that seen in full-blown Alzheimer's disease,

but it is an indicator that the symptoms of Alzheimer's disease are progressing. The combined presence of mild cognitive impairment and certain biological changes can lead to a fairly reliable diagnosis during this stage (see Dolgin, 2018). During the final stage of Alzheimer's, the *dementia stage*, there is initially a progressive decline in memory, deficits in attention, and personality changes; this is eventually followed by marked confusion, irritability, anxiety, and deterioration of speech. Eventually the patient deteriorates to the point that even simple responses such as swallowing and bladder control are difficult. Alzheimer's disease is a terminal illness.

The three defining neuropathological characteristics of the disease are neurofibrillary tangles, amyloid plaques, and neuron loss. *Neurofibrillary tangles* are threadlike tangles of **tau protein** in the neural cytoplasm. Tau protein normally plays a role in maintaining the overall structure of neurons (see Villemagne & Okamura, 2016; Wang & Mandelkow, 2016). *Amyloid plaques* are clumps of scar tissue composed of degenerating neurons and aggregates of another protein called **beta-amyloid** which is present in healthy brains in only small amounts. The presence of amyloid plaques in the brain of a patient who died of Alzheimer's disease is illustrated in Figure 10.13.

Although neurofibrillary tangles, amyloid plaques, and neuron loss tend to occur throughout the brains of

---

Figure 10.13 Amyloid plaques (stained blue) in the brain of a deceased patient who had Alzheimer's disease.

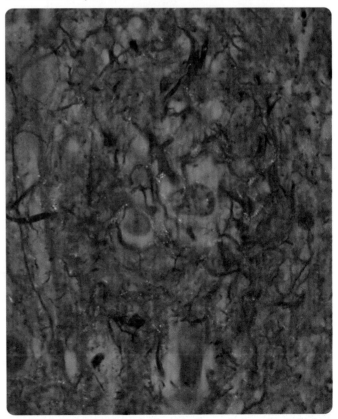

Dr. Cecil H. Fox/Science Source

Alzheimer's patients, they are more prevalent in some areas than in others. For example, neurofibrillary tangles are particularly prevalent in medial temporal lobe structures such as the *entorhinal cortex, amygdala,* and *hippocampus*—all structures involved in various aspects of memory. Figure 10.14 summarizes the various neuropathological changes seen in Alzheimer's disease.

Early attempts to identify the genes involved in Alzheimer's disease focused on a rare, early-onset form of the disorder that runs in a few families. Mutations to four different genes have been shown to contribute to the early-onset, familial form; however, these four gene mutations seem to contribute little (only about 1 percent) to the more common, late-onset form (see Kanekiyo, Xu, & Bu, 2014; Ulrich et al., 2017).

Subsequent research on the late-onset form of Alzheimer's disease has implicated other genes (see Tanzi, 2012). Attention has focused on one particular gene, the gene on chromosome 19 that codes for the protein *apolipoprotein E (APOE)* (see Belloy, M. E., Napolioni, V., & Greicius, 2019;

Yamazaki et al., 2019). Notably, the presence of a particular allele of the APOE gene, APOE4, has been shown to increase susceptibility to the late-onset form of Alzheimer's disease by approximately 50 percent (see Sala Frigerio & De Strooper, 2016; Karch, Cruchaga, & Goate, 2014; Yu, Tan, & Hardy, 2014). The exact cellular functions of APOE are not yet known (see Spinney, 2014). However, in Alzheimer's disease, it appears that APOE binds to beta-amyloid, and that such binding reduces beta-amyloid clearance from the brain, and it increases beta-amyloid clumping (see Ulrich et al., 2017)—leading to the development of amyloid plaques.

There is currently no cure for Alzheimer's disease. One factor complicating the search for a treatment or cure for Alzheimer's disease is that it is not entirely clear which symptom is primary (see O'Brien & Wong, 2011; Spires-Jones & Hyman, 2014). The *amyloid hypothesis* is currently the dominant view (see Castellani, Plascencia-Villa, & Perry, 2019). It proposes that amyloid plaques are the primary symptom of the disorder; that is, the plaques cause all the other symptoms (see Liu et al., 2019; Selkoe & Hardy, 2016).

Figure 10.14 A summary of the neuropathological changes that accompany Alzheimer's disease.

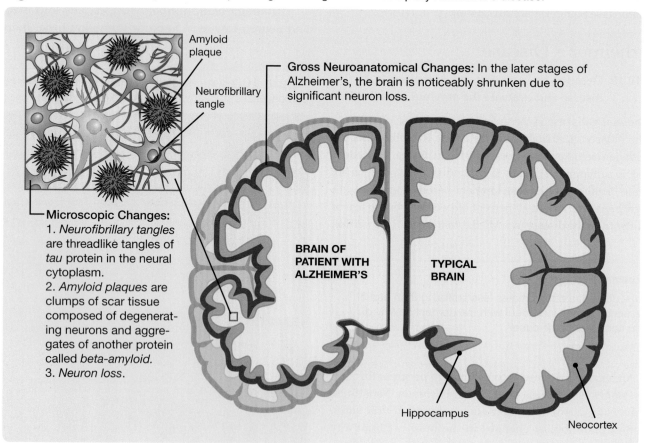

Base on Drew, L. (2018).

The main support for the amyloid hypothesis has come from the genetic analysis of families with early-onset Alzheimer's disease (see Herrup, 2015). All four different gene mutations that cause early-onset Alzheimer's disease influence the synthesis of beta-amyloid. Additional support for the amyloid hypothesis comes from the finding that the production of neurofibrillary tangles is downstream from alterations to beta-amyloid (Amar et al., 2017). One of the main arguments against the amyloid hypothesis is the fact that many people without observable dementia carry significant loads of amyloid plaques. These individuals are known as *high-plaque normals* (see Herrup, 2015; Spires-Jones & Hyman, 2014). However, these high-plaque normals also lacked the inflammation associated with Alzheimer's disease (see Abbott, 2018; Simon, Obst, & Gomez-Nicola, 2018; Villegas-Llerena et al., 2016). Thus, whether or not beta-amyloid initiates the cascade of events that lead to full-blown Alzheimer's disease is still a matter of debate (see Buxbaum, 2017; Reiman, 2016).

In the past decade, researchers trying to develop a cure for Alzheimer's disease by reducing the amyloid deposits have been taken on a roller-coaster ride: One treatment after another was found to be promising in animal models of the disease only to be found ineffective when applied to human patients (see Busche et al., 2015; Castellani, Plascencia-Villa, & Perry, 2019; Makin, 2018; Reiman, 2017). These failures have encouraged further challenges to the amyloid hypothesis (see Herrup, 2015; Karran, Mercken, & de Strooper, 2011).

It is widely believed that in order to be effective against Alzheimer's disease, treatments must be administered during the preclinical or prodromal stages (see Huang & Mucke, 2012; McDade & Bateman, 2017). The problem is that when patients first seek help for Alzheimer's symptoms, they typically are already in the dementia stage and have extensive brain pathology (see Selkoe, 2012). That is why recent advances in early diagnosis may prove to be critical in the development of effective treatments (see Hampel et al., 2018; Lee et al., 2019).

Attention has also focused on Down syndrome as a potential provider of insights into the neural mechanisms of Alzheimer's disease. This link stems from the fact that, by age 40, almost all individuals with Down syndrome have developed numerous amyloid plaques and neurofibrillary tangles, the core symptoms of Alzheimer's disease (see Lott & Head, 2019; Marshall, 2014); and that, by age 60, two-thirds of individuals with Down syndrome will have developed dementia (see Wiseman et al., 2015). It turns out that people with Down syndrome have three copies of chromosome 21, instead of two. The gene that codes for beta-amyloid resides on chromosome 21. Accordingly, scientists have hypothesized that the presence of an extra chromosome 21 leads to greater production of beta-amyloid in persons with Down syndrome (see Wiseman et al., 2015). Interestingly, recent research has shown that approximately 15 percent of the neurons in the brains of Alzheimer's patients without Down syndrome contain an extra copy of chromosome 21 (see Lott & Head, 2019; Marshall, 2014).

In recent years, increasing attention has been paid to a new hypothesis of Alzheimer's disease known as the *pathogenic spread hypothesis*. The pathogenic spread hypothesis proposes that many common neurodegenerative diseases (e.g., Alzheimer's disease, Parkinson's disease) result from the presence of misfolded proteins that initiate a chain reaction causing other proteins to misfold (see Collinge, 2016; Jucker & Walker, 2018). For example, according to this hypothesis, in Alzheimer's disease a misfolded tau or beta-amyloid protein "seeds" the misfolding of other tau or beta-amyloid proteins; these misfolded proteins subsequently spread to other neurons (see del Rio, Ferrer, & Gavín, 2018; DeVos et al., 2018; Wu et al., 2016) and then aggregate into the neurofibrillary tangles or amyloid plaques that are characteristic of Alzheimer's disease (see Goedert, Eisenberg, & Crowther, 2017; Walsh & Selkoe, 2016).

Scan Your Brain

This is a good place for you to pause and scan your brain to check your knowledge on neuroplasticity before moving on to the next module. Fill in the following blanks with the most appropriate terms. The correct answers are provided at the end of the exercise. Before proceeding, review material related to your errors and omissions.

1. An _____ is a pathological dilation at the wall of an artery and can be congenital or a result of an infection.

2. A bruise in an organ or a tissue caused by a localized blood clot is called a _____.

3. The _____ that accompanies a tonic-clonic seizure can cause brain damage.

4. _____ disease is a condition caused by severe degeneration of neurons in the substantia nigra that project to the basal ganglia.

5. Drug therapy for multiple sclerosis includes _____ that are only partly effective in delaying the symptoms of the disorder.

6. _____ plaques are one of the key characteristics of Alzheimer's disease.

7. Although neurofibrillary tangles, amyloid plaques, and neuron loss tend to occur throughout the brains of Alzheimer's patients, they are particularly prevalent in medial temporal lobe structures such as the _____, _____, and _____—all structures involved in various aspects of memory.

8. Genetic research suggests that one particular gene on chromosome 19 called _____ may be responsible for the late onset of Alzheimer's dementia.

Animal Models of Human Neurological Diseases

The first two modules of this chapter focused on neuropsychological diseases and their causes, but they also provided some glimpses into the ways in which researchers attempt to solve the puzzles of neurological dysfunction. This module focuses on one of these ways: the experimental investigation of animal models.

Because identifying the neuropathological bases of human neuropsychological diseases is seldom possible based on research on the patients themselves, research on animal models of the diseases often plays an important role. Unfortunately, using and interpreting animal models is far from straightforward: Even the best animal models of neuropsychological diseases display only some of the features of the diseases they are modeling (see Dawson, Golde, & Lagier-Tourenne, 2018; Wekerle et al., 2012), though researchers often treat animal models as if they duplicate in every respect the human conditions that they are claimed to model. That is why we have included this module on animal models in this chapter: We do not want you to fall into the trap of thinking about them in this way. This module discusses two widely used animal models: the kindling model of epilepsy and the MPTP model of Parkinson's disease.

Kindling Model of Epilepsy

LO 10.13 Describe the kindling model of epilepsy and explain the ways in which it models human epilepsy.

In the late 1960s, Goddard, McIntyre, and Leech (1969) delivered one mild electrical stimulation per day to rats through an implanted amygdalar electrode. There was no behavioral response to the first few stimulations, but soon each stimulation began to elicit a convulsive response. The first convulsions were mild, involving only a slight tremor of the face. However, with each subsequent stimulation, the elicited convulsions became more generalized, until each convulsion involved the entire body. The progressive development and intensification of convulsions elicited by

a series of periodic brain stimulations became known as the **kindling phenomenon**, one of the first neuroplastic phenomena to be widely studied.

Although kindling is most frequently studied in rats subjected to repeated amygdalar stimulation, it is a remarkably general phenomenon. For example, kindling has been reported in mice, rabbits, cats, dogs, and various primates. Moreover, kindling can be produced by the repeated stimulation of many brain sites other than the amygdala, and it can be produced by the repeated application of initially subconvulsive doses of convulsive chemicals.

There are many interesting features of kindling, but two warrant emphasis. The first is that the neuroplastic changes underlying kindling are permanent. A subject that has been kindled and then left unstimulated for several months still responds to each low-intensity stimulation with a generalized convulsion. The second is that kindling is produced by distributed, as opposed to massed, stimulations. If the intervals between successive stimulations are shorter than an hour or two, many more stimulations are usually required to kindle a subject, and under normal circumstances, no kindling at all occurs at intervals of less than about 20 minutes.

Journal Prompt 10.3

Kindling is produced by distributed, as opposed to massed, stimulations. Why is this an example of neuroplasticity? (Hint: Think about the effects of spaced vs. massed (i.e., cramming) study sessions when you prepare for an exam.)

Much of the interest in kindling stems from the fact that it models epilepsy in two ways (see Morimoto, Fahnestock, & Racine, 2004). First, the convulsions elicited in kindled animals are similar in many respects to those observed in some types of human epilepsy. Second, the kindling phenomenon itself is comparable to the **epileptogenesis** (the development, or genesis, of epilepsy) that can follow a head injury (see Goldberg & Coulter, 2013): Some individuals who at first appear to have escaped serious injury after a blow to the head begin to experience convulsions a few weeks later, and these convulsions often begin to recur more and more frequently and with greater and greater severity.

It must be stressed that the kindling model as it is applied in most laboratories does not model epilepsy in one important respect. You will recall from earlier in this chapter that epilepsy is a disease in which seizures recur spontaneously; in contrast, kindled convulsions are elicited. However, a model that overcomes this shortcoming has been developed in several species. If subjects are kindled for a very long time—about 300 stimulations in rats—a syndrome can be induced that is much more like epilepsy, in the sense that the subjects begin to display spontaneous seizures and continue to display them even after the regimen of stimulation is curtailed (see Grone & Baraban, 2015).

MPTP Model of Parkinson's Disease

LO 10.14 Describe the events that led to the discovery of the MPTP model of Parkinson's disease, and evaluate the utility of this animal model.

The preeminent animal model of Parkinson's disease grew out of an unfortunate accident, which resulted in the following cases (Langston, 1985).

The Case of the Frozen Drug Users

Parkinson's disease rarely occurs before the age of 50. Thus, it was surprising when a group of young drug-addicted individuals developed severe parkinsonism. The one link among these patients was their recent use of a new "synthetic heroin." They exhibited all of the typical symptoms of Parkinson's disease, including *bradykinesia* (slowness of movement), tremor, and muscle rigidity. Even subtle Parkinson's symptoms, such as *seborrhea* (oiliness of the skin) and *micrographia* (small handwriting), were present. After obtaining samples of the synthetic heroin, the offending agent was identified as 1-methyl-4-phenyl-1,2,3,6-tetrahydropyridine or **MPTP**. None of the patients recovered.

Researchers immediately turned the misfortune of these few individuals to the advantage of many by developing a much-needed animal model of Parkinson's disease. It was quickly established that nonhuman primates respond to MPTP in many of the same ways that humans do. They display Parkinsonian motor symptoms, cell loss in the substantia nigra, and a major reduction in brain dopamine. For unknown reasons, rats are relatively resistant to MPTP, and mice vary greatly from strain to strain in their response to it (see Blesa et al., 2012).

Although the MPTP model does not model all aspects of Parkinson's disease, the model has proven extremely useful (see Dawson, Golde, & Lagier-Tourenne, 2018; Porras, Li, & Bezard, 2011). It has been instrumental in the development of many of the treatments, mainly dopamine agonists, that are currently in use (see Fox & Brotchie, 2010).

Responses to Nervous System Damage: Degeneration, Regeneration, Reorganization, and Recovery

In the first three modules of this chapter, you have learned about three things: (1) causes of brain damage, (2) neurological diseases associated with brain damage, and (3) animal models of neurological diseases. This module focuses on four neuroplastic responses of the brain to damage: degeneration, regeneration, reorganization, and recovery of function.

Neural Degeneration

LO 10.15 Explain the various types of neural degeneration that ensue following axotomy.

Neural degeneration (neural deterioration and death) is a component of both brain development and disease. Neural degeneration, as it typically occurs, is a complex process: It is greatly influenced by nearby glial cells (see Burda & Sofroniew, 2014; Salter & Stevens, 2017; Tsuda & Inoue, 2016), by the activity of the degenerating neurons (see Bell & Hardingham, 2011), and by the particular cause of the degeneration (see Conforti, Adalbert, & Coleman, 2007). In the laboratory, however, neural degeneration is often induced in a simple, controlled way: by cutting axons (i.e., by *axotomy*). Two kinds of neural degeneration ensue: anterograde degeneration and retrograde degeneration. **Anterograde degeneration** is the degeneration of the **distal segment**—the segment of a cut axon from the cut to the synaptic terminals (see Neukomm & Freeman, 2014). **Retrograde degeneration** is the degeneration of the **proximal segment**—the segment of a cut axon from the cut back to the cell body.

Anterograde degeneration occurs quickly following axotomy because the cut separates the distal segment of the axon from the cell body, which is the metabolic center of the neuron. The entire distal segment becomes badly swollen within a few hours, and it breaks apart into fragments within a few days.

The course of retrograde degeneration is different; it progresses gradually back from the cut to the cell body. In about 2 or 3 days, major changes become apparent in the cell bodies of most axotomized neurons. These early cell body changes are either degenerative or regenerative in nature. Early degenerative changes to the cell body (e.g., a decrease in size) suggest that the neuron will ultimately die—usually by apoptosis but sometimes by necrosis or a combination of both. Early regenerative changes (e.g., an increase in size) indicate that the cell body is involved in a massive synthesis of the proteins that will be used to replace the degenerated axon. But early regenerative changes in the cell body do not guarantee the long-term survival of the neuron; if the regenerating axon does not manage to make synaptic contact with an appropriate target, the neuron eventually dies.

Sometimes, degeneration spreads from damaged neurons to neurons that are linked to them by synapses; this is called **transneuronal degeneration**. In some cases, transneuronal degeneration spreads from damaged neurons to the neurons on which they synapse; this is called *anterograde transneuronal degeneration*. And in some cases, it spreads from damaged neurons to the neurons that synapse on them; this

is called *retrograde transneuronal degeneration*. Neural and transneuronal degeneration are illustrated in Figure 10.15.

Neural Regeneration

LO 10.16 Compare neural regeneration within the CNS vs. the PNS.

Neural regeneration—the regrowth of damaged neurons—does not proceed as successfully in mammals and other higher vertebrates as it does in most invertebrates and lower vertebrates. For example, in Chapter 9, you learned about accurate regeneration in the frog visual system as demonstrated by Sperry's eye-rotation experiments. The capacity for accurate axonal growth, which higher

Figure 10.15 Neural and transneuronal degeneration following axotomy.

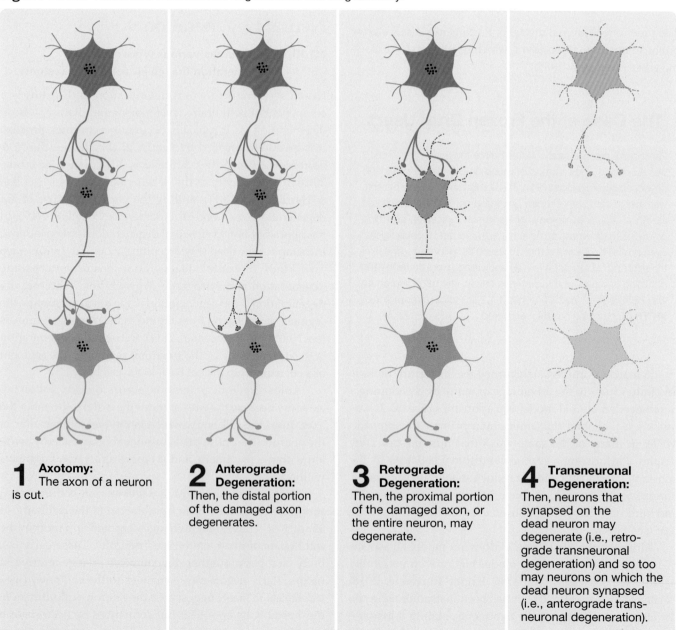

1 Axotomy: The axon of a neuron is cut.

2 Anterograde Degeneration: Then, the distal portion of the damaged axon degenerates.

3 Retrograde Degeneration: Then, the proximal portion of the damaged axon, or the entire neuron, may degenerate.

4 Transneuronal Degeneration: Then, neurons that synapsed on the dead neuron may degenerate (i.e., retrograde transneuronal degeneration) and so too may neurons on which the dead neuron synapsed (i.e., anterograde transneuronal degeneration).

vertebrates possess during their development, is mostly lost once they reach maturity. Regeneration of axons is virtually nonexistent in the CNS of adult mammals. When regeneration of axons does occur, it usually requires very specific conditions (e.g., Krol & Roska, 2016; Lim et al., 2016; Tedeschi & Bradke, 2017).

In the mammalian PNS, regrowth from the proximal stump of a damaged nerve usually begins 2 or 3 days after axonal damage, once new growth cones have formed (see Bradke, Fawcett, & Spira, 2012). What happens next depends on the nature of the injury; there are three possibilities. First, if the original Schwann cell myelin sheaths remain intact, the regenerating peripheral axons grow through them to their original targets at a rate of a few millimeters per day. Second, if the peripheral nerve is severed and the cut ends become separated by a few millimeters, regenerating axon tips often grow into incorrect sheaths and are guided by them to incorrect destinations; that is why it is often difficult to regain the coordinated use of a limb affected by nerve damage even if there has been substantial regeneration. And third, if the cut ends of a severed mammalian peripheral nerve become widely separated or if a lengthy section of the nerve is damaged, there may be no meaningful regeneration at all; regenerating axon tips grow in a tangled mass around the proximal stump. These three patterns of mammalian peripheral nerve regeneration are illustrated in Figure 10.16.

Why do mammalian PNS neurons regenerate but mammalian CNS neurons normally do not? The obvious answer is that adult PNS neurons are inherently capable of regeneration whereas adult CNS neurons are not (He & Jin, 2016; Mahar & Cavalli, 2018). However, this answer has proved to be only partially correct. Some CNS neurons are capable of regeneration if they are transplanted to the PNS, whereas some PNS neurons are not capable of regeneration if they are transplanted to the CNS. Clearly, something about the environment of the PNS promotes regeneration

Figure 10.16 Three patterns of axonal regeneration that have been observed in mammalian peripheral nerves.

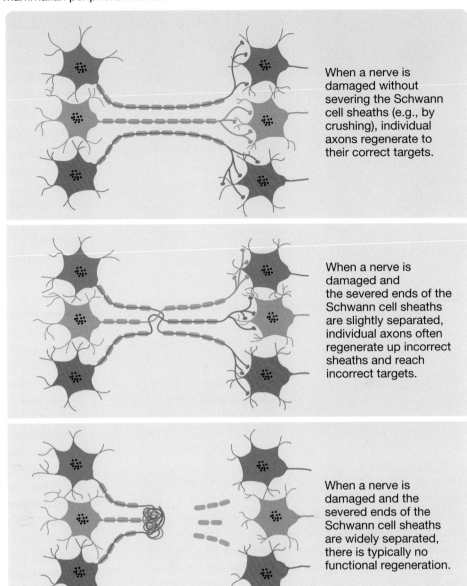

When a nerve is damaged without severing the Schwann cell sheaths (e.g., by crushing), individual axons regenerate to their correct targets.

When a nerve is damaged and the severed ends of the Schwann cell sheaths are slightly separated, individual axons often regenerate up incorrect sheaths and reach incorrect targets.

When a nerve is damaged and the severed ends of the Schwann cell sheaths are widely separated, there is typically no functional regeneration.

and something about the environment of the CNS does not. Schwann cells seem to be one factor (see Painter, 2017).

Schwann cells, which myelinate PNS axons, clear the debris and scar tissue resulting from degeneration, proliferate to produce more Schwann cells, and promote regeneration in the mammalian PNS both by producing neurotrophic factors and by forming physical tracks to guide the regrowth of axons (see Painter, 2017). In contrast, **oligodendroglia**, which myelinate CNS axons, do not clear debris or stimulate or guide regeneration; indeed, they release factors that actively block regeneration (see Geoffroy & Zheng, 2014). Moreover, in the CNS, astrocytes form a *glial scar* after injury that presents a physical barrier to axonal regrowth and also actively releases inhibitors of

Figure 10.17 Collateral sprouting after neural degeneration.

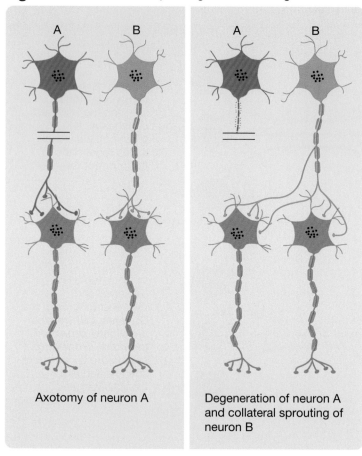

Axotomy of neuron A

Degeneration of neuron A and collateral sprouting of neuron B

axonal growth (see Ghibaudi, Boido, & Vercelli, 2017; but see Anderson et al., 2016).

When an axon degenerates, axon branches grow out from adjacent healthy axons and synapse at the sites vacated by the degenerating axon; this is called **collateral sprouting**. Collateral sprouts may grow out from the axon terminal branches or the nodes of Ranvier on adjacent neurons. Collateral sprouting is illustrated in Figure 10.17.

Neural Reorganization

LO 10.17 Describe three examples of cortical reorganization following damage to the brain, and discuss the mechanisms that might underlie such reorganization.

You learned in Chapter 9 that adult mammalian brains have the ability to reorganize themselves in response to experience. You will learn in this section that they can also reorganize themselves in response to damage.

CORTICAL REORGANIZATION FOLLOWING DAMAGE IN LABORATORY ANIMALS. Most studies of neural reorganization following damage have focused on the sensory and motor cortex of laboratory animals. Sensory and motor cortex are ideally suited to the study of neural reorganization

because of their topographic layout. Damage-induced reorganization of the primary sensory and motor cortices has been studied under two conditions: following damage to peripheral nerves and following damage to the cortical areas themselves.

Demonstrations of cortical reorganization following neural damage in laboratory animals started to be reported in substantial numbers in the early 1990s. The following three studies were particularly influential:

- Kaas and colleagues (1990) assessed the effect of making a small lesion in one retina and removing the other. Several months after the retinal lesion was made, primary visual cortex neurons that originally had receptive fields in the lesioned area of the retina were found to have receptive fields in the area of the retina next to the lesion; remarkably, this change began within minutes of the lesion (Gilbert & Wiesel, 1992).

- Pons and colleagues (1991) mapped the primary somatosensory cortex of monkeys whose contralateral arm sensory neurons had been cut 10 years before. They found that the cortical face representation had systematically expanded into the original arm area. This study created a stir because the scale of the reorganization was far greater than had been assumed to be possible: The primary somatosensory cortex face area had expanded its border by well over a centimeter over the 10-year interval between surgery and testing.

- Sanes, Suner, and Donoghue (1990) transected the motor neurons that controlled the muscles of rats' *vibrissae* (whiskers). A few weeks later, stimulation of the area of motor cortex that had previously elicited vibrissae movement now activated other muscles of the face. This result is illustrated in Figure 10.18.

CORTICAL REORGANIZATION FOLLOWING DAMAGE IN HUMANS. Demonstrations of cortical reorganization in controlled experiments on nonhumans provided an incentive to search for similar effects in human clinical populations. One such line of research has used brain-imaging technology to study the cortices of blind individuals. The findings are consistent with the hypothesis that there is continuous competition for cortical space by functional circuits. Without visual input to the cortex, there is an expansion of auditory and somatosensory cortex (see Elbert et al., 2002), and auditory and somatosensory input is processed in formerly visual areas (see Amedi et al., 2005). There seems to be a functional consequence to this reorganization: Blind volunteers have skills superior to those of sighted control volunteers on a variety of auditory and somatosensory tasks (see Merabet & Pascual-Leone, 2010).

Figure 10.18 Reorganization of the rat motor cortex following transection of the motor neurons that control movements of the vibrissae. The motor cortex was mapped by brain stimulation before transection and then again a few weeks after.

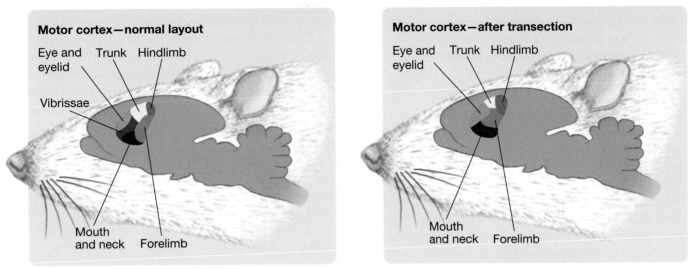

Based on Sanes, J. N., Suner, S., & Donoghue, J. P. (1990). Dynamic organization of primary motor cortex output to target muscles in adult rats.

MECHANISMS OF NEURAL RE-ORGANIZATION. Two kinds of mechanisms have been proposed to account for the reorganization of neural circuits: a strengthening of existing connections, possibly through release from inhibition, and the establishment of new connections by collateral sprouting (see Cafferty, McGee, & Strittmatter, 2008). Indirect support for the first mechanism comes from two observations: Reorganization often occurs too quickly to be explained by neural growth, and rapid reorganization never involves changes of more than 2 millimeters of cortical surface. Indirect support for the second mechanism comes from the observation that the magnitude of long-term reorganization can be too great to be explained by changes in existing connections. Figure 10.19 illustrates how these two mechanisms might account for the reorganization that occurs after damage to a peripheral somatosensory nerve.

Although sprouting and release from inhibition are considered to be the likely mechanisms of reorganization following brain damage, these are not the only possibilities. For example, neural degeneration, adjustment of dendritic trees, and adult neurogenesis may all be involved. It is also important to appreciate that reorganization following damage is not necessarily mediated by changes to the damaged area itself (see Grefkes & Fink, 2011; Rehme et al., 2011; Rossignol & Frigon, 2011).

Figure 10.19 Two proposed mechanisms for the reorganization of neural circuits: (1) strengthening of existing connections through release from inhibition and (2) establishment of new connections by collateral sprouting.

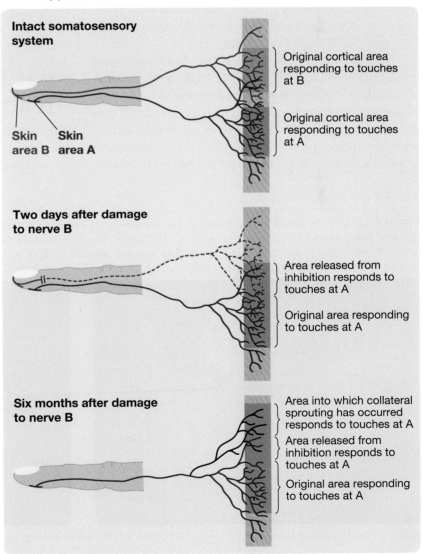

Recovery of Function after CNS Damage

LO 10.18 Describe the concept of "cognitive reserve," and discuss the potential role of adult neurogenesis in recovery following CNS damage.

Recovery of function after central nervous system damage in adult humans is poorly understood. In large part, this is so because improvements tend to be modest or nonexistent (see Teuber, 1975)—not the positive picture usually portrayed by the entertainment media, where all that is needed is lots of effort to guarantee a positive outcome. Another difficulty in studying recovery of function after CNS damage is that there are other compensatory changes that can easily be confused with it. For example, any improvement in the week or two after damage could reflect a decline in *cerebral edema* (brain swelling) rather than a recovery from the neural damage itself, and any gradual improvement in the months after damage could reflect the learning of new cognitive and behavioral strategies (i.e., substitution of functions) rather than the return of lost functions. Despite these difficulties, it is clear that recovery is most likely when lesions are small and patients are young (see Marquez de la Plata et al., 2008; Yager et al., 2006).

Cognitive reserve (roughly equivalent to education and intelligence) is thought to play a role in the improvements observed after brain damage that do not result from true recovery of brain function. Let us explain. Kapur (1997) conducted a biographical study of doctors and neuroscientists with brain damage, and he observed a surprising degree of what appeared to be cognitive recovery. His results suggested, however, that the observed improvement did not occur because these patients had actually recovered lost brain function but because their cognitive reserve allowed them to accomplish tasks in alternative ways.

Does adult neurogenesis contribute to recovery from brain damage? In adult laboratory animals, there is an increased migration of stem cells into nearby damaged areas (see Figure 10.20), and some of these develop into neurons that can survive for a few months (see Machado et al., 2015; Sun, 2015). This finding has been replicated in one human patient who died 1 week after suffering a stroke (see Minger et al., 2007). However, although stem cells can migrate short distances, there is no evidence that they can migrate from their usual sites of genesis in the hippocampus and subventricular zone to distant areas of damage in the adult human brain. In any case, there is no direct evidence that stem cells contribute to recovery, even when they do survive.

Journal Prompt 10.4

Given that brain damage seems to induce adult neurogenesis, what adaptive role, if any, might these new cells be playing?

Neuroplasticity and the Treatment of CNS Damage

This module reveals one reason for all the excitement about the phenomenon of neuroplasticity: the dream that recent discoveries about neuroplasticity can be applied to the

---

Figure 10.20 Increased neurogenesis in the dentate gyrus following damage. The left panel shows (1) an electrolytic lesion in the dentate gyrus (damaged neurons are stained turquoise) and (2) the resulting increase in the formation of new cells (stained red), many of which develop into mature neurons (stained dark blue). The right panel displays the control area in the unlesioned hemisphere, showing the normal number of new cells (stained red).

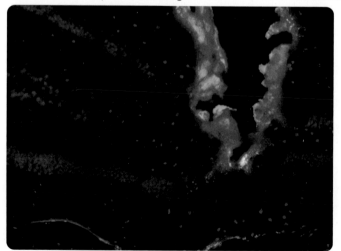 

Images courtesy of Carl Ernst and Brian Christie, Department of Psychology, University of British Columbia.

treatment of CNS damage in human patients. The following sections describe research on some major new treatment approaches that have been stimulated by the discovery of neuroplasticity.

Neurotransplantation as a Treatment for CNS Damage: Early Research

LO 10.19 Discuss early work on neurotransplantation for the treatment of CNS damage.

In the late 1980s, researchers, stimulated by research on neurodevelopment, began to explore the possibility of repairing damaged CNS tissue by implanting embryonic tissue near the damaged area. Their hope was that the embryonic tissue would mature and replace the damaged cells. Could the *donor* tissue develop and become integrated into the *host* brain and, in so doing, alleviate the symptoms? This approach focused on Parkinson's disease. Parkinson's patients lack the dopamine-releasing cells of the nigrostriatal pathway: Could they be cured by transplanting the appropriate fetal tissue into the substantia nigra?

Early signs were positive. Bilateral transplantation of fetal substantia nigra cells was successful in treating the MPTP monkey model of Parkinson's disease (see Bankiewicz et al., 1990; Sladek et al., 1987). Fetal substantia nigra transplants survived in the MPTP-treated monkeys; the transplanted cells innervated adjacent striatal tissue, released dopamine, and, most importantly, alleviated the severe poverty of movement, tremor, and rigidity produced by the MPTP.

Soon after the favorable effects of neurotransplants in the MPTP monkey model were reported, neurotransplantation was prematurely offered as a treatment for Parkinson's disease at major research hospitals. The results of the first case studies were promising. The fetal substantia nigra implants survived, and they released dopamine into the host striatum (see Sawle & Myers, 1993). More importantly, some of the patients improved.

These preliminary results triggered two double-blind evaluation studies of neurotransplants in patients suffering from advanced Parkinson's disease. The initial results were encouraging: The implants survived, and there was a modest reduction of symptoms. Unfortunately, however, some of the patients started to display a variety of uncontrollable writhing and chewing movements about a year after the surgery (see Greene et al., 1999).

The negative results of these two studies had two positive effects. First, it all but curtailed the premature clinical use of neurotransplantation to treat human patients with Parkinson's disease. Many had warned that it was inappropriate to use such an invasive treatment without a solid foundation of research identifying its mechanisms, its hazards, and the best ways of maximizing its effectiveness

(see Dunnett, Björkland, & Lindvall, 2001). Second, it stimulated researchers to take a more careful and systematic look at the effects of various kinds of neurotransplantation in animal models to answer fundamental questions.

In Chapter 4, you were introduced to Roberto Garcia d'Orta—the Lizard. D'Orta, who suffered from Parkinson's disease, initially responded to L-dopa therapy; but, after 3 years of therapy, his condition worsened. D'Orta was in a desperate state when he heard about *adrenal medulla autotransplantation* (transplanting a patient's own adrenal medulla cells into their striatum, usually for the treatment of Parkinson's disease). Adrenal medulla cells release small amounts of dopamine, and there were some early indications that adrenal medulla autotransplantation might alleviate the symptoms of Parkinson's disease.

D'Orta demanded adrenal medulla autotransplantation from his doctor. When his doctor refused, on the grounds that the effectiveness and safety of the treatment were unproven, d'Orta found another doctor—a neurosurgeon who was not so responsible (Klawans, 1990).

The Case of Roberto Garcia d'Orta: The Lizard Gets an Autotransplant

Roberto flew to Juarez for surgery. As long as Roberto would pay, the neurosurgeon would perform the autotransplantation.

Were there any risks? The neurosurgeon seemed insulted by the question. If Señor d'Orta didn't trust him, he could go elsewhere.

Roberto flew home 2 weeks after the surgery even though his condition had not improved. He was told to be patient and to wait for the cells to grow.

A few weeks later, Roberto died of a stroke. Was the stroke a complication of his surgery? It was more than a mere possibility.

Another approach to neurotransplantation research was developed in the late 1990s. Instead of implanting developing cells, researchers implanted nonneural cells to block neural degeneration or to stimulate and guide neural regeneration. For example, Xu and colleagues (1999) induced cerebral ischemia in rats by limiting blood flow to the brain. This had two major effects on rats in the control group: It produced damage to the hippocampus, a structure particularly susceptible to ischemic damage, and it produced deficits in the rats' spatial memory. The hippocampi of rats in the experimental group were treated with viruses genetically engineered to release *apoptosis inhibitor protein*. Amazingly, the apoptosis inhibitor protein reduced both the loss of hippocampal neurons and the deficits in Morris water maze performance. Many other neurotrophic factors have been shown to reduce degeneration—*brain-derived neurotrophic factor* (BDNF) and *glial-cell-line-derived factor* (GDNF) have been among the most widely studied (see Wang et al., 2011; Zhu et al., 2011).

A slightly different technique of this type involves implanting *Schwann cell sheathes*, which promote regeneration and guide axon growth. For example, Cheng, Cao, and Olson (1996) transected the spinal cords of rats, thus rendering them *paraplegic* (paralyzed in the posterior portion of their bodies). The researchers then transplanted sections of myelinated peripheral nerve across the transection. As a result, spinal cord neurons regenerated through the implanted Schwann cell myelin sheaths, and the regeneration allowed the rats to regain use of their hindquarters.

Modern Research on Neurotransplantation

LO 10.20 Discuss the methods and findings of modern research on neurotransplantation.

It was feared that the adverse consequences of prematurely rushing neurotransplantation to clinical practice for the treatment of Parkinson's disease might stifle further neurotransplantation research. It hasn't. In fact, since the failure of neurotransplantation to benefit human Parkinson's disease sufferers, neurotransplantation has become one of the most active areas of clinical neuroscientific research. The need for a method of repairing brain damage is simply too great for research on neurotransplantation to be deterred.

So, how active is the study of neurotransplantation? Hundreds of studies of neurotransplantation have been conducted since the turn of the century, involving animal models of virtually all common neurological disorders. And several of the protocols that have been developed in these studies have entered preliminary tests on human patients (see De Feo et al., 2012; Lemmens & Steinberg, 2013; Steinbeck & Studer, 2015). Given all this research activity, it is not surprising that there have been some major advances. Here are some of them.

At the turn of the century, stem cells were successfully isolated from the human embryo and maintained in tissue culture (see Hochedlinger, 2010). Because stem cells (see Chapter 9) divide indefinitely and are *pluripotent* (capable of developing into many, but not all, classes of adult cells), it seemed that stem-cell cultures would provide lasting sources of cells for transplantation, thus circumventing the ethical and technical difficulties in transplanting tissue obtained from human embryos. However, this did not prove to be the case: In practice, stem-cell cultures deteriorate or generate tumors as chromosomal abnormalities gradually accumulate during repeated cell division (see Kumar et al., 2017; Pires et al., 2017). That is why the report in 2007 that human skin cells could be induced to return to stem cells (i.e., to return to a pluripotent state) generated great excitement (Takahashi et al., 2007). Induced stem cells have been the focus of a massive research effort; however, the genetic manipulation required to create them (i.e., the insertion of four genes) makes it difficult to predict how they might behave in a human patient (see Pires et al., 2017). Still, there is a growing list of successful transplantation technologies in nonhuman animals (e.g., Falkner et al., 2016; Wuttke et al., 2018).

It is often assumed that the therapeutic effect of neurotransplantation results from the replacement of dead or dying neurons with healthy ones (see Gaillard & Jaber, 2011; Olson et al., 2012). However, in many cases this does not appear to be the mechanism (see Bonnamain, Neveu, & Naveilhan, 2012; Olson et al., 2012). In animal models, implants have been shown to have neuroprotective properties, and to stimulate regeneration and remyelination of injured axons (see Assinck et al., 2017). Any of these effects could be therapeutic.

Promoting Recovery from CNS Damage by Rehabilitative Training

LO 10.21 Discuss methods of promoting recovery from CNS damage through rehabilitative treatment.

Several demonstrations of the important role of experience in the organization of the developing and adult brain kindled a renewed interest in the use of rehabilitative training to promote recovery from CNS damage. The following innovative rehabilitative training programs were derived from such findings. Perhaps neurotransplants would be more effective if accompanied by the appropriate training?

TREATING STROKES. Small strokes produce a core of brain damage, which is often followed by a gradually expanding loss of neural function in the surrounding penumbra. Nudo and colleagues (1996) produced small *ischemic lesions* in the hand area of the motor cortex of monkeys. Then, 5 days later, a program of hand training and practice was initiated. During the ensuing 3 or 4 weeks, the monkeys plucked hundreds of tiny food pellets from food wells of different sizes. This practice substantially reduced the expansion of cortical damage into the surrounding penumbra. The monkeys that received the rehabilitative training also showed greater recovery in the use of their affected hand. Such rehabilitative training in humans has yielded comparable effects (see Sampaio-Baptista, Sanders, & Johansen-Berg, 2018).

One of the principles that has emerged from the study of neurodevelopment is that neurons seem to be in a competitive situation: They compete with other neurons for synaptic sites and neurotrophins, and the losers die. Weiller and Rijntjes (1999) designed a rehabilitative program based on this principle. Their procedure, called *constraint-induced therapy* (see Kwakkel et al., 2015; Jones, 2017), was to tie down the functioning arm for 2 weeks while the affected arm received intensive training. Performance with the affected arm improved markedly over the 2 weeks, and there was an increase in the area of motor cortex controlling that arm.

TREATING SPINAL INJURY. In one approach to treating spinal injuries (see Wolpaw & Tennissen, 2001), patients who were incapable of walking were supported by a harness over a moving treadmill. With most of their weight supported and the treadmill providing feedback, the patients gradually learned to make walking movements. Then, as they improved, the amount of support was gradually reduced. In one study using this technique, more than 90 percent of the trained patients eventually became independent walkers compared with only 50 percent of those receiving conventional physiotherapy. The effectiveness of this treatment has been confirmed and extended in human patients (e.g., Herman et al., 2002) and in nonhuman subjects (Frigon & Rossignol, 2008).

BENEFITS OF COGNITIVE AND PHYSICAL EXERCISE. Individuals who are cognitively and physically active are less likely to contract neurological disorders; and if they do, their symptoms tend to be less severe and their recovery better (see Stranahan & Mattson, 2012; Voss et al., 2013). However, in such correlational studies, there are always problems of causal interpretation: Do more active individuals tend to have better neurological outcomes because they are more active, or do they tend to be more active because they are more healthy? Because of these problems of causal interpretation, research in this area has relied heavily on controlled experiments using animal models (see Fryer et al., 2011; Gitler, 2011; Liu et al., 2019).

Journal Prompt 10.5

Design a carefully controlled longitudinal study to explore whether physical activity has protective effects on the onset of Alzheimer's dementia.

One experimental approach to studying the benefits of cognitive and physical activity has been to assess the neurological benefits of housing animals in enriched environments. **Enriched environments** are those designed to promote cognitive and physical activity—they typically involve group housing, toys, activity wheels, and changing stimulation (see Figure 10.21). The health-promoting effects of enriched environments have already been demonstrated in animal models of epilepsy, Huntington's disease, Alzheimer's disease, Parkinson's disease, Down syndrome, brain tumors, and various forms of stroke and traumatic brain injury (see Garofalo et al., 2015; Hannan, 2014; Mering & Jolkkonen, 2015). Although the mechanisms underlying the neurological benefits of enriched environments are

Figure 10.21 A rodent in an enriched laboratory environment.

Carolyn A. McKeone/Science Source

unclear, there are many possibilities: Enriched environments have been shown to increase dendritic branching, the size and number of dendritic spines, the size of synapses, the rate of adult neurogenesis, and the levels of various neurotrophic factors.

Physical exercise in the form of daily wheel running has also been shown to have a variety of beneficial effects on the rodent brain (see Cotman, Berchtold, & Christie, 2007): increased adult neurogenesis in the hippocampus, reduced age-related declines in the number of neurons in the hippocampus, and improved performance on tests of memory and navigation (two abilities linked to the hippocampus). Also, Adlard and colleagues (2005) found that wheel running reduced the development of amyloid plaques in mice genetically predisposed to develop a model of Alzheimer's disease.

TREATING PHANTOM LIMBS. Most amputees continue to experience the limbs that have been amputated—a condition referred to as **phantom limb**. Even some individuals (20 percent) born with a missing limb report experiencing a phantom limb (see Melzack et al., 1997).

The most striking feature of phantom limbs is their reality. Their existence is so compelling that a patient may try to jump out of bed onto a nonexistent leg or to lift a cup with a nonexistent hand. In most cases, the amputated limb behaves like a normal limb; for example, as an amputee walks, a phantom arm seems to swing back and forth in perfect coordination with the intact arm. However, sometimes an amputee feels that the amputated limb is stuck in a peculiar position.

About 50 percent of amputees experience severe chronic pain in their phantom limbs. A typical complaint is that an amputated hand is clenched so tightly that the fingernails are digging into the palm of the hand. Phantom limb pain can occasionally be treated by having the amputee concentrate on opening the amputated hand, but often surgical treatments are attempted. Based on the premise that phantom limb pain results from irritation at the stump, surgical efforts to control it have often involved cutting off the stump or various parts of the neural pathway between the stump and the cortex. Unfortunately, these treatments haven't worked (see Melzack, 1992).

Carlos and Philip experienced phantom limbs. Their neuropsychologist was the esteemed V. S. Ramachandran.

Cases of Carlos and Philip: Phantom Limbs and Ramachandran

Dr. Ramachandran read about the study of Pons and colleagues (1991), which you have already encountered in this chapter. In this study, severing the sensory neurons in the arms of monkeys led to a reorganization of somatosensory cortex: The area of the somatosensory cortex that originally received input from the damaged arm now received input from areas of the body normally mapped onto adjacent areas of somatosensory cortex. Ramachandran was struck by a sudden insight: Perhaps phantom limbs were not in the stump at all, but in the brain; perhaps the perception of a phantom arm originated from parts of the body that now innervated the original arm area of the somatosensory cortex (see Ramachandran & Blakeslee, 1998).

Excited by his hypothesis, Dr. Ramachandran asked one of his patients, Carlos, if he would participate in a simple test. He touched various parts of Carlos's body and asked Carlos what he felt. Remarkably, when he touched the side of Carlos's face on the same side as his amputated arm, Carlos felt sensations from various parts of his phantom hand as well as his face. A second map of his hand was found on his shoulder (see Figure 10.22).

Philip, another patient of Dr. Ramachandran, suffered from severe chronic pain in his phantom arm. For a decade, Philip's phantom arm had been frozen in an awkward position (Ramachandran & Rogers-Ramachandran, 2000), and Philip suffered great pain in his elbow.

Could Philip's pain be relieved by teaching him to move his phantom arm? Knowing how important feedback is in movement (see Chapter 8), Dr. Ramachandran constructed a special feedback apparatus for Philip. This was a box divided in two by a vertical mirror. Philip was instructed to put his good right hand into the box through a hole in the front and view it through a hole in the top. When he looked at his hand, he could see it and its mirror image. He was instructed to put his phantom limb in the box and try to position it, as best he could, so that it corresponded to the mirror image of his good hand. Then, he was instructed to make synchronous, bilaterally symmetrical movements of his arms—his actual right arm and

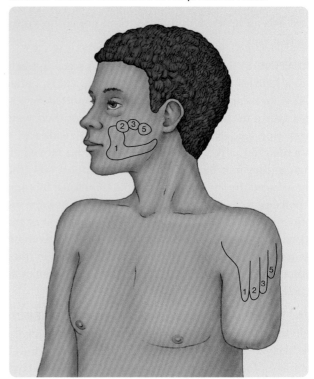

Figure 10.22 The places on Carlos's body where touches elicited sensations in his phantom hand.

Based on Ramachandran, V. S., & Blakeslee, S. (1998). *Phantoms in the Brain*. New York, NY: Morrow.

his phantom left arm—while viewing his good arm and its mirror image. Ramachandran sent Philip home with the box and instructed him to use it frequently. Three weeks later, Philip's pain had subsided.

The Ironic Case of Professor P.: Recovery

If you remember the chapter-opening case study, I (JP) am sure you will appreciate why this chapter has special meaning for me. Writing it played a part in my recovery.

When I was released from the hospital, I had many problems related to my neurosurgery. I knew that I would have to live with hearing and balance problems because I no longer had a right auditory-vestibular nerve. My other problems concerned me more. The right side of my face sagged, and making facial expressions was difficult. My right eye was often painful—likely because of inadequate tearing. I had difficulty talking, and I experienced debilitating attacks of fatigue. Unfortunately, neither my neurosurgeon nor GP seemed to know how to deal with these problems, and I was pretty much left to fend for myself.

I used information that I had learned from writing this chapter. Little about recovery from brain damage had been proven, but the results of experiments on animal models were suggestive. I like to think that the program I devised contributed to my current good health, but, of course, there is no way of knowing for sure.

I based my program of recovery on evidence that cognitive and physical exercise promotes recovery and other forms of neuroplasticity. My job constituted the cognitive part of my recovery program. Because I was influenced by recent evidence that the beneficial effects of exercise are greatest soon after the brain trauma, I returned to work 2 weeks after leaving the hospital.

Once back at work, I got more mental, oral, and facial exercise than I had anticipated. A few conversations were enough to make my throat and face ache and to totally exhaust me, at which point I would retreat to my office until I was fit to emerge once again for more "treatment."

Being a university professor is not physically demanding—perhaps you've noticed. I needed some physical exercise, but my balance problems limited my options. I turned to African hand drumming. I love the rhythms, and I found that learning and playing them could be a serious cognitive and physical challenge—particularly for somebody as enthusiastic and inept as I. So I began to practice, take lessons, and drum with my new friends at every opportunity. Gradually, I worked, talked, smiled, and drummed myself to recovery. Today, my face is reasonably symmetrical, my speech is good, I am fit, and my balance has improved.

Themes Revisited

This is the second chapter of the text to focus on neuroplasticity. It covered the neuroplastic changes associated with neurological disease and brain damage and the efforts to maximize various neuroplastic changes to promote recovery.

The clinical implications theme predominated. For example, there were many case studies of clinical syndromes: the ironic case of Professor P.; Junior Seau, the all-star American football player; the cases of complex partial epilepsy; the cases of MPTP poisoning; and Carlos and Philip, the amputees with phantom limbs.

The chapter stressed creative thinking in several places. Attention was drawn to thinking about the need to identify the primary symptom of Alzheimer's disease, about the applicability of animal models to humans, and about the correlation between exercise and recovery of function after nervous system damage. Particularly interesting was the creative approach that Dr. Ramachandran took in treating Philip, who suffered from phantom limb pain.

The evolutionary perspective theme was also highlighted at several points. You were introduced to the concept of animal models, which is based on the comparative approach, and you learned that most of the research on neural regeneration and reorganization following brain damage has been done with animal models. You also learned that research into the mechanisms of neural regeneration has been stimulated by the fact that this process occurs accurately in some species.

Finally, the emerging themes of consciousness and thinking about epigenetics were present in the discussion of epilepsy. In terms of consciousness, different forms of epilepsy induce different disruptions and alterations of consciousness (e.g., absence seizures involve momentary cessations of consciousness, and epileptic auras involve momentary alterations of consciousness). In terms of epigenetics, certain forms of epilepsy seem to arise as the result of epigenetic mechanisms.

Key Terms

Causes of Brain Damage

Tumor (neoplasm), p. 260
Meningiomas, p. 260
Encapsulated tumors, p. 260
Benign tumors, p. 260
Infiltrating tumors, p. 260
Malignant tumors, p. 260
Gliomas, p. 260
Metastatic tumors, p. 261
Strokes, p. 261
Penumbra, p. 261
Cerebral hemorrhage, p. 261
Aneurysm, p. 261
Congenital, p. 262
Cerebral ischemia, p. 262
Thrombosis, p. 262
Embolism, p. 262

Arteriosclerosis, p. 262
Glutamate, p. 262
NMDA (N-methyl-D-aspartate) receptors, p. 262
Traumatic brain injury (TBI), p. 262
Closed-head TBIs, p. 263
Contusions, p. 263
Subdural hematoma, p. 263
Contrecoup injuries, p. 263
Mild TBI (mTBI), p. 263
Chronic traumatic encephalopathy, (CTE), p. 263
Dementia, p. 263
Encephalitis, p. 264
Meningitis, p. 264
General paresis, p. 264
Toxic psychosis, p. 265

Tardive dyskinesia (TD), p. 265
Down syndrome, p. 265
Apoptosis, p. 265

Neurological Diseases

Epilepsy, p. 266
Convulsions, p. 266
Epileptic auras, p. 266
Focal seizure, p. 267
Simple seizures, p. 267
Complex seizures, p. 267
Generalized seizures, p. 267
Tonic-clonic seizure, p. 267
Hypoxia, p. 267
Absence seizure, p. 268
3-per-second spike-and-wave discharge, p. 268

Chapter 11
Learning, Memory, and Amnesia

How Your Brain Stores Information

Africa Studio/Shutterstock

 ## Chapter Overview and Learning Objectives

Amnesic Effects of
Bilateral Medial Temporal
Lobectomy

LO 11.1 Describe five specific memory tests that were used to assess H.M.'s anterograde amnesia.

LO 11.2 Describe three major scientific contributions of H.M.'s case.

LO 11.3 Discuss what research on medial temporal lobe amnesias has taught us about learning and memory.

LO 11.4 Describe the difference between semantic and episodic memories.

LO 11.5 Discuss two pieces of evidence that support the notion that selective hippocampal dysfunction can cause medial temporal lobe amnesia.

Amnesias of Korsakoff's Syndrome and Alzheimer's Disease	**LO 11.6**	Describe the etiology and symptoms of the amnesia of Korsakoff's syndrome.
	LO 11.7	Describe the symptoms of Alzheimer's disease that have been associated with amnesia.
Amnesia after Traumatic Brain Injury: Evidence for Consolidation	**LO 11.8**	Summarize the effects of a closed-head traumatic brain injury (TBI) on memory.
	LO 11.9	Describe the classic view of memory consolidation and some of the evidence it rests upon. Contrast that with current thinking about memory consolidation.
Evolving Perspective of the Role of the Hippocampus in Memory	**LO 11.10**	Describe the delayed nonmatching-to-sample tests for monkeys and rats.
	LO 11.11	Describe the neuroanatomical basis for the object-recognition deficits that result from bilateral medial temporal lobectomy.
Neurons of the Medial Temporal Lobes and Memory	**LO 11.12**	Describe hippocampal place cells and entorhinal grid cells and the relationship between these two cell types.
	LO 11.13	Explain what a concept cell is and describe the key properties of concept cells with reference to the experimental evidence.
	LO 11.14	Explain what an engram cell is and describe how these cells were identified using optogenetics.
Where Are Memories Stored?	**LO 11.15**	For each of the following brain structures, describe the type(s) of memory they have been implicated in: inferotemporal cortex, amygdala, prefrontal cortex, cerebellum, and striatum.
Cellular Mechanisms of Learning and Memory	**LO 11.16**	Describe the phenomenon known as long-term potentiation (LTP) and provide evidence for its role in learning and memory.
	LO 11.17	Describe the mechanisms underlying the induction of LTP.
	LO 11.18	Describe four findings that have emerged from the study of the maintenance and expression phases of LTP.
	LO 11.19	Define long-term depression (LTD) and metaplasticity.
	LO 11.20	Describe two sorts of neuroplastic changes that occur outside the synapse that may play a role in learning and memory.
Conclusion: Biopsychology of Memory and You	**LO 11.21**	Define infantile amnesia and describe two experiments that investigated whether infantile amnesia extends to implicit memories.
	LO 11.22	Discuss the findings on the efficacy of smart drugs.
	LO 11.23	Explain what the case of R.M. tells us about the relationship between posttraumatic amnesia and episodic memory.

Learning and memory are two ways of thinking about the same thing: Both are neuroplastic processes; they deal with the ability of the brain to change its functioning in response to experience. **Learning** deals with how experience changes the brain, and **memory** deals with how these changes are stored and subsequently reactivated. Without the ability to learn and remember, we would experience every moment as if waking from a lifelong sleep—each person would be a stranger, each act a new challenge, and each word incomprehensible.

This chapter focuses on the roles played by various brain structures in the processes of learning and memory. Our knowledge of these roles has come to a great extent from the study of neuropsychological patients with brain-damage-produced **amnesia** (any pathological loss of memory) and from research on animal models of the same memory problems.

Amnesic Effects of Bilateral Medial Temporal Lobectomy

Ironically, the person who contributed more than any other to our understanding of the neuropsychology of memory was not a neuropsychologist. In fact, although he collaborated on dozens of studies of memory, he had no formal research training and not a single degree to his name. He was H.M., a man who in 1953, at the age of 27, had the medial portions of his temporal lobes removed for the treatment of a severe case of epilepsy. Just as the Rosetta Stone provided archaeologists with important clues to the meaning of Egyptian hieroglyphics, H.M.'s memory deficits were instrumental in the achievement of our current understanding of the neural bases of memory.

The Case of H.M., the Man Who Changed the Study of Memory

During the 11 years preceding his surgery, H.M. suffered an average of one generalized seizure each week and many focal seizures each day, despite massive doses of anticonvulsant medication. Electroencephalography suggested that H.M.'s seizures arose from foci in the medial portions of both his left and right temporal lobes. Because the removal of one medial temporal lobe had proved to be an effective treatment for patients with a unilateral temporal lobe focus, the decision was made to perform a **bilateral medial temporal lobectomy**—the removal of the medial portions of both temporal lobes, including most of the **hippocampus**, **amygdala**, and adjacent cortex (see Figure 11.1). (A **lobectomy** is an operation in which a lobe, or a

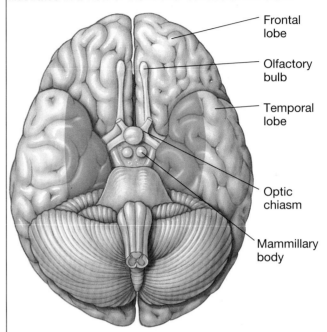

Figure 11.1 Medial temporal lobectomy. The portions of the medial temporal lobes removed from H.M.'s brain are illustrated in a view of the inferior surface of the brain.

Frontal lobe
Olfactory bulb
Temporal lobe
Optic chiasm
Mammillary body

Tissue typically excised in medial temporal lobectomy

major part of one, is removed from the brain; a **lobotomy** is an operation in which a lobe, or a major part of one, is separated from the rest of the brain by a large cut but is not removed.)

In several respects, H.M.'s bilateral medial temporal lobectomy was an unqualified success. His generalized seizures were all but eliminated, and the incidence of focal seizures was reduced to one or two per day, even though the level of his anticonvulsant medication was substantially reduced. Furthermore, H.M. entered surgery a reasonably well-adjusted individual with normal perceptual and motor abilities and normal intelligence, and he left it in nearly the same condition. Be that as it may, H.M. was the last patient to receive a bilateral medial temporal lobectomy—because of its devastating amnesic effects.

In assessing the amnesic effects of brain surgery, it is usual to administer tests of the patient's ability to remember things learned before the surgery and tests of the patient's ability to remember things learned after the surgery. Deficits on the former tests lead to a diagnosis of **retrograde** (backward-acting) **amnesia**; those on the latter tests lead to a diagnosis of **anterograde** (forward-acting) **amnesia**. If a patient is found to have anterograde amnesia, the next step is usually to determine whether the difficulty in storing new memories influences **short-term memory** (storage of new information for brief periods of time while a person attends to it), **long-term memory** (storage of new information once the person stops attending to it), or both.

Like his intellectual abilities, H.M.'s memory for events predating his surgery remained largely intact. Although he had a mild retrograde amnesia for those events that occurred in the 2 years before his surgery, his memory for more remote events (e.g., for the events of his childhood) was reasonably normal.

H.M.'s short-term anterograde memory also remained normal: For example, his **digit span**, the classic test of short-term memory (see Chapter 5), was six digits (see Wickelgren, 1968)—this means that if a list of six digits was read to him, he could usually repeat the list correctly, but he would have difficulty repeating longer lists.

In contrast, H.M. had an almost total inability to form new long-term memories: Once he stopped thinking about a new experience, it was lost forever. In effect, H.M. became suspended in time on that day in 1953 when he regained his health but lost his future. His family moved shortly after his surgery, but he was never able to remember his new address or where commonly used items were kept in his new residence. He never learned to recognize people (e.g., doctors and nurses) who he did not meet until after his surgery, and he read the same magazines over and over without finding them familiar. If you met H.M. at a party he could chat quite normally until he was distracted (e.g., by the phone); then he would not remember you, the conversation, or where he was. It was as if he was continually regaining consciousness.

Formal Assessment of H.M.'s Anterograde Amnesia: Discovery of Unconscious Memories

LO 11.1   **Describe five specific memory tests that were used to assess H.M.'s anterograde amnesia.**

In order to characterize H.M.'s anterograde memory problems, researchers began by measuring his performance on objective tests of various kinds of memory. This subsection describes five tests that were used to assess H.M.'s long-term memory. The results of the first two tests documented H.M.'s severe deficits in long-term memory, whereas the results of the last three indicated that H.M.'s brain was capable of storing long-term memories but that H.M. had no conscious awareness of those memories. This finding changed the way biopsychologists think about the brain and memory.

DIGIT-SPAN + 1 TEST.   H.M.'s inability to form certain long-term memories was objectively illustrated by his performance on the *digit-span + 1 test*, a classic test of verbal long-term memory. H.M. was asked to repeat 5 digits that were read to him at 1-second intervals. He repeated the sequence correctly. On the next trial, the same 5 digits were presented in the same sequence with 1 new digit added on the end. This same 6-digit sequence was presented a few times until he got it right, and then another digit was added to the end of it, and so on. After 25 trials, H.M. had not managed to repeat the 8-digit-sequence. Most people can correctly repeat about 15 digits after 25 trials of the digit-span + 1 test (see Drachman & Arbit, 1966).

BLOCK-TAPPING TEST.   H.M. had **global amnesia**—amnesia for information presented in all sensory modalities. Milner (1971) demonstrated that H.M.'s amnesia was not restricted to verbal material by assessing his performance on the *block-tapping test*. An array of 9 blocks was spread out on a board in front of H.M., and he was asked to watch the neuropsychologist touch a sequence of them and then to repeat the same sequence of touches. Whereas a typical person has a block-tapping span of 6, H.M. could not learn to correctly touch a sequence of 6 blocks—even when the same sequence was repeated 12 times.

MIRROR-DRAWING TEST.   The first indication that H.M.'s anterograde amnesia did not involve all long-term memories came from the results of a *mirror-drawing test* (see Milner, 1965). H.M.'s task was to draw a line within the boundaries of a star-shaped target by watching his hand in a mirror. H.M. was asked to trace the star 10 times on each of 3 consecutive days, and the number of times he went outside the boundaries on each trial was recorded. As Figure 11.2 shows,

---

Figure 11.2 The learning and retention of the mirror-drawing task by H.M. Despite his good retention of the task, H.M. had no conscious recollection of having performed it before.

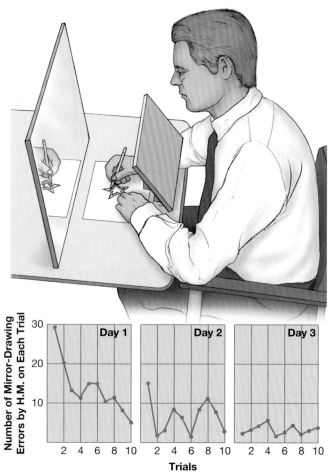

Based on Milner, B. 1965. Memory disturbances after bilateral hippocampal lesions. In P. Milner & S. Glickman (Eds.), *Cognitive Processes and the Brain* (pp. 104–105). Princeton, NJ: D. Van Nostrand.

H.M.'s performance improved over the 3 days, which indicates retention of the task. However, despite his improved performance, H.M. could not recall ever having completed the task before.

INCOMPLETE-PICTURES TEST.   The discovery that H.M. was capable of forming long-term memories for mirror drawing suggested that sensorimotor tasks were the one exception to his inability to form long-term memories. However, this view was challenged by the demonstration that H.M. could also form new long-term memories for the **incomplete-pictures test**—a nonsensorimotor test of memory that employs five sets of fragmented drawings. Each set contains drawings of the same 20 objects, but the sets differ in their degree of completeness: Set 1 contains the most fragmented drawings, and set 5 contains the complete drawings. The subject is asked to identify the 20 objects from the most fragmented set (set 1); then, those objects that go unrecognized are presented in their set 2 versions, and so on, until all 20 items have been identified. Figure 11.3 illustrates the performance of H.M. on this test and his improved performance 1 hour later (Milner et al., 1968). Despite his improved performance, H.M. could not recall previously performing the task.

PAVLOVIAN CONDITIONING.   H.M. learned an eye-blink Pavlovian conditioning task, albeit at a slower rate (Woodruff-Pak, 1993). A tone was sounded just before a puff of air was administered to his eye; these trials were repeated until the tone alone elicited an eye blink. Two years later, H.M. retained this conditioned response almost perfectly, although he had no conscious recollection of the training.

Three Major Scientific Contributions of H.M.'s Case

LO 11.2   Describe three major scientific contributions of H.M.'s case.

H.M.'s case is a story of personal tragedy, but his contributions to the study of the neural basis of memory were immense. The following three contributions proved to be particularly influential.

First, by showing that the medial temporal lobes play an especially important role in memory, H.M.'s case challenged the then prevalent view that memory functions are diffusely and equivalently distributed throughout the brain. In so doing, H.M.'s case renewed efforts to relate individual brain structures to specific *mnemonic* (memory-related) processes; in particular, H.M.'s case spawned a massive research effort aimed at clarifying the mnemonic functions of the hippocampus and other medial temporal lobe structures.

Figure 11.3 Two items from the incomplete-pictures test. H.M.'s memory for the 20 items on the test was indicated by his ability to recognize the more fragmented versions of them when he was retested. Nevertheless, he had no conscious awareness of having previously seen the items.

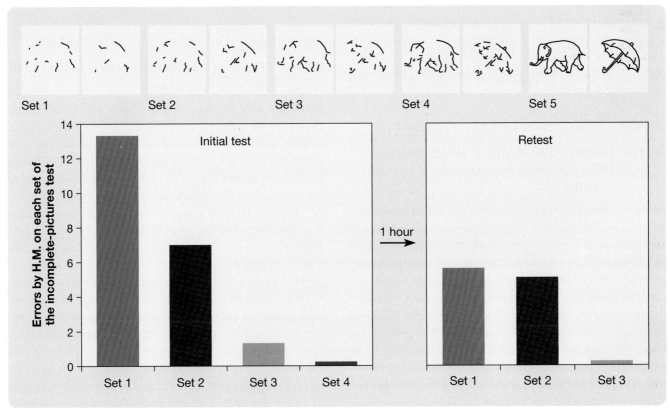

Second, the discovery that bilateral medial temporal lobectomy abolished H.M.'s ability to form certain kinds of long-term memories without disrupting his performance on tests of short-term memory or his **remote memory** (memory for experiences in the distant past) supported the theory that there are different modes of storage for short-term, long-term, and remote memory. H.M.'s specific problem appeared to be a difficulty in **memory consolidation** (the translation of short-term memories into long-term memories).

Third, H.M.'s case was the first to reveal that an amnesic patient might claim no recollection of a previous experience while demonstrating memory for it by improved performance (e.g., on the mirror-drawing and incomplete-pictures tests). This discovery led to the creation of two distinct categories of long-term memories: Conscious long-term memories became known as **explicit memories**, and long-term memories demonstrated by improved test performance without conscious awareness became known as **implicit memories**. As you will soon learn, this distinction is of general relevance: Many people with amnesia lose their ability to form explicit memories while maintaining their ability to form implicit memories.

Journal Prompt 11.1

Suppose one of your patients shows preserved ability to play the violin but has no recollection of what they had for lunch. How would you use implicit memory to help your patient learn new things, and will they be aware that they have gained this new knowledge?

H.M. died in 2008. Before his death, H.M. donated his brain to science, and his brain was sliced a few years after his death. Images of those brain slices are now archived online, where anyone can view them: H.M. continues to contribute to science, long after his death. H.M.'s real name was Henry Molaison.

Medial Temporal Lobe Amnesia

LO 11.3   Discuss what research on medial temporal lobe amnesias has taught us about learning and memory.

Neuropsychological patients with a profile of mnemonic deficits similar to those of H.M., with preserved intellectual functioning, and with evidence of medial temporal lobe damage are said to suffer from **medial temporal lobe amnesia**.

Research on medial temporal lobe amnesia has shown that H.M.'s difficulty in forming explicit long-term memories while retaining the ability to form implicit long-term memories of the same experiences is not unique to him. This problem has proved to be a symptom of medial temporal lobe amnesia, as well as many other amnesic disorders. As a result, the assessment of implicit long-term memories now plays an important role in the study of human memory (see Reber, 2013).

Tests that assess implicit memory are called **repetition priming tests**. The incomplete-pictures test and mirror-drawing task are two examples, but repetition priming tests that involve memory for words are more common. First, the participants are asked to examine a list of words; they are not asked to learn or remember anything. Later, they are shown a series of fragments (e.g., _ O B _ _ E R) of words from the original list and are simply asked to complete them. Controls who have seen the original words perform well. Surprisingly, participants with amnesia often perform equally well, even though they have no explicit memory of seeing the original list. (By the way, the correct answer to the repetition priming example is "lobster.")

The discovery that there are two memory systems—explicit and implicit—raises an important question: Why do we have two parallel memory systems, one conscious (explicit) and one unconscious (implicit)? Presumably, the implicit system was the first to evolve because it is more simple (it does not involve consciousness), so the question is actually this: What is the advantage in having a second, conscious system?

Two experiments, one with amnesic patients (Reber, Knowlton, & Squire, 1996) and one with amnesic monkeys with medial temporal lobe lesions (Buckley & Gaffan, 1998), suggest that the answer is "flexibility." In both experiments, the amnesic subjects learned an implicit learning task as well as control subjects did; however, if they were asked to use their implicit knowledge in a different way or in a different context, they failed miserably. Presumably, the evolution of explicit memory systems provided for the flexible use of information.

Semantic and Episodic Memories

LO 11.4   Describe the difference between semantic and episodic memories.

H.M. was able to form very few new explicit memories. However, most people with medial temporal lobe amnesia display memory deficits that are less complete. The study of these amnesics has found that explicit memories fall into two categories and that many of these amnesics tend to have far greater difficulties with one category than the other.

Explicit long-term memories come in two varieties: semantic and episodic (see Squire et al., 2015). **Semantic memories** are explicit memories for general facts or information; **episodic memories** are explicit memories for specific moments (i.e., episodes) in one's life (see Rugg & Vilberg, 2013). People with medial temporal lobe amnesia have particular difficulty with episodic memories. In other words, they have difficulty remembering specific events from their lives, even though their memory for general information is often normal. Although they can't remember having breakfast with an old friend in the morning, or going to a new movie in the afternoon, they often remember a language they learned, world events, and the sorts of things learned at school.

Endel Tulving has been a major force in research on the semantic-episodic dichotomy (Tulving, 2002). Following is a description of Tulving's patient K.C. Episodic memory (also called *autobiographical memory*) has been likened to traveling back in time mentally and experiencing one's past.

The Case of K.C., the Man Who Can't Time Travel

K.C. had a motorcycle accident in 1981. He suffered diffuse brain damage, including damage to the medial temporal lobes. Despite severe amnesia, K.C.'s other cognitive abilities remain remarkably normal. His general intelligence and use of language are normal; he has no difficulty concentrating; he plays the organ, chess, and various card games; and his reasoning abilities are good. His knowledge of mathematics, history, science, geography, and other school subjects is good.

Similarly, K.C. has good retention of many of the facts of his early life. He knows his birth date, where he lived as a youth, where his parents' summer cottage was located, the names of schools he attended, the makes and colors of cars that he has owned.

Still, in the midst of these normal memories, K.C. has severe amnesia for personal experiences. He cannot recall a single personal event for more than a minute or two. This inability to recall any episodes (events) at which he was present covers his entire life. Despite these serious memory problems, K.C. has no difficulty having a conversation, and his memory problems are far less obvious to others than one would expect. Basically, he does quite well using only his semantic memory.

K.C. understands the concept of time but he cannot "time travel" into either the past or the future. He cannot imagine future events any better than he can recall his past: He can't imagine what he will be doing for the rest of the day, the week, or his life.

Vargha-Khadem and colleagues (1997) followed the maturation of three patients with medial temporal lobe amnesia who experienced bilateral medial temporal lobe damage early in life. Remarkably, although they could remember few of the experiences they had during their daily lives (episodic memory), they progressed through mainstream schools and acquired reasonable levels of language ability and factual knowledge (semantic memory). However, despite their academic success, their episodic memory did not improve (de Haan et al., 2006).

It is difficult to spot episodic memory deficits, even when the deficits are extreme. This occurs in part because neuropsychologists usually have no way of knowing the true events of a patient's life and in part because the patients

become very effective at providing semantic answers to episodic questions. The following paraphrased exchange illustrates why neuropsychologists have difficulty spotting episodic memory problems.

The Case of the Clever Neuropsychologist: Spotting Episodic Memory Deficits

Neuropsychologist: I understand that you were a teacher.
Patient: That's right, I taught history.
Neuropsychologist: You must have given some good lectures in your time. Can you recall one of them that stands out?
Patient: Sure. I have given thousands of lectures. I especially liked Greek history.
Neuropsychologist: Was there any particular lecture that stood out—perhaps because it was very good or because something funny happened?
Patient: Oh, yes. Many stand out. My students liked my lectures—at least some of them—and sometimes I was quite funny.
Neuropsychologist: But is there one—just one—that you remember? And can you tell me something about it?
Patient: Oh yes, no problem. I didn't understand what you wanted. I can remember giving lectures and all my students were there watching and smiling.
Neuropsychologist: But can you describe a lecture where something happened that never happened in any other lecture? Perhaps something funny or disturbing.
Patient: That's hard.
Neuropsychologist: Before I go, I have some news for you that I think you will like. I understand that you are a hockey fan and follow the Toronto Maple Leafs.
Patient: Jeez, you guys know everything.
Neuropsychologist: Last night was a great night for Toronto. They beat New York 6–0. Do you think that you can remember that score for me? I will ask you about it a bit later.
Patient: That's great news. I will have no problem remembering that.
[Neuropsychologist leaves the room and returns an hour later.]
Neuropsychologist: I asked you to remember something the last time we chatted. Do you remember it?
Patient: I don't think so. I seem to have forgotten. It must have been a long time ago.
Neuropsychologist: That's strange. Do you remember anything specific about our last meeting, or even when it was?
Patient: Yes, I think we chatted about my memory.
Neuropsychologist: I understand that you are a Toronto Maple Leafs fan. Are they a good team?
Patient: Yes, they are very good. I used to go to every game with my father when I was a kid. They had great players; they were fast skaters and worked very hard. Did you know that they beat the New York Rangers 6–0? Now that's good.

Effects of Global Cerebral Ischemia on the Hippocampus and Memory

LO 11.5 Discuss two pieces of evidence that support the notion that selective hippocampal dysfunction can cause medial temporal lobe amnesia.

Patients who have experienced **global cerebral ischemia**—that is, have experienced an interruption of blood supply to their entire brain—often suffer from medial temporal lobe amnesia. R.B. is one such individual (Zola-Morgan, Squire, & Amaral, 1986).

The Case of R.B., Product of a Bungled Operation

At the age of 52, R.B. underwent cardiac bypass surgery. The surgery was bungled, and, as a consequence, R.B. suffered brain damage. The pump that was circulating R.B.'s blood to his body while his heart was disconnected broke down, and it was several minutes before a replacement arrived from another part of the hospital. R.B. lived, but the resulting ischemic brain damage left him amnesic.

Although R.B.'s amnesia was not as severe as H.M.'s, it was comparable in many aspects. R.B. died in 1983, and a detailed postmortem examination of his brain was carried out with the permission of his family. Obvious brain damage was restricted largely to the **pyramidal cell layer** of just one part of the hippocampus—the **CA1 subfield** (see Figure 11.4).

R.B.'s case suggested that hippocampal damage by itself can produce medial temporal lobe amnesia. However, in such cases of cerebral ischemia, it is difficult to rule out the possibility of subtle dysfunction to other areas of the brain.

Arguably, the strongest evidence that selective hippocampal damage can cause medial temporal lobe amnesia comes from cases of transient global amnesia. **Transient global amnesia** is defined by its sudden onset in the absence of any obvious cause in otherwise normal adults. As in other cases of medial temporal lobe amnesia, there is severe anterograde amnesia and moderate retrograde amnesia for explicit episodic memories (see Arena & Rabinstein, 2015; Bartsch & Butler, 2013). However, in the case of transient global amnesia, the amnesia is transient, typically lasting only 4 to 6 hours. Imagine the distress of the otherwise-healthy people who suddenly develop the symptoms of medial temporal lobe amnesia.

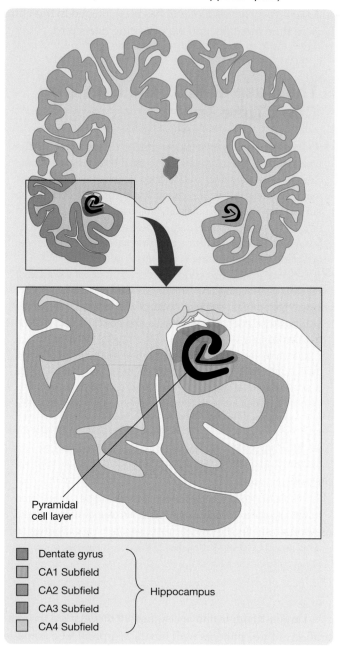

Figure 11.4 The major components of the hippocampus: CA1, CA2, CA3, and CA4 subfields and the dentate gyrus. R.B.'s brain damage appeared to be restricted largely to the pyramidal cell layer of the CA1 subfield. (*CA* stands for "cornu ammonis," another name for hippocampus.)

Pyramidal cell layer

- ▨ Dentate gyrus
- ▨ CA1 Subfield ⎫
- ▨ CA2 Subfield ⎬ Hippocampus
- ▨ CA3 Subfield ⎪
- ▨ CA4 Subfield ⎭

The sudden onset of transient global amnesias suggests they are caused by stroke; however, until recently no brain pathology could be linked to the disorder. But, in recent years, investigators have identified abnormalities in the CA1 subfield of the hippocampus (see Arena & Rabinstein, 2015; Bartsch & Butler, 2013). The time course of these abnormalities—they are not usually apparent for several hours after the beginning of the attack and have usually cleared up 10 days later—are suggestive of ischemia-induced damage (see Hunter, 2011).

Amnesias of Korsakoff's Syndrome and Alzheimer's Disease

The preceding module focused on amnesias associated with different sorts of brain dysfunction that both occur over a relatively brief period of time and are localized to specific brain areas. This module considers the amnesias associated with two syndromes that involve both a much slower progression of brain dysfunction and dysfunction that is more diffuse: Korsakoff's syndrome and Alzheimer's disease.

Amnesia of Korsakoff's Syndrome

LO 11.6   Describe the etiology and symptoms of the amnesia of Korsakoff's syndrome.

As you learned in Chapter 1, **Korsakoff's syndrome** is a disorder of memory that is most common in people who have consumed large amounts of alcohol; the disorder is largely attributable to the brain dysfunction associated with the thiamine deficiency that often accompanies heavy alcohol consumption (see Scalzo et al., 2015). In its advanced stages, it is characterized by a variety of sensory and motor problems, extreme confusion, personality changes, and a risk of death from liver, gastrointestinal, or heart disorders. Postmortem examination typically reveals lesions to the *medial diencephalon* (the medial-thalamus and the medial-hypothalamus) and diffuse damage to several other brain structures, most notably the neocortex, hippocampus, and cerebellum (see Fama, Pitel, & Sullivan, 2012; Kril & Harper, 2012; Savage, Hall, & Resende, 2012).

The amnesia of Korsakoff's syndrome is similar to medial temporal lobe amnesia in some respects. For example, during the early stages of the disorder, anterograde amnesia for explicit episodic memories is the most prominent symptom. However, as the disorder progresses, retrograde amnesia, which can eventually extend back into childhood, also develops. Deficits in implicit memory depend on the particular test used, but in general they are less severe than those in explicit memory (see Oudman et al., 2011; Van Tilborg et al., 2011).

Because the brain damage associated with Korsakoff's syndrome is diffuse, it has been difficult to identify which part of it is specifically responsible for the amnesia. Still, much attention has focused on one pair of medial diencephalic nuclei: the **mediodorsal nuclei** of the thalamus; this is because there is almost always damage to these nuclei in Korsakoff patients. However, it is unlikely that the memory deficits of Korsakoff patients are attributable to the damage of any single structure.

N.A. is a particularly well-known patient with **medial diencephalic amnesia** (amnesia, such as Korsakoff amnesia, associated with damage to the medial diencephalon). Although his memory deficits were conventional, their cause was not (Teuber, Milner, & Vaughan, 1968).

The Up-Your-Nose Case of N.A.

N.A. joined the U.S. Air Force after a year of college, serving as a radar technician until his accident. On the fateful day, N.A.'s roommate was playing with a fencing foil behind N.A.'s chair. N.A. turned unexpectedly and was stabbed up the right nostril. The foil punctured the *cribriform plate* (the thin bone around the base of the frontal lobes), taking an upward course into N.A.'s brain.

When tested a few weeks after his accident, N.A. was unable to recall any significant personal, national, or international events that had occurred in the 2 years preceding his accident. However, when retested 3 years later, his retrograde amnesia had decreased in duration, covering only those events that occurred in the 2 weeks before the accident.

N.A.'s recall of day-to-day events that occurred after the accident was extremely poor. On initial testing, he could not remember what he ate for breakfast, people whom he had recently met, or visits from his family. However, unpredictably he would sometimes recall specific experiences of no particular significance. Although his ability to remember new experiences has improved somewhat since he was first tested, he has not been able to function well enough to gain employment.

An MRI of N.A.'s brain was taken in the late 1980s (Squire et al., 1989). It revealed extensive medial diencephalic damage, including damage to the mediodorsal nuclei and mammillary bodies.

Journal Prompt 11.2
How have case studies played an important role in the study of memory?

Amnesia of Alzheimer's Disease

LO 11.7   Describe the symptoms of Alzheimer's disease that have been associated with amnesia.

Alzheimer's disease is another major cause of amnesia. The first sign of Alzheimer's disease is often a mild deterioration of memory. However, the disorder is progressive: Eventually, *dementia* develops and becomes so severe that the patient is incapable of even simple activities (e.g., eating, speaking, recognizing a spouse, or bladder control). Alzheimer's disease is terminal.

Efforts to understand the neural basis of Alzheimer's amnesia have focused on *predementia Alzheimer's patients* (Alzheimer's patients who have yet to develop dementia). The memory deficits of these patients are more general than those associated with medial temporal lobe damage, medial diencephalic damage, or Korsakoff's syndrome. In addition to major anterograde and retrograde deficits in tests of explicit memory, predementia Alzheimer's patients often display deficits in short-term memory and in some types of implicit memory: Their implicit memory for verbal and perceptual material is often deficient, whereas their implicit memory for sensorimotor learning is not (see Postle, Corkin, & Growdon, 1996).

The level of acetylcholine is greatly reduced in the brains of Alzheimer's patients. This reduction results from the degeneration of the **basal forebrain** (a midline area located just above the hypothalamus; see Figure 11.16), which is the brain's main source of acetylcholine. This finding, coupled with the finding that strokes in the basal forebrain area can cause amnesia, led to the view that acetylcholine depletion is the cause of Alzheimer's amnesia.

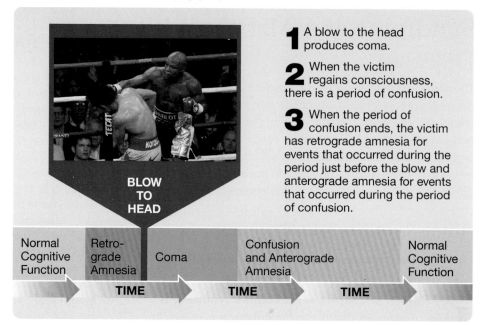

Figure 11.5 The retrograde amnesia and anterograde amnesia associated with a closed-head traumatic brain injury (TBI).

Craig Durling/ZUMA Wire/Alamy Live News

Although acetylcholine depletion resulting from damage to the basal forebrain may contribute to Alzheimer's amnesia, it is clearly not the only factor. The brain damage associated with Alzheimer's disease is extremely diffuse, involving many areas including the medial temporal lobes and the prefrontal cortex, which play major roles in memory (see Braskie & Thompson, 2013).

Amnesia after Traumatic Brain Injury: Evidence for Consolidation

Closed-head traumatic brain injuries (TBIs) (brain injuries produced by blows to the head that do not penetrate the skull; see Chapter 10) are the most common cause of amnesia. The amnesia following a closed-head TBI is called **posttraumatic amnesia**.

Posttraumatic Amnesia

LO 11.8 **Summarize the effects of a closed-head traumatic brain injury (TBI) on memory.**

The *coma* (pathological state of unconsciousness) following a severe blow to the head usually lasts a few seconds or minutes, but in severe cases it can last weeks. Once the patient regains consciousness, he or she experiences a period of confusion. Victims of closed-head TBIs are typically not tested by a neuropsychologist until after the period of confusion—if they are

tested at all. Testing usually reveals that the patient has permanent retrograde amnesia for the events that led up to the blow and permanent anterograde amnesia for many of the events that occurred during the subsequent period of confusion.

The anterograde memory deficits that follow a closed-head TBI are often quite puzzling to the friends and relatives who have talked to the patient during the period of confusion—for example, during a hospital visit. The patient may seem reasonably lucid at the time, because short-term memory is normal, but later may have no recollection whatsoever of the conversation.

Figure 11.5 summarizes the effects of a closed-head TBI on memory. Note that the duration of the period of confusion and anterograde amnesia is typically longer than that of the coma, which is typically longer than the period of retrograde amnesia. More severe blows to the head tend to produce longer comas, longer periods of confusion, and longer periods of amnesia. Not illustrated in Figure 11.5 are *islands of memory*—surviving memories for isolated events that occurred during periods for which other memories have been wiped out.

Gradients of Retrograde Amnesia and Memory Consolidation

LO 11.9 **Describe the classic view of memory consolidation and some of the evidence it rests upon. Contrast that with current thinking about memory consolidation.**

Gradients of retrograde amnesia after closed-head TBI seem to provide evidence for *memory consolidation*. The fact that closed-head TBIs preferentially disrupt recent memories suggests that the storage of older memories has been strengthened (i.e., consolidated).

The classic theory of memory consolidation is Hebb's theory. He argued that memories of experiences are stored in the short term by neural activity *reverberating* (circulating) in closed circuits (see Hanslmayr, Staresina, & Bowman, 2016). These reverberating patterns of neural activity are susceptible to disruption—for example, by a blow to the head—but eventually they induce structural changes in the involved synapses, which provide stable long-term storage.

Electroconvulsive shock seemed to provide a controlled method of studying memory consolidation. **Electroconvulsive shock (ECS)** is an intense, brief, diffuse, seizure-inducing current that is administered to the brain through large electrodes attached to the scalp. The rationale for using ECS to study memory consolidation was that by disrupting neural activity, ECS would erase from storage only those memories that had not yet been converted to structural synaptic changes; the length of the period of retrograde amnesia produced by an ECS would thus provide an estimate of the amount of time needed for memory consolidation.

Many studies have employed ECS to study the duration of consolidation. Some studies have been conducted on human patients receiving ECS for the treatment of depression; others have been conducted with laboratory animals. Since the 1950s, hundreds of studies have examined ECS-produced gradients of retrograde amnesia in order to estimate the duration of memory consolidation. Hebb's theory implies that memory consolidation is relatively brief, a few seconds or minutes, about as long as specific patterns of reverberatory neural activity could conceivably maintain a memory. However, many studies found evidence for much longer gradients.

The classic study of Squire, Slater, and Chace (1975) is an example of a study that found a long gradient of ECS-produced retrograde amnesia. They measured the memory of a group of ECS-treated patients for television shows that had played for only one season in different years prior to their electroconvulsive therapy. They tested each patient twice on different forms of the test: once before they received a series of five electroconvulsive shocks and once after. The difference between the before- and after-scores served as an estimate of memory loss for the events of each year. Figure 11.6 illustrates that five electroconvulsive shocks disrupted the retention of television shows that had played in the 3 years prior to treatment but not those that had played earlier.

The current view of memory consolidation is that it continues for a very long time if not indefinitely (see Dudai,

Figure 11.6 Demonstration of a long gradient of ECS-produced retrograde amnesia. A series of five electroconvulsive shocks produced retrograde amnesia for television shows that played for only one season in the 3 years before the shocks; however, the shocks did not produce amnesia for one-season shows that had played prior to that.

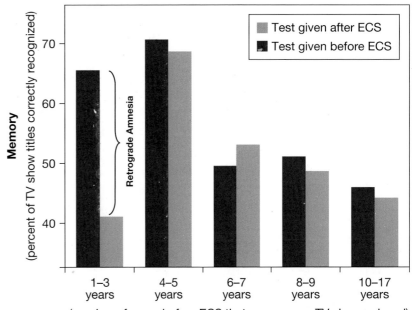

Based on Squire, L. R., Slater, P. C., & Chace, P. M. (1975). Retrograde amnesia: Temporal gradient in very long term memory following electroconvulsive therapy. *Science, 187*, 77–79.

Karni, & Born, 2015; Dudai & Morris, 2013). In other words, the evidence indicates that lasting memories become more and more resistant to disruption throughout a person's life. Each time a memory is activated, it is updated and linked to additional memories (see Sandrini, Cohen, & Censor, 2015). These additional links increase the memory's resistance to disruption by cerebral trauma, such as concussion or ECS.

HIPPOCAMPUS AND CONSOLIDATION. The case of H.M. provided evidence of memory consolidation, and it seemed to suggest that the hippocampus played a special role in it. To account for the fact that the bilateral medial temporal lobectomy disrupted only those retrograde memories acquired in the few years before H.M.'s surgery, Scoville and Milner (1957) suggested that memories are temporarily stored in the hippocampus until they can be transferred to a more stable cortical storage system. This theory has become known as the **standard consolidation theory** or **dual-trace theory** (see Clark & Maguire, 2016; Dudai, Karni, & Born 2015; Moscovitch et al., 2016).

Today, there are few adherents to standard consolidation theory. As you have just read, temporally graded retrograde amnesia is a feature of many forms of human amnesia (e.g., Alzheimer's amnesia, Korsakoff's amnesia); consequently, it seems unlikely that the hippocampus plays a special role in consolidation. It appears that when a conscious experience occurs, it is rapidly and sparsely encoded in a distributed fashion throughout the hippocampus and other involved structures.

According to Moscovitch and colleagues (e.g., Nadel and Moscovitch, 1997; Winocur and Moscovitch, 2011), retained memories become progressively more resistant to disruption by hippocampal dysfunction because each time a similar experience occurs or the original memory is recalled, a new **engram** (a change in the brain that stores a memory) is established and linked to the original engram. With the addition of each new engram, aspects of the original memory are progressively transformed into a semantic memory whose storage is less dependent on the hippocampus and more dependent on cortical structures (see Clarke & Maguire, 2016; Moscovitch et al., 2016). This makes the memory easier to recall and the original engram more difficult to disrupt.

RECONSOLIDATION.   One theoretical construct that has attracted significant attention is **reconsolidation** (see Bonin & De Koninck, 2015). The hypothesis is that each time a memory is retrieved from long-term storage, it is temporarily held in *labile* (changeable or unstable) short-term memory, where it is once again susceptible to posttraumatic amnesia until it is reconsolidated.

Interest in the process of reconsolidation originated with several studies in the 1960s, but then faded until a key study by Nader, Schafe, and LeDoux (2000) rekindled it. These researchers infused the protein-synthesis inhibitor *anisomycin* into the amygdalae of rats shortly after the rats had been required to recall a fear-conditioning trial. The infusion produced retrograde amnesia for the fear conditioning, even though the original conditioning trial had occurred many days before. Most research on reconsolidation has involved fear conditioning, but some evidence suggests that it may be a general phenomenon in the nervous system (see Bonin & De Koninck, 2015).

Scan Your Brain

This chapter is about to move from discussion of human memory disorders to consideration of animal models of human memory disorders. Are you ready? Scan your brain to assess your knowledge of human memory disorders by filling in the blanks in the following sentences. The correct answers are provided at the end of the exercise. Before proceeding, review the material related to your errors and omissions.

1. Improved test performance without conscious awareness, as in the case of H.M.'s mirror-drawing task, exemplifies long-term memories which became known as _____ memories.
2. Explicit long-term memory is broadly divided in two main categories: episodic and _____.
3. Ischemia in the _____ could result in transient global amnesia.
4. Amnesia due to Korsakoff's syndrome is typically associated with large lesions to the _____.

5. _____ amnesia refers to the loss of memories that took place before the brain injury.
6. _____ depletion resulting from damage to the basal forebrain may contribute to Alzheimer's amnesia.
7. Memory consolidation is studied using _____ shock, which induces seizures using large electrodes.
8. Each time a memory is recalled or a similar experience occurs, a new _____ is formed, making that memory more difficult to forget.
9. Amnesia caused by a nonpenetrative head injury is called _____.
10. Most research on reconsolidation has involved _____ conditioning using rats.

Scan Your Brain answers: (1) implicit, (2) semantic, (3) hippocampus, (4) medial diencephalon, (5) Anterograde, (6) Acetylcholine, (7) electroconvulsive, (8) engram, (9) posttraumatic amnesia (PTA), (10) fear.

Evolving Perspective of the Role of the Hippocampus in Memory

As interesting and informative as the study of patients with amnesia can be, it has major limitations. Many important questions about the neural bases of amnesia can be answered only by controlled experiments. For example, in order to identify the particular structures of the brain that participate in various kinds of memory, it is necessary to make precise lesions in various structures and to control what and when the subjects learn and how and when their retention is tested. Because such experiments are not feasible with humans, a major effort has been made to develop animal models of human brain-damage-produced amnesia.

The first reports of H.M.'s case in the 1950s triggered a massive effort to develop an animal model of his disorder so that it could be subjected to experimental analysis. In its early years, this effort was a dismal failure; lesions of medial temporal lobe structures did not produce severe anterograde amnesia in rats, monkeys, or other nonhuman species.

In retrospect, there were two reasons for the initial difficulty in developing an animal model of medial temporal lobe amnesia. First, it was not initially apparent that H.M.'s anterograde amnesia did not extend to all kinds of long-term memory—that is, it was specific to explicit long-term memories—and most animal memory tests widely used in the 1950s and 1960s were tests of implicit memory (e.g., Pavlovian and operant conditioning). Second, it was incorrectly assumed that the amnesic effects of medial temporal lobe lesions were largely, if not entirely, attributable to hippocampal damage; and most efforts to develop animal models of medial temporal lobe amnesia thus focused on hippocampal lesions.

Figure 11.7 An example of a delayed nonmatching-to-sample trial.

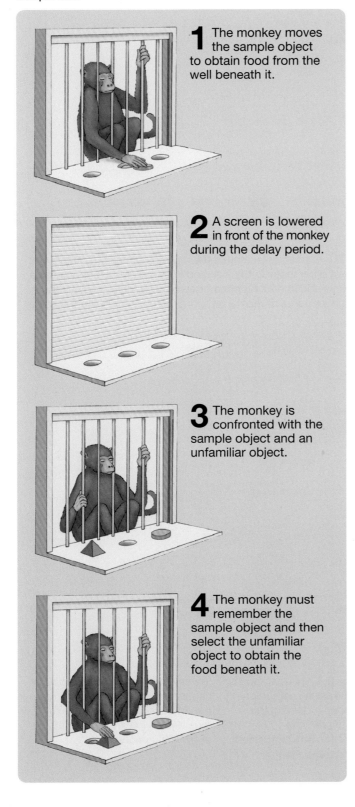

1 The monkey moves the sample object to obtain food from the well beneath it.

2 A screen is lowered in front of the monkey during the delay period.

3 The monkey is confronted with the sample object and an unfamiliar object.

4 The monkey must remember the sample object and then select the unfamiliar object to obtain the food beneath it.

Animal Models of Object-Recognition Amnesia: The Delayed Nonmatching-to-Sample Test

LO 11.10 Describe the delayed nonmatching-to-sample tests for monkeys and rats.

Finally, in the mid-1970s, more than two decades after the first reports of H.M.'s remarkable case, an animal model of his disorder was developed. It was hailed as a major breakthrough because it opened up the neuroanatomy of medial temporal lobe amnesia to experimental investigation.

MONKEY VERSION OF THE DELAYED NONMATCHING-TO-SAMPLE TEST. In separate laboratories, Gaffan (1974) and Mishkin and Delacour (1975) showed that monkeys with bilateral medial temporal lobectomies have major problems forming long-term memories for objects encountered in the **delayed nonmatching-to-sample test**. In this test, a monkey is presented with a distinctive object (the *sample object*), under which it finds food (e.g., a banana pellet). Then, after a delay, the monkey is presented with two test objects: the sample object and an unfamiliar object. The monkey must remember the sample object so that it can select the unfamiliar object to obtain food concealed beneath it. The correct performance of a trial is illustrated in Figure 11.7.

Intact, well-trained monkeys performed correctly on about 90 percent of delayed nonmatching-to-sample trials when the retention intervals were a few minutes or less. In contrast, monkeys with bilateral medial temporal lobe lesions had major object-recognition deficits (see Figure 11.8). These deficits modeled those of H.M. in key

Figure 11.8 The performance deficits of monkeys with large bilateral medial temporal lobe lesions on the delayed nonmatching-to-sample test. There were significant deficits at all but the shortest retention interval. These deficits parallel the memory deficits of humans with medial temporal lobe amnesia on the same task.

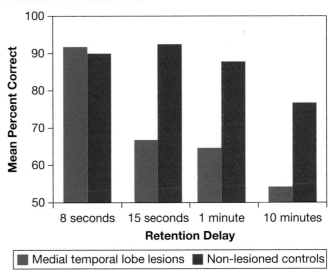

Based on Squire, L. R. & Zola-Morgan, S. (1991). The medial temporal lobe memory system. *Science, 253*, 1380–1386.

Figure 11.9 The three major structures of the medial temporal lobe, illustrated in the monkey brain: the hippocampus, the amygdala, and the medial temporal cortex.

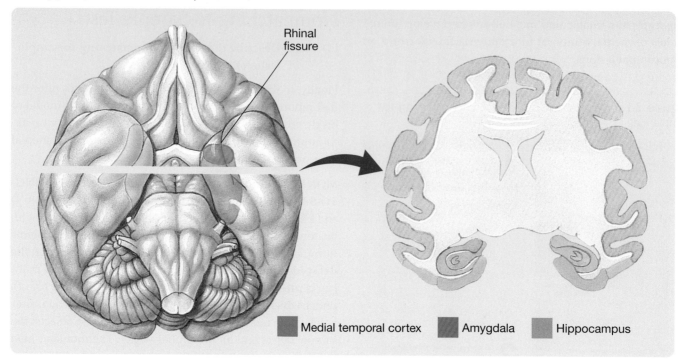

respects. For example, the monkeys' performance was normal at delays of a few seconds but fell off to near chance levels at delays of several minutes, and their performance was extremely susceptible to the disruptive effects of distraction (see Squire & Zola-Morgan, 1985). In fact, humans with medial temporal lobe amnesia have been tested on the delayed nonmatching-to-sample test—their rewards were coins rather than banana slices—and their performance mirrored that of monkeys with similar brain damage.

The development of the delayed nonmatching-to-sample test for monkeys provided a means of testing the assumption that the amnesia resulting from medial temporal lobe damage is entirely the consequence of hippocampal damage—Figure 11.9 illustrates the locations in the monkey brain of three major temporal lobe structures: hippocampus, amygdala, and adjacent **medial temporal cortex**. But before we consider this important line of research, we need to look at another important methodological development: the rat version of the delayed nonmatching-to-sample test.

RAT VERSION OF THE DELAYED NON-MATCHING-TO-SAMPLE TEST. In order to understand why the development of the rat version of the delayed nonmatching-to-sample test played an important role in assessing the specific role of hippocampal damage in medial temporal lobe amnesia, examine Figure 11.10,

Figure 11.10 Aspiration lesions of the hippocampus in monkeys and rats. Because of differences in the size and location of the hippocampus (pink) in monkeys and in rats, hippocampectomy typically involves the removal of large amounts of medial temporal cortex (red) in monkeys, but not in rats.

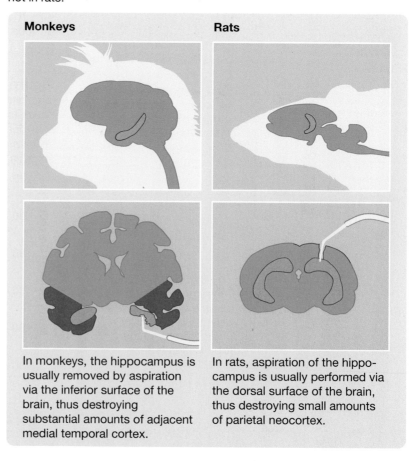

In monkeys, the hippocampus is usually removed by aspiration via the inferior surface of the brain, thus destroying substantial amounts of adjacent medial temporal cortex.

In rats, aspiration of the hippocampus is usually performed via the dorsal surface of the brain, thus destroying small amounts of parietal neocortex.

which illustrates the usual methods of making hippocampal lesions in monkeys and rats. Because of the size and location of the hippocampus, almost all studies of hippocampal lesions in monkeys have involved *aspiration* (suction) of large portions of the medial temporal cortex in addition to the hippocampus. However, in rats, the extraneous damage associated with aspiration lesions of the hippocampus is typically limited to a small area of parietal neocortex. Furthermore, the rat hippocampus is small enough that it can be lesioned electrolytically or with intracerebral neurotoxin injections—-methods that produce less extraneous damage.

The version of the delayed nonmatching-to-sample test for rats that most closely resembles that for monkeys was developed by David Mumby using an apparatus that has become known as the **Mumby box**. This rat version of the test is illustrated in Figure 11.11.

It was once assumed that rats could not perform a task as complex as that required for the delayed nonmatching-to-sample test; Figure 11.12 indicates otherwise. Rats perform almost as well as monkeys with delays of up to 1 minute (Mumby, Pinel, & Wood, 1989).

The validity of the rat version of the delayed nonmatching-to-sample test has been established by studies of the effects

Figure 11.11 The Mumby box and the rat version of the delayed nonmatching-to-sample test.

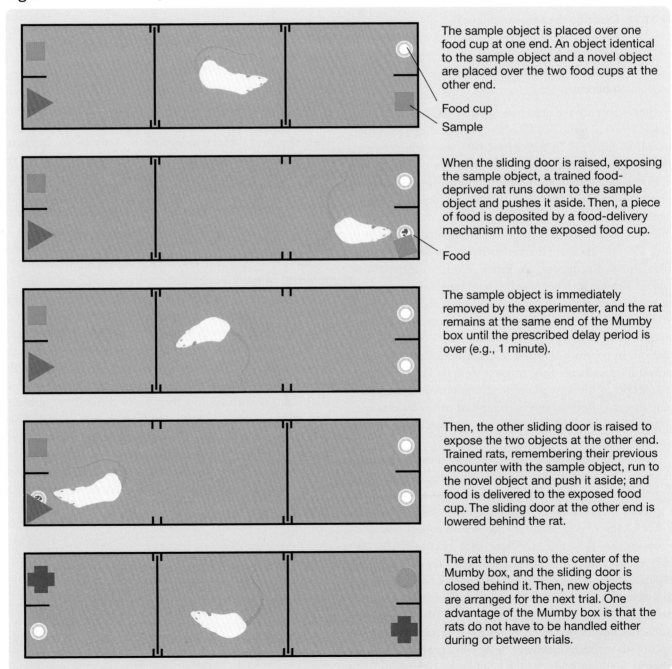

The sample object is placed over one food cup at one end. An object identical to the sample object and a novel object are placed over the two food cups at the other end.

Food cup

Sample

When the sliding door is raised, exposing the sample object, a trained food-deprived rat runs down to the sample object and pushes it aside. Then, a piece of food is deposited by a food-delivery mechanism into the exposed food cup.

Food

The sample object is immediately removed by the experimenter, and the rat remains at the same end of the Mumby box until the prescribed delay period is over (e.g., 1 minute).

Then, the other sliding door is raised to expose the two objects at the other end. Trained rats, remembering their previous encounter with the sample object, run to the novel object and push it aside; and food is delivered to the exposed food cup. The sliding door at the other end is lowered behind the rat.

The rat then runs to the center of the Mumby box, and the sliding door is closed behind it. Then, new objects are arranged for the next trial. One advantage of the Mumby box is that the rats do not have to be handled either during or between trials.

of medial temporal lobe lesions. As in humans and monkeys, bilateral lesions of the rats' hippocampus, amygdala, and medial temporal cortex combined produce major deficits at all but the shortest retention intervals (Mumby, Wood, & Pinel, 1992).

Neuroanatomical Basis of the Object-Recognition Deficits Resulting from Bilateral Medial Temporal Lobectomy

LO 11.11 Describe the neuroanatomical basis for the object-recognition deficits that result from bilateral medial temporal lobectomy.

To what extent are the object-recognition deficits following bilateral medial temporal lobectomy a consequence of hippocampal damage? In the early 1990s, researchers began assessing the relative effects of lesions to various medial temporal lobe structures on performance in the delayed nonmatching-to-sample test, in both monkeys and rats. Challenges to the view that hippocampal damage is the critical factor in medial temporal amnesia quickly accumulated. Most reviewers of this research (see Bussey & Saksida, 2005; Duva, Kornecook, & Pinel, 2000; Mumby, 2001) reached similar conclusions: Bilateral surgical removal of the medial temporal cortex consistently produces severe and permanent deficits in performance on the delayed nonmatching-to-sample test and other tests of object recognition. In contrast, bilateral surgical removal of the hippocampus produces only modest deficits, and bilateral destruction of the amygdala produces none. Figure 11.13 compares the effects of medial temporal cortex lesions and hippocampus-plus-amygdala lesions on object recognition memory in rats.

The reports that object-recognition memory is severely disrupted by medial temporal cortex lesions but only moderately by hippocampal lesions led to a resurgence of interest in the case of R.B. and others like it. Earlier in this chapter, you learned that R.B. was left amnesic following an ischemic accident that occurred during heart surgery and that subsequent analysis of his brain revealed that obvious cell loss was restricted largely to the pyramidal cell layer of his CA1 hippocampal subfield (see Figure 11.4). This result has been replicated in both monkeys (Zola-Morgan et al., 1992) and rats (Wood et al., 1993). In both monkeys and rats, global cerebral ischemia leads to a loss of CA1 hippocampal pyramidal

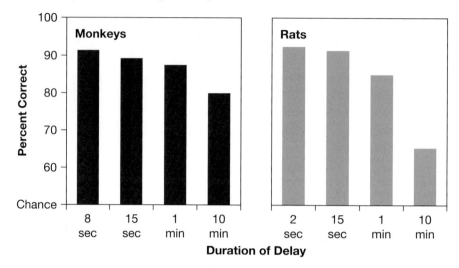

Figure 11.12 A comparison of the performance of intact monkeys (Zola-Morgan, Squire, & Mishkin, 1982) and intact rats (Mumby, Pinel, & Wood, 1989) on the delayed nonmatching-to-sample test.

cells and severe deficits in performance on the delayed nonmatching-to-sample test.

The relation between ischemia-produced hippocampal damage and object-recognition deficits in humans, monkeys, and rats seems to provide strong support for the theory that the hippocampus plays a key role in object-recognition memory. However, there is a gnawing problem with this conclusion: How can ischemia-produced lesions to one small part of the hippocampus be associated with severe deficits in performance on the delayed nonmatching-to-sample test when the deficits associated with total removal of the hippocampus are only modest? This line of evidence suggests that damage to brain structures other than the hippocampus contributes to the amnesia observed in patients following global cerebral ischemia (see Mumby et al., 1996). Indeed, although the most obvious damage following cerebral ischemia is in the CA1 subfield of the hippocampus, there is substantial damage to other areas that is more diffuse and thus more difficult to quantify (see Katsumata et al., 2006; van Groen et al., 2005). Allen and colleagues (2006) found that ischemic patients with a greatly reduced hippocampal volume were much more likely to suffer from anterograde amnesia; however, these same patients also tended to have extensive neocortical damage.

Journal Prompt 11.3

Before proceeding further, explain why ischemia-produced lesions to one small part of the hippocampus are associated with severe deficits in performance on the delayed nonmatching-to-sample test when the deficits associated with total removal of the hippocampus are only modest.

Figure 11.13 Effects of medial temporal cortex lesions and hippocampus-plus-amygdala lesions in rats. Lesions of the medial temporal cortex, but not of the hippocampus and amygdala combined, produced severe deficits in performance of the delayed nonmatching-to-sample test in rats.

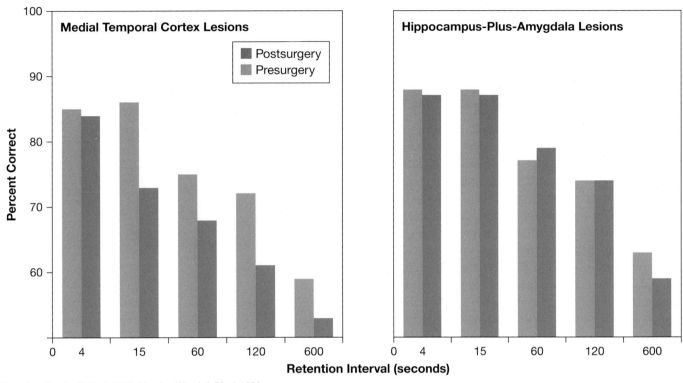

Based on Mumby & Pinel, 1994; Mumby, Wood, & Pinel, 1992.

So far in this chapter, you have seen that our perspective of the role of the hippocampus in memory has changed since the first published reports of H.M.'s case. Initially, the hippocampus was thought to be the site of temporary storage for all newly formed memories. However, it was soon discovered that the structures of the medial temporal lobes have a more specific function—they appear to play a major role only in explicit episodic memories. Then, as you just learned in this section, it was discovered that the role of the hippocampus in object-recognition memory is minor compared to the contribution of adjacent medial temporal cortex. Today, the hippocampus is considered to be just one of several brain structures that play important roles in memory. The next module considers its particular mnemonic function.

Neurons of the Medial Temporal Lobes and Memory

Although the first suggestion that the medial temporal lobes play a major role in memory came from the study of the effects of damage to that area, the study of the responses of medial temporal lobe neurons have also been enlightening.

The first major studies of the responses of medial temporal lobe neurons focused on the study of hippocampal neurons in rats. This research was stimulated by the finding that bilateral lesions of the hippocampus invariably disrupt the performance of tasks that involve memory for spatial location. For example, hippocampal lesions disrupt performance on the Morris water maze test and the radial arm maze test.

MORRIS WATER MAZE TEST. In the **Morris water maze test**, intact rats placed at various locations in a circular pool of murky water rapidly learn to swim to a stationary platform hidden just below the surface. Rats with hippocampal lesions learn this simple task with great difficulty.

RADIAL ARM MAZE TEST. In the **radial arm maze test**, several (e.g., eight) arms radiate out from a central starting chamber, and the same few arms are baited with food each day. Intact rats readily learn to visit only those arms that contain food and do not visit the same arm more than once each day. The ability to visit only the baited arms of the radial arm maze is a measure of **reference memory** (memory for the general principles and skills that are required to perform a task), and the ability to refrain from visiting an arm more than once in a given day is a measure of **working memory** (temporary memory that is necessary for the successful performance of a task on which one is currently working).

Rats with hippocampal lesions display major deficits on both the reference memory and the working memory measures of radial arm maze performance.

Hippocampal Place Cells and Entorhinal Grid Cells

LO 11.12 Describe hippocampal place cells and entorhinal grid cells and the relationship between these two cell types.

Consistent with the view that the hippocampus plays a role in spatial processing is the fact that many hippocampal neurons are **place cells** (see Moser & Moser, 2016; Moser, Moser, & McNaughton, 2017)—neurons that respond only when a subject is in specific locations (i.e., in the *place fields* of the neurons). For example, when a rat is first placed in an unfamiliar test environment, none of its hippocampal neurons have a place field in that environment; then, as the rat familiarizes itself with the environment, many hippocampal neurons acquire a place field in it (see Silva, Feng, & Foster, 2015)—that is, each cell fires only when the rat is in a particular part of the test environment. Accordingly, each place cell has a place field in a different part of the environment. Place cells have been identified in a variety of species (see Finkelstein, Las, & Ulanovsky, 2016; Geva-Sagiv et al., 2016; Moser et al., 2014)—including human patients (Miller et al., 2013).

By placing a rat in an ambiguous situation in a familiar test environment, it is possible to determine where the rat thinks it is located along the route that it takes to get to the location in the environment where it has previously been rewarded. Using this strategy, researchers have shown that the firing of a rat's place cells indicates where the rat "thinks" it is in the test environment, not necessarily where it actually is (see Moser, Moser, & McNaughton, 2017).

One line of research on spatial processing has focused on the **entorhinal cortex** (an area of the medial temporal cortex that is a major source of neural signals to the hippocampus)—see Figure 11.14. A possible answer to the question of how hippocampal place cells obtain their spatial information came from the discovery of so-called grid cells in the entorhinal cortex. **Grid cells** are entorhinal neurons that each have a repeating pattern of evenly spaced hexagon-shaped place fields that tile the surface of an environment (see Moser & Moser, 2013; Rowland & Moser, 2014; Underwood, 2014). Grid cells fire when an animal physically traverses a point of intersection in the hexagonal grid (see Hardcastle, Ganguli, & Giocomo, 2017), or when its gaze traverses a point of intersection while surveying the environment (see Killian & Buffalo, 2018; Nau et al., 2018). The hexagonal grid of place fields is

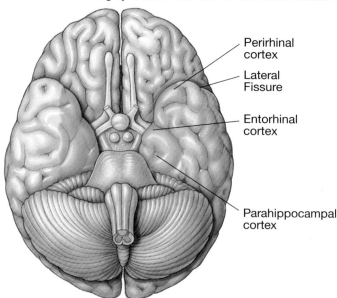

Figure 11.14 Areas of human medial temporal cortex. These areas are largely hidden from view in the lateral fissure.

Perirhinal cortex

Lateral Fissure

Entorhinal cortex

Parahippocampal cortex

flexible: When spatial cues are rotated or sheared, the grid pattern is also rotated or sheared, respectively (see Krupic et al., 2015; Rowland et al., 2016; Stensola et al., 2015). The even spacing of the place fields in grid cells could enable spatial computations in hippocampal place cells. Grid cells have also been identified in other species including humans (e.g., Nau et al., 2018). Other types of neurons in the entorhinal cortex are associated with spatial computations: For example, *head-direction cells* are tuned to the direction of head orientation, and *border cells* fire when the subject is near the borders of its immediate environment (see Rowland et al., 2016; Sanders et al., 2015; Winter, Clark, & Taube, 2015).

The nature of the relationship between entorhinal grid cells and hippocampal place cells is still a matter of ongoing debate (see Rowland et al., 2016; Sanders et al., 2015). Two lines of evidence suggested that the responses of hippocampal place cells depend on input from entorhinal grid cells (but see Bush, Barry, & Burgess, 2014). First, there is a major pathway from the entorhinal cortex to the hippocampus. Second, entorhinal grid cells respond in an ongoing fashion to an animal's location, whereas hippocampal place cells are only active in particular spatial locations (see Mallory et al., 2018). However, despite the appeal of such a one-way relationship between grid cells and place cells, the relationship between these cell types is more complex; three lines of evidence make this point. First, the properties of hippocampal place cells emerge in developing rat pups prior to the emergence of stable entorhinal grid cell firing (see Derdikman & Moser, 2010; Langston et al., 2010; Wills & Cacucci, 2014). Second, there is evidence that place cells can still function after entorhinal grid cells have been eliminated (see Sanders et al., 2015). Third, intact inputs from place cells are necessary for the reliable firing of grid cells (see Rowland et al., 2016).

THE HIPPOCAMPUS AS A COGNITIVE MAP.   Although the hippocampus plays a clear role in spatial learning and memory, hippocampal cells fire to more than just spatial cues. For example, certain cells in the hippocampus have been shown to code for the temporal aspects of an experience—so-called "time cells" (see Eichenbaum, 2014; Giacomo, 2015; Navratilova & Battaglia, 2015). Moreover, the hippocampus has been shown to play a role in learning about social organization ("social space") in humans (see Eichenbaum, 2015; Tavares et al., 2015) and in mice (see Hitti & Siegelbaum, 2014). Finally, as you will learn in the next section, cells in the hippocampus and its surrounding structures have been shown to play a role in the coding of concepts. Accordingly, it has been argued that it is better to think of the hippocampus as generating a "cognitive map" rather than just a "spatial map" (see Lisman et al., 2017; Epstein et al., 2017).

Jennifer Aniston Neurons: Concept Cells

LO 11.13 **Explain what a concept cell is and describe the key properties of concept cells with reference to the experimental evidence.**

Recording electrodes are sometimes implanted in the brains of patients with severe epilepsy, usually as a precursor to surgery. This provides an opportunity to record the activity of particular neurons in patients as they perform various tasks. Many of these electrodes are implanted in the structures of the medial temporal lobes because they are particularly susceptible to epileptic discharges.

As you have previously learned, the major structures of the medial temporal lobes are the hippocampus, amygdala, and medial temporal cortex. The medial temporal cortex (illustrated in Figure 11.14) is composed of entorhinal, perirhinal, and parahippocampal cortices.

In one neuron studied in this way, the neuron fired in response to images of the actor Jennifer Aniston, but not to 80 other images (see Quiroga, 2012). Other medial temporal lobe neurons were discovered that responded to other individuals (e.g., politicians, celebrities) or objects (e.g., famous buildings) known to the patients, but because the first neuron responded to Jennifer Aniston, they were initially called *Jennifer Aniston neurons* (see Quiroga, 2016).

Jennifer Aniston neurons are highly selective. Each neuron responded to only a small number of test objects or individuals—often only one could be found. Also, their responses are highly invariant: If a neuron responded to a particular

object on test 1, it tended to respond to that object on all subsequent tests. The Jennifer Aniston cells of the hippocampus were more selective and more invariant than those of the other medial temporal lobe structures (i.e., parahippocampal cortex, perirhinal cortex, entorhinal cortex, and amygdala).

Without question, the most remarkable feature of Jennifer Aniston neurons is that they respond to ideas or concepts rather than to particulars, which is why they have come to be known as **concept cells** (see Quiroga, Fried, & Koch, 2013). For example, a Halle Berry neuron responded to all photos of the actor (even when she was dressed in her Catwoman costume), to her printed name, and to the sound of her name (see Quiroga, 2016). In one case, a neuron that invariably responded to the Sydney Opera House responded to photos of the Bahá'í temple in India. When questioned about it later, the patient said she thought the Bahá'í temple photos were photos of the Sydney Opera House. Similarly, when participants are given ambiguous human faces (faces that are the average of two well-known faces, such as Whoopi Goldberg and Bob Marley), these concept cells respond only when the viewer perceives the concept to which the cells are attuned to—for example, only when the viewer perceives Whoopi Goldberg, as opposed to Bob Marley (see Quiroga et al., 2014; Reddy & Thorpe, 2014).

Interestingly, when concept cells have been found to respond to more than one concept, there is usually an obvious relationship between them. For example, on a second day of testing, it was discovered that the first Jennifer Aniston cell also responded to Lisa Kudrow, Jennifer Aniston's costar in the television series *Friends*. Another neuron responded to both Luke Skywalker and Yoda, both characters from the movie series *Star Wars*. Accordingly, it has been suggested that related concepts trigger activity in circuits of concept cells in the medial temporal lobes (see Quiroga, 2012; De Falco et al., 2016).

Figure 11.15 If researchers identified a "Harry Potter neuron" in a patient's brain, what other stimuli might it fire in response to?

WARNER BROS. PICTURES/Album/Newscom

Although it is not yet clear how concept cells contribute to the storage of memories, it is clear that they play a role, and their discovery is a major step forward (see De Falco et al., 2016).

Engram Cells

LO 11.14 Explain what an engram cell is and describe how these cells were identified using optogenetics.

As you learned in Chapter 5, one new approach in the toolkit available to biopsychologists is *optogenetics*. If you recall, in optogenetics, neuroscientists insert an opsin gene into particular neurons, after which they can then use light to either hyperpolarize or depolarize those neurons. In recent years, this tool has been used extensively in studies of learning and memory in mice (see Goshen, 2014). One line of research has been particularly interesting because it can shed light on the location of the neurons that maintain an engram: so-called engram cells (see Josselyn, Köhler, and Frankland, 2015; Tonegawa, Morrissey, & Kitamura, 2018). **Engram cells** are neurons that undergo a persistent change as the result of experience such that when they are subsequently activated or inhibited, the retrieval of the original experience is triggered or suppressed, respectively (see Bliss et al., 2018; Tonegawa, Morrissey, & Kitamura, 2018).

The identification of an engram cell via optogenetics is typically a two-stage process. In the first stage, the *tagging stage*, the neurons that are active during the learning task are induced to express opsins while an animal engages in a particular learning task. In the second stage, the *manipulate stage*, the previously active neurons are now either inhibited or excited by using light to influence the activity of the opsin-tagged neurons (see Josselyn, Köhler, & Frankland, 2015; Redondo et al., 2014; Ryan et al., 2015; Takeuchi & Morris, 2014).

In essence, researchers are now able to observe, suppress, or activate engram cells in different parts of the nervous system. For example, researchers have been able to reverse depressive-like behavior in mice by optogenetically reactivating hippocampal cells that had previously been active during the encoding of a positive experience (Ramirez et al., 2015). Moreover, researchers have shown that in mouse models of Alzheimer's disease, activating engram cells leads to the retrieval of memories that are otherwise inaccessible—suggesting that the memory deficits of Alzheimer's disease are retrieval deficits rather than encoding deficits (see Roy et al., 2016; Shrestha & Klann, 2016).

Journal Prompt 11.4

Do you think that optogenetics might have clinical implications for humans sometime in the future? If so, what do you think such interventions might look like?

Where Are Memories Stored?

In the mid-20th century, when modern biopsychology began, there was a major push to identify the areas in the brain where memories are stored. The search was largely championed by Karl Lashley, who wrote a famous review paper, *In Search of the Engram*, in which he described his fruitless efforts. Lashley and many who subsequently took up the search used the lesion method. If a particular structure were the storage site for all memories of a particular type, then destruction of that structure should eliminate all memories of that type that were acquired prior to the lesion. No brain structure has shown this result: Lesions of particular structures tend to produce either no retrograde amnesia at all or retrograde amnesia for only the experiences that occurred in the days or weeks just before the surgery. These findings have led to two major conclusions: (1) Memories are stored diffusely in the brain and thus can survive destruction of any single structure; and (2) memories become more resistant to disruption over time.

So far, this chapter has focused on four neural structures that appear to play some role in the storage of memories: (1) The hippocampus and (2) the medial temporal cortex play roles in episodic memory; and (3) the mediodorsal nucleus of the thalamus and (4) the basal forebrain have been implicated in the memory deficits of Korsakoff's and Alzheimer's diseases, respectively. In this module, we take a brief look at five other areas of the brain that have been implicated in memory: inferotemporal cortex, amygdala, prefrontal cortex, cerebellum, and striatum. See Figure 11.16.

Five Brain Areas Implicated in Memory

LO 11.15 For each of the following brain structures, describe the type(s) of memory they have been implicated in: inferotemporal cortex, amygdala, prefrontal cortex, cerebellum, and striatum.

INFEROTEMPORAL CORTEX. Numerous electrophysiological recording and functional brain-imaging studies of memory have led to the same important conclusion: Areas of the brain that are active during the retention of an experience tend to be the same ones active during the original experience. This has focused attention on the mnemonic functions of the sensory and motor areas of the brain. In particular, attention has focused on **inferotemporal cortex** (cortex of the inferior temporal cortex), which has complex visual functions—see Lehky and Tanaka (2016); Naya and Suzuki (2011); Suzuki (2010).

Bussey and Saksida (2005) have argued that the inferotemporal cortex, in concert with adjacent perirhinal cortex (see Miyashita, 2019; Suzuki & Naya, 2014), plays an important role in storing memories of visual input. In support of this view, Naya, Yoshida, and Miyashita (2001) recorded the responses of neurons in inferotemporal cortex and perirhinal cortex while monkeys learned the relation between the two items in pairs of visual images. When a pair was presented, responses were first recorded in inferotemporal neurons and then in perirhinal neurons; however, when the monkeys were required to recall that pair, activity was recorded in perirhinal neurons before inferotemporal neurons. Naya and colleagues concluded that this reversed pattern of activity reflected the retrieval of visual memories from inferotemporal cortex.

AMYGDALA. The amygdala is thought to play a special role in memory for the emotional significance of experiences (see Herry & Johansen, 2014; Paz & Pare, 2013). Rats with amygdalar lesions, unlike intact rats, do not respond with fear to a neutral stimulus that has previously been followed by electric foot shock (see Maren, 2015; Nader, 2015). Also, Bechara and colleagues (1995) reported the case of a neuropsychological patient with bilateral damage to the amygdala who could not acquire conditioned autonomic startle responses to various visual or auditory stimuli but had good explicit memory for the training. However, there is little evidence that the amygdala stores memories; it appears to be involved in strengthening emotionally significant memories stored in other structures (Do-Monte, Quiñones-Laracuente, & Quirk, 2015; Likhtik & Paz, 2015). The amygdala might be the reason why emotion-provoking events are remembered better than neutral events (see Dunsmoor et al., 2015; McGaugh, 2015; Yonelinas & Ritchey, 2015).

PREFRONTAL CORTEX. Patients with damage to the **prefrontal cortex** (the area of frontal cortex anterior to motor cortex) are not grossly amnesic; they often display no deficits at all on conventional tests of memory. This lack of reliable memory deficits in patients with prefrontal damage may in part result from the fact that different parts of the prefrontal cortex play different roles in memory and that patients with damage to different areas of prefrontal cortex are often combined for analysis.

Be that as it may, two episodic memory abilities are often lost by patients with large prefrontal lesions. Patients with large prefrontal lesions often display both anterograde

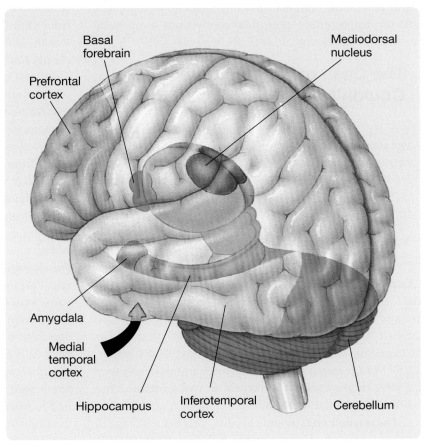

Figure 11.16 The structures of the brain that have been shown to play a role in memory. Because it would have blocked the view of other structures, the striatum is not included. (See Figure 3.27 on page 93.)

and retrograde deficits in memory for the temporal order of events, even when they can remember the events themselves. They also display deficits in *working memory* (the ability to maintain relevant memories while a task is being completed)—see D'Esposito and Postle (2015); Constantinidis & Klingberg (2016). As a result of these two deficits, patients with prefrontal cortex damage often have difficulty performing tasks that involve a series of responses (see Colvin, Dunbar, & Grafman, 2001).

Journal Prompt 11.5

The study of the anatomy of memory has come a long way since H.M.'s misfortune. What kind of advances do you think will be made in the next decade?

The prefrontal cortex is a large structure that is composed of many anatomically distinct areas that have different connections and, presumably, different functions. Functional brain-imaging studies are finding that specific complex patterns of prefrontal activity are associated with various memory functions. Some regions of prefrontal cortex seem to perform fundamental cognitive processes (e.g., attention and task management) during working memory tasks, and other regions of prefrontal cortex participate in other memory processes (e.g., Lehky & Tanaka, 2016; Morici, Bekinschtein, & Weisstaub, 2015). For example, recent studies in laboratory animals have highlighted an important role for hippocampal–prefrontal connections in episodic memory (see Barker et al., 2017; Eichenbaum, 2017).

The Case of the Cook Who Couldn't

The story of one patient with prefrontal cortex damage is very well known because she was the sister of Wilder Penfield, the famous Montreal neurosurgeon (Penfield & Evans, 1935). Before her brain damage, she had been an excellent cook; and afterward, she still remembered her favorite recipes and how to perform each individual cooking technique. However, she was incapable of preparing even simple meals because she could not carry out the various steps in proper sequence.

CEREBELLUM AND STRIATUM. Just as explicit memories of experiences are presumed to be stored in the circuits of the brain that mediated their original perception, implicit memories of sensorimotor learning are presumed to be stored in sensorimotor circuits (see Graybiel & Grafton, 2015). Most research on the neural mechanisms of memory for sensorimotor tasks has focused on two structures: the cerebellum and the striatum.

The **cerebellum** is thought to participate in the storage of memories of learned sensorimotor skills through its various neuroplastic mechanisms (see Gao, van Beugen, & De Zeeuw, 2012). Its role in the Pavlovian conditioning of the eye-blink response of rabbits has been most intensively investigated (see Freeman, 2015). In this paradigm, a tone (conditional stimulus) is sounded just before a puff of air (unconditional stimulus) is delivered to the eye. After several trials, the tone comes to elicit an eye blink. The convergence of evidence from stimulation, recording, and lesion studies suggests that the effects of this conditioning are stored in the form of changes in the way that cerebellar neurons respond to the tone (see De Zeeuw & Ten Brinke, 2015; Freeman, 2015).

The **striatum** is thought to store memories for consistent relationships between stimuli and responses—the type of memories that develop incrementally over many trials (see Graybiel & Grafton, 2015). Sometimes this striatum-based form of learning is referred to as *habit-formation* (see O'Tousa & Grahame, 2014).

Although few would disagree that the cerebellum and the striatum play a role in sensorimotor memory, there is growing evidence that these structures also play a role in certain types of memory with no obvious motor component. For example, Knowlton, Mangels, and Squire (1996) found that Parkinson's patients with striatal damage could not solve a probabilistic discrimination problem. The problem was a computer "weather forecasting" game, and the task was to correctly predict the weather by pressing one of two keys, rain or shine. The patients based their predictions on stimulus cards presented on the screen—each card had a different probability of leading to sunshine, which the patients had to learn and remember. The Parkinson's patients did not improve over 50 trials, although they displayed normal explicit (conscious) memory for the training episodes. In contrast, amnesic patients with medial temporal lobe or medial diencephalic damage displayed marked improvement in performance but had no explicit memory of their training.

Scan Your Brain

The preceding modules of this chapter have dealt with the gross neuroanatomy of memory—with the cells and structures of the brain that are involved in various aspects. Before you proceed to the next module, which deals with the synaptic mechanisms of learning and memory, test your knowledge by filling in the blanks in the following sentences. The correct answers are provided at the end of the exercise. Before proceeding, review the material related to your errors and omissions.

1. Monkeys with large lesions in the _____ _____ lobe show deficits on the delayed nonmatching-to-sample test.

2. Aspiration lesions of the _____ are more efficient in rats than monkeys because of the size and location of this brain structure.

3. Lesions of the medial temporal lobe in rats produce more pronounced deficits in object recognition compared to lesions of the hippocampus and the _____.

4. Rats with hippocampal lesions display major deficits in both _____ and working memory.

5. Neurons in the hippocampus that respond only when the subject is in specific locations are called _____.

6. _____ neurons are very selective and invariant. This means that they respond to concepts and ideas exclusively, which is why they are also called concept cells.

7. The _____ has been known for its role in Pavlovian conditioning of eye-blink response of rabbits.

8. One of the main structures involved in habit formation is the _____.

9. Patients with lesions in the _____ cortex display deficits in working memory.

10. The _____ cortex is involved in the storage and retention of visual memories.

Scan Your Brain answers: (1) medial temporal, (2) hippocampus, (3) amygdala, (4) reference, (5) place cells, (6) Jennifer Aniston, (7) cerebellum, (8) striatum, (9) prefrontal, (10) inferotemporal.

Cellular Mechanisms of Learning and Memory

So far, this chapter has focused on the particular structures of the human brain that are involved in learning and memory and on what happens when these structures are dysfunctional. In this module, the level of analysis changes: The focus shifts to the neuroplastic mechanisms within these structures that are thought to be the fundamental bases of learning and memory.

Most modern thinking about the neural mechanisms of memory began with Hebb (1949). Hebb argued so convincingly that enduring changes in the efficiency of synaptic transmission were the basis of long-term memory that the search for the neural bases of learning and memory has focused almost exclusively on the synapse.

Synaptic Mechanisms of Learning and Memory: Long-Term Potentiation

LO 11.16 Describe the phenomenon known as long-term potentiation (LTP) and provide evidence for its role in learning and memory.

Because Hebb's hypothesis that enduring facilitations of synaptic transmission are the neural bases of learning and memory was so influential, there was great excitement when such an effect was discovered. In 1973, Bliss and Lømo showed that there is a facilitation of synaptic transmission following high-frequency electrical stimulation applied to presynaptic neurons. This phenomenon has been termed **long-term potentiation (LTP)**.

LTP has been demonstrated in many species and in many parts of their brains, but it has been most frequently studied in the rodent hippocampus. Figure 11.17 illustrates

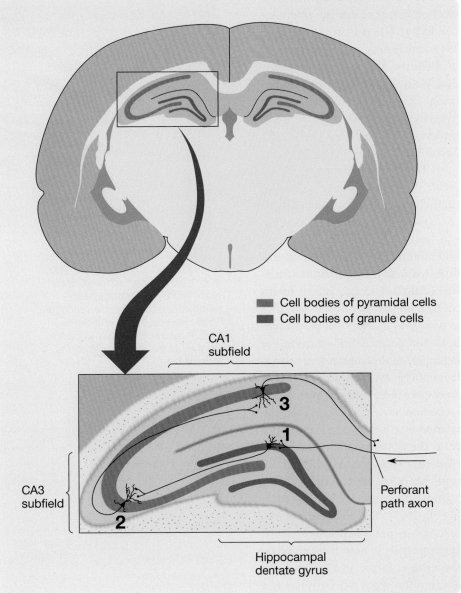

Figure 11.17 A slice of rat hippocampal tissue that illustrates the three synapses at which LTP is most commonly studied: (1) the dentate granule cell synapse, (2) the CA3 pyramidal cell synapse, and (3) the CA1 pyramidal cell synapse.

Cell bodies of pyramidal cells
Cell bodies of granule cells

CA1 subfield
CA3 subfield
Perforant path axon
Hippocampal dentate gyrus

three hippocampal synapses at which LTP has been commonly studied.

Figure 11.18 illustrates LTP in the granule cell layer of the rat hippocampal dentate gyrus. First, a single low-intensity pulse of current was delivered to the perforant path (the major input to the dentate gyrus), and the response was recorded through an extracellular multiple-unit-electrode in the granule cell layer of the hippocampal dentate gyrus; the purpose of this initial stimulation was to determine the initial response baseline. Second, high-frequency stimulation lasting 10 seconds was delivered to the perforant path to induce the LTP. Third, the granule cells' responses to single pulses of low-intensity current were measured again after various delays. Figure 11.18 shows that transmission at the granule cells' synapses was still potentiated 1 week after the high-frequency stimulation.

LTP is among the most widely studied neuroscientific phenomena. Why? The reason goes back to 1949 and Hebb's influential theory of memory. The synaptic changes that Hebb hypothesized as underlying long-term memory seemed to be the same kind of changes that underlie LTP (see Bliss et al., 2018; Nicoll, 2017).

LTP has two key properties that Hebb proposed as characteristics of the physiological mechanisms of learning and memory. First, LTP can last for a long time—for several months to a year after multiple high-frequency stimulations (Abraham, 2006). Second, many forms of LTP develop only if there is *co-occurrence of activity* in the presynaptic and postsynaptic neurons. For example, neurotransmitter release from the presynaptic neuron that co-occurs with postsynaptic depolarization (see Chapter 4), or the release of a retrograde transmitter from the postsynaptic neuron that co-occurs with presynaptic activity (see Monday & Castillo, 2017; Monday, Younts, & Castillo, 2018; Nicoll, 2017). The co-occurrence of activity in presynaptic and postsynaptic cells is now recognized as the critical factor in LTP, and the assumption that co-occurrence is a physiological necessity for learning and memory is often referred to as *Hebb's postulate for learning*.

Support for the idea that LTP is related to the neural mechanisms of learning and memory has come from several observations: (1) LTP can be elicited by low levels of stimulation that mimic normal neural activity; (2) LTP effects are most prominent in structures that have been implicated in learning and memory, such as the hippocampus; (3) learning can produce LTP-like changes in the hippocampus; (4) many drugs that influence learning and memory have parallel effects on LTP; (5) disruption of LTP impairs memory performance on many behavioral tasks, whereas enhancement of LTP improves memory performance (see Ricciarelli & Fedele, 2018); (6) behavioral changes that appear to be

Figure 11.18 Long-term potentiation in the granule cell layer of the rat hippocampal dentate gyrus.

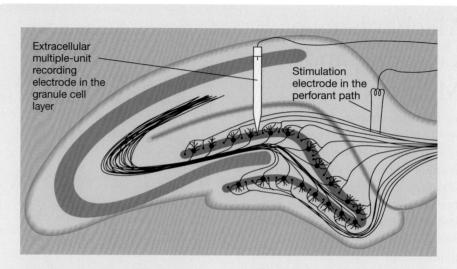

A single pulse of stimulation was administered to the perforant path, and the baseline response was recorded by an extracellular electrode in the granule cell layer. Then, several trains of intense high-frequency stimulation were applied to the perforant path to induce the LTP.

A single pulse of stimulation was administered 1 day later and again 1 week later to assess the magnitude of and duration of the potentiation. One measure of LTP is the increased amplitude of the *population spike*, in this case, the spike created by the firing of a greater number of granule cells.

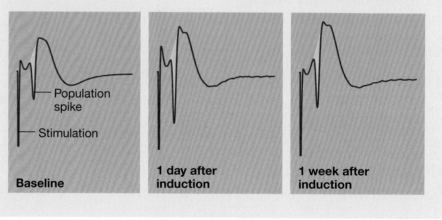

Traces courtesy of Michael Corcoran, Department of Anatomy, Physiology, and Pharmacology, University of Saskatchewan.

memories can be induced in mice via LTP (Nabavi et al., 2014); and (7) LTP occurs at specific synapses that have been shown to participate in learning and memory in simple invertebrate nervous systems. Still, it is important to keep in mind that much of this evidence is indirect and that LTP as induced in the laboratory by electrical stimulation is at best a caricature of the subtle cellular events that underlie learning and memory.

Conceiving of LTP as a three-part process, many researchers are investigating the mechanisms of *induction, maintenance,* and *expression*—that is, the processes by which high-frequency stimulations induce LTP (learning), the changes responsible for the maintenance of LTP (memory), and the changes that allow it to be expressed during the test (recall).

Induction of LTP: Learning

LO 11.17 Describe the mechanisms underlying the induction of LTP.

The NMDA (or N-methyl-D-aspartate) receptor is prominent at the synapses at which LTP is commonly studied. The **NMDA receptor** is a receptor for **glutamate**—the main excitatory neurotransmitter of the brain, as you learned in Chapter 4. An NMDA receptor does not respond maximally unless two events occur simultaneously: Glutamate must bind to it, and the postsynaptic neuron must already be partially depolarized. This dual requirement stems from the fact that the calcium channels associated with NMDA receptors only trigger a large influx of calcium ions if the postsynaptic neuron is already depolarized when glutamate binds to the receptors; it is the influx of calcium ions that triggers the cascade of events in the postsynaptic neuron that induces LTP.

An important characteristic of LTP induction at many glutamatergic synapses stems from the nature of the NMDA receptor and the requirement for co-occurrence of presynaptic glutamate release and postsynaptic depolarization for LTP to occur. This characteristic is not obvious under the usual, but unnatural, experimental condition in which LTP is induced by high-intensity, high-frequency stimulation, which always activates the postsynaptic neurons through massive temporal and spatial summation. However, when a more natural, low-intensity stimulation is applied, LTP is not induced unless the postsynaptic neurons are already partially depolarized so that their calcium channels open wide when glutamate binds to their NMDA receptors (see Bliss et al., 2018; Lisman, 2017).

The requirement for the postsynaptic neurons to be depolarized when glutamate binds to the NMDA receptors is an extremely important characteristic of conventional LTP because it permits neural networks to learn "associations." Let us explain. If one glutamatergic neuron were to fire and release its glutamate neurotransmitter without co-occurring depolarization at the postsynaptic neuron, there would be no LTP. However, if the postsynaptic neuron were depolarized by input from other neurons when the presynaptic neuron fired, the binding of the glutamate to the NMDA receptors would open wide the calcium channels, calcium ions would flow into the postsynaptic neuron, and transmission across the synapses between the presynaptic and postsynaptic neuron would be potentiated. Accordingly, the requirement for co-occurrence and the dependence of NMDA receptors on simultaneous binding and depolarization mean that, under natural conditions, LTP "records" the fact that there has been simultaneous activity in at least two converging inputs to the postsynaptic neuron—as would be produced, for example, by the simultaneous presentation of a conditional stimulus and an unconditional stimulus. Figure 11.19 summarizes the induction of NMDA-receptor–mediated LTP.

It is important to note that NMDA-receptor–mediated LTP, which is largely postsynaptic, is not the only type of LTP. For example, there are postsynaptic forms of LTP that do not rely on the NMDA receptor (Petrovic et al., 2017), presynaptic forms of LTP (see Bouvier et al., 2018; Monday & Castillo, 2017;

Figure 11.19 The induction of NMDA-receptor–mediated LTP.

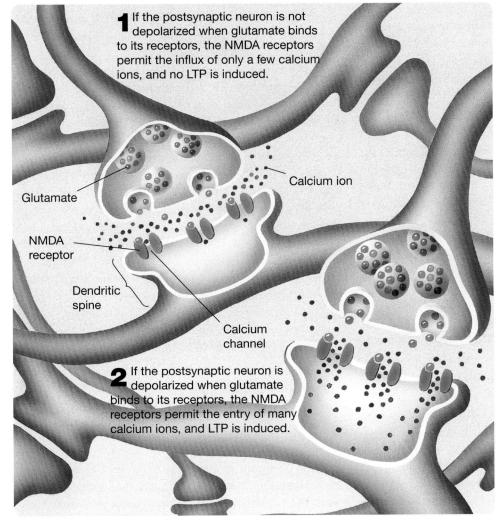

Monday, Younts, & Castillo, 2018), and LTP that relies on glial-cell activation (Kronschläger et al., 2016).

Maintenance and Expression of LTP: Storage and Recall

LO 11.18 Describe four findings that have emerged from the study of the maintenance and expression phases of LTP.

The search for the mechanisms underlying the maintenance and expression of LTP began with attempts to determine whether these mechanisms occur in presynaptic or postsynaptic neurons. This question has been answered: The maintenance and expression of LTP involve changes in both presynaptic and postsynaptic neurons. This discovery indicated that the mechanisms underlying the maintenance and expression of LTP are complex. Indeed, after more than four decades of research, we still do not have satisfactory answers. However, the quest for the neural mechanisms of LTP maintenance and expression has contributed to several important discoveries. Here are four of them:

- Once it became apparent that LTP occurs only at specific synapses on a postsynaptic neuron, it was clear that there must be a mechanism for keeping the events at one set of synapses on a postsynaptic neuron from affecting other synapses on the same neuron. This specificity is due to the **dendritic spines**; the calcium ions that enter a dendritic spine do not readily pass out of it, and thus they exert their effects locally (see Abraham, Jones, & Glanzman, 2019; Attardo, Fitzgerald, & Schnitzer, 2015; Colgan & Yasuda, 2014).

- It became apparent that maintenance of LTP involves structural changes, which depend on protein synthesis (see Abraham, Jones, & Glanzman, 2019; Bliss et al., 2018; Sweatt, 2016). The discovery that LTP causes structural changes was exciting because the structure of neurons and neural circuits had been assumed to be static. Many kinds of structural changes have been described (e.g., increases in number and size of synapses, increases in number and size of postsynaptic dendritic spines, changes in presynaptic and postsynaptic membranes, and changes in dendritic branching), and the changes have turned out to occur far more rapidly and more frequently than was once assumed (see Abraham, Jones, & Glanzman, 2019; Buonarati et al., 2019; Park et al., 2018; Wiegert & Oertner, 2015; Zhang et al., 2015).

- The discovery of structural changes in neurons following the induction of LTP stimulated a search for a mechanism by which a neuron's activity could change its structure. This led to the discovery of numerous **transcription factors** (intracellular proteins that bind to DNA and influence the operation of particular genes)

that were activated by neural activity (see Abraham, Jones, & Glanzman, 2019; Ryan et al., 2015; Sweatt, 2016).

- Francis Crick (co-discoverer of the DNA molecule) proposed in 1984 that DNA methylation (one sort of *epigenetic mechanism*; see Chapter 2) might serve an important role in memory storage. Since then, research has made it abundantly clear that many epigenetic mechanisms play a role in the maintenance of both LTP and memory (see Abraham, Jones, & Glanzman, 2019; Asok et al., 2019; Kim & Kaang, 2017; Leighton et al., 2017; Sweatt, 2016).

Variability of LTP

LO 11.19 Define long-term depression (LTD) and metaplasticity.

When we first learned about LTP, we were both excited. LTP seemed like a good model of learning and memory, and its simplicity suggested that its mechanisms could be identified. A generation of neuroscientists has shared our view, and LTP has become the focus of a massive research effort. However, it seems that researchers are further from ultimate answers than we naively thought they were about a quarter century ago. What has happened? Many important discoveries have been made, but rather than pointing to a simple mechanism, they have revealed that LTP is far more complex than first thought.

Most of the research on LTP has focused on NMDA-receptor–mediated LTP in the hippocampus. It is now clear that NMDA-receptor–mediated LTP involves a complex array of changes that are difficult to sort out. In addition, LTP has been documented in many other parts of the CNS, where it tends to be mediated by different mechanisms (see Grau, 2014; Gruart et al., 2015). There is also **long-term depression (LTD)**, which is the "flip side" of LTP and occurs in response to prolonged low-frequency stimulation of presynaptic neurons (see Atwood, Lovinger, & Mathur, 2014; Bliss et al., 2018; Connor & Wang, 2015). And then there is **metaplasticity**, which refers to the fact that LTP and/or LTD induction can be modulated by prior synaptic activity (see Abraham, Jones, & Glanzman, 2019; Hulme, Jones, & Abraham, 2013; Müller-Dahlhaus & Ziemann, 2015). Presumably, a full understanding of LTP and its role in memory will require an understanding of LTD and metaplasticity.

Journal Prompt 11.6

LTP is one of the most intensely studied of all neuro-scientific phenomena. Why do you think that is?

The dream of discovering the neural basis of learning and memory has attracted many neuroscientists to the study of LTP. Although this dream has not yet been fulfilled, the study of LTP has led to several important discoveries about the function and plasticity of neural systems. By this criterion, its study has been a great success.

Nonsynaptic Mechanisms of Learning and Memory

LO 11.20 Describe two sorts of neuroplastic changes that occur outside the synapse that may play a role in learning and memory.

Although LTP, a synaptic form of neuroplasticity, is the most widely studied cellular mechanism of learning and memory, there are others. For example, we now know that there are neuroplastic changes that occur outside the synapse that might also play a role in learning and memory. Such changes include epigenetic mechanisms in the cell nucleus (see Abraham, Jones, & Glanzman, 2019) and changes to the structure of axons. With respect to the latter, it has recently been observed that oligodendrocytes (a glial cell of the CNS; see Chapter 3) modulate their myelination of axons in response to experience (see Chang, Redmond, & Chan, 2016).

Contrary to popular belief, the myelin sheath is not a static part of an axon. Rather, myelination is an ongoing process throughout one's lifespan, and changes in the myelin sheath can affect the speed of transmission of information both within and between neurons. Changes in myelination occur through at least five mechanisms. First, unmyelinated axons can be myelinated by oligodendrocytes. Second, axons that were previously myelinated can have their myelin sheaths removed. Third, myelin can be replaced when old myelin sheaths degrade. Fourth, myelin sheaths can be thinned or thickened. Finally, myelin sheaths can be lengthened or shortened (see Chang, Redmond, & Chan, 2016; Forbes & Gallo, 2017).

Changes in myelination have been shown to occur following learning. For example, when rats learn a new spatial navigation task (e.g., the Morris water maze) there is an associated increase in myelination. In humans, both motor-skill learning and working-memory training lead to an increase in myelination in those brain areas associated with the particular type of learning (see Sampaio-Baptista & Johansen-Berg, 2017). In short, plasticity of myelin sheathing seems to play an important role in learning and memory (see Kaller et al., 2017; Monje, 2018; Mount & Monje, 2017).

Conclusion: Biopsychology of Memory and You

Because this chapter has emphasized the amnesic effects of brain dysfunction, you may have been left with the impression that the biopsychological study of memory has little

direct relevance to individuals, perhaps like you, with intact healthy brains. This final module shows that such is not the case. It makes this point by describing two interesting lines of research and one provocative case study.

Infantile Amnesia

LO 11.21 Define infantile amnesia and describe two experiments that investigated whether infantile amnesia extends to implicit memories.

We all experience **infantile amnesia**; that is, we remember virtually nothing of the events of our infancy (see Callaghan, Li, & Richardson, 2014; Sneed, 2014; Travaglia et al., 2016). Newcombe and her colleagues (2000) addressed the following question: Do normal children who fail to explicitly recall or recognize things from their early childhood display preserved implicit memories for these things? The results of two experiments indicate that the answer is "yes."

In one study of infantile amnesia (Newcombe & Fox, 1994), children were shown a series of photographs of preschool-aged children, some of whom had been their preschool classmates. The children recognized a few of their former preschool classmates. However, whether they explicitly remembered a former classmate or not, they consistently displayed a large skin conductance response to the photographs of their former classmates but not to the control photographs.

In a second study of infantile amnesia, Drummey and Newcombe (1995) used a version of the incomplete-pictures test. First, they showed a series of drawings to 3-year-olds, 5-year-olds, and adults. Three months later, the researchers assessed the implicit memories for these drawings by asking each participant to identify them ("It's a car," "It's a chair," etc.) and some control drawings as quickly as they could. During the test, the drawings were first presented badly out of focus, but became progressively sharper over time. Following this test of implicit memory, explicit memory was assessed by asking the participants which of the drawings they remembered seeing before. The 5-year-olds and adults showed better explicit memory than the 3-year-olds did; that is, they were more likely to recall seeing drawings from the original series. However, all three groups displayed substantial implicit memory: All participants were able to identify the drawings they had previously seen sooner, even when they had no conscious recollection of them.

Smart Drugs: Do They Work?

LO 11.22 Discuss the findings on the efficacy of smart drugs.

Nootropics, or **smart drugs**, are substances (drugs, supplements, herbal extracts, etc.) that are thought to improve memory. The shelves of health food stores are full of them, and even more are available on the Internet. Perhaps you have heard of, or even tried, some of them: *ginkgo biloba* extracts, ginseng extracts, multivitamins, glucose, cholinergic agonists, Piracetam, antioxidants, phospholipids, stimulants (e.g., amphetamine, methylphenidate), and many more. Those offering nootropics for sale claim that scientific evidence has proven that these substances improve the memories of healthy children and adults and block the adverse effects of aging on memory. Are these claims really supported by valid scientific evidence?

The evidence that nootropics enhance memory has been reviewed several times by independent scientists (e.g., Hussain & Mehta, 2011; Partridge et al., 2012). The following pattern of conclusions has emerged from these reviews:

- Although nootropics are often marketed to healthy adults, most research has been done on either non-humans or humans with memory difficulties (i.e., the elderly; e.g., Onaolapo, Obelawo, & Onaolapo, 2019; Uddin et al., 2019).
- The relevant research with humans tends to be of low quality, with few participants and poor controls.
- For each purported nootropic, there are typically a few positive findings on which the vendors focus; however, these findings are often difficult to replicate or represent very small effect sizes (see Farah, 2015).

In short, no purported nootropic has been convincingly shown to have memory-enhancing effects. There may be enough positive evidence to warrant continued investigation of some potential nootropics (e.g., Hinnebusch, 2015), but there is not nearly enough to justify the various claims that are made in advertisements for these substances. Why do you think there is such a huge gulf between the evidence and the claims?

Journal Prompt 11.7

Address the question posed to you in this paragraph: With respect to nootropics, why do you think there is such a huge gulf between the evidence and the claims?

Posttraumatic Amnesia and Episodic Memory

LO 11.23 Explain what the case of R.M. tells us about the relationship between posttraumatic amnesia and episodic memory.

This chapter began with the case of H.M.; it ends with the case of R.M. The case of R.M. is one of the most ironic that we have encountered. R.M. is a biopsychologist, and, as you will learn, his vocation played an important role in one of his symptoms.

The Case of R.M., the Biopsychologist Who Remembered H.M.

R.M. fell on his head while skiing; when he regained consciousness, he was suffering from both retrograde and anterograde amnesia. For several hours, he could recall few of the events of his previous life. He could not remember if he was married, where he lived, or where he worked. He had lost most of his episodic memory.

Also, many of the things that happened to him in the hours after his accident were forgotten as soon as his attention was diverted from them. For example, in the car on his way to the hospital, R.M. chatted with the person sitting next to him—a friend of a friend with whom he had skied all day. But each time his attention was drawn elsewhere—for example, by the mountain scenery—he completely forgot this person and reintroduced himself.

This was a classic case of posttraumatic amnesia. Like H.M., R.M. was trapped in the present, with only a cloudy past and seemingly no future. The irony of the situation was that during those few hours, when R.M. could recall few of the events of his own life, his thoughts repeatedly drifted to one semantic memory—his memory of a person he remembered learning about somewhere in his muddled past. Through the haze, he remembered H.M., his fellow prisoner of the present and wondered if the same fate lay in store for him.

R.M. recovered fully and looks back on what he can recall of his experience with relief and a feeling of empathy for H.M. Unlike H.M., R.M. received a reprieve, but his experience left him with a better appreciation for the situation of those like H.M., who served a life sentence.

Themes Revisited

All four of the book's major themes played roles in this chapter. The biopsychology of memory is a neuroplastic phenomenon—it focuses on the changes in neural function that store experiences. Thus, the neuroplasticity theme was implicit throughout the chapter.

Because the study of the neural mechanisms of memory is based largely on the study of humans with amnesia, the clinical implications theme played a significant role. The study of memory disorders has so far been a one-way street: We have learned much about memory and its neural mechanisms from studying amnesic patients, but we have not yet learned enough to treat their memory problems.

Animal models have also played a major role in the study of memory disorders. Only so much progress can be made studying human clinical cases; questions of causation must be addressed with animal models. The study of medial temporal lobe amnesia illustrates the comparative approach at its best.

Finally, the thinking creatively theme appeared at three points in the chapter where you were encouraged to think in unconventional ways. You were encouraged to consider: (1) why it proved so difficult to develop an animal model of medial temporal lobe amnesia; (2) why cases of cerebral ischemia do not provide strong evidence for involvement of the hippocampus in object recognition memory; and

(3) the gulf between the actual evidence for smart drugs and the advertised claims.

Both emerging themes appeared in this chapter. First, the thinking about epigenetics theme appeared when epigenetics was discussed in the context of both synaptic and nonsynaptic mechanisms of learning and memory. Second, the consciousness theme was present in the discussions of the types of memory we have. Indeed, the distinction between conscious and unconscious forms of memory (e.g., explicit vs. implicit memory) is fundamental to current memory research.

Key Terms

Learning, p. 289
Memory, p. 289
Amnesia, p. 289

Amnesic Effects of Bilateral Medial Temporal Lobectomy

Bilateral medial temporal lobectomy, p. 289
Hippocampus, p. 289
Amygdala, p. 289
Lobectomy, p. 289
Lobotomy, p. 289
Retrograde amnesia, p. 289
Anterograde amnesia, p. 289
Short-term memory, p. 289
Long-term memory, p. 289
Digit span, p. 290
Global amnesia, p. 290
Incomplete-pictures test, p. 291
Remote memory, p. 292
Memory consolidation, p. 292
Explicit memories, p. 292
Implicit memories, p. 292
Medial temporal lobe amnesia, p. 292
Repetition priming tests, p. 292
Semantic memories, p. 292
Episodic memories, p. 292
Global cerebral ischemia, p. 294
Pyramidal cell layer, p. 294

CA1 subfield, p. 294
Transient global amnesia, p. 294

Amnesias of Korsakoff's Syndrome and Alzheimer's Disease

Korsakoff's syndrome, p. 295
Mediodorsal nuclei, p. 295
Medial diencephalic amnesia, p. 295
Alzheimer's disease, p. 295
Basal forebrain, p. 296

Amnesia after Traumatic Brain Injury: Evidence for Consolidation

Posttraumatic amnesia, p. 296
Electroconvulsive shock (ECS), p. 297
Standard consolidation theory, p. 297
Dual-trace theory, p. 297
Engram, p. 298
Reconsolidation, p. 298

Evolving Perspective of the Role of the Hippocampus in Memory

Delayed nonmatching-to-sample test, p. 299
Medial temporal cortex, p. 300
Mumby box, p. 301

Neurons of the Medial Temporal Lobes and Memory

Morris water maze test, p. 303
Radial arm maze test, p. 303

Reference memory, p. 303
Working memory, p. 303
Place cells, p. 304
Entorhinal cortex, p. 304
Grid cells, p. 304
Concept cells, p. 305
Engram cells, p. 306

Where Are Memories Stored?

Inferotemporal cortex, p. 306
Prefrontal cortex, p. 307
Cerebellum, p. 308
Striatum, p. 308

Cellular Mechanisms of Learning and Memory

Long-term potentiation (LTP), p. 309
NMDA receptor, p. 311
Glutamate, p. 311
Dendritic spines, p. 312
Transcription factors, p. 312
Long-term depression (LTD), p. 312
Metaplasticity, p. 312

Conclusion: Biopsychology of Memory and You

Infantile amnesia, p. 313
Nootropics (smart drugs), p. 313

Chapter 12
Hunger, Eating, and Health

Why Do So Many People Eat Too Much?

filadendron/Getty Images

 Chapter Overview and Learning Objectives

Digestion, Energy Storage, and Energy Utilization

LO 12.1 Summarize the process of digestion and explain how energy is stored in the body.

LO 12.2 Explain the three phases of energy metabolism.

Theories of Hunger and Eating: Set Points versus Positive Incentives

LO 12.3 Describe the set-point assumption and describe the glucostatic and lipostatic set-point theories of hunger and eating. Also, outline three problems with set-point theories of hunger and eating.

LO 12.4 Describe the positive-incentive perspective on hunger and eating.

Factors That Determine What, When, and How Much We Eat

LO 12.5 Describe at least two factors that determine what we eat.

LO 12.6 Describe at least two factors that influence when we eat.

LO 12.7 Describe some of the major factors that influence how much we eat.

Physiological Research on Hunger and Satiety

LO 12.8 Explain the nature of the relationship between blood glucose levels and hunger and satiety.

LO 12.9 Describe the evolution of thinking about the role of various hypothalamic nuclei in hunger and satiety.

LO 12.10 Describe the role of the gastrointestinal tract in satiety.

LO 12.11 Describe the discovery of the role of hypothalamic circuits, peptides, and the gut in food consumption and metabolism.

LO 12.12 Describe the role of serotonin in satiety.

LO 12.13 Describe the symptoms and etiology of Prader-Willi syndrome.

Body-Weight Regulation: Set Points versus Settling Points

LO 12.14 Evaluate the evidence for set-point assumptions about body weight and eating.

LO 12.15 Compare and evaluate set-point and settling-point models of body-weight regulation.

Human Overeating: Causes, Mechanisms, and Treatments

LO 12.16 Explain why there is cause for concern surrounding the overeating epidemic.

LO 12.17 Describe, from an evolutionary perspective, why there is a current epidemic of overeating.

LO 12.18 Give some reasons as to why some people gain weight from overeating while others do not.

LO 12.19 Explain why weight-loss programs are typically ineffective.

LO 12.20 Explain how leptin and insulin are feedback signals for the regulation of body fat.

LO 12.21 Describe two sorts of treatments for overeating and/or high body-fat levels.

Anorexia and Bulimia Nervosa

LO 12.22 Describe the symptoms of anorexia nervosa and bulimia nervosa.

LO 12.23 Explain how anorexia and bulimia are, and are not, related.

LO 12.24 Explain why those starving due to anorexia do not appear to be as hungry as they should.

LO 12.25 Explain how anorexia might result from conditioned taste aversions.

Eating is a behavior that is of interest to virtually everyone. We all do it, and most of us derive great pleasure from it. But for many of us, it becomes a source of serious personal and health problems.

Most eating-related health problems in industrialized nations are associated with eating too much—the average American consumes 3,800 calories per day, about twice the average daily requirement. Such high levels of consumption lead to increased body weight in most individuals, which can in turn cause serious health problems (see Sharma & Campbell-Scherer, 2017; Frühbeck et al., 2019).

Increased body weight as a result of excessive consumption is so common that it is considered to be an epidemic (see Ogden et al., 2014). Indeed, a 2016 meta-analysis reported that for the first time in recorded history, overweight individuals outnumbered underweight individuals (Di Angelantonio et al., 2016). The resulting financial and personal costs of overeating are staggering. Each year, an estimated 400,000 U.S. citizens die unnecessarily from disorders caused by excessive eating (see Masters et al., 2013). Although the United States is a trendsetter when it comes to overeating, many other countries are not far behind (see Scully, 2014). Ironically, as overeating has reached epidemic proportions, there has been a related increase in disorders associated with eating too little. For example, about 1.2 percent of Americans will suffer from *anorexia* or *bulimia* at some point in their lives, and these conditions can be life-threatening in extreme cases (see Swanson et al., 2011).

The increases in eating-related disorders that have occurred over the past few decades in many countries stand in direct opposition to most people's thinking about hunger and eating. Many people—and we assume this includes you—believe that hunger and eating are normally triggered when the body's energy resources fall below a prescribed optimal level, or **set point**. They appreciate that many factors influence hunger and eating, but they assume that in general the hunger and eating system has evolved to supply the body with just the right amount of energy.

This chapter explores the incompatibility of the set-point assumption with the current epidemic of eating disorders. If we all have hunger and eating systems whose primary function is to maintain energy resources at optimal levels, then eating disorders should be rare. The fact that eating disorders are so prevalent suggests that hunger and eating are regulated in some other way. This chapter will repeatedly challenge you to think in new ways about issues that impact your health and longevity and will provide new insights of great personal relevance—we guarantee it.

Before you move on, we would like you to pause to consider a case study that links this chapter with the preceding one (Rozin et al., 1998). What would a severely amnesic patient do if offered a meal shortly after finishing one? If his hunger and eating were controlled by energy set points, he would refuse the second meal. Did he?

The Case of the Man Who Forgot Not to Eat

R.H. was a 48-year-old male whose progress in graduate school was interrupted by the development of a severe anterograde amnesia for long-term explicit memory. His amnesia was similar in pattern and severity to that of H.M., whom you met in Chapter 11, and an MRI examination revealed bilateral damage to the medial temporal lobes.

The meals offered to R.H. were selected on the basis of interviews with him about the foods he liked: veal parmigiana

(about 750 calories) plus all the apple juice he wanted. On one occasion, he was offered a second meal about 15 minutes after he had eaten the first, and he ate it. When offered a third meal 15 minutes later, he ate that, too. When offered a fourth meal he rejected it, claiming that his "stomach was a little tight."

Then, a few minutes later, R.H. announced that he was going out for a good walk and a meal. When asked what he was going to eat, his answer was "veal parmigiana."

Clearly, R.H.'s hunger (i.e., motivation to eat) did not result from an energy deficit. Other cases like that of R.H. have been reported by Higgs and colleagues (2008).

Digestion, Energy Storage, and Energy Utilization

The primary purpose of hunger is to increase the probability of eating, and the primary purpose of eating is to supply the body with the molecular building blocks and energy it needs to survive and function. This module provides the foundation for our consideration of hunger and eating by providing a brief overview of the processes by which food is digested, stored, and converted to energy.

Digestion and Energy Storage in the Body

LO 12.1 Summarize the process of digestion and explain how energy is stored in the body.

DIGESTION. The *gastrointestinal tract* and the process of digestion are illustrated in Figure 12.1. **Digestion** is the gastrointestinal process of breaking down food and absorbing its constituents into the body. In order to appreciate the basics of digestion, it is useful to consider the body without its protuberances, as a simple living tube with a hole at each end. To supply itself with energy and other nutrients, the tube puts food into one of its two holes—the one with teeth—and passes the food along its internal canal so that the food can be broken down and partially absorbed from the canal into the body. Much of the work of breaking down the food we ingest is done by the constituents of our **gut microbiome** (the bacteria and other organisms that live in our gastrointestinal tract). The leftovers of what we ingest are jettisoned from the other end. Although this is not a particularly appetizing description of eating, it does serve to illustrate that, strictly speaking, food has not been consumed until it has been digested.

ENERGY STORAGE IN THE BODY. As a consequence of digestion, energy is delivered to the body in three forms: (1) **lipids** (fats), (2) **amino acids** (the breakdown products of proteins), and (3) **glucose** (a simple sugar that

Figure 12.1 The gastrointestinal tract and the process of digestion. Not shown in the figure is the gut microbiome, which includes the bacteria and other organisms that live inside our gastrointestinal tract and help break down, store, and regulate the food we ingest—see Fetisov (2017), Martin et al. (2018), and Vuong et al. (2017).

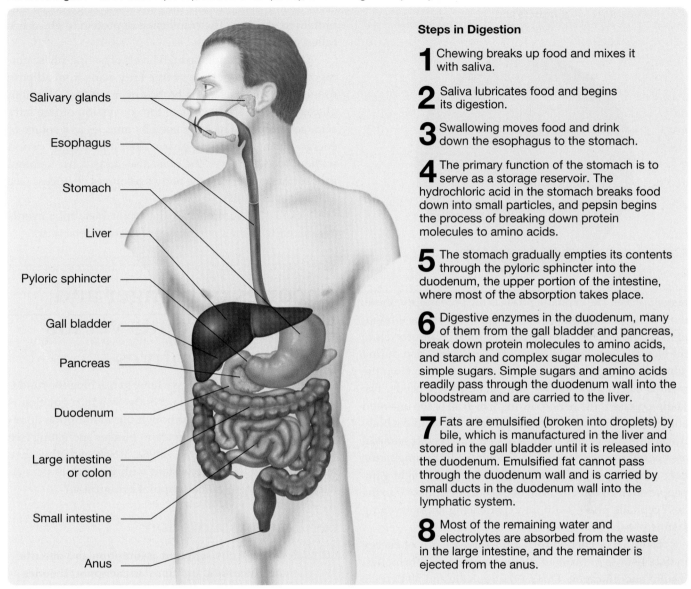

Steps in Digestion

1 Chewing breaks up food and mixes it with saliva.

2 Saliva lubricates food and begins its digestion.

3 Swallowing moves food and drink down the esophagus to the stomach.

4 The primary function of the stomach is to serve as a storage reservoir. The hydrochloric acid in the stomach breaks food down into small particles, and pepsin begins the process of breaking down protein molecules to amino acids.

5 The stomach gradually empties its contents through the pyloric sphincter into the duodenum, the upper portion of the intestine, where most of the absorption takes place.

6 Digestive enzymes in the duodenum, many of them from the gall bladder and pancreas, break down protein molecules to amino acids, and starch and complex sugar molecules to simple sugars. Simple sugars and amino acids readily pass through the duodenum wall into the bloodstream and are carried to the liver.

7 Fats are emulsified (broken into droplets) by bile, which is manufactured in the liver and stored in the gall bladder until it is released into the duodenum. Emulsified fat cannot pass through the duodenum wall and is carried by small ducts in the duodenum wall into the lymphatic system.

8 Most of the remaining water and electrolytes are absorbed from the waste in the large intestine, and the remainder is ejected from the anus.

is the breakdown product of complex *carbohydrates*, that is, starches and sugars).

The body uses energy continuously, but its consumption is intermittent; therefore, it must store energy for use in the intervals between meals. Energy is stored in three forms: *fats, glycogen,* and *proteins*. Most of the body's energy reserves are stored as fats, relatively little as glycogen and proteins (see Figure 12.2). Thus, changes in the body weights of adult humans are largely a consequence of changes in the amount of their stored body fat.

Why is fat the body's preferred way of storing energy? Glycogen, which is largely stored in the liver and muscles, might be expected to be the body's preferred mode of energy storage because it is so readily converted to glucose—the body's main directly utilizable source of energy. But there are two reasons why fat, rather than glycogen, is the primary mode of energy storage: (1) A gram of fat can store almost twice as much energy as a gram of glycogen, and (2) glycogen, unlike fat, attracts and holds substantial quantities of water. Consequently, if all your fat calories were stored as glycogen, you would likely weigh well over 275 kilograms (600 pounds).

Three Phases of Energy Metabolism

LO 12.2 Explain the three phases of energy metabolism.

There are three phases of *energy metabolism* (the chemical changes by which energy is made available for an organism's use): the cephalic phase, the absorptive phase, and the

Figure 12.2 Distribution of stored energy in an average person.

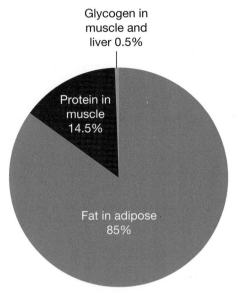

Glycogen in muscle and liver 0.5%

Protein in muscle 14.5%

Fat in adipose 85%

fasting phase. The **cephalic phase** is the preparatory phase; it often begins with the sight, smell, or even just the thought of food, and it ends when the food starts to be absorbed into the bloodstream. The **absorptive phase** is the period during which the energy absorbed into the bloodstream from the meal is meeting the body's immediate energy needs. The **fasting phase** is the period during which all of the unstored energy from the previous meal has been used and the body is withdrawing energy from its reserves to meet its immediate energy requirements; it ends with the beginning of the next cephalic phase. During periods of rapid weight gain, people often go directly from one absorptive phase into the next cephalic phase, without experiencing an intervening fasting phase.

The flow of energy during the three phases of energy metabolism is controlled by two pancreatic hormones: insulin and glucagon. During the cephalic and absorptive phases, the pancreas releases a great deal of insulin into the bloodstream and very little glucagon. **Insulin** does three things: (1) It promotes the use of glucose as the primary source of energy by the body; (2) it promotes the conversion of bloodborne fuels to forms that can be stored: glucose to glycogen and fat and amino acids to proteins; and (3) it promotes the storage of glycogen in liver and muscle, fat in adipose tissue, and proteins in muscle (see Figure 12.2). In short, the function of insulin during the cephalic phase is to lower the levels of bloodborne fuels, primarily glucose, in anticipation of the impending influx; and its function during the absorptive phase is to minimize the increasing levels of bloodborne fuels by utilizing and storing them.

In contrast to the cephalic and absorptive phases, the fasting phase is characterized by high blood levels of **glucagon** and low levels of insulin. Without high levels of insulin, glucose has difficulty entering most body cells; thus,

glucose stops being the body's primary fuel. In effect, this saves the body's glucose for the brain, because insulin is not required for glucose to enter most brain cells. The low levels of insulin also promote the conversion of glycogen and protein to glucose. (The conversion of protein to glucose is called **gluconeogenesis**.)

On the other hand, the high levels of fasting-phase glucagon promote the release of **free fatty acids** from adipose tissue and their use as the body's primary fuel. The high glucagon levels also stimulate the conversion of free fatty acids to **ketones**, which are used by muscles as a source of energy during the fasting phase. After a prolonged period without food, however, the brain also starts to use ketones, thus further conserving the body's resources of glucose (see Mattson et al., 2018).

Figure 12.3 summarizes the major metabolic events associated with the three phases of energy metabolism.

Theories of Hunger and Eating: Set Points versus Positive Incentives

One of the main difficulties we have in teaching the fundamentals of hunger, eating, and body-weight regulation is the set-point assumption. Although this assumption dominates most people's thinking about hunger and eating (see Assanand, Pinel, & Lehman, 1998a, 1998b), whether they realize it or not, it is inconsistent with the bulk of the evidence. What exactly is the set-point assumption?

Set-Point Assumption

LO 12.3 Describe the set-point assumption and describe the glucostatic and lipostatic set-point theories of hunger and eating. Also, outline three problems with set-point theories of hunger and eating.

Most people attribute *hunger* (the motivation to eat) to the presence of an energy deficit, and they view eating as the means by which the energy resources of the body are returned to their optimal level—that is, to the *energy set point*. Figure 12.4 summarizes this **set-point assumption**. After a *meal* (a bout of eating), a person's energy resources are assumed to be near their set point and to decline thereafter as the body uses energy to fuel its physiological processes. When the level of the body's energy resources falls far enough below the set point, a person becomes motivated by hunger to initiate another meal. The meal continues, according to the set-point assumption, until the energy level returns to its set point and the person feels *satiated* (not hungry).

Figure 12.3 The major events associated with the three phases of energy metabolism: the cephalic, absorptive, and fasting phases.

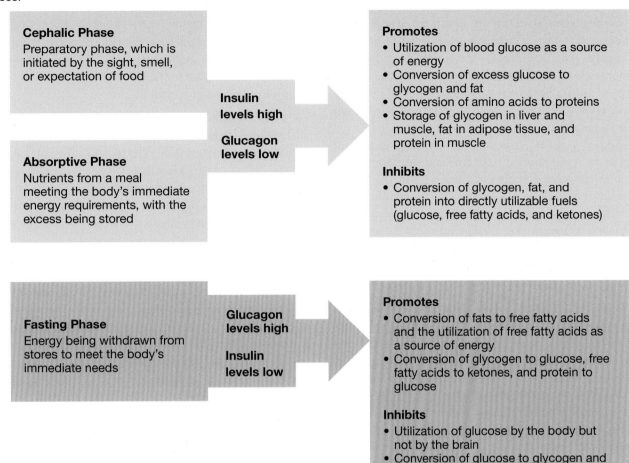

Cephalic Phase
Preparatory phase, which is initiated by the sight, smell, or expectation of food

Absorptive Phase
Nutrients from a meal meeting the body's immediate energy requirements, with the excess being stored

Insulin levels high

Glucagon levels low

Promotes
- Utilization of blood glucose as a source of energy
- Conversion of excess glucose to glycogen and fat
- Conversion of amino acids to proteins
- Storage of glycogen in liver and muscle, fat in adipose tissue, and protein in muscle

Inhibits
- Conversion of glycogen, fat, and protein into directly utilizable fuels (glucose, free fatty acids, and ketones)

Fasting Phase
Energy being withdrawn from stores to meet the body's immediate needs

Glucagon levels high

Insulin levels low

Promotes
- Conversion of fats to free fatty acids and the utilization of free fatty acids as a source of energy
- Conversion of glycogen to glucose, free fatty acids to ketones, and protein to glucose

Inhibits
- Utilization of glucose by the body but not by the brain
- Conversion of glucose to glycogen and fat, and amino acids to protein
- Storage of fat in adipose tissue

Figure 12.4 The energy set-point view that is the basis of many people's thinking about hunger and eating.

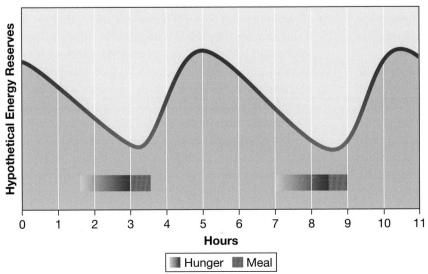

Hypothetical Energy Reserves

Hours

Hunger Meal

Set-point models assume that hunger and eating work in much the same way as a thermostat-regulated heating system in a cool climate. The heater increases the house temperature until it reaches its set point (the thermostat setting). The heater then shuts off, and the temperature of the house gradually declines until it becomes low enough to turn the heater back on. All set-point systems have three components: a set-point mechanism, a detector mechanism, and an effector mechanism. The *set-point mechanism* defines the set point, the *detector mechanism* detects deviations from the set point, and the *effector mechanism* acts to eliminate the deviations. For example, the set-point, detector, and effector mechanisms of a heating system are the thermostat, the thermometer, and the heater, respectively.

All set-point systems are **negative feedback systems**—systems in which feedback from changes in one direction elicit compensatory effects in the opposite direction. Negative feedback systems are common in mammals because they act to maintain **homeostasis**—a stable internal environment—which is critical for mammals' survival (see Clemmensen et al., 2017; Kotas & Medzhitov, 2015; Ramsay & Woods, 2014). Set-point systems combine negative feedback with a set point to keep an internal environment fixed at the prescribed point.

Set-point systems seemed necessary when the adult human brain was assumed to be immutable: Because the brain couldn't change, energy resources had to be highly regulated. However, we now know that the adult human brain is plastic and capable of considerable adaptation. Thus, there is no longer a logical imperative for the set-point regulation of eating. Throughout this chapter, you will need to put aside your preconceptions and base your thinking about hunger and eating entirely on the evidence.

GLUCOSTATIC THEORY. In the 1940s and 1950s, researchers working under the assumption that eating is regulated by some type of set-point system speculated about the nature of the regulation. Several researchers suggested that eating is regulated by a system designed to maintain a blood glucose set point—the idea being that we become hungry when our blood glucose levels drop significantly below their set point and that we become satiated when eating returns our blood glucose levels to their set point. The various versions of this theory are collectively referred to as the **glucostatic theory**. It seemed to make good sense that the main purpose of eating is to defend a blood glucose set point because glucose is the brain's primary fuel.

LIPOSTATIC THEORY. The **lipostatic theory** is another set-point theory proposed in various forms in the 1940s and 1950s. According to this theory, every person has a set point for body fat, and deviations from this set point produce compensatory adjustments in the level of eating that return levels of body fat to their set point. The most frequently cited support for the theory is the fact that the body weights of adults stay relatively constant.

The glucostatic and lipostatic theories were viewed as complementary, not mutually exclusive. The glucostatic theory was thought to account for meal initiation and termination, whereas the lipostatic theory was thought to account for long-term regulation. Thus, the dominant view in the 1950s was that eating is regulated by the interaction between two set-point systems: a short-term glucostatic system and a long-term lipostatic system. The simplicity of these 1950s theories is appealing. Remarkably, they are still being presented as well-established facts both in the popular media and in some scientific literature; perhaps you have encountered them.

PROBLEMS WITH SET-POINT THEORIES OF HUNGER AND EATING. Set-point theories of hunger and eating have several serious weaknesses. You have already learned one fact that undermines these theories: There is an epidemic of overeating, which should not occur if eating is regulated by a set point. Let's look at three more major weaknesses with glucostatic, lipostatic, and other set-point theories of hunger and eating.

- First, set-point theories of hunger and eating are inconsistent with basic eating-related evolutionary pressures as we understand them. The major eating-related problem faced by our ancestors was the inconsistency and unpredictability of the food supply. Thus, in order to survive, it was important for them to eat large quantities of good food when it was available so that calories could be banked in the form of body fat. For any warm-blooded species to survive under natural conditions, it needs a hunger and eating system that prevents energy deficits, rather than one that merely responds to them once they have developed. From this perspective, it is difficult to imagine how a hunger and feeding system based entirely on set points could have evolved in mammals (see Johnson, 2013).

- Second, major predictions of the set-point theories of hunger and eating have not been confirmed. Early studies seemed to support the set-point theories by showing that large reductions in body fat, produced by starvation, or large reductions in blood glucose, produced by insulin injections, induce increases in eating in laboratory animals. The problem is that reductions in blood glucose of the magnitude needed to reliably induce eating rarely occur naturally. Indeed, efforts to reduce meal size by having volunteers unknowingly consume a high-calorie drink before eating have been unsuccessful. Conversely, high levels of fat deposits at the time of eating are associated with increased, rather than decreased, hunger (see Ribeiro et al., 2018).

- Third, set-point theories of hunger and eating are deficient because they fail to recognize the major influences on hunger and eating of such important factors as taste, learning, and social influences. To convince yourself of the importance of these factors, pause for a minute and imagine the sight, smell, and taste of your favorite food. Perhaps it is a succulent morsel of lobster meat covered with melted garlic butter, a piece of cheesecake, sour gummy candies (SB's favorite), or a plate of sizzling homemade french fries (JP's favorite). Are you starting to feel a bit hungry? If a plate of fries was sitting in front of you right now, wouldn't you eat one, or maybe eat the whole plateful? Have you not on occasion felt discomfort after a large main course, only to polish off a substantial dessert? The usual positive answers to these questions lead unavoidably to the conclusion that hunger and eating are not rigidly controlled by deviations from energy set points.

Positive-Incentive Perspective

LO 12.4 Describe the positive-incentive perspective on hunger and eating.

The inability of set-point theories to account for the basic phenomena of eating and hunger led to the development of an alternative theoretical perspective (see Berridge, 2004). The central assertion of this perspective, commonly referred to as **positive-incentive theory**, is that humans and other animals are not normally driven to eat by internal energy deficits but are drawn to eat by the anticipated pleasure of eating—the anticipated pleasure of a behavior is called its **positive-incentive value**. There are several different positive-incentive theories, and we refer generally to all of them as the *positive-incentive perspective*.

The major tenet of the positive-incentive perspective on eating is that eating is controlled in much the same way as sexual behavior: We engage in sexual behavior not because we have an internal deficit but because we have evolved to crave it. The evolutionary pressures of unexpected food shortages have shaped us and all other warm-blooded animals (who need a continuous supply of energy to maintain their body temperatures) to take advantage of good food when it is present and eat it. According to the positive-incentive perspective, it is the presence of good food, or the anticipation of it, that normally makes us hungry, not an energy deficit.

According to the positive-incentive perspective, the degree of hunger you feel at any particular time depends on the interaction of all the factors that influence the positive-incentive value of eating (see Palmiter, 2007). These include the following: the flavor of the food you are likely to consume, what you have learned about the effects of this food either from eating it previously or from other people, the amount of time since you last ate, the type and quantity of food in your gut, whether or not other people are present and eating, whether or not your blood glucose levels are within the normal range. This partial list illustrates one strength of the positive-incentive perspective. Unlike set-point theories, positive-incentive theories do not single out one factor as the major determinant of hunger and ignore the others.

In this module, you learned that most people think about hunger and eating in terms of energy set points and were introduced to an alternative way of thinking—the positive-incentive perspective. Which way is correct? If you are like most people, you have an attachment to familiar ways of thinking and a resistance to new ones. Try to put this tendency aside and base your views about this important issue entirely on the evidence.

The next module describes some of the things that biopsychological research has taught us about hunger and eating. As you progress through the module, notice the superiority of the positive-incentive theories over set-point theories in accounting for the basic facts.

Factors That Determine What, When, and How Much We Eat

This module describes major factors that collectively determine what we eat, when we eat, and how much we eat. Notice that energy deficits are not included among these factors. Although major energy deficits clearly increase hunger and eating, they are not a common factor for those of us who can readily access food (see Kaviani & Cooper, 2017). If you have such easy access to food, you might believe your body is usually short of energy just before a meal. It is not.

Factors That Influence What We Eat

LO 12.5 Describe at least two factors that determine what we eat.

Certain tastes have a high positive-incentive value for virtually all members of a species. For example, most humans have a special fondness for sweet, fatty, and salty tastes. This species-typical pattern of human taste preferences is adaptive because in nature sweet and fatty tastes are typically characteristic of high-energy foods rich in vitamins and minerals, and salty tastes are characteristic of sodium-rich foods. In contrast, bitter tastes, for which most humans have an aversion, are often associated with toxins. Superimposed on our species-typical taste preferences and aversions, each of us has the ability to learn new specific taste preferences and aversions (see Clouard, Meunier-Salaün, & Val-Laillet, 2012).

LEARNED TASTE PREFERENCES AND AVERSIONS. Animals learn to prefer tastes that are followed by an infusion of calories, and they learn to avoid tastes that are followed by illness. In addition, humans and other animals learn what to eat from their conspecifics. For example, rats learn to prefer flavors they experience in mother's milk and those that they smell on the breath of other rats (see Galef, Whishkin, & Bielavska, 1997). Similarly, in humans, many food preferences are culturally specific—for example, in some cultures, various nontoxic insects are considered to be a delicacy.

LEARNING TO EAT VITAMINS AND MINERALS. How do animals select a diet that provides all of the vitamins and minerals they need? To answer this question, researchers have studied how dietary deficiencies influence diet selection. Two patterns of results have emerged: one for sodium and one for the other essential vitamins and minerals. When an animal is deficient in sodium, it develops an immediate and compelling preference for the taste of sodium salt (see Jarvie & Palmiter, 2017). In contrast, an animal deficient in some vitamin or mineral other than sodium must learn to consume foods that are rich in the missing nutrient by experiencing their positive effects; this is because vitamins and minerals other than sodium normally have no detectable taste in food. For example, rats maintained on a diet deficient in *thiamine* (vitamin B1) develop an aversion to the taste of that diet, and if they are offered two new diets, one deficient in thiamine and one rich in thiamine, they usually develop a preference for the taste of the thiamine-rich diet over the ensuing days, as it becomes associated with improved health.

If we, like rats, are capable of learning to select diets rich in the vitamins and minerals we need, why are dietary deficiencies so prevalent in our society? One reason is that, in order to maximize profits, manufacturers produce foods that have the tastes we prefer but lack many of the nutrients we need to maintain our health. (Even rats prefer chocolate chip cookies to nutritionally complete rat chow.) The second reason is illustrated by the classic study of Harris and associates (1933). When thiamine-deficient rats were offered two new diets, one with thiamine and one without, almost all of them learned to eat the complete diet and avoid the deficient one. However, when they were offered 10 new diets, only one of which contained the badly needed thiamine, few developed a preference for the complete diet. The number of different substances, both nutritious and not, consumed each day by most people in industrialized societies is immense, and this makes it difficult, if not impossible, for their bodies to learn which foods are beneficial and which are not.

Factors That Influence When We Eat

LO 12.6 Describe at least two factors that influence when we eat.

Collier and his colleagues (see Collier, 1986) found that most mammals choose to eat many small meals (snacks) each day if they have ready access to a continuous supply of food. In contrast to the usual mammalian preference, most people, particularly those living in family groups, tend to eat a few large meals each day at regular times. Interestingly, each person's regular mealtimes are the very same times at which that person is likely to feel most hungry; in fact, many people experience attacks of malaise (headache, nausea, and an inability to concentrate) when they miss a regularly scheduled meal.

PREMEAL HUNGER. We are sure you have experienced attacks of premeal hunger. Subjectively, they seem to provide compelling support for set-point theories. Your body seems to be crying out: "I need more energy. I cannot function without it. Please feed me." But things are not always as they seem. Woods straightened out the confusion (see Begg & Woods, 2013; Woods & Begg, 2015).

According to Woods, the key to understanding hunger is to appreciate that eating meals stresses the body. Before a meal, the body's energy reserves are in reasonable homeostatic balance; then, as a meal is consumed, there is a major homeostasis-disturbing influx of fuels into the bloodstream. The body does what it can to defend its homeostasis. At the first indication that a person will soon be eating—for example, when the usual mealtime approaches—the body enters the cephalic phase and takes steps to soften the impact of the impending homeostasis-disturbing influx by releasing insulin into the blood and thus reducing blood glucose. Woods's message is that the strong, unpleasant feelings of hunger you may experience at mealtimes are not cries from your body for food; they are the sensations of your body's preparations for the expected homeostasis-disturbing meal. Mealtime hunger is caused by the expectation of food, not by an energy deficit.

As a high school student, I (JP) ate lunch at exactly 12:05 every day and was overwhelmed by hunger as the time approached. Now, my eating schedule is different, and I never experience noontime hunger pangs; I now get hungry just before the time at which I usually eat. Have you had a similar experience?

PAVLOVIAN CONDITIONING OF HUNGER. In a classic series of Pavlovian conditioning experiments on laboratory rats, Weingarten (1983, 1984, 1985) provided strong support for the view that hunger is often caused by the expectation of food, not by an energy deficit. During the conditioning phase of one of his experiments, Weingarten presented rats with six meals per day at irregular intervals, and he signaled the impending delivery of each meal with a buzzer-and-light conditional stimulus. This conditioning procedure was continued for 11 days. Throughout the ensuing test phase of the experiment, the food was continuously available. Despite the fact that the subjects were never deprived during the test phase, they started to eat each time the buzzer and light were presented—even if they had recently completed a meal (see Johnson, 2013).

Factors That Influence How Much We Eat

LO 12.7 Describe some of the major factors that influence how much we eat.

The motivational state that causes us to stop eating a meal when there is food remaining is **satiety**. Satiety mechanisms play a major role in determining how much we eat.

SATIETY SIGNALS.   As you will learn in the next module of the chapter, food in the gut and glucose entering the blood can induce satiety signals, which inhibit subsequent consumption. These signals depend on both the volume and the **nutritive density** (calories per unit volume) of the food.

The effects of nutritive density have been demonstrated in studies in which laboratory rats have been maintained on a single diet. Once a stable baseline of consumption has been established, the nutritive density of the diet is changed. Some rats eventually learn to adjust the volume of food they consume to keep their caloric intake and body weights relatively stable. However, there are major limits to this adjustment: Rats rarely increase their intake sufficiently to maintain their body weights if the nutritive density of their conventional laboratory feed is reduced by more than 50 percent or if there are major changes in the diet's palatability.

SHAM EATING.   The study of **sham eating** indicates that satiety signals from the gut or blood are not necessary to terminate a meal. In sham-eating experiments, food is chewed and swallowed by the subject; but rather than passing down the subject's esophagus into the stomach, it passes out of the body through an implanted tube (see Figure 12.5).

Because sham eating adds no energy to the body, set-point theories predict that all sham-eaten meals should be huge. But this is not the case. The first sham meal of rats sham eating their usual diet is usually the same size as previous normal meals, thus indicating that satiety is a function of previous experience, not the current increases in the

Figure 12.5 The sham-eating preparation.

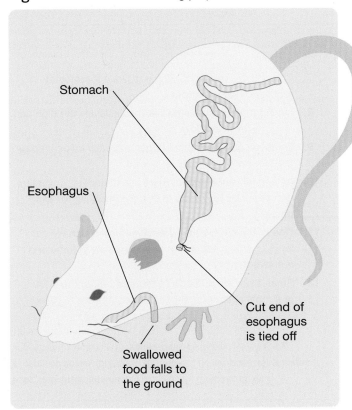

Stomach

Esophagus

Cut end of esophagus is tied off

Swallowed food falls to the ground

body's energy resources. However, after the first few sham meals, rats begin to sham eat larger meals.

APPETIZER EFFECT AND SATIETY.   The next time you attend a dinner party, you may experience a major weakness of the set-point theory of satiety. If appetizers are served, you will notice that small amounts of food consumed before a meal actually increase hunger rather than reduce it. This is the **appetizer effect**. Presumably, it occurs because the consumption of small amounts of food is particularly effective in eliciting cephalic-phase responses.

SERVING SIZE AND SATIETY.   Many experiments have shown that the amount of consumption is influenced by serving size (see Hollands et al., 2015). The larger the servings, the more we tend to eat.

SOCIAL INFLUENCES AND SATIETY.   Feelings of satiety also depend on whether we are eating alone or with others. People consume more when eating with others. Laboratory rats do the same.

SENSORY-SPECIFIC SATIETY.   The number of different tastes available at each meal has a major effect on meal size. For example, the effect of offering a laboratory rat a varied diet of highly palatable foods—a **cafeteria diet**—is dramatic. Adult rats that were offered bread and chocolate in addition to their usual laboratory diet increased their average intake of calories by 84 percent, and after 120 days they had increased their average body weights by 49 percent (Rogers & Blundell, 1980). The spectacular effects of cafeteria diets on consumption and body weight clearly run counter to the idea that eating is rigidly controlled by internal energy set points.

The effect on meal size of cafeteria diets results from the fact that satiety is to a large degree sensory-specific. As you eat one food, the positive-incentive value of all foods declines slightly, but the positive-incentive value of that particular food plummets. As a result, you soon become satiated on that food and stop eating it. However, if another food is offered to you, you will often begin eating again (see Yamada et al., 2017).

In one study of **sensory-specific satiety** (Rolls et al., 1981), human volunteers were asked to rate the palatability of eight different foods, and then they ate a meal of one of them. After the meal, they were asked to rate the palatability of the eight foods once again, and it was found that their rating of the food they had just eaten had declined substantially more than had their ratings of the other seven foods. Moreover, when the volunteers were offered an unexpected second meal, they consumed most of it unless it was the same as the first.

Booth (1981) asked volunteers to rate the momentary pleasure produced by the flavor, the smell, the sight, or just the thought of various foods at different times after consuming a large, high-calorie, high-carbohydrate liquid meal. There was an immediate sensory-specific decrease in the palatability of foods of the same or similar flavor

Thomas Barwick/Getty Images

Tooga/Getty Images

photographer and designer/Moment/Getty Images

Many different factors can affect satiety.

as soon as the liquid meal was consumed. This was followed by a general decrease in the palatability of all substances about 30 minutes later. Thus, it appears that signals from taste receptors produce an immediate decline in the positive-incentive value of similar tastes and that signals associated with the postingestive consequences of eating produce a general decrease in the positive-incentive value of all foods.

Rolls (1990) suggested that sensory-specific satiety has two kinds of effects: (1) relatively brief effects that influence the selection of foods within a single meal and (2) relatively enduring effects that influence the selection of foods from meal to meal. Some foods seem to be relatively immune to long-lasting sensory-specific satiety; foods such as rice, bread, potatoes, sweets, and green salads can be eaten almost every day with only a slight decline in their palatability.

The phenomenon of sensory-specific satiety has two adaptive consequences. First, it encourages the consumption of a varied diet. If there were no sensory-specific satiety, a person would tend to eat their preferred food and nothing else, and the result would be malnutrition. Second, sensory-specific satiety encourages animals that have access to a variety of foods to eat a lot; an animal that has eaten its fill of one food will often begin eating again if it encounters a different one (see Raynor & Epstein, 2001). This encourages animals to take full advantage of times of abundance, which are all too rare in nature.

This module has introduced you to several important properties of hunger and eating. How many of those properties support the set-point assumption, and how many are inconsistent with it?

> **Journal Prompt 12.1**
> Briefly summarize the evidence for and against the set-point theory of hunger and eating.

Scan Your Brain

Are you ready to move on to the discussion of the physiology of hunger and satiety in the following module? Find out by completing the following sentences with the most appropriate terms. The correct answers are provided at the end of the exercise. Before proceeding, review material related to your incorrect answers and omissions.

1. The gastrointestinal tract helps the process of _____, which is the breakdown of food and absorption of nutrients.

2. Fats, proteins, and sugars are the equivalents of _____, amino acids, and glucose, respectively.

3. The primary mode of energy storage is _____, which stores twice as much energy as glycogen.

4. The first phase of energy metabolism is the _____.

5. _____ promotes the storage of glycogen in the liver and muscles.

6. During the fasting phase, free fatty acids are converted to _____, which are used by muscles for energy.

7. The _____ model of hunger is like a thermostat-regulated heating system.

8. During the absorptive phase, insulin levels are high and _____ levels are low.

9. Evidence suggests that hunger is greatly influenced by the current _____ value of food.

10. As a meal is consumed, there is a major _____ influx and the body responds to this by preparing for each phase.

11. One of the factors that determine how much we eat is _____ signals, which depend on the volume and nutritive density of the food.

12. Small amounts of foods consumed before the start of a meal actually increase hunger, a process called _____.

Scan Your Brain answers: (1) digestion, (2) lipids, (3) fat, (4) cephalic phase, (5) Insulin, (6) ketones, (7) set-point, (8) glucagon, (9) positive-incentive, (10) homeostasis-disturbing, (11) satiety, (12) appetizer effect.

Physiological Research on Hunger and Satiety

Now that you have been introduced to set-point theories, the positive-incentive perspective, and some basic factors that affect why, when, and how much we eat, this module will introduce you to six prominent lines of research on the physiology of hunger and satiety.

Role of Blood Glucose Levels in Hunger and Satiety

LO 12.8 Explain the nature of the relationship between blood glucose levels and hunger and satiety.

As we have already explained, efforts to link blood glucose levels to eating have been largely unsuccessful. However, there was a renewed interest in the role of glucose in the regulation of eating in the 1990s, following the development of methods of continually monitoring blood glucose levels (see Grayson, Seeley, & Sandoval, 2013). In the classic experiment of Campfield and Smith (1995), rats were housed with free access to food and water, and their blood glucose levels were continually monitored. In this situation, baseline blood glucose levels rarely fluctuated more than 2 percent. However, about 10 minutes before a meal was initiated, the levels quickly dropped about 8 percent. Evidence does not support the glucostatic interpretation of this observation: that the premeal decline in blood glucose produces hunger and eating. Rather, evidence suggests that the causation goes in the opposite direction: that the intention to start eating triggers the decline in blood glucose. The following are four observations that support this view:

- The time course of the glucose decline is not consistent with the idea that it reflects a gradual decline in the body's energy—it occurs suddenly just before eating begins.
- Eliminating the premeal drop in blood glucose does not eliminate the meal.
- If an expected meal is not served, blood glucose soon returns to its previous level.
- The glucose levels in the extracellular fluids that surround CNS neurons stay relatively constant, even when blood glucose levels in general circulation drop (see Seeley & Woods, 2003).

Evolution of Research on the Role of Hypothalamic Nuclei in Hunger and Satiety

LO 12.9 Describe the evolution of thinking about the role of various hypothalamic nuclei in hunger and satiety.

In this section, we examine some of the research on the role of hypothalamic nuclei in hunger and satiety. As you will discover, early research was guided by an overly simplistic assumption: that one hypothalamic nucleus is necessary for the control of hunger, and that another is necessary for the control of satiety. More recent research has dispelled such simplistic thinking, and it has evolved considerably (see Clemmensen et al., 2017).

THE MYTH OF HYPOTHALAMIC HUNGER AND SATIETY CENTERS. In the 1950s, experiments on rats seemed to suggest that eating behavior is controlled by two different regions of the hypothalamus: satiety by the **ventromedial hypothalamus (VMH)** and feeding by the **lateral hypothalamus (LH)**—see Figure 12.6. This theory turned out to be wrong, but it stimulated several important discoveries.

In 1940, it was discovered that large bilateral electrolytic lesions to the VMH produce **hyperphagia** (excessive eating) and extreme obesity in rats (Hetherington & Ranson, 1940). This *VMH syndrome* has two different phases: dynamic and static. The **dynamic phase**, which begins as soon as the subject regains consciousness after the operation, is characterized by several weeks of grossly excessive eating and rapid weight gain. However, after that, consumption gradually declines to a level just sufficient to maintain the rat's new weight; this marks the beginning of the **static phase**.

The most important feature of the static phase of the VMH syndrome is that the animal maintains its new body weight. If a rat in the static phase is deprived of food until it has lost a substantial amount of weight, it will regain the lost weight once the deprivation ends; conversely, if it is made to gain weight by forced feeding, it will lose the excess weight once the forced feeding is curtailed.

In 1951, Anand and Brobeck reported that bilateral electrolytic lesions to the LH produce **aphagia**—a complete cessation of eating that ultimately leads to death. Even rats that were first made hyperphagic by VMH lesions were rendered aphagic by the addition of LH lesions. Anand and Brobeck concluded that the lateral region of the hypothalamus is a feeding center. Teitelbaum and Epstein (1962) subsequently discovered two important features of the *LH syndrome*. First, they found that the aphagia was accompanied by **adipsia**—a complete cessation of drinking. Second, they found that LH-lesioned rats partially recover if they are kept alive by tube feeding. Following the cessation of tube feeding, they begin to eat wet, palatable foods, such as chocolate chip cookies soaked in milk, and eventually they will eat dry food pellets if water is concurrently available.

The theory that the VMH and the LH are satiety and hunger centers, respectively, crumbled in the face of several lines of evidence. With respect to the theory that the VMH is a satiety center, two lines of evidence dispelled this myth. The first line showed that VMH-lesioned rats overeat because they gain weight, not that they gain weight because they overeat—as had initially been assumed. Bilateral VMH lesions increase blood insulin levels, which in turn increases **lipogenesis** (the production of body fat) and decreases **lipolysis** (the breakdown of body fat to utilizable forms of energy). Because the calories ingested by VMH-lesioned rats are converted to fat at such a high rate, the rats must keep eating to ensure they have enough calories in their blood to meet their immediate energy requirements.

The second line of evidence that undermined the theory of a VMH satiety center showed that many of the effects of VMH lesions are not directly attributable to VMH damage. A large bundle of axons that project from the nearby **paraventricular nuclei** courses past the VMH and is thus inevitably damaged by large electrolytic VMH lesions; in particular, bilateral lesions of that bundle of axons or the paraventricular nuclei itself produce hyperphagia and weight gain, just as VMH lesions do.

Most of the evidence against the notion that the LH is a hunger center has come from a thorough analysis of the effects of bilateral LH lesions. Early research focused exclusively on the aphagia and adipsia produced by LH lesions, but subsequent research demonstrated that LH lesions produce a wide range of severe motor disturbances and a general lack of responsiveness to sensory input (of which food and drink are but two examples). Consequently, the idea that the LH is a center specifically dedicated to hunger no longer warrants serious consideration.

Journal Prompt 12.2

Try to briefly summarize the evidence that undermined the theories of a VMH satiety center and an LH hunger center.

MODERN RESEARCH ON THE ROLE OF HYPOTHALAMIC NUCLEI IN HUNGER AND SATIETY. Despite the failure of the concept of the LH and VMH as nuclei that control hunger and satiety, respectively, recent evidence suggests that certain distinct cell populations within the hypothalamus can influence hunger and satiety. For example, certain neurons within the paraventricular nucleus of the hypothalamus (see Appendix VI) have been shown to act as

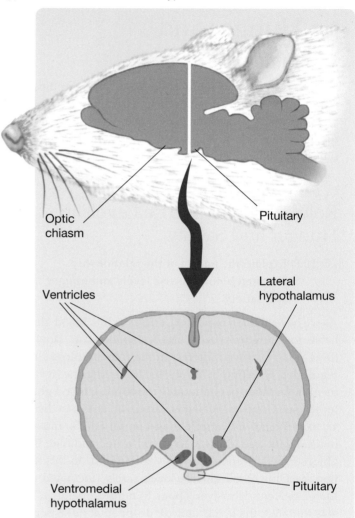

Figure 12.6 Locations in the rat brain of the ventromedial hypothalamus and the lateral hypothalamus.

nutrient sensors that can influence feeding and satiety (see Lagerlöf et al., 2016; Schwartz, 2016). Moreover, several distinct neuronal populations within the **arcuate nucleus** of the hypothalamus have been shown to influence the metabolism of energy resources, that is, consumed food (see Lagerlöf et al., 2016; Schwartz, 2016). Indeed, the arcuate nucleus (see Appendix VI) appears to be the center of a neural network that interacts with receptors in the blood and gut (see Clemmensen et al., 2017).

Role of the Gastrointestinal Tract in Satiety

LO 12.10 Describe the role of the gastrointestinal tract in satiety.

One of the most influential early studies of hunger was published by Cannon and Washburn in 1912. It was a perfect collaboration: Cannon had the ideas, and Washburn had the ability to swallow a balloon. First, Washburn swallowed an empty balloon tied to the end of a thin tube. Then, Cannon pumped some air into the balloon and connected the end of the tube to a water-filled glass U-tube so that Washburn's stomach contractions produced a momentary increase in the level of the water at the other end of the U-tube. Washburn reported a "pang" of hunger each time a large stomach contraction was recorded (see Figure 12.7).

Cannon and Washburn's finding led to the theory that hunger is the feeling of contractions caused by an empty stomach, whereas satiety is the feeling of stomach distention. However, support for this theory and interest in the role of the gastrointestinal tract in hunger and satiety quickly waned with the discovery that human patients whose stomach had been surgically removed and whose esophagus had been hooked up directly to their **duodenum** (the first segment of the small intestine, which normally carries food away from the stomach) continued to report feelings of hunger and satiety and continued to maintain their normal body weight by eating more meals of smaller size.

In the 1980s, there was a resurgence of interest in the role of the gastrointestinal tract in eating. It was stimulated by a series of experiments that indicated the gastrointestinal tract is the source of satiety signals. For example, Koopmans (1981) transplanted an extra stomach and length of intestine into rats and then joined the major arteries and veins of the implants to the recipients' circulatory systems (see Figure 12.8). Koopmans found that food injected into the transplanted stomach and kept there by a noose around the *pyloric sphincter* decreased eating in proportion to both its caloric content and volume. Because the transplanted stomach had no functional nerves, the gastrointestinal satiety signal had to be reaching the brain through the blood. And because nutrients are not absorbed from the stomach, the bloodborne satiety signal could not have been a nutrient. It had to be some chemical or chemicals that were released from the stomach in response to the caloric value and volume of the food—which leads us nicely into the next section.

Figure 12.7 The system developed by Cannon and Washburn in 1912 for measuring stomach contractions. They found that large stomach contractions were related to pangs of hunger.

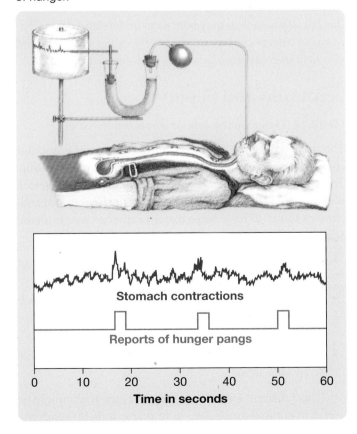

Figure 12.8 Transplantation of an extra stomach and length of intestine in a rat. Koopmans (1981) implanted an extra stomach and length of intestine in each of his experimental subjects. He then connected the major blood vessels of the implanted stomachs to the circulatory systems of the recipients. Food injected into the extra stomach and kept there by a noose around the pyloric sphincter decreased eating in proportion to its volume and caloric value.

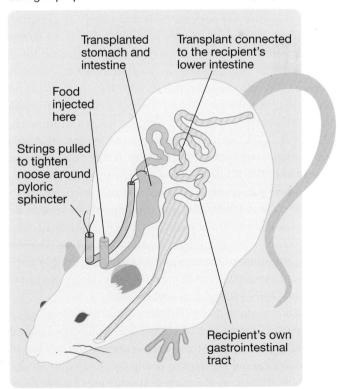

Hypothalamic Circuits, Peptides, and the Gut

LO 12.11 Describe the discovery of the role of hypothalamic circuits, peptides, and the gut in food consumption and metabolism.

Soon after the discovery that the stomach and other parts of the gastrointestinal tract release chemical signals to the brain, evidence began to accumulate that many of these chemicals were *peptides*, short chains of amino acids that can function as hormones and neurotransmitters. Ingested food interacts with receptors in the gastrointestinal tract and in so doing causes the tract to release peptides into the bloodstream. In 1973, Gibbs, Young, and Smith injected one of these gut peptides, **cholecystokinin (CCK)**, into hungry rats and found that they ate smaller meals. This led to the hypothesis that circulating gut peptides provide the brain with information about the quantity and nature of food in the gastrointestinal tract and that this information plays a role in satiety (see Dockray, 2014; Woods, 2013).

There has also been considerable evidence that some peptides can function as satiety signals (see Gao & Horvath, 2007). Many gut peptides have been shown to bind to receptors in the brain, particularly in those areas of the hypothalamus involved in energy metabolism (e.g., arcuate nucleus), and some of these gut peptides (e.g., *CCK, bombesin, glucagon,* and *somatostatin*) have been reported to produce metabolic changes that reduce food intake (see Crespo et al., 2014). These have become known as *satiety peptides* (peptides that decrease appetite).

In studying the appetite-reducing effects of peptides, researchers had to rule out the possibility that these effects are not merely the consequence of illness. Indeed, evidence suggests that one peptide in particular, CCK, induces illness: CCK administered to rats after they have eaten an unfamiliar substance induces a *conditioned taste aversion* for that substance, and CCK induces nausea in humans. However, CCK reduces appetite and eating at doses substantially below those required to induce taste aversion in rats, and thus it qualifies as a legitimate satiety peptide.

Several *hunger peptides* (peptides that increase appetite) have also been discovered. These peptides tend to be synthesized in the brain, particularly in the hypothalamus, and produce metabolic effects that increase eating. The following are some of the most widely studied hunger peptides: *neuropeptide Y, galanin, orexin-A,* and *ghrelin* (see Liu & Borgland, 2015; Wilson et al., 2014).

Clemmensen et al. (2017) have argued that the study of the effects of hunger and satiety peptides and related neural circuitry has been a particularly active and productive line of research on the mechanisms of eating and has led to several important discoveries. We agree, with one major qualification.

First, on the positive side, the discovery of so many hunger and satiety peptides has indicated that the neural system that controls eating likely reacts to many different signals, not just to one or two (e.g., not just to glucose and fat). Second, the discovery that many of the hunger and satiety peptides have receptors in the hypothalamus has led to the discovery of numerous hypothalamic neural circuits that are involved in food metabolism (see Al Massadi et al., 2017; Betley et al., 2015; Kim, Seeley, & Sandoval, 2018; Stuber & Wise, 2016). The third important contribution of research on hunger and satiety peptides is that it has led to an entirely different view of the gut: The gut is now seen as an important center for analysis and communication—much like the brain in many respects—not just a sack to hold consumed food. It is now clear that hypothalamic circuits are part of a two-way communication system between the brain and gut that influences eating, digestion, and the regulation of energy resources (see Grayson et al., 2013; Morton, Meek, & Schwartz, 2014; Sohn, Elmquist, & Williams, 2013; Trivedi, 2014).

What is the one major qualification that we have about this research? There is no question that research on the gut, hypothalamic circuits, and peptides has been remarkably successful. The one problem is that its focus has been on food metabolism. When eating effects were observed, they seemed to be secondary effects of the metabolic changes—we do not yet understand how the hypothalamic circuits feed into the circuits that directly control our eating. So we now have a lot of information about the neural and gut mechanisms of food metabolism but precious little about our major interest in this chapter: the control of eating. However, as progress is made in this line of research, it may help us to understand the mechanisms of current clinical treatments for eating disorders, and also help us to develop new ones (Clemmensen et al., 2017).

Serotonin and Satiety

LO 12.12 Describe the role of serotonin in satiety.

The monoaminergic neurotransmitter serotonin is another chemical that seems to play a role in metabolism and eating. The initial evidence for this role came from a line of research on satiety in rats. In these studies, serotonin-produced satiety was found to have three major properties (see Blundell & Halford, 1998):

- It caused the rats to resist the powerful attraction of highly palatable cafeteria diets.

- It reduced the amount of food consumed during each meal rather than reducing the number of meals (see Clifton, 2000).

- It was associated with a shift in food preferences away from fatty foods.

This profile of effects suggested that serotonin might be useful in combating obesity in humans. Indeed, serotonin agonists (e.g., fenfluramine, dexfenfluramine, fluoxetine) have been shown to reduce hunger and eating in some individuals who overeat under some conditions (see Voigt & Fink, 2015). We have mentioned serotonin here because later in this chapter, you will learn about the use of serotonin to treat overeating.

Prader-Willi Syndrome: Patients with Insatiable Hunger

LO 12.13 Describe the symptoms and etiology of Prader-Willi syndrome.

Prader-Willi syndrome could prove critical in the discovery of the neural mechanisms of hunger and satiety (see Tauber et al., 2014). Individuals with **Prader-Willi syndrome**, which results from an accident of chromosomal replication, experience insatiable hunger, little or no satiety, and an exceptionally slow metabolism (see Griggs, Sinnayah, & Mathai, 2015). In short, the Prader-Willi patient acts as though he or she is starving. Other common physical and neurological symptoms include weak muscles, small hands and feet, feeding difficulties in infancy, tantrums, compulsivity, and skin picking. If untreated, most patients accumulate an enormous amount of body fat, and they often die in early adulthood from diabetes, heart disease, or other disorders related to increased fat mass. Some have even died from gorging until their stomachs split open. Miss A. was diagnosed in infancy and received excellent care, which kept her from gaining too much weight (Martin et al., 1998).

Prader-Willi Syndrome: The Case of Miss A.

Miss A. was born with little muscle tone. Because her sucking reflex was so weak, she was tube fed. By the time she was 2 years old, her *hypotonia* (below-normal muscle tone) had resolved itself, but a number of characteristic deformities and developmental delays began to appear.

At 3 and a half years of age, Miss A. suddenly began to display a voracious appetite and quickly gained weight. Fortunately, her family maintained her on a low-calorie diet and kept all food locked away.

Miss A. is moderately intellectually disabled, and she suffers from psychiatric problems. Her major problem is her tendency to have tantrums any time something changes in her environment (e.g., a substitute teacher at school). Thanks largely to her family and pediatrician, she has received excellent care, which has minimized the complications that arise with Prader-Willi syndrome—most notably those related to increased fat deposits and its pathological effects.

Although the study of Prader-Willi syndrome has yet to provide any direct evidence about the neural mechanisms of hunger and eating, there has been a marked surge in its investigation. This increase has been stimulated by the recent identification of the genetic cause of the condition: an accident of reproduction that deletes or disrupts a section of chromosome 15 coming from the father.

Body-Weight Regulation: Set Points versus Settling Points

One strength of set-point theories of eating is that they also explain body-weight regulation. You have already learned that set-point theories are largely inconsistent with the facts of eating, but how well do they account for the regulation of body weight? Certainly, many people in our culture believe that body weight is regulated by a body-fat set point (see Assanand, Pinel, & Lehman, 1998a, 1998b). They believe that when fat deposits are below a person's set point, a person becomes hungrier and eats more, which results in a return of body-fat levels to that person's set point. And, conversely, they believe that when fat deposits are above a person's set point, a person becomes less hungry and eats less, which results in a return of body-fat levels to their set point.

Set-Point Assumptions about Body Weight and Eating

LO 12.14 Evaluate the evidence for set-point assumptions about body weight and eating.

You have already learned that set-point theories do a poor job of explaining the characteristics of hunger and eating. Do they do a better job of accounting for the facts of body-weight regulation? Let's begin by looking at three lines of evidence that challenge fundamental aspects of many set-point theories of body-weight regulation.

VARIABILITY OF BODY WEIGHT.   A set-point mechanism should make it virtually impossible for an adult to gain or lose large amounts of weight. Yet, many adults experience large and lasting changes in body weight. Set-point thinking crumbles in the face of the epidemic of overeating currently sweeping fast-food societies (see Morris et al., 2014).

Set-point theories of body-weight regulation suggest that the best method of maintaining a constant body weight is to eat each time there is a motivation to eat because, according to the theory, the main function of hunger is to defend the set point. However, many people avoid overeating only by resisting their urges to eat.

SET POINTS AND HEALTH.   One implication of set-point theories of body-weight regulation is that each person's set point is optimal for that person's health—or at least not incompatible with good health. This is why pop psychologists commonly advise people to "listen to the wisdom of their bodies" and eat as much as they need to satisfy their hunger. Experimental results indicate that this common prescription for good health could not be further from the truth.

Three kinds of evidence suggest that typical *ad libitum* (free-feeding) levels of consumption are unhealthy. First are the results of nonexperimental studies of humans who consume fewer calories than others. For example, people living on the Japanese island of Okinawa eat so few calories that their eating habits became a concern of health officials. When the health officials took a closer look, here is what they found (see Kagawa, 1978). Adult Okinawans were found to consume, on average, 20 percent fewer calories than other adult Japanese, and Okinawan schoolchildren were found to consume 38 percent fewer calories than recommended by public health officials. It was somewhat surprising then that rates of mortality and of all aging-related diseases were found to be substantially lower in Okinawa than in other parts of Japan, a country in which overall levels of caloric intake and obesity are far below Western norms. For example, the death rates from stroke, cancer, and heart disease in Okinawa were only 59 percent, 69 percent, and 59 percent, respectively, of those in the rest of Japan. Indeed, the proportion of Okinawans living to be over 100 years of age was up to four times greater than for inhabitants of the United States. In short, low-calorie diets seem to slow down the aging process (see Fontana & Partridge, 2015).

Because the Okinawan study and other nonexperimental studies of humans who eat less are not controlled, they must be interpreted with caution. For example, perhaps it is not simply the consumption of fewer calories that leads to health and longevity; perhaps people who eat less tend to eat healthier diets.

Controlled experiments of calorie restriction in more than a dozen different mammalian species, including monkeys and humans (see Colman et al., 2014; Mattison et al., 2017; Ravussin et al., 2015), constitute the second kind of evidence that *ad libitum* levels of consumption are unhealthy. In typical *calorie-restriction experiments*, one group of subjects is allowed to eat as much as they choose, while other groups of subjects have their caloric intake of the same diets substantially reduced (by between 25 and 65 percent in various studies). Results of such experiments have been remarkably consistent: In experiment after experiment, substantial reductions in the caloric intake of balanced diets have improved numerous indices of health and increased longevity. For example, in one experiment by Weindruch et al. (1996), groups of mice had their caloric intake of a well-balanced commercial diet reduced below free-feeding levels by either 25 percent, 55 percent, or 65 percent after weaning. All levels of dietary restriction substantially improved health and increased longevity, but the benefits were greatest in the mice whose intake was reduced the most. These mice had the lowest incidence of cancer, the best immune responses, and the greatest maximum life span—they lived 67 percent longer than mice that ate as much as they liked.

Evidence suggests that dietary restriction can have beneficial effects even if it is not initiated until later in life (see Mair et al., 2003; Vaupel, Carey, & Christensen, 2003). The few experiments of calorie restriction in humans have been conducted over shorter periods of time (e.g., 2 years), but they support the findings in nonhumans (see Ravussin et al., 2015).

Remarkably, there is evidence that dietary restriction can be used to treat some neurological conditions. For example, caloric restriction has been shown to reduce seizure susceptibility in humans with epilepsy (see Maalouf, Rho, & Mattson, 2008) and to improve memory in the elderly (see Witte et al., 2009).

Please stop and think about the implications of all these findings about calorie restriction. How much do you eat?

REGULATION OF BODY WEIGHT BY CHANGES IN THE EFFICIENCY OF ENERGY UTILIZATION. Of course, how much someone eats plays a role in his or her body weight, but it is now clear that the body controls its fat levels, to a large degree, by changing the efficiency with which it uses energy. As a person's level of body fat declines, that person starts to use energy resources more efficiently, which limits further weight loss (see Tremblay et al., 2013); conversely, weight gain is limited by a progressive decrease in the efficiency of energy utilization. Rothwell and Stock (1982) created a group of rats of increased weight by maintaining them on a cafeteria diet, and they found that the resting level of energy expenditure in these rats was 45 percent greater than in control rats.

This point is illustrated by the progressively declining effectiveness of weight-loss programs. Initially, low-calorie diets produce substantial weight loss, but the rate of weight loss diminishes with each successive week on the diet, until an equilibrium is achieved and little or no further weight loss occurs. Most dieters are familiar with this disappointing trend. A similar but opposite effect occurs with weight-gain programs (see Figure 12.9).

The mechanism by which the body adjusts the efficiency of its energy utilization in response to its levels of body fat has been termed **diet-induced thermogenesis**. Increases in the levels of body fat produce increases in body temperature, which require additional energy to maintain them—and decreases in the level of body fat have the opposite effects (see Jun et al., 2014).

There are major differences among humans both in **basal metabolic rate** (the rate at which energy is utilized to maintain bodily processes when resting) and in the ability to adjust the metabolic rate in response to changes in the levels of body fat. We all know people who remain slim even though they eat gluttonously. However, the research on calorie-restricted diets suggests that these people may not eat with impunity: There seems to be a health cost to pay for overeating even in the absence of significant weight gain (e.g., Colman et al., 2014; Mattison et al., 2017).

Set Points and Settling Points in Weight Control

LO 12.15   Compare and evaluate set-point and settling-point models of body-weight regulation.

Several prominent reviews of research on hunger and weight regulation generally acknowledge that a strict set-point model cannot account for the facts of weight regulation, and they argue for a more flexible model (see Berthoud, 2002; Mercer & Speakman, 2001; Woods et al., 2000). Because the body-fat set-point model still dominates the thinking of many people, we want to review the main advantages of an alternative and more flexible regulatory model: the settling-point model. Can you change your thinking?

According to the settling-point model, body weight tends to drift around a natural **settling point**—the level at which the various factors that influence body weight achieve an equilibrium. The idea is that as body-fat levels increase, changes occur that tend to limit further increases until a balance is achieved between all factors that encourage weight gain and all those that discourage it.

The settling-point model provides a loose kind of homeostatic regulation, without a set-point mechanism or mechanisms to return body weight to a set point. According to the settling-point model, body weight remains stable as long as there are no long-term changes in the factors that influence it; and if there are such changes, their impact is limited by negative feedback (see Figure 12.9). In the settling-point model, the negative feedback merely limits further changes in the same direction, whereas in the set-point model, negative feedback triggers a return to the set point. A neuron's resting potential is another well-known biological settling point—see Chapter 4.

Figure 12.9 The diminishing effects on body weight of a low-calorie diet and a high-calorie diet.

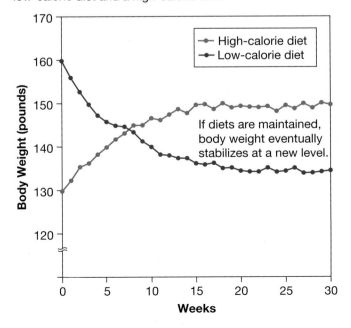

The seductiveness of the set-point mechanism is attributable in no small part to the existence of the thermostat model, which provides a vivid means of thinking about it. Figure 12.10 presents an analogy we like to use to think about the settling-point mechanism. We call it the **leaky-barrel model**: (1) The amount of water entering the hose is analogous to the amount of food available to the subject; (2) the water pressure at the nozzle is analogous to the positive-incentive value of the available food; (3) the amount of water entering the barrel is analogous to the amount of energy consumed; (4) the water level in the barrel is analogous to the level of body fat; (5) the amount of water leaking from the barrel is analogous to the amount of energy being expended; and (6) the weight of the barrel on the hose is analogous to the strength of the satiety signal.

The main advantage of the settling-point model of body-weight regulation over the body-fat set-point model is that it is more consistent with the data. Another advantage is that in those cases in which both models make the same prediction, the settling-point model does so more parsimoniously—that is, with a simpler mechanism that requires fewer assumptions. Let's use the leaky-barrel analogy to see how the two models account for four key facts of weight regulation.

- Body weight remains relatively constant in many adults. On the basis of this fact, it has been argued that body fat must be regulated around a set point. However, constant body weight does not require, or even imply, a set point. Consider the leaky-barrel model. As water from the tap begins to fill the barrel, the weight of the water in the barrel increases. This increases the amount of water leaking out of the barrel and decreases the amount of water entering the barrel by increasing the pressure of the barrel on the hose. Eventually, this system settles into an equilibrium where the water level stays constant; but because this level is neither predetermined nor actively defended, it is a settling point, not a set point.

- Many adults experience enduring changes in body weight. Set-point systems are designed to maintain internal constancy in the face of fluctuations of the external environment. Thus, the fact that many adults experience long-term changes in body weight is a strong argument against the set-point model. In contrast, the settling-point model predicts that when there is an enduring change in one of the parameters that affect body weight—for example, a major increase in the positive-incentive value of available food—body weight will drift to a new settling point.

- If a person's intake of food is reduced, metabolic changes that limit the loss of weight occur; the opposite happens when the subject overeats. This fact is often cited as evidence for set-point regulation of body weight; however, because the metabolic changes

Figure 12.10 The leaky-barrel model: a settling-point model of eating and body-weight homeostasis.

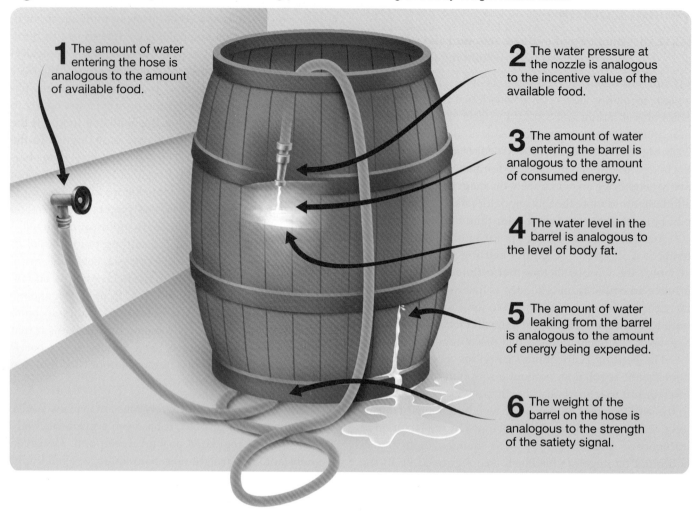

1 The amount of water entering the hose is analogous to the amount of available food.

2 The water pressure at the nozzle is analogous to the incentive value of the available food.

3 The amount of water entering the barrel is analogous to the amount of consumed energy.

4 The water level in the barrel is analogous to the level of body fat.

5 The amount of water leaking from the barrel is analogous to the amount of energy being expended.

6 The weight of the barrel on the hose is analogous to the strength of the satiety signal.

merely limit further weight changes rather than eliminating those that have occurred, they are more consistent with a settling-point model. For example, when water intake in the leaky-barrel model is reduced, the water level in the barrel begins to drop; but the drop is limited by a decrease in leakage and an increase in inflow attributable to the falling water pressure in the barrel. Eventually, a new settling point is achieved, but the reduction in water level is not as great as one might expect because of the loss-limiting changes.

- After an individual has lost a substantial amount of weight (by dieting, exercise, or the surgical removal of fat), there is a tendency for the original weight to be regained once he or she returns to the previous eating- and energy-related lifestyle. Although this finding is often offered as irrefutable evidence of a body-weight set point, the settling-point model readily accounts for it. When the water level in the leaky-barrel model is reduced—by temporarily decreasing input (dieting), by temporarily increasing output (exercising), or by

scooping out some of the water (surgical removal of fat)—only a temporary drop in the settling point is produced. When the original conditions are reinstated, the water level inexorably drifts back to the original settling point.

Does it really matter whether we think about body-weight regulation in terms of set points or settling points—or is making such a distinction just splitting hairs? It certainly matters to biopsychologists: Understanding that body weight is regulated by a settling-point system helps them better understand, and more accurately predict, the changes in body weight likely to occur in various situations; it also indicates the kinds of physiological mechanisms that are likely to mediate these changes. And it should matter to you. If the set-point model is correct, attempting to change your body weight would be a waste of time; you would inevitably be drawn back to your body-weight set point. On the other hand, the leaky-barrel model suggests that it is possible to permanently change your body weight by permanently changing any of the factors that influence energy intake or output.

The leaky-barrel model is the best analogy we can think of that captures the various factors that contribute to a settling point. Perhaps you can think of a better one that helps you think clearly about body weight regulation? In either case, be sure to ask your instructor if you are responsible for knowing the leaky-barrel analogy.

Scan Your Brain

Are you ready to move on to the final two modules of the chapter, which deal with eating disorders? This is a good place to pause and scan your brain to see if you understand the physiological bases of hunger and satiety. Complete the following sentences by filling in the blanks. The correct answers are provided at the end of the exercise. Before proceeding, review material related to your incorrect answers and omissions.

1. Early studies reported that bilateral electrolytic lesions in the _____ _____ produce aphagia, which is a complete cessation of eating.

2. Lesions in the ventromedial hypothalamus can cause insulin levels to increase, which increases _____ and decreases lipolysis.

3. More recent research suggests that distinct neural populations in the _____ nucleus of the hypothalamus may control metabolism.

4. Short chains of amino acids called _____ are released into the blood stream and function as hormones alerting the brain about the nature of food in the gastrointestinal system.

5. _____ is particularly known to induce conditioned taste aversion in rats and nausea in humans.

6. Individuals with a genetic disorder called _____ syndrome show insatiable hunger, little or no satiety, and an exceptionally slow metabolism.

7. _____ is the monoaminergic neurotransmitter that seems to play a role in satiety.

8. Okinawans eat less and live _____.

9. The _____ model suggests that people drift around a certain body weight to achieve equilibrium.

10. In the _____ model, the amount of water entering the hose is analogous to the amount of food available to the subject.

11. The fact that many adults have difficulty maintaining their body weight over time is evidence against the _____ model.

12. As an individual gains more weight, further weight gain is minimized by diet-induced _____.

Scan Your Brain answers: (1) lateral hypothalamus, (2) lipogenesis, (3) arcuate, (4) peptides, (5) Cholecystokinin (CCK), (6) Prader-Willi, (7) Serotonin, (8) longer, (9) settling-point, (10) leaky-barrel, (11) set-point, (12) thermogenesis.

Human Overeating: Causes, Mechanisms, and Treatments

This is an important point in this chapter. The chapter opened by describing the current epidemic of overeating and its adverse effects on health and longevity. Then, as the chapter progressed, you learned that many common beliefs about eating and weight regulation are incompatible with the evidence. Most importantly, you were challenged to think about eating and weight regulation in new ways that are more consistent with current evidence. Now that you are armed with these new ways of thinking, the chapter concludes with a discussion of overeating, anorexia, and bulimia and their treatment.

Overeating: Who Needs to Be Concerned?

LO 12.16 Explain why there is cause for concern surrounding the overeating epidemic.

Almost everyone needs to be concerned about the problem of overeating. If you currently overeat, the reason for concern is obvious: The relation between the accumulation of body fat and poor health has been repeatedly documented (see Flegal et al., 2013; Oestrich & Moley, 2017; Simonds & Cowley, 2013). Moreover, some studies have shown that even individuals who have higher levels of body fat but are metabolically healthy run a greater risk of developing health problems (see Oestrich & Moley, 2017). And the risk is not only to one's own health: Women with high levels of body fat are at an increased risk of having infants with health problems (see Avci et al., 2015; Bell, 2017; Oestrich & Moley, 2017; Zeltser, 2018). Even if you overeat but are currently slim, there is cause for concern; many people who are slim in their youth accumulate significant fat deposits as they age.

Overeating: Why Is There An Epidemic?

LO 12.17 Describe, from an evolutionary perspective, why there is a current epidemic of overeating.

Let's begin our analysis of overeating by considering the pressures that are likely to have led to the evolution of our eating and weight-regulation systems (see Genné-Bacon, 2014). It is believed that during the course of evolution, inconsistent food supplies were one of the main threats to survival. As a result, the fittest individuals were likely to have been those who preferred high-calorie foods, ate to capacity when food was available, stored as many excess calories as possible in the form of body fat, and used their stores of calories as efficiently as possible. Individuals who did not have these characteristics were unlikely to survive a food shortage or a harsh winter, and so these characteristics were passed on to future generations (see also Genné-Bacon, 2014; Sellayah, Cagampang, & Cox, 2014; Speakman, 2013).

The development of numerous cultural practices and beliefs that promote consumption has augmented the effects of evolution. For example, in many cultures, it is commonly believed that one should eat multiple meals per day at regular times, whether one is hungry or not; that food should be the focus of most social gatherings; that meals should be served in courses of progressively increasing palatability; and that salt, sweets (e.g., sugar), and fats (e.g., butter or cream) should be added to foods to improve their flavor and thus increase their consumption. Moreover, the tendency to make unhealthy food choices by parents tends to be passed on to their offspring (see Campbell et al., 2007).

Each of us possesses an eating and weight-regulation system that evolved to deal effectively with periodic food shortages, and many of us live in cultures whose eating-related practices evolved for the same purpose. However, our current environment differs from our "natural" environment in critical food-related ways. We live in an environment in which an endless variety of foods of the highest positive-incentive and caloric value are readily and continuously available. The consequence is an appallingly high level of consumption.

Why Do Some People Gain Weight from Overeating While Others Do Not?

LO 12.18 Give some reasons as to why some people gain weight from overeating while others do not.

Many factors lead some people to eat more than others who have comparable access to food. For example, some people consume more energy because they have strong preferences for the taste of high-calorie foods (see Blundell & Finlayson, 2004; Epstein et al., 2007), some consume more because they were raised in families and/or cultures that promote excessive eating, and some consume more because they have particularly large cephalic-phase responses to the sight or smell of food (see Rodin, 1985).

This raises an important question: Why do some people who overeat accumulate significant fat deposits while others do not? At a superficial level, the answer is obvious: Those who have high levels of body fat are those whose energy intake has exceeded their energy output; those who are slim are those whose energy intake has not exceeded

their energy output (see Drenowatz, 2015). Although this answer provides little insight, it does serve to emphasize that two kinds of individual differences play a role in high levels of body fat: those that lead to differences in energy input and those that lead to differences in energy output. Other differences also play a role.

DIFFERENCES IN ENERGY EXPENDITURE. With respect to energy output, people differ markedly from one another in the degree to which they can dissipate excess consumed energy. The most obvious difference is that people differ substantially in the amount of exercise they get; however, there are others. You have already learned about two of them: differences in *basal metabolic rate* and in the ability to react to fat increases by *diet-induced thermogenesis*. The third factor is called **nonexercise activity thermogenesis (NEAT)**, which is generated by activities such as fidgeting and the maintenance of posture and muscle tone; NEAT plays a small role in dissipating excess energy (see Villablanca et al., 2015).

DIFFERENCES IN GUT MICROBIOME COMPOSITION. Our gastrointestinal tract is replete with microbes, such as bacteria, that help us digest the food we eat—collectively known as our **gut microbiome**. Indeed, these microbes are so numerous that they outnumber our own bodily cells by 10 to 1 (see Ackerman, 2012; Wallis, 2014). In recent years, there has been a growing appreciation that the microbes that reside inside us can have major influences on brain and behavior (see Walker & Parkhill, 2013). For example, they can influence neurodevelopment, the blood–brain barrier, and even myelination of certain CNS axons (see Flight, 2014; Reardon, 2014; Smith, 2015).

Several recent findings have raised the question of whether our personal gut microbiome might protect us from, or predispose us to, obesity (see Deweerdt, 2014). For example, Ridaura and colleagues (2013) reported on the effects of taking mice raised in a germ-free environment and colonizing them with the fecal microbiota of human twin pairs that were discordant for body-fat levels (i.e., one twin had high levels of body fat, whereas the other did not). That is, half the mice were colonized by microbes from the lean twins, and the other half were colonized by microbes from the co-twins that had high levels of body fat. Those mice that were colonized with microbes from the twins with high body-fat levels gained more weight and had greater amounts of body fat compared with those colonized with microbes from the lean co-twins.

GENETIC AND EPIGENETIC FACTORS. Given the number of factors that can influence food consumption and energy metabolism, it is not surprising that many genes can influence body weight. Indeed, about 100 human chromosome loci (regions) have already been linked to it (see Locke et al., 2015). Interestingly, some of these genes seem to influence the likelihood of being overweight by affecting one's gut microbiome (see Pennisi, 2014; van Opstal & Bordenstein, 2015). Although it is proving difficult to unravel the interactions among the various genetic factors that influence variations in body weight among healthy people, single gene mutations have been linked to pathological conditions that involve excessive weight gain. You will encounter an example of such a condition later in this module. In addition, there is evidence that transgenerational epigenetic effects (see Chapter 2) can predispose subsequent generations to increased body-fat levels (see Banik et al., 2017; Lopomo, Burgio, & Migliore, 2016; Rohde et al., 2019).

Why Are Weight-Loss Programs Often Ineffective?

LO 12.19 Explain why weight-loss programs are typically ineffective.

Figure 12.11 describes the course of the typical weight-loss program. Most weight-loss programs are unsuccessful in the sense that, as predicted by the settling-point model, most of the lost weight is regained once the dieter stops following the program and the original conditions are reestablished. The key to permanent weight loss is a permanent lifestyle change—which is easier said than done in most cases (see Nymo et al., 2018).

Exercise has many health-promoting effects; however, despite the general belief that exercise is the most effective method of losing weight, several studies have shown that it contributes less than believed (see Dhurandhar et al., 2015; Riou et al., 2015). One reason is that physical exercise normally accounts for only a small proportion of total energy expenditure: Most of the energy you expend is used to maintain the resting physiological processes of your body (e.g., body temperature, brain function) and to digest your food (see Hills, Mokhtar, & Byrne, 2014). Another reason is that our bodies are efficient machines, burning only a small number of calories during a typical workout. Moreover, after exercise, many people believe that they can consume extra drinks and foods that contain more calories than the relatively small number that were expended during the exercise (see Freedman, 2011).

Leptin and the Regulation of Body Fat

LO 12.20 Explain how leptin and insulin are feedback signals for the regulation of body fat.

Fat is more than a passive storehouse of energy; it actively releases a peptide hormone called **leptin**. The following three subsections describe (1) the discovery of leptin, (2) how its discovery has fueled the development of a new approach to the treatment of human overeating, and

Figure 12.11 The five stages of a typical weight-loss program.

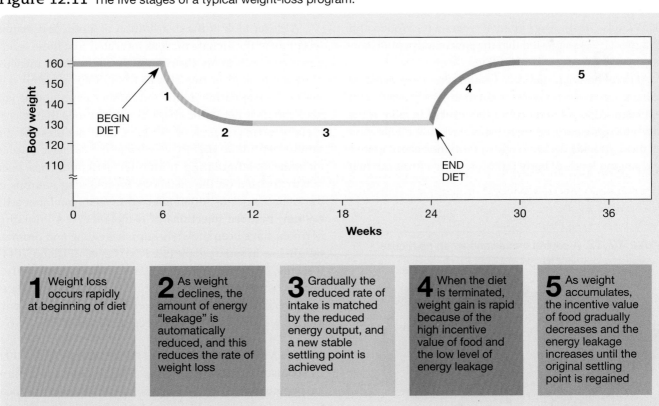

1 Weight loss occurs rapidly at beginning of diet

2 As weight declines, the amount of energy "leakage" is automatically reduced, and this reduces the rate of weight loss

3 Gradually the reduced rate of intake is matched by the reduced energy output, and a new stable settling point is achieved

4 When the diet is terminated, weight gain is rapid because of the high incentive value of food and the low level of energy leakage

5 As weight accumulates, the incentive value of food gradually decreases and the energy leakage increases until the original settling point is regained

(3) how the understanding that leptin and insulin are feedback signals led to the discovery of a hypothalamic nucleus that plays an important role in the regulation of body fat.

THE DISCOVERY OF LEPTIN. In 1950, a spontaneous genetic mutation occurred in the mouse colony being maintained in the Jackson Laboratory at Bar Harbor, Maine. The mutant mice were *homozygous* for the gene (ob), and they had very high levels of body fat, weighing up to three times as much as typical mice. These mutant mice are commonly referred to as **ob/ob mice**. See Figure 12.12.

Ob/ob mice eat more than control mice; they convert calories to fat more efficiently; and they use their calories more efficiently. Coleman (1979) hypothesized that ob/ob mice lack a critical hormone that normally inhibits fat production and maintenance.

In 1994, Friedman and his colleagues characterized and cloned the gene that is mutated in ob/ob mice. They found that this gene is *expressed* only in fat cells, and they characterized the protein it normally encodes, a peptide hormone they named *leptin*. Because of their mutation, ob/ob mice lack leptin. This finding led to an exciting hypothesis: Perhaps leptin is a negative feedback signal normally released from fat stores to decrease appetite and increase fat metabolism. Could leptin be administered to humans who overeat and/or have excessive body fat?

LEPTIN, INSULIN, AND THE ARCUATE MELANOCORTIN SYSTEM. There was great fanfare when leptin was discovered. However, it was not the first peptide hormone to be discovered that seems to function as a negative feedback signal in the regulation of body fat (see Schwartz, 2000; Woods, 2004). More than 40 years ago, Woods and colleagues (1979) suggested that the pancreatic peptide hormone insulin serves such a function.

At first, the suggestion that insulin serves as a negative feedback signal for body-fat regulation was viewed with skepticism. After all, how could the level of insulin in the body, which goes up and then comes back down following each meal, provide the brain with information about gradually changing levels of body fat? Moreover, it turns out that insulin does not readily penetrate the blood–brain barrier (e.g., Rhea & Banks, 2019). However, the following findings supported the hypothesis that insulin serves as a negative feedback signal in the regulation of body fat:

- Receptors for insulin were found in the brain (see Baura et al., 1993).
- Brain levels of insulin were found to be positively correlated with levels of body fat (see Seeley et al., 1996).
- Genetically modified mice that have lower levels of brain insulin also display higher levels of body fat (see Heni et al., 2015).
- Humans with high levels of body fat have been shown to have lower levels of insulin in their brains (see Heni et al., 2015).

Why are there two fat feedback signals? One reason may be that leptin levels are more closely correlated with **subcutaneous fat**, fat stored under the skin, whereas insulin levels are more closely correlated with **visceral fat**, fat stored around the internal organs of the body cavity (see Heni et al., 2015). Thus, each fat signal could provide different information. Visceral fat is more common in males than females and poses the greater threat to health (see Palmer & Clegg, 2015).

The discovery that leptin and insulin are signals that provide information to the brain about fat levels in the body provided a means for discovering parts of the neural circuits that participate in fat regulation. Receptors for both peptide hormones are located in many parts of the nervous system, but most are in the hypothalamus, particularly in the arcuate nucleus.

A closer look at the distribution of leptin and insulin receptors in the arcuate nucleus indicated that these receptors are not randomly distributed throughout the nucleus. They are located in two classes of neurons: neurons that release **neuropeptide Y** (the gut hunger peptide that you read about earlier in the chapter) and neurons that release **melanocortins**, a class of peptides that includes the gut satiety peptide α-*melanocyte-stimulating hormone* (alpha-melanocyte-stimulating hormone). Attention has been mostly focused on the melanocortin-releasing neurons in the arcuate nucleus (often referred to as the **melanocortin system**) because injections of α-melanocyte-stimulating hormone have been shown to suppress eating and promote weight loss in some people (see Kim, Leyva, & Diano, 2014). It seems, however, that the melanocortin system is only a minor component of a much larger system: Elimination of leptin receptors in the melanocortin system produces only a slight weight gain (see Münzberg & Myers, 2005).

LEPTIN AS A TREATMENT FOR HIGH BODY-FAT LEVELS IN HUMANS. The early studies of leptin seemed to confirm the hypothesis that it could function as an effective treatment for excessive body fat. Receptors for leptin were found in the brain, and injecting it into ob/ob mice reduced both their

Figure 12.12 A control mouse and an ob'ob mouse.

printed/digital with permission from © The Jackson Laboratory, Photographer Jennifer Torrance.

eating and their body fat (see Seeley & Woods, 2003). All that remained was to prove leptin's effectiveness in human patients.

However, when research on leptin turned from ob/ob mice to humans with high body-fat levels, the program ran into two major snags. First, humans who were overweight—unlike ob/ob mice—were found to have high, rather than low, levels of leptin (see Münzberg & Myers, 2005). Second, injections of leptin did not reduce either the eating or the body weight of overweight humans (see Heymsfield et al., 1999). The current thinking is that such individuals might be resistant to the effects of leptin through several possible mechanisms (see Pan & Myers, 2018), including a reduced ability for leptin to cross the blood–brain barrier (see Cui, López, & Rahmouni, 2017).

The exact reason as to why the actions of leptin are different in humans and ob/ob mice has yet to be explained. Nevertheless, efforts to use leptin in the treatment of humans with high levels of body fat have not been a total failure. Although few of those individuals have low leptin levels (see Blüher & Mantzoros, 2015), leptin may be a panacea for those who do. Consider the following case.

The Case of the Child with No Leptin

The patient was of normal weight at birth, but her weight soon began to increase at a rapid rate. She demanded food continually and was disruptive when denied food. As a result of her extreme levels of body fat, deformities of her legs developed, and surgery was required.

She was 9 when she was referred for treatment. At this point, she weighed 94.4 kilograms (about 210 pounds), and her weight was still increasing at an alarming rate. She was found to be homozygous for the ob gene and had no detectable leptin. Thus, leptin therapy was commenced.

The leptin therapy immediately curtailed the weight gain. She began to eat less, and she lost weight steadily over the 12-month period of the study, a total of 16.5 kilograms (about 36 pounds), almost all in the form of fat. There were no obvious side effects (Farooqi et al., 1999).

Treatment of Overeating and High Body-Fat Levels

LO 12.21 Describe two sorts of treatments for overeating and/or high body-fat levels.

Because overeating and high body-fat levels constitute such serious health problems, there have been many efforts to develop effective treatments. Some of these—such as the leptin treatment you just read about—have worked for some individuals, but certainly not the majority. The following two subsections discuss two treatments that are at different stages of development: serotonergic agonists and gastric surgery.

SEROTONERGIC AGONISTS. Because—as you have already learned—serotonin agonists have been shown to reduce food consumption in both human and nonhuman subjects, they have considerable potential in the treatment of overeating (Voigt & Fink, 2015). Serotonin agonists seem to act by a mechanism different from that for leptin and insulin, which produce long-term satiety signals based on fat stores. Serotonin agonists seem to increase short-term satiety signals associated with the consumption of a meal (Halford & Blundell, 2000).

Serotonin agonists have been found in various studies of overweight patients to reduce the following: the urge to eat high-calorie foods, the consumption of fat, the subjective intensity of hunger, the size of meals, the number of between-meal snacks, and bingeing. Because of this extremely positive profile of effects and the severity of the overeating problem, serotonin agonists (fenfluramine and dexfenfluramine) were rushed into clinical use. However, they were subsequently withdrawn from the market because chronic use was found to be associated with heart disease in a small but significant number of users.

Currently, there is only one approved serotonin agonist for the treatment of individuals who are overweight that has a more favorable side-effect profile: lorcaserin (see Halpern & Halpern, 2015; Nigro, Luon, & Baker, 2013). However, the efficacy of lorcaserin for the treatment of excessive weight is only modest (see Adan, 2013). Still there is hope for pharmacological interventions: Drugs that increase the levels of multiple monoamines (i.e., serotonin, norepinephrine, and dopamine) appear to be more effective than serotonin agonists (see González-Muniesa et al., 2017).

GASTRIC SURGERY. Individuals who are extremely overweight are sometimes recommended an extreme treatment. **Gastric bypass** is a surgical treatment for those who are extremely overweight that involves short-circuiting the normal path of food through the digestive tract so that its absorption is reduced; this, in turn, produces downstream effects that collectively lead to a decrease in weight (see Sinclair, Brennan, & le Roux, 2018; Sandoval & Seeley, 2017).

The first gastric bypass was done in 1967, and it is currently the most commonly prescribed surgical treatment for individuals who are extremely overweight (see Berthoud, 2013; Sinclair, Brennan, & le Roux, 2018). An alternative is the **adjustable gastric band procedure**, which involves surgically positioning a hollow silicone band around the stomach to reduce the flow of food through it; the circumference of the band can be adjusted by injecting saline into the band through a port that is implanted in the skin. One advantage of the gastric band over the gastric bypass is that the band can readily be removed.

The gastric bypass and adjustable gastric band are illustrated in Figure 12.13. A meta-analysis of studies comparing the two procedures found both to be highly effective (see Chang et al., 2014). In general, gastric bypass was found to be more effective than the adjustable gastric band procedure

Figure 12.13 Two surgical methods for treating individuals who are extremely overweight: gastric bypass and adjustable gastric band. The gastric band can be tightened by injecting saline into the access port implanted just beneath the skin.

Gastric Bypass

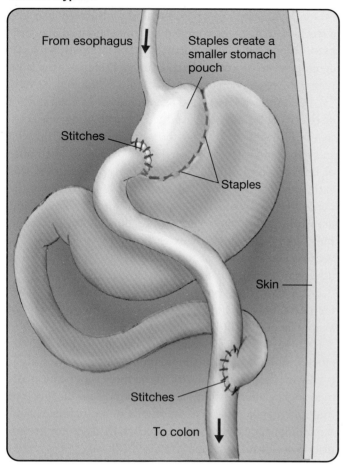

Adjustable Gastric Band

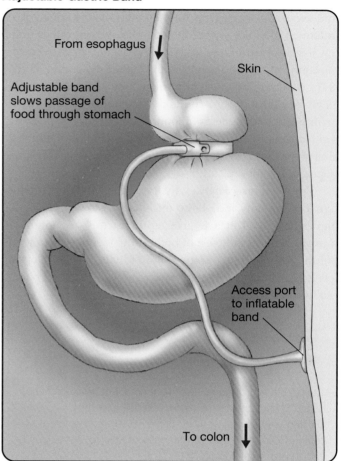

(see also Hughes, 2014) but was associated with more surgery-related complications.

Anorexia and Bulimia Nervosa

This chapter concludes with a discussion of two eating disorders of consumption: anorexia nervosa and bulimia nervosa.

Anorexia and Bulimia Nervosa

LO 12.22 Describe the symptoms of anorexia nervosa and bulimia nervosa.

ANOREXIA NERVOSA. Anorexia nervosa is a disorder of underconsumption (see Kaye et al., 2013). Individuals with anorexia eat so little that they experience health-threatening weight loss, and despite their emaciated appearance, they often perceive themselves as fat (see Gardner & Brown, 2014). Anorexia nervosa is a serious condition: In

approximately 4 percent of diagnosed cases, complications from starvation result in death, and there is a high rate of suicide among persons with anorexia.

BULIMIA NERVOSA. Bulimia nervosa is a disorder characterized by periods of not eating interrupted by *bingeing* (eating large amounts of food in short periods of time) followed by efforts to immediately eliminate the consumed calories from the body by *purging* (via vomiting or excessive use of laxatives, enemas, or diuretics) or by excessive exercise. Persons with bulimia may be overweight or of normal weight. Those who are underweight are usually diagnosed with *binge-eating/purging anorexia*. Bulimia nervosa is a serious condition: The mortality rate for individuals with bulimia is about 4 percent.

Relation between Anorexia and Bulimia

LO 12.23 Explain how anorexia and bulimia are, and are not, related.

Are anorexia nervosa and bulimia nervosa really different disorders, as current convention dictates? The answer to this question depends on one's perspective. From the perspective of a physician, it is important to distinguish between these disorders because starvation produces different health problems than does repeated bingeing and purging. For example, persons with anorexia often require treatment for reduced metabolism, *bradycardia* (slow heart rate), *hypotension* (low blood pressure), *hypothermia* (low body temperature), and *anemia* (deficiency of red blood cells). In contrast, persons with bulimia often require treatment for irritation and inflammation of the esophagus, vitamin and mineral deficiencies, electrolyte imbalance, dehydration, and acid reflux (see Westmoreland, Krantz, & Mehler, 2015).

Although anorexia and bulimia nervosa may seem like very different disorders from a physician's perspective, scientists often find it more appropriate to view them as variations of the same disorder. According to this view, both anorexia and bulimia begin with an obsession about body image and slimness and extreme efforts to lose weight. Persons with anorexia or bulimia both attempt to lose weight by strict dieting, but those with bulimia are less capable of controlling their appetites and thus enter into a cycle of starvation, bingeing, and purging. The following are two other similarities that support the view that anorexia and bulimia are variants of the same disorder (see Kaye et al., 2005):

- Individuals with anorexia or bulimia both tend to have distorted body images, seeing themselves as much heavier and less attractive than they are in reality (see Tabri et al., 2015).

- In practice, many patients seem to straddle the two diagnoses and cannot readily be assigned to one or the other categories, and many patients flip-flop between the two diagnoses as their circumstances change (see Kaye et al., 2013).

Anorexia and Positive Incentives

LO 12.24   Explain why those starving due to anorexia do not appear to be as hungry as they should.

The positive-incentive perspective on eating suggests that the decline in eating that defines both anorexia and bulimia is likely a consequence of a corresponding decline in the positive-incentive value of food. However, the positive-incentive value of food for anorexia patients has received little attention—in part, because anorexia patients often display substantial interest in food. The fact that many anorexia patients seem obsessed with food—continually talking about it, thinking about it, and preparing it for others—seems to suggest that food still holds a high positive-incentive value for them. However, to avoid confusion, it is necessary to keep in mind that the positive-incentive value of *interacting with food* is not necessarily the same as the positive-incentive value of *eating food*—and it is the positive-incentive value of eating food that is critical when considering anorexia nervosa.

A few studies have examined the positive-incentive value of various tastes in anorexia patients (see Drewnowski et al., 1987; Roefs et al., 2006; Sunday & Halmi, 1990). In general, these studies have found that the positive-incentive value of various tastes is lower in anorexia patients than in controls. However, these studies typically underestimate the importance of reductions in the positive-incentive value of food in the etiology of anorexia nervosa.

We can gain insight into the effects of anorexia nervosa on the positive-incentive value of food only by comparing individuals with anorexia to starving people of the same weight. Consider the behavior of volunteers undergoing semistarvation and compare it to people with anorexia. When asked how it felt to starve, one starving volunteer replied:

> I wait for mealtime. When it comes I eat slowly and make the food last as long as possible. The menu never gets monotonous even if it is the same each day or is of poor quality. It is food and all food tastes good. Even dirty crusts of bread in the street look appetizing. (Keys et al., 1950, p. 852)

Anorexia Nervosa: A Hypothesis

LO 12.25   Explain how anorexia might result from conditioned taste aversions.

The dominance of set-point theories in research into the regulation of hunger and eating has resulted in widespread inattention to one of the major puzzles of anorexia: Why does the adaptive massive increase in the positive-incentive value of eating that occurs in victims of starvation not occur in persons starving due to anorexia? Under conditions of starvation, the positive-incentive value of eating normally increases to such high levels that it is difficult to imagine how anybody who was starving—no matter how controlled, rigid, obsessive, and motivated that person was—could refrain from eating in the presence of palatable food. Why this protective mechanism is not activated in severe anorexia is a pressing question about the etiology of anorexia nervosa.

We believe part of the answer lies in the research of Woods and his colleagues on the aversive physiological effects of meals. At the beginning of meals, people are normally in reasonably homeostatic balance, and this homeostasis is disrupted by the sudden infusion of calories. The other part of the answer lies in the finding that the aversive effects of meals are much greater in people who have been eating little (see Brooks & Melnik, 1995). Meals, which produce adverse, but tolerable, effects in healthy individuals, may be extremely aversive for individuals who have undergone food deprivation. Evidence for the extremely noxious effects that eating meals has on starving humans is found in the reactions of World War II concentration camp victims to refeeding—many were rendered ill, and some were even killed by the food given to them by their liberators (see Keys et al., 1950; Solomon & Kirby, 1990).

So why do individuals with severe anorexia not experience a massive increase in the positive-incentive value of eating, similar to the increase experienced by other starving individuals? The answer may be *meals*. Meals consumed by a person with anorexia may produce a variety of conditioned taste aversions that reduce the motivation to eat. This hypothesis needs to be addressed because of its implication for treatment: Patients with anorexia—or anybody else who is severely undernourished—should not be encouraged, or even permitted, to eat meals. They should be fed—or infused with—small amounts of food intermittently throughout the day.

We have described the preceding hypothesis to show you the value of the new ideas you have encountered in this chapter: The major test of a new theory is whether it leads to innovative hypotheses. A few years ago, one of us (JP) was perusing an article on global famine and malnutrition and noticed an intriguing comment: One clinical complication that results from feeding meals to famine victims is anorexia (Blackburn, 2001). What do you make of this?

Journal Prompt 12.3

What do you think causes anorexia nervosa? Summarize the evidence that supports your view.

The Case of the Student with Anorexia

In societies where overeating is an epidemic, individuals with anorexia are out of step. People who are struggling to eat less have difficulty understanding those who have to struggle to eat. Still, when you stare anorexia in the face, it is difficult not to be touched by it.

She began by telling me (JP) how much she had been enjoying the course and how sorry she was to be dropping out of university. She was articulate and personable, and her grades were high—very high. Her problem was anorexia; she weighed only 82 pounds, and she was about to be hospitalized.

"But don't you want to eat?" I asked naively. "Don't you see that your plan to go to medical school will go up in smoke if you don't eat?"

"Of course I want to eat. I know I am terribly thin—my friends tell me I am. Believe me, I know this is wrecking my life. I try to eat, but I just can't force myself. In a strange way, I am pleased with my thinness."

She was upset, and I was embarrassed by my insensitivity. "It's too bad you're dropping out of the course before we cover the chapter on eating," I said, groping for safer ground.

"Oh, I've read it already," she responded. "The bit about positive incentives and learning was really good. I think my problem began when eating started to lose its positive-incentive value for me—in my mind, I kind of associated eating with being fat and all the boyfriend problems I was having. This made it easy to diet, but every once in a while I would get hungry and binge, or my parents would force me to eat a big meal. I would eat so much that I would feel ill. So I would put my finger down my throat and make myself throw up. This kept me from gaining weight, but I think it also taught my body to associate my favorite foods with illness—kind of a conditioned taste aversion. What do you think of my theory?"

After a lengthy chat, she got up to leave, and I walked her to the door of my office. I wished her luck and made her promise to come back for a visit. I never saw her again, but the image of her emaciated body walking down the hallway from my office has stayed with me.

Themes Revisited

Three of the text's four themes played prominent roles in this chapter. The thinking creatively theme was prevalent as you were challenged to critically evaluate your own beliefs and ambiguous research findings, to consider the scientific implications of your own experiences, and to think in new ways about phenomena with major personal and clinical implications. The chapter ended by using these new ideas to develop a potentially important hypothesis about the etiology of anorexia nervosa. Because of its emphasis on thinking, this chapter is our personal favorite.

Both aspects of the evolutionary perspective theme were emphasized repeatedly. First, you saw how thinking about hunger and eating from an evolutionary perspective leads to important insights. Second, you saw how controlled research on nonhuman species has contributed to our current understanding of human hunger and eating.

Finally, the clinical implications theme pervaded the chapter, but it was featured in the cases of the man who forgot not to eat, the child with Prader-Willi syndrome, the child with no leptin, and the student with anorexia.

One of the emerging themes was also touched upon in this chapter: The thinking about epigenetics theme appeared when transgenerational epigenetic effects on body-fat levels were discussed.

Key Terms

Set point, p. 318

Digestion, Energy Storage, and Energy Utilization
Digestion, p. 318
Gut microbiome, p. 318
Lipids, p. 318
Amino acids, p. 318
Glucose, p. 318
Cephalic phase, p. 320
Absorptive phase, p. 320
Fasting phase, p. 320
Insulin, p. 320
Glucagon, p. 320
Gluconeogenesis, p. 320
Free fatty acids, p. 320
Ketones, p. 320

Theories of Hunger and Eating: Set Points versus Positive Incentives
Set-point assumption, p. 320
Negative feedback systems, p. 322
Homeostasis, p. 322
Glucostatic theory, p. 322
Lipostatic theory, p. 322
Positive-incentive theory, p. 323
Positive-incentive value, p. 323

Factors That Determine What, When, and How Much We Eat
Satiety, p. 324
Nutritive density, p. 325
Sham eating, p. 325
Appetizer effect, p. 325
Cafeteria diet, p. 325
Sensory-specific satiety, p. 325

Physiological Research on Hunger and Satiety
Ventromedial hypothalamus (VMH), p. 327
Lateral hypothalamus (LH), p. 327
Hyperphagia, p. 327
Dynamic phase, p. 327
Static phase, p. 327
Aphagia, p. 327
Adipsia, p. 327
Lipogenesis, p. 328
Lipolysis, p. 328
Paraventricular nuclei, p. 328
arcuate nucleus, p. 328
Duodenum, p. 329
Cholecystokinin (CCK), p. 330
Prader-Willi syndrome, p. 331

Body-Weight Regulation: Set Points versus Settling Points
Diet-induced thermogenesis, p. 332
Basal metabolic rate, p. 332
Settling point, p. 333
Leaky-barrel model, p. 333

Human Overeating: Causes, Mechanisms, and Treatments
nonexercise activity thermogenesis (NEAT), p. 336
Gut microbiome, p. 336
Leptin, p. 337
Ob/ob mice, p. 338
Subcutaneous fat, p. 338
Visceral fat, p. 338
Neuropeptide Y, p. 338
Melanocortins, p. 338
Melanocortin system, p. 338
Gastric bypass, p. 339
Adjustable gastric band procedure, p. 339

Anorexia and Bulimia Nervosa
Anorexia nervosa, p. 340
Bulimia nervosa, p. 340

Chapter 13
Hormones and Sex

What's Wrong with the Mamawawa?

Hinterhaus Productions/Getty Images

 Chapter Overview and Learning Objectives

Neuroendocrine System

LO 13.1 Explain the distinction between exocrine glands and endocrine glands, describe the functions of the gonads, and distinguish between the X chromosome and the Y chromosome.

LO 13.2 Describe the three classes of hormones and then describe three classes of gonadal hormones.

LO 13.3 Explain why the pituitary is sometimes called the *master gland* and describe its anatomy. Discuss the female vs. male patterns of gonadal and gonadotropic hormone release and explain the evidence that discounted a role for the anterior pituitary in controlling those patterns of release.

LO 13.4 Explain how the anterior and posterior pituitary are controlled.

LO 13.5 Describe the research that led to the discovery of the hypothalamic releasing hormones.

LO 13.6 Describe three different types of signals that regulate hormone release. Also, describe how hormones are released over time and the effect this pattern of release has on levels of circulating hormones.

This chapter is about hormones and sex, a topic that some regard as unfit for conversation but one that fascinates many others. Perhaps the topic of hormones and sex is so fascinating because we are intrigued by the fact that our sex is so greatly influenced by the secretions of a small pair of glands. Because many of us think that our sex is fundamental and immutable, it could be disturbing to think it could be altered with a few surgical snips and some hormone injections. And there is something intriguing about the idea that our sex lives might be enhanced by the application of a few hormones. For whatever reason, the topic of hormones and sex is always a hit with our students. Some remarkable things await you in this chapter.

MEN-ARE-MEN-AND-WOMEN-ARE-WOMEN ASSUMPTION. Let's start with the fact that many students bring a piece of excess baggage to the topic of hormones and sex: the men-are-men-and-women-are-women assumption—or "mamawawa." This assumption is seductive; it seems so right that we are continually drawn to it without considering alternative views. Unfortunately, it is fundamentally flawed.

The men-are-men-and-women-are-women assumption is the tendency to think about femaleness and maleness as discrete, mutually exclusive, opposite categories. In thinking about hormones and sex, this general attitude leads one to assume that females have female sex hormones that give them female bodies and make them do female things and that males have male sex hormones that give them male bodies and make them do male things. Despite the fact that this approach to hormones and sex is inconsistent with the evidence, its simplicity, symmetry, and comfortable social implications draw us to it (see Carothers & Reis, 2013; Jordan-Young & Rumiati, 2012). That's why this chapter grapples with hormones and sex throughout and encourages you to think about them in new ways—ways that are more consistent with the evidence.

DEVELOPMENTAL AND ACTIVATIONAL EFFECTS OF SEX HORMONES. Before we begin discussing particular hormones and sex, you need to know that hormones influence sex in two fundamentally different ways (see Bale & Epperson, 2015): (1) by influencing the *development* from conception to sexual maturity of the anatomical, physiological, and behavioral characteristics that distinguish one as female or male; and (2) by *activating* the reproduction-related behavior of sexually mature adults (see Wu & Shah, 2011). The *developmental* (also called *organizational*) and *activational* effects of sex hormones are discussed in different parts of this chapter. However, in real life, these effects can occur simultaneously. For example, because the brain continues to develop into the late teens, adolescent hormone surges can have both effects (see Bell, 2018).

Neuroendocrine System

This module starts off by introducing the general principles of neuroendocrine function by describing the glands and hormones directly involved in sexual development and behavior.

The endocrine glands are illustrated in Figure 13.1. By convention, only the organs whose primary function appears to be the release of hormones are referred to as *endocrine glands*. However, other organs (e.g., the stomach, liver, and intestine) and body fat also release hormones into general circulation (see Chapter 12), and they are thus, strictly speaking, also part of the endocrine system.

Glands

LO 13.1 Explain the distinction between exocrine glands and endocrine glands, describe the functions of the gonads, and distinguish between the X chromosome and the Y chromosome.

There are two types of glands: exocrine glands and endocrine glands. **Exocrine glands** (e.g., sweat glands) release their chemicals into ducts, which carry them to their targets, mostly on the surface of the body. **Endocrine glands** (ductless glands) release their chemicals, which are called **hormones**, directly into the circulatory system. Once released by an endocrine gland, a hormone travels via the circulatory system until it

Figure 13.1 The endocrine glands.

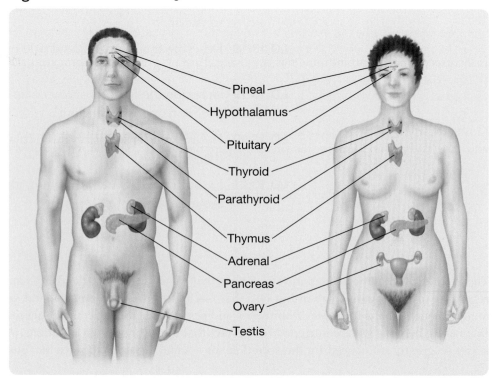

Pineal
Hypothalamus
Pituitary
Thyroid
Parathyroid
Thymus
Adrenal
Pancreas
Ovary
Testis

reaches the targets on which it normally exerts its effect (e.g., other endocrine glands, sites in the nervous system).

GONADS.   Central to any discussion of hormones and sex are the **gonads**—the male **testes** (pronounced TEST-eez) and the female **ovaries** (see Figure 13.1). As you learned in Chapter 2, the primary function of the testes and ovaries is the production of *sperm cells* and *ova*, respectively. After **copulation** (sexual intercourse), a single sperm cell may *fertilize* an *ovum* to form one cell called a **zygote**, which contains all of the information necessary for the typical growth of a complete adult organism in its natural environment. With the exception of ova and sperm cells, each cell of the human body has 23 pairs of chromosomes. In contrast, the ova and sperm cells contain only half that number, one member of each of the 23 pairs. Thus, when a sperm cell fertilizes an ovum, the resulting zygote ends up with the full complement of 23 pairs of chromosomes—one of each pair from the father and one of each pair from the mother.

Of particular interest in the context of this chapter is the pair of chromosomes called the **sex chromosomes**, so named because they contain the genetic programs that direct sexual development. The cells of females have two large sex chromosomes, which are called *X chromosomes*. In males, one sex chromosome is an X chromosome, and the other is called a *Y chromosome*. Consequently, the sex chromosome of every ovum is an X chromosome, whereas half the sperm cells have X chromosomes, and half have Y chromosomes. Your sex with all its social, economic, and personal ramifications was determined by which of your father's sperm cells won the dash to your mother's ovum. If a sperm cell with an X sex chromosome won, you are a female; if one with a Y sex chromosome won, you are a male.

You might reasonably assume that X chromosomes are X-shaped and Y chromosomes are Y-shaped, but this is incorrect. Once a chromosome has duplicated, the two products remain joined at one point, producing an X shape. This is true of all chromosomes, including Y chromosomes. Because the Y chromosome is much smaller than the X chromosome, early investigators failed to discern one small arm and thus saw a Y. In humans, the smaller Y-chromosome genes appear to control the synthesis of only 66 proteins (see Rengaraj, Kwon, & Pang, 2015)—in comparison to 615 for the larger X-chromosome genes (see Yamamoto et al., 2013).

Writing this section reminded me (JP) of my seventh-grade basketball team, the "Nads." The name puzzled our teacher because it was not at all like the names usually favored by pubescent boys—names such as the "Avengers," the "Marauders," and the "Vikings." Her puzzlement ended abruptly at our first game as our fans began to chant their support. You guessed it: "Go Nads, Go! Go Nads, Go!" My 14-year-old, spotted-faced teammates and I considered this to be humor of the most mature and sophisticated sort. The teacher didn't.

Hormones

LO 13.2   Describe the three classes of hormones and then describe three classes of gonadal hormones.

Vertebrate hormones fall into one of three classes: (1) amino acid derivatives, (2) peptides and proteins, and (3) steroids. **Amino acid derivative hormones** are hormones that are synthesized in a few simple steps from an amino acid molecule; an example is *epinephrine*, which is released from the *adrenal medulla* and synthesized from *tyrosine*. **Peptide hormones** and **protein hormones** are chains of amino acids; peptide hormones are short chains, and protein hormones are long chains. **Steroid hormones** are hormones that are synthesized from *cholesterol*, a type of fat molecule (see Abruzzese et al., 2018).

The hormones that influence sexual development and the activation of adult sexual behavior (i.e., the sex hormones) are all steroid hormones. Most other hormones produce their effects by binding to receptors in cell membranes. Steroid hormones can influence cells in this fashion; however, because they are small and fat-soluble, they can readily penetrate cell membranes and often affect cells in a second way. Once inside a cell, the steroid molecules can bind to receptors in the cytoplasm or nucleus and, by so doing, directly influence gene expression (amino acid derivative hormones and peptide hormones affect gene expression less commonly and by less direct mechanisms). Most importantly and most relevant to this chapter is that, of all the hormones, steroid hormones tend to have the most diverse and long-lasting effects on cellular function.

SEX STEROIDS.   The gonads do more than create sperm and egg cells; they also produce and release steroid hormones. Most people are surprised to learn that the testes and ovaries release the very same hormones. The two main classes of gonadal hormones are **androgens** and **estrogens; testosterone** is the most common androgen, and **estradiol** is the most common estrogen. The fact that adult ovaries tend to release more estrogens than androgens and that adult testes release more androgens than estrogens has led to the common, but misleading, practice of referring to androgens as "the *male* sex hormones" and to estrogens as "the *female* sex hormones." This practice should be avoided because of its men-are-men-and-women-are-women implication that androgens produce maleness and estrogens produce femaleness. They don't.

The ovaries and testes also release a third class of steroid hormones called **progestins**. The most common progestin is **progesterone**, which in females prepares the uterus and the breasts for pregnancy. Whether it serves a function in males is unclear, but in both males and females it seems to be involved in various forms of neuroplasticity (see González-Orozco & Camacho-Arroyo, 2019).

Because the primary function of the **adrenal cortex**—the outer layer of the *adrenal glands* (see Figure 13.1)—is the regulation of glucose and salt levels in the blood, it is not generally thought of as a sex gland. However, in addition to its principal steroid hormones, it does release small amounts of all of the sex steroids released by the gonads.

The Pituitary

LO 13.3 Explain why the pituitary is sometimes called the *master gland* and describe its anatomy. Discuss the female vs. male patterns of gonadal and gonadotropic hormone release and explain the evidence that discounted a role for the anterior pituitary in controlling those patterns of release.

The pituitary gland is frequently referred to as the *master gland* because most of its hormones are tropic hormones. *Tropic hormones'* primary function is to influence the release of hormones from other glands (*tropic* means "able to stimulate or change something"). For example, **gonadotropin** is a pituitary tropic hormone that travels through the circulatory system to the gonads, where it stimulates the release of gonadal hormones.

The pituitary gland is really two glands: the posterior pituitary and the anterior pituitary, which fuse during the course of embryological development. The **posterior pituitary** (see Figure 13.2) develops from a small outgrowth of hypothalamic tissue that eventually comes to dangle from the *hypothalamus* on the end of the **pituitary stalk** (see Figure 13.3). In contrast, the **anterior pituitary** begins as

part of the same embryonic tissue that eventually develops into the roof of the mouth; during the course of development, it pinches off and migrates upward to assume its position next to the posterior pituitary. It is the anterior pituitary that releases tropic hormones; thus, it is the anterior pituitary in particular, rather than the pituitary in general, that qualifies as the master gland.

FEMALE GONADAL HORMONE LEVELS ARE CYCLIC; MALE GONADAL HORMONE LEVELS ARE STEADY. Although males and females release exactly the same hormones, these hormones are not present at the same levels, and they do not necessarily perform the same functions. The major difference between the endocrine function of females and males is that in human females, the levels of gonadal and gonadotropic hormones go through a cycle that repeats itself every 28 days or so. It is these more-or-less regular hormone fluctuations that control the female **menstrual cycle**. In contrast—although we hate to admit it—human males are, from a neuroendocrine perspective, rather dull creatures: Males' levels of gonadal and gonadotropic hormones change little from day to day.

Because the anterior pituitary is the master gland, many early scientists assumed that an inherent difference between the male and female anterior pituitary was the basis for the difference in male and female patterns of gonadotropic and gonadal hormone release. However, this hypothesis was discounted by a series of clever transplant studies conducted by Geoffrey Harris in the 1950s (see Raisman, 2015). In these studies, a cycling pituitary removed from a mature female rat became a steady-state pituitary when transplanted at the appropriate site in a male, and a steady-state pituitary removed from a mature male rat began to cycle once transplanted into a female. What these studies established was that anterior pituitaries are not inherently female (cyclical) or male (steady-state); their patterns of hormone release are controlled by some other part of the body. The master gland seemed to have its own master. Where was it?

Control of the Pituitary

LO 13.4 Explain how the anterior and posterior pituitary are controlled.

The nervous system was implicated in the control of the anterior pituitary by behavioral research on birds and other animals that breed only during a specific time of the year. It was found that the seasonal variations in the light–dark cycle triggered many of the

Figure 13.2 A midline view of the posterior and anterior pituitary and surrounding structures.

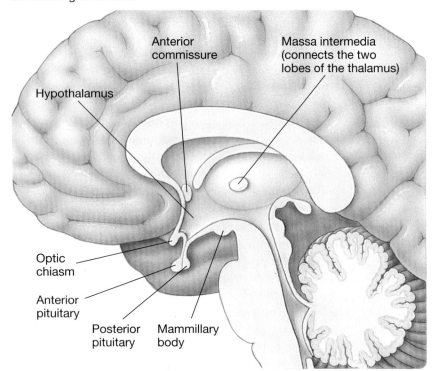

Figure 13.3 The neural connections between the hypothalamus and the pituitary. Notice the neural input to the pituitary all goes to the posterior pituitary; the anterior pituitary has no neural connections.

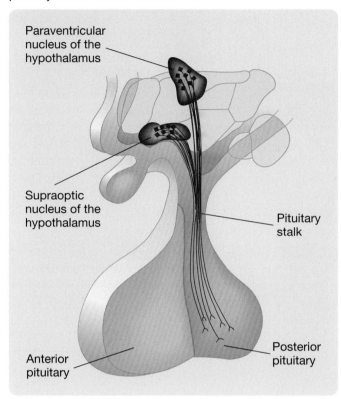

Paraventricular nucleus of the hypothalamus

Supraoptic nucleus of the hypothalamus

Pituitary stalk

Anterior pituitary

Posterior pituitary

breeding-related changes in hormone release. If the lighting conditions under which the animals lived were reversed, for example, by having the animals transported across the equator, the breeding seasons were also reversed. Somehow, visual input to the nervous system was controlling the release of tropic hormones from the anterior pituitary.

The search for the particular neural structure that controlled the anterior pituitary turned, naturally enough, to the hypothalamus, the structure from which the pituitary is suspended. Hypothalamic stimulation and lesion experiments quickly established that the hypothalamus is the regulator of the anterior pituitary, but how the hypothalamus carries out this role remained a mystery. You see, the anterior pituitary, unlike the posterior pituitary, receives no neural input whatsoever from the hypothalamus, or from any other neural structure (see Figure 13.3).

CONTROL OF THE ANTERIOR AND POSTERIOR PITUITARY BY THE HYPOTHALAMUS. There are two different mechanisms by which the hypothalamus controls the pituitary: one for the posterior pituitary and one for the anterior pituitary. The two major hormones of the posterior pituitary, **vasopressin** and **oxytocin**, are peptide hormones that are synthesized in the cell bodies of neurons in the **paraventricular nuclei** and **supraoptic nuclei** on each side of the hypothalamus (see Figure 13.3 and Appendix VI).

They are then transported along the axons of these neurons to their terminals in the posterior pituitary and are stored there until the arrival of action potentials causes them to be released into the bloodstream. (Neurons that release hormones into general circulation are called *neurosecretory cells*.) Oxytocin stimulates contractions of the uterus during labor and the ejection of milk during suckling; vasopressin (also called *antidiuretic hormone*) facilitates the reabsorption of water by the kidneys; and both seem to influence stress-coping and social responses (see Benarroch, 2013; Hammock, 2015; Shen, 2015).

The means by which the hypothalamus controls the release of hormones from the neuron-free anterior pituitary was more difficult to explain. Harris (1955) suggested that the release of hormones from the anterior pituitary was itself regulated by hormones released from the hypothalamus. Two findings provided early support for this hypothesis. The first was the discovery of a vascular network, the **hypothalamopituitary portal system**, that seemed well suited to the task of carrying hormones from the hypothalamus to the anterior pituitary. As Figure 13.4 illustrates, a network of hypothalamic capillaries feeds a bundle of portal veins that carries blood down the pituitary stalk into another network of capillaries in the anterior pituitary. (A *portal vein* is a vein that connects one capillary network with another.) The second finding was the discovery that cutting the portal veins of the pituitary stalk disrupts the release of anterior pituitary hormones until the damaged veins regenerate (Harris, 1955).

Discovery of Hypothalamic Releasing Hormones

LO 13.5 Describe the research that led to the discovery of the hypothalamic releasing hormones.

It was hypothesized that the release of each anterior pituitary hormone is controlled by a different hypothalamic hormone. Each hypothalamic hormone that was thought to stimulate the release of an anterior pituitary hormone was referred to as a **releasing hormone**. In contrast, each hormone thought to inhibit the release of an anterior pituitary hormone was referred to as a **release-inhibiting hormone**.

Efforts to isolate the putative (hypothesized) hypothalamic releasing and release-inhibiting hormones led to a major breakthrough in the late 1960s. Guillemin and his colleagues isolated **thyrotropin-releasing hormone** from the hypothalamus of sheep, and Schally and his colleagues isolated the same hormone from the hypothalamus of pigs. Thyrotropin-releasing hormone triggers the release of **thyrotropin** from the anterior pituitary, which in turn stimulates the release of hormones from the *thyroid gland*. For their efforts, Guillemin and Schally were awarded Nobel Prizes in 1977.

Schally's and Guillemin's isolation of thyrotropin-releasing hormone confirmed that hypothalamic releasing

Figure 13.4 Control of the anterior and posterior pituitary by the hypothalamus.

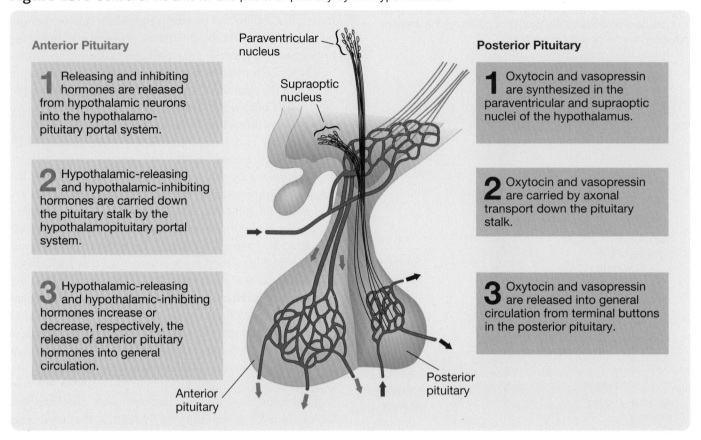

Anterior Pituitary

1 Releasing and inhibiting hormones are released from hypothalamic neurons into the hypothalamo-pituitary portal system.

2 Hypothalamic-releasing and hypothalamic-inhibiting hormones are carried down the pituitary stalk by the hypothalamopituitary portal system.

3 Hypothalamic-releasing and hypothalamic-inhibiting hormones increase or decrease, respectively, the release of anterior pituitary hormones into general circulation.

Paraventricular nucleus

Supraoptic nucleus

Posterior Pituitary

1 Oxytocin and vasopressin are synthesized in the paraventricular and supraoptic nuclei of the hypothalamus.

2 Oxytocin and vasopressin are carried by axonal transport down the pituitary stalk.

3 Oxytocin and vasopressin are released into general circulation from terminal buttons in the posterior pituitary.

Posterior pituitary

Anterior pituitary

hormones control the release of hormones from the anterior pituitary and thus provided the major impetus for the isolation and synthesis of other releasing and release-inhibiting hormones. Of direct relevance to the study of sex hormones was the subsequent isolation of **gonadotropin-releasing hormone** by Schally and his group (Schally, Kastin, & Arimura, 1971). This releasing hormone stimulates the release of both of the anterior pituitary's gonadotropins: **follicle-stimulating hormone (FSH)** and **luteinizing hormone (LH)**. All hypothalamic-releasing hormones, like all tropic hormones, have proven to be peptides.

Regulation of Hormone Levels

LO 13.6 Describe three different types of signals that regulate hormone release. Also, describe how hormones are released over time and the effect this pattern of release has on levels of circulating hormones.

Hormone release is regulated by three different kinds of signals: signals from the nervous system, signals from circulating hormones, and signals from circulating nonhormonal chemicals.

REGULATION BY NEURAL SIGNALS. All endocrine glands, with the exception of the anterior pituitary, are directly regulated by signals from the nervous system.

Endocrine glands located in the brain (i.e., the pituitary and pineal glands) are regulated by cerebral neurons. In contrast, those endocrine glands located outside the CNS are innervated by the *autonomic nervous system*—usually by both the *sympathetic* and *parasympathetic* branches, which often have opposite effects on hormone release.

It is extremely important to remember that hormone release can be influenced by experience—for example, many species that breed only in the spring are often prepared for reproduction by the release of sex hormones triggered by the increasing daily duration of daylight. This means that an explanation of any behavioral phenomenon in terms of a hormonal mechanism does not necessarily rule out an explanation in terms of an experiential mechanism.

Journal Prompt 13.1

Given what you have just learned about the hormonal differences between males and females, comment on the following statement: "Men are from Mars and women are from Venus."

REGULATION BY HORMONAL SIGNALS. The hormones themselves also influence hormone release. You have already learned, for example, that the tropic hormones of the anterior pituitary influence the release of hormones from their respective target glands. However, the regulation of endocrine function by the anterior pituitary is not a one-way street. Circulating hormones often provide feedback to the very structures that influence their release: the pituitary gland, the hypothalamus, and other sites in the brain. The function of most hormonal feedback is the maintenance of stable blood levels of the hormones. Thus, high gonadal hormone levels usually have effects on the hypothalamus and pituitary that decrease subsequent gonadal hormone release, and low levels usually have effects that increase hormone release.

REGULATION BY NONHORMONAL CHEMICALS.
Circulating chemicals other than hormones can play a role in regulating hormone levels. Glucose, calcium, and sodium levels in the blood all influence the release of particular hormones. For example, you learned in Chapter 12 that increases in blood glucose increase the release of *insulin* from the *pancreas*; in turn, insulin reduces blood glucose levels.

PULSATILE HORMONE RELEASE. Hormones tend to be released in pulses (see Plant, 2015); they are discharged several times per day in large surges, which typically last no more than a few minutes. Hormone levels in the blood are regulated by changes in the frequency and duration of the hormone pulses. One consequence of **pulsatile hormone release** is that there are often large minute-to-minute fluctuations in the levels of circulating hormones (see Lightman & Conway-Campbell, 2010). Accordingly, when the pattern of human male gonadal hormone release is referred to as "steady," it means that there are no major systematic changes in circulating gonadal hormone levels from day to day, not that the levels never vary.

Summary Model of Gonadal Endocrine Regulation

LO 13.7 Summarize the model of gonadal endocrine regulation.

Figure 13.5 is a summary model of the regulation of gonadal hormones. According to this model, the brain controls the release of gonadotropin-releasing hormone from the hypothalamus into the hypothalamopituitary portal system, which carries it to the anterior pituitary. In the anterior pituitary, the gonadotropin-releasing hormone stimulates the release of gonadotropins, which are carried by the circulatory system to the gonads. In response to the gonadotropins, the gonads release androgens, estrogens, and progestins, which feed back into the pituitary and hypothalamus to regulate subsequent gonadal hormone release. Armed with this general perspective of neuroendocrine function, you are now ready to consider how gonadal hormones direct sexual development and activate adult sexual behavior.

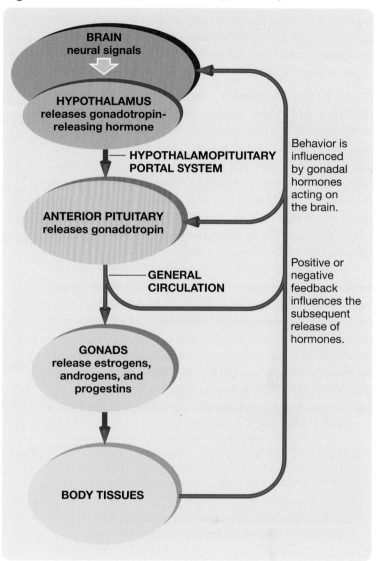

Figure 13.5 A summary model of the regulation of gonadal hormones.

Hormones and Sexual Development of the Body

You have undoubtedly noticed that humans are *dimorphic*—that is, most come in one of two models: female and male. This module describes how the development of female and male bodily characteristics is directed by hormones.

Sexual Differentiation

LO 13.8 Describe the development of the internal and external reproductive organs.

Sexual differentiation in mammals begins at fertilization with the production of one of two different kinds of zygotes: either one with an XX (female) pair of sex chromosomes or one with an XY (male) pair. It is the genetic information on the sex chromosomes that usually determines whether development will occur along female or male lines. But be cautious here: Do not fall into the seductive embrace of the men-are-men-and-women-are-women assumption. Do not begin by assuming that there are two parallel but opposite genetic programs for sexual development, one for female development and one for male development. As you are about to learn, sexual development is far more complex.

FETAL HORMONES AND DEVELOPMENT OF REPRODUCTIVE ORGANS. Figure 13.6 illustrates the structure of the gonads as they appear 6 weeks after fertilization. Notice that at this stage of development, each fetus, regardless of its genetic sex, has the same pair of gonadal structures, called *primordial gonads* (*primordial* means "existing at the beginning"). Each primordial gonad has an outer covering, or *cortex*, which has the potential to develop into an ovary; and each has an internal core, or *medulla*, which has the potential to develop into a testis.

In the seventh week after conception, the **Sry gene** on the Y chromosome of a male triggers the synthesis of **Sry protein** (see Arnold, 2017; Sekido & Lovell-Badge, 2013; Wu et al., 2012), and this protein causes the medulla of each primordial gonad to grow and to develop into a testis. In the absence of Sry protein, the cortical cells of the primordial gonads develop into ovaries (see Lin & Capel, 2015). So, if Sry protein is injected into a genetic female fetus 6 weeks after conception, the result is a genetic female with testes; or if drugs that

block the effects of Sry protein are injected into a genetic male fetus, the result is a genetic male with ovaries. Such examples of **intersexed persons** expose in a dramatic fashion the weakness of mamawawa thinking (thinking of "male" and "female" as mutually exclusive, opposite categories).

INTERNAL REPRODUCTIVE DUCTS. Six weeks after fertilization, all fetuses have two complete sets of reproductive ducts. They have a **Wolffian system**, which has the capacity to develop into male reproductive ducts (e.g., the *seminal vesicles*, which hold the fluid in which sperm cells are ejaculated; and the *vas deferens*, through which the sperm cells travel to the seminal vesicles). And they have a **Müllerian system**, which has the capacity to develop into female ducts (e.g., the *uterus*; the upper part of the *vagina*; and the *fallopian tubes*, through which ova travel from the ovaries to the uterus).

In the third month of male fetal development, the testes secrete testosterone and **Müllerian-inhibiting substance**. As Figure 13.7 illustrates, the testosterone stimulates the

Figure 13.6 The development of an ovary and a testis from the cortex and the medulla, respectively, of the primordial gonadal structure that is present 6 weeks after conception.

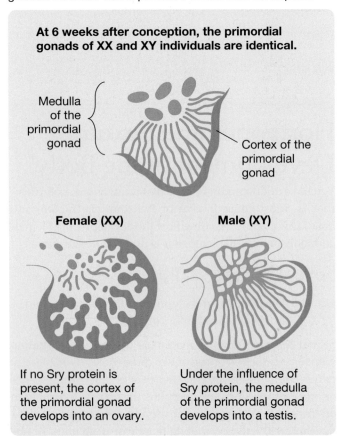

At 6 weeks after conception, the primordial gonads of XX and XY individuals are identical.

Medulla of the primordial gonad

Cortex of the primordial gonad

Female (XX)

Male (XY)

If no Sry protein is present, the cortex of the primordial gonad develops into an ovary.

Under the influence of Sry protein, the medulla of the primordial gonad develops into a testis.

Figure 13.7 The development of the internal ducts of the male and female reproductive systems from the Wolffian and Müllerian systems, respectively.

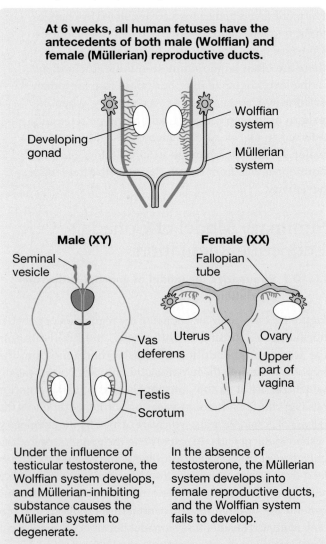

At 6 weeks, all human fetuses have the antecedents of both male (Wolffian) and female (Müllerian) reproductive ducts.

Developing gonad

Wolffian system

Müllerian system

Male (XY)

Seminal vesicle

Vas deferens

Testis

Scrotum

Female (XX)

Fallopian tube

Uterus

Ovary

Upper part of vagina

Under the influence of testicular testosterone, the Wolffian system develops, and Müllerian-inhibiting substance causes the Müllerian system to degenerate.

In the absence of testosterone, the Müllerian system develops into female reproductive ducts, and the Wolffian system fails to develop.

development of the Wolffian system, and the Müllerian-inhibiting substance causes the Müllerian system to degenerate and the testes to descend into the **scrotum**—the ball sac that holds the testes outside the body cavity (see Sajjad, 2010). Because it is testosterone—not the sex chromosomes—that triggers Wolffian development, genetic females who are injected with testosterone during the appropriate fetal period develop male reproductive ducts along with their female ones.

The differentiation of the internal ducts of the female reproductive system (see Figure 13.7) is not under the control of ovarian hormones; the ovaries are almost completely inactive during fetal development. The development of the Müllerian system occurs in any fetus that is not exposed to testicular hormones during the critical fetal period. Accordingly, female fetuses, ovariectomized female fetuses, and orchidectomized male fetuses all develop female reproductive ducts (Jost, 1972). **Ovariectomy** is the removal of the ovaries, and **orchidectomy** is the removal of the testes (the Greek word *orchis* means "testicle"). **Gonadectomy**, or *castration*, is the surgical removal of gonads—either ovaries or testes.

EXTERNAL REPRODUCTIVE ORGANS. There is a basic difference between the differentiation of the external reproductive organs and the differentiation of the internal reproductive organs (i.e., the gonads and reproductive ducts).

As you have just read, every typical fetus develops separate precursors for the male (medulla) and female (cortex) gonads and for the male (Wolffian system) and female (Müllerian system) reproductive ducts; then, only one set, male or female, develops. In contrast, both male and female **genitals** develop from the very same precursor (see Sajjad, 2010). This *bipotential precursor* and its subsequent differentiation are illustrated in Figure 13.8.

At the end of the third month of pregnancy, the bipotential precursor of the external reproductive organs consists of four parts: the glans, the urethral folds, the lateral bodies, and the labioscrotal swellings. Then it begins to differentiate. The *glans* grows into the head of the *penis* in the male or the *clitoris* in the female; the *urethral folds* fuse in the male or enlarge to become the *labia minora* in the female; the *lateral bodies* form the shaft of the penis in the male or the hood of the clitoris in the female; and the *labioscrotal swellings* form the *scrotum* in the male or the *labia majora* in the female.

Like the development of the internal reproductive ducts, the development of the external genitals is controlled

Figure 13.8 The development of male and female external reproductive organs from the same bipotential precursor.

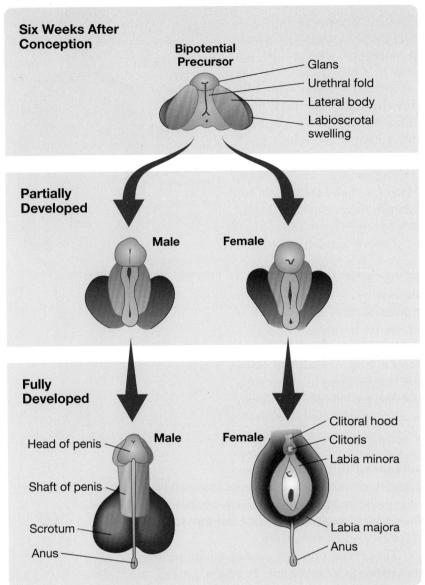

by the presence or absence of testosterone. If testosterone is present at the appropriate stage of fetal development, male external genitals develop from the bipotential precursor. Conversely, if testosterone is not present, development of the external genitals proceeds along female lines (see Matsushita et al., 2018).

Puberty: Hormones and Development of Secondary Sex Characteristics

LO 13.9 Describe the male and female secondary sex characteristics and the role of hormones in their development.

During childhood, levels of circulating gonadal hormones are low, reproductive organs are immature, and males and females differ little in general appearance. This period of developmental quiescence ends abruptly with the onset of *puberty*—the transitional period between childhood and adulthood during which fertility is achieved, the adolescent growth spurt occurs, and the secondary sex characteristics develop. **Secondary sex characteristics** are those features other than the reproductive organs that distinguish sexually mature males and females. Many of these secondary sex characteristics develop during puberty (see Figure 13.9).

Puberty is associated with an increase in the release of hormones by the anterior pituitary. The increase in the release of **growth hormone**—the only anterior pituitary hormone that does not have a gland as its primary target—acts directly on bone and muscle tissue to produce the pubertal growth spurt (see Colvin & Abdullatif, 2013; Russell et al., 2011). Increases in the release of gonadotropic hormone and **adrenocorticotropic hormone** cause the gonads and adrenal cortex to increase their release of gonadal and adrenal hormones, which in turn initiate the maturation of the genitals and the development of secondary sex characteristics.

The general principle guiding typical pubertal sexual maturation is a simple one: In pubertal males, androgen levels are higher than estrogen levels, and masculinization is the result; in pubertal females, the estrogens predominate, and the result is feminization (see Colvin & Abdullatif, 2013). Individuals castrated prior to puberty do not become sexually mature unless they receive replacement injections of androgens or estrogens.

But even during puberty, the men-are-men-and-women-are-women assumption stumbles badly. You see, **androstenedione**, an androgen that is released primarily by the adrenal cortex, is typically responsible for the growth of pubic hair and *axillary hair* (underarm hair) in females. It is hard to take seriously the practice of referring to androgens as "male hormones" when one of them is responsible for the development of the female pattern of pubic hair growth. The male pattern is a pyramid, and the female pattern is an inverted pyramid (see Figure 13.9).

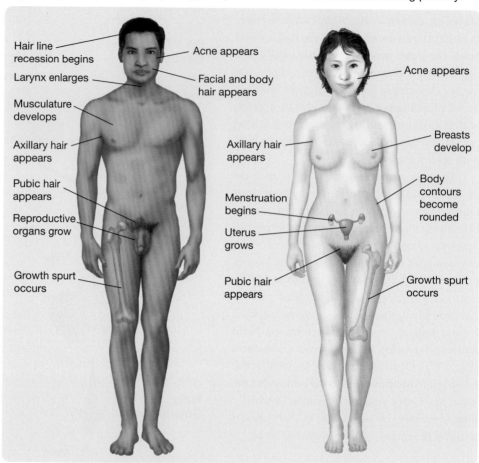

Figure 13.9 The changes that typically occur in males and females during puberty.

Hair line recession begins
Larynx enlarges
Musculature develops
Axillary hair appears
Pubic hair appears
Reproductive organs grow
Growth spurt occurs

Acne appears
Facial and body hair appears

Acne appears
Breasts develop
Body contours become rounded
Axillary hair appears
Menstruation begins
Uterus grows
Pubic hair appears
Growth spurt occurs

Sexual Development of Brain and Behavior

As you have just learned, the principles of the differentiation of the human body are well understood. But now, current research is focusing on a more difficult problem, the sexual differentiation of brain and behavior. This module reveals how seminal studies conducted in the 1930s generated theories that have morphed, under the influence of subsequent research, into our current views.

Sex differences in brain and behavior has been the identification of a systematic problem in neuroscience research. Until recently, male-only studies or studies where both males and females were lumped together as one group were the norm (see Fischer & Riddle, 2018). However, because of the growing body of research on sex-related brain differences, this practice is changing (see Choleris et al., 2018). Indeed, many funding agencies now require that their researchers always assess for potential sex differences in brain and behavior.

Sex Differences in the Brain

LO 13.10 Describe the evolution of research and thinking about sex differences in the brain.

The brains of males and females may look the same on casual inspection, and it may be politically correct to believe that they are—but they are not. The brains of males tend to be about 15 percent larger than those of females on average, and many other anatomical differences between male and female brains have been documented. There are differences in the average volumes of various cortical areas, nuclei and fiber tracts, in the numbers and types of neural and glial cells that compose various structures, in the plasticity of certain brain structures, and in the numbers and types of synapses that connect the cells in various structures (see Brecht, Lenschow, & Rao, 2018; Choleris et al., 2018; Grabowska, 2017; Hausmann, 2017; McCarthy, Nugent, & Lenz, 2017; Mottron et al., 2015).

Let's begin with the first functional sex difference to be identified in mammalian brains.

FIRST DISCOVERY OF A SEX DIFFERENCE IN MAMMALIAN BRAIN FUNCTION. The first attempts to discover sex differences in the mammalian brain focused on the factors that control the development of the steady and cyclic patterns of gonadotropin release in males and females, respectively. The seminal experiments were conducted by Pfeiffer in 1936. In his experiments, some neonatal rats (males and females) were gonadectomized and some were not, and some received gonad transplants (ovaries or testes) and some did not.

Remarkably, Pfeiffer found that gonadectomizing neonatal rats of either genetic sex caused them to develop into adults with the female cyclic pattern of gonadotropin release. In contrast, transplantation of testes into gonadectomized or intact female neonatal rats caused them to develop into adults with the steady male pattern of gonadotropin release. Transplantation of ovaries had no effect on the pattern of hormone release. Pfeiffer concluded that the female cyclic pattern of gonadotropin release develops unless the preprogrammed female cyclicity is overridden by testosterone during perinatal development.

Pfeiffer incorrectly concluded that the presence or absence of testicular hormones in neonatal rats influenced the development of the pituitary because he was not aware of something we know today: The release of gonadotropins from the anterior pituitary is controlled by the hypothalamus. Once this was discovered, it became apparent that Pfeiffer's experiments had provided the first evidence of the role of *perinatal* (around the time of birth) androgens in overriding the preprogrammed cyclic female pattern of gonadotropin release from the hypothalamus and initiating the development of the steady male pattern. This 1960s modification of Pfeiffer's theory of brain differentiation to include the hypothalamus was consistent with the facts of brain differentiation as understood at that time, but subsequent research necessitated major revisions. The first of these major revisions became known as the aromatization hypothesis.

AROMATIZATION HYPOTHESIS. What is aromatization? All gonadal and adrenal sex hormones are steroid hormones, and because all steroid hormones are derived from cholesterol, they have similar structures and are readily converted from one to the other. For example, a slight change to the testosterone molecule that occurs under the influence of the *enzyme* (a protein that influences a biochemical reaction without participating in it) **aromatase** converts testosterone to estradiol. This process is called **aromatization**.

According to the **aromatization hypothesis**, perinatal testosterone does not directly masculinize the brain; the brain is masculinized by estradiol that has been aromatized from perinatal testosterone. Although the idea that estradiol— the alleged female hormone—masculinizes the brain may seem counterintuitive, there is strong evidence for it in rats and mice: (1) findings demonstrating masculinizing effects on the brain of early estradiol injections, and (2) findings showing masculinization of the brain does not occur in response to testosterone administered with agents that block aromatization or in response to androgens that cannot be aromatized (e.g., dihydrotestosterone).

How do genetic females of species whose brains are masculinized by estradiol keep from being masculinized by their mothers' estradiol, which circulates through the fetal blood supply? Alpha fetoprotein is the answer. **Alpha fetoprotein** is present in the blood of rodents during the perinatal period, and it deactivates circulating estradiol by binding to it (see Yang & Shah, 2014). Although the role of alpha fetoprotein in deactivating estradiol is firmly established in rodents, its function in humans remains controversial (see Koebele & Bimonte-Nelson, 2015).

How, then, does estradiol masculinize the brain of the male rodent fetus in the presence of the deactivating effects of alpha fetoprotein? Because testosterone is immune to alpha fetoprotein, it can travel unaffected from the testes to the brain cells where it is converted to estradiol. Estradiol is not broken down in the rodent brain because alpha fetoprotein does not readily penetrate the blood–brain barrier.

How does the aromatization hypothesis fare when it comes to explaining human masculinization? The aromatization hypothesis was initially considered generalizable to humans. However, current research tells us that any masculinizing effects of testosterone are due to its direct effects on the human brain (see Motta-Mena & Puts, 2017; Puts & Motta-Mena, 2018).

SEX DIFFERENCES IN THE BRAIN: THE MODERN PERSPECTIVE. So far, our discussion of the development of sex differences has focused on the reproductive organs, secondary sex characteristics, and the hypothalamus. You have learned from this discussion that one theory accounts for the development of many sex differences: The default program of development is the female program, which is overridden in genetic males by perinatal exposure to testosterone. The initial assumption was that this same mechanism would prove to be the sole mechanism responsible for the development of other differences between male and female brains. However, this has not proven to be the case (see Arnold, 2017; Lenz, Nugent, & McCarthy, 2012).

Before considering the mechanisms by which sex differences in the brain develop, it is important to understand the nature of the differences. The vast majority of sex differences in the brain are not all-or-none, men-are-men-and-women-are-women differences. Most of the sex differences in the brain that have been documented are slight, variable, and statistical. In short, many differences exist between average male and female brains, but there is usually plenty of overlap. Indeed, in humans there is so much overlap that some researchers have argued it would be better for us not to think in terms of male brains versus female brains (see Joel & Fausto-Sterling, 2016).

At the same time, it would be a big mistake to ignore the study of differences between the brains of men and women. Even if most documented differences are slight, variable, and statistical, collectively those differences are sufficient to have significant health implications for the different sexes (see Ball, Balthazart, & McCarthy, 2014; Choleris et al., 2018; Geary, 2019). For example, there are marked sex differences in the incidence of neurological disorders: Parkinson's disease and autism spectrum disorder are far more common in males, whereas Alzheimer's disease is far more common in females (see Gurvich et al., 2018). Of major clinical relevance is the discovery that males do not recover from traumatic brain injury as well as females (see Mollayeva, Mollayeva, & Colantonio, 2018).

Although research on the development of sex differences in the brain is still in its infancy, one important principle has emerged. Brains are not masculinized or feminized as a whole: Sex differences develop independently in different parts of the brain at different points in time and by different mechanisms (see Joel & Fausto-Sterling, 2016). For example, aromatase is found in only a few areas of the rat brain (e.g., the hypothalamus), and it is only in these areas that aromatization is critical for testosterone's masculinizing effects (see de Bournonville et al., 2019; Lentini et al., 2012). Also, some sex differences in the brain are not manifested until puberty, and these differences are unlikely to be the product of perinatal hormones (see Ingalhalikar et al., 2014), which, as you have just learned, play a role in the development of some sex differences in the brain.

Further complicating the study of the development of sex differences in the brain is the fact that three factors that have proven to play little or no role in the sexual differentiation of the reproductive organs *do* play a role in the sexual differentiation of the brain. First, sex chromosomes have been found to influence brain development independent of their effect on hormones (see Arnold, 2017; Khramtsova et al., 2019; Maekawa et al., 2014); for example, different patterns of gene expression (see Chapter 2) exist in the brains of male and female mice even before the gonads become functional (e.g., Wolstenholme, Rissman, & Bekiranov, 2013). Second, epigenetic effects influence the emergence of brain sex differences. For example, epigenetic mechanisms (see Chapter 2) can interact with gonadal hormones to produce sex-specific effects on brain development (see Forger, 2018; McCarthy, Nugent, & Lenz, 2017). Third, although the female program of reproductive organ development typically proceeds in the absence of gonadal steroids, recent evidence suggests that estradiol plays an active role; knockout mice without the gene that forms estradiol receptors do not display a female pattern of brain development (see Maekawa et al., 2014). Thus, although the conventional view that a female program of development is the default program does a good job of explaining differentiation of the reproductive organs and hypothalamus, it falters badly when it comes to differentiation of the brain in general.

The mechanisms of brain differentiation appear to be much more complex and selective. Complicating their study is the fact that these complex mechanisms are different in different mammalian species (see Sekido, 2014). For example, as discussed earlier, although aromatization plays a role in the masculinization of the brains of rats and mice, it doesn't in humans.

Development of Sex Differences in Behavior

LO 13.11 Describe the results of studies of sex differences in behavior in humans and nonhumans.

Because it is not ethical to conduct experiments on the development of sex differences in humans, most of the research on this topic has focused on the development of reproductive behavior in laboratory animals. Until recently, this research has focused on the effects of perinatal hormones.

DEVELOPMENT OF REPRODUCTIVE BEHAVIORS IN LABORATORY ANIMALS.   Phoenix and colleagues (1959) were among the first to demonstrate that the perinatal injection of testosterone **masculinizes** and **defeminizes** a genetic female's adult reproductive behavior. First, they injected pregnant guinea pigs with testosterone. Then, when the litters were born, the researchers ovariectomized the female offspring. Finally, when these ovariectomized female guinea pigs reached maturity, the researchers injected them with testosterone and assessed their copulatory behavior. Phoenix and his colleagues found that the females exposed to perinatal testosterone displayed more male-like mounting behavior in response to testosterone injections in adulthood than adult females who had not been exposed to perinatal testosterone. And when as adults the female guinea pigs were injected with progesterone and estradiol and mounted by males, they displayed less **lordosis** (the intromission-facilitating arched-back posture that signals female rodent receptivity).

In a study complementary to that of Phoenix and colleagues, Grady, Phoenix, and Young (1965) found that the lack of early exposure of male rats to testosterone both **feminizes** and **demasculinizes** their reproductive behavior as adults. Male rats castrated shortly after birth failed to display the typical male copulatory pattern of mounting, **intromission** (penis insertion), and **ejaculation** (ejection of sperm) when as adults they were treated with testosterone and given access to a sexually receptive female. Moreover, when they were injected with estrogen and progesterone as adults, they exhibited more lordosis than uncastrated controls.

When it comes to the effects of perinatal testosterone on rat reproductive behavior development, timing is critical. The ability of single injections of testosterone to masculinize and defeminize rat reproductive behavior seems to be restricted to the first 5 days after birth (see McCarthy, Herold, & Stockman, 2018).

Because much of the research on hormones and the development of reproductive behavior has focused on the copulatory act, we know less about the role of hormones in the development of **proceptive behaviors** (solicitation behaviors) and in the development of sex-related behaviors that are not directly related to reproduction. However, perinatal testosterone has been found to disrupt the proceptive hopping, darting, and ear wiggling of receptive female rats.

Before you finish this section, we want to clarify an important point. If you are like many of our students, you may be wondering why biopsychologists who study the development of male–female behavioral differences always measure *masculinization* separately from *defeminization* and measure *feminization* separately from *demasculinization*. If you think that masculinization and defeminization are the same thing and that feminization and demasculinization are the same thing, you have likely fallen into the trap of the men-are-men-and-women-are-women assumption—that is, into the trap of thinking of maleness and femaleness as discrete, mutually exclusive, opposite categories. Indeed, male behaviors and female behaviors can coexist in the same individual, and they do not necessarily change in opposite directions if the individual receives physiological treatment such as hormones or brain lesions. For example, "male" behaviors (e.g., mounting receptive females) have been observed in the females of many different mammalian species, and "female" behaviors (e.g., lordosis) have been observed in males (see Reznikov et al., 2016). Furthermore, lesions in medial preoptic areas have been shown to abolish male reproductive behaviors in both male and female rats without affecting female behaviors (see Singer, 1968; Will, Hull, & Dominguez, 2014). Think about this idea carefully; it plays an important role in later parts of the chapter.

DEVELOPMENT OF SEX DIFFERENCES IN THE BEHAVIOR OF HUMANS.   There is much research on the development of behavioral differences in human females and males. However, because experimental investigations of this process are not ethical, virtually all of the research is based on case studies and correlational studies, which are difficult to interpret (see Chapter 1). Still, three general conclusions have emerged.

First, some sex differences in human behavior appear to be **sexual dimorphisms** (see McCarthy et al., 2012; Rigby & Kulathinal, 2015; Yang & Shaw, 2014). Sexual dimorphisms are instances where a behavior (or a structure) typically comes in two distinctive classes (male or female) into which most individuals can be unambiguously assigned. In the case of humans, it appears to be only reproduction-related behaviors that clearly fall into this category. The presence or absence of prenatal testosterone appears to be a major factor in the development of these behaviors (see Balthazart, 2011; Kreukels & Cohen-Kettenis, 2012).

Second, most differences between the behavior of average human males and average human females are small and characterized by substantial overlap between individuals of the two groups (see Hines, 2011; McCarthy et al., 2012; Miller & Halpern, 2014). For example, such human behavioral sex differences occur in play behavior, social interaction, reaction to pain, language, cognition, emotionality, drug sensitivity, and responses to stress (see Bale & Epperson, 2015; Eisenegger, Haushofer, & Fehr, 2011; Loyd & Murphy, 2014; Miller & Halpern, 2014; Mogil, 2012). The presence or absence of prenatal testosterone exposure has been shown to contribute to the development of these kinds of sex differences, but in general they account for only a portion of each difference.

The third conclusion that has emerged from the study of human behavioral sex differences is that there are often differences in the susceptibility of human males and females to behavioral disorders (see McCarthy et al., 2012; ter Horst et al., 2012; Zhao, Woody, & Chhibber, 2015). For example, *dyslexia* (reading difficulties), early-onset schizophrenia, stuttering, and autism spectrum disorders are each about three times more prevalent in males; and attention deficit hyperactivity disorder is 10 times more likely in males. In contrast, females are twice as likely to be diagnosed with depression, anxiety disorders, and Alzheimer's disease; and about 10 times as many females are diagnosed with certain eating disorders. The mechanisms leading to the development of any of these sex differences in susceptibility to behavioral disorders is unclear (see Cahill, 2014; ter Horst et al., 2012).

Before leaving the topic of sex differences in brain and behavior, we want to emphasize that the frequent finding that prenatal testosterone exposure influences the development of sex differences does not preclude other factors (see Hines, 2011). For example, cultural factors have been shown to play a major role in the development of many sex differences, perhaps by acting on the same brain mechanisms influenced by prenatal hormones.

Scan Your Brain

Before you proceed to a consideration of three cases of exceptional human sexual development, scan your brain to see whether you understand the basics of typical sexual development. Fill in the blanks in the following sentences. The correct answers are provided at the end of the exercise. Review material related to your errors and omissions before proceeding.

1. Under the influence of _____ protein, the medulla develops into a testis.

2. The increase in the release of _____ hormone acts directly on bone and muscle tissue to produce the pubertal growth spurt.

3. Increases in the release of _____ hormone and _____ hormone cause the gonads and adrenal cortex to increase their release of gonadal and adrenal hormones, which in turn initiate the maturation of the genitals and the development of secondary sex characteristics.

4. The hormonal factor that triggers the development of the human Müllerian system is the lack of _____ around the third month of fetal development.

5. The scrotum and the _____ develop from the same bipotential precursor.

6. In rodents, the presence of _____ in the prenatal period blocks the circulation of estradiol by binding to it.

7. _____ chromosomes have been found to influence brain development independent of their effect on hormones.

8. _____ are instances where a behavior (or structure) comes in two distinct classes (male or female) into which most individuals can be unambiguously assigned.

9. Certain conditions such as early-onset schizophrenia, autism spectrum disorders, stuttering, and _____ are three times more prevalent in males.

Scan Your Brain answers: (1) Sry, (2) growth, (3) gonadotropic, adrenocorticotropic, (4) androgens (or testosterone), (5) labia majora, (6) alpha fetoprotein, (7) Sex, (8) Sexual dimorphisms (9) dyslexia.

Three Cases of Exceptional Human Sexual Development

This module discusses three cases of exceptional sexual development. We are sure you will be intrigued by these three cases, but that is not the only reason we have chosen to present them. Our main reason is expressed by a proverb: The exception proves the rule. Most people think this proverb means that the exception "proves" the rule in the sense that it establishes its truth, but this is clearly wrong: The truth of a rule is challenged by, not confirmed by, exceptions to it. The word *proof* in this usage comes from the Latin *probare*, which means "to test"—as in *proving ground*—and this is the sense in which it is used in the proverb. Hence, the proverb means that the explanation of exceptional cases is a testing ground for any theory.

Exceptional Cases of Human Sexual Development

LO 13.12 **Explain what androgen insensitivity syndrome, adrenogenital syndrome, and ablatio penis have taught us about human sexual development.**

So far in this chapter, you have learned the rules according to which hormones seem to influence typical sexual development. Now, three exceptional cases are offered to prove (to test) these rules.

The Case of Anne S., the Woman with Testes

Anne S., a 26-year-old female, sought treatment for two sex-related disorders: lack of menstruation and pain during sexual intercourse (Jones & Park, 1971). She sought help because she and her husband of 4 years had been trying without success to have children, and she correctly surmised that her lack of a menstrual cycle was part of the problem. A physical examination revealed that Anne was a healthy young female. Her only readily apparent peculiarity was the sparseness and fineness of her pubic and axillary hair. Examination of her external genitals revealed nothing atypical; however, there were some differences with her internal genitals. Her vagina was only 4 centimeters long, and her uterus was underdeveloped.

At the start of this chapter, we said that you would encounter some remarkable things, and the diagnosis of Anne's case certainly qualifies as one of them. Anne's doctors concluded that her sex chromosomes were XY; they concluded that Anne had the genes of a genetic male. Three lines of evidence supported their diagnosis. First, analysis of cells scraped from the inside of Anne's mouth revealed that they had both an X and a Y chromosome. Second, a tiny incision in Anne's abdomen, which enabled Anne's physicians to look inside, revealed a pair of internalized testes but no ovaries. Finally, hormone tests revealed that Anne's hormone levels were more typical of a male.

Anne suffers from **complete androgen insensitivity syndrome**; all her symptoms stem from a mutation of the androgen receptor gene that rendered her androgen receptors totally unresponsive (see Bramble et al., 2017; Chen et al., 2015; Mongan et al., 2015). Complete androgen insensitivity syndrome is rare, occurring in about 1 of 60,000 genetic male births (see Bramble et al., 2017; Walia et al., 2018).

During development, Anne's testes released male-typical levels of androgens, but her body could not respond to them

because of her unresponsive androgen receptors; thus, her development proceeded as if no androgens had been released. Her external genitals, her brain, and her behavior developed along female lines, without the effects of androgens to override the female program, and her testes could not descend from her body cavity with no scrotum for them to descend into. Furthermore, Anne did not develop typical internal female reproductive ducts because, like other genetic males, her testes released Müllerian-inhibiting substance; that is why her vagina was short and her uterus underdeveloped. At puberty, Anne's testes released enough estrogens to feminize her body in the absence of the counteracting effects of androgens; however, adrenal androstenedione was not able to stimulate the growth of pubic and axillary hair.

An interesting issue of medical ethics is raised by the androgen insensitivity syndrome. Many people believe that physicians should always disclose all relevant findings to their patients. If you were Anne's physician, would you tell her that she is a genetic male? Would you tell her husband? Her doctor did not. Anne's vagina was surgically enlarged, she was counseled to consider adoption, and, as far as we know, she is still happily married and unaware of her genetic sex. On the other hand, we have heard from several genetic females who have *partial androgen insensitivity*, and they recommended full disclosure. They had faced a variety of sexual ambiguities throughout their lives, and learning the cause helped them.

The Case of the Little Girl Who Grew into a Boy

The patient—let's call her Elaine—sought treatment in 1972. Elaine was born with somewhat ambiguous external genitals, but she was raised by her parents as a girl without incident until the onset of puberty, when she suddenly began to develop male secondary sex characteristics. This was extremely distressing for her. Her treatment had two aspects: surgical and hormonal. Surgical treatment was used to increase the size of her vagina and decrease the size of her clitoris; hormonal treatment was used to suppress androgen release so that her own estrogen could feminize her body. Following treatment, Elaine developed into a young female—narrow hips and a husky voice being the only signs of her brush with masculinity (see Money & Ehrhardt, 1972).

Elaine suffered from adrenogenital syndrome, an atypical form of sexual development, affecting about 1 in 10,000. **Adrenogenital syndrome** is caused by **congenital adrenal hyperplasia**—a congenital deficiency in the release of the hormone *cortisol* from the adrenal cortex, which results in compensatory adrenal hyperactivity and the excessive release of adrenal androgens (see Bramble et al., 2017; Gondim, Teles, & Barroso, 2018). This produces an array of serious health problems (see Turcu & Auchus, 2015), but its most widely studied effect is on sexual development. There is little effect on the sexual development of males, other than accelerating the onset of puberty, but it has major effects on the development of genetic females. Females with adrenogenital syndrome are usually born with an enlarged clitoris and partially fused labia. Their gonads and internal ducts are usually typical because the adrenal androgens are released too late to stimulate the development of the Wolffian system.

Most female cases of adrenogenital syndrome are identified at birth. In such cases, the external genitals are altered to be more typically female, and cortisol is administered to reduce the levels of circulating adrenal androgens. Following early treatment, adrenogenital females develop as a typical female except that the onset of menstruation is likely to be later than usual. This makes them good participants for studies of the effects of fetal androgen exposure on psychosexual development.

Adrenogenital teenage girls who have received early treatment tend to display more tomboyishness, greater strength, and more aggression than most teenage girls. Moreover, they tend to prefer boys' clothes and toys, and play mainly with boys (e.g., Pasterski et al., 2011). However, it is important not to lose sight of the fact that many teenage girls display similar characteristics—and why not? Accordingly, the behavior of treated adrenogenital females, although tending toward what is stereotypically masculine, is usually within the typical range for female behavior.

The most interesting questions about the development of females with adrenogenital syndrome concern their romantic and sexual preferences as adults. They seem to lag behind typical females in dating and sex—perhaps because of the delayed onset of their menstrual cycle. Most are heterosexual, although a few studies have found an increased tendency for these females to express an interest in other females and a tendency to be less involved in heterosexual relationships (see Bramble et al., 2017; Piaggio, 2014). Complicating the situation further is the fact that these slight differences may not be direct consequences of early androgen exposure but may arise from the fact that some adrenogenital girls have ambiguous genitalia and certain male characteristics (e.g., body hair), which may result in different experiential influences (see Jordan-Young, 2012).

Prior to the development of cortisol therapy in 1950, genetic females with adrenogenital syndrome were left untreated. Some were raised as boys and some as girls, but the direction of their pubertal development was unpredictable. In some cases, adrenal androgens predominated and masculinized their bodies; in others, ovarian estrogens predominated and feminized their bodies. Thus, some who were raised as boys were transformed at puberty into females and some who were raised as girls were transformed into males—sometimes with devastating emotional consequences.

The Case of the Twin Who Lost His Penis

One of the most famous cases in the literature on sexual development is David Reimer, a monozygotic twin whose penis was accidentally destroyed during circumcision at the age of 7 months. Because there was no satisfactory way of surgically replacing the lost penis, an expert in such matters, John Money, recommended that he be castrated and raised as a girl, that an artificial vagina be created, and that estrogen be administered at puberty to feminize the body. After a great deal of anguish, the parents followed Money's advice.

Money's (1975) report of this case of **ablatio penis** was influential. It was seen by some as the ultimate test of the *nature–nurture controversy* (see Chapter 2) with respect to the development of gender identity and behavior. It seemed to pit the masculinizing effects of male genes and male prenatal hormones against the effects of being reared as a girl. And the availability of a genetically identical control, the twin brother, made the case all the more interesting. According to Money, the outcome strongly supported his *social-learning theory* of gender identity. Money reported in 1975, when the patient was 12, that "she" had developed as a typical female, thus confirming his prediction that being gonadectomized, having the genitals surgically altered, and being raised as a girl would override the masculinizing effects of male genes and early androgens.

A long-term follow-up to this study published by impartial experts tells an entirely different story (see Diamond & Sigmundson, 1997). Despite having female genitalia and being raised as a girl, he/she developed along male lines. Apparently, the organ that determines the course of psychosocial development is the brain, not the genitals (see Reiner, 1997). The following description gives you a glimpse of her/his life:

> From an early age, she tended to act in a masculine way, preferring boys' activities and games and displaying little interest in dolls, sewing, or other conventional female activities. At the age of four, she refused to put on mother's make-up, demanding instead to shave like dad. And by seven, she felt like a boy. Despite the absence of a penis, she often tried to urinate while standing and would sometimes go to the boys' lavatory.
>
> Her appearance was typical of a girl, but when she moved or talked her masculinity became apparent. She was teased by the other girls, and she often retaliated violently, which resulted in her expulsion from school. Put on an estrogen regimen at the age of 12, she rebelled. She did not want to be feminized; she hated her developing breasts and refused to wear a bra.
>
> At 14, she changed her name to David and decided to live as a man. At that time, David's father revealed David's entire early history to him. All of a sudden everything clicked; for the first time, he understood who and what he was.

Figure 13.10 David Reimer, the twin whose penis was accidentally destroyed.

STR OLD/Reuters

David requested androgen treatment, a *mastectomy* (surgical removal of breasts), and *phalloplasty* (surgical creation of a penis). He became a popular young man. He married at the age of 25 and adopted his wife's children. His ability to ejaculate and experience orgasm returned following his androgen treatments. However, his early castration permanently eliminated his reproductive capacity.

David remained bitter about his early treatment and his inability to produce offspring. To save others from his experience, he cooperated in writing his biography, *As Nature Made Him* (Colapinto, 2000). However, David never recovered from his emotional scars, and he committed suicide. His case suggests that the clinical practice of surgically modifying a person's sex at birth should be curtailed—irrevocable treatments should await early puberty and the emergence of the person's gender identity. At that stage, a compatible course of treatment can be selected.

Journal Prompt 13.2

What treatment(s) do you think should be given to infants born with ambiguous external genitals? Why?

DO THE EXCEPTIONAL CASES PROVE THE RULE?   Do current theories of hormones and sexual development pass the test of the three preceding cases of exceptional sexual development? In our view, the answer is "yes." Although current theories do not supply all of the answers, especially when it comes to brain and behavior, they have contributed greatly to the understanding of exceptional patterns of sexual differentiation of the body.

Notice one more thing about the three cases: Each of the three was male in some respects and female in others.

Accordingly, each case is a serious challenge to the men-are-men-and-women-are-women assumption: Male and female are not opposite, mutually exclusive categories (see Ainsworth, 2015; Carothers & Reis, 2013).

Effects of Gonadal Hormones on Adults

Once an individual reaches sexual maturity, gonadal hormones begin to play a role in activating reproductive behavior. These activational effects are the focus of the first two sections of this module. They deal with the role of hormones in activating the sexual behavior of males and females, respectively. The third section of this module deals with anabolic steroids.

Male Sexual Behavior and Gonadal Hormones

LO 13.13   Describe the role of gonadal hormones in male sexual behavior.

The important role played by gonadal hormones in the activation of male sexual behavior is clearly demonstrated by the asexualizing effects of orchidectomy. Bremer (1959) reviewed the cases of 215 orchidectomized Norwegians. About half had committed sex-related offenses and had agreed to castration to reduce the length of their prison terms.

Two important generalizations can be drawn from Bremer's study. The first is that orchidectomy leads to a reduction in sexual interest and behavior; the second is that the rate and the degree of the loss are variable. About half the males became completely asexual within a few weeks of the operation; others quickly lost their ability to achieve an erection but continued to experience some sexual interest and pleasure; and a few continued to copulate successfully, although somewhat less enthusiastically, for the duration of the study. There were also body changes: a reduction of hair on the torso, limbs, and face; an increase in fat on the hips and chest; a softening of the skin; and a reduction in muscle mass.

Of the 102 sex offenders in Bremer's study, only three were reconvicted of sex offenses. Accordingly, he recommended castration as an effective treatment of last resort for male sex offenders.

Why do some males remain sexually active for months after orchidectomy, despite the fact that testicular hormones are cleared from their bodies within days? It has been suggested that adrenal androgens may play some role in the maintenance of sexual activity in some castrated males, but there is no direct evidence for this hypothesis.

Orchidectomy removes, in one fell swoop—or, to put it more precisely, in two fell swoops—a pair of glands that release many hormones. Because testosterone is the major testicular hormone, the major symptoms of orchidectomy have been generally attributed to the loss of testosterone rather than to the loss of some other testicular hormone or to some nonhormonal consequence of the surgery. The therapeutic effects of **replacement injections** of testosterone have confirmed this assumption.

The Case of the Man Who Lost and Regained His Manhood

The very first case report of the effects of testosterone replacement therapy concerned an unfortunate 38-year-old World War I veteran, who was castrated in 1918 at the age of 19 by a shell fragment that removed his testes but left his penis undamaged (de Kruif, 1945).

His body was soft, as if he had little muscle, and his hips had grown wider and his shoulders narrower.

He got married—though the doctors had told him he would surely be **impotent** (unable to achieve an erection). Indeed, he made attempts at having sex with his partner, but as his doctors had warned, he was unable to achieve an erection.

He began to receive injections of testosterone into his muscles, and after the fifth injection, erections became rapid and prolonged. Moreover, after 12 weeks of treatment he had gained 18 pounds, mainly of muscle, and all his clothes had become too small. Testosterone had resurrected a broken man to the "manhood" that had been taken from him by the shell fragment. Since this first clinical trial, testosterone has breathed sexuality into the lives of many men. Testosterone does not, however, eliminate the *sterility* (inability to reproduce) of males who lack functional testes.

The fact that testosterone is necessary for male sexual behavior has led to two widespread assumptions: (1) that the level of a man's sexuality is a function of the amount of testosterone he has in his blood, and (2) that a man's sex drive can be increased by increasing his testosterone levels. Both assumptions are incorrect. Sex drive and testosterone levels are uncorrelated in healthy men, and testosterone injections do not increase their sex drive.

It seems that each healthy male has far more testosterone than required to activate the neural circuits that produce his sexual behavior, and having more than the minimum is of no advantage in this respect. A classic experiment by Grunt and Young (1952) clearly illustrates this point.

First, Grunt and Young rated the sexual behavior of each of the male guinea pigs in their experiment. Then, on the basis of the ratings, the researchers divided the male guinea pigs into three experimental groups: low, medium, and high sex drive. Following castration, the sexual behavior of all of

the guinea pigs fell to negligible levels within a few weeks (see Figure 13.11), but it recovered after the initiation of a series of testosterone replacement injections. The important point is that although each subject received the same large replacement injections of testosterone, the injections simply returned each to its previous level of copulatory activity. The conclusion is clear: With respect to the effects of testosterone on sexual behavior, more is not necessarily better.

Dihydrotestosterone, a nonaromatizable androgen, restores the copulatory behavior of castrated male primates; however, it fails to restore the copulatory behavior of castrated male rodents (see Hull & Dominguez, 2007). These findings indicate that the restoration of copulatory behavior by testosterone occurs by different mechanisms in rodents and primates: It appears to be a direct effect of testosterone in primates, but it appears to be produced by estradiol aromatized from testosterone in rodents.

Female Sexual Behavior and Gonadal Hormones

LO 13.14 Describe the role of gonadal hormones in female sexual behavior.

Sexually mature female rats and guinea pigs display 4-day cycles of gonadal hormone release. There is a gradual increase in the secretion of estrogens by the developing *follicle* (ovarian structure in which eggs mature) in the 2 days prior to ovulation, followed by a sudden surge in progesterone as the egg is released. These surges of estrogens and progesterone initiate **estrus**—a period of 12 to 18 hours during which the female is *fertile, receptive* (likely to assume the lordosis posture when mounted), *proceptive* (likely to engage in behaviors that serve to attract the male), and *sexually attractive* (smelling of chemicals that attract males).

The close relation between the cycle of hormone release and the **estrous cycle**—the cycle of sexual receptivity—in female rats and guinea pigs and in many other mammalian species suggests that female sexual behavior in these species is under hormonal control. The effects of ovariectomy confirm this conclusion: Ovariectomy of female rats and guinea pigs produces a rapid decline of both proceptive and receptive behaviors. Furthermore, estrus can be induced in ovariectomized rats and guinea pigs by an injection of estradiol followed a day and a half later by an injection of progesterone.

Female primates, including human females, are different from female rats, guinea pigs, and other mammals when it comes to the hormonal control of their sexual behavior: Female primates are the only female mammals motivated to copulate during periods of nonfertility (Ziegler, 2007). However, human females have been shown to display an even greater level of sexual desire, experience more sexual fantasies, and an even greater propensity to initiate sex during periods when they are fertile (see Shimoda, Campbell, & Barton, 2018).

Figure 13.11 The sexual behavior of male guinea pigs with low, medium, and high sex drive. Sexual behavior was disrupted by castration and returned to its original level by very large replacement injections of testosterone.

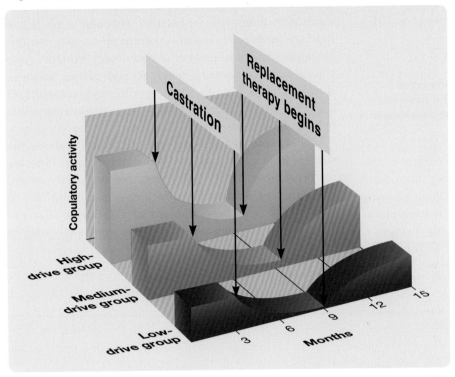

(Based on Grunt, J. A., & Young, W. C. (1952). Differential reactivity of individuals and the response of the male guinea pig to testosterone propionate. *Endocrinology*, *51*, 237–248.)

Conversely, ovariectomy has been shown to reduce sexual desire and frequency of sexual fantasies in human females (see Cappelletti & Wallen, 2016).

So which particular ovarian hormone is responsible for sexual desire in females?

- In ovariectomized human females, estradiol supplements reinstated their sexual desire and pleasure. Likewise, testosterone supplements also reinstated their sexual desire and pleasure (see Motta-Mena & Puts, 2017).

- Experiments on nonhuman ovariectomized and adrenalectomized rhesus monkeys have shown that replacement injections of testosterone, but not estradiol, increased their proceptivity (see Everitt & Herbert, 1972).

- Some studies of postmenopausal human females found that estrogen therapy renewed sexual interest, but other studies have not. Similarly, only some studies have found that testosterone renewed the sexual interest of postmenopausal females (see Cappelletti & Wallen, 2016; Motta-Mena & Puts, 2017).

- Still other studies have found that testosterone increased sexual desire in postmenopausal human females but only when administered at unnaturally high levels (see Cappalletti & Wallen, 2016; Davis & Braunstein, 2012).

In short, the evidence suggests that both estradiol and testosterone can influence female sexual behavior, at least in some cases (see Motta-Mena & Puts, 2017).

Anabolic Steroid Abuse

LO 13.15 Describe the dangers associated with anabolic steroid use.

Anabolic steroids are steroids, such as testosterone, that have *anabolic* (growth-promoting) effects. Testosterone itself is not very useful as an anabolic drug because it is broken down soon after injection and because it has undesirable side effects. Chemists have managed to synthesize a number of potent anabolic steroids that are long-acting, but they have not yet managed to synthesize one that does not have side effects.

We are currently in the midst of an epidemic of anabolic steroid abuse. Many competitive athletes and bodybuilders are self-administering appallingly large doses, and many others use them for cosmetic purposes. Because steroids are illegal in many countries, estimates of the numbers who use them are likely underestimates. Still, the results of numerous surveys have been disturbing. Elite athletes began to use anabolic steroids in the 1950s to improve their athletic performance, but by the 1980s they were in widespread use by the general population, often by adolescents (e.g., Calatayud et al., 2019) and often for cosmetic reasons (see Figueiredo & Silva, 2014). Particularly troubling is the scale of anabolic steroid use: Global estimates of lifetime prevalence rates for anabolic steroids are 6.4 percent for males and 1.6 percent for females (see Sagoe et al., 2014).

Although it is about 40 years since anabolic steroids began to be used by the general public, we still do not fully understand all the associated risks. This is because anabolic steroid use is illegal and thus concealed, because there are more than 100 different testosterone derivatives, and because users vary greatly in their patterns and doses of administration. Hazardous effects, although well documented, typically occur in only a small proportion of users (see van Amsterdam, Opperhuizen, & Hartgens, 2010).

Anabolic steroids have been shown to have a variety of cardiovascular effects that have been linked to premature death (see Angell et al., 2012; Kanayama et al., 2010). Also, oral anabolic steroids have been shown to have adverse effects on the liver, including the growth of liver tumors. Less dangerous, but still disturbing, are the muscle spasms, muscle pains, blood in the urine, acne, general swelling from the retention of water, bleeding of the tongue, nausea, vomiting, and fits of depression and anger (see Kraus et al., 2012).

Journal Prompt 13.3

What should be done about the current epidemic of anabolic steroid abuse? Would you make the same recommendation if a safe anabolic steroid were developed?

The main question of relevance in the context of this chapter is the following: Do the outrageously large doses of anabolic steroids routinely administered by many users enhance their sexual function? The answer is an emphatic "no." In fact, the effects appear to be uniformly disruptive.

In males, the negative feedback from high levels of anabolic steroids reduces gonadotropin release; this leads to a reduction in testicular activity, which can result in *testicular atrophy* (wasting away of the testes) and sterility. *Gynecomastia* (breast growth in males) can also occur, presumably as the result of the aromatization of anabolic steroids to estrogens. In females, anabolic steroids can produce *amenorrhea* (cessation of menstruation), sterility, *hirsutism* (excessive growth of body hair), growth of the clitoris, development of a masculine body shape, baldness, shrinking of the breasts, and deepening and coarsening of the voice. Unfortunately, some of the masculinizing effects of anabolic steroids on females appear to be irreversible.

There are two major points of concern about the adverse health consequences of anabolic steroids. First, the use of anabolic steroids in puberty, before developmental programs of sexual differentiation are complete, is particularly risky. Second, many of the adverse effects of anabolic steroids may take years to be manifested—steroid users who experience few immediate adverse effects may pay the price later.

Scan Your Brain

You encountered many forms of atypical sexual development and also many clinical problems in the preceding two modules of the chapter. Do you remember them? Write the name of the appropriate condition or syndrome in each blank based on the clues provided. The answers appear at the end of the exercise. Before proceeding, review material related to your errors and omissions.

Name of condition	Clues or syndrome
1. _____	Genetic male, sparse pubic hair, short vagina
2. _____	Congenital adrenal hyperplasia, elevated androgen levels
3. _____	David Reimer, destruction of penis

4. _____	Castrated males, gonadectomized males
5. _____	Castrated females, gonadectomized females
6. _____	Unable to achieve erection
7. _____	Enlargement of a male's breasts
8. _____	Cessation of menstruation
9. _____	Excessive body hair in females

Scan Your Brain answers: (1) androgen insensitivity syndrome, (2) adrenogenital syndrome, (3) ablatio penis, (4) orchidectomized (5) ovariectomized, (6) impotent, (7) gynecomastia, (8) amenorrhea, (9) hirsutism

Brain Mechanisms of Sexual Behavior

Human sexual behavior is complex and varied. Sexual practices vary from culture to culture, and from person to person within each culture. Furthermore, behavioral preferences of individuals are often changed by experience (see Hoffman, Peterson, & Garner, 2012). However, there are four brain structures whose role in sexual behavior has been well established: cortex, hypothalamus, amygdala, and ventral striatum. This module focuses on these four structures.

Four Brain Structures Associated with Sexual Activity

LO 13.16 Describe the roles of the cortex, hypothalamus, amygdala, and ventral striatum in sexual activity.

Our current knowledge of the brain mechanisms of sexual behavior is based on the study of both human volunteers and nonhuman subjects (see Georgiadis, Kringelbach, & Pfaus, 2012). Because functional brain imaging is the main technology used to study the relation between brain activity and sexual behavior in human volunteers, there are some major limitations. Specifically, it is virtually impossible using current imaging technology to investigate the neural activity associated with copulation in humans given cultural constraints, the requirement to remain motionless during brain scans, and the difficulty of squeezing two active adults into a conventional brain scanner. Consequently, research on humans has focused on sexual arousal, occasionally to orgasm, triggered by sexually provocative visual images or masturbation.

The study of the brain mechanisms of sexual behavior in laboratory animals—most commonly rats—circumvents the problems associated with studies of human volunteers. First, it is possible to study brain mechanisms in more detail by using invasive techniques. Second, it is possible to study natural patterns of copulatory activity. Conversely, several important aspects of sexual activity are difficult or impossible to study in laboratory animals: for example, sexual imagery, delayed orgasm, female orgasm, and feelings of sexual attraction. The important point here is that both humans and nonhumans have major weaknesses as subjects in the investigation of the brain mechanisms of sexual behavior, but the weaknesses tend to be complementary. Thus, knowledge in this area often depends on an amalgamation of both kinds of research (see Georgiadis et al., 2012).

CORTEX AND SEXUAL ACTIVITY. Because of its fundamental role in reproduction, and thus the very survival of our species, sexual behavior was once assumed to be regulated by archaic circuits in the brain stem of early evolutionary origin. This assumption is no longer tenable. Widespread cortical activation has been routinely recorded during functional brain imaging studies of volunteers exposed to sexually arousing stimuli (see Georgiadis, 2012; Stoléru et al., 2012). In both males and females, the following areas are often activated: occipitotemporal, inferotemporal, parietal, orbitofrontal, medial prefrontal, insular, cingulate, and premotor cortices. Interestingly, the activity in secondary visual cortex (occipitotemporal and inferotemporal cortices) occurs during sexual arousal even when eyes are closed (see Georgiadis, 2012), and the activity in prefrontal cortex is suppressed during orgasm (see Stoléru et al., 2012).

Presumably cortical activation mediates the most complex aspects of sexual experience. These may include feelings of release and loss of control, changes in self-awareness, disturbances of awareness of space and time, and feelings of love.

HYPOTHALAMUS AND SEXUAL ACTIVITY. Interest in the role of the hypothalamus in sexual behavior was driven by the discovery of a specific structural difference in the male and female hypothalamus. In 1978, Gorski and his colleagues discovered a nucleus in the **medial preoptic area** of the rat hypothalamus that was several times larger in males. They called this nucleus the **sexually dimorphic nucleus**.

At birth, the sexually dimorphic nuclei of male and female rats are the same size. In the first few days after birth, the male sexually dimorphic nuclei grow at a high rate and the female sexually dimorphic nuclei do not. The growth of the male sexually dimorphic nuclei is normally triggered by estradiol that has been aromatized from testosterone (Gorski, 1980)—see Figure 13.12. Since the discovery of the sexually dimorphic nuclei in rats, many other sex differences in hypothalamic anatomy have been identified in rats and in other species. These sex differences in the hypothalamus include differences in volume of various nuclei, cell number, connectivity, cell morphology, neural activity, and neurotransmitter type—all of which are influenced by perinatal exposure to estradiol (see Lenz & McCarthy, 2010).

The medial preoptic area (which includes the sexually dimorphic nucleus) is one area of the hypothalamus that plays a key role in male sexual behavior (see Will, Hull, & Dominguez, 2014). Destruction of the entire area abolishes sexual behavior in the males of all mammalian species that

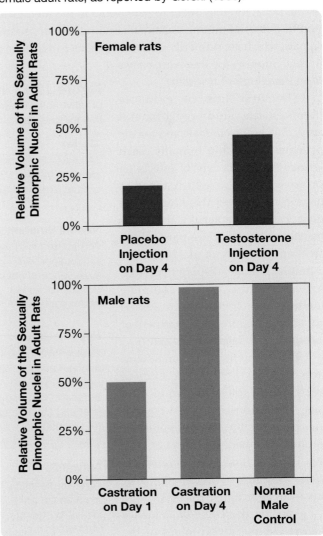

Figure 13.12 Effects of neonatal testosterone exposure on the size of the sexually dimorphic nuclei in male and female adult rats, as reported by Gorski (1980).

have been studied (see Calabrò et al., 2019). In contrast, medial preoptic area lesions do not eliminate the sexual behaviors of females, but they do eliminate the male sexual behaviors (e.g., mounting) that are sometimes observed in females. Conversely, electrical stimulation of the medial preoptic area elicits copulatory behavior in male rats (see Guadarrama-Bazante & Rodríguez-Manzo, 2019), and copulatory behavior can be reinstated in castrated male rats by medial preoptic implants of testosterone.

In female rats, the **ventromedial nuclei (VMN)** of the hypothalamus contain circuits that appear to be critical for sexual behavior. Female rats with bilateral lesions of the VMN do not display lordosis. This effect of VMN lesions seems to be mediated by estradiol receptors in the VMN—female rats that have had their VMN estrogen receptors selectively inactivated also do not display lordosis (see Ventura-Aquino & Paredes, 2020).

AMYGDALA AND SEXUAL ACTIVITY. The amygdalas, located in the left and right medial temporal lobes, play a general role in the experience of emotions and social cognition (see Olson et al., 2013; Rolls, 2015). With respect to sexual behavior, they seem to play a role in the identification of potential mating partners based on sensory social signals, which are primarily visual in humans and olfactory in rats. Support for this view comes from three lines of research.

The first line of evidence involves the study of bilateral amygdala lesions in male and female primates, including humans. Such lesions have a variety of effects on primate behavior, which together have been termed the *Kluver-Bucy syndrome* (see Chapter 17). The effects on sexual behavior are particularly striking (see Clay et al., 2019). For example, humans display flat affect, hypersexuality, and a complete inability to focus their sexual advances to appropriate partners or locations (see Calabró et al., 2019).

The second line of research stems from Everitt's (1990) classic study of bilateral amygdalar lesions in male rats. The lesions disrupted copulatory behavior, but this did not occur because they had difficulty copulating. The deficit occurred because the males were incapable of limiting their advances to receptive females.

And the third line of evidence suggesting the involvement of the amygdalas in sexual behavior comes from studies of the reactions of human males and females to erotic images. Males have been found to be more likely than females to be sexually aroused by erotic images (see Stark et al., 2019), and this difference is reflected in differences in their amygdalar responses to such images. In several studies, erotic images presented to human males and females undergoing functional brain scans produced greater amygdalar activation in men (see Stoléru, 2012; Calabró et al., 2019).

VENTRAL STRIATUM AND SEXUAL ACTIVITY. Sexual activity is clearly among the most pleasurable human activities. Accordingly, it is not surprising that sexual activity can serve as a powerful reinforcer. Indeed, Pfaus and colleagues (2012) have shown that both male and female rats learn to prefer partners, locations, and odors associated with copulation.

Because orgasm is associated with pleasure, it comes as no surprise that the ventral striatum is activated in human volunteers by sexually provocative visual images (see Stoléru et al., 2012). Research on rats has shown that the activity in this area is associated with the anticipation and experience of sex and other forms of pleasure (see Matsumoto et al., 2012).

Figure 13.13 summarizes this module. It illustrates the location of the four brain structures that were discussed and summarizes their putative roles in sexual activity.

Figure 13.13 The cortex, hypothalamus, amygdala, and ventral striatum: their putative roles in sexual activity. The amygdala and ventral striatum are not visible in this midline view.

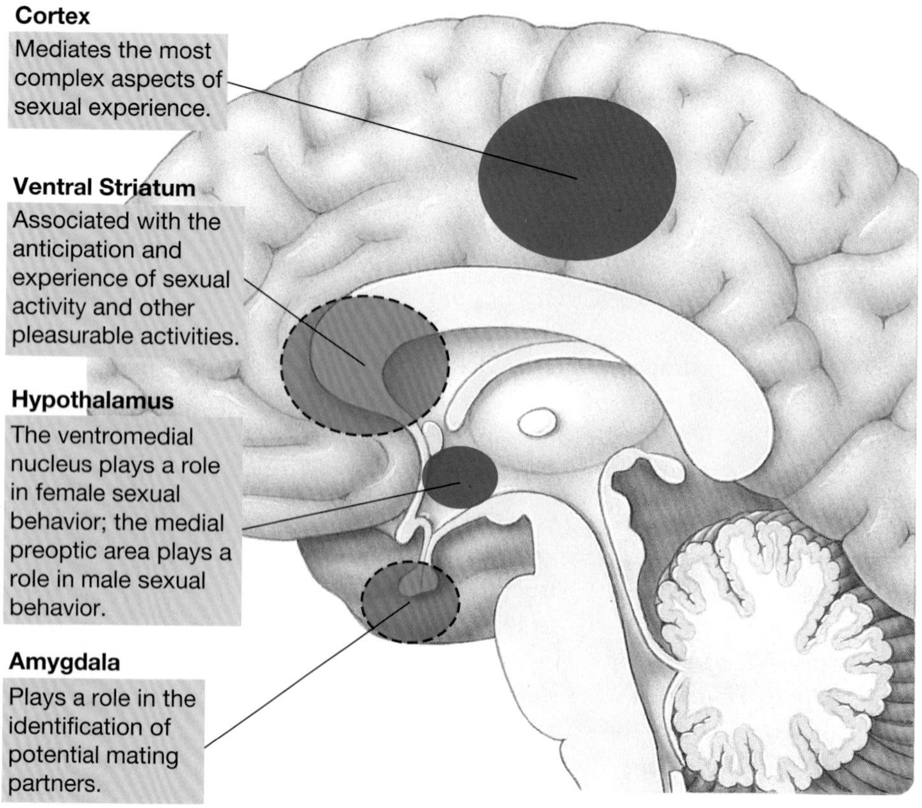

Cortex
Mediates the most complex aspects of sexual experience.

Ventral Striatum
Associated with the anticipation and experience of sexual activity and other pleasurable activities.

Hypothalamus
The ventromedial nucleus plays a role in female sexual behavior; the medial preoptic area plays a role in male sexual behavior.

Amygdala
Plays a role in the identification of potential mating partners.

Sexual Orientation and Gender Identity

So far, this chapter has not systematically addressed the topic of sexual orientation. As you know, people differ in their sexual orientation: Some people are **heterosexual** (women or men who are attracted to men or women, respectively), some are **gay** or **lesbian** (men who are attracted to men and women who are attracted to women, respectively), some are **bisexual** (sexually attracted to both men and women), and some are **asexual** (not sexually attracted to others). Other sorts of sexual orientations exist of course, such as a preference for two or more simultaneous sexual partners versus just one (see Fausto-Sterling, 2019).

Nor has this chapter yet addressed the topic of gender identity. People vary in their **gender identity**. Gender identity is the gender that a person most identifies with, including woman, man, **transgender** (an individual who identifies as a man, a woman, or some intersection thereof), or another gender category. A discussion of sexual orientation and gender identity is a fitting conclusion to this chapter because it further underscores its anti-mamawawa theme.

Sexual Orientation

LO 13.17 Describe the results of the two studies on the genetics of sexual orientation by Bailey and Pillard (1991, 1993) and describe the fraternal birth order effect and why it is thought to occur.

For a variety of historical reasons, heterosexual, gay, and lesbian individuals have dominated the study of sexual orientation (see Bogaert & Skorska, 2020). Accordingly, this section focuses exclusively on research done with heterosexual, gay, and lesbian persons.

SEXUAL ORIENTATION AND GENES. Research has shown that differences in sexual orientation are influenced by genes. For example, Bailey and Pillard (1991) studied a group of gay males who had twin brothers, and they found that 52 percent of the monozygotic twin brothers and 22 percent of the dizygotic twin brothers were gay. In a comparable study of female twins by the same research group (Bailey et al., 1993), it was found that 48 percent of the monozygotic twin sisters and 16 percent of the dizygotic twin sisters were lesbians.

SEXUAL ORIENTATION AND EARLY HORMONES. Many people mistakenly assume that gay males and lesbians have lower levels of sex hormones. They don't. For example, heterosexuals and gay males do not differ in their levels of circulating hormones. Moreover, orchidectomy reduces the sexual behavior of both heterosexual and gay males, but it

does not redirect it; and replacement injections simply reactivate the preferences that existed prior to surgery.

Some people also assume that sexual orientation is a matter of choice. It isn't. People discover their sexual orientation; they don't choose it. Sexual orientation seems to develop very early, and a child's first indication of their sexual orientation usually does not change as they mature. Could perinatal hormone exposure be the early event that shapes sexual orientation?

Because experiments involving levels of perinatal hormone exposure are not feasible with humans, efforts to determine whether perinatal hormone levels influence the development of sexual orientation have focused on nonhuman species. A consistent pattern of findings has emerged. In those species whose exposure to early sex hormones has been modified (e.g., rats, hamsters, ferrets, pigs, zebra finches, and dogs), it has not been uncommon to see males engaging in female-typical sexual behavior such as being mounted by other males. Nor has it been uncommon to see females engaging in male-typical sexual behavior such as mounting other females. However, because the defining feature of sexual orientation is sexual preference, the key studies have examined the effect of early hormone exposure on the sex of preferred sexual partners. In general, the perinatal castration of males has increased their preference as adults for sex with males; similarly, perinatal testosterone exposure in females has increased their preference as adults for sex with females (see Henley, Nunez, & Clemens, 2011).

On the one hand, we need to exercise prudence in drawing conclusions about the development of sexual orientation in humans based on the results of experiments on laboratory species; it would be a mistake to ignore the profound cognitive and emotional components of human sexuality, which have no counterpart in laboratory animals. On the other hand, it would also be a mistake to think that a pattern of results that runs so consistently through so many species has no relevance to humans.

There are indications that perinatal hormones do influence sexual orientation in humans, although the evidence is controversial (see Fausto-Sterling, 2019; Bogaert & Skorska, 2020). One example is the well-known quasiexperimental study of Ehrhardt and her colleagues (1985). They interviewed adult females whose mothers had been exposed to *diethylstilbestrol* (a synthetic estrogen) during pregnancy. The females' responses indicated that they were significantly more sexually attracted to females than was a group of matched controls. Ehrhardt and her colleagues concluded that perinatal estrogen exposure does encourage sexual attraction to other females but noted that its effect is relatively weak: The sexual behavior of all but 1 of the 30 participants in this study was primarily heterosexual. Even though this effect could not be replicated (see Puts & Motta-Mena, 2018), it is still widely cited as evidence of the influence of perinatal hormones on sexual orientation.

A more promising line of research on sexual orientation focuses on the **fraternal birth order effect**, the reliable finding that the probability of a male being gay increases as a function of the number of older brothers he has (see Alexander et al., 2011; Blanchard, 2018; Bogaert & Skorska, 2020). A recent study of blended families (families in which biologically related siblings were raised with adopted siblings or step-siblings) found that the effect is related to the number of males previously born to the mother, not the number of older males one is reared with (see Bogaert & Skorska, 2020). The effect is relatively large: The probability of a male being gay increases by 33.3 percent for every older brother he has (see Balthazart, 2018).

The **maternal immune hypothesis** has been proposed to explain the fraternal birth order effect. Currently, the evidence favors the notion that mothers develop an immune response to masculinizing hormones from male fetuses (see Balthazart, 2018), such that a mother's immune system will increasingly suppress the effects of masculinizing hormones in her younger sons (see Bogaert & Skorska, 2020). Indeed, one recent study found that the mothers of gay men with older brothers had a more pronounced immune response to one particular protein that is important during brain development (see Balthazart, 2018; Bogaert et al., 2018).

What Triggers the Development of Sexual Attraction?

LO 13.18 Describe the hypothesized role of adrenal cortex steroids in the emergence of sexual attraction.

The evidence indicates that most females and males living in Western countries experience their first intense feelings of sexual attraction at about 10 years of age, whether they are heterosexual, bisexual, gay, or lesbian (see Fausto-Sterling, 2019). This finding is at odds with the usual assumption that sexual interest is triggered by puberty, which currently tends to occur at 10.5 years of age in females and at 11.5 years in males.

McClintock and Herdt (1996) have suggested that the emergence of sexual attraction may be stimulated by adrenal cortex steroids. Unlike gonadal maturation, adrenal maturation occurs at about the age of 10.

What Differences in the Brain Can Account for Differences in Sexual Attraction?

LO 13.19 Describe the famous study of LeVay (1991) and two major problems with its finding.

The brains of gay, lesbian, and bisexual persons and heterosexuals must differ in some way, but how? Many studies have attempted to identify neuroanatomical, neuropsychological, neurophysiological, and hormonal response differences between gay men and heterosexual men.

Unfortunately, there has been little research on other sexual orientations (e.g., lesbian, bisexual), mostly due to political and cultural factors (see Byne, 2017). This lack of research on lesbian, gay, and bisexual individuals has major implications for the provision of appropriate health care to them (e.g., see Boehmer & Elk, 2016; O'Hanlan, Gordon, & Sullivan, 2018).

In a highly publicized study, LeVay (1991) found that the structure of one hypothalamic nucleus in gay males was intermediate in size to that of female heterosexuals and male heterosexuals. However, this influential study has not been consistently replicated, and interpretations of its results have been controversial (see Byne et al., 2001). For example, a common misinterpretation of the study has been that the brains of gay men lie near the middle of a continuum of "female-like" and "male-like" brains (see Fausto-Sterling, 2019; Roselli, 2018). This sort of thinking is a prime example of the mamawawa.

Gender Identity

LO 13.20 Define gender identity, and explain what is meant by "gender dysphoria."

Gender identity is the gender that a person most identifies with, including woman, man, transgender (an individual who identifies as a man, a woman, or some intersection thereof), or another gender category. Gender identity often coincides with a person's anatomical sex, but not always.

A transgender person has a gender identity that is inconsistent with their anatomical sex (see Kreukels & Cohen-Kettenis, 2012). Many transgender persons face a strong conflict known as **gender dysphoria**. For example: "I am a woman trapped in a male body. Help!" It is important to appreciate the desperation of these individuals; they do not merely think that life might be better if their gender were different. Their desperation is indicated by how some of them had dealt with their problem before surgical sexual reassignment was an option: Some biological males with female gender identities attempted self-castration, and others consumed copious quantities of estrogen-containing face creams in an attempt to feminize their bodies.

Independence of Sexual Orientation and Gender Identity

LO 13.21 Explain how sexual attraction, gender identity, and body type are independent.

To complete this chapter, we would like to remind you of two of its main themes and show you how useful they are in thinking about one of the puzzles of human sexuality. One of the two themes is that the exception proves the rule: that a powerful test of any theory is its ability to explain exceptional cases. The second is that the mamawawa is seriously flawed: We have seen that males and females are similar in some ways and different in others, but they are certainly not opposites, and their programs of development are neither parallel nor opposite. Moreover, some people do not fall unambiguously into one or either of these two categories.

Here, we want to focus on the puzzling fact that sexual orientation, gender identity, and body type are sometimes unrelated (see Fausto-Sterling, 2019). For example, consider transgender persons: They, by definition, have a body type that is inconsistent with their gender identity, but their sexual orientation is an independent matter. Some transgender persons are sexually attracted to females, others are sexually attracted to males, and others are sexually attracted to neither—and this is not changed by sexual reassignment.

Obviously, the mere existence of individuals who have non-heterosexual sexual orientations and/or a gender that is neither man or woman is a challenge to the mamawawa, the assumption that males and females belong to distinct and opposite categories. Many people tend to think of "femaleness" and "maleness" as being at opposite ends of a continuum, with a few atypical cases somewhere between the two. Perhaps this is what you think. However, the fact that body type, sexual orientation, and gender identity are often independent constitutes a serious attack on any assumption that femaleness and maleness lie at opposite ends of a single continuum. Clearly, femaleness or maleness is a combination of many different attributes (e.g., body type, sexual orientation, and gender identity), each of which can develop quite independently. This is a real puzzle for many people, including scientists, but what you have already learned in this chapter suggests a solution.

Think back to the section on brain differentiation. Until recently, it was assumed that the differentiation of the human brain into its typical female and male forms occurred through a single testosterone-based mechanism. However, a different notion has developed from recent evidence. Now, it is clear that male and female brains differ in many ways and that the differences develop at different times and by different mechanisms. If you keep this developmental principle in mind, you will have no difficulty understanding how it is possible for some individuals to be female-like in some ways, male-like in others, and neither male- nor female-like in others.

This analysis exemplifies a point we make many times in this text. The study of biopsychology often has important personal and social implications: The search for the neural basis of a behavior frequently provides us with a greater understanding of that behavior. We hope that you now have a greater understanding of, and acceptance of, differences in human sexuality.

Journal Prompt 13.4

Somebody tells you that all transgender persons with a female gender identity prefer males as their sexual partners. What would your response be?

Themes Revisited

Three of the book's four major themes were repeatedly emphasized in this chapter: the evolutionary perspective, clinical implications, and thinking creatively themes.

The evolutionary perspective theme was pervasive. It received frequent attention because most experimental studies of hormones and sex have been conducted in nonhuman species. The other major source of information about hormones and sex has been the study of human clinical cases, which is why the clinical implications theme was prominent in the cases of the woman who wasn't, the little girl who grew into a boy, the twin who lost his penis, and the man who lost and regained his manhood.

The thinking creatively theme was emphasized throughout the chapter because conventional ways of thinking about hormones and sex have often been at odds with the results of biopsychological research. If you are now better able to resist the seductive appeal of the men-are-men-and-women-are-women assumption, you are a more broadminded and understanding person than when you began this chapter. We hope you have gained an abiding appreciation of the fact that maleness and femaleness are multidimensional and often ambiguous variations of each other.

The fourth major theme of the book, neuroplasticity, arose during the discussions of the effects of hormones on the development of sex differences in the brain.

Of the two emerging themes in the book, only the epigenetics theme appeared. Epigenetic mechanisms are one of the factors that contribute to the emergence of brain sex differences.

Key Terms

Neuroendocrine System

Exocrine glands, p. 346
Endocrine glands, p. 346
Hormones, p. 346
Gonads, p. 347
Testes, p. 347
Ovaries, p. 347
Copulation, p. 347
Zygote, p. 347
Sex chromosomes, p. 347
Amino acid derivative
  hormones, p. 347
Peptide hormones, p. 347
Protein hormones, p. 347
Steroid hormones, p. 347
Androgens, p. 347
Estrogens, p. 347
Testosterone, p. 347
Estradiol, p. 347
Progestins, p. 347
Progesterone, p. 347
Adrenal cortex, p. 347
Gonadotropin, p. 348
Posterior pituitary, p. 348
Pituitary stalk, p. 348
Anterior pituitary, p. 348
Menstrual cycle, p. 348
Vasopressin, p. 349
Oxytocin, p. 349
Paraventricular nuclei, p. 349
Supraoptic nuclei, p. 349
Hypothalamopituitary portal
  system, p. 349
Releasing hormone, p. 349
Release-inhibiting hormone, p. 349
Thyrotropin-releasing hormone, p. 349
Thyrotropin, p. 349
Gonadotropin-releasing
  hormone, p. 350

Follicle-stimulating hormone
  (FSH), p. 350
Luteinizing hormone (LH), p. 350
Pulsatile hormone release, p. 351

**Hormones and Sexual Development
of the Body**

Sry gene, p. 352
Sry protein, p. 352
Intersexed person, p. 352
Wolffian system, p. 352
Müllerian system, p. 352
Müllerian-inhibiting
  substance, p. 352
Scrotum, p. 353
Ovariectomy, p. 353
Orchidectomy, p. 353
Gonadectomy, p. 353
Genitals, p. 353
Secondary sex characteristics, p. 354
Growth hormone, p. 354
Adrenocorticotropic hormone, p. 354
Androstenedione, p. 354

**Sexual Development of Brain
and Behavior**

Aromatase, p. 355
Aromatization, p. 355
Aromatization hypothesis, p. 355
Alpha fetoprotein, p. 355
Masculinizes, p. 357
Defeminizes, p. 357
Lordosis, p. 357
Feminizes, p. 357
Demasculinizes, p. 357
Intromission, p. 357
Ejaculation, p. 357
Proceptive behaviors, p. 357
Sexual dimorphisms, p. 357

**Three Cases of Exceptional Human
Sexual Development**

Complete androgen insensitivity
  syndrome, p. 359
Adrenogenital syndrome, p. 359
Congenital adrenal hyperplasia, p. 359
Ablatio penis, p. 360

**Effects of Gonadal Hormones on
Adults**

Replacement injections, p. 362
Impotent, p. 362
Estrus, p. 362
Estrous cycle, p. 362
Anabolic steroids, p. 363

**Brain Mechanisms of Sexual
Behavior**

Medial preoptic area, p. 365
Sexually dimorphic nucleus, p. 365
Ventromedial nucleus (VMN), p. 366

**Sexual Orientation and Gender
Identity**

Heterosexual, p. 367
Gay, p. 367
Lesbian, p. 367
Bisexual, p. 367
Asexual, p. 367
Gender identity, p. 367
Transgender, p. 367
Fraternal birth order effect, p. 368
Maternal immune hypothesis, p. 368
Gender dysphoria, p. 368

Chapter 14
Sleep, Dreaming, and Circadian Rhythms

How Much Do You Need to Sleep?

fizkes/Shutterstock

 Chapter Overview and Learning Objectives

Stages of Sleep

LO 14.1 Describe the three standard physiological measures of sleep.

LO 14.2 Describe the three stages of the sleep EEG and explain the difference between REM and non-REM sleep.

Dreaming

LO 14.3 Describe the discovery of the relationship between REM sleep and dreaming.

LO 14.4 Describe five common beliefs about dreaming and assess their validity.

LO 14.5 Understand the relationship between REM sleep, NREM sleep, and dreaming.

LO 14.6 Define lucid dreaming and discuss the research that supports its existence.

LO 14.7 Discuss the issue associated with studying dream content and describe findings related to what can influence dream content.

LO 14.8 Compare and contrast three different theories of why we dream.

LO 14.9 Identify three brain areas that have been implicated in dreaming.

Why Do We Sleep, and Why Do We Sleep When We Do?

LO 14.10 Describe the two kinds of theories of sleep.

LO 14.11 Explain four conclusions that have resulted from the comparative analysis of sleep.

Effects of Sleep Deprivation

LO 14.12 Explain how stress can often be a confounding variable when considering the effects of sleep deprivation.

LO 14.13 List the three predictions that recuperation theories make about the effects of sleep deprivation.

LO 14.14 Describe two classic sleep-deprivation case studies.

LO 14.15 Describe the major effects of sleep deprivation in humans.

LO 14.16 Describe the key studies of sleep deprivation in laboratory animals. Provide a critique of the carousel apparatus as a method of sleep deprivation.

LO 14.17 Describe the effects of REM-sleep deprivation.

LO 14.18 Describe six pieces of evidence that indicate that less sleep is associated with more efficient sleep.

Circadian Sleep Cycles

LO 14.19 Describe the circadian sleep–wake cycle and the role of zeitgebers in maintaining circadian rhythms.

LO 14.20 Describe free-running rhythms and internal desynchronization and explain why they are incompatible with recuperation theories of sleep.

LO 14.21 Describe the disruptive effects of jet lag and shift work on circadian rhythmicity and how one can minimize such effects.

LO 14.22 Describe the research that led to the discovery of a circadian clock in the suprachiasmatic nucleus (SCN) of the hypothalamus.

LO 14.23 Explain the mechanism by which SCN neurons are entrained by the 24-hour light–dark cycles.

LO 14.24 Understand the genetics of circadian rhythms and the important discoveries that have resulted from the discovery of circadian genes.

Four Areas of the Brain Involved in Sleep

LO 14.25 Describe the research that led to the identification of the anterior and posterior hypothalamus as brain regions involved in the regulation of sleep and wakefulness.

LO 14.26 Describe the research that led to the identification of the reticular formation as a brain region involved in the regulation of sleep and wakefulness.

LO 14.27 Discuss how REM sleep is controlled by the reticular formation and what implications this has for understanding the neural mechanisms of behavior.

Drugs That Affect Sleep

LO 14.28 Describe three classes of hypnotic drugs. Compare and contrast them in terms of their efficacy and side effects.

LO 14.29 Describe three classes of antihypnotic drugs.

LO 14.30 Understand the relationship between the pineal gland and melatonin, and how melatonin affects sleep.

Sleep Disorders

LO 14.31 Describe four causes of insomnia.

LO 14.32 Describe the symptoms of narcolepsy and the role of orexin (hypocretin) in this disorder.

LO 14.33 Describe one REM-sleep-related disorder and its presumed neural mechanisms.

Effects of Long-Term Sleep Reduction

LO 14.34 List the main differences between short and long sleepers.

LO 14.35 Describe the results of studies of long-term reduction of nightly sleep.

LO 14.36 Describe the results of studies of long-term reduction of sleep by napping.

LO 14.37 Recognize how shorter sleep times relate to longevity.

Most of us have a fondness for eating and sex—the two highly esteemed motivated behaviors discussed in Chapters 12 and 13. But the amount of time devoted to these behaviors by even the most amorous gourmands pales in comparison to the amount of time spent sleeping: Most of us will sleep for well over 175,000 hours in our lifetimes. This extraordinary commitment of time implies that sleep fulfills a critical biological function. But what is it? And what about dreaming: Why do we spend so much time dreaming? And why do we tend to get sleepy at about the same time every day? Answers to these questions await you in this chapter.

Almost every time we lecture about sleep, somebody asks, "How much sleep do we need?" Each time, we provide the same unsatisfying answer: We explain that there are two fundamentally different answers to this question, but neither has emerged as a clear winner. One answer stresses the presumed health-promoting and recuperative powers of sleep and suggests that people need as much sleep as they can comfortably get—the usual prescription being at least 8 hours per night. The other answer is that many of us sleep more than we need to and are consequently sleeping part of our life away. Just think how your life could change if you slept 5 hours per night instead of 8. You would have an extra 21 waking hours each week, a mind-boggling 10,952 hours each decade.

As we prepared to write this chapter, I (JP) began to think of the personal implications of the idea that we get more sleep than we need. That is when I decided to do something a bit unconventional. I am going to try to get no more than 5 hours of sleep per night—11:00 p.m. to 4:00 a.m.—until this chapter is written. As I begin, I am excited by the prospect of having more time to write, but a little worried that this extra time might cost me too dearly.

It is now the next day—4:50 Saturday morning, to be exact—and I am just sitting down to write. There was a party last night, and I didn't make it to bed by 11:00; but considering that I slept for only 2 hours and 15 minutes, I feel quite good. I wonder what I will feel like later in the day. In any case, I will report my experiences to you at the end of the chapter.

The following case study challenges several common beliefs about sleep (Meddis, 1977). Ponder its implications before proceeding to the body of the chapter.

The Case of the Woman Who Wouldn't Sleep

Miss M. usually sleeps only 1 hour per night. Although she is retired, she keeps busy painting, writing, and volunteering in the community. Although she becomes tired physically, she never reports feeling sleepy. During the night she sits on her bed reading, writing, crocheting, or painting. At about 2:00 a.m., she often falls asleep without any preceding drowsiness, and when she wakes about 1 hour later, she feels wide awake.

She came to the laboratory for some sleep tests, but on the first night we ran into a snag. She told us she did not sleep at all if she had interesting things to do, and she found her visit to a sleep laboratory very interesting. She had someone to talk to for the whole night.

In the morning, we broke into shifts so that some could sleep while at least one person stayed with her. The second night was the same as the first.

Finally, on the third night, she promised to try to sleep, and she did. The only abnormal thing about her sleep was that it was brief. After 99 minutes, she could sleep no more.

Stages of Sleep

Many changes occur in your body during sleep. This module introduces you to three standard measures of those physiological changes, and how they change over the course of a night's sleep.

Three Standard Psychophysiological Measures of Sleep

LO 14.1 Describe the three standard physiological measures of sleep.

There are major changes in the human EEG during the course of a night's sleep. Although the EEG waves that accompany sleep are generally high-voltage and slow, there are periods throughout the night that are dominated by low-voltage, fast waves similar to those in nonsleeping individuals. In the 1950s, it was discovered that *rapid eye movements* (*REMs*) occur under the closed eyelids of sleepers during these periods of low-voltage, fast EEG activity. And in 1962, Berger and Oswald discovered that there is also a loss of electromyographic activity in the neck

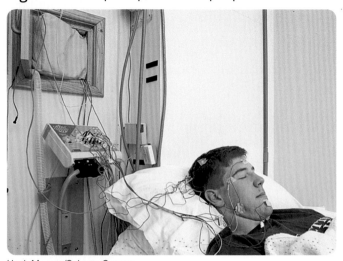

Figure 14.1 A participant in a sleep experiment.

Hank Morgan/Science Source

muscles during these same sleep periods. Subsequently, the **electroencephalogram (EEG)**, the **electrooculogram (EOG)**, and the neck **electromyogram (EMG)** became the three standard psychophysiological bases for defining the stages of sleep.

Figure 14.1 depicts a volunteer participating in a sleep experiment. A participant's first night of sleep in a laboratory is often fitful. That's why the usual practice is to have each participant sleep several nights in the laboratory before commencing a sleep study. The disturbance of sleep observed during the first night in a sleep laboratory is called the *first-night phenomenon*. It is well known to graders of introductory psychology examinations because of the creative definitions of it that are offered by students who forget that it is a sleep-related, rather than a sex-related, phenomenon.

Three Stages of Sleep EEG

LO 14.2 Describe the three stages of the sleep EEG and explain the difference between REM and non-REM sleep.

According to the American Academy of Sleep Medicine, there are three stages of sleep EEG: stage 1, stage 2, and stage 3 (see Grigg-Damberger, 2012). Examples of these three stages are presented in Figure 14.2.

After the eyes are shut and a person prepares to go to sleep, **alpha waves**—waxing and waning bursts of 8- to 12-Hz EEG waves—begin to punctuate the low-voltage, high-frequency waves of alert wakefulness. Then, as the person falls asleep, there is a sudden transition to a period of stage 1 sleep EEG. The stage 1 sleep EEG is a low-voltage, high-frequency signal that is similar to, but slower than, that of alert wakefulness.

There is a gradual increase in EEG voltage and a decrease in EEG frequency as the person progresses from stage 1 sleep through stages 2 and 3. Accordingly, the stage 2 sleep EEG has a slightly higher amplitude and a lower frequency than the stage 1 EEG; in addition, it is punctuated by two characteristic wave forms: K complexes and sleep spindles. Each *K complex* is a single large negative wave (upward deflection) followed immediately by a single large positive wave (downward deflection)—see Mak-McCully and colleagues (2014). Each *sleep spindle* is a 0.5- to 2-second waxing and waning burst of 11- to 15-Hz waves (see Purcell et al., 2017). The stage 3 sleep EEG is defined by a predominance of **delta waves**—the largest and slowest EEG waves, with a frequency of 1 to 2 Hz.

Once sleepers reach stage 3 EEG sleep, they stay there for a time, and then they retreat back through the stages of sleep to stage 1. However, when they return to stage 1, things are not at all the same as they were the first time through. The first period of stage 1 EEG during a night's sleep (**initial stage 1 EEG**) is not marked by any striking electromyographic or electrooculographic changes, whereas subsequent periods of stage 1 sleep EEG (**emergent stage 1 EEG**) are accompanied by rapid eye movements (REMs) and by a loss of tone in the muscles of the body core.

After the first cycle of sleep EEG—from initial stage 1 to stage 3 and back to emergent stage 1—the rest of the night is spent going back and forth through the stages. Figure 14.3 illustrates the EEG cycles of a typical night's sleep and the close relationship between emergent stage 1 sleep, REMs, and the loss of tone in core muscles. Notice that each cycle tends to be about 90 minutes long and that, as the night progresses, more and more time is spent in emergent stage 1 sleep, and less and less time is spent in the other stages, particularly stage 3. Notice also that there are brief periods during the night when the person is awake, although he or she usually does not remember these periods of wakefulness in the morning.

Let's pause here to get some sleep-stage terms straight. The sleep associated with emergent stage 1 EEG is often called **REM sleep** (pronounced "rehm"), after the associated rapid eye movements; whereas all other stages of sleep together are called *NREM sleep* (non-REM sleep). Accordingly, initial stage 1, stage 2, and stage 3 sleep are sometimes referred to as NREM 1, NREM 2, and NREM 3, respectively (see Grigg-Damberger, 2012). NREM 3 is often referred to as **slow-wave sleep (SWS)**, after the delta waves that characterize it. Table 14.1 summarizes the various sleep-stage terms.

REMs, loss of core-muscle tone, and a low-amplitude, high-frequency EEG are not the only physiological correlates

Figure 14.3 The course of EEG stages during a typical night's sleep and the relation of emergent stage 1 EEG to REMs and lack of tone in core muscles.

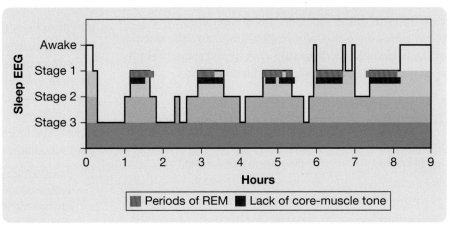

Figure 14.2 The EEG of alert wakefulness, the EEG that precedes sleep onset, and the three stages of sleep EEG. Each trace is about 10 seconds long.

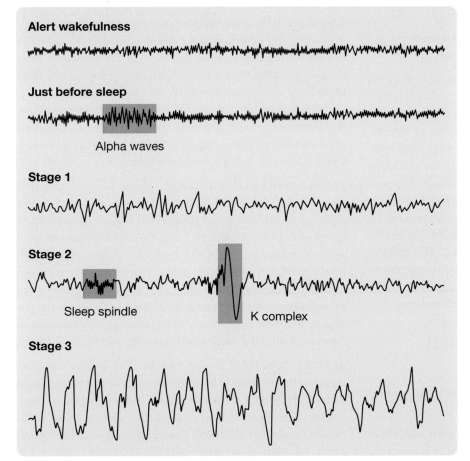

of REM sleep. Cerebral activity (e.g., oxygen consumption, blood flow, and neural firing) increases to waking levels in many brain structures, and there is a general increase in the variability of autonomic nervous system activity (e.g., in blood pressure, pulse, and respiration). Also, the muscles of the extremities occasionally twitch, and there is almost always some degree of penile or clitoral erection.

Table 14.1   Summary of Various Sleep-Stage Terms

REM sleep	Emergent stage 1 EEG
NREM sleep	Initial stage 1 EEG (NREM 1)
	Stage 2 EEG (NREM 2)
	Stage 3 EEG (NREM 3)
Slow-wave sleep	Stage 3 EEG (NREM 3)

Dreaming

Of all the topics I (SB) teach, few inspire more discussion in class than the topic of dreaming. This is in part because people have so many beliefs and theories about dreaming. A major goal of this module is to assess those beliefs and explore current theories and research related to dreaming.

Discovery of the Relationship between REM Sleep and Dreaming

LO 14.3   Describe the discovery of the relationship between REM sleep and dreaming.

Nathaniel Kleitman's laboratory was an exciting place in 1953. REM sleep had just been discovered, and Kleitman and his students were driven by the fascinating implication of the discovery. With the exception of the loss of tone in the core muscles, all of the other measures suggested that REM sleep episodes were emotion-charged. Could REM sleep be the physiological correlate of dreaming? Support for the theory that REM sleep is the physiological correlate of dreaming came from the observation that 80 percent of awakenings from REM sleep, but only 7 percent of awakenings from NREM sleep, led to dream recall. The dreams recalled from NREM sleep tended to be isolated experiences (e.g., "I was falling"), while those associated with REM sleep tended to take the form of stories, or narratives.

Testing Common Beliefs About Dreaming

LO 14.4   Describe five common beliefs about dreaming and assess their validity.

The relationship between REM sleep and dream recall that Kleitman and colleagues discovered provided them with an opportunity to test some common beliefs about dreaming. The following five beliefs were among the first to be addressed.

EXTERNAL STIMULI AND DREAMS.   Many people believe that external stimuli can become incorporated into their dreams. Dement and Wolpert (1958) sprayed water on sleeping volunteers after they had been in REM sleep for a few minutes and then awakened them a few seconds later. In 14 of 33 cases, the water was incorporated into the dream report. The following narrative was reported by one participant who had been dreaming that he was acting in a play:

> I was walking behind the leading lady when she suddenly collapsed and water was dripping on her. I ran over to her and water was dripping on my back and head. The roof was leaking . . . . I looked up and there was a hole in the roof. I dragged her over to the side of the stage and began pulling the curtains. Then I woke up. (Dement, W. C., & Wolpert, E. A. (1958))

We now know that some external stimuli are more likely to be incorporated into a dream than others (see Broughton, 1982; Schredl et al., 2009). Water droplets happen to be one of those. The stimulus that is most readily incorporated into a dream is the feeling of pressure on a limb (see Schredl et al., 2009).

DREAM DURATION.   Some people believe dreams last only an instant, but early research on the topic suggested that dreams run in "real time." In one study (Dement & Kleitman, 1957), participants were awakened 5 or 15 minutes after the beginning of a REM episode and asked to decide on the basis of the duration of the events in their dreams whether they had been dreaming for 5 or 15 minutes. They were correct in 92 of 111 cases.

Current research on dream duration supports the notion that dreams run slightly slower than real time. For example, physical movement while in a dream can take up to 40% longer than in a waking state (see Erlacher et al., 2014).

PEOPLE WHO DON'T DREAM.   Some people claim that they do not dream. However, studies conducted in the 1950s showed that these people have just as much REM sleep as normal dreamers: Most reported dreams if they were awoken during REM episodes (see Goodenough et al., 1959), although they did so less frequently than did normal dreamers. The results of more recent studies are mixed: Some studies have found these individuals do not appear to dream (e.g., Pagel, 2003), whereas others are more consistent with the earlier findings (e.g., Herlin et al., 2015).

SEXUAL CONTENT IN DREAMS.   Penile erections are commonly assumed to be indicative of dreams with sexual content. However, erections are no more complete during dreams with frank sexual content than during those without it (Karacan et al., 1966). Even babies have REM-related penile erections (see Fanni et al., 2018).

SLEEPTALKING AND SLEEPWALKING. Many people believe sleeptalking (*somniloquy*) and sleepwalking (*somnambulism*) occur during REM sleep and are therefore associated with dreaming. This is not so (see Alfonsi et al., 2019). Sleeptalking has no special association with REM sleep—it can occur during any stage but often occurs during the transition to wakefulness. Sleepwalking usually occurs during slow-wave sleep, and it never occurs during REM sleep, when core muscles tend to be totally relaxed (see Castelnovo et al., 2018; Januszko et al., 2015).

Does REM Sleep = Dreaming?

LO 14.5   Understand the relationship between REM sleep, NREM sleep, and dreaming.

While there are still adherents to the theory that REM sleep and dreaming can be equated, the situation has been complicated by three key discoveries. First, it is now well established that dreaming occurs during NREM sleep (see Siclari & Tononi, 2017). Second, the qualities of many NREM dreams are comparable to REM dreams (see McNamara et al., 2010). This is especially true for NREM dreams that occur later in the night: They are hallucinatory and have a storyline, just like REM dreams (see Siclari & Tononi, 2017). Finally, REM sleep and dreaming can be dissociated (see Siegel, 2011). For example, antidepressants greatly reduce or abolish REM sleep without affecting aspects of dream recall (see Oudiette et al., 2012). Conversely, specific cortical lesions can abolish dreaming without affecting REM sleep (see Nir & Tononi, 2010).

Lucid Dreaming

LO 14.6   Define lucid dreaming and discuss the research that supports its existence.

Have you ever been aware that you are dreaming? If so, have you ever taken control of your dream and made it unfold as you so desire? **Lucid dreaming** is the ability to be consciously aware that one is dreaming and, in some cases, be able to control the content of one's dream (see LaBerge, LaMarca, & Baird, 2018; Ribeiro, Gounden, & Quaglino, 2016). Think of the joy of being able to control one's dream. Want to have great sex? Want to be a superhero? Want to travel to the International Space Station? Having a lucid dream allows you to take control of your dream and do any of these things. I (SB) experience lucid dreams from time to time.

The Case of the Levitating Teenager

It was a recurring dream that plagued me (SB) as a teenager: I am in an otherwise empty movie theatre, apart from myself and two other individuals, one seated on either side of me. We are watching a film that I know quite well. The person to the right of me starts talking loudly to the person on my left, arguing over the title of the film.

"It's *Ferris Bueller's Day Off*."

"No, you idiot, it's *The Breakfast Club*," says the one on my left.

"Like you know anything about films," says the one on my right, voice rising further.

"More than you ever will," yells the other.

Meanwhile I am stuck in the center of this escalating argument, trying my best to enjoy the film and growing increasingly frustrated to the point of yelling "shut up!"

Then I often wake up. Not infrequently, I wake up on a bus: turning bright red with the realization that I had just told the whole bus to shut up; the other riders looking quite amused by my embarrassment.

Then, one night I had that same dream. But something was different this time. When the argument began as usual, I suddenly realized I was in a dream and decided to take control of things. I slowly levitated up from my seat. Then, floating high above my two nemeses, I smirked, leaned back, and enjoyed the film.

For a long time, sleep researchers believed it was impossible to have a lucid dream (see Baird, Mota-Rolim, & Dresler, 2019). In the 1980s, LaBerge conducted a series of studies that left little doubt that lucid dreaming was a real phenomenon (see Schredl, 2018). Indeed, previous surveys had indicated that many people have at least one lucid dream during their lives; some have them regularly; and a select few have them every night (see Ribeiro, Gounden, & Quaglino, 2016; Saunders et al., 2016).

LaBerge's technique for demonstrating lucid dreaming in his studies involved having self-proclaimed lucid dreamers sleep in his laboratory. Before allowing them to fall asleep, LaBerge would instruct them to carry out a specific bodily movement (e.g., repeated left-right-left-right eye movements while in their dream when given a specific signal, such as flashes of light on their eyelids; see Baird, Mota-Rolim, & Dresler, 2019; LaBerge, Baird, & Zimbardo, 2018). The following case study is an example of such a procedure, and it illustrates just how much control some lucid dreamers have over their dreams.

The Case of the Artistic Dreamer

A sleep researcher was at a dinner party with a large group. The topic of dreaming came up during the meal, and one of the guests, M.S., claimed that they always had lucid dreams. The sleep researcher was skeptical. She decided to put M.S.'s claim to the test. She invited M.S. to sleep a few nights in her sleep lab. When M.S. arrived, she sat them down on the bed, attached EEG electrodes to their scalp, attached EOG electrodes around their eyes, attached some EMG electrodes to their arm, and then asked them to lie down.

The sleep researcher gave M.S. the following instructions: "When you see a flashing light in your dream use your right arm to draw a star."

M.S. chuckled: "You could at least give me something interesting to draw." Then they fell asleep.

The sleep researcher watched and waited for M.S. to enter REM sleep. After an hour and a half, M.S. was showing the signs: Their EEG was low amplitude and high frequency, and they were displaying rapid eye movements. The researcher took her flashlight and pointed it at M.S.'s eyes. She rapidly switched the flashlight off and on, watching the EMG carefully as she did so. What she saw next made her believe in lucid dreams: M.S.'s arm muscles, in the absence of overt movement, displayed a series of contractions identical to what you would see if some-one were drawing a star.

"Careful what you wish for," G.D. said from her usual spot in the back row of the classroom. It was towards the end of the term and it was the first time that she had spoken.

"What do you mean?" I asked.

"I lucid dream every night," G.D. said. "I have as long as I can remember. I only rarely have non-lucid dreams and they are wonderful when I do."

"How so?" I said in a tone dripping with puzzlement.

"Imagine you knew the entire plot line of a movie before you begin watch it. Now, imagine if that were true for every movie you ever watched."

"I get it. If you always lucid dream it's like always knowing how your dream will unfold."

"Yep! For me, lucid dreaming is about as boring as a lecture," said G.D.

The class chuckled. I turned bright red.

Most students get excited about the prospect of lucid dreaming. I am often asked how one might become a lucid dreamer (see Aviram & Soffer-Dudek, 2018). There are several methods that have been shown to increase the likelihood that one will have a lucid dream (see Bazzari, 2018). Some examples include transcranial electrical stimulation (see Chapter 5) applied to the frontal and temporal areas of the skull (see Bray, 2014; Voss et al., 2014), certain cognitive training techniques (see LaBerge, LaMarca, & Baird, 2018), and the administration of acetylcholine agonists (see LaBerge, LaMarca, & Baird, 2018).

Don't worry. We have not forgotten you. We know that many of you were hoping that you might get some pointers in this module on how you might experience lucid dreaming yourself, and we also know that you are unlikely to have your own brain stimulator or stash of acetylcholine agonists or to be an expert in cognitive training techniques. Unfortunately, we do not have enough space here to instruct you ourselves, but fortunately there are several websites and books that can instruct you. The problem is that most of the techniques promoted are ineffective, or worse, dangerous. What we can do is send you to a good source: *Exploring the World of Lucid Dreaming,* a book by Stephen LaBerge, the most respected researcher in lucid dreaming. If you have success, we would like to hear about your experiences.

One would think everyone would want to be a lucid dreamer. However, as the following case study illustrates, it can be a mixed blessing.

The Case of the Bored Lucid Dreamer

I (SB) was giving a lecture about dreaming. I told my students about my own experiences with lucid dreaming and how wonderful it was. At the end of my story, I said, "I wish I could lucid dream every night."

Another reason lucid dreams might be considered boring is that they tend to be less bizarre than nonlucid dreams (see Yu & Shen, 2020). Also, concerns have been raised that lucid dreaming might have a negative impact on sleep quality (see Soffer-Dudek, 2020; but see Schredl, Dyck, & Kühnel, 2020).

Lucid dreamers have been a boon to sleep researchers. Because they are aware in their dreams and are better able to recall their dreams when they awaken (see Aspy et al., 2017), they are particularly well equipped to describe the sensory qualities of their dreams. Two examples:

- Lucid dreamers report that gustatory, olfactory, and somatosensory stimuli are uncommon in their dreams (Kahan & LaBerge, 2011).
- Lucid dreamers report being unable to perceive fine-grained visual details in their dreams (Kahan & LaBerge, 2011).

To date, only two studies have examined the brain changes associated with lucid dreaming—one case study (Dresler et al., 2012) and one quasiexperimental study (Baird et al., 2018). Unfortunately, there was little consistency in their results (see Baird, Mota-Rolim, & Dresler, 2019).

Why Do We Dream What We Do?

LO 14.7   Discuss the issue associated with studying dream content and describe findings related to what can influence dream content.

Most of us are fascinated by the content of our dreams. People often seek out therapists who are self-proclaimed dream interpreters. Is it possible to interpret dreams?

Sigmund Freud believed so (Freud, 1913). Freud believed that dreams are triggered by unacceptable repressed wishes, often of a sexual nature. He argued that because dreams represent unacceptable wishes, the dreams we experience (our *manifest dreams*) are merely disguised versions of our real dreams (our *latent dreams*): He hypothesized an unconscious censor that disguises and subtracts information from our real dreams so that we can endure them. Freud thus concluded that one of the keys to understanding people and dealing with their psychological problems is to expose the meaning of their latent dreams through the interpretation of their manifest dreams.

There is no convincing scientific evidence for the Freudian theory of dreams; indeed, the brain science of the 1890s, which served as its foundation, is now obsolete. Yet many people still believe that dreams bubble up from a troubled unconscious and that they represent repressed thoughts and wishes. Although dream interpretation is no longer a focus of scientific research, biopsychologists are increasingly interested in understanding why we dream the content that we do.

If a researcher wants to study dream content, they must rely on the detailed self-reports of dreamers. This reliance on self-reports of dreams has been viewed as a problem by some, who question whether we can trust these self-reports of dreams given that dream recall is often difficult if not impossible for many (see Mangiaruga et al., 2018; Nir & Tononi, 2010). Conversely, others have argued that self-reports of dreams are a legitimate window into the world of dream content (see Cipolli et al., 2017; Windt, 2013).

Despite the widely acknowledged limitations of dream recall and dream self-reports, studies of dream content have flourished. Two key findings have emerged. First, there is a general consensus that dream content is influenced by what we have experienced in the prior period of wakefulness (see Cipolli et al., 2017; Fogel et al., 2018)—even the most mundane of experiences, like taking out the garbage, can enter a subsequent dream (see Vallat et al., 2017). Second, the amount of anxiety experienced prior to a bout of dreaming affects the emotional content of dreams (see Sikka, Pesonen, & Revonsuo, 2018; Vallat et al., 2017).

Why Do We Dream?

LO 14.8   Compare and contrast three different theories of why we dream.

Rather than asking why we dream what we dream, many researchers are more interested in why we dream at all. What is dreaming good for, if anything? There are many theories as to why we dream. In this section, we present three of the most prominent.

HOBSON'S ACTIVATION-SYNTHESIS HYPOTHESIS. One theory of why we dream is Hobson's (1989) *activation-synthesis hypothesis* (see Palagini & Rosenlicht, 2011). It is based on the observation that, during sleep, many brain-stem circuits become active and bombard the cerebral cortex with neural signals. The essence of Hobson's hypothesis is that the information supplied to the cortex during sleep is

paul prescott/Shutterstock
What do you see when you look at these clouds?

largely random and that the resulting dream is the cortex's effort to make sense of these random signals. You might liken this process to what happens when you stare up at the clouds and see faces or figures in them: The clouds are randomly patterned, but your brain is trying its best to make sense of that random pattern.

REVONSUO'S EVOLUTIONARY THEORY OF DREAMS. A second theory of why we dream is Revonsuo's (2000) *evolutionary theory of dreams*. According to this theory, dreams serve an important biological function, and this function has implications for the fitness of an organism in the Darwinian sense (see Chapter 2). Specifically, Revonsuo (2000) proposed that we dream to simulate threatening events, such as physical attack, threats to social relationships, or threats to one's livelihood, and that such simulation allows us to better predict and respond to such threats when we are awake—see Gauchat et al. (2015).

HOBSON'S PROTOCONSCIOUSNESS HYPOTHESIS. In the years following the publication of his activation-synthesis hypothesis, Hobson began to believe there was a purpose for dreaming and became increasingly critical of his own activation-synthesis hypothesis.

Like Revonsuo (2000), Hobson now believed that dreaming conferred an evolutionary advantage. Unlike Revonsuo, his new *protoconsciousness hypothesis* proposed that dreaming conferred that advantage by simulating everything, not just threatening situations. According to his hypothesis, dreaming is important (1) during early development, when sensory input is limited by underdeveloped sensory systems—the visual system in particular, and (2) throughout one's life, by anticipating and predicting how events will unfold while awake.

In short, according to Hobson's (2009) new theory, dreaming is a training mechanism, with each dream representing a virtual real-life scenario (see Boag, 2017; Hobson & Friston, 2012; Llewellyn, 2016; Scarpelli et al., 2019). Hence the term "protoconsciousness": a virtual prototype of our conscious experiences.

Hobson's protoconsciousness hypothesis stands in stark contrast to his earlier activation-synthesis hypothesis, which proposed that dreams are, in essence, leftovers from a brain built for a day job that serves no particular function. It might strike you as odd that a researcher would propose a theory that contradicts his own theory. However, this is a sign of good critical thinking: In order to think critically, one must be just as critical of one's own theories as one is of others'.

The Dreaming Brain

LO 14.9 **Identify three brain areas that have been implicated in dreaming.**

What brain areas are involved in dreaming? Studies of patients with brain lesions and neuroimaging studies of individuals dreaming tell a clear story.

There are two types of brain lesions that produce a cessation of dreaming in patients (as determined either by self report, by in-lab sleep studies, or both): (1) bilateral lesions of the *temporo-parieto junction* (that location in the cerebral cortex where the temporal lobes and the parietal lobes meet), and (2) bilateral lesions of the *medial prefrontal cortex*; see Figure 14.4 (see Domhoff & Fox, 2015; Vallat et al., 2018). In addition, lesions to those areas of secondary visual cortex in the medial occipital lobe lead to a loss of visual imagery in dreams (see Figure 14.4).

There have been many fMRI (see Chapter 5) studies of healthy participants engaged in REM sleep (see Domhoff & Fox, 2015). Consistent with the lesion studies, the temporo-parieto junction, the medial prefrontal cortex, and medial occipital cortex all display increased neural activity during REM sleep (see Figure 14.4; Domhoff & Fox, 2015). Whether NREM dreams are associated with comparable patterns of brain activation remains to be

Figure 14.4 Two areas of the brain implicated in dreaming and one implicated in visual imagery. Both lesions studies and brain imaging studies have implicated the medial prefrontal cortex and the tempero-parieto junction in dreaming. Both lesion studies and brain imaging studies have implicated the medial occipital lobe in the visual imagery within dreams.

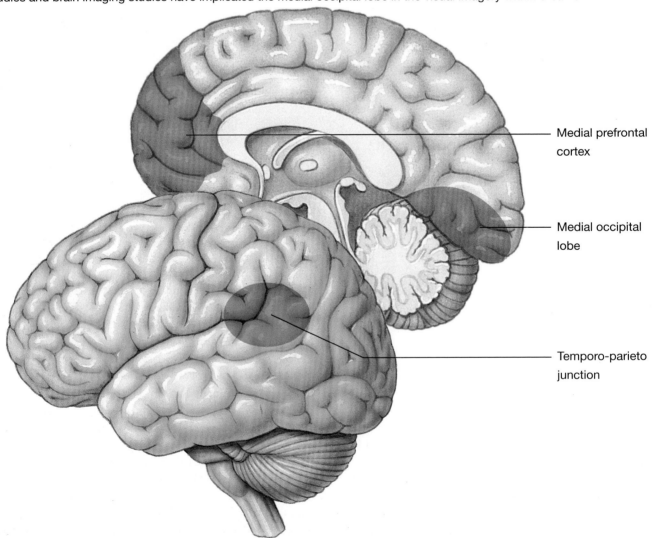

Medial prefrontal cortex

Medial occipital lobe

Temporo-parieto junction

Adapted from Figure 1 of Domhoff, G. W., & Fox, K. C. (2015). Dreaming and the default network: A review, synthesis, and counterintuitive research proposal. *Consciousness and Cognition, 33,* 342–353.

seen (see Mutz & Javadi, 2017). We know of only one study that has addressed the neuroanatomical correlates of NREM dreaming. The study by Siclari et al. (2017) demonstrated that activity in the tempero-parieto junction was associated with both REM and NREM dreaming, suggesting that this brain region is critical for dreaming in general.

Why Do We Sleep, and Why Do We Sleep When We Do?

You have been introduced to the properties of sleep and its various stages, and you have also learned about the topic of dreaming. The focus of this chapter now shifts to a consideration of two fundamental questions about sleep: Why do we sleep? And why do we sleep when we do?

Two Kinds of Theories of Sleep

LO 14.10 Describe the two kinds of theories of sleep.

Two kinds of theories of sleep have been proposed: *recuperation theories* and *adaptation theories*. The differences between these two theoretical approaches are revealed by the answers they offer to the two fundamental questions about sleep.

The essence of **recuperation theories of sleep** is that being awake disrupts the *homeostasis* (internal physiological stability) of the body in some way and sleep is required to restore it. Various recuperation theories differ in terms of the particular physiological disruption they propose as the trigger for sleep. For example, the three most common recuperation theories of sleep are that the function of sleep is to (1) restore energy levels that decline during wakefulness (see Porkka-Heiskanen, 2013), (2) clear toxins (e.g., beta-amyloid—see Chapter 10) from the brain and other tissues that accumulate during wakefulness (see Da Mesquita et al., 2018; Holth et al., 2019; Lei et al., 2017; Sun et al., 2018; Sweeney & Zlokovic, 2018), or (3) restore the synaptic plasticity that might dissipate during wakefulness (see Cirelli & Tononi, 2015; de Vivo et al., 2017; Tononi & Cirelli, 2019). However, regardless of the particular function postulated by recuperation theories of sleep, they all imply that sleepiness is triggered by a deviation from homeostasis caused by wakefulness and that sleep is terminated by a return to homeostasis.

The essence of **adaptation theories of sleep** is that sleep is not a reaction to the disruptive effects of being awake but the result of an internal 24-hour timing mechanism—that is, we humans are programmed to sleep at night regardless of what happens to us during the day. Adaptation theories of sleep focus more on when we sleep than on the function of sleep. Some of these theories even propose that sleep plays

no role in the efficient physiological functioning of the body. According to these theories, early humans had enough time to get their eating, drinking, and reproducing out of the way during the daytime, and their strong motivation to sleep at night evolved to conserve their energy resources, make them less susceptible to mishaps in the dark (e.g., predation) (Joiner, 2016), and to carry out certain brain functions that aren't possible during wakefulness (see Chen & Wilson, 2017; Frank, 2017; Nagai et al., 2017; Stickgold, 2015). Adaptation theories suggest that sleep is like reproductive behavior in the sense that we are highly motivated to engage in it, but we don't need it to stay healthy.

It should be noted that this distinction between recuperation theories and adaptation theories of sleep does not mean that a specific theory cannot incorporate elements of both. For example, a theory of sleep might propose that we have evolved to sleep at night, but that the duration of our nightly sleep is determined by recuperative mechanisms (e.g., Eban-Rothschild, Giardino, & de Lecea, 2017; Yadav et al., 2017).

Comparative Analysis of Sleep

LO 14.11 Explain four conclusions that have resulted from the comparative analysis of sleep.

Sleep has been studied in only a small number of species, but the evidence so far suggests that all mammals and most birds sleep (see Manger & Siegel, 2020). Furthermore, the sleep of mammals and birds, like ours, is characterized by high-amplitude, low-frequency EEG waves punctuated by periods of low-amplitude, high-frequency waves (see Manger & Siegel, 2020). The evidence for sleep in amphibians, reptiles, fish, and insects is less clear: Some display periods of inactivity and unresponsiveness, but the relation of these periods to mammalian sleep has not been established (see Siegel, 2008; Zimmerman et al., 2008). Table 14.2 gives the average number of hours per day that various mammalian species spend sleeping.

> **Journal Prompt 14.1**
> If you were a sleep researcher studying sleep in various organisms, how would you define sleep?

The comparative investigation of sleep has led to several important conclusions. Let's consider four of these.

First, the fact that most mammals and birds sleep (see Siegel, 2012) suggests that sleep serves some important physiological function, rather than merely protecting animals from mishap and conserving energy. The evidence is strongest in species that are at increased risk of predation when they sleep (e.g., antelopes) and in species that have evolved complex mechanisms that enable them to sleep. For

Table 14.2 Average Number of Hours Slept per Day by Various Mammalian Species

Mammalian Species	Hours of Sleep per Day
Giant sloth	20
Opossum, brown bat	19
Giant armadillo	18
Owl monkey, nine-banded armadillo	17
Arctic ground squirrel	16
Tree shrew	15
Cat, golden hamster	14
Mouse, rat, gray wolf, ground squirrel	13
Arctic fox, chinchilla, gorilla, raccoon	12
Mountain beaver	11
Jaguar, vervet monkey, hedgehog	10
Rhesus monkey, chimpanzee, baboon, red fox	9
Human, rabbit, guinea pig, pig	8
Gray seal, gray hyrax, Brazilian tapir	6
Tree hyrax, rock hyrax	5
Cow, goat, elephant, donkey, sheep	3
Roe deer, horse, zebra	2

Figure 14.5 After gorging themselves on a kill, African lions often sleep almost continuously for 2 or 3 days. And where do they sleep? Anywhere they want!

CRSuber/Getty Images

example, some birds and marine mammals sleep with only half of their brain at a time so that the other half can control resurfacing for air (e.g., Lyamin et al., 2018; Lyamin & Siegel, 2019; Mascetti, 2016; Yadav et al., 2017). It is against the logic of natural selection for some animals to risk predation while sleeping and for others to have evolved complex mechanisms to permit them to safely sleep, unless sleep itself serves some critical function.

Second, the fact that most mammals and birds sleep suggests that the primary function of sleep is not some special, higher-order human function. For example, suggestions that sleep helps humans reprogram our complex brains or that it permits some kind of emotional release to maintain our mental health are improbable in view of the comparative evidence.

Third, the large between-species differences in sleep time suggest that although sleep may be essential for survival, it is not necessarily needed in large quantities (see Table 14.2). Horses and many other animals get by quite nicely on 2 or 3 hours of sleep per day. Moreover, it is important to realize that the sleep patterns of mammals and birds in their natural environments can vary substantially from their patterns in captivity, which is where they are typically studied. For example, some animals that sleep a great deal in captivity sleep little in the wild when food is in short supply or during periods of migration or mating (see Lesku et al., 2012; Siegel, 2012).

Fourth, many studies have tried to identify the reasons why some species are long sleepers and others are short sleepers. Why do cats tend to sleep about 14 hours a day and horses only about 2? Under the influence of recuperation theories, researchers have focused on energy-related factors in their efforts. However, there is no strong relationship between a species' sleep time and its level of activity, its body size, or its body temperature (see Siegel, 2005). The fact that giant sloths sleep 20 hours per day is a strong argument against the theory that sleep is a compensatory reaction to energy expenditure—similarly, energy expenditure has been shown to have little effect on subsequent sleep in humans (see Kelley & Kelley, 2017). In contrast, adaptation theories correctly predict that the daily sleep time of each species is related to how vulnerable it is while it is asleep and how much time it must spend each day to feed itself and to take care of its other survival requirements. For example, zebras must graze almost continuously to get enough to eat and are extremely vulnerable to predatory attack when they are asleep—and they sleep only about 2 hours per day. In contrast, African lions often sleep more or less continuously for 2 or 3 days after they have gorged themselves on a kill. Figure 14.5 says it all.

Effects of Sleep Deprivation

One way to identify the functions of sleep is to determine what happens when a person is deprived of sleep. This module begins with a cautionary note about the interpretation of the effects of sleep deprivation, a description of the predictions that recuperation theories make about sleep deprivation, and two classic case studies of sleep deprivation. Then, it summarizes the results of sleep-deprivation research.

Interpretation of the Effects of Sleep Deprivation: The Stress Problem

LO 14.12 Explain how stress can often be a confounding variable when considering the effects of sleep deprivation.

When you sleep substantially less than you are used to, the next day you feel out of sorts and unable to function as well as you usually do. Although such experiences of sleep deprivation are compelling, you need to be cautious in interpreting them. In Western cultures, most people who sleep little or irregularly do so because they are under stress (e.g., from illness, excessive work, shift work, drugs, or examinations), which could have adverse effects independent of any sleep loss. Even when sleep-deprivation studies are conducted on healthy volunteers in controlled laboratory environments, stress can be a contributing factor because many of the volunteers will find the sleep-deprivation procedure itself stressful. Accordingly, because it is difficult to separate the effects of sleep loss from the effects of stressful conditions that may have induced the loss, results of sleep-deprivation studies must be interpreted with caution.

Be that as it may, almost every week we read a news article decrying the effects of sleep loss in the general population. Such articles will typically point out that many people who are pressured by the demands of their work schedule sleep little and experience a variety of health and accident problems. There is a place for this kind of research because it identifies a public health issue that requires attention (see Chattu et al., 2018); however, because the low levels of sleep are hopelessly confounded with high levels of stress (see Anafi, Kayser, & Raizen, 2019), most sleep-deprivation studies tell us little about the functions of sleep and how much we need.

Predictions of Recuperation Theories about Sleep Deprivation

LO 14.13 List the three predictions that recuperation theories make about the effects of sleep deprivation.

Because recuperation theories of sleep are based on the premise that sleep is a response to the accumulation of some debilitating effect of wakefulness, they make the following three predictions about sleep deprivation:

- Long periods of wakefulness will produce physiological and behavioral disturbances.

- These disturbances will grow worse as the sleep deprivation continues.

- After a period of deprivation has ended, much of the missed sleep will be regained.

Have these predictions been confirmed?

Two Classic Sleep-Deprivation Case Studies

LO 14.14 Describe two classic sleep-deprivation case studies.

Let's look at two widely cited sleep-deprivation case studies. First is the groundbreaking study of a group of sleep-deprived students (Kleitman, 1963); second is the bizarre case of Randy Gardner (Dement, 1978).

The Case of the Sleep-Deprived Students

Most of the volunteer students subjected to total sleep deprivation by Kleitman experienced the same effects. During the first night, they read or studied with little difficulty until after 3:00 a.m., when they experienced an attack of sleepiness. At this point, their watchers had to be particularly careful that they did not sleep. The next day the students felt alert as long as they were active. During the second night, reading or studying was next to impossible because sleepiness was so severe, and as on the first night, there came a time after 3:00 a.m. when sleepiness became overpowering. However, as before, later in the morning, there was a decrease in sleepiness, and the students could perform tasks around the lab during the day as long as they were standing and moving.

The cycle of sleepiness on the third and fourth nights resembled that on the second, but the sleepiness became even more severe. Surprisingly, things did not grow worse after the fourth night, and those students who persisted repeatedly went through the same daily cycle.

The Case of Randy Gardner

Randy Gardner and two classmates, who were entrusted with keeping him awake, planned to break the then world record of 260 hours of consecutive wakefulness. Dement learned about the project and, seeing an opportunity to collect some important data, joined the team, much to the comfort of Randy's parents. Randy did complain vigorously when his team would not permit him to close his eyes. However, in no sense could Randy's behavior be considered disturbed. Near the end of his vigil, Randy held a press conference, and he conducted himself impeccably. Randy went to sleep exactly 264 hours and 12 minutes after he had awakened 11 days before. And how long did he sleep? Only 14 hours the first night, and thereafter he returned to his usual 8-hour schedule.

Although you may be surprised that Randy did not have to sleep longer to "catch up" on his lost sleep, the lack of substantial recovery sleep is typical of such cases.

Studies of Sleep Deprivation in Humans

LO 14.15 Describe the major effects of sleep deprivation in humans.

Since the first studies of sleep deprivation by Dement and Kleitman in the mid-20th century, there have been hundreds of studies assessing the effects on humans of sleep-deprivation schedules ranging from a slightly reduced amount of sleep during one night to total sleep deprivation for several nights (see Krause et al., 2017). The studies have assessed the effects of these schedules on many different measures of sleepiness, mood, cognition, motor performance, physiological function, and even molecular function (see Krause et al., 2017).

Even moderate amounts of sleep deprivation—for example, sleeping 3 or 4 hours less than normal for one night—have been found to have three consistent effects. First, sleep-deprived individuals display an increase in sleepiness: They report being more sleepy, and they fall asleep more quickly if given the opportunity. Second, sleep-deprived individuals display negative affect on various tests of mood. And third, they perform poorly on tests of sustained attention (e.g., watching for a moving light on a computer screen)—see Kirszenblat and van Swinderen (2015).

The effects of sleep deprivation on complex cognitive functions have been less consistent (see Basner et al., 2013) and show marked variability between individuals—ranging from virtually no effects to severe effects (see Krause et al., 2017; Satterfield & Killgore, 2019). Consequently, researchers have preferred to assess performance on the simple, dull, monotonous tasks most sensitive to the effects of sleep deprivation, such as tasks that require sustained attention (see Hudson, Van Dongen, & Honn, 2019; Satterfield & Killgore, 2019). Nevertheless, a large number of studies have been able to demonstrate disruption of the performance of complex cognitive tasks by sleep deprivation (see Basner et al., 2013; Krause et al., 2017).

The inconsistent effects of sleep deprivation on cognitive function in various studies was clarified by the discovery that only some cognitive functions appear to be susceptible. Many early studies of the effects of sleep deprivation on cognitive function used tests of logical deduction or critical thinking, and these tests proved to be largely immune to disruption. In contrast, performance on tests of executive function proved to be more susceptible to disruption by sleep loss. **Executive function** is a group of cognitive abilities, such as problem solving, working memory, decision making, and assimilating new information to update plans and strategies (see Miller & Wallis, 2009).

The hypothesis that only some cognitive processes are susceptible to disruption by sleep loss clearly requires more systematic investigation. For example, several researchers have pointed out the need to determine the degree to which the deficits in vigilance and motivation produced by sleep loss can be mistaken for cognitive deficits (see Axelsson et al., 2019; Basner et al., 2013; Engle-Friedman, 2014; Massar, Lim, & Huettel, 2019).

Journal Prompt 14.2

Why would deficits in vigilance and motivation lead to cognitive deficits? Give an example as part of your answer.

The adverse effects of sleep deprivation on physical performance have been surprisingly inconsistent considering the general belief that a good night's sleep is essential for optimal motor performance. Only a few measures tend to be affected (e.g., Patrick et al., 2017), even after lengthy periods of deprivation, and there are indications that the effects on these measures are unreliable (see Fullager et al., 2015; Knufinke et al., 2018; Vaara et al., 2018). Altogether, the study of the effects of sleep deprivation on physical performance remains controversial (see Vaara et al., 2018).

Sleep deprivation has been found to have a variety of physiological consequences such as reduced body temperature, increases in blood pressure, decreases in some aspects of immune function, hormonal changes, and metabolic changes (see Cedernaes, Schiöth, & Benedict, 2015; Irwin, Olmstead, & Carroll, 2016; Maggio et al., 2013). The problem is that there is little evidence that these changes have any consequences for health or performance. For example, the fact that a decline in immune function was discovered in sleep-deprived volunteers does not necessarily mean that they would be more susceptible to infection—the immune system is extremely complicated and a decline in one aspect is often compensated for by other changes. This is why we want to single out a study by Prather and colleagues (2015) for commendation: Rather than studying immune function, these researchers focused directly on susceptibility to infection and illness.

Prather and colleagues exposed 164 healthy volunteers to a cold virus. Those who slept less than 6 hours a night were not less likely to become infected, but they were more likely to develop a cold. This is a *correlational study* (see Chapter 1), and thus it cannot directly implicate sleep duration as the causal factor. Still, the suggestion that there may be a causal relation between sleep and susceptibility to infection warrants further research.

After 2 or 3 days of continuous sleep deprivation, most volunteers experience microsleeps, unless they are in a laboratory environment where the microsleeps can be interrupted as soon as they begin. **Microsleeps** are brief periods of sleep, typically about 2 or 3 seconds long, during which the eyelids droop and the volunteers become less responsive to external stimuli, even though they remain sitting or standing (see Toppi et al., 2016). As one would expect, microsleeps severely disrupt the performance of tests of vigilance, but even sleep-deprived individuals not experiencing microsleeps experience some vigilance problems (see Hudson, Van Dongen, & Honn, 2019).

It can be useful to compare sleep deprivation with the deprivation of the motivated behaviors discussed in Chapters 12 and 13. If people were deprived of the opportunity to eat or engage in sexual activity, the effects would be severe and unavoidable: In the first case, starvation and death would ensue; in the second, there would be a total loss of reproductive capacity. Despite our powerful drive to sleep, the effects of sleep deprivation tend to be comparatively subtle, selective, and variable (see Satterfield & Killgore, 2019; Vaara et al., 2018). This is puzzling. Another puzzling thing is that performance deficits observed after extended periods of sleep deprivation disappear so readily—for example, in one study, 4 hours of sleep eliminated the performance deficits produced by 64 hours of sleep deprivation (Rosa, Bonnet, & Warm, 2007).

Sleep-Deprivation Studies of Laboratory Animals

LO 14.16 Describe the key studies of sleep deprivation in laboratory animals. Provide a critique of the carousel apparatus as a method of sleep deprivation.

The **carousel apparatus** (see Figure 14.6) has been used to deprive rats of sleep. Two rats, an experimental rat and its *yoked control*, are placed in separate chambers of the apparatus. Each time the EEG activity of the experimental rat indicates that it is sleeping, the disk, which serves as the floor of half of both chambers, starts to slowly rotate. As a result, if the sleeping experimental rat does not awaken immediately, it gets shoved off the disk into a shallow pool of water. The yoked control is exposed to exactly the same pattern of disk rotations; but if it is not sleeping, it can easily avoid getting dunked by walking in the direction opposite to the direction of disk rotation. The experimental rats typically died after about 12 days, while the yoked controls stayed reasonably healthy (see Rechtschaffen & Bergmann, 1995).

The fact that humans and rats have been sleep-deprived by other means for similar periods of time without dire consequences argues for caution in interpreting the results of the carousel sleep-deprivation experiments (see Rial et al., 2007; Siegel, 2009, 2012). It may be that repeatedly being awakened by this apparatus kills the experimental rats not because it keeps them from sleeping but because it is stressful. This interpretation is consistent with the pathological problems in the experimental rats that were revealed by

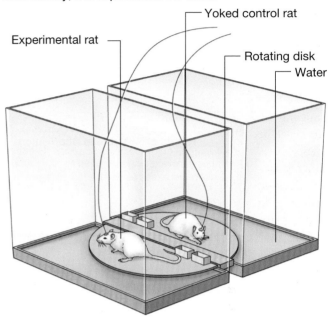

Figure 14.6 The carousel apparatus used to deprive an experimental rat of sleep while a yoked control rat is exposed to the same number and pattern of disk rotations. The disk on which both rats stand rotates every time the experimental rat displays sleep EEG. If the sleeping rat does not awaken immediately, it is deposited in the water.

Based on Rechtschaffen, A., Gilliland, M. A., Bergmann, B. M., & Winter, J. B. (1983). Physiological correlates of prolonged sleep deprivation in rats. *Science, 221,* 182–184.

postmortem examination: swollen adrenal glands, gastric ulcers, and internal bleeding (see Geissmann, Beckwith, & Gilestro, 2019).

You have already encountered many examples in this text of the value of the comparative approach. However, sleep deprivation may be one phenomenon that cannot be productively studied in nonhumans because of the unavoidable confounding effects of extreme stress (see McEwen & Karatsoreos, 2015; Minkel et al., 2014).

REM-Sleep Deprivation

LO 14.17 Describe the effects of REM-sleep deprivation.

Because of its early association with dreaming, REM sleep has been the subject of intensive investigation. In an effort to reveal the particular functions of REM sleep, sleep researchers have specifically deprived sleeping volunteers of REM sleep by waking them up each time a bout of REM sleep begins.

REM-sleep deprivation has been shown to have two consistent effects (see Figure 14.7). First, following REM-sleep deprivation, participants display a *REM rebound*; that is, they have more than their usual amount of REM sleep for the first two or three nights (Lyamin et al., 2018; McCarthy et al., 2016). Second, with each successive night of deprivation, there is a greater tendency for participants

Figure 14.7 The two effects of REM-sleep deprivation.

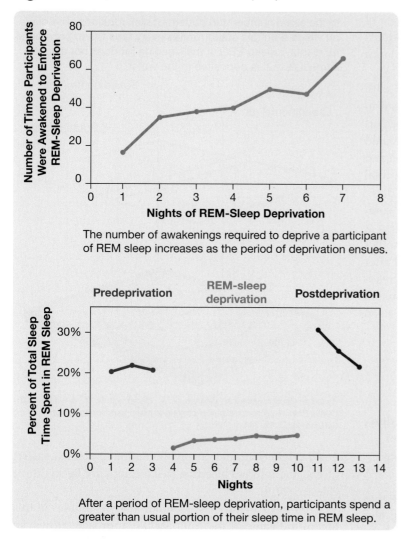

The number of awakenings required to deprive a participant of REM sleep increases as the period of deprivation ensues.

After a period of REM-sleep deprivation, participants spend a greater than usual portion of their sleep time in REM sleep.

to initiate REM sleep. Thus, as REM-sleep deprivation proceeds, participants have to be awakened more and more frequently to keep them from accumulating significant amounts of REM sleep (see Maisuradze, 2019; McCarthy et al., 2016). For example, during the first night of REM-sleep deprivation in one experiment (Webb & Agnew, 1967), the participants had to be awakened 17 times to keep them from having extended periods of REM sleep; but during the seventh night of deprivation, they had to be awakened 67 times.

The compensatory increase in REM sleep following a period of REM-sleep deprivation suggests that the amount of REM sleep is regulated separately from the amount of slow-wave sleep and that REM sleep serves a special function (see Hayashi et al., 2015; Vyazovskiy, 2015). This finding, coupled with the array of interesting physiological and psychological events that define REM sleep, has led to much speculation about its function.

Considerable attention has focused on the role of sleep in strengthening memory. For example, some researchers believe that REM sleep strengthens explicit memories (see Diekelmann & Born, 2010)—particularly those with emotional content (see Morgenthaler et al., 2014; Wiesner et al., 2015), other researchers believe that slow-wave sleep promotes memory consolidation (see Euston & Steenland, 2014; Inostroza & Born, 2013; Tononi & Cirelli, 2013), and still others believe that the memories of our daily experiences are processed (e.g., modified) prior to consolidation during sleep (see Chatburn, Lushington, & Kohler, 2014; Dudai, Karni, & Born, 2015; Stickgold & Walker, 2013). However, overall the results are still not convincing: Some studies have not observed any relationship between memory and sleep (see Ackermann et al., 2015).

The sleep of depressed patients constitutes a serious challenge to the hypothesis that sleep influences memory. Treated with antidepressant drugs, they may experience no REM sleep for years, again with no experience of memory deficits (see Genzel et al., 2014; Tribl, Wetter, & Schredl, 2013).

The *default theory* of REM sleep is a different approach to understanding the functions of REM sleep (see Horne, 2013). According to this theory, it is difficult to stay continuously in NREM sleep, so the brain periodically switches to one of two other states. If there are any immediate bodily needs to be taken care of (e.g., the need for food or water), the brain switches to wakefulness; if there are no immediate needs, it switches to REM sleep. According to the default theory, REM sleep is more adaptive than wakefulness when there are no immediate bodily needs. In addition, according to the default theory, REM sleep functions to prepare organisms for wakefulness in natural environments where immediate effective activity may be required upon wakening (see Horne, 2013; Klemm, 2011).

Most support for the default theory of REM sleep is indirect, coming from the many similarities between REM sleep and wakefulness. However, the surprising results of Nykamp and colleagues (1998) provide more direct support. They awakened young adults every time they entered REM sleep, but instead of letting them go back to sleep immediately, they substituted a 15-minute period of wakefulness for each lost REM period. Under these conditions, the participants were not tired the next day, despite getting only 5 hours of sleep, and they displayed no REM rebound. In other words, there seemed to be no need for REM sleep if periods of wakefulness were substituted for it. This finding is consistent with the finding that as antidepressants reduce REM sleep, the number of nighttime awakenings increases (see Lia, 2019; Wichniak et al., 2017).

Sleep Deprivation Increases the Efficiency of Sleep

LO 14.18 Describe six pieces of evidence that indicate that less sleep is associated with more efficient sleep.

One of the most important findings of human sleep-deprivation research is that individuals who are deprived of sleep become more efficient sleepers. In particular, their sleep has a higher proportion of slow-wave sleep (NREM 3), which seems to serve the main restorative function. Because this is such an important finding, let's look at six major pieces of evidence that support it:

- Although people regain only a small proportion of their total lost sleep after a period of sleep deprivation, they regain most of their lost slow-wave sleep (see Léger et al., 2018).

- After sleep deprivation, the slow-wave sleep EEG of humans is characterized by an even higher proportion of slow waves than usual (see Léger et al., 2018).

- People who sleep 6 hours or less per night normally get as much or more slow-wave sleep as people who sleep 8 hours or more (see Åkerstedt et al., 2019).

- Even during REM sleep, slow-wave activity occurs at different times and at different locations in the brain (see Bernardi et al., 2019; Funk et al., 2016; Siclari & Tononi, 2017).

- People who reduce their usual sleep time get less NREM 1 and NREM 2 sleep, but the duration of their slow-wave sleep remains about the same as before (see Mullaney et al., 1977; Webb & Agnew, 1975).

- Repeatedly waking individuals during REM sleep produces little increase in the sleepiness they experience the next day, whereas repeatedly waking individuals during slow-wave sleep has major effects (see Nykamp et al., 1998).

The fact that sleep becomes more efficient in people who sleep less means that conventional sleep-deprivation studies are virtually useless for discovering how much sleep people need. Certainly, our bodies respond negatively when we get less sleep than we are used to getting. However, the negative consequences of sleep loss in inefficient sleepers do not indicate whether the lost sleep was really needed.

The true need for sleep can be assessed only by experiments in which sleep is regularly reduced for many weeks, to give the participants the opportunity to adapt to getting less sleep by maximizing their sleep efficiency. Only when people are sleeping at their maximum efficiency is it possible to determine how much sleep they really need. Such sleep-reduction studies are discussed later in the chapter, but please pause here to think about this point—it is extremely important.

This is an appropriate time, here at the end of the module on sleep deprivation, for me (JP) to file a brief progress report. It has now been 2 weeks since I began my 5-hours-per-night sleep schedule. Generally, things are going well. My progress on this chapter has been faster than usual. I am not having any difficulty getting up on time or getting my work done, but I am finding that it takes a major effort to stay awake in the evening. If I try to read or watch a bit of television after 10:30, I experience microsleeps. My so-called friends delight in making sure that my transgressions are quickly punished.

Scan Your Brain

Before continuing with this chapter, scan your brain by completing the following exercise to make sure you understand the fundamentals of sleep. The correct answers appear at the end of the exercise. Before proceeding, review material related to your errors and omissions.

1. After the eyes are shut and a person prepares to go to sleep, _____ waves begin to punctuate the range between 8 and 12 Hz.

2. Sleep spindles and _____ are both observed in stage 2 sleep.

3. According to the _____ hypothesis, dreaming is the attempt of the cortex to make sense of the random signals generated during REM sleep.

4. While adaptation theories of sleep focus on when we sleep, _____ theories focus largely on the restorative function of sleep.

5. _____ is a confounding factor when interpreting research on sleep deprivation.

6. Sleep deprivation or even moderate sleep difficulties over consecutive nights can cause _____ and behavioral disturbances.

7. The effects of sleep loss on cognitive functions are largely observed when using tasks that assess _____ functions.

8. REM-sleep-deprived participants exhibit REM _____ over the next two or three nights, suggesting that REM sleep is regulated separately from the amount of slow-wave sleep.

Scan Your Brain answers: (1) alpha, (2) K complexes, (3) activation-synthesis, (4) recuperation, (5) Stress, (6) physiological, (7) executive, (8) rebound.

Circadian Sleep Cycles

This module explores the topic of circadian rhythms. We begin by describing what circadian rhythms are, then we move on to examples of what happens when circadian rhythms are disrupted. The module concludes with a discussion of the neural mechanisms and genetics of circadian rhythms.

Circadian Rhythms

LO 14.19 Describe the circadian sleep–wake cycle and the role of zeitgebers in maintaining circadian rhythms.

The world in which we live cycles from light to dark and back again once every 24 hours. Most surface-dwelling species have adapted to this regular change in their environment with a variety of **circadian rhythms** (*circadian* means "lasting about a day"). For example, most species display a regular circadian sleep–wake cycle. Humans take advantage of the light of day to take care of their biological needs, and then they sleep for much of the night; in contrast, *nocturnal animals*, such as rats, sleep for much of the day and stay awake at night.

Although the sleep–wake cycle is the most obvious circadian rhythm, it is difficult to find a physiological, biochemical, or behavioral process in animals that does not display some measure of circadian rhythmicity (see McGinnis & Young, 2016; Todd et al., 2018; Turek, 2016; Sassone-Corsi, 2016). Each day, our bodies adjust themselves in a variety of ways to meet the demands of the two environments in which we live: light and dark.

Our circadian cycles are kept right on their every-24-hours schedule by temporal cues in the environment. The most important of these cues for the regulation of mammalian circadian rhythms is the daily cycle of light and dark. Environmental cues, such as the light–dark cycle, that can *entrain* (control the timing of) circadian rhythms are called **zeitgebers** (pronounced "ZITE-gay-bers"), a German word that means "time givers." In controlled laboratory environments, it is possible to lengthen or shorten circadian cycles somewhat by adjusting the duration of the light–dark cycle; for example, when exposed to alternating 11.5-hour periods of light and 11.5-hour periods of dark, subjects' circadian cycles begin to conform to a 23-hour day. In a world without 24-hour cycles of light and dark, other *zeitgebers* can entrain circadian cycles. For example, the circadian sleep–wake cycles of hamsters living in continuous darkness or in continuous light can be entrained by regular daily bouts of social interaction, hoarding, eating, or exercise (see Mistlberger, 2011). Hamsters display particularly clear circadian cycles and thus are frequent subjects of research on circadian rhythms.

Free-Running Circadian Sleep–Wake Cycles

LO 14.20 Describe free-running rhythms and internal desynchronization and explain why they are incompatible with recuperation theories of sleep.

What happens to sleep–wake cycles and other circadian rhythms in an environment that is devoid of *zeitgebers*? Remarkably, under conditions in which there are absolutely no temporal cues, humans and other animals maintain all of their circadian rhythms. Circadian rhythms in constant environments are said to be **free-running rhythms**, and their duration is called the **free-running period**. Free-running periods vary in length from individual to individual, are of relatively constant duration within a given individual, and are usually longer than 24 hours—about 24.2 hours is typical in humans living under constant moderate illumination. It seems that we all have an internal *biological clock* that habitually runs a little slow unless it is entrained by time-related cues in the environment.

A typical free-running circadian sleep–wake cycle is illustrated in Figure 14.8. Notice its regularity. Without any external cues, this man fell asleep at intervals of approximately 25.3 hours for an entire month. The regularity of free-running sleep–wake cycles despite variations in physical and mental activity provides support for the dominance of circadian factors over recuperative factors in the regulation of sleep.

Free-running circadian cycles do not have to be learned. Even rats that are born and raised in an unchanging laboratory environment (in continuous light or in continuous darkness) display regular free-running sleep–wake cycles that are slightly longer than 24 hours (Richter, 1971).

Many animals display a circadian cycle of body temperature that is related to their circadian sleep–wake cycle: They tend to sleep during the falling phase of their circadian body temperature cycle and awaken during its rising phase. However, when subjects are housed in constant laboratory environments, their sleep–wake and body temperature cycles sometimes break away from one another. This phenomenon is called **internal desynchronization** (see Daan, Honma & Honma, 2013). For example, in one human volunteer, the free-running periods of *both* the sleep–wake and body temperature cycles were initially 25.7 hours; then, for some unknown reason, there was an increase in the free-running period of the sleep–wake cycle to 33.4 hours and a decrease in the free-running period of the body temperature cycle to 25.1 hours. The potential for the simultaneous existence of two different free-running periods was the first evidence that there is more than one circadian timing mechanism, and that sleep is not causally related to the decreases in body temperature normally associated with it.

Figure 14.8 A free-running circadian sleep–wake cycle 25.3 hours in duration. Despite living in an unchanging environment with no time cues, the man went to sleep each day approximately 1.3 hours later than he had the day before.

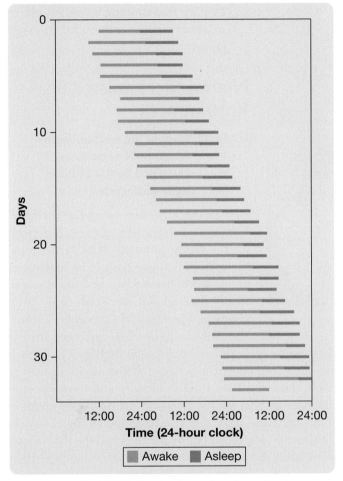

Based on Wever, R. A. (1979). The circadian system of man. Seewiesen-Andechs, Germany: Max-Planck-Institut für Verhaltensphysiologie.

There is another point about free-running circadian sleep–wake cycles that is incompatible with recuperation theories of sleep. On occasions when volunteers stay awake longer than usual, the following sleep time is shorter rather than longer (Wever, 1979). Humans and other animals are programmed to have sleep–wake cycles of approximately 24 hours; hence, the more wakefulness there is during a cycle, the less time there is for sleep.

Jet Lag and Shift Work

LO 14.21 Describe the disruptive effects of jet lag and shift work on circadian rhythmicity and how one can minimize such effects.

People in modern industrialized societies are faced with two different disruptions of circadian rhythmicity: jet lag and shift work. **Jet lag** occurs when the *zeitgebers* that control the phases of various circadian rhythms are accelerated during east-bound flights (*phase advances*) or decelerated during

west-bound flights (*phase delays*). In *shift work*, the *zeitgebers* stay the same, but workers are forced to adjust their natural sleep–wake cycles in order to meet the demands of changing work schedules. Both of these disruptions produce sleep disturbances, fatigue, general malaise, and deficits on tests of physical and cognitive function (see Lewis, 2015; van Ee et al., 2016). The disturbances can last for many days; for example, it typically takes about 8 days to completely adjust to the phase advance of 8 hours that one experiences on a Seattle-to-London flight.

What can be done to reduce the disruptive effects of jet lag and shift work? Two general approaches have been proposed for the reduction of jet lag: pharmacological and nonpharmacological. Pharmacological interventions, especially melatonin or melatonin agonists taken before bedtime at the new location, have been shown effective (see Reid & Abbott, 2015). Nonpharmacological interventions, such as light exposure prior to travel that is set to match the desired wake time at the destination, have limited support for their effectiveness at this time (see Bin, Postnova, & Cistulli, 2019).

Companies that employ shift workers have had success in improving the productivity and job satisfaction of those workers by scheduling phase delays rather than phase advances; whenever possible, shift workers are transferred from their current schedule to one that begins later in the day. It is much more difficult to go to sleep 4 hours earlier and get up 4 hours earlier (a phase advance) than it is to go to sleep 4 hours later and get up 4 hours later (a phase delay).

A Circadian Clock in the Suprachiasmatic Nuclei

LO 14.22 Describe the research that led to the discovery of a circadian clock in the suprachiasmatic nucleus (SCN) of the hypothalamus.

The fact that circadian sleep–wake cycles persist in the absence of temporal cues from the environment indicates that the physiological systems that regulate sleep are controlled by an internal timing mechanism—the **circadian clock**.

The first breakthrough in the search for the circadian clock was Richter's 1967 discovery that large medial hypothalamic lesions disrupt circadian cycles of eating, drinking, and activity in rats. Next, specific lesions of the **suprachiasmatic nuclei (SCN)** of the medial hypothalamus were shown to disrupt various circadian cycles, including sleep–wake cycles (see Figure 14.9). Although SCN lesions do not greatly affect the amount of time mammals spend sleeping, they do abolish the circadian periodicity of sleep cycles. Further support for the conclusion that the suprachiasmatic nuclei contain a circadian timing mechanism comes from the observation that the nuclei display circadian cycles of electrical, metabolic, and biochemical

Figure 14.9 Location of suprachiasmatic nuclei (SCN).

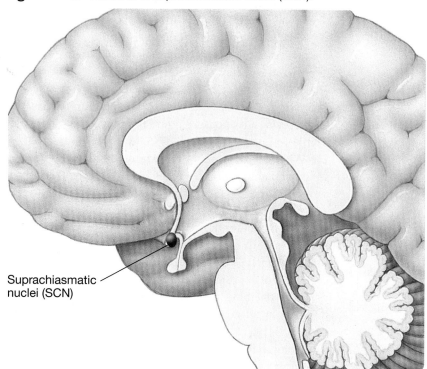

Suprachiasmatic nuclei (SCN)

activity that can be entrained by the light–dark cycle (see Pauls et al., 2016; Sollars & Pickard, 2015; Zelinski, Deibel, & McDonald, 2014).

If there was any lingering doubt about the location of the circadian clock, it was eliminated by the brilliant experiment of Ralph and colleagues (1990). They removed the SCN from the fetuses of a strain of mutant hamsters that had an abnormally short (20-hour) free-running sleep–wake cycle. Then, they transplanted the SCN into normal adult hamsters whose free-running sleep–wake cycles of 25 hours had been abolished by SCN lesions. These transplants restored free-running sleep–wake cycles in the recipients; but, remarkably, the cycles were about 20 hours long rather than the original 25 hours. Transplants in the other direction—that is, from normal hamster fetuses to SCN-lesioned adult mutants—had the complementary effect: They restored free-running sleep–wake cycles to about 25 hours long rather than the original 20 hours.

Although the suprachiasmatic nuclei are unquestionably the major circadian clocks in mammals, they are not the only ones (see Rosenwasser & Turek, 2015). Three lines of experiments, largely conducted in the 1980s and 1990s, pointed to the existence of other circadian timing mechanisms:

- Under certain conditions, bilateral SCN lesions have been shown to leave some circadian rhythms unaffected while abolishing others.
- Bilateral SCN lesions do not eliminate the ability of all environmental stimuli to entrain circadian rhythms (see Saper, 2013); for example, SCN lesions can block entrainment by light but not by regular food or water availability (see Mistlberger, 2011).
- Just like suprachiasmatic neurons, cells from other parts of the body often display free-running circadian cycles of activity when maintained in tissue culture.

Neural Mechanisms of Entrainment

LO 14.23 Explain the mechanism by which SCN neurons are entrained by the 24-hour light–dark cycles.

How do the SCN control circadian rhythms? The timing mechanisms of the SCN depend on the firing patterns of SCN neurons. Many SCN neurons tend to be inactive at night, start to fire at dawn, and fire at a slow, steady pace all day (see Pauls et al., 2016; Hastings, Maywood, & Brancaccio, 2018).

How does the 24-hour light–dark cycle entrain the sleep–wake cycle and other circadian rhythms? To answer this question, researchers began at the obvious starting point: the eyes. They tried to identify and track the specific neurons that left the eyes and carried the information about light and dark that entrained the biological clock. Cutting the *optic nerves* before they reached the *optic chiasm* eliminated the ability of the light–dark cycle to entrain circadian rhythms; however, when the *optic tracts* were cut at the point where they left the optic chiasm, the ability of the light–dark cycle to entrain circadian rhythms was unaffected. As Figure 14.10 illustrates, these two findings indicated that visual axons critical for the entrainment of circadian rhythms branch off from the optic nerve in the vicinity of the optic chiasm. This finding led to the discovery of the *retinohypothalamic tracts*, which leave the optic chiasm and project to the adjacent suprachiasmatic nuclei (see Junko et al., 2019).

Surprisingly, although the retinohypothalamic tracts mediate the ability of light to entrain circadian rhythms, neither rods nor cones are necessary for the entrainment. The critical photoreceptors have proven to be neurons, *retinal ganglion cells* with distinctive functional properties (see Lazzerini Ospri, Prusky, & Hattar, 2017; Roecklein et al., 2013). During the course of evolution, these photoreceptors have sacrificed the ability to respond quickly and briefly to rapid changes of light in favor of the ability to respond consistently to slowly changing levels of background illumination. Their photopigment is **melanopsin** (see Lazzerini Ospri, Prusky, & Hattar, 2017; Renna et al., 2015).

Figure 14.10 The discovery of the retinohypothalamic tracts. Neurons from each retina project to both suprachiasmatic nuclei.

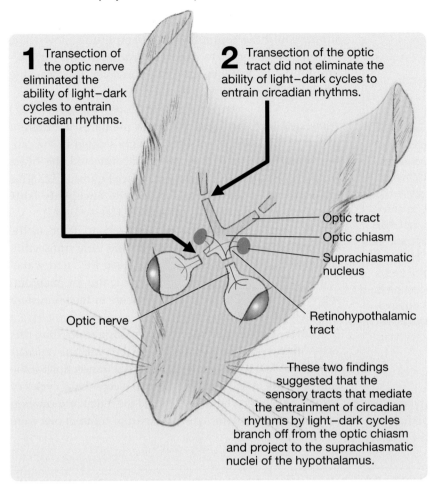

1 Transection of the optic nerve eliminated the ability of light–dark cycles to entrain circadian rhythms.

2 Transection of the optic tract did not eliminate the ability of light–dark cycles to entrain circadian rhythms.

Optic tract
Optic chiasm
Suprachiasmatic nucleus
Optic nerve
Retinohypothalamic tract

These two findings suggested that the sensory tracts that mediate the entrainment of circadian rhythms by light–dark cycles branch off from the optic chiasm and project to the suprachiasmatic nuclei of the hypothalamus.

The identification of circadian genes has led to three important discoveries:

- The same or similar circadian genes have been found in many species of different evolutionary ages (e.g., bacteria, flies, fish, frogs, mice, and humans), indicating that circadian genes evolved early in evolutionary history and have been conserved in various descendant species (Sun et al., 2019).

- Once the circadian genes were discovered, the fundamental molecular mechanism of circadian rhythms was quickly clarified. The key mechanism seems to be gene expression; the transcription of proteins by the circadian genes displays a circadian cycle (see Brancaccio et al., 2014; Musiek & Holtzman, 2016).

- The identification of circadian genes provided a more direct method of exploring the circadian timing capacities of parts of the body other than the SCN. Molecular circadian timing mechanisms similar to those in the SCN exist in most cells of the body (see Masri & Sassone-Corsi, 2013; Patke, Young, & Axelrod, 2019). Although most cells contain potential circadian timing mechanisms, these cellular clocks are normally regulated by neural and hormonal signals from the SCN.

Genetics of Circadian Rhythms

LO 14.24 Understand the genetics of circadian rhythms and the important discoveries that have resulted from the discovery of circadian genes.

An important breakthrough in the study of circadian rhythms came in 1988 when routine screening of a shipment of hamsters revealed that some of them had abnormally short 20-hour free-running circadian rhythms. Subsequent breeding experiments showed that the abnormality was the result of a genetic mutation, and the gene that was mutated was named **tau** (Ralph & Menaker, 1988; see Arnes et al., 2019).

Although tau was the first mammalian circadian gene to be identified, it was not the first to have its molecular structure characterized. This honor went to *clock*, a mammalian circadian gene discovered in mice. The structure of the clock gene was characterized in 1997, and that of the tau gene was characterized in 2000 (Lowrey et al., 2000). The molecular structures of several other mammalian circadian genes have since been specified (see Brancaccio et al., 2014; Cox & Takahashi, 2019; FitzGerald, 2014).

Four Areas of the Brain Involved in Sleep

You have just learned about the neural structures involved in controlling the circadian timing of sleep. This module describes four areas of the brain that are directly involved in producing or reducing sleep. You will learn more about their effects in the later module on sleep disorders.

Two Areas of the Hypothalamus Involved in Sleep

LO 14.25 Describe the research that led to the identification of the anterior and posterior hypothalamus as brain regions involved in the regulation of sleep and wakefulness.

It is remarkable that two areas of the brain involved in the regulation of sleep were discovered early in the 20th century—long before the advent of modern behavioral neuroscience. The discovery was made by Baron Constantin von Economo, a Viennese neurologist (see Scammell, Arrigoni, & Lipton, 2017; Weber & Dan, 2016).

The Case of Constantin von Economo, the Insightful Neurologist

During World War I, the world was swept by a serious viral infection of the brain: *encephalitis lethargica*. Many of its victims slept almost continuously. Baron Constantin von Economo discovered that the brains of deceased victims who had problems with excessive sleep all had damage in the *posterior hypothalamus* and adjacent parts of the midbrain. He then turned his attention to the brains of a small group of victims of encephalitis lethargica who had had the opposite sleep-related problem: In contrast to most victims, they had difficulty sleeping. He found that the brains of the deceased victims in this minority always had damage in the *anterior hypothalamus* and adjacent parts of the basal forebrain. On the basis of these clinical observations, von Economo concluded that the posterior hypothalamus promotes wakefulness, whereas the anterior hypothalamus promotes sleep.

Since von Economo's discovery of the involvement of the posterior hypothalamus and the anterior hypothalamus in human wakefulness and sleep, respectively, that involvement has been confirmed by lesion and recording studies in experimental animals (see Schwartz & Kilduff, 2015). The locations of the posterior and anterior hypothalamus are shown in Figure 14.11.

Reticular Formation and Sleep

LO 14.26 Describe the research that led to the identification of the reticular formation as a brain region involved in the regulation of sleep and wakefulness.

Another area involved in sleep was discovered through the comparison of the effects of two different brain-stem transections in cats. First, in 1936, Bremer severed the brain stems of cats between their *inferior colliculi* and *superior colliculi* in order to disconnect their forebrains from ascending sensory input (see Figure 14.12). This surgical preparation is called a **cerveau isolé preparation** (pronounced "ser-VOE ees-o-LAY"— literally, "isolated forebrain").

Bremer found that the cortical EEG of the isolated cat forebrains was indicative of almost continuous slow-wave sleep. Only when strong visual or olfactory stimuli were presented (the cerveau isolé has intact visual and olfactory input) could the continuous high-amplitude, slow-wave activity be changed to a **desynchronized EEG**—a low-amplitude, high-frequency EEG. However, this arousing effect barely outlasted the stimuli.

Next, for comparison purposes, Bremer (1937) *transected* (cut through) the brain stems of a different group of cats. These transections were located in the caudal brain stem, and thus, they disconnected the brain from the rest of the nervous system (see Figure 14.12). This experimental preparation is called the **encéphale isolé preparation** (pronounced "on-say-FELL ees-o-LAY").

Although it cut most of the same sensory fibers as the cerveau isolé transection, the encéphale isolé transection did not disrupt the normal cycle of sleep EEG and wakefulness EEG. This suggested that a structure for maintaining wakefulness was located somewhere in the brain stem between the two transections.

Later, two important findings suggested that this wakefulness structure in the brain stem was the *reticular formation*. First, it was shown that partial transections at the cerveau isolé level disrupted normal sleep–wake cycles of cortical EEG only when they severed the reticular formation core of the brain stem; when the partial transections were

Figure 14.11 Two regions of the brain involved in sleep. The anterior hypothalamus and adjacent basal forebrain are thought to promote sleep; the posterior hypothalamus and adjacent midbrain are thought to promote wakefulness.

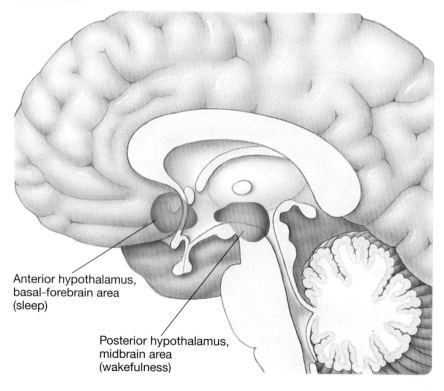

Anterior hypothalamus, basal-forebrain area (sleep)

Posterior hypothalamus, midbrain area (wakefulness)

restricted to more lateral areas, which contain the ascending sensory tracts, they had little effect on the cortical EEG (Lindsey, Bowden, & Magoun, 1949). Second, it was shown that electrical stimulation of the reticular formation of sleeping cats awakened them and produced a lengthy period of EEG desynchronization (Moruzzi & Magoun, 1949).

In 1949, Moruzzi and Magoun considered these four findings together: (1) the effects on cortical EEG of the cerveau isolé preparation, (2) the effects on cortical EEG of the encéphale isolé preparation, (3) the effects of reticular formation lesions, and (4) the effects on sleep of stimulation of the reticular formation. From these four key findings, Moruzzi and Magoun proposed that low levels of activity in the reticular formation produce sleep and that high levels produce wakefulness (see Larson-Prior, Ju, & Galvin, 2014). Indeed, this theory is so widely accepted that the reticular formation is commonly referred to as the **reticular activating system**, even though maintaining wakefulness is only one of the functions of the many nuclei it comprises.

Figure 14.12 Four pieces of evidence that the reticular formation is involved in sleep.

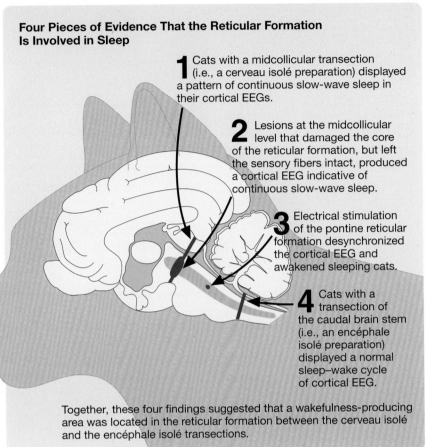

Four Pieces of Evidence That the Reticular Formation Is Involved in Sleep

1 Cats with a midcollicular transection (i.e., a cerveau isolé preparation) displayed a pattern of continuous slow-wave sleep in their cortical EEGs.

2 Lesions at the midcollicular level that damaged the core of the reticular formation, but left the sensory fibers intact, produced a cortical EEG indicative of continuous slow-wave sleep.

3 Electrical stimulation of the pontine reticular formation desynchronized the cortical EEG and awakened sleeping cats.

4 Cats with a transection of the caudal brain stem (i.e., an encéphale isolé preparation) displayed a normal sleep–wake cycle of cortical EEG.

Together, these four findings suggested that a wakefulness-producing area was located in the reticular formation between the cerveau isolé and the encéphale isolé transections.

Reticular REM-Sleep Nuclei

LO 14.27 Discuss how REM sleep is controlled by the reticular formation and what implications this has for understanding the neural mechanisms of behavior.

The fourth area of the brain involved in sleep controls REM sleep and is included in the brain area we have just described—it is part of the caudal reticular formation. It makes sense that an area of the brain involved in maintaining wakefulness would also be involved in the production of REM sleep because of the similarities between the two states. Indeed, REM sleep is controlled by a variety of nuclei scattered throughout the caudal reticular formation. Each site is responsible for controlling one of the major indices of REM sleep (see Liu & Dan, 2019; Peever, Luppi, & Montplaisir, 2014; Weber et al., 2015)—a site for the reduction of core-muscle tone, a site for EEG desynchronization, a site for rapid eye movements, and so on. The approximate location in the caudal brain stem of each of these REM-sleep nuclei is illustrated in Figure 14.13.

Please think for a moment about the broad implications of these various REM-sleep nuclei. In thinking about the brain mechanisms of behavior, many people assume that if there is one name for a behavior, there must be a single

structure for it in the brain; in other words, they assume that evolutionary pressures have acted to shape the human brain according to our current language and theories. Here we see the weakness of this assumption: The brain is organized along different principles, and REM sleep occurs only when a network of independent structures becomes active together. Relevant to this is the fact that the physiological changes that go together to define REM sleep sometimes break apart and go their separate ways—and the same is true of the changes that define slow-wave sleep. For example, during REM-sleep deprivation, penile erections, which normally occur during REM sleep, begin to occur during slow-wave sleep. And during total sleep deprivation, slow waves, which normally occur only during slow-wave sleep, begin to occur during wakefulness. This suggests that REM sleep, slow-wave sleep, and wakefulness are not each controlled by a single mechanism. Each state seems to result from the interaction of several mechanisms that are capable under certain conditions of operating independently of one another.

Journal Prompt 14.3

Someone tells you that the prefrontal cortex is the brain structure responsible for higher cognition. What would your response be?

Figure 14.13 A sagittal section of the brain stem of the cat illustrating the areas that control the various physiological indices of REM sleep.

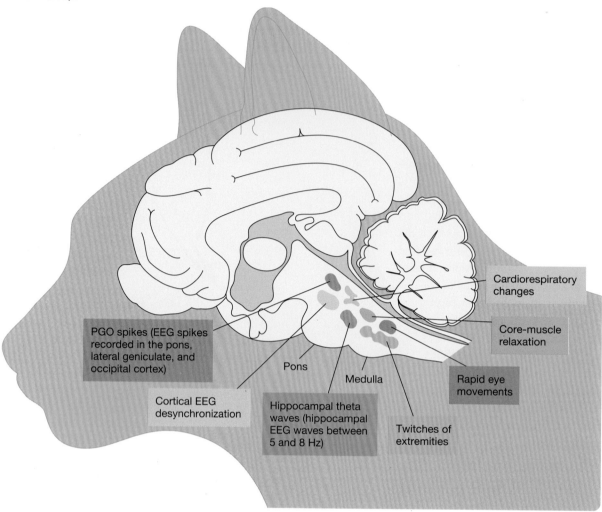

Vertes, R. P. (1983). Brainstem control of the events of REM sleep. *Progress in Neurobiology*, 22, 241–288.

Scan Your Brain

Before continuing with this chapter, scan your brain by completing the following exercise. The correct answers appear at the end of the exercise. Before proceeding, review material related to your errors and omissions.

1. _____ means "lasting about a day" and is largely determined by the light–dark cycle.

2. Cues that control the timing of light–dark cycles are called _____. These can be artificially manipulated in a laboratory setting to prolong or reduce the circadian rhythm.

3. When subjects are housed in constant laboratory environments, their sleep–wake and body temperature cycles sometimes break away from one another. This phenomenon is called _____.

4. Jet lag occurs when light–dark cues that control circadian rhythms are accelerated during _____ flights and decelerated during _____ flights.

5. Circadian rhythms exist in the absence of environmental cues, suggesting that there is an internal timing mechanism commonly referred to as the _____.

6. Lesions of the _____ disrupt sleep–wake cycles.

7. Research found that cutting the optic nerves before they reached the _____ eliminated the ability of light–dark cycle.

8. Critical photoreceptors called _____ are necessary for the entrainment of circadian rhythms.

9. Electrical stimulation of the _____ desynchronized cortical EEG and awakened sleeping cats.

10. According to Bremer's experiments, the _____ preparation did not disturb the normal sleep cycle in cats.

Scan Your Brain answers: (1) Circadian, (2) zeitgebers, (3) internal desynchronization, (4) east-bound, west-bound, (5) circadian clock, (6) suprachiasmatic nuclei, (7) optic chiasm, (8) retinal ganglion cells, (9) reticular formation, (10) encéphale isolé.

Drugs That Affect Sleep

Most drugs that influence sleep fall into two different classes: hypnotic and antihypnotic. **Hypnotic drugs** are drugs that increase sleep; **antihypnotic drugs** are drugs that reduce sleep. A third class of sleep-influencing drugs comprises those that influence its circadian rhythmicity; the main drug of this class is **melatonin**.

Hypnotic Drugs

LO 14.28 Describe three classes of hypnotic drugs. Compare and contrast them in terms of their efficacy and side effects.

The **benzodiazepines** (e.g., diazepam, clonazepam), which are GABA$_A$ agonists, were developed and tested for the treatment of anxiety, yet they are some of the most commonly prescribed hypnotic medications. In the short term, they increase drowsiness, decrease the time it takes to fall asleep, reduce the number of awakenings during a night's sleep, and increase total sleep time (DeKosky & Williamson, 2020). Thus, they can be effective in the treatment of occasional difficulties in sleeping.

Although benzodiazepines can be effective therapeutic hypnotic agents in the short term, their prescription for the treatment of chronic sleep difficulties, though common, is ill-advised (see Atkin, Comai, & Gobbi, 2018; Bourgeois et al., 2014; Mignot, 2013). Five complications are associated with the chronic use of benzodiazepines as hypnotic agents:

- Tolerance develops to the hypnotic effects of benzodiazepines; thus, patients must take larger and larger doses to maintain the drugs' efficacy (see Atkin, Comai, & Gobbi, 2018; Frase et al., 2018).

- Cessation of benzodiazepine therapy after chronic use causes *insomnia* (sleeplessness), which can exacerbate the very problem that the benzodiazepines were intended to correct (see Guina & Merrill, 2018).

- Benzodiazepines distort the normal pattern of sleep; they increase the duration of NREM 2 sleep while decreasing the duration of both slow-wave and REM sleep.

- Benzodiazepines lead to next-day drowsiness (Ware, 2008) and are associated with an increased risk of traffic accidents (see Atkin, Comai, & Gobbi, 2018).

- Most troubling is that chronic use of benzodiazepines has been shown to substantially reduce life expectancy (see Votaw et al., 2019).

Journal Prompt 14.4

Given all the problems associated with the long-term use of benzodiazepines, why do you think they are they so commonly prescribed for the treatment of insomnia?

Problems with the benzodiazepines have led to a search for other pharmacological agents with the same hypnotic effects as the benzodiazepines but fewer side effects. In the early 1990s, a new class of GABA$_A$ agonists, the **imidazopyridines**, was marketed for the treatment of insomnia. It was claimed that they have fewer adverse side effects and less potential for addiction. One of the most widely prescribed imidazopyridines is Zolpidem (see Krystal, 2015; Neubauer, 2014), yet Zolpidem has been found to be no safer or more effective than benzodiazepines (see Arbon, Knurowska, & Dijk, 2015).

Evidence that the raphé nuclei, which are serotonergic, play a role in sleep suggested that serotonergic drugs might be effective hypnotics. Efforts to demonstrate the hypnotic effects of such drugs have focused on **5-hydroxytryptophan (5-HTP)**—the precursor of serotonin—because 5-HTP, but not serotonin, readily passes through the blood–brain barrier. Injections of 5-HTP do reverse the insomnia produced in both cats and rats by the serotonin antagonist PCPA; however, they appear to be of no therapeutic benefit in the treatment of human insomnia.

Antihypnotic Drugs

LO 14.29 Describe three classes of antihypnotic drugs.

The mechanisms of the following three classes of antihypnotic drugs are well understood: *cocaine-derived stimulants, amphetamine-derived stimulants,* and *tricyclic antidepressants.* The drugs in these three classes seem to promote wakefulness by boosting the activity of catecholamines (norepinephrine, epinephrine, and dopamine)—by increasing their release into synapses, by blocking their reuptake from synapses, or both.

The regular use of antihypnotic drugs is risky. Antihypnotics tend to produce a variety of adverse side effects, such as loss of appetite, anxiety, tremor, addiction, and disturbance of normal sleep patterns. Moreover, they may mask the pathology that is causing the excessive sleepiness.

Melatonin

LO 14.30 Understand the relationship between the pineal gland and melatonin, and how melatonin affects sleep.

Melatonin is a hormone synthesized from the neurotransmitter serotonin in the **pineal gland** (see Alghamdi, 2018; Cecon et al., 2015). The pineal gland is an inconspicuous gland that René Descartes, whose dualistic philosophy was discussed in Chapter 2, once believed to be the seat of the soul. The pineal gland is located on the midline of the brain just ventral to the rear portion of the corpus callosum (see Figure 14.14).

Figure 14.14 The location of the pineal gland, the source of melatonin.

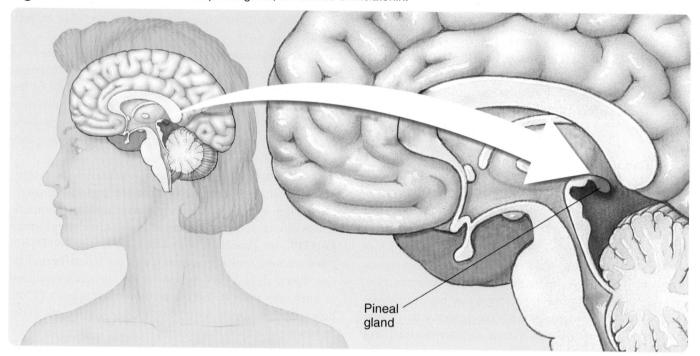

Pineal gland

The pineal gland has important functions in birds, reptiles, amphibians, and fish (see Sapède & Cau, 2013). The pineal gland of these species has inherent timing properties and regulates circadian rhythms and seasonal changes in reproductive behavior through its release of melatonin. In humans and other mammals, however, the functions of the pineal gland and melatonin are not as apparent.

In humans and other mammals, circulating levels of melatonin display circadian rhythms under control of the suprachiasmatic nuclei (see Videnovic et al., 2014; Zisapel, 2018), with the highest levels being associated with darkness and sleep (see Morris, Aeschbach, & Scheer, 2012). On the basis of this correlation, it has long been assumed that melatonin plays a role in promoting sleep or in regulating its timing in mammals.

In order to put the facts about melatonin in perspective, it is important to keep one significant point firmly in mind. In adult mammals, pinealectomy and the consequent elimination of melatonin appear to have little effect. The pineal gland plays a role in the development of mammalian sexual maturity, but its functions after puberty are not at all obvious.

Does *exogenous* (externally produced) melatonin improve sleep, as widely believed? A *meta-analysis* (a combined analysis of results of more than one study) of 19 studies indicated that exogenous melatonin has a slight, but statistically significant, *soporific* (sleep-promoting) effect (Ferracioli-Oda, Qawasmi, & Bloch, 2013).

There is also good evidence that melatonin can shift the timing of mammalian circadian cycles. Indeed, several researchers have argued that melatonin is better classified as a **chronobiotic** (a substance that adjusts the timing of internal biological rhythms) than as a soporific (see Golombek et al., 2015).

Table 14.3 offers a summary of the drugs that affect sleep.

Table 14.3 Summary of Drugs That Affect Sleep

Effects on Sleep	Drug Class or Name
Hypnotic	Benzodiazepines
	Imidazopyridines
	5-hydroxytryptophan (5-HTP)
Antihypnotic	Cocaine-derived stimulants
	Amphetamine-derived stimulants
	Tricyclic antidepressants
Chronobiotic	Melatonin

Sleep Disorders

Many sleep disorders fall into one of two complementary categories: insomnia and hypersomnia. **Insomnia** includes all disorders of initiating and maintaining sleep.

tab62/Fotolia

Hypersomnia includes disorders of excessive sleep or sleepiness. A third major class of sleep disorders includes all those disorders that are specifically related to REM-sleep dysfunction. Both insomnia and hypersomnia are common symptoms of many psychiatric disorders (see Rumble, White, & Benca, 2015).

In various surveys, approximately 30 percent of respondents report significant sleep-related problems. However, it is important to recognize that complaints of sleep problems often come from people whose sleep appears normal in laboratory sleep tests. For example, many people normally sleep 6 hours or less a night and seem to do well sleeping that amount, but they are pressured by their doctors, their friends, and their own expectations to sleep more (e.g., at least 8 hours). As a result, they spend more time in bed than they should and have difficulty getting to sleep. Often, the anxiety associated with their inability to sleep makes it even more difficult for them to sleep (see Harvey & Tang, 2012; Khoury & Doghramji, 2015). Such patients can often be helped by counseling that persuades them to go to bed only when they are very sleepy. Others with disturbed sleep have more serious problems (see Khoury & Doghramji, 2015).

Insomnia

LO 14.31 Describe four causes of insomnia.

Many cases of insomnia are **iatrogenic** (physician-created)—in large part because sleeping pills (e.g., benzodiazepines), which are usually prescribed by physicians, are a major cause of insomnia. At first, hypnotic drugs may be effective in increasing sleep, but soon the patient may become trapped in a rising spiral of drug use, as *tolerance* to the drug develops and progressively more of it is required to produce its original hypnotic effect. Soon, the patient cannot stop taking the drug without running the risk of experiencing *withdrawal symptoms*, which include insomnia. The case of Mr. B. illustrates this problem (Dement, 1978).

Mr. B., the Case of Iatrogenic Insomnia

Mr. B. was studying for an exam, the outcome of which would affect his life. He was under stress and found it difficult to sleep. He consulted his physician who prescribed a moderate dose of barbiturate (i.e., sodium amytal) at bedtime. Mr. B. found the medication to be effective for a few nights, but after a week, he started having trouble sleeping again. So, he decided to take two sleeping pills each night. Twice more this cycle was repeated, so that on the night before his exam he was taking four times the original dose.

The next night, with the pressure off, Mr. B. took no pills, but he couldn't sleep. Accordingly, Mr. B. decided he had a serious case of insomnia and returned to the pills. By the time he consulted a sleep clinic several years later, he was taking approximately 1,000 mg sodium amytal every night, and his sleep was more disturbed than ever.

Patients may go on for years with ever-increasing doses of one medication after another, never realizing that their troubles are caused by the pills.

In one study, individuals with insomnia claimed to take an average of 1 hour to fall asleep and to sleep an average of only 4.5 hours per night; but when they were tested in a sleep laboratory, they were found to have an average *sleep latency* (time to fall asleep) of only 15 minutes and an average nightly sleep duration of 6.5 hours. It used to be common medical practice to assume that people who claimed to suffer from insomnia but slept more than 6.5 hours per night were neurotic. However, this practice stopped when some of those diagnosed with *neurotic pseudoinsomnia* were subsequently found to be suffering from sleep apnea, nocturnal myoclonus, or other sleep-disturbing problems. Insomnia is not necessarily a problem of too little sleep; it is often a problem of too little undisturbed sleep.

The insomnia associated with **sleep apnea** is well documented. The patient with sleep apnea stops breathing many times each night. Each time, the patient awakens, begins to breathe again, and drifts back to sleep. Sleep apnea usually leads to a sense of having slept poorly and is thus often diagnosed as insomnia. However, some patients are totally unaware of their multiple awakenings and instead complain of excessive sleepiness during the day, which can lead to a diagnosis of *hypersomnia*. Sleep apnea disorders are of two types: (1) *obstructive sleep apnea* results from obstruction of the respiratory passages by muscle spasms or *atonia* (lack of muscle tone) and often occurs in individuals who are vigorous snorers (see Yenigun et al., 2020); (2) *central sleep apnea* results from the failure of the central nervous system to stimulate respiration (see Dempsey, 2019). Sleep apnea is more common in males, in people who are overweight, and in the elderly (see Badran et al., 2015; Mansur et al., 2019).

Two other specific causes of insomnia are related to the legs: periodic limb movement disorder and restless legs syndrome. **Periodic limb movement disorder** is characterized by periodic, involuntary movements of the limbs, often involving twitches of the legs during sleep. Most patients suffering from this disorder complain of poor sleep and daytime sleepiness but are unaware of the nature of their problem. In contrast, people with **restless legs syndrome** are all too aware of their problem. They complain of a hard-to-describe tension or uneasiness in their legs that keeps them from falling asleep (see Lai et al., 2017). Once established, both of these disorders are chronic (see Ondo, 2019; Stevens, 2015). There are no effective treatments for these disorders, although L-dopa (see Chapter 10) can help some patients (see Ondo, 2019).

One of the most effective treatments for insomnia is *sleep restriction therapy* (see Kay-Stacey & Attarian, 2016; Trauer et al., 2015). First, the amount of time that an individual with insomnia is allowed to spend in bed is substantially reduced. Then, after a period of sleep restriction, the amount of time spent in bed is gradually increased in small increments, as long as sleep latency remains in the normal range. Even people with severe insomnia can benefit from this treatment.

Hypersomnia

LO 14.32 Describe the symptoms of narcolepsy and the role of orexin (hypocretin) in this disorder.

Narcolepsy is the most widely studied disorder of hypersomnia. It occurs in about 1 out of 2,000 individuals (see Arango, Kivity, & Schoenfeld, 2015; Mahoney et al., 2019) and has two prominent symptoms. First, persons with narcolepsy experience severe daytime sleepiness and repeated, brief (10- to 15-minute) daytime sleep episodes. Individuals with narcolepsy typically sleep only about an hour per day more than average; it is the inappropriateness of their sleep episodes that most clearly defines their condition. Most of us occasionally fall asleep on the beach, in front of the television, or in that most soporific of all daytime sites—the large, stuffy, dimly lit lecture hall. But individuals with narcolepsy fall asleep in the middle of a conversation, while eating, while scuba diving, or even while having sex.

The second prominent symptom of narcolepsy is cataplexy (see Dauvilliers et al., 2014). **Cataplexy** is characterized by recurring losses of muscle tone during wakefulness, often triggered by an emotional experience. In its mild form, it may simply force the patient to sit down for a few seconds until it passes. In its extreme form, the patient drops to the ground as if shot and remains there for a minute or two, fully conscious.

In addition to the two prominent symptoms of narcolepsy (daytime sleep attacks and cataplexy), people with narcolepsy often experience two other symptoms: sleep paralysis and hypnagogic hallucinations. **Sleep paralysis** is the inability to move just as one is falling asleep or waking up (see Sharpless & Klíková, 2019). **Hypnagogic hallucinations** are dreamlike experiences during wakefulness. Many healthy people occasionally experience sleep paralysis and hypnagogic hallucinations (see Denis, French, & Gregory, 2018; Larøi et al., 2019). Have you experienced them?

Three lines of evidence suggested to early researchers that narcolepsy results from an abnormality in the mechanisms that trigger REM sleep. First, unlike people without narcolepsy, those with narcolepsy often go directly into REM sleep when they fall asleep. Second and third, as you have already learned, narcoleptics often experience two REM-sleep characteristics (dreamlike states and loss of muscle tone) during wakefulness (see Mahoney et al., 2019).

Some of the most exciting research on the neural mechanisms of sleep in general and narcolepsy in particular began with the study of a strain of narcoleptic dogs. After 10 years of studying the genetics of these narcoleptic dogs, Lin and colleagues (1999) finally isolated the gene that causes the disorder. The gene encodes a receptor protein that binds to a neuropeptide called **orexin** (sometimes called *hypocretin*; Mieda, 2017; Richter, Woods, & Schier, 2014), which exists in two forms: orexin-A and orexin-B. Although the discovery of the orexin gene has drawn attention to genetic factors in narcolepsy, the concordance rate for narcolepsy in monozygotic twins is only about 25 percent (see Liblau et al., 2015).

Several studies have documented reduced levels of orexin in the cerebrospinal fluid of individuals living with narcolepsy and in the brains of deceased individuals who had narcolepsy (see Nishino & Kanbayashi, 2005). Also, the number of orexin-releasing neurons has been found to be reduced in the brains of persons with narcolepsy (see Mahoney et al., 2019; Shan, Dauvilliers, & Siegel, 2015). One popular explanation for the loss of orexin-releasing neurons in narcolepsy is that it is the result of an autoimmune response (see Arango, Kivity, & Shoenfeld, 2015; Liblau et al., 2015; Mahoney et al., 2019).

Where is orexin synthesized in the brain? Orexin is synthesized by neurons in the region of the hypothalamus that has been linked to the promotion of wakefulness: the posterior hypothalamus. The orexin-producing neurons project diffusely throughout the brain, but they show many connections with neurons in the other wakefulness-promoting area of the brain: the reticular formation. Currently, there is considerable interest in understanding the role of the orexin circuits in wakefulness and REM sleep (see Mignot, 2013; Mahoney et al., 2019).

Narcolepsy has traditionally been treated with stimulants (e.g., amphetamine, methylphenidate), but these have a potential for addiction and produce many undesirable side effects. The antihypnotic stimulant modafinil has been shown to be effective in the treatment of some cases of narcolepsy, and certain antidepressants can be effective against cataplexy (see Scammell, 2015).

REM-Sleep-Related Disorders

LO 14.33 Describe one REM-sleep-related disorder and its presumed neural mechanisms.

Several sleep disorders are specific to REM sleep; these are classified as *REM-sleep-related disorders*. Even narcolepsy, which is usually classified as a hypersomnic disorder, can be considered to be a REM-sleep-related disorder—for reasons you have just encountered.

Occasionally, patients who have little or no REM sleep are discovered. Although this disorder is rare, it is important because of its theoretical implications. Lavie and colleagues (1984) described a patient who had suffered a brain injury that presumably involved damage to the REM-sleep controllers in the caudal reticular formation. The most important finding of this case study was that the patient did not appear to be adversely affected by his lack of REM sleep. After receiving his injury, he completed high school, college, and law school and established a thriving law practice.

Some patients experience REM sleep without core-muscle atonia. This condition is known as **REM-sleep behavior disorder** (see Schenck et al., 2013; Högl, Stefani, & Videnovic, 2017) and is common in individuals with Parkinson's disease (see Högl, Stefani, & Videnovic, 2017; Neikrug et al., 2014; Tekriwal et al., 2016). It has been suggested that the function of REM-sleep atonia is to prevent the acting out of dreams. This theory receives support from case studies of people who suffer from this disorder—case studies such as the following one (Schenck et al., 1986).

The Case of the Sleeper Who Ran Over Tackle

(The patient was dreaming about football.) The quarterback lateraled the ball to me and I was supposed to cut back over tackle and—this is very vivid—as I cut back there is this 280-pound tackle waiting, so I gave him my shoulder and knocked him out of the way. When I awoke I was standing in front of our dresser. I had knocked lamps, mirrors, and everything off the dresser; hit my head against the wall; and banged my knee against the dresser.

Effects of Long-Term Sleep Reduction

When people sleep less than they are used to sleeping, they do not feel or function well. We are sure you have experienced these effects. But what do they mean? Most people—nonexperts and experts alike—believe the adverse effects of sleep loss indicate that we need the sleep we typically get. However, there is an alternative interpretation—one that is more consistent with the plasticity of the adult human brain. Perhaps the brain needs a small amount of sleep each day but will sleep much more under ideal conditions because of sleep's high positive incentive value. The brain then slowly adapts to the amount of sleep it is getting—even though this amount may be far more than it needs—and is disturbed when there is a sudden reduction.

Fortunately, there are ways to determine which of these two interpretations of the effects of sleep loss is correct. The key is to study individuals who sleep little, either because they have always done so or because they have purposefully reduced their sleep times. If people need at least 8 hours of sleep each night, short sleepers should be suffering from a variety of health and performance problems. Before we summarize the results of this key area of research, we want to emphasize one point: Because they are so time-consuming, few studies of long-term sleep patterns have been conducted, and some of those that have been conducted are not sufficiently thorough. Nevertheless, there have been enough of them for a clear pattern of results to have emerged. We think they will surprise you.

This final module begins with a comparison of short and long sleepers. Then, it discusses two kinds of long-term sleep-reduction studies: studies in which volunteers reduced the amount they slept each night and studies in which volunteers reduced their sleep by restricting it to naps. Next comes a discussion of studies that have examined the relation between sleep duration and health. Finally, I (JP) relate my own experience of long-term sleep reduction.

Differences between Short and Long Sleepers

LO 14.34 List the main differences between short and long sleepers.

Numerous studies have compared short sleepers (those who sleep 6 hours or less per night) and long sleepers (those who sleep 8 hours or more per night). We focus here on the 2004 study of Fichten and colleagues because it is the most thorough. The study had three strong features:

- It included a relatively large sample (239) of adult short sleepers and long sleepers.

- It compared short and long sleepers in terms of 48 different measures, including daytime sleepiness, daytime naps, regularity of sleep times, busyness, regularity of meal times, stress, anxiety, depression, life satisfaction, and worrying.

- Before the study began, the researchers carefully screened out volunteers who were ill or under various kinds of stress or pressure; thus, the study was conducted with a group of healthy volunteers who slept the amount that they felt was right for them.

The findings of Fichten and colleagues are nicely captured by the title of their paper, "Long sleepers sleep more and short sleepers sleep less." In other words, other than the differences in sleep time, there were no differences between the two groups on any of the other measures—no indication that the short sleepers were suffering in any way from their shorter sleep time. Fichten and colleagues report that these results are consistent with most previous comparisons of short and long sleepers (e.g., Monk et al., 2001), except for a few studies that did not screen out participants who were sleeping little because they were under stress (e.g., from worry, illness, or a demanding work schedule). Those studies did report some negative characteristics in the short-sleep group, which likely reflected the stress experienced by some in that group.

Long-Term Reduction of Nightly Sleep

LO 14.35 Describe the results of studies of long-term reduction of nightly sleep.

Are short sleepers able to live happy, productive lives because they are genetically predisposed to be short sleepers, or is it possible for average people to adapt to a short sleep schedule? There have been only two published studies in which healthy volunteers have reduced their nightly sleep for several weeks or longer. In one (Webb & Agnew, 1974), a group of 16 volunteers slept for only 5.5 hours per night for 60 days, with only one detectable deficit on an extensive battery of mood, medical, and performance tests: a slight deficit on a test of vigilance.

In the other systematic study of long-term nightly sleep reduction (Friedman et al., 1977; Mullaney et al., 1977), eight volunteers reduced their nightly sleep by 30 minutes every 2 weeks until they reached 6.5 hours per night, then by 30 minutes every 3 weeks until they reached 5 hours, and then by 30 minutes every 4 weeks thereafter. After a participant indicated a lack of desire to reduce sleep further, the person spent 1 month sleeping the shortest duration of nightly sleep that had been achieved. Finally, each participant slept at the shortest duration plus 30 minutes for 1 year.

The minimum duration of nightly sleep achieved during this experiment was 5.5 hours for 2 participants, 5.0 hours for 4 participants, and an impressive 4.5 hours for 2 participants. In each participant, a reduction in sleep time was associated with an increase in sleep efficiency: a decrease in the amount of time it took to fall asleep after going to bed, a decrease in the number of nighttime awakenings, and an increase in the proportion of slow-wave sleep. After the participants had reduced their sleep to 6 hours per night, they began to experience daytime sleepiness, and this became a problem as sleep time was further reduced. Nevertheless, there were no deficits on any of the mood, medical, or performance tests administered throughout the experiment. The most encouraging result was that an unexpected follow-up 1 year after the end of the study found that all participants were sleeping less than they had before the study—between 7 and 18 hours less each week—with no excessive sleepiness.

Long-Term Sleep Reduction by Napping

LO 14.36 Describe the results of studies of long-term reduction of sleep by napping.

Most mammals and human infants display **polyphasic sleep cycles**; that is, they regularly sleep more than once per day. In contrast, most adult humans display **monophasic sleep cycles**; that is, they sleep once per day. Nevertheless, most adult humans do display polyphasic cycles of sleepiness, with periods of sleepiness occurring in late afternoon and late morning. Have you ever experienced them?

Do adult humans need to sleep in one continuous period per day, or can they sleep effectively in several naps as human infants and other mammals do? Which of the two sleep patterns is more efficient? Research has shown that naps have recuperative powers out of proportion with their brevity (see Ru et al., 2019), suggesting that polyphasic sleep might be particularly efficient.

Interest in the value of polyphasic sleep was stimulated by the legend that Leonardo da Vinci managed to generate a steady stream of artistic and engineering accomplishments during his life by napping for 15 minutes every 4 hours, thereby limiting his sleep to 1.5 hours per day. As unbelievable as this sleep schedule may seem, it has been replicated in several studies (see Stampi, 1992). Here are the main findings of these truly mind-boggling studies: First, participants required several weeks to adapt to a polyphasic sleep schedule. Second, once adapted to polyphasic sleep, participants were content and displayed no deficits on the performance tests they were given. Third, Leonardo's 4-hour schedule worked quite well, but in unstructured working situations (e.g., around-the-world solo sailboat races), individuals often varied the duration of the cycle without feeling negative consequences. Fourth, most people displayed a strong preference for particular sleep durations (e.g., 25 minutes) and refrained from sleeping too little, which left them unrefreshed, or too much, which left them groggy for several minutes when they awoke—an effect called **sleep inertia** (see Hilditch & McHill, 2019). Fifth, when individuals first adopted a polyphasic sleep cycle, most of their sleep was slow-wave sleep, but eventually they returned to a mix of REM and slow-wave sleep.

The following are the paraphrased words of artist Giancarlo Sbragia (1992), who adopted Leonardo's purported sleep schedule:

> At first, the schedule was difficult to follow. It took about 3 weeks to get used to. But I soon reached a point of comfort, and it turned out to be a thrilling experience. How beautiful my life became: I discovered dawns, silence, and concentration. I had far more time for myself, for painting, and for developing my career.

Effects of Shorter Sleep Times on Health

LO 14.37 Recognize how shorter sleep times relate to longevity.

For decades, it was believed that sleeping 8 hours or more per night is ideal for promoting optimal health and longevity. Then, a series of large-scale epidemiological studies conducted around the world challenged this belief (see Liu et al., 2017). These studies did not include participants who were a potential source of bias, for example, people who slept little because they were ill, depressed, or under stress. The studies started with a sample of healthy volunteers and followed their health for several years.

Liu et al. (2017) conducted a meta-analysis of 40 studies that collectively followed 2,200,425 participants for an average duration of 12 years. Figure 14.15 presents a summary of their results. You will immediately see that sleeping less than 8 hours per night does not increase the risk of death. However, sleeping more than 8 hours per night does, and dramatically.

Because these epidemiological data are correlational, it is important not to interpret them causally. They do not prove that sleeping 9 or more hours a night causes health problems: Perhaps there is something about people who sleep 9 hours or more per night that leads them to die sooner than people who sleep less. Thus, these studies do not prove that reducing your sleep will cause you to live longer. These studies do, however, provide strong evidence that sleeping less than 8 hours is not the risk to life and health that it is often made out to be. Consistent with this idea is the recent finding that adults who live in pre-industrial hunter-gatherer societies tend to sleep less than 8 hours per day—the sleep times in these societies range from 5.7 to 7.1 hours per day—yet they have a lower incidence

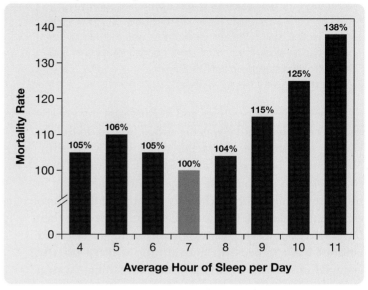

Figure 14.15 The mortality rates associated with different amounts of sleep, based on the data of 2,200,425 volunteers over an average duration of 12 years. The mortality rate at 7 hours of sleep per night has been arbitrarily set at 100 percent, and the other mortality rates are presented in relation to it.

Liu, T. Z., Xu, C., Rota, M., Cai, H., Zhang, C., Shi, M. J., ... & Sun, X. (2017). Sleep duration and risk of all-cause mortality: a flexible, non-linear, meta-regression of 40 prospective cohort studies. *Sleep Medicine Reviews*, 32, 28–36.

of many illnesses and higher levels of physical fitness than people living in industrial societies (Yetish et al., 2015).

Long-Term Sleep Reduction: A Personal Case Study

We began this chapter 4 weeks ago with both zeal and trepidation. One of us (JP) was fascinated by the idea that one could wring 2 or 3 extra hours of living out of each day by sleeping less, and I hoped that adhering to a sleep-reduction program while writing about sleep would create an enthusiasm for the subject that would color our writing and be passed on to you. I began with a positive attitude because I was aware of the relevant evidence; still, I was more than a little concerned about the negative effect that reducing my sleep by 3 hours per night might have on my writing.

The Case of the Author Who Reduced His Sleep

Rather than using a gradual stepwise reduction method, I jumped directly into my 5-hours-per-night sleep schedule. This proved to be less difficult than you might think. I took advantage of a trip to the East Coast from my home on the West Coast to reset my circadian clock. While I was in the East, I got up at 7:00 a.m., which is 4:00 a.m. on the West Coast, and I just kept on the same schedule when I got home. I decided to add my extra waking hours to the beginning of my day rather than to the end so there would be no temptation for me to waste them—there are not too many distractions around my university at 5:00 a.m.

Figure 14.16 is a record of my sleep times for the 4-week period that it took us to write a first draft of this chapter. I didn't quite meet my goal of sleeping less than 5 hours every night, but I didn't miss by much: My overall mean was 5.05 hours per night. Notice that in the last week, there was a tendency for my circadian clock to run a bit slow; I began sleeping in until 4:30 a.m. and staying up until 11:30 p.m.

What were the positives and negatives of my experience? The main positive was the added time to do things: Having an extra 21 hours per week was wonderful. Furthermore, because my daily routine was out of synchrony with everybody else's, I spent little time sitting in traffic or waiting in lines. The only negative of the experience was sleepiness. It was no problem during the day, when I was active. However, staying awake during the last hour before I went to bed—an hour during which I usually engaged in sedentary activities, such as reading—was at times a problem. This is when I became personally familiar with the phenomenon of microsleeps, and it was then that I required some assistance in order to stay awake. Each night of sleep became a highly satisfying but all too brief experience.

We began this chapter with this question: How much sleep do we need? Then, we gave you our best professorial it-could-be-this, it-could-be-that answer. However, that was a month ago. Now, after one of us experienced sleep reduction firsthand and we have reviewed the evidence, we are less inclined toward wishy-washiness

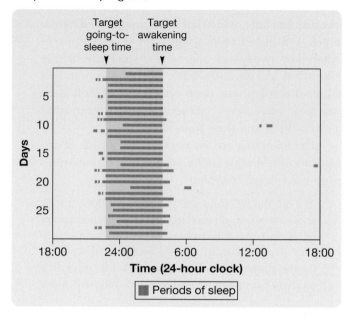

Figure 14.16 Sleep record of Pinel during a 4-week sleep-reduction program.

on the topic of sleep. The fact that most committed volunteers who are active during the day can reduce their sleep to about 5.5 hours per night without great difficulty or major adverse consequences suggested to us that the answer is 5.5 hours of sleep—not substantially different from the 6-hour daily sleep requirement advocated by one esteemed sleep expert (see Horne, 2010, 2011). However, we reached this conclusion before we learned about polyphasic sleep schedules. Now, we must revise our estimate downward.

Themes Revisited

The thinking creatively theme pervaded this chapter. In this chapter, you learned that many people sleep little with no apparent ill effects and that people who are average sleepers can reduce their sleep time substantially, again with few apparent ill effects. You also learned that epidemiological studies indicate that people who sleep between 5 and 7 hours a night live the longest. Together, this evidence challenges the widely held belief that humans have a fundamental need for at least 8 hours of sleep per night. Has this chapter changed your thinking about sleep? Writing it changed ours.

The evolutionary perspective theme also played a prominent role in this chapter. You learned how thinking about the adaptive function of sleep and comparing sleep in different species have led to interesting insights. Also, you saw how research into the physiology and genetics of sleep has been conducted on nonhuman species.

The clinical implications theme received emphasis in the module on sleep disorders. Perhaps most exciting and interesting were the recent breakthroughs in understanding the genetics and physiology of narcolepsy.

Finally, the neuroplasticity theme arose in a fundamental way. The fact that the adult human brain has the capacity to change and adapt raises the possibility that it might successfully adapt to a consistent long-term schedule of sleep that is of shorter duration than most people currently choose.

Only one of the two emerging themes appeared in this chapter, and it was pervasive: Consciousness. You learned that sleep is a complex set of altered states of consciousness (e.g., REM sleep, slow-wave sleep), and that dreaming is a state of consciousness that is of considerable interest to researchers.

Key Terms

Stages of Sleep

Electroencephalogram (EEG), p. 374
Electrooculogram (EOG), p. 374
Electromyogram (EMG), p. 374
Alpha waves, p. 374
Delta waves, p. 375
Initial stage 1 EEG, p. 375
Emergent stage 1 EEG, p. 375
REM sleep, p. 375
Slow-wave sleep (SWS), p. 375

Dreaming

Lucid dreaming, p. 377

Why Do We Sleep, and Why Do We Sleep When We Do?

Recuperation theories of sleep, p. 381
Adaptation theories of sleep, p. 381

Effects of Sleep Deprivation

Executive function, p. 384
Microsleeps, p. 384
Carousel apparatus, p. 385

Circadian Sleep Cycles

Circadian rhythms, p. 388
Zeitgebers, p. 388
Free-running rhythms, p. 388
Free-running period, p. 388
Internal desynchronization, p. 388
Jet lag, p. 389
Circadian clock, p. 389
Suprachiasmatic nuclei (SCN), p. 389
Melanopsin, p. 390
Tau, p. 391

Four Areas of the Brain Involved in Sleep

Cerveau isolé preparation, p. 392
Desynchronized EEG, p. 392
Encéphale isolé preparation, p. 392
Reticular activating system, p. 393

Drugs That Affect Sleep

Hypnotic drugs, p. 395
Antihypnotic drugs, p. 395
Melatonin, p. 395
Benzodiazepines, p. 395

Imidazopyridines, p. 395
5-hydroxytryptophan (5-HTP), p. 395
Pineal gland, p. 395
Chronobiotic, p. 396

Sleep Disorders

Insomnia, p. 396
Hypersomnia, p. 397
Iatrogenic, p. 397
Sleep apnea, p. 397
Periodic limb movement disorder, p. 398
Restless legs syndrome, p. 398
Narcolepsy, p. 398
Cataplexy, p. 398
Sleep paralysis, p. 398
Hypnagogic hallucinations, p. 398
Orexin, p. 398
REM-sleep behavior disorder, p. 399

Effects of Long-Term Sleep Reduction

Polyphasic sleep cycles, p. 400
Monophasic sleep cycles, p. 400
Sleep inertia, p. 401

Chapter 15
Drug Use, Drug Addiction, and the Brain's Reward Circuits

Chemicals That Harm with Pleasure

Phanie/Alamy Stock Photo

 ## Chapter Overview and Learning Objectives

Five Commonly Used Drugs	**LO 15.7**	Describe the health hazards associated with nicotine consumption.
	LO 15.8	Describe the health hazards associated with alcohol consumption and the various stages of the full-blown alcohol withdrawal syndrome.
	LO 15.9	Explain the health effects of marijuana and the mechanism of action of THC.
	LO 15.10	Describe the health hazards associated with the consumption of cocaine and other stimulants.
	LO 15.11	Describe the health hazards associated with the consumption of opioids and the opioid withdrawal syndrome.
Comparing the Health Hazards of Commonly Used Drugs	**LO 15.12**	Explain why it is difficult to determine causality in studies of the health hazards of drugs.
	LO 15.13	Compare the direct health hazards of nicotine, alcohol, marijuana, cocaine, and heroin.
Early Biopsychological Research on Addiction	**LO 15.14**	Explain the physical-dependence and positive-incentive perspectives of addiction.
	LO 15.15	Describe the intracranial self-stimulation (ICSS) paradigm.
	LO 15.16	Describe two methods for measuring the rewarding effects of drugs.
	LO 15.17	Explain the role of the nucleus accumbens in drug addiction.
Current Approaches to the Mechanisms of Addiction	**LO 15.18**	Describe the three stages in the development of a drug addiction.
	LO 15.19	Describe two sets of findings that have challenged the relevance of drug self-administration studies.
	LO 15.20	Explain the significance of the case of Sigmund Freud.

Drug addiction is a serious problem in most parts of the world. Globally, more than 1 billion people are addicted to nicotine; more than 100 million are addicted to alcohol; more than 5 million are addicted to marijuana; more than 28 million are addicted to illegal drugs; and tens of millions are addicted to prescription drugs (Degenhardt et al., 2018; Sinha et al., 2018; Farrell et al., 2019). Pause for a moment and think about the sheer magnitude of the problem represented by such figures—more than a billion addicted people worldwide. The incidence of drug addiction is so high that it is almost certain that you or somebody dear to you will be adversely affected by drugs.

This chapter introduces you to some basic **pharmacological** (pertaining to the scientific study of drugs) principles and concepts, compares the effects of five commonly used drugs, and reviews the research on the neural mechanisms of addiction. You likely already have strong views about drug addiction; thus, as you progress through this chapter, it is particularly important that you do not let your thinking be clouded by preconceptions. In particular, it is important that you do not fall into the trap of assuming that a drug's legal status has much to say about its safety (see Nutt, King, & Nichols, 2013). You will be less likely to assume that legal drugs are safe and illegal drugs are dangerous if you remember that most laws governing drug use in various parts of the world were enacted in the early part of the 20th century, long before there was any scientific research on the topic.

The Case of the Drugged High School Teachers

People's tendency to equate drug legality with drug safety was once conveyed to me (JP) in a particularly ironic fashion: I was invited to address a convention of high school teachers on the topic of drug misuse. When I arrived at the convention center to give my talk, I was escorted to a special suite, where I was encouraged to join the executive committee in a round of drug taking—the drug being a special single-malt whiskey. Later, the irony of the situation had its full impact. As I stepped to the podium under the influence of a psychoactive drug (the whiskey), I looked out through the haze of cigarette smoke at an audience of educators who had invited me to speak to them because they were concerned about the unhealthy impact of drugs on their students. The welcoming applause gradually gave way to the melodic tinkling of ice cubes in liquor glasses, and I began. They did not like what I had to say.

Basic Principles of Drug Action

This module focuses on the basic principles of drug action, with an emphasis on **psychoactive drugs**—drugs that influence subjective experience and behavior by acting on the nervous system.

Drug Administration, Absorption, and Penetration of the Central Nervous System

LO 15.1 Compare the various routes of drug administration.

Drugs are usually administered in one of four ways: oral ingestion, injection, inhalation, or absorption through the mucous membranes of the nose, mouth, or rectum. The route of administration influences the rate at which and the degree to which the drug reaches its sites of action in the body.

ORAL INGESTION. The oral route is the preferred route of administration for many drugs. Once they are swallowed, drugs dissolve in the fluids of the stomach and are carried to the intestine, where they are absorbed into the bloodstream. However, some drugs readily pass through the stomach wall (e.g., alcohol), and these take effect sooner because they do not have to reach the intestine to be absorbed. Drugs that are not readily absorbed from the digestive tract or that are broken down into inactive *metabolites* (breakdown products of the body's chemical reactions) before they can be absorbed must be taken by some other route.

The two main advantages of the oral route of administration over other routes are its ease and relative safety. Its main disadvantage is its unpredictability: Absorption from the digestive tract into the bloodstream can be greatly influenced by such difficult-to-gauge factors as the amount and type of food in the stomach.

INJECTION. Drug injection is common in medical practice because the effects of injected drugs are strong, fast, and predictable. Drug injections are typically made *subcutaneously* (*SC*), into the fatty tissue just beneath the skin; *intramuscularly* (*IM*), into the large muscles; or *intravenously* (*IV*), directly into veins at points where they run just beneath the skin. Many drug-addicted persons prefer the intravenous route because the bloodstream delivers the drug directly to the brain. However, the speed and directness of the intravenous route are mixed blessings; after an intravenous injection, there is little or no opportunity to counteract the effects of an overdose, an impurity, or an allergic reaction. Furthermore, many drug users develop scar tissue, infections, and collapsed veins at the few sites on their bodies where there are large, accessible veins.

INHALATION. Some drugs can be absorbed into the bloodstream through the rich network of capillaries in the lungs. Many anesthetics are typically administered by *inhalation*, as are tobacco and marijuana. The two main shortcomings of this route are that it is difficult to precisely regulate the dose of inhaled drugs, and many substances damage the lungs if they are inhaled chronically.

ABSORPTION THROUGH MUCOUS MEMBRANES. Some drugs can be administered through the mucous membranes of the nose, mouth, and rectum. Cocaine, for example, is commonly self-administered through the nasal membranes (snorted)—but not without damaging them (see Walker, Joshi, & D'Souza, 2017).

Drug Action, Metabolism, and Elimination

LO 15.2 Explain the ways in which drugs can influence the nervous system and how they are eliminated from the body.

DRUG PENETRATION OF THE CENTRAL NERVOUS SYSTEM. Once a drug enters the bloodstream, it is carried to the blood vessels of the central nervous system. Fortunately, a protective filter, the *blood–brain barrier* (see Chapter 3), makes it difficult for many potentially dangerous bloodborne-chemicals to pass from the blood vessels of the CNS into the extracellular space around CNS neurons and glia (see Sweeney et al., 2019).

MECHANISMS OF DRUG ACTION. Psychoactive drugs influence the nervous system in many ways. Some drugs (e.g., alcohol and many of the general anesthetics) act diffusely on neural membranes throughout the CNS. Others act in a more specific way: by binding to particular synaptic receptors; by influencing the synthesis, transport, release, or deactivation of particular neurotransmitters; or by influencing the chain of chemical reactions elicited in postsynaptic neurons by the activation of their receptors (see Chapter 4).

DRUG METABOLISM AND ELIMINATION. The actions of most drugs are terminated by enzymes synthesized by the *liver*. These liver enzymes stimulate the conversion of active drugs to nonactive forms—a process referred to as **drug metabolism**. In many cases, drug metabolism eliminates a drug's ability to pass through lipid membranes of cells so that it can no longer penetrate the blood–brain barrier. In addition, small amounts of some psychoactive drugs are passed from the body in urine, sweat, feces, breath, and mother's milk.

Drug Tolerance, Drug Withdrawal Effects, and Physical Dependence

LO 15.3 Describe how the body becomes tolerant to drugs and the process of drug withdrawal. Explain what it means to be physically dependent on a drug.

DRUG TOLERANCE. **Drug tolerance** is a state of decreased sensitivity to a drug that develops as a result of exposure to it. Drug tolerance can be demonstrated in two ways: by showing that a given dose of the drug has less effect than it had before drug exposure or by showing that it takes more of the drug to produce the same effect. In essence, what this means is that drug tolerance is a shift in the *dose-response curve* (a graph of the magnitude of the effect of different doses of the drug) to the right (see Figure 15.1) (see Bespalov et al., 2016).

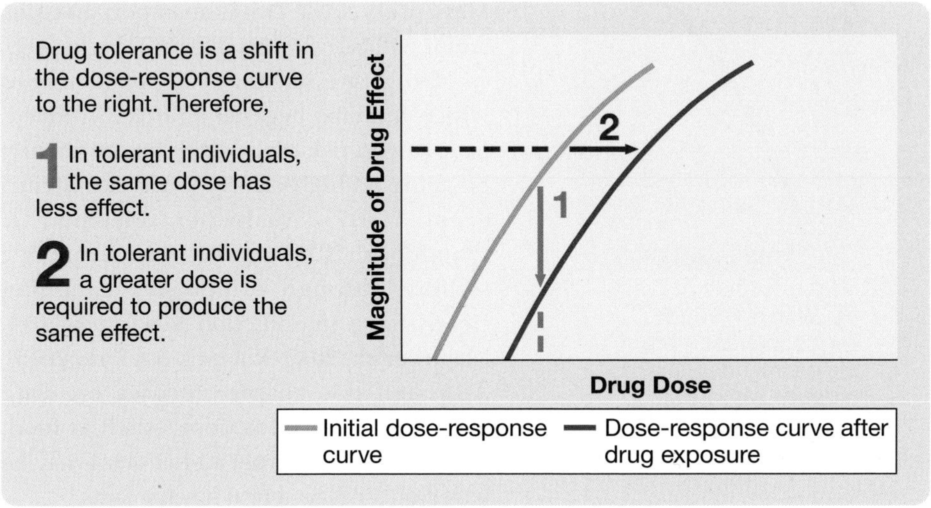

There are three important points to remember about the specificity of drug tolerance:

- One drug can produce tolerance to other drugs that act by the same mechanism; this is known as **cross tolerance** (e.g., Schiavi et al., 2019).
- Drug tolerance often develops to some effects of a drug but not to others (e.g., Castelló et al., 2014). Failure to understand this second point can have tragic consequences for people who think that because they have become tolerant to some effects of a drug (e.g., to the nauseating effects of alcohol), they are tolerant to all of them. In fact, tolerance may develop to some effects of a drug while sensitivity to other effects of the same drug increases. Increasing sensitivity to a drug is called **drug sensitization**.
- Drug tolerance is not a unitary phenomenon; that is, there is no single mechanism that underlies all examples of it (Koshimizu et al., 2018; Siciliano et al., 2016). When a drug is administered at doses that affect nervous system function, many kinds of adaptive changes can occur to reduce its effects.

Two categories of changes underlie drug tolerance: metabolic and functional. Drug tolerance that results from changes that reduce the amount of the drug getting to its sites of action is called **metabolic tolerance**. Drug tolerance that results from changes that reduce the reactivity of the sites of action to the drug is called **functional tolerance** (see Bespalov et al., 2016).

Tolerance to psychoactive drugs is largely functional. Functional tolerance to psychoactive drugs can result from several different types of adaptive neural changes (see Martyn, Mao, & Bittner, 2019). For example, exposure to a psychoactive drug can reduce the number of receptors for it, decrease the efficiency with which it binds to existing receptors, or diminish the impact of receptor binding on the activity of the cell. At least some of these adaptive neural changes are the result of epigenetic mechanisms (e.g., Ghezzi et al., 2013; Liang et al., 2013).

DRUG WITHDRAWAL EFFECTS AND PHYSICAL DEPENDENCE. After significant amounts of a drug have been in the body for a period of time (e.g., several days), its sudden elimination can trigger an adverse physiological reaction called a **withdrawal syndrome**. The effects of drug withdrawal are virtually always opposite to the initial effects of the drug. For example, the withdrawal of anticonvulsant drugs

Figure 15.1 Drug tolerance is a shift in the dose-response curve to the right as a result of exposure to the drug.

Drug tolerance is a shift in the dose-response curve to the right. Therefore,

1 In tolerant individuals, the same dose has less effect.

2 In tolerant individuals, a greater dose is required to produce the same effect.

Magnitude of Drug Effect / Drug Dose

—— Initial dose-response curve —— Dose-response curve after drug exposure often triggers convulsions, and the withdrawal of sleeping pills often produces insomnia. Individuals who suffer withdrawal reactions when they stop taking a drug are said to be **physically dependent** on that drug.

Journal Prompt 15.1

What do you think the withdrawal reaction might be when one suddenly stops taking an antidepressant medication after having taken it for many years?

The fact that withdrawal effects are frequently opposite to the initial effects of the drug suggests that withdrawal effects may be produced by the same neural changes that produce drug tolerance (see Figure 15.2). According to this theory, exposure to a drug produces compensatory changes in the nervous system that offset the drug's effects and produce tolerance. Then, when the drug is eliminated from the body, these compensatory neural changes—without the

drug to offset them—manifest themselves as withdrawal symptoms that are opposite to the initial effects of the drug.

The severity of withdrawal symptoms depends on the particular drug in question, on the duration and degree of the preceding drug exposure, and on the speed with which the drug is eliminated from the body. In general, longer exposure to greater doses followed by more rapid elimination produces greater withdrawal effects.

Drug Addiction: What Is It?

LO 15.4 Define drug addiction.

Drug-addicted individuals are habitual drug users, but not all habitual drug users are drug-addicted individuals. **Drug-addicted individuals** are those habitual drug users who continue to use a drug despite its adverse effects on their health and social life (see Heilig et al., 2016; Zakiniaeiz & Potenza, 2018), and despite their repeated efforts to stop using it (see Keramati, Ahmed, & Gutkin, 2017).

The greatest confusion about the nature of drug addiction concerns its relation to physical dependence. Many people equate the two: They see addicted persons as people who are trapped on a merry-go-round of drug taking, withdrawal symptoms, and further drug taking to combat the withdrawal symptoms. Although appealing in its simplicity, this conception of drug addiction is inconsistent with the evidence. Addicted individuals sometimes take drugs to prevent or alleviate their withdrawal symptoms, but this is often not the major motivating factor in their addiction. If it were, drug-addicted individuals could be easily cured by hospitalizing them for a few days, until their withdrawal symptoms subsided. However, most addicted individuals renew their drug taking even after months of enforced abstinence (see Meye et al., 2017). This is an important issue, and it will be revisited later in this chapter.

Drugs are not the only substances to which humans become addicted. Indeed, people who risk their health by continually bingeing on high-calorie foods (see Majuri et al., 2017; Westwater, Fletcher, & Ziauddeen, 2016) or risk their economic stability through compulsive gambling clearly have an addiction (see Clark, 2014; Majuri et al., 2017; Robbins & Clark, 2015). Although this chapter focuses on drug addiction, other addictions—such as food, gambling, and Internet addictions—may be based on similar neural mechanisms.

Figure 15.2 The relation between drug tolerance and withdrawal effects. The same adaptive neurophysiological changes that develop in response to drug exposure and produce drug tolerance manifest themselves as withdrawal effects once the drug is removed. As the neurophysiological changes develop, tolerance increases; as they subside, the severity of the withdrawal effects decreases.

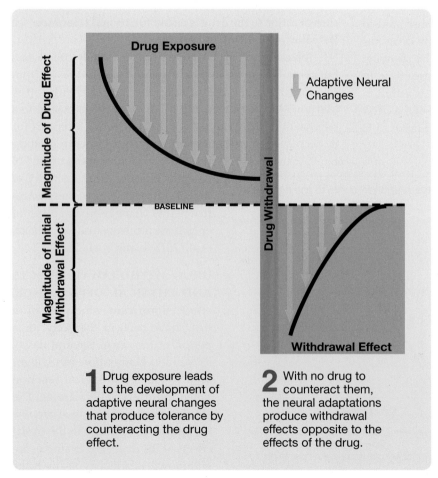

1 Drug exposure leads to the development of adaptive neural changes that produce tolerance by counteracting the drug effect.

2 With no drug to counteract them, the neural adaptations produce withdrawal effects opposite to the effects of the drug.

Role of Learning in Drug Tolerance

An important line of psychopharmacologic research has shown that learning plays a major role in drug tolerance. In addition to contributing to our understanding of drug tolerance, this research has established that efforts to understand the effects of psychoactive drugs without considering the experience and behavior of the subjects can provide only partial answers.

Research on the role of learning in drug tolerance has focused on two phenomena: contingent drug tolerance and conditioned drug tolerance. These two phenomena are discussed in the following sections.

Contingent Drug Tolerance

LO 15.5 Explain contingent drug tolerance.

Contingent drug tolerance refers to demonstrations that tolerance develops only to drug effects that are actually experienced. Most studies of contingent drug tolerance employ the **before-and-after design**. In before-and-after experiments, two groups of subjects receive the same series of drug injections and the same series of repeated tests, but the subjects in one group receive the drug before each test of the series and those in the other group receive the drug after each test of the series. At the end of the experiment, all subjects receive the same dose of the drug followed by a final test so that the degree to which the drug disrupts test performance in the two groups can be compared.

My colleagues and I (Pinel, Mana, & Kim, 1989) used the before-and-after design to study contingent tolerance to the anticonvulsant effect of alcohol. In one study, two groups of rats received exactly the same regimen of alcohol injections: one injection every 2 days for the duration of the experiment. During the tolerance development phase, the rats in one group received each alcohol injection 1 hour before a mild convulsive amygdala stimulation so that the anticonvulsant effect of the alcohol could be experienced on each trial. The rats in the other group received their injections 1 hour after each convulsive stimulation so that the anticonvulsant effect of the alcohol could not be experienced. At the end of the experiment, all of the subjects

received a test injection of alcohol, followed 1 hour later by a convulsive stimulation so that the amount of tolerance to the anticonvulsant effect of alcohol could be compared in the two groups. As Figure 15.3 illustrates, the rats that received alcohol on each trial before a convulsive stimulation became almost completely tolerant to alcohol's anticonvulsant effect, whereas those that received the same injections and stimulations in the reverse order developed no tolerance whatsoever to alcohol's anticonvulsant effect. Contingent drug tolerance has been demonstrated to many other drug effects and in many species, including humans (see Wolgin & Jakubow, 2003).

Conditioned Drug Tolerance

LO 15.6 Describe conditioned drug tolerance and conditioned compensatory responses.

Whereas studies of contingent drug tolerance focus on what subjects do while they are under the influence of drugs, studies of conditioned drug tolerance focus on the situations in which drugs are taken. **Conditioned drug tolerance** refers to demonstrations that tolerance effects are maximally expressed only when a drug is administered in the same situation in which it has previously been administered (see Castelló, Molina, & Arias, 2017; Siegel, 2011).

In one demonstration of conditioned drug tolerance (Crowell, Hinson, & Siegel, 1981), two groups of rats received

---

Figure 15.3 Contingent tolerance to the anticonvulsant effect of alcohol. The rats that received alcohol before each convulsive stimulation became tolerant to its anticonvulsant effect; those that received alcohol after each convulsive stimulation did not become tolerant.

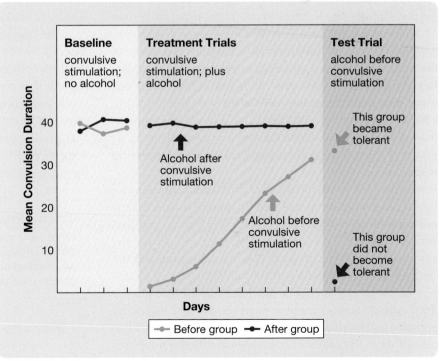

Based on Pinel, J. P. J., Mana, M. J., & Kim, C. K. (1989). Effect-dependent tolerance to ethanol's anticonvulsant effect on kindled seizures.

20 alcohol and 20 saline injections in an alternating sequence, one injection every other day. The only difference between the two groups was that the rats in one group received all 20 alcohol injections in a distinctive test room and the 20 saline injections in their colony room, while the rats in the other group received the alcohol in the colony room and the saline in the distinctive test room. At the end of the injection period, the tolerance of all rats to the *hypothermic* (temperature-reducing) effects of alcohol was assessed in both environments. As Figure 15.4 illustrates, tolerance was observed only when the rats were injected in the environment that had previously been paired with alcohol administration. There have been dozens of other demonstrations of the *situational specificity of drug tolerance*: The effects are large, reliable, and general.

The situational specificity of drug tolerance led Siegel and his colleagues to propose that drug users may be particularly susceptible to the lethal effects of a drug overdose when the drug is administered in a new context. Their hypothesis is that drug users become tolerant when they repeatedly self-administer their drug in the same environment and, as a result, begin taking larger and larger doses to counteract the diminution of drug effects. Then, if the drug user administers the usual massive dose in an unusual situation, tolerance effects are not present to counteract the effects of the drug, and there is a greater risk of death from overdose. In support of this hypothesis, Siegel and colleagues (1982) found that many more heroin-tolerant rats died following a high dose of heroin administered in a novel environment than died in the usual injection environment. (Heroin, as you will learn later in this chapter, kills by suppressing respiration.)

Siegel views each incidence of drug administration as a Pavlovian conditioning trial (see Chapter 5) in which various environmental stimuli (e.g., particular rooms, drug paraphernalia, or other drug users) that regularly predict the administration of the drug are conditional stimuli and the drug effects are unconditional stimuli. The central assumption of the theory is that conditional stimuli that predict drug administration come to elicit conditional responses opposite to the unconditional effects of the drug. Siegel has termed these hypothetical opposing conditional responses **conditioned compensatory responses**. The theory is that conditional stimuli that repeatedly predict the effects of a

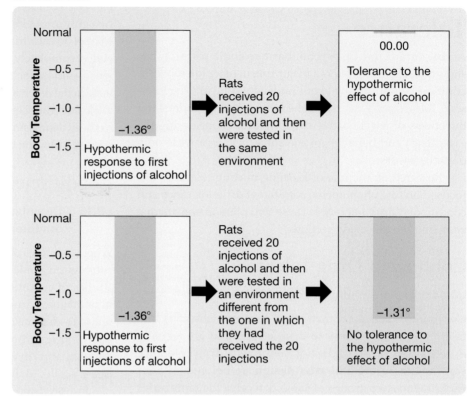

Figure 15.4 The situational specificity of tolerance to the hypothermic effects of alcohol in rats.

Based on Crowell, C. R., Hinson, R. E., & Siegel, S. (1981). The role of conditional drug responses in tolerance to the hypothermic effects of ethanol. *Psychopharmacology, 73*, 51–54.

drug come to elicit greater and greater conditioned compensatory responses; and those conditioned compensatory responses increasingly counteract the unconditional effects of the drug and produce situationally specific tolerance (see González et al., 2019).

Alert readers will have recognized the relation between Siegel's theory of drug tolerance and Woods's theory of mealtime hunger, which you learned about in Chapter 12. Stimuli that predict the homeostasis-disrupting effects of meals trigger conditioned compensatory responses to minimize a meal's disruptive effects in the same way that stimuli that predict the homeostasis-disrupting effects of a drug trigger conditioned compensatory responses to minimize the drug's disruptive effects.

Journal Prompt 15.2

What other external stimuli, besides the drug-administration environment, do you think might serve as effective conditional stimuli for the development of conditioned drug tolerance?

Most demonstrations of conditioned drug tolerance have employed **exteroceptive stimuli** (external, public stimuli, such as the drug-administration environment) as the conditional stimuli. However, **interoceptive stimuli** (internal, private stimuli) are just as effective in this role. For example, both the thoughts and feelings produced by the drug-taking ritual and the drug effects experienced soon after administration can, through conditioning, come to reduce the full impact of a drug (Siegel, 2008). This point about interoceptive stimuli is important because it indicates that just thinking about a drug can evoke conditioned compensatory responses.

Drug withdrawal effects and conditioned compensatory responses are similar: They are both responses that are opposite to the unconditioned effect of the drug. The difference is that drug withdrawal effects are produced by elimination of the drug from the body, whereas conditioned compensatory responses are elicited by drug-predictive cues in the absence of the drug. In complex, real-life situations, it is nearly impossible to tell them apart.

Although tolerance develops to many drug effects, sometimes the opposite occurs, that is, drug sensitization. *Drug sensitization*, like drug tolerance, can be situationally specific (e.g., Carey, 2020; Singer et al., 2014). For example, Anagnostaras and Robinson (1996) demonstrated the situational specificity of sensitization to the motor stimulant effects of amphetamine. They found that 10 amphetamine injections, one every 3 or 4 days, greatly increased the ability of amphetamine to activate the motor activity of rats—but only when the rats were injected and tested in the same environment in which they had experienced the previous amphetamine injections.

THINKING ABOUT DRUG CONDITIONING. In any situation in which drugs are repeatedly administered, conditioned effects are inevitable. That is why it is particularly important to understand them. However, most theories of drug conditioning have a serious problem: They have difficulty predicting the direction of the conditioned effects. For example, Siegel's conditioned compensatory response theory predicts that conditioned drug effects will always be opposite to the unconditioned effects of the drug, but there are many documented instances in which conditional stimuli elicit responses similar to those of the drug (e.g., conditioned drug sensitization).

Ramsay and Woods (1997) contend that much of the confusion about conditioned drug effects stems from a misunderstanding of Pavlovian conditioning. In particular, they criticize the common assumption that the unconditional stimulus (i.e., the stimulus to which the subject reflexively reacts) in a drug-tolerance experiment is the drug and that the unconditional responses are whatever changes in physiology or behavior the experimenter happens to be recording. They argue instead that the unconditional stimulus is the disruption of neural functioning that has been directly produced by the drug and that the unconditional responses are the various neurally mediated compensatory reactions to the unconditional stimulus, which the experimenter may or may not be recording.

This change in perspective makes a big difference. For example, in the previously described alcohol tolerance experiment by Crowell and colleagues (1981), alcohol was designated as the unconditional stimulus and the resulting hypothermia as the unconditional response. Instead, Ramsay and Woods would argue that the unconditional stimulus was the hypothermia directly produced by the exposure to alcohol, and the compensatory changes that tended to counteract the reductions in body temperature were the unconditional responses. The important point about all of this is that once one determines the unconditional stimulus and unconditional response, it is easy to predict the direction of the conditional response in any drug-conditioning experiment: The conditional response is always similar to the unconditional response.

Five Commonly Used Drugs

This module focuses on the health hazards associated with the chronic use of five commonly used drugs: nicotine, alcohol, marijuana, cocaine, and the opioids.

Nicotine

LO 15.7 Describe the health hazards associated with nicotine consumption.

The drug **nicotine**—the major psychoactive ingredient of tobacco—is most commonly administered through inhalation, though other methods of administration are possible (e.g., orally, as in the case of nicotine gum). There are two common methods of nicotine inhalation: (1) **smoking**—inhaling the smoke from the burning of tobacco (e.g., *cigarettes, cigars*) and (2) **vaping**—inhaling a vapor that contains nicotine (e.g., *e-cigarettes*). Of the two, smoking is the most common method, although vaping is quickly catching up.

Angela Hampton Picture Library/Alamy Stock Photo

TOBACCO SMOKING. When a cigarette is smoked, nicotine and some 4,000 other chemicals, collectively referred to as *tar*, are absorbed through the lungs. Nicotine acts on nicotinic cholinergic receptors in the brain (see Nees, 2014; Pistillo et al., 2014). Tobacco is the leading cause of preventable death and disease in developed nations (see Tuesta et al., 2017). It contributes to more than 6.5 million deaths a year across the globe—about 1 in every 10 deaths (Britton, 2017; Reitsma et al., 2017).

Because considerable tolerance develops to some of the immediate effects of tobacco, the effects of smoking a cigarette on nonsmokers and smokers can be quite different. Nonsmokers often respond to a few puffs of a cigarette with various combinations of nausea, vomiting, coughing, sweating, abdominal cramps, dizziness, flushing, and diarrhea. In contrast, smokers are better able to tolerate nicotine, and they report that they are more relaxed, more alert, and less hungry after a cigarette (see Tuesta et al., 2017).

The consequences of long-term tobacco use are alarming. **Smoker's syndrome** is characterized by chest pain, labored breathing, wheezing, coughing, and a heightened susceptibility to infections of the respiratory tract. Chronic smokers are highly susceptible to a variety of potentially lethal lung disorders, including pneumonia, *bronchitis* (chronic inflammation of the bronchioles of the lungs), *emphysema* (loss of elasticity of the lung from chronic irritation), and lung cancer. Although the increased risk of lung cancer receives the greatest publicity, smoking also increases the risk of cancer of the larynx (voice box), mouth, esophagus, kidneys, pancreas, bladder, and stomach. Smokers also run a greater risk of developing a variety of cardiovascular diseases, which may culminate in heart attack or stroke.

Sufferers from Buerger's disease provide a shocking illustration of the addictive power of nicotine. In **Buerger's disease**—which occurs in about 15 of 100,000 individuals, mostly in male smokers—the blood vessels, especially those supplying the legs, become constricted.

> If a patient with this condition continues to smoke, gangrene may eventually set in. First a few toes may have to be amputated, then the foot at the ankle, then the leg at the knee, and ultimately at the hip. Somewhere along this gruesome progression gangrene may also attack the other leg. Patients are strongly advised that if they will only stop smoking, it is virtually certain that the otherwise inexorable march of gangrene up the legs will be curbed. Yet surgeons report that it is not at all uncommon to find a patient with Buerger's disease vigorously puffing away in his hospital bed following a second or third amputation operation. (Brecher, 1972, pp. 215–216)

The adverse effects of tobacco smoke are unfortunately not restricted to those who smoke. Individuals who live or work with smokers are more likely to develop heart disease and cancer than those who don't. Even the unborn are vulnerable (see Abbott & Winzer-Serhan, 2012). Nicotine is a **teratogen** (an agent that can disturb the normal development of the fetus; see Jung et al., 2016): Smoking during pregnancy increases the likelihood of miscarriage, stillbirth, early death of the child, psychiatric disorders during adolescence (e.g., Hunter et al., 2020), and other health consequences (e.g., Baskaran et al., 2019).

Even the consumption of nicotine by the father can affect their children. For example, there is evidence that the administration of nicotine to male rats can increase depression and anxiety in their offspring animal models of those conditions (see Goldberg & Gould, 2019). This is an example of *transgenerational epigenetics*: the transmission of epigenetic changes to subsequent generations (see Chapter 2).

If you or a loved one is a cigarette smoker, we have some good news and some bad news. First the bad news: Treatments for nicotine addiction are only marginally effective (see Zwar, Mendelsohn, & Richmond, 2014). The good news: Many people do stop smoking, and they experience major health benefits. Indeed, even the replacement of tobacco with another form of nicotine administration (e.g., nicotine gum, nicotine skin patch, nicotine mouth spray) is likely to lead to major health benefits (see Fagerström & Bridgman, 2014; Kupferschmidt, 2014).

NICOTINE VAPING. In 2003, the pharmacist Hon Lik presented the world with the first e-cigarette (see Löhler & Wallenberg, 2019). For over a decade it was marketed as a safer alternative to tobacco for consuming nicotine. Current evidence supports the notion that e-cigarettes can help people quit smoking (see Hajek et al., 2019; Liu et al., 2018). At the same time, many who start using e-cigarettes will have difficulty quitting (see Liu et al., 2018), and some will move on to the smoking of tobacco (see Chapman, Bareham, & Maziak, 2019; Liu et al., 2018).

Is vaping a safer alternative to smoking? In general, there haven't been enough studies yet to answer that question. However, there have been an increasing number of reports of respiratory ailments amongst heavy vapers. We know of no controlled studies that have looked at long-term health outcomes associated with e-cigarette use (see Pisinger, Godtfredsen, & Bender, 2019).

Particularly disturbing has been the rapid rise in their use by youth. For example, in 2016, 11% of American high school students and 4% of American middle school students reporting using e-cigarettes in the past month (see Warner & Mendez, 2019).

ADDICTION AND NICOTINE. There is no question that heavy smokers and vapers are addicted in every sense of the word. Can you think of any other psychoactive drug that is self-administered almost continually—even while the addicted person is walking along the street? The compulsive **drug craving** (an affective state in which there is a strong desire for the drug), which is a major defining feature of addiction, is readily apparent in any habitual smoker who has run out of cigarettes, or any habitual vaper, who is forced by circumstance to refrain from nicotine inhalation for several hours. Furthermore, habitual smokers and vapers who stop their nicotine inhalation experience a variety of withdrawal effects, such as depression, anxiety, restlessness, irritability, constipation, and difficulties in sleeping and concentrating.

About 68 percent of all people who experiment with smoking become addicted—this figure compares unfavorably with 23 percent for alcohol and 9 percent for marijuana (see Curran et al., 2016). Moreover, nicotine addiction typically develops quickly, within a few weeks, and only about 20 percent of all attempts to stop smoking are successful for 2 years or more.

Alcohol

LO 15.8 Describe the health hazards associated with alcohol consumption and the various stages of the full-blown alcohol withdrawal syndrome.

This section discusses another commonly used drug: alcohol. Alcohol is involved in more than 3 million deaths each year across the globe, including deaths from birth defects, ill health, accidents, and violence (see Degenhardt et al., 2108; Parrott & Eckhardt, 2018). Worldwide, approximately 100 million people are heavy users of alcohol. Because alcohol molecules are small and soluble in both fat and water, they invade all parts of the body. Alcohol is classified as a **depressant** because at moderate-to-high doses it depresses neural firing; however, at low doses, it can stimulate neural firing (see Alasmari et al., 2018) and facilitate social interaction. Alcohol addiction has a major genetic component.

With moderate doses, the alcohol drinker experiences various degrees of cognitive, perceptual, verbal, and motor impairment, as well as a loss of control that can lead to a

Marcos Mesa Sam Wordley/Shutterstock

variety of socially unacceptable actions. High doses result in unconsciousness; and if blood levels reach 0.5 percent, there is a risk of death from respiratory depression. The telltale red facial flush of alcohol intoxication is produced by the dilation of blood vessels in the skin; this dilation increases the amount of heat lost from the blood to the air and leads to a decrease in body temperature (*hypothermia*). Alcohol is also a *diuretic*; that is, it increases the production of urine by the kidneys.

Alcohol, like many addictive drugs, produces both tolerance and physical dependence. The livers of heavy drinkers metabolize alcohol more quickly than the livers of nondrinkers, but this increase in metabolic efficiency contributes only slightly to overall alcohol tolerance; most alcohol tolerance is functional. Withdrawal from alcohol, even after a single bout of drinking, can produce a withdrawal syndrome of headache, nausea, vomiting, and tremors, which is euphemistically referred to as a *hangover*.

Withdrawal from alcohol after a long bout of heavy drinking produces a full-blown alcohol withdrawal syndrome comprising four phases (see Perry, 2014). The first phase begins 6 to 8 hours after the cessation of alcohol consumption and is characterized by anxiety, tremor, nausea, and *tachycardia* (rapid heartbeat). The second phase begins 10 to 30 hours after cessation of drinking and is characterized by hyperactivity, insomnia, and hallucinations. The defining feature of the third phase, which typically occurs between 12 and 48 hours after cessation of drinking, is convulsive activity. The fourth phase, which usually begins 3 to 5 days after the cessation of drinking and lasts up to a week, is called **delirium tremens (DTs)**. The DTs are characterized by disturbing hallucinations, bizarre delusions, disorientation, agitation, confusion, *hyperthermia* (high body temperature), and tachycardia. The convulsions and the DTs produced by alcohol withdrawal can be lethal.

Alcohol attacks almost every tissue in the body (see González-Reimers et al., 2014). Chronic alcohol consumption produces extensive brain damage (see Zahr & Pfefferbaum, 2017). This damage is produced both directly (see Zahr, Kaufman, & Harper, 2011) and indirectly (see Weil, Corrigan, & Karelina, 2018). For example, you learned in Chapter 1 that alcohol indirectly causes **Korsakoff's syndrome** (a neuropsychological disorder characterized by memory loss, sensory and motor dysfunction, and, in its advanced stages, severe dementia) by inducing and interacting with thiamine deficiency (see Moretti et al., 2017). But even heavy alcohol users without Korsakoff's syndrome display changes in brain structures: The most common finding is a general loss of cortical white and gray matter (see Everitt & Robbins, 2016).

Alcohol affects the brain function of drinkers in other ways as well. For example, it interferes with the function of second messengers inside neurons; it disrupts GABAergic and glutaminergic transmission; it leads to

DNA methylation; and it triggers apoptosis (see ali Shah et al., 2013; Berkel & Pandey, 2017)—it is a neurotoxin (see Henriques et al., 2018).

Chronic alcohol consumption also causes extensive scarring, or **cirrhosis**, of the liver, which is the major cause of death among heavy alcohol users. Alcohol erodes the muscles of the heart and thus increases the risk of heart attack. It irritates the lining of the digestive tract and, in so doing, increases the risk of oral and liver cancer, stomach ulcers, *pancreatitis* (inflammation of the pancreas), and *gastritis* (inflammation of the stomach). And not to be forgotten is the carnage that alcohol produces from accidents on our roads, in our homes, in our workplaces, and at recreational sites—in the United States, more than 10,000 people die each year in alcohol-related traffic accidents alone.

Many people assume the adverse effects of alcohol occur only in people who drink a lot—they tend to define "a lot" as "much more than they themselves consume." But they are wrong. Several large-scale studies have shown that even low-to-moderate regular drinking (a drink or two per day) is associated with elevated levels of many cancers, including breast, oral cavity, and colorectal cancer (Bagnardi et al., 2013; Castro & Castro, 2014; Huber & Tantiwongkosi, 2014; Stone et al., 2014).

The offspring of mothers who consume substantial quantities of alcohol during pregnancy can develop **fetal alcohol syndrome (FAS)**, which has a worldwide prevalence of 0.7% (see Lange et al., 2017; Valenzuela et al., 2012). A child with FAS suffers from some or all of the following symptoms: brain damage, intellectual disability, poor coordination, poor muscle tone, low birth weight, delayed growth, and/or physical deformity (see Landgraf et al., 2014; Subramoney et al., 2018). Because alcohol can disrupt brain development in so many ways (e.g., by disrupting the production of cell-adhesion molecules or by disrupting normal patterns of apoptosis), there is no time during pregnancy when alcohol consumption is safe (see Paintner, Williams, & Burd, 2012). Moreover, there seems to be no safe amount (see Charness, Riley, & Sowell, 2016). Although full-blown FAS is rarely seen in the babies of mothers who never had more than one drink a day during pregnancy, children of mothers who drank only moderately while pregnant are sometimes found to have a variety of cognitive problems, even though they are not diagnosed with FAS (see Hendricks et al., 2019; Koren et al., 2014).

There is evidence that alcohol consumption might have effects on subsequent generations, even when consumed by the male parent; that is, alcohol consumption has been shown to produce *transgenerational epigenetic effects* (see Chapter 2)—see Kippin (2014); Vassoler, Byrnes, and Pierce (2014). For example, the offspring of alcohol-consuming male rats display impairments on various cognitive tasks (see Goldberg & Gould, 2019; Vassoler, Byrnes, & Pierce, 2014). Moreover, there have been reports of human children born with characteristics of FAS whose mothers did not drink but whose fathers were alcoholics (see Vassoler, Byrnes, & Pierce, 2014; Chastain & Sarkar, 2017).

One of the most widely publicized findings about alcohol is that moderate drinking reduces the risk of coronary heart disease. This conclusion is based on the finding that the incidence of coronary heart disease is less among moderate drinkers than among abstainers. You learned in Chapter 1 about the difficulty in basing causal interpretations on correlational data, and researchers worked diligently to identify and rule out factors other than the alcohol that might protect moderate drinkers from coronary heart disease. They seemed to rule out every other possibility. However, a thoughtful analysis led to a different conclusion. Let us explain. In a culture in which alcohol consumption is the norm, any large group of abstainers will always include some people who have stopped drinking because they are ill—perhaps this is why abstainers have more heart attacks than moderate drinkers (see Roerecke & Rehm, 2011). This hypothesis was tested by including in a meta-analysis only those studies that used an abstainers control group consisting of individuals who had never consumed alcohol. This meta-analysis indicated that alcohol in moderate amounts does not prevent coronary heart disease; that is, moderate drinkers did not suffer less coronary heart disease than lifelong abstainers (Fillmore et al., 2006; Stockwell, 2012). Likewise, a more recent meta-analysis found that moderate alcohol consumption did not reduce the risk of mortality (see Stockwell et al., 2016).

Journal Prompt 15.3

Before reading this section, what were your views on the effects of moderate drinking? Have your views changed at all now? Why or why not?

Marijuana

LO 15.9 Explain the health effects of marijuana and the mechanism of action of THC.

Marijuana is the name commonly given to the dried flower buds of female *Cannabis* plants. Cannabis, the common hemp plant, has three species: *Cannabis sativa*, *Cannabis indica*, and *Cannabis ruderalis*. The usual mode of consumption is to smoke these flowers in a *joint* (a cigarette of marijuana) or a pipe, but marijuana is also effective when ingested orally if first baked into an oil-rich substrate, such as a chocolate brownie, to promote absorption from the gastrointestinal tract.

The psychoactive effects of marijuana are largely attributable to a constituent called **THC** (delta-9-tetrahydrocannabinol). However, marijuana contains more than 80 *cannabinoids* (chemicals of the same chemical class as THC), which may also be psychoactive. For example, one cannabinoid, THCV (delta-9-tetrahydrocannabivarin), has been shown to have antipsychotic effects (see Chapter 18)—see Cascio et al. (2014). Most of the cannabinoids are found in a sticky resin covering the leaves and flowers of the plant; this resin can be extracted and dried to form a dark corklike material called **hashish**.

Findings from archeological sites suggests that *Cannabis Sativa* was used for its psychological effects as early as 11,000 years ago (see Bonini et al., 2018; Wei et al., 2017). Written records of cannabis use go back 6,000 years in China, where its stems were used to make rope, its seeds were used as a grain, and its leaves and flowers were used for their psychoactive and medicinal effects (see Bonini et al., 2018). In the Middle Ages, cannabis cultivation spread into Europe, where it was grown primarily for the manufacture of rope. During the period of European imperialism, rope was in high demand for sailing vessels, and the American colonies responded to this demand by growing cannabis as a cash crop. George Washington, the first president of the United States, was one of the more notable cannabis growers.

The practice of smoking the flower buds of the cannabis plant and the word *marijuana* itself seem to have been introduced to the southern United States in the early part of the 20th century. In 1926, an article appeared in a New Orleans newspaper exposing the "menace of marijuana," and soon similar stories were appearing in newspapers all over the United States claiming that marijuana turns people into violent, drug-crazed criminals. The misrepresentation of the effects of marijuana by the news media led to the rapid enactment of laws against the drug. In many states, marijuana was legally classified a **narcotic** (a legal term generally used to refer to opioids), and punishment for its use was dealt out accordingly. Marijuana bears no resemblance to opioid narcotics.

Stanimir G.Stoev/Shutterstock

Popularization of marijuana smoking among the middle and upper classes in the 1960s stimulated a massive program of research. One of the difficulties in studying the effects of marijuana is that they are subtle, difficult to measure, and greatly influenced by the social situation:

> At low, usual "social" doses, the intoxicated individual may experience an increased sense of well-being: initial restlessness and hilarity followed by a dreamy, carefree state of relaxation; alteration of sensory perceptions including expansion of space and time; and a more vivid sense of touch, sight, smell, taste, and sound; a feeling of hunger, especially a craving for sweets; and subtle changes in thought formation and expression. To an unknowing observer, an individual in this state of consciousness would not appear noticeably different. (National Commission on Marijuana and Drug Abuse, 1972, p. 68)

Although the effects of typical social doses of marijuana are subtle, high doses do impair psychological functioning. At high doses, short-term memory is impaired, and the ability to carry out tasks involving multiple steps to reach a specific goal declines. Speech becomes slurred, and meaningful conversation becomes difficult. A sense of unreality, emotional intensification, sensory distortion, feelings of paranoia, and motor impairment are also common. Driving under the influence of marijuana is obviously dangerous (see Neavyn et al., 2014).

Some people do become addicted to marijuana, but its addiction potential is low. About 9% of all marijuana users become addicted—considerably less than for cocaine (21%), alcohol (23%), and tobacco (68%) (see Curran et al., 2016). Most people who use marijuana do so only occasionally, with only about 10 percent of them using daily; moreover, most people who try marijuana do so in their teens and curtail their use by their 30s or 40s. Tolerance to marijuana develops during periods of sustained use, and obvious withdrawal symptoms (e.g., irritability, nausea, nightmares, and other sleep disturbances) are present in about 50% of daily users who cease their marijuana use (see Curran et al., 2016).

What are the health hazards of marijuana use? There are respiratory symptoms associated with heavy marijuana smoking: bronchitis and coughing being the most common. There are also cardiovascular symptoms. For example, there is evidence of an increase in the likelihood of a heart attack in individuals of all ages, with older individuals showing the most pronounced increases (see Singh et al., 2018). There is no evidence that marijuana use increases the likelihood of cancer (see Cohen, Weizmann, & Weinstein, 2019).

You have likely heard that marijuana causes brain damage. This claim has been spread by governmental and social agencies attempting to discourage marijuana use. But what is the actual evidence?

Surprisingly, no damage that can reasonably be attributed to marijuana use has been found in the brains of living or deceased marijuana users (see Hall & Degenhardt, 2014). However, four lines of indirect correlational (see Chapter 1) evidence have a bearing on the question:

- Brain-imaging studies have found that hippocampal volumes tend to be slightly reduced in some heavy marijuana users (see Batalla et al., 2013; Cohen, Weizmann, & Weinstein, 2019). However, such findings might be the result of preexisting differences between users and nonusers (e.g., Goldman, 2015; Pagliaccio et al., 2015).

- Functional brain-imaging studies have found reliable differences in marijuana users when they are asked to perform any one of many different tasks while undergoing functional brain imaging (e.g., fMRI; see Chapter 5). Yet, these studies provided no indication as to whether the observed changes are harmful or beneficial (see Zehra et al., 2018).

- Heavy marijuana users tend to have memory problems (see Crane et al., 2013; Hall & Degenhardt, 2014). However, those memory deficits dissipate within about 72 hours after the cessation of marijuana use (see Scott et al., 2018). Accordingly, it is not clear if the memory problems are indicative of persistent brain damage (see Mechoulam & Parker, 2013; Scott et al., 2018; but see Berliner, Collins, & Coker, 2018).

- Heavy marijuana users are slightly more likely to be diagnosed with schizophrenia (see Chapter 18)—especially if they began using marijuana during adolescence (see Radhakrishnan et al., 2014; Cohen, Weizmann, & Weinstein, 2019). Until the reasons for this correlation are sorted out (see Hill, 2015; Pasman et al., 2018; Renard et al., 2014), youths with a history of schizophrenia in their families should refrain from marijuana use (see Burns, 2013).

In short, regardless of what you have heard to the contrary, there is no convincing evidence that marijuana causes brain damage. Complicating the situation further is that marijuana may actually have neuroprotective effects. For example, Nguyen and colleagues (2014) reviewed the data of adults who were treated for traumatic brain injury (see Chapter 10) and found that those individuals who tested positive for marijuana use were 80 percent less likely to die from the brain injury than nonusers of marijuana. Also suggestive is the finding that old mice display improved performance in a variety of learning paradigms after receiving a low dose of THC (see Bilkei-Gorzo et al., 2017).

Research on THC changed irrevocably in the early 1990s with the discovery of two receptors for it: CB_1 and CB_2 (see Scherma et al., 2019). CB_1 turned out to be one of the most prevalent metabotropic receptors in the brain (see Chapter 4)—see Parsons and Hurd (2015), and it is present in other parts of the body as well (e.g., Hedlund, 2014); CB_2 is found throughout the CNS and in the cells of the immune system (see Mechoulam & Parker, 2013; Parsons & Hurd, 2015; Wei et al., 2017). But why are there receptors for THC in the brain? They could hardly have evolved to mediate the effects of marijuana smoking. This puzzle was quickly solved with the discovery of a class of endogenous cannabinoid neurotransmitters: the endocannabinoids (see Lu et al., 2019; McPartland et al., 2014; Yin, Wang, & Zhang, 2019). The first endocannabinoid neurotransmitter to be isolated and characterized was named **anandamide**, from a word that means "internal bliss" (see Piomelli, 2014; Scherma et al., 2019).

THC has been shown to have several therapeutic effects (see Katz-Talmor et al., 2018; Noonan, 2015). Since the early 1990s, it has been used to suppress nausea and vomiting in cancer patients and to stimulate the appetite of patients with AIDS (see Robson, 2014). THC has also been shown to block seizures; to dilate the bronchioles of asthmatics; to decrease the severity of *glaucoma* (a disorder characterized by an increase in the pressure of the fluid inside the eye); and to reduce anxiety, some kinds of pain, and the symptoms of multiple sclerosis (see Koppel et al., 2014). In 2010, *Sativex*, a mouth spray that contains THC and other cannabinoids, was introduced into several countries for the treatment of multiple sclerosis symptoms (see Cressey, 2015; Zlebnik & Cheer, 2016), for which it seems to be effective (see Chiurchiù et al., 2018).

There is evidence that marijuana consumption might have effects on subsequent generations, even when consumed by the father; that is, marijuana consumption has been shown to produce transgenerational epigenetic effects (see Andaloussi, Taghzouti, & Abboussi, 2019; Meccariello et al., 2020; Szutorisz & Hurd, 2018). For example, Andaloussi, Taghzouti, & Abboussi (2019) found that the offspring of male rats who had been given a CB_1-receptor agonist for 20 days displayed increased vulnerability to stress-induced anxiety.

We cannot end this discussion of marijuana (e.g., *Cannabis Sativa*) without telling you the following story:

> You can imagine how surprised I (JP) was when my colleague went to his back door, opened it, and yelled, "Sativa, here Sativa, dinner time."
>
> "What was that you called your dog?" I asked as he returned to his beer.
>
> "Sativa," he said. "The kids picked the name. I think they learned about it at school; a Greek goddess or something. Pretty, isn't it? And catchy too: Every kid on the street seems to remember her name."
>
> "Yes," I said. "Very pretty."

Cocaine and Other Stimulants

LO 15.10 Describe the health hazards associated with the consumption of cocaine and other stimulants.

Stimulants are drugs whose primary effect is to produce general increases in neural and behavioral activity. Although stimulants all have a similar profile of effects, they differ greatly in their potency. Coca-Cola is a mild commercial stimulant preparation consumed by many people around the world. Today, its stimulant action is attributable to *caffeine*, but when it was first introduced, it packed a real wallop in the form of small amounts of cocaine. **Cocaine** and its derivatives are the focus of this section; but we will also be discussing other stimulants.

Cocaine is prepared from the leaves of the coca shrub, which grows primarily in western South America. For centuries, a crude extract called *coca paste* has been made directly from the leaves and eaten. Today, it is more common to treat the coca paste and extract *cocaine hydrochloride*, the nefarious white powder that is referred to simply as *cocaine* and typically consumed by snorting or by injection. Cocaine hydrochloride may be converted to its base form by boiling it in a solution of baking soda until the water has evaporated. The impure residue of this process is **crack**, a potent, cheap, smokeable form of cocaine (see Fukushima et al., 2019). However, because crack is impure and consumed by smoking, it is difficult to study, and most research on cocaine derivatives has thus focused on pure cocaine hydrochloride. More than 18 million people used cocaine in the past year across the globe (Farrell et al., 2019).

Cocaine hydrochloride is an effective local anesthetic and was once widely prescribed as such until it was supplanted by synthetic analogues such as *procaine* and *lidocaine* (see Farrell et al., 2019). It is not, however, cocaine's anesthetic actions that are of interest to users. People eat, smoke, snort, or inject cocaine or its derivatives

Jan Mika/model released people collection/Alamy

in order to experience its psychological effects. Users report being swept by a wave of well-being; they feel self-confident, alert, energetic, friendly, outgoing, fidgety, and talkative; and they have less than their usual desire for food and sleep.

Individuals who are addicted to cocaine tend to go on so-called **cocaine sprees**, binges in which extremely high levels of intake are maintained for periods of a day or two. During a cocaine spree, users become increasingly tolerant to the euphoria-producing effects of cocaine. Accordingly, larger and larger doses are often administered. The spree usually ends when the cocaine is gone or when it begins to have serious toxic effects. The effects of cocaine sprees include sleeplessness, tremors, nausea, hyperthermia, and, in rare cases, psychotic symptoms, which is called **cocaine psychosis** and has sometimes been mistakenly diagnosed as *schizophrenia* (see Chapter 18). During cocaine sprees, there is a risk of loss of consciousness, seizures, respiratory arrest, heart attack, or stroke (see Zimmerman, 2012; Stankowski, Kloner, & Rezkalla, 2014). Although tolerance develops to most effects of cocaine (e.g., to the euphoria), repeated cocaine exposure sensitizes subjects (i.e., makes them even more responsive) to its motor effects (see Li & Wolf, 2012). The withdrawal effects triggered by abrupt termination of a cocaine spree are relatively mild. Common cocaine withdrawal symptoms include a negative mood swing and insomnia.

Although cocaine and its derivatives are widely misused, **amphetamine** and its relatives are currently the most widely misused stimulants (UN Global ATS Assessment, 2011). Amphetamine has been in wide use since the 1960s. It is usually consumed orally in the potent form called *d-amphetamine* (dextroamphetamine). Some of the effects of *d*-amphetamine are comparable to those of cocaine; for example, it can also produce a syndrome of psychosis called *amphetamine psychosis*.

In the 1990s, *d*-amphetamine was supplanted as the favored amphetamine-like drug by several more potent relatives. One is *methamphetamine*, or "meth" (see Hsieh et al., 2014), which is commonly used in its even more potent, smokeable, crystalline form (crystal meth). Another potent relative of amphetamine is *3,4-methylenedioxymethamphetamine* (MDMA, or ecstasy), which is taken orally (see Cole, 2014). Besides being a stimulant, MDMA is also considered an empathogen. **Empathogens** are psychoactive drugs that produce feelings of empathy (see Bedi, Hyman, & de Wit, 2010; Bershad et al., 2016).

The primary mechanism by which cocaine and its derivatives exert their effects is by altering the activity of **dopamine transporters**, molecules in the presynaptic membrane that normally remove dopamine from synapses and transfer it back into presynaptic neurons. Other stimulants increase the release of monoamines into synapses (see Sitte & Freissmuth, 2015).

Do stimulants have long-term adverse effects on the health of habitual users? There is some evidence that they do. For example, many studies have reported cognitive impairments in both methamphetamine and MDMA users (see Marshall & O'Dell, 2012; Parrott, 2013; Moratalla et al., 2017), though the effects have often been small or difficult to reproduce (see Amoroso, 2015; Betzler, Viohl, & Romanczuk-Seiferth, 2017; Frazer, Richards, & Keith, 2018; Mueller et al., 2016; Müller et al., 2019). In addition, methamphetamine and amphetamine users, but not cocaine users, have a greater risk of developing Parkinson's disease (see Callaghan et al., 2012; Curtin et al., 2015). There is also evidence of heart pathology: Many cocaine-dependent, amphetamine-dependent, and methamphetamine-dependent patients have been found to have electrocardiographic abnormalities (see Carvalho et al., 2012; Maceira et al., 2014). However, because all these results are correlational, one cannot rule out other explanations (see Amoroso, 2015; Hart et al., 2012; Krebs & Johansen, 2012).

There is growing evidence that cocaine consumption can produce transgenerational epigenetic effects (see Goldberg & Gould, 2019). For example, Le et al., (2017) found that the offspring of male rats who had self-administered cocaine heavily were more likely to display increases in cocaine self-administration. Many other studies have found transgenerational epigenetic effects of cocaine consumption. In addition to an increased propensity for cocaine self-administration, the offspring of male rats given cocaine are more likely to display increases in anxiety- and depression-like behaviors (see Goldberg & Gould, 2019).

The Opioids: Heroin and Morphine

LO 15.11 Describe the health hazards associated with the consumption of opioids and the opioid withdrawal syndrome.

Opium—the dried form of sap exuded by the seedpods of the opium poppy—has several psychoactive ingredients. Most notable are **morphine** and **codeine**, its weaker relative. Morphine, codeine, and other drugs that have similar structures or effects are commonly referred to as **opioids**. The opioids exert their effects by binding to receptors whose normal function is to bind to endogenous opioids (see Stein, 2016; Darcq & Kieffer, 2018). The endogenous opioid neurotransmitters that bind to such receptors are of two classes: *endorphins* and *enkephalins* (see Chapter 4).

The opioids have a Jekyll-and-Hyde character. On their Dr. Jekyll side, the opioids are effective as **analgesics** (painkillers; see Weibel et al., 2013); they are also extremely effective in the treatment of cough and diarrhea. But, unfortunately, the kindly Dr. Jekyll brings with him the evil Mr. Hyde—the risk of addiction.

Opioids have been used for their euphoric effects and to relieve pain for over 4000 years (Darcq & Kieffer, 2018). Three historic events fanned the flame of opioid addiction. First, in 1644, the Emperor of China banned tobacco smoking, and this contributed to a gradual increase in opium smoking in China, spurred on by the smuggling of opium into China by the British East India Company. Because smoking opium has a greater effect on the brain than does eating it, many more people became addicted. Second, morphine, the most potent constituent of opium, was isolated in 1803, and it became available commercially in the 1830s (see Darcq & Kieffer, 2018; Volkow & Koroshetz, 2019). Third, the hypodermic needle was invented in 1856, and soon the injured were introduced to morphine through a needle.

Until the early part of the 20th century, opium was available legally in many parts of the world, including Europe and North America. Indeed, opium was an ingredient in cakes, candies, and wines, as well as in a variety of over-the-counter medicinal offerings. Opium potions such as *laudanum* (a very popular mixture of opium and alcohol), *Godfrey's Cordial*, and *Dalby's Carminative* were very popular. (The word *carminative* should win first prize for making a sow's ear at least sound like a silk purse: A carminative is a drug that expels gas from the digestive tract, thereby reducing stomach cramps and flatulence. *Flatulence* is the obvious pick for second prize.) There were even over-the-counter opium potions just for baby—such

threerocksimages/Shutterstock

as *Mrs. Winslow's Soothing Syrup* and the aptly labeled *Street's Infant Quietness*. Although pure morphine required a prescription at the time, physicians prescribed it for so many different maladies that morphine addiction was common among those who could afford a doctor.

The **Harrison Narcotics Act**, passed in 1914, made it illegal to sell or use opium, morphine, or cocaine in the United States—although morphine and its analogues are still legally prescribed for their medicinal properties. However, the act did not include the semisynthetic opioid **heroin**. Heroin was synthesized in 1870 by the addition of two acetyl groups to the morphine molecule, which greatly increased its ability to penetrate the blood–brain barrier. In 1898, heroin was marketed by the Bayer Drug Company (see Darcq & Kieffer, 2018); it was freely available without prescription and was widely advertised as a superior kind of aspirin. Tests showed that it was a more potent analgesic than morphine and that it was less likely to induce nausea and vomiting. Moreover, the Bayer Drug Company, on the basis of flimsy evidence, claimed that heroin was not addictive; this is why it was not covered by the Harrison Narcotics Act. The consequence of omitting heroin from the Harrison Narcotics Act was that opioid-dependent individuals in the United States, forbidden by law to use opium or morphine, turned to the readily available and much more potent heroin—and the flames of addiction were further fanned. In 1924, the U.S. Congress made it illegal for anybody to possess, sell, or use heroin. Unfortunately, the laws enacted to stamp out opioid use in the United States have been far from successful: In the United States, the number of deaths due to overdose on prescription or illegal opioids now exceeds that of car accidents (see Manchikanti et al., 2012). More than 53,000 people died of opioid overdose in 2016 (high doses of heroin kill by suppressing breathing). This is in part due to the development of synthetic opioids, such as *fentanyl* and *oxycodone*, that are both more potent and more addictive than heroin—even though many of these were marketed as having a very low risk of addiction (see Volkow & Koroshetz, 2019). Indeed, by all accounts we are currently in an epidemic of opioid abuse.

The effect of opioids most valued by users is the *rush* that follows intravenous injection. The *heroin rush* is a wave of intense abdominal, orgasmic pleasure that evolves into a state of serene, drowsy euphoria. Many opioid users, drawn by these pleasurable effects, begin to use the drug more and more frequently. Then, once they reach a point where they keep themselves drugged much of the time, tolerance and physical dependence develop and contribute to the problem. Opioid tolerance encourages users to progress to higher doses, to more potent drugs (e.g., heroin, fentanyl), and to more direct routes of administration (e.g., IV injection); and physical dependence adds to the already high motivation to take the drug.

The classic opioid withdrawal syndrome usually begins 6 to 12 hours after the last dose. The first withdrawal sign is typically an increase in restlessness; the opioid user begins to pace and fidget. Watering eyes, running nose, yawning, and sweating are also common during the early stages of opioid withdrawal. Then, the person often falls into a fitful sleep, which typically lasts for several hours. Once they wake up, the original symptoms may be joined in extreme cases by chills, shivering, profuse sweating, gooseflesh, nausea, vomiting, diarrhea, cramps, dilated pupils, tremor, and muscle pains and spasms. The gooseflesh skin and leg spasms of the opioid withdrawal syndrome are the basis for the expressions "going cold turkey" and "kicking the habit." The symptoms of opioid withdrawal are typically most severe in the second or third day after the last injection, and by the seventh day they have all but disappeared. In short, opioid withdrawal is about as serious as a bad case of the flu:

> [Opioid] withdrawal is probably one of the most misunderstood aspects of drug use. This is largely because of the image of withdrawal that has been portrayed in the movies and popular literature for many years….Few…take enough drug to cause the…severe withdrawal symptoms that are shown in the movies. Even in its most severe form, however, [opioid] withdrawal is not as dangerous or terrifying as withdrawal from barbiturates or alcohol. (McKim, 1986, p. 199)

Although many opioids are highly addictive, the direct health hazards of chronic exposure are surprisingly minor. The main direct risks are constipation, pupil constriction, menstrual irregularity, and reduced sex drive. Many opioid users have taken pure heroin or morphine for years with no serious ill-effects. In fact, opioid addiction is more prevalent among doctors, nurses, and dentists than among other professionals (e.g., Brewster, 1986):

> An individual tolerant to and dependent upon an [opioid] who is socially or financially capable of obtaining an adequate supply of good quality drug, sterile syringes and needles, and other paraphernalia may maintain his or her proper social and occupational functions, remain in fairly good health, and suffer little serious incapacitation as a result of the dependence. (Julien, 1981, p. 117)
>
> One such individual was Dr. William Steward Halsted, one of the founders of Johns Hopkins Medical School and one of the most brilliant surgeons of his day…known as "the father of modern surgery." And yet, during his career he was addicted to morphine, a fact that he was able to keep secret from all but his closest friends. In fact, the only time his habit caused him any trouble was when he was attempting to reduce his dosage. (McKim, 1986, p. 197)

Most medical risks of opioid addiction are indirect—that is, not entirely attributable to the drug itself. Many of the medical risks arise out of the battle between the relentless addictive power of opioids and the attempts of governments to eradicate addiction by making opioids illegal. The opioid user who cannot give up their habit—treatment programs report success rates of only about 10 percent—are caught in the middle. Because most opioid users must purchase their morphine and heroin from illicit dealers at greatly inflated prices, those who are not wealthy become trapped in a life of poverty and petty crime. They are poor, they are undernourished, they receive poor medical care, and they run great risk of contracting HIV and other infections from unsterile needles. Moreover, they never know for sure what they are injecting: Some street drugs are poorly processed, and virtually all have been *cut* (stretched by the addition of some other substance) to an unknown degree. The sad reality is that strict laws against opioid sales are associated with an increased number of overdose deaths—due in large part to the increase in cutting of heroin with more powerful synthetic opioids like fentanyl (see Csete et al., 2016).

Like every other drug presented in this module, opioids have also been shown to produce transgenerational epigenetic effects. Two general findings are that the offspring of male rats who have been administered opioids: (1) exhibit more severe opioid withdrawal symptoms; and (2) display decreases in synaptic plasticity (see Goldberg & Gould, 2019).

The primary treatments for heroin addiction in most countries are *methadone* and *buprenorphine*. Both methadone and buprenorphine have a high and long-lasting affinity for opioid receptors. Ironically, they are both opioids with many of the same adverse effects as heroin. However, because they produce less pleasure than heroin, the strategy has been to block heroin withdrawal effects with either methadone or buprenorphine and then maintain the individual on one of those drugs until they can be weaned from it. Methadone replacement has been shown to improve the success rate of some treatment programs, but its adverse effects are problematic (see Nutt, 2015). Buprenorphine has fewer adverse side effects than methadone (see Nutt, 2015) but is considered less effective than methadone (see Mattick et al., 2014).

In 1994, the Swiss government took an alternative approach to the problem of heroin addiction. It established a series of clinics in which, as part of a total treatment package, Swiss heroin users could receive heroin injections from a physician for a small fee. The Swiss government wisely funded a major research program to evaluate the clinics (see Khan et al., 2014). The results have been uniformly positive. Once they had a cheap, reliable source of heroin, most heroin users gave up their criminal lifestyles, and their health improved once they were exposed to the specialized medical and counseling staff at the clinics. Many heroin users returned to their family and

jobs, and many opted to reduce or curtail their heroin use. As a result, heroin use is no longer present in Swiss streets and parks; drug-related crime has substantially declined; and the physical and social well-being of the heroin users has greatly improved. Furthermore, the number of new cases of heroin addiction has declined (see Csete et al., 2016).

These positive results have led to the establishment of similar experimental programs in other countries (e.g., Canada, Germany, Netherlands, Spain, Switzerland, and United Kingdom) with similar success. Indeed, in a large randomized trial, supervised treatment with heroin was shown to be more effective and less costly than methadone (Nosyk et al., 2012). Furthermore, safe injection facilities have managed to reduce the spread of infection (see Southwell et al., 2019) and death from heroin overdose in many cities (e.g., Csete et al., 2016). Given the unqualified success of such programs in dealing with the drug problem, it is interesting to consider why some governments have not adopted them (e.g., Csete et al., 2016; Kupferschmidt, 2014). What do you think?

Comparing the Health Hazards of Commonly Used Drugs

The previous module described the health hazards of five commonly used drugs, some legal, some illegal. This module directly compares those health hazards for you.

Interpreting Studies of the Health Hazards of Drugs

LO 15.12 Explain why it is difficult to determine causality in studies of the health hazards of drugs.

You have probably noticed a repeated disclaimer in the previous module: Interpretation of the adverse effects observed in drug users is almost always complicated by the fact that the relevant research is correlational. Because most studies compare the health of known drug users with that of nonusers, one can never be certain that any observed differences in health are due to the drug or to some other difference between the two groups. An additional complication arises from the fact that most studies recruit drug users from addiction treatment clinics. Such users are typically the most severely addicted. Indeed, when studies use samples of drug users from the general population, the outcomes can be quite different (see Krebs & Johansen, 2012). For example, studies that recruit samples of drug users from addiction treatment clinics typically find that drug addiction is difficult to treat. In contrast, studies that recruit samples of drug users from the general population have found that many drug users can successfully treat their own addictions without professional help and live productive lives thereafter (see Blanco et al., 2013; Chapman & MacKenzie, 2010). The following quote from a former student of mine (SB), who was a heavy user of methamphetamine, illustrates this point:

> *"I dropped out of high school 17 years ago... Somewhere along the line, knowing full well I was playing with fire, I developed a methamphetamine habit. For 3 years I used almost daily, most often injecting. I never lost my job or my house, never experienced psychiatric symptoms, and never received any treatment for medical conditions related to my drug use—but it did permeate my life and interfere with my goals. Eventually I fell in love with the woman who would later become my wife, and that relationship provided all the motivation I needed to stop using regularly. Until we decided to get pregnant (5 years later) I continued to use meth at parties, but never more than twice a year... More than a decade after dropping out of high school I enrolled in university and found I was able to do my homework and get good grades."

Comparison of the Hazards of Nicotine, Alcohol, Marijuana, Cocaine, and Heroin

LO 15.13 Compare the direct health hazards of nicotine, alcohol, marijuana, cocaine, and heroin.

One way of comparing the adverse effects of tobacco, alcohol, marijuana, cocaine, and heroin is to compare the prevalence of their abuse in society as a whole. In terms of this criterion, it is clear that tobacco and alcohol have a

greater negative impact than marijuana, cocaine, and heroin (see Figure 15.5). Another method of comparison is based on global death rates: Tobacco has been implicated in approximately 6.5 million deaths per year, alcohol in approximately 3 million deaths per year, and all other drugs combined in about 450,000 deaths per year (see Britton, 2017; Degenhardt et al., 2018; Reitsma et al., 2017).

Journal Prompt 15.4

Do you or somebody you love use a hard drug such as nicotine or alcohol? How would you summarize the relevant data to someone who you wanted to stop using the drug?

But what about the individual drug user? Who is taking greater health risks: the cigarette smoker, the alcohol drinker, the marijuana smoker, the cocaine user, or the heroin user? You now have the information to answer this question. Would you have ranked the health risks of these drugs in the same way before you began this chapter? How have the laws, or lack thereof, influenced the hazards associated with the five drugs?

Figure 15.5 Global prevalence of addiction to each of six commonly used psychoactive drugs.

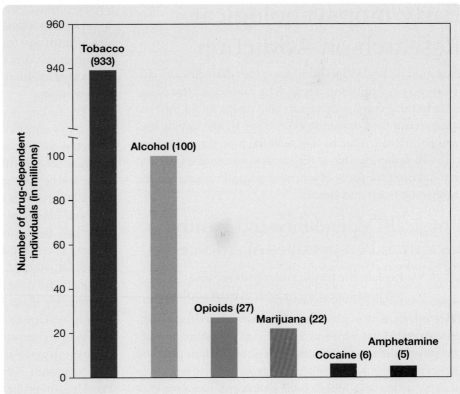

Based on Britton, 2017; Degenhardt et al., 2018; Reitsma et al., 2017.

*Student Excerpt. Anonymous.

Scan Your Brain

So far in this chapter, you have learned about drug action, addiction, and tolerance. This is a good place to pause and reinforce what you have learned. In each blank, write the appropriate term. The correct answers are provided at the end of the exercise. Review material related to your errors and omissions before proceeding.

1. The conversion of a drug from its active form to a nonactive form is called _____.

2. _____ is a state of decreased sensitivity to a drug that develops as a result of exposure to it.

3. People experience _____ symptoms after the sudden elimination of a drug.

4. _____ drug tolerance is demonstrated when tolerance effects are maximally expressed only when a drug is administered in the same situation in which it has previously been administered.

5. According to Siegel, just thinking about a drug can induce conditioned responses, which shows that _____ stimuli are just as effective in this process.

6. Drug users are particularly vulnerable to the lethal effects of drug overdose when administered in a new environment, a process known as _____.

7. During tobacco use, the psychoactive substance _____ acts on cholinergic receptors in the brain.

8. Long-term (mostly male) smokers may develop _____ syndrome, which causes blood vessels (particularly in the legs) to become restricted and may lead to gangrene.

9. A few days after alcohol cessation, people may experience _____, which is characterized by hallucinations, delusions, agitation, and disorientation.

10. Excessive alcohol consumption can cause liver _____ and _____, conditions that may both be lethal.

11. The psychoactive effects of cannabis are due to a constitutent called _____.

12. _____ is a semisynthetic opioid that penetrates the blood–brain barrier more effectively than morphine.

13. _____ and _____ are endogenous opioid neurotransmitters that bind to receptors and act as analgesics/painkillers.

Scan Your Brain answers: (1) drug metabolism, (2) Drug tolerance, (3) withdrawal, (4) Contingent, (5) interoceptive, (6) situational specificity, (7) nicotine, (8) Buerger's, (9) delirium tremens, (10) cirrhosis, pancreatitis, (11) delta-9-tetrahydrocannabinol (THC), (12) Heroin, (13) Endorphins, enkephalins.

Early Biopsychological Research on Addiction

This module begins by introducing two diametrically different ways of thinking about drug addiction: Are drug-addicted individuals driven to take drugs by an internal need, or are they drawn to take drugs by the anticipated positive effects? After having read the preceding chapters, you will recognize this is the same fundamental question that has been the focus of biopsychological research on the motivation to eat and sleep.

Physical-Dependence and Positive-Incentive Perspectives of Addiction

LO 15.14 Explain the physical-dependence and positive-incentive perspectives of addiction.

Early attempts to explain the phenomenon of drug addiction attributed it to physical dependence. According to various **physical-dependence theories of addiction**, physical dependence traps addicted individuals in a vicious circle of drug taking and withdrawal symptoms. The idea was that drug users whose intake has reached a level sufficient to induce physical dependence are driven by their withdrawal symptoms to self-administer the drug each time they attempt to curtail their intake.

Early drug addiction treatment programs were based on the physical-dependence perspective. They attempted to break the vicious cycle of drug taking by gradually withdrawing drugs from addicted individuals in a hospital environment. Unfortunately, once discharged, almost all detoxified habitual drug users return to their former drug-taking habits.

The failure of detoxification as a treatment for addiction is not surprising, for two reasons. First, some highly addictive drugs, such as cocaine and amphetamines, do not produce severe withdrawal distress. Second, the pattern of drug taking routinely displayed by many habitual drug users involves an alternating cycle of binges and detoxification. There are a variety of reasons for this pattern of drug use. For example, some addicted individuals adopt it because weekend binges are compatible with their work schedules, others adopt it because they do not have enough money to use drugs continuously, and others have it forced on them by their repeated unsuccessful efforts to shake their habit. However, whether detoxification is by choice or necessity, it does not stop addicted individuals from renewing their drug-taking habits.

As a result of these problems with physical-dependence theories of addiction, a different approach began to predominate in the 1970s and 1980s (see Higgins, Heil, & Lussier, 2004). This approach was based on the assumption that most addicted individuals take drugs not to escape or to avoid the unpleasant consequences of withdrawal, but rather to obtain the drugs' positive effects. Theories of addiction based on this premise are called **positive-incentive theories of addiction**. They hold that the primary factor in most cases of addiction is the craving for the positive-incentive (expected pleasure-producing) properties of the drug.

There is no question that physical dependence does play a role in addiction: Addicted individuals do sometimes consume the drug to alleviate their withdrawal symptoms. However, most researchers now assume that the more important factor in addiction is the drugs' *hedonic* (pleasurable) effects (see Cardinal & Everitt, 2004; Everitt, Dickinson, & Robbins, 2001).

The remainder of this module summarizes the early biopsychological research into the brain mechanisms of addiction. As you will learn, this research was largely based on the positive-incentive theory of addiction.

Intracranial Self-Stimulation and the Mesotelencephalic Dopamine System

LO 15.15 Describe the intracranial self-stimulation (ICSS) paradigm.

Rats, humans, and many other species will administer brief bursts of weak electrical stimulation to specific sites in their own brains (see Figure 15.6). This phenomenon is known as **intracranial self-stimulation (ICSS)**, and the brain sites capable of mediating the phenomenon are often called *pleasure centers*. When research on addiction turned to positive incentives in the 1970s and 1980s, what had been learned about the neural mechanisms of pleasure from studying intracranial self-stimulation served as a starting point for the study of the neural mechanisms of addiction.

Olds and Milner (1954), the discoverers of intracranial self-stimulation, argued that the specific brain sites that mediate self-stimulation are those that normally mediate the pleasurable effects of natural rewards (i.e., food, water, and sex). Accordingly, researchers studied the self-stimulation of various brain sites in order to map the neural circuits that mediate the experience of pleasure.

It was initially assumed that intracranial self-stimulation was a unitary phenomenon—that is, that its fundamental properties were the same regardless of the site of stimulation. Most early studies of intracranial self-stimulation involved septal or lateral hypothalamic

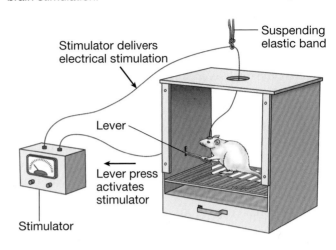

Figure 15.6 A rat pressing a lever to obtain rewarding brain stimulation.

stimulation because the rates of self-stimulation from these sites are spectacularly high: Rats typically press a lever thousands of times per hour for stimulation of these sites, stopping only when they become exhausted. However, self-stimulation of many other brain structures has been documented.

The mesotelencephalic dopamine system plays an important role in intracranial self-stimulation. The **mesotelencephalic dopamine system** is a system of dopaminergic neurons that projects from the mesencephalon (the midbrain) into various regions of the telencephalon. As Figure 15.7 indicates, the neurons that compose the

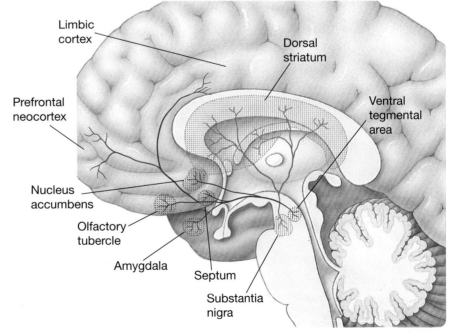

Figure 15.7 The mesotelencephalic dopamine system in the human brain, consisting of the nigrostriatal pathway (green) and the mesocorticolimbic pathway (red).

Based on Klivington, K. A. (Ed.). (1992). Gehirn und Geist. Heidelberg, Germany: Spektrum Akademischer Verlag.

mesotelencephalic dopamine system have their cell bodies in two midbrain nuclei—the **substantia nigra** and the **ventral tegmental area**. Their axons project to a variety of telencephalic sites, including specific regions of the prefrontal cortex, the limbic cortex, the olfactory tubercle, the amygdala, the septum, the dorsal striatum, and, in particular, the **nucleus accumbens** (a nucleus of the ventral striatum)—see Kringelbach and Berridge (2012).

Most of the axons of dopaminergic neurons that have their cell bodies in the substantia nigra project to the dorsal striatum; this component of the mesotelencephalic dopamine system is called the *nigrostriatal pathway*. It is degeneration in this pathway that is associated with Parkinson's disease (see Chapter 10).

Most of the axons of dopaminergic neurons that have their cell bodies in the ventral tegmental area project to various cortical and limbic sites. This component of the mesotelencephalic dopamine system is called the **mesocorticolimbic pathway**. Although there is some intermingling of the neurons between these two dopaminergic pathways, several pieces of evidence have supported the view that the mesocorticolimbic pathway plays an important role in mediating intracranial self-stimulation. The following are four of them:

- Many of the brain sites at which self-stimulation occurs are part of the mesocorticolimbic pathway.
- Intracranial self-stimulation is often associated with an increase in dopamine release in the mesocorticolimbic pathway.

- Dopamine agonists tend to increase intracranial self-stimulation, and dopamine antagonists tend to decrease it.
- Lesions of the mesocorticolimbic pathway tend to disrupt intracranial self-stimulation.

Early Evidence of the Involvement of Dopamine in Drug Addiction

LO 15.16 Describe two methods for measuring the rewarding effects of drugs.

In the 1970s, following much research on the role of dopamine in intracranial self-stimulation, experiments began to implicate dopamine in the rewarding effects of natural reinforcers and addictive drugs (mainly stimulants). These experiments, which were largely conducted in nonhumans, used one of two methods for measuring the rewarding effects of drugs: the drug self-administration paradigm and the conditioned place-preference paradigm (see Badiani et al., 2011; Kuhn, Kalivas, & Bobadilla, 2019). These two methods, still in use today, are illustrated in Figure 15.8.

> **Journal Prompt 15.5**
>
> What do you think are some of the limitations of using nonhuman animals to study the neural basis of drug addiction?

In the **drug self-administration paradigm**, nonhuman animals press a lever to inject drugs into themselves

Figure 15.8 Two behavioral paradigms that are used extensively in the study of the neural mechanisms of addiction: the drug self-administration paradigm and the conditioned place-preference paradigm.

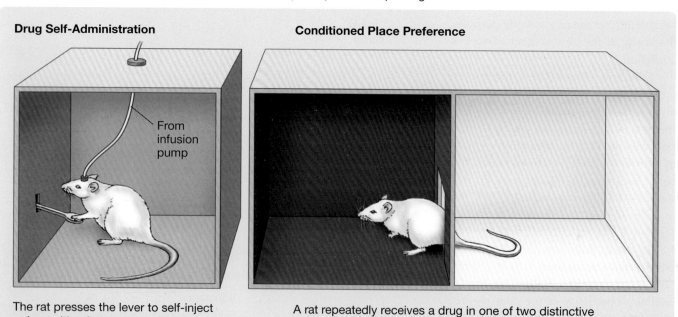

Drug Self-Administration

From infusion pump

The rat presses the lever to self-inject a drug, either into an area of its brain or into general circulation.

Conditioned Place Preference

A rat repeatedly receives a drug in one of two distinctive compartments. Then, on the test, the tendency of the rat, now drug-free, to prefer the drug compartment is assessed.

through implanted *cannulas* (thin tubes). They readily learn to self-administer intravenous injections of drugs to which humans become addicted (see Kuhn, Kalivas, & Bobadilla, 2019; O'Connor et al., 2011). Studies in which microinjections have been self-administered directly into particular brain structures have proved particularly enlightening.

In the **conditioned place-preference paradigm**, nonhuman animals repeatedly receive a drug in one compartment (the *drug compartment*) of a two-compartment box. Then, during the test phase, the drug-free rat is placed in the box, and the proportion of time it spends in the drug compartment, as opposed to the equal-sized but distinctive *control compartment,* is measured. Subjects usually prefer the drug compartment over the control compartment when the drug compartment has been associated with the effects of drugs to which humans become addicted. The main advantage of the conditioned place-preference paradigm is that the subjects are tested while they are drug-free, which means that the measure of the incentive value of a drug is not confounded by other effects the drug might have on behavior.

Experiments that used these methods showed that dopamine played an important role in the rewarding effects of addictive drugs and natural reinforcers. For example, in rats, dopamine antagonists blocked the self-administration of, or the conditioned preference for, several different addictive drugs, and they reduced the reinforcing effects of food. These findings suggested that dopamine signaled something akin to reward value or pleasure.

Nucleus Accumbens and Drug Addiction

LO 15.17 Explain the role of the nucleus accumbens in drug addiction.

Once evidence had accumulated linking dopamine to natural reinforcers and drug-induced reward, investigators began to explore particular sites in the mesocorticolimbic dopamine pathway by conducting experiments on laboratory animals. Their findings soon focused attention on the nucleus accumbens. Events occurring in the nucleus accumbens and in the dopaminergic input to it from the ventral tegmental area appeared to be most clearly related to the experience of reward and pleasure.

The following are four kinds of findings from research on laboratory animals that focused attention on the nucleus accumbens (see Deadwyler et al., 2004; Nestler, 2005; Pierce &

Kumaresan, 2006). Most of the early studies focused on stimulants.

- Laboratory animals self-administered microinjections of addictive drugs directly into the nucleus accumbens.
- Microinjections of addictive drugs into the nucleus accumbens produced a conditioned place preference for the compartment in which they were administered.
- Lesions to either the nucleus accumbens or the ventral tegmental area blocked the self-administration of addictive drugs into general circulation or the development of drug-associated conditioned place preferences.
- Both the self-administration of addictive drugs and the experience of natural reinforcers were found to be associated with elevated levels of extracellular dopamine in the nucleus accumbens.

Current Approaches to the Mechanisms of Addiction

The previous module brought us from the beginnings of research on the brain mechanisms of addiction to current research, which will be discussed in this module. Figure 15.9 summarizes the major shifts in thinking about the brain mechanisms of addiction that have occurred over time.

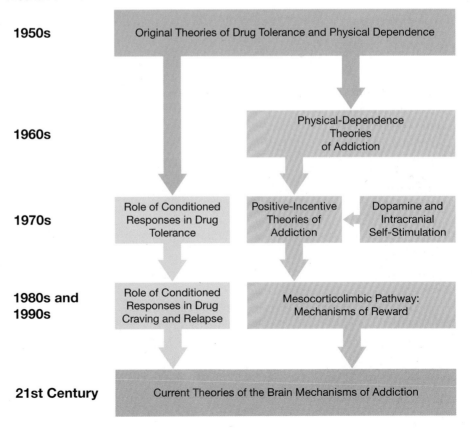

Figure 15.9 Historic influences that shaped current thinking about the brain mechanisms of addiction.

Figure 15.9 shows that two lines of thinking about the brain mechanisms of addiction both had their origins in classic research on drug tolerance and physical dependence. One line developed into physical-dependence theories of addiction, which, though appealing in their simplicity, proved to be inconsistent with the evidence, and these inconsistencies led to the emergence of positive-incentive theories. The positive-incentive approach to addiction, in combination with research on dopamine and intracranial self-stimulation, led to a focus on the mesocorticolimbic pathway and the mechanisms of reward. The second line of thinking about the brain mechanisms of addiction also began with early research on drug tolerance and physical dependence. This line moved ahead with the discovery that drug-associated cues come to elicit conditioned compensatory responses through a Pavlovian conditioning mechanism and that these conditioned responses are largely responsible for functional drug tolerance. This finding gained further prominence when researchers discovered that conditioned responses elicited by drug-associated cues were major factors in drug craving and **relapse** (the return to one's drug-taking habit after a period of voluntary abstinence).

These two lines of research together have shaped modern thinking about the brain mechanisms of addiction, but as you can see from Figure 15.9, this was not the end of the story. In this module you will learn about modern approaches to the study of drug addiction.

Three Stages in the Development of an Addiction

LO 15.18 Describe the three stages in the development of a drug addiction.

Modern approaches to drug addiction are increasingly concerned with modeling each of the three stages involved in the development of an addiction: (1) initial drug taking, (2) habitual drug taking, and (3) drug craving and repeated relapse (see Figure 15.10).

INITIAL DRUG TAKING. Not everyone given access to a drug will consume it, and of those that do, many will never take the drug more than once. Why does someone choose to take a drug those first few times? Price and availability of the drug, peer pressure, and prior life experiences are all well-known factors in initial drug taking; however, research has suggested a role for several others.

Experimental studies of drug self-administration in rats have pointed to a variety of factors that facilitate, or protect from, initial drug taking. Food restriction (Carroll & Meisch, 1984), social stress (see Ahmed, 2005), and environmental stress (Ambroggi et al., 2009; Lu et al., 2003) facilitate the acquisition of drug self-administration by rats. In contrast,

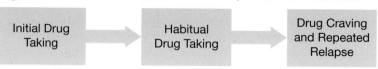

Figure 15.10 Three Stages in the Development of an Addiction.

Initial Drug Taking → Habitual Drug Taking → Drug Craving and Repeated Relapse

environmental enrichment (Puhl et al., 2012), social interaction (Raz & Berger, 2010), and access to nondrug reinforcers (Ahmed, Lenoir, & Guillem, 2013) all protect against the acquisition of drug self-administration.

The acquisition of drug self-administration can also be predicted from certain behavioral traits. For example, rats that prefer to drink sweetened water and rats that are more active in a novel environment are both more likely to self-administer cocaine. These two behavioral traits have been likened to *novelty seeking*, a behavioral trait commonly associated with initial drug taking in humans.

When drugs are viewed as tools, or *instruments* (see Müller & Schumann, 2011), the answer to the question of why people start taking them is simple: People first take a drug to determine if it will be useful to them in some way (or to confirm rumors of its usefulness); and their choice to continue taking that drug will depend on whether they find some use in it (Boys & Marsden, 2003). As an example, I (SB) often have a coffee in the morning before driving to campus to ensure that I am alert behind the wheel. In other words, I use caffeine as an instrument to increase my alertness. Other stimulants, and nicotine, are often used for this same reason. Alcohol, in low doses, can be used as an instrument to facilitate social interaction and sexual intercourse (see Patrick & Maggs, 2009). Alcohol is also commonly used as an instrument to relieve stress and anxiety, and the same is true for marijuana (Müller & Schumann, 2011). Furthermore, people with medical conditions often take specific drugs to self-medicate. For example, many individuals with schizophrenia use nicotine to alleviate any cognitive impairments and **anhedonia** (a general inability to experience pleasure)—see Rezvani and Levin (2001). Of course, many people use drugs simply for their pleasurable effects (see Hart, 2013).

HABITUAL DRUG TAKING. Habitual drug use, a necessary component of an addiction, is not the inevitable outcome of taking a drug. Many people periodically use addictive drugs and experience their hedonic effects without becoming habitual users (see Badiani et al., 2011). What is responsible for the transition from initial drug taking to habitual drug taking?

Positive-incentive theories of drug addiction have trouble explaining why some users become habitual users and others do not. Another challenge faced by positive-incentive theories is that they are unable to explain why addicted individuals often experience a big discrepancy between the hedonic value (*liking*) and the positive-incentive value (*wanting*) of their preferred drug. **Positive-incentive value** refers specifically to the anticipated pleasure associated with an action (e.g., taking a drug), whereas **hedonic value** refers to the amount of pleasure that is actually experienced. Addicted individuals often report that they are compulsively driven to take their drug by its positive-incentive value (they *want* the drug), although taking the drug is often not as pleasurable as it once was (they no longer *like* the drug)—see Everitt & Robbins (2016).

One more recent theory of drug addiction, the **incentive-sensitization theory**, is able to explain why some drug users become habitual users and others do not. It is also able to explain the discrepancy between the hedonic value and the positive-incentive value of drug taking in addicted persons (see Berridge, Robinson, & Aldridge, 2009). The central tenet of the incentive-sensitization theory is that the positive-incentive value of addictive drugs increases (i.e., becomes sensitized) with repeated drug use in addiction-prone individuals (see Miles et al., 2004; Robinson & Berridge, 2008). This renders addiction-prone individuals highly motivated to seek and consume the drug. A key point of the incentive-sensitization theory is that it isn't the pleasure (*liking*) of taking the drug that is the basis of habitual drug use and addiction; it is the anticipated pleasure of drug taking (i.e., the drug's positive-incentive value)—the *wanting* or *craving* for the drug. Initially, a drug's positive-incentive value is closely tied to its pleasurable effects; but tolerance often develops to the pleasurable effects, whereas the drug-addicted individual's craving for the drug is sensitized. Thus, in drug-addicted individuals, the positive-incentive value of the drug is often out of proportion with the pleasure actually derived from it: Many addicted individuals are miserable, their lives are in ruin, and the drug effects are not that great anymore; but they crave the drug more than ever.

Inspired by this important distinction between the wanting (positive-incentive value) vs. liking (hedonic value) of a drug, researchers have been trying to identify the brain circuitry responsible for each. There is now general agreement that there are differences in the circuitry for wanting vs. liking, and that dopamine release in the nucleus accumbens (via the mesocorticolimbic pathway) is more closely associated with the wanting of a drug, rather than the liking of it (see Berridge & Kringelbach, 2015; Kringelbach & Berridge, 2012). For example, studies have shown that neutral stimuli that signal the impending delivery of a reward (e.g., food or an addictive drug) are sufficient to trigger dopamine release in the nucleus accumbens (see Floresco, 2015). Researchers are still determining the brain circuitry responsible for the liking of a drug, but we do know that dopamine is less important for this aspect of

drug taking than was once believed and that the structures involved only partially overlap with those that mediate the wanting of a drug (see Kringelbach & Berridge, 2012; Liu et al., 2011).

Several changes in brain function have been associated with the transition from initial drug taking to habitual drug taking. First, there is a difference in how the striatum of drug-addicted individuals reacts to drugs and drug-associated cues. In addicted individuals, striatal control of drug taking is shifted from the nucleus accumbens (i.e., the ventral striatum) to the dorsal striatum (Pierce & Vanderschuren, 2010), an area that is known to play a role in habit formation and retention (see Chapter 11). Also, at the same time, there are impairments in the function of the prefrontal cortex, which likely relates to the loss of self-control that accompanies addiction (George & Koob, 2010; Goldstein & Volkow, 2011; Volkow & Warren, 2011). It was originally believed that these brain differences merely accompanied the transition to habitual drug taking, but Ersche and colleagues (2012) showed that these differences predispose an individual to that transition (Volkow & Baler, 2012).

There has been a growing appreciation that drug addiction is a specific expression of a more general behavioral problem: the inability to refrain from a behavior despite its adverse effects. Addicted individuals have been found to make poor decisions, to engage in excessive risk taking, and to have deficits in self-control (see Volkow et al., 2013). These behavioral problems are not limited to drug-addicted individuals. A lot of attention has recently been paid to overeating, compulsive gambling, kleptomania (compulsive shoplifting), and compulsive shopping as addictive behaviors (see Jabr, 2013; Potenza, 2015; Robbins & Clark, 2015).

Another change that habitual drug users often experience is *anhedonia* (a general inability to experience pleasure in response to natural reinforcers)—see Ahmed (2005); Robbins (2016). Those things that most of us find pleasurable (e.g., sex, eating, sleeping) are often less pleasurable to the habitual drug user. This devaluation of natural reinforcers poses a major problem for the treatment of addiction because it persists even after stopping the drug. Without the experience of pleasure from natural reinforcers, many abstinent drug users experience craving for their drug and then relapse.

DRUG CRAVING AND RELAPSE. Addiction is not necessarily a life sentence; many addicted individuals successfully treat their addiction by either stopping or reducing their drug intake (see Klingemann, Sobell, & Sobell, 2009). Others manage to stop or reduce their drug intake with the help of a treatment program. However, weeks, months, or even years later, some abstainers may experience drug craving. Such craving can lead to a relapse. This propensity to relapse, even after a long period of voluntary abstinence, is a hallmark of addiction.

Thus, understanding the causes of relapse is one key to understanding addiction and its treatment.

Three different causes of relapse in addicted individuals have been identified (see Badiani et al., 2011; Pickens et al., 2011):

- Many therapists and patients point to stress as a major factor in relapse (see Moschak et al., 2018). The impact of stress on drug taking was illustrated in a dramatic fashion by the marked increases in cigarette, alcohol, and marijuana consumption by New Yorkers following the terrorist attacks of September 11, 2001 (see Vlahov et al., 2004).

- **Drug priming** (a single exposure to the formerly misused drug) is another cause of relapse. Many addicted individuals who have managed to abstain feel they have their addiction under control. Reassured by this feeling, they sample their addictive drug just once and are immediately plunged back into full-blown addiction.

- Exposure to cues (e.g., people, times, places, or objects) that have previously been associated with drug taking has been shown to precipitate relapse (see Everitt & Robbins, 2016; Milton & Everitt, 2012; Steketee & Kalivas, 2011). The fact that the many U.S. soldiers who became addicted to heroin while fighting in the Vietnam War easily shed their addiction when they returned home has been attributed to removal from their drug-associated environment.

Explanation of the effects of drug-associated cues on relapse is related to our discussion of conditioned drug tolerance earlier. You may recall from earlier in this chapter that cues that predict drug exposure come to elicit conditioned compensatory responses through a Pavlovian conditioning mechanism, and because conditioned compensatory responses are usually opposite to the original drug effects, they produce tolerance. The point here is that conditioned compensatory responses seem to increase craving in abstinent drug-addicted individuals and, in so doing, trigger relapse. Indeed, just thinking about a drug is enough to induce craving and relapse.

It turns out that exactly when a drug-associated cue is presented is an important determinant of its effects: Cues presented soon after drug withdrawal are less likely to elicit craving and relapse than cues presented later. This time-dependent increase in cue-induced drug craving and relapse is known as the **incubation of drug craving** (Pickens et al., 2011). This phenomenon is illustrated in Figure 15.11. Notice in Figure 15.11 that shortly after cocaine withdrawal a drug-associated cue elicited few presses to a lever that had always delivered drug before withdrawal. However, as time passed, there was a gradual increase in lever pressing in response to presentations of a drug-associated cue. In addition to cocaine, the incubation of drug craving has also been demonstrated

Figure 15.11 Incubation of cocaine craving in rats that were previously self-administering cocaine. After cocaine withdrawal, there was a time-dependent increase in the number of lever presses the rats made in response to a drug-associated cue.

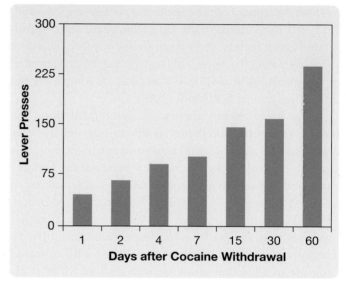

Data redrawn from Pickens, C. L., Airavaara, M., Theberge, F., Fanous, S., Hope, B. T., & Shaham, Y. (2011). Neurobiology of the incubation of drug craving. *Trends in Neurosciences, 34*, 411–420.

with heroin, methamphetamine, alcohol, and nicotine (see Pickens et al., 2011). This important phenomenon may help explain why some individuals who have seemingly recovered from an addiction experience craving and relapse even after years of abstinence (Bedi et al., 2011).

The animal literature on drug self-administration has suggested two additional factors that may play a role in drug craving and relapse. First, environmental enrichment after drug withdrawal has reduced cue- and stress-induced, but not drug-priming-induced, relapse of drug self-administration (Chauvet et al., 2009, 2012). Second, even a few brief exposures to nondrug reinforcers can reliably reduce relapse of cocaine self-administration (e.g., Liu & Grigson, 2005; Quick et al., 2011). Perhaps the best way of preventing relapse in recovered drug-addicted individuals is to improve their opportunities and surroundings. Indeed, such interventions have proven successful (e.g., Cao et al., 2011; Dutra et al., 2008; Friedmann et al., 2004).

Current Concerns about the Drug Self-Administration Paradigm

LO 15.19 Describe two sets of findings that have challenged the relevance of drug self-administration studies.

Much of what we now know about the neurobiology of drug addiction has been derived from animal studies of drug self-administration. Two sets of recent findings have challenged the relevance of drug self-administration studies and have opened new avenues of research.

UNNATURAL HOUSING AND TESTING CONDITIONS.
In conventional drug self-administration studies, rats are housed individually and tested in a barren test chamber where the only rewarding thing they can do is press a lever for a drug injection (see Figure 15.8). What would happen if the rats had more natural housing and testing conditions? When rats were either group housed (Raz & Berger, 2010), given access to enriched environments (Nader et al., 2008; Solinas et al., 2010), provided with the opportunity to obtain nondrug reinforcers (Ahmed, 2010, 2012), or housed in a large naturalistic environment (Alexander et al., 1981), they were much less likely to self-administer drugs. For example, only 10 percent of rats given a choice between consuming sucrose and self-administering cocaine displayed a preference for the cocaine (Lenoir et al., 2007).

The finding that more naturalistic housing and testing conditions reduce drug self-administration has led some researchers to question the relevance of conventional drug self-administration research to human addiction (Ahmed, Lenoir, & Guillem, 2013). However, when viewed another way, these findings suggest that human drug addiction might be prevented by improving the environment and life choices of those most vulnerable to addiction.

> **Journal Prompt 15.6**
> If you could change those policies related to how we treat drug-addicted individuals, what would be the first thing you would change?

EXCESSIVE FOCUS ON STIMULANTS.
It has long been assumed that the mechanisms of addiction are independent of the specific addictive drug. Thus, there has been little concern over the specific drug used in studies of addiction (see Badiani et al., 2011). Indeed, most drug self-administration studies have been done with stimulants, and most biopsychological theories of drug addiction have been built on the results of such studies.

Research on the self-administration of stimulants has led to two major conclusions about the mechanisms of drug addiction: (1) that all addictive drugs activate the mesocorticolimbic pathway and (2) that dopamine is important for the reinforcing properties of all addictive drugs. However, studies of opioid self-administration have led to different conclusions (see Nutt et al., 2015). For example, although mesocorticolimbic pathway lesions or dopamine antagonists disrupt habitual cocaine self-administration, they did not disrupt habitual heroin self-administration (see Badiani et al., 2011).

A Noteworthy Case of Addiction

LO 15.20   Explain the significance of the case of Sigmund Freud.

To illustrate in a more personal way some of the things you have learned about addiction, this chapter concludes with a case study of one drug-addicted individual: Sigmund Freud, a man of great significance to psychology.

Freud's case is particularly important for two reasons. First, it shows that nobody, no matter how powerful their intellect, is immune to the addictive effects of drugs.

Second, it allows comparisons between the two addictive drugs with which Freud had problems.

The Case of Sigmund Freud

In 1883, a German army physician prescribed cocaine, which had recently been isolated, to Bavarian soldiers to help them deal with the demands of military maneuvers. When Freud read about this, he decided to procure some of the drug.

In addition to taking cocaine himself, Freud pressed it on his friends and associates, both for themselves and for their patients. He even sent some to his fiancée. In short, by today's standards, Freud was a public menace.

Freud's famous essay "Song of Praise" was about cocaine and was published in July 1884. Freud wrote in such glowing terms about his own personal experiences with cocaine that he created a wave of interest in the drug. But within a year, there was a critical reaction to Freud's premature advocacy of the drug. As evidence accumulated that cocaine was highly addictive and produced a psychosis-like state at high doses, so too did published criticisms of Freud.

Freud continued to praise cocaine until the summer of 1887, but soon thereafter he suddenly stopped all use of cocaine—both personally and professionally. Despite the fact that he had used cocaine for 3 years, he seems to have had no difficulty stopping.

Some 7 years later, in 1894, when Freud was 38, his physician and close friend ordered him to stop smoking because it was causing a heart arrhythmia. Freud was a heavy smoker; he smoked approximately 20 cigars per day.

Freud did stop smoking, but 7 weeks later he started again. On another occasion, Freud stopped for 14 months, but at the age of 58, he was still smoking 20 cigars a day—and still struggling against his addiction. He wrote to friends that smoking was adversely affecting his heart and making it difficult for him to work…yet he kept smoking.

In 1923, at the age of 67, Freud developed sores in his mouth. They were cancerous. When he was recovering from oral surgery, he wrote to a friend that smoking was the cause of his cancer…yet he kept smoking.

In addition to the cancer, Freud began to experience severe heart pains (tobacco angina) whenever he smoked…still he kept smoking.

At 73, Freud was hospitalized for his heart condition and stopped smoking. He made an immediate recovery. But 23 days later, he started to smoke again.

In 1936, at the age of 79, Freud was experiencing more heart trouble, and he had had 33 operations to deal with his recurring oral cancer. His jaw had been entirely removed and replaced by an artificial one. He was in constant pain, and he could swallow, chew, and talk only with difficulty...yet he kept smoking.

Freud died of cancer in 1939 (see Sheth, Bhagwate, & Sharma, 2005).

Themes Revisited

Two of this text's themes—thinking creatively and clinical implications—received strong emphasis in this chapter because they are integral to its major objective: to sharpen your thinking about the effects of addiction on people's health. You were repeatedly challenged to think about drug addiction in ways that may have been new to you but are more consistent with the evidence.

The evolutionary perspective theme was also highlighted frequently in this chapter, largely because of the nature of biopsychological research into drug addiction. Because of the risks associated with the administration of addictive drugs and the direct manipulation of brain structures, the majority of biopsychological studies of drug addiction involve nonhumans—mostly rats and monkeys. Also, in studying the neural mechanisms of addiction, there is a need to maintain an evolutionary perspective. It is important not to lose sight of the fact that brain mechanisms did not evolve to support addiction; they evolved to serve natural-adaptive functions and have somehow been co-opted by addictive drugs.

Of the two emerging themes, the thinking about epigenetics theme appeared throughout this chapter. Most prominent was the consistent finding that drug taking can have effects on one's offspring via transgenerational epigenetic mechanisms.

Key Terms

Pharmacological, p. 405

Basic Principles of Drug Action

Psychoactive drugs, p. 406
Drug metabolism, p. 407
Drug tolerance, p. 407
Cross tolerance, p. 407
Drug sensitization, p. 407
Metabolic tolerance, p. 407
Functional tolerance, p. 407
Withdrawal syndrome, p. 407
Physically dependent, p. 408
Drug-addicted individual, p. 408

Role of Learning in Drug Tolerance

Contingent drug tolerance, p. 409
Before-and-after design, p. 409
Conditioned drug tolerance, p. 409
Conditioned compensatory responses, p. 410
Exteroceptive stimuli, p. 410
Interoceptive stimuli, p. 410

Five Commonly Used Drugs

Nicotine, p. 411
Smoking, p. 411
Vaping, p. 411
Smoker's syndrome, p. 412

Buerger's disease, p. 412
Teratogen, p. 412
Drug craving, p. 412
Depressant, p. 413
Delirium tremens (DTs), p. 413
Korsakoff's syndrome, p. 413
Cirrhosis, p. 414
Fetal alcohol syndrome (FAS), p. 414
Cannabis, p. 414
THC, p. 414
Hashish, p. 415
Narcotic, p. 415
Anandamide, p. 416
Stimulants, p. 417
Cocaine, p. 417
Crack, p. 417
Cocaine sprees, p. 417
Cocaine psychosis, p. 417
Amphetamine, p. 417
Empathogens, p. 417
Dopamine transporters, p. 417
Opium, p. 418
Morphine, p. 418
Codeine, p. 418
Opioids, p. 418
Analgesics, p. 418
Harrison Narcotics Act, p. 419
Heroin, p. 419

Early Biopsychological Research on Addiction

Physical-dependence theories of addiction, p. 422
Positive-incentive theories of addiction, p. 423
Intracranial self-stimulation (ICSS), p. 423
Mesotelencephalic dopamine system, p. 423
Substantia nigra, p. 424
Ventral tegmental area, p. 424
Nucleus accumbens, p. 424
Mesocorticolimbic pathway, p. 424
Drug self-administration paradigm, p. 424
Conditioned place-preference-paradigm, p. 425

Current Approaches to the Mechanisms of Addiction

Relapse, p. 426
Anhedonia, p. 426
Positive-incentive value, p. 427
Hedonic value, p. 427
Incentive-sensitization theory, p. 427
Drug priming, p. 428
Incubation of drug craving, p. 428

Chapter 16
Lateralization, Language, and the Split Brain

The Left Brain and Right Brain

Tom Wang/Shutterstock

Chapter Overview and Learning Objectives

Cerebral Lateralization of Function: Introduction

LO 16.1 Summarize early studies of the cerebral lateralization of function.

LO 16.2 Describe three techniques for assessing cerebral lateralization of function.

LO 16.3 Outline the discovery of the relationship between speech laterality and handedness.

LO 16.4 Describe and evaluate the hypothesis that male brains are more lateralized than female brains.

The Split Brain

LO 16.5 Outline the groundbreaking experiment of Myers and Sperry on split-brain cats.

LO 16.6 Describe the method used to demonstrate the hemispheric independence of visual experience in human split-brain patients.

LO 16.7 Describe the evidence that indicates that the hemispheres of split-brain patients can function independently.

LO 16.8 Outline the process of cross-cuing in split-brain patients.

LO 16.9 Describe the helping-hand phenomenon and the use of the chimeric figures test in experiments on split-brain patients.

LO 16.10 Describe a case where the right hemisphere tried to take control of a split-brain patient's everyday behavior.

LO 16.11 Explain how complete hemispheric independence is not an inevitable consequence of split-brain surgery.

Differences Between Left and Right Hemispheres

LO 16.12 Describe five examples of abilities that have been found to be lateralized.

LO 16.13 Discuss how we've come to understand that the lateralization of function is better understood in terms of individual cognitive processes rather than clusters of abilities.

LO 16.14 Describe three anatomical asymmetries in the human brain.

Evolution of Cerebral Lateralization and Language

LO 16.15 Describe and evaluate three theoretical explanations for why cerebral lateralization of function exists.

LO 16.16 List those species that display cerebral lateralization and explain what this tells us about when cerebral lateralization evolved.

LO 16.17 Describe what the study of nonhuman primates has suggested about the evolution of human language.

Cortical Localization of Language: Wernicke-Geschwind Model

LO 16.18 Describe the historical antecedents of the Wernicke-Geschwind model. Include descriptions of the following disorders: Broca's and Wernicke's aphasia, conduction aphasia, agraphia, and alexia.

LO 16.19 Describe the Wernicke-Geschwind model.

Wernicke-Geschwind Model: The Evidence

LO 16.20 Identify the effects of cortical damage and brain stimulation on language abilities, and evaluate the Wernicke-Geschwind model in light of these findings.

LO 16.21 Summarize the current status of the Wernicke-Geschwind model.

Cognitive Neuroscience of Language

LO 16.22 Describe the three premises that define the cognitive neuroscience approach to language, and compare them with the premises on which the Wernicke-Geschwind model is based.

LO 16.23 Describe two influential functional imaging studies of the localization of language, and explain what their findings indicate.

Cognitive Neuroscience of Dyslexia

LO 16.24 Describe the causes and neural mechanisms of developmental dyslexia.

LO 16.25 Describe the difference between the lexical procedure and the phonetic procedure for reading aloud. Then describe the difference between surface dyslexia and deep dyslexia.

With the exception of a few midline orifices, we humans have two of almost everything—one on the left and one on the right. Even the brain, which most people view as the unitary and indivisible basis of self, reflects this general principle of bilateral duplication. In its upper reaches, the brain comprises two structures—the left and right cerebral hemispheres—which are entirely separate except for the **cerebral commissures** connecting them. The fundamental duality of the human forebrain and the locations of the cerebral commissures are illustrated in Figure 16.1.

Although the left and right hemispheres are similar in appearance, there are major differences between them in function. This chapter is about these differences, a topic commonly referred to as **lateralization of function**. The study of **split-brain patients**—patients whose left and right hemispheres have been separated by **commissurotomy**—is a major focus of discussion. Another focus is the cortical localization of language abilities in the left hemisphere; language abilities are the most highly lateralized of all cognitive abilities.

You will learn in this chapter that your left and right hemispheres have different abilities and that they have the capacity to function independently—to have different thoughts, memories, and emotions. Thus, this chapter will challenge the concept you have of yourself as a unitary being. We hope you both enjoy it.

Figure 16.1 The cerebral hemispheres and cerebral commissures.

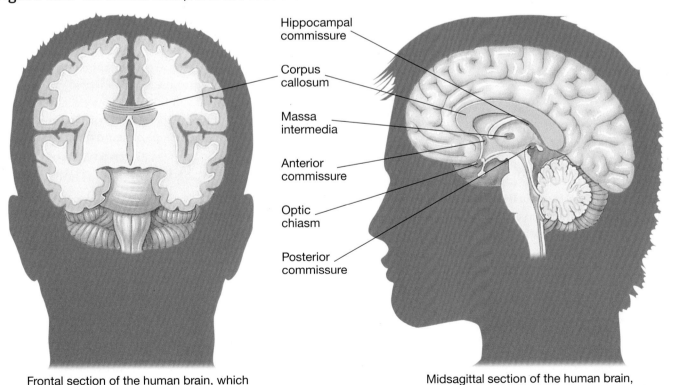

Hippocampal commissure

Corpus callosum

Massa intermedia

Anterior commissure

Optic chiasm

Posterior commissure

Frontal section of the human brain, which illustrates the fundamental duality of the human forebrain.

Midsagittal section of the human brain, which illustrates the corpus callosum and other commissures.

Cerebral Lateralization of Function: Introduction

In 1836, Marc Dax, an unknown country doctor, presented a short report at a medical society meeting in France. It was his first and only scientific presentation. Dax was struck by the fact that of the 40 or so brain-damaged patients with speech problems whom he had seen during his career, not a single one had damage restricted to the right hemisphere. His report aroused little interest, and Dax died the following year unaware that he had anticipated one of the most important areas of modern neuropsychological research.

Discovery of the Specific Contributions of Left-Hemisphere Damage to Aphasia and Apraxia

LO 16.1 Summarize early studies of the cerebral lateralization of function.

One reason Dax's important paper had so little impact was that most of his contemporaries believed that the brain acted as a whole and that specific functions could not be attributed to particular parts of it. This view began to change 25 years later in 1861, when Paul Broca reported his postmortem examination of two aphasic patients. **Aphasia** is a brain damage–produced deficit in the ability to produce or comprehend language.

Both of Broca's patients had a left-hemisphere lesion that involved an area in the frontal cortex just in front of the face area of the primary motor cortex. Broca at first did not recognize the relation between aphasia and the side of the brain damage; he did not know about Dax's report. However, by 1864, Broca had performed postmortem examinations on seven more aphasic patients, and he was struck by the fact that, like the first two, they all had damage to the *inferior prefrontal cortex* of the left hemisphere—which by then had become known as **Broca's area** (see Figure 16.2).

In the early 1900s, another example of *cerebral lateralization of function* was discovered. Hugo-Karl Liepmann found that **apraxia**, like aphasia, is almost always associated with left-hemisphere damage, despite the fact that its symptoms are *bilateral* (involving both sides of the body). Apraxic patients have difficulty performing movements when asked to perform them out of context, even though they often have no difficulty performing the same movements when they are not thinking about doing so (see Chapter 8).

The combined impact of the evidence that the left hemisphere plays a special role in both language and voluntary movement led to the theory of *cerebral dominance*. According to this theory, one hemisphere—usually the left—assumes the dominant role in the control of all complex behavioral

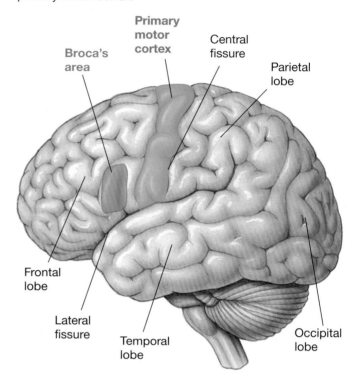

Figure 16.2 The location of Broca's area: In the inferior left prefrontal cortex, just anterior to the face area of the left primary motor cortex.

and cognitive processes, and the other plays only a minor role. This thinking led to the practice of referring to the left hemisphere as the **dominant hemisphere** and the right hemisphere as the **minor hemisphere**.

In addition, the discovery that language and motor abilities are lateralized to the left hemisphere triggered a search for other lateralized functions. In effect, the discovery of language and motor lateralization established *lateralization of function* as a major area of neuroscientific research.

Tests of Cerebral Lateralization

LO 16.2 Describe three techniques for assessing cerebral lateralization of function.

Early research on the cerebral lateralization of function compared the effects of left-hemisphere and right-hemisphere lesions. Now, however, other techniques are also used for this purpose. The sodium amytal test, the dichotic listening test, and functional brain imaging are three of them.

SODIUM AMYTAL TEST. The **sodium amytal test** of language lateralization (Wada, 1949) is often given to patients prior to neurosurgery (see Bauer et al., 2014). The neurosurgeon uses the results of the test to plan the surgery; every effort is made to avoid damaging areas of the cortex that are likely to be involved in language. The sodium amytal test involves the injection of a small amount of sodium amytal into the carotid artery on one side of the neck. The injection anesthetizes the hemisphere on that side for a few minutes, thus allowing the capacities of the other hemisphere to be assessed. During the test, the patient is asked to recite well-known series (e.g., letters of the alphabet, days of the week, months of the year) and to name pictures of common objects. Then, an injection is administered to the other side, and the test is repeated. When the hemisphere specialized for speech, usually the left hemisphere, is anesthetized, the patient is rendered completely mute for a minute or two; and once the ability to talk returns, there are errors of serial order and naming. In contrast, when the other hemisphere, usually the right, is anesthetized, mutism often does not occur at all, and errors are few.

DICHOTIC LISTENING TEST. Unlike the sodium amytal test, the **dichotic listening test** is noninvasive; thus, it can be administered to healthy individuals. In the standard dichotic listening test, three pairs of spoken digits are presented through earphones; the digits of each pair are presented simultaneously, one to each ear (see Blass et al., 2015; Kimura, 2011). For example, a person might hear the sequence 3, 9, 2 through one ear and at the same time 1, 6, 4 through the other. The person is then asked to report all of the digits. Kimura found that most people report slightly more of the digits presented to the right ear than the left, which is indicative of left-hemisphere specialization for language. In contrast, Kimura found that all the patients who had been identified by the sodium amytal test as having right-hemisphere specialization for language performed better with the left ear than the right. Kimura argued that although the sounds from each ear are projected to both hemispheres, the contralateral connections are stronger and take precedence when two different sounds are simultaneously competing for access to the same cortical auditory centers.

FUNCTIONAL BRAIN IMAGING. Lateralization of function has also been studied using functional brain-imaging techniques. While a volunteer engages in some activity, such as reading, the activity of the brain is monitored by positron emission tomography (PET) or functional magnetic resonance imaging (fMRI). On language tests, functional brain-imaging techniques typically reveal far greater activity in the left hemisphere than in the right hemisphere (see Bauer et al., 2014).

Discovery of the Relation Between Speech Laterality and Handedness

LO 16.3 Outline the discovery of the relationship between speech laterality and handedness.

Lesion studies have clarified the relation between the cerebral lateralization of speech and handedness. For example, one study involved military personnel who suffered brain damage in World War II (Russell & Espir, 1961), and another focused on neurological patients who underwent unilateral excisions for the treatment of neurological disorders (Penfield & Roberts, 1959). In both studies, approximately 60 percent of **dextrals** (right-handers) with left-hemisphere lesions and 2 percent of those with right-hemisphere lesions were diagnosed as aphasic; the comparable figures for **sinestrals** (left-handers) were about 30 and 24 percent, respectively. These results indicate that the left hemisphere is dominant for language-related abilities in almost all dextrals and in the majority of sinestrals. In effect, sinestrals are more variable (less predictable) than dextrals with respect to their hemisphere of language lateralization. This increased variability also holds for other aspects of brain function; for example, sinestrals show greater variability in their lateralization of attention and face recognition (see Vingerhoets, 2019).

Results of the sodium amytal test have confirmed the relation between handedness and language lateralization that was first observed in early lesion studies. For example, Milner (1974) found that almost all right-handed patients without early left-hemisphere damage had left-hemisphere specialization for speech (92 percent), most left-handed and ambidextrous patients without early left-hemisphere damage had left-hemisphere specialization for speech (69 percent), and early left-hemisphere damage decreased left-hemisphere specialization for speech in left-handed and ambidextrous patients (30 percent).

Sex Differences in Brain Lateralization

LO 16.4 Describe and evaluate the hypothesis that male brains are more lateralized than female brains.

In 1972, Levy proposed that the brains of females and males differ in their degree of lateralization. Interest in Levy's hypothesis was stimulated by McGlone's (1977, 1980) observation that male victims of unilateral strokes were three times more likely to suffer from aphasia than female victims. McGlone concluded that male brains are more lateralized than female brains.

Journal Prompt 16.1

What other clinical implications might this observed sex difference in brain lateralization have?

Levy's hypothesis of a sex difference in brain lateralization has been widely embraced, and it has been used to explain almost every imaginable cognitive and behavioral difference between the sexes. But there is currently limited support for Levy's hypothesis. For example, some researchers have failed to confirm McGlone's report of a sex difference in the effects of unilateral brain lesions (see Inglis & Lawson, 1982). Even more problematic, a meta-analysis of 17 functional brain-imaging studies did not find significant effects of sex on the lateralization of brain function (see Hirnstein, Hugdahl, & Hausmann, 2019). Accordingly, Levy's hypothesis of sex differences in brain lateralization is no longer tenable (see Hirnstein, Hugdahl, & Hausmann, 2019).

The Split Brain

In the previous module, you were introduced to early research on the lateralization of function, and you learned about four methods of studying cerebral lateralization of function: comparing the effects of unilateral left- and right-hemisphere brain lesions, the sodium amytal test, the dichotic listening test, and functional brain imaging. This module focuses on a fifth method.

In the early 1950s, the **corpus callosum**—the largest cerebral commissure—constituted a paradox of major proportions. Its size, an estimated 200 million axons, and its central position, right between the two cerebral hemispheres, implied that it performed an extremely important function; yet research in the 1930s and 1940s seemed to suggest that it did nothing at all. The corpus callosum had been cut in monkeys and in several other laboratory species, but the animals seemed no different after the surgery than they had been before. Similarly, human patients who were born without a corpus callosum or had it damaged seemed quite normal. In the early 1950s, Roger Sperry, whom you may remember for the eye-rotation experiments described in Chapter 9, and his colleagues were intrigued by this paradox (see Aboitiz, 2017).

Groundbreaking Experiment of Myers and Sperry

LO 16.5 Outline the ground-breaking experiment of Myers and Sperry on split-brain cats.

The solution to the puzzle of the corpus callosum was provided in 1953 by an experiment on cats by Myers and Sperry. The experiment made two astounding theoretical points. First, it showed that one function of the corpus callosum is

to transfer learned information from one hemisphere to the other. Second, it showed that when the corpus callosum is cut, each hemisphere can function independently; each split-brain cat appeared to have two brains. If you find the thought of a cat with two brains provocative, you will almost certainly be bowled over by similar observations about split-brain humans. But we are getting ahead of ourselves. Let's first consider the research on cats.

In their experiment, Myers and Sperry trained cats to perform a simple visual discrimination task. On each trial, each cat was confronted by two panels, one with a circle on it and one with a square on it. The relative positions of the circle and square (right or left) were varied randomly from trial to trial, and the cats had to learn which symbol to press in order to get a food reward. Myers and Sperry correctly surmised that the key to split-brain research was to develop procedures for teaching and testing one hemisphere at a time. Figure 16.3 illustrates the method they used to isolate visual-discrimination learning in one hemisphere of the cats. There are two routes by which visual information can cross from one eye to the contralateral hemisphere: via the corpus callosum or via the optic chiasm. Accordingly, in their key experimental group, Myers and Sperry *transected* (cut completely through) both the optic chiasm and the corpus callosum of each cat and put a patch on one eye. This restricted all incoming visual information to the hemisphere ipsilateral to the uncovered eye.

The results of Myers and Sperry's experiment are illustrated in Figure 16.4. In the first phase of the study, all cats

Figure 16.3 Restricting visual information to one hemisphere in cats. To restrict visual information to one hemisphere, Myers and Sperry (1) cut the corpus callosum, (2) cut the optic chiasm, and (3) blindfolded one eye. This restricted the visual information to the hemisphere ipsilateral to the uncovered eye.

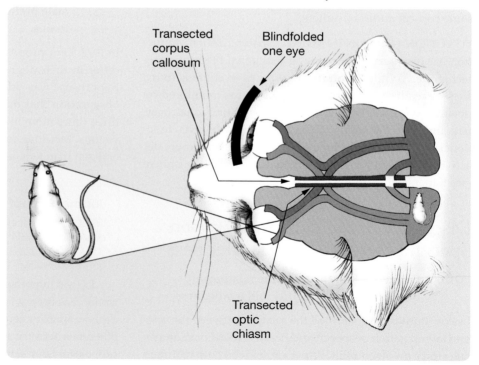

Figure 16.4 Schematic illustration of Myers and Sperry's (1953) groundbreaking split-brain experiment. There were four groups: (1) the key experimental group with both the optic chiasm and corpus callosum transected, (2) a control group with only the optic chiasm transected, (3) a control group with only the corpus callosum transected, and (4) an unlesioned control group. The performance of the three control groups did not differ, so they are illustrated together here.

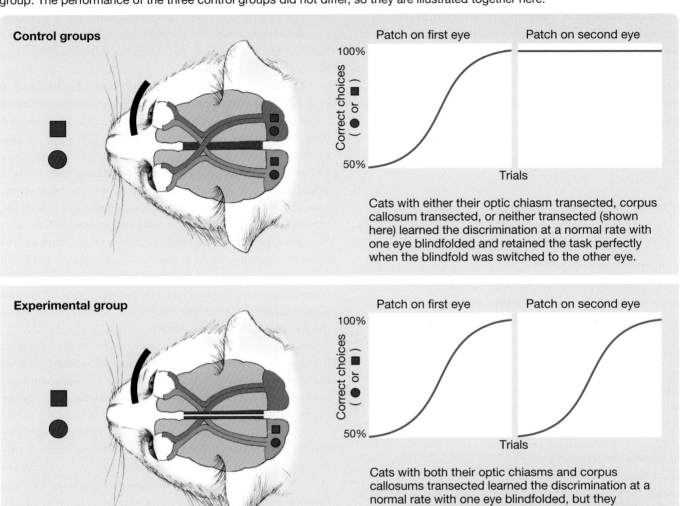

Control groups

Patch on first eye   Patch on second eye

Cats with either their optic chiasm transected, corpus callosum transected, or neither transected (shown here) learned the discrimination at a normal rate with one eye blindfolded and retained the task perfectly when the blindfold was switched to the other eye.

Experimental group

Patch on first eye   Patch on second eye

Cats with both their optic chiasms and corpus callosums transected learned the discrimination at a normal rate with one eye blindfolded, but they showed no retention whatsoever when the blindfold was switched to the other eye.

learned the task with a patch on one eye. The cats in the key experimental group (those with both the optic chiasm and the corpus callosum transected) learned the simple discrimination as rapidly as did unlesioned control cats or control cats with either the corpus callosum or the optic chiasm transected, despite the fact that cutting the optic chiasm produced a **scotoma**—an area of blindness—involving the entire medial half of each retina. This result suggested that one hemisphere working alone can learn simple tasks as rapidly as two hemispheres working together.

More surprising were the results of the second phase of Myers and Sperry's experiment, during which the patch was transferred to each cat's other eye. The transfer of the patch had no effect on the performance of the intact control cats or of the control cats with either the optic chiasm or the corpus callosum transected; these subjects continued to perform the task with close to 100 percent accuracy. In contrast,

transferring the eye patch had a devastating effect on the performance of the experimental cats. In effect, it blindfolded the hemisphere that had originally learned the task and tested the knowledge of the other hemisphere, which had been blindfolded during initial training. When the patch was transferred, the performance of the experimental cats dropped immediately to baseline (i.e., to 50 percent correct); and then the cats relearned the task with no savings whatsoever, as if they had never seen it before. Myers and Sperry concluded that the cat brain has the capacity to act as two separate brains and that the function of the corpus callosum is to transmit information between them.

Myers and Sperry's startling conclusions about the fundamental duality of the cat brain and the information-transfer function of the corpus callosum have been confirmed in a variety of species with a variety of test procedures. For example, split-brain monkeys cannot perform tasks requiring fine tactual discriminations (e.g., rough versus smooth) or fine motor responses (e.g., unlocking a puzzle) with one hand if they have learned them with the other—provided they are not allowed to watch their hands, which would allow the information to enter both hemispheres. This failure of intermanual transfer of fine tactual and motor information in split-brain monkeys occurs because the somatosensory and motor fibers involved in fine sensory and motor discriminations are all contralateral and because the hemispheres have lost their ability to communicate directly.

Commissurotomy in Humans with Epilepsy

LO 16.6 Describe the method used to demonstrate the hemispheric independence of visual experience in human split-brain patients.

In the first half of the 20th century, when the normal function of the corpus callosum was still a mystery, it was known that epileptic discharges often spread from one hemisphere to the other through the corpus callosum. This, along with the fact that cutting the corpus callosum had proven in numerous studies to have no obvious effect on performance outside the contrived conditions of Sperry's laboratory, led two neurosurgeons, Vogel and Bogen, to initiate a program of commissurotomy for the treatment of severe intractable cases of epilepsy—despite the fact that a previous similar attempt had failed, presumably because of incomplete transections (Van Wagenen & Herren, 1940). The rationale underlying therapeutic commissurotomy—which typically involves transecting the corpus callosum and leaving the smaller commissures intact—was that the severity of the patient's convulsions might be reduced if the discharges could be limited to the hemisphere of their origin (see Mancuso et al., 2019). The therapeutic benefits of commissurotomy turned out to be even greater than anticipated: Despite the fact that commissurotomy is performed in only the most severe cases, many commissurotomized patients do not experience another major convulsion.

Journal Prompt 16.2

The decision to perform commissurotomies on patients with epilepsy turned out to be a good one. In Chapter 1, you learned that the decision to perform prefrontal lobotomies on patients with mental illness turned out to be a bad one. Was this just the luck of the draw? (Hint: Make one list for commissurotomy, and one list for prefrontal lobotomy, of the evidence for and against each that existed when they were first adopted.)

Evaluation of the neuropsychological status of Vogel and Bogen's split-brain patients was conducted by Sperry

and his associate Gazzaniga, and this work was a major factor in Sperry receiving a Nobel Prize in 1981 (see Table 1.1). Sperry and Gazzaniga began by developing a battery of tests based on the same methodological strategy that had proved so informative in Sperry's studies of laboratory animals: delivering information to one hemisphere while keeping it out of the other (see Mancuso et al., 2019; Uddin, 2011).

They could not use the same visual-discrimination procedure that had been used in studies of split-brain laboratory animals (i.e., cutting the optic chiasm and blindfolding one eye) because cutting the optic chiasm produces a scotoma. Instead, they employed the procedure illustrated in Figure 16.5. Each split-brain patient was asked to fixate on the center of a display screen; then, visual stimuli were flashed onto the left or right side of the screen for 0.1 second. The 0.1-second exposure time was long enough for the subjects to perceive the stimuli but short enough to preclude the confounding effects of eye movement. All stimuli thus presented in the left visual field were transmitted to the right visual cortex, and all stimuli thus presented in the right visual field were transmitted to the left visual cortex (see Aboitiz, 2017).

Fine tactual and motor tasks were performed by each hand under a ledge. This procedure was used so that the nonperforming hemisphere—that is, the ipsilateral hemisphere—could not monitor the performance via the visual system.

The results of the tests on split-brain patients have confirmed the findings in split-brain laboratory animals in one major respect but not in another. Like split-brain laboratory animals, human split-brain patients seem to have in some respects two independent brains, each with its own stream of consciousness, abilities, memories, and emotions (e.g., Gazzaniga, 1967; Gazzaniga & Sperry, 1967; Sperry, 1964). But unlike the hemispheres of split-brain laboratory animals, the hemispheres of split-brain patients are far from equal in their ability to perform certain tasks. Most notably, the left hemisphere of most split-brain patients is capable of speech, whereas the right hemisphere is not.

Before we recount some of the key results of the tests on split-brain humans, let us give you some advice. Some students become confused by the results of these tests because their tendency to think of the human brain as a single unitary organ is deeply engrained. If you become confused, think of each split-brain patient as two separate individuals: Right Hemisphere Ray, who understands a few simple instructions but cannot speak, who receives sensory information from the left visual field and left hand, and who controls the fine motor responses of the left hand; and Left Hemisphere Logan, who is verbally adept, who receives sensory information from the right visual field and right hand, and who controls the fine motor responses of the right hand. In everyday life, the behavior of split-brain patients is

Figure 16.5 The testing procedure used to evaluate the neuropsychological status of split-brain patients. Visual input goes from each visual field to the contralateral hemisphere; fine tactile input goes from each hand to the contralateral hemisphere; and each hemisphere controls the fine motor movements of the contralateral hand.

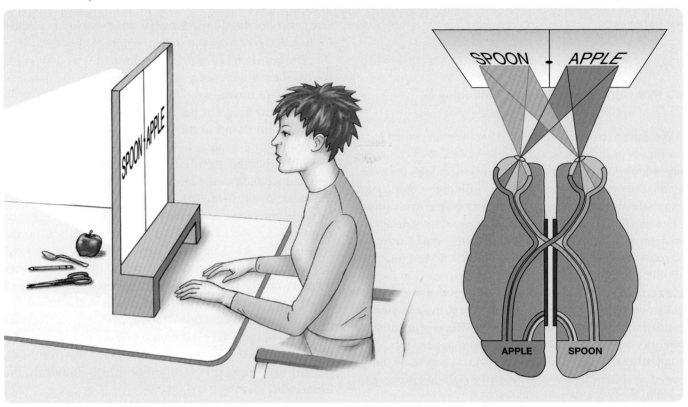

reasonably normal because their two brains go through life together and acquire much of the same information; however, in the neuropsychological laboratory, major discrepancies in what the two hemispheres learn can be created. As you are about to find out, this situation has interesting consequences.

Evidence That the Hemispheres of Split-Brain Patients Can Function Independently

LO 16.7 **Describe the evidence that indicates that the hemispheres of split-brain patients can function independently.**

If a picture of an apple were flashed in the right visual field of a split-brain patient, the left hemisphere could do one of two things to indicate that it had received and stored the information. Because it is the hemisphere that speaks, the left hemisphere could simply tell the experimenter that it saw a picture of an apple. Or the patient could reach under a ledge with the right hand, feel the test objects, and pick out the apple. Similarly, if the apple were presented to the left hemisphere by being placed in the patient's right hand, the left hemisphere could indicate to the experimenter that

it was an apple either by saying so or by putting the apple down and picking out another apple with the right hand from the test objects under the ledge. If, however, the non-speaking right hemisphere were asked to indicate the identity of an object that had previously been presented to the left hemisphere, it could not do so. Although objects that have been presented to the left hemisphere can be accurately identified with the right hand, performance is no better than chance with the left hand.

When test objects are presented to the right hemisphere either visually (in the left visual field) or tactually (in the left hand), the pattern of responses is entirely different. A split-brain patient asked to name an object flashed in the left visual field is likely to claim that nothing appeared on the screen. (Remember that it is the left hemisphere who is talking and the right hemisphere who has seen the stimulus.) A patient asked to name an object placed in the left hand is usually aware that something is there, presumably because of the crude tactual information carried by ipsilateral somatosensory fibers, but is unable to say what it is. Amazingly, all the while the patient is claiming (i.e., all the while the left hemisphere is claiming) the inability to identify a test object presented in the left visual field or left hand, the left hand (i.e., the right hemisphere) can identify the correct object. Imagine how confused the patient must become when, in trial after trial, the left hand can feel an object and then fetch another just like it from a collection of test items under the ledge, while the left hemisphere is vehemently claiming that it does not know the identity of the test object.

Cross-Cuing

LO 16.8 Outline the process of cross-cuing in split-brain patients.

Although the two hemispheres of a split-brain patient have no means of direct neural communication, they can communicate neurally via indirect pathways through the brain stem. They can also communicate with each other by an external route, by a process called **cross-cuing**. An example of cross-cuing occurred during a series of tests designed to determine whether the left hemisphere could respond to colors presented in the left visual field. To test this possibility, a red or a green stimulus was presented in the left visual field, and the split-brain patient was asked to verbally report the color: red or green. At first, the patient performed at a chance level on this task (50 percent correct); but after a time, performance improved appreciably, thus suggesting that the color information was somehow transferred over neural pathways from the right hemisphere to the left.

However, this proved not to be the case:

If a green light was presented and the patient happened to correctly guess green, she would be correct and the trial would end. However, if the green light was presented and the patient guessed red, after a pause, she would frown, shake her head, and then change her answer: "Oh no, I meant green." The right hemisphere saw the green light and heard the left hemisphere guess "red." Knowing that red was wrong, the right hemisphere initiated a frown and shook her head, no. This signaled to the left hemisphere that the answer was wrong and that it needed to be corrected.

This example demonstrates how neurological patients can use different cognitive strategies to perform the same task. The fact that neurological patients can perform the same task in different ways often clouds their deficits and greatly complicates the assessment of their neurological status.

Doing Two Things at Once

LO 16.9 Describe the helping-hand phenomenon and the use of the chimeric figures test in experiments on split-brain patients.

In many of the classes we teach, a student fits the following stereotype: He sits—or rather sprawls—near the back of the class; and despite good grades, he tries to create the

impression he is above it all by making sarcastic comments. Such a student inadvertently triggered an interesting discussion in one of our classes. His comment went something like this: "If getting my brain cut in two can create two separate brains, perhaps I should get it done so that I can study for two different exams at the same time."

The question raised by this comment is a good one. If the two hemispheres of a split-brain patient are capable of independent functioning, then they should be able to do two different things at the same time—in this case, learn two different things at the same time. Can they? Indeed they can. For example, in one test, two different visual stimuli appeared simultaneously on the test screen—let's say a pencil in the left visual field and an orange in the right visual field. The split-brain patient was asked to simultaneously reach into two bags—one with each hand—and grasp in each hand the object that was on the screen. After grasping the objects, but before withdrawing them, the patient was asked to tell the experimenter what was in the two hands; the patient (i.e., the left hemisphere) replied, "Two oranges." Much to the bewilderment of the verbal left hemisphere, when the hands were withdrawn, there was an orange in the right hand and a pencil in the left. The two hemispheres of the split-brain patient had learned two different things at exactly the same time.

In another test in which two visual stimuli were presented simultaneously—again, let's say a pencil to the left visual field and an orange to the right—the split-brain patient was asked to pick up the presented object from an assortment of objects on a table, this time in full view. As the right hand reached out to pick up the orange under the direction of the left hemisphere, the right hemisphere saw what was happening and thought an error was being made (remember that the right hemisphere saw a pencil). On some trials, the right hemisphere dealt with this problem in the only way that it could: The left hand shot out, grabbed the right hand away from the orange, and redirected it to the pencil. This response is called the **helping-hand phenomenon**.

The special ability of split brains to do two things at once has also been demonstrated on tests of attention. Each hemisphere of split-brain patients appears to be able to maintain an independent focus of attention (see Gazzaniga, 2005). This leads to an ironic pattern of results: Split-brain patients can search for, and identify, a visual target item in an array of similar items more quickly than healthy controls can (Luck et al., 1989)—presumably because the two split hemispheres are conducting two independent searches.

Yet another example of the split brain's special ability to do two things at once involves the phenomenon of **visual completion**. As you may recall from Chapter 6, individuals with scotomas are often unaware of them

Figure 16.6 The chimeric figures test. When a split-brain patient focuses on a chimeric face, the left hemisphere sees a single normal face that is a completed version of the half face on the right. At the same time, the right hemisphere sees a single normal face that is a completed version of the half face on the left.

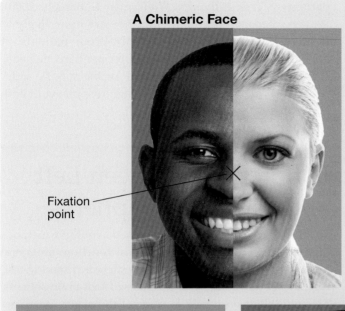

A Chimeric Face

Fixation point

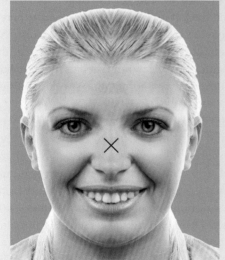

Left hemisphere of a split-brain patient sees this.

Right hemisphere of a split-brain patient sees this.

ASDF_MEDIA/Shutterstock; Jason Stitt/Shutterstock

split-brain patients—see Figure 16.6. The patients were then asked to describe what they saw or to indicate what they saw by pointing to it in a series of photographs of intact faces. Amazingly, each patient (i.e., each left hemisphere) reported seeing a complete, bilaterally symmetrical face, even when asked such leading questions as "Did you notice anything peculiar about what you just saw?" When the patients were asked to describe what they saw, they usually described a completed version of the half that had been presented to the right visual field (i.e., the left hemisphere).

Dual Mental Functioning and Conflict in Split-Brain Patients

LO 16.10 Describe a case where the right hemisphere tried to take control of a split-brain patient's everyday behavior.

In most split-brain patients, the right hemisphere does not seem to have a strong will of its own; the left hemisphere seems to control most everyday activities. However, in a few split-brain patients, the right hemisphere takes a more active role in controlling behavior, and in these cases, there can be serious conflicts between the left and right hemispheres. One patient (let's call him Peter) was such a case.

because their brains have the capacity to fill them in (to complete them) by using information from the surrounding areas of the visual field. In a sense, each hemisphere of a split-brain patient is a participant with a scotoma covering the entire ipsilateral visual field. The ability of the hemispheres of a split-brain patient to simultaneously and independently engage in completion has been demonstrated in studies using the **chimeric figures test**—named after *Chimera*, a mythical monster composed of parts of different animals. Levy, Trevarthen, and Sperry (1972) flashed photographs composed of fused-together half-faces of two different people onto the center of a screen in front of

The Case of Peter, the Split-Brain Patient Tormented by Conflict

At the age of 8, Peter began to suffer from complex partial seizures. Antiepileptic medication was ineffective, and at 20, he received a commissurotomy, which greatly improved his condition but did not completely block his seizures. A sodium amytal test administered prior to surgery showed that he was left-hemisphere dominant for language.

Following surgery, the independent mischievous behavior of Peter's right hemisphere often caused him (his left hemisphere) considerable frustration. He (his left hemisphere) complained that his left hand would turn off television shows that he was enjoying, that his left leg would not always walk in the intended direction, and that his left arm would sometimes perform embarrassing, socially unacceptable acts (e.g., striking a relative).

In the laboratory, Peter (his left hemisphere) sometimes became angry with his left hand, swearing at it, striking it, and trying to force it with his right hand to do what he (his left hemisphere) wanted (Joseph, 1988).

Independence of Split Hemispheres: Current Perspective

LO 16.11 Explain how complete hemispheric independence is not an inevitable consequence of split-brain surgery.

Discussions of split-brain patients tend to focus on cases in which there seems to be a complete separation of left-hemisphere and right-hemisphere function, and this is what we have done here. But complete hemispheric independence is not an inevitable consequence of split-brain surgery. In most split-brain patients, it is possible to demonstrate some communication of information between hemispheres, depending on the particular surgery, the time since surgery, the particular information, and the method of testing. For example, feelings of emotion appear to be readily passed between the hemispheres of most split-brain patients. This is easily demonstrated by presenting emotionally loaded images to the right hemisphere and asking patients to respond verbally to the images. Their verbal left hemisphere often responds with the appropriate emotion, even when the left hemisphere is unaware of the image (Sperry, Zaidel, & Zaidel, 1979).

Consider the following remarkable exchange (paraphrased from Sperry, Zaidel, & Zaidel, 1979, pp. 161–162). The patient's right hemisphere was presented with an array of photos, and the patient was asked if one was familiar. He pointed to the photo of his aunt.

Experimenter: "Is this a neutral, a thumbs-up, or a thumbs-down person?"
Patient: With a smile, he made a thumbs-up sign and said, "This is a happy person."
Experimenter: "Do you know him personally?"
Patient: "Oh, it's not a him, it's a her."
Experimenter: "Is she an entertainment personality or an historical figure?"
Patient: "No, just…"
Experimenter: "Someone you know personally?"
Patient: He traced something with his left index finger on the back of his right hand, and then he exclaimed, "My aunt, my Aunt Edie."

Experimenter: "How do you know?"
Patient: "By the E on the back of my hand."

Another factor that has been shown to contribute substantially to the hemispheric independence of split-brain patients is task difficulty (Weissman & Banich, 2000). As tasks become more difficult, they are more likely to involve both hemispheres of split-brain patients. It appears that simple tasks are best processed in one hemisphere, the hemisphere specialized for the specific activity, but complex tasks require the cognitive power of both hemispheres.

Differences Between Left and Right Hemispheres

So far in this chapter, you have learned about five methods of studying cerebral lateralization of function: unilateral lesions, the sodium amytal test, the dichotic listening test, functional brain imaging, and studies of split-brain patients. This module takes a look at some of the major functional differences between the left and right cerebral hemispheres that have been discovered using these methods. Because the verbal and motor abilities of the left hemisphere are readily apparent, most research on the lateralization of function has focused on uncovering the not-so-obvious special abilities of the right hemisphere.

Before we introduce you to some of the differences between the left and right hemispheres, we need to clear up a common misconception: For many functions, there are no substantial differences between the hemispheres; and when functional differences do exist, these tend to be slight biases in favor of one hemisphere or the other—not absolute differences. Disregarding these facts, the popular media inevitably portray left–right cerebral differences as absolute. As a result, it is widely believed that various abilities reside exclusively in one hemisphere or the other. For example, it is widely believed that the left hemisphere has exclusive control over language and the right hemisphere has exclusive control over emotion and creativity.

Journal Prompt 16.3

Prior to delving into this module, list your preconceptions about cerebral lateralization of function. What side of the brain does what?

Language-related abilities provide a particularly good illustration of the fact that lateralization of function is statistical rather than absolute. Language is the most lateralized of all cognitive abilities. Yet, even in this most extreme case, lateralization is far from total; there is substantial language-related activity in the right hemisphere. For example, on the dichotic listening test, people who are left-hemisphere dominant for language tend to identify more digits with the right ear than the left ear, but this right-ear advantage is only 55 to 45 percent. Furthermore, the right hemispheres of most left-hemisphere dominant split-brain patients can understand many spoken or written words and simple sentences (see Gazzaniga, 2013).

Examples of Cerebral Lateralization of Function

LO 16.12 Describe five examples of abilities that have been found to be lateralized.

Table 16.1 lists some of the abilities that are often found to be lateralized. They are arranged in two columns: those that seem to be controlled more by the left hemisphere and those that seem to be controlled more by the right hemisphere. Let's consider several examples of cerebral lateralization of function.

SUPERIORITY OF THE LEFT HEMISPHERE IN CONTROLLING IPSILATERAL MOVEMENT. One unexpected left-hemisphere specialization was revealed by

functional brain-imaging studies. When complex, cognitively driven movements are made by one hand, most of the activation is observed in the *contralateral* hemisphere, as expected. However, some activation is also observed in the *ipsilateral* hemisphere, and these ipsilateral effects are substantially greater in the left hemisphere than in the right (see Bundy & Leuthardt, 2019; Hervé et al., 2013). Consistent with this observation is the finding that left-hemisphere lesions are more likely than right-hemisphere lesions to produce ipsilateral motor problems—for example, left-hemisphere lesions are more likely to reduce the accuracy of left-hand movements than right-hemisphere lesions are to reduce the accuracy of right-hand movements.

SUPERIORITY OF THE RIGHT HEMISPHERE IN SPATIAL ABILITY. In a classic early study, Levy (1969) placed a three-dimensional block of a particular shape in either the right hand or the left hand of split-brain patients. Then, after they had thoroughly *palpated* (tactually investigated) it, she asked them to point to the two-dimensional test stimulus that best represented what the three-dimensional block would look like if it were made of cardboard and unfolded. She found a right-hemisphere superiority on this task, and she found that the two hemispheres seemed to go about the task in different ways. The performance of the left hand and right hemisphere was rapid and silent, whereas the performance of the right hand and left hemisphere was hesitant and often accompanied by a running verbal commentary that was difficult for the patients to inhibit. Levy concluded that the right hemisphere is superior to the left at spatial tasks. This conclusion has been frequently confirmed (see Dietz et al., 2014; Hirnstein, Hugdahl, & Hausmann, 2019; Zaidel, 2013), and it is consistent with the finding that disorders of spatial perception (e.g., contralateral neglect—see Chapters 7 and 8) tend to be associated with right-hemisphere damage.

SPECIALIZATION OF THE RIGHT HEMISPHERE FOR EMOTION. According to the old concept of left-hemisphere dominance, the minor right hemisphere is not involved in emotion. This presumption has been proven false. Indeed, analysis of the effects of unilateral brain lesions indicates that the right hemisphere may be superior to

Table 16.1 Abilities that display some degree of cerebral lateralization.

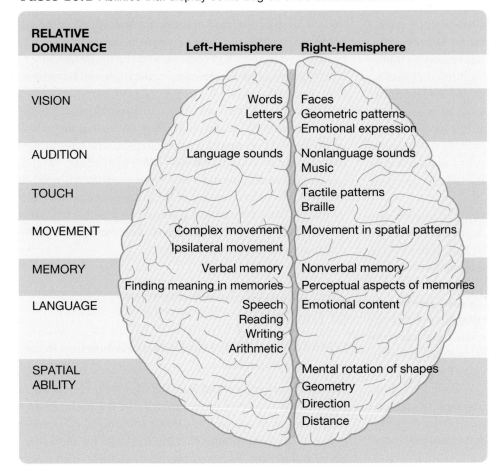

RELATIVE DOMINANCE	Left-Hemisphere	Right-Hemisphere
VISION	Words / Letters	Faces / Geometric patterns / Emotional expression
AUDITION	Language sounds	Nonlanguage sounds / Music
TOUCH		Tactile patterns / Braille
MOVEMENT	Complex movement / Ipsilateral movement	Movement in spatial patterns
MEMORY	Verbal memory / Finding meaning in memories	Nonverbal memory / Perceptual aspects of memories
LANGUAGE	Speech / Reading / Writing / Arithmetic	Emotional content
SPATIAL ABILITY		Mental rotation of shapes / Geometry / Direction / Distance

the left at performing some tests of emotion—for example, in accurately identifying facial expressions of emotion (see Bernard et al., 2018; Gainotti, 2018, 2019a, 2019b; Mitchell & Phillips, 2015; Prete et al., 2015). Although the study of unilateral brain lesions suggests a general right-hemisphere dominance for some aspects of emotional processing, functional brain-imaging studies have not provided unambiguous support for this view (see Costanzo et al., 2015; Gainotti, 2019a).

SUPERIOR MUSICAL ABILITY OF THE RIGHT HEMI-SPHERE. Kimura (1964) compared the performance of 20 right-handers on the standard digit version of the dichotic listening test with their performance on a version of the test involving the dichotic presentation of melodies. In the melody version of the test, Kimura simultaneously played two different melodies—one to each ear—and then asked the participants to identify the two they had just heard from four that were subsequently played to them through both ears. The right ear (i.e., the left hemisphere) was superior in the perception of digits, whereas the left ear (i.e., the right hemisphere) was superior in the perception of melodies. This is consistent with the observation that right temporal lobe lesions are more likely to disrupt music discriminations than are left temporal lobe lesions (see Casey, 2013).

HEMISPHERIC DIFFERENCES IN MEMORY. Early studies of the lateralization of cognitive function were premised on the assumption that particular cognitive abilities reside in one or the other of the two hemispheres. However, the results of research have led to an alternative way of thinking: The two hemispheres have similar abilities that tend to be expressed in different ways. The study of the lateralization of memory was one of the first areas of research on cerebral lateralization to lead to this modification in thinking. You see, both the left and right hemispheres have the ability to perform on tests of memory, but the left hemisphere is better on some tests, whereas the right hemisphere is better on others.

There are two approaches to studying the cerebral lateralization of memory. One approach is to try to link particular memory processes with particular hemispheres—for example, it has been argued that the left hemisphere is specialized for encoding episodic memory (see Chapter 11). The other approach is to link the memory processes of each hemisphere to specific materials rather than to specific processes. In general, the left hemisphere has been found to play the greater role in memory for verbal material, whereas the right hemisphere has been found to play the greater role in memory for nonverbal material (e.g., Willment & Golby, 2013). Whichever of these two approaches ultimately proves more fruitful, they represent an advance over the tendency to think that memory is totally lateralized to one hemisphere.

What Is Lateralized? Broad Clusters of Abilities or Individual Cognitive Processes?

LO 16.13 Discuss how we've come to understand that the lateralization of function is better understood in terms of individual cognitive processes rather than clusters of abilities.

Early theories of cerebral laterality tended to ascribe complex clusters of mental abilities to one hemisphere or the other. The left hemisphere tended to perform better on language tests, so it was presumed to be dominant for language-related abilities; the right hemisphere tended to perform better on some spatial tests, so it was presumed to be dominant for space-related abilities; and so on. Perhaps this was a reasonable first step, but now the consensus among researchers is that this conclusion is simplistic.

The problem is that categories such as language, emotion, musical ability, and spatial ability are each composed of dozens of different individual cognitive activities, and there is no reason to assume that all those activities associated with a general label (e.g., spatial ability) will necessarily be lateralized in the same hemisphere. Indeed, major exceptions to all broad categories of cerebral lateralization have emerged (see Cai & Van der Haegen, 2015; Crepaldi et al., 2013; Turner et al., 2015). How is it possible to argue that all language-related abilities are lateralized in the left hemisphere when the right hemisphere has been shown to be involved in speech perception and the understanding of word meaning (see Kreitewolf, Friederici, & von Kriegstein, 2014; Poeppel, 2014)?

Many researchers are taking a different approach to the study of cerebral lateralization. They are basing their studies on the work of cognitive psychologists, who have broken down complex cognitive tasks—such as reading, judging space, and remembering—into their *constituent cognitive processes*. Once the laterality of the individual cognitive elements has been determined, it should be possible to predict the laterality of cognitive tasks based on the specific cognitive processes that compose them.

Anatomical Asymmetries of the Brain

LO 16.14 Describe three anatomical asymmetries in the human brain.

The discovery of cerebral lateralization of function led to a search for anatomical asymmetries in the brain. In particular, it led to a search for those anatomical differences between the hemispheres that are the basis for their functional differences. For example, do anatomical differences between the left and right hemispheres make the left hemisphere more suited for the control of language?

Most efforts to identify interhemispheric differences in brain anatomy have focused on the size of three areas of cortex that are important for language, the most lateralized of our cognitive abilities: the frontal operculum, the planum temporale, and Heschl's gyrus (see Figure 16.7). The **frontal operculum** is the area of frontal lobe cortex that lies just in front of the face area of the primary motor cortex; in the left hemisphere, it is the location of Broca's area. The planum temporale and Heschl's gyrus are areas of temporal lobe cortex. The **planum temporale** lies in the posterior region of the lateral fissure; it is thought to play a role in the comprehension of language and is often referred to as *Wernicke's area*. **Heschl's gyrus** is located in the lateral fissure just anterior to the planum temporale in the temporal lobe; it is the location of primary auditory cortex.

Many anatomical differences between the average left and right hemispheres of the human brain have been reported, and the nature of those asymmetries is a function of sex and age (see Kong et al., 2018). There is no question that the average human cerebral hemispheres tend to be anatomically different, but the functional consequences of the differences have not been apparent (see Kong et al., 2018). Let's consider research on the three cortical language areas.

There are two serious difficulties in studying anatomical asymmetry of the language areas. First, their boundaries are unclear, with no consensus on how best to define them (see Hagoort, 2014; Poeppel, 2014). Second, there are large differences among healthy people in the structure of these cortical language areas (see Amunts & Zilles, 2012). Given these two difficulties, it is not surprising that reports of their anatomical asymmetry have been variable (see Amunts & Zilles, 2012; Kong et al., 2018). In many cases, the predicted size advantage of the left-hemisphere language areas is reported, but in other cases there is no asymmetry, or even a right-hemisphere size advantage.

Any report that one of the three cortical language areas tends to be larger in the left hemisphere typically leads to the suggestion that the anatomical asymmetry might have caused, or have been caused by, the lateralization of language to the left hemisphere. However, there is little support for such conjectures (see Bishop, 2013). The fact that a particular cortical area is on the average larger in the left hemisphere does not suggest that it is causally linked to language lateralization, even if the cortical area has been linked to language. At a bare minimum, it must be shown that the anatomical and functional asymmetries are correlated—that the degree of anatomical lateralization in a person reflects the degree of language lateralization in the same person. To our knowledge, no such correlations have been found.

We want to end this section by highlighting the results of an important recent study that bears directly on the question of whether there are anatomical asymmetries in language-related brain areas. In 2018, Kong et al. conducted a large-scale analysis of the MRI data from over 17,000 healthy individuals. They reported large asymmetries in the size of the frontal operculum and Heschl's gyrus. Heschl's gyrus was larger on the left, as predicted. However, different parts of the frontal operculum showed different directions of asymmetry: Its anterior portion was larger on the right, and its posterior portion was larger on the left. Clearly, if there is a relationship between anatomical asymmetries and language function, it is far more complex than previously thought.

In short, the search for anatomical differences between the two hemispheres has been only partially successful. Many anatomical asymmetries have been discovered, but few have been clearly related to functional asymmetries (see Kong et al., 2019). Several researchers have suggested that studies of differences in the microstructure (e.g., differences in cell type, synapses, and neural circuitry) between the two hemispheres may prove more informative than comparisons of differences in the size of vaguely defined areas (see Chance, 2014).

Figure 16.7 Three language areas of the cerebral cortex that have been the focus of studies on neuroanatomical asymmetry: The frontal operculum, the planum temporale (Wernicke's area), and Heschl's gyrus (primary auditory cortex).

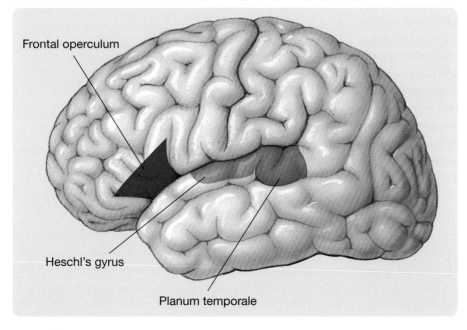

Evolution of Cerebral Lateralization and Language

You have already seen in this chapter how the discussion of cerebral lateralization inevitably leads to a discussion of language: Language is the most lateralized cognitive function. This module considers the evolution of cerebral lateralization and then the evolution of language.

Theories of the Evolution of Cerebral Lateralization

LO 16.15 Describe and evaluate three theoretical explanations for why cerebral lateralization of function exists.

Many theories have been proposed to explain why cerebral lateralization of function evolved. Most are based on one of two general premises: (1) that it is advantageous for areas of the brain that perform similar functions to be located in the same hemisphere, and (2) that it is advantageous to place some functions on one side of the brain and others on the other side so as to minimize redundancy of function between hemispheres (see Corballis, 2017). However, each theory of cerebral asymmetry postulates a different fundamental distinction between left and right hemisphere function. Consider the following three theories.

ANALYTIC–SYNTHETIC THEORY. The *analytic–synthetic theory of cerebral asymmetry* holds that there are two basic modes of thinking—an analytic mode and a synthetic mode—which have become segregated during the course of evolution in the left and right hemispheres, respectively. According to this theory, the left hemisphere operates in a logical, analytical, computerlike fashion, analyzing and abstracting stimulus input sequentially and attaching verbal labels; the right hemisphere is primarily a synthesizer, which organizes and processes information in terms of gestalts, or wholes.

Although the analytic–synthetic theory has been the darling of pop psychology (e.g., "You are so good with math—you must be left-brained!"), its vagueness is a problem (see Karolis, Corbetta, & de Schotten, 2019). Because it is not possible to specify the degree to which any task requires either analytic or synthetic processing, it has been difficult to subject the theory to empirical tests.

MOTOR THEORY. The *motor theory of cerebral asymmetry* (Kimura, 1979) holds that the left hemisphere is specialized not for the control of speech specifically but for the control of fine movements, of which speech is only one category (see Neubauer et al., 2020). Support for this theory comes from reports that lesions that produce aphasia often produce other motor deficits (see Kimura, 2011). One shortcoming of

the motor theory of cerebral asymmetry is that it does not suggest why motor function became lateralized in the first place (see Beaton, 2003).

LINGUISTIC THEORY. A third theory of cerebral asymmetry, the *linguistic theory of cerebral asymmetry*, posits that the primary role of the left hemisphere is language; this is in contrast to the analytic–synthetic and motor theories, which view language as a secondary specialization residing in the left hemisphere because of that hemisphere's primary specialization for analytic thought and skilled motor activity, respectively.

The linguistic theory of cerebral asymmetry is based to a large degree on the study of deaf people who use *American Sign Language* (a sign language with a structure similar to spoken language) and who suffer unilateral brain damage (see Campbell, MacSweeney, & Waters, 2008; Rogalsky et al., 2013). The fact that left-hemisphere damage can disrupt the use of sign language but not *pantomime gestures* (gestures that express meaning), as occurred in the case of W.L., suggests that the fundamental specialization of the left hemisphere may be language.

The Case of W.L., the Man Who Experienced Aphasia for Sign Language

W.L. is a congenitally deaf, right-handed male who grew up using American Sign Language. Seven months prior to testing, W.L. was admitted to hospital complaining of right-side weakness and motor problems. A CT scan revealed a large left frontotemporoparietal stroke. At that time, W.L.'s wife noticed he was making many uncharacteristic errors in signing and was having difficulty understanding the signs of others.

W.L.'s neuropsychologists managed to obtain a 2-hour videotape of an interview with him recorded 10 months before his stroke, which served as a valuable source of prestroke performance measures. Formal poststroke neuropsychological testing confirmed that W.L. had suffered a specific loss in his ability to use and understand sign language. The fact that he could produce and understand complex pantomime gestures suggested that his sign-language aphasia was specific to language (Corina et al., 1992).

When Did Cerebral Lateralization Evolve?

LO 16.16 List those species that display cerebral lateralization and explain what this tells us about when cerebral lateralization evolved.

Until recently, cerebral lateralization had been assumed to be an exclusive feature of the hominin brain. For example, one version of the motor theory of cerebral asymmetry is that left-hemisphere specialization for motor control evolved in early hominins in response to their use of tools, and then the capacity for vocal language subsequently evolved in the left hemisphere because of its greater motor dexterity. However, there is evidence of lateralization of function in many species (i.e., species from all five classes of vertebrates—fishes, reptiles, birds, amphibians, and mammals; and some invertebrate species) that evolved long before we humans were around (see Prieur et al., 2019).

The discovery of examples of cerebral lateralization in species from all five vertebrate classes and in certain invertebrate species suggests that cerebral lateralization must have survival advantages: But what are they? There seem to be two fundamental advantages. First, in some cases, it may be more efficient for the neurons performing a particular function to be concentrated in one hemisphere. For example, in most cases, it is advantageous to have one highly skilled hand rather than having two moderately skilled hands. Second, in some cases, two different kinds of cognitive processes may be more readily performed simultaneously if they are lateralized to different hemispheres (see Corballis, 2015).

As an example, right-handedness may have evolved from a preference for the use of the right side of the body for feeding—such a preference has been demonstrated in species of all five classes of vertebrates (fishes, reptiles, birds, amphibians, and mammals). Then, once hands evolved, those species with hands (i.e., species of monkeys and apes) displayed a right- or left-hand preference for feeding and other complex responses such as tool use and communication, which in turn led to a lateralization of those functions as well (see Prieur et al., 2019).

Evolution of Human Language

LO 16.17 Describe what the study of nonhuman primates has suggested about the evolution of human language.

Human communication is different from the communication of other species. Human language is a system allowing a virtually limitless number of ideas to be expressed by combining a finite set of elements (see Hauser et al., 2014). Other species do have language of sorts, but it can't compare with human language. For example, monkeys have distinct warning calls for different threats, but they do not combine the calls to express new ideas. Also, birds and whales sing complex songs, but there is no creative recombination of the songs to express new ideas (see Hage & Nieder, 2016).

Infants only babble prior to 10 months of age (see Laake & Bridgett, 2018), but 30-month-old infants speak in complete sentences and use more than 500 words. Also, over this same 20-month period, the plastic infant brain reorganizes itself to learn its parents' languages. At 10 months, human infants can distinguish the sounds of all human languages, but by 30 months, they can readily discriminate only those sounds that compose the languages to which they have been exposed. Once the ability to discriminate particular speech sounds is lost, it is difficult to regain, which is one reason why adults usually have difficulty learning to speak new languages without an accent.

Words do not leave fossils, and thus, insights into the evolution of human language can be obtained only through the comparative study of existing species. Naturally enough, researchers interested in the evolution of human language turned first to the vocal communications of our primate relatives.

VOCAL COMMUNICATION IN NONHUMAN PRIMATES. As you have just learned, no other species has a language that can compare with human language. However, each nonhuman primate species has a variety of calls, each with a specific meaning that is understood by conspecifics. Moreover, the calls are not simply reflexive reactions to particular situations: They are dependent on the social context (see Seyfarth & Cheney, 2014). For example, vervet monkeys do not make alarm calls unless other vervet monkeys are nearby, and the calls are most likely to be made if the nearby vervets are relatives (see Oller & Griebel, 2014). And chimpanzees vary the screams they produce during aggressive encounters depending on the severity of the encounter, their role in it, and which other chimpanzees can hear them (see Slocombe et al., 2010).

A consistent pattern has emerged from studies of nonhuman vocal language: There is typically a substantial difference between the capacity for vocal production and the capacity for auditory comprehension. Even the most vocal nonhumans can produce relatively few calls, yet they are capable of interpreting a wide range of other sounds in their environments (see Fitch, 2018; Hauser et al., 2014). This suggests that the ability of nonhumans to produce vocal language may be limited, not by their inability to interpret sounds, but by their inability to exert the fine motor control over their voices that is observed in humans (see Fitch, 2018; Ghazanfar & Takahashi, 2014). It also suggests that human language may have evolved from a competence in comprehension of sounds already existing in our primate ancestors.

MOTOR THEORY OF SPEECH PERCEPTION. It was reasonable for Broca to believe that Broca's area played a specific role in language expression (speech): After all, Broca's area is part of the left premotor cortex. However, the **motor theory of speech perception** goes one step further: It proposes that the perception and comprehension of speech depends on the words activating the same neural circuits in the motor system that would have been activated if the listener had said the words (see Cogan et al., 2014; but see Arsenault & Buchsbaum, 2015). General support for this theory has come from the discovery that just thinking about performing a particular action often activates the same areas of the brain as performing the action and from the discovery of *mirror neurons* (see Chapter 8), motor cortex neurons that fire when particular responses are either performed or observed (see Cook et al., 2014).

Further support for the motor theory of speech perception has come from many functional brain-imaging studies that have revealed activity in primary or secondary motor cortex during language comprehension tests that do not involve language expression (i.e., speaking or writing). Scott, McGettigan, and Eisner (2009) compiled and evaluated the results of studies that recorded activity in motor cortex during speech perception and concluded that the motor cortex is particularly active during the perception of conversational exchanges.

More direct support for the motor theory of speech perception has come from several studies that have applied transcranial magnetic stimulation (TMS) to areas of the motor cortex involved in speech articulation while volunteers listened to syllables and words. As hypothesized, the motor-cortex stimulation disrupted the perception of the syllables and words (see D'Ausilio et al., 2014; Schomers et al., 2015; Smalle, Rogers, & Möttönen, 2015).

However, there is some clinical evidence that contradicts the predictions of the motor theory of speech perception. Specifically, case studies of patients with damage to their motor cortex have failed to reveal the predicted deficits in speech perception (see Stasenko, Garcea, & Mahon, 2013; Stasenko et al., 2015).

Amelandfoto/Shutterstock

Each nonhuman primate species has a variety of calls, each with a specific meaning that is understood by conspecifics.

GESTURAL LANGUAGE. Because only humans are capable of a high degree of motor control over their vocal apparatus, language in nonhuman primates might be mainly gestural, rather than vocal. To test this hypothesis, Pollick and de Waal (2007) compared the gestures and the vocalizations of chimpanzees. They found a highly nuanced vocabulary of hand gestures that were used in many situations and in various combinations. In short, the chimpanzees' gestures were much more like human language than their vocalizations. Could primate gestures have been a critical stage in the evolution of human language (see de Boer, 2019; Gillespie-Lynch et al., 2014; Prieur et al., 2018)?

Scan Your Brain

So far in this chapter, you have learned about cerebral lateralization and split-brain patients. Before moving to the cerebral mechanisms of language and language disorder, pause and check your knowledge by filling in the blanks in the following sentences with the most appropriate terms. The correct answers appear at the end of the exercise. Be sure to review material related to your errors and omissions before proceeding.

1. Broca's discovery that patients with left-hemisphere damage exhibit _____ suggested that each hemisphere is dominant for specific cognitive functions.
2. An invasive technique of assessing cerebral lateralization involves the injection of a small amount of _____ in the carotid artery, which causes a brief paralysis to the injected hemisphere.
3. _____ is a noninvasive technique for assessing lateralization in which separate auditory messages are delivered to each ear through earphones.
4. Studies on the relationship between _____ and lateralization suggest that a majority of right-handers are left-hemisphere dominant for language while only 70 percent of left-handers show the same pattern.
5. Early research on sex differences in lateralization suggested that _____ are more lateralized than

_____. A more recent meta-analysis failed to find such effects.
6. The two hemispheres are connected by a bundle of neuronal axons that form the _____.
7. When a split-bran patient is presented with an object in the _____ visual field, they are unable to name the object.
8. In the _____ test, the left hemisphere of split-brain patients sees a single face that is the completed version of the half face on the right.
9. Motor functions of one side of the body are projected to the _____ hemisphere.
10. Kimura found that the right ear was superior in the perception of _____, while the left ear was superior in the perception of _____.
11. In musicians with perfect pitch, the _____ of the left hemisphere is larger than that of the right hemisphere.
12. According to the _____ of speech perception, auditory words activate the same mirror neurons that would be activated if the listener had spoken the words.

Scan Your Brain answers: (1) aphasia, (2) sodium amytal, (3) Dichotic listening test, (4) handedness, (5) males, females, (6) corpus callosum, (7) left, (8) chimeric figures, (9) contralateral, (10) digits, melodies, (11) planum temporale, (12) motor theory.

Cortical Localization of Language: Wernicke-Geschwind Model

This module focuses on the cerebral localization of language. In contrast to language lateralization, which refers to the relative control of language-related functions by the left and right hemispheres, *language localization* refers to the location within the hemispheres of the circuits that participate in language-related activities.

Like most introductions to language localization, the following discussion begins with the *Wernicke-Geschwind model*, the predominant theory of language localization. Because most of the research on the localization of language has been conducted and interpreted within the context of this model, reading about the localization of language without a basic understanding of the Wernicke-Geschwind model would be like watching a game of chess without knowing the rules.

Historical Antecedents of the Wernicke-Geschwind Model

LO 16.18 Describe the historical antecedents of the Wernicke-Geschwind model. Include descriptions of the following disorders: Broca's and Wernicke's aphasia, conduction aphasia, agraphia, and alexia.

The history of the localization of language and the history of the lateralization of function began at the same point, with Broca's assertion that a small area (Broca's area) in the inferior portion of the left prefrontal cortex is the center for speech production. Broca hypothesized that programs of articulation are stored within this area and that speech is produced when these programs activate the adjacent area of the precentral gyrus, which controls the muscles of the face and oral cavity. According to Broca, damage restricted to Broca's area should disrupt speech production without producing deficits in language comprehension.

The next major event in the study of the cerebral localization of language occurred in 1874, when Carl Wernicke (pronounced "VER-ni-key") concluded on the basis of 10 clinical cases that there is a language area in the left temporal lobe just posterior to the primary auditory cortex (i.e., in the left planum temporale). This second language area, which Wernicke argued was the cortical area of language comprehension, subsequently became known as **Wernicke's area**.

Wernicke suggested that selective lesions of Broca's area produce a syndrome of aphasia whose symptoms are primarily **expressive**—characterized by normal comprehension of both written and spoken language and by speech that retains its meaningfulness despite being slow, labored, disjointed, and poorly articulated. This hypothetical form of aphasia became known as **Broca's aphasia**. In contrast, Wernicke suggested that selective lesions of Wernicke's area produce a syndrome of aphasia whose deficits are primarily **receptive**—characterized by poor comprehension of both written and spoken language and speech that is meaningless but still retains the superficial structure, rhythm, and intonation of normal speech. This hypothetical form of aphasia became known as **Wernicke's aphasia**, and the normal-sounding but nonsensical speech of Wernicke's aphasia became known as *word salad*.

The following are examples of the kinds of speech presumed to be associated with selective damage to Broca's and Wernicke's areas (Geschwind, 1979, p. 183):

> *Broca's aphasia:* A patient who was asked about a dental appointment replied haltingly and indistinctly: "Yes…Monday…Dad and Dick…Wednesday nine o'clock…10 o'clock…doctors…and…teeth."

> *Wernicke's aphasia:* A patient who was asked to describe a picture that showed two boys stealing cookies reported smoothly: "Mother is away here working her work to get her better, but when she's looking the two boys looking in the other part. She's working another time."

Wernicke reasoned that damage to the pathway connecting Broca's and Wernicke's areas—the **arcuate fasciculus**—would produce a third type of aphasia, one he called **conduction aphasia**. He contended that comprehension and spontaneous speech would be largely intact in patients with damage to the arcuate fasciculus but that they would have difficulty repeating words they had just heard.

The left **angular gyrus**—the area of left temporal and parietal cortex just posterior to Wernicke's area—is another cortical area that has been implicated in language. Its role in language was recognized in 1892 by neurologist Joseph Jules Dejerine on the basis of the postmortem examination of one special patient. The patient suffered from **alexia** (the inability to read) and **agraphia** (the inability to write). What made this case special was that the alexia and agraphia were exceptionally pure: Although the patient could not read or write, he had no difficulty speaking or understanding speech. Dejerine's postmortem examination revealed damage in the pathways connecting the visual cortex with the left angular gyrus. He concluded that the left angular gyrus is responsible for comprehending language-related visual input, which is received directly from the adjacent left visual cortex and indirectly from the right visual cortex via the corpus callosum.

During the era of Broca, Wernicke, and Dejerine, many influential scholars (e.g., Freud, Head, and Marie) opposed

Figure 16.8 The seven components of the Wernicke-Geschwind model. All of the components are in the left hemisphere.

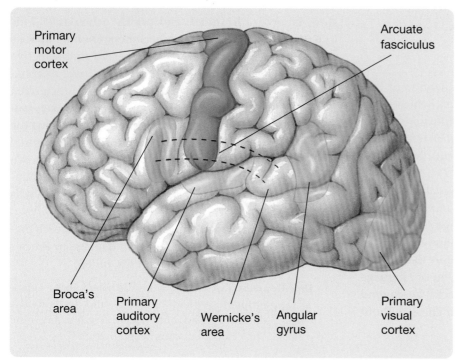

Primary motor cortex

Arcuate fasciculus

Broca's area

Primary auditory cortex

Wernicke's area

Angular gyrus

Primary visual cortex

their attempts to localize various language-related abilities to specific neocortical areas. In fact, advocates of the holistic approach to brain function gradually gained the upper hand, and interest in the cerebral localization of language waned. However, in the mid-1960s, Norman Geschwind (1970) revived the old localizationist ideas of Broca, Wernicke, and Dejerine, added some new data and insightful interpretation, and melded the mix into a powerful theory: the Wernicke-Geschwind model.

The Wernicke-Geschwind Model

LO 16.19 Describe the Wernicke-Geschwind model.

The following are the seven components of the **Wernicke-Geschwind model**: primary visual cortex, angular gyrus, primary auditory cortex, Wernicke's area, arcuate fasciculus, Broca's area, and primary motor cortex—all of which are in the left hemisphere. They are shown in Figure 16.8.

The following two examples illustrate how the Wernicke-Geschwind model is presumed to work (see Figure 16.9). First, when you are having a conversation, the auditory signals triggered by the speech of the other person are received by your primary auditory cortex and conducted to Wernicke's area, where they are comprehended. If a response is in order, Wernicke's area generates the neural representation of the thought underlying the reply, and it is transmitted to Broca's area via the left arcuate fasciculus. In Broca's area, this signal activates the appropriate program of articulation that drives the appropriate neurons of your primary motor cortex and ultimately your muscles of articulation. Second, when you are reading aloud, the signal received by your primary visual cortex is transmitted to your left angular gyrus, which translates the visual form of the word into its auditory code and transmits it to Wernicke's area for comprehension. Wernicke's area then triggers the appropriate responses in your arcuate fasciculus, Broca's area, and motor cortex, respectively, to elicit the appropriate speech sounds.

Figure 16.9 How the Wernicke-Geschwind model works in a person who is responding to a heard question and reading aloud. The hypothetical circuit that allows the person to respond to heard questions is in pink; the hypothetical circuit that allows the person to read aloud is in black.

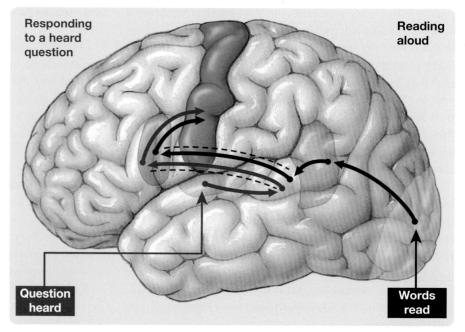

Responding to a heard question

Reading aloud

Question heard

Words read

Scan Your Brain

Before proceeding to the following evaluation of the Wernicke-Geschwind model, scan your brain to confirm that you understand its fundamentals. The correct answers are provided at the end of the exercise. Review material related to your errors and omissions before proceeding.

According to the Wernicke-Geschwind model, the following seven areas of the left cerebral cortex play a role in language-related activities:

1. The _____ gyrus translates the visual form of a read word into an auditory code.

2. The _____ cortex controls the muscles of articulation.

3. The _____ cortex perceives the written word.

4. _____ area is the center for language comprehension.

5. The _____ cortex perceives the spoken word.

6. _____ area contains the programs of articulation.

7. The left _____ carries signals from Wernicke's area to Broca's area.

Scan Your Brain answers: (1) angular, (2) primary motor, (3) primary visual, (4) Wernicke's, (5) primary auditory, (6) Broca's, (7) arcuate fasciculus.

Wernicke-Geschwind Model: The Evidence

Unless you are reading this text from back to front, you should have read the preceding description of the Wernicke-Geschwind model with some degree of skepticism. By this point in the text, you will almost certainly recognize that any model of a complex cognitive process that involves a few localized neocortical centers joined in a serial fashion by a few arrows is sure to have major shortcomings, and you will appreciate that the neocortex is not divided into neat compartments whose cognitive functions conform to vague concepts such as language comprehension, speech motor programs, and conversion of written language to auditory language (see Cahana-Amitay & Albert, 2014). Initial skepticism aside, the ultimate test of a theory's validity is the degree to which its predictions are consistent with the empirical evidence.

Before we examine this evidence, we want to emphasize one point. The Wernicke-Geschwind model was initially based on case studies of aphasic patients with strokes, tumors, and penetrating brain injuries. Damage in such cases is often diffuse, and it inevitably encroaches on subcortical nerve fibers that connect the lesion site to other areas of the brain. For example, Figure 16.10 shows the extent of the cortical damage in one of Broca's two original cases (see Mohr, 1976)—the damage is so diffuse that the case provides little evidence that Broca's area plays a role in speech.

Effects of Cortical Damage and Brain Stimulation on Language Abilities

LO 16.20 Identify the effects of cortical damage and brain stimulation on language abilities, and evaluate the Wernicke-Geschwind model in light of these findings.

In view of the fact that the Wernicke-Geschwind model grew out of the study of patients with cortical damage, it is appropriate to begin evaluating it by assessing its ability to

Figure 16.10 The extent of brain damage in one of Broca's two original patients. Like this patient, most aphasic patients have diffuse brain damage. It is thus difficult to determine from studying them the precise location of particular cortical language areas.

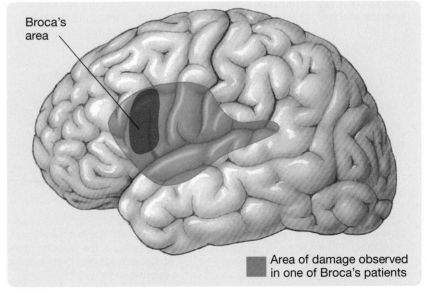

Broca's area

■ Area of damage observed in one of Broca's patients

Based on Mohr, J. P. (1976). Broca's area and Broca's aphasia. In H. Whitaker & H. A. Whitaker (Eds.), *Studies in Neurolinguistics* (Vol. 1, pp. 201–235). New York, NY: Academic Press.

predict the language-related deficits produced by damage to various parts of the cortex.

> **Journal Prompt 16.5**
>
> For each structure in the Wernicke-Geschwind model, write down your prediction of the effects of damage to it. Later, compare your predictions with the evidence you will encounter in this module.

EVIDENCE FROM STUDIES OF THE EFFECTS OF CORTICAL DAMAGE. Studies of patients in whom discrete areas of cortex have been surgically removed have been particularly informative about the cortical localization of language, because the location and extent of these patients' lesions can be derived with reasonable accuracy from the surgeons' reports. The study of neurosurgical patients has not confirmed the predictions of the Wernicke-Geschwind model by any stretch of the imagination. See the six cases summarized in Figure 16.11.

Surgery that destroys all of Broca's area but little surrounding tissue typically has no lasting effects on speech (Penfield & Roberts, 1959; Rasmussen & Milner, 1975; Zangwill, 1975). Some speech problems were observed after the removal of Broca's area, but their temporal course suggested that they were products of postsurgical *edema* (swelling) in the surrounding neural tissue rather than of the *excision* (cutting out) of Broca's area itself. Prior to the use of effective anti-inflammatory drugs, patients with excisions of Broca's area often regained consciousness with

Figure 16.11 The lack of permanent disruption of language-related abilities after surgical excision (indicated in orange) of the classic Wernicke-Geschwind language areas (outlined with dotted lines).

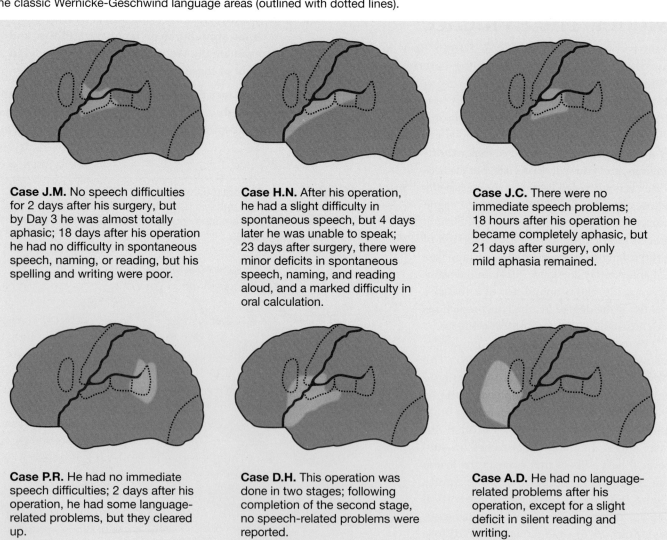

Case J.M. No speech difficulties for 2 days after his surgery, but by Day 3 he was almost totally aphasic; 18 days after his operation he had no difficulty in spontaneous speech, naming, or reading, but his spelling and writing were poor.

Case H.N. After his operation, he had a slight difficulty in spontaneous speech, but 4 days later he was unable to speak; 23 days after surgery, there were minor deficits in spontaneous speech, naming, and reading aloud, and a marked difficulty in oral calculation.

Case J.C. There were no immediate speech problems; 18 hours after his operation he became completely aphasic, but 21 days after surgery, only mild aphasia remained.

Case P.R. He had no immediate speech difficulties; 2 days after his operation, he had some language-related problems, but they cleared up.

Case D.H. This operation was done in two stages; following completion of the second stage, no speech-related problems were reported.

Case A.D. He had no language-related problems after his operation, except for a slight deficit in silent reading and writing.

▓ Area of Surgical Excision

Based on Penfield, W., & Roberts, L. (1959). *Speech and Brain Mechanisms*. Princeton, NJ: Princeton University Press.

their language abilities fully intact only to have serious language-related problems develop over the next few hours and then subside in the following weeks. Similarly, permanent speech difficulties were not produced by discrete surgical lesions to the arcuate fasciculus, and permanent alexia and agraphia were not produced by surgical lesions restricted to the cortex of the angular gyrus (Rasmussen & Milner, 1975).

The consequences of surgical removal of Wernicke's area are less well documented; surgeons have been hesitant to remove it in light of Wernicke's dire predictions. Nevertheless, in some cases, a good portion of Wernicke's area has been removed without lasting language-related deficits (e.g., Ojemann, 1979; Penfield & Roberts, 1959).

Hécaen and Angelergues (1964) published the first large-scale study of accidental or disease-related brain damage and aphasia. They rated the articulation, fluency, comprehension, naming ability, ability to repeat spoken sentences, reading, and writing of 214 right-handed patients with left-hemisphere damage. The extent and location of the damage in each case were estimated by either postmortem examination or visual inspection during subsequent surgery.

Hécaen and Angelergues found that small lesions to Broca's area seldom produced lasting language deficits and that small lesions restricted to Wernicke's area did not always produce lasting language deficits. Larger lesions did produce more lasting language deficits; but, in contrast to the predictions of the Wernicke-Geschwind model, problems of articulation were just as likely to occur following parietal or temporal lesions as they were following comparable lesions in the vicinity of Broca's area. It is noteworthy that none of the 214 patients studied by Hécaen and Angelergues displayed syndromes of aphasia that were either totally expressive (Broca's aphasia) or totally receptive (Wernicke's aphasia).

EVIDENCE FROM STRUCTURAL NEUROIMAGING STUDIES. Since their development in the 1970s, CT and MRI techniques have been used extensively to analyze the brain damage associated with aphasia. Several large studies have assessed the CT and structural MRI scans of aphasic patients with accidental or disease-related brain damage (e.g., Yourganov et al., 2015). In confirming and extending the results of earlier studies, they have not been kind to the Wernicke-Geschwind model. The following have been their major findings:

- No aphasic patients have damage restricted to Broca's area or Wernicke's area.
- Aphasic patients almost always have significant damage to subcortical white matter.
- Large anterior lesions are more likely to produce expressive symptoms, whereas large posterior lesions are more likely to produce receptive symptoms.

- **Global aphasia** (a severe disruption of all language-related abilities) is usually related to massive lesions of anterior cortex, posterior cortex, and underlying white matter.
- Aphasic patients sometimes have brain damage that does not encroach on the Wernicke-Geschwind areas—aphasia has been observed in patients with visible damage to only the medial frontal lobe, subcortical white matter, basal ganglia, or thalamus.

In summary, large-scale, objective studies of the relationship between language deficits and brain damage—whether utilizing autopsy, direct observation during surgery, or brain scans—have not confirmed the major predictions of the Wernicke-Geschwind model (see Nasios, Dardiotis, & Messinis, 2019; Rahimpour et al., 2019). Has the model been treated more favorably by studies of electrical brain stimulation?

EVIDENCE FROM STUDIES OF ELECTRICAL STIMULATION OF THE CORTEX. The first large-scale electrical brain-stimulation studies of humans were conducted by Wilder Penfield and his colleagues in the 1940s at the Montreal Neurological Institute (see Rahimpour et al., 2019). One purpose of the studies was to map the language areas of each patient's brain so that tissue involved in language could be avoided during the surgery. The mapping was done by assessing the responses of conscious patients, who were under local anesthetic, to stimulation applied to various points on the cortical surface. The description of the effects of each stimulation was dictated to a stenographer (this was before the days of tape recorders) and then a tiny numbered card was dropped on the stimulation site for subsequent photography.

Figure 16.12 illustrates the responses to stimulation of a 37-year-old right-handed patient with epilepsy. He had started to have seizures about 3 months after receiving a blow to the head; at the time of his operation, in 1948, he had been suffering from seizures for 6 years, despite efforts to control them with medication. In considering his responses, remember that the cortex just posterior to the central fissure is primary somatosensory cortex and that the cortex just anterior to the central fissure is primary motor cortex.

Because electrical stimulation of the cortex is much more localized than a brain lesion, it has been a useful method for testing predictions of the Wernicke-Geschwind model. Penfield and Roberts (1959) published the first large-scale study of the effects of cortical stimulation on speech. They found that sites at which stimulation blocked or disrupted speech in conscious neurosurgical patients were scattered throughout a large expanse of frontal, temporal, and parietal cortex, rather than being restricted to the Wernicke-Geschwind language areas (see Figure 16.13). They also found no tendency for particular kinds of speech

Figure 16.12 The responses of the left hemisphere of a 37-year-old right-handed person with epilepsy to electrical stimulation. Numbered cards were placed on the brain during surgery to mark the sites where brain stimulation had been applied.

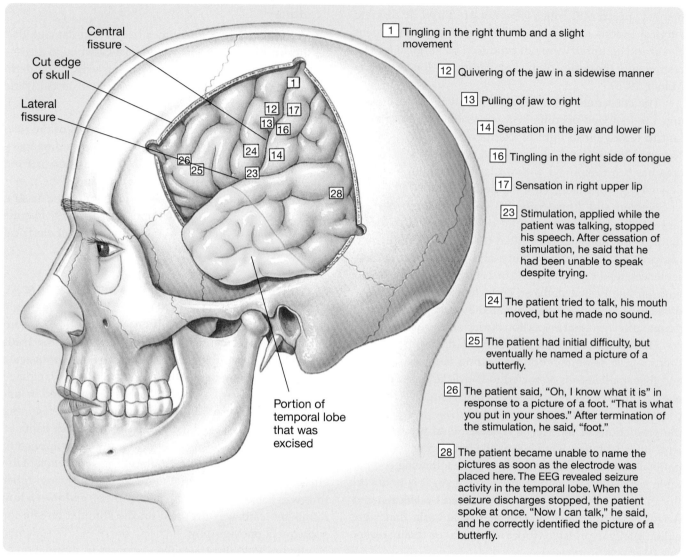

1 Tingling in the right thumb and a slight movement

12 Quivering of the jaw in a sidewise manner

13 Pulling of jaw to right

14 Sensation in the jaw and lower lip

16 Tingling in the right side of tongue

17 Sensation in right upper lip

23 Stimulation, applied while the patient was talking, stopped his speech. After cessation of stimulation, he said that he had been unable to speak despite trying.

24 The patient tried to talk, his mouth moved, but he made no sound.

25 The patient had initial difficulty, but eventually he named a picture of a butterfly.

26 The patient said, "Oh, I know what it is" in response to a picture of a foot. "That is what you put in your shoes." After termination of the stimulation, he said, "foot."

28 The patient became unable to name the pictures as soon as the electrode was placed here. The EEG revealed seizure activity in the temporal lobe. When the seizure discharges stopped, the patient spoke at once. "Now I can talk," he said, and he correctly identified the picture of a butterfly.

Based on Penfield & Roberts, *Speech and Brain Mechanisms* 1959. Princeton University Press.

Figure 16.13 The wide distribution of left hemisphere sites where cortical stimulation either blocked speech or disrupted it.

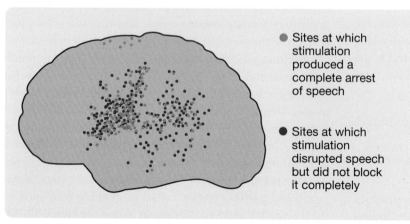

● Sites at which stimulation produced a complete arrest of speech

● Sites at which stimulation disrupted speech but did not block it completely

Based on Penfield & Roberts, *Speech and Brain Mechanisms* 1959. Princeton University Press.

disturbances to be elicited from particular areas of the cortex: Sites at which stimulation produced disturbances of pronunciation, confusion of counting, inability to name objects, or misnaming of objects were pretty much intermingled. Right-hemisphere stimulation almost never disrupted speech.

Ojemann and his colleagues (see Corina et al., 2010; McDermott, Watson, & Ojemann, 2005) assessed naming, reading of simple sentences, short-term verbal memory, ability to mimic movements of the face and mouth, and ability to recognize phonemes during cortical stimulation. A **phoneme** is the smallest unit of sound that distinguishes various words in a language; the pronunciation of each phoneme varies slightly, depending on the sounds next to it. The following are the findings of Ojemann and colleagues related to the Wernicke-Geschwind model:

- Stimulation to areas far beyond the boundaries of the Wernicke-Geschwind language areas was capable of disrupting the use of language.

- Each of the language tests was disrupted by stimulation at widely scattered sites.

- There were major differences among the patients in the organization of their language abilities.

Because the disruptive effects of stimulation at a particular site were frequently quite specific (i.e., disrupting only a single test), Ojemann suggested that the language cortex is organized like a *mosaic*, with the discrete columns of tissue that perform a particular function widely distributed throughout the language areas of cortex.

Current Status of the Wernicke-Geschwind Model

LO 16.21 **Summarize the current status of the Wernicke-Geschwind model.**

Evidence from studies of brain damage and from observations of electrical stimulation of the brain has supported the Wernicke-Geschwind model in two general respects. First, the evidence has confirmed that Broca's and Wernicke's areas play important roles in language; many aphasics have diffuse cortical damage that involves one or both of these areas. Second, there is a tendency for aphasias associated with anterior damage to involve deficits that are more expressive and those associated with posterior damage to involve deficits that are more receptive. However, other observations have not confirmed predictions of the Wernicke-Geschwind model:

- Damage restricted to the boundaries of the Wernicke-Geschwind cortical areas often has little lasting effect on the use of language—aphasia is typically associated with widespread damage.

- Brain damage that does not include any of the Wernicke-Geschwind areas can produce aphasia (see Lorch, 2019).

- Broca's and Wernicke's aphasias rarely exist in the pure forms implied by the Wernicke-Geschwind model; aphasia virtually always involves both expressive and receptive symptoms (see O'Sullivan, Brownsett, & Copland, 2019).

- There are major differences in the locations of cortical language areas in different people.

Despite these problems, the Wernicke-Geschwind model has been an extremely important theory. It guided the study and clinical diagnosis of aphasia for more than four decades. Indeed, clinical neuropsychologists still use *Broca's aphasia* and *Wernicke's aphasia* as diagnostic categories, but with an understanding that the syndromes are much less selective and the precipitating damage much more diffuse and variable than implied by the model (see Nasios, Dardiotis, & Messinis, 2019; Rahimpour et al., 2019). Because of the lack of empirical support for its major predictions, the Wernicke-Geschwind model has been largely abandoned by researchers, but it is still prominent in the classroom and clinic (see Dick, Bernal, & Tremblay, 2014; Nasios, Dardiotis, & Messinis, 2019; Poeppel et al., 2012).

Cognitive Neuroscience of Language

The *cognitive neuroscience approach*, which currently dominates research on language, is the focus of the final two modules of this chapter. We begin by examining the three premises that define the cognitive neuroscience approach to language and differentiate it from the premises on which the Wernicke-Geschwind model is based.

Three Premises That Define the Cognitive Neuroscience Approach to Language

LO 16.22 **Describe the three premises that define the cognitive neuroscience approach to language, and compare them with the premises on which the Wernicke-Geschwind model is based.**

- *Premise 1:* The use of language is mediated by widespread activity in all the areas of the brain that participate in the cognitive processes involved in the particular language-related behavior. As you have just learned, the Wernicke-Geschwind model is based on the assumption that particular areas of the brain involved in language are each dedicated to a complex process such as speech production, language comprehension, or reading. In contrast, cognitive neuroscience research is based on the premise that each of these complex processes results from a combination of several *constituent cognitive processes*, which may be organized separately in different parts of the brain (see Bouchard, 2015). Accordingly, the specific constituent cognitive processes, not the general Wernicke-Geschwind processes, are assumed to be the appropriate level at which to conduct cognitive neuroscientific analysis. Cognitive neuroscientists typically divide analysis of the constituent cognitive processes involved in language into three categories: **phonological analysis** (analysis of the sound of language), **grammatical analysis** (analysis of the structure of language), and **semantic analysis** (analysis of the meaning of language).

- *Premise 2:* The areas of the brain involved in language are not dedicated solely to that purpose (see Friederici & Singer, 2015). In the Wernicke-Geschwind model, large areas of left cerebral cortex were thought to be dedicated solely to language, whereas the cognitive neuroscience approach assumes that many of the constituent cognitive processes involved in language also play roles in other kinds of behavior.

- *Premise 3:* Because many of the areas of the brain that perform specific language functions are also parts of other functional systems, these areas are likely to be small, widely distributed, and specialized (see Friederici & Singer, 2015). In contrast, the language areas of the Wernicke-Geschwind model are assumed to be large, circumscribed, and homogeneous (see Friederici & Gierhan, 2013).

In addition to these three premises, the methodology of the cognitive neuroscience approach to language distinguishes it from previous approaches. The Wernicke-Geschwind model rests heavily on the analysis of brain-damaged patients, whereas researchers using the cognitive neuroscience approach also employ an array of other techniques—most notably, functional brain imaging—in studying the localization of language in healthy volunteers.

It is important to remember that functional brain-imaging studies cannot prove causation (see Cloutman et al., 2011). There is a tendency to assume that brain activity recorded during a particular cognitive process plays a causal role in that process, but, as you learned in Chapter 1, correlation cannot prove causation. For example, substantial right-hemisphere activity is virtually always recorded during various language-related cognitive tasks, and it is tempting to assume that this activity is thus crucial to language-related cognitions. However, lesions of the right hemisphere are only rarely associated with lasting language-related deficits (see Hickok et al., 2008).

Journal Prompt 16.6

Substantial right-hemisphere activity is virtually always recorded during various language-related cognitive tasks even though language tends to be lateralized to the left hemisphere. Why do you think this happens? (Hint: Think about the connections between the two hemispheres.)

Functional Brain Imaging and the Localization of Language

LO 16.23 Describe two influential functional imaging studies of the localization of language, and explain what their findings indicate.

Numerous PET and fMRI studies of volunteers engaging in various language-related tasks have been published (see Price, 2012). The following two have been particularly influential.

BAVELIER'S FMRI STUDY OF READING. Bavelier and colleagues (1997) used fMRI to measure the brain activity of healthy volunteers while they read silently. The methodology of these researchers was noteworthy in two respects. First, they used a particularly sensitive fMRI machine that allowed them to identify areas of activity with more accuracy than in most previous studies and without having to average the scores of several participants (see Chapter 5). Second, they recorded brain activity during the reading of sentences—rather than during simpler, controllable, unnatural tasks (e.g., listening to individual words) as in most functional brain-imaging studies of language.

The volunteers in Bavelier and colleagues' study viewed sentences displayed on a screen. Interposed between periods of silent reading were control periods, during which the participants were presented with strings of consonants. The differences in cortical activity during the reading and control periods served as the basis for determining those areas of cortical activity associated with reading.

Let's begin by considering the findings obtained for individual participants on individual trials before any averaging took place. Three important points emerged from this analysis:

- The areas of activity were patchy; that is, they were tiny areas of activity separated by areas of inactivity (see Leech & Saygin, 2011).

- The patches of activity were variable; that is, the areas of activity differed from participant to participant and even from trial to trial in the same participant.

- Although activity was often observed in parts of the classic Wernicke-Geschwind areas, it was widespread over the lateral surfaces of the brain.

Even though Bavelier and colleagues' fMRI machine was sensitive enough to render averaging unnecessary, they did average their data in the usual way to illustrate its misleading effects. Figure 16.14 illustrates the reading-related increases of activity averaged over all the trials and participants in the study by Bavelier and colleagues—as they are typically reported. The averaging creates the false impression that large, homogeneous expanses of tissue were active during reading, whereas the patches of activity induced on

Figure 16.14 The areas in which reading-associated increases in activity were observed in the fMRI study of Bavelier and colleagues (1997). These maps were derived by averaging the scores of all participants, each of whom displayed patchy increases of activity in 5 to 10 percent of the indicated areas on any particular trial.

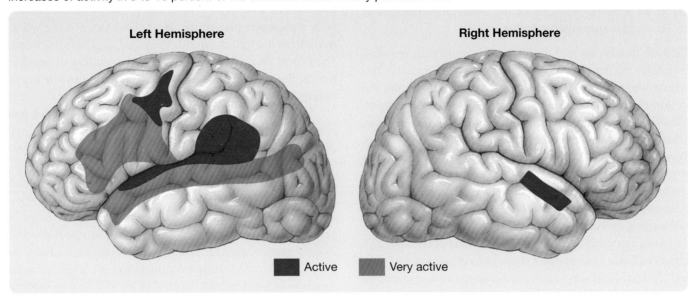

any given trial comprised only between 5 and 10 percent of the illustrated areas. Still, two points are clear from the averaged data: First, although there was significant activity in the right hemisphere, there was far more activity in the left hemisphere; second, the activity extended far beyond those areas predicted by the Wernicke-Geschwind model to be involved in silent reading (e.g., activity in Broca's area and the primary motor cortex would not have been predicted).

DAMASIO'S PET STUDY OF NAMING. The objective of the study of Damasio and colleagues (1996) was to look selectively at the temporal-lobe activity involved in naming objects within particular categories. PET activity was recorded from the left temporal lobes of healthy volunteers while they named images presented on a screen. The images were of three different types: famous faces, animals, and tools. To get a specific measure of the temporal-lobe activity involved in naming, the researchers subtracted from the activity recorded during this task the activity recorded while the volunteers judged the orientation of the images. The researchers limited the PET images to the left temporal lobe to permit a more fine-grained PET analysis.

Naming objects activated areas of the left temporal lobe outside the classic Wernicke's language area. Remarkably, the precise area activated by the naming depended on the category: Famous faces, animals, and tools each activated a slightly different area.

Other functional brain-imaging studies have confirmed category-specific encoding of words in the left temporal lobe (see Yi, Leonard, & Chang, 2019). Moreover, aphasic patients who have naming difficulties that are specific to famous faces, animals, or tools have been shown to have

damage in one of the three areas of the left temporal lobe identified by Damasio and colleagues.

Cognitive Neuroscience of Dyslexia

This final module of the chapter looks at how the cognitive neuroscience approach to language views dyslexia, one of the major subjects of cognitive neuroscience research.

Dyslexia is a reading disorder, one that does not result from general visual, motor, or intellectual deficits. There are two fundamentally different types of dyslexias: **developmental dyslexias**, which become apparent when a child is learning to read; and **acquired dyslexias**, which are caused by brain damage in individuals who were already capable of reading. Developmental dyslexia is widespread. Estimates of the overall incidence of developmental dyslexia among English-speaking children range from 5 to 12 percent (see Norton, Beach, & Gabrieli, 2015), depending on the criteria that are employed to define dyslexia, and the incidence is two to three times higher among boys than girls. In contrast, acquired dyslexias are relatively rare.

Journal Prompt 16.7

Do you or someone you know have dyslexia? What were some of the difficulties that you or the other person experienced?

Developmental Dyslexia: Causes and Neural Mechanisms

LO 16.24 Describe the causes and neural mechanisms of developmental dyslexia.

Because developmental dyslexia is far more common and its causes are less obvious, most research on dyslexia has focused on this form. However, the task of identifying the neural correlates of developmental dyslexia has been complicated by the fact that the disorder occurs in various forms, which likely have different neural correlates (see O'Brien, Wolf, & Lovett, 2012; Peyrin et al., 2012). Another problem is that the major reduction of reading may induce changes in the brain itself, making it difficult to determine whether a difference in the brain of a person with developmental dyslexia is the cause or the result of the disorder (see Bishop, 2013).

One approach to dealing with the issue of whether any observed brain changes are a cause or a consequence of having dyslexia has been to compare children with dyslexia to "ability-matched" children who are typically many years younger than them (see Norton, Beach, & Gabrieli, 2015). The rationale for such a comparison is that the two groups would be matched in terms of the amount of reading they have engaged in. In one such study by Hoeft et al. (2007), children with dyslexia exhibited reduced fMRI activity in their left parietal cortex and in their left occipitotemporo junction (the area of the cortex where the occipital lobe borders the temporal lobe) relative to their ability-matched controls.

Many researchers who study the neural mechanisms of dyslexia have focused on one kind of brain pathology and have tried to attribute developmental dyslexia to it. For example, developmental dyslexia has been attributed to attentional, auditory, visual, and sensorimotor deficits (see Goswami, 2014; Harrar et al., 2014; Stoodley & Stein, 2013). However, although many persons with dyslexia do experience a variety of subtle attentional, auditory, visual, and sensorimotor deficits, many do not. Moreover, even when these deficits are present in persons with dyslexia, they do not account for all aspects of the disorder. As a result, there is now fairly widespread agreement that dyslexia results most commonly from a specific disturbance of *phonological processing* (the representation and comprehension of speech sounds)—see Ramus and colleagues (2013); Hahn, Foxe, and Molholm (2014); but see Stein (2018).

Cognitive Neuroscience of Deep and Surface Dyslexia

LO 16.25 Describe the difference between the lexical procedure and the phonetic procedure for reading aloud. Then describe the difference between surface dyslexia and deep dyslexia.

Cognitive psychologists have long recognized that reading aloud can be accomplished in two entirely different ways.

One is by a **lexical procedure**, which is based on specific stored information that has been acquired about written words: The reader simply looks at the word, recognizes it, and says it. The other way reading can be accomplished is by a **phonetic procedure**: The reader looks at the word, recognizes the letters, sounds them out, and says the word. The lexical procedure dominates in the reading of familiar words; the phonetic procedure dominates in the reading of unfamiliar words.

This simple cognitive analysis of reading aloud has proven useful in understanding the symptoms of two kinds of dyslexia resulting from acquired brain damage: *surface dyslexia* and *deep dyslexia*. In cases of **surface dyslexia**, patients have lost their ability to pronounce words based on their specific memories of the words (i.e., they have lost the *lexical procedure*), but they can still apply rules of pronunciation in their reading (i.e., they can still use the *phonetic procedure*). Accordingly, they retain their ability to pronounce words whose pronunciation is consistent with common rules (e.g., *fish, river,* and *glass*) and their ability to pronounce nonwords according to common rules of pronunciation (e.g., *spleemer* and *twipple*); but they have great difficulty pronouncing words that do not follow common rules of pronunciation (e.g., *have, lose,* and *steak*). The errors they make often involve the misapplication of common rules of pronunciation; for example, *have, lose,* and *steak* are typically pronounced as if they rhymed with *cave, hose,* and *beak*.

In cases of **deep dyslexia** (also called *phonological dyslexia*), patients have lost their ability to apply rules of pronunciation in their reading (i.e., they have lost the *phonetic procedure*), but they can still pronounce familiar words based on their specific memories of them (i.e., they can still use the *lexical procedure*). Accordingly, they are completely incapable of pronouncing nonwords and have difficulty pronouncing uncommon words and words whose meaning is abstract. In attempting to pronounce words, patients with deep dyslexia try to react to them by using various lexical strategies, such as responding to the overall look of the word, the meaning of the word, or the derivation of the word. This leads to a characteristic pattern of errors. A patient with deep dyslexia might say "quill" for *quail* (responding to the overall look of the word), "hen" for *chicken* (responding to the meaning of the word), or "wise" for *wisdom* (responding to the derivation of the word).

Where are the lexical and phonetic procedures performed in the brain? Much of the research attempting to answer this question has focused on the study of deep dyslexia. Persons with deep dyslexia most often have extensive damage to the left-hemisphere language areas, suggesting that the disrupted phonetic procedure is widely distributed in the frontal and temporal areas of the left hemisphere. But which part of the brain maintains the lexical procedure in persons with deep dyslexia? There have been two theories, both of which have received some support. One theory is that the surviving lexical abilities of persons with deep dyslexia are mediated by activity in surviving parts of the left-hemisphere language areas. Evidence for this theory comes from the observation of neural activity in the surviving regions during reading (see Laine et al., 2000; Price et al., 1998). The other theory is that the surviving lexical abilities of persons with deep dyslexia are mediated by activity in the right hemisphere. The following remarkable case study provides support for this theory.

The Case of N.I., the Woman Who Read with Her Right Hemisphere

Prior to the onset of her illness, N.I. was a healthy girl. At the age of 13, she began to experience periods of aphasia, and several weeks later, she suffered a generalized convulsion. She subsequently had many convulsions, and her speech and motor abilities deteriorated badly. CT scans indicated ischemic brain damage to the left hemisphere.

Two years after the onset of her disorder, N.I. was experiencing continual seizures and blindness in her right visual field, and there was no meaningful movement or perception in her right limbs. In an attempt to relieve these symptoms, a total left **hemispherectomy** was performed; that is, her left hemisphere was totally removed. Her seizures were totally arrested by this surgery.

The reading performance of N.I. is poor, but she displays a pattern of retained abilities strikingly similar to those displayed by persons with deep dyslexia or split-brain patients reading with their right hemispheres. For example, she recognizes letters but is totally incapable of translating them into sounds; she can read concrete familiar words; she cannot pronounce even simple nonsense words (e.g., *neg*); and her reading errors indicate that she is reading on the basis of the meaning and appearance of words rather than by translating letters into sounds (e.g., when presented with the word *fruit*, she responded, "Juice...it's apples and pears and...fruit"). In other words, she suffers from a severe case of deep dyslexia (Patterson, Vargha-Khadem, & Polkey, 1989).

The case of N.I. completes the circle: The chapter began with a discussion of lateralization of function, and the case of N.I. concludes it on the same note.

Themes Revisited

This chapter contributes to all of the themes of the book. The clinical implications theme is the most prevalent because much of what we know about the lateralization of function and the localization of language in the brain comes from the study of neuropsychological patients.

Because lateralization of function and language localization are often covered by the popular media, they have become integrated into pop culture, and many widely held ideas about these subjects are overly simplistic. In this chapter, you were encouraged to think creatively about aspects of laterality and language that require unconventional ways of thinking.

The evolutionary perspective played a key role in trying to understand why cerebral lateralization of function evolved in the first place, and the major breakthrough in understanding the split-brain phenomenon came from comparative research.

The neuroplasticity theme arose during the discussion of the sensitive period for the learning of new languages and during the discussion of the neurodevelopmental bases of developmental dyslexia. Damage to the brain always triggers a series of neuroplastic changes, which can complicate the study of the behavioral effects of the damage.

Of the two emerging themes, the consciousness theme dominated: Studies of split-brain patients, which seem to produce two selves, question our assumptions of the unity of our conscious experience. Accordingly, the split-brain phenomenon has been prominent in theories of consciousness.

Key Terms

Cerebral commissures, p. 433
Lateralization of function, p. 433
Split-brain patients, p. 433
Commissurotomy, p. 433

Cerebral Lateralization of Function: Introduction
Aphasia, p. 434

Broca's area, p. 434
Apraxia, p. 434
Dominant hemisphere, p. 434
Minor hemisphere, p. 434
Sodium amytal test, p. 434
Dichotic listening test, p. 435
Dextrals, p. 435
Sinestrals, p. 435

The Split Brain
Corpus callosum, p. 436
Scotoma, p. 437
Cross-cuing, p. 440
Helping-hand phenomenon, p. 440
Visual completion, p. 440
Chimeric figures test, p. 441

Chapter 17
Biopsychology of Emotion, Stress, and Health

Fear, the Dark Side of Emotion

fizkes/Shutterstock

⌄ Chapter Overview and Learning Objectives

Biopsychology of Emotion: Introduction	**LO 17.1**	Summarize the major events in the history of research on the biopsychology of emotion.
	LO 17.2	Summarize the research on the relationship between the autonomic nervous system and emotions.
	LO 17.3	Describe some research on the facial expression of emotions.
Fear, Defense, and Aggression	**LO 17.4**	Describe the work that led to the distinction between aggressive and defensive behaviors in mammals.
	LO 17.5	Describe the relation between testosterone levels and aggression in males.

This chapter is about the biopsychology of emotion, stress, and health. It begins with a historical introduction to the biopsychology of emotion and then focuses in the next two modules on the dark end of the emotional spectrum: fear. Biopsychological research on emotions has concentrated on fear not because biopsychologists are a scary bunch, but because fear has three important qualities: It is the easiest emotion to infer from behavior in various species; it plays an important adaptive function in motivating the avoidance of threatening situations; and chronic fear is one common source of stress. In the final two modules of the chapter, you will learn how some brain structures have been implicated in human emotion, and how stress increases susceptibility to illness.

Biopsychology of Emotion: Introduction

To introduce the biopsychology of emotion, this module reviews several classic early discoveries and then discusses the role of the autonomic nervous system in emotional experience and the facial expression of emotion.

Early Landmarks in the Biopsychological Investigation of Emotion

LO 17.1 Summarize the major events in the history of research on the biopsychology of emotion.

This section describes, in chronological sequence, six early landmarks in the biopsychological investigation of emotion. It begins with the 1848 case of Phineas Gage.

The Mind-Blowing Case of Phineas Gage

In 1848, Phineas Gage, a 25-year-old construction foreman for the Rutland and Burlington Railroad, was the victim of a tragic accident. In order to lay new tracks, the terrain had to be leveled, and Gage was in charge of the blasting. His task involved drilling holes in the rock, pouring some gunpowder into each hole, covering it with sand, and tamping the material down with a large tamping iron before detonating it with a fuse. On the fateful day, the gunpowder exploded while Gage was tamping

Biopsychology of Emotion, Stress, and Health **463**

it, launching the 3-cm-thick, 90-cm-long tamping iron through his face, skull, and brain and out the other side.

Amazingly, Gage survived his accident, but he survived it a changed man. Before the accident, Gage had been a responsible, intelligent, socially well-adapted person, who was well liked by his friends and fellow workers. Once recovered, he appeared to be as able-bodied and intellectually capable as before, but his personality and emotional life had totally changed. Formerly a religious, respectful, reliable man, Gage became irreverent and impulsive. In particular, his abundant profanity offended many. He became so unreliable and undependable that he soon lost his job, and was never again able to hold a responsible position.

Gage became itinerant, roaming the country for a dozen years until his death in San Francisco. His bizarre accident and apparently successful recovery made headlines around the world, but his death went largely unnoticed and unacknowledged.

Gage was buried next to the offending tamping iron. Five years later, neurologist John Harlow was granted permission from Gage's family to exhume the body and tamping iron to study them. Since then, Gage's skull and the tamping iron have been on display in the Warren Anatomical Medical Museum at Harvard University.

In 1994, Damasio and her colleagues brought the power of computerized reconstruction to bear on Gage's classic case. They began by taking an x-ray of the skull and measuring it precisely, paying particular attention to the position of the entry and exit holes. From these measurements, they reconstructed the accident and determined the likely region of Gage's brain damage (see Figure 17.1). It was apparent that the damage to Gage's brain affected both medial prefrontal lobes, which we now know are involved in planning, decision making, and emotion (see Jin & Maren, 2015; Lee & Seo, 2016; Simon, Wood & Moghaddam, 2015).

DARWIN'S THEORY OF THE EVOLUTION OF EMOTION. The first major event in the study of the biopsychology of emotion was the publication in 1872 of Darwin's book *The Expression of Emotions in Man and Animals*. In it, Darwin argued that particular emotional responses, such as human facial expressions, tend to accompany the same emotional states in all members of a species.

Darwin believed that expressions of emotion, like other behaviors, are products of evolution; he therefore tried to understand them by comparing them in different species (see Brecht & Freiwald, 2012). From such interspecies comparisons, Darwin developed a theory of the evolution of emotional expression that was composed of three main ideas:

- Expressions of emotion evolve from behaviors that indicate what an animal is likely to do next.

- If the signals provided by such behaviors benefit the animal that displays them, they will evolve in ways that enhance their communicative function, and their original function may be lost.

- Opposite messages are often signaled by opposite movements and postures, an idea called the *principle of antithesis*.

Consider how Darwin's theory accounts for the evolution of *threat displays*. Originally, facing one's enemies, rising up, and exposing one's weapons were the components of the early stages of combat. But once enemies began to recognize these behaviors as signals of impending aggression, a survival advantage accrued to attackers that could communicate their aggression most effectively and intimidate their victims without actually fighting. As a result, elaborate threat displays evolved, and actual combat declined.

To be most effective, signals of aggression and submission must be clearly distinguishable; thus, they tended to evolve in opposite directions. For example, gulls signal aggression by pointing their beaks at one another and submission by pointing their beaks away from one another; primates signal aggression by staring and submission by

Figure 17.1 A reconstruction of the brain injury of Phineas Gage. The damage focused on the medial prefrontal lobes.

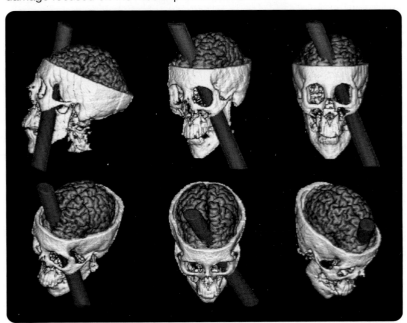

Patrick Landmann/Science Source

averting their gaze. Figure 17.2 reproduces the woodcuts Darwin used in his 1872 book to illustrate this principle of antithesis in dogs.

JAMES-LANGE AND CANNON-BARD THEORIES. The first physiological theory of emotion was proposed independently by James and Lange in 1884. According to the **James-Lange theory**, emotion-inducing sensory stimuli are received and interpreted by the cortex, which triggers changes in the visceral organs via the autonomic nervous system and in the skeletal muscles via the somatic nervous system. Then, the autonomic and somatic responses trigger the experience of emotion in the brain. In effect, what the James-Lange theory did was to reverse the

usual commonsense way of thinking about the causal relation between the experience of emotion and its expression (see Figure 17.3). James and Lange argued that the autonomic activity and behavior that are triggered by the emotional event (e.g., rapid heartbeat and running away) produce the feeling of emotion, not vice versa (see Figure 17.3).

Around 1915, Cannon proposed an alternative to the James-Lange theory of emotion, and it was subsequently extended and promoted by Bard. According to the **Cannon-Bard theory**, emotional stimuli have two independent excitatory effects: They excite both the feeling of emotion in the brain and the expression of emotion in the autonomic and somatic nervous systems. That is, the Cannon-Bard theory, in contrast to the James-Lange theory, views emotional experience and emotional expression as parallel processes that have no direct causal relation.

The James-Lange and Cannon-Bard theories make different predictions about the role of feedback from autonomic and somatic nervous system activity in emotional experience. According to the James-Lange theory, emotional experience depends entirely on feedback from autonomic and somatic nervous system activity; according to the Cannon-Bard theory, emotional experience is totally independent of such feedback. Both extreme positions have proved to be incorrect. On the one hand, it seems that the autonomic and somatic feedback is not necessary for the experience of emotion: Human patients whose autonomic and somatic feedback has been largely eliminated by a broken neck are capable of a full range of emotional experiences, though there does seem to be some dampening of fear and anger (see Pistoia et al., 2015). On the other hand, there have been numerous reports—some of which you will soon encounter—that autonomic and somatic responses to emotional stimuli can influence emotional experience.

Failure to find unqualified support for either the James-Lange or the Cannon-Bard theory led to the modern biopsychological view. According to this view, each of the three principal factors in an emotional response—the perception of the emotion-inducing stimulus, the autonomic and somatic responses to the stimulus, and the experience of the emotion—can influence the other two (e.g., Scherer & Moors, 2019; see Figure 17.3).

SHAM RAGE. In the late 1920s, Bard (1929) discovered that **decorticate** cats—cats whose cortex has been removed—respond aggressively to the slightest provocation: After a light touch, they arch their backs, erect their hair, hiss, and expose their teeth.

The aggressive responses of decorticate animals are abnormal in two respects: They are inappropriately severe,

Figure 17.2 Two woodcuts from Darwin's 1872 book, *The Expression of Emotions in Man and Animals,* that he used to illustrate the principle of antithesis. The aggressive posture of dogs features ears forward, back up, hair up, and tail up; the submissive posture features ears back, back down, hair down, and tail down.

Aggression

Submission

Figure 17.3 Four ways of thinking about the relations among the perception of emotion-inducing stimuli, the autonomic and somatic responses to the stimuli, and the emotional experience.

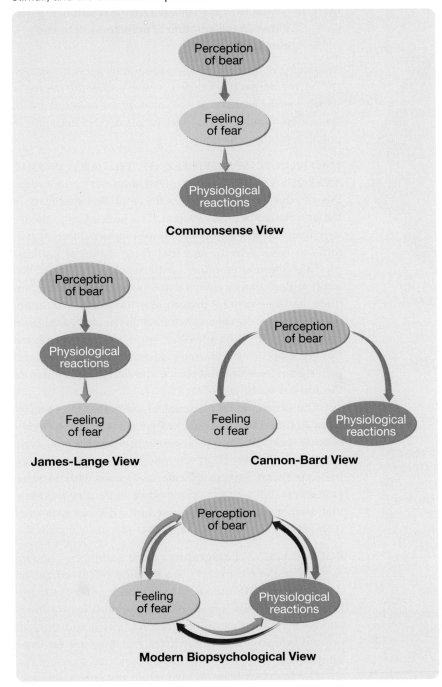

is critical for the expression of aggressive responses and that the function of the cortex is to inhibit and direct these responses.

LIMBIC SYSTEM AND EMOTION. In 1937, Papez (pronounced "Payps") proposed that emotional expression is controlled by several interconnected nuclei and tracts that ring the thalamus. Figure 17.4 illustrates some of the key structures in this circuit: the amygdala, mammillary body, hippocampus, fornix, cingulate cortex, septum, olfactory bulb, and hypothalamus. Papez proposed that emotional states are expressed through the action of the other structures of the circuit on the hypothalamus and that they are experienced through their action on the cortex. Papez's theory of emotion was revised and expanded by Paul MacLean in 1952 and became the influential *limbic system theory of emotion*. Indeed, many of the structures in Papez's circuit are part of what is now known as the **limbic system** (see Figure 3.27).

KLÜVER-BUCY SYNDROME. In 1939, Klüver and Bucy observed a striking *syndrome* (pattern of behavior) in monkeys whose anterior temporal lobes had been removed. This syndrome, which is commonly referred to as the **Klüver-Bucy syndrome**, includes the following behaviors: the consumption of almost anything that is edible, increased sexual activity often directed at inappropriate objects, a tendency to repeatedly investigate familiar objects, a tendency to investigate objects with the mouth, and a lack of fear. Monkeys that could not be handled before surgery were transformed by bilateral anterior temporal lobectomy into tame subjects that showed no fear whatsoever—even in response to snakes, which terrify normal monkeys. In primates, most of the symptoms of the Klüver-Bucy syndrome have been attributed to damage to the **amygdala** (see LeDoux, Michel, & Lau, 2020; Schröder, Moser, & Huggenberger, 2020), a structure that has played a major role in research on emotion, as you will learn later in this chapter.

The Klüver-Bucy syndrome has been observed in several species. Following is a description of the syndrome in a human patient with a brain infection.

and they are not directed at particular targets. Bard referred to the exaggerated, poorly directed aggressive responses of decorticate animals as **sham rage**.

Sham rage can be elicited in cats whose cerebral hemispheres have been removed down to, but not including, the hypothalamus; but it cannot be elicited if the hypothalamus is also removed. On the basis of this observation, Bard concluded that the hypothalamus

Figure 17.4 The location of the structures that Papez proposed controlled emotional expression.

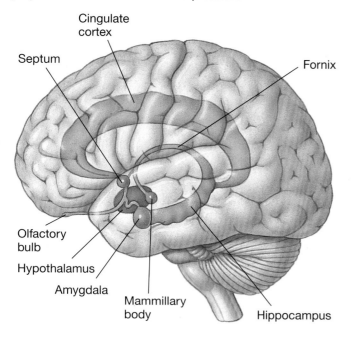

A Human Case of Klüver-Bucy Syndrome

At first he was listless, but eventually he became very placid with flat affect. He reacted little to people or to other aspects of his environment. He spent much time staring at the television, even when it was not turned on. On occasion he would become extremely silly, smiling inappropriately and mimicking the actions of others, and once he began copying the movements of another person, he would persist for extended periods of time. In addition, he tended to engage in oral exploration, sucking, licking, or chewing all small objects that he could reach.

The six early landmarks in the study of brain mechanisms of emotion just reviewed are listed in Table 17.1.

Table 17.1 Biopsychological Investigation of Emotion: Six Early Landmarks

Event	Date
Case of Phineas Gage	1848
Darwin's theory of the evolution of emotion	1872
James-Lange and Cannon-Bard theories	about 1900
Discovery of sham rage	1929
Discovery of Klüver-Bucy syndrome	1939
Limbic system theory of emotion	1952

Emotions and the Autonomic Nervous System

LO 17.2 Summarize the research on the relationship between the autonomic nervous system and emotions.

Research on the role of the autonomic nervous system (ANS) in emotion has focused on two issues: the degree to which specific patterns of ANS activity are associated with specific emotions and the effectiveness of ANS measures in polygraphy (lie detection).

EMOTIONAL SPECIFICITY OF THE AUTONOMIC NERVOUS SYSTEM. The James-Lange and Cannon-Bard theories differ in their views of the emotional specificity of the autonomic nervous system. The James-Lange theory says that different emotional stimuli induce different patterns of ANS activity and that these different patterns produce different emotional experiences. In contrast, the Cannon-Bard theory claims that all emotional stimuli produce the same general pattern of sympathetic activation, which prepares the organism for action (i.e., increased heart rate, increased blood pressure, pupil dilation, increased flow of blood to the muscles, increased respiration, and increased release of epinephrine and norepinephrine from the adrenal medulla).

The experimental evidence suggests that the specificity of ANS reactions lies somewhere between the extremes of total specificity and total generality (see Kreibig, 2010; Quigley & Barrett, 2014). On one hand, ample evidence indicates that not all emotions are associated with the same pattern of ANS activity; on the other, there is no evidence that each emotion is characterized by a distinct pattern of ANS activity (see Siegel et al., 2018).

POLYGRAPHY. Polygraphy (more commonly known as the "lie detector test") is a method of interrogation that employs ANS indexes of emotion to infer the truthfulness of a person's responses. Polygraph tests administered by skilled examiners can be useful additions to normal interrogation procedures, but they are far from infallible.

The main problem in evaluating the effectiveness of polygraphy is that it is rarely possible in real-life situations to know for certain whether a suspect is guilty or innocent. Consequently, many studies of polygraphy have employed the *mock-crime procedure*: Volunteers participate in a mock crime and are then subjected to a polygraph test by an examiner who is unaware of their "guilt" or "innocence." The usual interrogation method is the **control-question technique**, in which the physiological response to the target question (e.g., "Did you steal that purse?") is compared with the physiological responses to control questions whose answers are known (e.g., "Have you ever been in jail before?"). The assumption is that lying will be associated with greater sympathetic activation. A review of the use of the control-question technique in real-life crime settings led to an estimated success rate of about 55 percent—just slightly better than chance (i.e., 50%) (see Iacono & Ben-Shakhar, 2019).

Despite being commonly referred to as *lie detection*, polygraphy detects ANS activity, not lies. Consequently, it is less likely to successfully identify lies in real life than in experiments. In real-life situations, questions such as "Did you steal that purse?" are likely to elicit an emotional reaction from all suspects, regardless of their guilt or innocence, making it difficult to detect deception (see Ambach & Gamer, 2018). The **guilty-knowledge technique**, also known as the *concealed information test*, circumvents this problem. In order to use this technique, the polygrapher must have a piece of information concerning the crime that would be known only to the guilty person. Rather than attempting to catch the suspect in a lie, the polygrapher simply assesses the suspect's reaction to a list of actual and contrived details of the crime. Innocent suspects, because they have no knowledge of the crime, react to all such details in the same way; the guilty react differentially (see Ambach & Gamer, 2018).

In the classic study of the guilty-knowledge technique (Lykken, 1959), volunteers waited until the occupant of an office went to the washroom. Then, they entered her office, stole her purse from her desk, removed the money, and left the purse in a locker. The critical part of the interrogation went something like this: "Where do you think we found the purse? In the washroom? . . . In a locker? . . . Hanging on a coat rack?" Even though electrodermal activity was the only measure of ANS activity used in this study, 88 percent of the mock criminals were correctly identified; more importantly, none of the innocent control volunteers was judged guilty—see Ben-Shakhar (2012), Ambach & Gamer (2018).

Emotions and Facial Expression

LO 17.3 Describe some research on the facial expression of emotions.

Ekman and his colleagues have been preeminent in the study of facial expression (see Ekman, 2016). They began in the 1960s by analyzing hundreds of films and photographs of people experiencing various real emotions. From these, they compiled an atlas of the facial expressions that are normally associated with different emotions (Ekman & Friesen, 1975). For example, to produce the facial expression for surprise, models were instructed to pull their brows upward so as to wrinkle their forehead,

to open their eyes wide so as to reveal white above the iris, to slacken the muscles around their mouth, and to drop their jaw. Try it.

UNIVERSALITY OF FACIAL EXPRESSION. Several early studies found that people of different cultures make similar facial expressions in similar situations and that they can correctly identify the emotional significance of facial expressions displayed by people from cultures other than their own. The most convincing of these studies was a study of the members of an isolated New Guinea tribe who had had little or no contact with the outside world (see Ekman & Friesen, 1971).

PRIMARY FACIAL EXPRESSIONS. Ekman and Friesen concluded that the facial expressions of the following six emotions are primary: surprise, anger, sadness, disgust, fear, and happiness. They further concluded that all other facial expressions of genuine emotion are composed of mixtures of these six primaries. Figure 17.5 illustrates these six primary facial expressions.

FACIAL FEEDBACK HYPOTHESIS. Is there any truth to the old idea that putting on a happy face can make you feel better? Research suggests that there is. The hypothesis that our facial expressions influence our emotional experience is called the **facial feedback hypothesis**. In a test of the facial feedback hypothesis, Rutledge and Hupka (1985) instructed volunteers to assume one of two patterns of facial contractions while they viewed a series of slides; the patterns corresponded to happy or angry faces, although the volunteers were unaware of that. They reported that the slides made them feel more happy and less angry when they were making happy faces and less happy and more angry when they were making angry faces (see Figure 17.6). A recent meta-analysis of the facial feedback hypothesis confirmed the reliability of these and similar findings; however, the effects were smaller than originally believed (see Coles, Larsen, & Lench, 2019).

Check It Out
Experiencing Facial Feedback

Why don't you try the facial feedback hypothesis? Pull your eyebrows down and together; raise your upper eyelids and tighten your lower eyelids, and narrow your lips and press them together. Now, hold this expression for a few seconds. If it makes you feel slightly angry and uncomfortable, you have just experienced the effect of facial feedback.

Figure 17.5 Ekman's six primary facial expressions of emotion.

ANGER · DISGUST · FEAR

HAPPINESS · SADNESS · SURPRISE

Katerina Solovyeva/123RF

Figure 17.6 The effects of facial expression on the experience of emotion. Participants reported feeling more happy and less angry when they viewed slides while making a happy face and less happy and more angry when they viewed slides while making an angry face.

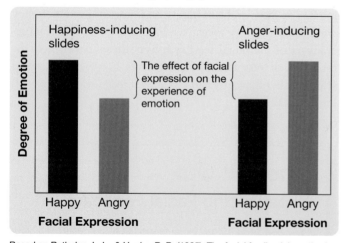

Based on Rutledge, L. L., & Hupka, R. B. (1985). The facial feedback hypothesis: Methodological concerns and new supporting evidence. *Motivation and Emotion*, 9, 219–240.

VOLUNTARY CONTROL OF FACIAL EXPRESSION. Because we can exert voluntary control over our facial muscles, it is possible to inhibit true facial expressions and to substitute false ones. There are many reasons for choosing to put on a false facial expression. Some of them are positive (e.g., putting on a false smile to reassure a worried friend), and some are negative (e.g., putting on a false smile to disguise a lie). In either case, it is difficult to fool an expert.

There are two ways of distinguishing true expressions from false ones (Ekman, 1985). First, *microexpressions* (brief facial expressions) of the real emotion often break through the false one (see Wang et al., 2015). Such microexpressions last only about 0.05 second, but with practice they can be detected without the aid of slow-motion photography. Second, there are often subtle differences between genuine facial expressions and false ones that can be detected by skilled observers.

The most widely studied difference between a genuine and a false facial expression was first described by the French anatomist Duchenne in 1862. Duchenne said that the smile of enjoyment could be distinguished from deliberately produced smiles by consideration of the two facial muscles that are contracted during genuine smiles: *orbicularis oculi*, which encircles the eye and pulls the skin from the cheeks and forehead toward the eyeball, and *zygomaticus major*, which pulls the lip corners up (see Figure 17.7). According to Duchenne, the zygomaticus major can be controlled voluntarily, whereas the orbicularis oculi is normally

Figure 17.7 A fake smile. The orbicularis oculi and the zygomaticus major are two muscles that contract during genuine (Duchenne) smiles. Because the lateral portion of the orbicularis oculi is difficult for most people to contract voluntarily, fake smiles usually lack this component. This young woman is faking a smile for the camera. Look at her eyes.

Steven J. Barnes

Figure 17.8 An expression of pride.

Reproduced with permission of Jessica Tracy, Department of Psychology, University of British Columbia.

contracted only by genuine pleasure. Thus, inertia of the orbicularis oculi in smiling unmasks a false friend—a fact you would do well to remember. Ekman named the genuine smile the **Duchenne smile**.

FACIAL EXPRESSIONS: CURRENT PERSPECTIVES. Ekman's work on facial expressions began before video recording became commonplace. Now, video recordings provide almost unlimited access to natural facial expressions made in response to real-life situations. This technology has contributed to four important qualifications to Ekman's original theory. First, it is now clear that Ekman's six primary facial expressions of emotion rarely occur in pure form—they are ideals with many subtle variations. Second, the existence of other primary emotions has been recognized (see Whalen et al., 2013). Third, body cues, not just facial expressions, are known to play a major role in expressions of emotion (see Sznycer, 2019). For example, pride is expressed through a small smile, with the head tilted back slightly and the hands on the hips, raised above the head, or clenched in fists with the arms crossed on the chest—see Figure 17.8 (see Witkower & Tracy, 2019). Fourth, there is evidence that Ekman's six primary facial expressions may not be as universal as originally believed. For example, there seem to be distinct differences, in terms of both the expression and recognition of facial expressions, between Western Caucasian and East Asian individuals (see Calvo & Nummenmaa, 2015; Jack et al., 2012; Wood et al., 2016). Moreover, recent studies of isolated tribes by Crivelli et al. (2016, 2017) indicate that facial expressions of emotion are not as universal as once thought.

Fear, Defense, and Aggression

Most biopsychological research on emotion has focused on fear and defensive behaviors. **Fear** is the emotional reaction to threat; it is the motivating force for defensive behaviors. **Defensive behaviors** are behaviors whose primary function is to protect the organism from threat or harm. In contrast, **aggressive behaviors** are behaviors whose primary function is to threaten or harm.

Although one purpose of this module is to discuss fear, defense, and aggression, it has another important purpose: to explain a common problem faced by biopsychologists and the way in which those who conduct research in this particular area have managed to circumvent it. Barrett (2006) pointed out that progress in the study of the neural basis of emotion has been limited because neuroscientists have often been guided by unsubstantiated cultural assumptions about emotion: Because we have words such as *fear, happiness,* and *anger* in our language, scientists have often assumed that these emotions exist as entities in the brain, and they have searched for them—often with little success. The following lines of research on fear, defense, and aggression illustrate how biopsychologists can overcome the problem of vague, subjective, everyday concepts by basing their search for neural mechanisms on the thorough descriptions of relevant behaviors, the environments in which they occur, and the putative adaptive functions of such behaviors (see Kasai et al., 2015; LeDoux & Hofmann, 2018).

Journal Prompt 17.1

Because we have a word for it, many people believe that "intelligence" is a real entity. Yet, it is a complex construct that was developed by psychologists. Treating a psychological construct (e.g., intelligence) as if it actually exists is a logical error, known as an *error of reification*. Have you ever encountered such errors in the popular media? Give an example.

Types of Aggressive and Defensive Behaviors

LO 17.4 Describe the work that led to the distinction between aggressive and defensive behaviors in mammals.

Considerable progress in the understanding of aggressive and defensive behaviors has come from the research of Blanchard and Blanchard (see Blanchard, Summers, & Blanchard, 2013; Koolhaas et al., 2013) on the *colony intruder model of aggression and defense* in rats. Blanchard and Blanchard have derived rich descriptions of rat intraspecific aggressive and defensive behaviors by studying the interactions between the **alpha male**—the dominant male—of an established mixed sex colony and a small male intruder: Upon encountering the intruder, the alpha male typically chases it away, repeatedly biting its back during the pursuit. The intruder eventually stops running and turns to face the alpha male. The intruder then rears up on its hind legs, still facing its attacker and using its forelimbs to ward off the attack. In response, the alpha male changes to a lateral orientation, with the side of its body perpendicular to the front of the defending intruder. Then, the alpha moves sideways toward the intruder, crowding and trying to push it off balance. If the defending intruder stands firm against this "lateral attack," the alpha often reacts by making a quick lunge around the defender's body in an attempt to bite its back. In response to such attacks, the defender pivots on its hind feet, in the same direction as the attacker is moving, continuing its frontal orientation to the attacker in an attempt to prevent the back bite.

Another excellent illustration of how careful observation of behavior has led to improved understanding of aggressive and defensive behaviors is provided by Pellis and colleagues' (1988) study of cats. They began by videotaping interactions between cats and mice. They found that different cats reacted to mice in different ways: Some were efficient mouse killers, some reacted defensively, and some seemed to play with the mice. Careful analysis of the "play"

sequences led to two important conclusions. The first conclusion was that, in contrast to the common belief, cats do not play with their prey; the cats that appeared to be playing with the mice were simply vacillating between attack and defense. The second conclusion was that one can best understand each cat's interactions with mice by locating the interactions on a linear scale, with total aggressiveness at one end, total defensiveness at the other, and various proportions of the two in between.

Pellis and colleagues tested their conclusions by reducing the defensiveness of the cats with an antianxiety drug. As predicted, the drug moved each cat along the scale toward more efficient killing. Cats that avoided mice before the injection "played with" them after the injection, those that "played with" them before the injection killed them after the injection, and those that killed them before the injection killed them more quickly after the injection.

Based on the numerous detailed descriptions of aggressive and defensive behaviors provided by the Blanchards, Pellis and colleagues, and other biopsychologists who have followed their example, most researchers now distinguish among different categories of such behaviors. These categories of aggressive and defensive behaviors are based on three criteria: (1) their *topography* (form), (2) the situations that elicit them, and (3) their apparent function. Several of these categories for rats are described in Table 17.2 (see Blanchard et al., 2011; Jager et al., 2017; Kim & Jung, 2018).

The analysis of aggressive and defensive behaviors has led to the development of the **target-site concept**—the idea that the aggressive and defensive behaviors of an animal are often designed to attack specific sites on the body of another animal while protecting specific sites on its own. For example, the behavior of a socially aggressive rat (e.g., lateral attack) appears to be designed to deliver bites to the defending rat's back and to protect its own face, the likely target of a defensive attack. Conversely, most of the maneuvers of the defending rat (e.g., boxing and pivoting) appear to be designed to protect the target site on its back.

The discovery that aggressive and defensive behaviors occur in a variety of stereotypical species-common forms was the necessary first step in the identification of their neural bases. Because the different categories of aggressive and defensive behaviors are mediated by different neural circuits, little progress was made in identifying these circuits before the categories were first delineated. For example, the lateral septum was once believed to inhibit all aggression, because lateral septal lesions rendered laboratory rats notoriously difficult to handle—the behavior of the lesioned rats was commonly referred to as *septal aggression* or *septal rage*. However, we now know that lateral septal lesions do not increase aggression: Rats with lateral septal lesions do not initiate more attacks, but they are hyperdefensive when threatened.

Table 17.2 Categories of Aggressive and Defensive Behaviors in Rats

Aggressive Behaviors	Predatory Aggression	The stalking and killing of members of other species for the purpose of eating them. Rats kill prey, such as mice and frogs, by delivering bites to the back of the neck.
	Social Aggression	Unprovoked aggressive behavior that is directed at a *conspecific* (member of the same species) for the purpose of establishing, altering, or maintaining a social hierarchy. In mammals, social aggression occurs primarily among males. In rats, it is characterized by piloerection, lateral attack, and bites directed at the defender's back.
Defensive Behaviors	Intraspecific Defense	Defense against social aggression. In rats, it is characterized by freezing and flight and by various behaviors, such as boxing, that are specifically designed to protect the back from bites.
	Defensive Attacks	Attacks that are launched by animals when they are cornered by threatening members of their own or other species. In rats, they include lunging, shrieking, and biting attacks that are usually directed at the face of the attacker.
	Freezing and Flight	Responses that many animals use to avoid attack. For example, if a human approaches a wild rat, it will often freeze until the human penetrates its safety zone, whereupon it will explode into flight.
	Maternal Defensive Behaviors	The behaviors by which mothers protect their young. Despite their defensive function, they are similar to male social aggression in appearance.
	Risk Assessment	Behaviors that are performed by animals in order to obtain specific information that helps them defend themselves more effectively. For example, rats that have been chased by a cat into their burrow do not emerge until they have spent considerable time at the entrance scanning the surrounding environment.
	Defensive Burying	Rats and other rodents spray sand and dirt ahead with their forepaws to bury dangerous objects in their environment, to drive off predators, and to construct barriers in burrows.

Aggression and Testosterone

LO 17.5 Describe the relation between testosterone levels and aggression in males.

The fact that social aggression in many species occurs more commonly among males than among females is usually explained with reference to the organizational and activational effects of testosterone (see Chapter 13). The brief period of testosterone release that occurs around birth in genetic males is thought to organize their nervous systems along masculine lines and hence to create the potential for male patterns of social aggression to be activated by the high testosterone levels that are present after puberty. These organizational and activational effects have been demonstrated in some mammalian species. For example, neonatal castration of male mice eliminates the ability of testosterone injections to induce social aggression in adulthood, and adult castration eliminates social aggression in male mice that do not receive testosterone replacement injections. Unfortunately, research on testosterone and aggression in other species has not been so straightforward (see Carré & Olmstead, 2015).

The extensive comparative research literature on testosterone and aggression has been reviewed several times (Demas et al., 2005; Munley, Rendon, & Demas, 2018; Soma, 2006). Here are the major conclusions:

- Testosterone increases social aggression in the males of many species; aggression is largely abolished by castration in these same species (see Hashikawa et al., 2018).

- In some species, castration has no effect on social aggression; in still others, castration reduces social aggression during the breeding season but not at other times.

- The relation between aggression and testosterone levels is difficult to interpret because engaging in aggressive activity can itself increase testosterone levels—for example, just playing with a gun increased the testosterone levels of male college students (Klinesmith, Kasser, & McAndrew, 2006).

- The blood level of testosterone, which is the only measure used in many studies, is not the best measure. What matters more are the testosterone levels in the relevant areas of the brain. Although studies focusing on brain levels of testosterone are rare, it has been shown that testosterone can be synthesized in particular brain sites and not in others.

It is unlikely that humans are an exception to the usual involvement of testosterone in mammalian social aggression. However, the evidence is far from clear. In human males, aggressive behavior does not increase at puberty as testosterone levels in the blood increase; aggressive behavior is not eliminated by castration; and it is not increased by testosterone injections that elevate blood levels of testosterone. A few studies have found that violent male criminals (see Fragkaki, Cima, & Granic, 2018) and aggressive male and female athletes tend to have higher testosterone levels than normal (see Batrinos, 2012; Denson et al., 2018); however, this correlation may indicate that aggressive behaviors increase testosterone, rather than vice versa.

The lack of strong evidence of the involvement of testosterone in human aggression could mean that hormonal and neural regulation of aggression in humans differs from that in many other mammalian species. Or, it could mean that the research on human aggression and testosterone is flawed. For example, human studies are typically based on blood levels of testosterone (often inferred from saliva levels because collecting saliva is safer and easier than collecting blood) rather than on brain levels. However, the blood levels of a hormone aren't necessarily indicative of how much hormone is reaching the brain. Also, the researchers who study human aggression have often failed to appreciate the difference between social aggression, which is related to testosterone in many species, and defensive attack, which is not (see Montoya et al., 2012; Sobolewski, Brown, & Mitani, 2013). Most seemingly aggressive outbursts in humans are overreactions to real or perceived threat, and thus they are more appropriately viewed as defensive attack, not social aggression.

Neural Mechanisms of Fear Conditioning

Much of what we know about the neural mechanisms of fear has come from the study of fear conditioning. **Fear conditioning** is the establishment of fear in response to a previously neutral stimulus (the *conditional stimulus*) by presenting it, usually several times, before the delivery of an aversive stimulus (the *unconditional stimulus*).

In a standard fear conditioning experiment, the subject, often a rat, hears a tone (conditional stimulus) and then receives a mild electric shock to its feet (unconditional stimulus). After several pairings of the tone and the shock, the rat responds to the tone with a variety of defensive behaviors (e.g., freezing and increased susceptibility to startle) and sympathetic nervous system responses (e.g., increased heart rate and blood pressure). LeDoux and his colleagues have mapped the neural mechanism that mediates this form of auditory fear conditioning (see Kim & Jung, 2018; Ledoux, 2014).

Amygdala and Fear Conditioning

LO 17.6 Describe the role of the amygdala in fear conditioning.

LeDoux and his colleagues began their search for the neural mechanisms of *auditory fear conditioning* (fear conditioning that uses a sound as a conditional stimulus) by making lesions in the auditory pathways of rats. They found that bilateral lesions to the *medial geniculate nucleus* (the auditory relay nucleus of the thalamus) blocked fear conditioning to

a tone, but bilateral lesions to the auditory cortex did not. This indicated that for auditory fear conditioning to occur, it is necessary for signals elicited by the tone to reach the medial geniculate nucleus but not the auditory cortex. It also indicated that a pathway from the medial geniculate nucleus to a structure other than the auditory cortex plays a key role in fear conditioning. This pathway proved to be the pathway from the medial geniculate nucleus to the amygdala. Lesions of the amygdala, like lesions of the medial geniculate nucleus, blocked auditory fear conditioning. The amygdala receives input from all sensory systems, and it is believed to be the structure in which the emotional significance of sensory signals is learned and retained.

Several pathways carry signals from the amygdala to brain stem structures that control the various emotional responses (see Dampney, 2015). For example, a pathway to the periaqueductal gray of the midbrain elicits appropriate defensive responses (see Kim et al., 2013), whereas another pathway to the lateral hypothalamus elicits appropriate sympathetic responses.

The fact that auditory cortex lesions do not disrupt fear conditioning to simple tones does not mean that the auditory cortex is not involved in auditory fear conditioning. There are two pathways from the medial geniculate nucleus to the amygdala: the direct one, which you have already learned about, and an indirect one that projects via the auditory cortex. Both routes are capable of mediating fear conditioning to simple sounds; if only one is destroyed, conditioning progresses normally. However, only the cortical route is capable of mediating fear conditioning to complex sounds (see Chang & Grace, 2015).

Figure 17.9 illustrates the circuit of the brain that is thought to mediate the effects of fear conditioning to an auditory conditional stimulus (see Calhoun & Tye, 2015; Herry & Johansen, 2014). The sound signal from an auditory conditional stimulus travels from the medial geniculate nucleus of the thalamus to reach the amygdala directly, or indirectly via the auditory cortex. The amygdala assesses the emotional significance of the sound on the basis of previous encounters with it, and then the amygdala activates the appropriate response circuits—for example, behavioral circuits in the periaqueductal gray and sympathetic circuits in the hypothalamus.

Contextual Fear Conditioning and the Hippocampus

LO 17.7 Describe the role of the hippocampus in contextual fear conditioning.

Environments, or *contexts*, in which fear-inducing stimuli are encountered can come to elicit fear. For example, if you repeatedly encountered a bear on a particular trail in the forest, the trail itself would begin to elicit fear. The process

Figure 17.9 The structures thought to mediate the sympathetic and behavioral responses conditioned to an auditory conditional stimulus.

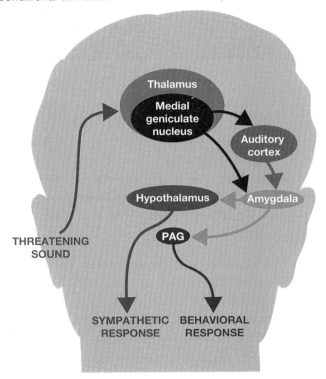

by which benign contexts come to elicit fear through their association with fear-inducing stimuli is called **contextual fear conditioning**.

Journal Prompt 17.2

Can you think of an instance where you have been subjected to contextual fear conditioning? Describe that instance.

Contextual fear conditioning has been produced in the laboratory in two ways. First, it has been produced by the conventional fear conditioning procedure, which we just discussed. For example, if a rat repeatedly receives an electric shock following a conditional stimulus, such as a tone, the rat will become fearful of the conditional context (the test chamber) as well as the tone. Second, contextual fear conditioning has been produced by delivering aversive stimuli in a particular context in the absence of any other

conditional stimulus. For example, if a rat receives shocks in a distinctive test chamber, the rat will become fearful of that chamber.

In view of the fact that the **hippocampus** plays a key role in memory for spatial location, it is reasonable to expect that it would be involved in contextual fear conditioning. This seems to be the case (see Chaaya, Battle, & Johnson, 2018; Maren, Phan, & Liberzon, 2013). Bilateral hippocampal lesions block the subsequent development of a fear response to the context without blocking the development of a fear response to the explicit conditional stimulus (e.g., a tone; see Moscarello & Maren, 2018).

Amygdala Complex and Fear Conditioning

LO 17.8 Describe the role of two specific amygdalar nuclei in fear conditioning.

The preceding discussion has probably left you with the impression that the amygdala is a single brain structure; it isn't. It is actually a cluster of many nuclei, often referred to as the *amygdala complex*. The amygdala is composed of a dozen or so major nuclei, which are themselves divided into subnuclei. Each of these subnuclei is structurally distinct, has different connections, and is thus likely to have different functions (see Duvarci & Pare, 2014; Janak & Tye, 2015).

The study of fear conditioning provides a compelling demonstration of the inadvisability of assuming that the amygdala is a single structure. Evidence has been accumulating that the **lateral nucleus of the amygdala**—not the entire amygdala—is critically involved in the acquisition, storage, and expression of conditioned fear (see Duvarci & Pare, 2014; Janak & Tye, 2015; Tovote, Fadok, & Lüthi, 2015). Both the prefrontal cortex and the hippocampus project to the lateral nucleus of the amygdala: The **prefrontal cortex** is thought to act on the lateral nucleus of the amygdala to suppress conditioned fear (see Gilmartin, Balderston, & Helmstetter, 2014), and the hippocampus is thought to interact with that part of the amygdala to mediate learning about the context of fear-related events. The amygdala is thought to control defensive behavior via outputs from the **central nucleus of the amygdala** (see Janak & Tye, 2015; Kim & Jung, 2018; Pellman & Kim, 2016; Ressler & Maren, 2019).

Scan Your Brain

This chapter is about to change direction: The remaining two modules focus on the neural mechanisms of human emotion and on the effects of stress on health. This is a good point for you to scan your brain to see whether it has retained the introductory material on emotion and fear. Fill in each of the following blanks with the most appropriate term. The correct answers are provided at the end of the exercise. Before continuing, review material related to your errors and omissions.

1. Bard discovered that the _____ is critical for the expression of aggressive responses.

2. Emotional processing is supported by a circuit in the brain called the _____.

3. In primates, most of the symptoms of Klüver–Bucy syndrome are the result of damage to the _____.

4. _____ theory suggests that an emotional event triggers autonomic activity and behavior, which then produce the feeling of emotion.

5. _____ is a method of interrogation that uses changes to the ANS to detect lies in a person's responses.

6. There are six primary emotions: happiness, anger, fear, surprise, sadness, and _____.

7. According to the _____ _____ hypothesis, facial expressions may alter our emotional state.

8. Two muscle groups, orbicularis oculi and _____, have been identified to depict a Duchenne smile.

9. The emotional reaction to threat is _____.

10. _____ increases aggression in the males of many species, and castration decreases it.

11. In a standard _____ task, a previously neutral stimulus is paired with an aversive stimulus.

12. The _____ plays a key role in memory for spatial locations.

13. Projections from the _____ _____ to the amygdala act to suppress conditioned fear.

Scan Your Brain answers: (1) hypothalamus, (2) limbic system, (3) amygdala, (4) James-Lange, (5) Polygraph, (6) disgust, (7) facial feedback, (8) zygomaticus major, (9) fear, (10) Testosterone, (11) fear-conditioning, (12) hippocampus, (13) prefrontal cortex.

Brain Mechanisms of Human Emotion

This module deals with the brain mechanisms of human emotion. We still do not know how the human brain controls the experience or expression of emotion, or how the brain interprets emotion in others, but progress has been made. Each of the following sections illustrates an area of progress.

Cognitive Neuroscience of Emotion

LO 17.9 Describe the current status of cognitive neuroscience research on emotion.

Cognitive neuroscience is currently the dominant approach being used to study the brain mechanisms of human emotion. There have been many functional brain imaging studies of people experiencing or imagining emotions or watching others experiencing them. These studies have established three points that have advanced our understanding of the brain mechanisms of emotion in fundamental ways (see Neumann et al., 2014; Wood et al., 2016):

• Brain activity associated with each human emotion is diffuse—there is not a center for each emotion (see Feinstein, 2013). Think "mosaic," not "center," for locations of brain mechanisms of emotion.

• There is virtually always activity in motor and sensory cortices when a person experiences an emotion.

• Similar patterns of brain activity tend to be recorded when a person experiences an emotion, imagines that emotion, or sees somebody else experience that emotion (see Figure 17.10).

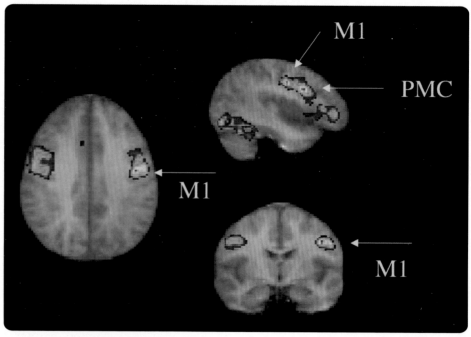

Figure 17.10 Horizontal, sagittal, and coronal functional MRIs show areas of increased activity in the primary motor cortex (M1) and the premotor cortex (PMC) when volunteers watched facial expressions of emotion. The same areas were active when the volunteers made the expressions themselves.

From Carr et al., 2003, Neural Mechanisms of Empathy, 100, pgs. 5497-5502. Figure 1 on page 550. Copyright (2003) National Academy of Sciences, U.S.A.

These three fundamental findings are influencing how researchers are thinking about the neural mechanisms of emotion. For example, the activity observed in sensory and motor cortex during the experience of human emotions is now believed to be an important part of the mechanism by which the emotions are experienced. The re-experiencing of related patterns of motor, autonomic, and sensory neural activity during emotional experiences is generally referred to as the *embodiment of emotions* (see Wang et al., 2016).

Amygdala and Human Emotion

LO 17.10 Describe the role of the amygdala in human emotion.

You have already learned that the amygdalae play an important role in fear conditioning in rats. Numerous functional brain-imaging studies have suggested that the function of the human amygdalae is more general. Although the human amygdalae appear to respond most robustly to fear, they also respond to other emotions (see Hsu et al., 2015; Koelsch & Skouras, 2014; Patin & Pause, 2015). Indeed, the amygdalae appear to play a role in the performance of any task with an emotional component, whether positive or negative (see Fastenrath et al., 2014; Stillman, Van Bavel, & Cunningham, 2015). This has led to the view that the amygdalae play a role in evaluating the emotional significance of situations.

Although the results of brain-imaging studies suggest that the amygdalae play a general role in emotions, the study of some patients with amygdalar damage suggests a specific role in fear. The following case illustrates this point.

The Case of S.P., the Woman Who Couldn't Perceive Fear

At the age of 48, S.P. had her right amygdala and adjacent tissues removed for the treatment of epilepsy. Because her left amygdala had been damaged, she in effect had a bilateral amygdalar lesion.

Journal Prompt 17.3

Before reading any further, based on the animal research on the amygdala that you read about in the previous module, try to predict the sorts of deficits you would expect to see in patient S.P.

Following her surgery, S.P. had an above average I.Q., and her perceptual abilities were generally normal. Of particular relevance was the fact that she had no difficulty in identifying faces or extracting information from them (e.g., information about age or gender). However, S.P. did have a severe postsurgical deficit in recognizing facial expressions of fear and less striking deficits in recognizing facial expressions of disgust, sadness, and happiness.

In contrast, S.P. had no difficulty specifying which emotion would go with particular sentences. Also, she had no difficulty using facial expressions upon request to express various emotions (see Anderson & Phelps, 2000).

The case of S.P. is similar to reported cases of Urbach-Wiethe disease (see Meletti et al., 2014). **Urbach-Wiethe disease** is a genetic disorder that often results in *calcification* (hardening by conversion to calcium carbonate, the main component of bone) of the amygdala and surrounding anterior medial temporal lobe structures in both hemispheres. One Urbach-Wiethe patient with bilateral amygdalar damage was found to have lost the ability to recognize facial expressions of fear (see Adolphs, 2006). Indeed, she could not describe fear-inducing situations or produce fearful expressions, although she had no difficulty on tests involving other emotions.

Medial Prefrontal Lobes and Human Emotion

LO 17.11 Describe the role of the medial prefrontal lobes in human emotion.

Emotion and cognition are often studied independently, but it is now believed that they are better studied as different components of the same system (see Barrett & Satpute, 2013). The medial portions of the prefrontal lobes (including the medial portions of the orbitofrontal cortex and anterior cingulate cortex) are the sites of emotion–cognition interaction that have received the most attention (e.g., Etkin, Büchel, & Gross, 2015; Hiser & Koenigs, 2017; Kragel et al., 2018). Functional brain-imaging studies have found evidence of activity in the medial prefrontal lobes when emotional reactions are being cognitively suppressed or re-evaluated (see Okon-Singer et al., 2015).

Many studies of medial prefrontal lobe activity employ suppression paradigms or reappraisal paradigms. In studies that use **suppression paradigms**, participants are directed to inhibit their emotional reactions to unpleasant films or pictures; in studies that use **reappraisal paradigms**, participants are instructed to reinterpret a picture to change their emotional reaction to it. The medial prefrontal lobes are active when both of these paradigms are used, and they seem to exert their cognitive control of emotion by interacting with the amygdala (see Whalen et al., 2013).

Many theories of the specific functions of the medial prefrontal lobes have been proposed. The medial prefrontal lobes have been hypothesized to monitor the difference between outcome and expectancy (see Diekhof et al., 2012), to encode stimulus value over time (Tsetsos et al., 2014), to predict the likelihood of error (see Hoffmann & Beste, 2015), to mediate the conscious awareness of emotional stimuli (see Mitchell & Greening, 2011), and to mediate social decision making (see Lee & Seo, 2016; Phelps, Lempert, & Sokol-Hessner, 2014). Which hypothesis is correct? Perhaps all are; the medial prefrontal cortex is large and complex, and it likely performs many functions. This point was made by the study of Kawasaki and colleagues (2005).

Kawasaki and colleagues used microelectrodes to record from 267 neurons in the anterior cingulate cortices (part of the medial prefrontal cortex) of four patients prior to surgery. They assessed the activity of the neurons when the patients viewed photographs with emotional content. Of these 267 neurons, 56 responded most strongly and consistently to negative emotional content. This confirms previous research linking the medial prefrontal lobes with negative emotional reactions, but it also shows that not all neurons in the area perform the same function—neurons directly involved in emotional processing appear to be sparse and widely distributed in the human medial prefrontal lobes.

Lateralization of Emotion

LO 17.12 Describe the research on the lateralization of emotion.

There is evidence suggesting that emotional functions are lateralized, that is, the left and right cerebral hemispheres are specialized to perform different emotional functions—as you learned in Chapter 16. This evidence has led to several theories of the cerebral lateralization of emotion; the following are the two most prominent (see Gainotti, 2019):

- The *right hemisphere model* of the cerebral lateralization of emotion holds that the right hemisphere is specialized for all aspects of emotional processing: perception, expression, and experience of emotion.

- The *valence model* proposes that the right hemisphere is specialized for processing negative emotion and the left hemisphere is specialized for processing positive emotion.

Which of the two theories does the evidence support? Most studies of the cerebral lateralization of emotion have employed functional brain-imaging methods, and the results have been complex and variable. Wager and colleagues (2003) performed a meta analysis of the data from 65 such studies.

The main conclusion of Wager and colleagues was that the current theories of lateralization of emotion are too general from a neuroanatomical perspective. Overall comparisons between left and right hemispheres revealed no interhemispheric differences in either the amount of emotional processing or the valence of the emotions being processed. However, when the comparisons were conducted on a structure-by-structure basis, they revealed substantial evidence of lateralization of emotional processing. Some kinds of emotional processing were lateralized to the left hemisphere in certain structures and to the right in others. Functional brain-imaging studies of emotion have commonly observed lateralization in the amygdalae—more activity is often observed in the left amygdala. Clearly, neither the right hemisphere model nor the valence model of the lateralization of emotion is supported by the evidence. The models are too general.

Another approach to studying the lateralization of emotions is based on observing the asymmetry of facial expressions. In most people, each facial expression begins on the left side of the face and, when fully expressed, is more pronounced there—which implies right hemisphere dominance for facial expressions (see Figure 17.11). Remarkably, the same asymmetry of facial expressions has been documented in monkeys (see Lindell, 2013).

Figure 17.11 The asymmetry of facial expressions. Notice that the expressions are more obvious on the left side of two well-known faces: those of Mona Lisa and Albert Einstein.

Dennis Hallinan/Alamy Stock Photo

Everett Collection Historical/Alamy Stock Photo

Neural Mechanisms of Human Emotion: Current Perspectives

LO 17.13   Describe the current perspective on the neural mechanisms of human emotion that has emerged from brain-imaging studies.

Although there is a general consensus that the amygdalae and medial prefrontal cortex play major roles in the perception and experience of human emotion, the results of brain-imaging studies have put this consensus into perspective (see Pessoa, 2018; Todd et al., 2020). Here are four important points:

- Emotional situations produce widespread increases in cerebral activity, not just in the amygdalae and prefrontal cortex.

- All brain areas activated by emotional stimuli are also activated during other psychological processes.

- No brain structure has been invariably linked to a particular emotion.

- The same emotional stimuli often activate different areas in different people.

Stress and Health

When the body is exposed to harm or threat, the result is a cluster of physiological changes generally referred to as the *stress response*—or just **stress**. All **stressors** (experiences that induce the stress response) produce the same core pattern of physiological changes, whether psychological (e.g., dismay at the loss of one's job) or physical (e.g., long-term exposure to cold). However, it is *chronic psychological stress* that has been most frequently implicated in ill health, which is the focus of this module.

The Stress Response

LO 17.14   Describe the components of the stress response.

Hans Selye (pronounced "SELL-yay") first described the stress response in the 1950s, and he emphasized its dual nature. In the short term, it produces adaptive changes that help the animal respond to the stressor (e.g., mobilization of energy resources); in the long term, however, it produces changes that are maladaptive (e.g., enlarged adrenal glands).

Selye attributed the stress response to the activation of the *anterior pituitary adrenal cortex system*. He concluded that stressors acting on neural circuits stimulate the release of **adrenocorticotropic hormone (ACTH)** from the anterior pituitary, that ACTH in turn triggers the release of **glucocorticoids** from the **adrenal cortex**, and that the glucocorticoids produce many of the components of the stress response (see

Russell & Lightman, 2019; Shirazi et al., 2015; Spiga et al., 2014). The level of circulating glucocorticoids is the most commonly employed physiological measure of stress.

Selye largely ignored the contributions of the sympathetic nervous system to the stress response. However, stressors activate the sympathetic nervous system, thereby increasing the amounts of epinephrine and norepinephrine released from the **adrenal medulla**. Most modern theories of stress acknowledge the roles of both the anterior pituitary adrenal cortex system and the sympathetic nervous system adrenal medulla system (see Carter & Goldstein, 2015). Figure 17.12 illustrates the two-system view.

The major feature of Selye's landmark theory is its assertion that both physical and psychological stressors induce the same general stress response. This assertion has proven to be partly correct. There is good evidence that all kinds of common psychological stressors—such as losing a job, taking a final exam, or ending a relationship—act like physical stressors. However, Selye's contention that there is only one stress response has proven to be a simplification. Stress responses are complex and varied, with the exact response depending on the stressor, its timing, the nature of the stressed person, and how the stressed person reacts to

Figure 17.12 The two-system view of the stress response.

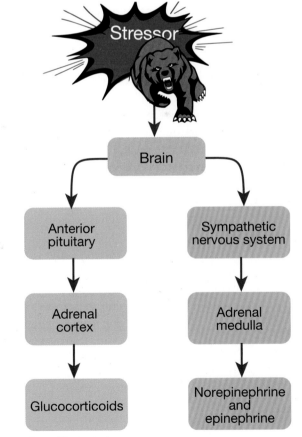

Evgeny Turaev/Shutterstock

the stressor (see Hostinar, Sullivan, & Gunnar, 2014; Oken, Chamine, & Wakeland, 2015). For example, in a study of women awaiting surgery for possible breast cancer, the levels of stress were lower in those who had convinced themselves that they could not possibly have cancer, that their prayers were certain to be answered, or that it was counterproductive to worry (see Katz et al., 1970).

In the 1990s, there was an important advance in the understanding of the stress response (see Grippo & Scotti, 2013). It was discovered that stressors produce physiological reactions that participate in the body's inflammatory responses. Most notably, it was found that stressors produce an increase in blood levels of **cytokines**, a group of peptide hormones that are released by many cells and participate in a variety of physiological and immunological responses, causing inflammation and fever (see Padro & Sanders, 2014).

Animal Models of Stress

LO 17.15 Describe research on animal models of stress, including that on subordination stress.

Most of the early research on stress was conducted with nonhumans, and even today most lines of stress research begin with controlled experiments involving nonhumans before moving to correlational studies of humans. Early stress research on nonhumans tended to involve extreme forms of stress such as repeated exposure to electric shock or long periods of physical restraint. There are two problems with this kind of research. First is the problem of ethics. Any research that involves creating stressful situations is going to be controversial, but many of the early stress studies were "over the top" and would not be permitted today in many countries. The second problem is that studies that use extreme, unnatural forms of stress are often of questionable scientific value. Responses to extreme stress tend to mask normal variations in the stress response, and it is difficult to relate the results of such studies to common human stressors.

Better animal models of stress involve the study of social threat from *conspecifics* (members of the same species). Virtually all mammals—particularly males—experience threats from conspecifics at certain points in their lives. When conspecific threat becomes an enduring feature of daily life, the result is **subordination stress** (e.g., Rodriguez-Arias et al., 2016).

Subordination stress is most readily studied in social species that form *dominance hierarchies* (pecking orders; see Chapter 2). What do you think happens to subordinate male rodents who are continually attacked by more dominant

males? They are more likely to attack juveniles, and they have smaller testes, shorter life spans, lower blood levels of testosterone, and higher blood levels of glucocorticoids (see Barik et al., 2013). If it has not already occurred to you, the chronic social threat that induces subordination stress in the members of many species is termed **bullying** in our own.

Psychosomatic Disorders: The Case of Gastric Ulcers

LO 17.16 Describe how our view of psychosomatic disorders has been refined by the results of research on gastric ulcers.

Interest in pathological effects of stress has increased as researchers have identified more and more **psychosomatic disorders** (medical disorders in which psychological factors play a causal role). So many adverse effects of stress on health (e.g., in heart disease, asthma, and skin disorders) have been documented that it is now more reasonable to think of most, if not all, medical disorders as psychosomatic.

Gastric ulcers were one of the first medical disorders to be classified as psychosomatic. **Gastric ulcers** are painful lesions to the lining of the stomach and duodenum, which in extreme cases can be life threatening. About 500,000 new cases are reported each year in the United States.

The view of gastric ulcers as the prototypical psychosomatic disorder changed with the discovery that they seemed to be caused by bacteria. It was claimed that the bacteria *Helicobacter pylori* (i.e., *H. pylori*) are responsible for all cases of gastric ulcers except those caused by nonsteroidal anti-inflammatory agents such as aspirin. This seemed to rule out stress as a causal factor, but a consideration of the evidence suggests otherwise.

There is no denying that *H. pylori* damage the stomach wall or that antibiotic treatment of gastric ulcers helps many sufferers. The facts do, however, suggest that *H. pylori* infection alone is insufficient to produce the disorder in most people. Although most patients with gastric ulcers display signs of *H. pylori* infection, so too do many healthy individuals (see Maixner et al., 2016; Testerman & Morris, 2014). Also, antibiotics improve the condition of many patients with gastric ulcers, but so do psychological treatments—and they do it without reducing signs of *H. pylori* infection. Apparently, another factor increases the susceptibility of the stomach wall to damage from *H. pylori*, and this factor appears to be stress. Gastric ulcers occur more commonly in people living in stressful situations, and stressors can produce gastric ulcers in laboratory animals.

Psychoneuroimmunology: Stress, the Immune System, and the Brain

LO 17.17 Define psychoneuroimmunology, and describe the four components that make up our bodies' defenses against foreign pathogens.

A major change in the study of psychosomatic disorders came in the 1970s with the discovery that stress can increase susceptibility to infectious diseases. Up to that point, infectious diseases had been regarded as "strictly physical." The discovery that stress can increase susceptibility to infection led to the emergence of a new field of research in the early 1980s: **psychoneuroimmunology**—the study of interactions among psychological factors, the nervous system, and the immune system. Psychoneuroimmunological research is the focus of this section. Let's begin with an introduction to the immune system.

Microorganisms of every description revel in the warm, damp, nutritive climate of your body. However, the body has four lines of defense to keep it from being overwhelmed. First is what has been termed the *behavioral immune systems:* Humans are motivated to avoid contact with individuals who are displaying symptoms of illness (see Murray & Schaller, 2016), and their bodies are primed to respond more aggressively to infection when they perceive signs of infection in others (see Schaller et al., 2010). Second are a variety of surface barriers that keep the body from being overwhelmed. The major surface barrier is skin, but there are other mechanisms that protect from invasions through bodily openings (e.g., respiratory tract, eyes, and gastrointestinal tract). These mechanisms include coughing, sneezing, tears, mucous, and numerous chemical barriers.

If microorganisms do manage to breach the surface barriers and enter the body, they are met by two additional lines of defense: the innate immune system and the adaptive immune system. Together, these two lines of defense constitute the **immune system** (see Kipnis, 2018; Pringle, 2013).

INNATE IMMUNE SYSTEM.   The **innate immune system** is the first component of the immune system to react. It reacts quickly and generally near points of entry of **pathogens** (disease-causing agents) to the body. It is triggered when receptors called **toll-like receptors** (because they are similar to *toll,* a receptor previously discovered in fruit flies) bind to molecules on the surface of the pathogens or when injured cells send out alarm signals (see De Nardo, 2015). The reaction of the innate immune system includes a complex, but general, array of chemical and cellular reactions—they are general in the sense that the reactions to all pathogens are the same.

One of the first reactions of the innate immune system to the invasion of pathogens is *inflammation* (swelling). Inflammation is triggered by the release of chemicals from damaged cells. Particularly influential are the cytokines, which attract **leukocytes** (white blood cells) and other **phagocytes** (cells that engulf and destroy pathogens) into the infected area. Microglia are phagocytes that are specific to the central nervous system (see Aguzzi, Barres, & Bennett, 2013; Su et al., 2016). Cytokines also promote healing of the damaged tissue once the pathogens are destroyed (see Kyritsis et al., 2012; Werneburg et al., 2017).

Phagocytosis (destruction of pathogens by phagocytes) is thought to be one of the first immune reactions to have evolved. Phagocytes have been identified in all vertebrates and invertebrates that have been examined. A phagocyte is shown attacking bacteria in Figure 17.13.

ADAPTIVE IMMUNE SYSTEM.   The **adaptive immune system** differs from the innate immune system in the following four respects:

- It evolved more recently, first appearing in early vertebrates.
- It is slower; its immune reaction to pathogens takes longer to be fully manifested.
- It is specific in the sense that it reacts against specific antigens.

Figure 17.13 Phagocytosis: A phagocyte about to ingest and destroy bacteria (red blobs).

- It has a memory; once it has reacted against a particular pathogen, it reacts more effectively against that same pathogen in the future.

The main cells of the adaptive immune system are specialized leukocytes called **lymphocytes**. Lymphocytes are produced in bone marrow and the thymus gland and are stored in the *lymphatic system* until they are activated. There are two major classes of lymphocytes: T cells and B cells (see Plesnila, 2016). **Cell-mediated immunity** is directed by **T cells** (T lymphocytes); **antibody-mediated immunity** is directed by **B cells** (B lymphocytes).

The cell-mediated immune reaction begins when a phagocyte ingests a foreign microorganism. The phagocyte then displays the microorganism's **antigens** (molecules, usually proteins, that can trigger an immune response) on the surface of its cell membrane, and this display attracts T cells. Each T cell has two kinds of receptors on its surface, one for molecules that are normally found on the surface of phagocytes and other body cells, and one for a specific foreign antigen. There are millions of different receptors for foreign antigens on T cells, but there is only one kind on each T cell, and there are only a few T cells with each kind of receptor. Once a T cell with a receptor for the foreign antigen binds to the surface of an infected macrophage, a series of reactions is initiated. Among these reactions is the multiplication of the bound T cell, creating more T cells with the specific receptor necessary to destroy all invaders that contain the target antigens and all body cells that have been infected by the invaders.

The antibody-mediated immune reaction begins when a B cell binds to a foreign antigen for which it contains an appropriate receptor. This causes the B cell to multiply and to synthesize a lethal form of its receptor molecules. These lethal receptor molecules, called **antibodies**, are released into the intracellular fluid, where they bind to the foreign antigens and destroy or deactivate the microorganisms that possess them. Memory B cells for the specific antigen are also produced during the process; these cells have a long life and accelerate antibody-mediated immunity if there is a subsequent infection by the same microorganism.

The memory of the adaptive immune system is the mechanism that gives vaccinations their *prophylactic* (preventive) effect—**vaccination** involves administering a weakened form of a virus so that if the virus later invades, the adaptive immune system is prepared to act against it. For example, smallpox has been largely eradicated by programs of vaccination with the weakened form of its largely benign relative, cowpox. The process of creating immunity through vaccination is termed **immunization**.

Until recently, most immunological research has focused on the adaptive immune system; however, the discovery of the role of cytokines in the innate immune system stimulated interest in that system.

WHAT EFFECT DOES STRESS HAVE ON IMMUNE FUNCTION: DISRUPTIVE OR BENEFICIAL? It is widely believed that the main effect of stress on immune function is disruptive. We are sure you have heard this from family members, friends, and even physicians. But is this true?

One of the logical problems with the view that stress always disrupts immune function is that it is inconsistent with the principles of evolution. Virtually every individual organism encounters many stressors during the course of its life, and it is difficult to see how a maladaptive response to stress, such as a disruption of immune function, could have evolved—or could have survived if it had been created by a genetic accident or as a *spandrel* (a nonadaptive byproduct of an adaptive evolutionary change; see Chapter 2).

Two events have helped clarify the relation between stress and immune function. The first was the meta-analysis of Segerstrom and Miller (2004), which reviewed about 300 previous studies of stress and immune function. Segerstrom and Miller found that the effects of stress on immune function depended on the kind of stress. They found that acute (brief) stressors (i.e., those lasting less than 100 minutes, such as public speaking, an athletic competition, or a musical performance) actually led to improvements in immune function. Not surprisingly, the improvements in immune function following acute stress occurred mainly in the innate immune system, whose components can be marshaled quickly. In contrast, chronic (long-lasting) stressors, such as caring for an ill relative or experiencing a period of unemployment, adversely affected the adaptive immune system. Stress that disrupts health or other aspects of functioning is called *distress*, and stress that improves health or other aspects of functioning is called *eustress*.

The second event that has helped clarify the relation between stress and immune function was the discovery of the bidirectional role played by the cytokines in the innate immune system. Short-term cytokine-induced inflammatory responses help the body combat infection, whereas long-term cytokine release is associated with a variety of adverse health consequences (see Dhabhar, 2014). This finding provided an explanation of the pattern of results discovered by Segerstrom and Miller's meta-analysis.

HOW DOES STRESS INFLUENCE IMMUNE FUNCTION? The mechanisms by which stress influences immune function have been difficult to specify because there are so many possibilities. Stress produces widespread changes in the body through its effects on the anterior-pituitary adrenal-cortex system and the sympathetic-nervous-system adrenal-medulla system, and there are innumerable mechanisms by which those systems can influence immune function. For example, both T cells and B cells have receptors for glucocorticoids; and lymphocytes have receptors for epinephrine, norepinephrine, and glucocorticoids. In addition, many of the neuropeptides that are released by neurons are also released by cells of the immune system. Conversely, cytokines, originally thought to be produced only by cells of the immune system, have been found to be produced by cells of the nervous system (see Jin & Yamashita, 2016).

It is important to appreciate that there are behavioral routes by which stress can affect immune function. For example, people under severe stress often change their diet, exercise, sleep, and drug use, any of which could influence immune function. Also, the behavior of a stressed or ill person can produce stress and illness in others. For example, Wolf and colleagues (2007) found that stress in mothers aggravates asthmatic symptoms in their children; conversely, asthma in the children increases measures of stress in their mothers.

DOES STRESS AFFECT SUSCEPTIBILITY TO INFECTIOUS DISEASE?   You have just learned that stress influences immune function. Most people assume that this means that stress increases susceptibility to infectious diseases. But it doesn't mean this at all, and it is important that you understand why.

> **Journal Prompt 17.4**
> Before reading further, try jotting down some reasons as to why it would be a mistake to think that stress increases susceptibility to infectious diseases.

There are at least three reasons why stress-produced decreases in immune function may not be reflected in an increased susceptibility to infectious disease:

- The immune system seems to have many redundant components; thus, disruption of one of them may have little or no effect on vulnerability to infection.

- Stress-produced changes in immune function may be too short-lived to have substantial effects on the probability of infection.

- Declines in some aspects of immune function may induce compensatory increases in others.

It has been difficult to prove that stress causes increases in susceptibility to infectious diseases in humans. One reason for this difficulty is that only correlational studies are possible. Numerous studies have reported *positive* correlations between stress and ill health in humans; for example, students in one study reported more respiratory infections during final exams (Glaser et al., 1987). However, interpretation of such correlations is never straightforward: People may report more illness during times of stress because they expect to be more ill, because their experience of illness during times of stress is more unpleasant, or because the stress changed their behavior in ways that increased their susceptibility to infection.

Despite the difficulties of proving a direct causal link between stress and susceptibility to infectious disease in humans, the evidence for such a link is strong. Three basic types of evidence, when considered together, are persuasive:

- Correlational studies in humans—as you have just learned—have found correlations between stress levels and numerous measures of health.

- Controlled experiments conducted with laboratory animals show that stress can increase susceptibility to infectious disease in these species.

- A few partially controlled studies of humans have added greatly to the weight of evidence.

One of the first partially controlled studies demonstrating stress-induced increases in the susceptibility of humans to infectious disease was conducted by Cohen and colleagues (1991). Using questionnaires, they assessed psychological stress levels in 394 healthy participants. Then, each participant randomly received saline nasal drops that contained a respiratory virus or only saline. Then, all of the participants were quarantined until the end of the study. A higher proportion of those participants who scored highly on the stress scales developed colds.

Early Experience of Stress

LO 17.18 Describe the effects of early exposure to severe stress.

Early exposure to severe stress can have a variety of adverse effects on subsequent development. Children subjected to maltreatment or other forms of severe stress display a variety of brain and endocrine system abnormalities (see Klengel & Binder, 2015). For example, early exposure to stress often increases the intensity of subsequent stress responses (e.g., increases the release of glucocorticoids in response to stressors).

It is important to understand that the developmental period during which early stress can adversely affect neural and endocrine development begins before birth. Many experiments have demonstrated the adverse effects of prenatal stress in laboratory animals; pregnant females have been exposed to stressors, and the adverse effects of that exposure on their offspring have subsequently been documented (e.g., Sowa et al., 2015).

One particularly interesting line of research on the role of early experience in the development of the stress response began with the observation that handling of rat pups by researchers for a few minutes per day during the first few weeks of the rats' lives has a variety of salutary (health-promoting) effects (see Raineki, Lucion, & Weinberg, 2014). The majority of these effects seemed to result from a decrease in the magnitude of the handled pups' responses to stressful events. As adults, rats that had been handled as pups displayed smaller increases in circulating glucocorticoids in response to stressors (see Francis & Meaney, 1999). It seemed remarkable that a few hours of handling early in life could have such a significant and lasting effect. However, evidence supports an alternative interpretation.

Liu and colleagues (1997) found that handled rat pups are groomed (licked) more by their mothers, and they hypothesized that the salutary effects of the early handling resulted from the extra grooming, rather than from the handling itself. They confirmed this hypothesis by showing that unhandled rat pups that received a lot of grooming from their mothers developed the same profile of less glucocorticoid release that was observed in handled pups (see Champagne et al., 2008).

Early separation of rat pups from their mothers seems to have effects opposite to those that result from high levels of early grooming (see Zhang et al., 2013). For example, rats that are separated from their mothers in infancy display elevated behavioral and hormonal responses to stress as adults.

Stress and the Hippocampus

LO 17.19 Describe the effects of stress on the hippocampus.

Exposure to stress affects the structure and function of the brain in a variety of ways (see Lupien et al., 2018; McEwen, Gray, & Nasca, 2015; Sandi & Haller, 2015). However, the hippocampus appears to be particularly susceptible to stress-induced effects (see Kim, Pellman, & Kim, 2015; McEwen, Nasca, & Gray, 2016). The reason for this susceptibility may be the particularly dense population of glucocorticoid receptors in the hippocampus.

Stress has been shown to reduce dendritic branching in the hippocampus, to reduce adult neurogenesis in the hippocampus (see Egeland, Zunszain, & Pariante, 2015), to modify the structure of some hippocampal synapses, and to disrupt the performance of hippocampus-dependent tasks (see Kim, Pellman, & Kim, 2015). These effects of stress on the hippocampus appear to be mediated by elevated-glucocorticoid levels: They can be induced by **corticosterone** (a major glucocorticoid)

and can be blocked by **adrenalectomy** (surgical removal of the adrenal glands)—see de Quervain, Schwabe, & Roozendaal (2017); Shirazi et al. (2015).

CONCLUSION.   In this chapter, you have learned that the amygdala plays a role in emotion. The chapter ends with a troubling case that reinforces this point. Fortunately, not everybody reacts in the same way to amygdalar damage.

The Case of Charles Whitman, the Texas Tower Sniper

After having lunch with his wife and his mother, Charles Whitman went home and typed a letter of farewell—perhaps as an explanation for what would soon happen.

He stated in his letter that he was having many compelling and bizarre ideas. Psychiatric care had been no help. He asked that his brain be autopsied after he was through; he was sure they would find the problem.

By all reports, Whitman had been a nice person. An Eagle Scout at 12 and a high school graduate at 17, he then enlisted in the Marine Corps, where he established himself as an expert marksman. After his discharge, he entered the University of Texas to study architectural engineering.

Nevertheless, in the evening of August 1, 1966, Whitman killed his wife and mother. He professed love for both of them, but he did not want them to face the aftermath of what was to follow.

The next morning, at about 11:30, Whitman went to the Tower of the University of Texas, carrying six guns, ammunition, several knives, food, and water. He clubbed the receptionist to death and shot four more people on his way to the observation deck. Once on the deck, he opened fire on people crossing the campus and on nearby streets. His accuracy was deadly: He killed people as far as 300 meters away—people who assumed they were out of range.

At 1:24 that afternoon, the police fought their way to the platform and shot Whitman to death. All told, 17 people, including Whitman, had been killed, and another 31 had been wounded (Helmer, 1986).

An autopsy was conducted. Whitman had been correct: They found a walnut-sized tumor in his right amygdala.

Scan Your Brain

Review the preceding module and fill in each of the following blanks. The correct answers are provided at the end of the exercise. Review material related to your errors or omissions before proceeding.

1. Functional MRI studies indicated similar activation in the _____ and prefrontal cortex when volunteers observed and executed facial expressions.

2. A genetic disorder that results in calcification of the amygdala and surrounding anterior medial temporal lobe structures in both hemispheres, can be linked with the inability to perceive fear, and is called _____.

3. In a typical _____ paradigm, participants are asked to inhibit their negative emotions when exposed to unpleasant stimuli.

4. Many studies have shown that one of the functions of the _____ is to exert cognitive control of emotions.

5. Hans Selye suggested that stress responses are attributed to the activation of the _____ system.

6. According to Hans Selye, neural circuits stimulate the release of _____, which in turn triggers the release of glucocorticoids.

7. Gastric ulcers, painful lesions to the lining of the stomach and duodenum that in extreme cases can be life-threatening, were one of the first medical disorders to be classified as _____.

8. The main cells of the adaptive immune system are _____.

9. _____ involves administering a weakened form of a virus so that if the virus later invades, the adaptive immune system is prepared to act against it.

10. Stress that disrupts health or other aspects of functioning is called _____, and stress that improves health or other aspects of functioning is called _____.

11. Stress has been shown to decrease the dendritic branching in the _____, causing memory difficulties.

12. T cells and B cells have receptors for _____, which is one of the ways that stress influences the immune system.

Themes Revisited

All four of the book's themes were prevalent in this chapter. The clinical implications theme appeared frequently, both because brain-damaged patients have taught us much about the neural mechanisms of emotion and because emotions have a major impact on health. The evolutionary perspective theme also occurred frequently because comparative research and the consideration of evolutionary pressures have also had a major impact on current thinking about the biopsychology of emotion.

The thinking creatively theme appeared where the text encouraged you to think in unconventional ways about the relation between testosterone and human aggression and the interpretation of reports of correlations between stress and ill health. Neuroplasticity was the major theme of the discussion of the effects of stress on the hippocampus.

The two emerging themes did not figure prominently in this chapter. However, thinking clearly about epigenetics has become important in modern research on the effects of stress: Epigenetic mechanisms appear to mediate many of the effects of stress on brain function.

Key Terms

Biopsychology of Emotion: Introduction
James-Lange theory, p. 464
Cannon-Bard theory, p. 464
Decorticate, p. 464
Sham rage, p. 465
Limbic system, p. 465
Klüver-Bucy syndrome, p. 465
Amygdala, p. 465
Polygraphy, p. 466
Control-question technique, p. 466
Guilty-knowledge technique, p. 467
Facial feedback hypothesis, p. 467
Duchenne smile, p. 469

Fear, Defense, and Aggression
Fear, p. 469
Defensive behaviors, p. 469
Aggressive behaviors, p. 469
Alpha male, p. 470
Target-site concept, p. 470

Neural Mechanisms of Fear Conditioning
Fear conditioning, p. 472

Contextual fear conditioning, p. 473
Hippocampus, p. 473
Lateral nucleus of the amygdala, p. 473
Prefrontal cortex, p. 473
Central nucleus of the amygdala, p. 473

Brain Mechanisms of Human Emotion
Urbach-Wiethe disease, p. 475
Suppression paradigms, p. 475
Reappraisal paradigms, p. 475

Stress and Health
Stress, p. 477
Stressors, p. 477
Adrenocorticotropic hormone (ACTH), p. 477
Glucocorticoids, p. 477
Adrenal cortex, p. 477
Adrenal medulla, p. 477
Cytokines, p. 478
Subordination stress, p. 478
Bullying, p. 478

Psychosomatic disorders, p. 478
Gastric ulcers, p. 478
Psychoneuroimmunology, p. 479
Immune system, p. 479
Innate immune system, p. 479
Pathogens, p. 479
Toll-like receptors, p. 479
Leukocytes, p. 479
Phagocytes, p. 479
Phagocytosis, p. 479
Adaptive immune system, p. 479
Lymphocytes, p. 480
Cell-mediated immunity, p. 480
T cells, p. 480
Antibody-mediated immunity, p. 480
B cells, p. 480
Antigens, p. 480
Antibodies, p. 480
Vaccination, p. 480
Immunization, p. 480
Corticosterone, p. 482
Adrenalectomy, p. 482

Chapter 18
Biopsychology of Psychiatric Disorders

The Brain Unhinged

Matthieu Spohn/ès/Corbis

 ## Chapter Overview and Learning Objectives

Schizophrenia

LO 18.1 Describe the positive and negative symptoms of schizophrenia, and provide specific examples of each.

LO 18.2 Describe the discovery of the first two widely prescribed antipsychotic drugs.

LO 18.3 Describe the evolution of the dopamine theory of schizophrenia.

LO 18.4 Describe two current lines of research on schizophrenia.

LO 18.5 Explain what is currently known about the genetics and epigenetics of schizophrenia.

LO 18.6 Describe the various brain changes associated with schizophrenia.

Depressive Disorders	**LO 18.7**	Explain what a clinical depression is.
	LO 18.8	Describe the early research that led to the discovery of antidepressant medications. Also, list each of the five major classes of antidepressant drugs, and provide one specific example of each.
	LO 18.9	Describe two forms of treatment for depression that utilize brain stimulation.
	LO 18.10	Describe two theories of the etiology of major depressive disorder.
	LO 18.11	Describe our current state of knowledge of the genetic and epigenetic mechanisms of depression.
	LO 18.12	Describe the various brain differences associated with major depressive disorder.
Bipolar Disorder	**LO 18.13**	Describe the symptoms associated with each of the two types of bipolar disorder.
	LO 18.14	Describe the discovery of the first mood stabilizer.
	LO 18.15	Describe some factors that may contribute to the onset and maintenance of bipolar disorder.
	LO 18.16	Describe our current state of knowledge of the genetic and epigenetic mechanisms of bipolar disorder.
	LO 18.17	Describe the brain differences associated with bipolar disorder.
Anxiety Disorders	**LO 18.18**	Describe four anxiety disorders.
	LO 18.19	Describe three sorts of drugs used in the treatment of anxiety disorders.
	LO 18.20	Describe three animal models of anxiety disorders.
	LO 18.21	Describe our current state of knowledge of the genetic and epigenetic mechanisms of anxiety disorders.
	LO 18.22	Describe the various brain differences associated with anxiety disorders.
Tourette's Disorder	**LO 18.23**	Describe the symptoms of Tourette's disorder.
	LO 18.24	Describe how Tourette's disorder is treated.
	LO 18.25	Describe our current state of knowledge of the genetic and epigenetic mechanisms of Tourette's disorder.
	LO 18.26	Describe the research findings related to the neural bases of Tourette's disorder.
Clinical Trials: Development of New Psychotherapeutic Drugs	**LO 18.27**	Describe the three phases of clinical trials.
	LO 18.28	Identify six controversial aspects of clinical trials.
	LO 18.29	Discuss the relative effectiveness of clinical trials.

This chapter is about the biopsychology of **psychiatric disorders** (disorders of psychological function sufficiently severe to require treatment). One of the main difficulties in studying or treating psychiatric disorders is that they are difficult to diagnose. The psychiatrist or clinical psychologist must first decide whether a patient's psychological function is pathological or merely an extreme of normal human variation: For example, does a patient with a poor memory suffer from a pathological condition, or is he merely a healthy person with a poor memory?

If a patient is judged to be suffering from a psychiatric disorder, then the particular disorder must be diagnosed. Because we cannot yet identify the specific brain pathology associated with various disorders, their diagnosis usually rests entirely on the patient's symptom profile. Currently, the diagnosis is guided by the **DSM-5** (the current edition of the *Diagnostic and Statistical Manual* of the American Psychiatric Association). There are two main difficulties in diagnosing particular psychiatric disorders: (1) patients suffering from the same disorder often display different symptoms, and (2) patients suffering from different disorders often display many of the same symptoms. Consequently, experts often disagree on the diagnosis of particular cases, and the guidelines provided by the DSM change with each new edition (see Blashfield et al., 2014). One purpose of this chapter is to help you understand why it is important to periodically revise the diagnosis of psychiatric disorders.

This chapter begins with discussions of five sorts of psychiatric disorders: schizophrenia, depressive disorders, bipolar disorder, anxiety disorders, and Tourette's disorder. It ends with a description of how new *psychotherapeutic* drugs are developed and tested.

Schizophrenia

Schizophrenia means "the splitting of psychic functions." The term was coined in the early years of the 20th century to describe what was assumed at the time to be the primary symptom of the disorder: the breakdown of integration among emotion, thought, and action.

Schizophrenia is considered to be a severe psychiatric disorder. It attacks about 1 percent of individuals of all races and cultural groups, typically beginning in adolescence or early adulthood (see Sikela & Quick, 2018). Schizophrenia occurs in many forms, but the case of Lena introduces you to some of its common features (Meyer & Salmon, 1988).

Schizophrenia: The Case of Lena

Lena's mother was hospitalized with schizophrenia when Lena was 2. As a child, Lena displayed periods of hyperactivity; as an adolescent, she was viewed as odd. She enjoyed her classes and got good grades, but she had few friends.

Shortly after their marriage, Lena's husband noticed that Lena was becoming more withdrawn. She would sit for hours barely moving a muscle, often having lengthy discussions with nonexistent people.

One day, Lena's husband found her sitting on the floor in an odd posture staring into space. She was totally unresponsive. When he tried to move her, Lena displayed *waxy flexibility*—that is, she reacted like a mannequin, not resisting movement and holding her new position until she was moved again. She was diagnosed with *schizophrenia with catatonia* (schizophrenia characterized by long periods of immobility and waxy flexibility).

In the hospital, Lena displayed a speech pattern exhibited by some individuals with schizophrenia: *echolalia* (vocalized repetition of some or all of what has just been heard).

Doctor: How are you feeling today?

Lena: I am feeling today, feeling the feelings today.

Doctor: Are you still hearing the voices?

Lena: Am I still hearing the voices, voices?

What Is Schizophrenia?

LO 18.1 Describe the positive and negative symptoms of schizophrenia, and provide specific examples of each.

The major difficulty in studying and treating schizophrenia is accurately defining it (see Bhati, 2013). Its symptoms are complex and diverse; they overlap greatly with those of other psychiatric disorders and frequently change during the progression of the disorder. Also, various neurological conditions (e.g., complex seizures; see Chapter 10) have symptoms that might suggest a diagnosis of schizophrenia. Because the current definition of schizophrenia overlaps with that of several different disorders, the DSM-5 prefers to use the label *schizophrenia spectrum disorders* to refer to schizophrenia and related disorders (see Bhati, 2013).

The following are some symptoms of schizophrenia, although none of them appears in all cases. In an effort to categorize cases of schizophrenia so that they can be studied and treated more effectively, it is common practice to consider **positive symptoms** (symptoms that seem to represent an excess of typical function) separately from **negative symptoms** (symptoms that seem to represent a reduction or loss of typical function)—see Smigielski et al. (2020).

Examples of positive symptoms include the following:

- *Delusions.* Delusions of being controlled (e.g., "Martians are making me steal"), delusions of persecution (e.g., "My mother is poisoning me"), or delusions of grandeur (e.g., "Steph Curry admires my jump shot").
- *Hallucinations.* Imaginary voices making critical comments or telling patients what to do.
- *Inappropriate affect.* Reacting with an inappropriate emotional response to positive or negative events.
- *Disorganized speech or thought.* Illogical thinking, peculiar associations among ideas, belief in supernatural forces.
- *Odd behavior.* Talking in rhymes, difficulty performing everyday tasks.

Examples of negative symptoms include the following:

- *Affective flattening.* Diminished emotional expression.
- *Avolition.* Reduction or absence of motivation.
- *Catatonia.* Remaining motionless, often in awkward positions for long periods.

The frequent recurrence of any two of these symptoms for 1 month is currently sufficient for the diagnosis of schizophrenia—provided that one of the symptoms is delusions, hallucinations, or disorganized speech.

Discovery of the First Antipsychotic Drugs

LO 18.2 Describe the discovery of the first two widely prescribed antipsychotic drugs.

The first major breakthrough in the study of the biochemistry of schizophrenia was the accidental discovery in the early 1950s of the first **antipsychotic** drug (a drug that is meant to treat certain symptoms of schizophrenia and bipolar disorder), **chlorpromazine**. Chlorpromazine was developed by a French drug company as an antihistamine. Then, in 1950, a French surgeon noticed that chlorpromazine given prior to surgery to counteract swelling had a calming effect on some of his patients, and he suggested that it might have a calming effect on difficult-to-handle patients with **psychosis** (a loss of touch with reality). His suggestion triggered research that led to the discovery that chlorpromazine alleviates the symptoms of schizophrenia: Agitated patients with schizophrenia were calmed by chlorpromazine, and emotionally blunted patients with schizophrenia were activated by it. Don't get the idea that chlorpromazine cures schizophrenia. It doesn't. But it often reduces the severity of symptoms enough to allow institutionalized patients to be discharged.

Shortly after the antipsychotic action of chlorpromazine was first documented, an American psychiatrist became interested in reports that the snakeroot plant had long been used in India for the treatment of mental illness. He gave **reserpine**—the active ingredient of the snakeroot plant—to his patients with schizophrenia and confirmed its antipsychotic action. Reserpine is no longer used in the treatment of schizophrenia because it produces a dangerous decline in blood pressure at the doses needed for successful treatment.

Although the chemical structures of chlorpromazine and reserpine are dissimilar, their antipsychotic effects are similar in two major respects. First, the antipsychotic effect of both drugs is manifested only after a patient has been medicated for 2 or 3 weeks. Second, the onset of this antipsychotic effect is usually associated with motor effects similar to the symptoms of Parkinson's disease (e.g., muscular rigidity, a general decrease in voluntary movement). These similarities suggested to researchers that chlorpromazine and reserpine were acting through the same mechanism—one that was related to Parkinson's disease.

The Dopamine Theory of Schizophrenia

LO 18.3 Describe the evolution of the dopamine theory of schizophrenia.

The next major breakthrough in the study of schizophrenia came from research on Parkinson's disease. In 1960, it was reported that the *striatums* (caudates plus putamens; see Figure 3.28) of persons with Parkinson's disease had been depleted of dopamine (see Goetz, 2011). This finding suggested that a disruption of dopaminergic transmission might produce both Parkinson's disease and the antipsychotic effects of chlorpromazine and reserpine. Thus was born the *dopamine theory of schizophrenia*—the theory that schizophrenia is caused by too much dopamine and, conversely, that antipsychotic drugs exert their effects by decreasing dopamine levels (see McCutcheon, Abi-Dargham, & Howes, 2019).

Lending instant support to the dopamine theory of schizophrenia were two already well-established facts. First, the antipsychotic drug reserpine was known to deplete the brain of dopamine and other monoamines by breaking down the synaptic vesicles in which these neurotransmitters are stored. Second, drugs such as amphetamine and cocaine, which can trigger episodes that resemble schizophrenia in healthy users, were known to increase the extracellular levels of dopamine and other monoamines in the brain.

An important step in the evolution of the dopamine theory of schizophrenia came in 1963, when Carlsson and Lindqvist assessed the effects of chlorpromazine on extracellular levels of dopamine and its *metabolites* (substances that are created by the breakdown of another substance in cells). Although they expected to find that chlorpromazine, like reserpine, depletes the brain of dopamine, they didn't. The extracellular levels of dopamine were unchanged by chlorpromazine, and the extracellular levels of its metabolites were increased. The researchers concluded that both chlorpromazine and reserpine antagonize transmission at dopamine synapses but that they do it in different ways: reserpine by depleting the brain of dopamine and chlorpromazine by binding to dopamine receptors.

Carlsson and Lindqvist argued that chlorpromazine is a *receptor blocker* at dopamine synapses—that is, it binds to dopamine receptors without activating them and, in so doing, keeps dopamine from activating them (see Figure 18.1). We now know that many psychoactive drugs are receptor blockers, but chlorpromazine was the first to be identified as such.

Carlsson and Lindqvist further postulated that the lack of activity at postsynaptic dopamine receptors sent a feedback signal to the presynaptic cells that increased their release of dopamine, which was broken down in the synapses. This explained why dopaminergic activity was reduced while extracellular levels of dopamine stayed about the same and extracellular levels of its metabolites were increased. Carlsson and Lindqvist's findings led to an important revision of the dopamine theory of schizophrenia: Rather than high dopamine levels, the main factor in schizophrenia was presumed to be high levels of activity at dopamine receptors.

In the mid-1970s, Snyder and his colleagues (see Creese, Burt, & Snyder, 1976; Madras, 2013) assessed the degree to which the various antipsychotic drugs that had been developed by that time bind to dopamine receptors. First, they added radioactively labeled dopamine to samples of dopamine-receptor-rich neural membrane obtained from calf striatums. Then, they rinsed away the unbound dopamine molecules from the samples and measured the amount of radioactivity left in them to obtain a measure of the number of dopamine receptors. Next, in other samples, they measured each drug's ability to block the binding of radioactive dopamine to the sample; the assumption was that the drugs with a high affinity for dopamine receptors would leave fewer sites available for the dopamine. In general, they found that chlorpromazine

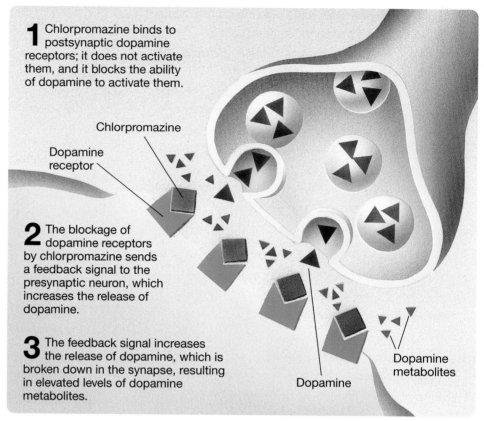

Figure 18.1 Chlorpromazine is a receptor blocker at dopamine synapses. Chlorpromazine was the first receptor blocker to be identified, and its discovery changed psychopharmacology.

1 Chlorpromazine binds to postsynaptic dopamine receptors; it does not activate them, and it blocks the ability of dopamine to activate them.

Chlorpromazine

Dopamine receptor

2 The blockage of dopamine receptors by chlorpromazine sends a feedback signal to the presynaptic neuron, which increases the release of dopamine.

3 The feedback signal increases the release of dopamine, which is broken down in the synapse, resulting in elevated levels of dopamine metabolites.

Dopamine metabolites

Dopamine

and the other effective antipsychotic drugs had a high affinity for dopamine receptors, whereas ineffective antipsychotic drugs had a low affinity. There were, however, several major exceptions, including haloperidol. Although **haloperidol** was one of the most potent antipsychotic drugs of its day, it had a relatively low affinity for dopamine receptors.

A solution to the haloperidol puzzle came with the discovery that dopamine binds to more than one dopamine receptor subtype—five have been identified (see Beaulieu, Espinoza, & Gainetdinov, 2015). It turns out that chlorpromazine and other antipsychotic drugs in the same chemical class (the **phenothiazines**) all bind effectively to both D_1 and D_2 receptors, whereas haloperidol and the other antipsychotic drugs in its chemical class (the **butyrophenones**) all bind effectively to D_2 receptors but not to D_1 receptors.

This discovery of the selective binding of butyrophenones to D_2 receptors led to an important revision in the dopamine theory of schizophrenia. It suggested that schizophrenia is caused by hyperactivity specifically at D_2 receptors, rather than at dopamine receptors in general. Snyder and his colleagues (see Madras, 2013; Snyder, 1978) subsequently confirmed that the degree to which **typical antipsychotics** (the first generation of antipsychotic

Figure 18.2 The positive correlation between the ability of various antipsychotics to bind to D_2 receptors and their clinical potency.

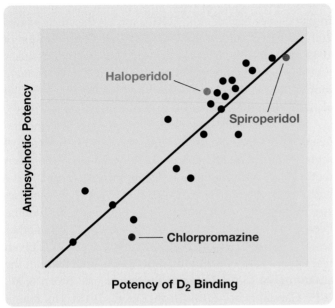

Based on Snyder, S. H. (1978). Neuroleptic drugs and neurotransmitter receptors. Journal of Clinical and Experimental Psychiatry, 133, 21–31.

drugs) bind to D_2 receptors is highly correlated with their effectiveness in suppressing the symptoms of schizophrenia (see Figure 18.2). For example, the butyrophenone *spiroperidol* had the greatest affinity for D_2 receptors and the most potent antipsychotic effect.

Although the evidence implicating D_2 receptors in schizophrenia is strong, it has become apparent that the D_2 version of the dopamine theory of schizophrenia could not explain two general findings:

- Although typical antipsychotics block activity at D_2 receptors within hours, their therapeutic effects are usually not apparent for several weeks.

- Almost all antipsychotics are only effective in the treatment of schizophrenia's positive symptoms, but not its negative symptoms (see Aleman et al., 2018; Osoegawa et al., 2018).

Appreciation of these limitations has led to the current version of the dopamine theory. This version holds that excessive activity at D_2 receptors is one factor in the disorder but that there are many other factors as well (see Poels et al., 2014; Sibley & Shi, 2018).

Those of you who read about the *mesocorticolimbic dopamine pathway* and the *nigrostriatal dopamine pathway* in Chapter 15 (see Figure 15.7) might be wondering which of those two pathways is affected in schizophrenia. The dopamine theory of schizophrenia proposes that schizophrenia is caused by excessive activity in the mesocorticolimbic pathway.

Schizophrenia: Beyond the Dopamine Theory

LO 18.4 Describe two current lines of research on schizophrenia.

Although the dopamine theory of schizophrenia is still influential, current lines of research into atypical antipsychotics and psychedelic drug effects are leading to interesting new perspectives. These two areas of research will be described in the following two subsections.

ATYPICAL ANTIPSYCHOTICS. Currently, atypical antipsychotics (also known as *second-generation antipsychotics*) are often the drugs of choice for the treatment of schizophrenia. **Atypical antipsychotics** are drugs that are effective against schizophrenia but yet do not bind strongly to D_2 receptors. For example, **clozapine**, the first atypical antipsychotic to be approved for clinical use, has an affinity for D_1 receptors, D_4 receptors, and several serotonin and histamine receptors, but only a slight affinity for D_2 receptors (see Humbert-Claude et al., 2012). *Aripiprazole, respiridone,* and *quetiapine* are but a few of the other commonly prescribed atypical antipsychotics.

RENEWED INTEREST IN HALLUCINOGENIC DRUGS. The study of **psychedelic drugs** (drugs whose primary action is to alter perception, emotion, and cognition) began in 1943 with the discovery of **lysergic acid diethylamide (LSD)** (see Garcia-Romeu & Richards, 2018; Nichols, 2016). In addition to *classical hallucinogens* (such as LSD, psilocybin, and mescaline), psychedelic drugs include a variety of other drugs such as the *dissociative hallucinogens* (e.g., ketamine and *phencyclidine*).

Researchers have pursued two lines of research on psychedelics. One line focused on those psychedelic drugs that produce effects similar to the symptoms of psychiatric disorders (e.g., illusions, hallucinations, paranoia, panic), and they used the drugs to model the disorders. The other line focused on the feelings of boundlessness, unity, and bliss reported by some users and attempted to use psychedelics in the treatment of psychiatric disorders. Unfortunately, these promising lines of research ground to a halt in the 1970s when many governments, troubled by the association of LSD and related drugs with various societal subcultures, made it extremely difficult for researchers to study their effects, particularly in humans (see Belouin & Henningfield, 2018; Rucker, Iliff, & Nutt, 2018; but see Oram, 2016).

In the 1990s, there was a gradual renewal of interest in utilizing psychedelic drugs to study the mechanisms of schizophrenia and other psychiatric disorders (see Belouin & Henningfield, 2018; Rucker, Iliff, & Nutt, 2018). This renewal was stimulated by the development of techniques for imaging the effects of drugs in the human brain and by an increased understanding of the mechanisms of psychedelic drug action (e.g., Tylš, Páleníček, & Horáček, 2014). This research led to three important conclusions:

- The psychedelic effects of classical hallucinogens, such as LSD, mimic the positive symptoms of schizophrenia (e.g., hallucinations and disorganized thought) by acting as an agonist of the serotonin type-2a receptor.

- That antagonists of the serotonin type-2a receptor are effective antipsychotics (e.g., the atypical antipsychotic *risperidone*) (see Girgis et al., 2018).

- Dissociative hallucinogens (e.g., ketamine) mimic the negative symptoms of schizophrenia by acting as antagonists of glutamate receptors (see Laruelle, 2014).

Genetic and Epigenetic Mechanisms of Schizophrenia

LO 18.5 Explain what is currently known about the genetics and epigenetics of schizophrenia.

It is clear that schizophrenia involves multiple genetic and epigenetic mechanisms. Many genes have been linked to the disorder (see Flint & Manufò, 2014; Reardon, 2014; Ripke et al., 2014), but no single gene seems capable of causing schizophrenia by itself, although certain genes have been more strongly implicated than others (see Dhindsa & Goldstein, 2016; Sikela & Quick, 2018; Sekar et al., 2016).

The study of schizophrenia-related genes and their expression (see Chapter 2; Figure 2.18) is still in its early stages, but it has already pointed to several physiological changes that could play important roles in development of the disorder (see Kotlar et al., 2015). For example, the expression of schizophrenia-related genes is associated with multiple aspects of brain development (see Birnbaum & Weinberger, 2017; Jaffe et al., 2018; Smigielski et al., 2020), myelination (see Voineskos et al., 2012), transmission at glutamatergic and GABAergic synapses (see Sacchetti et al., 2013), and changes in dopaminergic neuron physiology (see Dong et al., 2018; but see Gürel et al., 2020); and some genes that increase a person's susceptibility to schizophrenia have also been linked to other psychiatric and neurological disorders (see Rizzardi et al., 2019).

A variety of early experiential factors have been implicated in the development of schizophrenia—for example, birth complications, maternal stress, prenatal infections, socioeconomic factors, urban birth or residing in an urban setting, and childhood adversity (see Owen, Sawa, & Mortensen, 2016). Such early experiences are thought to alter the typical course of neurodevelopment leading to schizophrenia in individuals who have a genetic susceptibility (see Negrón-Oyarzo et al., 2016; Owen, Sawa, & Mortensen, 2016), presumably through *epigenetic mechanisms* (see Chapter 2)—see Birnbaum & Weinberger (2017), Hannon et al. (2016), Jaffe et al. (2016), and Sharp

& Akbarian (2016). Supporting this neurodevelopmental theory of schizophrenia are (1) the fact that schizophrenia and autism spectrum disorders share many of the same causal factors (e.g., genetic risk factors, environmental triggers)—see Millan et al. (2016), and (2) the study of two 20th-century famines: the Nazi-induced Dutch famine of 1944–1945 and the Chinese famine of 1959–1961. Fetuses whose pregnant mothers suffered in those famines were more likely to develop schizophrenia as adults (see Li et al., 2015; Schmitt et al., 2014).

Recent research has identified many epigenetic mechanisms that contribute to the emergence and persistence of schizophrenia (see Rizzardi et al., 2019). For example, DNA methylation and histone modifications (see Chapter 2) have both been implicated in the expression of genes for synapse-specific proteins in prefrontal cortex neurons (see Caldeira, Peça, & Carvalho, 2019; Rizzardi et al., 2019). This and other research on the role of epigenetic mechanisms has given researchers better insight into how genes interact with the environment to produce schizophrenia (see Breen et al., 2019; Gandal et al., 2018; Girdhar et al., 2018). The role of transgenerational epigenetic mechanisms (see Chapter 2) in psychiatric disorders like schizophrenia is also of great interest to researchers (see Yeshurun & Hannan, 2019).

Neural Bases of Schizophrenia

LO 18.6 Describe the various brain changes associated with schizophrenia.

There is a long history of research on the neural bases of schizophrenia. Many studies have assessed brain development in patients with, or at risk for, schizophrenia. Four important findings have emerged from various meta-analyses of those studies (see Fusar-Poli et al., 2011; Steen et al., 2006; Vita et al., 2006):

- Individuals who have not been diagnosed with schizophrenia but are at risk for the disorder (e.g., because they have close relatives with schizophrenia) display volume reductions in some parts of the brain—for example, in the hippocampus (see Haukvik et al., 2018).

- Extensive brain changes already exist when patients first seek medical treatment and receive their first brain scans.

- Subsequent brain scans reveal that the brain changes continue to develop after the initial diagnosis.

- Alterations to different areas of the brain develop at different rates (see Gogtay & Thompson, 2010).

One exciting line of research on the neural bases of schizophrenia has challenged the idea that the mesocorticolimbic dopamine pathway is involved in schizophrenia; if you remember, this was a tenet of the classic dopamine theory of schizophrenia. This line of research has shown that the neuropathological changes occur in the nigrostriatal dopamine pathway rather than in the mesocorticolimbic pathway (see McCutcheon, Abi-Dargham, & Howes, 2019).

Recent research on the neural bases of schizophrenia has used modern functional brain-imaging techniques to study *functional connectivity* (see Chapter 5) in the brains of individuals with schizophrenia. Research on functional connectivity in schizophrenia has been of two sorts. The first is the study of functional connectivity during hallucinations in individuals with schizophrenia. This line of research has shown that when participants with schizophrenia are hallucinating there is a change in the pattern of functional connectivity as compared to when they are not hallucinating (see Weber et al., 2020). The second line of research is examining whether patterns of *intrinsic functional connectivity* (see Chapter 5) might be used to predict treatment response to antipsychotic medications (see Chan et al., 2019).

CONCLUSION. Although there has been significant progress in our understanding of the mechanisms and treatment of schizophrenia, a careful reading of the research results suggests that ultimate answers will not be forthcoming until its diagnosis is "sharpened up." There is a general consensus that those patients diagnosed with schizophrenia under the current DSM-5 criteria do not suffer from a single unitary disorder resulting from the same neural pathology (see Kim et al., 2017; Krynicki et al., 2018): The current diagnosis seems to lump together a group of related disorders under one label. This is suggested by the variety and variability of psychological symptoms, neural pathology, and by the marked variability in patient response to particular antipsychotic drugs.

Depressive Disorders

We commonly use the word "depression" to refer to a reaction to grievous loss, such as the loss of a loved one, the loss of self-esteem, or the loss of health. However, "depression" is also used to refer to a psychiatric disorder; that disorder is the focus of this module.

What Are Depressive Disorders?

LO 18.7 Explain what a clinical depression is.

Some people experience deep depression and/or **anhedonia** (loss of the capacity to experience pleasure; see Husain & Roiser, 2018; Post & Warden, 2018), often for no apparent reason (see Treadway & Zald, 2011). Their depression can be so extreme that it can impair cognition (see Cambridge et al., 2018; Dillon & Pizzagalli, 2018) and makes it almost impossible for them to meet the essential requirements of

their daily lives: to keep a job, to eat, or to maintain social contacts and personal hygiene. Sleep disturbances and thoughts of suicide are common. When this condition lasts for 2 weeks or longer, these people are said to be suffering from a **clinical depression**, also known as **major depressive disorder**. The case of S.B. introduces you to some of the main features of clinical depression.

The Case of S.B., the Depressed Biopsychology Student

S.B. excelled during his first year at university—earning top marks in his program of study: biosychology. However, beginning in the second year of his studies, S.B. began to suffer from depression: He began to sleep excessively, he had trouble concentrating on his studies, and he thought about death and suicide frequently. He also suffered from delusions: He thought he was stupid and disliked, and he felt persecuted by his instructors and peers. Having seen these symptoms previously in certain members of his family, S.B. made the astute decision of seeking out help from a psychiatrist.

His psychiatrist offered him medications, but S.B. refused, as he was reminded of the many drug-related side effects he had seen in his family members. Rather, he attended psychotherapy for the remaining years of his degree. Although it helped to talk about his problems, S.B. saw little improvement in his condition during this period. Still, he was able to complete his degree with relatively high grades and subsequently applied and was admitted to graduate studies in biopsychology.

A few months after beginning graduate school, S.B.'s depression became so severe that he could no longer function. For example, S.B. had impairments in his memory and attention that affected his ability to read; delusional ideas and suicidal thoughts constantly plagued his mind. After seeing S.B. in this state, his psychiatrist immediately hospitalized him. While he was hospitalized, he was started on an antidepressant medication to calm his depression and an antipsychotic medication to help him deal with his delusional thoughts. Upon being released from the hospital, his psychiatrist advised him to take a leave of absence from his studies, which he did. S.B. returned to graduate school several months later. However, since his symptoms still persisted despite all the medications, albeit to a lesser degree, he was barely capable of keeping things together.

Don't forget S.B.; you will learn more about him later in this chapter.

Depression is often divided into two categories. Depression triggered by an obvious negative experience (e.g., the death of a friend, the loss of a job) is called **reactive depression**; depression with no apparent cause (as in the case of S.B.) is called **endogenous depression** (see Malki et al., 2014).

Depression affects 2–5 percent of the global population (see Han, Russo, & Nestler, 2019). Females are about twice as likely to a receive a diagnosis of depression during their lifetime (see Kuehner, 2017), with this sex difference being most pronounced during adolescence (see Salk, Hyde, & Abramson, 2017)—although the reasons for this sex difference are still unclear, gonadal-hormone-related explanations are currently popular (see Altemus, Sarvaiya, & Epperson, 2014; Bangasser & Valentino, 2014; Kokras & Dalla, 2014; Kuehner, 2017). For reasons that are still unclear, non-caucasian individuals are more likely to suffer from chronic and more debilitating depression than caucasians (see Bailey, Mokonogho, & Kumar, 2019). The lifetime risk of completed suicide in an individual diagnosed with clinical depression has been found to range between 4 and 15 percent in various studies (see Wang et al., 2015).

Clinical depressions attack children, adolescents, and adults. In adults, clinical depression is often **comorbid** (the tendency for two health conditions to occur together in the same individual) with one or more other health conditions—for example, anxiety disorders, coronary heart disease, and diabetes (see Scott, 2014).

There are two subtypes of major depressive disorder whose cause is more apparent because of the timing of the episodes. One is **seasonal affective disorder (SAD)**, in which episodes of depression and lethargy typically recur during particular seasons—usually during the winter months (see Tyrer et al., 2016). Two lines of evidence suggest that the episodes are triggered by the reduction in sunlight. One is that the incidence of the disorder is higher in Alaska (9 percent) than in Florida (1 percent), where the winter days are longer and brighter (see Melrose, 2015). The other is that *light therapy* (e.g., exposure to 15–30 minutes of very bright light each morning) is often effective in reducing the symptoms of SAD (see Oren, Koziorowski, & Desan, 2013; Song et al., 2015). The second subtype of major depressive disorder with an obvious cause is **peripartum depression**, the intense, sustained depression experienced by some women during pregnancy, after they give birth, or both (see Serati et al., 2016; Pawluski, Lonstein, & Fleming, 2017). Although estimates vary, the disorder seems to be associated with about 13 percent of pregnancies (see Silver et al., 2018).

Antidepressant Drugs

LO 18.8 Describe the early research that led to the discovery of antidepressant medications. Also, list each of the five major classes of antidepressant drugs, and provide one specific example of each.

Five major classes of drugs have been used for the treatment of depressive disorders (see Willner, Scheel-Krüger, & Belzung, 2013): monoamine oxidase inhibitors, tricyclic antidepressants, selective monoamine-reuptake inhibitors, atypical antidepressants, and NMDA-receptor antagonists.

MONOAMINE OXIDASE INHIBITORS. **Iproniazid**, the first antidepressant drug, was originally developed for the treatment of tuberculosis, for which it proved to be a dismal flop. However, interest in the antidepressant potential of the drug was kindled by the observation that it left patients with tuberculosis less concerned about their disorder. As a result, iproniazid was tested on a mixed group of psychiatric patients and seemed to act against clinical depression. It was first marketed as an antidepressant drug in 1957.

Iproniazid is a monoamine agonist; it increases the levels of monoamines (e.g., norepinephrine and serotonin) by inhibiting the activity of *monoamine oxidase (MAO)*, the enzyme that breaks down monoamine neurotransmitters in the *cytoplasm* (cellular fluid) of the neuron. **MAO inhibitors** have several side effects; the most dangerous is known as the **cheese effect** (see Finberg & Gillman, 2011). Foods such as cheese, wine, and pickles contain an amine called *tyramine*, which is a potent elevator of blood pressure. Normally, these foods have little effect on blood pressure because tyramine is rapidly metabolized in the liver by MAO. However, people who take MAO inhibitors and consume tyramine-rich foods run the risk of stroke caused by surges in blood pressure.

TRICYCLIC ANTIDEPRESSANTS. The **tricyclic antidepressants** are so named because of their antidepressant action and because their chemical structures include three rings of atoms. **Imipramine**, the first tricyclic antidepressant, was initially thought to be an antipsychotic drug. However, when its effects on a mixed sample of psychiatric patients were assessed, it had no effect against schizophrenia but seemed to help some depressed patients. Tricyclic antidepressants block the reuptake of both serotonin and norepinephrine, thus increasing their levels in the brain. They are a safer alternative to MAO inhibitors.

SELECTIVE MONOAMINE-REUPTAKE INHIBITORS. In the late 1980s, a new class of drugs—the selective serotonin-reuptake inhibitors—was introduced for treating clinical depression. **Selective serotonin-reuptake inhibitors (SSRIs)** are serotonin agonists that exert their agonistic effects by blocking the reuptake of serotonin from synapses—see Figure 18.3.

Fluoxetine (marketed as Prozac) was the first SSRI to be developed. Now there are many more (e.g., *paroxetine, sertraline, fluvoxamine*). Fluoxetine's structure is a slight variation of that of imipramine and other tricyclic antidepressants; in fact, fluoxetine is no more effective than imipramine in treating depression. Nevertheless, it was immediately embraced by the psychiatric community and has been prescribed in many millions of cases. The remarkable popularity of fluoxetine and other SSRIs is attributable to two things: First, they have fewer side effects than tricyclics and MAO inhibitors; second, they act against a wide range of psychological disorders in addition to depression.

Figure 18.3 Blocking of serotonin reuptake by *fluoxetine* (Prozac).

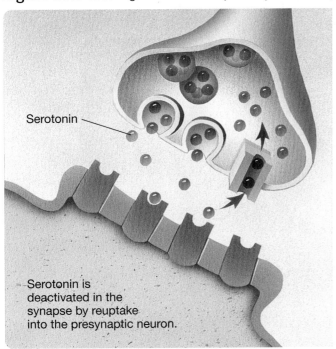

Serotonin

Serotonin is deactivated in the synapse by reuptake into the presynaptic neuron.

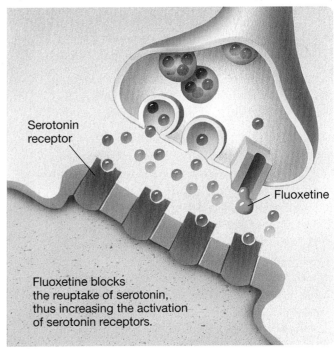

Serotonin receptor

Fluoxetine

Fluoxetine blocks the reuptake of serotonin, thus increasing the activation of serotonin receptors.

The success of the SSRIs spawned the introduction of a similar class of drugs, the **selective norepinephrine-reuptake inhibitors (SNRIs)**. These (e.g., *reboxetine*) have proven to be just as effective as the SSRIs in the treatment of depression. Also effective are drugs that block the reuptake of more than one monoamine neurotransmitter (e.g., *venlafaxine*) (see Harmer, Duman, & Cowen, 2017).

ATYPICAL ANTIDEPRESSANTS. Beginning in the 1980s, several new antidepressants began to appear on the market that did not neatly fit into the three aforementioned classes (i.e., MAO inhibitors, tricyclic antidepressants, and selective monoamine-reuptake inhibitors). Accordingly, a new class of antidepressant medications emerged that is really just a catch-all class comprising drugs that have many different modes of action: the **atypical antidepressants** (see Willner, Scheel-Krüger, & Belzung, 2013). For example, one of the drugs in this class, *bupropion*, has several effects on neurotransmission: It is a blocker of dopamine and norepinephrine reuptake, and it is also a blocker of nicotinic acetylcholine receptors (see Carroll et al., 2014). Another example of a drug in this class is *agomelatine*—a melatonin receptor agonist (see Taylor et al., 2014). There are many other drugs in this class, each with its own unique mechanism of action (see Willner, Scheel-Krüger, & Belzung, 2013).

NMDA-RECEPTOR ANTAGONISTS. Beginning in the early 1990s, several studies reported a positive effect of antagonizing the glutamate *NMDA receptor* on depressive disorders. In the early 2000s, one agent in particular was

shown to be remarkably effective: the dissociative hallucinogen **ketamine**. Remarkably, even a single low dose of ketamine rapidly reduces depression, even in patients who had been experiencing a severe episode (see Amit et al., 2015; Cui, Hu, & Hu, 2019; McGirr et al., 2015; Yang et al., 2018). However, because ketamine has undesirable side effects, researchers are now in the process of trying to identify more selective NMDA-receptor antagonists, and antagonists of other glutamate receptors, with fewer side effects (see Duman, Sanacora, & Krystal, 2019; Malinow, 2016).

EFFECTIVENESS OF DRUGS IN THE TREATMENT OF DEPRESSIVE DISORDERS. About $15 billion is spent in the United States each year on antidepressants. But how effective are antidepressants? Numerous studies have evaluated the effectiveness of antidepressant drugs against major depressive disorder. Many studies have reported that about 50 percent of clinically depressed patients improve with antidepressants (see Cipriani et al., 2018b). This rate seems quite good; however, control groups typically show a rate of improvement of about 35 percent, so only about 15 percent of depressed individuals are actually helped by the antidepressants (see Cipriani et al., 2018a). Still, one can argue that a 15 percent improvement over control conditions is certainly meaningful when one considers how debilitating and dangerous depression can be (see Cipriani et al., 2018b; Maslej et al., 2020). Recent meta-analyses have made it clear that some antidepressant drugs are more effective than others, and that their relative efficacy is a function of the sex and age of the depressed individual (e.g., Cipriani et al., 2018a).

Brain Stimulation to Treat Depression

LO 18.9 Describe two forms of treatment for depression that utilize brain stimulation.

Two treatments involving brain stimulation have been developed for depression: repetitive transcranial magnetic stimulation and deep brain stimulation.

REPETITIVE TRANSCRANIAL MAGNETIC STIMULATION. **Repetitive transcranial magnetic stimulation (rTMS)** is a form of transcranial magnetic stimulation (TMS; see Chapter 5) that involves the noninvasive delivery of repetitive magnetic pulses at either high frequencies (e.g., five pulses per second; *high-frequency rTMS*) or low frequencies (e.g., less than one pulse per second; *low-frequency rTMS*) to specific cortical areas—usually the prefrontal cortex (see Gaynes et al., 2014). High-frequency rTMS and low-frequency rTMS are believed to stimulate and inhibit, respectively, activity within those brain regions to which they are applied (see Berlim et al., 2013, 2014). Meta-analyses have shown reliable improvement of depressive symptoms after either low-frequency (see Berlim et al., 2013) or high-frequency (see Berlim et al., 2014) rTMS when compared with sham rTMS.

DEEP BRAIN STIMULATION. Chronic brain stimulation through an implanted electrode (see Figure 18.4) has been shown to have a therapeutic effect in some depressed patients who have failed to respond to other treatments. Lozano and colleagues (2008) implanted the tip of a stimulation electrode into an area of the white matter of the anterior cingulate gyrus in the medial prefrontal cortex (see Figure 18.5). The stimulator, which was implanted under the skin, delivered continual pulses of electrical stimulation that could not be detected by the patients. The 20 patients in this study were selected because they had repeatedly failed to respond to conventional treatments.

> **Journal Prompt 18.1**
>
> There are many other treatments for depression that aren't discussed in this module. Name a few that you know of. What is the evidence for their efficacy?

Considering that patients had failed to respond to other treatments, the results were strikingly positive: 60 percent showed substantial improvements, 35 percent were largely symptom free, and most of the patients were improved for at least 1 year (the duration of the study). These positive results

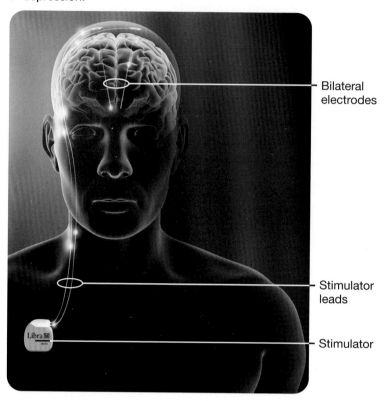

Figure 18.4 Implantation of bilateral anterior cingulate electrodes and a stimulator for chronic deep brain stimulation for the treatment of depression.

- Bilateral electrodes
- Stimulator leads
- Stimulator

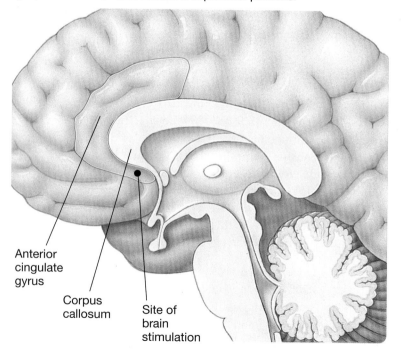

Figure 18.5 The site in the anterior cingulate gyrus at which chronic brain stimulation to subcortical white matter alleviated symptoms in treatment-resistant depressed patients.

- Anterior cingulate gyrus
- Corpus callosum
- Site of brain stimulation

have been replicated at other treatment centers (see Holtzheimer et al., 2011; Lozano et al., 2012; Lozano & Mayberg, 2015).

Theories of Depression

LO 18.10 Describe two theories of the etiology of major depressive disorder.

There are several theories of the etiology of major depressive disorder. As you will soon find out, most theories are based almost entirely on those therapies that have been found to be effective against depression.

MONOAMINE THEORY OF DEPRESSION. One prominent theory of clinical depression is the *monoamine theory*. The monoamine theory of depression holds that depression is associated with underactivity at serotonergic and noradrenergic synapses. The theory is largely based on the fact that monoamine oxidase inhibitors, tricyclic antidepressants, and selective monoamine-reuptake inhibitors are all agonists of serotonin, norepinephrine, or both.

Other support for the monoamine theory of depression has been provided by autopsy studies. Norepinephrine and serotonin receptors have been found to be more numerous in the brains of deceased depressed individuals who had not received pharmacological treatment. This implicates a deficit in monoamine release: When an insufficient amount of a neurotransmitter is released at a synapse, there is usually a compensatory increase in the number of receptors for that neurotransmitter—a process called **up-regulation**.

Three lines of evidence have challenged the monoamine theory of depression. First was the discovery that monoamine agonists, although widely prescribed, are not effective in the treatment of most depressed patients (see Malhi, Lingford-Hughes, & Young, 2016), and even when they are effective, they are only slightly better than placebo (see Linde et al., 2015). Second was the observation that, although monoamine activity is potentiated almost immediately after monoamine-agonist administration, it can take days to weeks for the antidepressant effects of the drug to emerge (see Harmer, Duman, & Cowen, 2017). Third was the discovery that other neurotransmitters (e.g., GABA, glutamate, acetylcholine) play a role in the development of depression (Murrough, Abdallah, & Mathew, 2017; Northoff, 2013; Pytka et al., 2016).

NEUROPLASTICITY THEORY OF DEPRESSION. Nearly all antidepressant drugs rapidly increase transmission at monoaminergic synapses, yet any therapeutic effects of those increases typically are not manifested until weeks after the beginning of drug therapy. Therefore, it is clear that the agonistic effects at monoaminergic synapses cannot be the critical therapeutic mechanism: There must be some change that occurs downstream from the synaptic changes. One theory is that the critical downstream change is an increase in neuroplasticity.

In a nutshell, the *neuroplasticity theory* of depression is that depression results from a decrease of neuroplastic processes in various brain structures (e.g., the hippocampus), which leads to neuron loss and other neural pathology (see Castrén & Hen, 2013; Miller & Hen, 2015). General support for the neuroplasticity theory of depression comes from two kinds of research: (1) research showing that stress and depression are associated with the disruption of various neuroplastic processes (e.g., a reduction in the synthesis of neurotrophins, a decrease in adult hippocampal neurogenesis; see Chapter 9) and (2) research showing that antidepressant treatments are associated with an enhancement of neuroplastic processes (e.g., an increase in the synthesis of neurotrophins (see Chapter 9), an increase in synaptogenesis, and an increase in adult hippocampal neurogenesis)—see Brandon and McKay (2015), Christian et al. (2014), Mahar et al. (2014), and Samuels et al. (2015).

One neurotrophin has been of great interest to researchers: *Brain-derived neurotropic factor (BDNF)*—because treatments that improve depression (both pharmacological and nonpharmacological) have been found to increase BDNF levels only in those patients who show improvement (see Homberg et al., 2014). Indeed, it has been proposed that decreased blood levels of BDNF might be a *biomarker* (a biological state that is predictive of a particular disorder) for depression, and that increased blood levels of BDNF might be a biomarker for the successful treatment of depression (see Polyakova et al., 2015). Moreover, it has been hypothesized that the antidepressant-induced increase in BDNF levels boost certain neuroplastic processes (e.g., increase adult hippocampal neurogenesis) that lead to the alleviation of depression (see Björkholm & Monteggia, 2016).

Genetic and Epigenetic Mechanisms of Depression

LO 18.11 Describe our current state of knowledge of the genetic and epigenetic mechanisms of depression.

Until very recently, most studies of the genetic contribution to clinical depression were not replicable. However, since 2015, a series of studies have reliably identified many genes as contributors to depression (see Ormel, Hartman, & Snieder, 2019).

Because so many different genes have been found to contribute to depression, it is considered to be a complex trait that involves many potential interactions with environmental factors through epigenetic mechanisms (see Heller et al., 2014; Nestler, 2014; Whalley, 2014). The general finding has been that several different epigenetic mechanisms contribute to depression (see Penner-Goeke & Binder, 2019) and that changes to the *epigenome* (see Chapter 2) of an individual might one day be used to predict a future clinical depression in an individual (see Barbu et al., 2020).

Neural Bases of Depression

LO 18.12 Describe the various brain differences associated with major depressive disorder.

Numerous structural and functional neuroimaging studies of the brains of depressed patients have been published. Structural neuroimaging studies have found consistent reductions in gray matter volumes in the prefrontal cortex, hippocampus, amygdala, and cingulate cortex (see Lener & Iosifescu, 2015). White matter reductions have also been noted in several brain regions—most reliably in the frontal cortex (see Russo & Nestler, 2013; Wang et al., 2014). Functional neuroimaging studies have found atypical activity in frontal, cingulate, and insular cortices as well as in the amygdala, thalamus, and striatum.

There are a growing number of studies that have examined alterations in *functional connectivity* (see Chapter 5) in depressed individuals (e.g., Goldstein-Piekarski et al., 2018). These studies have identified a large number of differences in the brains of individuals with depression (see Goldstein-Piekarski & Williams, 2019; Korgaonkar et al., 2019) indicating that the neural networks of depressed patients are markedly different from healthy participants. In most cases, these changes have not been clearly related to changes in behavior or cognition, but such linkages are beginning to emerge (e.g., Albert et al., 2019).

CONCLUSION. Although a number of promising lines of research currently focus on the mechanisms and treatment of clinical depression, treatments are not much better than they were 50 years ago. Many researchers contend that current diagnostic tools, such as those that generate large heterogeneous groups of patients—like the DSM-5—are misguiding research and that more precise approaches to diagnosis are necessary (see Insel & Cuthbert, 2015).

Bipolar Disorder

Some people experience periods of clinical depression and also periods of hypomania or mania. Those who do are said to suffer from **bipolar disorder**. Bipolar disorder affects about 3 percent of the global population (see Ashok et al., 2017; Vieta et al., 2018), with much intercultural variability in its presentation and course (see Subramanian, Sarkar, & Kattimani, 2017). It is a serious psychiatric disorder with one of the highest rates of attempted and completed suicide (see Lima, Peckham, & Johnson, 2018; Plans et al., 2019).

What Is Bipolar Disorder?

LO 18.13 Describe the symptoms associated with each of the two types of bipolar disorder.

Hypomania and mania are in some respects the opposite of depression. **Hypomania** is characterized by a reduced need for sleep, high energy, and positive affect. During periods of hypomania, people are talkative, energetic, impulsive, positive, and very confident. In this state, they can be very effective at certain jobs and can be great fun to be with. **Mania** has the same features as hypomania but taken to an extreme; it also has additional symptoms, such as delusions of grandeur, overconfidence, impulsivity, and distractibility. Mania usually involves psychosis. When mania is full-blown, the person often exhibits unbridled enthusiasm with an outflow of incessant chatter that hurtles from topic to topic. No task is too difficult. No goal is unattainable. This confidence and grandiosity, coupled with high energy, distractibility, and a leap-before-you-look impulsiveness, can result in a series of disasters. Mania often leaves behind a trail of unfinished projects, unpaid bills, and broken relationships.

Those persons who only experience bouts of depression and hypomania are said to have **bipolar disorder type II**; those who also experience bouts of mania, or only experience mania (see Makin, 2019), are said to have **bipolar disorder type I**. The case of S.B., previously introduced to you in the module on depressive disorders, will introduce you to the main features of both forms of bipolar disorder.

The Case of S.B. Revisited: The Biopsychology Student with Bipolar Disorder

S.B. continued his graduate studies in biopsychology while still suffering from a residual depression. Although he struggled through the remaining years of his program, he was successful in attaining a master's degree in biopsychology.

S.B. subsequently began a Ph.D. program in biopsychology. During the first summer after beginning this new program, S.B. started to feel exceptionally good: His mood was elevated; he was sleeping less than 3 hours per night; he became highly sociable and very charismatic; and he found he could read faster and understand materials that he had previously found difficult. Moreover, he became very productive. Indeed, many of the ideas that would later form the basis of his Ph.D. thesis came to him during this period of elation. S.B. was experiencing all the symptoms of a hypomania. Because S.B. had experienced both depression and hypomania, he now met criteria for a diagnosis of bipolar disorder type II. His hypomania persisted for several months, but things were about to take a turn for the worse.

S.B. began to sleep less than 2 hours per night, and sometimes he would simply not sleep at all for several days. He read incessantly—his living room was transformed into a labyrinthian pile of books and academic papers. S.B. also began to believe that he had some unique talents and insights. For example, he believed that he had developed a comprehensive theory that explained every aspect of how society functioned and he began to see linkages between everything that he thought about and experienced. He filled many notebooks with diagrams and writings that he thought summarized his theory quite well; however, nobody seemed to understand his theory except for him. But this didn't deter him. In short, he was displaying delusions of grandeur: He believed he had intellectual capacities that surpassed all those around him. Moreover, whenever he spoke with anyone, they always told him to slow down, or they simply gave him funny looks. At this point, S.B. was in a state of mania and now met criteria for a diagnosis of bipolar disorder type I.

However, his enthusiasm and positive affect began to slowly dwindle. Because nobody seemed to understand him and his theories, he felt increasingly alone and rejected. Soon, suicidal thoughts began to dominate his mind. After several weeks of experiencing intense suicidal ideation, barely sleeping, and often forgetting to eat, S.B. contacted his psychiatrist. He told his psychiatrist over the phone about his "brilliant" theory. Feeling suspicious, she asked S.B. to come see her at the hospital to talk about his theory further. When she saw S.B. and talked with him, she immediately realized he was in a **mixed state**: He was displaying symptoms of both severe depression (e.g., suicidal ideation) and mania (e.g., delusions of grandeur). She promptly committed S.B., and he was subsequently placed on a psychiatric ward where he stayed for 6 weeks.

So far, the case of S.B. has introduced you to the features of both depression and bipolar disorder. Here, his case makes an important point about bipolar disorder: Contrary to popular belief, bipolar disorder does not involve rapid alternations in mood (i.e., alternations occurring within hours to days); rather, the mood episodes often last weeks to months. To put things in context, there is a subtype of bipolar disorder known as *rapid cycling* bipolar disorder—defined as involving 4 or more mood episodes per year (see Cao et al., 2017; Carvalho et al., 2014).

S.B.'s case also makes an important point about psychiatric diagnoses in general: Psychiatric patients often careen from one diagnosis to another (see Phillips & Kupfer, 2013). For example, S.B.'s diagnosis changed from clinical depression to bipolar type II and then to bipolar type I. But stay tuned: There is more to the case of S.B.

Journal Prompt 18.2

Do you think that such careening between diagnoses represents an actual change in one's illness or a general problem with psychiatric diagnoses?

Mood Stabilizers

LO 18.14 Describe the discovery of the first mood stabilizer.

Ideally, **mood stabilizers** are drugs that effectively treat depression or mania without increasing the risk of mania or depression, respectively (see Karanti et al., 2016). Protection against the recurrence of mood episodes is important because mood episodes in bipolar disorder typically become both more severe and more frequent if left untreated (see Post, 2016; Post, Fleming, & Kapczinski, 2012). The mechanism by which mood stabilizers work is still a matter of debate (see Oruch et al., 2014; Rapoport, 2014), but for some reason many mood stabilizers are also effective in the treatment of epilepsy (see Prabhavalkar, Poovanpallil, & Bhatt, 2015; Yildiz et al., 2011) and schizophrenia (see Post, 2016).

Lithium, a simple metallic ion, was the first drug found to act as a mood stabilizer. The discovery of lithium's mood stabilizing action is yet another important pharmacological breakthrough that occurred largely by accident. John Cade, an Australian psychiatrist, injected guinea pigs with the urine of various psychiatric inpatients and found that the urine of manic patients was the most toxic (i.e., it killed the most guinea pigs). Next, he set about investigating which chemical constituent of the manic patients' urine caused the increased toxicity: He began by injecting guinea pigs with urea (one constituent of urine) and found that it was toxic, but not nearly so toxic as the urine of the patients with mania. Something else was contributing to the toxicity of the manic patients' urine. What was it? Cade thought it might be uric acid (another constituent of urine), so he injected both urea and lithium urate (a soluble form of uric acid that takes the form of a lithium salt) into a group of guinea pigs. Contrary to his hypothesis, he found that lithium urate protected the guinea pigs from the toxicity of the urea. Next, Cade injected some guinea pigs with both urea and lithium carbonate (another lithium salt) to check whether it was the lithium or the uric acid that was protective. He found that lithium carbonate also protected the guinea pigs from the toxicity of urea. Accordingly, Cade hypothesized that manic patients have lower levels of lithium than non-manic patients.

Cade wanted to test lithium carbonate on a group of manic patients, but he decided to first test it on a group of guinea pigs. The lithium carbonate seemed to calm the guinea pigs. However, we now know that at the doses he used, lithium carbonate produces extreme nausea; so, his subjects weren't calm—they were just sick. In any case, excited by what he thought was the success of his guinea pig experiments, in 1954 Cade gave lithium to a group of 10 manic patients, 6 patients with schizophrenia, and 3 depressed patients. The lithium had a dramatic effect, but only in the manic patients. The effect was so dramatic that some of the manic patients were even discharged from hospital (see Mitchell & Hadzi-Pavlovic, 2000), something unheard of in the pre-drug era of psychiatry.

Unfortunately, there was little immediate reaction to Cade's report—few scientists were impressed by his conference presentations, and few drug companies were interested in spending money to evaluate the therapeutic potential of a metallic ion that could not be protected by a patent. Consequently, it was not until the late 1960s that lithium was conclusively shown to be an effective mood stabilizer. Over the 50 years since its introduction, lithium is still considered by many to be the best mood stabilizer (see Oruch et al., 2014; Machado-Vieira, 2018), and it is now known to have neuroprotective effects (see Kerr, Bjedov, & Sofola-Adesakin, 2018; Sun et al., 2018) and anti-suicidal effects (see Baldessarini, Tondo, & Vásquez, 2019).

All mood stabilizers (i.e., lithium, certain anticonvulsants, and certain atypical antipsychotics) act against bouts of mania, some act against depression, and some act against both (see Prabhavalkar, Poovanpallil, & Bhatt, 2015; Yildiz et al., 2011), but they do not eliminate all symptoms. Moreover, many of them produce an array of adverse side effects (e.g., weight gain, tremor, blurred vision, dizziness; see Murru et al., 2015), which often leads patients to stop taking these medications (see Jann, 2014; Schloesser, Martinowich, & Manji, 2012).

Theories of Bipolar Disorder

LO 18.15 Describe some factors that may contribute to the onset and maintenance of bipolar disorder.

An understanding of the mechanisms underlying the development and maintenance of bipolar disorder has been hampered by the lack of a clear understanding of the mechanisms underlying the efficacy of various mood stabilizers (e.g., Can, Schulze, & Gould, 2014) and by the lack of an adequate animal model of bipolar disorder (see Logan & McClung, 2016). However, several physiological disturbances have been identified that might be contributing to the onset and maintenance of bipolar disorder. For example, there is evidence of hypothalamic–pituitary–adrenal (HPA) axis dysregulation in bipolar disorder; there are marked disruptions in the circadian rhythms in both patients with bipolar disorder and their nonbipolar relatives; and there are also alterations to GABA, glutamate, and dopamine neurotransmission in patients with bipolar disorder (see Ashok et al., 2017; Beyer & Freund, 2017; Maletic & Raison, 2014). Finally, there is evidence that BDNF (see Chapter 9) levels are lower in patients with bipolar disorder when they are either depressed or manic (see Maletic & Raison, 2014).

One theory of bipolar disorder, the *reward hypersensitivity theory* of bipolar disorder (see Alloy, Nusslock, & Boland, 2015), proposes that bipolar disorder results from a dysfunctional brain reward system that overreacts to rewards or the lack thereof. For example, individuals with bipolar disorder who are **euthymic** (they have no symptoms of depression, hypomania, or mania) make much riskier choices than individuals without bipolar disorder (see Johnson et al., 2019). The reward hypersensitivity theory of bipolar disorder explains both the manic and depressive phases of bipolar disorder by proposing that: (1) repeatedly rewarding individuals with bipolar disorder for their activities leads to excessive goal seeking and ultimately to hypomania or mania, and (2) when people with bipolar disorder fail to achieve their goals this leads to an excessive decrease in reward seeking and to depression. Consistent with this theory is the finding that individuals with bipolar disorder display increased activity in prefrontal and striatal reward circuits in both depressive and manic states (see Alloy, Nusslock, & Boland, 2015).

Many other theories of bipolar disorder exist. For example, there are emerging theories that propose a role for inflammation (see So et al., 2017), or a role for alterations in synaptic function (see Harrison, Geddes, & Tunbridge, 2017).

Genetic and Epigenetic Mechanisms of Bipolar Disorder

LO 18.16 Describe our current state of knowledge of the genetic and epigenetic mechanisms of bipolar disorder.

Many studies have been conducted on the genetics of bipolar disorder, based on the early observation that family members of individuals with bipolar disorder have a very high likelihood of developing bipolar disorder. The take-home message from all of the genetic studies of bipolar disorder is that there are hundreds to thousands of genes associated with bipolar disorder (see Gandal et al., 2018). Studies of these genes have yielded little in the way of understanding the mechanisms of bipolar disorder. Accordingly, current research efforts are focused on identifying epigenetic mechanisms that might affect transcription of the identified genes (see Duffy et al., 2019; Gandal et al., 2018).

Neural Bases of Bipolar Disorder

LO 18.17 Describe the brain differences associated with bipolar disorder.

Numerous MRI studies of the brains of patients with bipolar disorder have been published. In general, cognitive deficits are common in bipolar disorder and such deficits are associated with a variety of changes in brain function (see Lima, Peckham, & Johnson, 2019). Consistent overall reductions in gray matter volume have been reported (see Maletic & Raison, 2014). Moreover, the extent of gray matter volume reductions in frontal and limbic areas is correlated with clinical outcome (see Dusi et al., 2018). In addition, there have been reports of several specific brain structures being smaller in patients with bipolar disorder, including the medial prefrontal cortex, the left anterior cingulate, the left superior temporal gyrus, certain prefrontal regions, and the hippocampus (see Hanford et al., 2016; Knöchel et al., 2014; Otten & Meeter, 2015; Savitz, Price, & Drevets, 2014).

Meta-analyses of fMRI studies of patients with bipolar disorder have found atypical activation in the frontal cortex, medial temporal lobe structures, and basal ganglia, as well as atypical functional connectivity between some of these structures while in a variety of cognitive states (see Favre et al., 2014; Maletic & Raison, 2014). The functional connectivity disturbances seem to be centered around networks in the brain that are involved in emotional processing (see Hafeman et al., 2019; Perry et al., 2019).

Anxiety Disorders

Anxiety—chronic fear that persists in the absence of any direct threat—is a common psychological correlate of stress (see Mahan & Ressler, 2012). Anxiety is adaptive if it motivates effective coping behaviors; however, when it becomes so severe that it disrupts functioning, it is referred to as an **anxiety disorder**. All anxiety disorders are associated with feelings of anxiety (e.g., fear, worry) and with a variety of physiological stress reactions—for example, *tachycardia* (rapid heartbeat), *hypertension* (high blood pressure), nausea, breathing difficulties, sleep disturbances, and high glucocorticoid levels.

Anxiety disorders are the most prevalent of all psychiatric disorders. Estimates suggest that between 14 and 34 percent of people suffer from an anxiety disorder at some point in their lives, and the incidence seems to be almost twice as great in females as in males (see Bandelow & Michaelis, 2015; Gallo et al., 2018). M.R., a woman who was afraid to leave her home, suffered from one type of anxiety disorder.

The Case of M.R., the Woman Who Was Afraid to Go Out

M.R. was a 35-year-old woman who developed a pathological fear of leaving her house. The onset of her problem was sudden. Following an argument with her husband, she went out to mail a letter and cool off, but before she could accomplish her task, she was overwhelmed by dizziness and fear. She immediately struggled back to her house and rarely left it again, for about 2 years. Then, she gradually started to improve.

Her recovery was abruptly curtailed, however, by the death of her sister and another argument with her husband. Following the argument, she tried to go shopping, panicked, and had to be escorted home by a stranger. Following that episode, she was not able to leave her house by herself without experiencing an anxiety attack. Shortly after leaving home by herself, she would feel dizzy and sweaty, and her heart would start to pound; at that point, she would flee home to avoid a full-blown panic attack.

Although M.R. could manage to go out if she was escorted by her husband or one of her children, she felt anxious the entire time.

Four Anxiety Disorders

LO 18.18   Describe four anxiety disorders.

The following are four anxiety disorders:

- **Generalized anxiety disorder** is characterized by extreme feelings of anxiety and worry about a large number of different activities or events (see Mochcovitch et al., 2017).

- **Specific phobias** involve a strong fear or anxiety about particular objects (e.g., birds, spiders) or situations (e.g., enclosed spaces, darkness). A person with a phobia will usually try to avoid those specific objects or situations that are anxiety producing.

- **Agoraphobia** is the pathological fear of public places and open spaces. Although it might be considered as a specific phobia (see above), it is generally considered to be more incapacitating than most specific phobias and is, thus, treated as a separate diagnostic category in the DSM-5. M.R., the woman who was afraid to go out, suffered from agoraphobia.

- **Panic disorder** is characterized by recurrent rapid-onset attacks of extreme fear and severe symptoms of stress (e.g., choking, heart palpitations, shortness of breath). Such **panic attacks** also occur in certain cases of generalized anxiety disorder, specific phobia, and agoraphobia (see Cosci & Mansueto, 2019a; Johnson, Federici, & Shekhar, 2014). M.R., the woman who was afraid to go out, experienced panic attacks.

Pharmacological Treatment of Anxiety Disorders

LO 18.19   Describe three sorts of drugs used in the treatment of anxiety disorders.

Three sorts of drugs are commonly prescribed for the treatment of anxiety disorders: benzodiazepines, certain antidepressants, and pregabalin (see Quagliato, Freire, & Nardi, 2018).

BENZODIAZEPINES. Benzodiazepines such as *chlordiazepoxide* (marketed as Librium) and *diazepam* (marketed as Valium) are widely prescribed for the treatment of anxiety disorders. They are also prescribed as *hypnotics* (sleep-inducing drugs), anticonvulsants, and muscle relaxants. Indeed, benzodiazepines are the most widely prescribed psychoactive drugs; approximately 5 percent of adult North Americans are currently taking them (see Balon & Starcevic, 2020). The behavioral effects of benzodiazepines are thought to be mediated by their agonistic action on GABA$_A$ receptors.

The benzodiazepines have several adverse side effects: sedation, *ataxia* (disruption of motor activity), tremor, nausea, and a withdrawal reaction that includes rebound anxiety. Another serious problem with benzodiazepines is that they are addictive (see Quagliato, Freire, & Nardi, 2018). Consequently, they are typically prescribed for only short-term use.

ANTIDEPRESSANT DRUGS. One of the complications in studying anxiety disorders is their high comorbidity with other psychiatric disorders (e.g., Preti et al., 2018). For example, about 47 percent of individuals with a bipolar disorder and about 53 percent of individuals with major depressive disorder have a comorbid anxiety disorder (see Moscati, Flint, & Kendler, 2015; Vázquez, Baldessarini, & Tondo, 2014). Consistent with the comorbidity of anxiety disorders and clinical depression is the observation that antidepressants, such as monoamine agonists and tricyclics, are often effective against anxiety disorders (see Bandelow, 2020), and **anxiolytic drugs** (antianxiety drugs) are often effective against clinical depression.

PREGABALIN. Pregabalin is one of the newest drugs being prescribed for anxiety disorders—it is particularly effective for generalized anxiety disorder (see Generoso et al., 2017). The effects of pregabalin are believed to be due to its ability to modulate voltage-gated calcium channels, thus affecting calcium levels inside nervous system cells.

CONCLUSION. Although many drugs have been shown to produce slight, but statistically significant, improvements in groups of patients suffering from anxiety disorders, the treatment of anxiety disorders leaves a lot to be desired. Many patients are not helped at all by existing drug therapies, and many curtail therapy because of adverse side effects.

Animal Models of Anxiety Disorders

LO 18.20 Describe three animal models of anxiety disorders.

Animal models have played an important role in the study of anxiety disorders and in the assessment of the anxiolytic potential of new drugs. A weakness of these models is that they typically involve animal defensive behaviors, the implicit assumption being that defensive behaviors are motivated by fear and that fear and anxiety are similar states (see Dias et al., 2013; LeDoux, 2014). Three animal behaviors that model anxiety are elevated-plus-maze performance, defensive burying, and risk assessment.

Journal Prompt 18.3

What do you think is wrong with assuming that the defensive behaviors of nonhuman animals are representative of anxiety? What clinical implications might this assumption have for the development of new therapies for anxiety disorders?

In the **elevated-plus-maze test** (see Chapter 5), rats are placed on a four-armed plus-sign-shaped maze that rests about 50 centimeters above the floor. Two arms have sides and two arms have no sides, and the measure of anxiety is the proportion of time the rats spend in the enclosed arms, rather than venturing onto the exposed arms.

In the **defensive-burying test** (see Figure 5.25), rats are shocked by a wire-wrapped wooden dowel mounted on the wall of a familiar test chamber. The measure of anxiety is the amount of time the rats spend spraying bedding material from the floor of the chamber at the source of the shock with forward thrusting movements of their head and forepaws.

In the **risk-assessment test**, after a single brief exposure to a cat on the surface of a laboratory burrow system, rats flee to their burrows and freeze. Then, they engage in a variety of risk-assessment behaviors (e.g., scanning the surface from the mouth of the burrow or exploring the surface in a cautious stretched posture) before their behavior eventually returns to normal. The measures of anxiety in this test are the amounts of time that the rats spend in freezing and in risk assessment.

The elevated-plus-maze, defensive-burying, and risk-assessment tests of anxiety have all been validated by demonstrations that benzodiazepines reduce the various indices of anxiety used in the tests, whereas nonanxiolytic drugs usually do not. However, a potential problem with this line of evidence stems from the fact that many cases of anxiety do not respond well to benzodiazepine therapy. Therefore, existing animal models of anxiety may be models of benzodiazepine-sensitive anxiety rather than of anxiety in general, and, thus, the models may not be sensitive to anxiolytic drugs that act by a different (i.e., a nonGABAergic) mechanism.

Genetic and Epigenetic Mechanisms of Anxiety Disorders

LO 18.21 Describe our current state of knowledge of the genetic and epigenetic mechanisms of anxiety disorders.

Researchers have been unable to definitively locate genes related to anxiety disorders (see Meier & Deckert, 2019). Accordingly, research has shifted to trying to understand the environmental factors and epigenetic mechanisms that might be contributing to anxiety disorders. Many epigenetic mechanisms have been implicated in anxiety disorders, and researchers have a long way to go before they identify the critical ones (see Cosci & Mansueto, 2020b; Lin & Tsai, 2020). Part of the problem is the heterogeneity of anxiety disorders. There are so many different types of anxiety disorders and so many different presentations, that it will be more fruitful to study specific symptoms as opposed to focusing on diagnostic categories.

Neural Bases of Anxiety Disorders

LO 18.22 Describe the various brain differences associated with anxiety disorders.

Like current theories of the neural bases of schizophrenia, depression, and bipolar disorder, current theories of the neural bases of anxiety disorders rest heavily on the analysis of therapeutic drug effects. The fact that many anxiolytic drugs affect GABAergic neurotransmission (e.g., the benzodiazepines) or serotoninergic neurotransmission (e.g., fluoxetine) has focused attention on those two neurotransmitters.

There is substantial overlap between the brain structures involved in major depressive disorder and anxiety disorders. Indeed, the prefrontal cortex, hippocampus, and amygdala, which you have just learned are implicated in major depressive disorder, have also been implicated in anxiety disorders (see Calhoon & Tye, 2015). This is hardly surprising given the comorbidity of depression and anxiety disorders and the effectiveness of many drugs against both.

Although the prefrontal cortex, hippocampus, and amygdala have been implicated in both depression and anxiety disorders, the patterns of evidence differ. With major depressive disorder, you have already seen that there seems to be *atrophy* (shrinkage) of these structures; however, with anxiety disorders, there appears to be no significant atrophy. Most of the evidence linking these structures to anxiety disorders has come from functional brain-imaging studies in which atypical activity in these areas has been recorded during the performance of various emotional tasks (see Fonzo & Etkin, 2017; Maron & Nutt, 2017). Currently, much attention is focused on differences in *intrinsic functional connectivity* (see Chapter 5): Each anxiety disorder has a distinct pattern of intrinsic functional connectivity (see Northoff, 2020).

Tourette's Disorder

Tourette's disorder is the last of the psychiatric disorders discussed in this chapter. It differs from the others that have already been discussed (i.e., schizophrenia, depressive disorders, bipolar disorder, and anxiety disorders) in the specificity of its symptoms. And, as you are about to learn, they are as interesting as they are specific. The case of R.G. introduces you to Tourette's disorder.

The Case of R.G.—Barking Like a Dog

When R.G. was 15, he developed *tics* (involuntary, repetitive, stereotyped movements or vocalizations). For the first week, his tics took the form of involuntary blinking, but after that they started to involve other parts of the body, particularly his arms and legs (Spitzer et al., 1983).

R.G. and his family were religious, so it was particularly distressing when his tics became verbal. He began to curse repeatedly and involuntarily. (Involuntary cursing is a common symptom of Tourette's disorder and of several other psychiatric and neurological disorders.) R.G. also started to bark like a dog. Finally, he developed echolalia: When his mother said, "Dinner is ready," he responded, "Is ready, is ready."

Prior to the onset of R.G.'s symptoms, he was a top student, he was happy, and he had an outgoing, engaging personality. Once his symptoms developed, he was jeered at, imitated, and ridiculed by his schoolmates. He responded by becoming anxious, depressed, and withdrawn. His grades plummeted.

Once R.G. was taken to a psychiatrist by his parents, his condition was readily diagnosed—the symptoms of Tourette's disorder are unmistakable. Medication eliminated 99 percent of his symptoms, and then his anxiety and depression lifted and he returned to his former outgoing self.

Imagine how difficult it would be to get on with your life if you suffered from an extreme form of Tourette's disorder—for example, if you frequently made obscene gestures and barked like a dog. No matter how polite, intelligent, and kind you were inside, not many people would be willing to socialize with you or employ you (see Smith, Fox, & Trayner, 2015). However, if their friends, family members, and colleagues are understanding and supportive, people with Tourette's disorder can live happy, productive lives—for example, Tim Howard (shown in the photo on the next page—wearing the blue shirt) has Tourette's disorder and is the goalkeeper for a USL Championship team: the Memphis 901 FC.

What Is Tourette's Disorder?

LO 18.23 Describe the symptoms of Tourette's disorder.

Tourette's disorder is a disorder of **tics** (involuntary, repetitive, stereotyped movements or vocalizations). It typically begins early in life—usually in childhood or early

Shaun Clark/Getty Images

Pharmacological Treatment of Tourette's Disorder

LO 18.24 Describe how Tourette's disorder is treated.

Although tics are the defining feature of Tourette's disorder, treatment typically begins by focusing on other aspects of the disorder. First, the patient, family members, friends, and teachers are educated about the nature of the syndrome. Second, the treatment focuses on the ancillary emotional problems (e.g., anxiety and depression). Once these first two steps have been taken, attention turns to treating the tics.

adolescence—with simple motor tics, such as eye blinking or head movements, but the symptoms tend to become more complex and severe as the patient grows older. Common complex motor tics include hitting, touching objects, squatting, hopping, twirling, and sometimes even making lewd gestures. Common verbal tics include inarticulate sounds (e.g., barking, coughing, grunting), *coprolalia* (uttering obscenities), *echolalia* (repetition of another's words), and *palilalia* (repetition of one's own words). The symptoms usually reach a peak after a few years and gradually subside as the patient matures (see Cox, Seri, & Cavanna, 2018).

Tourette's disorder develops in 0.3–1 percent of the global population (see Serajee & Huq, 2015). It is four times more frequent in male children than in female children (see Hallett, 2015), but this sex difference is not as profound in adult patients (see Jackson et al., 2015).

Some patients with Tourette's disorder also display signs of *attention-deficit/hyperactivity disorder, obsessive-compulsive disorder*, or both (see Serajee & Huq, 2015). For example, R.G. was obsessed by odd numbers and refused to sit in even-numbered seats.

Although the tics of Tourette's disorder are involuntary, they can be temporarily suppressed with concentration and effort by the patient. The effect of suppression has been widely misunderstood. Many medical professionals believe that tic suppression is inevitably followed by a *rebound* (that the tics become even worse following a period of suppression; e.g., Novotny, Valis, & Klimova, 2018). However, this is not the case—see Figure 18.6.

> **Journal Prompt 18.4**
> Tourette's disorder is a disorder of onlookers. Explain.

The tics of Tourette's disorder are usually treated with *antipsychotics* (see Kim et al., 2018; Pandey & Dash, 2019; Quezada & Coffman, 2018), although behavioral interventions can also be effective (see Pringsheim et al., 2019). Antipsychotics can reduce tics by about 70 percent, but patients or their caregivers often refuse them because of the adverse side effects (e.g., weight gain, fatigue, dry mouth). The success of antipsychotics in blocking Tourette's tics is consistent with the hypothesis that the disorder is related

Figure 18.6 No rebound effect above baseline was observed following periods of tic suppression by children with Tourette's disorder.

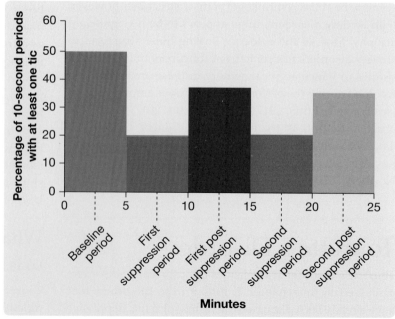

Based on Himle, M.B., & Woods, D.W., 2005. An experimental evaluation of tic suppression and the tic rebound effect. *Behaviour Research and Therapy, 43,* 1443–1451.

to changes in the cortical-striatal-thalamic-cortical circuit because that circuit relies heavily on dopaminergic signaling (see Jackson et al., 2015).

Genetic and Epigenetic Mechanisms of Tourette's Disorder

LO 18.25 Describe our current state of knowledge of the genetic and epigenetic mechanisms of Tourette's disorder.

The chance of monozygotic twins both having Tourette's disorder is about 80 percent, whereas it is 20 percent for dizogotic twins (see Burton et al., 2020). Thus, it would appear that Tourette's disorder is highly heritable. However, to date there have been no genes definitively linked to the disorder.

Studies of the epigenetic mechanisms of Tourette's disorder have been more fruitful. Several studies have found specific genes that are methylated (DNA methylation is one epigenetic mechanism; see Chapter 2) in individuals with Tourette's disorder, such as the methylation of genes for potassium channels. Interestingly, one gene has been identified whose degree of methylation is correlated with tic severity in individuals with Tourette's disorder (see Burton et al., 2020).

Neural Bases of Tourette's Disorder

LO 18.26 Describe the research findings related to the neural bases of Tourette's disorder.

Because Tourette's disorder is a well-defined disorder with clearly observable symptoms, its neural bases are more amenable to study than those of the other disorders that you have already encountered in this chapter. However, there are impediments to its study (e.g., the lack of a strong link to any particular gene is problematic). The greatest difficulty in studying Tourette's disorder is the fact that the symptoms usually subside as people age; because Tourette's patients are rarely under care for the disorder when they die, few postmortem studies of Tourette's disorder have been conducted. Consequently, the study of the disorder's neural bases is based almost exclusively on brain-imaging studies, which are difficult to conduct because of the requirement that the patients remain motionless.

Most research on the cerebral pathology associated with Tourette's disorder has focused on the striatum (caudate plus putamen). Patients with this disorder tend to have smaller striatal volumes (see Jackson et al., 2015), and when they suppress their tics, fMRI activity is recorded in both the prefrontal cortex and caudate nuclei (see Thomas & Cavanna, 2013). Presumably, the decision to suppress the tics comes from the prefrontal cortex, which initiates the suppression by acting on the caudate nuclei. Accordingly, Tourette's is sometimes viewed as the result of a dysfunctional caudate that is unable to suppress unwanted movements, like tics (see Gagné, 2019).

There is also evidence of dysfunctional dopaminergic and GABAergic signaling within the cortical-striatal-thalamic-cortical brain circuits in Tourette's disorder. This finding has been of particular interest because those brain circuits are implicated in motor learning—including habit formation (see Jackson et al., 2015).

Although most studies of the neural bases of Tourette's disorder have focused on the striatum, the brain differences appear to be more widespread. For example, an MRI study of children with Tourette's disorder (see Jackson et al., 2015) documented thinning in sensorimotor cortex gray matter that was particularly prominent in the areas that controlled the face, mouth, and *larynx* (voice box).

P.H. is a scientist who counsels Tourette's patients and their families. He also has Tourette's disorder, which provides him with a useful perspective (Hollenbeck, 2001).

The Case of P.H., the Neuroscientist with Tourette's Disorder

Tourette's disorder has been P.H.'s problem for more than three decades. Taking advantage of his position as a medical school faculty member, he regularly offers a series of lectures on the topic. Along with students, many other Tourette's patients and their families are attracted to his lectures.

Encounters with Tourette's patients of his own generation taught P.H. a real lesson. He was astounded to learn that most of them did not have his thick skin. About half of them were still receiving treatment for psychological wounds inflicted during childhood.

For the most part, these patients' deep-rooted pain and anxiety did not result from the tics themselves. They derived from being ridiculed and tormented by others and from the self-righteous advice repeatedly offered by well-meaning "clods." The malfunction may be in a patient's striatum, but in reality this is more a disorder of the onlooker than of the patient.

We received an e-mail from a professor of biological sciences at Purdue University. He came across this text because it was used in his department's behavioral neurobiology course. He thanked us for our coverage of Tourette's disorder but said that he found the case study "a bit eerie." The message began with "From one case study to another," and it ended "All the best, P.H."

Clinical Trials: Development of New Psychotherapeutic Drugs

Almost daily, there are news reports of exciting discoveries that appear to be pointing to effective new therapeutic drugs or treatments for psychiatric disorders. But most often, the promise does not materialize. For example, almost 50 years after the revolution in molecular biology began, not a single form of gene therapy is yet in widespread use for psychiatric disorders. The reason is that the journey of a drug or other medical treatment from promising basic research to useful reality is excruciatingly complex, time-consuming, and expensive. Research designed to translate basic scientific discoveries into effective clinical treatments is called **translational research**.

So far, the chapter has focused on early drug discoveries and their role in the development of theories of psychiatric disorders. In the early years, the development of psychotherapeutic drugs was largely a hit-or-miss process. New drugs were tested on patient populations with little justification and then quickly marketed to an unsuspecting public, often before it was discovered that they were dangerous or ineffective for their original purpose.

Things have changed. The testing of experimental drugs on human volunteers and their subsequent release for sale are now strictly regulated by government agencies. The process of gaining permission from the government to market a new psychotherapeutic drug begins with the synthesis of the drug, the development of economically efficient procedures for synthesizing the drug, and the collection of evidence from nonhuman subjects showing that the drug is likely safe for human consumption and has potential therapeutic benefits. These initial steps usually take a long time—at least 5 years—and only if the evidence is sufficiently promising is permission granted to proceed to clinical trials. **Clinical trials** are studies conducted on volunteers with the disorder to assess the therapeutic efficacy of an untested drug or other treatment. This final module of the chapter focuses on the process of conducting clinical trials—summarized in Table 18.1.

Clinical Trials: The Three Phases

LO 18.27 Describe the three phases of clinical trials.

Once approval has been obtained from the appropriate government agencies, clinical trials of a new drug with therapeutic potential can commence. Clinical trials are conducted in three separate phases: (1) screening for safety, (2) establishing the testing protocol, and (3) final testing (see Zivin, 2000). If any one of the three phases is unsuccessful, then study of the drug is usually curtailed.

Table 18.1 Phases of Drug Development

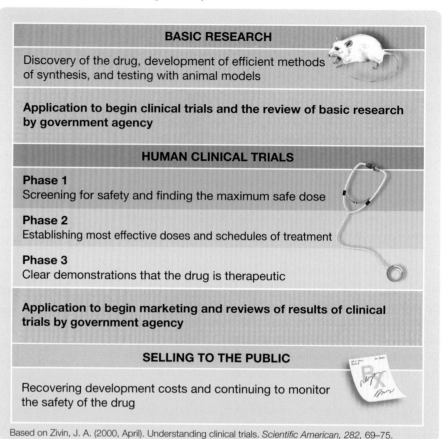

Based on Zivin, J. A. (2000, April). Understanding clinical trials. *Scientific American, 282,* 69–75.

PHASE 1: SCREENING FOR SAFETY. The purpose of the first phase of a clinical trial is to determine whether the drug is safe for human use and, if it is, to determine how much of the drug can be tolerated. Administering the drug to humans for the first time is always a risky process because there is no way of knowing for certain how they will respond. The subjects in phase 1 are typically healthy paid volunteers. Phase 1 clinical trials always begin with tiny injections, which are gradually increased as the tests proceed. The reactions of the volunteers are meticulously monitored, and if strong adverse reactions are observed, testing is curtailed.

PHASE 2: ESTABLISHING THE TESTING PROTOCOL. The purpose of the second phase of a clinical trial is to establish the *protocol* (the conditions) under which the final tests are likely to provide a clear result. For example, in phase 2, researchers hope to discover which doses are likely to be therapeutically effective, how frequently they should be administered, how long they need to be administered to have a therapeutic effect, what benefits are likely to occur, and which patients are likely to be helped. Phase 2 tests are conducted on volunteer patients suffering from the target disorder; the tests usually include *placebo-control groups* (groups of patients who receive a control substance rather than the drug), and their designs are usually *double-blind*—that is, the tests are conducted so that neither the patients nor the physicians interacting with them know which treatment (drug or placebo) each patient has received.

PHASE 3: FINAL TESTING.   Phase 3 of a clinical trial is typically a double-blind, placebo-control study on large numbers—often, many thousands—of patients suffering from the target disorder. The design of the phase 3 tests is based on the results of phase 2 so that the final tests are likely to demonstrate positive therapeutic effects if they exist. The first test of the final phase is often not conclusive, but if it is promising, a second test based on a redesigned protocol may be conducted. In most cases, two independent successful tests are required to convince government regulatory agencies. A test is typically deemed successful if the beneficial effects outweigh any adverse side effects.

Controversial Aspects of Clinical Trials

LO 18.28   Identify six controversial aspects of clinical trials.

The clinical trial process is not without controversy. The following are six points that have been focuses of criticism and debate (see Goldacre, 2013; London, Kimmelman, & Carlisle, 2012).

REQUIREMENT FOR DOUBLE-BLIND DESIGN AND PLACEBO CONTROLS.   In most clinical trials, patients are assigned to drug or placebo groups randomly and do not know for sure which treatment they are receiving. Thus, some patients whose only hope for recovery may be the latest experimental treatment will, without knowing it, receive the placebo. Drug companies and government agencies concede that this is true, but they argue that there can be no convincing evidence that the experimental treatment is effective until a double-blind, placebo-control trial has been completed. Because psychiatric disorders often improve after a placebo, a double-blind, placebo-control procedure is essential in the evaluation of any psychotherapeutic drug.

THE NEED FOR ACTIVE PLACEBOS.   Conventional wisdom has been that the double-blind, placebo-control procedure is the perfect control procedure to establish the effectiveness of new drugs, but it isn't (see Benedetti, 2014). Here is a new way to think about the double-blind placebo-control procedure. At therapeutic doses, many drugs have side effects that are obvious to people taking them, and, thus, the participants in double-blind, placebo-control studies who receive the drug can often accurately guess that they are not in the placebo group (see Geuter, Koban, & Wager, 2017). This knowledge may greatly contribute to the positive effects of the drug, independent of any real therapeutic effect. Accordingly, it is now widely recognized that an active placebo is better than an inert placebo as the control drug. **Active placebos** are control drugs that have no therapeutic effect but produce side effects similar to those produced by the drug under evaluation (see Kirsch, 2019; Shader, 2017).

LENGTH OF TIME REQUIRED.   Patients desperately seeking new treatments are frustrated by the amount of time needed for clinical trials. Therefore, researchers, drug companies, and government agencies are striving to speed up the evaluation process without sacrificing the quality of the procedures designed to protect patients from ineffective or dangerous treatments. It is imperative to strike the right compromise.

FINANCIAL ISSUES.   The drug companies pay the scientists, physicians, technicians, assistants, and patients involved in drug trials. Considering the billions these companies spend per candidate drug (see Fleming, 2018) and the fact that only about 22 percent of the candidate drugs entering phase 1 testing ever gain final approval (see Figure 18.7), it should come as no surprise that the companies are anxious to recoup their costs and that fewer are investing in novel drugs (see Grabb & Gobburu, 2017). In view of this pressure, many have questioned the impartiality of those conducting and reporting the trials (see Normile, 2014; Roest et al., 2015). The scientists themselves have often complained that the sponsoring drug company makes them sign an agreement that prohibits them from publishing or discussing negative findings without the company's consent. This is a serious ongoing problem: Any new drug will look promising if all negative evidence is suppressed (see Goldacre, 2013).

Figure 18.7 The probabilities that a drug that qualifies for testing in humans will reach each phase of testing and ultimately gain approval. Only 22 percent of the drugs that initially qualify for testing eventually gain approval.

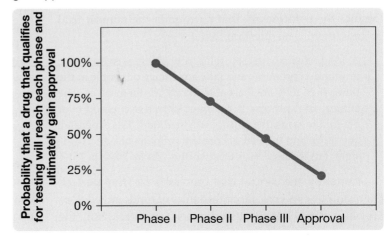

Another financial issue is profitability—drug companies seldom develop drugs to treat rare disorders because such treatments will not be profitable. Drugs for which the market is too small for them to be profitable are called **orphan drugs**. Some governments have passed laws intended to promote the development of orphan drugs. Also, the massive costs of clinical trials have contributed to a **translational bottleneck**—only a small proportion of potentially valuable ideas or treatments receive funding for translational research.

TARGETS OF PSYCHOPHARMACOLOGY. Hyman and Fenton (2003) have argued that a major impediment to the development of effective psychotherapeutic drugs is that the effort is often aimed at curing disease entities as currently conceived—for example, as defined by the DSM-5. The current characterizations of various psychiatric disorders are the best they can be given the existing evidence; however, it is clear that most psychiatric disorders, as currently conceived, are likely clusters of disorders, each with a different pattern of associated brain dysfunction. Thus, effective new drugs are likely to benefit only a proportion of those patients with a particular diagnosis, and thus their effectiveness might go unrecognized.

LACK OF DIVERSITY. Two related issues have come to light over the past decade. First is that there is a relative lack of diversity in the volunteers used in clinical trials (see Clark et al., 2019; Polo et al., 2019). If diverse groups (e.g., different ethnicities, sexes) aren't included in a clinical trial, then we cannot be sure that the drug will also work for individuals of different sexes or ethnicities. Second is that, even when there is diversity in a clinical trial, data are typically grouped and analyzed together—obscuring group differences and leading to potentially risky generalizations about the effectiveness of a drug across groups of individuals (see Eid, Gobinath, & Galea, 2019).

Effectiveness of Clinical Trials

LO 18.29 Discuss the relative effectiveness of clinical trials.

Despite the controversy that surrounds the clinical trial process, there is no question that it works.

> A long, dismal history tells of charlatans who make unfounded promises and take advantage of people at the time when they are least able to care for themselves. The clinical trial process is the most objective method ever devised to assess the efficacy of a treatment. It is expensive and slow, and in need of constant refinements, and oversight, but the process is trustworthy. (Zivin, 2000, p. 75)

Certainly, the clinical trial process is far from perfect. For example, concerns about the ethics of randomized double-blind, placebo-control studies are warranted. Still, the vast majority of those in the medical and research professions accept that these studies are the essential critical test of any new therapy. This is particularly true of psychotherapeutic drugs because psychiatric disorders often respond to placebo treatments (see Rutherford & Roose, 2013; Wager & Atlas, 2015) and because assessment of their severity can be greatly influenced by the expectations of the therapist.

Everybody agrees that clinical trials are too expensive and take too long. But Zivin (2000) responds to this concern in the following way: Clinical trials can be trustworthy, fast, or cheap; but in any one trial, only two of the three are possible. Think about it.

> **Journal Prompt 18.5**
> What do you think Zivin means when he suggests that clinical trials can be trustworthy, fast, or cheap; but that in any one trial, only two of the three are possible?

It is important to realize that every clinical trial is carefully monitored as it is being conducted. Any time the results warrant it, changes to the research protocol are made to reduce costs, improve safety, and reduce the time to get the drug to patients. We think this system would be greatly improved if there was a legal requirement for all partial or completed clinical trials to be published so that patients, doctors, and scientists would have access to all of the evidence. What do you think?

CONCLUSION. Ideally, patients in the same diagnostic category should display the same symptoms associated with the same underlying pathology caused by the same genetic and environmental factors. But more importantly, they should all respond to the same treatments. When it comes to disorders of the brain, our most complex organ, this ideal is rarely met. However, when it comes to the diagnosis of psychiatric disorders, we have not even been close.

As you progressed through this chapter, we hope you recognized the signs that the diagnosis of psychiatric disorders needs to be improved (with the possible exception of Tourette's disorder)—see Demkow & Wolańczyk (2017); Nusslock & Alloy (2017). The symptoms of schizophrenia, depressive disorders, bipolar disorder, and anxiety disorders are so variable that two individuals with the same diagnosis may share few of the same symptoms. Moreover, a wide range of genetic and environmental factors have been implicated in each diagnosis, and patients with the same diagnosis display substantial variability in any associated neural changes. Most problematic is that, despite the fact that many drugs acting by several mechanisms are used in the treatment of each disorder, many patients do not substantially improve. It does seem that each diagnosis contains several different disorders and that research and treatment could benefit from distinguishing among them.

The chapter, and indeed the book, ends with the case of S.B., who, if you recall from the module on bipolar disorder, suffers from bipolar disorder type I. So far, the case of S.B. has introduced you to the features of both depressive disorders and bipolar disorder. The conclusion of his case study demonstrates the value of a biopsychological education that stresses independent thinking and the importance of taking responsibility for one's own health. You see, S.B. took a course similar to the one you are currently taking, and the things that he learned in the course enabled him to steer his own treatment to a positive outcome.

Conclusion of the Case of S.B.: The Biopsychology Student Who Took Control

Heavily sedated on benzodiazepines, S.B. slept for much of the first week after he was committed to a psychiatric ward by his psychiatrist. When he came out of his stupor, the ward's resident psychiatrist informed S.B. that he would be placed on a particular mood stabilizer and that he would likely have to take it for the rest of his life. Two things made S.B. feel uncomfortable about this. First, many patients in the ward were taking this drug, and they looked like zombies; second, the resident psychiatrist seemed to know less about this drug and its mechanisms than S.B. did. So S.B. requested that he be given access to the hospital library so he could learn more about his disorder and the drugs used to treat it.

S.B. was amazed by what he found. The drug favored by the resident psychiatrist had been shown several months before to be no more effective in the long-term treatment of bipolar disorder than placebo. Moreover, a new drug that had recently cleared clinical trials was proving to be effective with fewer side effects. When S.B. confronted the resident psychiatrist with this evidence, he was surprised and agreed to prescribe the new drug.

Today, S.B. is feeling well and has finished graduate school. In fact, at the time of writing of this text, S.B. holds a faculty position at a leading university in Canada. In fact, S.B. is writing these very words. If it has not yet occurred to you, I (Steven Barnes) am S.B.

I still find it difficult to believe that I had enough nerve to question a psychiatrist and prescribe for myself. I never imagined that the lessons learned from biopsychology would have such a positive impact on my life. Although I still experience mood episodes associated with my bipolar disorder, they aren't nearly as severe as they were before I started taking a mood stabilizer, and I have learned to manage the residual symptoms using both pharmacological and psychosocial methods (see Geddes & Miklowitz, 2013; Murray et al., 2011). I am glad that I could tell you my story and hope that you or someone you care for will benefit from it.

Themes Revisited

This entire chapter focused on psychiatric disorders, so it should come as no surprise that the clinical implications theme was predominant. Nevertheless, the other three major themes of this text also received coverage. Moreover, both emerging themes received coverage.

The thinking creatively theme arose during the discussions of the following ideas: animal models of anxiety may be models of benzodiazepine effects, active placebos are needed to establish the clinical efficacy of psychotherapeutic drugs, scientists should try to focus on treatments for specific measurable symptoms rather than general diseases as currently conceived, and there needs to be a way to force drug companies to publish negative findings.

The evolutionary perspective theme came up twice: in the discussions of animal models of anxiety and of the important role played by research on nonhuman subjects in gaining official clearance to commence human clinical trials.

The neuroplasticity theme was discussed substantively only once, during the explanation of the neuroplasticity theory of depression.

Both emerging themes were prevalent in this final chapter of the text; one was explicit, the other implicit. The thinking about epigenetics theme was explicit throughout this chapter: It was present in each of the many discussions of the epigenetic factors that contribute to psychiatric disorders. The consciousness theme was implicit in this chapter: Many psychiatric disorders are associated with alterations of consciousness.

Key Terms

Psychiatric disorders, p. 486
DSM-5, p. 486

Schizophrenia
Positive symptoms, p. 486
Negative symptoms, p. 486

Antipsychotic drug, p. 487
Chlorpromazine, p. 487
Psychosis, p. 487

Epilogue

It is the summer of 2020, and it is a very different world from when we started writing this edition of *Biopsychology* almost two years ago. We are now living in the midst of the vast social and political changes brought on by the COVID-19 pandemic. We are also seeing great efforts to battle racism, and to address accessibility and equity issues faced by racialized persons and other marginalized groups in our society. The fact that such efforts are so prominent during a pandemic speaks to how pressing the need is to deal with these social injustices.

We feel relieved to be finishing *Biopsychology*, and we are excited by the prospect of being able to speak to so many students like you through this new edition. You must also feel relieved to be finishing this course; still, we hope that you feel a tiny bit of regret that our time together is over. Like good friends, we have shared good times and bad. We have shared the fun and wonder of Rhonelle, the dexterous cashier; the Nads basketball team; people who rarely sleep; the "mamawawa"; and split brains. But we have also been touched by many personal tragedies: for example, the victims of Alzheimer's disease and MPTP poisoning; Jimmie G.; H.M.; the man who mistook his wife for a hat; Professor P., the biopsychologist who experienced brain surgery from the other side of the knife; and S.B., the biopsychology student who guided the treatment of his own disease. Thank you for allowing us to share biopsychology with you. We hope you have found it to be an enriching experience.

As the founder and primary author of *Biopsychology* for eleven editions now, I (JP) am excited by the prospect that Steven Barnes will be carrying this book forward into future editions. Steven is a gifted writer, educator, researcher, and artist. His approach is eclectic. Although I will miss writing this book, I find comfort in the thought that I leave you in Steven's capable hands.

Right now, John Pinel is sitting in his home looking out over his garden and the Pacific Ocean, and Steven Barnes is sitting on the shores of a Rocky Mountain lake with his daughter. Our writing of this edition is complete. John's garden is calling for his attention, and the lake waters are calling for Steven to dive in. It is the afternoon of Friday, July 24, 2020.

Appendix I

The Autonomic Nervous System (ANS)

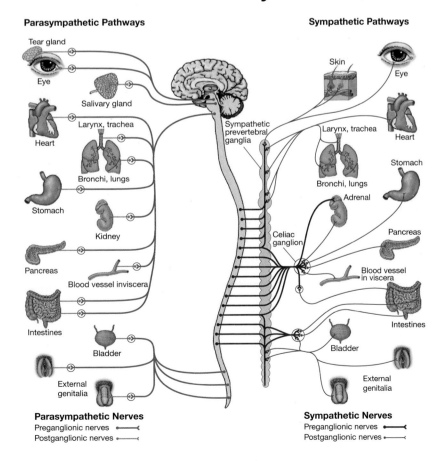

Parasympathetic Pathways

Tear gland
Eye
Salivary gland
Larynx, trachea
Heart
Bronchi, lungs
Stomach
Kidney
Pancreas
Blood vessel inviscera
Intestines
Bladder
External genitalia

Sympathetic Pathways

Skin
Eye
Sympathetic prevertebral ganglia
Larynx, trachea
Heart
Stomach
Bronchi, lungs
Adrenal
Celiac ganglion
Pancreas
Blood vessel in viscera
Intestines
Bladder
External genitalia

Parasympathetic Nerves
Preganglionic nerves
Postganglionic nerves

Sympathetic Nerves
Preganglionic nerves
Postganglionic nerves

Appendix II

Some Functions of Sympathetic and Parasympathetic Neurons

Organ	Sympathetic Effect	Parasympathetic Effect
Salivary gland	Decreases secretion	Increases secretion
Heart	Increases heart rate	Decreases heart rate
Blood vessels	Constricts blood vessels in most organs	Dilates blood vessels in a few organs
Penis	Ejaculation	Erection
Iris radial muscles	Dilates pupils	No effect
Iris sphincter muscles	No effect	Constricts pupils
Tear gland	No effect	Stimulates secretion
Sweat gland	Stimulates secretion	No effect
Stomach and intestine	No effect	Stimulates secretion
Lungs	Dilates bronchioles; inhibits mucous secretion	Constricts bronchioles; stimulates mucous secretion
Arrector pili muscles	Erects hair and creates gooseflesh	No effect

Appendix III

The Cranial Nerves

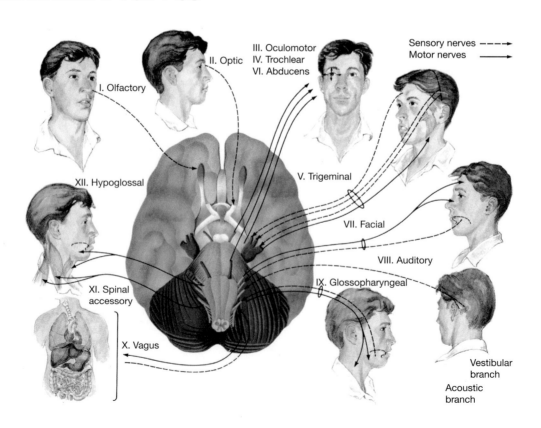

Appendix IV

Functions of the Cranial Nerves

Number	Name	General Function	Specific Functions
I	Olfactory	Sensory	Smell
II	Optic	Sensory	Vision
III	Oculomotor	Motor	Eye movement and pupillary constriction
		Sensory	Sensory signals from certain eye muscles
IV	Trochlear	Motor	Eye movement
		Sensory	Sensory signals from certain eye muscles
V	Trigeminal	Sensory	Facial sensations
		Motor	Chewing
VI	Abducens	Motor	Eye movement
		Sensory	Sensory signals from certain eye muscles
VII	Facial	Sensory	Taste from anterior two-thirds of tongue
		Motor	Facial expression, secretion of tears, salivation, cranial blood vessel dilation
VIII	Auditory-Vestibular	Sensory	Audition; sensory signals from the organs of balance in the inner ear
IX	Glossopharyngeal	Sensory	Taste from posterior third of tongue
		Motor	Salivation, swallowing
X	Vagus	Sensory	Sensations from abdominal and thoracic organs
		Motor	Control over abdominal and thoracic organs and muscles of the throat
XI	Spinal Accessory	Motor	Movement of neck, shoulders, and head
		Sensory	Sensory signals from muscles of the neck
XII	Hypoglossal	Motor	Tongue movements
		Sensory	Sensory signals from tongue muscles

NOTE: Some authors describe cranial nerves III, IV, VI, XI, and XII as purely motor. However, each of these cranial nerves contains a small proportion of sensory fibers that conduct information from receptors to the brain. This sensory information is necessary for directing the respective cranial nerve's motor responses. See the discussion of sensory feedback in Chapter 8.

Appendix V

Nuclei of the Thalamus

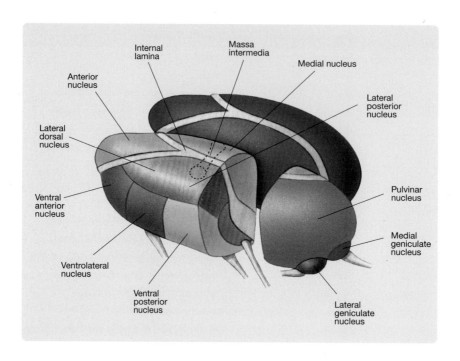

Appendix VI

Nuclei of the Hypothalamus

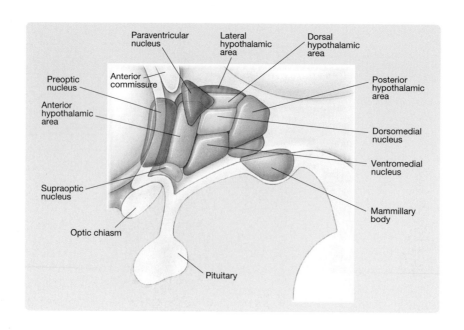

Glossary

3-per-second spike-and-wave discharge. The characteristic EEG pattern of the absence seizure.

5-hydroxytryptophan (5-HTP). The precursor of serotonin.

Ablatio penis. Accidental destruction of the penis via surgery.

Absence seizure. A type of generalized seizure whose primary behavioral symptom is a disruption of consciousness associated with a cessation of ongoing behavior, a vacant look, and sometimes fluttering eyelids.

Absolute refractory period. A brief period (typically 1 to 2 milliseconds) after the initiation of an action potential during which it is impossible to elicit another action potential in the same neuron.

Absorption spectrum. A graph of the ability of a substance to absorb light of different wavelengths.

Absorptive phase. The metabolic phase during which the body is operating on the energy from a recently consumed meal and is storing the excess as body fat, glycogen, and proteins.

Accommodation. The process of adjusting the configuration of the lenses to bring images into focus on the retina.

Acetylcholine. A neurotransmitter that is created by the addition of an acetyl group to a choline molecule.

Acetylcholinesterase. The enzyme that breaks down the neurotransmitter acetylcholine.

Acquired dyslexias. Dyslexias caused by brain damage in people previously capable of reading.

Action potential (AP). A massive momentary reversal of a neuron's membrane potential from about −70 mV to about +50 mV.

Activation-synthesis hypothesis. The theory that the information supplied to the cortex by the brain stem during REM sleep is largely random and that the resulting dream is the cortex's best effort to make sense of those random signals.

Activators. Proteins that bind to DNA and increase gene expression.

Active placebos. Control drugs that have no therapeutic effect but produce side effects similar to those produced by the drug under evaluation in a clinical trial.

Acuity. The ability to see the details of objects.

Adaptation theories of sleep. Theories of sleep based on the premise that sleep evolved to protect organisms from predation and accidents and to conserve their energy rather than to fulfill some particular physiological need.

Adaptive immune system. The division of the immune system that mounts targeted attacks on foreign pathogens by binding to antigens in their cell membranes.

Adipsia. Complete cessation of drinking.

Adjustable gastric band procedure. A surgical procedure for treating obesity in which an adjustable band is implanted around the stomach to reduce the flow of food.

Adrenal cortex. The outer layer of each adrenal gland, which releases glucocorticoids in response to stressors, as well as small amounts of steroid hormones.

Adrenal medulla. The core of each adrenal gland, which releases epinephrine and norepinephrine in response to stressors.

Adrenalectomy. Surgical removal of the adrenal glands.

Adrenocorticotropic hormone (ACTH). An anterior pituitary hormone that triggers the release of adrenal hormones from the adrenal cortices.

Adrenogenital syndrome. Caused by congenital adrenal hyperplasia, which results in the excessive release of adrenal androgens which have masculinizing effects in females.

Afferent nerves. Nerves that carry sensory signals to the central nervous system.

Ageusia. The inability to taste.

Aggregation. The alignment of neurons during the development of the nervous system.

Aggressive behaviors. Behaviors whose primary function is to threaten or harm other organisms.

Agnosia. A failure of recognition of sensory stimuli that is not attributable to a sensory or to verbal or intellectual impairment.

Agonists. Drugs that facilitate the effects of a particular neurotransmitter.

Agoraphobia. Pathological fear of public places and open spaces.

Agraphia. A specific inability to write; one that does not result from general visual, motor, or intellectual deficits.

Akinetopsia. A deficiency in the ability to see movement progress in a smooth fashion, which often results from damage to the MT area.

Alexia. A specific inability to read; one that does not result from general visual, motor, or intellectual deficits.

Alleles. The two genes that control the same trait.

All-or-none responses. Responses that are not graded; they either occur to their full extent or do not occur at all.

Alpha fetoprotein. A protein that is present in the blood of many mammals during the perinatal period and that deactivates circulating estradiol by binding to it.

Alpha male. The dominant male of a colony.

Alpha waves. Regular, 8- to 12-per-second, high-amplitude EEG waves that typically occur during relaxed wakefulness and just before falling asleep.

Alzheimer's disease. The most common form of dementia in the elderly. Its three defining characteristics are: neurofibrillary tangles, amyloid plaques, and neuron loss.

Amacrine cells. Retinal neurons that are specialized for lateral communication.

Amino acid derivative hormones. Hormones that are synthesized in a few simple steps from an amino acid molecule.

Amino acid neurotransmitters. A class of small-molecule neurotransmitters, which includes the amino acids glutamate, aspartate, glycine, and GABA.

Amino acids. The building blocks of proteins.

Amnesia. Any pathological loss of memory.

Amphetamine. A stimulant drug.

Amphibians. Species that must live in water during their larval phase; adult amphibians can survive on land.

Amygdala. A structure in the anterior temporal lobe, just anterior to the hippocampus; plays a role in emotion.

Anabolic steroids. Steroid drugs that are similar to testosterone and have powerful anabolic (growth-promoting) effects.

Analgesics. Drugs that reduce pain.

Analogous. Having a similar structure because of convergent evolution (e.g., a bird's wing and a bee's wing are analogous).

Anandamide. The first endogenous endocannabinoid to be discovered and characterized.

Androgen insensitivity syndrome. Results from a mutation to the androgen receptor gene that renders the androgen receptors unresponsive and leads to the development of a female body.

Androgens. The class of steroid hormones that includes testosterone.

Androstenedione. The adrenal androgen that is responsible for the growth of pubic hair and axillary hair in human females.

Aneurysm. A pathological balloonlike dilation that forms in the wall of an artery at a point where the elasticity of the artery wall is defective.

Angular gyrus. The gyrus of the posterior cortex at the boundary between the temporal and parietal lobes. According to the Wernicke-Geschwind model the left hemisphere angular gyrus plays a role in reading.

Anhedonia. A general inability to experience pleasure.

Anorexia nervosa. An eating disorder of underconsumption that results in health-threatening weight loss.

Anosmia. The inability to smell.

Anosognosia. The common failure of neuropsychological patients to recognize their own symptoms.

Antagonistic muscles. Pairs of muscles that act in opposition.

Antagonists. Drugs that inhibit the effects of a particular neurotransmitter.

Anterior. Toward the nose end of a vertebrate.

Anterior cingulate cortex. The cortex of the anterior cingulate gyrus.

Anterior pituitary. The part of the pituitary gland that releases tropic hormones.

Anterograde amnesia. Loss of memory for events occurring after the amnesia-inducing brain injury.

Anterograde degeneration. The degeneration of the distal segment of a cut axon.

Anterolateral system. A major somatosensory pathway that ascends in the anterolateral portion of the spinal cord and tends to carry information related to pain and temperature.

Antibodies. Proteins that bind to foreign antigens on the surface of microorganisms and in so doing promote the destruction of the microorganisms.

Antibody-mediated immunity. The immune reaction in which B cells destroy invading microorganisms via the production of antibodies.

Antidromic conduction. Axonal conduction opposite to the normal direction; conduction from axon terminals back toward the cell body.

Antigens. Molecules, usually proteins, that can trigger an immune response.

Antihypnotic drugs. Sleep-reducing drugs.

Anxiety. Chronic fear that persists in the absence of any direct threat.

Anxiety disorder. A psychiatric disorder that involves anxiety that is so extreme and so pervasive that it disrupts normal functioning.

Anxiolytic drugs. Drugs that have antianxiety effects.

Anxiolytics. Antianxiety drugs.

Aphagia. Complete cessation of eating.

Aphasia. A brain damage–produced deficit in the ability to produce or comprehend language.

Apoptosis. Cell death that is actively induced by genetic programs; programmed cell death.

Appetizer effect. The increase in hunger that is produced by the consumption of small amounts of food.

Applied research. Research that is intended to bring about some direct benefit to humankind.

Apraxia. A disorder in which patients have great difficulty performing movements when asked to do so out of context but can readily perform them spontaneously in natural situations.

Arachnoid membrane. The meninx that is located between the dura mater and the pia mater and has the appearance of a gauzelike spiderweb.

Arcuate fasciculus. The major neural pathway between Broca's area and Wernicke's area.

Arcuate nucleus. A nucleus of the hypothalamus that contains high concentrations of both leptin receptors and insulin receptors.

Area MT. An area of cortex, located near the junction of the temporal, parietal, and occipital lobes, whose function appears to be the perception of motion.

Aromatase. An enzyme that promotes the conversion of testosterone to estradiol.

Aromatization. The chemical process by which testosterone is converted to estradiol.

Aromatization hypothesis. The hypothesis that the brain is masculinized by estradiol that is produced from perinatal testosterone through a process called *aromatization*.

Arteriosclerosis. A condition in which blood vessels are narrowed or blocked by the accumulation of fat deposits on their walls.

Asomatognosia. A deficiency in the awareness of parts of one's own body that is typically produced by damage to the right parietal lobe.

Aspartate. An amino acid neurotransmitter.

Aspiration. A lesion technique in which tissue is drawn off by suction through the fine tip of a glass pipette.

Association cortex. An area of cortex that receives input from more than one sensory system.

Astereognosia. An inability to recognize objects by touch that is not attributable to a simple sensory deficit or to an intellectual impairment.

Astrocytes. Large, star-shaped glial cells that play multiple roles in the central nervous system.

Ataxia. A loss of motor coordination.

Atropine. A receptor blocker that exerts its antagonistic effect by binding to muscarinic receptors.

Attentional gaze. The shift in attention from one perceptual object to another.

Atypical antidepressants. A catch-all class for antidepressant drugs that do not fit into the other categories of antidepressants (e.g., monoamine oxidase inhibitors, tricyclic antidepressants). Each of the drugs in this class has its own unique mechanism of action.

Atypical antipsychotics. Drugs that are effective against schizophrenia but yet do not bind strongly to D_2 receptors. Also known as *second-generation antipsychotics*.

Auditory nerve. The branch of cranial nerve VIII that carries auditory signals from the hair cells in the basilar membrane.

Autism spectrum disorder (ASD). A complex neurodevelopmental disorder characterized by a reduced capacity for social interaction and communication and restricted and repetitive patterns of behavior, interests, or activities.

Autonomic nervous system (ANS). The part of the peripheral nervous system that participates in the regulation of the body's internal environment.

Autoradiography. The technique of photographically developing brain slices that have been exposed to a radioactively labeled substance (such as 2-deoxyglucose) so that regions of high uptake are made visible.

Autoreceptors. A type of metabotropic receptor located on the presynaptic membrane that bind to their neuron's own neurotransmitters.

Autosomal chromosomes. Chromosomes that come in matched pairs; in mammals, all of the chromosomes except the sex chromosomes are autosomal.

Axon hillock. The conical structure at the junction between the axon and cell body.

Axon initial segment. The segment of the axon where action potentials are generated—located immediately adjacent to the axon hillock.

B cells. B lymphocytes; lymphocytes that manufacture antibodies against antigens they encounter.

Basal forebrain. A midline area of the forebrain, which is located just in front of and above the hypothalamus and is the brain's main source of acetylcholine.

Basal ganglia. A collection of subcortical nuclei (e.g., striatum and globus pallidus).

Basal metabolic rate. The rate at which energy is utilized to maintain bodily processes when resting.

Basilar membrane. The membrane of the organ of Corti in which the hair cell receptors are embedded.

Before-and-after design. The experimental design used to demonstrate contingent drug tolerance; one group receives the drug before each of a series of behavioral tests and the other group receives the drug after each test.

Behavioral paradigm. A single set of procedures developed for the investigation of a particular behavioral phenomenon.

Benign tumors. Tumors that are surgically removable with little risk of further growth in the body.

Benzodiazepines. A class of $GABA_A$ agonists with anxiolytic, sedative, and anticonvulsant properties; drugs such as chlordiazepoxide (Librium) and diazepam (Valium).

Beta-amyloid. A protein that is present in normal brains in small amounts. Beta amyloid is a major constituent of the amyloid plaques of Alzheimer's disease.

Between-subjects design. An experimental design in which a different group of subjects is tested under each condition.

Bilateral medial temporal lobectomy. The removal of the medial portions of both temporal lobes, including the hippocampus, the amygdala, and the adjacent cortex.

Binding problem. When the brain combines individual sensory attributes to produce integrated perceptions.

Binocular. Cells in the visual system that are binocular respond to stimulation of either eye.

Binocular disparity. The difference in the position of the same image on the two retinas.

Biopsychology. The scientific study of the biology of behavior; a biological approach to the study of psychology.

Bipolar cells. Bipolar neurons that form the middle layer of the retina.

Bipolar disorders. A category of psychiatric disorders that involves alternate bouts of depression and mania or hypomania.

Bipolar disorder type I. A psychiatric disorder that involves alternate bouts of depression and mania.

Bipolar disorder type II. A psychiatric disorder that involves alternate bouts of depression and hypomania.

Bipolar neuron. A neuron with two processes extending from its cell body.

Bisexual. An individual who is sexually attracted to members of both sexes.

Bistable figures. A stimulus that produces two alternating perceptions.

Blind spot. The area on the retina where the bundle of axons from the retinal ganglion cells leave the eye as the optic nerve.

Blindsight. The ability to respond to visual stimuli in a scotoma without conscious awareness of those stimuli.

Blood–brain barrier. The mechanism that impedes the passage of toxic substances from the blood into the brain.

BOLD signal. The blood-oxygen-level-dependent signal that is recorded by functional MRI (fMRI).

Botox. *Botulinium toxin;* a neurotoxin released by bacterium often found in spoiled food. It blocks the release of acetylcholine at neuromuscular junctions and has applications in medicine and cosmetics.

Bottom-up. A sort of neural mechanism that involves activation of higher cortical areas by lower cortical areas.

Brain-derived neurotrophic factor. One type of neurotrophin.

Brain stem. The part of the brain on which the cerebral hemispheres rest; in general, it regulates reflex activities that are critical for survival (e.g., heart rate and respiration).

Bregma. The point on the surface of the skull where two of the major sutures intersect; commonly used as a reference point in stereotaxic surgery on rodents.

Broca's aphasia. A hypothetical disorder of speech production with no associated deficits in language comprehension.

Broca's area. The area of the inferior prefrontal cortex of the left hemisphere hypothesized by Broca to be the center of speech production.

Buerger's disease. A condition in which the blood vessels, especially those supplying the legs, are constricted whenever tobacco is smoked. The disease can progress to gangrene and amputation.

Bulimia nervosa. An eating disorder characterized by periods of not eating interrupted by bingeing followed by purging.

Bullying. A chronic social threat that induces subordination stress in members of our species.

Butyrophenones. A class of antipsychotic drugs that bind primarily to D_2 receptors.

CA1 subfield. A region of the hippocampus that is commonly damaged by cerebral ischemia.

Cafeteria diet. A diet offered to experimental animals that is composed of a wide variety of palatable foods.

Cannabis. The common hemp plant, which is the source of marijuana.

Cannon-Bard theory. The theory that emotional experience and emotional expression are parallel processes that have no direct causal relation.

Cannula. A fine, hollow tube that is implanted in the body for the purpose of introducing or extracting substances.

Carbon monoxide. A soluble-gas neurotransmitter.

Carousel apparatus. An apparatus used to study the effects of sleep deprivation in laboratory rats.

Cartesian dualism. The philosophical position of René Descartes, who argued that the universe is composed of two elements: physical matter and the human mind.

Case studies. Studies that focus on a single case, or subject.

Cataplexy. A disorder that is characterized by recurring losses of muscle tone during wakefulness and is often seen in cases of narcolepsy.

Catecholamines. The three monoamine neurotransmitters that are synthesized from the amino acid tyrosine: dopamine, epinephrine, and norepinephrine.

Caudate. The tail-like structure that is part of the striatum.

Cell-adhesion molecules (CAMs). Molecules on the surface of cells that have the ability to recognize specific molecules on the surface of other cells and adhere to them.

Cell-mediated immunity. The immune reaction by which T cells destroy invading microorganisms.

Central canal. The small cerebrospinal fluid-filled channel that runs the length of the spinal cord.

Central fissure. The large fissure that separates the frontal lobe from the parietal lobe.

Central nervous system (CNS). The portion of the nervous system within the skull and spine.

Central nucleus of the amygdala. A nucleus of the amygdala that is thought to control defensive behavior.

Central sensorimotor programs. Patterns of activity that are programmed into the sensorimotor system.

Cephalic phase. The metabolic phase during which the body prepares for food that is about to be absorbed.

Cerebellum. A metencephalic structure that is thought to participate in the storage of memories of learned sensorimotor skills.

Cerebral angiography. A contrast x-ray technique for visualizing the cerebral circulatory system by infusing a radio-opaque dye into a cerebral artery.

Cerebral aqueduct. A narrow channel that connects the third and fourth ventricles.

Cerebral commissures. Tracts that connect the left and right cerebral hemispheres.

Cerebral cortex. The layer of neural tissue covering the cerebral hemispheres of humans and other mammals.

Cerebral dialysis. A method for recording changes in brain chemistry in behaving animals in which a fine tube with a short semipermeable section is implanted in the brain and extracellular neurochemicals are continuously drawn off for analysis.

Cerebral hemorrhage. Bleeding in the brain.

Cerebral ischemia. An interruption of the blood supply to an area of the brain.

Cerebral ventricles. The four cerebrospinal fluid-filled internal chambers of the brain: the two lateral ventricles, the third ventricle, and the fourth ventricle.

Cerebrospinal fluid (CSF). The fluid that fills the subarachnoid space, the central canal, and the cerebral ventricles.

Cerebrum. The portion of the brain that sits above the brain stem; in general, it plays a role in complex adaptive processes (e.g., learning, perception, and motivation).

Cerveau isolé preparation. An experimental preparation in which the forebrain is disconnected from the rest of the brain by a midcollicular transection.

Change blindness. The difficulty perceiving major changes to unattended-to parts of a visual image when the changes are introduced during brief interruptions in the presentation of the image.

Charles Bonnet syndrome. A condition, most commonly seen in people with glaucoma, wherein affected individuals experience rich and complex hallucinations (e.g., people's faces, complex landscapes).

Cheese effect. The surges in blood pressure that occur when individuals taking MAO inhibitors consume tyramine-rich foods, such as cheese.

Chemoaffinity hypothesis. The hypothesis that growing axons are attracted to the correct targets by different chemicals released by the target sites.

Chemotopic. Organized, like the olfactory bulb, according to a map of various odors.

Chimeric figures test. A test of visual completion in split-brain subjects that uses pictures composed of the left and right halves of two different faces.

Chlorpromazine. The first antipsychotic drug.

Cholecystokinin (CCK). A peptide that is released by the gastrointestinal tract and is thought to function as a satiety signal.

Chordates. Animals with dorsal nerve cords.

Choroid plexuses. The networks of capillaries that protrude into the ventricles from the pia mater and produce cerebrospinal fluid.

Chromosomes. Threadlike structures in the cell nucleus that contain the genes; each chromosome is a DNA molecule.

Chronic traumatic encephalopathy. The dementia and cerebral scarring observed in boxers, rugby players, American football players, and other individuals who have experienced repeated concussive, or even subconcussive, blows to the head.

Chronobiotic. A substance that influences the timing of internal biological rhythms.

Ciliary muscles. The eye muscles that control the shape of the lenses.

Cingulate cortex. The cortex of the cingulate gyri, which are located on the medial surfaces of the frontal lobes.

Cingulate gyri. Large gyri located on the medial surfaces of the frontal lobes, just superior to the corpus callosum.

Cingulate motor areas. Two small areas of secondary motor cortex located in the cortex of the cingulate gyrus of each hemisphere.

Circadian clock. An internal timing mechanism that is capable of maintaining daily cycles of physiological functions.

Circadian rhythms. Daily cycles of bodily functions.

Cirrhosis. Scarring of the liver, which is a major cause of death among heavy alcohol users.

Clinical. Pertaining to illness or treatment.

Clinical depression (major depressive disorder). Depression that is so severe that it is difficult for the patient to meet the essential requirements of daily life.

Clinical trials. Studies conducted on human subjects to assess the therapeutic efficacy of an untested drug or other treatment.

Closed-head tramautic brain injuries (TBIs). Brain injuries produced by blows that do not penetrate the skull.

Clozapine. An atypical antipsychotic that is used to treat schizophrenia, does not produce Parkinsonian side effects, and has only a slight affinity for D_2 receptors.

Cocaine. A stimulant that exerts its effects by altering the activity of dopamine transporters.

Cocaine psychosis. Psychotic symptoms that are sometimes observed during cocaine sprees; similar in certain respects to schizophrenia.

Cocaine sprees. Binges of cocaine use.

Cochlea. The long, coiled tube in the inner ear that is filled with fluid and contains the organ of Corti and its auditory receptors.

Cocktail-party phenomenon. The ability to unconsciously monitor the contents of one conversation while consciously focusing on another.

Cocontraction. The simultaneous contraction of antagonistic muscles.

Codeine. A relatively weak psychoactive ingredient of opium.

Codon. A group of three consecutive nucleotide bases on a DNA or messenger RNA strand; each codon specifies the particular amino acid that is to be added to an amino acid chain during protein synthesis.

Coexistence. The presence of more than one neurotransmitter in the same neuron.

Cognition. Higher intellectual processes such as thought, memory, attention, and complex perceptual processes.

Cognitive neuroscience. A division of biopsychology that focuses on the use of functional brain imaging to study the neural mechanisms of human cognition.

Collateral sprouting. The growth of axon branches from mature neurons, usually to postsynaptic sites abandoned by adjacent axons that have degenerated.

Colony-intruder paradigm. A paradigm for the study of aggressive and defensive behaviors in male rats; a small male intruder rat is placed in an established colony in order to study the aggressive responses of the colony's alpha male and the defensive responses of the intruder.

Color constancy. The tendency of an object to appear the same color even when the wavelengths of light that it reflects change.

Columnar organization. The functional organization of the neocortex in vertical columns; the cells in each column form a mini-circuit that performs a single function.

Commissurotomy. Surgical severing of the cerebral commissures.

Comorbid. The tendency for two or more health conditions to occur together in the same individual.

Comparative approach. The study of biological processes by comparing different species—usually from the evolutionary perspective.

Comparative psychology. The division of biopsychology that studies the evolution, genetics, and adaptiveness of behavior, often by using the comparative approach.

Complementary colors. Pairs of colors that produce white or gray when combined in equal measure.

Completion. The visual system's automatic use of information obtained from receptors around the blind spot, or scotoma, to create a perception of the missing portion of the retinal image.

Complex cells. Neurons in the visual cortex that respond optimally to straight-edge stimuli in a certain orientation in any part of their receptive field.

Complex seizures. Seizures that are characterized by various complex psychological phenomena and are thought to originate in the temporal lobes.

Component theory. The theory that the relative amount of activity produced in three different classes of cones by light determines its perceived color (also called *trichromatic theory*).

Computed tomography (CT). A computer-assisted x-ray procedure that can be used to visualize the brain and other internal structures of the living body.

Concept cells. Cells, such as those found in the medial temporal lobe, that respond to ideas or concepts rather than to particulars. Also known as *Jennifer Aniston neurons*.

Conditioned compensatory responses. Hypothetical conditional physiological responses that are opposite to the effects of a drug that are thought to be elicited by stimuli that are regularly associated with experiencing the drug effects.

Conditioned defensive burying. The burial of a source of aversive stimulation by rats.

Conditioned drug tolerance. Tolerance effects that are maximally expressed only when a drug is administered in the same situation in which it has previously been administered.

Conditioned place-preference paradigm. A test that assesses a laboratory animal's preference for an environment in which it has previously experienced drug effects relative to a control environment.

Conditioned taste aversion. An avoidance response that develops to the taste of food whose consumption has been followed by illness.

Conduction aphasia. A hypothetical aphasia that is thought to result from damage to the arcuate fasciculus—the pathway between Broca's and Wernicke's areas.

Cones. The visual receptors in the retina that mediate high acuity color vision in good lighting.

Confounded variable. An unintended difference between the conditions of an experiment that could have affected the dependent variable.

Congenital. Present at birth.

Congenital adrenal hyperplasia. A congenital deficiency in the release of cortisol from the adrenal cortex, which leads to the excessive release of adrenal androgens.

Conscious awareness. The awareness of one's perceptions; typically inferred from the ability to verbally describe them.

Consciousness. The perception or awareness of some aspect of one's self or the world.

Conspecifics. Members of the same species.

Constituent cognitive processes. Simple cognitive processes that combine to produce complex cognitive processes.

Contextual fear conditioning. The process by which benign contexts (situations) come to elicit fear through their association with fear-inducing stimuli.

Contingent drug tolerance. Drug tolerance that develops as a reaction to the experience of the effects of drugs rather than to drug exposure alone.

Contralateral. Projecting from one side of the body to the other.

Contralateral neglect. A disturbance of the patient's ability to respond to stimuli on the side of the body opposite to a site of brain damage, usually the left side of the body following damage to the right parietal lobe.

Contrast enhancement. The intensification of the perception of edges.

Contrast x-ray techniques. X-ray techniques that involve the injection, into one compartment of the body, of a substance that absorbs x-rays either less than or more than surrounding tissues.

Contrecoup injuries. Contusions that occur on the side of the brain opposite to the side of a blow.

"Control of behavior" versus "conscious perception" theory. The theory that the dorsal stream mediates behavioral interactions with objects and the ventral stream mediates conscious perception of objects.

Control-question technique. A lie-detection interrogation method in which the polygrapher compares the physiological responses to target questions with the responses to control questions.

Contusions. Closed-head injuries that involve damage to the cerebral circulatory system, which produces internal hemorrhaging.

Convergent evolution. The evolution in unrelated species of similar solutions to the same environmental demands.

Converging operations. The use of several research approaches to solve a single problem.

Convolutions. Folds on the surface of the cerebral hemispheres.

Convulsions. Motor seizures.

Coolidge effect. The fact that a copulating male who becomes incapable of continuing to copulate with one sex partner can often recommence copulating with a new sex partner.

Copulation. Sexual intercourse.

Corpus callosum. The largest cerebral commissure.

Corticosterone. The predominant glucocorticoid in humans.

Crack. A potent, cheap, smokable form of cocaine.

Cranial nerves. The 12 pairs of nerves extending from the brain (e.g., optic nerves, olfactory nerves, and vagus nerves).

Critical period. A period during development in which a particular experience must occur for it to influence the course of subsequent development.

Critical thinking. The process of recognizing the weaknesses of existing ideas and the evidence on which they are based.

Cross-cuing. Communication between hemispheres that have been separated by commissurotomy via an external route.

Cross section. Section cut at a right angle to any long, narrow structure of the central nervous system.

Cross tolerance. Tolerance to the effects of one drug that develops as the result of exposure to another drug that acts by the same mechanism.

CRISPR/Cas9 method. A popular gene editing technique. It allows researchers to edit parts of the genome by removing from, adding to, or altering the DNA sequence.

Cytokines. A group of peptide hormones that are released by many cells and participate in a variety of physiological and immunological responses, causing inflammation and fever.

Decorticate. Lacking a cortex.

Decussate. To cross over to the other side of the brain.

Deep brain stimulation. A treatment in which low intensity electrical stimulation is continually applied to an area of the brain through an implanted electrode.

Deep dyslexia. A reading disorder in which the phonetic procedure is disrupted while the lexical procedure is not.

Default mode. The pattern of brain activity that is present when humans sit quietly and let their minds wander.

Default mode network. The network of brain structures that tends to be active when the brain is in default mode.

Defeminizes. Suppresses or disrupts female characteristics.

Defensive behaviors. Behaviors whose primary function is protection from threat or harm.

Defensive-burying test. An animal model of anxiety; anxious rats will bury objects that generate anxiety.

Delayed nonmatching-to-sample test. A test in which the subject is presented with an unfamiliar sample object and then, after a delay, is presented with a choice between the sample object and an unfamiliar object, where the correct choice is the unfamiliar object.

Delirium tremens (DTs). The phase of alcohol withdrawal syndrome characterized by hallucinations, delusions, disorientation, agitation, confusion, hyperthermia, and tachycardia.

Delta waves. The largest and slowest EEG waves.

Demasculinizes. Suppresses or disrupts male characteristics.

Dementia. General intellectual deterioration.

Dendritic spines. Tiny protrusions of various shapes that are located on the surfaces of many dendrites.

Deoxyribonucleic acid (DNA). The double-stranded, coiled molecule of genetic material.

Dependent variable. The variable measured by the experimenter to assess the effect of the independent variable.

Depolarize. To decrease the resting membrane potential.

Depressant. A drug that depresses neural activity.

Desynchronized EEG. Low-amplitude, high-frequency EEG.

Developmental dyslexias. Dyslexias that become apparent when a child is learning to read.

Dextrals. Right-handers.

Dichotic listening test. A test of language lateralization in which two different sequences of three spoken digits are presented simultaneously, one to each ear, and the subject is asked to report all of the digits heard.

Dichotomous traits. Traits that occur in one form or the other, never in combination.

Diencephalon. One of the five major divisions of the brain; it is composed of the thalamus and hypothalamus.

Diet-induced thermogenesis. The homeostasis-defending increases in body temperature that are associated with increases in body fat.

Diffusion tensor MRI. A magnetic resonance imaging (MRI) technique that is used for identifying major tracts.

Digestion. The process by which food is broken down and absorbed through the lining of the gastrointestinal tract.

Digit span. The longest sequence of random digits that can be repeated correctly 50 percent of the time—most people have a digit span of 7.

Directed synapses. Synapses at which the site of neurotransmitter release and the site of neurotransmitter reception are in close proximity.

Distal. Far from something.

Distal segment. The segment of a cut axon between the cut and the axon terminals.

Dizygotic twins. Twins that develop from two zygotes and thus tend to be as genetically similar as any pair of siblings.

DNA methylation. An epigenetic mechanism wherein a methyl group attaches to a DNA molecule, usually at cytosine sites in mammals. DNA methylation can either decrease or increase gene expression.

Dominant hemisphere. A term used in the past to refer to the left hemisphere, based on the incorrect assumption that the left hemisphere is dominant in all complex behavioral and cognitive activities.

Dominant trait. The trait of a dichotomous pair that is expressed in the phenotypes of heterozygous individuals.

Dopamine. One of the three catecholamine neurotransmitters.

Dopamine transporters. Molecules in the presynaptic membrane of dopaminergic neurons that attract dopamine molecules in the synaptic cleft and deposit them back inside the neuron.

Dorsal. Toward the surface of the back of a vertebrate or toward the top of the head.

Dorsal-column medial-lemniscus system. The division of the somatosensory system that ascends in the dorsal portion of the spinal white matter and tends to carry signals related to touch and proprioception.

Dorsal columns.   The somatosensory tracts that ascend in the dorsal portion of the spinal cord white matter.

Dorsal horns.   The two dorsal arms of the spinal gray matter.

Dorsal root ganglia.   Structures just outside the spinal cord that are composed of the cell bodies of dorsal root axons.

Dorsal stream.   The group of visual pathways that flows from the primary visual cortex to the dorsal prestriate cortex to the posterior parietal cortex.

Dorsolateral prefrontal association cortex.   An area of the prefrontal cortex that plays a role in the evaluation of external stimuli and the initiation of complex voluntary motor responses.

Down syndrome.   A disorder associated with the presence of an extra chromosome 21, resulting in disfigurement and intellectual impairment.

Drug-addicted individuals.   Those habitual drug users who continue to use a drug despite its adverse effects on their health and social life and despite their repeated efforts to stop using it.

Drug craving.   An affective state in which there is a strong desire for a particular drug.

Drug metabolism.   The conversion of a drug from its active form to a nonactive form.

Drug priming.   A single exposure to a formerly abused drug.

Drug self-administration paradigm.   A test of the addictive potential of drugs in which laboratory animals can inject drugs into themselves by pressing a lever.

Drug sensitization.   An increase in the sensitivity to a drug effect that develops as the result of exposure to the drug.

Drug tolerance.   A state of decreased sensitivity to a drug that develops as a result of exposure to the drug.

DSM-5.   The fifth and current edition of the *Diagnostic and Statistical Manual of Mental Disorders;* produced by the American Psychiatric Association.

Dual-trace theory.   The theory that memories are temporarily stored in the hippocampus until they can be transferred to a more stable cortical storage system. Also known as the *standard consolidation theory.*

Duchenne smile.   A genuine smile, one that includes contraction of the facial muscles called the *orbicularis oculi.*

Duodenum.   The upper portion of the intestine through which most of the glucose and amino acids are absorbed into the bloodstream.

Duplexity theory.   The theory that cones and rods mediate photopic and scotopic vision, respectively.

Dura mater.   The tough outer meninx.

Dynamic contraction.   Contraction of a muscle that causes the muscle to shorten.

Dynamic phase.   The first phase of the VMH syndrome, characterized by grossly excessive eating and rapid weight gain.

Dyslexia.   A reading disorder that does not result from general visual, motor, or intellectual deficits.

Efferent nerves.   Nerves that carry motor signals from the central nervous system to the skeletal muscles or internal organs.

Ejaculate.   To eject sperm from the penis.

Ejaculation.   Ejection of sperm.

Electrocardiogram (ECG or EKG).   A recording of the electrical signals associated with heartbeats.

Electroconvulsive shock (ECS).   An intense, brief, diffuse, seizure-inducing current administered to the brain via large electrodes attached to the scalp.

Electroencephalogram (EEG).   A measure of the gross electrical activity of the brain, commonly recorded through scalp electrodes.

Electroencephalography.   A technique for recording the gross electrical activity of the brain through electrodes, which in humans are usually attached to the surface of the scalp.

Electromyogram (EMG).   A record of muscle tension.

Electromyography.   A procedure for measuring muscle tension.

Electron microscopy.   A microscopy technique used to study the fine details of cellular structure.

Electrooculogram (EOG).   A measure of eye movement.

Electrooculography.   A technique for recording eye movements through electrodes placed around the eye.

Elevated plus maze.   An apparatus for recording defensiveness or anxiety in rats by assessing their tendency to avoid the two open arms of a plus sign–shaped maze mounted some distance above the floor.

Elevated-plus-maze test.   An animal model of anxiety; anxious rats tend to stay in the enclosed arms of the maze rather than venturing onto the open arms.

Embolism.   The blockage of blood flow in a smaller blood vessel by a plug that was formed in a larger blood vessel and carried by the bloodstream to the smaller one.

Emergent stage 1 EEG.   All periods of stage 1 sleep EEG except initial stage 1; each is associated with REMs.

Empathogens.   Psychoactive drugs that produce feelings of empathy.

Encapsulated tumors.   Tumors that grow within their own membrane.

Encéphale isolé preparation.   An experimental preparation in which the brain is separated from the rest of the nervous system by a transection of the caudal brain stem.

Encephalitis.   The inflammation associated with brain infection.

Endocannabinoids.   A class of unconventional neurotransmitters that are chemically similar to the active components of marijuana.

Endocrine glands.   Ductless glands that release chemicals called *hormones* directly into the circulatory system.

Endogenous.   Naturally occurring in the body (e.g., endogenous opioids).

Endogenous depression.   Depression that occurs with no apparent cause.

Endorphins.   A class of endogenous opioids.

Engram.   A change in the brain that stores a memory.

Engram cells.   Neurons that maintain an engram.

Enhancers.   Stretches of DNA that control the rate of expression of target genes.

Enkephalins.   The first class of endogenous opioids to be discovered.

Enriched environments.   Laboratory environments designed to promote cognitive and physical activity by providing opportunities for a greater variety of sensory and motor experiences than available in conventional laboratory environments; commonly used to study the effects of experience on development in rats and mice.

Entorhinal cortex.   An area of the medial temporal cortex that is a major source of neural signals to the hippocampus.

Enzymatic degradation.   The breakdown of chemicals by enzymes—one of the two mechanisms for deactivating released neurotransmitters.

Enzymes.   Proteins that stimulate or inhibit biochemical reactions without being affected by them.

Epidemiology. The study of the factors that influence the distribution of a disease in the general population.

Epigenetics. The study of all mechanisms of inheritance other than the genetic code and its expression.

Epigenetic. Not of the genes; refers to nongenetic means by which traits are passed from parents to offspring.

Epigenome. A catalogue of all the epigenetic mechanisms at play within a particular cell type.

Epilepsy. A neurological disorder characterized by spontaneously recurring seizures.

Epileptic auras. Psychological changes that precede the onset of a seizure.

Epileptogenesis. Development of epilepsy.

Epinephrine. One of the three catecholamine neurotransmitters.

Episodic memories. Explicit memories for the particular events and experiences of one's life.

Epitranscriptome. Refers to all those modifications of RNA that occur after transcription—that do not involve modifications to the RNA base sequence.

Estradiol. The most common estrogen.

Estrogens. The class of steroid hormones that are released in large amounts by the ovaries; an example is estradiol.

Estrous cycle. The cycle of sexual receptivity displayed by many female mammals.

Estrus. The portion of the estrous cycle characterized by proceptivity, sexual receptivity, and fertility (*estrus* is a noun and *estrous* an adjective).

Ethological research. The study of animal behavior in its natural environment.

Ethology. The study of the behavior of animals in their natural environments.

Euthymic. Individuals who are not currently displaying symptoms of depression, hypomania, or mania.

Event-related potentials (ERPs). The EEG waves that regularly accompany certain psychological events.

Evolutionary perspective. The approach that focuses on the environmental pressures that likely led to the evolution of the characteristics (e.g., of brain and behavior) of current species.

Evolve. To undergo gradual orderly change.

Exaptation. A characteristic that evolved to serve one function and was later co-opted to serve another function.

Excitatory postsynaptic potentials (EPSPs). Graded postsynaptic depolarizations, which increase the likelihood that an action potential will be generated.

Executive function. A collection of cognitive abilities (e.g., innovative thinking, lateral thinking, and insightful thinking) that appear to depend on the prefrontal cortex.

Exocrine glands. Glands that release chemicals into ducts that carry them to targets, mostly on the surface of the body.

Exocytosis. The process of releasing a neurotransmitter.

Explicit memories. Conscious memories.

Expressive. Pertaining to the generation of language; that is, pertaining to writing or talking.

Extensors. Muscles that act to straighten or extend a joint.

Exteroceptive stimuli. Stimuli that arise from outside the body.

Facial feedback hypothesis. The hypothesis that our facial expressions can influence the emotions we experience.

Far-field potentials. EEG signals recorded in attenuated form at the scalp because they originate far away—for example, in the brain stem.

Fasciculation. The tendency of developing axons to grow along the paths established by preceding axons.

Fasting phase. The metabolic phase that begins when energy from the preceding meal is no longer sufficient to meet the immediate needs of the body and during which energy is extracted from fat and glycogen stores.

Fear. The emotional reaction that is normally elicited by the presence or expectation of threatening stimuli.

Fear conditioning. Establishing fear of a previously neutral conditional stimulus by pairing it with an aversive unconditional stimulus.

Feminizes. Enhances or produces female characteristics.

Fetal alcohol syndrome (FAS). A syndrome produced by prenatal exposure to alcohol and characterized by brain damage, intellectual disability, poor coordination, poor muscle tone, low birth weight, retarded growth, and/or physical deformity.

Fissures. The large furrows in a convoluted cortex.

Fitness. According to Darwin, the ability of an organism to survive and contribute its genes to the next generation.

Fixational eye movements. Involuntary movements of the eyes (tremor, drifts, and saccades) that occur when a person tries to fix their gaze on (i.e., stare at) a point.

Flavor. The combined impression of taste and smell.

Flexors. Muscles that act to bend or flex a joint.

Fluorodeoxyglucose (FDG). A molecule that is similar to glucose, and is thus rapidly taken up by active cells. However, unlike glucose, fluorodeoxyglucose cannot be metabolized; it therefore accumulates in active cells until it is gradually broken down. A radioactive isotope of this molecule is commonly used in positron emission tomography (PET).

Fluoxetine. The first selective serotonin reuptake inhibitor (SSRI) to be developed. It was initially marketed under the tradename *Prozac*.

Focal seizure. A seizure that does not involve the entire brain.

Follicle-stimulating hormone (FSH). The gonadotropic hormone that stimulates development of ovarian follicles.

Fornix. The major tract of the limbic system; it connects the hippocampus with the septum and mammillary bodies.

Fourier analysis. A mathematical procedure for breaking down a complex wave form into component sine waves of various frequencies.

Fovea. The central indentation of the retina, which is specialized for high-acuity vision.

Fraternal birth order effect. The finding that the probability of a male being attracted to other males increases as a function of the number of older brothers he has.

Free fatty acids. The main source of the body's energy during the fasting phase; released from adipose tissue in response to high levels of glucagon.

Free nerve endings. Neuron endings that lack specialized structures on them and that detect cutaneous pain and changes in temperature.

Free-running period. The duration of one cycle of a free-running rhythm.

Free-running rhythms. Circadian rhythms that do not depend on environmental cues to keep them on a regular schedule.

Frontal eye field. A small area of prefrontal cortex that controls eye movements.

Frontal lobe. The most anterior of the four cerebral lobes.

Frontal operculum. The area of prefrontal cortex that in the left hemisphere is the location of Broca's area.

Frontal sections. Any slices of brain tissue cut in a plane that is parallel to the face; also termed *coronal sections*.

Functional connectivity. An approach used by cognitive neuroscientists that examines which brain regions have parallel activation patterns over time.

Functional connectome. A catalogue of the functional connectivity associated with each behavior and cognitive process.

Functional MRI (fMRI). A magnetic resonance imaging technique for inferring brain activity by measuring increased oxygen flow into particular areas.

Functional segregation. Organization into different areas, each of which performs a different function; for example, in sensory systems, different areas of secondary and association cortex analyze different aspects of the same sensory stimulus.

Functional tolerance. Drug tolerance that results from changes that reduce the reactivity of the sites of action to the drug.

Functional ultrasound imaging. A technique that uses ultrasound (sound waves of a higher frequency than we can hear) to measure changes in blood volume in particular brain regions.

Fusiform face area. An area of human cortex, located at the boundary between the occipital and temporal lobes, that is selectively activated by human faces.

G proteins. Proteins that are located inside neurons (and some other cells) and are attached to metabotropic receptors in the cell membrane.

Gametes. Egg cells and sperm cells.

Gamma-aminobutyric acid (GABA). The amino acid neurotransmitter that is synthesized from glutamate; the most prevalent inhibitory neurotransmitter in the mammalian central nervous system.

Ganglia. Clusters of neuronal cell bodies in the peripheral nervous system (singular *ganglion*).

Gap junctions. Narrow spaces between adjacent neurons that are bridged by fine tubular channels containing cytoplasm, through which electrical signals and small molecules can pass readily.

Gastric bypass. A surgical procedure for treating obesity in which the intestine is cut and connected to the upper portion of the stomach, which is isolated from the rest of the stomach by a row of staples.

Gastric ulcers. Painful lesions to the lining of the stomach or duodenum.

Gay. Sexually attracted to members of the same sex.

Gender dysphoria. The distress that can occur in people whose gender identity differs from their sex assigned at birth or sex-related physical characteristics.

Gender identity. The gender that a person most identifies with: female, male, some combination of male and female, neither female or male, or some other gender category.

Gene. A unit of inheritance; for example, the section of a chromosome that controls the synthesis of one protein.

Gene editing techniques. Techniques that allow researchers to edit genes at a particular time during development.

Gene expression. The production of the protein specified by a particular gene.

Gene knockin techniques. Procedures for creating organisms that have one or more additional genes.

Gene knockout techniques. Procedures for creating organisms that lack a particular gene.

General paresis. The mental illness and dementia resulting from a syphilitic infection.

Generalizability. The degree to which the results of a study can be applied to other individuals or situations.

Generalized anxiety disorder. An anxiety disorder characterized by stress responses and extreme feelings of anxiety and worry about a large number of different activities or events.

Generalized seizures. Seizures that involve the entire brain.

Genetic recombination. The meiotic process by which pairs of chromosomes cross over one another at random points, break apart, and exchange genes.

Genitals. The external reproductive organs.

Genotype. The traits that an organism can pass on to its offspring through its genetic material.

Glial cells. Several classes of nonneural cells of the nervous system.

Glia-mediated migration. One of two major modes of neural migration during development, by which immature neurons move away from the central canal along radial glial cells.

Gliomas. Brain tumors that develop from glial cells.

Global amnesia. Amnesia for information presented in all sensory modalities.

Global aphasia. Severe disruption of all language-related abilities.

Global cerebral ischemia. An interruption of blood supply to the entire brain.

Globus pallidus. A structure of the basal ganglia that is located between the putamen and thalamus.

Glucagon. A pancreatic hormone that promotes the release of free fatty acids from adipose tissue, their conversion to ketones, and the use of both as sources of energy.

Glucocorticoids. Steroid hormones that are released from the adrenal cortex in response to stressors.

Gluconeogenesis. The process by which protein is converted to glucose.

Glucose. A simple sugar that is the breakdown product of complex carbohydrates; it is the body's primary, directly utilizable source of energy.

Glucostatic theory. The theory that eating is controlled by deviations from a hypothetical blood glucose set point.

Glutamate. The most prevalent excitatory neurotransmitter in the central nervous system.

Glycine. An amino acid neurotransmitter.

Golgi complex. Structures in the cell bodies and terminal buttons of neurons that package neurotransmitters and other molecules in vesicles.

Golgi stain. A neural stain that completely darkens a few of the neurons in each slice of tissue, thereby revealing their silhouettes.

Golgi tendon organs. Receptors that are embedded in tendons and are sensitive to the amount of tension in the skeletal muscles to which their tendons are attached.

Gonadectomy. The surgical removal of the gonads (testes or ovaries); castration.

Gonadotropin. The pituitary tropic hormone that stimulates the release of hormones from the gonads.

Gonadotropin-releasing hormone. The hypothalamic releasing hormone that controls the release of the two gonadotropic hormones from the anterior pituitary.

Gonads. The testes and the ovaries.

Graded potentials. All postsynaptic potentials (i.e., both excitatory postsynaptic potentials and inhibitory postsynaptic potentials), are graded potentials, which means that the amplitudes of postsynaptic potentials are proportional to the intensity of the signals that elicit them: Weak signals elicit small PSPs, and strong signals elicit large ones.

Graded responses. Responses whose magnitude is proportional to the magnitude of the stimuli that elicit them.

Grammatical analysis. Analysis of the structure of language.

Gray matter. Portions of the nervous system that are gray because they are composed largely of cell bodies and unmyelinated interneurons.

Green fluorescent protein (GFP). A protein that is found in certain species of jellyfish and that fluoresces when exposed to blue light.

Grid cells. Entorhinal neurons that each have an extensive array of evenly spaced place fields, producing a pattern reminiscent of graph paper.

Growth cone. Amoebalike structure at the tip of each growing axon or dendrite that guides growth to the appropriate target.

Growth hormone. The anterior pituitary hormone that acts directly on bone and muscle tissue to produce the pubertal growth spurt.

Guilty-knowledge technique. A lie-detection method in which the polygrapher records autonomic nervous system responses to a list of control and crime-related information known only to the guilty person and the examiner; also known as the *concealed information test*.

Gut microbiome. The bacteria and other organisms that live inside our gastrointestinal tract.

Gyri. The cortical ridges that are located between fissures or sulci.

Hair cells. The receptors of the auditory system.

Haloperidol. A butyrophenone used as an antipsychotic drug.

Harrison Narcotics Act. The act passed in 1914 that made it illegal to sell or use opium, morphine, or cocaine in the United States.

Hashish. Dark corklike material extracted from the resin on the leaves and flowers of *Cannabis*.

Hedonic value. The amount of pleasure that is actually experienced as the result of some action.

Helping-hand phenomenon. The redirection of one hand of a split-brain patient by the other hand.

Hemianopsic. Having a scotoma that covers half of the visual field.

Hemispherectomy. The removal of one cerebral hemisphere.

Heritability estimate. A numerical estimate of the proportion of variability that occurred in a particular trait in a particular study and that resulted from the genetic variation among the subjects in that study.

Heroin. A semisynthetic opioid.

Heschl's gyrus. The temporal lobe gyrus that is the location of primary auditory cortex.

Heterosexual. Sexually attracted to members of the other sex.

Heterozygous. Possessing two different genes for a particular trait.

Hierarchical organization. Organization into a series of levels that can be ranked with respect to one another; for example, in sensory systems, primary cortex, secondary cortex, and association cortex perform progressively more detailed analyses.

Hippocampus. A structure of the medial temporal lobes that plays a role in various forms of memory.

Histone. A protein around which DNA is coiled.

Histone remodeling. An epigenetic mechanism wherein histones change their shape and in so doing influence the shape of the adjacent DNA. This can either increase or decrease gene expression.

Homeostasis. A stable internal environment.

Hominini. The tribe of primates that includes at least six genera: Australopithecus, Paranthropus, Sahelanthropus, Orrorin, Pan, and Homo.

Hominins. Primates of the same group that includes humans.

Homologous. Having a similar structure because of a common evolutionary origin (e.g., a human's arm and a bird's wing are homologous).

Homozygous. Possessing two identical genes for a particular trait.

Horizontal cells. Retinal neurons whose specialized function is lateral communication.

Horizontal sections. Any slices of brain tissue cut in a plane that are parallel to the top of the brain.

Hormones. Chemicals released by the endocrine system directly into the circulatory system.

Human Genome Project. The international research effort to construct a detailed map of the human chromosomes.

Human proteome. A map of the entire set of proteins encoded for by human genes.

Huntingtin. Dominant gene that is mutated in cases of Huntington's disease.

Huntingtin protein. Protein whose synthesis is controlled by the huntingtin gene and is thus abnormal in individuals with Huntington's disease.

Huntington's disease. A progressive terminal disorder of motor and intellectual function that is produced in adulthood by a dominant gene.

Hyperphagia. Excessive eating.

Hyperpolarize. To increase the resting membrane potential.

Hypersomnia. Disorders characterized by excessive sleep or sleepiness.

Hypertension. Chronically high blood pressure.

Hypnagogic hallucinations. Dreamlike experiences that occur during wakefulness.

Hypnotic drugs. Sleep-promoting drugs.

Hypomania. A state that is characterized by a reduced need for sleep, high energy, and positive affect. During periods of hypomania, people are talkative, energetic, impulsive, positive, and very confident.

Hypothalamic peptides. One of the five classes of neuropeptide transmitters; it consists of those first identified as hormones released by the hypothalamus.

Hypothalamopituitary portal system. The vascular network that carries hormones from the hypothalamus to the anterior pituitary.

Hypothalamus. The diencephalic structure that sits just below the anterior portion of the thalamus.

Hypoxia. Shortage of oxygen supply to tissue—for example, to the brain.

Iatrogenic. Physician-created.

Imidazopyridines. A class of $GABA_A$ agonists that were marketed for the treatment of insomnia.

Imipramine. The first tricyclic antidepressant drug.

Immune system. The system that protects the body against infectious microorganisms.

Immunization. The process of creating immunity through vaccination.

Immunocytochemistry. A procedure for locating particular proteins in the brain by labeling their antibodies with a dye or radioactive element and then exposing slices of brain tissue to the labeled antibodies.

Implicit memories. Memories that are expressed by improved performance without conscious recall or recognition.

Impotent. Unable to achieve a penile erection.

In situ hybridization. A technique for locating particular proteins in the brain; molecules that bind to the mRNA that directs the synthesis of the target protein are synthesized and labeled, and brain slices are exposed to them.

Incentive-sensitization theory. Theory that addictions develop when drug use sensitizes the neural circuits mediating wanting of the drug—not necessarily liking of the drug.

Incomplete-pictures test. A test of memory measuring the improved ability to identify fragmented figures that have been previously observed.

Incubation of drug craving. The time-dependent increase in cue-induced drug craving and relapse.

Independent variable. The difference between experimental conditions that is arranged by the experimenter.

Indolamines. The class of monoamine neurotransmitters that are synthesized from tryptophan; serotonin is the only member of this class found in the mammalian nervous system.

Infantile amnesia. The normal inability to recall events from early childhood.

Inferior. Toward the bottom of the primate head or brain.

Inferior colliculi. The structures of the tectum that receive auditory input from the superior olives.

Inferotemporal cortex. The cortex of the inferior temporal lobe, in which is located an area of secondary visual cortex.

Infiltrating tumors. Tumors that grow diffusely through surrounding tissue.

Inhibitory postsynaptic potentials (IPSPs). Graded postsynaptic hyperpolarizations, which decrease the likelihood that an action potential will be generated.

Initial stage 1 EEG. The period of the stage 1 EEG that occurs at the onset of sleep; it is not associated with REMs.

Innate immune system. The first component of the immune system to react. It reacts quickly and generally near points of entry of pathogens.

Insomnia. Sleeplessness.

Instinctive behaviors. Behaviors that occur in all like members of a species, even when there seems to have been no opportunity for them to have been learned.

Insulin. A pancreatic hormone that facilitates the entry of glucose into cells and the conversion of bloodborne fuels to forms that can be stored.

Integration. Adding or combining a number of individual signals into one overall signal.

Internal desynchronization. The cycling on different schedules of the free-running circadian rhythms of two or more different processes.

Interneurons. Neurons with short axons or no axons at all, whose function is to integrate neural activity within a single brain structure.

Interoceptive stimuli. Stimuli that arise from inside the body.

Intersexed person. A term used to refer to a person who is born with sexual anatomy that does not clearly fit into typical definitions of male and female sexual anatomy.

Intracranial self-stimulation (ICSS). The repeated performance of a response that delivers electrical stimulation to certain sites in the animal's brain.

Intrafusal motor neuron. A motor neuron that innervates an intrafusal muscle.

Intrafusal muscle. A threadlike muscle that adjusts the tension on a muscle spindle.

Intromission. Insertion of the penis into the vagina.

Ion channels. Pores in neural membranes through which specific ions pass.

Ionotropic receptors. Receptors that are associated with ligand-activated ion channels.

Ions. Positively or negatively charged particles.

Iproniazid. The first antidepressant drug; a monoamine oxidase inhibitor.

Ipsilateral. On the same side of the body.

Isometric contraction. Contraction of a muscle that increases the force of its pull but does not shorten the muscle.

James-Lange theory. The theory that emotion-inducing sensory stimuli are received and interpreted by the cortex, which triggers changes in the visceral organs via the autonomic nervous system and in the skeletal muscles via the somatic nervous system. Then, the autonomic and somatic responses trigger the experience of emotion in the brain.

Jet lag. The adverse effects on body function of the acceleration of zeitgebers during eastbound flights or their deceleration during westbound flights.

Ketamine. A drug that is a type of dissociative hallucinogen.

Ketones. Breakdown products of free fatty acids that are used by muscles as a source of energy during the fasting phase.

Kindling phenomenon. The progressive development and intensification of convulsions elicited by a series of periodic low-intensity brain stimulations—most commonly by daily electrical stimulations to the amygdala.

Kluver-Bucy syndrome. The syndrome of behavioral changes (e.g., lack of fear and hypersexuality) that is induced in primates by bilateral damage to the anterior temporal lobes.

Korsakoff's syndrome. A neuropsychological disorder that is common in alcoholics and whose primary symptoms include memory loss, sensory and motor dysfunction, and, in its advanced stages, severe dementia.

L-Dopa. The chemical precursor of dopamine, which is used in the treatment of Parkinson's disease.

Lateral. Away from the midline of the body of a vertebrate, toward the body's lateral surfaces.

Lateral fissure. The large fissure that separates the temporal lobe from the frontal lobe.

Lateral geniculate nuclei. The six-layered thalamic structures that receive input from the retinas and transmit their output to the primary visual cortex.

Lateral hypothalamus (LH). The area of the hypothalamus once thought to be the feeding center.

Lateral nucleus of the amygdala. The nucleus of the amygdala that plays the major role in the acquisition, storage, and expression of conditioned fear.

Lateralization of function. The unequal representation of various psychological functions in the two hemispheres of the brain.

Leaky-barrel model. An analogy for the settling-point model of body-fat regulation.

Learning. The brain's ability to change in response to experience.

Leptin. A protein normally synthesized in fat cells; it is thought to act as a negative feedback signal normally released by fat stores to decrease appetite and increase fat metabolism.

Lesbian. Women who are attracted to women.

Leucotome. A surgical device used in psychosurgery to cut out a core of brain tissue.

Leukocytes. White blood cells.

Lewy bodies. Clumps of proteins that can be found in the surviving dopaminergic neurons of the substantia nigra of Parkinson's patients.

Lexical procedure. A procedure for reading aloud that is based on specific stored information acquired about written words.

Ligand. A molecule that binds to another molecule; neurotransmitters are ligands of their receptors.

Limbic system. A collection of interconnected nuclei and tracts that ring the thalamus.

Lipids. Fats.

Lipogenesis. The production of body fat.

Lipolysis. The breakdown of body fat.

Lipostatic theory. The theory that eating is controlled by deviations from a hypothetical body-fat set point.

Lithium. A metallic ion that is a mood stabilizer; used in the treatment of bipolar disorders.

Lobectomy. An operation in which a lobe, or a major part of one, is removed from the brain.

Lobotomy. An operation in which a lobe, or a major part of one, is separated from the rest of the brain by a large cut but is not removed.

Longitudinal fissure. The large fissure that separates the two cerebral hemispheres.

Long-term depression (LTD). A long-lasting decrease in synaptic efficacy (the flip side of LTP) that occurs in response to prolonged low-frequency stimulation of presynaptic neurons.

Long-term memory. Memory for experiences that endures after the experiences are no longer the focus of attention.

Long-term potentiation (LTP). The enduring facilitation of synaptic transmission that occurs following activation of synapses by high-intensity, high-frequency stimulation of presynaptic neurons.

Lordosis. The arched-back, rump-up, tail-to-the-side posture of female rodent sexual receptivity.

Lordosis quotient. The proportion of mounts that elicit lordosis.

Luteinizing hormone (LH). The gonadotropic hormone that causes the developing ovum to be released from its follicle.

Lymphocytes. Specialized leukocytes that are produced in bone marrow and the thymus gland and play important roles in the body's immune reactions.

Lysergic acid diethylamide (LSD). Hallucinogenic drug that alters perception, emotion, and cognition.

Magnetic resonance imaging (MRI). A structural brain imaging procedure in which high-resolution images are constructed from the measurement of waves that hydrogen atoms emit when they are activated by radio-frequency waves in a magnetic field.

Magnetoencephalography (MEG). A technique for measuring changes in magnetic fields on the surface of the scalp that are produced by changes in underlying patterns of neural activity.

Magnocellular layers. The layers of the lateral geniculate nuclei that are composed of neurons with large cell bodies; the bottom two layers (also called *M layers*).

Malignant tumors. Tumors that are difficult to remove or destroy, and continue to grow after attempts to remove or destroy them.

Mammals. A class of animals whose young are fed from mammary glands.

Mammillary bodies. The pair of spherical nuclei that are located on the inferior surface of the hypothalamus.

Mania. A state that has the same features as hypomania but taken to an extreme; it also has additional symptoms, such as delusions of grandeur, overconfidence, and distractibility. Mania usually involves psychosis.

MAO inhibitors. Antidepressant drugs that increase the level of monoamine neurotransmitters by inhibiting the action of the enzyme monoamine oxidase.

Masculinizes. Enhances or produces male characteristics.

Massa intermedia. The neural structure located in the third ventricle that connects the two lobes of the thalamus.

Maternal immune hypothesis. The hypothesis that mothers become progressively more immune to some masculinizing hormone in their male fetuses; proposed to explain the fraternal birth order effect.

Mean difference image. In the context of functional neuroimaging, the average of the difference images (obtained via paired-image subtraction) obtained from multiple participants.

Medial. Toward the midline of the body.

Medial diencephalic amnesia. Amnesia that is associated with damage to the medial diencephalon (e.g., Korsakoff's amnesia).

Medial dorsal nuclei. The thalamic relay nuclei of the olfactory system.

Medial geniculate nuclei. The auditory thalamic nuclei that receive input from the inferior colliculi and project to primary auditory cortex.

Medial lemniscus. The somatosensory pathway between the dorsal column nuclei and the ventral posterior nucleus of the thalamus.

Medial preoptic area. The area of the hypothalamus that includes the sexually dimorphic nuclei and that plays a key role in the control of male sexual behavior.

Medial temporal cortex. Cortex in the medial temporal lobe that lies adjacent to the hippocampus and amygdala.

Medial temporal lobe amnesia. Amnesia associated with bilateral damage to the medial temporal lobes; its major features are anterograde and retrograde amnesia for explicit memories, with preserved intellectual functioning.

Mediodorsal nuclei. A pair of thalamic nuclei, damage to which is thought to be responsible for many of the memory deficits associated with Korsakoff's syndrome.

Meiosis. The process of cell division that produces cells (e.g., egg cells and sperm cells) with half the chromosomes of the parent cell.

Melanocortin system. Neurons in the arcuate nucleus that release melanocortins.

Melanocortins. A class of peptides that includes the gut satiety peptide α-melanocyte-stimulating hormone.

Melanopsin. Photopigment found in certain retinal ganglion cells that responds to changes in background illumination and plays a role in the entrainment of circadian rhythms.

Melatonin. A hormone that is synthesized from serotonin in the pineal gland, and is both a soporific and a chronobiotic.

Membrane potential. The difference in electrical charge between the inside and the outside of a cell.

Memory. The brain's ability to store and access the learned effects of experiences.

Memory consolidation. The transfer of short-term memories to long-term storage.

Meninges. The three protective membranes that cover the brain and spinal cord (singular *meninx*).

Meningiomas. Tumors that grow between the meninges.

Meningitis. Inflammation of the meninges, usually caused by bacterial infection.

Menstrual cycle. The hormone-regulated cycle in females of follicle growth, egg release, buildup of the uterus lining, and menstruation.

Mesencephalon. One of the five major divisions of the brain; it is composed of the tectum and tegmentum.

Mesocorticolimbic pathway. The component of the mesotelencephalic dopamine system that has cell bodies in the ventral tegmental area that project to various cortical and limbic sites.

Mesoderm layer. The middle of the three cell layers in the developing embryo.

Mesotelencephalic dopamine system. The ascending projections of dopamine-releasing neurons from the substantia nigra and ventral tegmental area of the mesencephalon into various regions of the telencephalon.

Messenger RNA. A strand of RNA that is transcribed from DNA and then moves out of the cell nucleus where it is translated into a protein.

Metabolic tolerance. Tolerance that results from a reduction in the amount of a drug getting to its sites of action.

Metabotropic receptors. Receptors that are associated with signal proteins and G proteins.

Metaplasticity. The modulation of long term potentiation (LTP) and/or long-term depression (LTD) induction by prior synaptic activity.

Metastatic tumors. Tumors that originate in one organ and spread to another.

Metencephalon. One of the five major divisions of the brain; it includes the pons and cerebellum.

Microelectrodes. Extremely fine recording electrodes, which are used for intracellular recording.

Microglia. Glial cells that respond to injury or disease by engulfing cellular debris and triggering inflammatory responses.

Microsleeps. Brief periods of sleep that occur in sleep-deprived subjects while they remain sitting or standing.

Migration. The movement of cells from their site of creation in the ventricular zone of the neural tube to their appropriate target location.

Mild TBI. When there is a disturbance of consciousness following a blow to the head and there is no evidence of contusion or other structural damage.

Minor hemisphere. A term used in the past to refer to the right hemisphere, based on the incorrect assumption that the left hemisphere is dominant.

Mirror neurons. Neurons that fire when an individual performs a particular goal-directed hand movement or when they observe the same goal-directed movement performed by another.

Mirror-like system. Areas of the cortex that are active both when a person performs a particular response and when the person perceives somebody else performing the same response.

Miscellaneous peptides. One of the five categories of neuropeptide transmitters; it include those neuropeptide transmitters that don't fit into one of the other four categories.

Mitosis. The process of cell division that produces cells with the same number of chromosomes as the parent cell.

Mixed state. A state that can occur in bipolar disorder type I, where the patient simultaneously displays symptoms of both depression and mania.

Monoamine neurotransmitters. Small-molecule neurotransmitters that are synthesized from monoamines and comprise two classes: catecholamines and indolamines.

Monocular. Involving only one eye.

Monogamy. A pattern of mate bonding in which one male and one female form an enduring bond.

Monophasic sleep cycles. Sleep cycles that regularly involve only one period of sleep per day, typically at night.

Monozygotic twins. Twins that develop from the same zygote and are thus genetically identical.

Mood stabilizers. Drugs that effectively treat depression or mania without increasing the risk of mania or depression, respectively.

Morgan's Canon. The rule that the simplest possible interpretation for a behavioral observation should be given precedence.

Morphine. The major psychoactive ingredient in opium.

Morris water maze. A pool of milky water that has a goal platform invisible just beneath its surface and is used to study the ability of rats to learn spatial locations.

Morris water maze test. A widely used test of spatial memory in which rats must learn to swim directly to a platform hidden just beneath the surface of a circular pool of murky water.

Motor end-plate. The receptive area on a muscle fiber at a neuromuscular junction.

Motor equivalence. The ability of the sensorimotor system to carry out the same basic movement in different ways that involve different muscles.

Motor homunculus. The somatotopic map of the human primary motor cortex.

Motor pool. All of the motor neurons that innervate the fibers of a given muscle.

Motor theory of speech perception. The theory that the perception of speech involves activation of the same areas of the brain that are involved in the production of speech.

Motor units. A single motor neuron and all of the skeletal muscle fibers that are innervated by it.

Movement vigor. The control of the speed and amplitude of movement based on motivational factors.

MPTP. A neurotoxin that produces a disorder in primates that is similar to Parkinson's disease.

Müllerian-inhibiting substance. The testicular hormone that causes the precursor of the female reproductive ducts (the Müllerian system) to degenerate and the testes to descend.

Müllerian system. The embryonic precursor of the female reproductive ducts.

Multiple sclerosis (MS). A progressive disease that attacks the myelin of axons in the central nervous system.

Multipolar neuron. A neuron with more than two processes extending from its cell body.

Multipotent. Capable of developing into different cells of only one class of cells (e.g., different kinds of blood cells).

Mumby box. An apparatus that is used in the rat version of the delayed nonmatching-to-sample test.

Muscle spindles. Receptors that are embedded in skeletal muscle tissue and are sensitive to changes in muscle length.

Mutations. Accidental alterations in individual genes.

Myelencephalon. The most posterior of the five major divisions of the brain; the medulla.

Myelin. A fatty insulating substance.

Myelin sheaths. Coverings on the axons of some neurons that are rich in myelin and increase the speed and efficiency of axonal conduction.

Narcolepsy. A disorder of hypersomnia that is characterized by repeated, brief daytime sleep attacks and cataplexy.

Narcotic. A legal term generally used to refer to opioids.

Nasal hemiretina. The half of each retina next to the nose.

Natural selection. The idea that those heritable traits that are associated with high rates of survival and reproduction are the most likely to be passed on to future generations.

Nature–nurture issue. The debate about the relative contributions of nature (genes) and nurture (experience) to the behavioral capacities of individuals.

NEAT. Nonexercise activity thermogenesis, which is generated by activities such as fidgeting and the maintenance of posture and muscle tone.

Necrosis. Passive cell death.

Negative feedback systems. Systems in which feedback from changes in one direction elicit compensatory effects in the opposite direction.

Negative symptoms. Symptoms of schizophrenia that seem to represent a reduction or loss of typical function.

Neocortex. Six-layered cerebral cortex of relatively recent evolution; it constitutes 90 percent of human cerebral cortex.

Neoplasm. Tumor; literally, "new growth."

Nerve growth factor (NGF). The first neurotrophin to be discovered.

Nerves. Bundles of axons in the peripheral nervous system.

Neural crest. A structure situated just dorsal to the neural tube. It is formed from cells that break off from the neural tube as it is being formed.

Neural plate. A small patch of ectodermal tissue on the dorsal surface of the vertebrate embryo, from which the neural groove, the neural tube, and, ultimately, the mature nervous system develop.

Neural proliferation. The rapid increase in the number of neurons that follows the formation of the neural tube.

Neural regeneration. The regrowth of damaged neurons.

Neural tube. The tube that is formed in the vertebrate embryo when the edges of the neural groove fuse and that develops into the central nervous system.

Neuroanatomy. The study of the structure of the nervous system.

Neurochemistry. The study of the chemical bases of neural activity.

Neuroendocrinology. The study of the interactions between the nervous system and the endocrine system.

Neurogenesis. The growth of new neurons.

Neuromuscular junctions. The synapses of a motor neuron on a muscle.

Neurons. Cells of the nervous system that are specialized for the reception, conduction, and transmission of electrochemical signals.

Neuropathic pain. Severe chronic pain in the absence of a recognizable pain stimulus.

Neuropathology. The study of nervous system disorders.

Neuropeptide. Short amino acid chains.

Neuropeptide transmitters. Peptides that function as neurotransmitters, of which about 100 have been identified; also called *neuropeptides*.

Neuropeptide Y. A gut hunger peptide.

Neuropharmacology. The study of the effects of drugs on neural activity.

Neurophysiology. The study of the functions and activities of the nervous system.

Neuroplasticity. The notion that the brain is a "plastic" (changeable) organ that continuously grows and changes in response to an individual's environment and experiences.

Neuropsychology. The division of biopsychology that studies the psychological effects of brain damage in human patients.

Neuroscience. The scientific study of the nervous system.

Neurotoxins. Neural poisons.

Neurotrophins. Chemicals that are supplied to developing neurons by their targets and that promote their survival.

Nicotine. The major psychoactive ingredient of tobacco.

Nigrostriatal pathway. The pathway along which axons from neurons in the substantia nigra project to the striatum.

Nissl stain. A neural stain that has an affinity for structures in neuron cell bodies.

Nitric oxide. A soluble-gas neurotransmitter.

NMDA (N-methyl-d-aspartate) receptors. Glutamate receptors that play key roles in the development of stroke-induced brain damage and long-term potentiation at glutaminergic synapses.

Nodes of Ranvier. The gaps between adjacent myelin sheaths on an axon.

Nondirected synapses. Synapses at which the site of neurotransmitter release and the site of neurotransmitter reception are not close together.

Nootropics (smart drugs). Drugs that purportedly improve memory.

Norepinephrine. One of the three catecholamine neurotransmitters.

Nuclei. The DNA-containing structures of cells; also, clusters of neuronal cell bodies in the central nervous system (singular *nucleus*).

Nucleotide bases. A class of chemical substances that includes adenine, thymine, guanine, and cytosine—constituents of DNA.

Nucleus accumbens. Nucleus of the ventral striatum and a major terminal of the mesocorticolimbic dopamine pathway.

Nucleus magnocellularis. The nucleus of the caudal reticular formation that promotes relaxation of the core muscles during REM sleep and during attacks of cataplexy.

Nutritive density. Calories per unit volume of a food.

Ob/ob mice. Mice that are homozygous for the mutant ob gene; their body fat produces no leptin, and they become very obese.

Occipital face area. An area in the occipital lobe that is implicated in the processing of faces.

Occipital lobe. The most posterior of the four cerebral lobes; its function is primarily visual.

Off-center cells. Visual neurons that respond to lights shone in the center of their receptive fields with "off" firing and to lights shone in the periphery of their receptive fields with "on" firing.

Olfactory bulbs. Their output goes primarily to the amygdala and piriform cortex.

Olfactory glomeruli. Discrete clusters of neurons that lie near the surface of the olfactory bulbs.

Olfactory mucosa. The mucous membrane that lines the upper nasal passages and contains the olfactory receptor cells.

Oligodendrocytes. Glial cells that myelinate axons of the central nervous system; also known as *oligodendroglia*.

Oligodendroglia. Glial cells that myelinate central nervous system axons; also known as *oligodendrocytes*.

On-center cells. Visual neurons that respond to lights shone in the center of their receptive fields with "on" firing and to lights shone in the periphery of their receptive fields with "off" firing.

Ontogeny. The development of individuals over their life span.

Open-field test. In this test an animal is placed in a large, barren chamber and its activity is recorded.

Operant conditioning paradigm. A paradigm in which the rate of a particular voluntary response is increased by reinforcement or decreased by punishment.

Opioids. Morphine, codeine, heroin, and other chemicals with similar structures or effects.

Opioid peptides. One of the five classes of neuropeptide transmitters; it consists of those with a structure similar to the active ingredients of opium.

Opium. The sap that exudes from the seed pods of the opium poppy.

Opponent-process theory. The theory that a visual receptor or a neuron signals one color when it responds in one way (e.g., by increasing its firing rate) and signals the complementary color when it responds in the opposite way (e.g., by decreasing its firing rate).

Opsins. Light-sensitive ion channels that are found in the cell membranes of certain bacteria and algae. When opsins are illuminated with light, they open and allow ions to enter the cell.

Optic chiasm. The X-shaped structure on the inferior surface of the diencephalon; the point where the optic nerves decussate.

Optic tectum. The main destination of retinal ganglion cells in non-mammalian vertebrates.

Optogenetics. A method that uses genetic engineering techniques to insert the opsin gene, or variants of the opsin gene, into particular types of neurons. By inserting an opsin gene into a particular type of neuron, a researcher can use light to hyperpolarize or depolarize those neurons.

Orbitofrontal cortex. The cortex of the inferior frontal lobes, adjacent to the orbits, which receives olfactory input from the thalamus.

Orchidectomy. The removal of the testes.

Orexin. A neuropeptide that has been implicated in narcolepsy; sometimes called *hypocretin*.

Organ of Corti. The auditory receptor organ, comprising the basilar membrane, the hair cells, and the tectorial membrane.

Orphan drugs. Drugs for which the market is too small for the necessary developmental research to be profitable.

Orthodromic conduction. Axonal conduction in the normal direction—from the cell body toward the terminal buttons.

Ossicles. The three small bones of the middle ear: the malleus, the incus, and the stapes.

Oval window. The membrane that transfers vibrations from the ossicles to the fluid of the cochlea.

Ovariectomy. The removal of the ovaries.

Ovaries. The female gonads.

Oxytocin. One of the two major peptide hormones of the posterior pituitary, which in females stimulates contractions of the uterus during labor and the ejection of milk during suckling.

Pacinian corpuscles. The largest and most deeply positioned cutaneous receptors, which are sensitive to sudden displacements of the skin.

Paired-image subtraction technique. The use of PET or fMRI to locate constituent cognitive processes in the brain by producing an image of the difference in brain activity associated with two cognitive tasks that differ in terms of a single constituent cognitive process.

Panic attacks. Rapid-onset attacks of extreme fear and severe symptoms of stress (e.g., choking, heart palpitations, shortness of breath).

Panic disorder. An anxiety disorder characterized by recurrent rapid-onset attacks of extreme fear and severe symptoms of stress (choking, heart palpitations, and shortness of breath).

Parallel processing. The simultaneous analysis of a signal in different ways by the multiple parallel pathways of a neural network.

Parasympathetic nerves. Those autonomic motor nerves that project from the brain to the sacral region of the spinal cord.

Paraventricular nuclei. Hypothalamic nuclei that play a role in eating and synthesize hormones released by the posterior pituitary.

Parietal lobe. One of the four cerebral lobes; it is located just posterior to the central fissure.

Parkinson's disease. A movement disorder that is associated with degeneration of dopaminergic neurons in the substantia nigra.

Parvocellular layers. The layers of the lateral geniculate nuclei that are composed of neurons with small cell bodies; the top four layers (also called *P layers*).

Patellar tendon reflex. The stretch reflex that is elicited when the patellar tendon is struck.

Pathogens. Disease-causing agents.

Pattern separation. The ability to separate distinct percepts into individual memories for storage.

Pavlovian conditioning paradigm. A paradigm in which the experimenter pairs an initially neutral stimulus (conditional stimulus) with a stimulus (unconditional stimulus) that elicits a reflexive response (unconditional response); after several pairings, the neutral stimulus elicits a conditional response.

Penumbra. The dysfunctional area of brain tissue around an infarct. The tissue in the penumbra may recover or die in the days following a stroke.

Peptide hormones. Hormones that are short chains of amino acids.

Percept. The outcome of perception.

Perception. The higher-order process of integrating, recognizing, and interpreting complete patterns of sensations.

Perceptual decision making. Decisions affecting perception that are based on prior experiences and current incoming sensory information.

Periaqueductal gray (PAG). The gray matter around the cerebral aqueduct, which contains opiate receptors and activates a descending analgesia circuit.

Perimetry test. The procedure used to map scotomas.

Periodic limb movement disorder. Characterized by periodic, involuntary movements of the limbs often involving twitches of the legs during sleep; one cause of insomnia.

Periodotopy. The notion that auditory cortex topography is linked to the temporal components of sound.

Peripartum depression. The intense, sustained depression experienced by some females during pregnancy, after they give birth, or both.

Peripheral nervous system (PNS). The portion of the nervous system outside the skull and spine.

Perseveration. The tendency to continue making a formerly correct response that is currently incorrect.

Phagocytes. Cells, such as macrophages and microglia, that destroy and ingest pathogens.

Phagocytosis. The destruction and ingestion of foreign matter by cells of the immune system.

Phantom limb. Phenomenon wherein amputees still perceive the presence of their missing limb.

Phantom percepts. Products of perception when there is an absence of sensory input.

Pharmacological. Pertaining to the scientific study of drugs.

Phenothiazines. A class of antipsychotic drugs that bind effectively to both D_1 and D_2 receptors.

Phenotype. An organism's observable traits.

Phenylketonuria (PKU). A neurological disorder whose symptoms are vomiting, seizures, hyperactivity, hyperirritability, intellectual disability, brain damage, and high levels of phenylpyruvic acid in the urine.

Phenylpyruvic acid. A substance that is found in abnormally high concentrations in the urine of those suffering from phenylketonuria.

Pheromones. Chemicals that are released by an animal and elicit through their odor specific patterns of behavior in its conspecifics.

Phoneme. The smallest unit of sound that distinguishes among various words in a language.

Phonetic procedure. A procedure for reading aloud that involves the recognition of letters and the application of a language's rules of pronunciation.

Phonological analysis. Analysis of the sound of language.

Photopic spectral sensitivity curve. The graph of the sensitivity of cone-mediated vision to different wavelengths of light.

Photopic vision. Cone-mediated vision, which predominates when lighting is good.

Phylogeny. The evolutionary development of species.

Physical-dependence theories of addiction. Theories holding that the main factor that motivates drug-addicted individuals to keep taking drugs is the prevention or termination of withdrawal symptoms.

Physically dependent. Being in a state in which the discontinuation of drug taking will induce withdrawal reactions.

Physiological psychology. The division of biopsychology that studies the neural mechanisms of behavior through direct manipulation of the brains of nonhuman animal subjects in controlled experiments.

Pia mater. The delicate, innermost meninx.

Pineal gland. The endocrine gland that is the human body's sole source of melatonin.

Pioneer growth cones. The first growth cones to travel along a particular route in the developing nervous system.

Piriform cortex. An area of medial temporal cortex that is adjacent to the amygdala and that receives direct olfactory input.

Pituitary gland. The gland that dangles from, and is controlled by, the hypothalamus.

Pituitary peptides. One of the five categories of neuropeptide transmitters; it contains neuropeptides that were first identified as hormones released by the pituitary.

Pituitary stalk. The structure connecting the hypothalamus and the pituitary gland.

Place cells. Neurons that respond only when the subject is in specific locations (i.e., in the place fields of the neurons).

Planum temporale. An area of temporal lobe cortex that lies in the posterior region of the lateral fissure and, in the left hemisphere, roughly corresponds to Wernicke's area.

Plethysmography. Any technique for measuring changes in the volume of blood in a part of the body.

Pluripotent. Cells that can develop into many, but not all, classes of body cells.

Polarized. In the context of membrane potentials, it is a membrane potential that is not zero.

Polyandry. A pattern of mate bonding in which one female bonds with more than one male.

Polygraphy. A method of interrogation that employs ANS indexes of emotion to infer the truthfulness of a person's responses.

Polygyny. A pattern of mate bonding in which one male bonds with more than one female; the most prevalent pattern of mate bonding in mammals.

Polyphasic sleep cycles. Sleep cycles that regularly involve more than one period of sleep per day.

Pons. The metencephalic structure that creates a bulge on the ventral surface of the brain stem.

Positive symptoms. Symptoms of schizophrenia that seem to represent an excess of typical function.

Positive-incentive theories of addiction. Theories holding that the primary factor in most cases of addiction is the craving for the positive-incentive (expected pleasure-producing) properties of the drug.

Positive-incentive theory. The idea that behaviors (e.g., eating and drinking) are motivated by their anticipated pleasurable effects.

Positive-incentive value. The anticipated pleasure associated with a particular action, such as taking a drug.

Positron emission tomography (PET). A technique for visualizing brain activity, usually by measuring the accumulation of radioactive fluorodeoxyglucose (FDG) in active areas of the brain.

Postcentral gyrus. The gyrus located just posterior to the central fissure; its function is primarily somatosensory.

Posterior. Toward the tail end of a vertebrate or toward the back of the head.

Posterior parietal association cortex. An area of association cortex that receives input from the visual, auditory, and somatosensory systems and is involved in the perception of spatial location and guidance of voluntary behavior.

Posterior parietal cortex. The posterior area of the parietal cortex.

Posterior pituitary. The part of the pituitary gland that contains the terminals of hypothalamic neurons.

Postnatal period. The period of development after birth.

Postsynaptic potentials. Potentials that move the postsynaptic cell's membrane potential away from the resting state.

Posttraumatic amnesia (PTA). Amnesia produced by a nonpenetrating head injury (a blow to the head that does not penetrate the skull).

Prader-Willi syndrome. A neurodevelopmental disorder that is characterized by insatiable appetite and exceptionally slow metabolism.

Precentral gyrus. The gyrus located just anterior to the central fissure; its function is primarily motor.

Prefrontal cortex. The areas of frontal cortex that are anterior to the frontal motor areas.

Prefrontal lobes. Areas of cortex, left and right, that are located at the very front of the brain—in the frontal lobes.

Prefrontal lobotomy. A surgical procedure in which the connections between the prefrontal lobes and the rest of the brain are cut, as a treatment for mental illness.

Premotor cortex. The area of secondary motor cortex that lies between the supplementary motor area and the lateral fissure.

Prenatal period. The period of development before birth.

Prestriate cortex. The band of tissue in the occipital lobe that surrounds the primary visual cortex and contains areas of secondary visual cortex.

Primary motor cortex. The cortex of the precentral gyrus, which is the major point of departure for motor signals descending from the cerebral cortex into lower levels of the sensorimotor system.

Primary sensory cortex. An area of sensory cortex that receives most of its input directly from the thalamic relay nuclei of one sensory system.

Primary visual cortex. The area of the cortex that receives direct input from the lateral geniculate nuclei (also called *striate cortex*).

Primates. One of 20 different orders of mammals; there are about a dozen families of primates.

Proceptive behaviors. Behaviors that solicit the sexual advances of members of the other sex.

Progesterone. A progestin that prepares the uterus and breasts for pregnancy.

Progestins. The class of steroid hormones that includes progesterone.

Promoters. Stretches of DNA whose function is to determine whether or not particular structural genes are converted into proteins through the process of gene expression.

Prosopagnosia. Visual agnosia for faces.

Protein hormones. Hormones that are long chains of amino acids.

Proteins. Long chains of amino acids.

Proximal. Close to something.

Proximal segment. The segment of a cut axon between the cut and the cell body.

Psychedelic drugs. Drugs whose primary action is to alter perception, emotion, and cognition.

Psychiatric disorder. A disorder of psychological function sufficiently severe to require treatment by a psychiatrist or clinical psychologist.

Psychoactive drugs. Drugs that influence subjective experience and behavior by acting on the nervous system.

Psychoneuroimmunology. The study of interactions among psychological factors, the nervous system, and the immune system.

Psychopharmacology. The division of biopsychology that studies the effects of drugs on the brain and behavior.

Psychophysiology. The division of biopsychology that studies the relation between physiological activity and psychological processes in human subjects by noninvasive methods.

Psychosis. A loss of touch with reality.

Psychosomatic disorder. Any physical disorder that can be caused or exacerbated by stress.

Psychosurgery. Any brain surgery performed for the treatment of a psychological problem (e.g., prefrontal lobotomy).

P300 wave. The positive EEG wave that usually occurs about 300 milliseconds after a momentary stimulus that has meaning for the subject.

Pulsatile hormone release. The typical pattern of hormone release: Hormones are discharged several times per day in large surges.

Pure research. Research motivated primarily by the curiosity of the researcher and done solely for the purpose of acquiring knowledge.

Purkinje effect. In intense light, red and yellow wavelengths look brighter than blue or green wavelengths of equal intensity; in dim light, blue and green wavelengths look brighter than red and yellow wavelengths of equal intensity.

Putamen. A structure that is joined to the caudate by a series of fiber bridges; together the putamen and caudate compose the striatum.

Pyramidal cell layer. One of the major layers of cell bodies in the hippocampus.

Pyramidal cells. Large multipolar cortical neurons with a pyramid-shaped cell body, an apical dendrite, and a very long axon.

Quasiexperimental studies. Studies of groups of subjects who have been exposed to the conditions of interest in the real world; such studies have the appearance of experiments but are not true experiments because potential confounded variables have not been controlled for.

Radial arm maze. A maze in which several arms radiate out from a central starting chamber; commonly used to study spatial learning in rats.

Radial arm maze test. A widely used test of rats' spatial ability in which the same arms are baited on each trial, and the rats must learn to visit only the baited arms once per trial.

Radial glial cells. Glial cells that exist in the neural tube during the period of neural migration and that form a network along which radial migration occurs. Some radial glial cells are stem cells.

Radial migration. Movement of cells in the developing neural tube from the ventricular zone in a straight line outward toward the tube's outer wall.

Reactive depression. Depression that is triggered by a negative experience.

Reappraisal paradigm. An experimental method for studying emotion; subjects are asked to reinterpret a film or photo to change their emotional reaction to it while their brain activity is recorded.

Receptive. Pertaining to the comprehension of language and speech.

Receptive field. The area of the visual field within which it is possible for the appropriate stimulus to influence the firing of a visual neuron.

Receptor blockers. Antagonistic drugs that bind to postsynaptic receptors without activating them and block the access of the usual neurotransmitter.

Receptor subtypes. The different types of receptors to which a particular neurotransmitter can bind.

Receptors. Cells that are specialized to receive chemical, mechanical, or radiant signals from the environment; also proteins that contain binding sites for particular neurotransmitters.

Recessive trait. The trait of a dichotomous pair that is not expressed in the phenotype of heterozygous individuals.

Reciprocal innervation. The principle of spinal cord circuitry that causes a muscle to automatically relax when a muscle that is antagonistic to it contracts.

Recuperation theories of sleep. Theories based on the premise that being awake disturbs the body's homeostasis and the function of sleep is to restore it.

Recurrent collateral inhibition. The inhibition of a neuron that is produced by its own activity via a collateral branch of its axon and an inhibitory interneuron.

Red nucleus. A structure of the sensorimotor system that is located in the tegmentum of the mesencephalon.

Reference memory. Memory for the general principles and skills that are required to perform a task.

Relapse. The return to one's drug taking habit after a period of voluntary abstinence.

Relative refractory period. A period after the absolute refractory period during which a higher-than-normal amount of stimulation is necessary to make a neuron fire.

Release-inhibiting hormones. Hypothalamic hormones that inhibit the release of hormones from the anterior pituitary.

Releasing hormones. Hypothalamic hormones that stimulate the release of hormones from the anterior pituitary.

REM sleep. The stage of sleep characterized by rapid eye movements, loss of core muscle tone, and emergent stage 1 EEG.

REM-sleep behavior disorder. A disorder where the individual experiences REM sleep without core-muscle atonia.

Remote memory. Memory for experiences in the distant past.

Repetition priming tests. Tests of implicit memory; in one example, a list of words is presented, then fragments of the original words are presented and the subject is asked to complete them.

Repetitive transcranial magnetic stimulation (rTMS). A form of transcranial magnetic stimulation (TMS) that involves the delivery of repetitive magnetic pulses at either high frequencies (e.g., five pulses per second; high-frequency rTMS) or low frequencies (e.g., less than one pulse per second; low-frequency rTMS) to specific cortical areas.

Replacement injections. Injections of a hormone whose natural release has been curtailed by the removal of the gland that normally releases it.

Replication. The process by which the DNA molecule duplicates itself.

Repressors. Proteins that bind to DNA and decrease gene expression.

Reserpine. The first monoamine antagonist to be used in the treatment of schizophrenia; the active ingredient of the snakeroot plant.

Response-chunking hypothesis. The idea that practice combines the central sensorimotor programs that control individual responses into programs that control sequences (chunks) of behavior.

Resting potential. The steady membrane potential of a neuron at rest, usually about –70 mV.

Resting state-fMRI. One application of functional magnetic resonance imaging (fMRI) wherein brain scans are carried out while the participant is not performing any explicit tasks.

Restless legs syndrome. Tension or uneasiness in the legs that keeps a person from falling asleep; one cause of insomnia.

Reticular activating system. The hypothetical arousal system in the reticular formation.

Retina-geniculate-striate pathway. The major visual pathway from each retina to the striate cortex (primary visual cortex) via the lateral geniculate nuclei of the thalamus.

Retinal ganglion cells. Retinal neurons whose axons leave the eyeball and form the optic nerve.

Retinex theory. Land's theory that the color of an object is determined by its reflectance, which the visual system calculates by comparing the ability of adjacent surfaces to reflect short, medium, and long wavelengths.

Retinotopic. Organized, like the primary visual cortex, according to a map of the retina.

Retrograde amnesia. Loss of memory for events or information learned before the amnesia-inducing brain injury.

Retrograde degeneration. Degeneration of the proximal segment of a cut axon.

Reuptake. The drawing back into the terminal button of neurotransmitter molecules after their release into the synapse; the most common mechanism for deactivating a released neurotransmitter.

Reversible lesions. Methods for temporarily eliminating the activity in a particular area of the brain while tests are being conducted.

Rhodopsin. The photopigment of rods.

Ribonucleic acid (RNA). A molecule that is similar to DNA except that it has the nucleotide base uracil and a phosphate and ribose backbone.

Ribosome. Intracellular structures found in large numbers in the cytoplasm of living cells. They are involved in the translation phase of gene expression.

Risk-assessment test. An animal model of anxiety. After a single brief exposure to a cat on the surface of a laboratory burrow system, rats flee to their burrows and freeze. Then they engage in a variety of risk-assessment behaviors.

RNA editing. An epigenetic mechanism wherein messenger RNA is modified through the actions of small RNA molecules and other proteins.

Rods. The visual receptors in the retina that mediate achromatic, low-acuity vision under dim light.

Rubber-hand illusion. The feeling that an extraneous object, usually a rubber hand, is actually part of one's own body.

Saccades. The rapid movements of the eyes between fixations.

Sagittal sections. Any slices of brain tissue cut in a plane that is parallel to the side of the brain.

Saltatory conduction. Conduction of an action potential from one node of Ranvier to the next along a myelinated axon.

Satiety. The motivational state that terminates a meal when there is food remaining.

Savants. Individuals with developmental disabilities who nevertheless display amazing and specific cognitive or artistic abilities; savant abilities are sometimes associated with autism spectrum disorder.

Schwann cells. The glial cells that compose the myelin sheaths of PNS axons and promote the regeneration of PNS axons.

Scientific inference. The logical process by which observable events are used to infer the properties of unobservable events.

Scotoma. An area of blindness produced by damage to, or disruption of, an area of the visual system.

Scotopic spectral sensitivity curve. The graph of the sensitivity of rod-mediated vision to different wavelengths of light.

Scotopic vision. Rod-mediated vision, which predominates in dim light.

Scrotum. The sac that holds the male testes outside the body cavity.

Seasonal affective disorder (SAD). Type of major depressive disorder in which episodes of depression typically recur during particular seasons—usually during the winter months.

Second messenger. A chemical synthesized in a neuron in response to the binding of a neurotransmitter to a metabotropic receptor in its cell membrane.

Secondary motor cortex. An area of the cerebral cortex that receives much of its input from association cortex and sends much of its output to primary motor cortex.

Secondary sensory cortex. An area of sensory cortex that receives most of its input from the primary sensory cortex of one sensory system or from other areas of secondary cortex of the same system.

Secondary sex characteristics. Body features, other than the reproductive organs, that distinguish males from females.

Secondary visual cortex. Areas of cerebral cortex that receive most of their input from primary visual cortex.

Selective attention. The ability to focus on a small subset of the multitude of stimuli that are being received at any one time.

Selective serotonin-reuptake inhibitors (SSRIs). Class of drugs that exert agonistic effects by blocking the reuptake of serotonin from synapses; typically used to treat depression.

Self-stimulation paradigm. A paradigm in which animals press a lever to administer reinforcing electrical stimulation to particular sites in their own brains.

Semantic analysis. Analysis of the meaning of language.

Semantic memories. Explicit memories for general facts or knowledge.

Semicircular canals. The receptive organs of the vestibular system.

Sensation. The process of detecting the presence of stimuli.

Sensitive period. An interval of time during development when an experience can have a greater effect on development if it occurs during that interval, as opposed to outside that interval.

Sensitivity. In vision, the ability to detect the presence of dimly lit objects.

Sensorimotor phase. The second of the two phases of birdsong development, during which juvenile birds progress from subsongs to adult songs.

Sensory evoked potential. A change in the electrical activity of the brain (e.g., in the cortical EEG) that is elicited by the momentary presentation of a sensory stimulus.

Sensory feedback. Sensory signals that are produced by a response and are often used to guide the continuation of the response.

Sensory phase. The first of the two phases of birdsong development, during which young birds do not sing but form memories of the adult songs they hear.

Sensory relay nuclei. Those nuclei of the thalamus whose main function is to relay sensory signals to the appropriate areas of cortex.

Sensory-specific satiety. The fact that the consumption of a particular food produces greater satiety for foods of the same taste than for other foods.

Septum. A midline nucleus of the limbic system, located near the anterior tip of the cingulate cortex.

Serotonin. An indolamine neurotransmitter; the only member of this class of monoamine neurotransmitters found in the mammalian nervous system.

Set point. The value of a physiological parameter that is maintained constantly by physiological or behavioral mechanisms; for example, the body's energy resources are often assumed to be maintained at a constant optimal level by compensatory changes in hunger.

Set-point assumption. The assumption that hunger is typically triggered by a decline in the body's energy reserves below their set point.

Settling point. The point at which various factors that influence the level of some regulated function (such as body weight) achieve an equilibrium.

Sex chromosomes. The pair of chromosomes that determine an individual's genetic sex: XX for a female and XY for a male.

Sex-linked traits. Traits that are influenced by genes on the sex chromosomes.

Sexual dimorphisms. Instances where a behavior (or structure) comes in two distinct classes (male or female) into which most individuals can be unambiguously assigned.

Sexually dimorphic nucleus. The nucleus in the medial preoptic area of rats that is larger in males than in females.

Sham eating. The experimental protocol in which an animal chews and swallows food, after which the food immediately exits its body through a tube implanted in its esophagus.

Sham rage. The exaggerated, poorly directed aggressive responses of decorticate animals.

Short-term memory. Storage of information for brief periods of time while a person attends to it.

Signal averaging. A method of increasing the signal-to-noise ratio by reducing background noise.

Simple cells. Neurons in the visual cortex that respond maximally to straight-edge stimuli of a particular width and orientation.

Simple seizures. Focal seizures in which the symptoms are primarily sensory or motor or both.

Simultanagnosia. A difficulty attending to more than one stimulus at a time.

Sinestrals. Left-handers.

Skeletal muscle (extrafusal muscle). Striated muscle that is attached to the skeleton and is usually under voluntary control.

Skin conductance level (SCL). A measure of the background level of skin conductance associated with a particular situation.

Skin conductance response (SCR). The transient change in skin conductance associated with discrete experiences.

Sleep apnea. A condition in which sleep is repeatedly disturbed by momentary interruptions in breathing.

Sleep inertia. The unpleasant feeling of grogginess that is sometimes experienced for a few minutes after awakening.

Sleep paralysis. A sleep disorder characterized by the inability to move (paralysis) just as a person is falling asleep or waking up.

Slow-wave sleep (SWS). Stage 3 sleep, which is characterized by the largest and slowest EEG waves.

Smoking. Inhaling the smoke from the burning of tobacco.

Smoker's syndrome. The chest pain, labored breathing, wheezing, coughing, and heightened susceptibility to infections of the respiratory tract commonly observed in tobacco smokers.

Sodium amytal test. A test involving the anesthetization of first one cerebral hemisphere and then the other to determine which hemisphere plays the dominant role in language.

Sodium–potassium pumps. An ion transporter that actively exchanges three Na^+ ions inside the neuron for two K^+ ions outside.

Solitary nucleus. The medullary relay nucleus of the gustatory system.

Soluble-gas neurotransmitters. A class of unconventional neurotransmitters that includes nitric oxide and carbon monoxide.

Somal translocation. One of two major modes of neural migration, in which an extension grows out from the undeveloped neuron and draws the cell body up into it.

Somatic nervous system (SNS). The part of the peripheral nervous system that interacts with the external environment.

Somatosensory homunculus. The somatotopic map in the primary somatosensory cortex.

Somatotopic. Organized, like the primary somatosensory cortex, according to a map of the surface of the body.

Spandrels. Incidental nonadaptive evolutionary by-products of some adaptive characteristic.

Spatial resolution. Ability of a recording technique to detect differences in spatial location (e.g., to pinpoint a location in the brain).

Spatial summation. The integration of signals that originate at different sites on the neuron's membrane.

Species. A group of organisms that is reproductively isolated from other organisms; the members of one species cannot produce fertile offspring by mating with members of other species.

Species-common behaviors. Behaviors that are displayed in the same manner by virtually all like members of a species.

Specific phobia. An anxiety disorder that involves strong fear or anxiety about particular objects (e.g., birds, spiders) or situations (e.g., enclosed spaces, darkness).

Spindle afferent neurons. Neurons that carry signals from muscle spindles into the spinal cord via the dorsal root.

Split-brain patients. Commissurotomized patients.

Sry gene. A gene on the Y chromosome that triggers the production of Sry protein.

Sry protein. A protein that causes the medulla of each primordial gonad to grow and develop into a testis.

Standard consolidation theory. The theory that memories are temporarily stored in the hippocampus until they can be transferred to a more stable cortical storage system. Also known as *dual-trace theory*.

Static phase. The second phase of the VMH syndrome, during which the obese animal maintains a stable level of obesity.

Stellate cells. Small star-shaped cortical interneurons.

Stem cells. Cells that have an almost unlimited capacity for self-renewal and the ability to develop into many different types of cells.

Stereognosis. The process of identifying objects by touch.

Stereotaxic atlas. A series of maps representing the three-dimensional structure of the brain that is used to determine coordinates for stereotaxic surgery.

Stereotaxic instrument. A device for performing stereotaxic surgery, composed of two parts: a head holder and an electrode holder.

Steroid hormones. Hormones that are synthesized from cholesterol.

Stimulants. Drugs that produce general increases in neural and behavioral activity.

Stress. The physiological changes that occur when the body is exposed to harm or threat.

Stressors. Experiences that induce a stress response.

Stretch reflex. A reflexive counteracting reaction to an unanticipated external stretching force on a muscle.

Striatum. A structure of the basal ganglia that is the terminal of the dopaminergic nigrostriatal pathway.

Strokes. Sudden-onset cerebrovascular disorders that cause brain damage.

Subarachnoid space. The space beneath the arachnoid membrane, which contains many large blood vessels and cerebrospinal fluid.

Subcutaneous fat. Fat stored under the skin.

Subordination stress. Stress experienced by animals, typically males, that are continually attacked by higher-ranking conspecifics.

Substantia nigra. The midbrain nucleus whose neurons project via the nigrostriatal pathway to the striatum of the basal ganglia; it is part of the mesotelencephalic dopamine system.

Subthalamic nucleus. A nucleus that lies just below the thalamus and is connected to the basal ganglia; deep brain stimulation applied to this site has been used to treat Parkinson's disease.

Subventricular zone. A region adjacent to the ventricular zone; the ventricular zone is adjacent to the ventricles.

Sulci. Small furrows in a convoluted cortex.

Superior. Toward the top of the primate head.

Superior colliculi. Two of the four nuclei that compose the tectum; they receive major visual input.

Superior olives. Medullary nuclei that play a role in sound localization.

Superior temporal gyri. The plural of superior temporal gyrus.

Superior temporal gyrus. The large gyrus of the temporal lobe adjacent to the lateral fissure; the location of auditory cortex.

Supplementary motor area. The area of secondary motor cortex that is within and adjacent to the longitudinal fissure.

Suppression paradigm. An experimental method for studying emotion; subjects are asked to inhibit their emotional reactions to unpleasant films or photos while their brain activity is recorded.

Suprachiasmatic nuclei (SCN). Nuclei of the medial hypothalamus that control the circadian cycles of various body functions.

Supraoptic nuclei. Hypothalamic nuclei in which the hormones of the posterior pituitary are synthesized.

Surface dyslexia. A reading disorder in which the lexical procedure is disrupted while the phonetic procedure is not.

Surface interpolation. The process by which we perceive surfaces; the visual system extracts information about edges and from it infers the appearance of large surfaces.

Sympathetic nerves. Those motor nerves of the autonomic nervous system that project from the central nervous system in the lumbar and thoracic region areas of the spinal cord.

Synaptic vesicles. Small spherical membranes that store neurotransmitter molecules and release them into the synaptic cleft.

Synaptogenesis. The formation of new synapses.

Synergistic muscles. Pairs of muscles whose contraction produces a movement in the same direction.

T cells. T lymphocytes; lymphocytes that bind to foreign micro-organisms and cells that contain them and, in so doing, destroy them.

Tangential migration. Movement of cells in the developing neural tube in a direction parallel to the tube's walls.

Tardive dyskinesia (TD). A motor disorder that results from chronic use of certain antipsychotic drugs.

Target-site concept. The idea that aggressive and defensive behaviors of an animal are often designed to attack specific sites on the body of another animal while protecting specific sites on its own.

Taste buds. Clusters of taste receptors found on the tongue and in parts of the oral cavity.

Tau. The first circadian gene to be identified in mammals.

Tau protein. Plays a role in maintaining the overall structure of neurons.

Tectorial membrane. The cochlear membrane that rests on the hair cells.

Tegmentum. The ventral division of the mesencephalon; it includes part of the reticular formation, substantia nigra, and red nucleus.

Telencephalon. The most superior of the brain's five major divisions.

Temporal hemiretina. The half of each retina next to the temple.

Temporal lobe. One of the four major cerebral lobes; it lies adjacent to the temples and contains the hippocampus and amygdala.

Temporal resolution. Ability of a recording technique to detect differences in time (i.e., to pinpoint when an event occurred).

Temporal summation. The integration of neural signals that occur at different times at the same synapse.

Teratogen. A drug or other chemical that causes birth defects.

Testes. The male gonads.

Testosterone. The most common androgen.

Thalamus. The large two-lobed diencephalic structure that constitutes the anterior end of the brain stem; many of its nuclei are sensory relay nuclei that project to the cortex.

THC. Delta-9-tetrahydrocannabinol, the main psychoactive constituent of marijuana.

Thermal grid illusion. The perception of pain that results from placing one's hand on a grid of metal rods that alternate between cool and warm.

Thigmotaxic. Tending to stay near the walls of an open space such as a test chamber.

Thinking creatively. Thinking in productive, unconventional ways.

Threshold of excitation. The level of depolarization necessary to generate an action potential; usually about −65 mV.

Thrombosis. The blockage of blood flow by a plug (a thrombus) at the site of its formation.

Thyrotropin. The anterior pituitary hormone that stimulates the release of hormones from the thyroid gland.

Thyrotropin-releasing hormone. The hypothalamic hormone that stimulates the release of thyrotropin from the anterior pituitary.

Tics. Involuntary, repetitive, stereotyped movements or vocalizations; the defining feature of Tourette's disorder.

Tinnitus. Ringing in the ears.

Token test. A preliminary test for language-related deficits that involves following verbal instructions to touch or move tokens of different shapes, sizes, and colors.

Toll-like receptors. Receptors found in the cell membranes of many cells of the innate immune system; they trigger phagocytosis and inflammatory responses.

Tonic-clonic seizure. A type of generalized seizure whose primary behavioral symptoms are loss of consciousness, loss of equilibrium, and a tonic-clonic convulsion—a convulsion involving both tonus and clonus.

Tonotopic. Organized, like the primary auditory cortex, according to the frequency of sound.

Top-down. A sort of neural mechanism that involves activation of lower cortical areas by higher cortical areas.

Topographic gradient hypothesis. The hypothesis that axonal growth is guided by the relative position of the cell bodies on intersecting gradients, rather than by point-to-point coding of neural connections.

Totipotent. Capable of developing into any type of body cell.

Tourette's disorder. A disorder of tics (involuntary, repetitive, stereotyped movements or vocalizations).

Toxic psychosis. A chronic psychiatric disorder produced by exposure to a neurotoxin.

Tracts. Bundles of axons in the central nervous system.

Transcranial electrical stimulation. A technique that can be used to stimulate ("turn on") an area of the cortex by applying an electrical current through two electrodes placed directly on the scalp.

Transcranial magnetic stimulation (TMS). A technique that can be used to stimulate ("turn on") or turn off an area of the cortex by creating a magnetic field under a coil positioned next to the skull.

Transcranial ultrasound stimulation. A technique that, like transcranial electrical stimulation and magnetic stimulation, can be used to activate particular brain structures.

Transcription. The first phase of gene expression, wherein a strand of messenger RNA (mRNA) is transcribed from one of the exposed DNA strands and carries the genetic code from the nucleus into the cytoplasm of the cell.

Transcription factors. Intracellular proteins that bind to DNA and influence the operation of particular genes.

Transduction. The conversion of one form of energy to another.

Transfer RNA. Molecules of RNA that carry amino acids to ribosomes during protein synthesis; each kind of amino acid is carried by a different kind of transfer RNA molecule.

Transgender. An individual who identifies as a man, a woman, or some intersection thereof.

Transgenerational epigenetics. A subfield of epigenetics that examines the transmission of experiences via epigenetic mechanisms across generations.

Transgenic mice. Mice into which the genetic material of another species has been introduced.

Translation. The second phase of gene expression, wherein the strand of messenger RNA (mRNA) is converted by a ribosome and transfer RNA (tRNA) into a protein.

Transient global amnesia. A sudden onset severe anterograde amnesia and moderate retrograde amnesia for explicit episodic memory that is transient—typically lasting only between 4 to 6 hours.

Translational bottleneck. A barrier keeping promising ideas and treatments from becoming the focus of translational research; largely created by the massive cost of such research.

Translational research. Research designed to translate basic scientific discoveries into effective applications (e.g., into clinical treatments).

Transneuronal degeneration. Degeneration of a neuron caused by damage to another neuron to which it is linked by a synapse.

Transorbital lobotomy. A prefrontal lobotomy performed with an instrument inserted through the eye socket.

Transporters. Mechanisms in the membrane of a cell that actively transport ions or molecules across the membrane.

Traumatic brain injury. Serious damage caused to the brain by a blow to the head.

Tricyclic antidepressants. Drugs with an antidepressant action and a three-ring molecular structure.

Tripartite synapse. A synapse that involves two neurons and an astroglia.

True-breeding lines. Breeding lines in which interbred members always produce offspring with the same trait, generation after generation.

Tumor (neoplasm). A mass of cells that grows independently of the rest of the body.

Tympanic membrane. The eardrum.

Typical antipsychotics. The first generation of antipsychotic drugs.

Unipolar neuron. A neuron with one process extending from its cell body.

Unipotent. Cells that can develop into only one type of cell.

Up-regulation. An increase in the number of receptors for a neurotransmitter in response to decreased release of that neurotransmitter.

Urbach-Wiethe disease. A genetic disorder that often results in the calcification of the amygdala and surrounding brain structures.

Vaccination. Administering a weakened form of a virus so that if the virus later invades, the adaptive immune system is prepared to deal with it.

Vaping. Inhaling a vapor that contains nicotine.

Vasopressin. One of the two major peptide hormones of the posterior pituitary; it facilitates reabsorption of water by kidneys and is thus also called *antidiuretic hormone*.

Ventral. Toward the chest surface of a vertebrate or toward the bottom of the head.

Ventral horns. The two ventral arms of the spinal gray matter.

Ventral posterior nucleus. A thalamic relay nucleus in both the somatosensory and gustatory systems.

Ventral stream. The group of visual pathways that flows from the primary visual cortex to the ventral prestriate cortex to the inferotemporal cortex.

Ventral tegmental area. The midbrain nucleus of the mesotelencephalic dopamine system that is the major source of the mesoscorticolimbic pathway.

Ventricular zone. The region adjacent to the ventricle in the developing neural tube.

Ventromedial hypothalamus (VMH). The area of the hypothalamus that was once thought to be a satiety center.

Ventromedial nucleus (VMN). A hypothalamic nucleus that is thought to be involved in female sexual behavior.

Vertebrates. Chordates that possess spinal bones.

Vestibular system. The sensory system that detects changes in the direction and intensity of head movements and that contributes to the maintenance of balance through its output to the motor system.

Visceral fat. Fat stored around the internal organs of the body cavity.

Visual agnosia. A failure to recognize visual stimuli that is not attributable to sensory, verbal, or intellectual impairment.

Visual association cortex. Areas of cerebral cortex that receive input from areas of secondary visual cortex as well as from secondary areas of other sensory systems.

Visual completion. The completion or filling in of a scotoma by the brain.

Voltage-activated ion channels. Ion channels that open and close in response to changes in the level of the membrane potential.

Wechsler Adult Intelligence Scale (WAIS). A widely used test of general intelligence that includes 11 subtests.

Wernicke-Geschwind model. An influential model of cortical language localization in the left hemisphere.

Wernicke's aphasia. A hypothetical disorder of language comprehension with no associated deficits in speech production.

Wernicke's area. The area of the left temporal cortex hypothesized by Wernicke to be the center of language comprehension.

"Where" versus "what" theory. The theory that the dorsal stream mediates the perception of where things are and the ventral stream mediates the perception of what things are.

White matter. Portions of the nervous system that are white because they are composed largely of myelinated axons.

Williams syndrome. A neurodevelopmental disorder characterized by intellectual disability, accompanied by preserved language and social skills.

Withdrawal reflex. The reflexive withdrawal of a limb when it comes in contact with a painful stimulus.

Withdrawal syndrome. The illness brought on by the elimination from the body of a drug on which the person is physically dependent.

Within-subjects design. An experimental design in which the same subjects are tested under each condition.

Wolffian system. The embryonic precursor of the male reproductive ducts.

Working memory. Temporary memory that is necessary for the successful performance of a task on which one is currently working.

Zeitgebers. Environmental cues, such as the light–dark cycle, that entrain circadian rhythms.

Zeitgeist. The general intellectual climate of a culture.

Zygote. The cell formed from the amalgamation of a sperm cell and an ovum.

References

Aarsland, D., Creese, B., Politis, M., & Chaudhuri, R. K. (2017). Cognitive decline in Parkinson disease. *Nature Reviews Neurology, 13*(4), 217–231.

Abbott, A. (2018). The brain inflamed. *Nature, 556,* 426–428.

Abbott, L. C., & Winzer-Serhan, U. H. (2012). Smoking during pregnancy: Lessons learned from epidemiological studies and experimental studies using animal models. *Clinical Reviews in Toxicology, 42,* 279–303.

Aboitiz, F. (2017). *A Brain for Speech: A View from Evolutionary Neuroanatomy.* Springer.

Abraham, W. C. (2006). Memory maintenance: The changing nature of neural mechanisms. *Current Directions in Psychological Science, 15,* 5–8.

Abraham, W. C., Jones, O. D., & Glanzman, D. L. (2019). Is plasticity of synapses the mechanism of long-term memory storage? *NPJ Science of Learning, 4*(1), 1–10.

Abruzzese, G. A., Crisosto, N., De Grava Kempinas, W., & Sotomayor-Zárate, R. (2018). Developmental programming of the female neuroendocrine system by steroids. *Journal of Neuroendocrinology, 30*(10), e12632.

Ackerman, J. (2012). The ultimate social network. *Scientific American, 306,* 36–43.

Ackermann, S., Hartmann, F., Papassotiropoulos, A., de Quervain, D. J., & Rasch, B. (2015). No associations between interindividual differences in sleep parameters and episodic memory consolidation. *Sleep, 38,* 951–959.

Ackman, J. B., & Crair, M. C. (2014). Role of emergent neural activity in visual map development. *Current Opinion in Neurobiology, 24,* 166–175.

Adair, J. C., & Barrett, A. M. (2008). Spatial neglect: Clinical and neuroscience review. *Annals of the New York Academy of Sciences, 1142*(1), 21–43.

Adams, C. P., and Brantner, V. V. (2010). Spending on new drug development. *Health Economics, 19,* 130–141.

Adams, K. V., & Morshead, C. M. (2018). Neural stem cell heterogeneity in the mammalian forebrain. *Progress in Neurobiology, 170,* 2–36.

Adan, R. A. H. (2013). Mechanisms underlying current and future anti-obesity drugs. *Trends in Neurosciences, 36,* 133–140.

Adlard, P. A., Perreau, V. M., Pop, V., & Cotman, C. W. (2005). Voluntary exercise decreases amyloid load in a transgenic model of Alzheimer's diseases. *Journal of Neuroscience, 25,* 4217–4221.

Adler, D. S., Wilkinson, K. N., Blockley, S., Mark, D. F., Pinhasi, R., et al. (2014). Early Levallois technology and the lower to middle paleolithic transition in the Southern Caucasus. *Science, 345,* 1609–1613.

Adli, M. (2018). The CRISPR tool kit for genome editing and beyond. *Nature Communications, 9*(1), 1–13.

Adolphs, R. (2006). Perception and emotion: How we recognize facial expressions. *Current Directions in Psychological Science, 15,* 222–226.

Adolphs, R. (2010). What does the amygdala contribute to social cognition? *Annals of the New York Academy of Sciences, 1191,* 42–61.

Adrian, M., Kusters, R., Wierenga, C. J., Storm, C., Hoogenraad, C. C., & Kapitein, L. C. (2014). Barriers in the brain: Resolving dendritic spine morphology and compartmentalization. *Frontiers in Neuroanatomy, 8,* 142.

Aflalo, T., Kellis, S., Klaes, C., Lee, B., Shi, Y., . . . & Andersen, R. A. (2015). Decoding motor imagery from the posterior parietal cortex of a tetraplegic human. *Science, 348,* 906–910.

Aguilera, O., Fernández, A. F., Muñoz, A., & Fraga, M. F. (2010). Epigenetics and environment: A complex relationship. *Journal of Applied Physiology, 109,* 243–251.

Aguzzi, A., Barres, B. A., & Bennett, M. L. (2013). Microglia: Scapegoat, saboteur, or something else? *Science, 339,* 156–161.

Ahmed, S. H. (2005). Imbalance between drug and non-drug reward availability: A major risk factor for addiction. *European Journal of Pharmacology, 526,* 9–20.

Ahmed, S. H. (2010). Validation crisis in animals models of drug addiction: Beyond non-disordered drug use toward drug addiction. *Neuroscience and Biobehavioral Reviews, 35,* 172–184.

Ahmed, S. H. (2012). The science of making drug addicted animals. *Neuroscience, 211,* 107–125.

Ahmed, S. H., Lenoir, M., & Guillem, K. (2013). Neurobiology of addiction versus drug use driven by lack of choice. *Current Opinion in Neurobiology, 23,* 1–7.

Aimone, J. B., Li, Y., Lee, S. W., Clemenson, G. D., Deng, W., & Gage, F. H. (2014). Regulation and function of adult neurogenesis: From genes to cognition. *Physiology Review, 94,* 991–1026.

Ainsworth, C. (2015). Sex redefined. *Nature, 518,* 288–291.

Akers, K. G., Martinez-Canabal, A., Restivo, L., Yiu, A. P., De Cristofaro, A., . . . & Frankland, P. W. (2014). Hippocampal neurogenesis regulates forgetting during adulthood and infancy. *Science, 344,* 598–602.

Åkerstedt, T., Schwarz, J., Gruber, G., Theorell-Haglöw, J., & Lindberg, E. (2019). Short sleep—poor sleep? A polysomnographic study in a large population-based sample of women. *Journal of Sleep Research, 28*(4), e12812.

Akil, H., Brenner, S., Kandel, E., Kendler, K. S., King, M-C., Scolnick, E., Watson, J. D., & Zoghbi, H. Y. (2010). The future of psychiatric research: Genomes and neural circuits. *Science, 327,* 1580–1581.

Al Massadi, O., López, M., Tschöp, M., Diéguez, C., & Nogueiras, R. (2017). Current understanding of the hypothalamic ghrelin pathways inducing appetite and adiposity. *Trends in Neurosciences, 40*(3), 167–180.

Alabi, A. A., & Tsien, R. W. (2012). Synaptic vesicle pools and dynamics. *Cold Spring Harbor Perspectives in Biology, 4,* a013680.

Alarcón, C. R., & Tavazoie, S. F. (2016). Cancer: Endothelial-cell killing promotes metastasis. *Nature, 536*(7615), 154–155.

Alasmari, F., Goodwani, S., McCullumsmith, R. E., & Sari, Y. (2018). Role of glutamatergic system and mesocorticolimbic circuits in alcohol dependence. *Progress in Neurobiology, 171,* 32–49.

Albert, K. M., Potter, G. G., Boyd, B. D., Kang, H., & Taylor, W. D. (2019). Brain network functional connectivity and cognitive performance in major depressive disorder. *Journal of Psychiatric Research, 110,* 51–56.

Albonico, A., & Barton, J. (2019). Progress in perceptual research: The case of prosopagnosia. *F1000Research, 8:* 765.

Albright, T. D., Kandel, E. R., & Posner, M. I. (2000). Cognitive neuroscience. *Current Opinion in Neurobiology, 10,* 612–624.

Aleman, A., Enriquez-Geppert, S., Knegtering, H., & Dlabac-de Lange, J. J. (2018). Moderate effects of noninvasive brain stimulation of the frontal cortex for improving negative symptoms in schizophrenia: Meta-analysis of controlled trials. *Neuroscience & Biobehavioral Reviews, 89,* 111–118.

Alexander, B. K., Beyerstein, B. L., Hadaway, P. F., & Coambs, R. B. (1981). Effects of early and later colony housing on oral ingestion of morphine in rats. *Psychopharmacology Biochemistry and Behavior, 58,* 176–179.

Alexander, B. M., Skinner, D. C., & Roselli, C. E. (2011). Wired on steroids: Sexual differentiation of the brain and its role in the expression of sexual partner preferences. *Frontiers in Endocrinology, 2,* 42.

Alfonsi, V., D'Atri, A., Scarpelli, S., Mangiaruga, A., & De Gennaro, L. (2019). Sleep talking: A viable access to mental processes during sleep. *Sleep Medicine Reviews, 44,* 12–22.

Alghamdi, B. S. (2018). The neuroprotective role of melatonin in neurological disorders. *Journal of Neuroscience Research, 96*(7), 1136–1149.

Algra, A., & Wermer, M. J. (2017). Stroke in 2016: Stroke is treatable, but prevention is the key. *Nature Reviews Neurology, 13*(2), 78–79.

ali Shah, S., Ullah, I., Lee, H. Y., & Kim, M. O. (2013). Anthocyanin's protect against ethanol-induced neuronal apoptosis via $GABA_{B1}$ receptors intracellular signaling in prenatal rat hippocampal neurons. *Molecular Neurobiology, 48,* 257–269.

Allen, J. S., Tranel, D., Bruss, J., & Damasio, H. (2006). Correlations between regional brain volumes and memory in anoxia. *Journal of Clinical and Experimental Neuropsychology, 28,* 457–476.

Allen, N. J. (2013). Role of glia in developmental synapse formation. *Current Opinion in Neurobiology, 23,* 1027–1033.

Allen, N. J., & Barres, B. A. (2005). Signaling between glia and neurons: Focus on synaptic plasticity. *Current Opinion in Neurobiology, 15,* 542–548.

Allen, N. J., & Eroglu, C. (2017). Cell biology of astrocyte-synapse interactions. *Neuron, 96*(3), 697–708.

Alloy, L. B., Nusslock, R., & Boland, E. M. (2015). The development and course of bipolar spectrum disorders: An integrated reward and circadian rhythm dysregulation model. *Annual Review of Clinical Psychology, 11,* 213–250.

Almad, A., & Maragakis, N. J. (2018). A stocked toolbox for understanding the role of astrocytes in disease. *Nature Reviews Neurology, 14*(14), 351–62.

Altemus, M., Sarvaiya, N., & Epperson, C. N. (2014). Sex differences in anxiety and depression clinical perspectives. *Frontiers in Neuroendocrinology, 35,* 320–330.

Alvarez, J. I., Katayama, T., & Prat, A. (2013). Glial influence on the blood brain barrier. *Glia, 61,* 1939–1958.

Alvarez, V. A., & Sabatini, B. L. (2007). Anatomical and physiological plasticity of dendritic spines. *Annual Review of Neuroscience, 30,* 79–97.

Amar, F., Sherman, M. A., Rush, T., Larson, M., Boyle, G., Chang, L., . . . & Lesné, S. E. (2017). The amyloid-β oligomer Aβ* 56 induces specific alterations in neuronal signaling that lead to tau phosphorylation and aggregation. *Science Signaling, 10*(478), eaal2021.

Amaral, D. G., Schumann, C. M., & Nordahl, C. W. (2008). Neuroanatomy of autism. *Trends in Neurosciences, 31*, 137–145.

Amato, D. (2015). Serotonin in antipsychotic drugs action. *Behavioural Brain Research, 277*, 125–135.

Ambach, W., & Gamer, M. (2018). Physiological measures in the detection of deception and concealed information. In *Detecting Concealed Information and Deception* (pp. 3–33). Academic Press.

Ambroggi, F., Turiault, M., Milet, A., Deroche-Gamonet, V., Parnaudeau, S., Balado, E., Barik, J., van der Veen, R., Maroteaux, G., Lemberger, T., Schütz, G., Lazar, M., Marinelli, M., Piazza, P. V., & Tronche, F. (2009). Stress and addiction: Glucocorticoid receptor in dopaminoceptive neurons facilitates cocaine seeking. *Nature Neuroscience, 12*, 247–249.

Amedi, A., Merabet, L. B., Bermpohl, F., & Pascual-Leone, A. (2005). The occipital cortex in the blind. *Current Directions in Psychological Science, 14*, 306–311.

American Psychiatric Association. (2013). *Diagnostic and statistical manual of mental disorders* (5th ed.). Washington, DC: Author.

Amit, B. H., Caddy, C., McCloud, T. L., Rendell, J. M., Hawton, K., Diamond, P. R.,…& Cipriani, A. (2015). Ketamine and other glutamate receptor modulators for depression in adults. *Cochrane Database of Systematic Reviews, 9*, CD011612.

Amoroso, T. (2015). The psychopharmacology of ± 3,4 methylenedioxymethamphetamine and its role in the treatment of posttraumatic stress disorder. *Journal of Psychoactive Drugs, 47*(5), 337–344.

Amso, D., & Scerif, G. (2015). The attentive brain: Insights from developmental cognitive neuroscience. *Nature Reviews Neuroscience, 16*, 606–619.

Amunts, K., & Zilles, K. (2012). Architecture and organizational principles of Broca's region. *Trends in Cognitive Sciences, 16*, 418–426.

Anacker, C., Luna, V. M., Stevens, G. S., Millette, A., Shores, R., Jimenez, J. C., . . . & Hen, R. (2018). Hippocampal neurogenesis confers stress resilience by inhibiting the ventral dentate gyrus. *Nature, 559*(7712), 98–102.

Anafi, R. C., Kayser, M. S., & Raizen, D. M. (2019). Exploring phylogeny to find the function of sleep. *Nature Reviews Neuroscience, 20*(2), 109–116.

Anagnostaras, S. G., & Robinson, T. E. (1996). Sensitization to the psychomotor stimulant effects of amphetamine: Modulation by associative learning. *Behavioral Neuroscience, 110*(6), 1397–1414.

Anand, B. K., & Brobeck, J. R. (1951). Localization of a "feeding center" in the hypothalamus of the rat. *Proceedings of the Society for Experimental Biology and Medicine, 77*, 323–324.

Andaloussi, Z. I. L., Taghzouti, K., & Abboussi, O. (2019). Behavioural and epigenetic effects of paternal exposure to cannabinoids during adolescence on offspring vulnerability to stress. *International Journal of Developmental Neuroscience, 72*, 48–54.

Andersen, R. A., Andersen, K. N., Hwang, E. J., & Hauschild, M. (2014). Optic ataxia: From Balint's syndrome to the parietal reach region. *Neuron, 81*, 967–983.

Anderson, A. K., & Phelps, E. A. (2000). Expression without recognition: Contributions of the human amygdala to emotional communication. *Psychological Science, 11*, 106–111.

Anderson, M. A., Burda, J. E., Ren, Y., Ao, Y., O'Shea, T. M., Kawaguchi, R.,…& Sofroniew, M. V. (2016). Astrocyte scar formation aids central nervous system axon regeneration. *Nature, 532*, 195–200.

Anderson, S., & Vanderhaeghen, P. (2014). Cortical neurogenesis from pluripotent stem cells: Complexity emerging from simplicity. *Current Opinion in Neurobiology, 27*, 151–157.

Andrade, C., & Rao, N. S. K. (2010). How antidepressant drugs act: A primer on neuroplasticity as the eventual mediator of antidepressant efficacy. *Indian Journal of Psychiatry, 52*, 378–386.

Andreae, L. C., & Burrone, J. (2014). The role of neuronal activity and transmitter release on synapse formation. *Current Opinion in Neurobiology, 24*, 47–52.

Andreae, L. C., & Burrone, J. (2018). The role of spontaneous neurotransmission in synapse and circuit development. *Journal of Neuroscience Research, 96*(3), 354–359.

Angell, P., Chester, N., Green, D., Somauroo, J., Whyte, G., & George, K. (2012). Anabolic steroids and cardiovascular risk. *Sports Medicine, 42*, 119–134.

Angeloni, C., & Geffen, M. N. (2018). Contextual modulation of sound processing in the auditory cortex. *Current Opinion in Neurobiology, 49*, 8–15.

Anson, J. A., & Kuhlman, D. T. (1993). Post-ictal Klüver-Bucy syndrome after temporal lobectomy. *Journal of Neurology, Neurosurgery, and Psychiatry, 56*, 311–313.

Anton-Erxleben, K., & Carrasco, M. (2013). Attentional enhancement of spatial resolution: Linking behavioral and neurophysiological evidence. *Nature Reviews Neuroscience, 14*, 188–200.

Antón, S. C., Potts, R., & Aiello, L. C. (2014). Evolution of early *Homo*: An integrated biological perspective. *Science, 345*, 45.

Antonini, A., & Stryker, M. P. (1993). Rapid remodeling of axonal arbors in the visual cortex. *Science, 260*, 1819–1821.

Apps, M. A. J., & Tsakiris, M. (2014). The free-energy self: A predictive coding account of self-recognition. *Neuroscience and Biobehavioral Reviews, 41*, 85–97.

Apps, R., & Hawkes, R. (2009). Cerebellar cortical organization: A one-map hypothesis. *Nature Reviews Neuroscience, 10*, 670–681.

Arango, C., Garibaldi, G., & Marder, S. R. (2013). Pharmacological approaches to treating negative symptoms: A review of clinical trials. *Schizophrenia Research, 150*, 346–352.

Arango, M.-T., Kivity, S., & Shoenfeld, Y. (2015). Is narcolepsy a classical autoimmune disease? *Pharmacological Research, 92*, 6–12.

Araque, A., Carmignoto, G., Haydon, P. G., Oliet, S. H. R., Robitaille, R., & Volterra, A. (2014). Gliotransmitters travel in time and space. *Neuron, 81*, 728–739.

Araya, C., Tawk, M., Girdler, G. C., Costa, M., Carmone-Fontaine, C., & Clarke, J. D. (2014). Mesoderm is required for coordinated cell movements within zebrafish neural plate in vivo. *Neural Development, 9*, 9.

Araya, R. (2014). Input transformation by dendritic spines of pyramidal neurons. *Frontiers in Neuroanatomy, 8*, 141.

Arbon, E. L., Knurowska, M., & Dijk, D. J. (2015). Randomised clinical trial of the effects of prolonged-release melatonin, temazepam, and zolpidem on slow-wave activity during sleep in healthy people. *Journal of Psychopharmacology, 29*, 764–776.

Ardila, A. (1992). Luria's approach to neuropsychological assessment. *International Journal of Neuroscience, 66*(1–2), 35–43.

Arena, J. E., & Rabinstein, A. A. (2015). Transient global amnesia. *Mayo Clinic Proceedings, 90*, 264–272.

Arguello, P. A., & Gogos, J. A. (2011). Genetic and cognitive windows into circuit mechanisms of psychiatric disease. *Trends in Neurosciences, 35*, 3–13.

Arlotta, P. (2018). Organoids required! A new path to understanding human brain development and disease. *Nature Methods, 15*(1), 27–29.

Arnegard, M. E., McGee, M. D., Matthews, B., Marchinko, K. B., Conte, G. L., et al. (2014). Genetics of ecological divergence during speciation. *Nature, 511*, 307–311.

Arnes, M., Alaniz, M. E., Karam, C. S., Cho, J. D., Lopez, G., Javitch, J. A., & Santa-Maria, I. (2019). Role of tau protein in remodeling of circadian neuronal circuits and sleep. *Frontiers in Aging Neuroscience, 11*, 320.

Arnold, A. P. (2017). A general theory of sexual differentiation. *Journal of neuroscience research, 95*(1–2), 291–300.

Arnott, S. R., & Alain, C. (2011). The auditory dorsal pathway: Orienting vision. *Neuroscience and Biobehavioral Reviews, 35*, 2162–2173.

Arsenault, J. S., & Buchsbaum, B. R. (2015). No evidence of somatotopic place of articulation feature mapping in motor cortex during passive speech perception. *Psychonomic Bulletin & Review*. Advance online publication. doi:10.3758/s13423-015-0988-z

Ascherio, A., & Munger, K. (2008). Epidemiology of multiple sclerosis: From risk factors to prevention. *Seminars in Neurology, 28*, 17–28.

Ashby, F. G., Turner, B. O., & Horvitz, J. C. (2010). Cortical and basal ganglia contributions to habit learning and automaticity. *Trends in Cognitive Sciences, 14*, 208–215.

Ashe, J., Lungu, O. V., Basford, A. T., & Lu, X. (2006). Cortical control of motor sequences. *Current Opinion in Neurobiology, 16*, 213–221.

Ashkan, K., Rogers, P., Bergman, H., & Ughratdar, I. (2017). Insights into the mechanisms of deep brain stimulation. *Nature Reviews Neurology, 13*(9), 548–554.

Ashok, A. H., Marques, T. R., Jauhar, S., Nour, M. M., Goodwin, G. M., Young, A. H., & Howes, O. D. (2017). The dopamine hypothesis of bipolar affective disorder: The state of the art and implications for treatment. *Molecular Psychiatry, 22*(5), 666–679.

Asok, A., Leroy, F., Rayman, J. B., & Kandel, E. R. (2019). Molecular mechanisms of the memory trace. *Trends in Neurosciences, 42*, 14–22.

Aspy, D. J., Delfabbro, P., Proeve, M., & Mohr, P. (2017). Reality testing and the mnemonic induction of lucid dreams: Findings from the national Australian lucid dream induction study. *Dreaming, 27*(3), 206.

Assali, A., Gaspar, P., & Rebsam, A. (2014). Activity dependent mechanisms of visual map formation—from retinal waves to molecular regulators. *Seminars in Cell and Developmental Biology, 35*, 136–146.

Assanand, S., Pinel, J. P. J., & Lehman, D. R. (1998a). Personal theories of hunger and eating. *Journal of Applied Social Psychology, 28*, 998–1015.

Assanand, S., Pinel, J. P. J., & Lehman, D. R. (1998b). Teaching theories of hunger and eating: Overcoming students' misconceptions. *Teaching Psychology, 25*, 44–46.

Assinck, P., Duncan, G. J., Hilton, B. J., Plemel, J. R., & Tetzlaff, W. (2017). Cell transplantation therapy for spinal cord injury. *Nature Neuroscience, 20*(5), 637–647.

Ast, G. (April, 2005). The alternative genome. *Scientific American, 292*, 58–65.

Asuelime, G. E., & Shi, Y. (2012). A case of cellular alchemy: Lineage reprogramming and its potential in regenerative medicine. *Journal of Molecular Cell Biology, 4*, 190–196.

Atkin, T., Comai, S., & Gobbi, G. (2018). Drugs for insomnia beyond benzodiazepines: Pharmacology, clinical applications, and discovery. *Pharmacological Reviews, 70*(2), 197–245.

Attardo, A., Fitzgerald, J. E., & Schnitzer, M. J. (2015). Impermanence of dendritic spines in live adult CA1 hippocampus. *Nature, 523*, 592–596.

Atwood, B. K., Lovinger, D. M., & Mathur, B. N. (2014). Presynaptic long-term depression mediated by Gi/o-coupled receptors. *Trends in Neurosciences, 37*, 663–673.

Aurora, R. N., Kristo, D. A., Bista, S. R., Rowley, J. A., Zak, R. S., . . . & American Academy of Sleep Medicine. The treatment of restless legs syndrome and periodic limb movement disorder in adults—an update for 2012: Practice parameters with an evidence-based systematic review and meta-analyses: An American academy of sleep medicine clinical practice guideline. *Sleep, 35*, 1039–1062.

Autism Genome Project. (2007). Mapping autism risk loci using genetic linkage and chromosomal rearrangements. *Nature Genetics, 39*, 319–328.

Avci, M. E., Sanlikan, F., Çelik, M., Avci, A., Kocaer, M., & Göçmen, A. (2015). Effects of maternal obesity on antenatal, perinatal, and neonatal outcomes. *Journal of Maternal-Fetal and Neonatal Medicine, 28*, 2080–2083.

Averbeck, B. B., & Costa, V. D. (2017). Motivational neural circuits underlying reinforcement learning. *Nature Neuroscience, 20*(4), 505.

Aviezer, H., Trope, Y., & Todorov, A. (2012). Body cues, not facial expressions, discriminate between intense positive and negative emotions. *Science, 338*, 1225–1228.

Aviram, L., & Soffer-Dudek, N. (2018). Lucid dreaming: Intensity, but not frequency, is inversely related to psychopathology. *Frontiers in Psychology, 9*, 384.

Axelsson, J., Ingre, M., Kecklund, G., Lekander, M., Wright, K. P., & Sundelin, T. (2019). Sleepiness as motivation: A potential mechanism for how sleep deprivation affects behavior. *Sleep,* zsz291.

Ayas, N. T., White, D. P., Manson, J. E., Stampfer, M. J., Speizer, F. E., Malhotra, A., & Hu, F. B. (2003). A prospective study of sleep duration and coronary heart disease in women. *Archives of Internal Medicine, 163*, 205–209.

Azad, T. D., Li, A., Pendharkar, A. V., Veeravagu, A., & Grant, G. A. (2016). Junior Seau: An illustrative case of chronic traumatic encephalopathy and update on chronic sports-related head injury. *World Neurosurgery, 86*, 515–e11.

Azevedo, F. A., Carvalho, L. R., Grinberg, L. T., Farfel, J. M., Ferretti, R. E., Leite, R. E., . . . & Herculano-Houzel, S. (2009). Equal numbers of neuronal and nonneuronal cells make the human brain an isometrically scaled-up primate brain. *Journal of Comparative Neurology, 513*, 532–541.

Azim, E., Fink, A. J., & Jessell, T. M. (2014). Internal and external feedback circuits for skilled forelimb movement. *Cold Spring Harbor Symposia on Quantitative Biology, 79*, 81–92.

Bachevalier, J., & Loveland, K. A. (2006). The orbitofrontal-amygdala circuit and self-regulation of social-emotional behavior in autism. *Neuroscience and Biobehavioural Reviews, 30*, 97–117.

Bachoud-Lévi, A. C., Ferreira, J., Massart, R., Youssov, K., Rosser, A., Busse, M., . . . & Squitieri, F. (2019). International Guidelines for the treatment of Huntington's Disease. *Frontiers in Neurology, 10*, 710.

Baden, T., Berens, P., Franke, K., Rosón, M. R., Bethge, M., & Euler, T. (2016). The functional diversity of retinal ganglion cells in the mouse. *Nature, 529*, 345–350.

Baden, T., Berens, P., Franke, K., Rosón, M. R., Bethge, M., & Euler, T. (2016). The functional diversity of retinal ganglion cells in the mouse. *Nature, 529*(7586), 345–350.

Baden, T., Euler, T., & Berens, P. (2019). Understanding the retinal basis of vision across species. *Nature Reviews Neuroscience, 21*, 5–20.

Badiani, A., Belin, D., Epstein, D., Calu, D., & Shaham, Y. (2011). Opiate versus psychostimulant addiction: The differences do matter. *Nature Reviews Neuroscience, 12*, 685–700.

Badran, M., Yassin, B. A., Fox, N., Laher, I., & Ayas, N. (2015). Epidemiology of sleep disturbances and cardiovascular consequences. *Canadian Journal of Cardiology, 31*, 873–879.

Bagnardi, V., Rota, M., Botteri, E., Tramacere, I., Islami, F., Fedirko, V., Scotti, L., Jenab, M., Turati, F., Pasquali, E., Pelucchi, C., Bellocco, R., Negri, E., Corrao, G., Rehm, J., Boffetta, P., & La Vecchia, C. (2013). Light alcohol drinking and cancer: A meta-analysis. *Annals of Oncology, 24*, 301–308.

Bagni, C., & Zukin, R. S. (2019). A synaptic perspective of fragile X syndrome and autism spectrum disorders. *Neuron, 101*(6), 1070–1088.

Bailey, J. M., Pillard, R. C., Neale, M. C., & Agyei, Y. (1993). Heritable factors influence sexual orientation in women. *Archives of General Psychiatry, 50*, 217–223.

Bailey, M. J., & Pillard, R. C. (1991). A genetic study of male sexual orientation. *Archives of General Psychiatry, 48*, 1089–1096.

Bailey, R. K., Mokonogho, J., & Kumar, A. (2019). Racial and ethnic differences in depression: Current perspectives. *Neuropsychiatric disease and treatment, 15*, 603.

Baillet, S. (2017). Magnetoencephalography for brain electrophysiology and imaging. *Nature Neuroscience, 20*(3), 327.

Baird, B., Castelnovo, A., Gosseries, O., & Tononi, G. (2018). Frequent lucid dreaming associated with increased functional connectivity between frontopolar cortex and temporoparietal association areas. *Scientific Reports, 8*(1), 1–15.

Baird, B., Mota-Rolim, S. A., & Dresler, M. (2019). The cognitive neuroscience of lucid dreaming. *Neuroscience and Biobehavioral Reviews, 100*, 305–323.

Bakhshandeh, B., Amin Kamaleddin, M., & Aalishah, K. (2017). A comprehensive review on exosomes and microvesicles as epigenetic factors. *Current Stem Cell Research & Therapy, 12*(1), 31–36.

Bakhshi, K., & Chance, S. A. (2015). The neuropathology of schizophrenia: A selective review of past studies and emerging themes in brain structure and cytoarchitecture. *Neuroscience, 303*, 82–102.

Bakken, T. E., Miller, J. A., Ding, S. L., Sunkin, S. M., Smith, K. A., Ng, L., . . . & Shapouri, S. (2016). A comprehensive transcriptional map of primate brain development. *Nature, 535*(7612), 367–375.

Baldessarini, R. J., Tondo, L., & Vázquez, G. H. (2019). Pharmacological treatment of adult bipolar disorder. *Molecular Psychiatry, 24*(2), 198–217.

Baldwin, K. T., & Eroglu, C. (2017). Molecular mechanisms of astrocyte-induced synaptogenesis. *Current Opinion in Neurobiology, 45*, 113–120.

Bale, T. L. (2015). Epigenetic and transgenerational reprogramming of brain development. *Nature Reviews Neuroscience, 16*(6), 332.

Bale, T. L., & Epperson, C. N. (2015). Sex differences and stress across the lifespan. *Nature Neuroscience, 18*, 1413–1420.

Ball, G. F., & Balthazart, J. (2006). Androgen metabolism and the activation of male sexual behavior: It's more complicated than you think! *Hormones and Behavior, 49*, 1–3.

Ball, G. F., Balthazart, J., & McCarthy, M. M. (2014). Is it useful to view the brain as a secondary sexual characteristic? *Neuroscience & Biobehavioral Reviews, 46*, 628–638.

Balon, R., & Starcevic, V. (2020). Role of benzodiazepines in anxiety disorders. In *Anxiety Disorders* (pp. 367–388). Springer, Singapore.

Balthazart, J. (2011). Minireview: Hormones and human sexual orientation. *Endocrinology, 152*, 2937–2947.

Balthazart, J. (2018). Fraternal birth order effect on sexual orientation explained. *Proceedings of the National Academy of Sciences, 115*(2), 234–236.

Baluch, F., & Itti, L. (2011). Mechanisms of top-down attention. *Trends in Neurosciences, 34*, 210–224.

Bandelow, B. (2020). Current and novel psychopharmacological drugs for anxiety disorders. In *Anxiety Disorders* (pp. 347–365). Springer, Singapore.

Bandelow, B., & Michaelis, S. (2015). Epidemiology of anxiety disorders in the 21st century. *Dialogues in Clinical Neuroscience, 17*, 327.

Bangasser, D. A., & Valentino, R. J. (2014). Sex differences in stress-related psychiatric disorders: Neurobiological perspectives. *Frontiers in Neuroendocrinology, 35*, 303–319.

Banik, A., Kandilya, D., Ramya, S., Stünkel, W., Chong, Y., & Dheen, S. (2017). Maternal factors that induce epigenetic changes contribute to neurological disorders in offspring. *Genes, 8*(6), 150.

Bankiewicz, K. S., Plunkett, R. J., Jaconowitz, D. M., Porrino, L., di Porzio, U., London, W. T., . . . & Oldfield, E. H. (1990). The effect of fetal mesencephalon implants on primate MPTP-induced Parkinsonism: Histochemical and behavioral studies. *Journal of Neuroscience, 72*, 231–244.

Barak, B., & Feng, G. (2016). Neurobiology of social behavior abnormalities in autism and Williams syndrome. *Nature Neuroscience, 19*(5), 647–655.

Barbu, M. C., Shen, X., Walker, R. M., Howard, D. M., Evans, K. L., Whalley, H. C., . . . & Marioni, R. E. (2020). Epigenetic prediction of major depressive disorder. *Molecular Psychiatry*, 1–12.

Bard P. (1929). The central representation of the sympathetic system. *Archives of Neurology and Psychiatry, 22*, 230–246.

Bardella, C., Al-Shammari, A. R., Soares, L., Tomlinson, I., O'Neill, E., & Szele, F. G. (2018). The role of inflammation in subventricular zone cancer. *Progress in Neurobiology, 170*, 37–52.

Barik, J., Marti, F., Morel, C., Fernandez, S. P., Lanteri, C., Godeheu, G., Tassin, J.-P., Mombereau, C., Faure, P., & Tronche, F. (2013). Chronic stress triggers social aversion via glucocorticoid receptor in dopaminoceptive neurons. *Science, 339*, 332–335.

Barker, G. R., Banks, P. J., Scott, H., Ralph, G. S., Mitrophanous, K. A., Wong, L. F., . . . & Warburton, E. C. (2017). Separate elements of episodic memory subserved by distinct hippocampal–prefrontal connections. *Nature Neuroscience, 20*(2), 242–252.

Baron, J. C. (2018). Protecting the ischaemic penumbra as an adjunct to thrombectomy for acute stroke. *Nature Reviews Neurology*, 325–337.

Barresi, V., Caffo, M., & Tuccari, G. (2016). Classification of human meningiomas: Lights, shadows, and future perspectives. *Journal of Neuroscience Research, 94*(12), 1604–1612.

Barrett, L. F. (2006). Are emotions natural kinds? *Perspectives on Psychological Science, 1*, 28–58.

Barrett, L. F., & Satpute, A. B. (2013). Large-scale brain networks in affective and social neuroscience: Towards an integrative functional architecture of the brain. *Current Opinion in Neurobiology, 23*, 361–372.

Barretto, R. P. J., Gillis-Smith, S., Chandrashekar, J., Yarmolinsky, D. A., . . . & Zuker, C. S. (2015). The neural representation of taste quality at the periphery. *Nature, 517*, 373–376.

Barrick, J. E., & Lenski, R. E. (2013). Genome dynamics during experimental evolution. *Nature Reviews Genetics, 14*, 827–839.

Bartal, I. B.-A., Decety, J., & Mason, P. (2011). Empathy and pro-social behavior in rats. *Science, 334*, 1427–1430.

Bartsch, T., & Butler, C. (2013). Transient amnesic syndromes. *Nature Reviews Neurology, 9*, 86–97.

Basbaum, A. I., & Fields, H. L. (1978). Endogenous pain control mechanisms: Review and hypothesis. *Annals of Neurology, 4*, 451–462.

Baskaran, V., Murray, R. L., Hunter, A., Lim, W. S., & McKeever, T. M. (2019). Effect of tobacco smoking on the risk of developing community acquired pneumonia: A systematic review and meta-analysis. *PLoS ONE, 14*(7), e0220204.

Basner, M., Rao, H., Goel, N., & Dinges, D. F. (2013). Sleep deprivation and neurobehavioral dynamics. *Current Opinion in Neurobiology, 23*, 854–863.

Bassett, D. S., Yang, M., Wymbs, N. F., & Grafton, S. T. (2015). Learning-induced autonomy of sensorimotor systems. *Nature Neuroscience, 5*, 744–751.

Bastian, A. J. (2006). Learning to predict the future: The cerebellum adapts feedforward movement control. *Current Opinion in Neurobiology, 16*, 645–649.

Batalla, A., Bhattacharyya, S., Yücel, M., Fusar-Poli, P., Crippa, J. A., Nogué, S., Torrens, M., Pujol, J., Farré, M., & Martin-Santos, R. (2013). Structural and functional imaging studies in chronic cannabis users: A systematic review of adolescent and adult findings. *PLoS ONE, 8*, e55821.

Batelaan, N. M., Van Balkom, A. J. L. M., & Stein, D. J. (2012). Evidence-based pharmacotherapy of panic disorder: An update. *International Journal of Neuropsychopharmacology, 15*, 403–415.

Batrinos, M. L. (2012). Testosterone and aggressive behavior in man. *Endocrinology and Metabolism, 10*, 563–568.

Bauer, P. R., Reitsma, J. B., Houweling, B. M., Ferrier, C. H., & Ramsey, N. F. (2014). Can fMRI safely replace the Wada test for preoperative assessment of language lateralisation? A meta-analysis and systematic review. *Journal of Neurology, Neurosurgery & Psychiatry, 85*, 581–588.

Baura, G., Foster, D., Porte, D., Kahn, S. E., Bergman, R. N., Cobelli, C., & Schwartz, M. W. (1993). Saturable transport of insulin from plasma into the central nervous system of dogs in vivo: A mechanism for regulated insulin delivery to the brain. *Journal of Clinical Investigations, 92*, 1824–1830.

Bavelier, D., Corina, D., Jessard, P., Padmanabhan, S., Clark, V. P., Karni, A., et al. (1997). Sentence reading: A functional MRI study at 4 tesla. *Journal of Cognitive Neuroscience, 9*, 664–686.

Bazargani, N., & Attwell, D. (2016). Astrocyte calcium signaling: The third wave. Nature Neuroscience, *19*(2), 182.

Bazzari, F. H. (2018). Can we induce lucid dreams? A pharmacological point of view. *International Journal of Dream Research*, 106–119.

Beaton, A. A. (2003). The nature and determinants of handedness. In K. Hugdahl and R. J. Davidson (Eds.), *The asymmetrical brain* (pp. 105–158). Cambridge, MA: Bradford Books/MIT Press.

Beaulieu, J. M., Espinoza, S., & Gainetdinov, R. R. (2015). Dopamine receptors–IUPHAR review 13. *British Journal of Pharmacology, 172*, 1–23.

Bechara, A., Tranel, D., Damasio, H., Adolphs, R., Rockland, C., & Damasio, A. R. (1995). Double dissociation of conditioning and declarative knowledge relative to the amygdala and hippocampus in humans. *Science, 269*, 1115–1118.

Becker, S. (2017). Neurogenesis and pattern separation: Time for a divorce. *Wiley Interdisciplinary Reviews: Cognitive Science, 8*(3), e1427.

Bedi, G., Hyman, D., & de Wit, H. (2010). Is ecstasy an "empathogen"? Effects of ±3,4—methylenedioxymethamphetamine on prosocial feelings and identification of emotional states in others. *Biological Psychiatry, 68*, 1134–1140.

Bedi, G., Preston, K. L., Epstein, D. H., Heishman, S. J., Marrone, G. F., Shaham, Y., & de Wit, H. (2011). Incubation of cue-induced cigarette craving during abstinence in human smokers. *Biological Psychiatry, 69*, 708–711.

Bedner, P., & Steinhäuser, C. (2016). Neuron–glia interactions in epilepsy. *Journal of Neuroscience Research, 94*, 779–780.

Begg, D. P., & Woods, S. C. (2013). The endocrinology of food intake. *Nature Reviews Endocrinology, 9*, 584–597.

Behrmann, M., Avidan, G., Marotta, J. J., & Kimchi, R. (2005). Detailed exploration of face-related processing in congenital prosopagnosia: 1. Behavioral findings. *Journal of Cognitive Neuroscience, 17*, 1130–1149.

Bekkers, J. M., & Suzuki, N. (2013). Neurons and circuits for odor processing in the piriform cortex. *Trends in Neurosciences, 36*, 429–438.

Bell, C. G. (2017). The epigenomic analysis of human obesity. *Obesity, 25*(9), 1471–1481.

Bell, C. C., Han, V., & Sawtell, N. B. (2008). Cerebellum-like structures and their implications for cerebellar function. *Annual Review of Neuroscience, 31*, 1–24.

Bell, J. T., & Saffery, R. (2012). The value of twins in epigenetic epidemiology. *International Journal of Epidemiology*, dyr179.

Bell, J. T., & Spector, T. D. (2011). A twin approach to unraveling epigenetics. *Trends in Genetics, 27*, 116–125.

Bell, K. F. S., & Hardingham, G. E. (2011). The influence of synaptic activity on neuronal health. *Current Opinion in Neurobiology, 21*, 299–305.

Bell, M. R. (2018). Comparing postnatal development of gonadal hormones and associated social behaviors in rats, mice, and humans. *Endocrinology, 159*(7), 2596–2613.

Bellono, N. W., Leitch, D. B., & Julius, D. (2017). Molecular basis of ancestral vertebrate electroreception. *Nature, 543*(7645), 391.

Belloy, M. E., Napolioni, V., & Greicius, M. D. (2019). A quarter century of APOE and Alzheimer's disease: Progress to date and the path forward. *Neuron, 101*(5), 820–838.

Belouin, S. J., & Henningfield, J. E. (2018). Psychedelics: Where we are now, why we got here, what we must do. *Neuropharmacology, 142*, 7–19.

Belousov, A. B. (2011). The regulation and role of neuronal gap junctions during development. *Communicative & Integrative Biology, 4*, 579–581.

Belousov, A. B., & Fontes, J. D. (2013). Neuronal gap junctions: Making and breaking connections during development and injury. *Trends in Neurosciences, 36*, 227–236.

Ben-Ari, Y. (2017). NKCC1 chloride importer antagonists attenuate many neurological and psychiatric disorders. *Trends in Neurosciences, 40*(9), 536–554.

Ben-Shakhar, G. (2012). Current research and potential applications of the concealed information test: An overview. *Frontiers in Psychology, 3*, 342.

Benarroch, E. E. (2013). Oxytocin and vasopressin: Social neuropeptides with complex neuromodulatory functions. *Neurology, 80*, 1521–1528.

Bendor, D., & Wang, X. (2005). The neuronal representation of pitch in primate auditory cortex. *Nature, 436*, 1161–1165.

Benedetti, F. (2014). Placebo effects: From the neurobiological paradigm to translational implications. *Neuron, 84*, 623–637.

Bennett, M. V. L., Contreras, J. E., Bukauskas, F. F., & Saez, J. C. (2003). New roles for astrocytes: Gap junction hemichannels have something to communicate. *Trends in Neurosciences, 26*, 610–617.

Bentivoglio, M., & Kristensson, K. (2014). Tryps and trips: Cell trafficking across the 100-year-old blood-brain barrier. *Trends in Neurosciences, 37*, 325–333.

Benton, A. L. (1994). Neuropsychological assessment. *Annual Review of Psychology, 45*, 1–23.

Berardi, N., Sale, A., & Maffei, L. (2014). Brain structural and functional development: Genetics and experience. *Developmental Medicine and Child Neurology, 57*(Suppl. 2), 4–9.

Berger, J. R., & Dean, D. (2014). Neurosyphilis. *Handbook of Clinical Neurology, 121*, 1461–1472.

Berger, R. J., & Oswald, I. (1962). Effects of sleep deprivation on behaviour, subsequent sleep, and dreaming. *Journal of Mental Science, 106*, 457–465.

Bergmann, O., & Frisén, J. (2013). Why adults need new brain cells. *Science, 340*, 695–696.

Berkel, T. D., & Pandey, S. C. (2017). Emerging role of epigenetic mechanisms in alcohol addiction. *Alcoholism: Clinical and Experimental Research, 41*(4), 666–680.

Berlim, M. T., Van den Eynde, F., & Daskalakis, Z. J. (2013). Clinically meaningful efficacy and acceptability of low-frequency repetitive transcranial magnetic stimulation (rTMS) for treating primary major depression: A meta-analysis of randomized, double-blind and sham-controlled trials. *Neuropsychopharmacology, 38*, 543–551.

Berlim, M. T., Van den Eynde, F., Tovar-Perdomo, S., & Daskalakis, Z. J. (2014). Response, remission and drop-out rates following high-frequency repetitive transcranial magnetic stimulation (rTMS) for treating major depression: A systematic review and meta-analysis of randomized, double-blind and sham-controlled trials. *Psychological Medicine, 44*, 225–239.

Berliner, J., Collins, K., & Coker, J. (2018). Cannabis conundrum. *Spinal Cord Series and Cases, 4*, 68.

Bernard, F., Lemée, J. M., Ter Minassian, A., & Menei, P. (2018). Right hemisphere cognitive functions: From clinical and anatomic bases to brain mapping during awake craniotomy part I: Clinical and functional anatomy. *World Neurosurgery, 118*, 348–359.

Bernardi, G., Betta, M., Ricciardi, E., Pietrini, P., Tononi, G., & Siclari, F. (2019). Regional delta waves in human rapid eye movement sleep. *Journal of Neuroscience, 39*(14), 2686–2697.

Berndt, A., & Deisseroth, K. (2015). Expanding the optogenetics toolkit. *Science, 349*, 590–591.

Bernstein, I. L., & Webster, M. M. (1980). Learned taste aversion in humans. *Physiology & Behavior, 25*, 363–366.

Berridge, K. C. (2004). Motivation concepts in behavioral neuroscience. *Physiology & Behavior, 81*, 179–209.

Berridge, K. C., & Kringelbach, M. L. (2015). Pleasure systems in the brain. *Neuron, 86*, 646–664.

Berridge, K. C., Robinson, T. E., & Aldridge, J. W. (2009). Dissecting components of reward: "Liking," "wanting," and learning. *Current Opinion in Pharmacology, 9*, 65–73.

Berry, K. P., & Nedivi, E. (2017). Spine dynamics: Are they all the same? *Neuron, 96*(1), 43–55.

Bershad, A. K., Miller, M. A., Baggott, M. J., & de Wit, H. (2016). The effects of MDMA on socio-emotional processing: Does MDMA differ from other stimulants? *Journal of Psychopharmacology, 30*(12), 1248–1258.

Berthoud, H.-R. (2002). Multiple neural systems controlling food intake and body weight. *Neuroscience and Biobehavioural Reviews, 26*, 393–428.

Berthoud, H.-R. (2013). Why does gastric bypass surgery work? *Science, 341*, 351–352.

Bertram, E. H. (2014). Extratemporal lobe circuits in temporal lobe epilepsy. *Epilepsy & Behavior, 38*, 13–18.

Bespalov, A., Müller, R., Relo, A. L., & Hudzik, T. (2016). Drug tolerance: A known unknown in translational neuroscience. *Trends in Pharmacological Sciences, 37*(5), 364–378.

Betley, J. N., Xu, S., Cao, Z. F. H., Gong, R., Magnus, C. J., Yu, Y., & Sternson, S. M. (2015). Neurons for hunger and thirst transmit a negative-valence teaching signal. *Nature, 521*, 180–185.

Betzler, F., Viohl, L., & Romanczuk-Seiferth, N. (2017). Decision-making in chronic ecstasy users: A systematic review. *European Journal of Neuroscience, 45*(1), 34–44.

Beyer, D. K., & Freund, N. (2017). Animal models for bipolar disorder: From bedside to the cage. *International Journal of Bipolar Disorders, 5*(1), 35.

Bhati, M. T. (2013). Defining psychosis: the evolution of DSM-5 schizophrenia spectrum disorders. *Current Psychiatry Reports, 15*, 1–7.

Bi, G.-Q., & Poo, M.-M. (2001). Synaptic modification by correlated activity: Hebb's postulate revisited. *Annual Review of Neuroscience, 24*, 139–166.

Bi, Y., Wang, X., & Caramazza, A. (2016). Object domain and modality in the ventral visual pathway. *Trends in Cognitive Sciences, 20*(4), 282–290.

Bidelman, G. M., & Grall, J. (2014). Functional organization for musical consonance and tonal pitch hierarchy in human auditory cortex. *NeuroImage, 101*, 204–214.

Bilder, D. A., Noel, J. K., Baker, E. R., Irish, W., Chen, Y., Merilainen, M. J., . . . & Winslow, B. J. (2016). Systematic review and meta-analysis of neuropsychiatric symptoms and executive functioning in adults with phenylketonuria. *Developmental Neuropsychology, 41*(4), 245–260.

Bilkei-Gorzo, A., Albayram, O., Draffehn, A., Michel, K., Piyanova, A., Oppenheimer, H., . . . & Bab, I. (2017). A chronic low dose of Δ 9-tetrahydrocannabinol (THC) restores cognitive function in old mice. *Nature Medicine, 23*(6), 782–787.

Billock, V. A., & Tsou, B. H. (2010, February). Seeing forbidden colors. *Scientific American, 302*, 72–77.

Bin, Y. S., Postnova, S., & Cistulli, P. A. (2019). What works for jetlag? A systematic review of non-pharmacological interventions. *Sleep Medicine Reviews, 43*, 47–59.

Bintu, L., Yong, J., Antebi, Y. E., McCue, K., Kazuki, Y., Uno, N., . . . & Elowitz, M. B. (2016). Dynamics of epigenetic regulation at the single-cell level. *Science, 351*, 720–724.

Birch, A. M., & Kelly, Á. M. (2019). Lifelong environmental enrichment in the absence of exercise protects the brain from age-related cognitive decline. *Neuropharmacology, 145*, 59–74.

Birklein, F., Ajit, S. K., Goebel, A., Perez, R. S., & Sommer, C. (2018). Complex regional pain syndrome—phenotypic characteristics and potential biomarkers. *Nature Reviews Neurology, 14*(5), 272–284.

Birnbaum, R., & Weinberger, D. R. (2017). Genetic insights into the neurodevelopmental origins of schizophrenia. *Nature Reviews Neuroscience, 18*(12), 727–740.

Birney, E. (2012). Journey to the genetic interior. Interview by Stephen S. Hall. *Scientific American, 307*, 80–82.

Bishop, D. V. M. (2013). Cerebral asymmetry and language development: Cause, correlate, or consequence? *Science, 340*, 1302.

Biyani, S., Morgan, P. S., Hotchkiss, K., Cecchini, M., & Derkay, C. S. (2015). Autism spectrum disorder 101: A primer for pediatric otolaryngologists. *International Journal of Pediatric Otorhinolaryngology, 79*, 798–802.

Bizley, J. K., Maddox, R. K., & Lee, A. K. (2016). Defining auditory-visual objects: Behavioral tests and physiological mechanisms. *Trends in Neurosciences, 39*(2), 74–85.

Björkholm, C., & Monteggia, L. M. (2016). BDNF—a key transducer of antidepressant effects. *Neuropharmacology, 102*, 72–79.

Black, J. E., Isaacs, K. R., Anderson, B. J., Alcantara, A. A., & Greenhough, W. T. (1990). Learning causes synaptogenesis, whereas motor activity causes angiogenesis, in cerebellar cortex of adult rats. *Proceedings of the National Academy of Science, USA, 87*, 5568–5572.

Blackburn, G. L. (2001). Pasteur's quadrant and malnutrition. *Nature, 409*, 397–401.

Blackmore, S. (2018). Decoding the puzzle of human consciousness: The hardest problem. *Scientific American, 319*(3), 48–53.

Blakemore, C., Clark, J. M., Nevalainen, T., Oberdorfer, M., & Sussman, A. (2012). Implementing the 3Rs in neuroscience research: A reasoned approach. *Neuron, 75*, 948–950.

Blakemore, S.-J. (2008). The social brain in adolescence. *Nature Reviews Neuroscience, 9*, 267–277.

Blanchard, D. C., & Blanchard, R. J. (1984). Affect and aggression: An animal model applied to human behavior. In D. C. Blanchard & R. J. Blanchard (Eds.), *Advances in the study of aggression* (pp. 1–62). Orlando, FL: Academic Press.

Blanchard, D. C., & Blanchard, R. J. (1988). Etho-experimental approaches to the biology of emotion. *Annual Review of Psychology, 39*, 43–68.

Blanchard, D. C., Griebel, G., Pobbe, R., & Blanchard, R. J. (2011). Risk assessment as an evolved threat detection and analysis process. *Neuroscience and Biobehavioral Reviews, 35*, 991–998.

Blanchard, D. C., Summers, C. H., & Blanchard, R. J. (2013). The role of behavior in translational models for psychopathology: Functionality and dysfunctional behaviors. *Neuroscience & Biobehavioral Reviews, 37*, 1567–1577.

Blanchard, R. (2018). Fraternal birth order, family size, and male homosexuality: Meta-analysis of studies spanning 25 years. *Archives of Sexual Behavior, 47*(1), 1–15.

Blanco, C., Secades-Villa, R., Garcia-Rodriguez, O., Labrador-Mendez, M., Wang, S., & Schwartz, R. P. (2013). Probability and predictors of remission from life-time prescription drug use disorders: Results from the national epidemiologic survey on alcohol and related conditions. *Journal of Psychiatric Research, 47*, 42–49.

Blanke, O., Landis, T., Safran, A. B., & Seeck, M. (2002). Direction-specific motion blindness induced by focal stimulation of human extrastriate cortex. *European Journal of Neuroscience, 15*, 2043–2048.

Blanke, O., Slater, M., & Serino, A. (2015). Behavioral, neural, and computational principles of bodily self-consciousness. *Neuron, 88*, 145–166.

Blankenship, A. G., & Feller, M. B. (2010). Mechanisms underlying spontaneous patterned activity in developing neural circuits. *Nature Reviews Neuroscience, 11*, 18–29.

Blashfield, R. K., Keeley, J. W., Flanagan, E. H., & Miles, S. R. (2014). The cycle of classification: DSM-I through DSM-5. *Annual Review of Clinical Psychology, 10*, 25–51.

Blesa, J., Phani, S., Jackson-Lewis, V., & Przedborski, S. (2012). Classic and new animal models of Parkinson's disease. *Journal of Biomedicine and Biotechnology, 2012*, article 845618.

Bless, J. J., Westerhausen, R., von Koss Torklidsen, J., Gudmundsen, M., Kompus, K., & Hughdahl, K. (2015). Laterality across languages: Results from a global dichotic listening study using a smartphone application. *Laterality, 20*, 434–452.

Bliss, T. V. P., & Lømo, T. (1973). Long-lasting potentiation of synaptic transmission in the dentate area of the anaesthetized rabbit following stimulation of the perforant path. *Journal of Physiology, 232*, 331–356.

Bliss, T. V., Collingridge, G. L., Kaang, B. K., & Zhuo, M. (2016). Synaptic plasticity in the anterior cingulate cortex in acute and chronic pain. *Nature Reviews Neuroscience, 17*(8), 485–496.

Bliss, T. V., Collingridge, G. L., Morris, R. G., & Reymann, K. G. (2018). Long-term potentiation in the hippocampus: Discovery, mechanisms and function. *Neuroforum, 24*, A103–A120.

Bloch, G. J., & Mills, R. (1995). Prepubertal testosterone treatment of neonatally gonadectomized male rats: Defeminization and masculinization of behavioral and endocrine function in adulthood. *Neuroscience and Behavioural Reviews, 19*, 187–200.

Bloch, G. J., Mills, R., & Gale, S. (1995). Prepubertal testosterone treatment of female rats: Defeminization of behavioral and endocrine function in adulthood. *Neuroscience and Biobehavioural Reviews, 19*, 177–186.

Bludau, A., Royer, M., Meister, G., Neumann, I. D., & Menon, R. (2019). Epigenetic Regulation of the Social Brain. *Trends in Neurosciences, 42*(7), 471–484.

Blüher, M., & Mantzoros, C. S. (2015). From leptin to other adipokines in health and disease: Facts and expectations at the beginning of the 21st century. *Metabolism, 61*, 131–145.

Blundell, J. E., & Finlayson, G. (2004). Is susceptibility to weight gain characterized by homeostatic or hedonic risk factors for overconsumption? *Physiology & Behavior, 82*, 21–25.

Blundell, J. E., & Halford, J. C. G. (1998). Serotonin and appetite regulation. *CNS Drugs, 9*, 473–495.

Boag, S. (2017). On dreams and motivation: Comparison of Freud's and Hobson's views. *Frontiers in Psychology, 7*, 2001.

Boehmer, U., & Elk, R. (2016). LGBT populations and cancer: Is it an ignored epidemic? *LGBT Health, 3*(1), 1–2.

Bogaert, A. F. (2006). Biological versus nonbiological older brothers and men's sexual orientation. *Proceedings of the National Academy of Sciences, USA, 103*, 10771–10774.

Bogaert, A. F., & Skorska, M. N. (2020). A short review of biological research on the development of sexual orientation. *Hormones and Behavior, 119*, 104659.

Bogaert, A. F., Skorska, M. N., Wang, C., Gabrie, J., MacNeil, A. J., Hoffarth, M. R., . . . & Blanchard, R. (2018). Male homosexuality and maternal immune responsivity to the Y-linked protein NLGN4Y. *Proceedings of the National Academy of Sciences, 115*(2), 302–306.

Böhm, U. L., & Wyart, C. (2016). Spinal sensory circuits in motion. *Current Opinion in Neurobiology, 41*, 38–43.

Boison, D. (2016). The biochemistry and epigenetics of epilepsy: Focus on adenosine and glycine. *Frontiers in Molecular Neuroscience, 9*, 26.

Bokiniec, P., Zampieri, N., Lewin, G. R., & Poulet, J. F. (2018). The neural circuits of thermal perception. *Current Opinion in Neurobiology, 52*, 98–106.

Bolton, M. M., & Eroglu, C. (2009). Look who is weaving the neural web: Glial control of synapse formation. *Current Opinion in Neurobiology, 19*, 491–497.

Bonin, R. P., & De Koninck, Y. (2015). Reconsolidation and the regulation of plasticity: Moving beyond memory. *Trends in Neurosciences, 38*, 336–344.

Bonini, L., & Ferrari, P. F. (2011). Evolution of mirror systems: A simple mechanism for complex cognitive functions. *Annals of the New York Academy of Sciences, 1225*, 166–175.

Bonini, S. A., Premoli, M., Tambaro, S., Kumar, A., Maccarinelli, G., Memo, M., & Mastinu, A. (2018). Cannabis sativa: A comprehensive ethnopharmacological review of a medicinal plant with a long history. *Journal of Ethnopharmacology, 227*, 300–315.

Bonnamain, V., Neveu, I., & Naveilhan, P. (2012). Neural stem/progenitor cells as promising candidates for regenerative therapy of the central nervous system. *Frontiers in Cellular Neuroscience, 6*, 17.

Bonnel, A., Mottron, L., Peretz, I., Trudel, M., Gallun, E., & Bonnel, A. M. (2003). Enhanced pitch sensitivity in individuals with autism: A signal detection analysis. *Journal of Cognitive Neuroscience, 15,* 226–235.

Boot, E., Hollak, C. E. M., Huijbregts, S. C. J., Jahja, R., van Vliet, D., Nederveen, A. J., . . . & Abeling, N. G. G. M. (2017). Cerebral dopamine deficiency, plasma monoamine alterations and neurocognitive deficits in adults with phenylketonuria. *Psychological Medicine, 47*(16), 2854–2865.

Booth, D. A. (1981). The physiology of appetite. *British Medical Bulletin, 37,* 135–140.

Borbély, A. A. (1983). Pharmacological approaches to sleep regulation. In A. R. Mayes (Ed.), *Sleep mechanisms and functions in humans and animals* (pp. 232–261). Wokingham, England: Van Nostrand Reinhold.

Borodinsky, L. N., Belgacem, Y. H., & Swapna, I. (2012). Electrical activitiy as a developmental regulator in the formation of spinal cord circuits. *Current Opinion in Neurobiology, 22,* 624–630.

Boroviak, T., & Nichols, J. (2014). The birth of embryonic pluripotency. *Philosophical Transactions of the Royal Society B, 369,* 2.

Borrell, V., & Götz, M. (2014). Role of radial glial cells in cerebral cortex folding. *Current Opinion in Neurobiology, 27,* 39–46.

Bostan, A. C., & Strick, P. L. (2018). The basal ganglia and the cerebellum: Nodes in an integrated network. *Nature Reviews Neuroscience, 19,* 338–350.

Bosworth, A. P., & Allen, N. J. (2017). The diverse actions of astrocytes during synaptic development. *Current Opinion in Neurobiology, 47,* 38–43.

Boto, E., Holmes, N., Leggett, J., Roberts, G., Shah, V., Meyer, S. S., . . . & Barnes, G. R. (2018). Moving magnetoencephalography towards real-world applications with a wearable system. *Nature, 555*(7698), 657.

Bouchard, D. (2015). Brain readiness and the nature of language. *Frontiers in Psychology, 6,* 1376.

Bouchard, T. J., Jr. (1998). Genetic and environmental influences on adult intelligence and special mental abilities. *Human Biology, 70,* 257–279.

Bouchard, T. J., Jr., & Pederson, N. L. (1999). Twins reared apart: Natures double experiment. In M. C. LaBuda & E. L. Grigorenko (Eds.), *Current Methodological Issues in Behavioral Genetics* (pp. 71–93). Commack, N. Y.: Nova Scientific.

Bourgeois, J., Elseviers, M. M., Van Bortel, L., Petrovic, M., & Vander Stichele, R. H. (2014). One-year evolution of sleep quality in older users of benzodiazepines: A longitudinal cohort study in Belgian nursing home residents. *Drugs & Aging, 31,* 677–682.

Bourne, J. N., & Harris, K. M. (2012). Nanoscale analysis of structural synaptic plasticity. *Current Opinion in Neurobiology, 22,* 372–382.

Bourzac, K. (2016). Bright sparks. *Nature, 531*(7592), S6–S8.

Bourzac, K. (2016). Diagnosis: Warning signs. *Nature, 538,* S5–S7.

Bouvier, G., Larsen, R. S., Rodríguez-Moreno, A., Paulsen, O., & Sjöström, P. J. (2018). Towards resolving the presynaptic NMDA receptor debate. *Current Opinion in Neurobiology, 51,* 1–7.

Bouzier-Sore, A-K., & Pellerin, L. (2013). Unraveling the complex metabolic nature of astrocytes. *Frontiers in Cellular Neuroscience, 7,* 179.

Bowler, P. J. (2009). Darwin's originality. *Science, 323,* 223–226.

Boyden, E. S. (2014, November/December). Let there be light. *Scientific American Mind,* 62–69.

Boyden, E. S. (2015). Optogenetics and the future of neuroscience. *Nature Neuroscience, 18,* 1200–1201.

Boys, A., & Marsden, J. (2003). Perceived functions predict intensity of use and problems in young polysubstance users. *Addiction, 98,* 951–963.

Bradke, F., Fawcett, J. W., & Spira, M. E. (2012). Assembly of a new growth cone after axotomy: The precursor to axon regeneration. *Nature Reviews Neuroscience, 13,* 183–192.

Bragatti, J. A., Torres, C. M., Abrahim, C., Leistner-Segal, S., & Bianchin, M. M. (2014). Is interictal EEG activity a biomarker for mood disorders in temporal lobe epilepsy? *Clinical Neurophysiology, 125,* 1952–1958.

Braidy, N., Poljak, A., Jayasena, T., Mansour, H., Inestrosa, N. C., & Sachdev, P. S. (2015). Accelerating Alzheimer's reserach through "natural" animal models. *Current Opinion in Psychiatry, 28,* 155–164.

Bramble, M. S., Lipson, A., Vashist, N., & Vilain, E. (2017). Effects of chromosomal sex and hormonal influences on shaping sex differences in brain and behavior: Lessons from cases of disorders of sex development. *Journal of Neuroscience Research, 95*(1–2), 65–74.

Brancaccio, M., Enoki, R., Mazuski, C. N., Jones, J., Evans, J. A., & Azzi, A. (2014). Network-mediated encoding of circadian time: The suprachiasmatic nucleus (SCN) from genes to neurons to circuits, and back. *Journal of Neuroscience, 34,* 15192–15199.

Brandon, N. J., & McKay, R. (2015). The cellular target of antidepressants. *Nature Neuroscience, 18,* 1537–1538.

Brandt, T., & Dieterich, M. (2017). The dizzy patient: Don't forget disorders of the central vestibular system. *Nature Reviews Neurology, 13*(6), 352–362.

Brang, D., Taich, Z. J., Hillyard, S. A., Grabowecky, M., & Ramachandran, V. S. (2013). Parietal connectivity mediates multisensory facilitation. *NeuroImage, 78,* 396–401.

Brann, J. H., & Firestein, S. J. (2014). A lifetime of neurogenesis in the olfactory system. *Frontiers in Neuroscience, 8,* 182.

Brascamp, J., Sterzer, P., Blake, R., & Knapen, T. (2018). Multistable perception and the role of the frontoparietal cortex in perceptual inference. *Annual Review of Psychology, 69,* 77–103.

Braskie, M. N., & Thompson, P. M. (2013). Understanding cognitive deficits in Alzheimer's disease based on neuroimaging findings. *Trends in Cognitive Sciences, 17,* 510–516.

Braver, T. S., Cole, M. W., & Yarkoni, T. (2010). Vive les differences! Individual variation in neural mechanisms of executive control. *Current Opinion in Neurobiology, 20,* 242–250.

Bray, N. (2014). Sleep: Inducing lucid dreams. *Nature Reviews Neuroscience, 15*(7), 428.

Brecher, E. M. (1972). *Licit and illicit drugs.* Boston, MA: Little, Brown.

Brecht, M. (2017). The body model theory of somatosensory cortex. *Neuron, 94*(5), 985–992.

Brecht, M., & Freiwald, W. A. (2012). The many facets of facial interactions in mammals. *Current Opinion in Neurobiology, 22,* 259–266.

Brecht, M., Lenschow, C., & Rao, R. (2018). Socio-sexual processing in cortical circuits. *Current Opinion in Neurobiology, 52,* 1–9.

Breen, M. S., Dobbyn, A., Li, Q., Roussos, P., Hoffman, G. E., Stahl, E., . . . & Buxbaum, J. D. (2019). Global landscape and genetic regulation of RNA editing in cortical samples from individuals with schizophrenia. *Nature Neuroscience, 22*(9), 1402–1412.

Bremen, P., & Middlebrooks, J. C. (2013). Weighting of spatial and spectro-temporal cues for auditory scene analysis by human listeners. *PLoS ONE, 8,* e59815.

Bremer, F. (1936). Nouvelles recherches sur le mécanisme du sommeil. *Comptes rendus de la Société de Biologie, 22,* 460–464.

Bremer, F. L. (1937). L'activité cérébrale au cours du sommeil et de la narcose. Contribution à l'étude du mécanisme du sommeil. *Bulletin de l'Académie Royale de Belgique, 4,* 68–86.

Bremer, J. (1959). *Asexualization.* New York, NY: Macmillan.

Brenner, S. (2012). The revolution in the life sciences. *Science, 338,* 1427–1428.

Bressler, S. L., & Richter, C. G. (2015). Interareal oscillatory synchronization in top-down neocortical processing. *Current Opinion in Neurobiology, 31,* 62–66.

Brewer, A. A., & Barton, B. (2016). Maps of the auditory cortex. *Annual Review of Neuroscience, 39,* 385–407.

Brewster, J. M. (1986). Prevalence of alcohol and other drug problems among physicians. *Journal of the American Medical Association, 255,* 1913–1920.

Brier, M. R., Gordon, B., Friedrichsen, K., McCarthy, J., Stern, A., Christensen, J., . . . & Cairns, N. J. (2016). Tau and Aβ imaging, CSF measures, and cognition in Alzheimer's disease. *Science Translational Medicine, 8*(338), 338ra66–338ra66.

Brinker, T., Stopa, E., Morrison, J., & Klinge, P. (2014). A new look at cerebrospinal fluid circulation. *Fluids and Barriers of the CNS, 11,* 10.

Britton, J. (2017). Death, disease, and tobacco. *The Lancet, 389*(10082), 1861–1862.

Broekman, M. L., Maas, S. L., Abels, E. R., Mempel, T. R., Krichevsky, A. M., & Breakefield, X. O. (2018). Multidimensional communication in the microenvirons of glioblastoma. *Nature Reviews Neurology, 14,* 482–495.

Brooks, M. J., & Melnik, G. (1995). The refeeding syndrome: An approach to understanding its complications and preventing its occurrence. *Pharmacotherapy, 15,* 713–726.

Brooks, S. P., & Dunnett, S. B. (2009). Tests to assess motor phenotype in mice: A user's guide. *Nature Reviews Neuroscience, 10,* 519–528.

Broughton, R. (1982). Human consciousness and sleep/waking rhythms: A review and some neuropsychological considerations. *Journal of Clinical and Experimental Neuropsychology, 4*(3), 193–218.

Brown, C. S., & Lichter-Konecki, U. (2016). Phenylketonuria (PKU): A problem solved? *Molecular Genetics and Metabolism Reports, 6,* 8–12.

Brown, G. C., & Neher, J. J. (2014). Microglial phagocytosis of live neurons. *Nature Reviews Neuroscience, 15,* 209–216.

Brown, J. W., & Braver, T. S. (2005). Learned predictions of error likelihood in the anterior cingulate cortex. *Science, 307,* 1118–1121.

Brown, R. D., & Broderick, J. P. (2014). Unruptured intracranial aneurysms: Epidemiology, natural history, management options, and familial screening. *The Lancet Neurology, 13,* 393–404.

Brown, R. E. (1994). *An introduction to neuroendocrinology.* Cambridge, England: Cambridge University Press.

Brown, R. E., & Milner, P. M. (2003). The legacy of Donald Hebb: More than the Hebb synapse. *Nature Reviews Neuroscience, 4,* 1013–1019.

Brown, R. E., & Milner, P. M. (2003). The legacy of Donald O. Hebb: More than the Hebb synapse. *Nature Reviews Neuroscience, 4*(12), 1013–1019.

Brunoni, A. R., Nitsche, M. A., Bolognini, N., Bikson, M., Wagner, T., Merabet, L.,. . . & Ferrucci, R. (2012). Clinical research with transcranial direct current stimulation (tDCS): Challenges and future directions. *Brain Stimulation, 5,* 175–195.

Buckingham-Howes, S., Berger, S. S., Scaletti, L. A., & Black, M. M. (2013). Systematic review of prenatal cocaine exposure and adolescent development. *Pediatrics, 131,* e1917.

Buckley, M. J., & Gaffan, D. (1998). Perirhinal cortex ablation impairs visual object identification. *Journal of Neuroscience, 18*(6), 2268–2275.

Buckner, R. L. (2013). The cerebellum and cognitive function: 25 years of insight from anatomy and neuroimaging. *Neuron, 80,* 807–815.

Budday, S., Steinmann, P., & Kuhl, E. (2015). Physical biology of human brain development. *Frontiers in Cellular Neuroscience, 9,* 257.

Buitrago-Delgado, E., Nordin, K., Rao, A., Geary, L., & LaBonne, C. (2015). Shared regulatory programs suggest retention of blastula-stage potential in neural crest cells. *Science, 348,* 1332–1335.

Bulters, D., Gaastra, B., Zolnourian, A., Alexander, S., Ren, D., Blackburn, S. L., . . . & Nyquist, P. (2018). Haemoglobin scavenging in intracranial bleeding: Biology and clinical implications. *Nature Reviews Neurology, 14*(7), 416–432.

Bundy, D. T., & Leuthardt, E. C. (2019). The cortical physiology of ipsilateral limb movements. *Trends in Neurosciences, 42*(11), 825–839.

Buonarati, O. R., Hammes, E. A., Watson, J. F., Greger, I. H., & Hell, J. W. (2019). Mechanisms of postsynaptic localization of AMPA-type glutamate receptors and their regulation during long-term potentiation. *Science Signaling, 12*(562), eaar6889.

Burda, J. E., & Sofroniew, M. V. (2014). Reactive gliosis and the multicellular response to CNS damage and disease. *Neuron, 81,* 229–248.

Burns, J. K. (2013). Pathways from cannabis to psychosis: A review of the evidence. *Frontiers in Psychiatry, 4,* 128.

Burton, C. L., Barta, C., Cath, D., Geller, D., van den Heuvel, O. A., Yao, Y., . . . & Zai, G. (2020). Genetics of obsessive-compulsive disorder and Tourette disorder. In *Personalized Psychiatry* (pp. 239–252). Academic Press.

Busche, M. A., Grienberger, C., Keskin, A. D., Song, B., Neumann, U., Staufenbiel, M., . . . & Konnerth, A. (2015). Decreased amyloid-beta and increased neuronal hyperactivity by immunotherapy in Alzheimer's models. *Nature Neuroscience, 18,* 1725–1728.

Buschman, T. J. (2015). Paying attention to the details of attention. *Neuron, 86,* 1111–1113.

Bush, D., Barry, C., & Burgess, N. (2014). What do grid cells contribute to place cell firing? *Trends in Neurosciences, 37,* 136–145.

Bushnell, M. C., Čeko, M., & Low, L. A. (2013). Cognitive and emotional control of pain and its disruption in chronic pain. *Nature Reviews Neuroscience, 14,* 502–511.

Bussey, T. J., & Saksida, L. M. (2005). Object memory and perception in the medial temporal lobe: An alternative approach. *Current Opinion in Neurobiology, 15,* 730–737.

Bussey, T. J., Warburton, E. C., Aggleton, J. P., & Muir, J. L. (1998). Fornix lesions can facilitate acquisition of the transverse patterning task: A challenge for "configural" theories of hippocampal function. *Journal of Neuroscience, 18*(4), 1622–1631.

Buxbaum, J. N. (2017). Alzheimer's Disease: It's More Than Aβ. *The FASEB Journal, 31*(1), 2–4.

Buzsáki, G. (2005). Similar is different in hippocampal networks. *Science, 309,* 568–569.

Byne, W. (1994). The biological evidence challenged. *Scientific American, 270*(5), 50–55.

Byne, W. (2017). Sustaining progress toward LGBT health equity: A time for vigilance, advocacy, and scientific inquiry. *LGBT Health, 4*(1), 1.

Byne, W., Karasic, D. H., Coleman, E., Eyler, A. E., Kidd, J. D., Meyer-Bahlburg, H. F., . . . & Pula, J. (2018). Gender dysphoria in adults: An overview and primer for psychiatrists. *Transgender Health, 3*(1), 57–A3.

Byne, W., Tobet, S., Mattiace, L. A., Lasco, M. S., Kemether, E., Edgar, M. A., . . . & Jones, L. B. (2001). The interstitial nuclei of the human anterior hypothalamus: An investigation of variation with sex, sexual orientation, and

HIV status. *Hormones and Behavior, 40*(2), 86–92.

Caballero, A., & Tseng, K. Y. (2016). GABAergic function as a limiting factor for prefrontal maturation during adolescence. *Trends in Neurosciences, 39*(7), 441–448.

Cabezas, R., Avila, M., Gonzalez, J., El-Bachá, R. S., Báez, E., et al. (2014). Astrocytic modulation of blood brain barrier: Perspectives on Parkinson's disease. *Frontiers in Cellular Neuroscience, 8,* 211.

Cacioppo, J. T., & Decety, J. (2009). What are the brain mechanisms on which psychological processes are based? *Perspectives on Psychological Science, 4,* 10–18.

Cafferty, W. B. J., McGee, A. W., & Strittmatter, S. M. (2008). Axonal growth therapeutics: Regeneration or sprouting or plasticity? *Trends in Neurosciences, 31,* 215–220.

Caggiula, A. R. (1970). Analysis of the copulation-reward properties of posterior hypothalamic stimulation in male rats. *Journal of Comparative and Physiological Psychology, 70,* 399–412.

Cahana-Amitay, D., & Albert, M. L. (2014). Brain and language: Evidence for neural multifunctionality. *Behavioural Neurology, 2014,* 260–381.

Cahill, L. (2006). Why sex matters for neuroscience. *Nature Reviews Neuroscience, 7,* 477–484.

Cahill, L. (2014). Equal ≠ the same: Sex differences in the human brain. *Cerebrum, April 2014,* 1–19.

Calabrò, R. S., Cacciola, A., Bruschetta, D., Milardi, D., Quattrini, F., Sciarrone, F., . . . & Anastasi, G. (2019). Neuroanatomy and function of human sexual behavior: A neglected or unknown issue? *Brain and Behavior, 9*(12), e01389.

Calatayud, V. A., Gras, T. S., Serrano, J. A., & Dols, S. T. (2019). The epidemiology of anabolic-androgenic steroids use among secondary students (Valencia-Spain). *Health and Addictions/Salud y Drogas, 19*(2), 1–7.

Caldeira, G. L., Peça, J., & Carvalho, A. L. (2019). New insights on synaptic dysfunction in neuropsychiatric disorders. *Current Opinion in Neurobiology, 57,* 62–70.

Calhoon, G. G., & Tye, K. M. (2015). Resolving the neural circuits of anxiety. *Nature Neuroscience, 18,* 1394–1404.

Calhoon, G. G., & Tye, K. M. (2015). Resolving the neural circuits of anxiety. *Nature Neuroscience, 18,* 1394–1404.

Callaghan, B. L., Li, S., & Richardson, R. (2014). The elusive engram: What can infantile amnesia tell us about memory? *Trends in Neurosciences, 37,* 47–53.

Callaghan, R. C., Cunningham, J. K., Sykes, J., & Kish, S. J. (2012). Increased risk of Parkinson's disease in individuals hospitalized with conditions related to the use of methamphetamine or other amphetamine-type drugs. *Drug and Alcohol Dependence, 120,* 35–40.

Callaway, E. M., & Garg, A. K. (2017). Brain technology: Neurons recorded en masse. *Nature, 551*(7679), 172.

Callaway, E., Sutikana, T., Roberts, R., Saptomo, W., Brown, P., et al. (2014). The discovery of *Homo floresiensis*: Tales of the hobbit. *Nature, 514,* 422–426.

Calmon, G., Roberts, N., Eldridge, P., & Thirion, J.-P. (1998). Automatic quantification of changes in the volume of brain structures. *Lecture Notes in Computer Science, 1496,* 761–769.

Calvo, M. G., & Nummenmaa, L. (2015). Perceptual and affective mechanisms in facial recognition: An integrative review. *Cognition and Emotion,* 1–26.

Camardese, G., Di Giuda, D., Di Nicola, M., Cocciolillo, F., Giordano, A., Janiri, L., & Guglielmo, R. (2014). Imaging studies on dopamine transporter and depression: A review of literature and suggestions for future research. *Journal of Psychiatric Research, 51,* 7–18.

Cambridge, O. R., Knight, M. J., Mills, N., & Baune, B. T. (2018). The clinical relationship between cognitive impairment and psychosocial functioning in major depressive disorder:

A systematic review. *Psychiatry Research, 269,* 157–171.

Cameron, H. A., Woolley, C. S., McEwen, B. S., & Gould, E. (1993). Differentiation of newly born neurons and glia in the dentate gyrus of the adult rat. *Neuroscience, 56,* 337–344.

Caminiti, R., Innocenti, G. M., & Battaglia-Mayer, A. (2015). Organization and evolution of parieto-frontal processing streams in macaque monkeys and humans. *Neuroscience and Biobehavioral Reviews, 56,* 73–96.

Campbell, K. J., Crawford, D. A., Salmon, J., Carver, A., Garnett, S. P., & Baur, L. A. (2007). Associations between the home food environment and obesity-promoting eating behaviors in adolescence. *Obesity, 15,* 719–730.

Campbell, R. R., & Wood, M. A. (2019). How the epigenome integrates information and reshapes the synapse. *Nature Reviews Neuroscience, 20*(3), 133–147.

Campbell, R., MacSweeney, M., & Waters, D. (2008). Sign language and the brain: A review. *Journal of Deaf Studies and Deaf Education, 13,* 3–20.

Campfield, L. A., & Smith, F. J. (1989). Transient declines in blood glucose signal meal initiation. *International Journal of Obesity, 14,* 15–31.

Campfield, L. A., Smith, F. J., Gulsez, Y., Devos, R., & Burn, P. (1995). Mouse Ob protein: Evidence for a peripheral signal linking adiposity and central neural networks. *Science, 269,* 546–550.

Can, A., Schulze, T. G., & Gould, T. D. (2014). Molecular actions and clinical pharmacogenetics of lithium therapy. *Pharmacology Biochemistry and Behavior, 123,* 3–16.

Candidi, M., Stienen, B. M., Aglioti, S. M., & de Gelder, B. (2015). Virtual lesion of right posterior superior temporal sulcus modulates conscious visual perception of fearful expressions in faces and bodies. *Cortex, 65,* 184–194.

Cang, J., & Feldheim, D. A. (2013). Developmental mechanisms of topographic map formation and alignment. *Annual Review of Neuroscience, 36,* 51–77.

Cannon, W. B., & Washburn, A. L. (1912). An explanation of hunger. *American Journal of Physiology, 29,* 441–454.

Canteras, N. S., Ribeiro-Barbosa, É. R., Goto, M., Cipolla-Nito, J., & Swanson, L. W. (2011). The retinohypothalamic tract: Comparison of axonal projection patterns from four major targets. *Brain Research Reviews, 65,* 150–183.

Cantor, J. M., Blanchard, R., Paterson, A. D., & Bogaert, A. F. (2004). How many gay men owe their sexual orientation to fraternal birth order? *Archives of Sexual Behavior, 33,* 63–71.

Cao, B., Stanley, J. A., Passos, I. C., Mwangi, B., Selvaraj, S., Zunta-Soares, G. B., & Soares, J. C. (2017). Elevated choline-containing compound levels in rapid cycling bipolar disorder. *Neuropsychopharmacology, 42*(11), 2252–2258.

Cao, D., Marsh, J. C., Shun, H. C., & Andrews, C. M. (2011). Improving health and social outcomes with targeted services in comprehensive substance abuse treatment. *American Journal of Drug and Alcohol Abuse, 37,* 250–258.

Cao, M., He, Y., Dai, Z., Liao, X., Jeon, T., Ouyang, M., . . . & Huang, H. (2017). Early development of functional network segregation revealed by connectomic analysis of the preterm human brain. *Cerebral Cortex, 27*(3), 1949–1963.

Capogrosso, M., Milekovic, T., Borton, D., Wagner, F., Moraud, E. M., Mignardot, J. B., . . . & Rey, E. (2016). A brain–spine interface alleviating gait deficits after spinal cord injury in primates. *Nature, 539*(7628), 284.

Cappelletti, M., & Wallen, K. (2016). Increasing women's sexual desire: The comparative effectiveness of estrogens and androgens. *Hormones and Behavior, 78,* 178–193.

Cappuccio, F. P., D'Elia, L., Strazzullo, P., & Miller, M. A. (2008). Sleep duration and all-cause mortality: A systematic review and meta-analysis of prospective studies. *Sleep, 33,* 585–592.

Caranci, F., Briganti, F., Cirillo, L., Leonardi, M., & Muto, M. (2013). Epidemiology and genetics of intracranial aneurysms. *European Journal of Radiology, 82*, 1598–1605.

Cardinal, R. N., & Everitt, B. J. (2004). Neural and psychological mechanisms underlying appetitive learning: Links to drug addiction. *Current Opinion in Neurobiology, 14*, 156–162.

Carey, R. J. (2020). Drugs and memory: Evidence that drug effects can become associated with contextual cues by being paired post-trial with consolidation/re-consolidation. Mini review. *Pharmacology Biochemistry and Behavior*, 172911.

Carlsson, A., & Lindqvist, M. (1963). Effect of chlorpromazine or haloperidol on formation of 3-methoxytyramine and normetanephrine in mouse brains. *Acta Pharmacologica et Toxicologica, 20*, 140–144.

Carlyle, B. C., Kitchen, R. R., Kanyo, J. E., Voss, E. Z., Pletikos, M., Sousa, A. M., . . . & Nairn, A. C. (2017). A multiregional proteomic survey of the postnatal human brain. *Nature Neuroscience, 20*(12), 1787–1795.

Carman, A. J., Ferguson, R., Cantu, R., Comstock, R. D., Dacks, P. A., . . . & Fillit, H. M. (2015). Mind the gaps—advancing research into short-term and long-term neuropsychological outcomes of youth sports-related concussions. *Nature Neurology, 11*, 230–244.

Caron, N. S., Dorsey, E. R., & Hayden, M. R. (2018). Therapeutic approaches to Huntington disease: From the bench to the clinic. *Nature Reviews Drug Discovery, 17*, 729–750.

Carothers, B. J., & Reis, H. T. (2013). Men and women are from Earth: Examining the latent structure of gender. *Journal of Personality and Social Psychology, 104*, 385.

Carpenter, A. M., Singh, I., Gandhi, C. D., & Prestigiacomo, C. J. (2016). Genetic risk factors for spontaneous intracerebral haemorrhage. *Nature Reviews Neurology, 12*(1), 40–49.

Carr, L., Iacoboni, M., Dubeau, M.-C., Mazziotta, J. C., & Lenzi, G. L. (2003). Neural mechanisms of empathy in humans: A relay from neural systems for imitation to limbic areas. *Proceedings of the National Academy of Sciences, USA, 100*, 5497–5502.

Carré, J. M., & Olmstead, N. A. (2015). Social neuroendocrinology of human aggression: Examining the role of competition-induced testosterone dynamics. *Neuroscience, 286*, 171–186.

Carroll, F., Blough, B., Mascarella, S., Navarro, H. A., Lukas, R. J., & Damaj, M. I. (2014). Bupropion and bupropion analogs as treatments for CNS disorders. *Advances in Pharmacology, 69*, 177–216.

Carroll, M. E., & Meisch, R. A. (1984). Increased drug-reinforced behavior due to food deprivation. *Advances in Behavioral Pharmacology, 4*, 47–88.

Carter, J. R., & Goldstein, D. S. (2015). Sympathoneural and adrenomedullary responses to mental stress. *Comprehensive Physiology, 5*, 119–146.

Carvalho, A. F., Dimellis, D., Gonda, X., Vieta, E., McIntyre, R. S., & Fountoulakis, K. N. (2014). Rapid cycling in bipolar disorder: A systematic review. *Journal of Clinical Psychiatry, 75*, 1–478.

Carvalho, M., Carmo, H., Costa, V. M., Capela, J. P., Pontes, H., Remião, F., Carvalho, F., & Bastos, M. D. (2012). Toxicity of amphetamines: An update. *Archives of Toxicology, 86*, 1167–1231.

Casadio, P., Fernandes, C., Murray, R. M., & Di Forti, M. (2011). Cannabis use in young people: The risk of schizophrenia. *Neuroscience and Biobehavioral Reviews, 35*, 1779–1787.

Cascio, M. G., Zamberletti, E., Marini, P., Parolaro, D., & Pertwee, R. G. (2014). The phytocannabinoid, delta-9-tetrahydrocannabivarin, can act through 5-HT1A receptors to produce antipsychotic effects. *British Journal of Pharmacology, 172*, 1305–1318.

Casey, D. A. J. (2013). Aetiology of auditory dysfunction in amusia: A systematic review. *International Archives of Medicine, 6*, 16.

Casey, L. (2013). Caring for children with phenylketonuria. *Canadian Family Physician, 59*, 837–840.

Castelblanco, R. L., Lee, M., & Hasbun, R. (2014). Epidemiology of bacterial meningitis in the USA from 1997 to 2010: A population-based observational study. *The Lancet Infectious Diseases, 14*, 813–819.

Castellani, R. J., Plascencia-Villa, G., & Perry, G. (2019). The amyloid cascade and Alzheimer's disease therapeutics: theory versus observation. *Laboratory Investigation, 99*(7), 958–970.

Castelló, S., Molina, J. C., & Arias, C. (2017). Long-term contextual memory in infant rats as evidenced by an ethanol conditioned tolerance procedure. *Behavioural Brain Research, 332*, 243–249.

Castelló, S., Revile, D. A., Molina, J. C., & Arias, C. (2014). Ethanol-induced tolerance and sex-dependent sensitization in preweanling rats. *Physiology & Behavior, 139*, 50–58.

Castelnovo, A., Lopez, R., Proserpio, P., Nobili, L., & Dauvilliers, Y. (2018). NREM sleep parasomnias as disorders of sleep-state dissociation. *Nature Reviews Neurology, 14*(8), 470–481.

Castrén, E., & Hen, R. (2013). Neuronal plasticity and antidepressant actions. *Trends in Neurosciences, 36*, 259–267.

Castro, G. D., & Castro, J. A. (2014). Alcohol drinking and mammary cancer: Pathogenesis and potential dietary preventive alternatives. *World Journal of Clinical Oncology, 5*, 713–729.

Catterall, W. A., Raman, I. M., Robinson, H. P. C., Sejnowski, T. J., & Paulsen, O. (2012). The Hodgkin-Huxley heritage: From channels to circuits. *Journal of Neuroscience, 32*, 14064–14073.

Cavalli, G., & Heard, E. (2019). Advances in epigenetics link genetics to the environment and disease. *Nature, 571*(7766), 489–499.

Cavanagh, S. E., Towers, J. P., Wallis, J. D., Hunt, L. T., & Kennerley, S. W. (2018). Reconciling persistent and dynamic hypotheses of working memory coding in prefrontal cortex. *Nature Communications, 9*(1), 1–16.

Cecon, E., Chen, M., Marçola, M., Fernandes, P. A., Jockers, R., & Markus, R. P. (2015). Amyloid β peptide directly impairs pineal gland melatonin synthesis and melatonin receptor signaling through the ERK pathway. *FASEB Journal, 29*, 2566–2582.

Cedernaes, J., Schiöth, H. B., & Benedict, C. (2015). Determinants of shortened, disrupted, and mistimed sleep and associated metabolic health consequences in healthy humans. *Diabetes, 64*, 1073–1080.

Cembrowski, M. S., & Menon, V. (2018). Continuous variation within cell types of the nervous system. *Trends in Neurosciences, 41*, 337–348.

Cepko, C. (2015). Intrinsically different retinal progenitor cells produce specific types of progeny. *Nature Reviews Neuroscience, 15*, 615–627.

Cermakian, N., & Sassone-Corsi, P. (2002). Environmental stimulus perception and control of circadian clocks. *Current Opinion in Neurobiology, 12*, 359–365.

Chaaya, N., Battle, A. R., & Johnson, L. R. (2018). An update on contextual fear memory mechanisms: Transition between amygdala and hippocampus. *Neuroscience & Biobehavioral Reviews, 92*, 43–54.

Chai, H., Diaz-Castro, B., Shigetomi, E., Monte, E., Octeau, J. C., Yu, X., . . . & Coppola, G. (2017). Neural circuit-specialized astrocytes: Transcriptomic, proteomic, morphological, and functional evidence. *Neuron, 95*(3), 531–549.

Chailangkarn, T., Trujillo, C. A., Freitas, B. C., Hrvoj-Mihic, B., Herai, R. H., Diana, X. Y., . . . & Stefanacci, L. (2016). A human neurodevelopmental model for Williams syndrome. *Nature, 536*(7616), 338–343.

Chalfie, M., Tu, Y., Euskirchen, G., Ward, W., & Prasher, D. (1994). Green fluorescent protein as a marker for gene expression. *Science, 263*, 802–805.

Champagne, D. L., Bagot, R. C., van Hasselt, F., Ramakers, G., Meaney, M. J., de Kloet, E. R., . . . & Krugers, H. (2008). Maternal care and hippocampal plasticity: Evidence for experience-dependent structural plasticity, altered synaptic functioning, and differential responsiveness to glucocorticoids and stress. *Journal of Neuroscience, 28*, 6037–6045.

Chan, N. K., Kim, J., Shah, P., Brown, E. E., Plitman, E., Carravaggio, F., . . . & Graff-Guerrero, A. (2019). Resting-state functional connectivity in treatment response and resistance in schizophrenia: A systematic review. *Schizophrenia Research, 211*, 10–20.

Chance, S. A. (2014). The cortical microstructural basis of lateralized cognition: A review. *Frontiers in Psychology, 5*, 820.

Chang, C. H., & Grace, A. A. (2015). Dopaminergic modulation of lateral amygdala neuronal activity: Differential D1 and D2 receptor effects on thalamic and cortical afferent inputs. *International Journal of Neuropsychopharmacology, 18*, pyv015.

Chang, C. Y., Esber, G. R., Marrero-Garcia, Y., Yau, H. J., Bonci, A., & Schoenbaum, G. (2016). Brief optogenetic inhibition of dopamine neurons mimics endogenous negative reward prediction errors. *Nature Neuroscience, 19*, 111–116.

Chang, K. J., Redmond, S. A., & Chan, J. R. (2016). Remodeling myelination: Implications for mechanisms of neural plasticity. *Nature Neuroscience, 19*(2), 190–197.

Chang, S. H., Stoll, C. R., Song, J., Varela, J. E., Eagon, C. J., & Colditz, G. A. (2014). The effectiveness and risks of bariatric surgery: An updated systematic review and meta-analysis, 2003–2012. *JAMA Surgery, 149*, 275–287.

Changeux, J.-P. (2013). The concept of allosteric interaction and its consequences for the chemistry of the brain. *Journal of Biological Chemistry, 288*, 26969–26986.

Chapman, S., & MacKenzie, R. (2010). The global research neglect of unassisted smoking cessation: Causes and consequences. *PLoS Medicine, 7*, e1000216.

Chapman, S., Bareham, D., & Maziak, W. (2019). The gateway effect of e-cigarettes: Reflections on main criticisms. *Nicotine and Tobacco Research, 21*(5), 695–698.

Charles, A. C., & Baca, S. M. (2013). Cortical spreading depression and migraine. *Nature Reviews Neurology, 9*, 637–644.

Charness, M. E., Riley, E. P., & Sowell, E. R. (2016). Drinking during pregnancy and the developing brain: Is any amount safe? *Trends in Cognitive Sciences, 20*, 80–82.

Charvin, D., Medori, R., Hauser, R. A., & Rascol, O. (2018). Therapeutic strategies for Parkinson disease: Beyond dopaminergic drugs. *Nature Reviews Drug Discovery, 17*(11), 804–822.

Chastain, L. G., & Sarkar, D. K. (2017). Alcohol effects on the epigenome in the germline: Role in the inheritance of alcohol-related pathology. *Alcohol, 60*, 53–66.

Chatburn, A., Lushington, K., & Kohler, M. J. (2014). Complex associative memory processing and sleep: A systematic review and meta-analysis of behavioural evidence and underlying EEG mechanisms. *Neuroscience & Biobehavioral Reviews, 47*, 646–655.

Chattarji, S., Tomar, A., Suvrathan, A., Ghosh, S., & Rahman, M. M. (2015). Neighborhood matters: Divergent patterns of stress-induced plasticity across the brain. *Nature Neuroscience, 18*, 1364–1375.

Chatterjee, A., & Morison, A. M. (2011). Monozygotic twins: Genes are not the destiny? *Bioinformation, 7*, 369–370.

Chattu, V. K., Sakhamuri, S. M., Kumar, R., Spence, D. W., BaHammam, A. S., & Pandi-Perumal, S. R. (2018). Insufficient sleep syndrome: Is it time to classify it as a major noncommunicable disease? *Sleep Science, 11*(2), 56–64.

Chauvet, C., Goldberg, S. R., Jaber, M., & Solinas, M. (2012). Effects of environmental enrichment on the incubation of drug craving. *Neuropharmacology, 63,* 635–641.

Chauvet, C., Lardeux, V., Goldberg, S. R., Jaber, M., & Solinas, M. (2009). Environmental enrichment reduces cocaine seeking and reinstatement induced by cues and stress but not by cocaine. *Neuropsychopharmacology, 34,* 2767–2778.

Cheetham, C. E., & Belluscio, L. (2014). An olfactory critical period. *Science, 344,* 157–158.

Chen, E., Hanson, M. D., Paterson, L. Q., Griffin, M. J., Walker, H. A., & Miller, G. E. (2006). Socioeconomic status and inflammatory processes in childhood asthma: The role of psychological stress. *Journal of Allergy and Clinical Immunology, 117,* 1014–1020.

Chen, I. (2013a). Lead's buried legacy. *Scientific American, 309,* 28–30.

Chen, J. L., Margolis, D. J., Stankov, A., Sumanovski, L. T., Schneider, B. L., & Helmchen, F. (2015). Pathway-specific reorganization of projection neurons in somatosensory cortex during learning. *Nature Neuroscience, 18,* 1101–1108.

Chen, M. J., Axelrad, M., Dietrich, J. E., Gargollo, P., Gunn, S., ... & Karaviti, L. P. (2015). Androgen insensitivity syndrome: Management considerations from infancy to adulthood. *Pediatric Endocrinology Reviews, 12,* 373–387.

Chen, R., Spencer, D. C., Weston, J., & Nolan, S. J. (2016). Transcranial magnetic stimulation for the treatment of epilepsy. *Cochrane Database of Systematic Reviews,* (8), CD011025.

Chen, T., Giri, M., Xia, Z., Subedi, Y. N., & Li, Y. (2017). Genetic and epigenetic mechanisms of epilepsy: A review. *Neuropsychiatric disease and treatment, 13,* 1841.

Chen, W. R., Midtgaard, J., & Shepherd, G. M. (1997). Forward and backward propagation of dendritic impulses and their synaptic control in mitral cells. *Science, 278,* 463–466.

Chen, X., Gabitto, M., Peng, Y., Ryba, N. J. P., & Zuker, C. S. (2011). A gustatopic map of taste qualities in the mammalian brain. *Science, 333,* 1262–1266.

Chen, Y., Kelton, C. M., Jing, Y., Guo, J. J., Li, X., & Patel, N. C. (2008). Utilization, price, and spending trends for antidepressants in the US medicaid program. *Research in Social and Administrative Pharmacy, 4,* 244–257.

Chen, Z., & Wilson, M. A. (2017). Deciphering neural codes of memory during sleep. *Trends in Neurosciences, 40*(5), 260–275.

Cheng, J., Cao, Y., & Olson, L. (1996). Spinal cord repair in adult paraplegic rats: Partial restoration of hind limb function. *Science, 273,* 510–513.

Cheung, K. J., & Ewald, A. J. (2016). A collective route to metastasis: Seeding by tumor cell clusters. *Science, 352,* 167–169.

Chica, A. B., Bartolomeo, P., & Lupiáñez, J. (2013). Two cognitive and neural systems for endogenous and exogenous spatial attention. *Behavioural Brain Research, 237,* 107–123.

Chiesa, P. A., Cavedo, E., Lista, S., Thompson, P. M., Hampel, H., & Alzheimer Precision Medicine Initiative. (2017). Revolution of resting-state functional neuroimaging genetics in Alzheimer's disease. *Trends in Neurosciences, 40*(8), 469–480.

Chisholm, N. C., & Sohrabji, F. (2015). Astrocytic response to cerebral ischemia is influenced by sex differences and impaired by aging. *Neurobiology of Disease, 85,* 245–253.

Chiurchiù, V., van der Stelt, M., Centonze, D., & Maccarone, M. (2018). The endocannabinoid system and its therapeutic exploitation in multiple sclerosis: Clues for other neuroinflammatory diseases. *Progress in Neurobiology, 160,* 82–100.

Choleris, E., Galea, L. A., Sohrabji, F., & Frick, K. M. (2018). Sex differences in the brain: Implications for behavioral and biomedical research. *Neuroscience & Biobehavioral Reviews, 85,* 126–145.

Chotard, C., & Salecker, I. (2004). Neurons and glia: Team players in axon guidance. *Trends in Neurosciences, 27,* 655–661.

Chow, B. W., & Gu, C. (2015). The molecular constituents of the blood–brain barrier. *Trends in Neurosciences, 38,* 598–608.

Christensen, R., Shao, Z., & Colón-Ramos, D. A. (2013). The cell biology of synaptic specificity during development. *Current Opinion in Neurobiology, 23,* 1018–1026.

Christison-Lagay, K. L., & Cohen, Y. E. (2014). Behavioral correlates of auditory streaming in rhesus macaques. *Hearing Research, 309,* 17–25.

Christison-Lagay, K. L., Gifford, A. M., & Cohen, Y. E. (2015). Neural correlates of auditory scene analysis and perception. *International Journal of Psychophysiology, 95,* 238–245.

Chun, J., Kihara, Y., Jonnalagadda, D., & Blaho, V. A. (2019). Fingolimod: Lessons learned and new opportunities for treating multiple sclerosis and other disorders. *Annual Review of Pharmacology and Toxicology, 59,* 149–170.

Churchland, P. C. (2002). Self-representation in nervous systems. *Science, 296,* 308–310.

Churchland, P. S. (2014). Looking-glass wars. *Nature, 511,* 532–533.

Cipolli, C., Ferrara, M., De Gennaro, L., & Plazzi, G. (2017). Beyond the neuropsychology of dreaming: insights into the neural basis of dreaming with new techniques of sleep recording and analysis. *Sleep Medicine Reviews, 35,* 8–20.

Cipriani, A., Furukawa, T. A., Salanti, G., Chaimani, A., Atkinson, L. Z., Ogawa, Y., ... & Egger, M. (2018a). Comparative efficacy and acceptability of 21 antidepressant drugs for the acute treatment of adults with major depressive disorder: a systematic review and network meta-analysis. *Focus, 16*(4), 420–429.

Cipriani, A., Salanti, G., Furukawa, T. A., Egger, M., Leucht, S., Ruhe, H. G., ... & Ogawa, Y. (2018b). Antidepressants might work for people with major depression: Where do we go from here? *The Lancet Psychiatry, 5*(6), 461–463.

Cirelli, C., & Tononi, G. (2015). Sleep and synaptic homeostasis. *Sleep, 38*(1), 161–162.

Citron, M. (2004). Strategies for disease modification in Alzheimer's disease. *Nature Reviews Neuroscience, 5,* 677–685.

Clark, I. A., & Maguire, E. A. (2016). Remembering preservation in hippocampal amnesia. *Annual Review of Psychology, 67,* 51–82.

Clark, L. (2014). Disordered gambling: The evolving concept of behavioural addiction. *Annals of the New York Academy of Sciences, 1327,* 46–61.

Clark, L. T., Watkins, L., Piña, I. L., Elmer, M., Akinboboye, O., Gorham, M., ... & Puckrein, G. (2019). Increasing diversity in clinical trials: Overcoming critical barriers. *Current Problems in Cardiology, 44*(5), 148–172.

Clarke, L. E., & Barres, B. A. (2013). Emerging roles of astrocytes in neural circuit development. *Nature Reviews Neuroscience, 14,* 311–321.

Clarke, L. E., & Liddelow, S. A. (2017). Neurobiology: Diversity reaches the stars. *Nature, 548*(7668), 396–397.

Clay, F. J., Kuriakose, A., Lesche, D., Hicks, A. J., Zaman, H., Azizi, E., ... & Hopwood, M. (2019). Klüver-Bucy syndrome following traumatic brain injury: A systematic synthesis and review of pharmacological treatment from cases in adolescents and adults. *The Journal of Neuropsychiatry and Clinical Neurosciences, 31*(1), 6–16.

Clemmensen, C., Müller, T. D., Woods, S. C., Berthoud, H. R., Seeley, R. J., & Tschöp, M. H. (2017). Gut-brain cross-talk in metabolic control. *Cell, 168*(5), 758–774.

Clifton, P. G. (2000). Meal patterning in rodents: Psychopharmacological and neuroanatomical studies. *Neuroscience and Biobehavioural Reviews, 24,* 213–222.

Clouard, C., Meunier-Salaün, M. C., & Val-Laillet, D. (2012). Food preferences and aversions in human health and nutrition: How can pigs help the biomedical research. *Animal, 6,* 118–136.

Cloud, L. J., Zutshi, D., & Factor, S. A. (2014). Tardive dyskinesia: Therapeutic options for an increasingly common disorder. *Neurotherapeutics, 11,* 166–176.

Cloutman, L. L., Newhart, M., Davis, C. L., Heidler-Gary, J., & Hillis, A. E. (2011). Neuroanatomical correlates of oral reading in acute left hemispheric stroke. *Brain and Language, 116,* 14–21.

Coen-Cagli, R., Kohn, A., & Schwartz, O. (2015). Flexible gating of contextual influences in natural vision. *Nature Neuroscience, 18,* 1648–1655.

Cogan, G. B., Thesen, T., Carlson, C., Doyle, W., Devinsky, O., & Pesaran, B. (2014). Sensory-motor transformations for speech occur bilaterally. *Nature, 507,* 94–98.

Cohen, J. (2007). Relative differences: The myth of 1%. *Science, 316,* 1836.

Cohen, K., Weizman, A., & Weinstein, A. (2019). Positive and negative effects of cannabis and cannabinoids on health. *Clinical Pharmacology & Therapeutics, 105*(5), 1139–1147.

Cohen, M. X. (2017). Where does EEG come from and what does it mean? *Trends in Neurosciences, 40*(4), 208–218.

Cohen, S., Tyrrell, D. A. J., & Smith, A. P. (1991). Psychological stress and susceptibility to the common cold. *New England Journal of Medicine, 325,* 606–612.

Cohen, Y. E. (2009). Multimodal activity in the parietal cortex. *Hearing Research, 258,* 100–105.

Colapinto, J. (2000). *As nature made him: The boy who was raised as a girl.* New York, NY: HarperCollins.

Cole, J. C. (2014). MDMA and the "Ecstasy Paradigm." *Journal of Psychoactive Drugs, 46,* 44–56.

Cole, M. W., Yeung, N., Freiwald, W. A., & Botvinick, M. (2009). Cingulate cortex: Diverging data from humans and monkeys. *Trends in Neurosciences, 32,* 566–574.

Coleman, D. L. (1979). Obesity genes: Beneficial effects in heterozygous mice. *Science, 203,* 663–665.

Coles, J. A., Myburgh, E., Brewer, J. M., & McMenamin, P. G. (2017). Where are we? The anatomy of the murine cortical meninges revisited for intravital imaging, immunology, and clearance of waste from the brain. *Progress in Neurobiology, 156,* 107–148.

Coles, N. A., Larsen, J. T., & Lench, H. C. (2019). A meta-analysis of the facial feedback literature: Effects of facial feedback on emotional experience are small and variable. *Psychological Bulletin, 145*(6), 610–651.

Colgan, L. A., & Yasuda, R. (2014). Plasticity of dendritic spines: Subcompartmentalization of signaling. *Annual Review of Physiology, 76,* 365–385.

Collaer, M. L., Brook, C. G. D., Conway, G. S., Hindmarsh, P. C., & Hines, M. (2008). Motor development in individuals with congenital adrenal hyperplasia: Strength, targeting, and fine motor skill. *Psychoneuroendocrinology, 34,* 249–258.

Collier, G. (1986). The dialogue between the house economist and the resident physiologist. *Nutrition and Behavior, 3,* 9–26.

Collin, T., Marty, A., & Llano, I. (2005). Presynaptic calcium stores and synaptic transmission. *Current Opinion in Neurobiology, 15,* 275–281.

Collinge, J. (2016). Mammalian prions and their wider relevance in neurodegenerative diseases. *Nature, 539*(7628), 217–226.

Collinger, J. L., Wodlinger, B., Downey, J. E., Wang, W., Tyler-Kabara, E. C., ... & Schwartz, A. B. (2013). High-performance neuroprosthetic control by an individual with tetraplegia. *Lancet, 381,* 557–564.

Collins, J. A., & Olson, I. R. (2014). Beyond the FFA: The role of the ventral anterior temporal lobes in face processing. *Neuropsychologia, 61,* 65–79.

Colman, R. J., Beasley, T. M., Kemnitz, J. W., Johnson, S. C., Weindruch, R., & Anderson, R. M. (2014). Caloric restriction reduces age-related and all-cause mortality in rhesus monkeys. *Nature Communications, 5,* 3557.

Colombo, M., & Broadbent, N. (2000). Is the avian hippocampus a functional homologue of the mammalian hippocampus? *Neuroscience and Biobehavioural Reviews, 24,* 465–484.

Colvin, C. W., & Abdullatif, H. (2013). Anatomy of female puberty: The clinical relevance of developmental changes in the reproductive system. *Clinical Anatomy, 26*(1), 115–129.

Colvin, M. K., Dunbar, K., & Grafman, J. (2001). The effects of frontal lobe lesions on goal achievement in the water jug task. *Journal of Cognitive Neuroscience, 13,* 1129–1147.

Conforti, L., Adalbert, R., & Coleman, M. P. (2007). Neuronal death: Where does the end begin? *Trends in Neurosciences, 30,* 159–166.

Conners, B. W., & Long, M. A. (2004). Electrical synapses in the mammalian brain. *Annual Review of Neuroscience, 27,* 393–418.

Connor, C. E., & Knierim, J. J. (2017). Integration of objects and space in perception and memory. *Nature Neuroscience, 20*(11), 1493–1503.

Connor, S. A., & Wang, Y. T. (2015). A place at the table: LTD as a mediator of memory genesis. *Neuroscientist,* 1073858415588498.

Constantinidis, C., & Klingberg, T. (2016). The neuroscience of working memory capacity and training. *Nature Reviews Neuroscience, 17*(7), 438–451.

Constantino, J. N., Kennon-McGill, S., Weichselbaum, C., Marrus, N., Haider, A., Glowinski, A. L., . . . & Jones, W. (2017). Infant viewing of social scenes is under genetic control and is atypical in autism. *Nature, 547*(7663), 340–344.

Cooper, J. A. (2013). Mechanisms of cell migration in the nervous system. *Journal of Cell Biology, 202,* 725–734.

Cooper, R. M., & Zubek, J. P. (1958). Effects of enriched and restricted early environments on the learning ability of bright and dull rats. *Canadian Journal of Psychology, 12,* 159–164.

Cooper, S. A., Joshi, A. C., Seenan, P. J., Hadley, D. M., . . . & Metcalfe, R. A. (2012). Akinetopsia: Acute presentation and evidence for persisting defects in motion vision. *Journal of Neurology, Neurosurgery and Psychiatry, 83,* 229–230.

Cope, E. C., & Gould, E. (2019). Adult Neurogenesis, glia, and the extracellular matrix. *Cell Stem Cell, 24,* 690–705.

Corballis, M. C. (2015). What's left in language? Beyond the classical model. *Annals of the New York Academy of Sciences, 1359,* 14–29.

Corballis, M. C. (2017). The evolution of lateralized brain circuits. *Frontiers in Psychology, 8,* 1021.

Corbetta, M., Miezin, F. M., Dobmeyer, S., Shulman, G. L., & Petersen, S. E. (1990). Attentional modulation of neural processing of shape, color, and velocity in humans. *Science, 248,* 1556–1559.

Corina, D. P., Loudermilk, B. C., Detwiler, L., Martin, R. F., Brinkley, J. F., & Ojemann, G. (2010). Analysis of naming errors during cortical stimulation mapping: Implications for models of language representation. *Brain and Language, 115,* 101–112.

Corina, D. P., Poizner, H., Bellugi, U., Feinberg, T., Dowd, D., & O'Grady-Batch, L. (1992). Dissociation between linguistic and nonlinguistic gestural systems: A case for compositionality. *Brain and Language, 43,* 414–447.

Corkin, S., Milner, B., & Rasmussen, R. (1970). Somatosensory thresholds. *Archives of Neurology, 23,* 41–59.

Corsi, P. I. (1991). *The enchanted loom: Chapters in the history of neuroscience.* New York, NY: Oxford University Press.

Cosci, F., & Mansueto, G. (2019a). Biological and clinical markers in panic disorder. *Psychiatry Investigation, 16*(1), 27–36.

Cosci, F., & Mansueto, G. (2020b). Biological and clinical markers to differentiate the type of anxiety disorders. In *Anxiety Disorders* (pp. 197–218). Springer, Singapore.

Cossu, G., & Pau, M. (2017). Subthalamic nucleus stimulation and gait in Parkinson's Disease: A not always fruitful relationship. *Gait & Posture, 52,* 205–210.

Costafreda, S. G., Brammer, M. J., David, A. S., & Fu, C. H. Y. (2008). Predictors of amygdala activation during the processing of emotional stimuli: A meta-analysis of 385 PET and fMRI studies. *Brain Research Reviews, 58,* 57–70.

Costanzo, E. Y., Villarreal, M., Drucaroff, L. J., Ortiz-Villafañe, M., Castro, M. N., . . . & Guinjoan, S. M. (2015). Hemispheric specialization in affective responses, cerebral dominance for language, and handedness: Lateralization of emotion, language, and dexterity. *Behavioural Brain Research, 288,* 11–19.

Cotman, C. W., Berchtold, N. C., & Christie, L.-A. (2007). Exercise builds brain health: Key roles of growth factor cascades and inflammation. *Trends in Neurosciences, 30,* 464–472.

Coulon, P., & Landisman, C. E. (2017). The potential role of gap junctional plasticity in the regulation of state. *Neuron, 93*(6), 1275–1295.

Coulthard, L. G., Hawksworth, O. A., & Woodruff, T. M. (2018). Complement: The emerging architect of the developing brain. *Trends in Neurosciences, 41*(6), 373–384.

Courchesne, E., & Pierce, K. (2005). Why the frontal cortex in autism might be talking only to itself: Local over-connectivity but long-distance disconnection. *Current Opinion in Neurobiology, 15,* 225–230.

Couzin-Frankel, J. (2013). When mice mislead. *Science, 342,* 922–925.

Cowan, W. M. (1979, September). The development of the brain. *Scientific American, 241,* 113–133.

Cox, J. H., Seri, S., & Cavanna, A. E. (2018). Sensory aspects of Tourette syndrome. *Neuroscience & Biobehavioral Reviews, 88,* 170–176.

Cox, J. J., Reimann, F., Nicholas, A. K., Thornton, G., Roberts, E., Springell, K., . . . & Woods, C. G. (2006). An SCN9A channelopathy causes congenital inability to experience pain. *Nature, 444,* 894–898.

Cox, K. H., & Takahashi, J. S. (2019). Circadian clock genes and the transcriptional architecture of the clock mechanism. *Journal of Molecular Endocrinology, 63*(4), R93–R102.

Craig, A. D., Reiman, E. M., Evans, A., & Bushnell, M. C. (1996). Functional imaging of an illusion of pain. *Nature, 384*(6606), 258–260.

Crane, N. A., Schuster, R. M., Fusar-Poli, P., & Gonzalez, R. (2013). Effects of cannabis on neurocognitive functioning: Recent advances, neurodevelopment influences, and sex differences. *Neuropsychology Review, 23,* 117–137.

Creese, I., Burt, D. R., & Snyder, S. H. (1976). Dopamine receptor binding predicts clinical and pharmacological potencies of anti-schizophrenic drugs. *Science, 192,* 481–483.

Crepaldi, D., Berlingeri, M., Cattinelli, L., Borghese, N. A., Luzzatti, C., & Paulesu, E. (2013). Clustering the lexicon in the brain: A meta-analysis of the neurofunctional evidence on noun and verb processing. *Frontiers in Human Neuroscience, 7,* 303.

Crespo, C. S., Cachero, A. P., Jiménez, L. P., Barrios, V., & Ferreiro, E. A. (2014). Peptides and food intake. *Frontiers in Endocrinology, 5,* 58.

Cressey, D. (2015). The cannabis experiment. *Nature, 524,* 280–283.

Crisp, J., & Ralph, M. A. L. (2006). Unlocking the nature of the phonological-deep dyslexia continuum: The keys to reading aloud are in phonology and semantics. *Journal of Cognitive Neuroscience, 18,* 348–362.

Crivelli, C., Jarillo, S., Russell, J. A., & Fernández-Dols, J. M. (2016). Reading emotions from faces in two indigenous societies. *Journal of Experimental Psychology: General, 145*(7), 830–843.

Crivelli, C., Russell, J. A., Jarillo, S., & Fernández-Dols, J. M. (2017). Recognizing spontaneous facial expressions of emotion in a small-scale society of Papua New Guinea. *Emotion, 17*(2), 337–347.

Crochet, S., Lee, S. H., & Petersen, C. C. (2019). Neural circuits for goal-directed sensorimotor transformations. *Trends in Neurosciences, 42,* 66–77.

Crone, E. A., & Dahl, R. E. (2012). Understanding adolescence as a period of social-affective engagement and goal flexibility. *Nature Reviews Neuroscience, 13,* 636–650.

Crowell, C. R., Hinson, R. E., & Siegel, S. (1981). The role of conditional drug responses in tolerance to the hypothermic effects of ethanol. *Psychopharmacology, 73,* 51–54.

Csete, J., Kamarulzaman, A., Kazatchkine, M., Altice, F., Balicki, M., Buxton, J., . . . & Hart, C. (2016). Public health and international drug policy. *The Lancet, 387*(10026), 1427–1480.

Cui, H., López, M., & Rahmouni, K. (2017). The cellular and molecular bases of leptin and ghrelin resistance in obesity. *Nature Reviews Endocrinology, 13*(6), 338–351.

Cui, Y., Hu, S., & Hu, H. (2019). Lateral habenular burst firing as a target of the rapid antidepressant effects of ketamine. *Trends in Neurosciences, 42*(3), 179–191.

Culbertson, F. M. (1997). Depression and gender. *American Psychologist, 52,* 25–31.

Culham, J. C., & Kanwisher, N. G. (2001). Neuroimaging of cognitive functions in human parietal cortex. *Current Opinion in Neurobiology, 11,* 157–163.

Cunniff, C., Frias, J., Kaye, C., Moeschler, J., Panny, S., Trotter, T., . . . & Lloyd-Puryear, M. (2001). Health care supervision for children with Williams syndrome. *Pediatrics, 107*(5), 1192–1204.

Curcio, G., Ferrara, M., & De Gennaro, L. (2006). Sleep loss, learning capacity and academic performance. *Sleep Medicine Reviews, 10,* 323–337.

Curran, H. V., Freeman, T. P., Mokrysz, C., Lewis, D. A., Morgan, C. J., & Parsons, L. H. (2016). Keep off the grass? Cannabis, cognition and addiction. *Nature Reviews Neuroscience, 17*(5), 293–306.

Curtin, K., Fleckenstein, A. E., Robison, R. J., Crookston, M. J., Smith, K. R., & Hanson, G. R. (2015). Methamphetamine/amphetamine abuse and risk of Parkinson's disease in Utah: A population-based assessment. *Drug and Alcohol Dependence, 146,* 30–38.

Curtiss, S. (1977). *Genie: A psycholinguistic study of a modern-day "wild child."* New York, NY: Academic Press.

Custers, R., & Aarts, H. (2010). The unconscious will: How the pursuit of goals operates outside of conscious awareness. *Science, 329,* 47–50.

Czeisler, C. A., Duffy, J. F., Shanahan, T. L., Brown, E. N., Mitchell, J. F., Rimmer, D. W., . . . & Kronauer, R. E. (1999). Stability, precision, and near-24-hour period of the human circadian pacemaker. *Science, 284,* 2177–2181.

Czyz, W., Morahan, J. M., Ebers, G. C., & Ramagopalan, S. V. (2012). Genetic, environmental and stochastic factors in monozygotic twin discordance with a focus on epigenetic differences. *BMC Medicine, 10,* 1.

D'Ausilio, A., Maffongelli, L., Bartoli, E., Campanella, M., Ferrari, E., . . . & Fadiga, L. (2014). Listening to speech recruits specific tongue motor synergies as revealed by transcranial magnetic stimulation and tissue-Doppler ultrasound imaging. *Philosophical Transactions of the Royal Society of London. Series B, Biological Sciences, 369,* 20130418.

D'Esposito, M., & Postle, B. R. (2015). The cognitive neuroscience of working memory. *Annual Review of Psychology, 66,* 115–142.

Da Mesquita, S., Fu, Z., & Kipnis, J. (2018). The meningeal lymphatic system: A new player in neurophysiology. *Neuron, 100*(2), 375–388.

Daan, S., Honma, S., & Honma, K. (2013). Body temperature predicts the direction of internal desynchronization in humans isolated from time cues. *Journal of Biological Rhythms, 28*, 403–411.

Dallérac, G., & Rouach, N. (2016). Astrocytes as new targets to improve cognitive functions. *Progress in Neurobiology, 144*, 48–67.

Dalva, M. B., McClelland, A. C., & Kayser, M. S. (2007). Cell adhesion molecules: Signalling functions at the synapse. *Nature Reviews Neuroscience, 8*, 206–220.

Daly, M., & Wilson, M. (1983). *Sex, evolution, and behavior*. Boston, MA: Allard Grant Press.

Damasio, H., Grabowski, T. J., Tranel, D., Hichwa, R. D., & Damasio, A. R. (1996). A neural basis for lexical retrieval. *Nature, 380*, 499–505.

Damasio, H., Grabowski, T., Frank, R., Galaburda, A. M., & Damasio, A. R. (1994). The return of Phineas Gage: Clues about the brain from the skull of a famous patient. *Science, 264*, 1102–1105.

Dampney, R. A. (2015). Central mechanisms regulating coordinated cardiovascular and respiratory function during stress and arousal. *American Journal of Physiology-Regulatory, Integrative and Comparative Physiology, 309*, R429–R443.

Dando, R., & Roper, S. D. (2009). Cell-to-cell communication in intact taste buds through ATP signalling from pannexin 1 gap junction hemichannels. *Journal of Physiology, 587*, 5899–5906.

Danna, J., & Velay, J. L. (2015). Basic and supplementary sensory feedback in handwriting. *Frontiers in Psychology, 6*, 169.

Dannemann, M., & Racimo, F. (2018). Something old, something borrowed: Admixture and adaptation in human evolution. *Current Opinion in Genetics and Development, 53*, 1–8.

Darcq, E., & Kieffer, B. L. (2018). Opioid receptors: Drivers to addiction? *Nature Reviews Neuroscience, 19*(8), 499–514.

Darlison, M. G., & Richter, D. (1999). Multiple genes for neuropeptides and their receptors: Co-evolution and physiology. *Trends in Neurosciences, 22*, 81–88.

Dasen, J. S. (2017). Master or servant? emerging roles for motor neuron subtypes in the construction and evolution of locomotor circuits. *Current Opinion in Neurobiology, 42*, 25–32.

Daubert, E. A., & Condron, B. G. (2010). Serotonin: A regulator of neuronal morphology and circuitry. *Trends in Neurosciences, 33*, 424–434.

Dauvilliers, Y., Siegel, J. M., Lopez, R., Torontali, Z. A., & Peever, J. H. (2014). Cataplexy—clinical aspects, pathophysiology and management strategy. *Nature Reviews Neurology, 10*, 386–395.

Davare, M., Kraskov, A., Rothwell, J. C., & Lemon, R. N. (2011). Interactions between areas of the cortical grasping network. *Current Opinion in Neurobiology, 21*, 565–570.

Davidson, A. G., Chan, V., O'Dell, R., & Schieber, M. H. (2007). Rapid changes in throughput from single motor cortex neurons to muscle activity. *Science, 318*, 1934–1937.

Davis, K. D., Flor, H., Greely, H. T., Iannetti, G. D., Mackey, S., Ploner, M., . . . & Wager, T. D. (2017). Brain imaging tests for chronic pain: Medical, legal and ethical issues and recommendations. *Nature Reviews Neurology, 13*(10), 624–638.

Davis, S. R., & Braunstein, G. D. (2012). Efficacy and safety of testosterone in the management of hypoactive sexual desire disorder in postmenopausal women. *Journal of Sexual Medicine, 9*, 1134–1148.

Davis, S. R., Davison, S. L., Donath, S., & Bell, R. J. (2005). Circulating androgen levels and self-reported sexual function in women. *Journal of the American Medical Association, 294*, 91–96.

Dawson, M., Soulières, I., Gernsbacher, M. A., & Mottron, L. (2007). The level and nature of autistic intelligence. *Psychological Science, 18*, 657–662.

Dawson, T. M., Golde, T. E., & Lagier-Tourenne, C. (2018). Animal models of neurodegenerative diseases. *Nature Neuroscience, 21*(10), 1370–1379.

de Boer, B. (2019). Evolution of speech: Anatomy and control. *Journal of Speech, Language, and Hearing Research, 62*(8S), 2932–2945.

de Bournonville, M. P., Vandries, L. M., Ball, G. F., Balthazart, J., & Cornil, C. A. (2019). Site-specific effects of aromatase inhibition on the activation of male sexual behavior in male Japanese quail (Coturnix japonica). *Hormones and Behavior, 108*, 42–49.

De Butte-Smith, M., Gulinello, M., Zukin, R. S., & Etgen, A. M. (2009). Chronic estradiol treatment increases CA1 cell survival but does not improve visual or spatial recognition memory after global ischemia in middle-aged female rats. *Hormones and Behavior, 55*, 442–453.

De Deurwaerdere, P., Di Giovanni, G., & Millan, M. J. (2017). Expanding the repertoire of L-DOPA's actions: A comprehensive review of its functional neurochemistry. *Progress in Neurobiology, 151*, 57–100.

De Falco, E., Ison, M. J., Fried, I., & Quiroga, R. Q. (2016). Long-term coding of personal and universal associations underlying the memory web in the human brain. *Nature Communications, 7*, 13408.

de Gelder, B. (2010). Uncanny sight in the blind. *Scientific American, 302*, 60–65.

De Groof, G., Poirier, C., George, I., Hausberger, M., & Van der Linden, A. (2013). Functional changes between seasons in the male songbird auditory forebrain. *Frontiers in Behavioral Neuroscience, 7*, 196.

de Haan, E. H. F., & Cowey, A. (2011). On the usefulness of 'what' and 'where' pathways in vision. *Trends in Cognitive Sciences, 15*, 460–466.

de Haan, M., Mishkin, M., Baldeweg, T., & Vargha-Khadem, F. (2006). Human memory development and its dysfunction after early hippocampal injury. *Trends in Neurosciences, 29*, 374–380.

de Kloet, E. R., Joëls, M., & Holsboer, F. (2005). Stress and the brain: From adaptation to disease. *Nature Reviews Neuroscience, 6*, 463–475.

de Knijff, P. (2014). How carrion and hooded crows defeat Linnaeus's curse. *Science, 344*, 1345.

de Kruif, P. (1945). *The male hormone*. New York, NY: Harcourt, Brace.

DeKosky, S. T., & Williamson, J. B. (2020). The long and the short of benzodiazepines and sleep medications: Short-term benefits, Long-term harms? Neurotherapeutics, 17, 153–155.

de Lange, F. P., Heilbron, M., & Kok, P. (2018). How do expectations shape perception? *Trends in Cognitive Sciences, in press*.

De Mees, C., Bakker, J., Szpirer, J., & Szpirer, C. (2006). Alpha-fetoprotein: From a diagnostic biomarker to a key role in female fertility. *Biomarker Insights, 1*, 82–85.

De Nardo, D. (2015). Toll-like receptors: Activation, signalling, and transcriptional modulation. *Cytokine, 74*, 181–189.

De Nicola, A. F., Brocca, M. E., Pietranera, L., & Garcia-Segura, L. M. (2012). Neuroprotection and sex steroids: Evidence of estradiol-mediated protection in hypertensive encephalopathy. *Mini Reviews in Medicinal Chemistry, 12*, 1081–1089.

de Quervain, D., Schwabe, L., & Roozendaal, B. (2017). Stress, glucocorticoids and memory: Implications for treating fear-related disorders. *Nature Reviews Neuroscience, 18*(1), 7–19.

DeValois, R. L., Cottaris, N. P., Mahon, L. E., Elfar, S. D., & Wilson, J. A. (2000). Spatial and temporal receptive fields of geniculate and cortical cells and directional selectivity. *Vision research, 40*(27), 3685–3702.

De Valois, R. L., Cottaris, N. P., Elfar, S. D., Mahon, L. E., & Wilson, J. A. (2000). Some transformations of color information from lateral geniculate nucleus to striate cortex. *Proceedings of the National Academy of Science, USA, 97*, 4997–5002.

de Vivo, L., Bellesi, M., Marshall, W., Bushong, E. A., Ellisman, M. H., Tononi, G., & Cirelli, C. (2017). Ultrastructural evidence for synaptic scaling across the wake/sleep cycle. *Science, 355*(6324), 507–510.

de Wit, J., & Ghosh, A. (2016). Specification of synaptic connectivity by cell surface interactions. *Nature Reviews Neuroscience, 17*, 22–35.

De Zeeuw, C. I., & Ten Brinke, M. M. (2015). Motor learning and the cerebellum. *Cold Spring Harbor Perspectives in Biology, 7*, a021683.

Deadwyler, S. A., Hayashizaki, S., Cheer, J., & Hampson, R. E. (2004). Reward, memory and substance abuse: Functional neuronal circuits in the nucleus accumbens. *Neuroscience and Biobehavioural Reviews, 27*, 703–711.

Dean, C., & Dresbach, T. (2006). Neuroligins and neurexins: Linking cell adhesion, synapse formation and cognitive function. *Trends in Neurosciences, 29*, 21–29.

Debanne, D. (2004). Information processing in the axon. *Nature Reviews Neuroscience, 5*, 304–316.

Deffieux, T., Demene, C., Pernot, M., & Tanter, M. (2018). Functional ultrasound neuroimaging: A review of the preclinical and clinical state of the art. *Current Opinion in Neurobiology, 50*, 128–135.

Degenhardt, L., & Hall, W. (2012). Extent of illicit drug use and dependence, and their contribution to the global burden of disease. *The Lancet, 379*, 55–70.

Degenhardt, L., Charlson, F., Ferrari, A., Santomauro, D., Erskine, H., Mantilla-Herrara, A., . . . & Rehm, J. (2018). The global burden of disease attributable to alcohol and drug use in 195 countries and territories, 1990–2016: A systematic analysis for the Global Burden of Disease Study 2016. *The Lancet Psychiatry, 5*(12), 987–1012.

DeGutis, J. M., Chiu, C., Grosso, M. E., & Cohan, S. (2014). Face processing improvements in prosopagnosia: Successes and failures over the last 50 years. *Frontiers in Human Neuroscience, 8*, 561.

Deisseroth, K. (2011). Optogenetics. *Nature Methods, 8*, 26–29.

Dekker, A. D., Strydom, A., Coppus, A. M. W., Nizetic, D., Vermeiren, Y., . . . & De Deyn, P. P. (2015). Behavioral and psychological symptoms of dementia in Down syndrome: Early indicators of clinical Alzheimer's disease? *Cortex, 73*, 36–61.

Dekkers, M. P., & Barde, Y. A. (2013). Programmed cell death in neuronal development. *Science, 340*, 39–41.

DeKosky, S. T., & Williamson, J. B. (2020). The long and the short of benzodiazepines and sleep medications: Short-term benefits, Long-term harms? *Neurotherapeutics, 17*, 153–155.

del Río, J. A., Ferrer, I., & Gavín, R. (2018). Role of cellular prion protein in interneuronal amyloid transmission. *Progress in Neurobiology, 165–167*, 87–102.

Deliagina, T. G., Zelenin, P. V., & Orlovsky, G. N. (2012). Physiological and circuit mechanisms of postural control. *Current Opinion in Neurobiology, 22*, 646–652.

Delmas, P., Hao, J., & Rodat-Despoix, L. (2011). Molecular mechanisms of mechanotransduction in mammalian sensory neurons. *Nature Reviews Neuroscience, 12*, 139–152.

Demas, G. E., Cooper, M. A., Albers, H. E., & Soma, K. K. (2005). Novel mechanisms underlying neuroendocrine regulation of aggression: A synthesis of rodent, avian, and primate studies. In J. D. Blaustein (Ed.), *Behavioral neurochemistry and neuroendocrinology* (pp. 2–25). New York, NY: Kluwer Press.

Dement, W. C. (1978). *Some must watch while some must sleep*. New York, NY: Norton.

Dement, W. C., & Kleitman, N. (1957). The relation of eye movement during sleep to dream activity: An objective method for the study of dreaming. *Journal of Experimental Psychology, 53,* 339–346.

Dement, W. C., & Wolpert, E. A. (1958). The relation of eye movements, body motility and external stimuli to dream content. *Journal of Experimental Psychology, 55,* 543–553.

Demkow, U., & Wolańczyk, T. (2017). Genetic tests in psychiatric disorders—integrating molecular medicine with clinical psychiatry— why is it so difficult? *Translational Psychiatry, 7*(6), e1151–e1151.

Dempsey, J. A. (2019). Central sleep apnea: Misunderstood and mistreated!. *F1000Research, 8*(981), 981.

Deng, W., Aimone, J. B., & Gage, F. H. (2010). New neurons and new memories: How does adult hippocampal neurogenesis affect learning and memory? *Nature Reviews Neuroscience, 11,* 339–350.

Denis, D., French, C. C., & Gregory, A. M. (2018). A systematic review of variables associated with sleep paralysis. *Sleep Medicine Reviews, 38,* 141–157.

Dennis, E. L., & Thompson, P. M. (2013). Typical and atypical brain development: A review of neuroimaging studies. *Dialogues in Clinical Neuroscience, 15,* 359–384.

Denson, T. F., O'Dean, S. M., Blake, K. R., & Beames, J. R. (2018). Aggression in women: Behavior, brain and hormones. *Frontiers in Behavioral Neuroscience, 12,* 81.

Deo, C., & Lavis, L. D. (2018). Synthetic and genetically encoded fluorescent neural activity indicators. *Current Opinion in Neurobiology, 50,* 101–108.

Deppmann, C. D., Mihalas, S., Shama, N., Lonze, B. E., Niebur, E., & Ginty, D. D. (2008). A model for neuronal competition during development. *Science, 320,* 369–373.

Derdikman, D., & Moser, E. I. (2010). A manifold of spatial maps in the brain. *Trends in Cognitive Sciences, 14,* 561–569.

Deshmukh, V. A., Tardiff, V., Lyssiotis, C. A., Green, C. C., Kerman, B.,... & Lairson, L. L. (2013). A regenerative approach to the treatment of multiple sclerosis. *Nature, 502,* 327–332.

Desmurget, M., & Sirigu, A. (2012). Conscious motor intention emerges in the inferior parietal lobule. *Current Opinion in Neurobiology, 22,* 1004–1011.

Desmurget, M., Reilly, K. T., Richard, N., Szathmari, A., Mottolese, C., & Sirigu, A. (2009). Movement intention after parietal cortex stimulation in humans. *Science, 324,* 811–813.

Dess, N. K., & Chapman, C. D. (1998). "Humans and animals"? On saying what we mean. *Psychological Science, 9*(2), 156–157.

Devinsky, O., Vezzani, A., Najjar, S., De Lanerolle, N. C., & Rogawski, M. A. (2013). Glia and epilepsy: Excitability and inflammation. *Trends in Neurosciences, 36,* 174–184.

DeVos, S. L., Corjuc, B. T., Oakley, D. H., Nobuhara, C. K., Bannon, R. N., Chase, A.,... & Hyman, B. T. (2018). Synaptic tau seeding precedes tau pathology in human Alzheimer's disease brain. *Frontiers in Neuroscience, 12,* 267.

Devreotes, P., & Horwitz, A. R. (2015). Signaling networks that regulate cell migration. *Cold Spring Harbor Perspectives in Biology, 7,* a005959.

Deweerdt, S. (2014). A complicated relationship status. *Nature, 508,* S61–S63.

Dewsbury, D. A. (1988). The comparative psychology of monogamy. In D. W. Leger (Ed.), *Comparative perspectives in modern psychology: Nebraska Symposium on Motivation* (Vol. 35, pp. 1–50). Lincoln: University of Nebraska Press.

Dewsbury, D. A. (1991). Psychobiology. *American Psychologist, 46,* 198–205.

Dhabhar, F. S. (2014). Effects of stress on immune function: The good, the bad, and the beautiful. *Immunologic Research, 58,* 193–210.

Dhurandhar, E. J., Kaiser, K. A., Dawson, J. A., Alcorn, A. S., Keating, K. D., & Allison, D. B. (2015). Predicting adult weight change in the real world: A systematic review and meta-analysis accounting for compensatory changes in energy intake or expenditure. *International Journal of Obesity, 39,* 1181–1187.

Di Angelantonio, E., Bhupathiraju, S. N., Wormser, D., Gao, P., Kaptoge, S., de Gonzalez, A. B.,... & Lewington, S. (2016). Body-mass index and all-cause mortality: Individual-participant-data meta-analysis of 239 prospective studies in four continents. *The Lancet, 388*(10046), 776–786.

Di Cristo, G., Awad, P. N., Hamidi, S., & Avoli, M. (2018). KCC2, epileptiform synchronization, and epileptic disorders. *Progress in Neurobiology, 162,* 1–16.

Di Lollo, V. (2012). The feature-binding problem is an ill-posed problem. *Trends in Cognitive Sciences, 16,* 317–321.

Di Lullo, E., & Kriegstein, A. R. (2017). The use of brain organoids to investigate neural development and disease. *Nature Reviews Neuroscience, 18*(10), 573.

Di Marzo, V., Stella, N., & Zimmer, A. (2015). Endocannabinoid signalling and the deteriorating brain. *Nature Reviews Neuroscience, 16,* 30–42.

Diamond, A. (1985). Development of the ability to use recall to guide action, as indicated by infants' performance on AB. *Child Development, 56,* 868–883.

Diamond, A. (1991). Neuropsychological insights into the meaning of object concept development. In S. Carey & R. Gelman (Eds.), *The epigenesis of mind: Essays on biology and cognition* (pp. 67–110). Hillsdale, NJ: Lawrence Erlbaum.

Diamond, J. (2004). The astonishing micropygmies. *Science, 306,* 2047–2048.

Diamond, M., & Sigmundson, H. K. (1997). Sex reassignment at birth: Long-term review and clinical implications. *Archives of Pediatric and Adolescent Medicine, 151,* 298–304.

Dias, B. G., & Ressler, K. J. (2014). Parental olfactory experience influences behavior and neural structure in subsequent generations. *Nature Neuroscience, 17,* 89–96.

Dias, B. G., Maddox, S. A., Klengel, T., & Ressler, K. J. (2015). Epigenetic mechanisms underlying learning and the inheritance of learned behaviors. *Trends in Neurosciences, 38,* 96–107.

DiCicco-Bloom, E. (2006). Neuron, know thy neighbor. *Science, 311,* 1560–1562.

Dick, A. S., Bernal, B., & Tremblay, P. (2014). The language connectome new pathways, new concepts. *The Neuroscientist, 20,* 453–467.

Diekelmann, S., & Born, J. (2010). The memory function of sleep. *Nature Reviews Neuroscience, 11,* 114–126.

Diekhof, E. K., Kaps, L., Falkai, P., & Gruber, O. (2012). The role of the human ventral striatum and the medial orbitofrontal cortex in the representation of reward magnitude—an activation likelihood estimation meta-analysis of neuroimaging studies of passive reward expectancy and outcome processing. *Neuropsychologia, 50,* 1252–1266.

Dieterich, D. C. (2010). Chemical reporters for the illumination of protein and cell dynamics. *Current Opinion in Neurobiology, 20,* 623–630.

Dietz, M. J., Friston, K. J., Mattingley, J. B., Roepstorff, A., & Garrido, M. I. (2014). Effective connectivity reveals right-hemisphere dominance in audiospatial perception: Implications for models of spatial neglect. *Journal of Neuroscience, 34,* 5003–5011.

DiFeo, G., & Shors, T. J. (2017). Mental and physical skill training increases neurogenesis via cell survival in the adolescent hippocampus. *Brain Research, 1654,* 95–101.

Dillon, D. G., & Pizzagalli, D. A. (2018). Mechanisms of memory disruption in depression. *Trends in Neurosciences, 41*(3), 137–149.

DiMaggio, E. N., Campisano, C. J., Rowan, J., Dupont-Nivet, G., Deino, A. L., Bibi, F.,... & Reed, K. E. (2015). Late Pliocene fossiliferous sedimentary record and the environmental context of early *Homo* from Afar, Ethiopia. *Science, 347,* 1355–1359.

Dimou, L., & Götz, M. (2014). Glial cells as progenitors and stem cells: New roles in the healthy and diseased brain. *Physiological Reviews, 94,* 709–737.

Dimou, L., & Simons, M. (2017). Diversity of oligodendrocytes and their progenitors. *Current opinion in neurobiology, 47,* 73–79.

Ding, H., Smith, R. G., Poleg-Polsky, A., Diamond, J. S., & Briggman, K. L. (2016). Species-specific wiring for direction selectivity in the mammalian retina. *Nature, 535*(7610), 105–110.

Dinges, D. F., Pack, F., Williams, K., Gillen, K. A., Powell, J. W., Ott, G. E., et al. (1997). Cumulative sleepiness, mood disturbance, and psychomotor vigilance performance decrements during a week of sleep restricted to 4–5 hours per night. *Sleep, 20,* 267–277.

Do-Monte, F. H., Quiñones-Laracuente, K., & Quirk, G. J. (2015). A temporal shift in the circuits mediating retrieval of fear memory. *Nature, 519,* 460–463.

Dobelle, W. H., Mladejovsky, M. G., & Girvin, J. P. (1974). Artificial vision for the blind: Electrical stimulation of visual cortex offers hope for a functional prosthesis. *Science, 183,* 440–444.

Dockray, G. J. (2014). Gastrointestinal hormones and the dialogue between gut and brain. *Journal of Physiology, 592,* 2927–2941.

Doetsch, F., & Hen, R. (2005). Young and excitable: The function of new neurons in the adult mammalian brain. *Current Opinion in Neurobiology, 15,* 121–128.

Dolgin, E. (2015). The elaborate architecture of RNA. *Nature, 523,* 398–399.

Dolgin, E. (2018). Alzheimer's disease: A tough spot. *Nature, 559,* S10–S12.

Domhoff, G. W., & Fox, K. C. (2015). Dreaming and the default network: A review, synthesis, and counterintuitive research proposal. *Consciousness and Cognition, 33,* 342–353.

Dominici, N., Ivanenko, Y. P., Cappellini, G., d'Avella, A., Mondì, V., Cicchese, M., Fabiano, A., Silei, T., Di Paolo, A., Giannini, C., Poppele, R. E., & Lacquaniti, F. (2011). Locomotor primitives in newborn babies and their development. *Science, 334,* 997–999.

Dong, X., Liao, Z., Gritsch, D., Hadzhiev, Y., Bai, Y., Locascio, J. J.,... & Adler, C. H. (2018). Enhancers active in dopamine neurons are a primary link between genetic variation and neuropsychiatric disease. *Nature Neuroscience, 21*(10), 1482–1492.

Dong, Y., & Yong, V. W. (2019). When encephalitogenic T cells collaborate with microglia in multiple sclerosis. *Nature Reviews Neurology, 15,* 704–717.

Dong, Z., Pan, K., Pan, J., Peng, Q., & Wang, Y. (2018). The possibility and molecular mechanisms of cell pyroptosis after cerebral ischemia. *Neuroscience Bulletin, 34*(6), 1131–1136.

Drachman, D. A., & Arbit, J. (1966). Memory and the hippocampal complex. *Archives of Neurology, 15,* 52–61.

Drenowatz, C. (2015). Reciprocal compensation to changes in dietary intake and energy expenditure within the concept of energy balance. *Advances in Nutrition, 6,* 592–599.

Dresler, M., Wehrle, R., Spoormaker, V. I., Koch, S. P., Holsboer, F., Steiger, A.,... & Czisch, M. (2012). Neural correlates of dream lucidity obtained from contrasting lucid versus non-lucid REM sleep: A combined EEG/fMRI case study. *Sleep, 35*(7), 1017–1020.

Drew, L., & Ashour, M. (2018). Alzheimer's disease: An age-old story. *Nature, 559,* S2–S3.

Drewnowski, A., Halmi, K. A., Pierce, B., Gibbs, J., & Smith, G. P. (1987). Taste and eating disorders. *American Journal of Clinical Nutrition, 46,* 442–450.

Drummey, A. B., & Newcombe, N. (1995). Remembering versus knowing the past: Children's explicit and implicit memories for pictures. *Journal of Experimental Child Psychology, 59*, 540–565.

Du, Y., Kong, L., Wang, Q., Wu, X., & Li, L. (2011). Auditory frequency-following response: A neurophysiological measure for studying the "cocktail-party problem." *Neuroscience and Biobehavioral Reviews, 35*, 2046–2057.

Duchaine, B. C., & Nakayama, K. (2006). Developmental prosopagnosia: A window to content-specific face processing. *Current Opinion in Neurobiology, 16*, 166–173.

Duchaine, B., & Nakayama, K. (2005). Dissociations of face and object recognition in developmental prosopagnosia. *Journal of Cognitive Neuroscience, 17*, 249–261.

Dudai, Y., & Morris, R. G. M. (2013). Memorable trends. *Neuron, 80*, 742–750.

Dudai, Y., Karni, A., & Born, J. (2015). The consolidation and transformation of memory. *Neuron, 88*, 20–32.

Dudanova, I., & Klein, R. (2013). Integration of guidance cues: Parallel signaling and crosstalk. *Trends in Neurosciences, 36*, 295–304.

Dudman, J. T., & Krakauer, J. W. (2016). The basal ganglia: From motor commands to the control of vigor. *Current Opinion in Neurobiology, 37*, 158–166.

Duffy, A., Goodday, S. M., Keown-Stoneman, C., Scotti, M., Maitra, M., Nagy, C., . . . & Turecki, G. (2019). Epigenetic markers in inflammation-related genes associated with mood disorder: A cross-sectional and longitudinal study in high-risk offspring of bipolar parents. *International Journal of Bipolar Disorders, 7*(1), 17.

Dugan, P., & Devinsky, O. (2013). Guidelines on vagus nerve stimulation for epilepsy. *Nature Reviews Neurology, 9*, 611–612.

Dully, H., & Fleming, C. (2007). *My lobotomy: A memoir.* New York, NY: Crown.

Duman, R. S., Sanacora, G., & Krystal, J. H. (2019). Altered connectivity in depression: GABA and glutamate neurotransmitter deficits and reversal by novel treatments. *Neuron, 102*(1), 75–90.

Dumoulin, S. O., & Knapen, T. (2018). How visual cortical organization is altered by ophthalmologic and neurologic disorders. *Annual Review of Vision Science, 4*, 357–379.

Dunnett, S. B., Björklund, A., & Lindvall, O. (2001). Cell therapy in Parkinson's disease—stop or go? *Nature Reviews Neuroscience, 2*, 365–368.

Dunsmoor, J. E., Murty, V. P., Davachi, L., & Phelps, E. A. (2015). Emotional learning selectively and retroactively strengthens memories for related events. *Nature, 520*, 345–348.

Duque, J., Greenhouse, I., Labruna, L., & Ivry, R. B. (2017). Physiological markers of motor inhibition during human behavior. *Trends in Neurosciences, 40*(4), 219–236.

Dusi, N., De Carlo, V., Delvecchio, G., Bellani, M., Soares, J. C., & Brambilla, P. (2019). MRI features of clinical outcome in bipolar disorder: A selected review: Special Section on "Translational and Neuroscience Studies in Affective Disorders". Section Editor, Maria Nobile MD, PhD. This Section of JAD focuses on the relevance of translational and neuroscience studies in providing a better understanding of the neural basis of affective disorders. The main aim is to briefly summaries relevant research findings in clinical neuroscience with particular regards to specific innovative topics in .... *Journal of Affective Disorders, 243*, 559–563.

Dutra, L., Stathopoulou, G., Basden, S. L., Leyro, T. M., Powers, M. B., & Otto, M. W. (2008). A meta-analytic review of psychosocial interventions for substance use disorders. *American Journal of Psychiatry, 165*, 179–187.

Duva, C. A., Kornecook, T. J., & Pinel, J. P. J. (2000). Animal models of medial temporal lobe amnesia: The myth of the hippocampus. In M. Haug & R. E. Whalen (Eds.), *Animal models of human*

emotion and cognition (pp. 197–214). Washington, DC: American Psychological Association.

Duvarci, S., & Pare, D. (2014). Amygdala microcircuits controlling learned fear. *Neuron, 82*, 966–980.

Eagleman, D. M. (2001). Visual illusions and neurobiology. *Nature Reviews Neuroscience, 2*, 920–926.

Eban-Rothschild, A., Giardino, W. J., & de Lecea, L. (2017). To sleep or not to sleep: Neuronal and ecological insights. *Current Opinion in Neurobiology, 44*, 132–138.

Ebbesen, C. L., & Brecht, M. (2017). Motor cortex—to act or not to act? *Nature Reviews Neuroscience, 18*(11), 694.

Egeland, M., Zunszain, P. A., & Pariante, C. M. (2015). Molecular mechanisms in the regulation of adult neurogenesis during stress. *Nature Reviews Neuroscience, 16*, 189–200.

Eggermont, J. J., & Tass, P. A. (2015). Maladaptive neural synchrony in tinnitus: Origin and restoration. *Frontiers in Neurology, 6*, 29.

Ehrhardt, A. A., Meyer-Bahlburg, H. F. L., Rosen, L. R., Feldman, J. F., Veridiano, N. P., Zimmerman, I., & McEwen, B. S. (1985). Sexual orientation after prenatal exposure to exogenous estrogen. *Archives of Sexual Behavior, 14*, 57–77.

Ehrlich, I., Humeau, Y., Grenier, F., Ciocchi, S., Herry, C., & Lüthi, A. (2009). Amygdala inhibitory circuits and the control of fear memory. *Neuron, 62*, 757–771.

Eichenbaum, H. (2014). Time cells in the hippocampus: A new dimension for mapping memories. *Nature Reviews Neuroscience, 15*, 732–744.

Eichenbaum, H. (2015). The hippocampus as a cognitive map...of social space. *Neuron, 87*, 9–11.

Eichenbaum, H. (2017). Prefrontal–hippocampal interactions in episodic memory. *Nature Reviews Neuroscience, 18*(9), 547–558.

Eid, R. S., Gobinath, A. R., & Galea, L. A. (2019). Sex differences in depression: Insights from clinical and preclinical studies. *Progress in Neurobiology, 176*, 86–102.

Eigsti, I. M., & Shapiro, T. (2004). A systems neuroscience approach to autism: Biological, cognitive, and clinical perspectives. *Mental Retardation and Developmental Disabilities Research Reviews, 9*, 205–215.

Eilat-Adar, S., Eldar, M., & Goldbourt, U. (2005). Association of intentional changes in body weight with coronary heart disease event rates in overweight subjects who have an additional coronary risk factor. *American Journal of Epidemiology, 161*, 352–358.

Einhäuser, W., & König, P. (2010). Getting real—sensory processing of natural stimuli. *Current Opinion in Neurobiology, 20*, 389–395.

Eisenegger, C., Haushofer, J., & Fehr, E. (2011). The role of testosterone in social interaction. *Trends in Cognitive Sciences, 15*, 263–271.

Eisinger, R. S., Urdaneta, M. E., Foote, K. D., Okun, M. S., & Gunduz, A. (2018). Non-motor characterization of the basal ganglia: Evidence from human and non-human primate electrophysiology. *Frontiers in Neuroscience, 12*, 385.

Ejaz, N., Hamada, M., & Diedrichsen, J. (2015). Hand use predicts the structure of representations in sensorimotor cortex. *Nature Neuroscience, 18*, 1034–1040.

Ekman, P. (1985). *Telling lies.* New York, NY: Norton.

Ekman, P. (2016). What scientists who study emotion agree about. *Perspectives on Psychological Science, 11*, 31–34.

Ekman, P., & Friesen, W. V. (1971). Constants across cultures in the face and emotion. *Journal of Personality and Social Psychology, 17*, 124–129.

Ekman, P., & Friesen, W. V. (1975). *Unmasking the face: A guide to recognizing emotions from facial clues.* Englewood Cliffs, NJ: Prentice-Hall.

Elbert, T., Pantev, C., Wienbruch, C., Rockstroh, B., & Taub, E. (1995). Increased cortical representation of the fingers of the left hand in string players. *Science, 270*, 305–307.

Elbert, T., Sterr, A., Rockstroh, B., Pantev, C., Müller, M. M., & Taub, E. (2002). Expansion of the tonotopic area in the auditory cortex of the blind. *Journal of Neuroscience, 22*, 9941–9944.

Elder, G. A., Gama Sosa, M. A., & De Gasperi, R. (2010). Transgenic mouse models of Alzheimer's disease. *Mount Sinai Journal of Medicine, 77*, 69–81.

Elgoyhen, A. B., Langguth, B., De Ridder, D., & Vanneste, S. (2015). Tinnitus: Perspectives from human neuroimaging. *Nature Reviews Neuroscience, 16*, 632–642.

Eling, P., Derckx, K., & Maes, R. (2008). On the historical and conceptual background of the Wisconsin Card Sorting Test. *Brain and Cognition, 67*, 247–253.

Elkouzi, A., Vedam-Mai, V., Eisinger, R. S., & Okun, M. S. (2019). Emerging therapies in Parkinson disease—repurposed drugs and new approaches. *Nature Reviews Neurology, 15*(4), 204–223.

Ellis, C. T., & Turk-Browne, N. B. (2018). Infant fMRI: A model system for cognitive neuroscience. *Trends in Cognitive Sciences, 22*(5), 375–387.

Ellis, J. G., Perlis, M. L., Neale, L. F., Espie, C. A., & Bastien, C. H. (2012). The natural history of insomnia: Focus on prevalence and incidence of acute insomnia. *Journal of Psychiatric Research, 46*, 1278–1285.

Ellis, L., & Ames, M. A. (1987). Neurohormonal functioning and sexual orientation: A theory of homosexuality-heterosexuality. *Psychological Bulletin, 101*, 233–258.

Elmenhorst, E.-M., Elmenhorst, D., Luks, N., Maass, H., Vejvoda, M., & Samel, A. (2008). Partial sleep deprivation: Impact on the architecture and quality of sleep. *Sleep Medicine, 9*, 840–850.

Elmer, S., Hänggi, J., Meyer, M., & Jäncke, L. (2013). Increased cortical surface area of the left planum temporale in musicians facilitates the categorization of phonetic and temporal speech sounds. *Cortex, 49*, 2812–2821.

Engle-Friedman, M. (2014). The effects of sleep loss on capacity and effort. *Sleep Science, 7*, 213–224.

Epstein, L. H., Temple, J. L., Neaderhiser, B. J., Salis, R. J., Erbe, R. W., & Leddy, J. J. (2007). Food reinforcement, the dopamine D_2 receptor genotype, and energy intake in obese and nonobese humans. *Behavioral Neuroscience, 121*, 877–886.

Epstein, R. A., Patai, E. Z., Julian, J. B., & Spiers, H. J. (2017). The cognitive map in humans: Spatial navigation and beyond. *Nature Neuroscience, 20*(11), 1504–1513.

Erikkson, P. S., Perfilieva, E., Björk-Eriksson, T., Alborn, A., Nordberg, C., Peterson, D. A., & Gage, F. H. (1998). Neurogenesis in the adult human hippocampus. *Nature Medicine, 4*, 1313–1317.

Erlacher, D., Schädlich, M., Stumbrys, T., & Schredl, M. (2014). Time for actions in lucid dreams: Effects of task modality, length, and complexity. *Frontiers in Psychology, 4*, 1013.

Ernst, A., Alkass, K., Bernard, S., Salehpour, M., Perl, S., . . . & Frisén, J. (2014). Neurogenesis in the striatum of the adult human brain. *Cell, 156*, 1072–1083.

Ersche, K. D., Jones, P. S., Williams, G. B., Turton, A. J., Robbins, T. W., & Bullmore, E. T. (2012). Abnormal brain structure implicated in stimulant drug addiction. *Science, 335*, 601–604.

Essa, M. M., Moghadas, M., Ba-Omar, T., Qoronfleh, M. W., Guillemin, G. J., Manivasagam, T., . . . & Fernandes, A. J. (2019). Protective effects of antioxidants in Huntington's disease: An extensive review. *Neurotoxicity Research, 35*, 739–774.

Esteller, M. (2018). The human epigenome—implications for the understanding of human disease. In W.B. Coleman & G. J. Tsongalis (Eds.) *Molecular Pathology (Second Edition)* (pp. 165–182). Elsevier: London.

Etkin, A., Büchel, C., & Gross, J. J. (2015). The neural bases of emotion regulation. *Nature Reviews Neuroscience, 16*, 693–700.

Euston, D. R., & Steenland, H. W. (2014). Memories—getting wired during sleep. *Science, 344*, 1087–1088.

Euston, D. R., Gruber, A. J., & McNaughton, B. L. (2012). The role of medial prefrontal cortex in memory and decision making. *Neuron, 76*, 1057–1070.

Evans, N. R., Tarkin, J. M., Buscombe, J. R., Markus, H. S., Rudd, J. H., & Warburton, E. A. (2017). PET imaging of the neurovascular interface in cerebrovascular disease. *Nature Reviews Neurology, 13*(11), 676–688.

Everitt, B. J. (1990). Sexual motivation: A neural and behavioral analysis of the mechanisms underlying appetitive and copulatory responses of male rats. *Neuroscience and Biobehavioral Reviews, 14*, 217–232.

Everitt, B. J., & Herbert, J. (1972). Hormonal correlates of sexual behavior in sub-human primates. *Danish Medical Bulletin, 19*, 246–258.

Everitt, B. J., & Robbins, T. W. (2016). Drug addiction: Updating actions to habits to compulsions ten years on. *Annual Review of Psychology, 67*, 23–50.

Everitt, B. J., Dickinson, A., & Robbins, T. W. (2001). The neuropsychological basis of addictive behaviour. *Brain Research Reviews, 36*, 129–138.

Everitt, B. J., Herbert, J., & Hamer, J. D. (1971). Sexual receptivity of bilaterally adrenalectomized female rhesus monkeys. *Physiology & Behavior, 8*, 409–415.

Evsyukova, I., Plestant, C., & Anton, E. S. (2013). Integrative mechanisms of oriented neuronal migration in the developing brain. *Annual Review of Cell and Developmental Biology, 29*, 299–353.

Fagerström, K. O., & Bridgman, K. (2014). Tobacco harm reduction: The need for new products that can compete with cigarettes. *Addictive Behaviors, 39*, 507–511.

Faggiani, E., & Benazzouz, A. (2017). Deep brain stimulation of the subthalamic nucleus in Parkinson's disease: From history to the interaction with the monoaminergic systems. *Progress in Neurobiology, 151*, 139–156.

Fagiolini, M., Jensen, C. L., & Champagne, F. A. (2009). Epigenetic influences on brain development and plasticity. *Current Opinion in Neurobiology, 19*, 207–212.

Faissner, S., Plemel, J. R., Gold, R., & Yong, V. W. (2019). Progressive multiple sclerosis: From pathophysiology to therapeutic strategies. *Nature Reviews Drug Discovery, 18*, 905–922.

Falasconi, M., Gutierrez-Galvez, A., Leon, M., Johnson, B. A., & Marco, S. (2012). Cluster analysis of rat olfactory bulb responses to diverse odorants. *Chemical Senses, 37*, 639–653.

Falco-Walter, J. J., Scheffer, I. E., & Fisher, R. S. (2018). The new definition and classification of seizures and epilepsy. *Epilepsy research, 139*, 73–79.

Falk, S., & Götz, M. (2017). Glial control of neurogenesis. *Current Opinion in Neurobiology, 47*, 188–195.

Falkner, S., Grade, S., Dimou, L., Conzelmann, K. K., Bonhoeffer, T., Götz, M., & Hübener, M. (2016). Transplanted embryonic neurons integrate into adult neocortical circuits. *Nature, 539*(7628), 248.

Fama, R., Pitel, A.-L., & Sullivan, E. V. (2012). Anterograde episodic memory in Korsakoff's syndrome. *Neuropsychology Review, 22*, 93–104.

Famulski, J. K., & Solecki, D. J. (2013). New spin on an old transition: Epithelial parallels in neuronal adhesion control. *Trends in Neurosciences, 36*, 163–173.

Fan, X., & Agid, Y. (2018). At the origin of the history of glia. *Neuroscience, in press.*

Fan, X., Dong, J., Zhong, S., Wei, Y., Wu, Q., Yan, L., . . . & Wang, W. (2018). Spatial transcriptomic survey of human embryonic cerebral cortex by single-cell RNA-seq analysis. *Cell Research, 28*(7), 730–745.

Fanni, C., Marcialis, M. A., Pintus, M. C., Loddo, C., & Fanos, V. (2018). The first case of neonatal priapism during hypothermia for hypoxic-ischemic encephalopathy and a literature review. *Italian Journal of Pediatrics, 44*(1), 85.

Farah, M. J. (2015). The unknowns of cognitive enhancement. *Science, 350*, 379–380.

Farah, M. J., & Murphy, N. (2009). Neuroscience and the soul [Letter to editor]. *Science, 323*, 1168.

Fareri, D. S., & Delgado, M. R. (2014). Social rewards and social networks in the human brain. *Neuroscientist, 20*, 387–402.

Farina, E., Borgnis, F., & Pozzo, T. (2020). Mirror neurons and their relationship with neurodegenerative disorders. *Journal of Neuroscience Research.*

Farmer, J., Zhao, X., Van Praag, H., Wodke, K., Gage, F. H., & Christie, B. R. (2004). Effects of voluntary exercise on synaptic plasticity and gene expression in the dentate gyrus of adult male Sprague-Dawley rats in vivo. *Neuroscience, 124*, 71–79.

Farooqi, I. S., Jebb, S. A., Langmack, G., Lawrence, E., Cheetham, C. H., Prentice, A. M., . . . & O'Rahilly, S. (1999). Effects of recombinant leptin therapy in a child with congenital leptin deficiency. *New England Journal of Medicine, 341*, 879–884.

Farrell, M., Martin, N. K., Stockings, E., Bórquez, A., Cepeda, J. A., Degenhardt, L., . . . & Shoptaw, S. (2019). Responding to global stimulant use: Challenges and opportunities. *The Lancet, 394*(10209), 1652–1667.

Fastenrath, M., Coynel, D., Spalek, K., Milnik, A., Gschwind, L., Roozendaal, B., . . . & de Quervain, D. J. (2014). Dynamic modulation of amygdala–hippocampal connectivity by emotional arousal. *Journal of Neuroscience, 34*, 13935–13947.

Fausto-Sterling, A. (2019). Gender/sex, sexual orientation, and identity are in the body: How did they get there? *The Journal of Sex Research, 56*(4–5), 529–555.

Fava, G. A., Ruini, C., Rafanelli, C., Finos, L., Conti, S., & Grandi, S. (2004). Six-year outcome of cognitive behavior therapy for prevention of recurrent depression. *American Journal of Psychiatry, 161*, 1872–1876.

Favre, P., Baciu, M., Pichat, C., Bougerol, T., & Polosan, M. (2014). fMRI evidence for abnormal resting-state functional connectivity in euthymic bipolar patients. *Journal of Affective Disorders, 165*, 182–189.

Fedirko, V., Tramacere, I., Bagnardi, V., Rota, M., Scotti, L., Islami, F., Negri, E., Straif, K., Romieu, I., La Vecchia, C., Boffetta, P., & Jenab, M. (2011). Alcohol drinking and colorectal cancer risk: An overall and dose-response meta-analysis of published studies. *Annals of Oncology, 22*, 1958–1972.

Feigin, V. L., Norrving, B., George, M. G., Foltz, J. L., Roth, G. A., & Mensah, G. A. (2016). Prevention of stroke: A strategic global imperative. *Nature Reviews Neurology, 12*, 501–512.

Feil, R., & Fraga, M. F. (2012). Epigenetics and the environment: Emerging patterns and implications. *Nature Reviews Genetics, 13*, 97–109.

Feinberg, T. E., Vennen, A., & Simone, A. M. (2009). The neuroanatomy of asomatognosia and somatoparaphrenia. *Journal of Neurology, Neurosurgery and Psychiatry, 81*, 276–281.

Feinstein, J. S. (2013). Lesion studies of human emotion and feeling. *Current Opinion in Neurobiology, 23*, 304–309.

Feldman, J. (2013). The neural binding problem(s). *Cognitive Neurodynamics, 7*, 1–11.

Feliciano, D. M., & Bordey, A. (2013). Newborn cortical neurons: Only for neonates? *Trends in Neurosciences, 36*, 51–61.

Feller, M. B., & Scanziani, M. (2005). A precritical period for plasticity in visual cortex. *Current Opinion in Neurobiology, 15*, 94–100.

Felsen, G., & Dan, Y. (2006). A natural approach to studying vision. *Nature Neuroscience, 8*, 1643–1646.

Fentress, J. C. (1973). Development of grooming in mice with amputated forelimbs. *Science, 179*, 704–705.

Ferracioli-Oda, E., Qawasmi, A., & Bloch, M. H. (2013). Meta-analysis: Melatonin for the treatment of primary sleep disorders. *PLoS One, 8*, e637773.

Ferrari, P. F., & Rizzolatti, G. (2014). Mirror neuron research: The past and the future. *Philosophical Transactions of the Royal Society B, 369*, 20130169.

Ferro, J. M., Caeiro, L., & Figueira, M. L. (2016). Neuropsychiatric sequelae of stroke. *Nature Reviews Neurology, 12*, 269–280.

Fetissov, S. O. (2017). Role of the gut microbiota in host appetite control: Bacterial growth to animal feeding behaviour. *Nature Reviews Endocrinology, 13*(1), 11–25.

Fichten, C. S., Libman, E., Creti, L., Bailes, S., & Sabourin, S. (2004). Long sleepers sleep more and short sleepers sleep less: A comparison of older adults who sleep well. *Behavioral Sleep Medicine, 2*, 2–23.

Fields, R. D. (2005, February). Making memories stick. *Scientific American, 292*, 75–80.

Fields, R. D., & Burnstock, G. (2006). Purinergic signalling in neuron-glia interactions. *Nature Reviews Neuroscience, 7*, 423–435.

Fiest, K. M., Sauro, K. M., Wiebe, S., Patten, S. B., Kwon, C. S., Dykeman, J., . . . & Jetté, N. (2017). Prevalence and incidence of epilepsy: A systematic review and meta-analysis of international studies. *Neurology, 88*(3), 296–303.

Figueiredo, V. C., & Silva, P. R. (2014). Cosmetic doping—when anabolic-androgenic steroids are not enough. *Substance Use and Misuse, 49*, 1163–1167.

Filippi, M., Bar-Or, A., Piehl, F., Preziosa, P., Solari, A., Vukusic, S., & Rocca, M. A. (2018). Multiple sclerosis. *Nature Reviews Disease Primers, 4*, 43.

Fillion, T. J., & Blass, E. M. (1986). Infantile experience with suckling odors determines adult sexual behavior in male rats. *Science, 231*, 729–731.

Fillmore, K. M., Kerr, W. C., Stockwell, T., Chikritzhs, T., & Bostrom, A. (2006). Moderate alcohol use and reduced mortality risk: Systematic error in prospective studies. *Addiction Research and Therapy, 14*, 101–132.

Finberg, J. P., & Gillman, K. (2011). Selective inhibitors of monoamine oxidase type B and the "cheese effect." *International Review of Neurobiology, 100*, 169–190.

Finkelstein, A., Las, L., & Ulanovsky, N. (2016). 3-D maps and compasses in the brain. *Annual Review of Neuroscience, 39*, 171–196.

Finn, R. (1991). Different minds. *Discover, 12*, 52–58.

Fischer, K. E., & Riddle, N. C. (2018). Sex differences in aging: Genomic instability. *The Journals of Gerontology: Series A, 73*(2), 166–174.

Fiser, A., Mahringer, D., Oyibo, H. K., Petersen, A. V., Leinweber, M., & Keller, G. B. (2016). Experience-dependent spatial expectations in mouse visual cortex. *Nature neuroscience, 19*(12), 1658–1664.

Fitch, W. T. (2017). A major blow to primate neonatal imitation and mirror neuron theory. *Behavioral and Brain Sciences, 40*.

Fitch, W. T. (2018). The biology and evolution of speech: A comparative analysis. *Annual Review of Linguistics, 4*, 255–279.

FitzGerald, G. A. (2014). Temporal targets of drug action. *Science, 346*, 921–922.

Fitzpatrick, S. M., & Rothman, D. L. (2000). Meeting report: Transcranial magnetic stimulation and studies of human cognition. *Journal of Cognitive Neuroscience, 12*, 704–709.

Flegal, K. M., Kit, B. K., Orpana, H., & Graubard, B. I. (2013). Association of all-cause mortality with overweight and obesity using standard body mass index categories. *JAMA, 309,* 71–82.

Fleming, N. (2018). Computer-calculated compounds. *Nature, 557*(7707), S55–S57.

Fleming, S. M., van der Putten, E. J., & Daw, N. D. (2018). Neural mediators of changes of mind about perceptual decisions. *Nature Neuroscience, 21,* 617–624.

Flight, M. H. (2014). The gut-microbiome-brain connection. *Nature Reviews Neuroscience, 15,* 65.

Flint, J., & Munafò, M. (2014). Schizophrenia: genesis of a complex disease. *Nature, 511*(7510), 412–413.

Flor, H., Nikolajsen, L., & Jensen, T. S. (2006). Phantom limb pain: A case of maladaptive CNS plasticity? *Nature Reviews Neuroscience, 7,* 873–881.

Floresco, S. B. (2015). The nucleus accumbens: An interface between cognition, emotion, and action. *Annual Review of Psychology, 66,* 25–52.

Floresco, S. B., Seamans, J. K., & Phillips, A. G. (1997). Selective roles for hippocampal, prefrontal cortical, and ventral striatal circuits in radial-arm maze tasks with or without a delay. *Journal of Neuroscience, 17*(5), 1880–1890.

Fogel, S. M., Ray, L. B., Sergeeva, V., De Koninck, J., & Owen, A. M. (2018). A novel approach to dream content analysis reveals links between learning-related dream incorporation and cognitive abilities. *Frontiers in Psychology, 9,* 1398.

Fomsgaard, L., Moreno, J. L., de la Fuente Revenga, M., Brudek, T., Adamsen, D., Rio-Alamos, C., . . . & Blazquez, G. (2018). Differences in 5-HT2A and mGlu2 receptor expression levels and repressive epigenetic modifications at the 5-HT2A promoter region in the Roman Low-(RLA-I) and High-(RHA-I) Avoidance rat strains. *Molecular Neurobiology, 55*(3), 1998–2012.

Fontana, L., & Partridge, L. (2015). Promoting health and longevity through diet: From model organisms to humans. *Cell, 161,* 106–118.

Fonzo, G. A., & Etkin, A. (2017). Affective neuroimaging in generalized anxiety disorder: An integrated review. *Dialogues in Clinical Neuroscience, 19*(2), 169–179.

Forbes, T. A., & Gallo, V. (2017). All wrapped up: Environmental effects on myelination. *Trends in Neurosciences, 40*(9), 572–587.

Forger, N. G. (2018). Past, present and future of epigenetics in brain sexual differentiation. *Journal of Neuroendocrinology, 30*(2), e12492.

Foster, D. H. (2011). Color constancy. *Vision Research, 51,* 674–700.

Fox, D. (2018). The brain, reimagined. *Scientific American, 318*(4), 60–67.

Fox, K. C. R., Spreng, R. N., Ellamil, M., Andrews-Hanna, J. R., & Christoff, K. (2015). The wandering brain: Meta-analysis of functional neuroimaging studies of mind wandering and related spontaneous thought processes. *NeuroImage, 111,* 611–621.

Fox, K. C., Foster, B. L., Kucyi, A., Daitch, A. L., & Parvizi, J. (2018). Intracranial electrophysiology of the human default network. *Trends in Cognitive Sciences, 22*(4), 307–324.

Fox, K., Glazewski, S., & Schulze, S. (2000). Plasticity and stability of somatosensory maps in thalamus and cortex. *Current Opinion in Neurobiology, 10,* 494–497.

Fox, S. H., & Brotchie, J. M. (2010). The MPTP-lesioned non-human primate models of Parkinson's disease. Past, present, and future. *Progress in Brain Research, 184,* 133–157.

Frackowiak, R., & Markram, H. (2015). The future of human cerebral cartography: a novel approach. *Philosophical Transactions of the Royal Society B: Biological Sciences, 370*(1668), 20140171.

Fraga, M. F., Ballestar, E., Paz, M. F., Ropero, S., Setien, F., Ballestar, M. L., Heine-Suñer, D.,

Cigudosa, J. C., Urioste, M., Benitez, J., Boix-Chornet, M., Sanchez-Aguilera, A., Ling, C., Carlsson, E., Poulsen, P., Vaag, A., Stephan, Z., Spector, T. D., Wu, Y. Z., Plass, C., & Esteller, M. (2005). Epigenetic differences arise during the lifetime of monozygotic twins. *Proceedings of the National Academy of Sciences USA, 102,* 10604–10609.

Fragkaki, I., Cima, M., & Granic, I. (2018). The role of trauma in the hormonal interplay of cortisol, testosterone, and oxytocin in adolescent aggression. *Psychoneuroendocrinology, 88,* 24–37.

Francis, D. D., & Meaney, M. J. (1999). Maternal care and the development of stress responses. *Current Opinion in Neurobiology, 9,* 128–134.

Franco, S. J., & Müller, U. (2013). Shaping our minds: Stem and progenitor cell diversity in the mammalian neocortex. *Neuron, 77,* 19–34.

Franco, S. J., Gil-Sanz, C., Martinez-Garay, I., Espinosa, A., Harkins-Perry, S. R., Ramos, C., & Müller, U. (2012). Fate-restricted neural progenitors in the mammalian cerebral cortex. *Science, 337,* 746–749.

Frank, M. G. (2017). Sleep and plasticity in the visual cortex: More than meets the eye. *Current Opinion in Neurobiology, 44,* 8–12.

Franklin, R. J. M., & Ffrench-Constant, C. (2017). Regenerating CNS myelin—from mechanisms to experimental medicines. *Nature Reviews Neuroscience, 18*(12), 753–769.

Franklin, T. B., & Mansuy, I. M. (2010). The prevalence of epigenetic mechanisms in the regulation of cognitive function and behaviour. *Current Opinion in Neurobiology, 20,* 441–449.

Frase, L., Nissen, C., Riemann, D., & Spiegelhalder, K. (2018). Making sleep easier: Pharmacological interventions for insomnia. *Expert Opinion on Pharmacotherapy, 19*(13), 1465–1473.

Frazer, K. M., Richards, Q., & Keith, D. R. (2018). The long-term effects of cocaine use on cognitive functioning: A systematic critical review. *Behavioural Brain Research, 348,* 241–262.

Freedman, D. H. (2011). How to fix the obesity crisis. *Scientific American, 304,* 40–47.

Freedman, D. J., & Assad, J. A. (2016). Neuronal mechanisms of visual categorization: An abstract view on decision making. *Annual Review of Neuroscience, 39,* 129–147.

Freedman, D. J., & Ibos, G. (2018). An integrative framework for sensory, motor, and cognitive functions of the posterior parietal cortex. *Neuron, 97*(6), 1219–1234.

Freeman, J. H. (2015). Cerebellar learning mechanisms. *Brain Research, 1621,* 260–269.

Freiwald, W., Duchaine, B., & Yovel, G. (2016). Face processing systems: From neurons to real-world social perception. *Annual Review of Neuroscience, 39,* 325–346.

Freud, E., Plaut, D. C., & Behrmann, M. (2016). 'What' is happening in the dorsal visual pathway. *Trends in Cognitive Sciences, 20*(10), 773–784.

Freud, S. (1913). The interpretation of dreams (AA Brill, Trans.). *New York: McMillan (Original work published 1900).*

Freund, J., Brandmaier, A. M., Lewejohann, L., Kirste, I., Kritzler, M., . . . & Kempermann, G. (2013). Emergence of individuality in genetically identical mice. *Science, 340,* 756–758.

Friederici, A. D., & Gierhan, S. M. E. (2013). The language network. *Current Opinion in Neurobiology, 23,* 250–254.

Friederici, A. D., & Singer, W. (2015). Grounding language processing on basic neurophysiological principles. *Trends in Cognitive Sciences, 19,* 329–338.

Friedman, J., Globus, G., Huntley, A., Mullaney, D., Naitoh, P., & Johnson, L. (1977). Performance and mood during and after gradual sleep reduction. *Psychophysiology, 14,* 245–250.

Friedmann, P. D., Hendrickson, J. C., Gerstein, D. R., & Zhang, Z. (2004). The effect of matching

comprehensive services to patients' needs on drug use improvement in addiction treatment. *Addiction, 99,* 962–972.

Friese, M. A., Schattling, B., & Fugger, L. (2014). Mechanisms of neurodegeneration and axonal dysfunction in multiple sclerosis. *Nature Reviews Neurology, 10,* 225–238.

Frigon, A., & Rossignol, S. (2008). Adaptive changes of the locomotor pattern and cutaneous reflexes during locomotion studied in the same cats before and after spinalization. *Journal of Physiology, 12,* 2927–2945.

Froemke, R. C., & Jones, B. J. (2011). Development of auditory cortical synaptic receptive fields. *Neuroscience and Biobehavioral Reviews, 35,* 2105–2113.

Frühbeck, G., Busetto, L., Dicker, D., Yumuk, V., Goossens, G. H., Hebebrand, J., . . . & Toplak, H. (2019). The ABCD of obesity: An EASO position statement on a diagnostic term with clinical and scientific implications. *Obesity Facts, 12*(2), 131–136.

Fryer, J. D., Yu, P., Kang, H., Mandel-Brehm, C., Carter, A. N., Crespo-Barreto, J., Gao, Y., Flora, A., Shaw, C., Orr, H. T., & Zoghbi, H. Y. (2011). Exercise and genetic rescue of SCA1 via the transcriptional repressor Capicua. *Science, 334,* 690–693.

Fuhrmann, D., Knoll, L. J., & Blakemore, S. J. (2015). Adolescence as a sensitive period of brain development. *Trends in Cognitive Sciences, 19,* 558–566.

Fujiwara, N., Imai, M., Nagamine, T., Mima, T., Oga, T., Takeshita, K., Toma, K., & Shibasaki, H. (2002). Second somatosensory area (SII) plays a role in selective somatosensory attention. *Cognitive Brain Research, 14,* 389–397.

Fukushima, A. R., Corrêa, L. T., Muñoz, J. W. P., Ricci, E. L., Carvalho, V. M., Carvalho, D. G. D., . . . & Chasin, A. A. D. M. (2019). Crack cocaine, a systematic literature review. *Forensic Research & Criminology International Journal, 7*(5), 247–253.

Fullager, H. H. K., Skorski, S., Duffield, R., Hammes, D., Coutts, A. J., & Meyer, T. (2015). Sleep and athletic performance: The effects of sleep loss on exercise performance, and physiological and cognitive responses to exercise. *Sports Medicine, 45,* 161–186.

Funk, C. M., Honjoh, S., Rodriguez, A. V., Cirelli, C., & Tononi, G. (2016). Local slow waves in superficial layers of primary cortical areas during REM sleep. *Current Biology, 26*(3), 396–403.

Fusar-Poli, P., Borgwardt, S., Crescini, A., Deste, G., Kempton, M. J., Lawrie, S., Mc Guire, P., & Sacchetti, E. (2011). Neuroanatomy of vulnerability to psychosis: A voxel-based meta-analysis. *Neuroscience and Biobehavioral Reviews, 35,* 1175–1185.

Fuster, J. M. (2000). The prefrontal cortex of the primate: A synopsis. *Psychobiology, 28,* 125–131.

Fyfe, I. (2018). Prediction of cognitive decline in PD. *Nature Reviews Neurology, 14,* 316–317.

Gaffan, D. (1974). Recognition impaired and association intact in the memory of monkeys after transection of the fornix. *Journal of Comparative and Physiological Psychology, 86,* 1100–1109.

Gage, F. H., & Temple, S. (2013). Neural stem cells: Generating and regenerating the brain. *Neuron, 80,* 588–601.

Gagné, J. P. (2019). The psychology of Tourette disorder: Revisiting the past and moving toward a cognitively-oriented future. *Clinical Psychology Review, 67,* 11–21.

Gaillard, A., & Jaber, M. (2011). Rewiring the brain with cell transplantation in Parkinson's disease. *Trends in Neurosciences, 34,* 124–133.

Gainotti, G. (2018). Anosognosia in degenerative brain diseases: The role of the right hemisphere and of its dominance for emotions. *Brain and Cognition, 127,* 13–22.

Gainotti, G. (2019a). The role of the right hemisphere in emotional and behavioural disorders of patients with Fronto-temporal degeneration:

An updated review. *Frontiers in Aging Neuroscience, 11*, 55.

Gainotti, G. (2019b). Emotions and the right hemisphere: Can new data clarify old models? *The Neuroscientist, 25*(3), 258–270.

Galef, B. G. (1989). Laboratory studies of naturally-occurring feeding behaviors: Pitfalls, progress and problems in ethoexperimental analysis. In R. J. Blanchard, P. F. Brain, D. C. Blanchard, & S. Parmigiani (Eds.), *Ethoexperimental approaches to the study of behavior* (pp. 51–77). Dordrecht, Netherlands: Kluwer Academic.

Galef, B. G., Whishkin, E. E., & Bielavska, E. (1997). Interaction with demonstrator rats changes observer rats' affective responses to flavors. *Journal of Comparative Psychology, 111*, 393–398.

Gallo, E. A. G., Munhoz, T. N., de Mola, C. L., & Murray, J. (2018). Gender differences in the effects of childhood maltreatment on adult depression and anxiety: A systematic review and meta-analysis. *Child Abuse & Neglect, 79*, 107–114.

Gallup, G. G., Jr. (1983). Toward a comparative psychology of mind. In R. L. Mellgren (Ed.), *Animal cognition and behavior* (pp. 473–505). New York, NY: North-Holland.

Gambino, F., Pagès, S., Kehayas, V., Baptista, D., Tatti, R., Carleton, A., & Holtmaat, A. (2014). Sensory-evoked LTP driven by dendritic plateau potentials in vivo. *Nature, 515*, 116–119.

Gandal, M. J., Zhang, P., Hadjimichael, E., Walker, R. L., Chen, C., Liu, S., . . . & Shieh, A. W. (2018). Transcriptome-wide isoform-level dysregulation in ASD, schizophrenia, and bipolar disorder. *Science, 362*(6420).

Gandhi, N. J., & Katnani, H. A. (2011). Motor functions of the superior colliculus. *Annual Review of Neuroscience, 34*, 205–231.

Ganel, T., Tanzer, M., & Goodale, M. A. (2008). A double dissociation between action and perception in the context of visual illusions. *Psychological Science, 19*, 221–225.

Gao, Q., & Horvath, T. L. (2007). Neurobiology of feeding and energy expenditure. *Annual Review of Neuroscience, 30*, 367–398.

Gao, Z., van Beugen, B. J., & De Zeeuw, C. I. (2012). Distributed synergistic plasticity and cerebellar learning. *Nature Reviews Neuroscience, 13*, 619–635.

Garcia-Romeu, A., & Richards, W. A. (2018). Current perspectives on psychedelic therapy: Use of serotonergic hallucinogens in clinical interventions. *International Review of Psychiatry, 30*, 291–316.

Garcia, J., & Koelling, R. A. (1966). Relation of cue to consequence in avoidance learning. *Psychonomic Science, 4*, 123–124.

Gardner, R. M., & Brown, D. L. (2014). Body size estimation in anorexia nervosa: A brief review of findings from 2003 through 2013. *Psychiatry Research, 219*, 407–410.

Garofalo, S., D'Alessandro, G., Chece, G., Brau, F., Maggi, L., . . . & Limatola, C. (2015). Enriched environment reduces glioma growth through immune and non-immune mechanisms in mice. *Nature Communications, 6*, 6623.

Gauchat, A., Séguin, J. R., McSween-Cadieux, E., & Zadra, A. (2015). The content of recurrent dreams in young adolescents. *Consciousness and Cognition, 37*, 103–111.

Gaynes, B. N., Lloyd, S. W., Lux, L., Gartlehner, G., Hansen, R. A., Brode, S., . . . & Lohr, K. N. (2014). Repetitive transcranial magnetic stimulation for treatment-resistant depression: A systematic review and meta-analysis. *Journal of Clinical Psychiatry, 75*, 1–478.

Gazzaniga, M. S. (1967, August). The split brain in man. *Scientific American, 217*, 24–29.

Gazzaniga, M. S. (2005). Forty-five years of split-brain research and still going strong. *Nature Reviews Neuroscience, 6*, 653–659.

Gazzaniga, M. S. (2010). Neuroscience and the correct level of explanation for understanding mind. *Trends in Cognitive Sciences, 14*, 291–292.

Gazzaniga, M. S. (2013). Shifting gears: Seeking new approaches for mind/brain mechanisms. *Annual Review of Psychology, 64*, 1–20.

Gazzaniga, M. S., & Sperry, R. W. (1967). Language after section of the cerebral commissure. *Brain, 90*, 131–148.

Geary, D. C. (2019). Evolutionary perspective on sex differences in the expression of neurological diseases. *Progress in Neurobiology, 176*, 33–53.

Geddes, J. R., & Miklowitz, D. J. (2013). Treatment of bipolar disorder. *Lancet, 381*, 1672–1682.

Geddes, J. R., Carney, S. M., Davies, C., Furukawa, T. A., Kupfer, D. J., Frank, E., & Goodwin, G. M. (2003). Relapse prevention with antidepressant drug treatment in depressive disorders: A systematic review. *The Lancet, 361*, 653–661.

Gehring, W. J. (2014). The evolution of vision. *WIREs Developmental Biology, 3*, 1–40.

Geissmann, Q., Beckwith, E. J., & Gilestro, G. F. (2019). Most sleep does not serve a vital function: Evidence from Drosophila melanogaster. *Science advances, 5*(2), eaau9253.

Generoso, M. B., Trevizol, A. P., Kasper, S., Cho, H. J., Cordeiro, Q., & Shiozawa, P. (2017). Pregabalin for generalized anxiety disorder: An updated systematic review and meta-analysis. *International Clinical Psychopharmacology, 32*(1), 49–55.

Geng, J. J., & Behrmann, M. (2002). Probability cuing of target location facilitates visual search implicitly in normal participants and patients with hemispatial neglect. *Psychological Science, 13*, 520–525.

Genné-Bacon, E. A. (2014). Thinking evolutionarily about obesity. *Yale Journal of Biology and Medicine, 87*, 99–112.

Genzel, L., Kroes, M. C. W., Dresler, M., & Battaglia, F. P. (2014). Light sleep versus slow wave sleep in memory consolidation: A question of global versus local processes? *Trends in Neurosciences, 37*, 10–19.

Geoffroy, C. G., & Zheng, B. (2014). Myelin-associated inhibitors in axonal growth after CNS injury. *Current Opinion in Neurobiology, 27*, 31–38.

George, O., & Koob, G. F. (2010). Individual differences in prefrontal cortex function and the transition from drug use to drug dependence. *Neuroscience and Biobehavioral Reviews, 35*, 232–247.

Georgiadis, J. R. (2012). Doing it . . . wild? On the role of the cerebral cortex in human sexual activity. *Socioaffective Neuroscience and Psychology, 2*, 17337.

Georgiadis, J. R., & Kringelbach, M. L. (2012). The human sexual response cycle: Brain imaging evidence linking sex to other pleasures. *Progress in Neurobiology, 98*, 49–81.

Georgiadis, J. R., Kringelbach, M. L., & Pfaus, J. G. (2012). Sex for fun: A synthesis of human and animal neurobiology. *Nature Reviews Urology, 9*, 486–498.

Georgopoulos, A. P., & Carpenter, A. F. (2015). Coding of movements in the motor cortex. *Current Opinion in Neurobiology, 33*, 34–39.

Geraldes, R., Ciccarelli, O., Barkhof, F., De Stefano, N., Enzinger, C., Filippi, M., . . . & DeLuca, G. C. (2018). The current role of MRI in differentiating multiple sclerosis from its imaging mimics. *Nature Reviews Neurology, 14*(4), 199–213.

Gervain, J., & Geffen, M. N. (2019). Efficient neural coding in auditory and speech perception. *Trends in Neurosciences, 42*(1), 56–65.

Geschwind, D. H., & Rakic, P. (2013). Cortical evolution: Judge the brain by its cover. *Neuron, 80*, 633–647.

Geschwind, N. (1970). The organization of language and the brain. *Science, 170*, 940–944.

Geschwind, N. (1979, September). Specializations of the human brain. *Scientific American, 241*, 180–199.

Geuter, S., Koban, L., & Wager, T. D. (2017). The cognitive neuroscience of placebo effects: Concepts, predictions, and physiology. *Annual Review of Neuroscience, 40*, 167–188.

Geva-Sagiv, M., Romani, S., Las, L., & Ulanovsky, N. (2016). Hippocampal global remapping for different sensory modalities in flying bats. *Nature Neuroscience, 19*(7), 952.

Gevins, A., Leong, H., Smith, M. E., Le, J., & Du, R. (1995). Mapping cognitive brain function with modern high-resolution electroencephalography. *Trends in Neurosciences, 18*, 429–436.

Ghazanfar, A. A., & Takahashi, D. Y. (2014). The evolution of speech: Vision, rhythm, cooperation. *Trends in Cognitive Sciences, 18*, 543–553.

Ghezzi, A., Krishna, H. R., Lew, L., Prado, F. J., Ong, D. S., & Atkinson, N. S. (2013). Alcohol-induced histone acetylation reveals a gene network involved in alcohol tolerance. *PLoS Genetics, 9*, e1003986.

Ghibaudi, M., Boido, M., & Vercelli, A. (2017). Functional integration of complex miRNA networks in central and peripheral lesion and axonal regeneration. *Progress in Neurobiology, 158*, 69–93.

Ghodrati, M., Khaligh-Razavi, S. M., & Lehky, S. R. (2017). Towards building a more complex view of the lateral geniculate nucleus: Recent advances in understanding its role. *Progress in Neurobiology, 156*, 214–255.

Ghose, D., & Wallace, M. T. (2014). Heterogeneity in the spatial receptive field architecture of multisensory neurons of the superior colliculus and its effects on multisensory integration. *Neuroscience, 256*, 147–162.

Giacobini, E., & Gold, G. (2013). Alzheimer disease therapy—moving from amyloid-β to tau. *Nature Reviews Neurology, 9*, 677–686.

Giacomo, L. M. (2015). Imagine a journey through time and space. *Nature Neuroscience, 18*, 163–164.

Giardino, N. D., Friedman, S. D., & Dager, S. R. (2007). Anxiety, respiration, and cerebral blood flow: Implications for functional brain imaging. *Comprehensive Psychiatry, 48*, 103–112.

Gibbons, A. (2009). A new kind of ancestor: *Ardipithecus* unveiled. *Science, 326*, 36–43.

Gibbons, A. (2014). Neanderthals and moderns made imperfect mates. *Science, 343*, 471–472.

Gibbons, A. (2015a). New human species discovered. *Science, 349*, 1149–1150.

Gibbs, J., Young, R. C., & Smith, G. P. (1973). Cholecystokinin decreases food intake in rats. *Journal of Comparative and Physiological Psychology, 84*, 488–495.

Giedd, J. N. (2015). The amazing teen brain. *Scientific American, 312*, 32–37.

Giese, M. A., & Rizzolatti, G. (2015). Neural and computational mechanisms of action processing: Interaction between visual and motor representations. *Neuron, 88*, 167–180.

Gilbert, C. D., & Li, W. (2013). Top-down influences on visual processing. *Nature Reviews Neuroscience, 14*, 350–363.

Gilbert, C. D., & Wiesel, T. N. (1992). Receptive field dynamics in adult primary visual cortex. *Nature, 356*, 150–152.

Gillebert, C. R., Mantini, D., Thijs, V., Sunaert, S., Dupont, P., & Vandenberghe, R. (2011). Lesion evidence for the critical role of the intraparietal sulcus in spatial attention. *Brain, 134*, 1694–1709.

Gillespie-Lynch, K., Greenfield, P. M., Lyn, H., & Savage-Rumbaugh, S. (2014). Gestural and symbolic development among apes and humans: Support for a multimodal theory of language development. *Frontiers in Psychology, 5*, 1228.

Gilmartin, M. R., Balderston, N. L., & Helmstetter, F. J. (2014). Prefrontal cortical regulation of fear learning. *Trends in Neurosciences, 37*, 455–464.

Gilmore, J. H., Knickmeyer, R. C., & Gao, W. (2018). Imaging structural and functional brain development in early childhood. *Nature Reviews Neuroscience, 19*(3), 123.

Gilmore, J. H., Shi, F., Woolson, S. L., Knickmeyer, R. C., Short, S. J., Lin, W., Zhu, H., Hamer, R. M.,

Styner, M., & Shen, D. (2012). Longitudinal development of cortical and subcortical gray matter from birth to 2 years. *Cerebral Cortex, 22*, 2478–2485.

Gingras, J., Smith, S., Matson, D. J., Johnson, D., Nye, K., . . . & McDonough, S. I. (2014). Global Nav1.7 knockout mice recapitulate the phenotype of human congenital indifference to pain. *PLoS ONE, 9*, e105895.

Ginsberg, M. D. (2016). Expanding the concept of neuroprotection for acute ischemic stroke: The pivotal roles of reperfusion and the collateral circulation. *Progress in Neurobiology, 145*, 46–77.

Girdhar, K., Hoffman, G. E., Jiang, Y., Brown, L., Kundakovic, M., Hauberg, M. E., . . . & Zharovsky, E. (2018). Cell-specific histone modification maps in the human frontal lobe link schizophrenia risk to the neuronal epigenome. *Nature Neuroscience, 21*(8), 1126–1136.

Girgis, R. R., Zoghbi, A. W., Javitt, D. C., & Lieberman, J. A. (2019). The past and future of novel, non-dopamine-2 receptor therapeutics for schizophrenia: A critical and comprehensive review. *Journal of Psychiatric Research, 108*, 57–83.

Giszter, S. F. (2015). Motor primitives—new data and future questions. *Current Opinion in Neurobiology, 33*, 156–165.

Gitler, A. D. (2011). Another reason to exercise. *Science, 334*, 606–607.

Gittis, A. H., & Brasier, D. J. (2015). Astrocytes tell neurons when to listen up. *Science, 349*, 690–691.

Gittis, A. H., Berke, J. D., Bevan, M. D., Chan, C. S., Mallet, N., Morrow, M. M., & Schmidt, R. (2014). New roles for the external globus pallidus in basal ganglia circuits and behavior. *Journal of Neuroscience, 34*, 15178–15183.

Glaser, R., Rice, J., Sheridan, J., Fertel, R., Stout, J., Speicher, C., . . . & Kiecolt-Glaser, J. (1987). Stress-related immune suppression: Health implications. *Brain, Behavior, and Immunity, 1*, 7–20.

Glasser, M. F., Smith, S. M., Marcus, D. S., Andersson, J. L., Auerbach, E. J., Behrens, T. E., . . . & Robinson, E. C. (2016). The human connectome project's neuroimaging approach. *Nature Neuroscience, 19*(9), 1175–1186.

Gliga, T., Jones, E. J. H., Bedford, R., Charman, T., & Johnson, M. H. (2014). From early markers to neuro-developmental mechanisms of autism. *Developmental Review, 34*, 189–207.

Goadsby, P. J. (2015). Decade in review—migraine: Incredible progress for an era of better migraine care. *Nature Reviews Neurology, 11*, 621–622.

Goddard, G. V., McIntyre, D. C., & Leech, C. K. (1969). A permanent change in brain function resulting from daily electrical stimulation. *Experimental Neurology, 25*, 295–330.

Godoy, M. D. C. L., Voegels, R. L., Pinna, F. D., Imamura, R., & Frafel, J. M. (2015). Olfaction in neurologic and neurodegenerative diseases: A literature review. *International Archives of Otorhinolaryngology, 19*, 176–179.

Godtfredsen, N. S., & Prescott, E. (2011). Benefits of smoking cessation with focus on cardiovascular and respiratory comorbidities. *Clinical Respiratory Journal, 5*, 187–194.

Goedert, M. (1993). Tau protein and the neurofibrillary pathology of Alzheimer's disease. *Trends in Neurosciences, 16*, 460–465.

Goedert, M., Eisenberg, D. S., & Crowther, R. A. (2017). Propagation of tau aggregates and neurodegeneration. *Annual Review of Neuroscience, 40*, 189–210.

Goense, J. B. M., & Logothetis, N. K. (2008). Neurophysiology of the BOLD fMRI signal in awake monkeys. *Current Biology, 18*, 631–640.

Goetz, C. G. (2011). The history of Parkinson's disease: Early clinical descriptions and neurological therapies. *Cold Spring Harbor Perspectives in Medicine, 1*, a008862.

Goetz, C. G., & Pal, G. (2014). Initial management of Parkinson's disease. *BMJ, 349*, g6258.

Gogolla, N., Galimberti, I., & Caroni, P. (2007). Structural plasticity of axon terminals in the adult. *Current Opinion in Neurobiology, 17*, 516–524.

Gogtay, N., & Thompson, P. M. (2010). Mapping gray matter development: Implications for typical development and vulnerability to psychopathology. *Brain and Cognition, 72*, 6–15.

Goldacre, B. (2013). *Bad pharma.* McClelland & Stewart: Toronto.

Goldberg, E. M., & Coulter, D. A. (2013). Mechanisms of epileptogenesis: A convergence on neural circuit dysfunction. *Nature Reviews Neuroscience, 14*, 337–348.

Goldberg, L. R., & Gould, T. J. (2019). Multigenerational and transgenerational effects of paternal exposure to drugs of abuse on behavioral and neural function. *European Journal of Neuroscience, 50*(3), 2453–2466.

Goldman, D. (2015). America's cannabis experiment. *JAMA Psychiatry, 72*, 969–970.

Goldman, S. A., & Nottebohm, F. (1983). Neuronal production, migration, and differentiation in a vocal control nucleus of the adult female canary brain. *Proceedings of the National Academy of Sciences, USA, 80*, 2390–2394.

Goldstein-Piekarski, A. N., & Williams, L. M. (2019). A neural circuit-based model for depression anchored in a synthesis of insights from functional neuroimaging. In *Neurobiology of Depression* (pp. 241–256). Academic Press.

Goldstein-Piekarski, A. N., Staveland, B. R., Ball, T. M., Yesavage, J., Korgaonkar, M. S., & Williams, L. M. (2018). Intrinsic functional connectivity predicts remission on antidepressants: A randomized controlled trial to identify clinically applicable imaging biomarkers. *Translational Psychiatry, 8*(1), 1–11.

Goldstein, R. Z., & Volkow, N. D. (2011). Dysfunction of the prefrontal cortex in addiction: Neuroimaging findings and clinical implications. *Nature Reviews Neuroscience, 12*, 652–669.

Goll, Y., Atlan, G., & Citri, A. (2015). Attention: The claustrum. *Trends in Neurosciences, 38*, 486–495.

Gollin, E. S. (1960). Developmental studies of visual recognition of incomplete objects. *Perceptual Motor Skills, 11*, 289–298.

Gollnick, P. D., & Hodgson, D. R. (1986). The identification of fiber types in skeletal muscle: A continual dilemma. *Exercise and Sport Sciences Reviews, 14*, 81–104.

Golombek, D. A., Pandi-Perumal, S. R., Brown, G. M., & Cardinali, D. P. (2015). Some implications of melatonin use in chronopharmacology of insomnia. *European Journal of Pharmacology, 762*, 42–48.

Golub, M. D., Chase, S. M., Batista, A. P., & Byron, M. Y. (2016). Brain—computer interfaces for dissecting cognitive processes underlying sensorimotor control. *Current Opinion in Neurobiology, 37*, 53–58.

Gomez-Marin, A., & Mainen, Z. F. (2016). Expanding perspectives on cognition in humans, animals, and machines. *Current Opinion in Neurobiology, 37*, 85–91.

Gomez, J. A., Beitnere, U., & Segal, D. J. (2019). Live-animal epigenome editing: Convergence of novel techniques. *Trends in Genetics, 35*, 527–541

Gonçalves, J. T., Schafer, S. T., & Gage, F. H. (2016). Adult neurogenesis in the hippocampus: From stem cells to behavior. *Cell, 167*(4), 897–914.

Gondim, R., Teles, F., & Barroso Jr, U. (2018). Sexual orientation of 46, XX patients with congenital adrenal hyperplasia: A descriptive review. *Journal of Pediatric Urology, 14*(6), 486–493.

González-Muniesa, P., Mártinez-González, M., Hu, F., Després, J., Matsuzawa, Y., Loos, R., & Martinez, J. A. (2017). Obesity. *Nature Reviews Disease Primers, 3*, 17034.

González-Orozco, J. C., & Camacho-Arroyo, I. (2019). Progesterone actions during central nervous system development. *Frontiers in Neuroscience, 13*, 503.

González-Reimers, E., Santolaria-Fernández, F., Martín-Gonzalaz, M. C., Fernández-Rodríguez, C. M., & Quintero-Platt, G. (2014). Alcoholism: A systemic proinflammatory condition. *World Journal of Gastroenterology, 20*, 14660–14671.

González, V., Miguez, G., Quezada, V., Mallea, J., & Laborda, M. (2019). Ethanol tolerance from a Pavlovian perspective. *Psychology & Neuroscience, 12*(4), 495–509.

Goodale, M. A., & Milner, A. D. (1992). Separate visual pathways for perception and action. *Trends in Neurosciences, 15*, 20–25.

Goodale, M. A., & Milner, A. D. (2004). *Sight unseen: An exploration of conscious and unconscious perception.* Oxford, England: Oxford University Press.

Goodale, M. A., & Westwood, D. A. (2004). An evolving view of duplex vision: Separate but interacting cortical pathways for perception and action. *Current Opinion in Neurobiology, 14*, 203–211.

Goodenough, D. R., Shapiro, A., Holden, M., & Steinschriber, L. (1959). A comparison of "dreamers" and "nondreamers": Eye movements, electroencephalograms, and the recall of dreams. *Journal of Abnormal and Social Psychology, 59*, 295–303.

Goodhill, G. J. (2016). Can molecular gradients wire the brain? *Trends in Neurosciences, 39*, 202–211.

Goodhill, G. J. (2016). Can molecular gradients wire the brain? *Trends in Neurosciences, 39*(4), 202–211.

Gorski, R. A. (1980). Sexual differentiation in the brain. In D. T. Krieger & J. C. Hughes (Eds.), *Neuroendocrinology* (pp. 215–222). Sunderland, MA: Sinauer.

Gorski, R. A., Gordon, J. H., Shryne, J. E., & Southam, A. M. (1978). Evidence for a morphological sex difference within the medial preoptic area of the rat brain. *Brain Research, 148*, 333–346.

Goshen, I. (2014). The optogenetic revolution in memory research. *Trends in Neurosciences, 37*, 511–522.

Goswami, U. (2014). Sensory theories of developmental dyslexia: Three challenges for research. *Nature Reviews Neuroscience, 16*, 43–54.

Gottesman, I. I., & Hanson, D. R. (2005). Human development: Biological and genetic processes. *Annual Review of Psychology, 56*, 263–286.

Gottfried, J. A. (2010). Central mechanisms of odour object perception. *Nature Reviews Neuroscience, 11*, 628–640.

Götz, J., Bodea, L. G., & Goedert, M. (2018). Rodent models for Alzheimer disease. *Nature Reviews Neuroscience, 19*(10), 583–598.

Goulas, A., Bastiani, M., Bezgin, G., Uylings, H. B. M., Roebroeck, A., & Stiers, P. (2014). Comparative analysis of the macroscale structural connectivity in the macaque and human brain. *PLoS Computational Biology, 10*, e1003529.

Grabb, M. C., & Gobburu, J. V. (2017). Challenges in developing drugs for pediatric CNS disorders: A focus on psychopharmacology. *Progress in Neurobiology, 152*, 38–57.

Grabowska, A. (2017). Sex on the brain: Are gender-dependent structural and functional differences associated with behavior? *Journal of Neuroscience Research, 95*(1–2), 200–212.

Grabska-Barwińska, A., Barthelmé, S., Beck, J., Mainen, Z. F., Pouget, A., & Latham, P. E. (2017). A probabilistic approach to demixing odors. *Nature Neuroscience, 20*(1), 98–106.

Grady, K. L., Phoenix, C. H., & Young, W. C. (1965). Role of the developing rat testis in differentiation of the neural tissues mediating mating behavior. *Journal of Comparative and Physiological Psychology, 59*, 176–182.

Granger, A. J., Wallace, M. L., & Sabatini, B. L. (2017). Multi-transmitter neurons in the mammalian central nervous system. *Current Opinion in Neurobiology, 45*, 85–91.

Grant, P. R. (1991, October). Natural selection and Darwin's finches. *Scientific American, 265,* 82–87.

Grasso, G., & Landi, A. (2016). Changing paradigm in mild traumatic brain injury research. *Journal of Neuroscience Research, 94,* 825–826.

Grau, J. W. (2014). Learning from the spinal cord: How the study of spinal cord plasticity informs our view of learning. *Neurobiology of Learning and Memory, 108,* 155–171.

Graybiel, A. M. (2000). The basal ganglia. *Current Biology, 10*(14), R509–R511.

Graybiel, A. M., & Grafton, S. T. (2015). The striatum: Where skills and habits meet. *Cold Spring Harbor Perspectives in Biology, 7,* a021691.

Grayson, B. E., Seeley, R. J., & Sandoval, D. A. (2013). Wired on sugar: The role of the CNS in the regulation of glucose homeostasis. *Nature Reviews Neuroscience, 14,* 24–37.

Graziano, M. S. (2016). Ethological action maps: A paradigm shift for the motor cortex. *Trends in Cognitive Sciences, 20*(2), 121–132.

Green, A. M., & Kalaska, J. F. (2010). Learning to move machines with the mind. *Trends in Neurosciences, 34,* 61–75.

Green, S. R., Kragel, P. A., Fecteau, M. E., & LaBar, K. S. (2014). Development and validation of an unsupervised scoring system (Autonomate) for skin conductance response analysis. *International Journal of Psychophysiology, 91,* 186–193.

Greene, N. D. E., & Copp, A. J. (2014). Neural tube defects. *Annual Review of Neuroscience, 37,* 221–242.

Greene, P. E., Fahn, S., Tsai, W. Y., Winfield, H., Dillon, S., Kao, R., et al. (1999). Double-blind controlled trial of human embryonic dopaminergic tissue transplants in advanced Parkinson's disease: Long-term unblinded follow-up phase. *Neurology, 52*(Suppl. 2).

Grefkes, C., & Fink, G. R. (2011). Reorganization of cerebral networks after stroke: New insights from neuroimaging with connectivity approaches. *Brain, 134,* 1264–1276.

Greig, L. C., Woodworth, M. B., Galazo, M. J., Padmanabhan, H., & Macklis, J. D. (2013). Molecular logic of neocortical projection neuron specification, development and diversity. *Nature Reviews Neuroscience, 14,* 755–769.

Grigg-Damberger, M. M. (2012). The AASM scoring manual four years later. *Journal of Clinical Sleep Medicine, 8,* 323–332.

Griggs, J., Sinnayah, P., & Mathai, M. L. (2015). Prader-Willi syndrome: From genetics to behaviour, with special focus on appetite treatments. *Neuroscience and Biobehavioral Reviews, 59,* 155–172.

Grill-Spector, K., & Malach, R. (2004). The human visual cortex. *Annual Review of Neuroscience, 27,* 649–677.

Grillner, S. (1985). Neurobiological bases of rhythmic motor acts in vertebrates. *Science, 228,* 143–149.

Grillner, S. (2011). Human locomotor circuits conform. *Science, 334,* 912–913.

Grillner, S., & Dickinson, M. (2002). Motor systems. *Current Opinion in Neurobiology, 12,* 629–632.

Grillner, S., & Jessell, T. M. (2009). Measured motion: Searching for simplicity in spinal locomotor networks. *Current Opinion in Neurobiology, 19,* 572–586.

Grippo, A. J., & Scotti, M.-A. L. (2013). Stress and neuroinflammation. In A. Halaris & B. E. Leonard (Eds.), *Inflammation in psychiatry* (pp. 20–32). Basel: Karger.

Grone, B. P., & Baraban, S. C. (2015). Animal models in epilepsy research: Legacies and new directions. *Nature Neuroscience, 18,* 339–343.

Grosche, A., & Reichenbach, A. (2013). Developmental refining of neuroglial signaling? *Science, 339,* 152–153.

Gross, C. G., Moore, T., & Rodman, H. R. (2004). Visually guided behavior after V1 lesions in young and adult monkeys and its relation to blindsight in humans. *Progress in Brain Research, 144,* 279–294.

Gruart, A., Leal-Campanario, R., López-Ramos, J. C., & Delgado-Garcia, J. M. (2015). Functional basis of associative learning and their relationships with long-term potentiation evoked in the involved neural circuits: Lessons from studies in behaving mammals. *Neurobiology of Learning and Memory, 124,* 3–18.

Grubbe, M. S., & Thompson, I. D. (2004). The influence of early experience on the development of sensory systems. *Current Opinion in Neurobiology, 14,* 503–512.

Grunt, J. A., & Young, W. C. (1952). Differential reactivity of individuals and the response of the male guinea pig to testosterone propionate. *Endocrinology, 51,* 237–248.

Gu, Y. (2018). Vestibular signals in primate cortex for self-motion perception. *Current Opinion in Neurobiology, 52,* 10–17.

Guadarrama-Bazante, I. L., & Rodríguez-Manzo, G. (2019). Nucleus accumbens dopamine increases sexual motivation in sexually satiated male rats. *Psychopharmacology, 236*(4), 1303–1312.

Guerrini, R., Dobyns, W. B., & Barkovich, A. J. (2008). Abnormal development of the human cerebral cortex: Genetics, functional consequences and treatment options. *Trends in Neurosciences, 31,* 154–162.

Guilhoto, L. M. (2017). Absence epilepsy: Continuum of clinical presentation and epigenetics? *Seizure, 44,* 53–57.

Guillamón, A., & Segovia, S. (1996). Sexual dimorphism in the CNS and the role of steroids. In T. W. Stone (Ed.), *CNS neurotransmitters and neuromodulators. Neuroactive steroids* (pp. 127–152). Boca Raton, FL: CRC Press.

Guillon, Q., Hadjikhani, N., Baduel, S., & Rogé, B. (2014). Visual social attention in autism spectrum disorder: Insights from eye tracking studies. *Neuroscience & Biobehavioral Reviews, 42,* 279–297.

Guina, J., & Merrill, B. (2018). Benzodiazepines I: Upping the care on downers: The evidence of risks, benefits and alternatives. *Journal of Clinical Medicine, 7*(2), 17.

Guo, Y., & Udin, S. B. (2000). The development of abnormal axon trajectories after rotation of one eye in *Xenopus. Journal of Neuroscience, 20,* 4189–4197.

Gupta, P., Albeanu, D. F., & Bhalla, U. S. (2015). Olfactory bulb coding of odors, mixtures and sniffs is a linear sum of odor time profiles. *Nature Neuroscience, 18,* 272–281.

Gürel, Ç., Kuşçu, G. C., Yavaşoğlu, A., & Avcı, B. (2020). The clues in solving the mystery of major psychosis: The epigenetic basis of schizophrenia and bipolar disorder. *Neuroscience and Biobehavioral Reviews, 113,* 51–61.

Gurvich, C., Hoy, K., Thomas, N., & Kulkarni, J. (2018). Sex differences and the influence of sex hormones on cognition through adulthood and the aging process. *Brain Sciences, 8*(9), 163.

Gusyatiner, O., & Hegi, M. E. (2018). Glioma epigenetics: From subclassification to novel treatment options. *Seminars in Cancer Biology, 51,* 50–58.

Gutchess, A. (2014). Plasticity of the aging brain: New directions in cognitive neuroscience. *Science, 346,* 579–582.

Gutfreund, Y., Zheng, W., & Knudsen, E. I. (2002). Gated visual input to the central auditory system. *Science, 297,* 1556–1559.

Gutruf, P., & Rogers, J. A. (2018). Implantable, wireless device platforms for neuroscience research. *Current Opinion in Neurobiology, 50,* 42–49.

Guyenet, P. G. (2006). The sympathetic control of blood pressure. *Nature Reviews Neuroscience, 7,* 335–346.

Hafeman, D. M., Chase, H. W., Monk, K., Bonar, L., Hickey, M. B., McCaffrey, A., . . . & Axelson, D. A. (2019). Intrinsic functional connectivity correlates of person-level risk for bipolar disorder in offspring of affected parents. *Neuropsychopharmacology, 44*(3), 629–634.

Haffenden, A. M., & Goodale, M. A. (1998). The effect of pictorial illusion on prehension and perception. *Journal of Cognitive Neuroscience, 10,* 122–136.

Hage, S. R., & Nieder, A. (2016). Dual neural network model for the evolution of speech and language. *Trends in Neurosciences, 39*(12), 813–829.

Hagoort, P. (2014). Nodes and networks in the neural architecture for language: Broca's region and beyond. *Current Opinion in Neurobiology, 28,* 136–141.

Hahn, N., Foxe, J. J., & Molholm, S. (2014). Impairments of multisensory integration and cross-sensory learning as pathways to dyslexia. *Neuroscience & Biobehavioral Reviews, 47,* 384–392.

Haim, L. B., & Rowitch, D. H. (2017). Functional diversity of astrocytes in neural circuit regulation. *Nature Reviews Neuroscience, 18*(1), 31–41.

Hajek, P., Phillips-Waller, A., Przulj, D., Pesola, F., Myers Smith, K., Bisal, N., . . . & Ross, L. (2019). A randomized trial of e-cigarettes versus nicotine-replacement therapy. *New England Journal of Medicine, 380*(7), 629–637.

Halford, J. C., & Blundell, J. E. (2000). Separate systems for serotonin and leptin in appetite control. *Annals of Medicine, 32,* 222–232.

Hall, S. S. (2010). Revolution postponed. *Scientific American, 303*(4), 60–67.

Hall, W., & Degenhardt, L. (2014). The adverse health effects of chronic cannabis use. *Drug Testing and Analysis, 6,* 39–45.

Hallett, M. (2015). Tourette syndrome: Update. *Brain and Development, 37,* 651–655.

Halpern, B., & Halpern, A. (2015). Safety assessment of FDA-approved (orlistat and lorcaserin) anti-obesity medications. *Expert Opinion on Drug Safety, 14,* 305–315.

Hamann, S., Herman, R. A., Nolan, C. L., & Wallen, K. (2004). Men and women differ in amygdala response to visual sexual stimuli. *Nature Neuroscience, 7,* 411–416.

Hammock, E. A. (2015). Developmental perspective on oxytocin and vasopressin. *Neuropsychopharmacology, 40,* 24–42.

Hammond, P. H., Merton, P. A., & Sutton, G. G. (1956). Nervous gradation of muscular contraction. *British Medical Bulletin, 12,* 214–218.

Hampel, H., O'Bryant, S. E., Molinuevo, J. L., Zetterberg, H., Masters, C. L., Lista, S., . . . & Blennow, K. (2018). Blood-based biomarkers for Alzheimer disease: Mapping the road to the clinic. *Nature Reviews Neurology, 14*(11), 639–652.

Han, M. H., Russo, S. J., & Nestler, E. J. (2019). Molecular, cellular, and circuit basis of depression susceptibility and resilience. In *Neurobiology of Depression* (pp. 123–136). Academic Press.

Handel, A. E., & Ramagoplan, S. V. (2010). Is Lamarckian evolution relevant to medicine? *BMC Medical Genetics, 11,* 73.

Hanford, L. C., Nazarov, A., Hall, G. B., & Sassi, R. B. (2016). Cortical thickness in bipolar disorder: A systematic review. *Bipolar Disorders, 18,* 4–18.

Hannan, A. J. (2014). Environmental enrichment and brain repair: Harnessing the therapeutic effects of cognitive stimulation and physical activity to enhance experience-dependent plasticity. *Neuropathology and Applied Neurobiology, 40,* 13–25.

Hannon, E., Spiers, H., Viana, J., Pidsley, R., Burrage, J., Murphy, T. M., . . . & Bray, N. J. (2016). Methylation QTLs in the developing brain and their enrichment in schizophrenia risk loci. *Nature Neuroscience, 19,* 48–54.

Hanslmayr, S., Staresina, B. P., & Bowman, H. (2016). Oscillations and episodic memory: Addressing the synchronization/desynchronization conundrum. *Trends in Neurosciences, 39*(1), 16–25.

Haque, F. N., Gottesman, I. I., & Wong, A. H. C. (2009). Not really identical: Epigenetic differences in monozygotic twins and implications for twin studies in psychiatry. *American Journal of Medical Genetics, 151C,* 136–141.

Haque, S., Vaphiades, M. S., & Lueck, C. J. (2018). The visual agnosias and related disorders. *Journal of Neuro-Ophthalmology, 38,* 379–392.

Hardcastle, K., Ganguli, S., & Giocomo, L. M. (2017). Cell types for our sense of location: Where we are and where we are going. *Nature Neuroscience, 20*(11), 1474–1482.

Harden, K. P., Turkheimer, E., & Loehlin, J. C. (2007). Genotype by environment interaction in adolescents' cognitive aptitude. *Behavioral Genetics, 37,* 273–283.

Hardy, J., & Selkoe, D. J. (2002). The amyloid hypothesis of Alzheimer's disease: Progress and problems on the road to therapeutics. *Science, 297,* 353–356.

Hare, D. J., Arora, M., Jenkins, N. L., Finkelstein, D. I., Doble, P. A., & Bush, A. I. (2015). Is early-life iron exposure critical in neurodegeneration? *Nature Reviews Neurology, 11,* 536–544.

Hari, R., & Parkkonen, L. (2015). The brain time-wise: How timing shapes and supports brain function. *Philosophical Transactions of the Royal Society B, 370,* 20140170.

Harmer, C. J., Duman, R. S., & Cowen, P. J. (2017). How do antidepressants work? New perspectives for refining future treatment approaches. *The Lancet Psychiatry, 4*(5), 409–418.

Harmon, K. (2013). Shattered ancestry. *Scientific American, 308,* 42–49.

Harrar, V., Tammam, J., Pérez-Bellido, A., Pitt, A., Stein, J., & Spence, C. (2014). Multisensory integration and attention in developmental dyslexia. *Current Biology, 24,* 531–535.

Harris, G. W. (1955). *Neural control of the pituitary gland.* London, England: Edward Arnold.

Harris, J. P., Burrell, J. C., Struzyna, L. A., Chen, H. I., Serruya, M. D., Wolf, J. A., . . . & Cullen, D. K. (2020). Emerging regenerative medicine and tissue engineering strategies for Parkinson's disease. *NJP Parkinson's Disease, 6*(1), 1–14.

Harris, K. D., Quiroga, R. Q., Freeman, J., & Smith, S. L. (2016). Improving data quality in neuronal population recordings. *Nature Neuroscience, 19*(9), 1165.

Harris, L. J., Clay, J., Hargreaves, F. J., & Ward, A. (1933). Appetite and choice of diet: The ability of the vitamin B deficient rat to discriminate between diets containing and lacking the vitamin. *Proceedings of the Royal Society of London (B), 113,* 161–190.

Harrison, P. J., Geddes, J. R., & Tunbridge, E. M. (2018). The emerging neurobiology of bipolar disorder. *Trends in Neurosciences, 41*(1), 18–30.

Harrison, T. C., & Murphy, T. H. (2014). Motor maps and the cortical control of movement. *Current Opinion in Neurobiology, 24,* 88–94.

Hart, C. (2013). *High price.* New York: Harper Collins.

Hart, C. L., Marvin, C. B., Silver, R., & Smith, E. E. (2012). Is cognitive functioning impaired in methamphetamine users? A critical review. *Neuropsychopharmacology, 37,* 586–608.

Harvey, A. G., & Tang, N. K. Y. (2012). (Mis)perception of sleep in insomnia: A puzzle and a resolution. *Psychological Bulletin, 138,* 77–101.

Hashikawa, K., Hashikawa, Y., Lischinsky, J., & Lin, D. (2018). The neural mechanisms of sexually dimorphic aggressive behaviors. *Trends in Genetics, 34*(10), 755–776.

Haslinger, R., Pipa, G., Lima, B., Singer, W., Brown, E. N., & Neuenschwander, S. (2012). Context matters: The illusive simplicity of macaque V1 receptive fields. *PLoS ONE, 7,* e39699.

Hastings, M. H., Maywood, E. S., & Brancaccio, M. (2018). Generation of circadian rhythms in the suprachiasmatic nucleus. *Nature Reviews Neuroscience, 19*(8), 453–469.

Hata, Y., & Stryker, M. P. (1994). Control of thalamocortical afferent rearrangement by postsynaptic activity in developing visual cortex. *Science, 265,* 1732–1735.

Hattar, S., Lucas, R. J., Mrosovsky, N., Thompson, S., Douglas, R. H., Hankins, M. W., . . . & Yau, K.-W. (2003). Melanopsin and rod-cone photoreceptive systems account for all major accessory visual functions in mice. *Nature, 424,* 75–81.

Haukvik, U. K., Tamnes, C. K., Söderman, E., & Agartz, I. (2018). Neuroimaging hippocampal subfields in schizophrenia and bipolar disorder: A systematic review and meta-analysis. *Journal of Psychiatric Research, 104,* 217–226.

Hauser, M. D., Yang, C., Berwick, R. C., Tattersall, I., Ryan, M. J., . . . & Lewontin, R. C. (2014). The mystery of language evolution. *Frontiers in Psychology, 5,* 401.

Hauser, S. L., Chan, J. R., & Oksenberg, J. R. (2013). Multiple sclerosis: Prospects and promise. *Annals of Neurology, 74,* 317–327.

Hausmann, M. (2017). Why sex hormones matter for neuroscience: A very short review on sex, sex hormones, and functional brain asymmetries. *Journal of Neuroscience Research, 95*(1–2), 40–49.

Hawkins, L. J., Al-attar, R., & Storey, K. B. (2018). Transcriptional regulation of metabolism in disease: From transcription factors to epigenetics. *PeerJ, 6,* e5062.

Haxby, J. V. (2006). Fine structure in representations of faces and objects. *Nature Neuroscience, 9,* 1084–1085.

Hayashi, Y., Kashiwagi, M., Yasuda, K., Ando, R., Kanuka, M., Sakai, K., & Itohara, S. (2015). Cells of a common developmental origin regulate REM/non-REM sleep and wakefulness in mice. *Science, 350,* 957–961.

Haynes, J.-D., & Rees, G. (2006). Decoding mental states from brain activity in humans. *Nature Reviews Neuroscience, 7,* 523–534.

He, Z., & Jin, Y. (2016). Intrinsic control of axon regeneration. *Neuron, 90*(3), 437–451.

Hebb, D. O. (1949). *The organization of behavior.* New York, NY: Wiley.

Hécaen, H., & Angelergues, R. (1964). Localization of symptoms in aphasia. In A. V. S. de Reuck & M. O'Connor (Eds.), *CIBA foundation symposium on the disorders of language* (pp. 222–256). London: Churchill Press.

Hedlund, P. (2014). Cannabinoids and the endocannabinoid system in lower urinary tract function and dysfunction. *Neurourology and Urodynamics, 33,* 46–53.

Heidenreich, M., & Zhang, F. (2016). Applications of CRISPR–Cas systems in neuroscience. *Nature Reviews Neuroscience, 17*(1), 36.

Heilig, M., Epstein, D. H., Nader, M. A., & Shaham, Y. (2016). Time to connect: Bringing social context into addiction neuroscience. *Nature Reviews Neuroscience, 17*(9), 592–599.

Heller, A. C., Amar, A. P., Liu, C. Y., & Apuzzo, M. L. J. (2006). Surgery of the mind and mood: A mosaic of issues in time and evolution. *Neurosurgery, 59*(4), 720–739.

Heller, E. A., Cates, H. M., Peña, C. J., Sun, H., Shao, N., Feng, J., . . . & Ferguson, D. (2014). Locus-specific epigenetic remodeling controls addiction- and depression-related behaviors. *Nature Neuroscience, 17,* 1720–1727.

Helm, M., & Motorin, Y. (2017). Detecting RNA modifications in the epitranscriptome: Predict and validate. *Nature Reviews Genetics, 18*(5), 275.

Helmer, W. J. (1986, February). The madman in the tower. *Texas Monthly.*

Hendricks, G., Malcolm-Smith, S., Adnams, C., Stein, D. J., & Donald, K. A. M. (2019). Effects of prenatal alcohol exposure on language, speech and communication outcomes: A review longitudinal studies. *Acta Neuropsychiatrica, 31*(2), 74–83.

Heni, M., Kullmann, S., Preissl, H., Fritsche, A., & Häring, H. U. (2015). Impaired insulin action in the human brain: Causes and metabolic consequences. *Nature Reviews Endocrinology, 11*(12), 701–711.

Henley, C. L., Nunez, A. A., & Clemens, L. G. (2011). Hormones of choice: The neuroendocrinology of partner preference in animals. *Frontiers in Neuroendocrinology, 32,* 146–154.

Henriques, J. F., Portugal, C. C., Canedo, T., Relvas, J. B., Summavielle, T., & Socodato, R. (2018). Microglia and alcohol meet at the crossroads: Microglia as critical modulators of alcohol neurotoxicity. *Toxicology Letters, 283,* 21–31.

Herlin, B., Leu-Semenescu, S., Chaumereuil, C., & Arnulf, I. (2015). Evidence that non-dreamers do dream: A REM sleep behaviour disorder model. *Journal of Sleep Research, 24*(6), 602–609.

Herman, J. P., McKlveen, J. M., Solomon, M. B., Carvalho-Netto, E., & Myers, B. (2011). Neural regulation of the stress response: Glucocorticoid feedback mechanisms. *Brazilian Journal of Medical and Biological Research, 45,* 292–298.

Herman, R., He, J., D'Luzansky, S., Willis, W., & Dilli, S. (2002). Spinal cord stimulation facilitates functional walking in a chronic, incomplete spinal cord injured. *Spinal Cord, 2,* 65–68.

Hernández-Fonseca, K., Massieu, L., de la Cadena, S. G., Guzmán, C., & Carnacho-Arroyo, I. (2012). Neuroprotective role of estradiol against neuronal death induced by glucose deprivation in cultured rat hippocampal neurons. *Neuroendocrinology, 96,* 41–50.

Hernandez, G., Hamdani, S., Rajabi, H., Conover, K., Stewart, J., Arvanitogiannis, A., & Shizgal, P. (2006). Prolonged rewarding stimulation of the rat medial forebrain bundle: Neurochemical and behavioral consequences. *Behavioral Neuroscience, 120,* 888–904.

Herrup, K. (2015). The case for rejecting the amyloid cascade hypothesis. *Nature Neuroscience, 18,* 794–798.

Herry, C., & Johansen, J. P. (2013). Encoding of fear learning and memory in distributed neuronal circuits. *Nature Neuroscience, 17,* 1644–1654.

Herry, C., & Johansen, J. P. (2014). Encoding of fear learning and memory in distributed neuronal circuits. *Nature Neuroscience, 17,* 1644–1654.

Hertzog, C., Kramer, A. F., Wilson, R. S., & Lindenberger, U. (2008). Enrichment effects on adult cognitive development: Can the functional capacity of older adults be preserved or enhanced? *Psychological Science in the Public Interest, 9,* 1–65.

Hervé, P.-Y., Zago, L., Petit, L., Mazoyer, B., & Tzourio-Mazoyer, N. (2013). Revisiting human hemispheric specialization with neuroimaging. *Trends in Cognitive Sciences, 17,* 69–80.

Herzfeld, D. J., & Shadmehr, R. (2014). Cerebellum estimates the sensory state of the body. *Trends in Cognitive Sciences, 18,* 66–67.

Herzfeld, D. J., Kojima, Y., Soetedjo, R., & Shadmehr, R. (2018). Encoding of error and learning to correct that error by the Purkinje cells of the cerebellum. *Nature Neuroscience, 21*(5), 736–743.

Hestrin, S., & Galarreta, M. (2005). Electrical synapses define networks of neocortical GABAergic neurons. *Trends in Neurosciences, 28,* 304–309.

Hetherington, A. W., & Ranson, S. W. (1940). Hypothalamic lesions and adiposity in the rat. *Anatomical Record, 78,* 149–172.

Heuser, J. E., Reese, T. S., Dennis, M. J., Jan, Y., Jan, L., & Evans, L. (1979). Synaptic vesicle exocytosis captured by quick freezing and correlated with quantal transmitter release. *Journal of Cell Biology, 81,* 275–300.

Heyes, C. (2010). Where do mirror neurons come from? *Neuroscience and Biobehavioral Reviews, 34,* 575–583.

Heymsfield, S. B., Greenberg, A. S., Fujioka, K., Dixon, R. M., Kushner, R., Hunt, T., . . . & McCamish, M. (1999). Recombinant leptin for weight loss in obese and lean adults. *Journal of the American Medical Association, 282*, 1568–1575.

Hickman, R. A., Faustin, A., & Wisniewski, T. (2016). Alzheimer disease and its growing epidemic: Risk factors, biomarkers, and the urgent need for therapeutics. *Neurologic Clinics, 34*(4), 941–953.

Hickok, G., Okada, K., Barr, W., Pa, J., Rogalsky, C., Donnelly, K., . . . & Grant, A. (2008). Bilateral capacity for speech sound processing in auditory comprehension: Evidence from Wada procedures. *Brain and Language, 107*, 179–184.

Higgins, S. T., Heil, S. H., & Lussier, J. P. (2004). Clinical implications of reinforcement as a determinant of substance use disorders. *Annual Review of Psychology, 55*, 431–461.

Higgs, S., Williamson, A. C., Rotshtein, P., & Humphreys, G. W. (2008). Sensory-specific satiety is intact in amnesics who eat multiple meals. *Psychological Science, 19*, 623–628.

Hikosaka, O., Kim, H. F., Yasuda, M., & Yamamoto, S. (2014). Basal ganglia circuits for reward value-guided behavior. *Annual Review of Neuroscience, 37*, 289–306.

Hilditch, C. J., & McHill, A. W. (2019). Sleep inertia: Current insights. *Nature and Science of Sleep, 11*, 155–165.

Hill, M. (2015). Perspective: Be clear about the real risks. *Nature, 525*, S14.

Hillier, D., Fiscella, M., Drinnenberg, A., Trenholm, S., Rompani, S. B., Raics, Z., . . . & Roska, B. (2017). Causal evidence for retina-dependent and-independent visual motion computations in mouse cortex. *Nature Neuroscience, 20*(7), 960–968.

Hillman, E. M. C. (2014). Coupling mechanism and significance of the BOLD signal: A status report. *Annual Review of Neuroscience, 37*, 161–181.

Hills, A. P., Mokhtar, N., & Byrne, N. M. (2014). Assessment of physical activity and energy expenditure: An overview of objective measures. *Frontiers in Nutrition, 1*, 5.

Himmelbach, M., & Karnath, H.-O. (2005). Dorsal and ventral stream interaction: Contributions from optic ataxia. *Journal of Cognitive Neuroscience, 17*, 632–640.

Hines, M. (2011). Gender development and the human brain. *Annual Review of Neuroscience, 34*, 69–88.

Hinnebusch, A. G. (2015). Blocking stress response for better memory? *Science, 348*, 967–968.

Hirnstein, M., Hugdahl, K., & Hausmann, M. (2019). Cognitive sex differences and hemispheric asymmetry: A critical review of 40 years of research. *Laterality: Asymmetries of Body, Brain and Cognition, 24*(2), 204–252.

Hiser, J., & Koenigs, M. (2018). The multifaceted role of the ventromedial prefrontal cortex in emotion, decision making, social cognition, and psychopathology. *Biological Psychiatry, 83*(8), 638–647.

Hitti, F. L., & Siegelbaum, S. A. (2014). The hippocampal CA2 region is essential for social memory. *Nature, 508*, 88–92.

Hobson, J. A. (1989). *Sleep.* New York, NY: Scientific American Library.

Hobson, J. A. (2009). REM sleep and dreaming: Towards a theory of protoconsciousness. *Nature Reviews Neuroscience, 10*(11), 803–813.

Hobson, J. A., & Friston, K. J. (2012). Waking and dreaming consciousness: Neurobiological and functional considerations. *Progress in Neurobiology, 98*(1), 82–98.

Hobson, J. A., & Pace-Schott, E. F. (2002). The cognitive neuroscience of sleep: Neuronal systems, consciousness and learning. *Nature Reviews Neuroscience, 3*, 679–693.

Hoeft, F., Meyler, A., Hernandez, A., Juel, C., Taylor-Hill, H., Martindale, J. L., . . . & Deutsch, G. K. (2007). Functional and morphometric brain dissociation between dyslexia and reading ability. *Proceedings of the National Academy of Sciences, 104*, 4234–4239.

Hoeren, M., Kümmerer, D., Bormann, T., Beume, L., Ludwig, V. M., . . . & Weiller, C. (2014). Neural bases of imitation and pantomime in acute stroke patients: Distinct streams for praxis. *Brain, 137*, 2796–2810.

Hofer, S. B., Mrsic-Flogel, T. D., Bonhoeffer, T., & Hübener, M. (2005). Prior experience enhances plasticity in adult visual cortex. *Nature Neuroscience, 9*, 127–132.

Hofer, S. B., Mrsic-Flogel, T. D., Bonhoeffer, T., & Hübener, M. (2006). Lifelong learning: Ocular dominance plasticity in mouse visual cortex. *Current Opinion in Neurobiology, 16*, 451–459.

Hoffman, H., Peterson, K., & Garner, H. (2012). Field conditioning of sexual arousal in humans. *Socioaffective Neuroscience and Psychology, 2*, 17336.

Hoffmann, S., & Beste, C. (2015). A perspective on neural and cognitive mechanisms of error commission. *Frontiers in Behavioral Neuroscience, 9*, 50.

Hofman, M. A. (2014). Evolution of the human brain: When bigger is better. *Frontiers in Neuroanatomy, 8*, 1–12.

Högl, B., Stefani, A., & Videnovic, A. (2018). Idiopathic REM sleep behaviour disorder and neurodegeneration—an update. Nature Reviews Neurology, 14(1), 40.

Holland, L., De Regt, H. W., & Drukarch, B. (2019). Thinking about the nerve impulse: The prospects for the development of a comprehensive account of nerve impulse propagation. *Frontiers in Cellular Neuroscience, 13*, 208.

Hollands, G. J., Shemilt, I., Marteau, T. M., Jebb, S. A., Lewis, H. B., Wei, Y., Higgins, J. P., & Ogilvie, D. (2015). Portion, package or tableware size for changing selection and consumption of food, alcohol, and tobacco. *Cochrane Database of Systematic Reviews, 9*, CD011045.

Hollenbeck, P. J. (2001). Insight and hindsight into Tourette syndrome. In D. J. Cohen, C. G. Goetz, & J. Jankovic (Eds.), *Tourette syndrome* (pp. 363–367). Philadelphia, PA: Lippincott Williams & Wilkins.

Hollins, M. (2010). Somesthetic senses. *Annual Review of Psychology, 61*, 243–271.

Holm, M. M., Kaiser, J., & Schwab, M. E. (2018). Extracellular vesicles: Multimodal envoys in neural maintenance and repair. *Trends in Neurosciences, 41*, 360–370.

Holth, J. K., Fritschi, S. K., Wang, C., Pedersen, N. P., Cirrito, J. R., Mahan, T. E., . . . & Lucey, B. P. (2019). The sleep-wake cycle regulates brain interstitial fluid tau in mice and CSF tau in humans. *Science, 363*(6429), 880–884.

Holtmaat, A., & Svoboda, K. (2009). Experience-dependent structural synaptic plasticity in the mammalian brain. *Nature Reviews Neuroscience, 10*, 647–658.

Holtzheimer, P. E., Kelley, M. E., Gross, R. E., Filkowski, M. M., Garlow, S. J., Barrocas, A., Wint, D., Craighead, M. C., Kozarsky, J., Chismar, R., Moreines, J. L., Mewes, K., Posse, P. R., Gutman, D. A., & Mayberg, H. S. (2011). Subcallosal cingulate deep brain stimulation for treatment-resistant unipolar and bipolar depression. *JAMA Psychiatry, 70*, 373–382.

Homberg, J. R., Molteni, R., Calabrese, F., & Riva, M. A. (2014). The serotonin–BDNF duo: Developmental implications for the vulnerability to psychopathology. *Neuroscience & Biobehavioral Reviews, 43*, 35–47.

Hoon, M., Okawa, H., Della Santina, L., & Wong, R. O. L. (2014). Functional architecture of the retina: Development and disease. *Progress in Retinal and Eye Research, 42*, 44–84.

Hoops, D., & Flores, C. (2017). Making dopamine connections in adolescence. *Trends in Neurosciences, 40*(12), 709–719.

Horikawa, T., Tamaki, M., Miyawaki, Y., & Kamitani, Y. (2013). Neural decoding of visual imagery during sleep. *Science, 340*, 639–642.

Horne, J. (2010). The end of sleep: 'Sleep debt' versus biological adaptation of human sleep to waking needs. *Biological Psychology, 87*, 1–14.

Horne, J. (2011). Sleepiness as a need for sleep: When is enough, enough? *Neuroscience and Biobehavioral Reviews, 34*, 108–118.

Horne, J. (2013). Why REM sleep? Clues beyond the laboratory in a more challenging world. *Biological Psychology, 92*, 152–168.

Horton, J. (2009). Akinetopsia from nefazodone toxicity. *American Journal of Ophthalmology, 128*, 530–531.

Horton, J. C., & Sincich, L. C. (2004). A new foundation for the visual cortical hierarchy. In M. Gazzaniga (Ed.), *The cognitive neurosciences* (pp. 233–244). Cambridge, MA: MIT Press.

Hoshi, E., & Tanji, J. (2007). Distinctions between dorsal and ventral premotor areas: Anatomical connectivity and functional properties. *Current Opinion in Neurobiology, 17*, 234–242.

Hostinar, C. E., Sullivan, R. M., & Gunnar, M. R. (2014). Psychobiological mechanisms underlying the social buffering of the hypothalamic–pituitary–adrenocortical axis: A review of animal models and human studies across development. *Psychological Bulletin, 140*, 256.

Howald, C., Merla, G., Digilio, M. C., Amenta, S., Lyle, R., Deutsch, S., . . . & Reymond, A. (2006). Two high throughput technologies to detect segmental aneuploidies identify new Williams-Beuren syndrome patients with atypical deletions. *Journal of Medical Genetics, 43*, 266–273.

Howard-Jones, P. A. (2014). Neuroscience and education: Myths and messages. *Nature Reviews Neuroscience, 15*, 817–824.

Howarth, C. (2014). The contribution of astrocytes to the regulation of cerebral blood flow. *Frontiers in Neuroscience, 8*, 103.

Howells, D. W., Sena, E. S., & MacLeod, M. R. (2014). Bringing rigour to translational medicine. *Nature Reviews Neurology, 10*, 37–43.

Hrabovszky, Z., & Hutson, J. M. (2002). Androgen imprinting of the brain in animal models and humans with intersex disorders: Review and recommendations. *The Journal of Urology, 168*, 2142–2148.

Hrvatin, S., Hochbaum, D. R., Nagy, M. A., Cicconet, M., Robertson, K., Cheadle, L., . . . & Sabatini, B. L. (2018). Single-cell analysis of experience-dependent transcriptomic states in the mouse visual cortex. *Nature Neuroscience, 21*(1), 120–129.

Hsieh, J. H., Stein, D. J., & Howells, F. M. (2014). The neurobiology of methamphetamine induced psychosis. *Frontiers in Human Neuroscience, 8*, 537.

Hsieh, J., & Schneider, J. W. (2013). Neural stem cells, excited. *Science, 339*, 1533–1534.

Hsu, C. T., Jacobs, A. M., Altmann, U., & Conrad, M. (2015). The magical activation of left amygdala when reading Harry Potter: An fMRI study on how descriptions of supra-natural events entertain and enchant. *PLoS One, 10*, e0118179.

Huang, Y., & Mucke, L. (2012). Alzheimer mechanisms and therapeutic strategies. *Cell, 148*, 1204–1222.

Hubel, D. H., & Wiesel, T. N. (1979, September). Brain mechanisms of vision. *Scientific American, 241*, 150–162.

Hubel, D. H., & Wiesel, T. N. (2004). *Brain and visual perception: The story of a 25-year collaboration.* New York: Oxford University Press.

Hubel, D. H., Wiesel, T. N., & LeVay, S. (1977). Plasticity of ocular dominance columns in the monkey striate cortex. *Philosophical Transactions of the Royal Society of London, 278*, 377–409.

Huber, D., Gutnisky, D. A., Peron, S., O'Connor, D. H., Wiegert, J. S., Tian, L., Oertner, T. G., Looger, L. L., & Svoboda, K. (2012). Multiple dynamic representations in the motor cortex during sensorimotor learning. *Nature, 484,* 473–478.

Huber, M. A., & Tantiwongkosi, B. (2014). Oral and oropharyngeal cancer. *Medical Clinics of North America, 6,* 1299–1321.

Hudson, A. N., Van Dongen, H. P., & Honn, K. A. (2020). Sleep deprivation, vigilant attention, and brain function: A review. *Neuropsychopharmacology, 45*(1), 21–30.

Hudson, J. I., Hiripi, E., Pope, H. G., & Kessler, R. C. (2007). The prevalence and correlates of eating disorders in the National Comorbidity Survey Replication. *Biological Psychiatry, 61,* 348–358.

Hudspeth, A. J. (2014). Integrating the active process of hair cells with cochlear function. *Nature Reviews Neuroscience, 15,* 600–614.

Huffman, K. M., Redman, L. M., Landerman, L. R., Pieper, C. F., Stevens, R. D., Muehlbauer, M. J., Wenner, B. R., Bain, J. R., Kraus, V. B., Newgard, C. B., Ravussin, E., & Kraus, W. E. (2012). Caloric restriction alters the metabolic response to a mixed-meal: Results from a randomized, controlled trial. *PLoS ONE, 7,* e28190.

Hughes, V. (2014). A gut-wrenching question. *Nature, 511,* 282–284.

Hughes, V. (2014). The sins of the father. *Nature, 507,* 22–24.

Hull, E. M., & Dominguez, J. M. (2007). Sexual behavior in male rodents. *Hormones and Behavior, 52,* 45–55.

Hulme, S. R., Jones, O. D., & Abraham, W. C. (2013). Emerging role of metaplasticity in behaviour and disease. *Trends in Neurosciences, 36,* 353–362.

Humbert-Claude, M., Davenas, E., Gbahou, F., Vincent, L., & Arrang, J.-M. (2012). Involvement of histamine receptors in the atypical antipsychotic profile of clozapine: A reassessment in vitro and in vivo. *Psychopharmacology, 220,* 225–241.

Hunter, A., Murray, R., Asher, L., & Leonardi-Bee, J. (2020). The effects of tobacco smoking, and prenatal tobacco smoke exposure, on risk of schizophrenia: A systematic review and meta-analysis. *Nicotine and Tobacco Research, 22*(1), 3–10.

Hunter, G. (2011). Transient global amnesia. *Neurologic Clinics, 29,* 1045–1054.

Hurlemann, R., Wagner, M., Hawellek, B., Reich, H., Pieperhoff, P., Amunts, K.,...& Dolan, R. J. (2007). Amygdala control of emotion-induced forgetting and remembering: Evidence from Urbach-Wiethe disease. *Neuropsychologia, 45,* 877–884.

Husain, M., & Roiser, J. P. (2018). Neuroscience of apathy and anhedonia: A transdiagnostic approach. *Nature Reviews Neuroscience, 19*(8), 470–484.

Hussain, M., & Mehta, M. A. (2011). Cognitive enhancement by drugs in health and disease. *Trends in Cognitive Science, 15,* 28–36.

Hutchinson, J. B., Uncapher, M. R., Weiner, K. S., Bressler, D. W., Silver, M. A.,...& Wagner, A. D. (2014). Functional heterogeneity in posterior parietal cortex across attention and episodic memory retrieval. *Cerebral Cortex, 24,* 49–66.

Hyde, J. S. (2005). The gender similarities hypothesis. *American Psychologist, 60,* 581–592.

Hyman, S. E., & Fenton, W. S. (2003). What are the right targets for psychopharmacology? *Science, 299,* 350–351.

Iacaruso, M. F., Gasler, I. T., & Hofer, S. B. (2017). Synaptic organization of visual space in primary visual cortex. *Nature, 547*(7664), 449–452.

Iacono, W. G., & Ben-Shakhar, G. (2019). Current status of forensic lie detection with the comparison question technique: An update of the 2003 National Academy of Sciences report on polygraph testing. *Law and Human Behavior, 43*(1), 86–98.

Iacono, W. G., & Koenig, W. G. R. (1983). Features that distinguish the smooth-pursuit eye-tracking performance of schizophrenic, affective-disorder, and normal individuals. *Journal of Abnormal Psychology, 92,* 29–41.

Iakoucheva, L. M., Muotri, A. R., & Sebat, J. (2019). Getting to the cores of autism. *Cell, 178*(6), 1287–1298.

Ibbotson, M., & Krekelberg, B. (2011). Visual perception and saccadic eye movements. *Current Opinion in Neurobiology, 21,* 553–558.

Iino, M., Goto, K., Kakegawa, W., Okado, H., Sudo, M., Ishiuchi, S.,...& Ozawa, S. (2001). Glia-synapse interaction through calcium-permeable AMPA receptors in Bergmann glia. *Science, 292,* 926–929.

Illert, M., & Kümmel, H. (1999). Reflex pathways from large muscle spindle afferents and recurrent axon collaterals to motoneurones of wrist and digit muscles: A comparison in cats, monkeys and humans. *Experimental Brain Research, 128,* 13–19.

Inestrosa, N. C., & Arenas, E. (2010). Emerging roles of Wnts in the adult nervous system. *Nature Reviews Neuroscience, 11,* 77–86.

Ingalhalikar, M., Smith, A., Parker, D., Satterthwaite, T. D., Elliott, M. A.,...& Verma, R. (2014). Sex differences in the structural connectome of the human brain. *PNAS, 111,* 823–828.

Inglis, J., & Lawson, J. S. (1982). A meta-analysis of sex differences in the effects of unilateral brain damage on intelligence test results. *Canadian Journal of Psychology, 36,* 670–683.

Innocenti, G. M., & Price, D. J. (2005). Exuberance in the development of cortical networks. *Nature Reviews Neuroscience, 6,* 955–965.

Inostroza, M., & Born, J. (2013). Sleep for preserving and transforming episodic memory. *Annual Review of Neuroscience, 36,* 79–102.

Inoue, K., & Tsuda, M. (2018). Microglia in neuropathic pain: Cellular and molecular mechanisms and therapeutic potential. *Nature Reviews Neuroscience, 19,* 138–152.

Insel, T. R., & Cuthbert, B. N. (2015). Brain disorders? Precisely. *Science, 348,* 499–500.

Inta, D., Cameron, H. A., & Gass, P. (2015). New neurons in the adult striatum: From rodents to humans. *Trends in Neurosciences, 38,* 517–523.

Interlandi, J. (2013). Breaking the brain barrier. *Scientific American, 308,* 52–57.

Iqbal, K., Liu, F., & Gong, C. X. (2016). Tau and neurodegenerative disease: The story so far. *Nature Reviews Neurology, 12,* 15–27.

Iremonger, K. J., Wamsteeker Cusulin, J. I., & Bains, J. S. (2013). Changing the tune: Plasticity and adaptation of retrograde signals. *Trends in Neurosciences, 8,* 471–479.

Irwin, M. R., Olmstead, R., & Carroll, J. E. (2016). Sleep disturbance, sleep duration, and inflammation: A systematic review and meta-analysis of cohort studies and experimental sleep deprivation. *Biological Psychiatry, 80,* 40–52.

Isoda, M., & Noritake, A. (2013). What makes the dorsomedial frontal cortex active during reading the mental states of others? *Frontiers in Neuroscience, 7,* 232.

Ito, K., & Suda, T. (2014). Metabolic requirements for the maintenance of self-renewing stem cells. *Nature Reviews Molecular Cell Biology, 15,* 243–256.

Itoh, Y., Tyssowski, K., & Gotoh, Y. (2013). Transcriptional coupling of neuronal fate commitment and the onset of migration. *Current Opinion in Neurobiology, 23,* 957–964.

Iwaniuk, A. N., & Whishaw, I. Q. (2000). On the origin of skilled forelimb movements. *Trends in Neurosciences, 23,* 372–376.

Iyer, A., & Marson, A. (2014). Pharmacotherapy for focal epilepsy. *Expert Opinion on Pharmacotherapy, 15,* 1543–1551.

Izaurralde, E. (2015). Breakers and blockers—miRNAs at work. *Science, 349,* 380–382.

Jabr, F. (2013). Gambling on the brain. *Scientific American, 309,* 28–30.

Jack, R. E., & Schyns, P. G. (2017). Toward a social psychophysics of face communication. *Annual Review of Psychology, 68,* 269–297.

Jack, R. E., Garrod, O. G., Yu, H., Caldara, R., & Schyns, P. G. (2012). Facial expressions of emotion are not culturally universal. *Proceedings of the National Academy of Sciences, 109,* 7241–7244.

Jackson, A. (2016). Spinal-cord injury: Neural interfaces take another step forward. *Nature, 539*(7628), 177.

Jackson, G. M., Draper, A., Dyke, K., Pépés, S. E., & Jackson, S. R. (2015). Inhibition, disinhibition, and the control of action in Tourette syndrome. *Trends in Cognitive Sciences, 19,* 655–665.

Jacobs, G. H., Williams, G. A., Cahill, H., & Nathans, J. (2007). Emergence of novel color vision in mice engineered to express a human cone photopigment. *Science, 315,* 1723–1725.

Jaeger, J.-J., & Marivaux, L. (2005). Shaking the earliest branches of anthropoid primate evolution. *Science, 310,* 244–245.

Jaepel, J., Hübener, M., Bonhoeffer, T., & Rose, T. (2017). Lateral geniculate neurons projecting to primary visual cortex show ocular dominance plasticity in adult mice. *Nature Neuroscience, 20*(12), 1708–1714.

Jaffe, A. E., Gao, Y., Deep-Soboslay, A., Tao, R., Hyde, T. M., Weinberger, D. R., & Kleinman, J. E. (2016). Mapping DNA methylation across development, genotype and schizophrenia in the human frontal cortex. *Nature Neuroscience, 19,* 40–47.

Jaffe, A. E., Straub, R. E., Shin, J. H., Tao, R., Gao, Y., Collado-Torres, L., . . . & Colantuoni, C. (2018). Developmental and genetic regulation of the human cortex transcriptome illuminate schizophrenia pathogenesis. *Nature Neuroscience, 21*(8), 1117–1125.

Jager, A., Maas, D. A., Fricke, K., de Vries, R. B., Poelmans, G., & Glennon, J. C. (2018). Aggressive behavior in transgenic animal models: A systematic review. *Neuroscience & Biobehavioral Reviews, 91,* 198–217.

Jameson, K. A., Highnote, S. M., & Wasserman, L. M. (2001). Richer color experience in observers with multiple photopigment opsin genes. *Psychonomic Bulletin & Review, 8*(2), 244–261.

Janak, P. H., & Tye, K. M. (2015). From circuits to behaviour in the amygdala. *Nature, 517,* 284–292.

Jankovic, J. (2017). Movement disorders in 2016: Progress in Parkinson disease and other movement disorders. *Nature Reviews Neurology, 13*(2), 76–77.

Jann, M. W. (2014). Diagnosis and treatment of bipolar disorders in adults: A review of the evidence on pharmacologic treatments. *American Health & Drug Benefits, 7,* 489.

Januszko, P., Niemcewicz, S., Gajda, T., Wolyńczyk-Gmaj, D., Piotrowska, A. J.,...& Szelenberger, W. (2015). Sleepwalking episodes are preceded by arousal-related activation in the cingulate motor area: EEG current density imaging. *Clinical Neurophysiology, 127,* 530–536.

Jarvie, B. C., & Palmiter, R. D. (2017). HSD2 neurons in the hindbrain drive sodium appetite. *Nature Neuroscience, 20*(2), 167–169.

Järvinen, A., Korenberg, J. R., & Bellugi, U. (2013). The social phenotype of Williams syndrome. *Current Opinion in Neurobiology, 23,* 414–422.

Järvinen, A., Ng, R., Crivelli, D., Neumann, D., Arnold, A. J.,...& Bellugi, U. (2015). Social functioning and autonomic nervous system sensitivity across vocal and musical emotion in Williams syndrome and autism spectrum disorder. *Developmental Psychobiology, 58,* 17–26.

Jasmin, K., Lima, C. F., & Scott, S. K. (2019). Understanding rostral–caudal auditory cortex contributions to auditory perception. *Nature Reviews Neuroscience, 20,* 425–434.

Jbabdi, S., Sotiropoulos, S. N., Haber, S. N., Van Essen, D. C., & Behrens, T. E. (2015). Measuring macroscopic brain connections in vivo. *Nature Neuroscience, 18,* 1546–1555.

Jeannerod, M., Arbib, M. A., Rizzolatti, G., & Sakarta, H. (1995). Grasping objects: The cortical mechanisms of visuomotor transformation. *Trends in Neurosciences, 18*(7), 314–327.

Jenkins, I. H., Brooks, D. J., Bixon, P. D., Frackowiak, R. S. J., & Passingham, R. E. (1994). Motor sequence learning: A study with positron emission tomography. *Journal of Neuroscience, 14*(6), 3775–3790.

Jensen, S. K., & Yong, V. W. (2016). Activity-dependent and experience-driven myelination provide new directions for the management of multiple sclerosis. *Trends in Neurosciences, 39*(6), 356–365.

Jerath, R., & Crawford, M. W. (2014). Neural correlates of visuospatial consciousness in 3D default space: Insights from contralateral neglect syndrome. *Consciousness and Cognition, 28*, 81–93.

Jernigan, T. L., Baaré, W. F., Stiles, J., & Madsen, K. S. (2011). Postnatal brain development: Structural imaging of dynamic neurodevelopmental process. *Progress in Brain Research, 189*, 77–92.

Jetté, N., Sander, J. W., & Keezer, M. R. (2016). Surgical treatment for epilepsy: The potential gap between evidence and practice. *The Lancet Neurology, 15*(9), 982–994.

Jiang, X., Andjelkovic, A. V., Zhu, L., Yang, T., Bennett, M. V., Chen, J., . . . & Shi, Y. (2018). Blood-brain barrier dysfunction and recovery after ischemic stroke. *Progress in Neurobiology, 163*, 144–171.

Jin, J., & Maren, S. (2015). Prefrontal-hippocampal interactions in memory and emotion. *Frontiers in Systems Neuroscience, 9*, 170.

Jin, X., & Yamashita, T. (2016). Microglia in central nervous system repair after injury. *Journal of Biochemistry, 159*, 491–496.

Joel, D., & Fausto-Sterling, A. (2016). Beyond sex differences: New approaches for thinking about variation in brain structure and function. *Philosophical Transactions of the Royal Society B: Biological Sciences, 371*(1688), 20150451.

Joesch, M., & Meister, M. (2016). A neuronal circuit for colour vision based on rod–cone opponency. *Nature, 532*(7598), 236–239.

Johnson, A. W. (2013). Eating beyond metabolic need: How environmental cues influence feeding behavior. *Trends in Neurosciences, 36*, 101–109.

Johnson, P. L., Federici, L. M., & Shekhar, A. (2014). Etiology, triggers and neurochemical circuits associated with unexpected, expected, and laboratory-induced panic attacks. *Neuroscience & Biobehavioral Reviews, 46*, 429–454.

Johnson, S. L., Mehta, H., Ketter, T. A., Gotlib, I. H., & Knutson, B. (2019). Neural responses to monetary incentives in bipolar disorder. *NeuroImage: Clinical, 24*, 102018.

Johnson, W., Turkheimer, E., Gottesmann, I. I., & Bouchard, T. J. (2009). Beyond heritability: Twin studies in behavioral research. *Current Directions in Psychological Science, 18*, 217–220.

Johnston, T. D. (1987). The persistence of dichotomies in the study of behavioral development. *Developmental Review, 7*, 149–182.

Joiner, W. J. (2016). Unraveling the evolutionary determinants of sleep. *Current Biology, 26*(20), R1073–R1087.

Joiner, W. M., Cavanaugh, J., FitzGibbon, E. J., & Wurtz, R. H. (2013). Corollary discharge contributes to perceived eye location in monkeys. *Journal of Neurophysiology, 110*, 2402–2413.

Jonas, J., Maillard, L., Frismand, S., Colnat-Coulbois, S., Vespignani, H., Rossion, B., & Vignal, J. P. (2014). Self-face hallucinations evoked by electrical stimulation of the human brain. *Neurology, 83*, 336–338.

Jones, H. W., & Park, I. J. (1971). A classification of special problems in sex differentiation. In D. Bergsma (Ed.), *The clinical delineation of birth defects. Part X: The endocrine system* (pp. 113–121). Baltimore, MD: Williams and Wilkins.

Jones, T. A. (2017). Motor compensation and its effects on neural reorganization after stroke. *Nature Reviews Neuroscience, 18*(5), 267.

Jones, W., & Klin, A. (2014). Attention to eyes is present but in decline in 2–6-month-old infants later diagnosed with autism. *Nature, 504*, 427–431.

Jordan-Young, R. M. (2012). Hormones, context, and "brain gender": A review of evidence from congenital adrenal hyperplasia. *Social Science & Medicine, 74*, 1738–1744.

Jordan-Young, R., & Rumiati. R. I. (2012). Hardwired for sexism? Approaches to sex/gender in neuroscience. *Neuroethics, 5*, 305–315.

Joseph, R. (1988). Dual mental functioning in a split-brain patient. *Journal of Clinical Psychology, 44*, 771–779.

Josselyn, S. A., Köhler, S., & Frankland, P. W. (2015). Finding the engram. *Nature Reviews Neuroscience, 16*, 521–534.

Jost, A. (1972). A new look at the mechanisms controlling sex differentiation in mammals. *Johns Hopkins Medical Journal, 130*, 38–53.

Jucker, M., & Walker, L. C. (2018). Propagation and spread of pathogenic protein assemblies in neurodegenerative diseases. *Nature Neuroscience, 21*(10), 1341–1349.

Julien, R. M. (1981). *A primer of drug action.* San Francisco, CA: W. H. Freeman.

Jun, H. J., Joshi, Y., Patil, Y., Noland, R. C., & Chang, J. S. (2014). NT-PGC-1 activation attenuates high-fat diet-induced obesity by enhancing brown fat thermogenesis and adipose tissue oxidative metabolism. *Diabetes, 63*, 3615–3625.

Jun, J. J., Steinmetz, N. A., Siegle, J. H., Denman, D. J., Bauza, M., Barbarits, B., . . . & Barbic, M. (2017). Fully integrated silicon probes for high-density recording of neural activity. *Nature, 551*(7679), 232.

Jung, Y., Hsieh, L. S., Lee, A. M., Zhou, Z., Coman, D., Heath, C. J., . . . & Bordey, A. (2016). An epigenetic mechanism mediates developmental nicotine effects on neuronal structure and behavior. *Nature Neuroscience, 19*(7), 905–914.

Junko, I., Okamoto-Uchida, Y., Nishimura, A., & Hirayama, J. (2019). Light-dependent regulation of circadian clocks in vertebrates. In *Chronobiology-The Science of Biological Time Structure.* IntechOpen.

Kaas, J. H., & Collins, C. E. (2001). The organization of sensory cortex. *Current Opinion in Neurobiology, 11*, 498–504.

Kaas, J. H., Krubtzer, L. A., Chino, Y. M., Langston, A. L., Polley, E. H., & Blair, N. (1990). Reorganization of retinotopic cortical maps in adult mammals after lesions of the retina. *Science, 248*, 229–231.

Kagawa, Y. (1978). Impact of Westernization on the nutrition of Japanese: Changes in physique, cancer, longevity, and centenarians. *Preventive Medicine, 7*, 205–217.

Kahan, T. L., & LaBerge, S. P. (2011). Dreaming and waking: Similarities and differences revisited. *Consciousness and Cognition, 20*(3), 494–514.

Kahn, O. I., & Baas, P. W. (2016). Microtubules and growth cones: Motors drive the turn. *Trends in Neurosciences, 39*(7), 433–440.

Kaila, K., Ruusuvuori, E., Seja, P., Voipio, J., & Puskarjov, M. (2014). GABA actions and ionic plasticity in epilepsy. *Current Opinion in Neurobiology, 26*, 34–41.

Kaiser, J. (2010). Cancer's circulation problem. *Science, 327*, 1072–1074.

Kalia, L. V., & Lang, A. E. (2015). Parkinson's disease. *Lancet, 386*, 896–912.

Kalil, K., & Dent, E. W. (2005). Touch and go: Guidance cues signal to the growth cone cytoskeleton. *Current Opinion in Neurobiology, 15*, 521–526.

Kalivas, P. W. (2005). How do we determine which drug-induced neuroplastic changes are important? *Nature Neuroscience, 8*, 1440–1441.

Kaller, C. P., Rahm, B., Spreer, J., Weiller, C., & Unterrainer, J. M. (2011). Dissociable contributions of left and right dorsolateral prefrontal cortex in planning. *Cerebral Cortex, 21*, 307–317.

Kaller, M. S., Lazari, A., Blanco-Duque, C., Sampaio-Baptista, C., & Johansen-Berg, H. (2017). Myelin plasticity and behaviour—connecting the dots. *Current Opinion in Neurobiology, 47*, 86–92.

Kalynchuk, L. E., Pinel, J. P. J., Treit, D., & Kippin, T. E. (1997). Changes in emotional behavior produced by long-term amygdala kindling in rats. *Biological Psychiatry, 41*, 438–451.

Kanai, R., & Rees, G. (2011). The structural basis of inter-individual differences in human behaviour and cognition. *Nature Reviews Neuroscience, 12*, 231–242.

Kanayama, G., Hudson, J. I., & Pope, H. G. (2010). Illicit anabolic-androgenic steroid use. *Hormones and Behavior, 58*, 111–121.

Kandel, E. R., & Squire, L. R. (2000). Neuroscience: Breaking down barriers to the study of brain and mind. *Science, 290*, 1113–1120.

Kanekiyo, T., Xu, H., & Bu, G. (2014). ApoE and Aβ in Alzheimer's disease: Accidental encounters or partners? *Neuron, 81*, 740–754.

Kanold, P. O., & Luhmann, H. J. (2010). The subplate and early cortical circuits. *Annual Review of Neuroscience, 33*, 23–48.

Kantarci, O. H. (2008). Genetics and natural history of multiple sclerosis. *Seminars in Neurology, 28*, 7–16.

Kappers, A. M. L. (2011). Human perception of shape from touch. *Philosophical Transactions of the Royal Society B, 366*, 3106–3114.

Kapur, N. (1997). *Injured brains of medical minds.* Oxford, England: Oxford University Press.

Karacan, I., Goodenough, D. R., Shapiro, A., & Starker, S. (1966). Erection cycle during sleep in relation to dream anxiety. *Archives of General Psychiatry, 15*, 183–189.

Karanti, A., Kardell, M., Lundberg, U., & Landén, M. (2016). Changes in mood stabilizer prescription patterns in bipolar disorder. *Journal of Affective Disorders, 195*, 50–56.

Karnath, H. O., & Rorden, C. (2012). The anatomy of spatial neglect. *Neuropsychologia, 50*(6), 1010–1017.

Karnath, H.-O. (2015). Spatial attention systems in spatial neglect. *Neuropsychologia, 75*, 61–73.

Karnath, H.-O., & Rorden, C. (2012). The anatomy of spatial neglect. *Neuropsychologia, 50*, 1010–1017.

Karolis, V. R., Corbetta, M., & De Schotten, M. T. (2019). The architecture of functional lateralisation and its relationship to callosal connectivity in the human brain. *Nature Communications, 10*(1), 1–9.

Karran, E., Mercken, M., & De Strooper, B. (2011). The amyloid cascade hypothesis for Alzheimer's disease: An appraisal for the development of therapeutics. *Nature Reviews Drug Discovery, 10*, 698–712.

Kasai, H., Fukuda, M., Watanabe, S., Hayashi-Takagi, A., & Noguchi, J. (2010). Structural dynamics of dendritic spines in memory and cognition. *Trends in Neurosciences, 33*, 121–130.

Kasai, K., Fukuda, M., Yahata, N., Morita, K., & Fujii, N. (2015). The future of real-world neuroscience: Imaging techniques to assess active brains in social environments. *Neuroscience Research, 90*, 65–71.

Katsumata, N., Kuroiwa, T., Ishibashi, S., Li, S., Endo, S., & Ohno, K. (2006). Heterogeneous hyperactivity and distribution of ischemic lesions after focal cerebral ischemia in Mongolian gerbils. *Neuropathology, 26*, 283–292.

Katz-Talmor, D., Katz, I., Porat-Katz, B. S., & Shoenfeld, Y. (2018). Cannabinoids for the treatment of rheumatic diseases—where do we stand? *Nature Reviews Rheumatology, 14*(8), 488–498.

Katz, J. L., Ackman, P., Rothwax, Y., Sachar, E. J., Weiner, H., Hellman, L., & Gallagher, T. F. (1970). Psychoendocrine aspects of cancer of the breast. *Psychosomatic Medicine, 32*, 1–18.

Katzner, S., & Weigelt, S. (2013). Visual cortical networks: Of mice and men. *Current Opinion in Neurobiology, 23*, 202–206.

Kaviani, S., & Cooper, J. A. (2017). Appetite responses to high-fat meals or diets of varying fatty acid composition: A comprehensive review. *European Journal of Clinical Nutrition, 71*(10), 1154–1165.

Kawai, R., Markman, T., Poddar, R., Ko, R., Fantana, A. L., Dhawale, A. K.,...& Ölveczky, B. P. (2015). Motor cortex is required for learning but not for executing a motor skill. *Neuron, 86*, 800–812.

Kawasaki, H., Adolphs, R., Oya, H., Kovach, C., Damasio, H., Kaufman, O., & Howard, M. (2005). Analysis of single-unit responses to emotional scenes in human ventromedial prefrontal cortex. *Journal of Cognitive Neuroscience, 17*, 1509–1518.

Kay-Stacey, M., & Attarian, H. (2016). Advances in the management of chronic insomnia. *BMJ, 353*, i2123.

Kaye, W. H., Bailer, U. F., Frank, G. K., Wagner, A., & Henry, S. E. (2005). Brain imaging of serotonin after recovery from anorexia and bulimia nervosa. *Physiology & Behavior, 86*, 15–17.

Kaye, W. H., Wierenga, C. E., Bailer, U. F., Simmons, A. N., & Bischoff-Grethe, A. (2013). Nothing tastes as good as skinny feels: The neurobiology of anorexia nervosa. *Trends in Neurosciences, 36*, 110–120.

Kaye, W. H., Wierenga, C. E., Bailer, U. F., Simmons, A. N., Wagner, A., & Bischoff-Grethe, A. (2013). Does a shared neurobiology for foods and drugs of abuse contribute to extremes of food ingestion in anorexia and bulimia nervosa? *Biological Psychiatry, 73*, 836–842.

Kelava, I., & Lancaster, M. A. (2016). Stem cell models of human brain development. *Cell Stem Cell, 18*(6), 736–748.

Kelber, A. (2016). Colour in the eye of the beholder: Receptor sensitivities and neural circuits underlying colour opponency and colour perception. *Current Opinion in Neurobiology, 41*, 106–112.

Kelley, G. A., & Kelley, K. S. (2017). Exercise and sleep: A systematic review of previous meta-analyses. *Journal of Evidence-Based Medicine, 10*(1), 26–36.

Kelly, C., & Castellanos, F. X. (2014). Strengthening connections: Functional connectivity and brain plasticity. *Neuropsychology Review, 24*(1), 63–76.

Kelsch, W., Sim, S., & Lois, C. (2010). Watching synaptogenesis in the adult brain. *Annual Review of Neuroscience, 33*, 131–149.

Keltner, D., Kring, A. M., & Bonanno, G. A. (1999). Fleeting signs of the course of life: Facial expression and personal adjustment. *Current Directions in Psychological Science, 8*, 18–22.

Kempermann, G. (2013). What the bomb said about the brain. *Science, 340*, 1180–1182.

Kempermann, G., & Gage, F. H. (1999, May). New nerve cells for the adult brain. *Scientific American, 282*, 48–53.

Kenakin, T. (2005, October). New bull's-eyes for drugs. *Scientific American, 293*, 51–57.

Kennedy-Costantini, S., Oostenbroek, J., Suddendorf, T., Nielsen, M., Redshaw, J., Davis, J., . . . & Slaughter, V. (2017). There is no compelling evidence that human neonates imitate. *Behavioral and Brain Sciences, 40*.

Keramati, M., Ahmed, S. H., & Gutkin, B. S. (2017). Misdeed of the need: Towards computational accounts of transition to addiction. *Current Opinion in Neurobiology, 46*, 142–153.

Kerr, F., Bjedov, I., & Sofola-Adesakin, O. (2018). Molecular mechanisms of lithium action: Switching the light on multiple targets for dementia using animal models. *Frontiers in Molecular Neuroscience, 11*, 297.

Kerstein, P. C., Nichol, R. H., IV, & Gomez, T. M. (2015). Mechanochemical regulation of growth cone motility. *Frontiers in Cellular Neuroscience, 9*, 244.

Keung, A. J., & Khalil, A. S. (2016). A unifying model of epigenetic regulation. *Science, 351*, 661–662.

Keys, A., Broz, J., Henschel, A., Mickelsen, O., & Taylor H. L. (1950). *The biology of human starvation.* Minneapolis: University of Minnesota Press.

Khadra, A., & Li, Y.-X. (2006). A model for the pulsatile secretion of gonadotropin-releasing hormone from synchronized hypothalamic neurons. *Biophysical Journal, 91*, 74–83.

Khakh, B. S., & Sofroniew, M. V. (2015). Diversity of astrocyte functions and phenotypes in neural circuits. *Nature Neuroscience, 18*, 942–952.

Khan, A. G., & Hofer, S. B. (2018). Contextual signals in visual cortex. *Current Opinion in Neurobiology, 52*, 131–138.

Khan, A. G., Poort, J., Chadwick, A., Blot, A., Sahani, M., Mrsic-Flogel, T. D., & Hofer, S. B. (2018). Distinct learning-induced changes in stimulus selectivity and interactions of GABAergic interneuron classes in visual cortex. *Nature Neuroscience, 21*, 851–859.

Khan, R., Khazaal, Y., Thorens, G., Zullino, D., & Uchtenhagen, A. (2014). Understanding Swiss drug policy change and the introduction of heroin maintenance treatment. *European Addiction Research, 20*, 200–207.

Khan, S. H. (2019). Genome-editing technologies: Concept, pros, and cons of various genome-editing techniques and bioethical concerns for clinical application. *Molecular Therapy: Nucleic Acids, 16*, 326–334.

Khang, P., & Dhand, A. (2015). Teaching video neuroimages: Movement of a paralyzed arm with yawning. *Neurology, 84*, e118.

Kheirbek, M. A., & Hen, R. (2013). (Radio)active neurogenesis in the human hippocampus. *Cell, 153*, 1183–1184.

Khoury, J., & Doghramji, K. (2015). Primary sleep disorders. *Psychiatric Clinics of North America, 38*, 683–704.

Khramtsova, E. A., Davis, L. K., & Stranger, B. E. (2019). The role of sex in the genomics of human complex traits. *Nature Reviews Genetics, 20*(3), 173–190.

Kiecker, C., & Lumsden, A. (2012). The role of organizers in patterning the nervous system. *Annual Review of Neuroscience, 35*, 347–367.

Kiehl, K. A., Liddle, P. F., Smith, A. M., Mendrek, A., Forster, B. B., & Hare, R. D. (1999). Neural pathways involved in the processing of concrete and abstract words. *Human Brain Mapping, 7*, 225–233.

Kiehn, O. (2016). Decoding the organization of spinal circuits that control locomotion. *Nature Reviews Neuroscience, 17*, 224–238.

Kiehn, O. (2016). Decoding the organization of spinal circuits that control locomotion. *Nature Reviews Neuroscience, 17*(4), 224–238.

Killian, N. J., & Buffalo, E. A. (2018). Grid cells map the visual world. *Nature Neuroscience, 21*(2), 161–162.

Kilteni, K., Maselli, A., Kording, K. P., & Slater, M. (2015). Over my fake body: Body ownership illusions for studying the multisensory basis of own-body perception. *Frontiers in Human Neuroscience, 9*, 141.

Kim, C. K., Adhikari, A., & Deisseroth, K. (2017). Integration of optogenetics with complementary methodologies in systems neuroscience. *Nature Reviews Neuroscience, 18*(4), 222.

Kim, D. D., Barr, A. M., Chung, Y., Yuen, J. W., Etminan, M., Carleton, B. C., . . . & Procyshyn, R. M. (2018). Antipsychotic-associated symptoms of Tourette syndrome: A systematic review. *CNS Drugs, 32*(10), 917–938.

Kim, E. J., Horovitz, O., Pellman, B. A., Tan, L. M., Li, Q., Richter-Levin, G., & Kim, J. J. (2013). Dorsal periaqueductal gray-amygdala pathway conveys both innate and learned fear responses in rats. *Proceedings of the National Academy of Sciences, 110*, 14795–14800.

Kim, E. J., Pellman, B., & Kim, J. J. (2015). Stress effects on the hippocampus: A critical review. *Learning & Memory, 22*, 411–416.

Kim, E., Howes, O. D., Veronese, M., Beck, K., Seo, S., Park, J. W., . . . & Kwon, J. S. (2017). Presynaptic dopamine capacity in patients with treatment-resistant schizophrenia taking clozapine: An [18 F] DOPA PET study. *Neuropsychopharmacology, 42*, 941–950.

Kim, J. J., & Jung, M. W. (2018). Fear paradigms: The times they are a-changin'. *Current Opinion in Behavioral Sciences, 24*, 38–43.

Kim, K. S., Seeley, R. J., & Sandoval, D. A. (2018). Signalling from the periphery to the brain that regulates energy homeostasis. *Nature Reviews Neuroscience, 19*(4), 185–196.

Kim, M.-S., Pinto, S. M., Getnet, D., Nirujogi, R. S., Manda, S. S., et al. (2014) A draft map of the human proteome. *Nature, 509*, 575–581.

Kim, S., & Kaang, B. K. (2017). Epigenetic regulation and chromatin remodeling in learning and memory. *Experimental & Molecular Medicine, 49*(1), e281.

Kim, Y. S., & Leventhal, B. L. (2015). Genetic epidemiology and insights into interactive genetic and environmental effects in autism spectrum disorders. *Biological Psychiatry, 77*, 66–74.

Kimble, G. A. (1989). Psychology from the standpoint of a generalist. *American Psychologist, 44*, 491–499.

Kimura, D. (1964). Left-right differences in the perception of melodies. *Quarterly Journal of Experimental Psychology, 16*, 355–358.

Kimura, D. (1973, March). The asymmetry of the human brain. *Scientific American, 228*, 70–78.

Kimura, D. (1979). Neuromotor mechanisms in the evolution of human communication. In H. E. Steklis & M. J. Raleigh (Eds.), *Neurobiology of social communication in primates* (pp. 197–219). New York, NY: Academic Press.

Kimura, D. (2011). From ear to brain. *Brain and Cognition, 76*, 214–217.

Kingsley, D. M. (2009, January). From atoms to traits. *Scientific American, 300*, 52–59.

Kipnis, J. (2018). The seventh sense. *Scientific American, 319*(2), 28–35.

Kippin, T. E. (2014). Adaptations underlying the development of excessive alcohol intake in selectively bred mice. *Alcoholism: Clinical and Experimental Research, 38*, 36–39.

Kirsch, I. (2019). The placebo effect. *Cambridge Handbook of Psychology, Health and Medicine, 93.*

Kirszenblat, L., & van Swinderen, B. (2015). The yin and yang of sleep and attention. *Trends in Neurosciences, 38*, 776–786.

Klawans, H. L. (1990). *Newton's madness: Further tales of clinical neurology.* New York: Harper & Row.

Klein, D. A., & Walsh, T. B. (2004). Eating disorders: Clinical features and pathophysiology. *Physiology & Behavior, 81*, 359–374.

Klein, R., & Kania, A. (2014). Ephrin signalling in the developing nervous system. *Current Opinion in Neurobiology, 27*, 16–24.

Kleiner-Fisman, G., Fisman, D. N., Sime, E., Saint-Cyr, J. A., Lozano, A. M., & Lang, A. E. (2003). Long-term follow up of bilateral deep brain stimulation of the subthalamic nucleus in patients with advanced Parkinson disease. *Journal of Neuroscience, 99*, 489–495.

Kleitman, N. (1963). *Sleep and wakefulness.* Chicago: University of Chicago Press.

Klemm, W. R. (2011). Why does REM sleep occur? A wake-up hypothesis. *Frontiers in Systems Neuroscience, 5*, 73.

Klengel, T., & Binder, E. B. (2015). Epigenetics of stress-related psychiatric disorders and gene × environment interactions. *Neuron, 86*, 1343–1357.

Klinesmith, J., Kasser, T., & McAndrew, F. T. (2006). Guns, testosterone, and aggression: An experimental test of a mediational hypothesis. *Psychological Science, 17*, 568–571.

Klivington, K. A. (Ed.). (1992). *Gehirn und Geist.* Heidelberg, Germany: Spektrum Akademischer Verlag.

Kluver, H., & Bucy, P. C. (1939). Preliminary analysis of the temporal lobes in monkeys. *Archives of Neurology and Psychiatry, 42*, 979–1000.

Knöchel, C., Stäblein, M., Storchak, H., Reinke, B., Jurcoane, A., Prvulovic, D.,...& Alves, G. (2014). Multimodal assessments of the hippocampal formation in schizophrenia and bipolar disorder: Evidences from neurobehavioral measures and functional and structural MRI. *NeuroImage: Clinical, 6*, 134–144.

Knowlton, B. J., Mangels, J. A., & Squire, L. R. (1996). A neostriatal habit learning system in humans. *Science, 273,* 1399–1402.

Knudsen, E. I., & Brainard, M. S. (1991). Visual instruction of the neural map of auditory space in the developing optic tectum. *Science, 253,* 85–87.

Knufinke, M., Nieuwenhuys, A., Maase, K., Moen, M. H., Geurts, S. A., Coenen, A. M., & Kompier, M. A. (2018). Effects of natural between-days variation in sleep on elite athletes' psychomotor vigilance and sport-specific measures of performance. *Journal of Sports Science & Medicine, 17*(4), 515.

Ko, C.-H., Liu, G.-C., Yen, J.-Y., Yen, C.-F., Chen, C.-S., & Lin, W.-C. (2013). The brain activations for both cue-induced gaming urge and smoking craving among subjects comorbid with internet gaming addiction and nicotine dependence. *Journal of Psychiatric Research, 47,* 486–493.

Koebele, S. V., & Bimonte-Nelson, H. A. (2015). Trajectories and phenotypes with estrogen exposures across the lifespan: What does Goldilocks have to do with it? *Hormones and Behavior, 74,* 86–104.

Koelsch, S., & Skouras, S. (2014). Functional centrality of amygdala, striatum and hypothalamus in a "small-world" network underlying joy: An fMRI study with music. *Human Brain Mapping, 35,* 3485–3498.

Kohwi, M., & Doe, C. Q. (2013). Temporal fate specification and neural progenitor competence during development. *Nature Reviews Neuroscience, 14,* 823–838.

Kokras, N., & Dalla, C. (2014). Sex differences in animal models of psychiatric disorders. *British Journal of Pharmacology, 171,* 4595–4619.

Kolb, B., & Whishaw, I. Q. (1990). *Fundamentals of human neuropsychology* (3rd ed.). New York, NY: Freeman.

Koleske, A. J. (2013). Molecular mechanisms of dendrite stability. *Nature Reviews Neuroscience, 14,* 536–550.

Kong, X. Z., Mathias, S. R., Guadalupe, T., Glahn, D. C., Franke, B., Crivello, F., . . . & ENIGMA Laterality Working Group. (2018). Mapping cortical brain asymmetry in 17,141 healthy individuals worldwide via the ENIGMA Consortium. *Proceedings of the National Academy of Sciences, 115*(22), E5154–E5163.

Kong, X. Z., Postema, M., Castillo, A. C., Pepe, A., Crivello, F., Joliot, M., . . . & Francks, C. (2019). Handedness and other variables associated with human brain asymmetrical skew. *bioRxiv,* 756395.

Koolhaas, J. M., Coppens, C. M., de Boer, S. F., Buwalda, B., Meerlo, P., & Timmermans, P. J. (2013). The resident-intruder paradigm: A standardized test for aggression, violence, and social stress. *Journal of Visualized Experiments, 77,* e4367.

Koopmans, H. S. (1981). The role of the gastrointestinal tract in the satiation of hunger. In L. A. Cioffi, W. B. T. James, & T. B. Van Italie (Eds.), *The body weight regulatory system: Normal and disturbed mechanisms* (pp. 45–55). New York, NY: Raven Press.

Kopelman, P. G. (2000). Obesity as a medical problem. *Nature, 404,* 635–648.

Kopp-Scheinpflug, C., Sinclair, J. L., & Linden, J. F. (2018). When sound stops: Offset responses in the auditory system. *Trends in Neurosciences, 41,* 712–728.

Koppel, B. S., Brust, J. C., Fife, T., Bronstein, J., Youssof, S., Gronseth, G., & Gloss, D. (2014). Systematic review: Efficacy and safety of medical marijuana in selected neurologic disorders: Report of the Guideline Development Subcommittee of the American Academy of Neurology. *Neurology, 82,* 1556–1563.

Koren, G., Zelner, I., Nash, K., & Koren, G. (2014). Foetal alcohol spectrum disorder: Identifying the neurobehavioral phenotype and effective interventions. *Current Opinion in Psychiatry, 27,* 98–104.

Korgaonkar, M. S., Goldstein-Piekarski, A. N., Fornito, A., & Williams, L. M. (2019). Intrinsic connectomes are a predictive biomarker of remission in major depressive disorder. *Molecular Psychiatry,* 1–13.

Kornack, D. R., & Rakic, P. (1999). Continuation of neurogenesis in the hippocampus of the adult macaque monkey. *Proceedings of the National Academy of Sciences, USA, 96,* 5768–5773.

Kornblith, S., & Tsao, D. Y. (2017). How thoughts arise from sights: Inferotemporal and prefrontal contributions to vision. *Current Opinion in Neurobiology, 46,* 208–218.

Kornell, N. (2009). Metacognition in humans and animals. *Current Directions in Psychological Science, 18,* 11–15.

Koropouli, E., & Kolodkin, A. L. (2014). Semaphorins and the dynamic regulation of synapse assembly, refinement, and function. *Current Opinion in Neurobiology, 27,* 1–7.

Koser, D. E., Thompson, A. J., Foster, S. K., Dwivedy, A., Pillai, E. K., Sheridan, G. K., . . . & Holt, C. E. (2016). Mechanosensing is critical for axon growth in the developing brain. *Nature Neuroscience, 19*(12), 1592–1598.

Koshimizu, T. A., Honda, K., Nagaoka-Uozumi, S., Ichimura, A., Kimura, I., Nakaya, M., . . . & Hirasawa, A. (2018). Complex formation between the vasopressin 1b receptor, β-arrestin-2, and the μ-opioid receptor underlies morphine tolerance. *Nature Neuroscience, 21*(6), 820–833.

Kosslyn, S. M., & Andersen, R. A. (1992). *Frontiers in cognitive neuroscience.* Cambridge, MA: MIT Press.

Kotas, M. E., & Medzhitov, R. (2015). Homeostasis, inflammation, and disease susceptibility. *Cell, 160,* 816–827.

Koziol, L. F., Budding, D., Andreasen, N., D'Arrigo, S., Bulgheroni, S., . . . & Yamazaki, T. (2014). Consensus paper: The cerebellum's role in movement and cognition. *Cerebellum, 13,* 151–177.

Kragel, P. A., Kano, M., Van Oudenhove, L., Ly, H. G., Dupont, P., Rubio, A., . . . & Ceko, M. (2018). Generalizable representations of pain, cognitive control, and negative emotion in medial frontal cortex. *Nature Neuroscience, 21*(2), 283–289.

Kral, A., & Sharma, A. (2012). Developmental neuroplasticity after cochlear implantation. *Trends in Neurosciences, 35,* 111–122.

Kraus, S. L., Emmert, S., Schön, M. P., & Haenssle, H. A. (2012). The dark side of beauty: Acne fulminans induced by anabolic steroids in a male bodybuilder. *Archives of Dermatology, 148,* 1210–1212.

Krause, A. J., Simon, E. B., Mander, B. A., Greer, S. M., Saletin, J. M., Goldstein-Piekarski, A. N., & Walker, M. P. (2017). The sleep-deprived human brain. *Nature Reviews Neuroscience, 18*(7), 404–418.

Krauzlis, R. J., Lovejoy, L. P., & Zénon, A. (2013). Superior colliculus and visual spatial attention. *Annual Review of Neuroscience, 36,* 165–182.

Krebs, T. S., & Johansen, P-Ø. (2012). Methodological weaknesses in non-randomized studies of ecstasy (MDMA) use: A cautionary note to readers and reviewers. *Neuropsychopharmacology, 37,* 1070–1071.

Kreibig, S. D. (2010). Autonomic nervous system activity in emotion: A review. *Biological Psychology, 84,* 394–421.

Kreitewolf, J., Friederici, A. D., & von Kriegstein, K. (2014). Hemispheric lateralization of linguistic prosody recognition in comparison to speech and speaker recognition. *Neuroimage, 102,* 332–344.

Kremer, D., Göttle, P., Hartung, H. P., & Küry, P. (2016). Pushing forward: Remyelination as the new frontier in CNS diseases. *Trends in Neurosciences, 39*(4), 246–263.

Kremkow, J., Jin, J., Wang, Y., & Alonso, J. M. (2016). Principles underlying sensory map topography in primary visual cortex. *Nature, 533*(7601), 52–56.

Kreukels, B. P. C., & Cohen-Kettenis, P. T. (2012). Male gender identity and masculine behavior: The role of sex hormones in brain development. In M. Maggi (Ed.), *Hormonal therapy for male sexual dysfunction* (1st ed.). New York, NY: John Wiley & Sons.

Kriegeskorte, N. (2011). Pattern-information analysis: From stimulus decoding to computational-model testing. *Neuroimage, 56,* 411–421.

Kriegstein, A., & Alvarez-Buylla, A. (2009). The glial nature of embryonic and adult neural stem cells. *Annual Review of Neuroscience, 32,* 149–184.

Kril, J. J., & Harper, C. G. (2012). Neuroanatomy and neuropathology associated with Korsakoff's syndrome. *Neuropsychology Review, 22,* 72–80.

Kring, A. M. (1999). Emotion in schizophrenia: Old mystery, new understanding. *Current Directions in Psychological Science, 8,* 160–163.

Kringelbach, M. L., & Berridge, K. C. (2012). The joyful mind. *Scientific American, 307,* 40–45.

Krishnan, A., & Schiöth, H. B. (2015). The role of G protein-coupled receptors in the early evolution of neurotransmission and the nervous system. *Journal of Experimental Biology, 218,* 562–571.

Krol, J., & Roska, B. (2016). Treatment synergy in axon regeneration. *Nature Neuroscience, 19*(8), 983–984.

Kronschläger, M. T., Drdla-Schutting, R., Gassner, M., Honsek, S. D., Teuchmann, H. L., & Sandkühler, J. (2016). Gliogenic LTP spreads widely in nociceptive pathways. *Science, 354*(6316), 1144–1148.

Krubitzer, L., & Stolzenberg, D. S. (2014). The evolutionary masquerade: Genetic and epigenetic contributions to the neocortex. *Current Opinion in Neurobiology, 24,* 157–165.

Krueger, D. D., Tuffy, L. P., Papadopoulos, T., & Brose, N. (2012). The role of neurexins and neuroligins in the formation, maturation, and function of vertebrate synapses. *Current Opinion in Neurobiology, 22,* 412–422.

Krupic, J., Bauza, M., Burton, S., Barry, C., & O'Keefe, J. (2015). Grid cell symmetry is shaped by environmental geometry. *Nature, 518,* 232–235.

Krynicki, C. R., Upthegrove, R., Deakin, J. F. W., & Barnes, T. R. (2018). The relationship between negative symptoms and depression in schizophrenia: A systematic review. *Acta Psychiatrica Scandinavica, 137,* 380–390.

Krystal, A. D. (2015). Current, emerging, and newly available insomnia medications. *Journal of Clinical Psychiatry, 76,* e1045.

Kuang, S., Morel, P., & Gail, A. (2015). Planning movements in visual and physical space in monkey posterior parietal cortex. *Cerebral Cortex, 26,* 731–747.

Kuba, H., Adachi, R., & Ohmori, H. (2014). Activity-dependent and activity-independent development of the axon initial segment. *Journal of Neuroscience, 34,* 3443–3453.

Kuchibhotla, K., & Bathellier, B. (2018). Neural encoding of sensory and behavioral complexity in the auditory cortex. *Current Opinion in Neurobiology, 52,* 65–71.

Kuehner, C. (2017). Why is depression more common among women than among men? *The Lancet Psychiatry, 4*(2), 146–158.

Kuhn, B. N., Kalivas, P. W., & Bobadilla, A. C. (2019). Understanding addiction using animal models. *Frontiers in Behavioral Neuroscience, 13,* 262.

Kuhn, H. G., Toda, T., & Gage, F. H. (2018). Adult hippocampal neurogenesis: A coming-of-age story. *Journal of Neuroscience, 38*(49), 10401–10410.

Kumar, A., Narayanan, K., Chaudhary, R. K., Mishra, S., Kumar, S., Vinoth, K. J., . . . & Gulyás, B. (2017). Current perspective of stem cell therapy in neurodegenerative and metabolic diseases. *Molecular Neurobiology, 54*(9), 7276–7296.

Kuner, R., & Flor, H. (2017). Structural plasticity and reorganisation in chronic pain. *Nature Reviews Neuroscience, 18*(1), 20–30.

Kupferschmidt, K. (2014). The dangerous professor. *Science, 343*, 478–481.

Kuratani, S., Kusakabe, R., & Hirasawa, T. (2018). The neural crest and evolution of the head/trunk interface in vertebrates. *Developmental Biology, 444*, S60–S66.

Kurbat, M. A., & Farah, M. J. (1998). Is the category-specific deficit for living things spurious? *Journal of Cognitive Neuroscience, 10*, 355–361.

Kwakkel, G., Veerbeek, J. M., van Wegen, E. E. H., & Wolf, S. L. (2015). Constraint-induced movement therapy after stroke. *The Lancet Neurology, 14*, 224–234.

Kyritsis, N., Kizil, C., Zocher, S., Kroehne, V., Kaslin, J., Freudenreich, D., Iltzsche, A., & Brand, M. (2012). Acute inflammation initiates the regenerative response in the adult zebrafish brain. *Science, 338*, 1353–1356.

Laake, L. M., & Bridgett, D. J. (2018). Early language development in context: Interactions between infant temperament and parenting characteristics. *Early Education and Development, 29*(5), 730–746.

LaBerge, S., Baird, B., & Zimbardo, P. G. (2018). Smooth tracking of visual targets distinguishes lucid REM sleep dreaming and waking perception from imagination. *Nature Communications, 9*(1), 1–8.

LaBerge, S., LaMarca, K., & Baird, B. (2018). Pre-sleep treatment with galantamine stimulates lucid dreaming: A double-blind, placebo-controlled, crossover study. *PLoS one, 13*(8).

Laeng, B., & Caviness, V. S. (2001). Prosopagnosia as a deficit in encoding curved surface. *Journal of Cognitive Neuroscience, 13*, 556–576.

LaFerla, F. M., & Green, K. N. (2012). Animal models of Alzheimer disease. *Cold Spring Harbor Perspectives in Medicine, 2*, a006320.

Lagerlöf, O., Slocomb, J. E., Hong, I., Aponte, Y., Blackshaw, S., Hart, G. W., & Huganir, R. L. (2016). The nutrient sensor OGT in PVN neurons regulates feeding. *Science, 351*, 1293–1296.

Lai, M. C., Lerch, J. P., Floris, D. L., Ruigrok, A. N., Pohl, A., Lombardo, M. V., & Baron-Cohen, S. (2017). Imaging sex/gender and autism in the brain: Etiological implications. *Journal of Neuroscience Research, 95*(1–2), 380–397.

Lai, Y. Y., Cheng, Y. H., Hsieh, K. C., Nguyen, D., Chew, K. T., Ramanathan, L., & Siegel, J. M. (2017). Motor hyperactivity of the iron-deficient rat—an animal model of restless legs syndrome. *Movement Disorders, 32*(12), 1687–1693.

Lai,. T. W., Zhang, S., & Wang, Y. T. (2014). Excitotoxicity and stroke: Identifying novel targets for neuroprotection. *Progress in Neurobiology, 115*, 157–188.

Laine, M., Salmelin, R., Helenius, P., & Marttila, R. (2000). Brain activation during reading in deep dyslexia: An MEG study. *Journal of Cognitive Neuroscience, 12*, 622–634.

Lamb, T. D., Collin, S. P., & Pugh, E. N. (2007). Evolution of the vertebrate eye: Opsins, photoreceptors, retina and eye cup. *Nature Reviews Neuroscience, 8*, 960–975.

Lamichhaney, S., Berglund, J., Almén, M. S., Maqbool, K., Grabherr, M., Martinez-Barrio, A., . . . & Grant, B. R. (2015). Evolution of Darwin's finches and their beaks revealed by genome sequencing. *Nature, 518*, 371–375.

Land, E. H. (1977, April). The retinex theory of color vision. *Scientific American, 237*, 108–128.

Land, M. F. (2014). Do we have an internal model of the outside world? *Philosophical Transactions of the Royal Society B, 369*, 20130045.

Landgraf, M. N., Nothacker, M., & Heinen, F. (2013). Diagnosis of fetal alcohol syndrome (FAS): German guideline version 2013. *European Journal of Paediatric Neurology, 17*, 437–446.

Landgraf, M., & Heinen, F. (2014). Diagnosis of fetal alcohol spectrum disorders. *Neuropediatrics, 45*, fp041.

Landhuis, E. (2017). Ultrasound for the brain. *Nature, 551*(7679), 257–259.

Lang, D.-Y., Li, X., & Clark, J. D. (2013). Epigenetic regulation of opioid-induced hyperalgesia, dependence, and tolerance in mice. *The Journal of Pain, 14*, 36–47.

Lange, N., & McDougle, C. J. (2013). Help for the child with autism. *Scientific American, 309*, 72–77.

Lange, S., Probst, C., Gmel, G., Rehm, J., Burd, L., & Popova, S. (2017). Global prevalence of fetal alcohol spectrum disorder among children and youth: A systematic review and meta-analysis. *JAMA Pediatrics, 171*(10), 948–956.

Langston, J. W. (1985). MPTP and Parkinson's disease. *Trends in Neurosciences, 8*, 79–83.

Langston, R. F., Ainge, J. A., Couey, J. J., Canto, C. B., Bjerknes, T. L., Witter, M. P., Moser, E. I., & Moser, M.-B. (2010). Development of the spatial representation system in the rat. *Science, 328*, 1576–1580.

Lappin, J. S. (2014). What is binocular disparity? *Frontiers in Psychology, 5*, 870.

Larochelle, C., Uphaus, T., Prat, A., & Zipp, F. (2016). Secondary progression in multiple sclerosis: Neuronal exhaustion or distinct pathology? *Trends in Neurosciences, 39*(5), 325–339.

Larøi, F., Bless, J. J., Laloyaux, J., Kråkvik, B., Vedul-Kjelsås, E., Kalhovde, A. M., . . . & Hugdahl, K. (2019). An epidemiological study on the prevalence of hallucinations in a general-population sample: Effects of age and sensory modality. *Psychiatry research, 272*, 707–714.

Larson-Prior, L. J., Ju, Y.-E., & Galvin, J. E. (2014). Cortical-subcortical interactions in hypersomnia disorders: Mechanisms underlying cognitive and behavioral aspects of the sleep-wake cycle. *Frontiers in Neurology, 5*, 165.

Larson, G., Piperno, D. R., Allaby, R. G., Purugganan, M. D., Andersson, L., et al. (2014). Current perspectives and the future of domestication studies. *PNAS, 111*, 6139–6146.

Lashley, K. S. (1941). Patterns of cerebral integration indicated by the scotomas of migraine. *Archives of Neurology and Psychiatry, 46*, 331–339.

Lashuel, H. A., Overk, C. R., Oueslati, A., & Masliah, E. (2013). The many faces of 24a-synuclein: From structure and toxicity to therapeutic target. *Nature Reviews Neuroscience, 14*, 38–48.

Lask, B., & Bryant-Waugh, R. (Eds.). (2000). *Anorexia nervosa and related eating disorders in childhood and adolescence.* Hove, England: Psychology Press.

Lau, B. K., & Vaughan, C. W. (2014). Descending modulation of pain: The GABA disinhibition hypothesis of analgesia. *Current Opinion in Neurobiology, 29*, 159–164.

Lau, B. K., & Werner, L. A. (2014). Perception of the pitch of unresolved harmonics by 3- and 7-month-old human infants. *Journal of the Acoustical Society of America, 136*, 760–767.

Laug, D., Glasgow, S. M., & Deneen, B. (2018). A glial blueprint for gliomagenesis. *Nature Reviews Neuroscience, 19*, 393–403.

Lavie, P., Pratt, H., Scharf, B., Peled, R., & Brown, J. (1984). Localized pontine lesion: Nearly total absence of REM sleep. *Neurology, 34*, 1118–1120.

Laviolette, S. R., & van der Kooy, D. (2004). The neurobiology of nicotine addiction: Bridging the gap from molecules to behavior. *Nature Reviews Neuroscience, 5*, 55–65.

Lavrov, I., Courtine, G., Dy, C. J., van den Brand, R., Fong, A. J., . . . & Edgerton, V. R. (2008). Facilitation of stepping with epidural stimulation in spinal rats: Role of sensory input. *Journal of Neuroscience, 28*, 7774–7780.

Law, S. W., & Levine, S. R. (2015). Stroke: Support for IV tPA in ischaemic stroke in elderly people. *Nature Reviews Neurology, 12*, 8–9.

Lawrence, D. G., & Kuypers, H. G. J. M. (1968). The functional organization of the motor system in the monkey: II. The effects of lesions of the descending brain-stem pathways. *Brain, 91*, 15–36.

Lazzerini Ospri, L., Prusky, G., & Hattar, S. (2017). Mood, the circadian system, and melanopsin retinal ganglion cells. *Annual Review of Neuroscience, 40*, 539–556.

Le, Q., Yan, B., Yu, X., Li, Y., Song, H., Zhu, H., . . . & Ma, L. (2017). Drug-seeking motivation level in male rats determines offspring susceptibility or resistance to cocaine-seeking behaviour. *Nature Communications, 8*(1), 1–13.

Learney, K., Van Wart, A., & Sur, M. (2009). Intrinsic patterning and experience-dependent mechanisms that generate eye-specific projections and binocular circuits in the visual pathway. *Current Opinion in Neurobiology, 19*, 181–187.

LeDoux, J. E. (2014). Coming to terms with fear. *Proceedings of the National Academy of Sciences, 111*, 2871–2878.

LeDoux, J. E., & Hofmann, S. G. (2018). The subjective experience of emotion: A fearful view. *Current Opinion in Behavioral Sciences, 19*, 67–72.

LeDoux, J. E., Michel, M., & Lau, H. (2020). A little history goes a long way toward understanding why we study consciousness the way we do today. *Proceedings of the National Academy of Sciences, 117*(13), 6976–6984.

Lee, A. K., & Brecht, M. (2018). Elucidating neuronal mechanisms using intracellular recordings during behavior. *Trends in Neurosciences, 41*(6), 385–403.

Lee, B. H., Smith, T., & Paciorkowski, A. R. (2015). Autism spectrum disorder and epilepsy: Disorders with a shared biology. *Epilepsy & Behavior, 47*, 191–201.

Lee, D., & Seo, H. (2016). Neural basis of strategic decision making. *Trends in Neurosciences, 39*, 40–48.

Lee, D., Seo, H., & Jung, M. W. (2012). Neural basis of reinforcement learning and decision making. *Annual Review of Neuroscience, 35*, 287–308.

Lee, H. S., Ghetti, A., Pinto-Duarte, A., Wang, X., Dziewczapolski, G., et al. (2014). Astrocytes contribute to gamma oscillations and recognition memory. *PNAS, 111*, e3343–e3352.

Lee, J. C., Kim, S. J., Hong, S., & Kim, Y. (2019). Diagnosis of Alzheimer's disease utilizing amyloid and tau as fluid biomarkers. *Experimental & Molecular Medicine, 51*(5), 1–10.

Lee, J., Kim, K., Chung, S., & Lee, C. (2013). Suppression of spontaneous activity before visual response in the primate V1 neurons during a visually guided saccade task. *Journal of Neuroscience, 33*, 3760–3764.

Lee, L. J., Hughes, T. R., & Frey, B. J. (2006). How many new genes are there? *Science, 3*, 1709.

Lee, S.-C., Patrick, S. L., Richardson, K. A., & Connors, B. W. (2014). Two functionally distinct networks of gap junction-coupled inhibitory neurons in the thalamic reticular nucleus. *Journal of Neuroscience, 34*, 13170–13182.

Lee, T.-S. (2008). Contextual influences in visual processing. In M. D. Binder, N. Hirokawa, & U. Windurst (Eds.), *Encyclopedia of Neuroscience* (pp. 867–871). Berlin: Springer.

Leech, R., & Saygin, A. P. (2011). Distributed processing and cortical specialization for speech and environmental sounds in human temporal cortex. *Brain and Language, 116*, 83–90.

Léger, D., Debellemaniere, E., Rabat, A., Bayon, V., Benchenane, K., & Chennaoui, M. (2018). Slow-wave sleep: From the cell to the clinic. *Sleep Medicine Reviews, 41*, 113–132.

Lehky, S. R., & Tanaka, K. (2016). Neural representation for object recognition in inferotemporal cortex. *Current Opinion in Neurobiology, 37*, 23–35.

Lei, Y., Han, H., Yuan, F., Javeed, A., & Zhao, Y. (2017). The brain interstitial system: Anatomy, modeling, in vivo measurement, and applications. *Progress in Neurobiology, 157*, 230–246.

Leighton, D. H., & Sternberg, P. W. (2016). Mating pheromones of Nematoda: Olfactory signaling with physiological consequences. *Current Opinion in Neurobiology, 38*, 119–124.

Leighton, L. J., Ke, K., Zajaczkowski, E. L., Edmunds, J., Spitale, R. C., & Bredy, T. W. (2018). Experience-dependent neural plasticity, learning, and memory in the era of epitranscriptomics. *Genes, Brain and Behavior, 17*(3), e12426.

Lemmens, R., & Steinberg, G. K. (2013). Stem cell therapy for acute cerebral injury: What do we know and what will the future bring? *Current Opinion in Neurology, 26*, 617–625.

Lener, M. S., & Iosifescu, D. V. (2015). In pursuit of neuroimaging biomarkers to guide treatment selection in major depressive disorder: A review of the literature. *Annals of the New York Academy of Sciences, 1344*, 50–65.

Leng, T., Shi, Y., Xiong, Z.-G., & Sun, D. (2014). Proton-sensitive cation channels and ion exchangers in ischemic brain injury: New therapeutic targets for stroke? *Progress in Neurobiology, 115*, 189–209.

Lennox, W. G. (1960). *Epilepsy and related disorders.* Boston, MA: Little, Brown.

Lenoir, M., Serre, F., Cantin, L., & Ahmed, S. H. (2007). Intense sweetness surpasses cocaine reward. *PLoS ONE, 8*, e698.

Lenroot, R. K., & Giedd, J. N. (2006). Brain development in children and adolescents: Insights from anatomical magnetic resonance imaging. *Neuroscience and Biobehavioural Reviews, 30*, 718–729.

Lense, M. D., Key, A. P., & Dykens, E. M. (2011). Attentional disengagement in adults with William syndrome. *Brain and Cognition, 77*, 201–207.

Lenz, K. M., & McCarthy, M. M. (2010). Organized for sex-steroid hormones and the developing hypothalamus. *European Journal of Neuroscience, 32*, 2096–2104.

Lenz, K. M., Nugent, B. M., & McCarthy, M. M. (2012). Sexual differentiation of the rodent brain: Dogma and beyond. *Frontiers in Neuroscience, 6*, 26.

Leopold, D. A. (2012). Primary visual cortex: Awareness and blindsight. *Annual Review of Neuroscience, 35*, 91–109.

Lerch, J. P., van der Kouwe, A. J., Raznahan, A., Paus, T., Johansen-Berg, H., Miller, K. L., . . . & Sotiropoulos, S. N. (2017). Studying neuroanatomy using MRI. *Nature Neuroscience, 20*(3), 314–326.

Lesica, N. A. (2018). Why do hearing aids fail to restore normal auditory perception? *Trends in Neurosciences, 41*(4), 174–185.

Lesku, J. A., Rattenborg, N. C., Valcu, M., Vyssotski, A. L., Kuhn, S., Kuemmeth, F., Heidrich, W., & Kempenaers, B. (2012). Adaptive sleep loss in polygynous pectoral sandpipers. *Science, 337*, 1652–1658.

Lester, G. L. L., & Gorzalka, B. B. (1988). Effect of novel and familiar mating partners on the duration of sexual receptivity in the female hamster. *Behavioral and Neural Biology, 49*, 398–405.

LeVay, S. (1991). A difference in hypothalamic structure between heterosexual and homosexual men. *Science, 253*, 1034–1037.

Levelt, C. N., & Hübener, M. (2012). Critical-period plasticity in the visual cortex. *Annual Review of Neuroscience, 35*, 309–330.

Levi-Montalcini, R. (1952). Effects of mouse motor transplantation on the nervous system. *Annals of the New York Academy of Sciences, 55*, 330–344.

Levi-Montalcini, R. (1975). NGF: An uncharted route. In F. G. Worden, J. P. Swazey, & G. Adelman (Eds.), *The neurosciences: Paths of discovery* (pp. 245–265). Cambridge, MA: MIT Press.

Levine, M. S., Cepeda, C., Hickey, M. A., Fleming, S. M., & Chesselet, M.-F. (2004). Genetic mouse models of Huntington's and Parkinson's diseases: Illuminating but imperfect. *Trends in Neurosciences, 27*, 691–697.

Levy, E. I., & Mokin, M. (2017). Stroke: Stroke thrombolysis and thrombectomy—not stronger together? *Nature Reviews Neurology, 13*, 198–199.

Levy, J. (1969). Possible basis for the evolution of lateral specialization of the human brain. *Nature, 224*, 614–615.

Levy, J. (1972). Lateral specialization of the human brain: Behavioral manifestations and possible evolutionary basis. In J. A. J. Kiger (Ed.), *The Biology of Behavior.* Oregon State University Press, Oregon, USA.

Levy, J., Trevarthen, C., & Sperry, R. W. (1972). Perception of bilateral chimeric figures following hemispheric deconnection. *Brain, 95*, 61–78.

Levy, R. G., Cooper, P. N., & Giri, P. (2012). Ketogenic diet and other dietary treatments for epilepsy. *Cochrane Database of Systematic Reviews, 3*, CD001903.

Lewis, D. A., Hashimoto, T., & Volk, D. W. (2005). Cortical inhibitory neurons and schizophrenia. *Nature Reviews Neuroscience, 6*, 312–324.

Lewis, S. (2015). Circadian rhythms: Remembering night and day. *Nature Reviews Neuroscience, 16*, 3.

Lezak, M. D. (1997). Principles of neuropsychological assessment. In T. E. Feinberg & M. J. Farah (Eds.), *Behavioral neurology and neuropsychology* (pp. 43–54). New York, NY: McGraw-Hill.

Li, G., Lin, W., Gilmore, J. H., & Shen, D. (2015). Spatial patterns, longitudinal development, and hemispheric asymmetries of cortical thickness in infants from birth to 2 years of age. *Journal of Neuroscience, 35*, 9150–9162.

Li, J., Na, L., Ma, H., Zhang, Z., Li, T., Lin, L., . . . & Li, Y. (2015). Multigenerational effects of parental prenatal exposure to famine on adult offspring cognitive function. *Scientific Reports, 5*, 13792.

Li, K., & Malhotra, P. A. (2015). Spatial neglect. *Practical Neurology, 15*, 333–339.

Li, X., & Wolf, M. E. (2014). Multiple faces of BDNF in cocaine addiction. *Behavioural Brain Research, 279*, 240–254.

Lia, M. M. (2019). The advantages of substitution of REM sleep stages with waking episodes to perform REM sleep reduction. *EC Neurology, 11*, 400–407.

Liang, D. Y., Li, X., & Clark, J. D. (2013). Epigenetic regulation of opioid-induced hyperalgesia, dependence, and tolerance in mice. *The Journal of Pain, 14*, 36–47.

Liblau, R. S., Vassalli, A., Seifinejad, A., & Tafti, M. (2015). Hypocretin (orexin) biology and the pathophysiology of narcolepsy with cataplexy. *Lancet Neurology, 14*, 318–328.

Lichtman, J. W., & Denk, W. (2011). The big and the small: Challenges of imaging the brain's circuits. *Science, 334*, 618–623.

Lieberwirth, C., Pan, Y., Liu, Y., Zhang, Z., & Wang, Z. (2016). Hippocampal adult neurogenesis: Its regulation and potential role in spatial learning and memory. *Brain Research, 1644*, 127–140.

Lien, W.-H., Klezovitch, O., Fernandez, T. E., Delrow, J., & Vasioukhin, V. (2006). AlphaE-catenin controls cerebral cortical size by regulating the hedgehog signaling pathway. *Science, 311*, 1609–1612.

Ligaard, J., Sannæs, J., & Pihlstrøm, L. (2019). Deep brain stimulation and genetic variability in Parkinson's disease: A review of the literature. *NJP Parkinson's Disease, 5*(1), 1–10.

Lightman, S. L., & Conway-Campbell, B. L. (2010). The crucial role of pulsatile activity of the HPA axis for continuous dynamic equilibration. *Nature Reviews Neuroscience, 11*, 710–717.

Likhtik, E., & Paz, R. (2015). Amygdala-prefrontal interactions in (mal)adaptive learning. *Trends in Neurosciences, 38*, 158–166.

Lim, J. H. A., Stafford, B. K., Nguyen, P. L., Lien, B. V., Wang, C., Zukor, K., . . . & Huberman, A. D. (2016). Neural activity promotes long-distance, target-specific regeneration of adult retinal axons. *Nature Neuroscience, 19*(8), 1073–1084.

Lim, S. J., Fiez, J. A., & Holt, L. L. (2014). How may the basal ganglia contribute to auditory categorization and speech perception. *Frontiers in Neuroscience, 8*, 230.

Lima, C. F., Krishnan, S., & Scott, S. K. (2016). Roles of supplementary motor areas in auditory processing and auditory imagery. *Trends in neurosciences, 39*(8), 527–542.

Lima, I. M., Peckham, A. D., & Johnson, S. L. (2018). Cognitive deficits in bipolar disorders: Implications for emotion. *Clinical Psychology Review, 59*, 126–136.

Limanowski, J., & Blankenburg, F. (2015). That's not quite me: Limb ownership encoding in the brain. *Social Cognitive and Affective Neuroscience*. Advance online publication. doi: 10.1093/scan/nsv079.

Limousin, P., & Foltynie, T. (2019). Long-term outcomes of deep brain stimulation in Parkinson disease. *Nature Reviews Neurology, 15*(4), 234–242.

Lin, C. C. J., Yu, K., Hatcher, A., Huang, T. W., Lee, H. K., Carlson, J., . . . & Mohila, C. A. (2017). Identification of diverse astrocyte populations and their malignant analogs. *Nature Neuroscience, 20*(3), 396–405.

Lin, E., & Tsai, S. J. (2020). Gene-Environment Interactions and Role of Epigenetics in Anxiety Disorders. In *Anxiety Disorders* (pp. 93–102). Springer, Singapore.

Lin, J. Y., Arthurs, J., & Reilly, S. (2014). Conditioned taste aversion, drugs of abuse and palatability. *Neuroscience & Biobehavioral Reviews, 45*, 28–45.

Lin, L., Faraco, J., Li, R., Kadotani, H., Rogers, W., Lin, X., . . . & Mignot, E. (1999). The sleep disorder canine narcolepsy is caused by a mutation in the hypocretin (orexin) receptor 2 gene. *Cell, 98*, 365–376.

Lin, M. Z., & Schnitzer, M. J. (2016). Genetically encoded indicators of neuronal activity. *Nature Neuroscience, 19*(9), 1142.

Lin, Y. T., & Capel, B. (2015). Cell fate commitment during mammalian sex determination. *Current Opinion in Genetics & Development, 32*, 144–152.

Linde, K., Kriston, L., Rücker, G., Jamil, S., Schumann, I., Meissner, K., . . . & Schneider, A. (2015). Efficacy and acceptability of pharmacological treatments for depressive disorders in primary care: Systematic review and network meta-analysis. *Annals of Family Medicine, 13*, 69–79.

Lindell, A. K. (2013). Continuities in emotion lateralization in human and non-human primates. *Frontiers in Human Neuroscience, 7*, 464.

Lindquist, K. A., Wager, T. D., Kober, H., Bliss-Moreau, E., & Barrett, L. F. (2012). The brain basis of emotion: A meta-analytic review. *Behavioral and Brain Sciences, 35*, 121–143.

Lindsay, P. H., & Norman, D. A. (1977). *Human information processing* (2nd ed.). New York, NY: Academic Press.

Lindsey, D. B., Bowden, J., & Magoun, H. W. (1949). Effect upon the EEG of acute injury to the brain stem activating system. *Electroencephalography and Clinical Neurophysiology, 1*, 475–486.

Linster, C., & Fontanini, A. (2014). Functional neuromodulation of chemosensation in vertebrates. *Current Opinion in Neurobiology, 29*, 82–87.

Lisman, J. (2017). Glutamatergic synapses are structurally and biochemically complex because of multiple plasticity processes: Long-term potentiation, long-term depression, short-term potentiation and scaling. *Philosophical Transactions of the Royal Society B: Biological Sciences, 372*(1715), 20160260.

Lisman, J., Buzsáki, G., Eichenbaum, H., Nadel, L., Ranganath, C., & Redish, A. D. (2017). Viewpoints: How the hippocampus contributes to memory, navigation and cognition. *Nature Neuroscience, 20*(11), 1434–1447.

Liu, C., & Grigson, P. S. (2005). Brief access to sweets protect against relapse to cocaine-seeking. *Brain Research, 1049*, 128–131.

Liu, D., & Dan, Y. (2019). A motor theory of sleep-wake control: Arousal-action circuit. *Annual Review of Neuroscience, 42*, 27–46.

Liu, D., Dorio, J., Tannenbaum, B., Caldji, C., Francis, D., Freedman, A., . . . & Meaney, M. J. (1997). Maternal care, hippocampal glucocorticoid receptors, and hypothalamic-pituitary-adrenal responses to stress. *Science, 277*, 1659–1662.

Liu, P. P., Xie, Y., Meng, X. Y., & Kang, J. S. (2019). History and progress of hypotheses and clinical trials for Alzheimer's disease. *Signal Transduction and Targeted Therapy, 4*(1), 1–22.

Liu, S., & Borgland, S. L. (2015). Regulation of the mesolimbic dopamine circuit by feeding peptides. *Neuroscience, 289*, 19–42.

Liu, T. Z., Xu, C., Rota, M., Cai, H., Zhang, C., Shi, M. J., . . . & Sun, X. (2017). Sleep duration and risk of all-cause mortality: A flexible, non-linear, meta-regression of 40 prospective cohort studies. *Sleep Medicine Reviews, 32*, 28–36.

Liu, X., Hairston, J., Schrier, M., & Fan, J. (2011). Common and distinct networks underlying reward valence and processing stages: A meta-analysis of functional neuroimaging studies. *Neuroscience and Biobehavioral Reviews, 35*, 1219–1236.

Liu, X., Lu, W., Liao, S., Deng, Z., Zhang, Z., Liu, Y., & Lu, W. (2018). Efficiency and adverse events of electronic cigarettes: A systematic review and meta-analysis (PRISMA-compliant article). *Medicine, 97*(19).

Liu, Y., Yan, T., Chu, J. M. T., Chen, Y., Dunnett, S., Ho, Y. S., . . . & Chang, R. C. C. (2019). The beneficial effects of physical exercise in the brain and related pathophysiological mechanisms in neurodegenerative diseases. *Laboratory Investigation, 99*, 943–957.

Livet, J., Weissman, T. A., Kang, H., Draft, R. W., Lu, J., Bennis, R. A., . . . & Lichtman, J. W. (2007). Transgenic strategies for combinatorial expression of fluorescent proteins in the nervous system. *Nature, 450*, 56–62.

Livingstone, M. S., & Hubel, D. H. (1988). Segregation of form, color, movement, and depth: Anatomy, physiology, and perception. *Science, 240*, 740–749.

Livingstone, M. S., & Tsao, D. Y. (1999). Receptive fields of disparity-selective neurons in macaque striate cortex. *Nature Reviews Neuroscience, 2*, 825–832.

Lleras, A., Cronin, D. A., Madison, A. M., Wang, M., & Buetti, S. (2017). Oh, the number of things you will process (in parallel)! *Behavioral and Brain Sciences, 40*.

Llewellyn, S. (2016). Dream to predict? REM dreaming as prospective coding. *Frontiers in Psychology, 6*, 1961.

Locke, A. E., Kahali, B., Berndt, S. I., Justice, A. E., Pers, T. H., . . . & Speliotes, E. K. (2015). Genetic studies of body mass index yield new insights for obesity biology. *Nature, 518*, 197–206.

Lodatto, S., Shetty, A. S., & Arlotta, P. (2015). Cerebral cortex assembly: Generating and reprogramming projection neuron diversity. *Trends in Neurosciences, 38*, 117–125.

Logan, R. W., & McClung, C. A. (2016). Animal models of bipolar mania: The past, present and future. *Neuroscience, 321*, 163–188.

Logothetis, N. K., & Sheinberg, D. L. (1996). Visual object recognition. *Annual Review of Neuroscience, 19*, 577–621.

Löhler, J., & Wollenberg, B. (2019). Are electronic cigarettes a healthier alternative to conventional tobacco smoking? *European Archives of Oto-Rhino-Laryngology, 276*(1), 17–25.

Lohse, A., Kjaer, T. W., Sabers, A., & Wolf, P. (2015). Epileptic aura and perception of self-control. *Epilepsy & Behavior, 45*, 191–194.

London, A. J., Kimmelman, J., & Carlisle, B. (2012). Rethinking research ethics: The case of postmarketing trials. *Science, 336*, 544–545.

Long, M. A., & Lee, A. K. (2012). Intracellular recording in behaving animals. *Current Opinion in Neurobiology, 22*, 34–44.

López-Hidalgo, M., & Schummers, J. (2014). Cortical maps: A role for astrocytes? *Current Opinion in Neurobiology, 24*, 176–189.

Lopomo, A., Burgio, E., & Migliore, L. (2016). Epigenetics of obesity. In *Progress in molecular biology and translational science* (Vol. 140, pp. 151–184). Academic Press.

Lorch, M. P. (2019). The long view of language localization. *Frontiers in Neuroanatomy, 13*, 52.

Lordkipanidze, D., de León, M. S. P., Margvelashvili, A., Rak, Y., Rightmire, G. P., Vekua, A., & Zollikofer, C. P. (2013). A complete skull from Dmanisi, Georgia, and the evolutionary biology of early *Homo. Science, 342*, 326–331.

Lott, I. T., & Head, E. (2019). Dementia in Down syndrome: Unique insights for Alzheimer disease research. *Nature Reviews Neurology, 15*(3), 135–147.

Lowdon, R. F., Jang, H. S., & Wang, T. (2016). Evolution of epigenetic regulation in vertebrate genomes. *Trends in Genetics, 32*(5), 269–283.

Lowenstein, D. H. (2015). Decade in review—epilepsy: Edging toward breakthroughs in epilepsy diagnostics and care. *Nature Reviews Neurology, 11*, 616–617.

Lowrey, P. L., Shimomura, K., Antoch, M. P., Yamazaki, S., Zemenides, P. D., Ralph, M. R., . . . & Takahashi, J. S. (2000). Positional syntenic cloning and functional characterization of the mammalian circadian mutation tau. *Science, 288*, 483–491.

Loyd, D. R., & Murphy, A. Z. (2014). The neuroanatomy of sexual dimorphism in opioid analgesia. *Experimental Neurology, 259*, 57–63.

Lozano, A. M., & Kalia, S. K. (2005, July). New movement in Parkinson's. *Scientific American, 293*, 68–75.

Lozano, A. M., & Mayberg, H. S. (2015). Treating depression at the source. *Scientific American, 312*, 68–73.

Lozano, A. M., Giacobbe, P., Hamani, C., Rizvi, S. J., Kennedy, S. H., Kolivakis, T. T., Debonnel, G., Sadikot, A. F., Lam, R. W., Howard, A. K., Ilcewicz-Klimek, M., Honey, C. R., & Mayberg, H. S. (2012). A multicenter pilot study of subcallosal cingulate area deep brain stimulation for treatment-resistant depression. *Journal of Neurosurgery, 116*, 315–322.

Lozano, A. M., Hutchison, W. D., & Kalia, S. K. (2017). What have we learned about movement disorders from functional neurosurgery? *Annual Review of Neuroscience, 40*, 453–477.

Lozano, A. M., Mayberg, H. S., Giacobbe, P., Hamani, C., Craddock, R. C., & Kennedy, S. H. (2008). Subcallosal cingulate gyrus deep brain stimulation for treatment-resistant depression. *Biological Psychiatry, 64*, 461–467.

Lu, D., Immadi, S. S., Wu, Z., & Kendall, D. A. (2019). Translational potential of allosteric modulators targeting the cannabinoid CB1 receptor. *Acta Pharmacologica Sinica, 40*(3), 324–335.

Lu, L., Shepard, J. D., Scott Hall, F., & Shaham, Y. (2003). Effect of environmental stressors on opiate and psychostimulant reinforcement, reinstatement and discrimination in rats: A review. *Neuroscience and Biobehavioral Reviews, 27*, 457–491.

Lucio, R. A., Rodríguez-Piedracruz, V., Tlachi-López, J. L., García-Lorenzana, M., & Fernández-Guasti, A. (2014). Copulation without seminal expulsion: The consequence of sexual satiation and the Coolidge effect. *Andrology, 2*, 450–457.

Luck, S. J., Hillyard, S. A., Mangoun, G. R., & Gazzaniga, M. S. (1989). Independent hemispheric attentional systems mediate visual search in split-brain patients. *Nature, 343*, 543–545.

Lumpkin, E. A., & Caterina, M. (2007). Mechanisms of sensory transduction in the skin. *Nature, 445*, 858–865.

Luo, C., Kuner, T., & Kuner, R. (2014). Synaptic plasticity in pathological pain. *Trends in Neurosciences, 37*, 343–354.

Luo, L., Callaway, E. M., & Svoboda, K. (2018). Genetic dissection of neural circuits: A decade of progress. *Neuron, 98*(2), 256–281.

Luo, T. Z., & Maunsell, J. H. (2015). Neuronal modulations in visual cortex are associated with only one of multiple components of attention. *Neuron, 86*, 1182–1188.

Lupien, S. J., Juster, R. P., Raymond, C., & Marin, M. F. (2018). The effects of chronic stress on the human brain: From neurotoxicity, to vulnerability, to opportunity. *Frontiers in Neuroendocrinology, 49*, 91–105.

Luppi, P.-H., Clément, O., Sapin, E., Gervasoni, D., Peyron, C., Léger, L., Salvert, D., & Fort, P. (2011). The neuronal network responsible for paradoxical sleep and its dysfunctions causing narcolepsy and rapid eye movement (REM) behavior disorder. *Sleep Medicine Reviews, 15*, 153–163.

Luria, A. R., & Majovski, L. V. (1977). Basic approaches used in American and Soviet clinical neuropsychology. *American Psychologist, 32*(11), 959.

Lyamin, O. I., Kosenko, P. O., Korneva, S. M., Vyssotski, A. L., Mukhametov, L. M., & Siegel, J. M. (2018). Fur seals suppress REM sleep for very long periods without subsequent rebound. *Current Biology, 28*(12), 2000–2005.

Lyamin, O. I., & Siegel, J. M. (2019). Sleep in aquatic mammals. *Handbook of Sleep Research, 30*, 375–393.

Lykken, D. T. (1959). The GSR in the detection of guilt. *Journal of Applied Psychology, 43*, 385–388.

Lynn, A. C., Padmanabhan, A., Simmonds, D., Foran, W., Hallquist, M. N., Luna, B., & O'Hearn, K. (2018). Functional connectivity differences in autism during face and car recognition: Underconnectivity and atypical age-related changes. *Developmental Science, 21*(1), e12508.

Lyon, A. M., Taylor, V. G., & Tesmer, J. J. G. (2014). Strike a pose: G_q complexes at the membrane. *Trends in Pharmacological Sciences, 35*, 23–30.

Ma, C. L., Ma, X. T., Wang, J. J., Liu, H., Chen, Y. F., & Yang, Y. (2017). Physical exercise induces hippocampal neurogenesis and prevents cognitive decline. *Behavioural Brain Research, 317*, 332–339.

Maalouf, M., Rho, J. M., & Mattson, M. P. (2008). The neuroprotective properties of caloric restriction, the ketogenic diet, and ketone bodies. *Brain Research Reviews, 59*, 293–315.

Macaluso, E., & Doricchi, F. (2013). Attention and predictions: Control of spatial attention beyond the endogenous-exogenous dichotomy. *Frontiers in Human Neuroscience, 7*, 685.

Macchi, G. (1989). Anatomical substrate of emotional reactions. In F. Boller & J. Grafman (Eds.), *Handbook of neuropsychology* (Vol. 3, pp. 283–304). New York, NY: Elsevier.

Maceira, A. M., Ripoll, C., Cosin-Sales, J., Igual, B., Gavilan, M., Salazar, J., . . . & Pennell, D. J. (2014). Long term effects of cocaine on the heart assessed by cardiovascular magnetic resonance at 3T. *Journal of Cardiovascular Magnetic Resonance, 16*, 26.

Machado-Vieira, R. (2018). Lithium, stress, and resilience in bipolar disorder: Deciphering this key homeostatic synaptic plasticity regulator. *Journal of Affective Disorders, 233*, 92–99.

Machado, V. M., Morte, M. I., Carreira, B. P., Azevedo, M. M., Takano, J.,...& Araújo, I. M. (2015). Involvement of calpains in adult neurogenesis: Implications for stroke. *Frontiers in Cellular Neuroscience, 9*, 22.

MacLean, P. D. (1952). Some psychiatric implications of physiological studies on frontotemporal portion of limbic system (visceral brain). *Electroencephalography and Clinical Neurophysiology, 4*, 407–418.

Madras, B. K. (2013). History of the discovery of the antipsychotic dopamine D2 receptor: A basis for the dopamine hypothesis of schizophrenia. *Journal of the History of the Neurosciences, 22*, 62–78.

Maeda, N. (2015). Proteoglycans and neuronal migration in the cerebral cortex during development and disease. *Frontiers in Neuroscience, 9*, 98.

Maekawa, F., Tsukahara, S., Kawashima, T., Nohara, K., & Ohki-Hamazaki, H. (2014). The mechanisms underlying sexual differentiation of behavior and physiology in mammals and birds: Relative contributions of sex steroids and sex chromosomes. *Frontiers in Neuroscience, 8*, 242.

Maggio, M., Colizzi, E., Fisichella, A., Valenti, G., Ceresini, G.,...& Ceda, G. P. (2013). Stress hormones, sleep deprivation and cognition in older adults. *Maturitas, 76*, 22–44.

Mahan, A. L., & Ressler, K. J. (2012). Fear conditioning, synaptic plasticity and the amygdala: Implications for posttraumatic stress disorder. *Trends in Neurosciences, 35*, 24–34.

Mahar, I., Bambico, F. R., Mechawar, N., & Nobrega, J. N. (2014). Stress, serotonin, and hippocampal neurogenesis in relation to depression and antidepressant effects. *Neuroscience & Biobehavioral Reviews, 38*, 173–192.

Mahar, M., & Cavalli, V. (2018). Intrinsic mechanisms of neuronal axon regeneration. *Nature Reviews Neuroscience, 19*, 323–337.

Mahoney, C. E., Cogswell, A., Koralnik, I. J., & Scammell, T. E. (2019). The neurobiological basis of narcolepsy. *Nature Reviews Neuroscience, 20*(2), 83–93.

Mahowald, M. W., & Schenck, C. H. (2005). Insights from studying human sleep disorders. *Nature, 437*, 1279–1285.

Mainland, J. D., Lundström, J. N., Reisert, J., & Lowe, G. (2014). From molecule to mind: An integrative perspective on odor intensity. *Trends in Neurosciences, 37*, 443–454.

Mair, W., Goymer, P., Pletcher, S. D., & Partridge, L. (2003). Demography of dietary restriction and death in *Drosophila. Science, 301*, 1731–1733.

Maisuradze. L. M. (2019). The Advantages of Substitution of REM Sleep Stages with Waking Episodes to Perform REM Sleep Reduction. *EC Neurology, 11*(6), 400–407.

Maixner, F., Krause-Kyora, B., Turaev, D., Herbig, A., Hoopmann, M. R., Hallows, J. L.,...& O'Sullivan, N. (2016). The 5300-year-old *Helicobacter pylori* genome of the Iceman. *Science, 351*, 162–165.

Majdan, M., & Shatz, C. J. (2006). Effects of visual experience on activity-dependent gene regulation in cortex. *Nature Neuroscience, 9*, 650–659.

Majewska, A. K., & Sur, M. (2006). Plasticity and specificity of cortical processing networks. *Trends in Neurosciences, 29*, 323–329.

Majuri, J., Joutsa, J., Johansson, J., Voon, V., Alakurtti, K., Parkkola, R., . . . & Forsback, S. (2017). Dopamine and opioid neurotransmission in behavioral addictions: A comparative PET study in pathological gambling and binge eating. *Neuropsychopharmacology, 42*(5), 1169–1177.

Mak-McCully, R. A., Deiss, S. R., Rosen, B. Q., Jung, K.-Y., Sejnowski, T. J.,...& Halgren, E. (2014). Synchronization of isolated downstates (k-complexes) may be caused by cortically-induced disruption of thalamic spindling. *PLoS Computational Biology, 10*, e1003855.

Makin, S. (2018). Alzheimer's disease: The amyloid hypothesis on trial. *Nature, 559*, S4–S7.

Makin, S. (2019). The undiscovered illness. *Scientific American, 320*(3), 36–41.

Makino, H., Hwang, E. J., Hedrick, N. G., & Komiyama, T. (2016). Circuit mechanisms of sensorimotor learning. *Neuron, 92*(4), 705–721.

Makinodan, M., Rosen, K. M., Ito, S., & Corfas, G. (2012). A critical period for social experience-dependent oligodendrocyte maturation and myelination. *Science, 337*, 1357–1360.

Malhi, G. S., Lingford-Hughes, A. R., & Young, A. H. (2016). Antidepressant treatment response: "I want it all, and I want it now!" *British Journal of Psychiatry, 208*, 101–103.

Malinow, R. (2016). Depression: Ketamine steps out of the darkness. *Nature, 533*(7604), 477–478.

Malki, K., Keers, R., Tosto, M. G., Lourdusamy, A., Carboni, L., Domenici, E.,...& Schalkwyk, L. C. (2014). The endogenous and reactive depression subtypes revisited: Integrative animal and human studies implicate multiple distinct molecular mechanisms underlying major depressive disorder. *BMC Medicine, 12*, 1.

Mallory, C. S., Hardcastle, K., Bant, J. S., & Giocomo, L. M. (2018). Grid scale drives the scale and long-term stability of place maps. *Nature Neuroscience, 21*(2), 270–282.

Malsbury, C. W. (1971). Facilitation of male rat copulatory behavior by electrical stimulation of the medial preoptic area. *Physiology & Behavior, 7*, 797–805.

Man, K., Damasio, A., Meyer, K., & Kaplan, J. T. (2015). Convergent and invariant object representations for sight, sound, and touch. *Human Brain Mapping, 36*, 3629–3640.

Man, K., Kaplan, J., Damasio, H., & Damasio, A. (2013). Neural convergence and divergence in the mammalian cerebral cortex: From experimental neuroanatomy to functional neuroimaging. *Journal of Comparative Neurology, 521*, 4097–4111.

Manchikanti, L., Fellows, B., Janata, J. W., Pampati, V., Grider, J. S., & Boswell, M. V. (2012). Opioid epidemic in the United States. *Pain physician, 15*(3 Suppl), ES9–38.

Mancuso, L., Uddin, L. Q., Nani, A., Costa, T., & Cauda, F. (2019). Brain functional connectivity in individuals with callosotomy and agenesis of the corpus callosum: A systematic review. *Neuroscience and Biobehavioral Reviews, 105*, 231–248.

Manger, P. R. (2013). Questioning the interpretations of behavioral observations of cetaceans: Is there really support for a special intellectual status for this mammalian order? *Neuroscience, 250*, 664–696.

Manger, P. R., & Siegel, J. M. (2020). Do all mammals dream? *Journal of Comparative Neurology, 1*–8.

Mangiaruga, A., Scarpelli, S., Bartolacci, C., & De Gennaro, L. (2018). Spotlight on dream recall: The ages of dreams. *Nature and Science of Sleep, 10*, 1–12.

Manglik, A., & Kobilka, B. (2014). The role of protein dynamics in GPCR function: Insights from the beta2AR and rhodopsin. *Current Opinion in Cell Biology, 27*, 136–143.

Manolio, T. A., Collins, F. S., Cox, N. J., Goldstein, D. B., Hindorff, L. A., Hunter, D. J., McCarthy, M. I., Ramos, E. M., Cardon, L. R., Chakravarti, A., Cho, J. H., Guttmacher, A. E., Kong A., Kruglyak, L., Mardis, E., Rotimi, C. N., Slatkin, M., Valle, D., Whittemore, A. S., Boehnke, M., Clark, A. G., Eichler, E. E., Gibson, G., Haines, J. L., Mackay, T. F., McCarroll, S. A., & Visscher, P. M. (2009). Finding the missing heritability of complex diseases. *Nature, 461*, 747–53.

Mansur, A., Castillo, P. R., Cabrero, F. R., & Bokhari, S. R. A. (2019). Restless leg syndrome. In *StatPearls [Internet]*. StatPearls Publishing.

Marchi, N., Granata, T., & Janigro, D. (2014). Inflammatory pathways of seizure disorders. *Trends in Neurosciences, 37*, 55–65.

Maren, S. (2015). Out with the old and in with the new: Synaptic mechanisms of extinction in the amygdala. *Brain Research, 1621*, 231–238.

Maren, S., Phan, K. L., & Liberzan, I. (2013). The contextual brain: Implications for fear conditioning, extinction, and psychopathology. *Nature Reviews Neuroscience, 23*, 417–428.

Marino, F., & Cosentino, M. (2016). Multiple sclerosis: Repurposing dopaminergic drugs for MS—the evidence mounts. *Nature Reviews Neurology, 12*(4), 191–192.

Markov, N. T., & Kennedy, H. (2013). The importance of being hierarchical. *Current Opinion in Neurobiology, 23*, 187–194.

Markram, H., Toledo-Rodriguez, M., Wang, Y., Gupta, A., Silberberg, G., & Wu, C. (2004). Interneurons of the neocortical inhibitory system. *Nature Reviews Neuroscience, 5*, 793–807.

Marlowe, W. B., Mancall, E. L., & Thomas, J. J. (1985). Complete Kluver-Bucy syndrome in man. *Cortex, 11*, 53–59.

Maron, E., & Nutt, D. (2017). Biological markers of generalized anxiety disorder. *Dialogues in Clinical Neuroscience, 19*(2), 147–157.

Maroon, J. C., Winkelman, R., Bost, J., Amos, A., Mathyssek, C., & Miele, V. (2015). Chronic traumatic encephalopathy in contact sports: A systematic review of all reported pathological cases. *PloS One, 10*, e0117338.

Marques, T., Nguyen, J., Fioreze, G., & Petreanu, L. (2018). The functional organization of cortical feedback inputs to primary visual cortex. *Nature Neuroscience, 21*(5), 757–764.

Marquez de la Plata, M., Hart, T., Hammond, F. M., Frol, A. B., Hudak, A., Harper, C. R., O'Neill-Pirozzi, T. M., Whyte, J., Carlile, M., & Diaz-Arrastia, R. (2008). Impact of age on long-term recovery from traumatic brain injury. *Archives of Physical Medicine and Rehabilitation, 89*, 896–903.

Marshall, J. F., & O'Dell, S. J. (2012). Methamphetamine influences on brain and behavior: Unsafe at any speed? *Trends in Neurosciences, 35*, 536–545.

Marshall, L. (2014). Two of a kind. *Scientific American, 311*, 22.

Martens, M. (2013). Developmental and cognitive troubles in Williams syndrome. *Handbook of Clinical Neurology, 111*, 291–293.

Martin-Fernandez, M., Jamison, S., Robin, L. M., Zhao, Z., Martin, E. D., Aguilar, J.,...& Araque, A. (2017). Synapse-specific astrocyte gating of amygdala-related behavior. *Nature Neuroscience, 20*(11), 1540–1548.

Martin, A., State, M., Koenig, K., Schultz, R., Dykens, E. M., Cassidy, S. B., & Leckman, J. F. (1998). Prader-Willi syndrome. *American Journal of Psychiatry, 155*, 1265–1272.

Martin, B. J. (1981). Effect of sleep deprivation on tolerance of prolonged exercise. *European Journal of Applied Physiology, 47*, 345–354.

Martin, C. R., Osadchiy, V., Kalani, A., & Mayer, E. A. (2018). The brain-gut-microbiome axis. *Cellular and Molecular Gastroenterology and Hepatology, 6*(2), 133–148.

Martinez-Conde, S., Macknik, S. L., & Hubel, D. H. (2004). The role of fixational eye movements in visual perception. *Nature Reviews Neuroscience, 5*, 229–240.

Martyn, J. J., Mao, J., & Bittner, E. A. (2019). Opioid tolerance in critical illness. *New England Journal of Medicine, 380*(4), 365–378.

Mascetti, G. G. (2016). Unihemispheric sleep and asymmetrical sleep: Behavioral, neurophysiological, and functional perspectives. *Nature and Science of Sleep, 8*, 221–238.

Mashour, G. A., Walker, E. E., & Martuza, R. L. (2005). Psychosurgery: Past, present, and future. *Brain Research Reviews, 48*, 409–419.

Maslej, M. M., Furukawa, T. A., Cipriani, A., Andrews, P. W., & Mulsant, B. H. (2020). Individual differences in response to antidepressants: A meta-analysis of placebo-controlled randomized clinical trials. *JAMA Psychiatry.*

Mason, M. F., Norton, M. I., Van Horn, J. D., Wegner, D. M., Grafton, S. T., & Macrae, C. N. (2007). Wandering minds: The default network and stimulus-independent thought. *Science, 315*, 393–395.

Massar, S. A., Lim, J., & Huettel, S. A. (2019). Sleep deprivation, effort allocation and performance. *Progress in Brain Research, 246*, 1–26.

Masters, R. K., Reither, E. N., Powers, D. A., Yang, Y. C., Burger, A. E., & Link, B. G. (2013). The impact of obesity on US mortality levels: The importance of age and cohort factors in population estimates. *American Journal of Public Health, 103*, 1895–1901.

Matosin, N., Fernandez-Enright, F., Lum, J. S., & Newell, K. A. (2017). Shifting towards a model of mGluR5 dysregulation in schizophrenia: Consequences for future schizophrenia treatment. *Neuropharmacology, 115*, 73–91.

Matsumoto, J., Urakawa, S., Hori, E., de Araujo, M. F. P., Sakuma, Y., Ono, T., & Nishijo, H. (2012). Neuronal responses in the nucleus accumbens shell during sexual behavior in male rats. *Journal of Neuroscience, 32*, 1672–1686.

Matsushita, S., Suzuki, K., Murashima, A., Kajioka, D., Acebedo, A. R., Miyagawa, S., . . . & Yamada, G. (2018). Regulation of masculinization: Androgen signalling for external genitalia development. *Nature Reviews Urology, 15*(6), 358–368.

Matsuzawa, T. (2013). Evolution of the brain and social behavior in chimpanzees. *Current Opinion in Neurobiology, 23*, 443–449.

Matthews, G., & Fuchs, P. (2010). The diverse roles of ribbon synapses in sensory neurotransmission. *Nature Reviews Neuroscience, 11*(12), 812–822.

Matthews, P. M. (2019). Chronic inflammation in multiple sclerosis—seeing what was always there. *Nature Reviews Neurology, 15*(10), 582–593.

Matthews, P. M., & Hampshire, A. (2016). Clinical concepts emerging from fMRI functional connectomics. *Neuron, 91*(3), 511–528.

Mattick, R. P., Breen, C., Kimber, J., & Davoli, M. (2014). Buprenorphine maintenance versus placebo or methadone maintenance for opioid dependence. *Cochrane Database of Systematic Reviews, 6*, CD002207.

Mattis, V. B., & Svendsen, C. N. (2018). Huntington modeling improves with age. *Nature Neuroscience, 21*(3), 301–302.

Mattison, J. A., Colman, R. J., Beasley, T. M., Allison, D. B., Kemnitz, J. W., Roth, G. S., . . . & Anderson, R. M. (2017). Caloric restriction improves health and survival of rhesus monkeys. *Nature Communications, 8*, 14063.

Mattson, M. P., Moehl, K., Ghena, N., Schmaedick, M., & Cheng, A. (2018). Intermittent metabolic switching, neuroplasticity and brain health. *Nature Reviews Neuroscience, 19*(2), 81–94.

Matzinger, P. (2002). The danger model: A renewed sense of self. *Science, 296*, 301–305.

Maurano, M. T., Humbert, R., Rynes, E., Thurman, R. E., Haugen, E., Wang, H., Reynolds, A. P., Sandstrom, R., Qu, H., Brody, J., Shafer, A., Neri, F., Lee, K., Kutyavin, T., Stehling-Sun, S., Johnson, A. K., Canfield, T. K., Giste, E., Diegel, M., Bates, D., Hansen, R. S., Neph, S., Sabo, P. J., Heimfeld, S., Raubitschek, A., Ziegler, S., Cotsapas, C., Sotoodehnia, N., Glass, I., Sunyaev, S. R., Kaul, R., & Stamatoyannopoulos, J. A. (2012). Systematic localization of common disease-associated variation in regulatory DNA. *Science, 337*, 1190–1195.

Maurer, D. (2017). Critical periods re-examined: Evidence from children treated for dense cataracts. *Cognitive Development, 42*, 27–36.

Maurer, D., & Lewis, T. L. (2018). Visual systems. In *The Neurobiology of Brain and Behavioral Development* (pp. 213–233). Academic Press.

Maurer, D., Ellemberg, D., & Lewis, T. L. (2006). Repeated measurements of contrast sensitivity reveal limits to visual plasticity after early binocular deprivation in humans. *Neuropsychologia, 44*, 2104–2112.

Mauss, A. S., Vlasits, A., Borst, A., & Feller, M. (2017). Visual circuits for direction selectivity. *Annual Review of Neuroscience, 40*, 211–230.

McCann, T. S. (1981). Aggression and sexual activity of male southern elephant seals, *Mirounga leonina. Journal of Zoology, 195*, 295–310.

McCarthy, A., Wafford, K., Shanks, E., Ligocki, M., Edgar, D. M., & Dijk, D. J. (2016). REM sleep homeostasis in the absence of REM sleep: Effects of antidepressants. *Neuropharmacology, 108*, 415–425.

McCarthy, M. M., Arnold, A. P., Ball, G. F., Blaustein, J. D., & De Vries, G. J. (2012). Sex differences in the brain: The not so inconvenient truth. *Journal of Neuroscience, 32*, 2241–2247.

McCarthy, M. M., Auger, A. P., & Perrot-Sinal, T. S. (2002). Getting excited about GABA and sex differences in the brain. *Trends in Neurosciences, 25*, 307–313.

McCarthy, M. M., Herold, K., & Stockman, S. L. (2018). Fast, furious and enduring: Sensitive versus critical periods in sexual differentiation of the brain. *Physiology & Behavior, 187*, 13–19.

McCarthy, M. M., Nugent, B. M., & Lenz, K. M. (2017). Neuroimmunology and neuroepigenetics in the establishment of sex differences in the brain. *Nature Reviews Neuroscience, 18*(8), 471–484.

McClintock, M. K., & Herdt, G. (1996). Rethinking puberty: The development of sexual attraction. *Current Directions in Psychological Sciences, 5*, 178–183.

McCoy, P. A., Huang, H-S., & Philpot, B. D. (2009). Advances in understanding visual cortex plasticity. *Current Opinion in Neurobiology, 19*, 298–304.

McCutcheon, R. A., Abi-Dargham, A., & Howes, O. D. (2019). Schizophrenia, dopamine and the striatum: From biology to symptoms. *Trends in Neurosciences, 42*, 205–220.

McDade, E., & Bateman, R. J. (2017). Stop Alzheimer's before it starts. *Nature, 547*(7662), 153–155.

McDermott, K. B., Watson, J. M., & Ojemann, J. G. (2005). Presurgical language mapping. *Current Directions in Psychological Science, 14*, 291–295.

McDougle, S. D., Ivry, R. B., & Taylor, J. A. (2016). Taking aim at the cognitive side of learning in sensorimotor adaptation tasks. *Trends in Cognitive Sciences, 20*(7), 535–544.

McEwen, B. S. (1983). Gonadal steroid influences on brain development and sexual differentiation. In R. O. Greep (Ed.), *Reproductive physiology IV*. Baltimore: University Park Press.

McEwen, B. S., & Karatsoreos, I. N. (2015). Sleep deprivation and circadian disruption: Stress, allostasis, and allostatic load. *Sleep Medicine Clinics, 10*, 1–10.

McEwen, B. S., Gray, J. D., & Nasca, C. (2015). Recognizing resilience: Learning from the effects of stress on the brain. *Neurobiology of Stress, 1*, 1–11.

McEwen, B. S., Nasca, C., & Gray, J. D. (2016). Stress effects on neuronal structure: Hippocampus, amygdala, and prefrontal cortex. *Neuropsychopharmacology, 41*(1), 3–23.

McGaugh, J. L. (2015). Consolidating memories. *Annual Review of Psychology, 66*, 1–24.

McGinnis, G. R., & Young, M. E. (2016). Circadian regulation of metabolic homeostasis: Causes and consequences. *Nature and Science of Sleep, 8*, 163–180.

McGirr, A., Berlim, M. T., Bond, D. J., Fleck, M. P., Yatham, L. N., & Lam, R. W. (2015). A systematic review and meta-analysis of randomized, double-blind, placebo-controlled trials of ketamine in the rapid treatment of major depressive episodes. *Psychological Medicine, 45*, 693–704.

McGlone, F., & Reilly, D. (2010). The cutaneous sensory system. *Neuroscience and Biobehavioral Reviews, 34*, 148–159.

McGlone, F., Wessberg, J., & Olausson, H. (2014). Discriminative and affective touch: Sensing and feeling. *Neuron, 82*, 737–755.

McGlone, J. (1977). Sex differences in the cerebral organization of verbal functions in patients with unilateral brain lesions. *Brain, 100*, 775–793.

McGlone, J. (1980). Sex differences in human brain asymmetry: A critical survey. *Behavioral and Brain Sciences, 3*, 215–263.

McGregor, M. M., & Nelson, A. B. (2019). Circuit mechanisms of Parkinson's disease. *Neuron, 101*(6), 1042–1056.

McKim, W. A. (1986). *Drugs and behavior: An introduction to behavioral pharmacology*. Englewood Cliffs, NJ: Prentice-Hall.

McKone, E., Crookes, K., Jeffery, L., & Dilks, D. D. (2012). A critical review of the development of face recognition: Experience is less important than previously believed. *Cognitive Neuropsychology, 29*, 174–212.

McLaughlin, T., Hindges, R., & O'Leary, D. D. M. (2003). Regulation of axial patterning of the retina and its topographic mapping in the brain. *Current Opinion in Neurobiology, 13*, 57–69.

McNamara, P., Johnson, P., McLaren, D., Harris, E., Beauharnais, C., & Auerbach, S. (2010). REM and NREM sleep mentation. *International Review of Neurobiology, 92*, 69–86.

McNaughton, N., & Zangrossi, H. H., Jr. (2008). Theoretical approaches to the modeling of anxiety in animals. In R. J. Blanchard, D. C. Blanchard, G. Griebel, & D. J. Nutt (Eds.), *Handbook of anxiety and fear* (Vol. 17, pp. 11–27). Oxford, England: Elsevier.

McPartland, J. M., Duncan, M., Di Marzo, V., & Pertwee, R. (2014). Are cannabidiol and delta-9-tetrahydrocannabivarin negative modulators of the endocannabinoid system? A systematic review. *British Journal of Pharmacology, 172*, 737–753.

Meaney, M. J., & Szyf, M. (2005). Maternal care as a model for experience-dependent chromatin plasticity? *Trends in Neurosciences, 28*, 456–463.

Meccariello, R., Santoro, A., D'Angelo, S., Morrone, R., Fasano, S., Viggiano, A., & Pierantoni, R. (2020). The epigenetics of the endocannabinoid system. *International Journal of Molecular Sciences, 21*(3), 1113.

Mechoulam, R., & Parker, L. A. (2013). The endocannabinoid system and the brain. *Annual Review of Psychology, 64*, 21–47.

Meddis, R. (1977). *The sleep instinct*. London, England: Routledge & Kegan Paul.

Medzhitov, R., & Janeway, C. A., Jr. (2002). Decoding the patterns of self and nonself by the innate immune system. *Science, 296*, 298–301.

Meier, M. H., Caspi, A., Ambler, A., Harrington, H., Houts, R., Keefe, R. S. E., McDonald, K., Ward, A., Poulton, R., & Moffitt, T. E. (2012). Persistent cannabis users show neuropsychological decline from childhood to midlife. *PNAS, 109*, E2657–E2664.

Meier, S. M., & Deckert, J. (2019). Genetics of anxiety disorders. *Current Psychiatry Reports, 21*(3), 16.

Mel, B. W. (2002). What the synapse tells the neuron. *Science, 295*, 1845–1846.

Meletti, S., Cantalupo, G., Santoro, F., Benuzzi, F., Marliani, A. F., Tassinari, C. A., & Rubboli, G. (2014). Temporal lobe epilepsy and emotion recognition without amygdala: A case study of Urbach-Wiethe disease and review of the literature. *Epileptic Disorders, 16*, 518–527.

Mello, N. K., & Negus, S. S. (1996). Preclinical evaluation of pharmacotherapies for treatment of cocaine and opioid abuse using drug self-administration procedures. *Neuropsychopharmacology, 14*, 375–424.

Melrose, S. (2015). Seasonal affective disorder: An overview of assessment and treatment approaches. *Depression Research and Treatment, 2015*, 178564.

Melzack, R. (1992, April). Phantom limbs. *Scientific American, 266*, 120–126.

Melzack, R., & Wall, P. D. (1982). *The challenge of pain*. London, England: Penguin.

Melzack, R., Israel, R., Lacroix, R., & Schultz, G. (1997). Phantom limbs in people with congenital limb deficiency or amputation in early childhood. *Brain, 120,* 1603–1620.

Melzer, N., Meuth, S. G., & Wiendl, H. (2012). Neuron-directed autoimmunity in the central nervous system: Entities, mechanisms, diagnostic clues, and therapeutic options. *Current Opinion in Neurology, 25,* 341–348.

Meney, I., Waterhouse, J., Atkinson, G., Reilly, T., & Davenne, D. (1988). The effect of one night's sleep deprivation on temperature, mood, and physical performance in subjects with different amounts of habitual physical activity. *Chronobiology International, 15,* 349–363.

Merabet, L. F., & Pascual-Leone, A. (2010). Neural reorganization following sensory loss: The opportunity of change. *Nature Reviews Neuroscience, 11,* 44–52.

Mercer, J. G., & Speakman, J. R. (2001). Hypothalamic neuropeptide mechanisms for regulating energy balance: From rodent models to human obesity. *Neuroscience and Biobehavioural Reviews, 25,* 101–116.

Mering, S., & Jolkkonen, J. (2015). Proper housing conditions in experimental stroke studies—special emphasis on environmental enrichment. *Frontiers in Neuroscience, 9,* 106.

Meye, F. J., Trusel, M., Soiza-Reilly, M., & Mameli, M. (2017). Neural circuit adaptations during drug withdrawal—spotlight on the lateral habenula. *Pharmacology Biochemistry and Behavior, 162,* 87–93.

Meyer-Lindenberg, A., Mervis, C. B., & Berman, K. F. (2006). Neural mechanisms in Williams syndrome: A unique window to genetic influences on cognition and behaviour. *Nature Reviews Neuroscience, 7,* 380–393.

Meyer, R. G., & Salmon, P. (1988). *Abnormal psychology* (2nd ed.). Boston, MA: Allyn & Bacon.

Meyhöfer, I., Steffens, M., Kasparbauer, A., Grant, P., Weber, B., & Ettinger, U. (2014). Neural mechanisms of smooth pursuit eye movements in schizotypy. *Human Brain Mapping,* Sept 5.

Mez, J., Daneshvar, D. H., Kiernan, P. T., Abdolmohammadi, B., Alvarez, V. E., Huber, B. R., . . . & Cormier, K. A. (2017). Clinicopathological evaluation of chronic traumatic encephalopathy in players of American football. *JAMA, 318,* 360–370.

Michel, P. P., Hirsch, E. C., & Hunot, S. (2016). Understanding dopaminergic cell death pathways in Parkinson disease. *Neuron, 90*(4), 675–691.

Micu, I., Plemel, J. R., Caprariello, A. V., Nave, K. A., & Stys, P. K. (2017). Axo-myelinic neurotransmission: a novel mode of cell signalling in the central nervous system. *Nature Reviews Neuroscience, 19*(1), 49–57.

Mieda, M. (2017). The roles of orexins in sleep/wake regulation. *Neuroscience Research, 118,* 56–65.

Mielke, M. M., Vemuri, P., & Rocca, W. A. (2014). Clinical epidemiology of Alzheimer's disease: Assessing sex and gender differences. *Clinical Epidemiology, 6,* 37–48.

Miendlarzewska, E. A., & Trost, W. J. (2014). How musical training affects cognitive development: Rhythm, reward and other modulating variables. *Frontiers in Neuroscience, 7,* 279.

Mignot, E. (2013). The perfect hypnotic? *Science, 340,* 36–38.

Miles, F. J., Everitt, B. J., Dalley, J. W., & Dickinson, A. (2004). Conditioned activity and instrumental reinforcement following long-term oral consumption of cocaine by rats. *Behavioral Neuroscience, 118,* 1331–1339.

Millan, M. J., Andrieux, A., Bartzokis, G., Cadenhead, K., Dazzan, P., Fusar-Poli, P., . . . & Kahn, R. (2016). Altering the course of schizophrenia: Progress and perspectives. *Nature Reviews Drug Discovery.*

Miller, D. I., & Halpern, D. F. (2014). The new science of cognitive sex differences. *Trends in Cognitive Sciences, 18,* 37–45.

Miller, E. K., & Buschman, T. J. (2013). Cortical circuits for the control of attention. *Current Opinion in Neurobiology, 23,* 216–222.

Miller, E. K., & Wallis, J. D. (2009). Executive function and higher-order cognition: Definition and neural substrates. *Encyclopedia of Neuroscience, 4*(99–104).

Miller, G. E., & Chen, E. (2006). Life stress and diminished expression of genes encoding glucocorticoid receptor and β_2-adrenergic receptor in children with asthma. *Proceedings of the National Academy of Sciences, USA, 103,* 5496–5501.

Miller, G. L., & Knudsen, E. I. (1999). Early visual experience shapes the representation of auditory space in the forebrain gaze fields of the barn owl. *Journal of Neuroscience, 19,* 2326–2336.

Miller, J. F., Neufang, M., Solway, A., Brandt, A., Trippel, . . . & Schulze-Bonhage, A. (2013). Neural activity in human hippocampal formation reveals the spatial context of retrieved memories. *Science, 342,* 111–1120.

Miller, K. D., Schnell, M. J., & Rall, G. F. (2016). Keeping it in check: chronic viral infection and antiviral immunity in the brain. *Nature Reviews Neuroscience, 17*(12), 766–776.

Miller, S. M., & Sahay, A. (2019). Functions of adult-born neurons in hippocampal memory interference and indexing. *Nature Neuroscience, 22,* 1565–1575.

Miller, T. M., & Cleveland, D. W. (2005). Treating neurodegenerative diseases with antibiotics. *Science, 307,* 361–362.

Milner, B. (1965). Memory disturbances after bilateral hippocampal lesions. In P. Milner & S. Glickman (Eds.), *Cognitive processes and the brain* (pp. 104–105). Princeton, NJ: D. Van Nostrand.

Milner, B. (1971). Interhemispheric differences in the localization of psychological processes in man. *British Medical Bulletin, 27,* 272–277.

Milner, B. (1974). Hemispheric specialization: Scope and limits. In F. O. Schmitt & F. G. Worden (Eds.), *The neurosciences: Third study program* (pp. 75–89). Cambridge, MA: MIT Press.

Milner, B., Corkin, S., & Teuber, H. L. (1968). Further analysis of the hippocampal amnesic syndrome: 14-year follow-up study of H.M. *Neuropsychologia, 6,* 317–338.

Milner, P. M. (1993, January). The mind and Donald O. Hebb. *Scientific American, 268,* 124–129.

Milton, A. L., & Everitt, B. J. (2012). The persistence of maladaptive memory: Addiction, drug memories and anti-relapse treatments. *Neuroscience and Biobehavioral Reviews, 36,* 1119–1139.

Mindell, D. P. (2009, January). Evolution in the everyday world. *Scientific American, 300,* 82–89.

Minger, S. L., Ekonomou, A., Carta, E. M., Chinoy, A., Perry, R. H., & Ballard, C. G. (2007). Endogenous neurogenesis in the human brain following cerebral infarction. *Regenerative Medicine, 2,* 69–74.

Minkel, J., Moreta, M., Muto, J., Htaik, O., Jones, C., . . . & Dinges, D. (2014). Sleep deprivation potentiates HPA axis stress reactivity in healthy adults. *Health Psychology, 33,* 1430–1434.

Mishkin, M., & Delacour, J. (1975). An analysis of short-term visual memory in the monkey. *Journal of Experimental Psychology: Animal Behavior Processes, 1,* 326–334.

Mishra, A., Reynolds, J. P., Chen, Y., Gourine, A. V., Rusakov, D. A., & Attwell, D. (2016). Astrocytes mediate neurovascular signaling to capillary pericytes but not to arterioles. *Nature Neuroscience, 19*(12), 1619–1627.

Mistlberger, R. E. (2011). Neurobiology of food anticipatory circadian rhythms. *Physiology & Behavior, 104,* 535–545.

Mistlberger, R. E., de Groot, M. H. M., Bossert, J. M., & Marchant, E. G. (1996). Discrimination of circadian phase in intact and suprachiasmatic nuclei-ablated rats. *Brain Research, 739,* 12–18.

Mitchell, D. G. V., & Greening, S. G. (2011). Conscious perception of emotional stimuli: Brain mechanisms. *The Neuroscientist, 18,* 386–398.

Mitchell, P. B., & Hadzi-Pavlovic, D. (2000). Lithium treatment for bipolar disorder. *Bulletin of the World Health Organization, 78,* 515–517.

Mitchell, R. L. C., & Phillips, L. H. (2015). The overlapping relationship between emotion perception and theory of mind. *Neuropsychologia, 70,* 1–10.

Miyashita, Y. (2004). Cognitive memory: Cellular and network machineries and their top-down control. *Science, 306,* 435–440.

Miyashita, Y. (2019). Perirhinal circuits for memory processing. *Nature Reviews Neuroscience, 20*(10), 577–592.

Mochcovitch, M. D., da Rocha Freire, R. C., Garcia, R. F., & Nardi, A. E. (2017). Can long-term pharmacotherapy prevent relapses in generalized anxiety disorder? A systematic review. *Clinical Drug Investigation, 37*(8), 737–743.

Moerel, M., De Martino, F., & Formisano, E. (2014). An anatomical and functional topography of human auditory cortical areas. *Frontiers in Neuroscience, 8,* 225.

Mogil, J. S. (2012). Sex differentiation in pain and pain inhibition: Multiple explanations of a controversial phenomenon. *Nature Reviews Neuroscience, 13,* 859–866.

Mohan, A., & Vanneste, S. (2017). Adaptive and maladaptive neural compensatory consequences of sensory deprivation—From a phantom percept perspective. *Progress in Neurobiology, 153,* 1–17.

Mohr, J. P. (1976). Broca's area and Broca's aphasia. In H. Whitaker & H. A. Whitaker (Eds.), *Studies in neurolinguistics* (Vol. 1, pp. 201–235). New York, NY: Academic Press.

Mok, V. C., Lam, B. Y., Wong, A., Ko, H., Markus, H. S., & Wong, L. K. (2017). Early-onset and delayed-onset poststroke dementia—revisiting the mechanisms. *Nature Reviews Neurology, 13,* 148–159.

Mokin, M., Rojas, H., & Levy, E. I. (2016). Randomized trials of endovascular therapy for stroke—impact on stroke care. *Nature Reviews Neurology, 12,* 86–94.

Moldovan, A.-S., Groiss, S. J., Elben, S., Südmeyer, M., Schnitzler, A., & Wojtecki, L. (2015). The treatment of Parkinson's disease with deep brain stimulation: Current issues. *Neural Regeneration Research, 10,* 1018–1022.

Molenberghs, P., Cunnington, R., & Mattingley, J. B. (2012). Brain regions with mirror properties: A meta-analysis of 125 human fMRI studies. *Neuroscience and Biobehavioral Reviews, 36,* 341–349.

Mollayeva, T., Mollayeva, S., & Colantonio, A. (2018). Traumatic brain injury: Sex, gender and intersecting vulnerabilities. *Nature Reviews Neurology, 14*(12), 711–722.

Monday, H. R., & Castillo, P. E. (2017). Closing the gap: Long-term presynaptic plasticity in brain function and disease. *Current Opinion in Neurobiology, 45,* 106–112.

Monday, H. R., Younts, T. J., & Castillo, P. E. (2018). Long-term plasticity of neurotransmitter release: Emerging mechanisms and contributions to brain function and disease. *Annual Review of Neuroscience, 41,* 299–322.

Money, J. (1975). Ablatio penis: Normal male infant sex-reassigned as a girl. *Archives of Sexual Behavior, 4*(1), 65–71.

Money, J., & Ehrhardt, A. A. (1972). *Man & woman, boy & girl.* Baltimore, MD: Johns Hopkins University Press.

Mongan, N. P., Tadokoro-Cuccaro, R., Bunch, T., & Hughes, I. A. (2015). Androgen insensitivity syndrome. *Best Practice & Research Clinical Endocrinology & Metabolism, 4,* 569–580.

Mongiat, L. A., & Schinder, A. F. (2014). A price to pay for adult neurogenesis. *Science, 344,* 594–595.

Monje, M. (2018). Myelin plasticity and nervous system function. *Annual Review of Neuroscience, 41,* 61–76.

Monk, T. H., Buysse, D. J., Welsh, D. K., Kennedy, K. S., & Rose, L. R. (2001). A sleep diary and questionnaire study of naturally short sleepers. *Journal of Sleep Research, 10*, 173–179.

Montero, T. D., & Orellana, J. A. (2015). Hemichannels: New pathways for gliotransmitter release. *Neuroscience, 286*, 45–59.

Montoya, E. R., Terburg, D., Bos, P. A., & van Honk, J. (2012). Testosterone, cortisol, and serotonin as key regulators of social aggression: A review and theoretical perspective. *Motivation and Emotion, 36*, 65–73.

Moore, D. J., West, A. B., Dawson, V. L., & Dawson, T. M. (2005). Molecular pathophysiology of Parkinson's disease. *Annual Review of Neuroscience, 28*, 57–87.

Moore, T., & Zirnsak, M. (2017). Neural mechanisms of selective visual attention. *Annual Review of Psychology, 68*, 47–72.

Moore, Y. E., Kelley, M. R., Brandon, N. J., Deeb, T. Z., & Moss, S. J. (2017). Seizing control of KCC2: A new therapeutic target for epilepsy. *Trends in Neurosciences, 40*(9), 555–571.

Mor, D. E., Tsika, E., Mazzulli, J. R., Gould, N. S., Kim, H., Daniels, M. J., . . . & Kalb, R. G. (2017). Dopamine induces soluble α-synuclein oligomers and nigrostriatal degeneration. *Nature Neuroscience, 20*(11), 1560–1568.

Moran, J., & Desimone, R. (1985). Selective attention gates visual processing in the extrastriate cortex. *Science, 229*, 782–784.

Moran, Y., Barzilai, M. G., Liebeskind, B. J., & Zakon, H. H. (2015). Evolution of voltage-gated ion channels at the emergence of the Metazoa. *Journal of Experimental Biology, 218*, 515–525.

Moratalla, R., Khairnar, A., Simola, N., Granado, N., García-Montes, J. R., Porceddu, P. F., . . . & Morelli, M. (2017). Amphetamine-related drugs neurotoxicity in humans and in experimental animals: Main mechanisms. *Progress in Neurobiology, 155*, 149–170.

Moretti, R., Caruso, P., Dal Ben, M., Gazzin, S., & Tiribelli, C. (2017). Thiamine and alcohol for brain pathology: Super-imposing or different causative factors for brain damage? *Current Drug Abuse Reviews, 10*(1), 44–51.

Morey, L., Santanach, A., & Di Croce, L. (2015). Pluripotency and epigenetic factors in mouse embyonic stem cell fate regulation. *Molecular and Cellular Biology, 35*, 2716–2728.

Morgan, T. H., Sturtevant, A. H., Muller, H. J., & Bridges, C. B. (1915). *The mechanism of Mendelian heredity*. New York, NY: Holt.

Morgenthaler, J., Wiesner, C. D., Hinze, K., Abels, L. C., Prehn-Kristensen, A., & Göder, R. (2014). Selective REM-sleep deprivation does not diminish emotional memory consolidation in young healthy subjects. *PLoS ONE, 9*, e89849.

Mori, K., & Sakano, H. (2011). How is the olfactory map formed and interpreted in the mammalian brain? *Annual Review of Neuroscience, 34*, 467–499.

Mori, M., Rikitake, Y., Mandai, K., & Takai, Y. (2014). Roles of nectins and nectin-like molecules in the nervous system. *Advances in Neurobiology, 8*, 91–116.

Morici, J. F., Bekinschtein, P., & Weisstaub, N. V. (2015). Medial prefrontal cortex in recognition memory in rodents. *Behavioural Brain Research, 292*, 241–251.

Morimoto, K., Fahnestock, M., & Racine, R. J. (2004). Kindling and status epilepticus models of epilepsy: Rewiring the brain. *Progress in Neurobiology, 73*, 1–60.

Morrie, R. D., & Feller, M. B. (2016). Development of synaptic connectivity in the retinal direction selective circuit. *Current Opinion in Neurobiology, 40*, 45–52.

Morris, C. J., Aeschbach, D., & Scheer, F. A. J. L. (2012). Circadian system, sleep and endocrinology. *Molecular and Cellular Endocrinology, 349*, 91–104.

Morris, M. J., Beilharz, J. E., Maniam, J., Reichelt, A. C., & Westbrook, R. F. (2014). Why is obesity such a problem in the 21st century? The intersection of palatable food, cues and reward pathways, stress, and cognition. *Neuroscience and Biobehavioral Reviews, 58*, 36–45.

Morris, N. M., Udry, J. R., Khan-Dawood, F., & Dawood, M. Y. (1987). Marital sex frequency and midcycle female testosterone. *Archives of Sexual Behavior, 16*, 27–37.

Morris, R. G. M. (1981). Spatial localization does not require the presence of local cues. *Learning and Motivation, 12*, 239–260.

Morton, G. J., Meek, T. H., & Schwartz, M. W. (2014). Neurobiology of food intake in health and disease. *Nature Reviews Neuroscience, 15*, 367–377.

Moruzzi, G., & Magoun, H. W. (1949). Brain stem reticular formation and activation of the EEG. *Electroencephalography and Clinical Neurophysiology, 1*, 455–473.

Moscarello, J. M., & Maren, S. (2018). Flexibility in the face of fear: Hippocampal–prefrontal regulation of fear and avoidance. *Current Opinion in Behavioral Sciences, 19*, 44–49.

Moscati, A., Flint, J., & Kendler, K. S. (2015). Classification of anxiety disorders comorbid with major depression: Common or distinct influences on risk? *Depression and Anxiety, 33*, 120–127.

Moschak, T. M., Terry, D. R., Daughters, S. B., & Carelli, R. M. (2018). Low distress tolerance predicts heightened drug seeking and taking after extended abstinence from cocaine self-administration. *Addiction Biology, 23*(1), 130–141.

Moscovitch, M., Cabeza, R., Winocur, G., & Nadel, L. (2016). Episodic memory and beyond: The hippocampus and neocortex in transformation. *Annual Review of Psychology, 67*, 105–134.

Moseley, G. L., Gallace, A., & Spence, C. (2012). Bodily illusions in health and disease: Physiological and clinical perspectives and the concept of a cortical 'body matrix.' *Neuroscience and Biobehavioral Reviews, 36*, 34–46.

Moseley, G. L., Zalucki, N., Birklein, F., Marinus, J., van Hilten, J. J., & Luomajoki, H. (2008). Thinking about movement hurts: The effect of motor imagery on pain and swelling in people with chronic arm pain. *Arthritis Care Research, 59*, 623–631.

Moser, E. I., & Moser, M.-B. (2013). Grid cells and neural coding in high-end cortices. *Neuron, 80*, 765–774.

Moser, E. I., Moser, M. B., & McNaughton, B. L. (2017). Spatial representation in the hippocampal formation: A history. *Nature Neuroscience, 20*(11), 1448–1464.

Moser, E. I., Roudi, Y., Witter, M. P., Kentros, C., Bonhoeffer, T., & Moser, M.-B. (2014). Grid cells and cortical representation. *Nature Reviews Neuroscience, 15*, 466–481.

Moser, M. B., & Moser, E. I. (2016). Where am I? Where am I going? *Scientific American, 314*, 26–33.

Motta-Mena, N. V., & Puts, D. A. (2017). Endocrinology of human female sexuality, mating, and reproductive behavior. *Hormones and Behavior, 91*, 19–35.

Mottron, L., Duret, P., Mueller, S., Moore, R. D., d'Arc B. F., . . . & Xiong, L. (2015). Sex differences in brain plasticity: A new hypothesis for sex ratio bias in autism. *Molecular Autism, 6*, 33.

Mount, C. W., & Monje, M. (2017). Wrapped to adapt: Experience-dependent myelination. *Neuron, 95*(4), 743–756.

Moye, L. S., & Pradhan, A. A. (2017). From blast to bench: A translational mini-review of posttraumatic headache. *Journal of Neuroscience Research, 95*(6), 1347–1354.

Mu, Y., Lee, S. W., & Gage, F. H. (2010). Signaling in adult neurogenesis. *Current Opinion in Neurobiology, 20*, 416–423.

Muckli, L., & Petro, L. S. (2013). Network interactions: Non-geniculate input to V1. *Current Opinion in Neurobiology, 23*, 195–201.

Mueller, F., Lenz, C., Steiner, M., Dolder, P. C., Walter, M., Lang, U. E., . . . & Borgwardt, S. (2016). Neuroimaging in moderate MDMA use: A systematic review. *Neuroscience & Biobehavioral Reviews, 62*, 21–34.

Mühlnickel, W., Elbert, T., Taub, E., & Flor, H. (1998). Reorganization of auditory cortex in tinnitus. *Proceedings of the National Academy of Sciences, USA, 95*, 10340–10343.

Mukamel, R., Ekstrom, A. D., Kaplan, J., Iacoboni, M., & Fried, I. (2010). Single neuron responses in humans during execution and observation of actions. *Current Biology, 20*, 750–756.

Mullette-Gillman, O. A., Cohen, Y. E., & Groh, J. M. (2005). Eye-centered, head-centered, and complex coding of visual and auditory targets in the intraparietal sulcus. *Journal of Neurophysiology, 94*, 2331–2352.

Mullaney, D. J., Johnson, L. C., Naitoh, P., Friedman, J. K., & Globus, G. G. (1977). Sleep during and after gradual sleep reduction. *Psychophysiology, 14*, 237–244.

Müller-Dahlhaus, F., & Ziemann, U. (2015). Metaplasticity in human cortex. *Neuroscientist, 21*, 185–202.

Müller, C. P., & Schumann, G. (2011). Drugs as instruments: A new framework for non-addictive psychoactive drug use. *Behavioral and Brain Sciences, 34*, 293–347.

Müller, F., Brändle, R., Liechti, M. E., & Borgwardt, S. (2019). Neuroimaging of chronic MDMA ("ecstasy") effects: A meta-analysis. *Neuroscience & Biobehavioral Reviews, 96*, 10–20.

Mullins, C., Fishell, G., & Tsien, R. W. (2016). Unifying views of autism spectrum disorders: A consideration of autoregulatory feedback loops. *Neuron, 89*, 1131–1156.

Mumby, D. G. (2001). Perspectives on object-recognition memory following hippocampal damage: Lessons from studies in rats. *Behavioural Brain Research, 14*, 159–181.

Mumby, D. G., & Pinel, J. P. J. (1994). Rhinal cortex lesions impair object recognition in rats. *Behavioral Neuroscience, 108*, 11–18.

Mumby, D. G., Cameli, L., & Glenn, M. J. (1999). Impaired allocentric spatial working memory and intact retrograde memory after thalamic damage caused by thiamine deficiency in rats. *Behavioral Neuroscience, 113*, 42–50.

Mumby, D. G., Pinel, J. P. J., & Wood, E. R. (1989). Nonrecurring items delayed nonmatching-to-sample in rats: A new paradigm for testing nonspatial working memory. *Psychobiology, 18*, 321–326.

Mumby, D. G., Wood, E. R., & Pinel, J. P. J. (1992). Object-recognition memory is only mildly impaired in rats with lesions of the hippocampus and amygdala. *Psychobiology, 20*, 18–27.

Mumby, D. G., Wood, E. R., Duva, C. A., Kornecook, T. J., Pinel, J. P. J., & Phillips, A. G. (1996). Ischemia-induced object-recognition deficits in rats are attenuated by hippocampal ablation before or soon after ischemia. *Behavioral Neuroscience, 110*(2), 266–281.

Munley, K. M., Rendon, N. M., & Demas, G. E. (2018). Neural androgen synthesis and aggression: Insights from a seasonally breeding rodent. *Frontiers in Endocrinology, 9*, 136.

Münzberg, H., & Myers, M. G. (2005). Molecular and anatomical determinants of central leptin resistance. *Nature Neuroscience, 8*, 566–570.

Muoio, V., Persson, P. B., & Sendeski, M. M. (2014). The neurovascular unit—concept review. *Acta Physiologica, 210*, 790–798.

Murdoch, B. E. (2010). The cerebellum and language: Historical perspective and review. *Cortex, 46*, 858–868.

Murphy, M. R., & Schneider, G. E. (1970). Olfactory bulb removal eliminates mating behavior in the male golden hamster. *Science, 157*, 302–304.

Murray, D. R., & Schaller, M. (2016). The behavioral immune system: Implications for social cognition, social interaction, and social influence. *Advances in Experimental Social Psychology, 53,* 75–129.

Murray, G., Suto, M., Hole, R., Hale, S., Amari, E., & Michalak, E. E. (2011). Self-management strategies used by "high functioning" individuals with bipolar disorder: From research to clinical practice. *Clinical Psychology & Psychotherapy, 18,* 95–109.

Murray, M. M., & Herrmann, C. S. (2013). Illusory contours: A window onto the neurophysiology of constructing perception. *Trends in Cognitive Sciences, 17,* 471–481.

Murray, M. M., Lewkowicz, D. J., Amedi, A., & Wallace, M. T. (2016). Multisensory processes: A balancing act across the lifespan. *Trends in Neurosciences, 39*(8), 567–579.

Murrough, J. W., Abdallah, C. G., & Mathew, S. J. (2017). Targeting glutamate signalling in depression: Progress and prospects. *Nature Reviews Drug Discovery, 16*(7), 472–486.

Murru, A., Popovic, D., Pacchiarotti, I., Hidalgo, D., León-Caballero, J., & Vieta, E. (2015). Management of adverse effects of mood stabilizers. *Current Psychiatry Reports, 17,* 1–10.

Murthy, V. N. (2011). Olfactory maps in the brain. *Annual Review of Neuroscience, 34,* 233–258.

Musiek, E. S., & Holtzman, D. M. (2016). Mechanisms linking circadian clocks, sleep, and neurodegeneration. *Science, 354*(6315), 1004–1008.

Mutz, J., & Javadi, A. H. (2017). Exploring the neural correlates of dream phenomenology and altered states of consciousness during sleep. *Neuroscience of Consciousness, 2017*(1), nix009.

Myers, R. E., & Sperry, R. W. (1953). Interocular transfer of a visual form discrimination habit in cats after section of the optic chiasma and corpus callosum. *Anatomical Record, 115,* 351–352.

Mysore, S. P., & Knudsen, E. I. (2014). Descending control of neural bias and selectivity in a spatial attention network: Rules and mechanisms. *Neuron, 84,* 214–226.

Nabavi, S., Fox, R., Proulx, C. D., Lin, J. Y., Tsien, R. Y., & Malinow, R. (2014). Engineering a memory with LTD and LTP. *Nature, 511,* 348–352.

Nachev, P., Kennard, C., & Husain, M. (2008). Functional role of the supplementary and pre-supplementary motor areas. *Nature Reviews Neuroscience, 9,* 856–869.

Nadeau, J. H. (2009). Transgenerational genetic effects on phenotypic variation and disease risk. *Human Molecular Genetics, 18*(R2), R202–R210.

Nadel, L., & Moscovitch, M. (1997). Memory consolidation, retrograde amnesia and the hippocampal complex. *Current Opinion in Neurobiology, 7*(2), 217–227.

Nader, K. (2015). Reconsolidation and the dynamic nature of memory. *Cold Spring Harbor Perspectives in Biology, 7,* a021782.

Nader, K., Schafe, G. E., & LeDoux, J. E. (2000). Fear memories require protein synthesis in the amygdala for reconsolidation after retrieval. *Nature, 406,* 722–726.

Nader, M. A., Czoty, P. W., Gould, R. W., & Riddick, N. V. (2008). Positron emission tomography imaging studies of dopamine receptors in primate models of addiction. *Philosophical Transactions of the Royal Society B, 363,* 3223–3232.

Nagai, C., Inui, T., & Iwata, M. (2011). Fading-figure tracing in William syndrome. *Brain and Cognition, 75,* 10–17.

Nagy, J. I., Dudek, F. E., & Rash, J. E. (2004). Update on connexins and gap junctions in neurons and glia in the mammalian nervous system. *Brain Research Reviews, 47,* 191–215.

Nambu, A. (2008). Seven problems on the basal ganglia. *Current Opinion in Neurobiology, 18,* 595–604.

Nasios, G., Dardiotis, E., & Messinis, L. (2019). From Broca and Wernicke to the neuromodulation era: Insights of brain language networks for neurorehabilitation. *Behavioural Neurology, 2019,* 9894571.

National Commission on Marijuana and Drug Abuse. (1972). *Marijuana: A signal of misunderstanding.* New York, NY: New American Library.

Nau, M., Schröder, T. N., Bellmund, J. L., & Doeller, C. F. (2018). Hexadirectional coding of visual space in human entorhinal cortex. *Nature Neuroscience, 21*(2), 188–190.

Navarrete, M., & Araque, A. (2014). The Cajal school and the physiological role of astrocytes: A way of thinking. *Frontiers in Neuroanatomy, 8,* 33.

Nave, K. A., & Ehrenreich, H. (2018). A bloody brake on myelin repair. *Nature, 553*(7686), 31–32.

Navratilova, E., & Porreca, F. (2014). Reward and motivation in pain and pain relief. *Nature Neuroscience, 17,* 1304–1312.

Navratilova, Z., & Battaglia, F. P. (2015). CA2: It's about time—and episodes. *Neuron, 85,* 8–10.

Naya, Y., & Suzuki, W. A. (2011). Integrating what and when across the primate medial temporal lobe. *Science, 333,* 773–775.

Naya, Y., Yoshida, M., & Miyashita, Y. (2001). Backward spreading of memory-retrieval signal in the primate temporal cortex. *Science, 291,* 661–664.

Neary, J. T., & Zimmerman, H. (2009). Trophic functions of nucleotides in the central nervous system. *Trends in Neurosciences, 32,* 189–198.

Neavyn, M. J., Blohm, E., Babu, K. M., & Bird, S. B. (2014). Medical marijuana and driving: A review. *Journal of Medical Toxicology, 10,* 269–279.

Nedergaard, M., & Goldman, S. A. (2016). Brain Drain. *Scientific American, 314,* 44–49.

Nees, F. (2014). The nicotinic cholinergic system function in the human brain. *Neuropharmacology, 96,* 289–301.

Negrón-Oyarzo, I., Lara-Vásquez, A., Palacios-García, I., Fuentealba, P., & Aboitiz, F. (2016). Schizophrenia and reelin: A model based on prenatal stress to study epigenetics, brain development and behavior. *Biological Research, 49,* 1.

Neikrug, A. B., Avanzino, J. A., Liu, L., Maglione, J. E., Natarajan, L.,…& Ancoli-Israel, S. (2014). Parkinson's disease and REM sleep behavior disorder result in increased non-motor symptoms. *Sleep Medicine, 15,* 959–966.

Nelson, A. B., & Kreitzer, A. C. (2014). Reassessing models of basal ganglia function and dysfunction. *Annual Review of Neuroscience, 37,* 117–135.

Nelson, V. R., Nadeau, J. H. (2010). Transgenerational genetic effects. *Epigenomics, 2,* 797–800.

Neniskyte, U., & Gross, C. T. (2017). Errant gardeners: Glial-cell-dependent synaptic pruning and neurodevelopmental disorders. *Nature Reviews Neuroscience, 18*(11), 658–670.

Nestler, E. J. (2005). Is there a common molecular pathway for addiction? *Nature neuroscience, 8,* 1445–1449.

Nestler, E. J. (2011). Hidden switches in the mind. *Scientific American, 305*(6), 76–83.

Nestler, E. J. (2014). Epigenetic mechanisms of depression. *JAMA Psychiatry, 71,* 454–456.

Nestler, E. J., Barrot, M., DiLeone, R. J., Eisch, A. J., Gold, S. J., & Monteggia, L. M. (2002). Neurobiology of depression. *Neuron, 34,* 13–25.

Netter, F. H. (1962). *The CIBA collection of medical illustrations. Vol. 1, The nervous system.* New York, NY: CIBA.

Neubauer, D. N. (2014). New and emerging pharmacotherapeutic approaches for insomnia. *International Review of Psychiatry, 26,* 214–224.

Neubauer, S., Gunz, P., Scott, N. A., Hublin, J. J., & Mitteroecker, P. (2020). Evolution of brain lateralization: A shared hominid pattern of endocranial asymmetry is much more variable in humans than in great apes. *Science Advances, 6*(7), eaax9935.

Neukomm, L. J., & Freeman, M. R. (2014). Diverse cellular and molecular modes of axon degeneration. *Trends in Cell Biology, 24,* 515–523.

Neumann, D., Keiski, M. A., McDonald, B. C., & Wang, Y. (2014). Neuroimaging and facial affect processing: Implications for traumatic brain injury. *Brain Imaging and Behavior, 8,* 460–473.

Neves, S. R., Ram, P. T., & Iyengar, R. (2002). G protein pathways. *Science, 296,* 1636–1639.

Newcombe, N. S., Drummey, A. B., Fox, N. A., Lie, E., & Ottinger Alberts, W. (2000). Remembering early childhood: How much, how, and why (or why not). *Current Directions in Psychological Science, 9,* 55–58.

Newcombe, N., & Fox, N. (1994). Infantile amnesia: Through a glass darkly. *Child Development, 65,* 31–40.

Newman, E. A. (2003). New roles for astrocytes: Regulation of synaptic transmission. *Trends in Neurosciences, 26,* 536–542.

Nguyen, B. M., Kim, D., Bricker, S., Bongard, F., Neville, A., Putnam, B., Smith, J., & Plurad, D. (2014). Effect of marijuana use on outcomes in traumatic brain injury. *The American Surgeon, 80,* 979–983.

Nichols, D. E. (2016). Psychedelics. *Pharmacological Reviews, 68,* 264–355.

Nicolesis, M. A. L., & Chapin, J. K. (October, 2002). People with nerve or limb injuries may one day be able to command wheelchairs, prosthetics and even paralyzed arms and legs by "thinking them through" the motions. *Scientific American, 287,* 47–53.

Nicoll, R. A. (2017). A brief history of long-term potentiation. *Neuron, 93*(2), 281–290.

Niculescu, D., & Lohmann, C. (2014). Gap junctions in developing thalamic and neocortical networks. *Cerebral Cortex, 24,* 3097–3106.

Niederberger, E., Resch, E., Parnham, M. J., & Geisslinger, G. (2017). Drugging the pain epigenome. *Nature Reviews Neurology, 13*(7), 434–447.

Niego, A., & Benítez-Burraco, A. (2019). Williams Syndrome, human self-domestication, and language evolution. *Frontiers in Psychology, 10,* 521–521.

Niego, A., & Benítez-Burraco, A. (2019). Williams syndrome, human self-domestication, and language evolution. *Frontiers in Psychology, 10,* 521.

Nielsen, J. B. (2016). Human spinal motor control. *Annual Review of Neuroscience, 39,* 81–101.

Niessen, E., Fink, G. R., & Weiss, P. H. (2014). Apraxia, pantomime and the parietal cortex. *Neuroimage: Clinical, 5,* 42–52.

Nigro, S. C., Luon, D., & Baker, W. L. (2013). Lorcaserin: A novel serotonin 2c agonist for the treatment of obesity. *Current Medical Research and Opinion, 29,* 839–848.

Nimmerjahn, A., & Bergles, D. E. (2015). Large-scale recording of astrocyte activity. *Current Opinion in Neurobiology, 32,* 95–106.

Nir, Y., & Tononi, G. (2010). Dreaming and the brain: From phenomenology to neurophysiology. *Trends in Cognitive Sciences, 14,* 88–100.

Nisbett, R. E., Aronson, J., Blair, C., Dickens, W., Flynn, J., Halpern, D. F., & Turkheimer, E. (2012). Intelligence: New findings and theoretical developments. *American Psychologist, 67,* 130–159.

Nishino, S., & Kanbayashi, T. (2005). Symptomatic narcolepsy, cataplexy and hypersomnia, and their implications in the hypothalamic hypocretin/orexin system. *Sleep Medicine Reviews, 9,* 269–310.

Nobre, A. C., & van Ede, F. (2018). Anticipated moments: Temporal structure in attention. *Nature Reviews Neuroscience, 19*(1), 34.

Nolden, A. A., & Feeney, E. L. (2020). Genetic Differences in Taste Receptors: Implications for the Food Industry. *Annual Review of Food Science and Technology, 11.*

Nomi, J. S., & Uddin, L. Q. (2015). Face processing in autism spectrum disorders: From brain regions to brain networks. *Neuropsychologia, 71*, 201–216.

Noonan, D. (2015). Marijuana's medical future. *Scientific American, 312*, 32–34.

Nordmann, G. C., Hochstoeger, T., & Keays, D. A. (2017). Magnetoreception—A sense without a receptor. *PLoS Biology, 15*(10), e2003234.

Normile, D. (2014). Faulty drug trials tarnish Japan's clinical research. *Science, 345*, 17.

Northoff, G. (2013). Gene, brains, and environment—genetic neuroimaging of depression. *Current Opinion in Neurobiology, 23*, 133–142.

Northoff, G. (2020). Anxiety disorders and the brain's resting state networks: from altered spatiotemporal synchronization to psychopathological symptoms. In *Anxiety Disorders* (pp. 71–90). Springer, Singapore.

Norton, E. S., Beach, S. D., & Gabrieli, J. D. E. (2015). Neurobiology of dyslexia. *Current Opinion in Neurobiology, 30*, 73–78.

Nosyk, B., Guh, D. P., Bansback, N. J., Oviedo-Joekes, E., Brissette, S., Marsh, D. C., Meikleham, E., Schecter, M. T., & Anis, A. H. (2012). Cost-effectiveness of diacetylmorphine versus methadone for chronic opioid dependence refractory to treatment. *CMAJ, 184*, E317–E328.

Noudoost, B., Chang, M. H., Steinmetz, N. A., & Moore, T. (2010). Top-down control of visual attention. *Current Opinion in Neurobiology, 20*, 183–190.

Novotny, M., Valis, M., & Klimova, B. (2018). Tourette syndrome: A mini-review. *Frontiers in neurology, 9*, 139.

Nowak, M. A. (2012). Why we help: Far from being a nagging exception to the rule of evolution, cooperation has been one of its primary architects. *Scientific American, 307*, 34–39.

Nudo, R. J., Jenkins, W. M., & Merzenich, M. M. (1996). Repetitive microstimulation alters the cortical representation of movements in adult rats. *Somatosensory Motor Research, 7*, 463–483.

Nusslock, R., & Alloy, L. B. (2017). Reward processing and mood-related symptoms: An RDoC and translational neuroscience perspective. *Journal of Affective Disorders, 216*, 3–16.

Nutt, D. J. (2015). Consideration on the role of buprenorphine in recovery from heroin addiction from a UK perspective. *Journal of Psychopharmacology, 29*, 43–49.

Nutt, D. J., King, L. A., & Nichols, D. E. (2013). Effects of Schedule I drug laws on neuroscience research and treatment innovation. *Nature Reviews Neuroscience, 14*, 577–585.

Nutt, D. J., Lingford-Hughes, A., Erritzoe, D., & Stokes, P. R. (2015). The dopamine theory of addiction: 40 years of highs and lows. *Nature Reviews Neuroscience, 16*, 305–312.

Nykamp, K., Rosenthal, L., Folkerts, M., Roehrs, T., Guido, P., & Roth, T. (1998). The effects of REM sleep deprivation on the level of sleepiness/alertness. *Sleep, 21*, 609–614.

Nymo, S., Coutinho, S. R., Eknes, P. H., Vestbostad, I., Rehfeld, J. F., Truby, H., . . . & Martins, C. (2018). Investigation of the long-term sustainability of changes in appetite after weight loss. *International Journal of Obesity, 42*(8), 1489–1499.

O'Brien, B. A., Wolf, M., & Lovett, M. W. (2012). A taxometric investigation of developmental dyslexia subtypes. *Dyslexia, 18*, 16–39.

O'Brien, R. J., & Wong, P. C. (2011). Amyloid precursor protein processing and Alzheimer's disease. *Annual Review of Neuroscience, 34*, 185–204.

O'Connor, E. C., Chapman, K., Butler, P., & Mead, A. N. (2011). The predictive validity of the rat self-administration model for abuse liability. *Neuroscience and Biobehavioral Reviews, 35*, 912–938.

O'Hanlan, K. A., Gordon, J. C., & Sullivan, M. W. (2018). Biological origins of sexual orientation and gender identity: Impact on health. *Gynecologic Oncology, 149*(1), 33–42.

O'Leary, C. M. (2004). Fetal alcohol syndrome: Diagnosis, epidemiology, and developmental outcomes. *Journal of Paediatrics and Child Health, 40*, 2–7.

O'Reilly, R. C. (2010). The *what* and *how* of prefrontal cortical organization. *Trends in Neurosciences, 33*, 355–361.

O'Sullivan, M., Brownsett, S., & Copland, D. (2019). Language and language disorders: Neuroscience to clinical practice. *Practical Neurology, 19*(5), 380–388.

O'Tousa, D., & Grahame, N. (2014). Habit formation: Implications for alcoholism research. *Alcohol, 48*, 327–335.

Oesch, N. W., Kothmann, W. W., & Diamond, J. S. (2011). Illuminating synapses and circuitry in the retina. *Current Opinion in Neurobiology, 21*, 238–244.

Oestreich, A. K., & Moley, K. H. (2017). Developmental and transmittable origins of obesity-associated health disorders. *Trends in Genetics, 33*(6), 399–407.

Ogden, C. L., Carroll, M. D., Kit, B. K., & Flegal, K. M. (2014). Prevalence of childhood and adult obesity in the United States, 2011–2012. *JAMA, 311*, 806–814.

Oh, S. W., Harris, J. A., Ng, L., Winslow, B., Cain, N., Mihalas, S., . . . & Mortrud, M. T. (2014). A mesoscale connectome of the mouse brain. *Nature, 508*, 207–214.

Ohno-Shosaku, T., & Kano, M. (2014). Endocannabinoid-mediated retrograde modulation of synaptic transmission. *Current Opinion in Neurobiology, 29*, 1–8.

Ohshima, T. (2015). Neuronal migration and protein kinases. *Frontiers in Neuroscience, 8*, 458.

Ohzawa, I. (1998). Mechanisms of stereoscopic vision: The disparity energy model. *Current Opinion in Neurobiology, 8*, 509–515.

Ojemann, G. A. (1979). Individual variability in cortical localization of language. *Journal of Neurosurgery, 50*, 164–169.

Oken, B. S., Chamine, I., & Wakeland, W. (2015). A systems approach to stress, stressors and resilience in humans. *Behavioural Brain Research, 282*, 144–154.

Okon-Singer, H., Hendler, T., Pessoa, L., & Shackman, A. J. (2015). The neurobiology of emotion–cognition interactions: Fundamental questions and strategies for future research. *Frontiers in Human Neuroscience, 9*, 58.

Oldenburg, I. A., & Sabatini, B. L. (2015). Antagonistic but not symmetric regulation of primary motor cortex by basal ganglia direct and indirect pathways. *Neuron, 86*, 1174–1181.

Olds, J., & Milner, P. (1954). Positive reinforcement produced by electrical stimulation of septal area and other regions of rat brain. *Journal of Comparative and Physiological Psychology, 47*, 419–427.

Oller, D. K., & Griebel, U. (2014). On quantitative comparative research in communication and language evolution. *Biological Theory, 9*, 296–308.

Olshansky, M. P., Bar, R. J., Fogarty, M., & DeSouza, J. F. (2015). Supplementary motor area and primary auditory cortex activation in an expert break-dancer during kinesthetic motor imagery of dance to music. *Neurocase, 21*, 607–617.

Olson, I. R., McCoy, D., Klobusicky, E., & Ross, L. A. (2013). Social cognition and the anterior temporal lobes: A review and theoretical framework. *Social Cognitive and Affective Neuroscience, 8*, 123–133.

Olson, S. D., Pollock, K., Kambal, A., Cary, W., Mitchell, G.-M., Tempkin, J., Stewart, H., McGee, J., Bauer, G., Kim, H. S., Tempkin, T., Wheelock, V., Annett, G., Dunbar, G., & Nolta, J. A. (2012). Genetically engineered mesenchymal stem cells as a proposed therapeutic for Huntington's disease. *Molecular Neurobiology, 45*, 87–98.

Olton, D. S., & Samuelson, R. J. (1976). Remembrance of places: Spatial memory in rats. *Journal of Experimental Psychology: Animal Behavior Processes, 2*, 97–116.

Omenn, G. S., Lane, L., Overall, C. M., Corrales, F. J., Schwenk, J. M., Paik, Y. K., . . . & Deutsch, E. W. (2018). Progress on identifying and characterizing the Human proteome: 2018 metrics from the HUPO Human proteome project. *Journal of Proteome Research.* 10.1021/acs.jproteome.8b00441.

Onaolapo, A. Y., Obelawo, A. Y., & Onaolapo, O. J. (2019). Brain ageing, cognition and diet: A review of the emerging roles of food-based nootropics in mitigating age-related memory decline. *Current Aging Science, 12*(1), 2–14.

Ondo, W. G. (2019). Treatment of restless legs syndrome and periodic limb movements. In *Therapy of Movement Disorders* (pp. 329–331). Humana, Cham.

Onishi, K., Hollis, E., & Zou, Y. (2014). Axon guidance and injury—lessons from Wnts and Wnt signaling. *Current Opinion in Neurobiology, 27*, 232–240.

Oram, M. (2016). Prohibited or regulated? LSD psychotherapy and the United States Food and Drug Administration. *History of Psychiatry, 27*, 290–306.

Orban, G. A., & Caruana, F. (2014). The neural basis of human tool use. *Frontiers in Psychology, 5*, 1–12.

Oren, D. A., Koziorowski, M., & Desan, P. H. (2013). SAD and the not-so-single photoreceptors. *American Journal of Psychiatry, 170*, 1403–1412.

Ormel, J., Hartman, C. A., & Snieder, H. (2019). The genetics of depression: Successful genome-wide association studies introduce new challenges. *Translational Psychiatry, 9*(1), 1–10.

Orser, B. A. (2007, June). Lifting the fog around anesthesia. *Scientific American, 296*, 54–61.

Oruch, R., Elderbi, M. A., Khattab, H. A., Pryme, I. F., & Lund, A. (2014). Lithium: A review of pharmacology, clinical uses, and toxicity. *European Journal of Pharmacology, 740*, 464–473.

Osoegawa, C., Gomes, J. S., Grigolon, R. B., Brietzke, E., Gadelha, A., Lacerda, A. L., . . . & Daskalakis, Z. J. (2018). Non-invasive brain stimulation for negative symptoms in schizophrenia: An updated systematic review and meta-analysis. *Schizophrenia Research, 197*, 34–44.

Ostry, D. J., & Gribble, P. L. (2016). Sensory plasticity in human motor learning. *Trends in Neurosciences, 39*(2), 114–123.

Otten, M., & Meeter, M. (2015). Hippocampal structure and function in individuals with bipolar disorder: A systematic review. *Journal of Affective Disorders, 174*, 113–125.

Ou, C.-Y., & Shen, K. (2010). Setting up presynaptic structures at specific positions. *Current Opinion in Neurobiology, 20*, 489–493.

Oudiette, D., Dealberto, M.-J., Uguccioni, G., Golmard, J.-L., Milagros, M.-L., Tafti, M., Garma, L., Schwartz, S., & Arnulf, I. (2012). Dreaming without REM sleep. *Consciousness and Cognition, 21*, 1129–1140.

Oudman, E., Van der Stigchel, S., Wester, A. J., Kessels, R. P. C., & Postma, A. (2011). Intact memory for implicit contextual information in Korsakoff's amnesia. *Neuropsychologia, 49*, 2848–2855.

Overmier, J. B., & Murison, R. (1997). Animal models reveal the "psych" in the psychosomatics of peptic ulcers. *Current Directions in Psychological Science, 6*, 180–184.

Owen, M. J., Sawa, A., & Mortensen, P. B. (2016). Schizophrenia. *Lancet.* Advance online publication. doi:10.1016/S0140-6736(15)01121-6.

Owens, D. M., & Lumpkin, E. A. (2014). Diversification and specialization of touch receptors in skin. *Cold Spring Harbor Perspectives in Medicine, 4*, a013656.

Oxenham, A. J. (2018). How we hear: The perception and neural coding of sound. *Annual Review of Psychology, 69*, 27–50.

Padro, C. J., & Sanders, V. M. (2014). Neuroendocrine regulation of inflammation. *Seminars in Immunology, 26*, 357–368.

Pagel, J. F. (2003). Non-dreamers. *Sleep Medicine, 4*(3), 235–241.

Pagliaccio, D., Barch, D. M., Bogdan, R., Wood, P. K., Lynskey, M. T., . . . & Agrawal, A. (2015). Shared predisposition in the association between cannabis use and subcortical brain structure. *JAMA Psychiatry, 72*, 994–1001.

Painter, M. W. (2017). Aging Schwann cells: Mechanisms, implications, future directions. *Current Opinion in Neurobiology, 47*, 203–208.

Paintner, A., Williams, A. D., & Burd, L. (2012). Fetal alcohol spectrum disorders—implications for child neurology, part I: Prenatal exposure and dosimetry. *Journal of Child Neurology, 27*, 258–263.

Paixão, S., & Klein, R. (2010). Neuron-astrocyte communication and synaptic plasticity. *Current Opinion in Neurobiology, 20*, 466–473.

Palagini, L., & Rosenlicht, N. (2011). Sleep, dreaming, and mental health: A review of historical and neurobiological perspectives. *Sleep Medicine Reviews, 15*, 179–186.

Palmer, B. F., & Clegg, D. J. (2015). The sexual dimorphism of obesity. *Molecular and Cellular Endocrinology, 402*, 113–119.

Palmer, L. M., & Stuart, G. J. (2006). Site of action potential initiation in layer 5 pyramidal neurons. *Journal of Neuroscience, 26*, 1854–1863.

Palmiter, R. D. (2007). Is dopamine a physiologically relevant mediator of feeding behavior? *Trends in Neurosciences, 30*, 375–381.

Pan, W. W., & Myers Jr, M. G. (2018). Leptin and the maintenance of elevated body weight. *Nature Reviews Neuroscience, 19*(2), 95–105.

Pandey, S., & Dash, D. (2019). Progress in pharmacological and surgical management of Tourette syndrome and other chronic tic disorders. *The Neurologist, 24*(3), 93–108.

Panksepp, J., & Trowill, J. A. (1967). Intraoral self-injection: II. The simulation of self-stimulation phenomena with a conventional reward. *Psychonomic Science, 9*, 407–408.

Pannasch U., & Rouach, N. (2013). Emerging role for astroglial networks in information processing: From synapse to behavior. *Trends in Neuroscience, 36*, 405–417.

Paoletti, P., Ellis-Davies, G. C., & Mourot, A. (2019). Optical control of neuronal ion channels and receptors. *Nature Reviews Neuroscience, 20*, 514–532.

Paolicelli, R. C., Bergamini, G., & Rajendran, L. (2018). Cell-to-cell communication by extracellular vesicles: Focus on microglia. *Neuroscience*, in press.

Papke, R. L. (2014). Merging old and new perspectives on nicotinic acetylcholine receptors. *Biochemical Pharmacology, 89*, 1–11.

Parikh, S. V., LeBlanc, S. R., & Ovanessian, M. M. (2010). Advancing bipolar disorder: Key lessons from the Systematic Treatment Enhancement Program for Bipolar Disorder (STEP-BD). *Canadian Journal of Psychiatry, 55*, 136–143.

Park, C. H., Chang, W. H., Lee, M., Kwon, G. H., Kim, L., . . . & Kim, Y. H. (2015). Which motor cortical region best predicts imagined movement? *Neuroimage, 113*, 101–110.

Park, H.-J., & Friston, K. (2013). Structural and functional brain networks: From connectomics to cognition. *Science, 342*, 579–584.

Park, H., & Poo, M. (2013). Neurotrophin regulation of neural circuit development and function. *Nature Reviews Neuroscience, 14*, 7–23.

Park, P., Kang, H., Sanderson, T. M., Bortolotto, Z. A., Georgiou, J., Zhuo, M., . . . & Collingridge, G. L. (2018). The role of calcium-permeable AMPARs in long-term potentiation at principal neurons in the rodent hippocampus. *Frontiers in Synaptic Neuroscience, 10*, 42.

Park, T. J., Brand, A., Koch, U., Ikebuchi, M., & Grothe, B. (2008). Dynamic changes in level influence spatial coding in the lateral superior olive. *Hearing Research, 238*, 58–67.

Parker, S. T., Mitchell, R. W., & Boccia, M. L. (1994). *Self-awareness in animals and humans: Developmental perspectives*. New York, NY: Cambridge University Press.

Parkhurst, C. N., Yang, G., Ninan, I., Savas, J. N., Yates, J. R. 3rd, et al. (2013). Microglia promote learning-dependent synapse formation through brain-derived neurotrophic factor. *Cell, 155*, 1596–1609.

Parrott, A. C. (2013). Human psychobiology of MDMA or "Ecstasy": An overview of 25 years of empirical research. *Human Psychopharmacology, 28*, 289–307.

Parrott, D. J., & Eckhardt, C. I. (2018). Effects of alcohol on human aggression. *Current Opinion in Psychology, 19*, 1–5.

Parsons, L. H., & Hurd, Y. L. (2015). Endocannabinoid signalling in reward and addiction. *Nature Reviews Neuroscience, 16*, 579–594.

Parsons, M. P., & Raymond, L. A. (2014). Extrasynaptic NMDA receptor involvement in central nervous system disorders. *Neuron, 82*, 279–293.

Partridge, B. J., Bell, S. K., Lucke, J. C., Yeates, S., & Hall, W. D. (2012). Smart drugs "as common as coffee": Media hype about neuroenhancement. *PLoS One, 6*, e28416.

Paşca, S. P. (2018). The rise of three-dimensional human brain cultures. *Nature, 553*, 437–445.

Pascual-Leone, A., Amedi, A., Fregni, F., & Merabet, L. B. (2005). The plastic human brain cortex. *Annual Review of Neuroscience, 28*, 377–401.

Pascual-Leone, A., Walsh, V., & Rothwell, J. (2000). Transcranial magnetic stimulation in cognitive neuroscience—virtual lesion, chronometry, and functional connectivity. *Current Opinion in Neurobiology, 10*, 232–237.

Pasman, J. A., Verweij, K. J., Gerring, Z., Stringer, S., Sanchez-Roige, S., Treur, J. L., . . . & Ip, H. F. (2018). GWAS of lifetime cannabis use reveals new risk loci, genetic overlap with psychiatric traits, and a causal effect of schizophrenia liability. *Nature Neuroscience, 21*(9), 1161–1170.

Pasterkamp, R. J. (2012). Getting neural circuits into shape with semaphorins. *Nature Reviews Neuroscience, 13*, 605–618.

Pasterski, V., Geffner, M. E., Brain, C., Hindmarsh, P., Brook, C., & Hines, M. (2011). Prenatal hormones and childhood sex-segregation: Playmate and play style preferences in girls with congenital adrenal hyperplasia. *Hormones & Behavior, 59*, 549–555.

Patel, G. H., Kaplan, D. M., & Snyder, L. H. (2014). Topographic organization in the brain: Searching for general principles. *Trends in Cognitive Sciences, 18*, 351–363.

Patin, A., & Pause, B. M. (2015). Human amygdala activations during nasal chemoreception. *Neuropsychologia, 78*, 171–194.

Patke, A., Young, M. W., & Axelrod, S. (2019). Molecular mechanisms and physiological importance of circadian rhythms. *Nature Reviews Molecular Cell Biology*, 1–18.

Patrick M. E., & Maggs, J. L. (2009). Does drinking lead to sex? Daily alcohol-sex behaviors and expectancies among college students. *Psychology of Addictive Behaviors, 23*, 472–481.

Patrick, Y., Lee, A., Raha, O., Pillai, K., Gupta, S., Sethi, S., . . . & Smith, S. F. (2017). Effects of sleep deprivation on cognitive and physical performance in university students. *Sleep and Biological Rhythms, 15*(3), 217–225.

Patterson, K., & Ralph, M. A. L. (1999). Selective disorders of reading? *Current Opinion in Neurobiology, 9*, 235–239.

Patterson, K., Vargha-Khadem, F., & Polkey, C. E. (1989). Reading with one hemisphere. *Brain, 112*, 39–63.

Patton, G. C., Olsson, C. A., Skirbekk, V., Saffery, R., Wlodek, M. E., Azzopardi, P. S., . . . & Bhutta, Z. A. (2018). Adolescence and the next generation. *Nature, 554*(7693), 458–466.

Patzke, N., Olaleye, O., Haagensen, M., Hof, P. R., Ihunwo, A. O., & Manger, P. R. (2014). Organization and chemical neuroanatomy of the African elephant (*Loxodonta africana*) hippocampus. *Brain Structure and Function, 219*, 1587–1601.

Paulesu, E., McCrory, E., Fazio, F., Menoncello, L., Brunswick, N., Cappa, S. F., . . . & Frith, U. (2000). A cultural effect on brain function. *Nature Neuroscience, 3*, 91–96.

Pauls, S. D., Honma, K. I., Honma, S., & Silver, R. (2016). Deconstructing circadian rhythmicity with models and manipulations. *Trends in Neurosciences, 39*(6), 405–419.

Paus, T. (2010). Growth of white matter in the adolescent brain: Myelin or axon? *Brain and Cognition, 72*, 26–35.

Pawluski, J. L., Lonstein, J. S., & Fleming, A. S. (2017). The neurobiology of postpartum anxiety and depression. *Trends in Neurosciences, 40*, 106–120.

Paz, R., & Pare, D. (2013). Physiological basis for emotional modulation of memory circuits by the amygdala. *Current Opinion in Neurobiology, 23*, 381–386.

Pearce, T. M., & Moran, D. W. (2012). Strategy-dependent encoding of planned arm movements in the dorsal premotor cortex. *Science, 337*, 984–988.

Peelen, M. V., & Kastner, S. (2014). Attention in the real world: Toward understanding its neural basis. *Trends in Cognitive Sciences, 18*, 242–250.

Peelle, J. E., & Wingfield, A. (2016). The neural consequences of age-related hearing loss. *Trends in Neurosciences, 39*(7), 486–497.

Peever, J., Luppi, P.-H., & Montplaisir, J. (2014). Breakdown in REM sleep circuitry underlies REM sleep behavior disorder. *Trends in Neurosciences, 37*, 279–288.

Pellis, S. M., O'Brien, D. P., Pellis, V. C., Teitelbaum, P., Wolgin, D. L., & Kennedy, S. (1988). Escalation of feline predation along a gradient from avoidance through "play" to killing. *Behavioral Neuroscience, 102*, 760–777.

Pellman, B. A., & Kim, J. J. (2016). What can ethobehavioral studies tell us about the brain's fear system? *Trends in Neurosciences, 39*(6), 420–431.

Pellow, S., Chopin, P., File, S. E., & Briley, M. (1985). Validation of open:closed arm entries in an elevated plus-maze as a measure of anxiety in the rat. *Journal of Neuroscience Methods, 14*, 149–167.

Pena, J. L., & Gutfreund, Y. (2014). New perspectives on the owl's map of auditory space. *Current Opinion in Neurobiology, 24*, 55–62.

Penfield, W., & Boldrey, E. (1937). Somatic motor and sensory representations in cerebral cortex of man as studied by electrical stimulation. *Brain, 60*, 389–443.

Penfield, W., & Evans, J. (1935). The frontal lobe in man: A clinical study of maximum removals. *Brain, 58*, 115–133.

Penfield, W., & Rasmussen, T. (1950). *The cerebral cortex of man: A clinical study of the localization of function*. New York, NY: Macmillan.

Penfield, W., & Roberts, L. (1959). *Speech and brain mechanisms*. Princeton, NJ: Princeton University Press.

Peng, K., Steele, S. C., Becerra, L., & Borsook, D. (2017). Brodmann area 10: Collating, integrating and high level processing of nociception and pain. *Progress in Neurobiology, 161*, 1–22.

Peng, Y., Gillis-Smith, S., Jin, H., Tränkner, D., Ryba, N. J., & Zuker, C. S. (2015). Sweet and bitter taste in the brain of awake behaving animals. *Nature, 527*, 512–515.

Penner-Goeke, S., & Binder, E. B. (2019). Epigenetics and depression. *Dialogues in Clinical Neuroscience, 21*(4), 397.

Pennisi, E. (2014). Genetics may foster bugs that keep you thin. *Science, 346*, 687.

Pennisi, E. (2014). Lengthy RNAs earn respect as cellular players. *Science, 344*, 1072.

Pereda, A. E. (2014). Electrical synapses and their functional interactions with chemical synapses. *Nature Reviews Neuroscience, 15*(4), 250–263.

Perry, A., Roberts, G., Mitchell, P. B., & Breakspear, M. (2019). Connectomics of bipolar disorder: a critical review, and evidence for dynamic instabilities within interoceptive networks. *Molecular Psychiatry, 24*(9), 1296–1318.

Perry, E. C. (2014). Inpatient management of acute alcohol withdrawal syndrome. *CNS Drugs, 28*, 401–410.

Persico, A. M., & Bourgeron, T. (2006). Searching for ways out of the autism maze: Genetic, epigenetic and environmental clues. *Trends in Neurosciences, 29*, 349–358.

Perugini, A., Ditterich, J., Shaikh, A. G., Knowlton, B. J., & Basso, M. A. (2018). Paradoxical decision-making: A framework for understanding Cognition in Parkinson's disease. *Trends in Neurosciences, 41*(8), 512–525.

Pessoa, L. (2018). Understanding emotion with brain networks. *Current Opinion in Behavioral Sciences, 19*, 19–25.

Peterburs, J., & Desmond, J. E. (2016). The role of the human cerebellum in performance monitoring. *Current Opinion in Neurobiology, 40*, 38–44.

Peters, A. J., Lee, J., Hedrick, N. G., O'Neil, K., & Komiyama, T. (2017). Reorganization of corticospinal output during motor learning. *Nature Neuroscience, 20*(8), 1133–1141.

Petersen, S. E., Fox, P. T., Posner, M. I., Mintun, M., & Raichle, M. E. (1988). Positron emission tomographic studies of the cortical anatomy of single-word processing. *Nature, 331*, 585–589.

Petronis, A. (2010). Epigenetics as a unifying principle in the aetiology of complex traits and diseases. *Nature, 465*, 721–727.

Petrovic, M. M., da Silva, S. V., Clement, J. P., Vyklicky, L., Mulle, C., González-González, I. M., & Henley, J. M. (2017). Metabotropic action of postsynaptic kainate receptors triggers hippocampal long-term potentiation. *Nature Neuroscience, 20*(4), 529–539.

Petrovska, S., Dejanova, B., & Jurisic, V. (2012). Estrogens: Mechanisms of neuroprotective effects. *Journal of Physiology and Biochemistry, 68*, 455–460.

Peyrin, C., Lallier, M., Démonet, J. F., Pernet, C., Baciu, M., Le Bas, J. F., & Valdois, S. (2012). Neural dissociation of phonological and visual attention span disorders in developmental dyslexia: FMRI evidence from two case reports. *Brain and Language, 120*, 381–394.

Pezawas, L., Meyer-Lindenberg, A., Drabant, E. M., Verchinski, B. A., Munoz, K. E., Kolachana, B. S., ... & Weinberger, D. R. (2005). 5-HTTLPR polymorphism impacts human cingulate-amygdala interactions: A genetic susceptibility mechanism for depression. *Nature Neuroscience, 8*, 828–834.

Pezzulo, G., van der Meer, M. A. A., Lansink, C. S., & Pennartz, C. M. A. (2014). Internally generated sequences in learning and executing goal-directed behavior. *Trends in Cognitive Sciences, 18*, 647–657.

Pfaus, J. G., Kippin, T. E., Coria-Avila, G. A., Gelez, H., Afonso, V. M., Ismail, N., & Parada, M. (2012). Who, what, where, when (and maybe even why)? How the experience of sexual reward connects sexual desire, preference, and performance. *Archives of Sexual Behavior, 41*, 31–62.

Pfeiffer, C. A. (1936). Sexual differences of the hypophyses and their determination by the gonads. *American Journal of Anatomy, 58*, 195–225.

Pfrieger, F. W. (2010). Role of glial cells in the formation and maintenance of synapses. *Brain Research Reviews, 63*, 39–46.

Phelps, E. A., Lempert, K. M., & Sokol-Hessner, P. (2014). Emotion and decision making: Multiple modulatory neural circuits. *Annual Review of Neuroscience, 37*, 263–287.

Phelps, M. E., & Mazziotta, J. (1985). Positron tomography: Human brain function and biochemistry. *Science, 228*, 804.

Philip, R. C. M., Dauvermann, M. R., Whalley, H. C., Baynham, K., Lawrie, S. M., & Stanfield, A. C. (2012). A systematic review and meta-analysis of the fMRI investigation of autism spectrum disorders. *Neuroscience and Biobehavioral Reviews, 36*, 901–942.

Phillips, A. G., Coury, A., Fiorino, D., LePiane, F. G., Brown, E., & Fibiger, H. C. (1992). Self-stimulation of the ventral tegmental area enhances dopamine release in the nucleus accumbens: A microdialysis study. *Annals of the New York Academy of Sciences, 654*, 199–206.

Phillips, D. R., Quinlan, C. K., & Dingle, R. N. (2012). Stability of central binaural sound localization mechanisms in mammals, and the Heffner hypothesis. *Neuroscience and Biobehavioral Reviews, 36*, 889–900.

Phillips, M. L., & Kupfer, D. J. (2013). Bipolar disorder diagnosis: Challenges and future directions. *Lancet, 381*, 1663–1671.

Phoenix, C. H., Goy, R. W., Gerall, A. A., & Young, W. C. (1959). Organizing action of prenatally administered testosterone proprionate on the tissues mediating mating behavior in the female guinea pig. *Endocrinology, 65*, 369–382.

Piaggio, L. A. (2014). Congenital adrenal hyperplasia: Review from a surgeon's perspective in the beginning of the twenty-first century. *Frontiers in Pediatrics, 1*, 50.

Pickens, C. L., Airavaara, M., Theberge, F., Fanous, S., Hope, B. T., & Shaham, Y. (2011). Neurobiology of the incubation of drug craving. *Trends in Neurosciences, 34*, 411–420.

Pickering, R., Dirks, P. H. G. M., Jinnah, Z., de Ruiter, D. J., Churchil, S. E., Herries, A. I. R., Woodhead, J. D., & Hellstrom, J. C., & Berger, L. R. (2011). *Australopethicus sediba* at 1.977 Ma and implications for the origins of the genus *Homo. Science, 333*, 1421–1423.

Pierce, R. C., & Kumaresan, V. (2006). The mesolimbic dopamine system: The final common pathway for the reinforcing effect of drugs of abuse? *Neuroscience and Behaviourial Reviews, 30*, 215–238.

Pierce, R. C., & Vanderschuren, L. J. M. J. (2010). Kicking the habit: The neural basis of ingrained behaviors in cocaine addiction. *Neuroscience and Biobehavioral Reviews, 35*, 212–219.

Pierce, S. E., Hutchinson, J. R., & Clack, J. A. (2013). Historical perspectives on the evolution of tetrapodomorph movement. *Integrative and Comparative Biology, 53*, 209–223.

Pinel, J. P. J. (1969). A short gradient of ECS-produced amnesia in a one-trial appetitive learning situation. *Journal of Comparative and Physiological Psychology, 68*, 650–655.

Pinel, J. P. J., & Mana, M. J. (1989). Adaptive interactions of rats with dangerous inanimate objects: Support for a cognitive theory of defensive behavior. In R. J. Blanchard, P. F. Brain, D. C. Blanchard, & S. Parmigiani (Eds.), *Ethoexperimental approaches to the study of behavior* (pp. 137–150). Dordrecht, Netherlands: Kluwer Academic Publishers.

Pinel, J. P. J., & Treit, D. (1978). Burying as a defensive response in rats. *Journal of Comparative and Physiological Psychology, 92*, 708–712.

Pinel, J. P. J., Assanand, S., & Lehman, D. R. (2000). Hunger, eating, and ill health. *American Psychologist, 55*, 1105–1116.

Pinel, J. P. J., Mana, M. J., & Kim, C. K. (1989). Effect-dependent tolerance to ethanol's anticonvulsant effect on kindled seizures. In R. J. Porter, R. H. Mattson, J. A. Cramer, & I. Diamond (Eds.), *Alcohol and seizures: Basic mechanisms and clinical implications* (pp. 115–125). Philadelphia, PA: F. A. Davis.

Piomelli, D. (2014). More surprises lying ahead. The endocannabinoids keep us guessing. *Neuropharmacology, 76*, 228–234.

Pires, A. O., Teixeira, F. G., Mendes-Pinheiro, B., Serra, S. C., Sousa, N., & Salgado, A. J. (2017). Old and new challenges in Parkinson's disease therapeutics. *Progress in Neurobiology, 156*, 69–89.

Pisanello, F., Mandelbaum, G., Pisanello, M., Oldenburg, I. A., Sileo, L., Markowitz, J. E., ... & Spagnolo, B. (2017). Dynamic illumination of spatially restricted or large brain volumes via a single tapered optical fiber. *Nature Neuroscience, 20*(8), 1180.

Pisinger, C., Godtfredsen, N., & Bender, A. M. (2019). A conflict of interest is strongly associated with tobacco industry–favourable results, indicating no harm of e-cigarettes. *Preventive Medicine, 119*, 124–131.

Pistillo, F., Clementi, F., Zoli, M., & Gotti, C. (2014). Nicotinic, glutamatergic, and dopaminergic synaptic transmission and plasticity in the mesocorticolimbic system: Focus on nicotine effects. *Progress in Neurobiology, 124*, 1–27.

Pistoia, F., Carolei, A., Sacco, S., Conson, M., Pistarini, C., ... & Sarà, M. (2015). Contribution of interoceptive information to emotional processing: Evidence from individuals with spinal cord injury. *Journal of Neurotrauma, 32*, 1981–1986.

Plans, L., Barrot, C., Nieto, E., Rios, J., Schulze, T. G., Papiol, S., ... & Benabarre, A. (2019). Association between completed suicide and bipolar disorder: A systematic review of the literature. *Journal of Affective Disorders, 242*, 111–122.

Plant, T. M. (2015). 60 years of neuroendocrinology: The hypothalamo-pituitary-gonadal axis. *Journal of Endocrinology, 226*, T41–T54.

Platel, J.-C., Stamboulian, S., Nguyen, I., & Bordey, A. (2010). Neurotransmitter signaling in postnatal neurogenesis: The first leg. *Brain Research, 63*, 60–71.

Plesnila, N. (2016). The immune system in traumatic brain injury. *Current Opinion in Pharmacology, 26*, 110–117.

Ploegh, H. L. (1998). Viral strategies of immune evasion. *Science, 280*, 248–252.

Plomin, R., & McGuffin, P. (2005). Psychopathology in the postgenomic era. *Annual Review of Psychology, 54*, 205–228.

Poels, E. M. P., Kegeles, L. S., Kantrowitz, J. T., Slifstein, M., Javitt, D. C., Lieberman, J. A., ... & Girgis, R. R. (2014). Imaging glutamate in schizophrenia: Review of findings and implications for drug discovery. *Molecular Psychiatry, 19*, 20–29.

Poeppel, D. (2014). The neuroanatomic and neurophysiological infrastructure for speech and language. *Current Opinion in Neurobiology, 28*, 142–149.

Poeppel, D., Emmorey, K., Hickock, G., & Pylkkänen, L. (2012). Towards a new neurobiology of language. *Journal of Neuroscience, 32*, 14125–14131.

Polanco, J. C., Li, C., Bodea, L. G., Martinez-marmol, R., Meunier, F. A., & Götz, J. (2018). Amyloid-β and tau complexity—towards improved biomarkers and targeted therapies. *Nature Reviews Neurology, 14*(1), 22–39.

Polania, R., Nitsche, M. A., & Ruff, C. C. (2018). Studying and modifying brain function with non-invasive brain stimulation. Nature Neuroscience, 21(2), 174–187.

Pollack, G. D. (2012). Circuits for processing dynamic interaural intensity disparities in the inferior colliculus. *Hearing Research, 288*, 47–57.

Pollick, A. S., & De Waal, F. B. (2007). Ape gestures and language evolution. *Proceedings of the National Academy of Sciences, 104*, 8184–8189.

Polo, A. J., Makol, B. A., Castro, A. S., Colón-Quintana, N., Wagstaff, A. E., & Guo, S. (2019). Diversity in randomized clinical trials of depression: A 36-year review. *Clinical Psychology Review, 67*, 22–35.

Polyakova, M., Stuke, K., Schuemberg, K., Mueller, K., Schoenknecht, P., & Schroeter, M. L. (2015). BDNF as a biomarker for successful treatment of mood disorders: A systematic & quantitative meta-analysis. *Journal of Affective Disorders, 174*, 432–440.

Pons, T. P., Garraghty, P. E., Ommaya, A. K., Kaas, J. H., Taub, E., & Mishkin, M. (1991). Massive cortical reorganization after sensory deafferentation in adult macaques. *Science, 252*, 1857–1860.

Porkka-Heiskanen, T. (2013). Sleep homeostasis. *Current Opinion in Neurobiology, 23*, 799–805.

Porras, G., Li, Q., & Bezard, E. (2012). Modeling Parkinson's disease in primates: The MPTP model. *Cold Spring Harbor Perspectives in Medicine, 2*, a009308.

Porter, R., & Lemon, R. N. (1993). Corticospinal function and voluntary movement. *Monographs of the Physiological Society, No. 45*. Oxford, England: Oxford University Press.

Pósfai, B., Cserép, C., Orsolits, B., & Dénes, Á. (2018). New insights into microglia–neuron interactions: A neuron's perspective. *Neuroscience, in press.*

Posner, M. I., & Raichle, M. E. (1994). *Images of the mind.* New York, NY: Scientific American Library.

Post, R. J., & Warden, M. R. (2018). Melancholy, anhedonia, apathy: The search for separable behaviors and neural circuits in depression. *Current Opinion in Neurobiology, 49*, 192–200.

Post, R. M. (2016). Treatment of bipolar depression: Evolving recommendations. *Psychiatric Clinics of North America, 39*, 11–33.

Post, R. M., Fleming, J., & Kapczinski, F. (2012). Neurobiological correlates of illness progression in the recurrent affective disorders. *Journal of Psychiatric Research, 46*, 561–573.

Postle, B. R., Corkin, S., & Growdon, J. H. (1996). Intact implicit memory for novel patterns in Alzheimer's disease. *Learning & Memory, 3*, 305–312.

Postuma, R. B., & Berg, D. (2016). Advances in markers of prodromal Parkinson disease. *Nature Reviews Neurology, 12*(11), 622–634.

Potenza, M. (2015). Perspective: Behavioural addictions matter. *Nature, 522*, S62.

Powell, L. J., Kosakowski, H. L., & Saxe, R. (2018). Social origins of cortical face areas. *Trends in Cognitive Sciences, in press.*

Prabhavalkar, K. S., Poovanpallil, N. B., & Bhatt, L. K. (2015). Management of bipolar depression with lamotrigine: An antiepileptic mood stabilizer. *Frontiers in Pharmacology, 6*, 242.

Prather, A. A., Janicki-Deverts, D., Hall, M. H., & Cohen, S. (2015). Behaviorally assessed sleep and susceptibility to the common cold. *Sleep, 38*, 1353–1359.

Prete, G., D'Ascenzo, S., Laeng, B., Fabri, M., Foschi, N., & Tommasi, L. (2015). Conscious and unconscious processing of facial expressions: Evidence from two split-brain patients. *Journal of Neuropsychology, 9*, 45–63.

Preti, A. (2007). New developments in the pharmacotherapy of cocaine abuse. *Addiction Biology, 12*, 133–151.

Preti, A., Vrublevska, J., Veroniki, A. A., Huedo-Medina, T. B., Kyriazis, O., & Fountoulakis, K. N. (2018). Prevalence and treatment of panic disorder in bipolar disorder: Systematic review and meta-analysis. *Evidence-Based Mental Health, 21*(2), 53–60.

Price, C. J. (2012). A review and synthesis of the first 20 years of PET and fMRI studies of heard speech, spoken language and reading. *Neuroimage, 62*, 816–847.

Price, C. J., Howard, D., Patterson, K., Warburton, E. A., Friston, K. J., & Frackowiak, R. S. J. (1998). A functional neuroimaging description of two deep dyslexic patients. *Journal of Cognitive Neuroscience, 10*, 303–315.

Prieur, J., Lemasson, A., Barbu, S., & Blois-Heulin, C. (2018). Challenges facing the study of the evolutionary origins of human right-handedness and language. *International Journal of Primatology, 39*(2), 183–207.

Prieur, J., Lemasson, A., Barbu, S., & Blois-Heulin, C. (2019). History, development and current advances concerning the evolutionary roots of human right-handedness and language: Brain lateralisation and manual laterality in non-human primates. *Ethology, 125*(1), 1–28.

Primakoff, P., & Myles, D. G. (2002). Penetration, adhesion, and fusion in mammalian sperm-egg interaction. *Science, 296*, 2183–2185.

Pringle, H. (2013). Long live the humans. *Scientific American, 309*, 48–55.

Pringle, H. (2013). The origins of creativity. *Scientific American, 308*, 36–43.

Pringsheim, T., Holler-Managan, Y., Okun, M. S., Jankovic, J., Piacentini, J., Cavanna, A. E., . . . & Jarvie, E. (2019). Comprehensive systematic review summary: Treatment of tics in people with Tourette syndrome and chronic tic disorders. *Neurology, 92*(19), 907–915.

Pringsheim, T., Jette, N., Frolkis, A., & Steeves, T. D. (2014). The prevalence of Parkinson's disease: A systematic review and meta-analysis. *Movement Disorders, 29*, 1583–1590.

Pritchard, J. K. (2010). How we are evolving. *Scientific American, 303*(4), 41–47.

Pritchett, D. L., & Carey, M. R. (2014). A matter of trial and error for motor learning. *Trends in Neurosciences, 37*, 465–466.

Pruszynski, J. A., & Diedrichsen, J. (2015). Reading the mind to move the body. *Science, 348*, 860–861.

Przedborski, S. (2017). The two-century journey of Parkinson disease research. *Nature Reviews Neuroscience, 18*(4), 251–259.

Ptak, R., Schnider, A., & Fellrath, J. (2017). The dorsal frontoparietal network: A core system for emulated action. *Trends in Cognitive Sciences, 21*(8), 589–599.

Puelles, L., Harrison, M., Paxinos, G., & Watson, C. (2013). A developmental ontology for the mammalian brain based on the prosomeric model. *Trends in Neurosciences, 36*, 570–578.

Puhl, M. D., Blum, J. S., Acosta-Torres, S., & Grigson, P. S. (2012). Environmental enrichment protects against the acquisition of cocaine self-administration in adult male rate, but does not eliminate avoidance of a drug-associated saccharin cue. *Behavioural Pharmacology, 23*, 43–53.

Purcell, S. M., Manoach, D. S., Demanuele, C., Cade, B. E., Mariani, S., Cox, R., . . . & Redline, S. (2017). Characterizing sleep spindles in 11,630 individuals from the National Sleep Research Resource. *Nature Communications, 8*(1), 1–16.

Purger, D., Gibson, E. M., & Monje, M. (2015). Myelin plasticity in the central nervous system. *Neuropharmacology.* Advance online publication. doi:10.1016/j.neuropharm.2015.08.001.

Purves, D., Augustine, G. J., Fitzpatrick, D., Hall, W. C., LaMantia, A-S., McNamara, J. O., & Williams, S. M. (2004). *Neuroscience* (3rd ed.). Sunderland, MA: Sinauer Associates.

Pusey, A. E., & Schroepfer-Walker, K. (2013). Female competition in chimpanzees. *Philosophical Transactions of the Royal Society B, 368*, 20130077.

Pusey, A., Williams, J., & Goodall, J. (1997). The influence of dominance rank on the reproductive success of female chimpanzees. *Science, 277*, 828–830.

Puts, D. A., Jordan, C. L., & Breedlove, S. M. (2006). O brother, where art thou? The fraternal birth-order effect on male sexual orientation. *Proceedings of the National Academy of Sciences, USA, 103*, 10531–10532.

Puts, D., & Motta-Mena, N. V. (2018). Is human brain masculinization estrogen receptor-mediated? Reply to Luoto and Rantala. *Hormones and Behavior, 97*, 3–4.

Pytka, K., Dziubina, A., Młyniec, K., Dziedziczak, A., Żmudzka, E., Furgała, A., . . . & Filipek, B. (2016). The role of glutamatergic, GABA-ergic, and cholinergic receptors in depression and antidepressant-like effect. *Pharmacological Reports, 68*, 443–450.

Qu, C., Ligneul, R., Van der Henst, J. B., & Dreher, J. C. (2017). An integrative interdisciplinary perspective on social dominance hierarchies. *Trends in Cognitive Sciences, 21*, 893–908.

Quadrato, G., Nguyen, T., Macosko, E. Z., Sherwood, J. L., Yang, S. M., Berger, D. R., . . . & Boyden, E. S. (2017). Cell diversity and network dynamics in photosensitive human brain organoids. *Nature, 545*(7652), 48–53.

Quagliato, L. A., Freire, R. C., & Nardi, A. E. (2018). Risks and benefits of medications for panic disorder: A comparison of SSRIs and benzodiazepines. *Expert Opinion on Drug Safety, 17*(3), 315–324.

Quammen, D. (2004, May). Was Darwin wrong? No. The evidence for evolution is overwhelming. *National Geographic, 206*, 2–35.

Quast, K. B., Ung, K., Froudarakis, E., Huang, L., Herman, I., Addison, A. P., . . . & Arenkiel, B. R. (2017). Developmental broadening of inhibitory sensory maps. *Nature Neuroscience, 20*, 189–199.

Quesnel-Vallieres, M., Weatheritt, R. J., Cordes, S. P., & Blencowe, B. J. (2019). Autism spectrum disorder: Insights into convergent mechanisms from transcriptomics. *Nature Reviews Genetics, 20*, 51–63.

Quezada, J., & Coffman, K. A. (2018). Current approaches and new developments in the pharmacological management of Tourette syndrome. *CNS Drugs, 32*(1), 33–45.

Quick, S. L., Pyszczynski, A. D., Colston, K. A., & Shahan, T. A. (2011). Loss of alternative non-drug reinforcement induces relapse of cocaine-seeking in rats: Role of dopamine D(1) receptors. *Neuropsychopharmacology, 36*, 1015–1020.

Quigley, K. S., & Barrett, L. F. (2014). Is there consistency and specificity of autonomic changes during emotional episodes? Guidance from the conceptual act theory and psychophysiology. *Biological Psychology, 98*, 82–94.

Quiroga, R. Q. (2012). Concept cells: The building blocks of declarative memory functions. *Nature Reviews Neuroscience, 13*, 587–597.

Quiroga, R. Q. (2016). Neuronal codes for visual perception and memory. *Neuropsychologia, 83*, 227–241.

Quiroga, R. Q., Fried, I., & Koch, C. (2013). Brain cells for grandmother. *Scientific American, 308*, 30–35.

Quiroga, R. Q., Kraskov, A., Mormann, F., Fried, I., & Koch, C. (2014). Single-cell responses to face adaptation in the human medial temporal lobe. *Neuron, 84*, 363–369.

Qureshi, I. A., & Mehler, M. F. (2012). Emerging roles of non-coding RNAs in brain evolution, development, plasticity and disease. *Nature Reviews Neuroscience, 13*, 528–541.

Radhakrishnan, R., Wilkinson, S. T., & D'Souza, D. C. (2014). Gone to pot —a review of the association between cannabis and psychosis. *Frontiers in Psychiatry, 5*, 54.

Rahimpour, S., Haglund, M. M., Friedman, A. H., & Duffau, H. (2019). History of awake mapping and speech and language localization: From modules to networks. *Neurosurgical Focus, 47*(3), E4.

Raichle, M. E. (2008). A brief history of human brain mapping. *Trends in Neurosciences, 32*, 118–126.

Raichle, M. E. (2010, March). The brain's dark energy. *Scientific American, 302*, 44–49.

Raichlen, D. A., Gordon, A. D., Harcourt-Smith, W. E. H., Foster, A. D., & Haas, Jr., W. R. (2010). Laetoli footprints preserve earliest direct evidence of human-like bipedal biomechanics. *PLoS One, 5,* e9769.

Raineki, C., Lucion, A. B., & Weinberg, J. (2014). Neonatal handling: An overview of the positive and negative effects. *Developmental Psychobiology, 56*(8), 1613–1625.

Raisman, G. (2015). 60 years of neuroendocrinology: Memoir: Geoffrey Harris and my brush with his unit. *Journal of Endocrinology, 226,* T1–T11.

Rakic, P. (1979). Genetic and epigenetic determinants of local neuronal circuits in the mammalian central nervous system. In F. O. Schmitt & F. G. Worden (Eds.), *The neurosciences: Fourth study program.* Cambridge, MA: MIT Press.

Ralph, M. R., & Menaker, M. (1988). A mutation of the circadian system in golden hamsters. *Science, 241,* 1225–1227.

Ralph, M. R., Foster, T. G., Davis, F. C., & Menaker, M. (1990). Transplanted suprachiasmatic nucleus determines circadian period. *Science, 247,* 975–978.

Ramachandran, V. S., & Blakeslee, S. (1998). *Phantoms in the brain.* New York, NY: Morrow.

Ramachandran, V. S., & Rogers-Ramachandran, D. (2000). Phantom limbs and neuronal plasticity. *Archives of Neurology, 57,* 317–320.

Ramirez, S., Liu, X., MacDonald, C. J., Moffa, A., Zhou, J.,…& Tonegawa, S. (2015). Activating positive memory engrams suppresses depression-like behaviour. *Nature, 522,* 335–339.

Ramsay, D. S., & Woods, S. C. (1997). Biological consequences of drug administration: Implications for acute and chronic tolerance. *Psychological Review, 104*(1), 170–193.

Ramsay, D. S., & Woods, S. C. (2014). Clarifying the roles of homeostasis and allostasis in physiological regulation. *Psychological Review, 121,* 225–247.

Ramus, F., Marshall, C. R., Rosen, S., & van der Lely, H. K. J. (2013). Phonological deficits in specific language impairment and developmental dyslexia: Towards a multidimensional model. *Brain, 136,* 630–645.

Ran, C., Hoon, M. A., & Chen, X. (2016). The coding of cutaneous temperature in the spinal cord. *Nature Neuroscience, 19*(9), 1201–1209.

Rana, A., & Dolmetsch, R. E. (2010). Using light to control signaling cascades in live neurons. *Current Opinion in Neurobiology, 20,* 617–622.

Rangarajan, V., Hermes, D., Foster, B. L., Weiner, K. S., Jacques, C., Grill-Spector, K., & Parvizi, J. (2014). Electrical stimulation of the left and right human fusiform gyrus causes different effects in conscious face pereception. *Journal of Neuroscience, 34,* 12828–12836.

Ransahoff, R. M., Hafler, D. A., & Lucchinetti, C. F. (2015). Multiple sclerosis—a quiet revolution. *Nature Reviews Neurology, 11,* 134–142.

Rao, S. C., Rainer, G., & Miller, E. K. (1997). Integration of what and where in the primate prefrontal cortex. *Science, 276,* 821–824.

Rapoport, S. I. (2014). Lithium and the other mood stabilizers effective in bipolar disorder target the rat brain arachidonic acid cascade. *ACS Chemical Neuroscience, 5,* 459–467.

Rasch, B., Büchel, C., Gais, S., & Born, J. (2007). Odor cues during slow-wave sleep prompt declarative memory consolidation. *Science, 315,* 1426–1429.

Rasch, B., Pommer, J., Diekelmann, S., & Born, J. (2008). Pharmacological REM sleep suppression paradoxically improves rather than impairs skill memory. *Nature Neuroscience, 12,* 396–397.

Rasmussen, T., & Milner, B. (1975). Clinical and surgical studies of the cerebral speech areas in man. In K. J. Zulch, O. Creutzfeldt, & G. C. Galbraith (Eds.), *Cerebral localization* (pp. 238–257). New York, NY: Springer-Verlag.

Rassnick, S., Stinus, L., & Koob, G. F. (1993). The effects of 6-hydroxydopamine lesions of the nucleus accumbens and the mesolimbic dopamine system on oral self-administration of ethanol in the rat. *Brain Research, 623,* 16–24.

Raushchecker, J. P. (2015). Auditory and visual cortex of primates: A comparison of two sensory systems. *European Journal of Neuroscience, 41,* 579–585.

Ravussin, E., Redman, L. M., Rochon, J., Das, S. K., Fontana, L.,…& Roberts, S. B. (2015). A 2-year randomized controlled trial of human caloric restriction: Feasibility and effects of predictors of health span and longevity. *Journals of Gerontology. Series A, Biological Sciences and Medical Sciences, 70,* 1097–1104.

Raynor, H. A., & Epstein, L. H. (2001). Dietary variety, energy regulation, and obesity. *Psychological Bulletin, 127,* 325–341.

Raz, A. (2012). From neuroimaging to tea leaves in the bottom of a cup. In: S. Choudhury, & J. Slaby (Eds.), *Critical neuroscience: A handbook of the social and cultural contexts of neuroscience, first edition.* Blackwell: Oxford.

Raz, S., & Berger, B. D. (2010). Social isolation increases morphine intake: Behavioral and psychopharmacological aspects. *Behavioural Pharmacology, 21,* 39–46.

Reardon, S. (2014). Gut-brain link grabs neuroscientists. *Nature, 515,* 175–177.

Reardon, S. (2016). A mouse's house may ruin experiments. *Nature, 530,* 264.

Reber, P. J. (2013). The neural basis of implicit learning and memory: A review of neuropsychological and neuroimaging research. *Neuropsychologia, 51,* 2026–2042.

Reber, P. J., Knowlton, B. J., & Squire, L. R. (1996). Dissociable properties of memory system: Differences in the flexibility of declarative and non-declarative knowledge. *Behavioral Neuroscience, 110*(5), 861–871.

Recanzone, G. H. (2004). Acoustic stimulus processing and multimodal interactions in primates. In M. S. Gazzaniga (Ed.), *The cognitive neurosciences* (pp. 359–368). Cambridge, MA: MIT Press.

Rechtschaffen, A., & Bergmann, B. M. (1995). Sleep deprivation in the rat by the disk-over-water method. *Behavioural Brain Research, 69,* 55–63.

Rechtschaffen, A., Gilliland, M. A., Bergmann, B. M., & Winter, J. B. (1983). Physiological correlates of prolonged sleep deprivation in rats. *Science, 221,* 182–184.

Reddy, L., & Kanwisher, N. (2006). Coding of visual objects in the ventral stream. *Current Opinion in Neurobiology, 16,* 408–414.

Reddy, L., & Thorpe, S. J. (2014). Concept cells through associative learning of high-level representations. *Neuron, 84,* 248–251.

Redondo, R. L., Kim, J., Arons, A. L., Ramirez, S., & Tonegawa, S. (2014). Bidirectional switch of the valence associated with a hippocampal contextual memory engram. *Nature, 513,* 426–430.

Rees, G., Kreiman, G., & Koch, C. (2002). Neural correlates of consciousness in humans. *Nature Reviews Neuroscience, 3,* 261–270.

Regan, P. C. (1996). Rhythms of desire: The association between menstrual cycle phases and female sexual desire. *Canadian Journal of Human Sexuality, 5*(3), 145–156.

Rehme, A. K., Fink, G. R., von Cramon, D. Y., & Grefkes, C. (2011). The role of the contralesional motor cortex for motor recovery in the early days after stroke assessed with longitudinal fMRI. *Cerebral Cortex, 21,* 756–768.

Reid, K. J., & Abbott, S. M. (2015). Jet lag and shift work disorder. *Sleep Medicine Clinics, 10,* 523–535.

Reiman, E. M. (2016). Alzheimer's disease: Attack on amyloid-β protein. *Nature, 537*(7618), 36–37.

Reiman, E. M. (2017). Alzheimer disease in 2016: Putting AD treatments and biomarkers to the test. *Nature Reviews Neurology, 13*(2), 74.

Reinarman, C., Cohen, P. D., & Kaal, H. L. (2004). The limited relevance of drug policy: Cannabis in Amsterdam and in San Francisco. *American Journal of Public Health, 94,* 836–842.

Reiner, A., & Levitz, J. (2018). Glutamatergic signaling in the central nervous system: ionotropic and metabotropic receptors in concert. *Neuron, 98*(6), 1080–1098.

Reiner, W. (1997). To be male or female—that is the question. *Archives of Pediatrics and Adolescent Medicine, 151,* 224–225.

Reitsma, M. B., Fullman, N., Ng, M., Salama, J. S., Abajobir, A., Abate, K. H.,. . . & Adebiyi, A. O. (2017). Smoking prevalence and attributable disease burden in 195 countries and territories, 1990–2015: A systematic analysis from the Global Burden of Disease Study 2015. *The Lancet, 389*(10082), 1885–1906.

Rektor, I., Bočková, M., Chrastina, J., Rektorová, I., & Baláz, M. (2015). The modulatory role of subthalamic nucleus in cognitive functions—a viewpoint. *Clinical Neurophysiology, 126,* 653–658.

Renard, J., Krebs, M.-O., Le Pen, G., & Jay, T. M. (2014). Long-term consequences of adolescent cannabinoid exposure in adult psychopathology. *Frontiers in Neuroscience, 8,* 361.

Rengaraj, D., Kwon, W.-S., & Pang, M.-G. (2015). Bioinformatics annotation of human y chromosome-encoded protein pathways and interactions. *Journal of Proteome Research, 14,* 3503–3518.

Renna, J. M., Chellappa, D. K., Ross, C. L., Stabio, M. E., & Berson, D. M. (2015). Melanopsin ganglion cells extend dendrites into the outer retina during early postnatal development. *Developmental Neurobiology, 75,* 935–946.

Represa, A., & Ben-Ari, Y. (2006). Trophic actions of GABA on neuronal development. *Trends in Neurosciences, 28,* 279–283.

Ressler, R. L., & Maren, S. (2019). Synaptic encoding of fear memories in the amygdala. *Current Opinion in Neurobiology, 54,* 54–59.

Revonsuo, A. (2000). The reinterpretation of dreams: An evolutionary hypothesis of the function of dreaming. *Behavioral and Brain Sciences, 23*(6), 877–901.

Revusky, S. H., & Garcia, J. (1970). Learned associations over long delays. In G. H. Bower & J. T. Spence (Eds.), *The psychology of learning and motivation* (Vol. 4, pp. 1–85). New York, NY: Academic Press.

Reynolds, D. V. (1969). Surgery in the rat during electrical analgesia induced by focal brain stimulation. *Science, 164,* 444–445.

Reznikov, A. G., Nosenko, N. D., Tarasenko, L. V., & Limareva, A. A. (2016). Changes in the brain testosterone metabolism and sexual behavior of male rats prenatally exposed to methyldopa and stress. *International Journal of Physiology and Pathophysiology, 7*(3).

Rezvani, A. H., & Levin, E. D. (2001). Cognitive effects of nicotine. *Biological Psychiatry, 49,* 258–267.

Rhea, E. M., & Banks, W. A. (2019). Role of the blood-brain barrier in central nervous system insulin resistance. *Frontiers in Neuroscience, 13,* 521.

Rheims, S., & Ryvlin, P. (2014). Pharmacotherapy for tonic-clonic seizures. *Expert Opinion on Pharmacotherapy, 15,* 1417–1426.

Rhodes, G., Byatt, G., Michie, P. T., & Puce, A. (2004). Is the fusiform face area specialized for faces, individuation, or expert individuation? *Journal of Cognitive Neuroscience, 16,* 189–203.

Rhodes, S. M., Riby, D. M., Fraser, E., & Campbell, L. E. (2011). The extent of working memory deficits associated with Williams syndrome: Exploration of verbal and spatial domains and executively controlled processes. *Brain and Cognition, 77,* 208–214.

Rial, R. V., Nicolau, M. C., Gamundi, A., Akaârir, M., Aparicio, S., Garau, C.,…& Estaban, S. (2007). The trivial function of sleep. *Sleep Medicine Reviews, 11,* 311–325.

Ribeiro, G., Camacho, M., Santos, O., Pontes, C., Torres, S., & Oliveira-Maia, A. J. (2018). Association between hedonic hunger and body-mass index versus obesity status. *Scientific Reports*, *8*(1), 5857.

Ribeiro, N., Gounden, Y., & Quaglino, V. (2016). Investigating on the methodology effect when evaluating lucid dream. *Frontiers in Psychology*, *7*, 1306.

Ricciarelli, R., & Fedele, E. (2018). cAMP, cGMP and amyloid β: Three ideal partners for memory formation. *Trends in Neurosciences*, *41*(5), 255–266.

Riccio, A. (2010). Dynamic epigenetic regulation in neurons: Enzymes, stimuli and signaling pathways. *Nature Neuroscience*, *11*, 1330–1337.

Richards, W. (1971, May). The fortification illusions of migraines. *Scientific American*, *224*, 89–97.

Richmond, L. L., & Zacks, J. M. (2017). Constructing experience: Event models from perception to action. *Trends in Cognitive Sciences*, *21*(12), 962–980.

Richter, C. P. (1967). Sleep and activity: Their relation to the 24-hour clock. *Proceedings of the Association for Research on Nervous and Mental Disorders*, *45*, 8–27.

Richter, C. P. (1971). Inborn nature of the rat's 24-hour clock. *Journal of Comparative and Physiological Psychology*, *75*, 1–14.

Richter, C., Woods, I. G., & Schier, A. F. (2014). Neuropeptidergic control of sleep and wakefulness. *Annual Review of Neuroscience*, *37*, 503–531.

Ridaura, V. K., Faith, J. J., Rey, F. E., Cheng, J., Duncan, A. E., . . . & Gordon, J. I. (2013). Gut microbiota from twins discordant for obesity modulate metabolism in mice. *Science*, *341*, 1079–1089.

Ridley, N. J., Draper, B., & Withall, A. (2013). Alcohol-related dementia: An update of the evidence. *Alzheimer's Research and Therapy*, *5*, 3.

Rigato, J., Murakami, M., & Mainen, Z. (2014). Spontaneous decisions and free will: Empirical results and philosophical considerations. *Cold Spring Harbor Symposia on Quantitative Biology*, *79*, 177–184.

Rigby, N., & Kulathinal, R. J. (2015). Genetic architecture of sexual dimorphism in humans. *Journal of Cellular Physiology*, *230*, 2304–2310.

Rigby, S. N., Stoesz, B. M., & Jakobson, L. S. (2018). Empathy and face processing in adults with and without autism spectrum disorder. *Autism Research*, *11*(6), 942–955.

Rijntjes, M., Dettmers, C., Büchel, C., Kiebel, S., Frackowiak, R. S. J., & Weiller, C. (1999). A blueprint for movement: Functional and anatomical representations in the human motor system. *Journal of Neuroscience*, *19*, 8043–8048.

Riou, M. È., Jomphe-Tremblay, S., Lamothe, G., Stacey, D., Szczotka, A., & Doucet, É. (2015). Predictors of energy compensation during exercise interventions: A systematic review. *Nutrients*, *7*, 3677–3704.

Rivlin-Etzion, M., Grimes, W. N., & Rieke, F. (2018). Flexible neural hardware supports dynamic computations in retina. *Trends in Neurosciences*, *41*, 224–237.

Rizzardi, L. F., Hickey, P. F., DiBlasi, V. R., Tryggvadóttir, R., Callahan, C. M., Idrizi, A., . . . & Feinberg, A. P. (2019). Neuronal brain-region-specific DNA methylation and chromatin accessibility are associated with neuropsychiatric trait heritability. *Nature Neuroscience*, *22*(2), 307–316.

Rizzolatti, G., & Fogassi, L. (2014). The mirror mechanism: Recent findings and perspectives. *Philosophical Transactions of the Royal Society B*, *369*, 20130420.

Rizzolatti, G., & Sinigaglia, C. (2016). The mirror mechanism: A basic principle of brain function. *Nature Reviews Neuroscience*, *17*, 757–765.

Rizzoli, S. O., & Betz, W. (2005). Synaptic vesicle pools. *Nature Reviews Neuroscience*, *6*, 57–69.

Robbins, T. W. (2016). Illuminating anhedonia. *Science*, *351*, 24–25.

Robbins, T. W., & Clark, L. (2015). Behavioural addictions. *Current Opinion in Neurobiology*, *30*, 66–72.

Robel, S., & Sontheimer, H. (2016). Glia as drivers of abnormal neuronal activity. *Nature Neuroscience*, *19*(1), 28–33.

Robertson, C. E., & Baron-Cohen, S. (2017). Sensory perception in autism. *Nature Reviews Neuroscience*, *18*(11), 671.

Robinson, T. E., & Berridge, K. C. (2008). The incentive sensitization theory of addiction: Some current issues. *Philosophical Transactions of the Royal Society B*, *363*, 3137–3146.

Robotham, R. J., & Starrfelt, R. (2018). Tests of whole upright face processing in prosopagnosia: A literature review. *Neuropsychologia*, *121*, 106–121.

Robson, P. J. (2014). Therapeutic potential of cannabinoid medicines. *Drug Testing and Analysis*, *6*, 24–30.

Rocchetti, M., Crescini, A., Borgwardt, S., Caverzasi, E., Politi, P., Atakan, Z., & Fusar-Poli, P. (2013). Is cannabis neurotoxic for the healthy brain? A meta-analytical review of structural brain alterations in non-psychotic users. *Psychiatry and Clinical Neurosciences*, *67*, 483–492.

Rodin, J. (1985). Insulin levels, hunger, and food intake: An example of feedback loops in body weight regulation. *Health Psychology*, *4*, 1–24.

Rodriguez-Arias, M., Navarrete, F., Blanco-Gandia, M. C., Arenas, M. C., Bartoll-Andrés, A., Aguilar, M. A., . . . & Manzanares, J. (2016). Social defeat in adolescent mice increases vulnerability to alcohol consumption. *Addiction Biology*, *21*, 87–97.

Roe, A. W., Pallas, S. L., Hahm, J.-O., & Sur, M. (1990). A map of visual space induced in primary auditory cortex. *Science*, *250*, 818–820.

Roecklein, K. A., Wong, P. M., Miller, M. A., Donofry, S. D., Kamarck, M. L., & Brainard, G. C. (2013). Melanopsin, photosensitive ganglion cells, and seasonal affective disorder. *Neuroscience and Biobehavioral Reviews*, *37*, 229–239.

Roefs, A., Werrij, M. Q., Smulders, F. T. Y., & Jansen, A. (2006). The value of indirect measures for assessing food preferences in abnormal eating. *Journal für Verbraucherschutz und Lebensmittelsicherheit*, *1*, 180–186.

Roerecke, M., & Rehm, J. (2011). Ischemic heart disease mortality and morbidity rates in former drinkers: A meta-analysis. *American Journal of Epidemiology*, *173*, 245–258.

Roest, A. M., de Jonge, P., Williams, C. D., de Vries, Y. A., Schoevers, R. A., & Turner, E. H. (2015). Reporting bias in clinical trials investigating the efficacy of second-generation antidepressants in the treatment of anxiety disorders: A report of 2 meta-analyses. *JAMA Psychiatry*, *72*, 500–510.

Roesti, M., & Salzburger, W. (2014). Natural selection: It's a many-small world after all. *Current Biology*, *24*, 959–960.

Rogalsky, C., Raphel, K., Tomkovicz, V., O'Grady, L., Damasio, H., . . . & Hickok, G. (2013). Neural basis of action understanding: Evidence from sign language aphasia. *Aphasiology*, *27*, 1147–1158.

Rogers, J., & Gibbs, R. A. (2014). Comparative primate genomics: Emerging patterns of genome content and dynamics. *Nature Reviews Genetics*, *15*, 347–359.

Rogers, P. J., & Blundell, J. E. (1980). Investigation of food selection and meal parameters during the development of dietary induced obesity. *Appetite*, *1*, 85–88.

Rohde, C., Mütze, U., Schulz, S., Thiele, A. G., Ceglarek, U., et al. (2014). Unrestricted fruits and vegetables in the PKU diet: A 1-year follow-up. *European Journal of Clinical Nutrition*, *68*, 401–403.

Rohde, K., Keller, M., la Cour Poulsen, L., Blüher, M., Kovacs, P., & Böttcher, Y. (2019). Genetics and epigenetics in obesity. *Metabolism*, *92*, 37–50.

Rolls, B. J. (1990). The role of sensory-specific satiety in food intake and food selection. In E. D. Capaldi & T. L. Powley (Eds.), *Taste, experience, and feeding* (pp. 28–42). Washington, DC: American Psychological Association.

Rolls, B. J., Rolls, E. T., Rowe, E. A., & Sweeney, K. (1981). Sensory specific satiety in man. *Physiology & Behavior*, *27*, 137–142.

Rolls, E. T. (1981). Central nervous mechanisms related to feeding and appetite. *British Medical Bulletin*, *37*, 131–134.

Rolls, E. T. (2008). Top—down control of visual perception: Attention in natural vision. *Perception*, *37*, 333–354.

Rolls, E. T. (2015). Limbic systems for emotion and for memory, but no single limbic system. *Cortex*, *62*, 119–157.

Romei, V., Thut, G., & Silvanto, J. (2016). Information-based approaches of noninvasive transcranial brain stimulation. *Trends in Neurosciences*, *39*(11), 782–795.

Romanoski, C. E., Glass, C. K., Stunnenberg, H. G., Wilson, L., & Almouzni, G. (2015). Roadmap for regulation. *Nature*, *518*(7539), 314–31.

Romo, R., & Rossi-Pool, R. (2020). Turning touch into perception. *Neuron*, *105*, 16–33.

Roney, J. R., & Simmons, Z. L. (2013). Hormonal predictors of sexual motivation in natural menstrual cycles. *Hormones and Behavior*, *63*, 636–645.

Rorden, C., & Karnath, H.-O. (2004). Using human brain lesions to infer function: A relic from a past era in the fMRI age? *Nature Reviews Neuroscience*, *5*, 813–819.

Rosa, M. G. P., Tweedale, R., & Elston, G. N. (2000). Visual responses of neurons in the middle temporal area of New World monkeys after lesions of striate cortex. *Journal of Neuroscience*, *20*, 5552–5563.

Rosa, R. R., Bonnet, M. H., & Warm, J. S. (2007). Recovery of performance during sleep following sleep deprivation. *Psychophysiology*, *20*, 152–159.

Rose, T., & Bonhoeffer, T. (2018). Experience-dependent plasticity in the lateral geniculate nucleus. *Current Opinion in Neurobiology*, *53*, 22–28.

Roselli, C. E. (2018). Neurobiology of gender identity and sexual orientation. *Journal of neuroendocrinology*, *30*(7), e12562.

Rosenwasser, A. M., & Turek, F. W. (2015). Neurobiology of circadian rhythm regulation. *Sleep Medicine Clinics*, *10*, 403–412.

Roska, B., & Sahel, J. A. (2018). Restoring vision. *Nature*, *557*(7705), 359–367.

Rossignol, S., & Frigon, A. (2011). Recovery of locomotion after spinal cord injury: Some facts and mechanisms. *Annual Review of Neuroscience*, *34*, 413–440.

Rossini, P. M., Burke, D., Chen, R., Cohen, L. G., Daskalakis, Z., Di Iorio, R., . . . & Hallett, M. (2015). Non-invasive electrical and magnetic stimulation of the brain, spinal cord, roots and peripheral nerves: basic principles and procedures for routine clinical and research application. An updated report from an IFCN Committee. *Clinical Neurophysiology*, *126*, 1071–1107.

Rossini, P. M., Martino, G., Narici, L., Pasquarelli, A., Peresson, M., Pizzella, V., . . . & Romani, G. L. (1994). Short-term brain "plasticity" in humans: Transient finger representation changes in sensory cortex somatotopy following ischemic anesthesia. *Brain Research*, *642*, 169–177.

Roth, M. M., Dahmen, J. C., Muir, D. R., Imhof, F., Martini, F. J., & Hofer, S. B. (2016). Thalamic nuclei convey diverse contextual information to layer 1 of visual cortex. *Nature Neuroscience*, *19*, 299–307.

Rothwell, J. C., Traub, M. M., Day, B. L., Obeso, J. A., Thomas, P. K., & Marsden, C. D. (1982). Manual motor performance in a deafferented man. *Brain*, *105*, 515–542.

Rothwell, N. J., & Stock, M. J. (1982). Energy expenditure derived from measurements of oxygen consumption and energy balance in hyperphagic, "cafeteria"-fed rats. *Journal of Physiology*, *324*, 59–60.

Rotstein, D., & Montalban, X. (2019). Reaching an evidence-based prognosis for personalized treatment of multiple sclerosis. *Nature Reviews Neurology, 15*(5), 287–300.

Rousseau, D. M., & Gunia, B. C. (2016). Evidence-based practice: The psychology of EBP implementation. *Annual Review of Psychology, 67*, 667–692.

Rowland, D. C., & Moser, M-B. (2014). From cortical modules to memories. *Current Opinion in Neurobiology, 24*, 22–27.

Rowland, D. C., & Moser, M.-B. (2015). A three-dimensional neural compass. *Nature, 517*, 156–157.

Rowland, D. C., Roudi, Y., Moser, M. B., & Moser, E. I. (2016). Ten years of grid cells. *Annual Review of Neuroscience, 39*, 19–40.

Roy, D. S., Arons, A., Mitchell, T. I., Pignatelli, M., Ryan, T. J., & Tonegawa, S. (2016). Memory retrieval by activating engram cells in mouse models of early Alzheimer's disease. *Nature, 531*, 508–512.

Roy, S. (2017). Synuclein and dopamine: The Bonnie and Clyde of Parkinson's disease. *Nature Neuroscience, 20*(11), 1514–1515.

Rozin, P., Dow, S., Moscovitch, M., & Rajaram, S. (1998). What causes humans to begin and end a meal? A role for memory for what has been eaten, as evidenced by a study of multiple meal eating in amnesic patients. *Psychological Science, 9*, 392–396.

Ru, T., Chen, Q., You, J., & Zhou, G. (2019). Effects of a short midday nap on habitual nappers' alertness, mood and mental performance across cognitive domains. *Journal of Sleep Research, 28*(3), e12638.

Ruby, N. F., Brennan, T. J., Xie, X., Cao, V., Franken, P., Heller, H. C., & O'Hara, B. F. (2002). Role of melanopsin in circadian responses to light. *Science, 298*, 2211–2213.

Rucker, J. J., Iliff, J., & Nutt, D. J. (2018). Psychiatry & the psychedelic drugs. Past, present & future. *Neuropharmacology, 142*, 200–218.

Ruder, L., & Arber, S. (2019). Brainstem circuits controlling action diversification. *Annual Review of Neuroscience, 42*, 485–504.

Rudy, J. W., & Sutherland, R. J. (1992). Configural and elemental associations and the memory coherence problem. *Journal of Cognitive Neuroscience, 4*, 208–216.

Ruff, C. C. (2013). Sensory processing: Who's in (top-down) control? *Annals of the New York Academy of Sciences, 1296*, 88–107.

Rugg, M. D., & Vilberg, K. L. (2013). Brain networks underlying memory retrieval. *Current Opinion in Neurobiology, 23*, 255–260.

Rumble, M. E., White, K. H., & Benca, R. M. (2015). Sleep disturbances in mood disorders. *Psychiatric Clinics of North America, 38*, 743–759.

Rusakov, D. A., Bard, L., Stewart, M. G., & Henneberger, C. (2014). Diversity of astroglial functions alludes to subcellular specialisation. *Trends in Neurosciences, 37*, 228–242.

Rushworth, M. F. S., Mars, R. B., & Sallet, J. (2013). Are there specialized circuits for social cognition and are they unique to humans? *Current Opinion in Neurobiology, 23*, 436–442.

Russell, G., & Lightman, S. (2019). The human stress response. *Nature Reviews Endocrinology, 15*(9), 525–534.

Russell, J. A., Bachorowski, J.-A., & Fernandez-Dols, J.-M. (2003). Facial and vocal expressions of emotion. *Annual Review of Psychology, 54*, 329–349.

Russell, M., Breggia, A., Mendes, N., Klibanski, A., & Misra, M. (2011). Growth hormone is positively associated with surrogate markers of bone turnover during puberty. *Clinical Endocrinology, 75*, 482–488.

Russell, W. R., & Espir, M. I. E. (1961). *Traumatic aphasia—a study of aphasia in war wounds of the brain.* London, England: Oxford University Press.

Russo, S. J., & Nestler, E. J. (2013). The brain reward circuitry in mood disorders. *Nature Reviews Neuroscience, 14*, 609–625.

Ruthazer, E. S., & Aizenman, C. D. (2010). Learning to see: Patterned visual activity and the development of visual function. *Trends in Neurosciences, 33*, 183–192.

Rutherford, B. R., & Roose, S. P. (2013). A model of placebo response in antidepressant clinical trials. *American Journal of Psychiatry, 170*, 723–733.

Rutledge, L. L., & Hupka, R. B. (1985). The facial feedback hypothesis: Methodological concerns and new supporting evidence. *Motivation and Emotion, 9*, 219–240.

Rutter, M. L. (1997). Nature-nurture integration: The example of antisocial behavior. *American Psychologist, 52*, 390–398.

Ryan, M. M., Guévremont, D., Luxmanan, C., Abraham, W. C., & WIlliams, J. M. (2015). Aging alters long-term potentiation-related gene networks and impairs synaptic protein synthesis in the rat hippocampus. *Neurobiology of Aging, 36*, 1868–1880.

Ryan, T. J., Roy, D. S., Pignatelli, M., Arons, A., & Tonegawa, S. (2015). Engram cells retain memory under retrograde amnesia. *Science, 348*, 1007–1013.

Rymer, R. (1993). *Genie.* New York, NY: HarperCollins

Sacchetti, E., Scassellati, C., Minelli, A., Valsecchi, P., Bonvicini, C., Pasqualetti, P., Galluzzo, A., Pioli, R., & Gennarelli, M. (2013). Schizophrenia susceptibility and NMDA-receptor mediated signalling: An association study involving 32 tagSNPs of DAO, DAOA, PPP3CC, and DTNBP1 genes. *BMC Biomedical Genetics, 14*, 33.

Sack, A. T. (2006). Transcranial magnetic stimulation, causal structure–function mapping and networks of functional relevance. *Current Opinion in Neurobiology, 16*, 593–599.

Sack, R. L. (2010). Jet lag. *New England Journal of Medicine, 362*, 440–447.

Sacks, O. (1985). *The man who mistook his wife for a hat and other clinical tales.* New York, NY: Summit Books.

Sacks, O. (1999). The man who mistook his wife for a hat. In *Summit Books.* Simon & Schuster New York.

Sacks, O. (2012). *Hallucinations.* Pan Macmillan.

Sacks, O. (2013). Hallucinations of musical notation. *Brain, 136*(7), 2318–2322.

Sadnicka, A., Kornysheva, K., Rothwell, J. C., & Edwards, M. J. (2018). A unifying motor control framework for task-specific dystonia. *Nature Reviews Neurology, 14*(2), 116.

Saenz, M., & Langers, D. R. M. (2014). Tonotopic mapping of human auditory cortex. *Hearing Research, 307*, 42–52.

Sagi, I., & Benvenisty, N. (2017). Haploidy in humans: An evolutionary and developmental perspective. *Developmental Cell, 41*(6), 581–589.

Sagoe, D., Molde, H., Andreassen, C. S., Torsheim, T., & Pallesen, S. (2014). The global epidemiology of anabolic-androgenic steroid use: A meta-analysis and meta-regression analysis. *Annals of Epidemiology, 24*, 383–398.

Sajjad, Y. (2010). Development of the genital ducts and external genitalia in the early human embryo. *Journal of Obstetrics and Gynaecology Research, 36*(5), 929–937.

Sakamoto, M., Kageyama, R., & Imayoshi, I. (2015). The functional significance of newly born neurons integrated into olfactory bulb circuits. *Frontiers in Neuroscience, 8*, 121.

Sala Frigerio, C., & De Strooper, B. (2016). Alzheimer's disease mechanisms and emerging roads to novel therapeutics. *Annual Review of Neuroscience, 39*, 57–79.

Sale, A. (2018). A systematic look at environmental modulation and its impact in brain development. *Trends in Neurosciences, 41*(1), 4–17.

Salk, R. H., Hyde, J. S., & Abramson, L. Y. (2017). Gender differences in depression in representative national samples: Meta-analyses of diagnoses and symptoms. *Psychological Bulletin, 143*(8), 783–822.

Salter, M. W., & Stevens, B. (2017). Microglia emerge as central players in brain disease. *Nature Medicine, 23*(9), 1018.

Sampaio-Baptista, C., & Johansen-Berg, H. (2017). White matter plasticity in the adult brain. *Neuron, 96*(6), 1239–1251.

Sampaio-Baptista, C., Sanders, Z. B., & Johansen-Berg, H. (2018). Structural plasticity in adulthood with motor learning and stroke rehabilitation. *Annual review of neuroscience, 41*, 25–40.

Samuels, B. A., Anacker, C., Hu, A., Levinstein, M. R., Pickenhagen, A., Tsetsenis, T.,…& Gross, C. T. (2015). 5-HT1A receptors on mature dentate gyrus granule cells are critical for the antidepressant response. *Nature Neuroscience, 18*, 1606–1616.

Sanders, D., & Bancroft, J. (1982). Hormones and the sexuality of women—the menstrual cycle. *Clinics in Endocrinology and Metabolism, 11*, 639–659.

Sanders, H., Rennó-Costa, C., Idiart, M., & Lisman, J. (2015). Grid cells and place cells: An integrated view of their navigational and memory function. *Trends in Neurosciences, 38*, 763–775.

Sandi, C., & Haller, J. (2015). Stress and the social brain: Behavioural effects and neurobiological mechanisms. *Nature Reviews Neuroscience, 16*, 290–304.

Sandoval, D. A., & Seeley, R. J. (2017). Physiology: Gut feeling for food choice. *Nature, 542*(7641), 302–303.

Sandrini, M., Cohen, L. G., & Censor, N. (2015). Modulating reconsolidation: A link to causal systems-level dynamics of human memories. *Trends in Cognitive Sciences, 19*, 475–482.

Sanes, D. H., & Bao, S. (2009). Tuning up the developing auditory CNS. *Current Opinion in Neurobiology, 19*, 188–199.

Sanes, J. N., Donoghue, J. P., Thangaraj, V., Edelman, R. R., & Warach, S. (1995). Shared neural substrates controlling hand movements in human motor cortex. *Science, 268*, 1775–1777.

Sanes, J. N., Suner, S., & Donoghue, J. P. (1990). Dynamic organization of primary motor cortex output to target muscles in adult rats. I. Long-term patterns of reorganization following motor or mixed peripheral nerve lesions. *Experimental Brain Research, 79*, 479–491.

Sapède, D., & Cau, E. (2013). The pineal gland from development to function. *Current Topics in Developmental Biology, 106*, 171–215.

Saper, C. B. (2013). The central circadian timing system. *Current Opinion in Neurobiology, 23*, 747–751.

Sassone-Corsi, P. (2016). The epigenetic and metabolic language of the circadian clock. In *A Time for Metabolism and Hormones* (pp. 1–11). Springer, Cham.

Sati, P., Oh, J., Constable, R. T., Evangelou, N., Guttmann, C. R., Henry, R. G., . . . & Nelson, F. (2016). The central vein sign and its clinical evaluation for the diagnosis of multiple sclerosis: A consensus statement from the North American Imaging in Multiple Sclerosis Cooperative. *Nature Reviews Neurology, 12*(12), 714–722.

Satoh, N. (2016). *Chordate origins and evolution: The molecular evolutionary road to vertebrates.* Academic Press, London.

Satterfield, B. C., & Killgore, W. D. (2019). Sleep loss, executive function, and decision-making. In *Sleep and Health* (pp. 339–358). Academic Press.

Saunders, D. T., Roe, C. A., Smith, G., & Clegg, H. (2016). Lucid dreaming incidence: A quality effects meta-analysis of 50 years of research. *Consciousness and Cognition, 43*, 197–215.

Savage, L. M., Hall, J. M., & Resende, L. S. (2012). Translational rodent models of Korsakoff syndrome reveal the critical neuroanatomical substrates of memory dysfunction and recovery. *Neuropsychology Review, 22*, 195–209.

Savitz, J. B., Price, J. L., & Drevets, W. C. (2014). Neuropathological and neuromorphometric abnormalities in bipolar disorder: View from the medial prefrontal cortical network. *Neuroscience & Biobehavioral Reviews, 42*, 132–147.

Sawle, G. V., & Myers, R. (1993) The role of positron emission tomography in the assessment of human neurotransplantation. *Trends in Neurosciences, 16,* 172–176.

Sbragia, G. (1992). Leonardo da Vinci and ultrashort sleep: Personal experience of an eclectic artist. In C. Stampi (Ed.), *Why we nap: Evolution, chronobiology, and functions of polyphasic and ultra-short sleep.* Boston, MA: Birkhaüser.

Scalzo, S. J., Bowden, S. C., Ambrose, M. L., Whelan, G., & Cook, M. J. (2015). Wernicke-Korsakoff syndrome not related to alcohol use: A systematic review. *Journal of Neurology, Neurosurgery, and Psychiatry, 86*(12), 1362–1368.

Scammell, T. E. (2015). Narcolepsy. *New England Journal of Medicine, 373,* 2654–2662.

Scammell, T. E., Arrigoni, E., & Lipton, J. O. (2017). Neural circuitry of wakefulness and sleep. *Neuron, 93*(4), 747–765.

Scarpelli, S., Bartolacci, C., D'Atri, A., Gorgoni, M., & De Gennaro, L. (2019). The functional role of dreaming in emotional processes. *Frontiers in Psychology, 10,* 459.

Schaller, M., Miller, G. E., Gervais, W. M., Yager, S., & Chen, E. (2010). Mere visual perception of other people's disease symptoms facilitates a more aggressive immune response. *Psychological Science, 21,* 649–652.

Schally, A. V., Kastin, A. J., & Arimura, A. (1971). Hypothalamic follicle-stimulating hormone (FSH) and luteinizing hormone (LH)–regulating hormone: Structure, physiology, and clinical studies. *Fertility and Sterility, 22,* 703–721.

Scharfman, H. E. (2015). Metabolic control of epilepsy. *Science, 347,* 1312–1313.

Schärli, H., Harman, A. M., & Hogben, J. H. (1999a). Blindsight in subjects with homonymous visual field defects. *Journal of Cognitive Neuroscience, 11,* 52–66.

Schärli, H., Harman, A. M., & Hogben, J. H. (1999b). Residual vision in a subject with damaged visual cortex. *Journal of Cognitive Neuroscience, 11,* 502–510.

Scheiber, M. H. (1999). Somatotopic gradients in the distributed organization of the human primary motor cortex hand area: Evidence from small infarcts. *Experimental Brain Research, 128,* 139–148.

Scheiber, M. H., & Poliakov, A. V. (1998). Partial inactivation of the primary motor cortex hand area: Effects on individuated finger movements. *Journal of Neuroscience, 18,* 9038–9054.

Schenck, C. H., Bundlie, S. R., Ettinger, M. G., & Mahowald, M. W. (1986). Chronic behavioral disorders of human REM sleep: A new category of parasomnia. *Sleep, 9,* 293–308.

Schenck, C. H., Montplaisir, J. Y., Frauscher, B., Hogl, B., Gagnon, J. F., . . . & Oertel, W. (2013). Rapid eye movement sleep behavior disorder: Devising controlled active treatment studies for symptomatic and neuroprotective therapy—a consensus statement from the international rapid eye movement sleep behavior disorder study group. *Sleep Medicine, 14,* 795–806.

Schendel, K., & Robertson, L. C. (2004). Reaching out to see: Arm position can attenuate human visual loss. *Journal of Cognitive Neuroscience, 16,* 935–943.

Scherer, K. R., & Moors, A. (2019). The emotion process: Event appraisal and component differentiation. *Annual Review of Psychology, 70,* 719–745.

Scherma, M., Masia, P., Satta, V., Fratta, W., Fadda, P., & Tanda, G. (2019). Brain activity of anandamide: A rewarding bliss? *Acta Pharmacologica Sinica, 40*(3), 309–323.

Schiavi, S., Manduca, A., Segatto, M., Campolongo, P., Pallottini, V., Vanderschuren, L. J., & Trezza, V. (2019). Unidirectional opioid-cannabinoid cross-tolerance in the modulation of social play behavior in rats. *Psychopharmacology, 236*(9), 2557–2568.

Schienle, A., Schäfer, A., Walter, B., Stark, R., & Vaitl, D. (2005). Brain activation of spider phobics towards disorder-relevant, generally disgust- and fear-inducing pictures. *Neuroscience Letters, 388,* 1–6.

Schloesser, R. J., Martinowich, K., & Manji, H. K. (2012). Mood-stabilizing drugs: Mechanisms of action. *Trends in Neurosciences, 35,* 36–46.

Schmid, M. C., & Maier, A. (2015). To see or not to see—Thalamo-cortical networks during blindsight and perceptual suppression. *Progress in Neurobiology, 126,* 36–48.

Schmidt-Kastner, R. (2015). Genomic approach to selective vulnerability of the hippocampus in brain ischemia-hypoxia. *Neuroscience, 309,* 259–279.

Schmithorst, V. J., & Yuan, W. (2010). White matter development during adolescence as shown by diffusion MRI. *Brain and Cognition, 72,* 16–25.

Schmitt, A., Malchow, B., Hasan, A., & Falkai, P. (2014). The impact of environmental factors in severe psychiatric disorders. *Frontiers in Neuroscience, 8,* 19.

Schmitz, T. W., & Duncan, J. (2018). Normalization and the cholinergic microcircuit: A unified basis for attention. *Trends in Cognitive Sciences, 22,* 422–437.

Schneggenburger, R., & Neher, E. (2005). Presynaptic calcium and control of vesicle fusion. *Current Opinion in Neurobiology, 15,* 266–274.

Schneider, D. M., & Mooney, R. (2018). How movement modulates hearing. *Annual Review of Neuroscience, 41,* 553–572.

Schnell, M. J., McGettigan, J. P., Wirblich, C., & Papaneri, A. (2010). The cell biology of rabies virus: Using stealth to reach the brain. *Nature Reviews Microbiology, 8,* 51–61.

Schoenfeld, M. A., Neuer, G., Tempelmann, C., Schüßler, K., Noesselt, T., Hopf, J.-M., & Heinze, H.-J. (2004). Functional magnetic resonance tomography correlates of taste perception in the human primary taste cortex. *Neuroscience, 127,* 347–353.

Schomers, M. R., Kirilina, E., Weigand, A., Bajbouj, M., & Pulvermüller, F. (2015). Causal influence of articulatory motor cortex on comprehending single spoken words: TMS evidence. *Cerebral Cortex, 25,* 3894–3902.

Schoppa, N. E. (2009). Making scents out of how olfactory neurons are ordered in space. *Nature Neuroscience, 12,* 103–104.

Schredl, M. (2018). Lucid dreaming. In *Researching Dreams* (pp. 163–173). Palgrave Macmillan, Cham.

Schredl, M., Atanasova, D., Hörmann, K., Maurer, J. T., Hummel, T., & Stuck, B. A. (2009). Information processing during sleep: The effect of olfactory stimuli on dream content and dream emotions. *Journal of Sleep Research, 18*(3), 285–290.

Schredl, M., Dyck, S., & Kühnel, A. (2020). Lucid dreaming and the feeling of being refreshed in the morning: A Diary Study. *Clocks & Sleep, 2*(1), 54–60.

Schreiner, C. E., & Polley, D. B. (2014). Auditory map plasticity: Diversity in causes and consequences. *Current Opinion in Neurobiology, 24,* 143–156.

Schröder, H., Moser, N., & Huggenberger, S. (2020). The mouse amygdaloid body. In *Neuroanatomy of the Mouse* (pp. 289–304). Springer, Cham.

Schroeder, L., Roseman, C. C., Cheverud, J. M., & Ackerman, R. R. (2014). Characterizing the evolutionary path(s) to early homo. *PLoS One, 9,* e114307.

Schübeler, D. (2012). Epigenetic islands in a genetic ocean. *Science, 338,* 756–757.

Schultz, C., & Engelhardt, M. (2014). Anatomy of the hippocampal formation. In K. Szabo, & M. G. Hennerici (Eds.), *The hippocampus in clinical neuroscience* (6–17). Basel: Karger AG.

Schultz, M. D., He, Y., Whitaker, J. W., Hariharan, M., Mukamel, E. A., Leung, D., . . . & Lin, S. (2015). Human body epigenome maps reveal noncanonical DNA methylation variation. *Nature, 523,* 212–216.

Schultz, W. (2002). Getting formal with dopamine and reward. *Neuron, 36,* 241–263.

Schultz, W. (2016). Reward functions of the basal ganglia. *Journal of Neural Transmission, 123*(7), 679–693.

Schwartz, A. B. (2004). Cortical neural prosthetics. *Annual Review of Neuroscience, 27,* 487–507.

Schwartz, G. J. (2016). A satiating signal. *Science, 351,* 1268–1269.

Schwartz, J. H., & Tattersall, I. (2015). Defining the genus *Homo. Science, 349,* 931–932.

Schwartz, M. D., & Kilduff, T. S. (2015). The neurobiology of sleep and wakefulness. *Psychiatric Clinics of North America, 38,* 615–644.

Schwartz, M. W. (2000). Staying slim with insulin in mind. *Science, 289,* 2066–2067.

Schweinsburg, A. D., Brown, S. A., & Tapert, S. F. (2008). The influence of marijuana use on neurocognitive functioning in adolescents. *Current Drug Abuse Reviews, 1,* 99–111.

Schweizer, F. E., & Ryan, T. A. (2006). The synaptic vesicle: Cycle of exocytosis and endocytosis. *Current Opinion in Neurobiology, 16,* 298–304.

Scott, J. C., Slomiak, S. T., Jones, J. D., Rosen, A. F., Moore, T. M., & Gur, R. C. (2018). Association of cannabis with cognitive functioning in adolescents and young adults: A systematic review and meta-analysis. *JAMA Psychiatry, 75*(6), 585–595.

Scott, K. M. (2014). Depression, anxiety, and incident cardiometabolic diseases. *Current Opinion in Psychiatry, 27,* 289–293.

Scott, S. H. (2016). A functional taxonomy of bottom-up sensory feedback processing for motor actions. *Trends in Neurosciences, 39*(8), 512–526.

Scott, S. K., McGettigan, C., & Eisner, F. (2009). A little more conversation, a little less action—candidate roles for the motor cortex in speech perception. *Nature Reviews Neuroscience, 10,* 295–302.

Scoville, W. B., & Milner, B. (1957). Loss of recent memory after bilateral hippocampal lesions. *Journal of Neurology, Neurosurgery and Psychiatry, 20,* 11–21.

Scully, T. (2014). Society at large. *Nature, 508,* S50–S51.

Seabrook, T. A., Burbridge, T. J., Crair, M. C., & Huberman, A. D. (2017). Architecture, function, and assembly of the mouse visual system. *Annual Review of Neuroscience, 40,* 499–538.

Searle, L. V. (1949). The organization of hereditary maze-brightness and maze-dullness. *Genetic Psychology Monographs, 39,* 279–325.

Sedley, W., Friston, K. J., Gander, P. E., Kumar, S., & Griffiths, T. D. (2016). An integrative tinnitus model based on sensory precision. *Trends in Neurosciences, 39*(12), 799–812.

Seeley, R. J., & Woods, S. C. (2003). Monitoring of stored and available fuel by the CNS: Implications for obesity. *Nature Reviews Neuroscience, 4,* 901–909.

Seeley, R. J., van Dijk, G., Campfield, L. A., Smith, F. J., Nelligan, J. A., Bell, S. M., et al. (1996). The effect of intraventricular administration of leptin (Ob protein) on food intake and body weight in the rat. *Hormone and Metabolic Research, 28,* 664–668.

Segerstrom, S. C., & Miller, G. E. (2004). Psychological stress and the human immune system: A meta-analytic study of 30 years of inquiry. *Psychological Bulletin, 130,* 601–630.

Segovia, S., Guillamón, A., del Cerro, M. C. R., Ortega, E., Pérez-Laso, C., Rodriguez-Zafra, M., & Beyer, C. (1999). The development of brain sex differences: A multisignaling process. *Behavioural Brain Research, 105,* 69–80.

Sekido, R. (2014). The potential role of SRY in epigenetic gene regulation during brain sexual differentiation in mammals. *Advances in Genetics, 86,* 135–165.

Sekido, R., & Lovell-Badge, R. (2013). Genetic control of testis development. *Sexual Development, 7,* 21–32.

Selkoe, D. J. (1991, November). Amyloid protein and Alzheimer's. *Scientific American, 265*, 68–78.

Selkoe, D. J. (2012). Preventing Alzheimer's disease. *Science, 337*, 1488–1494.

Selkoe, D. J., & Hardy, J. (2016). The amyloid hypothesis of Alzheimer's disease at 25 years. *EMBO Molecular Medicine, 8*(6), 595–608.

Sellayah, D., Cagampang, F. R., & Cox, R. D. (2014). On the evolutionary origins of obesity: A new hypothesis. *Endocrinology, 155*, 1573–1588.

Sellers, K. K., Mellin, J. M., Lustenberger, C. M., Boyle, M. R., Lee, W. H., Peterchev, A. V., & Fröhlich, F. (2015). Transcranial direct current stimulation (tDCS) of frontal cortex decreases performance on the WAIS-IV intelligence test. *Behavioural Brain Research, 290*, 32–44.

Senkowski, D., Höfle, M., & Engel, A. K. (2014). Crossmodal shaping of pain: A multisensory approach to nociception. *Trends in Cognitive Sciences, 18*, 319–327.

Serajee, F. J., & Huq, A. M. (2015). Advances in Tourette syndrome: Diagnoses and treatment. *Pediatric Clinics of North America, 62*, 687–701.

Serati, M., Redaelli, M., Buoli, M., & Altamura, A. C. (2016). Perinatal major depression biomarkers: A systematic review. *Journal of Affective Disorders, 193*, 391–404.

Sereno, M. I., & Huang, R. S. (2014). Multisensory maps in parietal cortex. *Current Opinion in Neurobiology, 24*, 39–46.

Servos, P., Engel, S. A., Gati, J., & Menon, R. (1999). fMRI evidence for an inverted face representation in human somatosensory cortex. *NeuroReport, 10*(7), 1393–1395.

Seung, H. S., & Sümbül, U. (2014). Neuronal cell types and connectivity: Lessons from the retina. *Neuron, 83*, 1262–1272.

Seyfarth, R. M., & Cheney, D. L. (2014). The evolution of language from social cognition. *Current Opinion in Neurobiology, 28*, 5–9.

Shader, R. I. (2017). Placebos, active placebos, and clinical trials. *Clinical Therapeutics, 39*(3), 451–454.

Shallice, T., & Cipolotti, L. (2018). The prefrontal cortex and neurological impairments of active thought. *Annual Review of Psychology, 69*, 157–180.

Shamma, S. A., Elhilali, M., & Micheyl, C. (2011). Temporal coherence and attention in auditory scene analysis. *Trends in Neurosciences, 34*, 114–122.

Shan, L., Dauvilliers, Y., & Siegel, J. M. (2015). Interactions of the histamine and hypocretin systems in CNS disorders. *Nature Reviews Neurology, 11*, 401–413.

Shapiro, B., & Hofreiter, M. (2014). A paleogenomic perspective on evolution and gene function: New insights from ancient DNA. *Science, 343*, 1236573.

Sharma, A. M., & Campbell-Scherer, D. L. (2017). Redefining obesity: Beyond the numbers. *Obesity, 25*(4), 660–661.

Sharp, A. J., & Akbarian, S. (2016). Back to the past in schizophrenia genomics. *Nature Neuroscience, 19*, 1–2.

Sharpee, T. (2014). Toward functional classification of neuronal types. *Neuron, 83*, 1329–1334.

Sharpee, T. O., Atencio, C. A., & Schreiner, C. E. (2011). Hierarchical representations in the auditory cortex. *Current Opinion in Neurobiology, 21*, 761–767.

Sharpless, B. A., & Klíková, M. (2019). Clinical features of isolated sleep paralysis. *Sleep medicine, 58*, 102–106.

Sheffer-Collins, S. I., & Dalva, M. B. (2012). EphBs: An integral link between synaptic function and synaptopathies. *Trends in Neurosciences, 35*, 293–304.

Shelton, J. F., Tancredi, D. J., & Hertz-Picciotto, I. (2010). Independent and dependent contributions of advanced maternal and paternal ages to autism risk. *Autism Research, 3*, 30–39.

Shen, H. (2015). Neuron encyclopaedia fires up to reveal brain secrets. *Nature, 520*, 13–14.

Shen, H. (2015). The hard science of oxytocin. *Nature, 522*, 410–412.

Shen, H. (2018). Embryo assembly 101. *Nature, 559*(7712), 19–22.

Shen, S., Lang, B., Nakamoto, C., Zhang, F., Pu, J., Kuan, S.-L., . . . & St. Clair, D. (2008). Schizophrenia-related neural and behavioral phenotypes in transgenic mice expressing truncated Disc1. *Journal of Neuroscience, 28*, 10893–10904.

Shepherd, R. K., Shivdasani, M. N., Nayagam, D. A. X., Williams, C. E., & Blamey, P. J. (2013). *Trends in biotechnology, 31*, 562–570.

Sherman, D. L., & Brophy, P. J. (2005). Mechanisms of axon ensheathment and myelin growth. *Nature Reviews Neuroscience, 6*, 683–690.

Sherman, M. (2007). The thalamus is more than just a relay. *Current Opinion in Neurobiology, 17*, 417–422.

Sheth, D. N., Bhagwate, M. R., & Sharma, N. (2005). Curious clicks—Sigmund Freud. *Journal of Postgraduate Medicine, 51*, 240–241.

Shevell, S. K., & Kingdom, F. A. A. (2008). Color in complex scenes. *Annual Review of Psychology, 59*, 143–166.

Shi, K., Zhang, J., Dong, J. F., & Shi, F. D. (2019). Dissemination of brain inflammation in traumatic brain injury. *Cellular & Molecular Immunology, 16*(6), 523–530.

Shimoda, R., Campbell, A., & Barton, R. A. (2018). Women's emotional and sexual attraction to men across the menstrual cycle. *Behavioral Ecology, 29*(1), 51–59.

Shimomura, O., Johnson, F., & Saiga, Y. (1962). Extraction, purification and properties of aequorin, a bioluminescent protein from the luminous hydromedusan, *Aequorea. Journal of Cellular and Comparative Physiology, 59*, 223–239.

Shin, O.-H. (2014). Exocytosis and synaptic vesicle function. *Comprehensive Physiology, 4*, 149–175.

Shirazi, S. N., Friedman, A. R., Kaufer, D., & Sakhai, S. A. (2015). Glucocorticoids and the brain: Neural mechanisms regulating the stress response in glucocorticoid signaling. *Advances in Experimental Medicine and Biology, 872*, 235–252.

Shirazi, S. N., Friedman, A. R., Kaufer, D., & Sakhai, S. A. (2015). Glucocorticoids and the brain: Neural mechanisms regulating the stress response. In *Glucocorticoid signaling* (pp. 235–252). New York: Springer.

Shore, S. E., Roberts, L. E., & Langguth, B. (2016). Maladaptive plasticity in tinnitus—triggers, mechanisms and treatment. *Nature Reviews Neurology, 12*(3), 150–160.

Short, S. D., & Hawley, P. H. (2015). The effects of evolution education: Examining attitudes toward and knowledge of evolution in college courses. *Evolutionary Psychology, 13*, 67–88.

Shrestha, P., & Klann, E. (2016). Alzheimer's disease: Lost memories found. *Nature, 531*, 450–451.

Shu, D. G., Luo, H. L., Morris, S. C., Zhang, X. L., Hu, S. X., Chen, L., . . . & Chen, L. Z. (1999). Lower Cambrian vertebrates from south China. *Nature, 402*(6757), 42.

Shultz, S., Klin, A., & Jones, W. (2018). Neonatal transitions in social behavior and their implications for autism. *Trends in Cognitive Sciences, 22*(5), 452–469.

Sibley, D. R., & Shi, L. (2018). A new era of rationally designed antipsychotics. *Nature, 555*, 170–172.

Siciliano, C. A., Fordahl, S. C., & Jones, S. R. (2016). Cocaine self-administration produces long-lasting alterations in dopamine transporter responses to cocaine. *Journal of Neuroscience, 36*(30), 7807–7816.

Siclari, F., Baird, B., Perogamvros, L., Bernardi, G., LaRocque, J. J., Riedner, B., . . . & Tononi, G. (2017). The neural correlates of dreaming. *Nature Neuroscience, 20*(6), 872.

Siclari, F., & Tononi, G. (2017). Local aspects of sleep and wakefulness. *Current Opinion in Neurobiology, 44*, 222–227.

Siegel, E. H., Sands, M. K., Van den Noortgate, W., Condon, P., Chang, Y., Dy, J., . . . & Barrett, L. F. (2018). Emotion fingerprints or emotion populations? A meta-analytic investigation of autonomic features of emotion categories. *Psychological Bulletin, 144*(4), 343–393.

Siegel, J. M. (2005). Clues to the functions of mammalian sleep. *Nature 437*, 1264–1271.

Siegel, J. M. (2008). Do all animals sleep?. *Trends in Neurosciences, 31*(4), 208–213.

Siegel, J. M. (2012). Suppression of sleep for mating. *Science, 337*, 1610–1611.

Siegel, M., Buschman, T. J., & Miller, E. K. (2015). Cortical information flow during flexible sensorimotor decisions. *Science, 348*, 1352–1355.

Siegel, S. (2011). The Four-Loko effect. *Perspectives on Psychological Science, 6*, 357–362.

Siegel, S., Hinson, R. E., Krank, M. D., & McCully, J. (1982). Heroin "overdose" death: Contribution of drug-associated environmental cues. *Science, 216*, 436–437.

Siegenthaler, J. A., Sohet, F., & Daneman, R. (2013). 'Sealing off the CNS': Cellular and molecular regulation of blood-brain barriergenesis. *Current Opinion in Neurobiology, 23*, 1057–1064.

Sikela, J. M., & Quick, V. S. (2018). Genomic trade-offs: Are autism and schizophrenia the steep price of the human brain? *Human Genetics, 137*(1), 1–13.

Sikka, P., Pesonen, H., & Revonsuo, A. (2018). Peace of mind and anxiety in the waking state are related to the affective content of dreams. *Scientific Reports, 8*(1), 12762.

Silbereis, J. C., Pochareddy, S., Zhu, Y., Li, M., & Sestan, N. (2016). The cellular and molecular landscapes of the developing human central nervous system. *Neuron, 89*, 248–268.

Silbereis, J. C., Pochareddy, S., Zhu, Y., Li, M., & Sestan, N. (2016). The cellular and molecular landscapes of the developing human central nervous system. *Neuron, 89*(2), 248–268.

Silva, D., Feng, T., & Foster, D. J. (2015). Trajectory events across hippocampal place cells require previous experience. *Nature Neuroscience, 18*, 1772–1779.

Silva, S., Martins, Y., Matias, A., & Blickstein, I. (2011). Why are monozygotic twins different? *Journal of Perinatal Medicine, 39*, 195–202.

Silvanto, J. (2014). Is primary visual cortex necessary for visual awareness? *Trends in Neurosciences, 37*, 618–619.

Silver, M., Moore, C. M., Villamarin, V., Jaitly, N., Hall, J. E., Rothschild, A. J., & Deligiannidis, K. M. (2018). White matter integrity in medication-free women with peripartum depression: A tract-based spatial statistics study. *Neuropsychopharmacology, 43*, 1573–1580.

Simms, B. A., & Zamponi, G. W. (2014). Neuronal voltage-gated calcium channels: Structure, function, and dysfunction. *Neuron, 82*, 24–45.

Simon, E., Obst, J., & Gomez-Nicola, D. (2019). The evolving dialogue of microglia and neurons in Alzheimer's disease: Microglia as necessary transducers of pathology. *Neuroscience, 405*, 24–34.

Simon, N. W., Wood, J., & Moghaddam, B. (2015). Action-outcome relationships are represented differently by medial prefrontal and orbitofrontal cortex neurons during action execution. *Journal of Neurophysiology, 114*, 3374–3385.

Simonds, S. E., & Cowley, M. A. (2013). Hypertension in obesity: Is leptin the culprit? *Trends in Neurosciences, 36*, 121–132.

Sinclair, P., Brennan, D. J., & le Roux, C. W. (2018). Gut adaptation after metabolic surgery and its influences on the brain, liver and cancer. *Nature Reviews Gastroenterology & Hepatology, 15*(10), 606–624.

Sinclair, S. V., & Mistlberger, R. E. (1997). Scheduled activity reorganizes circadian phase of Syrian hamsters under full and skeleton photoperiods. *Behavioural Brain Research, 87,* 127–137.

Singer, B. F., Neugebauer, N. M., Forneris, J., Rodvelt, K. R., Li, D., Bubula, N., & Vezina, P. (2014). Locomotor conditioning by amphetamine requires cyclin-dependent kinase 5 signaling in the nucleus accumbens. *Neuropharmacology, 85,* 243–252.

Singer, J. J. (1968). Hypothalamic control of male and female sexual behavior in female rats. *Journal of Comparative and Physiological Psychology, 66,* 738–742.

Singh, A., Saluja, S., Kumar, A., Agrawal, S., Thind, M., Nanda, S., & Shirani, J. (2018). Cardiovascular complications of marijuana and related substances: A review. *Cardiology and Therapy, 7*(1), 45–59.

Singh, J., Hallmayer, J., & Illes, J. (2007). Interacting and paradoxical forces in neuroscience and society. *Nature Reviews Neuroscience, 8,* 153–160.

Singleton, A., & Hardy, J. (2016). The evolution of genetics: Alzheimer's and Parkinson's diseases. *Neuron, 90*(6), 1154–1163.

Sinha, D. N., Gupta, P. C., Kumar, A., Bhartiya, D., Agarwal, N., Sharma, S., . . . & Mehrotra, R. (2018). The poorest of poor suffer the greatest burden from smokeless tobacco use: A study from 140 countries. *Nicotine and Tobacco Research, 20*(12), 1529–1532.

Sisk, C. L., & Zehr, J. L. (2005). Pubertal hormones organize the adolescent brain and behavior. *Frontiers in Neuroendocrinology, 26,* 163–174.

Sitte, H. H., & Freissmuth, M. (2015). Amphetamines, new psychoactive drugs, and the monoamine transporter cycle. *Trends in Pharmacological Sciences, 36,* 41–50.

Skinner, M. M., Stephens, N. B., Tsegai, Z. J., Foote, A. C., Nguyen, N. H., et al. (2015). Human-like hand use in *Australopithecus africanus. Science, 347,* 395–399.

Sladek, J. R., Jr., Redmond, D. E., Jr., Collier, T. J., Haber, S. N., Elsworth, J. D., Deutch, A. Y., & Roth, R. H. (1987). Transplantation of fetal dopamine neurons in primate brain reverses MPTP induced Parkinsonism. In F. J. Seil, E. Herbert, & B. M. Carlson (Eds.), *Progress in brain research* (Vol. 71, pp. 309–323). New York, NY: Elsevier.

Slocombe, K. E., Kaller, T., Call, J., & Zuberbühler, K. (2010). Chimpanzees extract social information from agonistic screams. *PLoS One, 5,* e11473.

Smalle, E. H., Rogers, J., & Möttönen, R. (2015). Dissociating contributions of the motor cortex to speech perception and response bias by using transcranial magnetic stimulation. *Cerebral Cortex, 25,* 3690–3698.

Smigielski, L., Jagannath, V., Rössler, W., Walitza, S., & Grünblatt, E. (2020). Epigenetic mechanisms in schizophrenia and other psychotic disorders: A systematic review of empirical human findings. *Molecular Psychiatry,* 1–31.

Smith, A. M., & Dragunow, M. (2014). The human side of microglia. *Trends in Neurosciences, 37,* 125–135.

Smith, D. H., Johnson, V. E., Trojanowski, J. Q., & Stewart, W. (2019). Chronic traumatic encephalopathy—confusion and controversies. *Nature Reviews Neurology, 15*(3), 179–183.

Smith, F. W., & Goodale, M. A. (2015). Decoding visual object categories in early somatosensory cortex. *Cerebral Cortex, 25,* 1020–1031.

Smith, H., Fox, J. R., & Trayner, P. (2015). The lived experiences of individuals with Tourette syndrome or tic disorders: A meta-synthesis of qualitative studies. *British Journal of Psychology, 106,* 609–634.

Smith, P. A. (2015). Brain, meet gut. *Nature, 526,* 312–314.

Sneed, A. (2014). Why babies forget. *Scientific American, 311,* 28.

Snyder, J. S. (2019). Recalibrating the relevance of adult neurogenesis. *Trends in Neurosciences, 42*(3), 164–178.

Snyder, S. H. (1978). Neuroleptic drugs and neurotransmitter receptors. *Journal of Clinical and Experimental Psychiatry, 133,* 21–31.

So, H. C., Chau, C. K. L., Chiu, W. T., Ho, K. S., Lo, C. P., Yim, S. H. Y., & Sham, P. C. (2017). Analysis of genome-wide association data highlights candidates for drug repositioning in psychiatry. *Nature neuroscience, 20*(10), 1342.

Sobolewski, M. E., Brown, J. L., & Mitani, J. C. (2013). Female parity, male aggression, and the challenge hypothesis in wild chimpanzees. *Primates, 54,* 81–88.

Soffer-Dudek, N. (2020). Are lucid dreams good for us? Are we asking the right question? A call for caution in lucid dream research. *Frontiers in Neuroscience, 13,* 1423.

Sofroniew, M. V. (2009). Molecular dissection of reactive astrogliosis and glial scar formation. *Trends in Neurosciences, 32,* 638–647.

Sohn, J.-W., Elmquist, J. K., & Williams, K. W. (2013). Neuronal circuits that regulate feeding behavior and metabolism. *Trends in Neurosciences, 36,* 504–512.

Sokolov, A. A., Miall, R. C., & Ivry, R. B. (2017). The cerebellum: Adaptive prediction for movement and cognition. *Trends in Cognitive Sciences, 21*(5), 313–332.

Solinas, M., Thiriet, N., Chauvet, C., & Jaber, M. (2010). Prevention and treatment of drug addiction by environmental enrichment. *Progress in Neurobiology, 92,* 572–592.

Sollars, P. J., & Pickard, G. E. (2015). The neurobiology of circadian rhythms. *Psychiatric Clinics of North America, 38,* 645–665.

Soloman, S. M., & Kirby, D. F. (1990). The refeeding syndrome: A review. *Journal of Parenteral and Enteral Nutrition, 14,* 90–97.

Somers, J. M., Goldner, E. M., Waraich, P., & Hsu, L. (2006). Prevalence and incidence studies of anxiety disorders: A systematic review of the literature. *Canadian Journal of Psychiatry, 51,* 100–112.

Song, C., Luchtman, D., Kang, Z., Tam, E. M., Yatham, L. N., Su, K. P., & Lam, R. W. (2015). Enhanced inflammatory and T-helper-1 type responses but suppressed lymphocyte proliferation in patients with seasonal affective disorder and treated by light therapy. *Journal of Affective Disorders, 185,* 90–96.

Sorrells, S. F., Paredes, M. F., Cebrian-Silla, A., Sandoval, K., Qi, D., Kelley, K. W., . . . & Chang, E. F. (2018). Human hippocampal neurogenesis drops sharply in children to undetectable levels in adults. *Nature, 555*(7696), 377–381.

Sousa-Ferreira, L., de Almeida, L. P., & Cavadas, C. (2014). Role of hypothalamic neurogenesis in feeding regulation. *Trends in Endocrinology and Metabolism, 25,* 80–88.

Southwell, M., Shelly, S., MacDonald, V., Verster, A., & Maher, L. (2019). Transforming lives and empowering communities: evidence, harm reduction and a holistic approach to people who use drugs. *Current Opinion in HIV and AIDS, 14*(5), 409–414.

Sowa, J., Bobula, B., Glombik, K., Slusarczyk, J., Basta-Kaim, A., & Hess, G. (2015). Prenatal stress enhances excitatory synaptic transmission and impairs long-term potentiation in the frontal cortex of adult offspring rats. *PLoS One, 10,* e0119407.

Soykan, T., Maritzen, T., & Haucke, V. (2016). Modes and mechanisms of synaptic vesicle recycling. *Current Opinion in Neurobiology, 39,* 17–23.

Spadoni, A. D., McGee, C. L., Fryer, S. L., & Riley, E. P. (2007). Neuroimaging and fetal alcohol spectrum disorders. *Neuroscience and Biobehavioral Reviews, 31,* 239–245.

Spalding, K. L., Bergmann, O., Alkass, K., Bernard, S., Salehpour, M., . . . & Frisén, J. (2013). Dynamics of hippocampal neurogenesis in adult humans. *Cell, 153,* 1219–1227.

Speakman, J. R. (2013). Evolutionary perspectives on the obesity epidemic: Adaptive, maladaptive, and neutral viewpoints. *Annual Review of Nutrition, 33,* 289–317.

Spering, M., & Carrasco, M. (2015). Acting without seeing: Eye movements reveal visual processing without awareness. *Trends in Neurosciences, 38,* 247–258.

Sperry, R. W. (1963). Chemoaffinity in the orderly growth of nerve fiber patterns and connections. *Proceedings of the National Academy of Sciences, USA, 50,* 703–710.

Sperry, R. W. (1964, January). The great cerebral commissure. *Scientific American, 210,* 42–52.

Sperry, R. W., Zaidel, E., & Zaidel, D. (1979). Self recognition and social awareness in the deconnected minor hemisphere. *Neuropsychologia, 17,* 153–166.

Spiga, F., Walker, J. J., Terry, J. R., & Lightman, S. L. (2014). HPA axis-rhythms. *Comprehensive Physiology, 4,* 1273–1298.

Spinney, L. (2014). The forgetting gene. *Nature, 510,* 26–27.

Spires-Jones, T. L., & Hyman, B. T. (2014). The intersection of amyloid beta and tau at synapses in Alzheimer's disease. *Neuron, 82,* 756–771.

Spitzer, N. C. (2017). Neurotransmitter switching in the developing and adult brain. *Annual Review of Neuroscience, 40,* 1–19.

Spitzer, R. L., Skodol, A. E., Gibbon, M., & Williams, J. B. W. (1983). *Psychopathology: A case book.* New York, NY: McGraw-Hill.

Sporns, O., & Betzel, R. F. (2016). Modular brain networks. *Annual Review of Psychology, 67,* 613–640.

Sprague, T. C., Saproo, S., & Serences, J. T. (2015). Visual attention mitigates information loss in small- and large-scale neural codes. *Trends in Cognitive Sciences, 19,* 215–226.

Squarzoni, P., Thion, M. S., & Garel, S. (2015). Neuronal and microglial regulators of cortical wiring: Usual and novel guideposts. *Frontiers in Neuroscience, 9,* 248.

Squire, L. R. (1987). *Memory and brain.* New York, NY: Oxford University Press.

Squire, L. R. & Zola-Morgan, S. (1991). The medial temporal lobe memory system. *Science, 253,* 1380–1386.

Squire, L. R., & Dede, A. J. O. (2015). Conscious and unconscious memory systems. *Cold Spring Harbor Perspectives in Biology, 7,* a021667.

Squire, L. R., & Zola-Morgan, S. (1985). The neuropsychology of memory: New links between humans and experimental animals. *Annals of the New York Academy of Sciences, 444,* 137–149.

Squire, L. R., Amaral, D. G., Zola-Morgan, S., Kritchevsky, M., & Press, G. (1989). Description of brain injury in the amnesic patient N.A. based on magnetic resonance imaging. *Experimental Neurology, 105,* 23–35.

Squire, L. R., Genzel, L., Wixted, J. T., & Morris, R. G. (2015). Memory consolidation. *Cold Spring Harbor Perspectives in Biology, 7,* a021766.

Squire, L. R., Slater, P. C., & Chace, P. M. (1975). Retrograde amnesia: Temporal gradient in very long term memory following electroconvulsive therapy. *Science, 187,* 77–79.

Squire, R. F., Noudoost, B., Schafer, R. J., & Moore, T. (2013). Prefrontal contributions to visual selective attention. *Annual Review of Neuroscience, 36,* 451–466.

St. George-Hyslop, P. H. (2000, December). Piecing together Alzheimer's. *Scientific American, 283,* 76–83.

Stampi, C. (Ed.). (1992). *Why we nap: Evolution, chronobiology, and functions of polyphasic and ultrashort sleep.* Boston, MA: Birkhäuser.

Stangel, M., Kuhlmann, T., Matthews, P. M., & Kilpatrick, T. J. (2017). Achievements and obstacles of remyelinating therapies in multiple sclerosis. *Nature Reviews Neurology, 13*(12), 742–754.

Stankowski, R. V., Kloner, R. A., & Rezkalla, S. H. (2015). Cardiovascular consequences of cocaine use. *Trends in Cardiovascular Medicine, 25,* 517–526.

Stark, R., Klein, S., Kruse, O., Weygandt, M., Leufgens, L. K., Schweckendiek, J., & Strahler, J. (2019). No sex difference found: Cues of sexual stimuli activate the reward system in both sexes. *Neuroscience, 416,* 63–73.

Stasenko, A., Bonn, C., Teghipco, A., Garcea, F. E., Sweet, C.,... & Mahon, B. Z. (2015). A causal test of the motor theory of speech perception: A case of impaired speech production and spared speech perception. *Cognitive Neuropsychology, 32,* 38–57.

Stasenko, A., Garcea, F. E., & Mahon, B. Z. (2013). What happens to the motor theory of perception when the motor system is damaged? *Language and Cognition, 5,* 225–238.

State, M. W., & Levitt, P. (2011). The conundrums of understanding genetic risks for autism spectrum disorders. *Nature Neuroscience, 14,* 1499–1506.

Steen, R. G., Mull, C., McClure, R., Hamer, R. M., & Lieberman, J. A. (2006). Brain volume in first-episode schizophrenia: Systematic review and meta-analysis of magnetic resonance imaging studies. *British Journal of Psychiatry, 188,* 510–518.

Stefani, A., & Videnovic, A. (2018). Idiopathic REM sleep behaviour disorder and neurodegeneration—an update. *Nature Reviews Neurology, 14*(1), 40–56.

Stefani, A., Trendafilov, V., Liguori, C., Fedele, E., & Galati, S. (2017). Subthalamic nucleus deep brain stimulation on motor-symptoms of Parkinson's disease: Focus on neurochemistry. *Progress in Neurobiology, 151,* 157–174.

Steimer, T. (2011). Animal models of anxiety disorders in rats and mice: Some conceptual issues. *Dialogues in Clinical Neuroscience, 13,* 495–506.

Stein, C. (2016). Opioid receptors. *Annual Review of Medicine, 67,* 433–451.

Stein, J. (2018). What is developmental dyslexia? *Brain Sciences, 8*(2), 26.

Steinbeck, J. A., & Studer, L. (2015). Moving stem cells to the clinic: Potential and limitations for brain repair. *Neuron, 86,* 187–206.

Steketee, J. D., & Kalivas, P. W. (2011). Drug wanting: Behavioral sensitization and relapse to drug seeking behavior. *Pharmacological Reviews, 63,* 348–365.

Stensola, T., Stensola, H., Moser, M.-B., & Moser, E. I. (2015). Shearing-induced asymmetry in entorhinal grid cells. *Nature, 518,* 207–212.

Stephenson-Jones, M., Yu, K., Ahrens, S., Tucciarone, J. M., van Huijstee, A. N., Mejia, L. A., ... & Li, B. (2016). A basal ganglia circuit for evaluating action outcomes. *Nature, 539*(7628), 289.

Stern, D. L. (2013). The genetic causes of convergent evolution. *Nature Reviews Genetics, 14,* 751–764.

Sternson, S. M., Betley, J. N., & Cao, Z. F. H. (2013). Neural circuits and motivational processes for hunger. *Current Opinion in Neurobiology, 23,* 353–360.

Stevens, C., & Bavelier, D. (2012). The role of selective attention on academic foundations: A cognitive neuroscience perspective. *Developmental Cognitive Neuroscience, 25,* S30–S48.

Stevens, M. S. (2015). Restless leg syndrome/Willis-Ekbom disease morbidity: Burden, quality of life, cardiovascular aspects, and sleep. *Sleep Medicine Clinics, 10,* 369–373.

Stevenson, R. J. (2012). The role of attention in flavour perception. *Flavour, 1,* 2.

Stickgold, R. (2005). Sleep-dependent memory consolidation. *Nature, 437,* 1272–1278.

Stickgold, R. (2015). Sleep on it!. *Scientific American, 313,* 52–57.

Stickgold, R., & Walker, M. P. (2005). Memory consolidation and reconsolidation: What is the role of sleep? *Trends in Neurosciences, 28,* 408–415.

Stickgold, R., & Walker, M. P. (2013). Sleep-dependent memory triage: Evolving generalization through selective processing. *Nature Neuroscience, 16,* 139–145.

Stillman, P. E., Van Bavel, J. J., & Cunningham, W. A. (2015). Valence asymmetries in the human amygdala: Task relevance modulates amygdala responses to positive more than negative affective cues. *Journal of Cognitive Neuroscience, 27,* 842–851.

Stockwell, T. (2012). Commentary on Roerecke & Rehm (2012): The state of the science on moderate drinking and health—a case of heterogeneity in and heterogeneity out? *Addiction, 107,* 1261–1262.

Stockwell, T., Zhao, J., Panwar, S., Roemer, A., Naimi, T., & Chikritzhs, T. (2016). Do "moderate" drinkers have reduced mortality risk? A systematic review and meta-analysis of alcohol consumption and all-cause mortality. *Journal of Studies on Alcohol and Drugs, 77,* 185–198.

Stogsdill, J. A., & Eroglu, C. (2017). The interplay between neurons and glia in synapse development and plasticity. *Current Opinion in Neurobiology, 42,* 1–8.

Stoléru, S., Fonteille, V., Cornélis, C., Joyal, C., & Moulier, V. (2012). Functional neuroimaging studies of sexual arousal and orgasm in healthy men and women: A review and meta-analysis. *Neuroscience and Biobehavioral Reviews, 36,* 1481–1509.

Stone, W. L., Krishnan, K., Campbell, S. E., & Palau, V. E. (2014). The role of antioxidants and pro-oxidants in colon cancer. *World Journal of Gastrointestinal Oncology, 6,* 55–66.

Stoodley, C. J., & Stein, J. F. (2013). Cerebellar function in developmental dyslexia. *Cerebellum, 12,* 267–276.

Storace, D., Rad, M. S., Kang, B., Cohen, L. B., Hughes, T., & Baker, B. J. (2016). Toward better genetically encoded sensors of membrane potential. *Trends in Neurosciences, 39*(5), 277–289.

Storrs, C. (2017). How poverty affects the brain. *Nature, 547*(7662), 150–152.

Stowers, L., & Kuo, T.-H. (2015). Mammalian pheromones: Emerging properties and mechanisms of detection. *Current Opinion in Neurobiology, 34,* 103–109.

Stranahan, A. M., & Mattson, M. P. (2012). Recruiting adaptive cellular stress responses for successful brain ageing. *Nature Reviews Neuroscience, 13,* 209–216.

Strathdee, S. A., & Pollini, R. A. (2007). A 21st-century Lazarus: The role of safer injection sites in harm reduction and recovery. *Addiction, 102,* 848–849.

Stringer, C. (2014). Human evolution: Small remains still pose big problems. *Nature, 514,* 427–429.

Strisciuglio, P., & Concolino, D. (2014). New strategies for the treatment of phenylketonuria (PKU). *Metabolites, 4,* 1007–1017.

Strub, R. L., & Black, F. W. (1997). The mental status exam. In T. E. Feinberg & M. J. Farah (Eds.), *Behavioral neurology and neuropsychology* (pp. 25–42). New York, NY: McGraw-Hill.

Stuart, G. J., & Spruston, N. (2015). Dendritic integration: 60 years of progress. *Nature Neuroscience, 18*(12), 1713.

Stuber, G. D., & Wise, R. A. (2016). Lateral hypothalamic circuits for feeding and reward. *Nature Neuroscience, 19*(2), 198–203.

Stuchlik, A. (2014). Dynamic learning and memory, synaptic plasticity and neurogenesis: An update. *Frontiers in Behavioral Neuroscience, 8,* 108.

Stumvoli, M., Goldstein, B. J., & van Haeften, T. W. (2005). Type 2 diabetes: Principles of pathogenesis and therapy. *The Lancet, 365,* 1333–1346.

Stuss, D. T., & Alexander, M. P. (2005). Does damage to the frontal lobes produce impairment in memory? *Current Directions in Psychological Science, 14,* 84–88.

Su, P., Zhang, J., Wang, D., Zhao, F., Cao, Z., Aschner, M., & Luo, W. (2016). The role of autophagy

in modulation of neuroinflammation in microglia. *Neuroscience, 319,* 155–167.

Subramanian, K., Sarkar, S., & Kattimani, S. (2017). Bipolar disorder in Asia: Illness course and contributing factors. *Asian Journal of Psychiatry, 29,* 16–29.

Subramoney, S., Eastman, E., Adnams, C., Stein, D. J., & Donald, K. A. (2018). The early developmental outcomes of prenatal alcohol exposure: a review. *Frontiers in Neurology, 9,* 1108.

Substance Abuse and Mental Health Services Administration. (2012). Results from the 2011 National Survey on Drug Use and Health: Summary of National Findings, NSDUH Series H-44, HHS Publication No. (SMA) 12–4713. Rockville, MD: Substance Abuse and Mental Health Services Administration.

Suchowersky, O. (2015). Decade in review—movement disorders: Tracking the pathogenesis of movement disorders. *Nature Reviews Neurology, 11,* 618–619.

Südhof, T. C. (2017). Molecular neuroscience in the 21st century: A personal perspective. *Neuron, 96*(3), 536–541.

Sugita, M., & Shipa, Y. (2005). Genetic tracing shows segregation of taste neuronal circuitries for bitter and sweet. *Science, 309,* 781–785.

Sulzer, D. (2007). Multiple hit hypotheses for dopamine neuron loss in Parkinson's disease. *Trends in Neurosciences, 30,* 244–250.

Sun, B. L., Wang, L. H., Yang, T., Sun, J. Y., Mao, L. L., Yang, M. F., . . . & Yang, X. Y. (2018). Lymphatic drainage system of the brain: A novel target for intervention of neurological diseases. *Progress in Neurobiology, 163,* 118–143.

Sun, D. (2015). Endogenous neurogenic cell response in the mature mammalian brain following traumatic injury. *Experimental Neurology, 275,* 405–410.

Sun, T., & Hevner, R. F. (2014). Growth and folding of the mammalian cerebral cortex: From molecules to malformations. *Nature Reviews Neuroscience, 15,* 217–232.

Sun, W., McConnell, E., Pare, J.-F., Xu, Q., Chen, M., Peng, W., Lovatt, D., Han, X., Smith, Y., & Nedergard, M. (2013). Glutamate-dependent neuroglial calcium signaling differs between young and adult brain. *Science, 339,* 197–200.

Sun, W., Tan, Z., Mensh, B. D., & Ji, N. (2016). Thalamus provides layer 4 of primary visual cortex with orientation-and direction-tuned inputs. *Nature Neuroscience, 19*(2), 308–315.

Sun, Y. R., Herrmann, N., Scott, C. J., Black, S. E., Khan, M. M., & Lanctôt, K. L. (2018). Global grey matter volume in adult bipolar patients with and without lithium treatment: A meta-analysis. *Journal of Affective Disorders, 225,* 599–606.

Sun, Y., Liu, C., Huang, M., Huang, J., Liu, C., Zhang, J., . . . & Wang, H. (2019). The molecular evolution of circadian clock genes in spotted gar (*Lepisosteus oculatus*). *Genes, 10*(8), 622.

Sunday, S. R., & Halmi, K. A. (1990). Taste perceptions and hedonics in eating disorders. *Physiology & Behavior, 48,* 587–594.

Sung, Y. J., Simino, J., Kume, R., Basson, J., Schwander, K., & Rao, D. C. (2014). Comparison of two methods for analysis of gene-environment interactions in longitudinal family data: The Framingham heart study. *Frontiers in Genetics, 5,* 9.

Sur, M., & Rubenstein, J. L. R. (2005). Patterning and plasticity of the cerebral cortex. *Science, 310,* 805–810.

Surmeier, D. J., Mercer, J. N., & Chan, C. S. (2005). Autonomous pacemakers in the basal ganglia: Who needs excitatory synapses anyway? *Current Opinion in Neurobiology, 15,* 312–318.

Surmeier, D. J., Obeso, J. A., & Halliday, G. M. (2017). Selective neuronal vulnerability in Parkinson disease. *Nature Reviews Neuroscience, 18*(2), 101–113.

Susilo, T., & Duchaine, B. (2013). Advances in developmental prosopagnosia research. *Current Opinion in Neurobiology, 23,* 423–429.

Suwa, G., Asfaw, B., Kono, R. T., Kubo, D., Lovejoy, C. O., & White, T. D. (2011). The Ardipithecus ramidus skull and its implications for hominid origins. *Science, 326*, 68.

Suwa, G., Kono, R. T., Simpson, S. W., Asfaw, B., Lovejoy, C. O., & White, T. D. (2011). Paleobiological implications of the *Ardipithecus ramidus* dentition. *Science, 326*, 94–99.

Suwanwela, N., & Koroshetz, W. J. (2007). Acute ischemic stroke: Overview of recent therapeutic developments. *Annual Review of Medicine, 58*, 89–106.

Suzuki, W. A. (2010). Untangling memory from perception in the medial temporal lobe. *Trends in Cognitive Sciences, 14*, 195–200.

Suzuki, W. A., & Naya, Y. (2014). The perirhinal cortex. *Annual Review of Neuroscience, 37*, 39–53.

Swaab, D. F. (2004). Sexual differentiation of the human brain: Relevance for gender identity, transsexualism and sexual orientation. *Gynecological Endocrinology, 19*, 301–312.

Swanson, L. W. (2000). What is the brain? *Trends in Neurosciences, 23*, 519–527.

Swanson, L. W., & Lichtman, J. W. (2016). From Cajal to connectome and beyond. *Annual Review of Neuroscience, 39*, 197–216.

Swanson, S. A., Crow, S. J., Le Grange, D., Swendsen, J., & Merikangas, K. R. (2011). Prevalence and correlates of eating disorders in adolescents: Results from the national comorbidity survey replication adolescent supplement. *Archives of General Psychiatry, 68*, 714–723.

Sweatt, J. D. (2013). The emerging field of neuroepigenetics. *Neuron, 80*, 624–632.

Sweatt, J. D. (2016). Neural plasticity and behavior–sixty years of conceptual advances. *Journal of Neurochemistry, 139*, 179–199.

Sweeney, M. D., & Zlokovic, B. V. (2018). A lymphatic waste-disposal system implicated in Alzheimer's disease. *Nature, 560*, 172–174.

Sweeney, M. D., Zhao, Z., Montagne, A., Nelson, A. R., & Zlokovic, B. V. (2019). Blood-brain barrier: From physiology to disease and back. *Physiological Reviews, 99*(1), 21–78.

Syntichaki, P., & Tavernarakis, N. (2003). The biochemistry of neuronal necrosis: Rogue biology? *Nature Reviews Neuroscience, 4*, 672–684.

Sytnyk, V., Leshchyns'ka, I., & Schachner, M. (2017). Neural cell adhesion molecules of the immunoglobulin superfamily regulate synapse formation, maintenance, and function. *Trends in Neurosciences, 40*(5), 295–308.

Szczupak, L. (2016). Functional contributions of electrical synapses in sensory and motor networks. *Current Opinion in Neurobiology, 41*, 99–105.

Sznycer, D. (2019). Forms and functions of the self-conscious emotions. *Trends in Cognitive Sciences, 23*(2), 143–157.

Sztainberg, Y., & Zoghbi, H. Y. (2016). Lessons learned from studying syndromic autism spectrum disorders. *Nature Neuroscience, 19*(11), 1408–1418.

Szutorisz, H., & Hurd, Y. L. (2018). High times for cannabis: Epigenetic imprint and its legacy on brain and behavior. *Neuroscience & Biobehavioral Reviews, 85*, 93–101.

Szyf, M. (2014). Lamarck revisited: Epigenetic inheritance of ancestral odor fear conditioning. *Nature Neuroscience, 17*, 2–4.

Szyf, M. (2014). Nongenetic inheritance and transgenerational epigenetics. *Trends in Molecular Medicine, 21*, 134–144.

Tabri, N., Murray, H. B., Thomas, J. J., Franko, D. L., Herzog, D. B., & Eddy, K. T. (2015). Overvaluation of body shape/weight and engagement in non-compensatory weight-control behaviors in eating disorders: Is there a reciprocal relationship? *Psychological Medicine, 45*, 2951–2958.

Takahashi, K., Tanabe, K., Ohnuki, M., Narita, M., Ichisaka, T., Tomoda, K., & Yamanaka, S. (2007). Induction of pluripotent stem cells from adult human fibroblasts by defined factors. *Cell, 131*, 861–872.

Takahashi, Y., Sipp, D., & Enomoto, H. (2013). Tissue interactions in neural crest cell development and disease. *Science, 341*, 860–863.

Takeichi, M. (2007). The cadherin superfamily in neuronal connections and interactions. *Nature Reviews Neuroscience, 8*, 11–20.

Takeuchi, T., & Morris, R. G. M. (2014). Shedding light on a change of mind. *Nature, 513*, 323–324.

Takumi, T., & Tamada, K. (2018). CNV biology in neurodevelopmental disorders. *Current Opinion in Neurobiology, 48*, 183–192.

Tamariz, E., & Varela-Echavarría, A. (2015). The discovery of the growth cone and its influence on the study of axon guidance. *Frontiers in Neuroanatomy, 9*, 51.

Tan, L. L., Pelzer, P., Heinl, C., Tang, W., Gangadharan, V., Flor, H., . . . & Kuner, R. (2017). A pathway from midcingulate cortex to posterior insula gates nociceptive hypersensitivity. *Nature Neuroscience, 20*(11), 1591–1601.

Tan, Q., Frost, M., Heijmans, B. T., von Bornemann Hjelmborg, J., Tobi, E. W., Christensen, K., & Christiansen, L. (2014). Epigenetic signature of birth weight discordance in adult twins. *BMC Genomics, 15*, 1.

Tank, D. W., Sugimori, M., Connoer, J. A., & Llinás, R. R. (1998). Spatially resolved calcium dynamics of mammalian Purkinje cells in cerebellar slice. *Science, 242*, 7733–7777.

Tanzi, R. E. (2012). The genetics of Alzheimer disease. *Cold Spring Harbor Perspectives in Medicine, 2*, a006296.

Tattersall, I., & Matternes, J. H. (2000, January). Once we were not alone. *Scientific American, 282*, 56–62.

Tauber, M., Diene, G., Mimoun, E., Çabal-Berthoumieu, S., Mantoulan, C., . . . & Salles, J. P. (2014). Prader-Willi syndrome as a model of human hyperphagia. *Frontiers of Hormone Research, 42*, 93–106.

Tavares, R. M., Mendelsohn, A., Grossman, Y., Williams, C. H., Shapiro, M., Trope, Y., & Schiller, D. (2015). A map for social navigation in the human brain. *Neuron, 87*, 231–243.

Taylor, D., Sparshatt, A., Varma, S., & Olofinjana, O. (2014). Antidepressant efficacy of agomelatine: Meta-analysis of published and unpublished studies. *British Medical Journal, 348*, g1888.

Tedeschi, A., & Bradke, F. (2017). Spatial and temporal arrangement of neuronal intrinsic and extrinsic mechanisms controlling axon regeneration. *Current Opinion in Neurobiology, 42*, 118–127.

Teicher, M. H., Samson, J. A., Anderson, C. M., & Ohashi, K. (2016). The effects of childhood maltreatment on brain structure, function and connectivity. *Nature Reviews Neuroscience, 17*(10), 652–666.

Teitelbaum, P. (1961). Disturbances in feeding and drinking behavior after hypothalamic lesions. In M. R. Jones (Ed.), *Nebraska symposium on motivation* (pp. 39–69). Lincoln: University of Nebraska Press.

Teitelbaum, P., & Epstein, A. N. (1962). The lateral hypothalamic syndrome: Recovery of feeding and drinking after lateral hypothalamic lesions. *Psychological Review, 69*, 74–90.

Teixeira, S., Machado, S., Velasques, B., Sanfim, A., Minc, D., . . . & Silva, J. G. (2014). Integrative parietal cortex processes: Neurological and psychiatric aspects. *Journal of Neurological Sciences, 338*, 12–22.

Tekriwal, A., Kern, D. S., Tsai, J., Ince, N. F., Wu, J., Thompson, J. A., & Abosch, A. (2017). REM sleep behaviour disorder: Prodromal and mechanistic insights for Parkinson's disease. *Journal of Neurology, Neurosurgery & Psychiatry, 88*(5), 445–451.

ter Horst, J. P., de Kloet, E. R., Schächinger, H., & Oitzl, M. S. (2012). Relevance of stress and female sex hormones for emotion and cognition. *Cellular and Molecular Neurobiology, 32*, 725–735.

Testerman, T. L., & Morris, J. (2014). Beyond the stomach: An updated view of *Helicobacter pylori* pathogenesis, diagnosis, and treatment. *World Journal of Gastroenterology, 20*, 12781–12808.

Teuber, H.-L. (1975). Recovery of function after brain injury in man. In *Outcomes of severe damage to the nervous system.* CIBA Foundation Symposium 34. Amsterdam, Netherlands: Elsevier North-Holland.

Teuber, H.-L., Battersby, W. S., & Bender, M. B. (1960). Recovery of function after brain injury in man. In *Outcomes of severe damage to the nervous system.* CIBA Foundation Symposium 34. Amsterdam, Netherlands: Elsevier North-Holland.

Teuber, H.-L., Milner, B., & Vaughan, H. G., Jr. (1968). Persistent anterograde amnesia after stab wound of the basal brain. *Neuropsychologia, 6*, 267–282.

Thach, W. T., & Bastian, A. J. (2004). Role of the cerebellum in the control and adaptation of gait in health and disease. *Progress in Brain Research, 143*, 353–366.

Thion, M. S., & Garel, S. (2017). On place and time: Microglia in embryonic and perinatal brain development. *Current Opinion in Neurobiology, 47*, 121–130.

Thissen, S., Vos, I. G., Schreuder, T. H., Schreurs, W. M. J., Postma, L. A., & Koehler, P. J. (2014). Persistent migraine aura: New cases, a literature review, and ideas about pathophysiology. *Headache, 54*, 1290–1309.

Thomas, R., & Cavanna, A. E. (2013). The pharmacology of Tourette syndrome. *Journal of Neural Transmission, 120*, 689–694.

Thomason, M. E., Scheinost, D., Manning, J. H., Grove, L. E., Hect, J., Marshall, N., . . . & Hassan, S. S. (2017). Weak functional connectivity in the human fetal brain prior to preterm birth. *Scientific Reports, 7*(1), 1–10.

Thompson, A., Gribizis, A., Chen, C., & Crair, M. C. (2017). Activity-dependent development of visual receptive fields. *Current Opinion in Neurobiology, 42*, 136–143.

Thompson, P. M., Vidal, C., Giedd, J. N., Gochman, P., Blumenthal, J., Nicolson, R., . . . & Rapoport, J. L. (2001). Mapping adolescent brain change reveals dynamic wave of accelerated gray matter loss in very early-onset schizophrenia. *Proceedings of the National Academy of Sciences, USA, 98*, 11650–11655.

Thompson, R. F. (2005). In search of memory traces. *Annual Review of Psychology, 56*, 1–23.

Tian, C., Wang, K., Ke, W., Guo, H., & Shu, Y. (2014). Molecular identity of axonal sodium channels in human cortical pyramidal cells. *Frontiers in Cellular Neuroscience, 8*, 297.

Tian, X. J., Zhang, H., Sannerud, J., & Xing, J. (2016). Achieving diverse and monoallelic olfactory receptor selection through dual-objective optimization design. *Proceedings of the National Academy of Sciences, 113*(21), E2889–E2898.

Tikidji-Hamburyan, A., Reinhard, K., Seitter, H., Hovhannisyan, A., Procyk, C. A., . . . & Münch, T. A., (2015). Retinal output changes qualitatively with every change in ambient luminance. *Nature Neuroscience, 18*, 66–74.

Tintore, M., Vidal-Jordana, A., & Sastre-Garriga, J. (2019). Treatment of multiple sclerosis—success from bench to bedside. *Nature Reviews Neurology, 15*(1), 53–58.

Tison, F., Meissner, W. G. (2014). Diagnosing and treating PD—the earlier the better? *Nature Reviews Neurology, 10*, 65–66.

Tlachi-López, J. L., Eguibar, J. R., Fernández-Guasti, A., & Lucio, R. A. (2012). Copulation and ejaculation in male rats under sexual satiety and the Coolidge effect. *Physiology & Behavior, 106*, 626–630.

Todd, R. M., Miskovic, V., Chikazoe, J., & Anderson, A. K. (2020). Emotional objectivity: Neural representations of emotions and their interaction with cognition. *Annual Review of Psychology, 71*, 25–48.

Todd, W. D., Fenselau, H., Wang, J. L., Zhang, R., Machado, N. L., Venner, A., . . . & Lowell, B. B. (2018). A hypothalamic circuit for the circadian control of aggression. *Nature Neuroscience, 21*(5), 717–724.

Toga, A. W., & Thompson, P. M. (2005). Genetics of brain structure and intelligence. *Annual Review of Neuroscience, 28*, 1–23.

Toga, A. W., Thompson, P. M., & Sowell, E. R. (2006). Mapping brain maturation. *Trends in Neurosciences, 29*, 148–159.

Toledano, R., & Gil-Nagel, A. (2008). Adverse effects of antiepileptic drugs. *Seminars in Neurology, 28*, 317–327.

Tomaszczyk, J. C., Green, N. L., Frasca, D., Colella, B., Turner, G. R., Christensen, B. K., & Green, R. E. (2014). Negative neuroplasticity in chronic traumatic brain injury and implications for neurorehabilitation. *Neuropsychology Review, 24*(4), 409–427.

Tomlinson, D. R., & Gardiner, N. J. (2008). Glucose neurotoxicity. *Nature Reviews Neuroscience, 9*, 36–45.

Tompa, T., & Sáry, G. (2010). A review on the inferior temporal cortex of the macaque. *Brain Research Reviews, 62*, 165–182.

Tonegawa, S., Morrissey, M. D., & Kitamura, T. (2018). The role of engram cells in the systems consolidation of memory. *Nature Reviews Neuroscience, 19*(8), 485–498.

Tong, F. (2003). Primary visual cortex and visual awareness. *Nature Reviews Neuroscience, 4*, 219–229.

Toni, N., Laplagne, D. A., Zhao, C., Lombardi, G., Ribak, C. E., Gage, F. H., & Schinder, A. F. (2008). Neurons born in the adult dentate gyrus form functional synapses with target cells. *Nature Neuroscience, 11*, 901–907.

Tononi, G., & Cirelli, C. (2013). Perchance to prune. *Scientific American, 309*, 34–39.

Tononi, G., & Cirelli, C. (2019). Sleep and synaptic down-selection. *The European Journal of Neuroscience*, 14355.

Tootell, R. B. H., Dale, A. M., Sereno, M. I., & Malach, R. (1996). New images from human visual cortex. *Trends in Neurosciences, 19*, 481–489.

Toppi, J., Astolfi, L., Poudel, G. R., Innes, C. R. H., Babiloni, F., & Jones, R. D. (2016). Time-varying effective connectivity of the cortical neuroelectric activity associated with behavioural microsleeps. *NeuroImage, 124*, 421–432.

Tordjman, S., Somogyi, E., Coulon, N., Kermarrec, S., Cohen, D., . . . & Xavier, J. (2014). Gene x environment interactions in autism spectrum disorders: Role of epigenetic mechanisms. *Frontiers in Psychiatry, 5*, 53.

Tovote, P., Fadok, J. P., & Lüthi, A. (2015). Neuronal circuits for fear and anxiety. *Nature Reviews Neuroscience, 16*, 317–331.

Tragante, V., Moore, J. H., & Asselbergs, F. W. (2014). The ENCODE project and perspectives on pathways. *Genetics and Epidemiology, 38*, 275–280.

Tramontin, A. D., & Brenowitz, E. A. (2000). Seasonal plasticity in the adult brain. *Trends in Neurosciences, 23*(6), 251–258.

Tranel, D., & Damasio, A. R. (1985). Knowledge without awareness: An autonomic index of facial recognition by prosopagnosics. *Science, 228*, 1453–1454.

Trauer, J. M., Qian, M. Y., Doyle, J. S., Rajaratnam, S. M. W., & Cunnington, D. (2015). Cognitive behavioral therapy for chronic insomnia: A systematic review and meta-analysis. *Annals of Internal Medicine, 163*, 191–204.

Travaglia, A., Bisaz, R., Sweet, E. S., Blitzer, R. D., & Alberini, C. M. (2016). Infantile amnesia reflects

a developmental critical period for hippocampal learning. *Nature Neuroscience, 19*(9), 1225–1233.

Treadway, M. T., & Zald, D. H. (2011). Reconsidering anhedonia in depression: Lessons from translational neuroscience. *Neuroscience and Biobehavioral Reviews, 35*, 537–555.

Treffert, D. A. (2014a). Savant syndrome: Realities, myths and misconceptions. *Journal of Autism and Developmental Disorders, 44*, 564–571.

Treffert, D. A. (2014b). Accidental genius. *Scientific American, 311*, 52–57.

Treit, D., Robinson, A., Rotzinger, S., & Pesold, C. (1993). Anxiolytic effects of serotonergic interventions in the shock-probe burying test and the elevated plus-maze test. *Behavioural Brain Research, 54*, 23–34.

Tremblay, A., Royer, M. M., Chaput, J. P., & Doucet, E. (2013). Adaptive thermogenesis can make a difference in the ability of obese individuals to lose body weight. *International Journal of Obesity, 37*, 759–764.

Tremblay, M. W., & Jiang, Y. H. (2019). DNA methylation and susceptibility to autism spectrum disorder. *Annual Review of Medicine, 70*, 151–166.

Trevelyan, A. J. (2016). Do cortical circuits need protecting from themselves? *Trends in Neurosciences, 39*(8), 502–511.

Tribl, G. G., Wetter, T. C., & Schredl, M. (2013). Dreaming under antidepressants: A systematic review on evidence in depressive patients and healthy volunteers. *Sleep Medicine Reviews, 17*, 133–142.

Triplett, J. W. (2014). Molecular guidance of retinotopic map development in the midbrain. *Current Opinion in Neurobiology, 24*, 7–12.

Trivedi, B. P. (2014). Dissecting appetite. *Nature, 508*, S64–S65.

Trottier, G., Srivastava, L., & Walker, C. D. (1999). Etiology of infantile autism: A review of recent advances in genetic and neurobiological research. *Journal of Psychiatry and Neuroscience, 24*, 103–115.

Tryon, R. C. (1934). Individual differences. In F. A. Moss (Ed.), *Comparative psychology* (pp. 409–448). New York, NY: Prentice-Hall.

Tsai, L., & Barnea, G. (2014). A critical period defined by axon-targeting mechanisms in the murine olfactory bulb. *Science, 344*, 197–200.

Tsakiris, M., Hesse, M. D., Boy, C., Haggard, P., & Fink, G. R. (2007). Neural signatures of body ownership: A sensory network for bodily self-consciousness. *Cerebral Cortex, 17*, 2235–2244.

Tsetsos, K., Wyart, V., Shorkey, S. P., & Summerfield, C. (2014). Neural mechanisms of economic commitment in the human medial prefrontal cortex. *eLife, 3*, e03701.

Tshala-Katumbay, D., Mwanza, J. C., Rohlman, D. S., Maestre, G., & Oriá, R. B. (2015). A global perspective on the influence of environmental exposures on the nervous system. *Nature, 527*, S187–S192.

Tsien, R. (1998). The green fluorescent protein. *Annual Review of Biochemistry, 67*, 509–544.

Tsuda, M., & Inoue, K. (2016). Neuron-microglia interaction by purinergic signaling in neuropathic pain following neurodegeneration. *Neuropharmacology, 104*, 76–81.

Tsunada, J., Liu, A. S., Gold, J. I., & Cohen, Y. E. (2016). Causal contribution of primate auditory cortex to auditory perceptual decision-making. *Nature Neuroscience, 19*(1), 135.

Tsunemoto, R., Lee, S., Szűcs, A., Chubukov, P., Sokolova, I., Blanchard, J. W., . . . & Sanna, P. P. (2018). Diverse reprogramming codes for neuronal identity. *Nature, 557*(7705), 375–380.

Tucker-Drob, E. M., Rhemtulla, M., Harden, K. P., Turkheimer, E., & Fask, D. (2011). Emergence of a gene x socioeconomic status interaction on infant mental ability between 10 months and 2 years. *Psychological Science, 22*, 125–133.

Tuesta, L. M., Chen, Z., Duncan, A., Fowler, C. D., Ishikawa, M., Lee, B. R., . . . & Kamenecka, T. M. (2017). GLP-1 acts on habenular avoidance

circuits to control nicotine intake. *Nature Neuroscience, 20*(5), 708–716.

Tulving, E. (2002). Episodic memory: From mind to brain. *Annual Review of Neuroscience, 53*, 1–25.

Turcu, A. F., & Auchus, R. J. (2015). Adrenal steroidogenesis and congenital adrenal hyperplasia. *Endocrinology and Metabolism Clinics of North America, 44*, 275–296.

Turek, F. W. (2016). Circadian clocks: Not your grandfather's clock. *Science, 354*(6315), 992–993.

Turella, L., & Lingnau, A. (2014). Neural correlates of grasping. *Frontiers in Human Neuroscience, 8*, 686.

Turkheimer, E. (2000). Three laws of behavior genetics and what they mean. *Current Directions in Psychological Science, 9*(5), 160–164.

Turkheimer, E., Haley, A., Waldron, M., D'Onofrio, B., & Gottesman, I. (2003). Socioeconomic status modifies heritability of IQ in young children. *Psychological Science, 14*(6), 623–628.

Turkheimer, E., Pettersson, E., & Horn, E. E. (2014). A phenotypic null hypothesis for the genetics of personality. *Annual Review of Psychology, 65*, 515–540.

Turner, B. O., Marinsek, N., Ryhal, E., & Miller, M. B. (2015). Hemispheric lateralization in reasoning. *Annals of the New York Academy of Sciences, 1359*, 47–64.

Tyebji, S., & Hannan, A. J. (2017). Synaptopathic mechanisms of neurodegeneration and dementia: Insights from Huntington's disease. *Progress in Neurobiology, 153*, 18–45.

Tyler, W. J., Lani, S. W., & Hwang, G. M. (2018). Ultrasonic modulation of neural circuit activity. *Current Opinion in Neurobiology, 50*, 222–231.

Tylš, F., Páleníček, T., & Horáček, J. (2014). Psilocybin—summary of knowledge and new perspectives. *European Neuropsychopharmacology, 24*, 342–356.

Tyrer, A. E., Levitan, R. D., Houle, S., Wilson, A. A., Nobrega, J. N., & Meyer, J. H. (2016). Increased seasonal variation in serotonin transporter binding in seasonal affective disorder. *Neuropsychopharmacology, 41*, 2447–2454.

U.S. Department of Health and Human Services, Substance Abuse and Mental Health Services Administration. (2006). *2006 National Survey on Drug Use and Health.* Retrieved from http://www.oas.samhsa.gov/nsduhLatest.htm

U.S. Department of Health and Human Services. Substance Abuse and Mental Health Services Administration. Center for Behavioral Health Statistics and Quality. (2010). National survey on drug use and health. ICPSR32722-v3. Ann Arbor, MI: Interuniversity Consortium for Political and Social Research.

Uchida, N., Poo, C., & Haddad, R. (2014). Coding and transformations in the olfactory system. *Annual Review of Neuroscience, 37*, 363–385.

Uddin, L. Q. (2011). Brain connectivity and the self: The case of cerebral disconnection. *Consciousness and Cognition, 20*, 94–98.

Uddin, M. S., Al Mamun, A., Kabir, M. T., Jakaria, M., Mathew, B., Barreto, G. E., & Ashraf, G. M. (2019). Nootropic and anti-Alzheimer's actions of medicinal plants: Molecular insight into therapeutic potential to alleviate Alzheimer's neuropathology. *Molecular Neurobiology, 56*(7), 4925–4944.

Ueno, M., & Yamashita, T. (2014). Bidirectional tuning of microglia in the developing brain: From neurogenesis to neural circuit formation. *Current Opinion in Neurobiology, 27*, 8–15.

Ulrich, J. D., Ulland, T. K., Colonna, M., & Holtzman, D. M. (2017). Elucidating the role of TREM2 in Alzheimer's disease. *Neuron, 94*(2), 237–248.

Ulrich, R. E. (1991). Commentary: Animal rights, animal wrongs and the question of balance. *Psychological Science, 2*, 197–201.

UN Global ATS Assessment. (2011). *Amphetamines and ecstasy.* United Nations Office on Drug and Crime.

Underwood, E. (2013). How to build a dream-reading machine. *Science, 340*, 21.

Underwood, E. (2013). Why do so many neurons commit suicide during brain development? *Science, 340*, 1157–1158.

Underwood, E. (2014). Brain's GPS finds top honor. *Science, 346*, 149–149.

Underwood, E. (2015). The brain's identity crisis. *Science, 349*, 575–577.

Underwood, E. (2015a). Neuroscience: Seeking tests for a contested brain disease. *Science, 348*, 378–379.

Ungerleider, L. G., & Haxby, J. V. (1994). "What" and "where" in the human brain. *Current Opinion in Neurobiology, 4*, 157–165.

Ungerleider, L. G., & Mishkin, M. (1982). Two cortical visual systems. In D. J. Ingle, M. A. Goodale, & R. J. W. Mansfield (Eds.), *Analysis of visual behavior* (pp. 549–586). Cambridge, MA: MIT Press.

Urban, N. N., & Castro, J. B. (2010). Functional polarity in neurons: what can we learn from studying an exception?. *Current opinion in neurobiology, 20*(5), 538–542.

Uzzi, B., Mukherjee, S., Stringer, M., & Jones, B. (2013). Atypical combinations and scientific impact. *Science, 342*, 468–472.

Vaara, J. P., Oksanen, H., Kyröläinen, H., Virmavirta, M., Koski, H., & Finni, T. (2018). 60-hour sleep deprivation affects submaximal but not maximal physical performance. *Frontiers in Physiology, 9*, 1437.

Valbuena, S., & Lerma, J. (2016). Non-canonical signaling, the hidden life of ligand-gated ion channels. *Neuron, 92*(2), 316–329.

Valenstein, E. S. (1973). *Brain control*. New York, NY: John Wiley & Sons.

Valenzuela, C. F., Morton, R. A., Diaz, M. R., & Topper, L. (2012). Does moderate drinking harm the fetal brain? Insights from animal models. *Trends in Neurosciences, 35*, 284–292.

Vallat, R., Chatard, B., Blagrove, M., & Ruby, P. (2017). Characteristics of the memory sources of dreams: A new version of the content-matching paradigm to take mundane and remote memories into account. *PloS One, 12*(10), e0185262.

Vallat, R., Eichenlaub, J. B., Nicolas, A., & Ruby, P. (2018). Dream recall frequency is associated with medial prefrontal cortex white-matter density. *Frontiers in Psychology, 9*, 1856.

Van Acker, G. M., Amundsen, S. L., Messamore, W. G., Zhang, H. Y., Luchies, C. W., & Cheney, P. D. (2014). Equilibrium-based movement endpoints elicited from primary motor cortex using repetitive microstimulation. *Journal of Neuroscience, 34*, 15722–15734.

van Amsterdam, J., Opperhuizen, A., & Hartgens, F. (2010). Adverse health effects of anabolic-androgenic steroids. *Regulatory Toxicology and Pharmacology, 57*, 117–123.

van den Hurk, J., Pegado, F., Martens, F., & de Beeck, H. P. O. (2015). The search for the face of the visual homunculus. *Trends in Cognitive Sciences, 19*, 638–641.

van der Heijden, K., Rauschecker, J. P., de Gelder, B., & Formisano, E. (2019). Cortical mechanisms of spatial hearing. *Nature Reviews Neuroscience, 20*, 609–623.

van Ee, R., Van de Cruys, S., Schlangen, L. J., & Vlaskamp, B. N. (2016). Circadian-time sickness: Time-of-day cue-conflicts directly affect health. *Trends in Neurosciences, 39*(11), 738–749.

van Erp, T. G. M., Hibar, D. P., Rasmussen, J. M., Glahn, D. C., Pearlson, G. D., Andreassen, O. A.,…& Melle, I. (2016). Subcortical brain volume abnormalities in 2028 individuals with schizophrenia and 2540 healthy controls via the ENIGMA consortium. *Molecular Psychiatry, 21*, 547–553.

van Groen, T., Puurunen, K., Mäki, H.-M., Sivenius, J., & Jolkkonen, J. (2005). Transformation of diffuse β-amyloid precursor protein and β-amyloid deposits to plaques in the thalamus after transient occlusion of the middle cerebral artery in rats. *Stroke, 36*, 1551.

Van Herwegen, J. (2015). Williams syndrome and its cognitive profile: The importance of eye movements. *Psychology Research and Behavior Management, 8*, 143–151.

van Opstal, E. J., & Bordenstein, S. R. (2015). Rethinking heritability of the microbiome. *Science, 349*, 1172–1173.

Van Praag, H., Christie, B. R., Sejnowski, T. J., & Gage, F. H. (1999). Running enhances neurogenesis, learning, and long-term potentiation in mice. *Proceedings of the National Academy of Sciences, USA, 19*, 13427–13431.

Van Praag, H., Schinder, A. F., Christie, B. R., Toni, N., Palmer, T. D., & Gage, F. H. (2002). Functional neurogenesis in the adult hippocampus. *Nature, 415*, 1030–1034.

Van Praag, H., Shubert, T., Zhao, C., & Gage, F. H. (2005). Exercise enhances learning and hippocampal neurogenesis in aged mice. *Journal of Neuroscience, 25*, 8680–8685.

Van Tilborg, I. A. D. A., Kessels, R. P. C., Krujit, P., Wester, A. J., & Hulstijn, W. (2011). Spatial and nonspatial implicit motor learning in Korsakoff's amnesia: Evidence for selective deficits. *Experimental Brain Research, 214*, 427–435.

Van Wagenen, W. P., & Herren, R. Y. (1940). Surgical division of commissural pathways in the corpus callosum: Relation to spread of an epileptic attack. *Archives of Neurology and Psychiatry, 44*, 740–759.

Vargha-Khadem, F., Gadian, D. G., Watkins, K. E., Connelly, A., van Paesschen, W., & Mishkin, M. (1997). Differential effects of early hippocampal pathology on episodic and semantic memory. *Science, 277*, 376–380.

Vassoler, F. M., & Sadri-Vakili, G. (2014). Mechanisms of transgenerational inheritance of addictive-like behaviours. *Neuroscience, 264*, 198–206.

Vassoler, F. M., Byrnes, E. M., & Pierce, R. C. (2014). The impact of exposure to addictive drugs on future generations: Physiological and behavioural effects. *Neuropharmacology, 76*, 269–275.

Vaupel, J. W., Carey, J. R., & Christensen, K. (2003). It's never too late. *Science, 301*, 1679–1681.

Vavvas, D. G., Dryja, T. P., Wilson, M. E., Olsen, T. W., Shah, A., Jurkunas, U.,…& Moreno-Montañés, J. (2018). Lens regeneration in children. *Nature, 556*(7699), E2.

Vázquez, G. H., Baldessarini, R. J., & Tondo, L. (2014). Co-occurrence of anxiety and bipolar disorders: Clinical and therapeutic overview. *Depression and Anxiety, 31*, 196–206.

Vecchia, D., & Pietrobon, D. (2012). Migraine: A disorder of brain excitatory-inhibitory balance? *Trends in Neurosciences, 35*, 507–520.

Veldhuizen, M. G., & Small, D. M. (2011). Modality-specific neural effects of selective attention to taste and odor. *Chemical Senses, 36*, 747–760.

Veldhuizen, M. G., Gitelman, D. R., & Small, D. M. (2012). An fMRI study of the interactions between the attention and the gustatory networks. *Chemosensory Perception, 5*, 117–127.

Veldman, M. B., & Yang, X. W. (2018). Molecular insights into cortico-striatal miscommunications in Huntington's disease. *Current Opinion in Neurobiology, 48*, 79–89.

Ventura-Aquino, E., & Paredes, R. G. (2020). Sexual behavior in rodents: Where do we go from here? *Hormones and Behavior, 118*, 104678.

Verkhratsky, A., Parpura, V., & Rodríguez, J. J. (2010). Where the thought dwells: The physiology of neuronal-glial "diffuse neural net." *Brain Research Reviews, 66*, 133–151.

Verstraeten, A., Theuns, J., & Van Broeckhoven, C. (2015). Progress in unraveling the genetic etiology of Parkinson's disease in a genomic era. *Trends in Genetics, 31*, 140–149.

Vertes, R. P., & Eastman, K. E. (2000). The case against memory consolidation in REM sleep. *Behavioral and Brain Sciences, 23*, 867–876.

Vetter, P., Grosbras, M.-H., & Muckli, L. (2015). TMS over V5 disrupts motion prediction. *Cerebral Cortex, 25*, 1052–1059.

Videnovic, A., Noble, C., Reid, K. J., Peng, J., Turek, F. W.,…& Zee, P. C. (2014). Circadian melatonin rhythm and excessive daytime sleepiness in Parkinson disease. *JAMA Neurology, 71*, 463–469.

Vidyasagar, T. R., & Eysel, U. T. (2015). Origins of feature selectivities and maps in the mammalian primary visual cortex. *Trends in Neurosciences, 38*, 475–485.

Vieta, E., Berk, M., Schulze, T. G., Carvalho, A. F., Suppes, T., Calabrese, J. R., … & Grande, I. (2018). Bipolar disorders. *Nature Reviews Disease Primers, 4*(1), 1–16.

Vijayakumar, N., de Macks, Z. O., Shirtcliff, E. A., & Pfeifer, J. H. (2018). Puberty and the human brain: Insights into adolescent development. *Neuroscience & Biobehavioral Reviews, 92*, 417–436.

Villablanca, P. A., Alegria, J. R., Mookadam, F., Holmes D. R., Wright, R. S., & Levine, J. A. (2015). Nonexercise activity thermogenesis in obesity management. *Mayo Clinic Proceedings, 90*, 509–519.

Villegas-Llerena, C., Phillips, A., Garcia-Reitboeck, P., Hardy, J., & Pocock, J. M. (2016). Microglial genes regulating neuroinflammation in the progression of Alzheimer's disease. *Current Opinion in Neurobiology, 36*, 74–81.

Villemagne, V. L., & Okamura, N. (2016). Tau imaging in the study of ageing, Alzheimer's disease, and other neurodegenerative conditions. *Current Opinion in Neurobiology, 36*, 43–51.

Villemagne, V. L., Doré, V., Burnham, S. C., Masters, C. L., & Rowe, C. C. (2018). Imaging tau and amyloid-β proteinopathies in Alzheimer disease and other conditions. *Nature Reviews Neurology, 14*(4), 225–236.

Villmoore, B., Kimbel, W. H., Seyoum, C., Campisano, C. J., DiMaggio, E. N., Rowan, J.,…& Reed, K. E. (2015). Early *Homo* at 2.8 Ma from Ledi-Geraru, Afar, Ethiopia. *Science, 347*, 1352–1355.

Vincis, R., & Fontanini, A. (2016). A gustocentric perspective to understanding primary sensory cortices. *Current Opinion in Neurobiology, 40*, 118–124.

Vingerhoets, G. (2019). Phenotypes in hemispheric functional segregation? Perspectives and challenges. *Physics of Life Reviews*. doi: https://doi.org/10.1016/j.plrev.2019.06.002

Visanji, N. P., Brotchie, J. M., Kalia, L. V., Koprich, J. B., Tandon, A., Watts, J. C., & Lang, A. E. (2016). α-Synuclein-based animal models of Parkinson's disease: Challenges and opportunities in a new era. *Trends in Neurosciences, 39*(11), 750–762.

Viscomi, M. T., & Molinari, M. (2014). Remote neurodegeneration: Multiple actors at play. *Molecular Neurobiology, 50*, 368–389.

Vita, A., De Peri, L., Silenzi, C., & Dieci, M. (2006). Brain morphology in first-episode schizophrenia: A meta-analysis of quantitative magnetic resonance imaging studies. *Schizophrenia Research, 82*, 75–88.

Vlahov, D., Galea, S., Ahern, J., Resnick, H., Boscarino, J. A.,…& Kilpatrick, D. (2004). Consumption of cigarettes, alcohol, and marijuana among New York City residents six months after the September 11 terrorist attacks. *American Journal of Drug and Alcohol Abuse, 30*, 385–407.

Vo, N. K., Cambronne, X. A., & Goodman, R. H. (2010). MicroRNA pathways in neural development and plasticity. *Current Opinion in Neurobiology, 20*, 457–465.

Vogeley, K. (2017). Two social brains: Neural mechanisms of intersubjectivity. *Philosophical Transactions of the Royal Society B: Biological Sciences, 372*(1727), 20160245.

Voigt, J.-P., & Fink, H. (2015). Serotonin controlling feeding and satiety. *Behavioural Brain Research, 277*, 14–31.

Voineskos, A. N., Felsky, D., Kovacevic, N., Tiwari, A. K., Zai, C., Chakravarty, M. M., Lobaugh, N. J., Shenton, M. E., Rajji, T. K., Miranda, D., Pollock, B. G., Mulsant, B. H., McIntosh, A. R., & Kennedy, J. L. (2012). Oligodendrocyte genes, white matter tract integrity, and cognition in schizophrenia. Cerebral Cortex. doi:10.1093/cercor/bhs188

Volkman, F. R., & McPartland, J. C. (2014). From Kanner to DSM-5: Autism as an evolving diagnostic concept. *Annual Review of Clinical Psychology, 10,* 193–212.

Volkow, N. D., & Baler, R. D. (2012). To stop or not to stop? *Science, 335,* 546–548.

Volkow, N. D., & Koroshetz, W. J. (2019). The role of neurologists in tackling the opioid epidemic. *Nature Reviews Neurology, 15*(5), 301–305.

Volkow, N. D., & Warren, K. R. (2011). The science of addiction; untangling complex systems. *Nature Reviews Neuroscience, 12,* 620.

Volkow, N. D., Wang, G.-J., Tomasi, D., & Baler, R. D. (2013). Unbalanced neuronal circuits in addiction. *Current Opinion in Neurobiology, 23,* 639–648.

von Bartheld, C. S. (2018). Myths and truths about the cellular composition of the human brain: A review of influential concepts. *Journal of Chemical Neuroanatomy, 93,* 2–15.

Vonck, K., Raedt, R., Naulaerts, J., De Vogelaere, F., Thiery, E., … & Boon, P. (2014). Vagus nerve stimulation…25 years later! What do we know about the effects on cognition? *Neuroscience and Biobehavioral Reviews, 45,* 63–71.

Vonderschen, K., & Wagner, H. (2014). Detecting interaural time differences and remodeling their representation. *Trends in Neurosciences, 37,* 289–300.

Voss, M. W., Vivar, C., Kramer, A. F., & van Praag, H. (2013). Bridging animal and human models of exercise-induced brain plasticity. *Trends in Cognitive Sciences, 17,* 525–544.

Voss, U., Holzmann, R., Hobson, A., Paulus, W., Koppehele-Gossel, J., Klimke, A., & Nitsche, M. A. (2014). Induction of self awareness in dreams through frontal low current stimulation of gamma activity. *Nature Neuroscience, 17*(6), 810–812.

Votaw, V. R., Geyer, R., Rieselbach, M. M., & McHugh, R. K. (2019). The epidemiology of benzodiazepine misuse: A systematic review. *Drug and alcohol dependence, 200,* 95.

Vuilleumier, P., Schwartz, S., Clarke, K., Husain, M., & Driver, J. (2002). Testing memory for unseen visual stimuli in patients with extinction and spatial neglect. *Journal of Cognitive Neuroscience, 14,* 875–886.

Vuong, H. E., Yano, J. M., Fung, T. C., & Hsiao, E. Y. (2017). The microbiome and host behavior. *Annual Review of Neuroscience, 40,* 21–49.

Vyazovskiy, V. V. (2015). Mapping the birth of the sleep connectome. *Science, 350,* 909–910.

Wada, J. A. (1949). A new method for the determination of the side of cerebral speech dominance. *Igaku to Seibutsugaku, 14,* 221–222.

Wager, T. D., & Atlas, L. Y. (2015). The neuroscience of placebo effects: Connecting context, learning and health. *Nature Reviews Neuroscience, 16,* 403–418.

Wager, T. D., Phan, K. L., Liberzon, I., & Taylor, S. F. (2003). Valence, gender, and lateralization of functional brain anatomy in emotion: A meta-analysis of findings from neuroimaging. *Neuroimage, 19,* 513–531.

Wake, H., Moorhouse, A. J., Miyamoto, A., & Nabekura, J. (2013). Microglia: Actively surveying and shaping neuronal circuit structure and function. *Trends in Neurosciences, 36,* 209–217.

Wald, G. (1964). The receptors of human color vision. *Science, 145,* 1007–1016.

Walia, R., Singla, M., Vaiphei, K., Kumar, S., & Bhansali, A. (2018). Disorders of sex development: A study of 194 cases. *Endocrine Connections, 7*(2), 364–371.

Walker, A. W., & Parkhill, J. (2013). Fighting obesity with bacteria. *Science, 341,* 1069–1070.

Walker, A., Joshi, A., & D'Souza, A. (2017). Care of the cocaine user with nasal deformity. *Facial Plastic Surgery, 33*(04), 411–418.

Wallis, C. (2014). Gut reactions. *Scientific American, 310,* 30–33.

Wallis, C. (2018). Ba-Boom! There goes your hearing. *Scientific American, 319*(2), 24.

Walløe, S., Pakkenberg, B., & Fabricius, K. (2014). Stereological estimation of total cell numbers in the human cerebral and cerebellar cortex. *Frontiers in Human Neuroscience, 8,* 508.

Walsh, D. M., & Selkoe, D. J. (2016). A critical appraisal of the pathogenic protein spread hypothesis of neurodegeneration. *Nature Reviews Neuroscience, 17,* 251–260.

Walsh, P., Elsabbagh, M., Bolton, P., & Singh, I. (2011). In search of biomarkers for autism: Scientific, social and ethical challenges. *Nature Reviews Neuroscience, 12,* 603–612.

Wan, H., Aggleton, J. P., & Brown, M. W. (1999). Different contributions of the hippocampus and perirhinal cortex to recognition memory. *Journal of Neuroscience, 19,* 1142–1148.

Wang, F., Zhu, J., Zhu, H., Lin, Z., & Hu, H. (2011). Bidirectional control of social hierarchy by synaptic efficacy in medial prefrontal cortex. *Science, 334,* 693–697.

Wang, H., Jing, M., & Li, Y. (2018). Lighting up the brain: Genetically encoded fluorescent sensors for imaging neurotransmitters and neuromodulators. *Current Opinion in Neurobiology, 50,* 171–178.

Wang, J., Yang, Y., Fan. L., Xu, J., Li, C., … & Jiang, T. (2015). Convergent functional architecture of the superior parietal lobule unraveled with multimodal neuroimaging approaches. *Human Brain Mapping, 36,* 238–257.

Wang, M., Alexanderson, K., Runeson, B., & Mittendorfer-Rutz, E. (2015). Sick-leave measures, socio-demographic factors and health care as risk indicators for suicidal behavior in patients with depressive disorders—a nationwide prospective cohort study in Sweden. *Journal of Affective Disorders, 173,* 201–210.

Wang, S. S.-H., Kloth, A. D., & Badura, A. (2014). The cerebellum, sensitive periods, and autism. *Neuron, 83,* 518–532.

Wang, X. (2018). Cortical coding of auditory features. *Annual Review of Neuroscience, 41,* 527–552.

Wang, Y., & Mandelkow, E. (2016). Tau in physiology and pathology. *Nature Reviews Neuroscience, 17*(1), 5–23.

Wang, Y., See, J., Phan, R. C., & Oh, Y. H. (2015). Efficient spatio-temporal local binary patterns for spontaneous facial micro-expression recognition. *PLoS One, 10,* e0124676.

Wang, Y., Zhang, D., Zou, F., Li, H., Luo, Y., Zhang, M., & Liu, Y. (2016). Gender differences in emotion experience perception under different facial muscle manipulations. *Consciousness and Cognition, 41,* 24–30.

Wang, Z., Sa, Y.-L., Ye, X.-X., Zhang, J., & Xu, Y.-M. (2014). Complete androgen insensitivity syndrome in juveniles and adults with female phenotypes. *Journal of Obstetrics and Gynaecology Research, 40,* 2044–2050.

Wanner, A. A., Genoud, C., Masudi, T., Siksou, L., & Friedrich, R. W. (2016). Dense EM-based reconstruction of the interglomerular projectome in the zebrafish olfactory bulb. *Nature Neuroscience, 19*(6), 816–825.

Ward, L. M. (2013). The thalamus: Gateway to the mind. *Wiley Interdisciplinary Reviews: Cognitive Science, 4*(6), 609–622.

Ware, J. C. (2008). Will sleeping pills ever wake us up? *Sleep Medicine, 9,* 811–812.

Warner, K. E., & Mendez, D. (2019). E-cigarettes: Comparing the possible risks of increasing smoking initiation with the potential benefits of increasing smoking cessation. *Nicotine and Tobacco Research, 21*(1), 41–47.

Watanabe, M., Fukuda, A., & Nabekura, J. (2014). The role of GABA in the regulation of GnRH neurons. *Frontiers in Neuroscience, 8,* 387.

Watanabe, N., & Yamamoto, M. (2015). Neural mechanisms of social dominance. *Frontiers in Neuroscience, 9,* 154.

Watson, J. B. (1930). *Behaviorism.* New York, NY: Norton.

Webb, W. B., & Agnew, H. W. (1967). Sleep cycling within the twenty-four hour period. *Journal of Experimental Psychology, 74,* 167–169.

Webb, W. B., & Agnew, H. W. (1974). The effects of a chronic limitation of sleep length. *Psychophysiology, 11,* 265–274.

Webb, W. B., & Agnew, H. W. (1975). The effects on subsequent sleep of an acute restriction of sleep length. *Psychophysiology, 12,* 367–370.

Weber, F., & Dan, Y. (2016). Circuit-based interrogation of sleep control. *Nature, 538*(7623), 51–59.

Weber, F., Chung, S., Beier, K. T., Xu, M., Luo, L., & Dan, Y. (2015). Control of REM sleep by ventral medulla GABAergic neurons. *Nature, 526,* 435–438.

Weber, S., Johnsen, E., Kroken, R. A., Løberg, E. M., Kandilarova, S., Stoyanov, D., … & Hugdahl, K. (2020). Dynamic functional connectivity patterns in schizophrenia and the relationship with hallucinations. *Frontiers in Psychiatry, 11,* 227.

Wei, D., Allsop, S., Tye, K., & Piomelli, D. (2017). Endocannabinoid signaling in the control of social behavior. *Trends in Neurosciences, 40*(7), 385–396.

Weibel, R., Reiss, D., Karchewski, L., Gardon, O., Matifas, A., Filliol, D., Becker, J. A. J., Wood, J. N., Kieffer, B. L., & Gaveriaux-Ruff, C. (2013). Mu opioid receptors on primary afferent Nav1.8 neurons contribute to opiate-induced analgesia: Insight from conditional knockout mice. *PLoS ONE, 8,* e74706.

Weigelt, S., Koldewyn, K., & Kanwisher, N. (2012). Face identity recognition in autism spectrum disorder: A review of behavioral studies. *Neuroscience and Biobehavioral Reviews, 36,* 1060–1084.

Weil, R. S., & Rees, G. (2011). A new taxonomy for perceptual filling-in. *Brain Research Reviews, 67,* 40–55.

Weil, Z. M., Corrigan, J. D., & Karelina, K. (2018). Alcohol use disorder and traumatic brain injury. *Alcohol Research: Current Reviews, 39*(2), 171–180.

Weiller, C., & Rijntjes, M. (1999). Learning, plasticity, and recovery in the central nervous system. *Experimental Brain Research, 128,* 134–138.

Weindruch, R. (1996). The retardation of aging by caloric restriction: Studies in rodents and primates. Toxicologic Pathology, 24, 742–745.

Weingarten, H. P. (1983). Conditioned cues elicit feeding in sated rats: A role for learning in meal initiation. *Science, 220,* 431–433.

Weingarten, H. P. (1984). Meal initiation controlled by learned cues: Basic behavioral properties. *Appetite, 5,* 147–158.

Weingarten, H. P. (1985). Stimulus control of eating: Implications for a two-factor theory of hunger. *Appetite, 6,* 387–401.

Weingarten, H. P. (1990). Learning, homeostasis, and the control of feeding behavior. In E. D. Capaldi & T. L. Powley (Eds.), *Taste, experience, and feeding* (pp. 14–27). Washington, DC: American Psychological Association.

Weiskrantz, L., Warrington, E. K., Sanders, M. D., & Marshall, J. (1974). Visual capacity in the hemianopic field following a restricted occipital ablation. *Brain, 97,* 709–728.

Weissman, D. H., & Banich, M. T. (2000). The cerebral hemispheres cooperate to perform complex but not simple tasks. *Neuropsychology, 14*, 41–59.

Weissman, M. M., & Olfson, M. (1995). Depression in women: Implications for health care research. *Science, 269*, 799–801.

Wekerle, H., Flügel, L., Schett, G., & Serreze, D. (2012). Autoimmunity's next top models. *Nature Medicine, 18*, 66–70.

Welberg, L. (2014). Epigenetics: A lingering smell? *Nature Reviews Neuroscience, 15*, 1.

Welberg, L. (2014). Synaptic plasticity: A synaptic role for microglia. *Nature Reviews Neuroscience, 15*, 69.

Welberg, L. (2014a). Neurogenesis: A striatal supply of new neurons. *Nature Reviews Neuroscience, 15*, 203.

Welberg, L. (2014b). Learning and memory: Neurogenesis erases existing memories. *Nature Reviews Neuroscience, 15*, 428–429.

Welberg, L. (2014c). Glia: A critical mass of microglia? *Nature Reviews Neuroscience, 15*, 133.

Weledji, E. P., & Assob, J. C. (2014). The ubiquitous neural cell adhesion molecule (N-CAM). *Annals of Medicine and Surgery, 3*, 77–81.

Werneburg, S., Feinberg, P. A., Johnson, K. M., & Schafer, D. P. (2017). A microglia-cytokine axis to modulate synaptic connectivity and function. *Current Opinion in Neurobiology, 47*, 138–145.

Westmoreland, P., Krantz, M. J., & Mehler, P. S. (2015). Medical complications of anorexia nervosa and bulimia. *American Journal of Medicine, 129*, 30–37.

Westwater, M. L., Fletcher, P. C., & Ziauddeen, H. (2016). Sugar addiction: The state of the science. *European Journal of Nutrition, 55*(2), 55–69.

Weth, F., Fiederling, F., Gebhardt, C., & Bastmeyer, M. (2014). Chemoaffinity in topographic mapping revisited—is it more about fiber-fiber than fiber-target interactions? *Seminars in Cell and Developmental Biology, 35*, 126–135.

Wever, R. A. (1979). *The circadian system of man.* Seewiesen-Andechs, Germany: Max-Planck-Institut für Verhaltensphysiologie.

Whalen, P. J., Raila, H., Bennett, R., Mattek, A., Brown, A., Taylor, J., . . . & Palmer, A. (2013). Neuroscience and facial expressions of emotion: The role of amygdala–prefrontal interactions. *Emotion Review, 5*, 78–83.

White, T. D., Asfaw, B., Beyene, Y., Haile-Selassie, Y., Lovejoy, C. O., Suwa, G., & WoldeGabriel, G. (2009). Ardipithecus ramidus and the paleobiology of early hominids. *Science, 326*, 64–86.

Whiten, A., & Boesch, C. (2001, January). The cultures of chimpanzees. *Scientific American, 284*, 61–67.

Whitney, D., & Leib, A. Y. (2018). Ensemble perception. *Annual Review of Psychology, 69*, 105–129.

Wichniak, A., Wierzbicka, A., Wałęcka, M., & Jernajczyk, W. (2017). Effects of antidepressants on sleep. *Current Psychiatry Reports, 19*(9), 63.

Wichterle, H., Gifford, D., & Mazzoni, E. (2013). Mapping neuronal diversity one cell at a time. *Science, 341*, 726–727.

Wickelgren, W. A. (1968). Sparing of short-term memory in an amnesic patient: Implications for strength theory of memory. *Neuropsychologia, 6*, 31–45.

Wiedemann, K. (2014). Where we came from. *Scientific American, 311*, 40–41.

Wiegert, J. S., & Oertner, T. G. (2015). Neighborly synapses help each other out. *Nature Neuroscience, 18*, 326–327.

Wiegert, J. S., Mahn, M., Prigge, M., Printz, Y., & Yizhar, O. (2017). Silencing neurons: Tools, applications, and experimental constraints. *Neuron, 95*(3), 504–529.

Wiesner, C. D., Pulst, J., Krause, F., Elsner, M., Baving, L., . . . & Göder, R. (2015). The effect of selective REM-sleep deprivation on the consolidation and affective evaluation of emotional memories. *Neurobiology of Learning and Memory, 122*, 131–141.

Wilber, A. A., Clark, B. J., Forster, T. C., Tatsuno, M., & McNaughton, B. L. (2014). Interaction of egocentric and world-centered reference frames in the rat posterior parietal cortex. *Journal of Neuroscience, 34*, 5431–5446.

Will, R. G., Hull, E. M., & Dominguez, J. M. (2014). Influences of dopamine and glutamate in the medial preoptic area on male sexual behavior. *Pharmacology, Biochemistry, and Behavior, 121*, 115–123.

Williams, M. (1970). *Brain damage and the mind.* Baltimore, MD: Penguin Books.

Williams, S. R., & Stuart, G. J. (2002). Dependence of EPSP efficacy on synapse location in neocortical pyramidal neurons. *Science, 295*, 1907–1910.

Williams, S. R., & Stuart, G. J. (2003). Role of dendritic synapse location in the control of action potential output. *Trends in Neurosciences, 26*(3), 147–154.

Willment, K. C., & Golby, A. (2013). Hemispheric lateralization interrupted: Material-specific memory deficits in temporal lobe epilepsy. *Frontiers in Human Neuroscience, 7*, 546.

Willner, P., Scheel-Krüger, J., & Belzung, C. (2013). The neurobiology of depression and antidepressant action. *Neuroscience & Biobehavioral Reviews, 37*, 2331–2371.

Wills, T. J., & Cacucci, F. (2014). The development of the hippocampal neural representation. *Current Opinion in Neurobiology, 24*, 111–119.

Willsey, A. J., & State, M. W. (2015). Autism spectrum disorders: From genes to neurobiology. *Current Opinion in Neurobiology, 30*, 92–99.

Wilson, C. R. E., Gaffan, D., Browning, P. G. F., & Baxter, M. G. (2010). Functional localization within the prefrontal cortex: Missing the forest for the trees? *Trends in Neurosciences, 33*, 533–540.

Wilson, M. E., Moore, C. J., Ethun, K. F., & Johnson, Z. P. (2014). Understanding the control of ingestive behavior in primates. *Hormones and Behavior, 66*, 86–94.

Wilson, R. I., & Mainen, Z. F. (2006). Early events in olfactory processing. *Annual Review of Neuroscience, 29*, 163–201.

Wiltstein, N. (1995, October 26). Quarry KO'd by dementia. *Vancouver Sun*, p. 135.

Wilusz, J. E., & Sharp, P. A. (2013). A circuitous route to noncoding RNA. *Science, 340*, 440–441.

Windt, J. M. (2013). Reporting dream experience: Why (not) to be skeptical about dream reports. *Frontiers in Human Neuroscience, 7*, 708.

Wingfield, A., Tun, P. A., & McCoy, S. L. (2005). Hearing loss in older adulthood: What it is and how it interacts with cognitive performance. *American Psychological Society, 14*, 144–148.

Winocur, G., & Moscovitch, M. (2011). Memory transformation and systems consolidation. *Journal of the International Neuropsychological Society, 17*(5), 766–780.

Winter, S. S., Clark, B. J., & Taube, J. S. (2015). Disruption of the head direction cell network impairs the parahippocampal grid cell signal. *Science, 347*, 870–874.

Wirths, O., & Bayer, T. A. (2010). Neuron loss in transgenic mouse models of Alzheimer's disease. *International Journal of Alzheimer's Disease, 2010*, article 723782.

Wiseman, F. K., Al-Janabi, T., Hardy, J., Karmiloff-Smith, A., Nizetic, D., Tybulewicz, V. L., . . . & Strydom, A. (2015). A genetic cause of Alzheimer disease: Mechanistic insights from Down syndrome. *Nature Reviews Neuroscience, 16*, 564–574.

Witkower, Z., & Tracy, J. L. (2019). Bodily communication of emotion: Evidence for extrafacial behavioral expressions and available coding systems. *Emotion Review, 11*(2), 184–193.

Witte, A. V., Fobker, M., Gellner, R., Knecht, S., & Flöel, A. (2009). Caloric restriction improves memory in elderly humans. *Proceedings of the National Academy of Sciences, USA, 106*, 1255–1260.

Wolf, J. M., Walker, D. A., Cochrane, K. L., & Chen, E. (2007, May). *Parental perceived stress influences inflammatory markers in children with asthma.* Paper delivered at the Western Psychological Association Convention, Vancouver, BC.

Wolgin, D. L., & Jakubow, J. J. (2003). Tolerance to amphetamine hypophagia: A microstructural analysis of licking behavior in the rat. *Behavioral Neuroscience, 117*, 95–104.

Wolpaw, J. R., & Tennissen, A. (2001). Activity-dependent spinal cord plasticity in health and disease. *Annual Review of Neuroscience, 24*, 807–843.

Wolstenholme, J. T., Rissman, E. F., & Bekiranov, S. (2013). Sexual differentiation in the developing mouse brain: Contributions of sex chromosome genes. *Genes, Brain and Behavior, 12*, 166–180.

Wommelsdorf, T., Anton-Erxleben, K., Pieper, F., & Treue, S. (2006). Dynamic shifts of visual receptive fields in cortical area MT by spatial attention. *Nature Neuroscience, 9*, 1156–1160.

Wong, A. C. Y., & Ryan, A. F. (2015). Mechanisms of sensorineural cell damage, death and survival in the cochlea. *Frontiers in Aging Neuroscience, 7*, 58.

Wong, C. C. Y., Caspi, A., Williams, B., Craig, I. W., Houts, R., Ambler, A., Moffitt, T. E., & Mill, J. (2010). A longitudinal study of epigenetic variation in twins. *Epigenetics, 5*, 1–11.

Wong, K. (2005a, February). The littlest human. *Scientific American, 292*, 56–65.

Wong, K. (2005b, June). The morning of the modern mind. *Scientific American, 292*, 86–95.

Wong, K. (2012). First of our kind. Sensational fossils from South Africa spark debate over how we came to be human. *Scientific American, 306*(4), 30–39.

Wong, K. (Feb 2015). Neandertal minds. *Scientific American, 312*, 36–43.

Wong, W. T., & Wong, R. O. L. (2000). Rapid dendritic movements during synapse formation and rearrangement. *Current Opinion in Neurobiology, 10*, 118–124.

Wood, A., Rychlowska, M., Korb, S., & Niedenthal, P. (2016). Fashioning the face: Sensorimotor simulation contributes to facial expression recognition. *Trends in Cognitive Sciences, 20*(3), 227–240.

Wood, B. (2010). Reconstructing human evolution: Achievements, challenges, and opportunities. *PNAS, 107*, 8902–8909.

Wood, B. (2011). Did early *Homo* migrate "out of" or "in to" Africa? *PNAS, 108*, 10375–10376.

Wood, B. (2014). Welcome to the family. *Scientific American, 311*, 42–47.

Wood, E. R., Mumby, D. G., Pinel, J. P. J., & Phillips, A. G. (1993). Impaired object recognition memory in rats following ischemia-induced damage to the hippocampus. *Behavioral Neuroscience, 107*, 51–62.

Wood, H. (2018). Retinal implants show promise in macular degeneration. *Nature Reviews Neurology, 14*, 317.

Woodruff-Pak, D. S. (1993). Eyeblink classical conditioning in H.M.: Delay and trace paradigms. *Behavioral Neuroscience, 107*, 911–925.

Woods, S. C. (2004). Lessons in the interactions of hormones and ingestive behavior. *Physiology & Behavior, 82*(1), 187–190.

Woods, S. C. (2013). Metabolic signals and food intake. Forty years of progress. *Appetite, 71*, 440–444.

Woods, S. C., & Begg, D. P. (2015). Food for thought: Revisiting the complexity of food intake. *Cell Metabolism, 22*, 348–351.

Woods, S. C., & Begg, D. P. (2016). Regulation of the motivation to eat. *Current Topics in Behavioural Neuroscience, 27*, 15–34.

Woods, S. C., Lotter, E. C., McKay, L. D., & Porte, D., Jr. (1979). Chronic intracerebroventricular infusion of insulin reduces food intake and body weight of baboons. *Nature, 282*, 503–505.

Woods, S. C., Schwartz, M. W., Baskin, D. G., & Seeley, R. J. (2000). Food intake and the regulation of body weight. *Annual Review of Neuroscience, 51*, 255–277.

Woodson, J. C., & Gorski, R. A. (2000). Structural sex differences in the mammalian brain: Reconsidering the male/female dichotomy. In A. Matsumoto (Ed.), *Sexual differentiation of the brain* (pp. 229–255). Boca Raton, FL: CRC Press.

Wu, J. W., Hussaini, S. A., Bastille, I. M., Rodriguez, G. A., Mrejeru, A., Rilett, K., . . . & Herman, M. (2016). Neuronal activity enhances tau propagation and tau pathology in vivo. *Nature Neuroscience, 19*(8), 1085–1092.

Wu, M. V., & Shah, N. M. (2011). Control of masculinization of the brain and behavior. *Current Opinion in Neurobiology, 21*, 116–123.

Wu, N., Yu, A.-B., Zhu, H.-B., & Lin, X-K. (2012). Effective silencing of Sry gene with RNA interference in developing mouse embryos resulted in feminization of XY gonad. *Journal of Biomedicine and Biotechnology, 2012*, 343891.

Wu, Y. E., Parikshak, N. N., Belgard, T. G., & Geschwind, D. H. (2016). Genome-wide, integrative analysis implicates microRNA dysregulation in autism spectrum disorder. *Nature Neuroscience, 19*(11), 1463.

Wu, Z., Grillet, N., Zhao, B., Cunningham, C., Harkins-Perry, S., Coste, B., . . . & Patapoutian, A. (2017). Mechanosensory hair cells express two molecularly distinct mechanotransduction channels. *Nature Neuroscience, 20*(1), 24–33.

Wurtz, R. H., McAlonan, K., Cavanaugh, J., & Berman, R. A. (2011). Thalamic pathways for active vision. *Trends in Cognitive Sciences, 15*, 177–184.

Wüst, S., Kasten, E., & Sabel, A. (2002). Blindsight after optic nerve injury indicates functionality of spared fibers. *Journal of Cognitive Neuroscience, 14*, 243–253.

Wuttke, T. V., Markopoulos, F., Padmanabhan, H., Wheeler, A. P., Murthy, V. N., & Macklis, J. D. (2018). Developmentally primed cortical neurons maintain fidelity of differentiation and establish appropriate functional connectivity after transplantation. *Nature Neuroscience, 21*(4), 517–529.

Wyatt, T. D. (2017). Pheromones. *Current Biology, 27*(15), R739–R743.

Xiao, L., Ohayon, D., McKenzie, I. A., Sinclair-Wilson, A., Wright, J. L., Fudge, A. D., . . . & Richardson, W. D. (2016). Rapid production of new oligodendrocytes is required in the earliest stages of motor-skill learning. *Nature Neuroscience, 19*(9), 1210–1217.

Xu, D., Bureau, Y., McIntyre, D. C., Nicholson, D. W., Liston, P., Zhu, Y., . . . & Robertson, G. S. (1999). Attenuation of ischemia-induced cellular and behavioral deficits by X chromosome–linked inhibitor of apoptosis protein overexpression in the rat hippocampus. *Journal of Neuroscience, 19*, 5026–5033.

Yadav, A., Kumar, R., Tiwari, J., Kumar, V., & Rani, S. (2017). Sleep in birds: Lying on the continuum of activity and rest. *Biological Rhythm Research, 48*(5), 805–814.

Yager, J. Y., Wright, S., Armstrong, E. A., Jahraus, C. M., & Saucier, D. M. (2006). The influence of aging on recovery following ischemic brain damage. *Behavioural Brain Research, 173*, 171–180.

Yagi, S., & Galea, L. A. (2019). Sex differences in hippocampal cognition and neurogenesis. *Neuropsychopharmacology, 44*(1), 200–213.

Yamada, H. (2017). Hunger enhances consistent economic choices in non-human primates. *Scientific Reports, 7*(1), 2394.

Yamamoto, T., Nakayama, K., Hirano, H., Tomonagai, T., Ishihama, Y., Yamada, T., . . . & Goshima, N. (2013). Integrated view of the human x-centric proteome project. *Journal of Proteome Research, 12*, 58–61.

Yamazaki, Y., Zhao, N., Caulfield, T. R., Liu, C. C., & Bu, G. (2019). Apolipoprotein E and Alzheimer disease: Pathobiology and targeting strategies. *Nature Reviews Neurology, 15*(9), 501–518.

Yang, C. F., & Shah, N. M. (2014). Representing sex in the brain, one module at a time. *Neuron, 82*, 261–278.

Yang, S. H., Liu, R., Wu, S. S., & Simpkins, J. W. (2003). The use of estrogens and related compounds in the treatment of damage from cerebral ischemia. *Annals of the New York Academy of Sciences, 1007*, 101–107.

Yang, Y., Cui, Y., Sang, K., Dong, Y., Ni, Z., Ma, S., & Hu, H. (2018). Ketamine blocks bursting in the lateral habenula to rapidly relieve depression. *Nature, 554*(7692), 317–322.

Yang, Y., Tang, B. S., & Guo, J. F. (2016). Parkinson's disease and cognitive impairment. *Parkinson's Disease, 2016*, 6734678.

Yao, B., Christian, K. M., He, C., Jin, P., Ming, G. L., & Song, H. (2016). Epigenetic mechanisms in neurogenesis. *Nature Reviews Neuroscience, 17*(9), 537.

Yap, Q. J., Teh, I., Fusar-Poli, P., Sum, M. Y., Kuswanto, C., & Sim, K. (2013). Tracking cerebral white matter changes across the lifespan: Insights from diffusion tensor imaging studies. *Journal of Neural Transmission, 120*, 1369–1395.

Yarmolinsky, D. A., Zuker, C. S., & Ryba, N. J. P. (2009). Common sense about taste: From mammals to insects. *Cell, 139*, 234–244.

Yau, J. M., Connor, C. E., & Hsiao, S. S. (2013). Representation of tactile curvature in macaque somatosensory area 2. *Journal of Neurophysiology, 109*, 2999–3012.

Yenigun, A., Tugrul, S., Dogan, R., Aksoy, F., & Ozturan, O. (2020). A feasibility study in the treatment of obstructive sleep apnea syndrome and snoring: Nasopharyngeal stent. *American Journal of Otolaryngology*, 102460.

Yesalis, C. E., & Bahrke, M. S. (2005). Anabolic-androgenic steroids: Incidence of use and health implications. *President's Council on Physical Fitness and Sports Research Digest, 5*, 1–6.

Yeshurun, S., & Hannan, A. J. (2018). Transgenerational epigenetic influences of paternal environmental exposures on brain function and predisposition to psychiatric disorders. *Molecular Psychiatry*, doi: 10.1038/s41380-018-0039-z.

Yeshurun, S., & Hannan, A. J. (2019). Transgenerational epigenetic influences of paternal environmental exposures on brain function and predisposition to psychiatric disorders. *Molecular Psychiatry, 24*(4), 536–548.

Yetish, G., Kaplan, H., Gurven, M., Wood, B., Pontzer, H., . . . & Siegel, J. M. (2015). Natural sleep and its seasonal variations in three pre-industrial societies. *Current Biology, 25*, 2862–2868.

Yi, H. G., Leonard, M. K., & Chang, E. F. (2019). The encoding of speech sounds in the superior temporal gyrus. *Neuron, 102*(6), 1096–1110.

Yildiz, A., Vieta, E., Leucht, S., & Baldessarini, R. J. (2011). Efficacy of antimanic treatments: Meta-analysis of randomized, controlled trials. *Neuropsychopharmacology, 36*, 375–389.

Yin, A. Q., Wang, F., & Zhang, X. (2019). Integrating endocannabinoid signaling in the regulation of anxiety and depression. *Acta Pharmacologica Sinica, 40*(3), 336–341.

Yogev, S., & Shen, K. (2017). Establishing neuronal polarity with environmental and intrinsic mechanisms. *Neuron, 96*(3), 638–650.

Yonelinas, A. P., & Ritchey, M. (2015). The slow forgetting of emotional episodic memories: An emotional binding account. *Trends in Cognitive Sciences, 19*, 259–267.

Yoo, S., & Blackshaw, S. (2018). Regulation and function of neurogenesis in the adult mammalian hypothalamus. *Progress in Neurobiology, 170*, 53–66.

Yoon, B.-E., & Lee, C. J. (2014). GABA as a rising gliotransmitter. *Frontiers in Neural Circuits, 8*, 141.

Yoon, J. M. D., & Vouloumanos, A. (2014). When and how does autism begin? *Trends in Cognitive Sciences, 18*, 272–273.

Younts, T. J., & Castillo, P. E. (2014). Endogenous cannabinoid signaling at inhibitory interneurons. *Current Opinion in Neurobiology, 26*, 42–50.

Yournagov, G., Smith, K. G., Fridriksson, J., & Rorden, C. (2015). Predicting aphasia type from brain damage measured with structural MRI. *Cortex, 73*, 203–215.

Yu, C., & Shen, H. (2020). Bizarreness of lucid and non-lucid dream: Effects of metacognition. *Frontiers in Psychology, 10*, 2946.

Yu, J.-T., Tan, L., & Hardy, J. (2014). Apolipoprotein E in Alzheimer's disease: An update. *Annual Review of Neuroscience, 37*, 79–100.

Yücel, M., Solowij, N., Respondek, C., Whittle, S., Fornito, A., Pantelis, C., & Lubman, D. I. (2008). Regional brain abnormalities associated with long-term heavy cannabis use. *Archives of General Psychiatry, 65*, 694–701.

Yuste, R., & Church, G. M. (2014). The new century of the brain. *Scientific American, 310*, 38–45.

Zahr, N. M., & Pfefferbaum, A. (2017). Alcohol's effects on the brain: Neuroimaging results in humans and animal models. *Alcohol Research: Current Reviews, 38*(2), 183–206.

Zahr, N. M., Kaufman, K. L., & Harper, C. G. (2011). Clinical and pathological features of alcohol-related brain damage. *Nature Reviews Neurology, 7*, 284–294.

Zaidel, D. W. (2013). Split-brain, the right hemisphere, and art: Fact and fiction. *Progress in Brain Research, 204*, 3–17.

Zakiniaeiz, Y., & Potenza, M. N. (2018). Gender-related differences in addiction: A review of human studies. *Current Opinion in Behavioral Sciences, 23*, 171–175.

Zangwill, O. L. (1975). Excision of Broca's area without persistent aphasia. In K. J. Zulch, O. Creutzfeldt, & G. C. Galbraith (Eds.), *Cerebral localization* (pp. 258–263). New York, NY: Springer-Verlag.

Zehra, A., Burns, J., Liu, C. K., Manza, P., Wiers, C. E., Volkow, N. D., & Wang, G. J. (2018). Cannabis addiction and the brain: A review. *Journal of Neuroimmune Pharmacology, 13*(4), 438–452.

Zeidán-Chuliá, F., Salmina, A. B., Malinovskaya, N. A., Noda, M., Verkhratsky, A., & Moreira, J. C. F. (2014). The glial perspective of autism spectrum disorders. *Neuroscience and Biobehavioral Reviews, 38*, 160–172.

Zeki, S. (2015). Area V5—a microcosm of the visual brain. *Frontiers in Integrative Neuroscience, 9*, 21.

Zelinski, E. L., Deibel, S.-J., & McDonald, R. J. (2014). The trouble with circadian clock dysfunction: Multiple deleterious effects on the brain and body. *Neuroscience & Biobehavioral Reviews, 40*, 80–101.

Zeltser, L. M. (2018). Feeding circuit development and early-life influences on future feeding behaviour. *Nature Reviews Neuroscience, 19*(5), 302–316.

Zembrzycki, A., Chou, S.-J., Ashery-Padan, R., Stoykova, A., & O'Leary, D. D. M. (2013). Sensory cortex limits cortical maps and drives top-down plasticity in thalamocortical circuits. *Nature Neuroscience, 8*, 1060–1068.

Zeng, H., & Sanes, J. R. (2017). Neuronal cell-type classification: Challenges, opportunities and the path forward. *Nature Reviews Neuroscience, 18*(9), 530.

Zetterberg, H., & Blennow, K. (2016). Fluid biomarkers for mild traumatic brain injury and related conditions. *Nature Reviews Neurology, 12*(10), 563–574.

Zhang, C., Kolodkin, A. L., Wong, R. O., & James, R. E. (2017). Establishing wiring specificity in visual system circuits: From the retina to the brain. *Annual Review of Neuroscience, 40*, 395–424.

Zhang, D., & Raichle, M. E. (2010). Disease and the brain's dark energy. *Nature Reviews Neurology, 6*, 15–28.

Zhang, D., Xia, H., Xu, L., Zhang, C., Yao, W., Wang, Y., Ren, J., Wu, J., Tian, Y., Liu, W., & Wang, X. (2012). Neuroprotective effects of 17beta-estradiol associate with KATP in rat brain. *Neuroreport, 23*, 952–957.

Zhang, N., Yacoub, E., Zhu, X.-H., Ugurbil, K., & Chen, W. (2009). Linearity of blood-oxygenation-level dependent signal at microvasculature. *NeuroImage, 48*, 313–318.

Zhang, T. Y., Labonté, B., Wen, X. L., Turecki, G., & Meaney, M. J. (2013). Epigenetic mechanisms for the early environmental regulation of hippocampal glucocorticoid receptor gene expression in rodents and humans. *Neuropsychopharmacology, 38*(1), 111–123.

Zhang, T.-Y., & Meaney, M. J. (2010). Epigenetics and the environmental regulation of the genome and its function. *Annual Review of Psychology, 61*, 439–466.

Zhang, Y., Chen, S. R., Laumet, G., Chen, H., & Pan, H. L. (2016). Nerve injury diminishes opioid analgesia through lysine methyltransferase-mediated transcriptional repression of μ-opioid receptors in primary sensory neurons. *Journal of Biological Chemistry, 291*, 8475–8485.

Zhang, Y., Cudmore, R. H., Lin, D. T., Linden, D. J., & Huganir, R. L. (2015). Visualization of NMDA receptor-dependent AMPA receptor synaptic plasticity in vivo. *Nature Neuroscience, 18*, 402–407.

Zhang, Z. Q. (2013). Animal biodiversity: An update of classification and diversity in 2013. *Zootaxa, 3703*(1), 5–11.

Zhao, L., Woody, S. K., & Chhibber, A. (2015). Estrogen receptor β in Alzheimer's disease: From mechanisms to therapeutics. *Ageing Research Reviews, 24*, 178–190.

Zhong, S., Zhang, S., Fan, X., Wu, Q., Yan, L., Dong, J.,. . . & Xu, X. (2018). A single-cell RNA-seq survey of the developmental landscape of the human prefrontal cortex. *Nature, 555*(7697), 524–528.

Zhou, Q., Zhou, P., Wang, A. L., Wu, D., Zhao, M., Südhof, T. C., & Brunger, A. T. (2017). The primed SNARE–complexin–synaptotagmin complex for neuronal exocytosis. *Nature, 548*(7668), 420.

Zhou, W., & Yuan, J. (2014). Necroptosis in health and diseases. *Seminars in Cell and Developmental Biology, 35*, 14–23.

Zhu, J. M., Zhao, Y. Y., Chen, S. D., Zhang, W. H., Lou, L., & Jin, X. (2011). Functional recovery after transplantation of neural stem cells modified by brain-derived neurotrophic factor in rats with cerebral ischaemia. *Journal of International Medical Research, 39*, 488–498.

Zhuo, M. (2008). Cortical excitation and chronic pain. *Trends in Neurosciences, 31*, 199–207.

Zhuo, M. (2016). Neural mechanisms underlying anxiety–chronic pain interactions. *Trends in Neurosciences, 39*(3), 136–145.

Ziauddeen, H., Farooqi, I. S., & Fletcher, P. C. (2012). Obesity and the brain: How convincing is the addiction model? *Nature Reviews Neuroscience, 13*, 279–286.

Ziegler, T. E. (2007). Female sexual motivation during non-fertile periods: A primate phenomenon. *Hormones and Behavior, 51*, 1–2.

Zilles, K., & Amunts, K. (2010). Centenary of Brodmann's map—conception and fate. *Nature Reviews Neuroscience, 11*, 139–145.

Zilles, K., Palermo-Gallagher, N., & Amunts, K. (2013). Development of cortical folding during evolution and ontogeny. *Trends in Neurosciences, 36*, 275–284.

Zimmer, C. (2011). 100 trillion connections. *Scientific American, 304*(1), 58–63.

Zimmer, E. R., Parent, M. J., Souza, D. G., Leuzy, A., Lecrux, C., Kim, H. I., . . . & Rosa-Neto, P. (2017). [18 F] FDG PET signal is driven by astroglial glutamate transport. *Nature Neuroscience, 20*, 393.

Zimmerman, A., Bai, L., & Ginty, D. D. (2014). The gentle touch receptors of mammalian skin. *Science, 346*, 950–954.

Zimmerman, J. E., Naidoo, N., Raizen, D. M., & Pack, A. I. (2008). Conservation of sleep: Insights from non-mammalian model systems. *Trends in Neurosciences, 31*, 371–376.

Zimmerman, J. L. (2012). Cocaine intoxication. *Critical Care Clinics, 28*, 517–526.

Zirnsak, M., & Moore, T. (2014). Saccades and shifting receptive fields: Anticipating consequences or selecting targets? *Trends in Cognitive Sciences, 18*, 621–628.

Zisapel, N. (2018). New perspectives on the role of melatonin in human sleep, circadian rhythms and their regulation. *British Journal of Pharmacology, 175*(16), 3190–3199.

Zivin, J. A. (2000). Understanding clinical trials. *Scientific American, 282*, 69–75.

Zlebnik, N. E., & Cheer, J. F. (2016). Beyond the CB1 receptor: Is cannabidiol the answer for disorders of motivation? *Annual Review of Neuroscience, 39*, 1–17.

Zoghbi, H. Y. (2003). Postnatal neurodevelopmental disorders: Meeting at the synapse? *Science, 302*, 826–830.

Zola-Morgan, S., Squire, L. R., & Amaral, D. G. (1986). Human amnesia and the medial temporal region: Enduring memory impairment following a bilateral lesion limited to field CA1 of the hippocampus. *Journal of Neuroscience, 6*, 2950–2967.

Zola-Morgan, S., Squire, L. R., & Mishkin, M. (1982). The neuroanatomy of amnesia: Amygdala-hippocampus versus temporal stem. *Science, 218*, 1337–1339.

Zola-Morgan, S., Squire, L. R., Rempel, N. L., Clower, R. P., & Amaral, D. G. (1992). Enduring memory impairment in monkeys after ischemic damage to the hippocampus. *Journal of Neuroscience, 12*, 2582–2596.

Zong, C., Lu, S., Chapman, A. R., & Xie, X. S. (2012). Genome-wide detection of single-nucleotide and copy-number variations of a single human cell. *Science, 338*, 1622–1626.

Zou, Y. (2004). Wnt signaling in axon guidance. *Trends in Neurosciences, 27*, 529–532.

Zuberi, S. M., & Brunklaus, A. (2018). Epilepsy in 2017: Precision medicine drives epilepsy classification and therapy. *Nature Reviews Neurology, 14*(2), 67–68.

Zwar, N. A., Mendelsohn, C. P., & Richmond, R. L. (2014). Supporting smoking cessation. *British Medical Journal, 348*, f7535.

Credits

Text Credits

p. 27: Based on Sacks, O. (1970) The Man Who Mistook His Wife for a Hat and Other Clinical Tales. Harper & Row; **p. 30:** Ulrich, R. E. (1991). Commentary: Animal rights, animal wrongs and the question of balance. *Psychological Science*, 2, 197–201; **p. 31:** Johnson, R.N. (1972) Aggression in man and animals. London: Saunders. **p. 32:** Based on Lester, G. L. L., & Gorzalka, B. B. (1988). Effect of novel and familiar mating partners on the duration of sexual receptivity in the female hamster. Behavioral and Neural Biology, 49, 398–405; **p. 34–35:** Kolb, B., & Whishaw, I.Q. (1980) Fundamentals of Human Neuropsychology, 3e. New York: W.H. Freeman; **p. 35:** Based on Iacono, W. G., & Koenig, W. G. (1983). Features that distinguish the smooth-pursuit eye-tracking performance of schizophrenic, affective-disorder, and normal individuals. Journal of Abnormal Psychology, 92(1), 29–41; **p. 38:** Based on Snyder, S. H. (1978). "Neuroleptic Drugs and Neurotransmitter Receptors." *Journal of Clinical and Experimental Psychiatry*, 133, 21–31; **p. 40:** Valenstein, E. S. (1973). Brain control. New York, NY: John Wiley & Sons; Based on Barry Blackwell, M.A, M.Phil, MB, BChir., M.D., FRCPsych, (2014) "JOSE DELGADO: A CASE STUDY" Science, Hubris, Nemesis and Redemption. International Network for the History of Neuropsychopharmacology Retrieved from http://www.inhn.org/fileadmin/previews/Case_Study_Delgado.pdf; **p. 46:** Corsi, P.I. (1991) The Enchanted Loom: Chapters in the History of Neuroscience. New York: Oxford University Press; Watson, J.B. (1930) Behaviorism. New York: W.W. Norton & Company, Inc.; **p. 47:** Based on Sacks, O. (1970) The Man Who Mistook His Wife for a Hat and Other Clinical Tales. Harper & Row; Gallup, G. G. Jr. (1983) "Toward a Comparative Psychology of Mind," *American Journal of Primatology*, 13, 473–510; **p. 48:** Mark Twain in an interview by White, F.M. "Mark Twain Amused," *The New York Journal* (2 June 1897); **p. 50:** Quammen, D. (2004, May). "Was Darwin Wrong? No. The Evidence for Evolution is Overwhelming." *National Geographic*, 206, 2–35; **p. 51:** Based on McCann, T. S. (1981). "Aggression and Sexual Activity of Male Southern Elephant Seals, Mirounga Leonina." *Journal of Zoology*, 195, 295–310; **p. 66:** Data from Cooper, R. M., & Zubek, J. P. (1958). "Effects of Enriched and Restricted Early Environments on the Learning Ability of Bright and Dull Rats." *Canadian Journal of Psychology*, 12, 159–164; **p. 67:** Data from Cooper, R. M., & Zubek, J. P. (1958). "Effects of Enriched and Restricted Early Environments on the Learning Ability of Bright and Dull Rats." *Canadian Journal of Psychology*, 12, 159–164; **p. 85:** Based on Himle, M.B., & Woods, D.W., (2005) "An Experimental Evaluation of Tic Suppression and the Tic Rebound Effect." *Behaviour Research and Therapy*, 43, 1443–1451; **p. 89:** Hollenbeck, P. J. (2001). "Insight And Hindsight Into Tourette Syndrome." *Advances In Neurology*, 85, 363–367; **p. 91:** Based on Zivin, J. A. (2000, April). "Understanding Clinical Trials." *Scientific American*, 282, 69–75; **p. 96:** Based on Rakic, P. (1979). "Genetic and Epigenetic Determinants of Local Neuronal Circuits in the Mammalian Central Nervous System." In F. O. Schmitt & F. G. Worden (eds.), The Neurosciences: Fourth Study Program. Cambridge, MA: MIT Press; **p. 97–98:** Zivin, J. A. (2000, April). "Understanding Clinical Trials." *Scientific American*, 282, 69–75; **p. 98:** Based on Klawans, H. L. (1990) Newton's Madness: Further Tales of Clinical Neurology. New York: Harper & Row; **p. 155:** Based on Lamb, T. D., Collin, S. P., & Pugh, E. N. (2007). "Evolution of the Vertebrate Eye: Opsins, Photoreceptors, Retina and Eye Cup." *Nature Reviews Neuroscience*, 8, 960–975; **p. 160:** Based on Lindsay, P. H., & Norman, D. A. (1977). Human Information Processing (2nd ed.). New York, NY: Academic Press; **p. 164:** Based on Netter, F. H. (1962). The CIBA Collection of Medical Illustrations. Vol. 1, The Nervous System. New York, NY: CIBA; **p. 165:** Eagleman, D.M. (2001) "Visual Illusions and Neurobiology." *Nature Reviews Neuroscience*. 2(12): 920–926; **p. 172:** Land, E. H. (1977). "The Retinex Theory of Color Vision." *Scientific American*, 237(6), 108–129; **p. 174:** Based on Teuber, H.-L., Battersby, W. S., & Bender, M. B. (1960). "Recovery of Function After Brain Injury in Man." In Outcomes of Severe Damage to the Nervous System. CIBA Foundation Symposium 34. Amsterdam, Netherlands: Elsevier North-Holland; **p. 175:** Lashley, K. S. (1941). "Patterns of Cerebral Integration Indicated by the Scotomas of Migraine." Archives of Neurology and Psychiatry, 46, 331–339, p. 338; Based on Lashley, K. S. (1941). "Patterns of Cerebral Integration Indicated by the Scotomas of Migraine." Archives of Neurology and Psychiatry, 46, 331–339, p. 338; Based on Lashley, K. S. (1941). "Patterns of Cerebral Integration Indicated by the Scotomas of Migraine." Archives of Neurology and Psychiatry, 46, 331–339, p. 338; **p. 178:** Based on Goodale, M. A. (2004). "Perceiving the World and Grasping it: Dissociations Between Conscious and Unconscious Visual Processing." In M. Gazzaniga (ed.), The Cognitive Neurosciences (pp. 1159–1172). Cambridge, MA: MIT Press; Based on Jeannerod, M., Arbib, M. A., Rizzolatti, G., & Sakarta, H. (1995). "Grasping Objects: The Cortical Mechanisms Of Visuomotor Transformation." Trends in Neurosciences, 18(7), 314–327; **p. 185:** Williams, M. (1970). Brain Damage and the Mind. Baltimore, MD: Penguin Books; **p. 197:** Based on Servos, P., Engel, S. A., Gati, J., & Menon, R. (1999). "fMRI Evidence for an Inverted Face Representation in Human Somatosensory Cortex." *NeuroReport*, 10(7), 1393–1395; **p. 198:** Based on Klawans, H.L. (1990) Newton's Madness: Further Tales of Clinical Neurology. New York: Harper & Row; **p. 201:** Based on Basbaum A. I., & Fields H. L. (1978) Endogenous Pain Control Mechanisms: Review and Hypothesis. *Ann Neurol.* 4(5):451–462; **p. 207:** Based on de Lange, F. P., Heilbron, M., & Kok, P. (2018). "How Do Expectations Shape Perception?" Trends in Cognitive Sciences, in press; **p. 221:** Based on Penfield, W., & Rasmussen, T. (1950). The Cerebral Cortex of Man: A Clinical Study of the Localization of Function. New York, NY: Macmillan; **p. 233:** Based on Jenkins, I. H., Brooks, D. J., Bixon, P. D., Frackowiak, R. S. J., & Passingham, R. E. (1994). "Motor Sequence Learning: A Study With Positron Emission Tomography." *Journal of Neuroscience*, 14(6), 3775–3790; **p. 240:** Based on Cowan, W. M. (1979, September). "The Development of the Brain." *Scientific American*, 241, 113–133; **p. 241:** Based on Kriegstein, A., & Alvarez-Buylla, A. (2009). "The Glial Nature of Embryonic and Adult Neural Stem Cells." *Annual Review of Neuroscience*, 32, 149–184; **p. 249:** Based on Antonini, A., & Stryker, M. P. (1993). "Rapid Remodeling of Axonal Arbors in the Visual Cortex." *Science*, 260, 1819–1821; **p. 253:** Used with Permission of Emma Smith; Finn, R. (1991 June) "Different Minds," *Discover*. Retrieved from http://www.bobfinn.net/williams+syndrome.htm.; **p. 264:** "Junior Seau: An Illustrative Case of Chronic Traumatic Encephalopathy and Update on Chronic Sports-Related Head Injury." Tej D. Azad, Amy Li, Arjun V. Pendharkar, Anand Veeravagu, Gerald A. Gran. *World Neuroscience*. February 2016 Volume 86, Pages 515.e11–515.e16. Elsevier; **p. 267:** Lennox, W. G. (1960). Epilepsy and Related Disorders. Boston, MA: Little, Brown; **p. 275:** Langston, J. W. (1985). "MPTP and Parkinson's Disease." *Trends in Neurosciences*, 8, 79–83; **p. 279:** Based on Sanes, J. N., Suner, S., & Donoghue, J. P. (1990). "Dynamic Organization of Primary Motor Cortex Output to Target Muscles in Adult Rats I. Long-term Patterns of Reorganization Following Motor or Mixed Peripheral Nerve Lesions." *Experimental Brain Research*, 79(3), 479–491; **p. 281:** Klawans, H. L. (1990). Newton's Madness: Further Tales of Clinical Neurology. New York: Harper & Row; **p. 284:** Based on V. S. Ramachandran & Sandra Blakeslee (1998) Phantoms in the Brain: Probing the Mysteries of the Human Mind. New York: Morrow.; Based on V. S. Ramachandran & Sandra Blakeslee (1998) Phantoms in the Brain: Probing the Mysteries of the Human Mind. New York: Morrow; **p. 290:** Based on Milner, B. (1965). "Memory Disturbances After Bilateral Hippocampal Lesions." In P. Milner & S. Glickman (eds.), Cognitive Processes and the Brain (pp. 104–105). Princeton, NJ: D. Van Nostrand; **p. 293:** Tulving, E. (2002). "Episodic Memory: From Mind to Brain." *Annual Review of Neuroscience*, 53, 1–25; **p. 294:** Zola-Morgan, S., Squire, L. R., & Amaral, D. G. (1986). "Human Amnesia and the Medial Temporal Region: Enduring Memory Impairment Following a Bilateral Lesion Limited to Field CA1 of the Hippocampus." *Journal of Neuroscience*, 6, 2950–2967; **p. 295:** Teuber, H.-L., Milner, B., & Vaughan, H. G., Jr. (1968). "Persistent Anterograde Amnesia After Stab Wound of the Basal Brain." *Neuropsychologia*, 6, 267–282; **p. 297:** Based on Squire, L. R., Slater, P. C., & Chace, P. M. (1975). "Retrograde Amnesia: Temporal Gradient in Very Long Term Memory Following Electroconvulsive Therapy. Science, 187, 77–79; **p. 299:** Based on Squire, L. R., & Zola-Morgan, S. (1991). 'The Medial Temporal Lobe Memory System." *Science*, 253(5026), 1380–1386; **p. 310:** Traces courtesy of Michael Corcoran, Department of Psychology, University of Saskatchewan; **p. 318:** Rozin, P., Dow, S., Moscovitch, M., & Rajaram, S. (1998). "What Causes Humans to Begin and End a Meal? A Role for Memory For What Has Been Eaten, As Evidenced by a Study of Multiple Meal Eating in Amnesic Patients." *Psychological Science*, 9, 392–396;

p. 331: Martin, A., State, M., Koenig, K., Schultz, R., Dykens, E. M., Cassidy, S. B., & Leckman, J. F. (1998). "Prader-Willi Syndrome." *American Journal of Psychiatry*, 155, 1265–1272; **p. 339:** Farooqi, I. S., Jebb, S. A., Langmack, G., Lawrence, E., Cheetham, C. H., Prentice, A. M., O'Rahilly, S. (1999). Effects of Recombinant Leptin Therapy in a Child with Congenital Leptin Deficiency." *New England Journal of Medicine*, 341, 879–884; **p. 341:** Keys, A., Broz, J., Henschel, A., Mickelsen, O., & Taylor H. L. (1950). The Biology of Human Starvation. Minneapolis: University of Minnesota Press; **p. 359:** Money, J., & Ehrhardt, A. A. (1972). Man & Woman, Boy & Girl. Baltimore, MD: Johns Hopkins University Press; **p. 362:** de Kruif, P. (1945). The Male Hormone. New York, NY: Harcourt, Brace; **p. 363:** Based on Grunt, J. A., & Young, W. C. (1952). "Differential Reactivity of Individuals and the Response of the Male Guinea Pig to Testosterone Propionate." *Endocrinology*, 51, 237–248; **p. 365:** Gorski, R. A. (1980) "Sexual Differentiation in the Brain." In D.T. Krieger & J. C. Hughes (eds.), *Neuroendocrinology*. Sunderland, MA: Sinauer. **p. 376:** Dement, W. C., & Wolpert, E. A. (1958). "The Relation of Eye Movements, Body Motility and External Stimuli to Dream Content." *Journal of Experimental Psychology*, 55, 543–553; **p. 380:** Adapted from Figure 1 of Domhoff, G. W., & Fox, K. C. (2015). "Dreaming and the Default Network: A Review, Synthesis, and Counterintuitive Research Proposal." *Consciousness and Cognition*, 33, 342–353; **p. 383:** Kleitman N. Sleep and Wakefulness. Chicago, IL: University of Chicago Press; 1963; Dement, W. C. (1978). Some Must Watch While Some Must Sleep. New York, NY: Norton; **p. 385:** Based on Rechtschaffen, A., Gilliland, M. A., Bergmann, B. M., & Winter, J. B. (1983). "Physiological Correlates of Prolonged Sleep Deprivation in Rats." *Science*, 221, 182–184; **p. 389:** Based on Wever, R. A. (1979). The Circadian System of Man. Seewiesen-Andechs, Germany: Max-Planck-Institut für Verhaltensphysiologie; **p. 392:** Saper, C. B., Scammell, T. E., & Lu, J. (2005). Hypothalamic Regulation of Sleep and Circadian Rhythms. *Nature*, 437, 1257–1263; **p. 394:** Vertes, R. P. (1983). Brainstem Control of the Events of REM Sleep. *Progress in Neurobiology*, 22, 241–288; **p. 397:** Dement, W. C. (1978). Some Must Watch While Some Must Sleep. New York, NY: Norton; **p. 399:** Schenck, C. H., Bundlie, S. R., Ettinger, M. G., & Mahowald, M. W. (1986). "Chronic Behavioral Disorders of Human REM Sleep: A New Category of Parasomnia." *Sleep*, 9, 293–308; **p. 401:** Sbragia G., "Leonardo da Vinci and Ultrashort Sleep: Personal Experience of an Eclectic Artist." In Why we Nap: Evolution Chronobiology and Functions of Polyphasic and Ultrashort Sleep, edited by C. A. Stampi (Boston: Birkhäuser, 1992), 180–185; Liu, T. Z., Xu, C., Rota, M., Cai, H., Zhang, C., Shi, M. J., . . . & Sun, X. (2017). "Sleep Duration and Risk of All-Cause Mortality: A Flexible, Non-Linear, Meta-Regression of 40 Prospective Cohort Studies." *Sleep Medicine Reviews*, 32, 28–36; **p. 409:** Based on Pinel, J. P. J., Mana, M. J., & Kim, C. K. (1989). "Effect-Dependent Tolerance to Ethanol's Anticonvulsant Effect on Kindled Seizures." In R. J. Porter, R. H. Mattson, J. A. Cramer, & I. Diamond (eds.), Alcohol and Seizures: Basic Mechanisms and Clinical Implications (pp. 115–125). Philadelphia, PA: F. A. Davis; **p. 410:** Based on Crowell, C. R., Hinson, R. E., & Siegel, S. (1981). "The Role of Conditional Drug Responses in Tolerance to the Hypothermic Effects of Ethanol." *Psychopharmacology*, 73, 51–54; **p. 412:** Brecher, E. M. (1972). Licit and Illicit Drugs. Boston, MA: Little, Brown; **p. 415:** The Schafer Commission, FKA, National Commission on Marijuana and Drug Abuse, 1972, p. 68 **p. 419:** McKim, W. A. (1986). Drugs and Behavior: An Introduction to Behavioral Pharmacology. Englewood Cliffs, NJ: Prentice-Hall.; McKim, W. A. (1986). Drugs and Behavior: An Introduction to Behavioral Pharmacology. Englewood Cliffs, NJ: Prentice-Hall; Julien, R. M. (1981). A Primer of Drug Action. San Francisco, CA: W. H. Freeman; **p. 421:** Based on the National Survey on Drug Use and Health, 2011. U.S. Department of Health and Human Services; Student Excerpt. Anonymous; **p. 423:** Based on Klivington, K. A. (Ed.). (1992). Gehirn und Geist. Heidelberg, Germany: Spektrum Akademischer Verlag; **p. 428:** Data redrawn from Pickens, C. L., Airavaara, M., Theberge, F., Fanous, S., Hope, B. T., & Shaham, Y. (2011). Neurobiology of the Incubation of Drug Craving. Trends in Neurosciences, 34(8), 411–420; **p. 441–442:** Joseph, R. (1988). "Dual Mental Functioning in a Split-Brain Patient." *Journal of Clinical Psychology*, 44, 771–779; **p. 442:** Paraphrased from Sperry, R. W., Zaidel, E., & Zaidel, D. W. (1979). "Self-Recognition and Social Awareness in the Disconnected Minor Hemisphere." *Neuropsychologia*, 17, 153–166; **p. 446:** Corina, D. P., Poizner, H., Bellugi, U., Feinberg, T., Dowd, D., & O'Grady-Batch, L. (1992). "Dissociation Between Linguistic and Nonlinguistic Gestural Systems: A Case For Compositionality." *Brain and Language*, 43, 414–447; **p. 449:** Geschwind, N. (1979). "Specializations of the Human Brain." *Scientific American*, 241(3), 180–201, p. 183; **p. 451:** Based on Mohr, J. P. (1976). "Broca's Area and Broca's Aphasia." In H. Whitaker & H. A. Whitaker (eds.), Studies in Neurolinguistics (Vol. 1, pp. 201–235). New York, NY: Academic Press; **p. 452:** Based on Penfield, W., & Roberts, L. (1959). Speech and Brain Mechanisms. Princeton, NJ: Princeton University Press;

p. 454: Based on Penfield, W., & Roberts, L. (1959). Speech and Brain Mechanisms. Princeton, NJ: Princeton University Press.; Based on Penfield, W., & Roberts, L. (1959). Speech and Brain Mechanisms. Princeton, NJ: Princeton University Press; **p. 459:** Patterson, Vargha-Khadem, & Polkey (1989) "Reading with One Hemisphere Brain," *A Journal of Neurology*. 112, 39–63; **p. 463:** Based on Damasio, H., Grabowski, T., Frank, R., Galaburda, A. M., & Damasio, A. R. (1994). "The Return of Phineas Gage: Clues about the Brain From the Skull of a Famous Patient." *Science*, 264, 1102–1105; **p. 468:** Based on Rutledge, L. L., & Hupka, R. B. (1985). "The Facial Feedback Hypothesis: Methodological Concerns and New Supporting Evidence." *Motivation and Emotion*, 9, 219–240.

Image Credits

p. 25: Image Source/Alamy Stock Photo; PhotoAlto/Alamy Stock Photo; **p. 26:** UHB Trust/The Image Bank/Getty Images; **p. 31:** Bettmann/Getty Images; **p. 36:** Todd C. Handy/University of British Columbia Department of Psychology; **p. 42:** Shaun Clark/Getty Images; **p. 44:** pyrozhenka/Shutterstock; **p. 48:** Photograph by Donna Bierschwale, courtesy of the New Iberia Research Center; **p. 51:** Francois Gohier/Science Source; **p. 52:** Beth Rooney Photography; **p. 53:** Vladimir Sazonov/Shutterstock; Anatoliy Lukich/Shutterstock; Michael Krabs/Alamy Stock Photo; Daniel Frauchiger, Switzerland/Moment/Getty Images; Kevin Schafer/Photolibrary/Getty Images; **p. 54:** Ryan McVay/Getty Images; Lealisa Westerhoff/AFP/Getty Images; **p. 55:** John Reader/Science Source; **p. 60:** David M. Phillips/Science Source; **p. 70:** J. Lee/Getty Images; **p. 72:** Tetra Images/Alamy Stock Photo; **p. 82:** GUNILLA ELAM/Science Source; **p. 83:** Martin M. Rotker/Science Source; Ernst/Brian Christie/University of British Columbia Department of Psychology; Carl Ernst/Brian Christie/University of British Columbia Department of Psychology; Science History Images/Alamy Stock Photo; **p. 94:** National Institute of Mental Health; **p. 95:** Mark Sykes/Alamy Stock Photo; **p. 97:** KTSDESIGN/SCIENCE PHOTO LIBRARY; **p. 110:** Don W. Fawcett/Science Source; **p. 115:** DR DAVID JACOBOWITZ/Science Source; **p. 119:** Viacheslav Iakobchuk/Alamy Stock Photo; **p. 121:** The Photolibrary Wales/Alamy Stock Photo; **p. 124:** CNRI/Science Source; **p. 125:** NIH/Science Source; **p. 126:** Image Source/Alamy Stock Photo; Scott Camazine/Science Source; Simon Fraser/Science Source; **p. 127:** Kent Kiehl/Peter Liddle/University of British Columbia Department of Psychiatry; **p. 130:** Image Source/Alamy Stock Photo; **p. 132:** David Kopf Instruments; **p. 136:** Rod Cooper/University of Calgary Department of Psychology; **p. 137:** Richard Mooney/University of Toledo College of Medicine Department of Neurosciences; **p. 139:** American Association for the Advancement of Science (AAAS); **p. 140:** Jeff Lichtman; **p. 149:** Dr. John Pinel; Dr. John Pinel; Dr. John Pinel; Dr. John Pinel; **p. 151:** Indiapicture/Alamy Stock Photo; **p. 154:** tarapong srichaiyos/Shutterstock; **p. 156:** Matthew Cuda/Alamy Stock Photo; Guiziou Franck/Hemis/Alamy Stock Photo; Naomi Engela Le Roux/123RF; Vasiliy Vishnevskiy/123RF; Colin Varndell/Nature Picture Library; C.K. Lorenz/Science Source; **p. 158:** Ralph C. Eagle, Jr./Science Source; **p. 159:** Ralph C. Eagle, Jr./Science Source; **p. 183:** Barry Diomede/Alamy Stock Photo; **p. 193:** AJPhoto/Science Source; GENE J. PUSKAR/AP Images; **p. 197:** Steven J. Barnes; **p. 201:** VanVang/Alamy Stock Photo **p. 204:** Omikron/Science Source; **p. 207:** Mala Iryna/Shuttertock; **p. 209:** Creativa Images/Shutterstock; **p. 212:** Matteo Carta/Alamy Stock Photo; **p. 226:** CNRI/Science Source; **p. 236:** Robbi Akbari Kamaruddin/Alamy Stock Photo; **p. 243:** Naweed Syed, PhD/Departments of Anatomy and Medical Physiology, the University of Calgary; **p. 251:** Carl Ernst/Brian Christie/Department of Psychology, University of British Columbia; **p. 258:** Michael Ventura/Alamy Stock Photo; **p. 260:** NEIL BORDEN/Living Art Enterprises/Science Source; **p. 261:** SMC Images/Photodisc/Getty Images; John Pinel; **p. 262:** ZEPHYR/SPL/Science Photo Library/Alamy Stock Photo; **p. 264:** Scott Camazine/Alamy Images; epa european pressphoto agency b.v./Alamy Stock Photo; **p. 271:** Dr. Cecil H. Fox/Science Source; **p. 271:** University of Rochester Medical Center; **p. 280:** Images courtesy of Carl Ernst and Brian Christie, Department of Psychology, University of British Columbia; **p. 283:** Carolyn A. McKeone/Science Source; **p. 287:** Africa Studio/Shutterstock; **p. 296:** Craig Durling/ZUMA Wire/Alamy Live News; **p. 305:** WARNER BROS. PICTURES/Album/Newscom; **p. 316:** filadendron/Getty Images; **p. 326:** Thomas Barwick/Getty Images; photographer and designer/Moment/Getty Images; Tooga/Getty Images; **p. 338:** printed/digital with permission from © The Jackson Laboratory, Photographer Jennifer Torrance; **p. 344:** Hinterhaus Productions/Getty Images; **p. 361:** Str Old/Reuters Pictures; **p. 371:** fizkes/Shutterstock; **p. 374:** Hank Morgan/Science Source; **p. 379:** paul prescott/Shutterstock; **p. 382:** CRSuber/Getty Images; **p. 397:** tab62/Fotolia; **p. 404:** Phanie/Alamy Stock Photo; **p. 411:** Angela Hampton/Angela Hampton Picture Library/Alamy Stock Photo;

Name Index

Begg, D. P., 324
Beggs, S., 208
Behrmann, M., 178, 179, 218
Bekinschtein, P., 308
Bekiranov, S., 356
Bekkers, J. M., 203
Bell, C. C., 223
Bell, J. T., 69, 70
Bell, K. F. S., 275
Bell, M. R., 346
Bellono, N. W., 185
Belloy, M. E., 272
Bellugi, U., 256
Belluscio, L., 203
Belouin, S. J., 489
Belousov, A. B., 242
Belzung, C., 492, 493
Ben-Ari, Y., 266
Ben-Shakhar, G., 467
Benarroch, E. E., 349
Benazzouz, A., 269
Benca, R. M., 397
Bender, A. M., 412
Bender, M. B., 174
Bendor, D., 192
Benedetti, F., 505
Benedict, C., 384
Benítez-Burraco, A., 255, 256
Bennett, M. L., 479
Bentivoglio, M., 77
Benton, A. L., 34
Benvenisty, N., 59
Berardi, N., 249
Berchtold, N. C., 283
Berens, P., 155, 170
Bergamini, G., 109
Berger, B. D., 426, 428
Berger, J. R., 264
Berger, R. J., 374
Bergles, D. E., 80
Bergmann, B. M., 385
Bergmann, O., 251, 252
Berkel, T. D., 414
Berlim, M. T., 494
Berliner, J., 416
Bernal, B., 455
Bernard, F., 444
Berndt, A., 140
Bernstein, I. L., 147
Bernstein, R. A., 261
Berridge, K. C., 323, 424, 427
Berry, K. P., 247
Bershad, A. K., 417
Berthoud, H. R., 333, 339
Bertram, E. H., 267
Bespalov, A., 407
Beste, C., 476
Betley, J. N., 330
Betzel, R. F., 144
Betzler, F., 418
Beyer, D. K., 498
Bezard, E., 275
Bhagwate, M. R., 430
Bhalla, U. S., 203
Bhati, M. T., 486

Bhatt, L. K., 497, 498
Bi, Y., 178
Bidelman, G. M., 187
Bielavska, E., 323
Bilder, D. A., 67
Bilkei-Gorzo, A., 416
Billock, V. A., 171
Bimonte-Nelson, H. A., 355
Bin, Y. S., 389
Binder, E. B., 481, 495
Bintu, L., 64
Birch, A. M., 251
Birklein, F., 202
Birnbaum, R., 490
Bishop, D. V. M., 445, 458
Bittner, E. A., 407
Bixon, P. D., 233
Biyani, S., 254
Bizley, J. K., 208
Bjedov, I., 498
Björkholm, C., 495
Björklund, A., 281
Blackburn, G. L., 342
Blackmore, S., 28
Blackshaw, S., 250
Blakemore, C., 30
Blakemore, S. J., 246
Blakeslee, S., 254, 284
Blanchard, D. C., 146, 470
Blanchard, R., 368
Blanchard, R. J., 146, 470
Blanco, C., 421
Blanke, O., 180, 199
Blankenburg, F., 199
Blashfield, R. K., 486
Blass, E. M., 435
Blennow, K., 263
Blesa, J., 275
Bliss, T. V., 306
Bliss, T. V. P., 202, 309, 310, 311, 312
Bloch, G. J., 396
Bludau, A., 65
Blüher, M., 339
Blundell, J. E., 325, 330, 336, 339
Boag, S., 379
Bobadilla, A. C., 424, 425
Boccia, M. L., 47
Bodea, L. G., 271
Bogaert, A. F., 367, 368
Bogen, J. G., 438
Böhm, U. L., 230
Boido, M., 278
Boisson, D., 266
Bokiniec, P., 200
Boland, E. M., 498
Bonhoeffer, T., 170
Bonin, R. P., 298
Bonini, L., 220
Bonini, S. A., 415
Bonnamain, B., 282
Bonnel, A., 254
Bonnet, M. H., 385
Boot, E., 67

Booth, D. A., 325
Bordenstein, S. R., 337
Bordey, A., 250
Borgland, S. L., 330
Borgnis, F., 220
Born, J., 297, 386
Boroviak, T., 238
Borsook, D., 208
Bosman, A. C., 356
Bosworth, A. P., 245
Boto, E., 130
Bouchard, T. J., Jr., 68, 455
Bourgeois, J., 395
Bourzac, K., 128, 268
Bouvier, G., 311
Bouzier-Sore, A. K., 81
Bowden, J., 393
Bowler, P. J., 50
Bowman, H., 297
Boyden, E. S., 139
Boys, A., 426
Bradke, F., 277
Bragatti, J. A., 266
Bramble, M. S., 359, 360
Branaccio, M., 391
Brancaccio, M., 390
Brandon, N. J., 495
Brandt, T., 189
Brang, D., 191
Brann, J. H., 251
Brascamp, J., 206, 207
Brasier, D. J., 81
Braskie, M. N., 296
Braunstein, G. D., 363
Braver, T. S., 145
Bray, N., 378
Brecher, E. M., 412
Brecht, M., 199, 223, 463
Brecht, M. C., 355
Breen, M. S., 490
Bremen, P., 189
Bremer, F. L., 392
Bremer, J., 361
Brennan, D. J., 339
Brenner, S., 59
Bressler, S. L., 187
Brewer, A. A., 190
Brewster, J. M., 419
Bridgett, D. J., 447
Bridgman, K., 412
Brier, M. R., 271
Brinker, T., 75
Britton, J., 412, 421
Brobeck, J. R., 327
Broca, P., 434, 449, 450
Broderick, J. P., 262
Broekman, M. L., 261
Brooks, M. J., 341
Brooks, S. P., 146
Brooks, V. B., 233
Brotchie, J. M., 275
Broughton, R., 376
Brown, C. S., 67
Brown, D. L., 340
Brown, G. C., 81

Brown, J. L., 472
Brown, R. D., 262
Brown, R. E., 28
Brownsett, S., 455
Brunklaus, A., 267
Bu, G., 272
Büchel, C., 475
Buchsbaum, B. R., 448
Buck, L., 33
Buckley, M. J., 292
Buckner, R. L., 223
Bucy, P. C., 465–466
Budday, S., 241
Buffalo, E. A., 304
Buitrago-Delgado, E., 242
Bulters, D., 261
Bundy, D. T., 443
Buonarati, O. R., 312
Burd, L., 414
Burda, J. E., 275
Burgess, N., 304
Burgio, E., 337
Burns, J. K., 416
Burrone, J., 244, 245
Burt, D. R., 488
Burton, C. L., 503
Busche, M. A., 273
Buschman, T. J., 208, 210, 218
Bush, D., 304
Bushnell, M. C., 200
Bussey, T. J., 302, 307
Butler, C., 294
Buxbaum, J. N., 273
Byne, W., 368
Byrne, N. M., 337
Byrnes, E. M., 414

C

Caballero, A., 248
Cabezas, R., 81
Cacioppo, J. T., 29
Cacucci, F., 304
Cade, J., 497–498
Caeiro, L., 261
Cafferty, W. B. J., 279
Caffo, M., 260
Cagampang, F. R., 336
Cahana-Amitay, D., 451
Cahill, L., 358
Cai, H., 401
Cai, Q., 444
Cajal, S. R., 33
Calabrò, R. S., 366
Calatayud, V. A., 363
Caldeira, G. L., 490
Calhoon, G. G., 501
Calhoun, V. D., 472
Callaghan, B. L., 313, 418
Callaway, E., 53
Callaway, E. M., 135, 138
Calvo, M. G., 469
Camacho-Arroyo, I., 347
Camardese, G., 125
Cambridge, O. R., 491
Cameli, L., 38

Subject Index